WITHDRAWN

D1605413

Handbook of Natural Language Processing

Handbook of Natural Language Processing

edited by

Robert Dale
Macquarie University
Sydney, Australia

Hermann Moisl
University of Newcastle
Newcastle-upon-Tyne, England

Harold Somers
UMIST
Manchester, England

MARCEL DEKKER, INC. NEW YORK • BASEL

ISBN: 0-8247-9000-6

This book is printed on acid-free paper.

Headquarters
Marcel Dekker, Inc.
270 Madison Avenue, New York, NY 10016
tel: 212-696-9000; fax: 212-685-4540

Eastern Hemisphere Distribution
Marcel Dekker AG
Hutgasse 4, Postfach 812, CH-4001 Basel, Switzerland
tel: 41-61-261-8482; fax: 41-61-261-8896

World Wide Web
http://www.dekker.com

The publisher offers discounts on this book when ordered in bulk quantities. For more information, write to Special Sales/Professional Marketing at the headquarters address above.

Current printing (last digit)
10 9 8 7 6 5 4 3 2 1

PRINTED IN THE UNITED STATES OF AMERICA

Preface

The discipline of Natural Language Processing (NLP) concerns itself with the design and implementation of computational machinery that communicates with humans using natural language. Why is it important to pursue such an endeavor? Given the self-evident observation that humans communicate most easily and effectively with one another using natural language, it follows that, in principle, it is the easiest and most effective way for humans and machines to interact; as technology proliferates around us, that interaction will be increasingly important. At its most ambitious, NLP research aims to design the language input–output components of artificially intelligent systems that are capable of using language as fluently and flexibly as humans do. The robots of science fiction are archetypical: stronger and more intelligent than their creators and having access to vastly greater knowledge, but human in their mastery of language. Even the most ardent exponent of artificial intelligence research would have to admit that the likes of HAL in Kubrick's *2001: A Space Odyssey* remain firmly in the realms of science fiction. Some success, however, has been achieved in less ambitious domains, where the research problems are more precisely definable and therefore more tractable. Machine translation is such a domain, and one that finds ready application in the internationalism of contemporary economic, political, and cultural life. Other successful areas of application are message-understanding systems, which extract useful elements of the propositional content of textual data sources such as newswire reports or banking telexes, and front ends for information systems such as databases, in which queries can be framed and replies given in natural language. This handbook is about the design of these and other sorts of NLP systems. Throughout, the emphasis is on practical tools and techniques for implementable systems; speculative research is minimized, and polemic excluded.

The rest of this preface is in three sections. In Section 1 we provide a sketch of the development of NLP, and of its relationship with allied disciplines such as linguistics and cognitive science. Then, in Section 2, we go on to delineate the scope of the handbook in terms of the range of topics covered. Finally, in Section 3, we provide an overview of how the handbook's content is structured.

1. THE DEVELOPMENT OF NATURAL LANGUAGE PROCESSING

A. History

In the 1930s and 1940s, mathematical logicians formalized the intuitive notion of an effective procedure as a way of determining the class of functions that can be computed algorithmically. A variety of formalisms were proposed—recursive functions, lambda calculus, rewrite systems, automata, artificial neural networks—all equivalent in terms of the functions they can compute. The development of these formalisms was to have profound effects far beyond mathematics. Most importantly, it led to modern computer science and to the computer technology that has transformed our world in so many ways. But it also generated a range of new research disciplines that applied computational ideas to the study and simulation of human cognitive functions: cognitive science, generative linguistics, computational linguistics, artificial intelligence, and natural language processing. Since the 1950s, three main approaches to natural language processing have emerged. We consider them briefly in turn.

- Much of the NLP work in the first 30 years or so of the field's development can be characterized as "NLP based on generative linguistics." Ideas from linguistic theory, particularly relative to syntactic description along the lines proposed by Chomsky and others, were absorbed by researchers in the field, and refashioned in various ways to play a role in working computational systems; in many ways, research in linguistics and the philosophy of language set the agenda for explorations in NLP. Work in syntactic description has always been the most thoroughly detailed and worked-out aspect of linguistic inquiry, so that at this level a great deal has been borrowed by NLP researchers. During the late 1980s and the early 1990s, this led to a convergence of interest in the two communities to the extent that at least some linguistic theorizing is now undertaken with explicit regard to computational issues. Head-driven Phase Structure Grammar and Tree Adjoining Grammar are probably the most visible results of this interaction. Linguistic theoreticians, in general, have paid less attention to the formal treatment of phenomena beyond syntax, and approaches to semantics and pragmatics within NLP have consequently tended to be somewhat more ad hoc, although ideas from formal semantics and speech act theory have found their way into NLP.
- The generative linguistics-based approach to NLP is sometimes constrasted with "empirical" approaches based on statistical and other data-driven analyses of raw data in the form of text corpora, that is, collections of machine-readable text. The empirical approach has been around since NLP began in the early 1950s, when the availability of computer technology made analysis of reasonably large text corpora increasingly viable, but it was soon pushed into the background by the well-known and highly influential opinions of Chomsky and his followers, who were strongly opposed to empirical methods in linguistics. A few researchers resisted the trend, however, and work on corpus collections, such as the Brown and LOB corpora, continued, but it is only in the last 10 years or so that empirical NLP has reemerged as a major alternative to "rationalist" lin-

guistics-based NLP. This resurgence is mainly attributable to the huge data storage capacity and extremely fast access and processing afforded by modern computer technology, together with the ease of generating large text corpora using word processors and optical character readers.

Corpora are primarily used as a source of information about language, and a number of techniques have emerged to enable the analysis of corpus data. Using these techniques, new approaches to traditional problems have also been developed. For example, syntactic analysis can be achieved on the basis of statistical probabilities estimated from a training corpus (as opposed to rules written by a linguist), lexical ambiguities can be resolved by considering the likelihood of one or another interpretation on the basis of context both near and distant, and measures of style can be computed in a more rigorous manner. "Parallel" corpora—equivalent texts in two or more languages—offer a source of contrastive linguistic information about different languages, and can be used, once they have been aligned, to extract bilingual knowledge useful for translation either in the form of traditional lexicons and transfer rules or in a wholly empirical way in example-based approaches to machine translation.

- Artificial neural network (ANN)–based NLP is the most recent of the three approaches. ANNs were proposed as a computational formalism in the early 1940s, and were developed alongside the equivalent automata theory and rewrite systems throughout the 1950s and 1960s. Because of their analogy with the physical structure of biological brains, much of this was strongly oriented toward cognitive modeling and simulation. Development slowed drastically when, in the late 1960s, it was shown that the ANN architectures known at the time were not universal computers, and that there were some very practical and cognitively relevant problems that these architectures could not solve. Throughout the 1970s relatively few researchers persevered with ANNs, but interest in them began to revive in the early 1980s, and for the first time some language-oriented ANN papers appeared. Then, in the mid-1980s, discovery of a way to overcome the previously demonstrated limitations of ANNs proved to be the catalyst for an explosion of interest in ANNs both as an object of study in their own right and as a technology for a range of application areas. One of these application areas has been NLP: since 1986 the volume of NL-oriented—and specifically NLP—research has grown very rapidly.

B. NLP and Allied Disciplines

The development of NLP is intertwined with that of several other language-related disciplines. This subsection specifies how, for the purposes of the handbook, these relate to NLP. This is not done merely as a matter of interest. Each of the disciplines has its own agenda and methodology, and if one fails to keep them distinct, confusion in one's own work readily ensues. Moreover, the literature associated with these various disciplines ranges from very large to huge, and much time can be wasted engaging with issues that are simply irrelevant to NLP. The handbook takes NLP to be exclusively concerned with the design and implementation of effective natural language input and output components for computational systems. On the basis of this definition, we can compare NLP with each of the related disciplines in the following ways.

1. *Cognitive Science*

Cognitive science is concerned with the development of psychological theories; the human language faculty has always been central to cognitive theorizing and is, in fact, widely seen as paradigmatic for cognition generally. In other words, cognitive science aims in a scientific sense to explain human language. NLP as understood by this book does not attempt such an explanation. It is interested in designing devices that implement some linguistically relevant mapping. No claims about cognitive explanation or plausibility are made for these devices. What matters is whether they actually implement the desired mapping, and how efficiently they do so.

2. *Generative Linguistics*

Like cognitive science, generative linguistics aims to develop scientific theories, although about human language in particular. For Chomsky and his adherents, linguistics is, in fact, part of cognitive science; the implication of this for present purposes has already been stated. Other schools of generative linguistics make no cognitive claims, and regard their theories as formal systems, the adequacy of which is measured in terms of the completeness, internal consistency, and economy by which all scientific theories are judged. The rigor inherent in these formal systems means that, in many cases, they characterize facts and hypotheses about language in such a way that these characterizations can fairly straightforwardly be used in computational processing. Such transfers from the theoretical domain to the context of NLP applications are quite widespread. At the same time, however, there is always scope for tension between the theoretical linguist's desire for a maximally economical and expressive formal system and the NLP system designer's desire for broad coverage, robustness, and efficiency. It is this tension that has in part provoked interest in methods other than the strictly symbolic.

3. *Artificial Intelligence*

Research in artificial intelligence (AI) aims to design computational systems that simulate aspects of cognitive behavior. It differs from cognitive science in that no claim to cognitive explanation or plausibility is necessarily made with respect to the architectures and algorithms used. AI is thus close in orientation to NLP, but not identical: in the view we have chosen to adopt in this book, NLP aims not to simulate intelligent behavior per se, but simply to design NL input–output components for arbitrary computational applications; these might be components of AI systems, but need not be.

4. *Computational Linguistics*

Computational linguistics (CL) is a term that different researchers interpret in different ways; for many the term is synonymous with NLP. Historically founded as a discipline on the back of research into Machine Translation, and attracting researchers from a variety of neighboring fields, it can be seen as a branch of linguistics, computer science, or literary studies.

- As a branch of linguistics, CL is concerned with the computational implementation of linguistic theory: the computer is seen as a device for testing the completeness, internal consistency, and economy of generative linguistic theories.
- As a branch of computer science, CL is concerned with the relationship between natural and formal languages: interest here focuses on such issues as language recognition and parsing, data structures, and the relationship between facts about language and procedures that make use of those facts.
- As a branch of literary studies, CL involves the use of computers to process large corpora of literary data in, for example, author attribution of anonymous texts.

Within whatever interpretation of CL one adopts, however, there is a clear demarcation between domain-specific theory on the one hand and practical development of computational language processing systems on the other. It is this latter aspect—practical development of computational language processing systems—that, for the purposes of this book, we take to be central to NLP. In recent years, this approach has often been referred to as "language technology" or "language engineering."

Our view of NLP can, then, be seen as the least ambitious in a hierarchy of disciplines concerned with NL. It does not aim to explain or even to simulate the human language faculty. What it offers, however, is a variety of practical tools for the design of input–output modules in applications that can benefit from the use of natural language. And, because this handbook is aimed at language-engineering professionals, that is exactly what is required.

2. THE SCOPE OF THIS BOOK

Taken at face value, the expression "natural language processing" encompasses a great deal. Ultimately, applications and technologies such as word processing, desktop publishing, and hypertext authoring are all intimately involved with the handling of natural language, and so in a broad view could be seen as part of the remit of this book. In fact, it is our belief that, in the longer term, researchers in the field will increasingly look to integration of these technologies into mainstream NLP research. But there are limits to what can usefully be packed between the covers of a book, even of a large book such as this, and there is much else to be covered, so a degree of selection is inevitable. Our particular selection of topics was motivated by the following considerations.

Any computational system that uses natural language input and output can be seen in terms of a five-step processing sequence (Fig. 1):

1. The system receives a physical signal from the external world, where that signal encodes some linguistic behavior. Examples of such signals are speech waveforms, bitmaps produced by an optical character recognition or handwriting recognition system, and ASCII text streams received from an electronic source. This signal is converted by a transducer into a linguistic representation amenable to computational manipulation.
2. The natural language analysis component takes the linguistic representation from step 1 as input and transforms it into an internal representation appropriate to the application in question.

3. The application takes the output from step 2 as one of its inputs, carries out a computation, and outputs an internal representation of the result.
4. The natural language generation component takes part of the output from step 3 as input, transforms it into a representation of a linguistic expression, and outputs that representation.
5. A transducer takes the output representation from step 4 and transforms it into a physical signal that, in the external world, is interpretable by humans as language. This physical signal might be a text stream, the glyphs in a printed document, or synthesized speech.

Ultimately, the input to a language-processing system will be in graphemic or phonetic form, as will the outputs. Phonetic form corresponds to speech input. Traditionally, however, the only kind of graphemic input dealt with in language-processing systems is digitally encoded text. In the case of such graphemic input it is generally assumed either that this will be the native form of the input (an assumption that is indeed true for a great deal of the data we might wish to process) or that some independent process will map the native form into digitally encoded text. This assumption is also made in much—although certainly not all—work on speech: a great deal of this research proceeds on the basis that mapping speech signals into a textual representation consisting of sequences of words can be achieved independently of any process that subsequently derives some other representation from this text. It is for these reasons that we have chosen in the present work to focus on the processing techniques that can be used with textual representations, and thus exclude from direct consideration techniques that have been developed for speech processing.

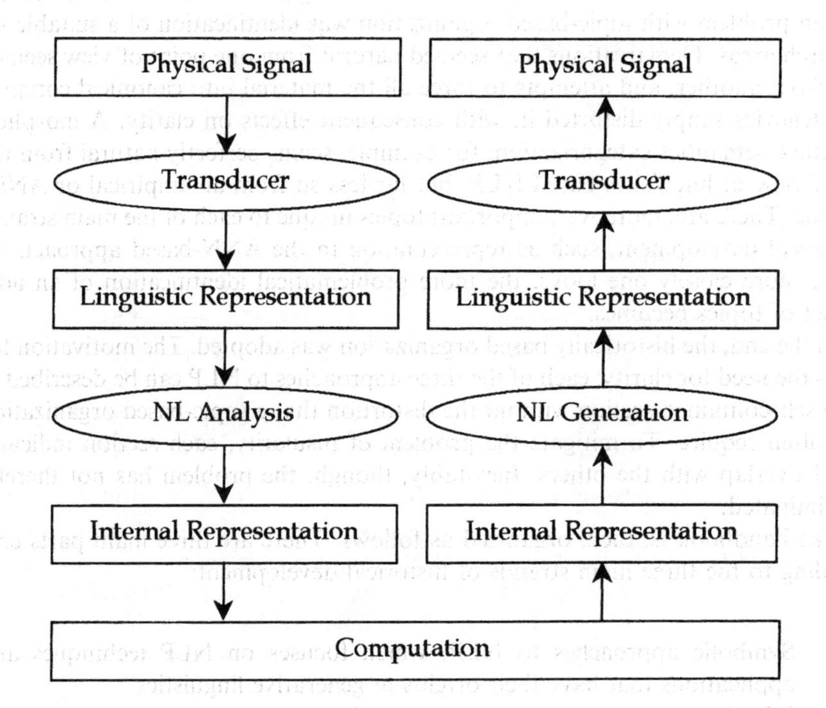

Fig. 1 The five steps in computational processing of natural language input and output.

Subject, then, to the foregoing restrictions of scope, coverage of NLP tools and applications in this handbook attempts to strike a balance between comprehensiveness and depth of coverage with the aim of providing a practical guide for practical applications.

3. THE STRUCTURE OF THIS BOOK

After some five decades of NLP research, and of NLP-related work in the aforementioned disciplines, a huge research literature has accumulated. Because the primary function of this handbook is to present that material as accessibly as possible, clarity of organization is paramount. After much discussion, we realized that no organization was going to be ideal, but two reasonable possibilities did emerge: a historically based organization and a topic-based one. The first of these would partition the material into three sections corresponding to the three main strands of development described in the foregoing historical sketch, and the second would attempt to identify a set of research topics that would comprehensively describe NLP and NLP-related research to date by presenting the relevant work in each topic area. There are drawbacks to both. The main problem with the historically based organization is that it might be seen to reinforce the distinctions between the three approaches to NLP: although these distinctions have historical validity, they are increasingly being broken down, and for the handbook to partition the material in this way would distort the current state of increasingly ecumenical NLP research. The main problem with topic-based organization was identification of a suitable set of research areas. Demarcations that seemed natural from one point of view seemed less so from another, and attempts to force all the material into canonical compromise categories simply distorted it, with consequent effects on clarity. A morphology–syntax–semantics categorization, for example, seems perfectly natural from the point of view of linguistics-based NLP, but far less so from an empirical or ANN-based one. There are, moreover, important topics unique to each of the main strands of historical development, such as representation in the ANN-based approach. In fact, the more closely one looks, the more problematical identification of an adequate set of topics becomes.

In the end, the historically based organization was adopted. The motivation for this was the need for clarity: each of the three approaches to NLP can be described in its own self-contained section without the distortion that a topic-based organization would often require. To mitigate the problem of insularity, each section indicates areas of overlap with the others. Inevitably, though, the problem has not thereby been eliminated.

The handbook is, then, organized as follows. There are three main parts corresponding to the three main strands of historical development:

1. Symbolic approaches to NLP, which focuses on NLP techniques and applications that have their origins in generative linguistics
2. NLP based on empirical corpus analysis
3. NLP based on artificial neural networks

Again for clarity, each section is subdivided into the following subsections:

1. An introduction that gives an overview of the approach in question
2. A set of chapters describing the fundamental concepts and tools appropriate to that approach
3. A set of chapters describing particular applications in which the approach has been successfully used

Beyond this basic format, no attempt has been made to impose uniformity on the individual parts of the handbook. Each is the responsibility of a different editor, and each editor has presented his material in a way that he considers most appropriate. In particular, given that work in ANN–based NLP will be relatively new to many readers of this book, the introduction to Part III includes a detailed overview of how these techniques have developed and found a place in NLP research.

Some acknowledgments are appropriate. First and foremost, we express our thanks to the many contributors for their excellent efforts in putting together what we hope will be a collection of long-standing usefulness. We would also like to thank the editorial staff of Marcel Dekker, Inc., for their considerable patience in response to the succession of delays that is inevitable in the marshaling of a book of this size and authorial complexity. Finally, Debbie Whittington of the Microsoft Research Institute at Macquarie University and Rowena Bryson of the Centre for Research in Linguistics at Newcastle University deserve special mention for their administrative help.

Robert Dale
Hermann Moisl
Harold Somers

Contents

Contributors

James Allen Department of Computer Science, University of Rochester, Rochester, New York

Elisabeth André German Research Center for Artificial Intelligence, Saarbrücken, Germany

Ion Androutsopoulos Institute of Informatics and Telecommunications, National Centre for Scientific Research "Demokritos," Athens, Greece

Eric Brill Microsoft Research, Redmond, Washington

John A. Carroll School of Cognitive and Computing Sciences, University of Sussex, Brighton, England

Chun-Hsien Chen Department of Information Management, Chang Gung University, Kwei-Shan, Tao-Yuan, Taiwan, Republic of China

Jim Cowie Computing Research Laboratory, New Mexico State University, Las Cruces, New Mexico

Ido Dagan Department of Mathematics and Computer Science, Bar-Ilan University, Ramat Gan, Israel

Robert Dale Department of Computing, Macquarie University, Sydney, Australia

Georg Dorffner Austrian Research Institute for Artificial Intelligence, and Department of Medical Cybernetics and Artificial Intelligence, University of Vienna, Vienna, Austria

George Ferguson Department of Computer Science, University of Rochester, Rochester, New York

Barbara J. Grosz Division of Engineering and Applied Sciences, Harvard University, Cambridge, Massachusetts

Simon Haykin Communications Research Laboratory, McMaster University, Hamilton, Ontario, Canada

George E. Heidorn Microsoft Research, Redmond, Washington

Stefan Heil Technical University of Munich, Munich, Germany

Vasant Honavar Department of Computer Science, Iowa State University, Ames, Iowa

Richard I. Kittredge* Department of Linguistics and Translation, University of Montreal, Montreal, Quebec, Canada

Ludovic Lebart Economics and Social Sciences Department, Ecole Nationale Supérieure des Télécommunications, Paris, France

Karen E. Lochbaum U S WEST Advanced Technologies, Boulder, Colorado

Simon Lucas Department of Electronic Systems Engineering, University of Essex, Colchester, England

Yuji Matsumoto Graduate School of Information Science, Nara Institute of Science and Technology, Ikoma, Nara, Japan

David D. McDonald Brandeis University, Arlington, Massachusetts

Tony McEnery Department of Linguistics, Bowland College, Lancaster University, Lancaster, England

Kathleen R. McKeown Department of Computer Science, Columbia University, New York, New York

Dieter Merkl Department of Software Technology, Vienna University of Technology, Vienna, Austria

Risto Miikkulainen Department of Computer Sciences, The University of Texas at Austin, Austin, Texas

Bradford W. Miller Cycorp, Austin, Texas

Hermann Moisl Centre for Research in Linguistics, University of Newcastle, Newcastle-upon-Tyne, England

---

**Also affiliated with*: CoGenTex, Inc., Ithaca, New York.

Michael Oakes Department of Linguistics, Lancaster University, Lancaster, England

David D. Palmer The MITRE Corporation, Bedford, Massachusetts

Rajesh G. Parekh Allstate Research and Planning Center, Menlo Park, California

Massimo Poesio Human Communication Research Centre and Division of Informatics, University of Edinburgh, Edinburgh, Scotland

Alain Polguère Department of Linguistics and Translation, University of Montreal, Montreal, Quebec, Canada

Dragomir R. Radev School of Information, University of Michigan, Ann Arbor, Michigan

Martin Rajman Artificial Intelligence Laboratory (LIA), EPFL, Swiss Federal Institute of Technology, Lausanne, Switzerland

Eric K. Ringger* Department of Computer Science, University of Rochester, Rochester, New York

Graeme Ritchie Division of Informatics, University of Edinburgh, Edinburgh, Scotland

Christer Samuelsson Xerox Research Centre Europe, Grenoble, France

Jürgen Schmidhuber IDSIA, Lugano, Switzerland

Candace L. Sidner Lotus Research, Lotus Development Corporation, Cambridge, Massachusetts

Teresa Sikorski Zollo Department of Computer Science, University of Rochester, Rochester, New York

Harold Somers Centre for Computational Linguistics, UMIST, Manchester, England

Richard Sproat Department of Human/Computer Interaction Research, AT&T Labs–Research, Florham Park, New Jersey

Henry S. Thompson Division of Informatics, University of Edinburgh, Edinburgh, Scotland

---

**Current affiliation*: Microsoft Research, Redmond, Washington

Takehito Utsuro Graduate School of Information Science, Nara Institute of Science and Technology, Ikoma, Nara, Japan

Stefan Wermter Centre for Informatics, SCET, University of Sunderland, Sunderland, England

Yorick Wilks Department of Computer Science, University of Sheffield, Sheffield, England

Mats Wirén Telia Research, Stockholm, Sweden

Michael Witbrock Lycos Inc., Waltham, Massachusetts

Dekai Wu Department of Computer Science, The Hong Kong University of Science and Technology, Kowloon, Hong Kong

David Yarowsky Department of Computer Science, Johns Hopkins University, Baltimore, Maryland

1

Symbolic Approaches to Natural Language Processing

ROBERT DALE

Macquarie University, Sydney, Australia

1. INTRODUCTION

Ever since the early days of machine translation in the 1950s, work in natural language processing (NLP) has attempted to use symbolic methods—where knowledge about language is explicitly encoded in rules or other forms of representation—as a means to solving the problem of how to automatically process human languages. In this first part of the handbook, we look at the body of techniques that have developed in this area over the past forty years, and we examine some categories of applications that make use of these techniques. In more recent years, the availability of large corpora and datasets has reminded us of the difficulty of scaling-up many of these symbolic techniques to deal with real-world problems; many of the techniques described in the second and third parts of this handbook are a reaction to this difficulty. Nonetheless, the symbolic techniques described here remain important when the processing of abstract representations is required, and it is now widely recognised that the key to automatically processing human languages lies in the appropriate combination of symbolic and nonsymbolic techniques. Indeed, some of the chapters in this part of the book already demonstrate, both in terms of techniques and the use of those techniques within specific applications, that the first steps toward hybrid solutions are being taken.

In this introduction, we provide an overview of the contents of this part of the handbook. The overview separates the chapters into two clusters, corresponding to the major subdivision we have adopted in all three parts of the book: the first six chapters contained here focus on specific techniques that can be used in building natural language processing applications, and the remaining seven chapters describe applications that all use some combination of the techniques described earlier.

2. TECHNIQUES

Traditionally, work in natural language processing has tended to view the process of language analysis as being decomposable into a number of stages, mirroring the theoretical linguistic distinctions drawn between SYNTAX, SEMANTICS, and PRAGMATICS. The simple view is that the sentences of a text are first analysed in terms of their syntax; this provides an order and structure that is more amenable to an analysis in terms of semantics, or literal meaning; and this is followed by a stage of pragmatic analysis whereby the meaning of the utterance or text in context is determined. This last stage is often seen as being concerned with DISCOURSE, whereas the previous two are generally concerned with sentential matters. This attempt at a correlation between a stratificational distinction (syntax, semantics, and pragmatics) and a distinction in terms of granularity (sentence versus discourse) sometimes causes some confusion in thinking about the issues involved in natural language processing; and it is widely recognised that in real terms, it is not so easy to separate the processing of language neatly into boxes corresponding to each of the strata. However, such a separation serves as a useful pedagogical aid, and also constitutes the basis for architectural models that make the task of natural language analysis more manageable from a software-engineering point of view.

Nonetheless, the tripartite distinction into syntax, semantics, and pragmatics serves, at best, only as a starting point when we consider the processing of real natural language texts. A finer-grained decomposition of the process is useful when we take into account the current state of the art in combination with the need to deal with real language data; this is reflected in Fig. 1.

Unusually for books on natural language processing, we identify here the stage of tokenisation and sentence segmentation as a crucial first step. Natural language text is generally not made up of the short, neat, well-formed and well-delimited sentences we find in textbooks; and for languages such as Chinese, Japanese, or Thai, which do not share the apparently easy space-delimited tokenisation we might believe to be a property of languages such as English, the ability to address issues of tokenisation is essential to getting off the ground at all. We also treat lexical analysis as a separate step in the process. To some degree this finer-grained decomposition reflects our current state of knowledge about language processing: we know quite a lot about general techniques for tokenisation, lexical analysis, and syntactic analysis, but much less about semantics and discourse-level processing. But it also reflects the fact that the known is the surface text, and anything deeper is a representational abstraction that is harder to pin down; so it is not so surprising that we have better-developed techniques at the more concrete end of the processing spectrum.

Natural language analysis is only one half of the story. We also have to consider natural language generation, when we are concerned with mapping from some (typically nonlinguistic) internal representation to a surface text. In the history of the field so far, there has been much less work on natural language generation than there has been on natural language analysis. One sometimes hears the suggestion that this is because natural language generation is easier, so that there is less to be said. As

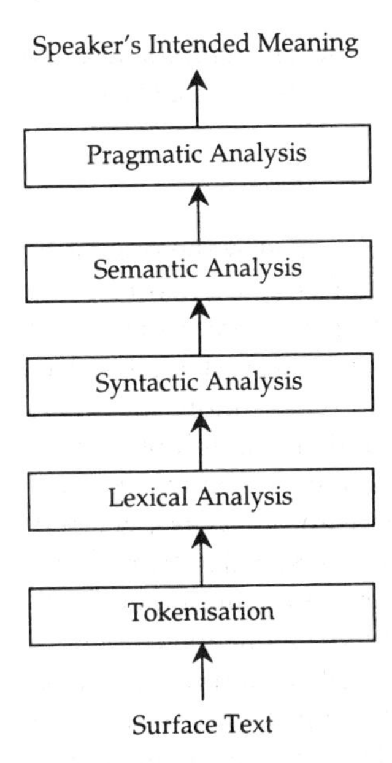

Fig. 1 The stages of analysis in processing natural language.

David McDonald demonstrates in Chap. 7, this is far from the truth: there are a great many complexities to be addressed in generating fluent and coherent multi-sentential texts from an underlying source of information. A more likely reason for the relative lack of work in generation is precisely the correlate of the observation made at the end of the previous paragraph: it is relatively straightforward to build theories around the processing of something known (such as a sequence of words), but much harder when the input to the process is more or less left to the imagination. This is the question that causes researchers in natural language generation to wake in the middle of the night in a cold sweat: what does generation start from? Without a clear idea of where to start, and without independently motivated input representations, it can be hard to say very much that is not idiosyncratic. McDonald offers some suggestions here, although many of the problems disappear when we face the need to construct real applications of natural language generation, two of which are described in the second half of this part of the handbook.

A. Chapter 2: Tokenisation and Sentence Segmentation

As we have already noted, not all languages deliver text in the form of words neatly delimited by spaces. Languages such as Chinese, Japanese, and Thai require first that a segmentation process be applied, analogous to the segmentation process that must first be applied to a continuous speech stream to identify the words that make up an utterance. As Palmer demonstrates in Chap. 2, there are also significant segmentation and tokenisation issues in apparently easier-to-segment languages—such as English. Fundamentally, the issue here is that of what constitutes a word; as Palmer shows, there is no easy answer here. This chapter also looks at the problem of sentence segmentation: because so much work in natural language processing views the sentence as the unit of analysis, clearly it is of crucial importance to ensure that, given a text, we can break it into sentence-sized pieces. This turns out not to be so trivial either. Palmer offers a catalog of tips and techniques that will be useful to anyone faced with dealing with real raw text as the input to an analysis process, and provides a healthy reminder that these problems have tended to be idealised away in much earlier, laboratory-based work in natural language processing.

B. Chapter 3: Lexical Analysis

The previous chapter addressed the problem of breaking a stream of input text into the words and sentences that will be subject to subsequent processing. The words are not atomic and are themselves open to further analysis. Here we enter the realms of computational morphology, the focus of Richard Sproat's chapter. By taking words apart, we can uncover information that will be useful at later stages of processing. The combinatorics also mean that decomposing words into their parts, and maintaining rules for how combinations are formed, is much more efficient in terms of storage space than would be true if we simply listed every word as an atomic element in a huge inventory. And, once more returning to our concern with the handling of real texts, there will always be words missing from any such inventory; morphological processing can go some way toward handling such unrecognised words. Sproat provides a wide-ranging and detailed review of the techniques that can be used to carry out morphological processing, drawing on examples from languages other than English to demonstrate the need for sophisticated processing methods; along the way he provides some background in the relevant theoretical aspects of phonology and morphology, and draws connections with questions of tokenisation discussed in the previous chapter.

C. Chapter 4: Parsing Techniques

A presupposition in most work in natural language processing is that the basic unit of meaning analysis is the sentence: a sentence expresses a proposition, idea, or thought, and says something about some real or imaginary world. Extracting the meaning from a sentence is thus a key issue. Sentences are not, however, just linear sequences of words, and so it is widely recognised that to carry out this task requires an analysis of each sentence which determines its structure in one way or another. In NLP approaches based on generative linguistics, this is generally taken to involve determination of the syntactic or grammatical structure of each sentence. In their chapter, Samuelsson and Wirén present a range of techniques that can be used to achieve this end. This area is probably the most well-established in the field of NLP, enabling the authors here to provide an inventory of basic concepts in parsing, followed by a detailed catalog of parsing techniques that have been explored in the literature. Samuelsson and Wirén end by pointing to some of the issues that arise in applying these techniques to real data, with the concomitant problems of undergeneration (where one's grammar is unable to provide an analysis for a specific sentence) and ambiguity (where multiple analyses are provided).

D. Chapter 5: Semantic Analysis

Identifying the underlying syntactic structure of a sequence of words is only one step in determining the meaning of a sentence; it provides a structured object that is more amenable to further manipulation and subsequent interpretation. It is these subsequent steps that derive a meaning for the sentence in question. Massimo Poesio's chapter turns to these deeper issues. It is here that we begin to reach the bounds of what has managed to move from the research laboratory to practical application. As pointed out earlier in this introduction, the semantics of natural language have been less studied than syntactic issues, and so the techniques described here are not yet developed to the extent that they can easily be applied in a broad-coverage fashion.

After setting the scene by identifying some of the issues that arise in semantic interpretation, Poesio begins with a brief primer on the variety of knowledge representation formalisms that have been used in semantic interpretation. He then goes on to show how semantic interpretation can be integrated with the earlier stages of linguistic processing, relying on some of the machinery introduced in Samuelsson and Wirén's chapter. A substantial proportion of the chapter is given over to a discussion of how semantic interpretation plays a role in one particularly ubiquitous problem in language processing: that of the interpretation of anaphoric elements, such as pronouns and definite noun phrases, the meaning of which relies on an interpretation of prior elements of a text.

E. Chapter 6: Discourse Structure and Intention Recognition

The chapter by Lochbaum, Grosz, and Sidner picks up where Poesio's leaves off. Just as a sentence is more than a linear sequence of words, a discourse is more than a linear sequence of sentences. The relations between the meanings of individual sentences mean that discourses can have quite complex hierarchical structures; and these structures are closely tied to the intentions of the speaker or speakers in creating the resulting discourse.

Lochbaum, Grosz, and Sidner consider these two interrelated aspects of discourse processing, showing how the speaker's intentions play a role in determining the structure of the discourse, and how it is important to recognise the speaker's intentions when attempting to determine the role of an individual utterance in a larger discourse. The chapter first surveys a number of approaches to analysing the structure of discourse, and then goes on to look at what is involved in recognising the intention or intentions underlying a speaker's utterance. These two aspects are then drawn together using the notion of SharedPlans.

The techniques presented in this chapter are currently beyond what we can reasonably achieve in the broad-coverage analysis of unrestricted text; however, they can serve a useful purpose in restricted domains of discourse, and their discussion raises awareness of the issues that can have ramifications for shallower approaches to multisentential text processing.

F. Chapter 7: Natural Language Generation

At the end of the day, determining the speaker's intentions underlying an utterance is only really one-half of the story of natural language processing: in many situations, a response then needs to be generated, either in natural language alone or in combination with other modalities. For many of today's applications, what is required here is rather trivial and can be handled by means of canned responses; increasingly, however, we are seeing natural language generation techniques applied in the context of more sophisticated back-end systems, where the need to be able to custom-create fluent multisentential texts on demand becomes a priority. Chapters 11 and 12 bear testimony to the scope here.

In Chap. 7, David McDonald provides a far reaching survey of work in the field of natural language generation. McDonald begins by lucidly characterising the differences between natural language analysis and natural language generation. He goes on to show what can be achieved using natural language generation techniques, drawing examples from systems developed over the last 25 years. The bulk of the chapter is then concerned with laying out a picture of the component processes and representations required to generate fluent multisentential or multiparagraph texts, built around the now-standard distinction between text planning and linguistic realisation.

3. APPLICATIONS

In one of the first widely available published reports on the scope for practical applications of natural language processing, the Ovum report on *Natural Language Computing: The Commercial Applications* (Johnson, 1985) identified a number of specific application types:

- Text critiquing
- Natural language database query
- Message understanding
- Machine translation
- Speech recognition

At the time, enthusiasm for the technology was high. With the benefit of hindsight, it is clear that the predictions made at that time regarding the size of the market that would develop have been undermined by slower progress in developing the technology than was then thought likely. However, significant advances have been made in the nearly 15 years since that report appeared; and, perhaps surprisingly, the perceived application types have remained relatively constant in broad terms. An application area not predicted by the Ovum report is that of automatic text generation: this has arisen as an area of significant promise because of the scope for dynamic document tailoring on the World Wide Web and the scope for the use of large existing databases of information as an input to a text generation process.

In this second half of this part of the handbook, we present a number of reports on working NLP systems from these application areas.

A. Chapter 8: Intelligent Writing Assistance

In this chapter, George Heidorn describes the workings of the grammar checker first introduced in Microsoft Word 97, probably the first natural language processing application to reach desktops in the millions. Heidorn goes through the various steps of processing used in the grammar checker in considerable detail, demonstrating how issues discussed in Chaps. 2–4, in particular, have a bearing in constructing a real-world application of broad-coverage NLP technology. The chapter also contains useful and interesting details on a variety of issues in the development and testing of the grammar-checking technology; the chapter is a valuable look behind the scenes at what it takes to make a piece of successful commercial-strength NLP technology.

B. Chapter 9: Database Interfaces

The idea of being able to replace the manual construction of complex database queries in SQL or some other query language has long held an appeal for researchers in natural language processing; this chapter provides an extensive survey of the issues involved in building such systems.

After sketching some history, Androutsopoulos and Ritchie present a detailed analysis of the architecture of natural language database interfaces (NLDBIs). The model developed here draws on ideas presented in Chaps. 2–5. The authors go on to address the important question of the portability of these kinds of systems—we want to be able to avoid having to rebuild from scratch for each new database—and then consider a number of more advanced issues that arise in trying to build truly flexible and robust systems. The chapter ends with a discussion of variants on true natural language interfaces, where controlled languages or menu-based interfaces are used.

C. Chapter 10: Information Extraction

An area of significant interest in the late 1980s and the 1990s has been that commonly referred to as INFORMATION EXTRACTION or MESSAGE UNDERSTANDING. In opposition to the more traditionally conceived task of natural language processing, where the aim is to understand a complete text to a sophisticated degree of analysis, work in this area adopts a much simpler notion of what it is to extract meaning from a text. Specifically, if we know we have a collection of texts that are of essentially the same kind—for example, they may all be reports about terrorist incidents in some part of the world—then in many contexts it can be useful to extract from each of these texts a predefined set of elements; in essence, who did what to whom, when, and where.

The chapter by Cowie and Wilks provides an overview of the techniques involved in building such systems. The authors also provide some views on the role of this kind of work in the field of NLP as a whole.

D. Chapter 11: The Generation of Reports from Databases

In the first of two chapters focussing on practical applications of natural language generation technology, Richard Kittredge and Alain Polguère look at how textual reports can be automatically generated from databases. They begin by providing a useful characterisation of the properties a dataset needs to have to be amenable to the fruitful use of this technology; they then go on to describe the kinds of knowledge that need to be built into such systems, before describing a large number of example systems that carry out report generation. The second half of their chapter focusses on the computational techniques required for report generation, providing a usefully detailed analysis of many of the core ideas introduced in McDonald's earlier chapter (see Chap. 7). The chapter ends by contrasting report generation from databases with the generation of text from richer knowledge bases.

E. Chapter 12: The Generation of Multimedia Presentations

Text is often not the best way to convey information: as the oft-cited adage reminds us, 'a picture is worth a thousand words.' Generally, however, those pictures need to be supported by words, or words need to be supported by pictures; neither alone is sufficient.

Following on from the previous chapter, Elisabeth André describes work that aims to generate documents that contain both text and graphic elements. After presenting a general architecture for the development of such systems, André first shows how ideas from notions of discourse structure developed in linguistics can be applied to the generation of documents containing pictoral elements; she then discusses a variety of issues that arise in building real working systems that integrate multiple modalities.

F. Chapter 13: Machine Translation

Machine translation is the oldest application of natural language processing technology, having been around for 50 years. Appropriately, Harold Somers begins this chapter by presenting a historical overview of the field. He then shows how various problems in natural language processing more generally manifest themselves in machine translation; the discussion here echoes points made in Chaps. 3–5 of this book. Somers goes on to describe the two principal approaches taken to the construction of machine translation systems: interlingua-based approaches, in which an intermediate language-independent representational level is used, and transfer approaches, in which mappings are carried out more directly between language pairs. The chapter ends by discussing the user's perspective on the technology, highlighting how a consideration of how the technology can actually be used in practice has an influence on how we build systems.

G. Chapter 14: Dialogue Systems: From Theory to Practice in TRAINS-96

Full-blown natural language dialogue systems are probably the ultimate dream of many of those working in natural language processing. Allen and his colleagues describe here a system that takes advantage, in one way or another, of all the techniques described in the first half of this part of the handbook: the TRAINS system, an end-to-end spoken dialogue system, interacts with the user to solve route planning problems. As well as using the symbolic approaches to syntax, semantics, and pragmatics, outlined earlier, the TRAINS system also makes use of some of the statistical techniques outlined in the second part of this handbook; and contrary to our focus in this book on the processing of text, the TRAINS system takes spoken input and produces spoken output. It is thus an excellent example of what can be achieved by integrating the results of the last several decades of research into NLP; although there are many outstanding issues to be addressed in scaling-up a system of the sophistication of TRAINS to a real-world application, the system serves as a demonstration of where we can expect developments to take us in the next 5–10 years.

4. FURTHER READING

Each of the authors of the chapters presented here has provided an extensive bibliography for material relevant to their specific subject matter. We provide here some pointers to more general background literature in symbolic approaches to natural language processing.

At the time of writing, the best widely available introduction to techniques in natural language understanding is James Allen's book (Allen, 1995); another excellent starting point is Gazdar and Mellish's book (Gazdar and Mellish, 1989). Reiter and Dale (2000) provide an extensive introduction to the issues involved in building natural language generation systems.

For a broad overview of seminal research in the field, see Grosz et al. (1986); although now a little dated, this collects together a large number of papers that were influential in setting directions for the field. A more recent collection of papers can be found in Pereira and Grosz (1993). Work in speech recognition is not addressed in this handbook; but see Waibel and Lee (1990) for a useful collection of important papers. For a broad overview of natural language processing that covers the territory exhaustively without going into technical detail, see the Cambridge Human Language Technology Survey (Cole et al., 1998).

REFERENCES

Allen J. 1995. Natural Language Understanding. 2nd ed. Redwood City, CA: Benjamin Cummings.

Cole R, Mariani J, Uszkoreit H, Varile GB, Zaenen A, Zampolli A. 1998. Survey of the State of the Art in Human Language Technology. Cambridge, UK: Cambridge University Press.

Gazdar G, Mellish C. 1989. Natural Language Processing in Lisp: An Introduction to Computational Linguistics. Wokingham, UK: Addison–Wesley.

Grosz BJ, Sparck Jones K, Webber B, eds. 1986. Readings in Natural Language Processing. San Francisco, CA: Morgan-Kaufmann.

Johnson T. 1985. Natural Language Computing: The Commercial Applications. London: Ovum Ltd.

Pereira F, Grosz BJ, eds. 1993. Artif Intell 63(1–2). Special Issue on Natural Language Processing.

Reiter E, Dale R. 2000. Building Natural Language Generation Systems. Cambridge, UK: Cambridge University Press.

Waibel A, Lee K-F. 1990. Readings in Speech Recognition. San Francisco, CA: Morgan Kaufmann.

2

Tokenisation and Sentence Segmentation

DAVID D. PALMER

The MITRE Corporation, Bedford, Massachusetts

1. INTRODUCTION

In linguistic analysis of a natural language text, it is necessary to clearly define what constitutes a word and a sentence. Defining these units presents different challenges depending on the language being processed, but neither task is trivial, especially when considering the variety of human languages and writing systems. Natural languages contain inherent ambiguities, and much of the challenge of Natural Language Processing (NLP) involves resolving these ambiguities.

In this chapter we will discuss the challenges posed by **text segmentation**, the task of dividing a text into linguistically meaningful units—at the lowest level characters representing the individual graphemes in a language's written system, words consisting of one or more characters, and sentences consisting of one or more words. Text segmentation is a frequently overlooked yet essential part of any NLP system, because the words and sentences identified at this stage are the fundamental units passed to further processing stages, such as morphological analyzers, part-of-speech taggers, parsers, and information retrieval systems.

Tokenisation is the process of breaking up the sequence of characters in a text by locating the **word boundaries**, the points where one word ends and another begins. For computational linguistics purposes, the words thus identified are frequently referred to as **tokens**. In written languages where no word boundaries are explicitly marked in the writing system, tokenisation is also known as **word segmentation**, and this term is frequently used synonymously with tokenisation.

Sentence segmentation is the process of determining the longer processing units consisting of one or more words. This task involves identifying **sentence boundaries** between words in different sentences. Since most written languages have punctuation marks that occur at sentence boundaries, sentence segmentation is frequently referred to as **sentence boundary detection**, **sentence boundary disambiguation**, or **sentence boundary recognition**. All these terms refer to the same task: determining how a text should be divided into sentences for further processing.

In practice, sentence and word segmentation cannot be performed successfully independent from one another. For example, an essential subtask in both word and sentence segmentation for English is identifying abbreviations, because a period can be used in English to mark an abbreviation as well as to mark the end of a sentence. When a period marks an abbreviation, the period is usually considered a part of the abbreviation token, whereas a period at the end of a sentence is usually considered a token in and of itself. Tokenising abbreviations is complicated further when an abbreviation occurs at the end of a sentence, and the period marks *both* the abbreviation and the sentence boundary.

This chapter will provide an introduction to word and sentence segmentation in a variety of languages. We will begin in Sec. 2 with a discussion of the challenges posed by text segmentation, and emphasize the issues that must be considered before implementing a tokenisation or sentence segmentation algorithm. The section will describe the dependency of text segmentation algorithms on the language being processed and the character set in which the language is encoded. It will also discuss the dependency on the application that uses the output of the segmentation and the dependency on the characteristics of the specific corpus being processed.

In Sec. 3 we will introduce some common techniques currently used for tokenisation. The first part of the section will focus on issues that arise in tokenising languages in which words are separated by whitespace. The second part of the section will discuss tokenisation techniques in languages for which no such whitespace word boundaries exist. In Sec. 4 we will discuss the problem of sentence segmentation and introduce some common techniques currently used to identify sentences boundaries in texts.

2. WHY IS TEXT SEGMENTATION CHALLENGING?

There are many issues that arise in tokenisation and sentence segmentation that need to be addressed when designing NLP systems. In our discussion of tokenisation and sentence segmentation, we will emphasize the main types of dependencies that must be addressed in developing algorithms for text segmentation: **language dependence** (Sec. 2.A), **character-set dependence** (Sec. 2.B), **application dependence** (Sec. C), and **corpus dependence** (Sec. 2.D). In Sec. 2.E, we will briefly discuss the evaluation of the text segmentation algorithms.

A. Language Dependence

We focus here on written language, rather than transcriptions of spoken language. This is an important distinction, for it limits the discussion to those languages that have established writing systems. Because every human society uses language, there are thousands of distinct languages and dialects; yet only a small minority of the languages and dialects currently have a system of visual symbols to represent the elements of the language.

Just as the spoken languages of the world contain a multitude of different features, the written adaptations of languages have a diverse set of features. Writing systems can be **logographic**, in which a large number (often thousands) of individual symbols represent words. In contrast, writing systems can be **syllabic**, in which individual symbols represent syllables, or **alphabetic**, in which individual symbols (more or less) represent sounds; unlike logographic systems, syllabic and alphabetic systems typically have fewer than 100 symbols. In practice, no modern writing system employs symbols of only one kind, so no natural language writing system can be classified as purely logographic, syllabic, or alphabetic. Even English, with its relatively simple writing system based on the Roman alphabet, utilizes logographic symbols including Arabic numerals (0–9), currency symbols ($, £), and other symbols (%, &, #). English is nevertheless predominately alphabetic, and most other writing systems comprise symbols that are mainly of one type.

In addition to the variety of symbol types used in writing systems, there is a range of orthographic conventions used in written languages to denote the boundaries between linguistic units such as syllables, words, or sentences. In many written Amharic texts, for example, both word and sentence boundaries are explicitly marked, whereas in written Thai texts neither is marked. In the latter example in which no boundaries are explicitly indicated in the written language, written Thai is similar to spoken language, for which there are no explicit boundaries and few cues to indicate segments at any level. Between the two extremes are languages that mark boundaries to different degrees. English employs whitespace between most words and punctuation marks at sentence boundaries, but neither feature is sufficient to segment the text completely and unambiguously. Tibetan and Korean both explicitly mark syllable boundaries, either through layout or by punctuation, but neither marks word boundaries. Written Chinese and Japanese have adopted punctuation marks for sentence boundaries, but neither denotes word boundaries. For a very thorough description of the various writing systems employed to represent natural languages, including detailed examples of all languages and features discussed in this chapter, we recommend Daniels and Bright (1996).

The wide range of writing systems used by the languages of the world result in language-specific as well as orthography-specific features that must be taken into account for successful text segmentation. The first essential step in text segmentation is thus to understand the writing system for the language being processed. In this chapter we will provide general techniques applicable to a variety of different writing systems. As many segmentation issues are language-specific, we will also highlight the challenges faced by robust, broad-coverage tokenisation efforts.

B. Character-Set Dependence

At its lowest level, a computer-based text or document is merely a sequence of digital bits stored in a computer's memory. The first essential task is to interpret these bits as characters of a writing system of a natural language. Historically, this was trivial, because nearly all texts were encoded in the 7-bit ASCII character set, which allowed only 128 characters and included only the Roman alphabet and essential characters for writing English. This limitation required the "asciification" or "romanization" of many texts, in which ASCII equivalents were defined for characters not defined in the character set. An example of this asciifiation is the adaptation of many European languages containing umlauts and accents, in which the umlauts are replaced by a double quotation mark or the letter **e** and accents are denoted by a single quotation mark or even a number code. In this system, the German word *über* would be written as *u''ber* or *ueber*, and the French word *déjà* would be written *de'ja'* or *de1ja2*. Languages less similar to English, such as Russian and Hebrew, required much more elaborate romanization systems, to allow ASCII processing. In fact, such adaptations are still common because many current e-mail systems are limited to this 7-bit encoding.

Extended character sets have been defined to allow 8-bit encodings, but an 8-bit computer byte can still encode just 256 distinct characters, and there are tens of thousands of distinct characters in all the writing systems of the world. Consequently, characters in different languages are currently encoded in a large number of overlapping character sets. Most alphabetic and syllabic writing systems can be encoded in a single byte, but larger character sets, such as those of written Chinese and Japanese, which have several thousand distinct characters, require a two-byte system in which a single character is represented by a pair of 8-bit bytes. It is further complicated by the fact that multiple encodings currently exist for the same character set; for example, the Chinese character set is encoded in two widely used variants: GB and Big-5.[1] The fact that the same range of numeric values represents different characters in different encodings can be a problem for tokenisation, because tokenisers are usually targeted to a specific language in a specific encoding. For example, a tokeniser for a text in English or Spanish, which are normally stored in the common encoding Latin-1 (or ISO 8859-1), would need to be aware that bytes in the (decimal) range 161–191 in Latin-1 represent punctuation marks and other symbols (such as '¡', '¿', '£', and '©'); tokenisation rules would be required to handle each symbol (and thus its byte code) appropriately for that language. However, the same byte range 161–191 in the common Thai alphabet encoding TIS620 corresponds to a set of Thai consonants, which would naturally be treated differently from punctuation or other symbols. A completely different set of tokenisation rules would be necessary to successfully tokenise a Thai text in TIS620 owing to the use of overlapping character ranges.[2]

Determining characters in two-byte encodings involves locating pairs of bytes representing a single character. This process can be complicated by the tokenisation equivalent of code-switching, in which characters from many different writing systems occur within the same text. It is very common in texts to encounter multiple writing systems and thus multiple encodings. In Chinese and Japanese newswire texts, which are usually encoded in a two-byte system, one byte (usually Latin-1) letters, spaces, punctuation marks (e.g., periods, quotation marks, and parentheses), and Arabic numerals are commonly interspersed with the Chinese and Japanese characters. Such texts also frequently contain Latin-1 SGML headers.

---

[1] The Unicode standard (Consortium, 1996) specifies a single 2-byte encoding system that includes 38,885 distinct coded characters derived from 25 supported scripts. However, until Unicode becomes universally accepted and implemented, the many different character sets in use will continue to be a problem.

[2] Actually, due to the nature of written Thai, there are more serious issues than character set encoding which prevent the use of an English tokeniser on Thai texts, and we will discuss these issues in Sec. 3.B.

C. Application Dependence

Although word and sentence segmentation are necessary, in reality, there is no absolute definition for what constitutes a word or a sentence. Both are relatively arbitrary distinctions that vary greatly across written languages. However, for the purposes of computational linguistics we need to define exactly what we need for further processing; usually, the language and task at hand determine the necessary conventions. For example, the English words *I am* are frequently contracted to *I'm*, and a tokeniser frequently expands the contraction to recover the essential grammatical features of the pronoun and verb. A tokeniser that does not expand this contraction to the component words would pass the single token *I'm* to later processing stages. Unless these processors, which may include morphological analysers, part-of-speech taggers, lexical lookup routines, or parsers, were aware of both the contracted and uncontracted forms, the token may be treated as an unknown word.

Another example of the dependence of tokenisation output on later-processing stages is the treatment of the English possessive *'s* in various tagged corpora.[3] In the Brown corpus (Francis and Kucera, 1982), the word *governor's* is considered one token and is tagged as a possessive noun. In the Susanne corpus (Sampson, 1995), on the other hand, the same word is treated as two tokens, *governor* and *'s*, tagged singular noun and possessive, respectively.

Because of this essential dependence, the tasks of word and sentence segmentation overlap with the techniques discussed in other chapters in this handbook: *Lexical Analysis* in Chap. 3, *Parsing Techniques* in Chap. 4, and *Semantic Analysis* in Chap. 5, as well as the practical applications discussed in later chapters.

D. Corpus Dependence

Until recently, the problem of robustness was rarely addressed by NLP systems, which normally could process only well-formed input conforming to their hand-built grammars. The increasing availability of large corpora in multiple languages that encompass a wide range of data types (e.g., newswire texts, e-mail messages, closed captioning data, optical character recognition [OCR] data, multimedia documents) has required the development of robust NLP approaches, as these corpora frequently contain misspellings, erratic punctuation and spacing, and other irregular features. It has become increasingly clear that algorithms that rely on input texts to be well formed are much less successful on these different types of texts.

Similarly, algorithms that expect a corpus to follow a set of conventions for a written language are frequently not robust enough to handle a variety of corpora. It is notoriously difficult to prescribe rules governing the use of a written language; it is even more difficult to get people to "follow the rules." This is in large part due to the nature of written language, in which the conventions are determined by publishing houses and national academies and are arbitrarily subject to change.[4] So, although punctuation roughly corresponds to the use of suprasegmental features in spoken

[3] This example is taken from Grefenstette and Tapanainen (1994).

[4] This is evidenced by the numerous spelling "reforms" that have taken place in various languages, most recently in the German language, or the attempts by governments (such as the French) to "purify" the language or to outlaw foreign words.

language, reliance on well-formed sentences delimited by predictable punctuation can be very problematic. In many corpora, traditional prescriptive rules are commonly ignored. This fact is particularly important to our discussion of both word and sentence segmentation, which largely depend on the regularity of spacing and punctuation. Most existing segmentation algorithms for natural languages are both language-specific and corpus-dependent, developed to handle the predictable ambiguities in a well-formed text. Depending on the origin and purpose of a text, capitalization and punctuation rules may be followed very closely (as in most works of literature), erratically (as in various newspaper texts), or not at all (as in e-mail messages or closed captioning). For example, an area of current research is summarization, filtering, and sorting techniques for e-mail messages and usenet news articles. Corpora containing e-mail and usenet articles can be ill-formed, such as Example {1}, an actual posting to a usenet newsgroup, which shows the erratic use of capitalization and punctuation, "creative" spelling, and domain-specific terminology inherent in such texts.

ive just loaded pcl onto my akcl. when i do an 'in-package' to load pcl, ill get the prompt but im not able to use functions like defclass, etc... is there womething basic im missing or am i just left hanging, twisting in the breeze? {1}

Many online corpora, such as those from OCR or handwriting recognition, contain substitutions, insertions, and deletions of characters and words, which affect both tokenisation and sentence segmentation. Example {2} shows a line taken from a corpus of OCR data in which large portions of several sentences (including the punctuation) were clearly elided by the OCR algorithm. In this extreme case, even accurate sentence segmentation would not allow for full processing of the partial sentences that remain.

newsprint. Furthermore, shoe presses have Using rock for granite roll Two years ago we reported on advances in {2}

Robust text segmentation algorithms designed for use with such corpora must, therefore, have the capability to handle the range of irregularities that distinguish these texts from well-formed newswire documents frequently processed.

E. Evaluation of Text Segmentation Algorithms

Because of the dependencies discussed in the foregoing, evaluation and comparison of text segmentation algorithms is very difficult. Owing to the variety of corpora for which the algorithms are developed, an algorithm that performs very well on a specific corpus may not be successful on another corpus. Certainly, an algorithm fine-tuned for a particular language will most likely be completely inadequate for processing another language. Nevertheless, evaluating algorithms provides important information about their efficacy, and there are common measures of performance for both tokenisation and sentence segmentation.

The performance of word segmentation algorithms is usually measured using *recall* and *precision*, where recall is defined as the percent of words in the manually segmented text identified by the segmentation algorithm, and precision is defined as the percentage of words returned by the algorithm that also occurred in the hand-segmented text in the same position. For a language such as English, in which a great deal of the initial tokenisation can be done by relying on whitespace, tokenisation is rarely scored. However, for unsegmented languages (see Sec. 3.B), evaluation of word segmentation is critical for improving all aspects of NLP systems.

A sentence segmentation algorithm's performance is usually reported in terms of a single score equal to the number of punctuation marks correctly classified divided by the total number of punctuation marks. It is also possible to evaluate sentence segmentation algorithms using recall and precision, given the numbers of **false positives**, punctuation marks erroneously labeled as a sentence boundary, and **false negatives**, actual sentence boundaries not labeled as such.

3. TOKENISATION

Section 2 discussed the many challenges inherent in segmenting freely occurring text. In this section we will focus on the specific technical issues that arise in tokenisation.

Tokenisation is well established and well understood for artificial languages such as programming languages.[5] However, such artificial languages can be strictly defined to eliminate lexical and structural ambiguities; we do not have this luxury with natural languages, in which the same character can serve many different purposes and in which the syntax is not strictly defined. Many factors can affect the difficulty of tokenising a particular natural language. One fundamental difference exists between tokenisation approaches for **space-delimited** languages and approaches for **unsegmented** languages. In space-delimited languages, such as most European languages, some word boundaries are indicated by the insertion of whitespace. The character sequences delimited are not necessarily the tokens required for further processing, however, so there are still many issues to resolve in tokenisation. In unsegmented languages, such as Chinese and Thai, words are written in succession with no indication of word boundaries. Tokenisation of unsegmented languages, therefore, requires additional lexical and morphological information.

In both unsegmented and space-delimited languages, the specific challenges posed by tokenisation are largely dependent on both the writing system (logographic, syllabic, or alphabetic, as discussed in Sec. 2.A) and the typographical structure of the words. There are three main categories into which word structures can be placed,[6] and each category exists in both unsegmented and space-delimited writing systems. The morphology of words in a language can be **isolating**, in which words do not divide into smaller units; **agglutinating** (or **agglutinative**), in which words divide into smaller units (morphemes) with clear boundaries between the morphemes; or **inflectional**, in which the boundaries between morphemes are not clear and the component morphemes can express more than one grammatical meaning. Although individual languages show tendencies toward one specific type (e.g., Mandarin Chinese is predominantly isolating, Japanese is strongly agglutinative, and Latin is largely inflectional), most languages exhibit traces of all three.[7]

---

[5] For a thorough introduction to the basic techniques of tokenisation in programming languages, see Aho et al. (1986).

[6] This classification comes from Comrie et al. (1996) and Crystal (1987).

[7] A fourth typological classification frequently studied by linguists, **polysynthetic**, can be considered an extreme case of agglutinative, where several morphemes are put together to form complex words that can function as a whole sentence. Chukchi and Inuit are examples of polysynthetic languages.

Because the techniques used in tokenising space-deliminated languages are very different from those used in tokenising unsegmented languages, we will discuss the techniques separately, in Sects. 3.A and 3.B, respectively.

A. Tokenisation in Space-Delimited Languages

In many alphabetic writing systems, including those that use the Latin alphabet, words are separated by whitespace. Yet even in a well-formed corpus of sentences, there are many issues to resolve in tokenisation. Most tokenisation ambiguity exists among uses of punctuation marks, such as periods, commas, quotation marks, apostrophes, and hyphens, since the same punctuation mark can serve many different functions in a single sentence, let alone a single text. Consider example sentence {3} from the *Wall Street Journal* (1988).

> *Clairson International Corp. said it expects to report a net loss for its second quarter ended March 26 and doesn't expect to meet analysts' profit estimates of $3.9 to $4 million, or 76 cents a share to 79 cents a share, for its year ending Sept. 24.* {3}

This sentence has several items of interest that are common for Latinate, alphabetic, space-delimited languages. First, it uses periods in three different ways—within numbers as a decimal point (*$3.9*), to mark abbreviations (*Corp.* and *Sept.*), and to mark the end of the sentence, in which the period following the number *24* is not a decimal point. The sentence uses apostrophes in two ways—to mark the genitive case (where the apostrophe denotes possession) in *analysts'* and to show contractions (places where letters have been left out of words) in *doesn't*. The tokeniser must thus be aware of the uses of punctuation marks and be able to determine when a punctuation mark is part of another token and when it is a separate token.

In addition to resolving these cases, we must make tokenisation decisions about a phrase such as *76 cents a share*, which on the surface consists of four tokens. However, when used adjectivally such as in the phrase *a 76-cents-a-share dividend*, it is normally hyphenated and appears as one. The semantic content is the same despite the orthographic differences, so it makes sense to treat the two identically, as the same number of tokens. Similarly, we must decide whether to treat the phrase *$3.9 to $4 million* differently than if it had been written *3.9 to 4 million dollars* or *$3,900,000 to $4,000,000*. We will discuss these ambiguities and other issues in the following sections.

A logical initial tokenisation of a space-delimited language would be to consider as a separate token any sequence of characters preceded and followed by space. This successfully tokenises words that are a sequence of alphabetic characters, but does not take into account punctuation characters. Frequently, characters such as commas, semicolons, and periods, should be treated as separate tokens, although they are not preceded by whitespace (such as with the comma after *$4 million* in Example {3}). Additionally, many texts contain certain classes of character sequences that should be filtered out before actual tokenisation; these include existing markup and headers (including SGML markup), extra whitespace, and extraneous control characters.

1. Tokenising Punctuation

Although punctuation characters are usually treated as separate tokens, there are many cases when they should be "attached" to another token. The specific cases vary from one language to the next, and the specific treatment of the punctuation characters need to be enumerated within the tokeniser for each language. In this section we will give examples of English tokenisation.

Abbreviations are used in written language to denote the shortened form of a word. In many cases abbreviations are written as a sequence of characters terminated with a period. For this reason, recognizing abbreviations is essential for both tokenisation and sentence segmentation, because abbreviations can occur at the end of a sentence, in which case the period serves two purposes. Compiling a list of abbreviations can help in recognizing them, but abbreviations are productive, and it is not possible to compile an exhaustive list of all abbreviations in any language. Additionally, many abbreviations can also occur as words elsewhere in a text (e.g., the word *Mass* is also the abbreviation for *Massachusetts*). An abbreviation can also represent several different words, as is the case for *St.* which can stand for *Saint*, *Street*, or *State*. However, as *Saint* it is less likely to occur at a sentence boundary than *Street* or *State*. Examples {4} and {5} from the *Wall Street Journal* (1991 and 1987, respectively) demonstrate the difficulties produced by such ambiguous cases, where the same abbreviation can represent different words and can occur both within and at the end of a sentence.

The contemporary viewer may simply ogle the vast wooded vistas rising up from the Saguenay River and Lac St. Jean, standing in for the St. Lawrence River. {4}

The firm said it plans to sublease its current headquarters at 55 Water St. A spokesman declined to elaborate. {5}

Recognizing an abbreviation is thus not sufficient for complete tokenisation, and we will discuss abbreviations at sentence boundaries fully in Sec. 4.B.

Quotation marks and apostrophes (" " ') are a major source of tokenisation ambiguity. Usually, single and double quotes indicate a quoted passage, and the extent of the tokenisation decision is to determine whether they open or close the passage. In many character sets, single quote and apostrophe are the same character; therefore, it is not always possible to immediately determine if the single quotation mark closes a quoted passage, or serves another purpose as an apostrophe. In addition, as discussed in Sec. 2.B, quotation marks are also commonly used when "romanizing" writing systems, in which umlauts are replaced by a double quotation mark and accents are denoted by a single quotation mark or apostrophe.

The apostrophe is a very ambiguous character. In English, the main uses of apostrophes are to mark the genitive form of a noun, to mark contractions, and to mark certain plural forms. In the genitive case, some applications require a separate token while some require a single token, as discussed in Sec. 2.C. How to treat the genitive case is important, as in other languages, the possessive form of a word is not marked with an apostrophe and cannot be as readily recognized. In German, for example, the possessive form of a noun is usually formed by adding the letter *s* to the word, without an apostrophe, as in *Peters Kopf* (*Peter's head*). However, in modern (informal) usage in German, *Peter's Kopf* would also be common; the apostrophe is also frequently omitted in modern (informal) English such that *Peters head* is a possible construction. Further, in English, *'s* also serves as a contraction for the verb *is*, as in *he's*, *it's*, and *she's*, as well as the plural form of some words, such as *I.D.'s* or *1980's* (although the apostrophe is also frequently omitted from such plurals). The tokenisation decision in these cases is context-dependent and is closely tied to syntactic analysis.

In the case of apostrophe as contraction, tokenisation may require the expansion of the word to eliminate the apostrophe, but the cases for which this is necessary are very language-dependent. The English contraction *I'm* could be tokenised as the two words *I am*, and *we've* could become *we have*. And depending on the application, *wouldn't* probably would expand to *would not*, although it has been argued by Zwicky and Pullum (1981) that the ending *n't* is an inflectional marker, rather than a clitic. Written French contains a completely different set of contractions, including contracted articles (*l'homme, c'etait*), as well as contracted pronouns (*j'ai, je l'ai*) and other forms such as *n'y*, *qu'ils*, *d'ailleurs*, and *aujourd'hui*. Clearly, recognizing the contractions to expand requires knowledge of the language, and the specific contractions to expand, as well as the expanded forms, must be enumerated. All other word-internal apostrophes are treated as a part of the token and not expanded, which allows the proper tokenisation of multiply contracted words such as *fo'c's'le* and *Pudd'n'head* as single words. In addition, because contractions are not always demarcated with apostrophes, as in the French *du*, which is a contraction of *de la*, or the Spanish del, contraction of *de el*, other words to expand must also be listed in the tokeniser.

2. *Multipart Words*

To different degrees, many written languages contain space-delimited words composed of multiple units, each expressing a particular grammatical meaning. For example, the single Turkish word *çöplüklerimizdekilerdenmiydi* means "was it from those that were in our garbate cans?"[8] This type of construction is particularly common in strongly agglutinative languages such as Swahili, Quechua, and most Altaic languages. It is also common in languages such as German, where noun–noun (*Lebensversicherung*, life insurance), adverb–noun (*Nichtraucher*, nonsmoker), and preposition–noun (*Nachkriegszeit*, postwar period) compounding are all possible. In fact, although it is not an agglutinative language, German compounding can be quite complex, as in *Feuerundlebensversicherung* (fire and life insurance) or *Kundenzufriedenheitsabfragen* (customer satisfaction survey).

To some extent, agglutinating constructions are present in nearly all languages, although this compounding can be marked by hyphenation, in which the use of hyphens can create a single word with multiple grammatical parts. In English it is commonly used to create single-token words such as *end-of-line*, as well as multi-token words, such as *Boston-based*. As with the apostrophe, the use of the hyphen is not uniform; for example, hyphen usage varies greatly between British and American English, as well as between different languages. However, as with apostrophes as

---

[8] This example is from Hankamer (1986).

contractions, many common language-specific uses of hyphens can be enumerated in the tokeniser.

Many languages use the hyphen to create essential grammatical structures. In French, for example, hyphenated compounds such as *va-t-il*, *c'est-à-dire*, and *celui-ci* need to be expanded during tokenisation, to recover necessary grammatical features of the sentence. In these cases, the tokeniser needs to contain an enumerated list of structures to be expanded, as with the contractions, discussed previously.

Another tokenisation difficulty involving hyphens stems from the practice, common in traditional typesetting, of using hyphens at the ends of lines to break a word too long to include on one line. Such end-of-line hyphens can thus occur within words that are not normally hyphenated. Removing these hyphens is necessary during tokenisation, yet it is difficult to distinguish between such incidental hyphenation and cases where naturally hyphenated words happen to occur at a line break. In an attempt to dehyphenate the artificial cases, it is possible to incorrectly remove necessary hyphens. Grefenstette and Tapanainen (1994) found that nearly 5% of the end-of-line hyphens in an English corpus were word-internal hyphens which happened to also occur as end-of-line hyphens.

In tokenising multipart words, such as hyphenated or agglutinative words, whitespace does not provide much useful information to further processing stages. In such cases, the problem of tokenisation is very closely related both to tokenisation in unsegmented languages, discussed in Sec. 3.B of this chapter, and to lexical analysis, discussed in Chap. 3.

3. *Multiword Expressions*

Spacing conventions in written languages do not always correspond to the desired tokenisation. For example, the three-word English expression *in spite of* is, for all intents and purposes, equivalent to the single word *despite*, and both could be treated as a single token. Similarly, many common English expressions, such as *au pair*, *de facto*, and *joie de vivre*, consist of foreign loan words that can be treated as a single token.

Multiword numerical expressions are also commonly identified in the tokenisation stage. Numbers are ubiquitous in all types of texts in every language. For most applications, sequences of digits and certain types of numerical expressions, such as dates and times, money expressions, and percents, can be treated as a single token. Several examples of such phrases can be seen in the foregoing Example {3}: *March 26*, *$3.9 to $4 million*, and *Sept. 24* could each be treated as a single token. Similarly, phrases such as *76 cents a share* and *$3-a-share* convey roughly the same meaning, despite the difference in hyphenation, and the tokeniser should normalize the two phrases to the same number of tokens (either one or four). Tokenising numeric expressions requires knowledge of the syntax of such expressions, as numerical expressions are written differently in different languages. Even within a language or in languages as similar as English and French, major differences exist in the syntax of numeric expressions, in addition to the obvious vocabulary differences. For example, the English date *November 18, 1989* could alternatively appear in English texts as any number of variations, such as *Nov. 18, 1989*, or *18 November 1989*, or *11/18/89*.

Closely related to hyphenation, treatment of multiword expressions is highly language-dependent and application-dependent, but can easily be handled in the tokenisation stage if necessary. We need to be careful, however, when combining words into a single token. The phrase *no one*, along with *noone* and *no-one*, is a commonly encountered English equivalent for *nobody* and should normally be treated as a single token. However, in a context such as *No one man can do it alone*, it needs to be treated as two words. The same is true of the two-word phrase *can not*, which is not always equivalent to the single word *cannot* or the contraction *can't*.[9] In such cases, it is safer to allow a later process (such as a parser) to make the decision.

4. A Flex Tokeniser

Tokenisation in space-delimited languages can usually be accomplished with a standard lexical analyzer, such as lex (Lesk and Schmidt, 1975) or flex (Nicol, 1993), or the UNIX tools awk, sed, or perl. Figure 1 shows an example of a basic English tokeniser written in flex, a lexical analyzer with a simple syntax that allows the user to write rules in a regular grammar. This flex example contains implementations of several issues discussed in the foregoing.

The first part of the tokeniser in Fig. 1 contains a series of character category definitions. For example, NUMBER_CHARACTER contains characters that may occur within a number, NUMBER_PREFIX contains characters that may occur at the beginning of a numeric expression, DIACRITIC_LETTER contains letters with diacritics (such as ñ), which are common in many European languages and may occur in some English words. Note that some category definitions contain octal byte codes (such as in NUMBER_PREFIX, where \243 and \245 represent the currency symbols for English pounds and Japanese yen) as discussed in Sec. 2.B.

The character categories defined in the first part of the tokeniser are combined in the second part in a series of rules explicitly defining the tokens to be output. The first rule, for example, states that if the tokeniser encounters an apostrophe followed by the letters 'v' and 'e', it should expand the contraction and output the word "have." The next rule states that any other sequence of letters following an apostrophe should (with the apostrophe) be tokenised separately and not expanded. Other rules in Fig. 1 show examples of tokenising multiword expressions (*au pair*), dates (*Jan. 23rd*), numeric expressions (*$4,347.12*), punctuation characters, abbreviations, and hyphenated words. Note that this basic tokeniser is by no means a complete tokeniser for English, but rather intended to show some simple examples of flex implementations of the issues discussed in previous sections.

B. Tokenisation in Unsegmented Languages

The nature of the tokenisation task in unsegmented languages, such as Chinese, Japanese, and Thai, is essentially different from tokenisation in space-delimited languages, such as English. The lack of any spaces between words necessitates a more informed approach than simple lexical analysis, and tools such as flex are not as successful. However, the specific approach to word segmentation for a particular unsegmented language is further limited by the writing system and orthography of the language, and a single general approach has not been developed. In Sec. 3.B.1, we will describe some general algorithms that have been

[9] For example, consider the following sentence: "Why is my soda can not where I left it?"

```
DIGIT                      [0-9]
NUMBER_CHARACTER           [0-9\.\,]
NUMBER_PREFIX              [\$\243\245]
NUMBER_SUFFIX              (\%|\242|th|st|rd)
ROMAN_LETTER               [a-zA-Z]
DIACRITIC_LETTER           [\300-\377]
INTERNAL_CHAR              [\-\/]
WORD_CHAR                  ({ROMAN_LETTER}|{DIACRITIC_LETTER})
ABBREVIATION               (mr|dr)
WHITESPACE                 [\040\t\n]
APOSTROPHE                 [\']
UNDEFINED                  [\200-\237]
PERIOD                     [\.]
SINGLE_CHARACTER           [\.\,\;\!\?\:\"\251]
MONTH                      (jan(uary)?|feb(ruary)?)

%%
{APOSTROPHE}(ve)           { printf("have\n", yytext); } ;
{APOSTROPHE}{ROMAN_LETTER}+    { printf("%s\n", yytext); } ;
(au){WHITESPACE}+(pair) { printf("au pair\n", yytext); } ;

{MONTH}{PERIOD}*{WHITESPACE}+{DIGIT}+{NUMBER_SUFFIX}*          |
{MONTH}{PERIOD}*{WHITESPACE}+{DIGIT}+[\,]{WHITESPACE}+{DIGIT}+ |
{NUMBER_PREFIX}*{NUMBER_CHARACTER}+{NUMBER_SUFFIX}*            |
{SINGLE_CHARACTER} | {ABBREVIATION}{PERIOD}                     |
{WORD_CHAR}+({INTERNAL_CHAR}{WORD_CHAR}+)*    { printf("%s\n",yytext); } ;

{WHITESPACE}+|{UNDEFINED}+                                   ;

%%
yywrap() {
  printf("\n");
  return(1);
```

Fig. 1 An (incomplete) basic flex tokeniser for English.

applied to the problem to obtain an initial approximation for a variety of languages. In Sec. 3.B.2, we will give details of some successful approaches to Chinese segmentation, and in Sec. 3.B.3, we will describe some approaches that have been applied to other languages.

1. *Common Approaches*

An extensive word list combined with an informed segmentation algorithm can help achieve a certain degree of accuracy in word segmentation, but the greatest barrier to accurate word segmentation is in recognizing unknown words, words not in the lexicon of the segmenter. This problem is dependent both on the source of the lexicon as well as the correspondence (in vocabulary) between the text in question and the lexicon. Wu and Fung (1994) reported that segmentation accuracy is significantly higher when the lexicon is constructed using the same type of corpus as the corpus on which it is tested.

Another obstacle to high-accuracy word segmentation is that there are no widely accepted guidelines for what constitutes a word; therefore, there is no agreement on how to "correctly" segment a text in an unsegmented language. Native speakers of a language do not always agree about the "correct" segmentation, and the same text could be segmented into several very different (and equally correct) sets of words by different native speakers. A simple example from English would be the hyphenated phrase *Boston-based*. If asked to "segment" this phrase into words, some native English speakers might say *Boston-based* is a single word and some might say *Boston* and *based* are two separate words; in this latter case there might also be disagreement about whether the hyphen "belongs" to one of the two words (and to which one) or whether it is a "word" by itself. Disagreement by native speakers of Chinese is much more prevalent; in fact, Sproat et al. (1996) give empirical results showing that native speakers of Chinese frequently agree on the correct segmentation less than 70% of the time. Such ambiguity in the definition of what constitutes a word makes it difficult to evaluate segmentation algorithms that follow different conventions, as it is nearly impossible to construct a "gold standard" against which to directly compare results.

A simple word segmentation algorithm is to consider each character a distinct word. This is practical for Chinese because the average word length is very short (usually between one and two characters, depending on the corpus[10]), and actual words can be recognized with this algorithm. Although it does not assist in tasks such as parsing, part-of-speech tagging, or text-to-speech systems (see Sproat et al., 1996), the character-as-word segmentation algorithm has been used to obtain good performance in Chinese information retrieval (Buckley et al., 1996), a task in which the words in a text play a major role in indexing.

A very common approach to word segmentation is to use a variation of the *maximum matching algorithm*, frequently referred to as the *greedy algorithm*. The greedy algorithm starts at the first character in a text and, using a word list for the language being segmented, attempts to find the longest word in the list starting with that character. If a word is found, the maximum-matching algorithm marks a boundary at the end of the longest word, then begins the same longest match search starting at the character following the match. If no match is found in the word list, the greedy algorithm simply segments that character as a word (as in the foregoing character-as-word algorithm) and begins the search starting at the next character. A variation of the greedy algorithm segments a sequence of unmatched characters as a single word; this variant is more likely to be successful in writing systems with longer average word lengths. In this manner, an initial segmentation can be obtained that is more informed than a simple character-as-word approach. The success of this algorithm is largely dependent on the word list.

As a demonstration of the application of the character-as-word and greedy algorithms, consider an example of "desegmented" English, in which all the white space has been removed: the desegmented version of the phrase, *the table down there*, would thus be *thetabledownthere*. Applying the character-as-word algorithm would result in the useless sequence of tokens *t h e t a b l e d o w n t h e r e*, which is why this algorithm only makes sense for languages such as Chinese. Applying the greedy

[10] As many as 95% of Chinese words consist of one or two characters, according to Fung and Wu (1994).

algorithm with a "perfect" word list containing all known English words would first identify the word *theta*, since that is the longest sequence of letters starting at the initial *t* which forms an actual word. Starting at the *b* following *theta*, the algorithm would then identify *bled* as the maximum match. Continuing in this manner, *thetabledownthere* would be segmented by the greedy algorithm as *theta bled own there*.

A variant of the maximum matching algorithm is the *reverse maximum matching* algorithm, in which the matching proceeds from the end of the string of characters, rather than the beginning. In the foregoing example, *thetabledownthere* would be correctly segmented as *the table down there* by the reverse maximum matching algorithm. Greedy matching from the beginning and the end of the string of characters enables an algorithm such as *forward–backward matching*, in which the results are compared and the segmentation optimized based on the two results. In addition to simple greedy matching, it is possible to encode language-specific heuristics to refine the matching as it progress.

2. *Chinese Segmentation*

The Chinese writing system is frequently described as logographic, although it is not entirely so because each character (known as *Hanzi*) does not always represent a single word. It has also been classified as morphosyllabic (DeFrancis, 1984), in that each Hanzi represents both a single lexical (and semantic) morpheme as well as a single phonological syllable. Regardless of its classification, the Chinese writing system has been the focus of a great deal of computational linguistics research in recent years.

Most previous work[11] in Chinese segmentation falls into one of three categories: statistical approaches, lexical rule-based approaches, and hybrid approaches that use both statistical and lexical information. Statistical approaches use data, such as the mutual information between characters, compiled from a training corpus, to determine which characters are most likely to form words. Lexical approaches use manually encoded features about the language, such as syntactic and semantic information, common phrasal structures, and morphological rules, to refine the segmentation. The hybrid approaches combine information from both statistical and lexical sources. Sproat et al. (1996) describe such an approach that uses a weighted finite-state transducer to identify both dictionary entries as well as unknown words derived by productive lexical processes. Palmer (1997) also describes a hybrid statistical–lexical approach in which the segmentation is incrementally improved by a trainable sequence of transformation rules.

3. *Other Segmentation Algorithms*

According to Comrie et al. (1996), the majority of all written languages use an alphabetic or syllabic system. Common unsegmented alphabetic and syllabic languages are Thai, Balinese, Javenese, and Khmer. Although such writing systems have fewer characters, they also have longer words. Localized optimization is thus not as practical as in Chinese segmentation. The richer morphology of such lan-

[11] A detailed treatment of Chinese word segmentation is beyond the scope of this chapter. Much of our summary is taken from Sproat et al. (1996). For a comprehensive recent summary of work in Chinese segmentation, we also recommend Wu and Tseng (1993).

guages often allows initial segmentations based on lists of words, names, and affixes, usually using some variation of the maximum-matching algorithm. Successful high-accuracy segmentation requires a thorough knowledge of the lexical and morphological features of the language. Very little research has been published on this type of word segmentation, but a recent discussion can be found in Kawtrakul et al. (1996), which describes a robust Thai segmenter and morphological analyzer.

Languages such as Japanese and Korean have writing systems that incorporate alphabetic, syllabic, and logographic symbols. Modern Japanese texts, for example, frequently consist of many different writing systems: Kanji (Chinese symbols), hiragana (a syllabary for grammatical markers and for words of Japanese origin), katakana (a syllabary for words of foreign origin), romanji (words written in the Roman alphabet), Arabic numerals, and various punctuation symbols. In some ways, the multiple character sets make tokenisation easier, as transitions between character sets give valuable information about word boundaries. However, character set transitions are not enough, for a single word may contain characters from multiple character sets, such as inflected verbs, which can contain a Kanji base and katakana inflectional ending. Company names also frequently contain a mix of Kanji and romanji. To some extent, Japanese can be segmented using the same techniques developed for Chinese. For example, Nagata (1994) describes an algorithm for Japanese segmentation similar to that used for Chinese segmentation by Sproat et al. (1996).

4. SENTENCE SEGMENTATION

Sentences in most written languages are delimited by punctuation marks, yet the specific usage rules for punctuation are not always coherently defined. Even when a strict set of rules exists, the adherence to the rules can vary dramatically, based on the origin of the text source and the type of text. Additionally, in different languages, sentences and subsentences are frequently delimited by different punctuation marks. Successful sentence segmentation for a given language thus requires an understanding of the various uses of punctuation characters in that language. In most languages, the problem of sentence segmentation reduces to disambiguating all instances of punctuation characters that may delimit sentences. The scope of this problem varies greatly by language, as does the number of different punctuation marks that need to be considered.

Written languages that do not use many punctuation marks present a very difficult challenge in recognizing sentence boundaries. Thai, for one, does not use a period (or any other punctuation mark) to mark sentence boundaries. A space is sometimes used at sentence breaks, but very often the space is indistinguishable from the carriage return, or there is no separation between sentences. Spaces are sometimes also used to separate phrases or clauses, where commas would be used in English, but this is also unreliable. In cases such as written Thai for which punctuation gives no reliable information about sentence boundaries, locating sentence boundaries is best treated as a special class of locating word boundaries.

Even languages with relatively rich punctuation systems, such as English, present surprising problems. Recognizing boundaries in such a written language involves determining the roles of all punctuation marks that can denote sentence boundaries: periods, question marks, exclamation points, and sometimes semicolons,

colons, dashes, and commas. In large document collections, each of these punctuation marks can serve several different purposes, in addition to marking sentence boundaries. A period, for example, can denote a decimal point or a thousands marker, an abbreviation, the end of a sentence, or even an abbreviation at the end of a sentence. Ellipsis (a series of periods [...]), can occur both within sentences and at sentence boundaries. Exclamation points and question marks can occur at the end of a sentence, but also within quotation marks or parentheses (really!) or even (albeit infrequently) within a word, such as in the band name *Therapy?* and the language name *!Xũ*. However, conventions for the use of these two punctuation marks also vary by language; in Spanish, both can be unambiguously recognized as sentence delimiters by the presence of '¡' or '¿' at the start of the sentence. In this section we will introduce the challenges posed by the range of corpora available and the variety of techniques that have been successfully applied to this problem and discuss their advantages and disadvantages.

A. Sentence Boundary Punctuation

Just as the definition of what constitutes a sentence is rather arbitrary, the use of certain punctuation marks to separate sentences depends largely on an author's adherence to changeable, and frequently ignored, conventions. In most NLP applications, the only sentence boundary punctuation marks considered are the period, question mark, and exclamation point, and the definition of what constitutes a sentence is limited to the *text-sentence*, as defined by Nunberg (1990). However, grammatical sentences can be delimited by many other punctuation marks, and restricting sentence boundary punctuation to these three can cause an application to overlook many meaningful sentences or unnecessarily complicate processing by allowing only longer, complex sentences. Consider Examples {6} and {7}, two English sentences that convey exactly the same meaning; yet, by the traditional definitions, the first would be classified as two sentences, the second as just one. The semicolon in Example {7} could likewise be replaced by a comma or a dash, retain the same meaning, but still be considered a single sentence. Replacing the semicolon with a colon is also possible, although the resulting meaning would be slightly different.

Here is a sentence. Here is another. {6}

Here is a sentence; here is another. {7}

The distinction is particularly important for an application such as part-of-speech tagging. Many taggers, such as the one described in Cutting et al. (1991), seek to optimize a tag sequence for a sentence, with the locations of sentence boundaries being provided to the tagger at the outset. The optimal sequence will usually be different depending on the definition of sentence boundary and how the tagger treats "sentence-internal" punctuation.

For an even more striking example of the problem of restricting sentence boundary punctuation, consider Example {8}, from Lewis Carroll's *Alice in Wonderland*, in which *.!?* are completely inadequate for segmenting the meaningful units of the passage:

There was nothing so VERY remarkable in that; nor did Alice think it so VERY much out of the way to hear the Rabbit say to itself, 'Oh dear! Oh dear! I shall be late!' (when she thought it over afterwards, it occurred to her that she ought to have wondered at this, but at the time it all seemed quite natural); but when the Rabbit actually TOOK A WATCH OUT OF ITS WAISTCOAT-POCKET, and looked at it, and then hurried on, Alice started to her feet, for it flashed across her mind that she had never before seen a rabbit with either a waistcoat-pocket, or a watch to take out of it, and burning with curiosity, she ran across the field after it, and fortunately was just in time to see it pop down a large rabbit-hole under the hedge. {8}

This example contains a single period at the end and three exclamation points within a quoted passage. However, if the semicolon and comma were allowed to end sentences, the example could be decomposed into as many as ten grammatical sentences. This decomposition could greatly assist in nearly all NLP tasks, because long sentences are more likely to produce (and compound) errors of analysis. For example, parsers consistently have difficulty with sentences longer than 15–25 words, and it is highly unlikely that any parser could ever successfully analyze this example in its entirety.

In addition to determining which punctuation marks delimit sentences, the sentence in parentheses as well as the quoted sentences *'Oh dear! Oh dear! I shall be late!'* suggest the possibility of a further decomposition of the sentence boundary problem into types of sentence boundaries, one of which would be "embedded sentence boundary." Treating embedded sentences and their punctuation differently could assist in the processing of the entire text-sentence. Of course, multiple levels of embedding would be possible, as in Example {9}, taken from Adams (1972). In this example, the main sentence contains an embedded sentence (delimited by dashes), and this embedded sentence also contains an embedded quoted sentence.

The holes certainly were rough—"Just right for a lot of vagabonds like us," said Bigwig—but the exhausted and those who wander in strange country are not particular about their quarters. {9}

It should be clear from these examples that true sentence segmentation, including treatment of embedded sentences, can be achieved only through an approach which integrates segmentation with parsing. Unfortunately there has been little research in integrating the two; in fact, little research in computational linguistics has focused on the role of punctuation in written language.[12] With the availability of a wide range of corpora and the resulting need for robust approaches to natural language processing, the problem of sentence segmentation has recently received a lot of attention. Unfortunately, nearly all published research in this area has focused on the problem of sentence boundary detection in English, and all this work has focused exclusively on disambiguating the occurrences of period, exclamation point, and question mark. A promising development in this area is the recent focus on trainable approaches to sentence segmentation, which we will discuss in Sec. 4.D.

---

[12] A notable exception is Nunberg (1990).

These new methods, which can be adapted to different languages and different text genres, should make a tighter coupling of sentence segmentation and parsing possible. While the remainder of this chapter will focus on published work that deals with the segmentation of a text into text-sentences, the foregoing discussion of sentence punctuation indicates the application of trainable techniques to broader problems may be possible.

B. The Importance of Context

In any attempt to disambiguate the various uses of punctuation marks, whether in text-sentences or embedded sentences, some amount of the context in which the punctuation occurs is essential. In many cases, the essential context can be limited to the character immediately following the punctuation mark. When analyzing well-formed English documents, for example, it is tempting to believe that sentence boundary detection is simply a matter of finding a period followed by one or more spaces followed by a word beginning with a capital letter, perhaps also with quotation marks before or after the space. A single rule[13] to represent this pattern would be:

```
IF (right context = period + space + capital letter
            OR period + quote + space + capital letter
            OR period + space + quote + capital letter)
THEN sentence boundary
```

Indeed, in some corpora (e.g., literary texts) this single pattern accounts for almost all sentence boundaries. In *The Call of the Wild* by Jack London, for example, which has 1640 periods as sentence boundaries, this single rule will correctly identify 1608 boundaries (98%). However, the results are different in journalistic texts, such as the *Wall Street Journal (WSJ)*. In a small corpus of the *WSJ* from 1989 that has 16,466 periods as sentence boundaries, this simple rule would detect only 14,562 (88.4%), while producing 2,900 **false positives**, placing a boundary where one does not exist.

Most of the errors resulting from this simple rule are when the period occurs immediately after an abbreviation. Expanding the context to consider the word preceding the period is thus a logical step. An improved rule would be:

```
IF ((right context = period + space + capital letter
            OR period + quote + space + capital letter
            OR period + space + quote + capital letter)
  AND (left context != abbreviation))
THEN sentence boundary
```

This can produce mixed results, for the use of abbreviations in a text depends on the particular text and text genre. The new rule improves performance on *The Call of the Wild* to 98.4% by eliminating 5 false positives (previously introduced by the phrase "St. Bernard" within a sentence). On the *WSJ* corpus, this new rule also eliminates all but 283 of the false positives introduced by the first rule. However, this rule also introduces 713 **false negatives**, erasing boundaries where they were pre-

[13] Parts of this discussion originally appeared in Bayer et al. (1998).

viously correctly placed, yet still improving the overall score. Recognizing an abbreviation, therefore, is not sufficient to disambiguate the period, because we also must determine if the abbreviation occurs at the end of a sentence.

The difficulty of disambiguating abbreviation-periods can vary depending on the corpus. Liberman and Church (1992) report that 47% of the periods in a *Wall Street Journal* corpus denote abbreviations, compared with only 10% in the Brown corpus (Francis and Kucera, 1982) as reported by Riley (1989). In contrast, Müller (1980) reports abbreviation-period statistics ranging from 54.7 to 92.8% within a corpus of English scientific abstracts. Such a range of figures suggests the need for a more informed treatment of the context that considers more than just the word preceding or following the punctuation mark. In difficult cases, such as an abbreviation that can occur at the end of a sentence, three or more words preceding and following must be considered. This is true in the following examples of "garden path sentence boundaries," the first consisting of a single sentence, the other of two sentences.

Two high-ranking positions were filled Friday by Penn St. University President Graham Spanier. {10}

Two high-ranking positions were filled Friday at Penn St. University President Graham Spanier announced the appointments. {11}

Many contextual factors assist sentence segmentation in difficult cases. These contextual factors include:

- **Case distinctions**—In languages and corpora where both upper-case and lower-case letters are consistently used, whether a word is capitalized provides information about sentence boundaries.
- **Part of speech**—Palmer and Hearst (1997) showed that the parts of speech of the words within three tokens of the punctuation mark can assist in sentence segmentation. Their results indicate that even an estimate of the *possible* parts of speech can produce good results.
- **Word length**—Riley (1989) used the length of the words before and after a period as one contextual feature.
- **Lexical endings**—Müller et al. (1980) used morphological analysis to recognize suffixes and thereby filter out words that were not likely to be abbreviations. The analysis made it possible to identify words that were not otherwise present in the extensive word lists used to identify abbreviations.
- **Prefixes and suffixes**—Reynar and Ratnaparkhi (1997) used both prefixes and suffixes of the words surrounding the punctuation mark as one contextual feature.
- **Abbreviation classes**—Riley (1989) and Reynar and Ratnaparkhi (1997) further divided abbreviations into categories, such as titles (which are not likely to occur at a sentence boundary) and corporate designators (which are more likely to occur at a boundary).

C. Traditional Rule-Based Approaches

The success of the few simple rules described in the previous section is a major reason sentence segmentation has been frequently overlooked or idealized away. In well-behaved corpora, simple rules relying on regular punctuation, spacing, and capitalization can be quickly written, and are usually quite successful. Traditionally, the method widely used for determining sentence boundaries is a regular grammar, usually with limited lookahead. More elaborate implementations include extensive word lists and exception lists to attempt to recognize abbreviations and proper nouns. Such systems are usually developed specifically for a text corpus in a single language and rely on special language-specific word lists; as a result they are not portable to other natural languages without repeating the effort of compiling extensive lists and rewriting rules. Although the regular grammar approach can be successful, it requires a large manual effort to compile the individual rules used to recognize the sentence boundaries. Nevertheless, since rule-based sentence segmentation algorithms can be very successful when an application does deal with well-behaved corpora, we provide a description of these techniques.

An example of a very successful regular–expression-based sentence segmentation algorithm is the text segmentation stage of the Alembic information extraction system (Aberdeen et al., 1995), which was created using the lexical scanner generator flex (Nicol, 1993). The Alembic system uses flex in a preprocess pipeline to perform tokenisation and sentence segmentation at the same time. Various modules in the pipeline attempt to classify all instances of punctuation marks by identifying periods in numbers, date, and time expressions, and abbreviations. The preprocess utilizes a list of 75 abbreviations and a series of over 100 hand-crafted rules and was developed over the course of more than 6 staff months. The Alembic system alone achieved a very high accuracy rate (99.1%) on a large *Wall Street Journal* corpus. However, the performance was improved when integrated with the trainable system Satz, described in Palmer and Hearst (1997) and summarized later in this chapter. In this hybrid system, the rule-based Alembic system was used to disambiguate the relatively unambiguous cases, whereas Satz was used to disambiguate difficult cases such as the five abbreviations *Co.*, *Corp.*, *Ltd.*, *Inc.*, and *U.S.*, which frequently occur in English texts both within sentences and at sentence boundaries. The hybrid system achieved an accuracy of 99.5%, higher than either of the two component systems alone.

D. Robustness and Trainability

Throughout this chapter we have emphasized the need for robustness in NLP systems, and sentence segmentation is no exception. The traditional rule-based systems, which rely on features such as spacing and capitalization, will not be as successful when processing texts where these features are not present, such as in Example {1}. Similarly, some important kinds of text consist solely of upper-case letters; closed captioning (CC) data is an example of such a corpus. In addition to being upper-case-only, CC data also has erratic spelling and punctuation, as can be seen from the following example of CC data from CNN:

> *THIS IS A DESPERATE ATTEMPT BY THE REPUBLICANS TO SPIN THEIR STORY THAT NOTHING SEAR WHYOUS— SERIOUS HAS BEEN DONE AND TRY TO SAVE THE SPEAKER'S SPEAKERSHIP AND THIS HAS BEEN A SERIOUS PROBLEM FOR THE SPEAKER, HE DID NOT TELL THE TRUTH TO THE COMMITTEE, NUMBER ONE.* {12}

The limitations of manually crafted rule-based approaches suggest the need for trainable approaches to sentence segmentation, to allow for variations between languages, applications, and genres. Trainable methods provide a means for addressing the problem of embedded sentence boundaries discussed earlier, as well as the capability of processing a range of corpora and the problems they present, such as erratic spacing, spelling errors, single-case, and OCR errors.

For each punctuation mark to be disambiguated, a typical trainable sentence segmentation algorithm will automatically encode the context using some or all of the features just described. A set of training data, in which the sentence boundaries have been manually labeled, is then used to train a machine-learning algorithm to recognize the salient features in the context. As we describe in the following, machine-learning algorithms that have been used in trainable sentence segmentation systems have included neural networks, decision trees, and maximum entropy calculation.

E. Trainable Algorithms

One of the first published works describing a trainable sentence segmentation algorithm was by Riley (1989). The method described used regression trees (Breiman et al., 1984) to classify periods according to contextual features describing the single word preceding and following the period. These contextual features included word length, punctuation after the period, abbreviation class, case of the word, and the probability of the word occurring at beginning or end of a sentence. Riley's method was trained using 25 million words from the Associated Press (AP) newswire, and he reported an accuracy of 99.8% when tested on the Brown corpus.

Reynar and Ratnaparkhi (1997) described a trainable approach to identifying English sentence boundaries (.!?) that used a statistical maximum entropy model. The system used a system of contextual templates that encoded one word of context preceding and following the punctuation mark, using such features as prefixes, suffixes, and abbreviation class. They also reported success in including an abbreviation list from the training data for use in the disambiguation. The algorithm, trained in less than 30 min on 40,000 manually annotated sentences, achieved a high accuracy rate (98%+) on the same test corpus used by Palmer and Hearst (1997), without requiring specific lexical information, word lists, or any domain-specific information. Although they reported results only for English, they indicated the ease of trainability should allow the algorithm to be used with other Roman alphabet languages, given adequate training data.

Palmer and Hearst (1997) developed a sentence segmentation system called Satz, which used a machine-learning algorithm to disambiguate all occurrences of periods, exclamation points, and question marks. The system defined a contextual feature array for three words preceding and three words following the punctuation mark; the feature array encoded the context as the parts of speech that can be attributed to each word in the context. Using the lexical feature arrays, both a neural network and a decision tree were trained to disambiguate the punctuation marks, and both achieved a high accuracy rate (98–99%) on a large corpus from the *Wall Street Journal*. They also demonstrated the algorithm, which was trainable in as little as one minute and required fewer than one thousand sentences of training data, to be rapidly portable to new languages. They adapted the system to French and German, in each case achieving a very high accuracy. Additionally, they demonstrated the trainable method to be extremely robust, as it was able to successfully disambiguate single-case texts and OCR data.

Trainable sentence segmentation algorithms, such as these, are clearly a step in the right direction toward enabling robust processing of a variety of texts and languages. Algorithms that offer rapid training while requiring small amounts of training data will permit systems to be retargeted in hours or minutes to new text genres and languages.

5. CONCLUSION

Until recently, the problem of text segmentation was overlooked or idealized away in most NLP systems; tokenisation and sentence segmentation were frequently dismissed as uninteresting "preprocessing" steps. This was possible because most systems were designed to process small texts in a single language. When processing texts in a single language with predictable orthographic conventions, it was possible to create and maintain hand-built algorithms to perform tokenisation and sentence segmentation. However, the recent explosion in availability of large unrestricted corpora in many different languages, and the resultant demand for tools to process such corpora, has forced researchers to examine the many challenges posed by processing unrestricted texts. The result has been a move toward developing robust algorithms that do not depend on the well-formedness of the texts being processed. Many of the hand-built techniques have been replaced by trainable corpus-based approaches that use machine learning to improve their performance.

Although the move toward trainable robust segmentation systems has been very promising, there is still room for improvement of tokenisation and sentence segmentation algorithms. Because errors at the text segmentation stage directly affect all later processing stages, it is essential to completely understand and address the issues involved in tokenisation and sentence segmentation and how they influence further processing. Many of these issues are language-dependent: the complexity of tokenisation and sentence segmentation and the specific implementation decisions depend largely on the language being processed and the characteristics of its writing system. For a corpus in a particular language, the corpus characteristics and the application requirements also affect the design and implementation of tokenisation and sentence segmentation algorithms. Frequently, because text segmentation is not the primary objective of NLP systems, it cannot be thought of as simply an independent "preprocessing" step, but rather, must be tightly integrated with the design and implementation of all other stages of the system.

REFERENCES

Aberdeen J, Burger J, Day D, Hirschman L, Robinson P, Vilain, M. (1995) MITRE: Description of the Alembic system used for MUC-6. Proceedings of the Sixth Message Understanding Conference (MUC-6), Columbia, MD, November 1995.

Adams R. (1972). *Watership Down*. New York: Macmillan Publishing.

Aho AV, Sethi R, Ullman JD. (1986). Compilers, Principles, Techniques, and Tools. Reading, MA: Addison–Wesley.

Bayer S, Aberdeen J, Burger J, Hirschman L, Palmer D, Vilain M. (1998). Theoretical and computational linguistics: toward a mutual understanding. In: Lawler J, Dry, HA, eds. Using Computers in Linguistics. London: Routledge.

Breiman L, Friedman J.H, Olshen R, Stone C.J (1984). Classification and Regression Trees. Belmont, CA: Wadsworth International Group.

Buckley C, Singhal A, Mitra M. (1996). Using query zoning and correlation within SMART: TREC 5. In: Harman DK, ed. Proceedings of the Fifth Text Retrieval Conference (TREC-5), Gaithersburg, MD, November 20–22.

Comrie B, Matthews S, Polinsky M. (1996). The Atlas of Languages. London: Quarto.

Consortium, Unicode. (1996). The Unicode Standard, Version 2.0. Reading, MA: Addison–Wesley.

Crystal D. (1987). The Cambridge Encyclopedia of Language. Cambridge, UK: Cambridge University Press.

Cutting D, Kupiec J, Pedersen J, Sibun P. (1991). A practical part-of-speech tagger. 3rd Conference on Applied Natural Processing, Trento, Italy.

Daniels PT, Bright W. (1996). The World's Writing Systems. New York: Oxford University Press.

DeFrancis J. (1984). The Chinese Language. Honolulu: The University of Hawaii Press.

Nelson FW, Kucera H. (1982). Frequency Analysis of English Usage. New York: Houghton Mifflin.

Fung, P, Wu D. (1994). Statistical augmentation of a Chinese machine-readable dictionary. Proceedings of Second Workshop on Very Large Corpora (WVLC-94).

Grefenstette G, Tapanainen P. (1994). What is a word, What is a sentence? Problems of tokenization. 3rd International Conference on Computational Lexicography (COMPLEX 1994).

Hankamer J. (1986) Finite state morphology and left to right phonology. Proceedings of the Fifth West Coast Conference on Formal Linguistics.

Harman DK. (1996). Proceedings of the Fifth Text Retrieval Conference (TREC-5). Gaithersburg, MD, November 20–22.

Kawtrakul A, Thumkanon C, Jamjanya T, Muangyunnan P, Poolwan K, Inagaki Y. (1996). A gradual refinement model for a robust Thai morphological analyzer. Proceedings of COLING96, Copenhagen, Denmark.

Lawler J, Dry HA. (1998). Using Computers in Linguistics. London: Routledge.

Lesk ME, Schmidt E. (1975). Lex—a lexical analyzer generator. Computing Science Technical Report 39, AT&T Bell Laboratories, Murray Hill, NJ.

Liberman MY, Church KW. (1992). Text analysis and word pronunciation in text-to-speech synthesis. In Furui S, Sondhi MM, eds, Advances in Speech Signal Processing. New York: Marcel Dekker, pp. 791–831.

Müller H, Amerl V, Natalis G. (1980). Worterkennungsverfahren als Grundlage einer Universalmethode zur automatischen Segmentierung von Texten in Sätze. Ein Verfahren zur maschinellen Satzgrenzenbestimmung im Englischen. Sprache und Datenverarbeitung, 1.

Nagata M. (1994). A stochastic Japanese morphological analyzer using a forward-dp backward A* n-best search algorithm. Proceedings of COLING94.

Nicol GT. (1993). Flex—The Lexical Scanner Generator. Cambridge, MA: Free Software Foundation.

Nunberg G. (1990). The Linguistics of Punctuation. CSLI Lecture Notes, Number 18. Center for the Study of Language and Information, Stanford, CA.

Palmer DD. (1997). A trainable rule-based algorithm for word segmentation. Proceedings of the 35th Annual Meeting of the Association for Computational Linguistics (ACL97), Madrid.

Palmer DD, Hearst MA. (1997). Adaptive multilingual sentence boundary disambiguation. Comput Linguist 23:241–267.

Reynar JC, Ratnaparkhi A. (1997). A maximum entropy approach to identifying sentence boundaries. Proceedings of the Fifth ACL Conference on Applied Natural Language Processing, Washington, DC.

Riley MD. (1989). Some applications of tree-based modelling to speech and language indexing. Proceedings of the DARPA Speech and Natural Language Workshop, Morgan Kaufmann, pp 339–352.

Sampson GR. (1995). English for the Computer, New York: Oxford University Press.

Sproat RW, Shih C, Gale W, Chang N. (1996). A stochastic finite-state word-segmentation algorithm for Chinese. Comput Linguist 22:377–404.

Wu D, Fung P. (1994). Improving Chinese tokenization with linguistic filters on statistical lexical acquisition. Proceedings of the Fourth ACL Conference on Applied Natural Language Processing, Stuttgart, Germany.

Wu Z, Tseng G. (1993). Chinese text segmentation for text retrieval: Achievements and problems. J Am Soc Inform Sci, 44:532–542.

Zwicky AM, Pullum GK. (1981). Cliticization vs. inflection: English n't. 1981 Annual Meeting of the Linguistics Society of America.

3

Lexical Analysis

RICHARD SPROAT

AT&T Labs—Research, Florham Park, New Jersey

1. INTRODUCTION

An important component of any natural language system is *lexical analysis*, which can be broadly defined as the determination of lexical features for each of the individual words of a text. Which precise set of features are identified largely depends on the broader application within which the lexical analysis component is functioning. For example, if the lexical analysis component is part of a 'preprocessor' for a part-of-speech tagger or a syntactic parser, then it is likely that the lexical features of most interest will be purely grammatical features, such as part of speech, or perhaps verbal subcategorization features. If the lexical analysis is being performed as part of a machine-translation system, then it is likely that at least some 'semantic' information (such as the gloss in the other language) would be desired. And if the purpose is text-to-speech synthesis (TTS), then the pronunciation of the word will be among the set of features to be extracted.

The simplest model is one in which all words are simply listed along with their lexical features: lexical analysis thus becomes a simple matter of table lookup. It is safe to say that there is probably no natural language for which such a strategy is practical for anything other than a limited task. The reason is that all natural languages exhibit at least some productive lexical processes by which a great many words can be derived. To give a concrete if somewhat standard example, regular verbs in Spanish can typically occur in any of approximately thirty five forms.[1] The

[1] This is quite small by some standards: in Finnish it is possible to derive thousands of forms from a lexical item.

forms are derivable by regular rules, and their grammatical features are perfectly predictable: it is predictable, for example, that the first-person plural subjunctive present of the first conjugation verb *hablar* 'speak' is *hablemos*, composed of the stem *habl-* the present subjunctive theme vowel *-e-* and the first-person plural ending *-mos*; it is similarly predictable that the masculine form of the past participle is *hablado*, formed from the stem *habl-* the theme vowel *-a-*, the participial affix *-d-* and the masculine ending *-o*. Although it would certainly be possible in Spanish to simply list all possible forms of all verbs (say all the verbs that one might find in a large Spanish desk dictionary), this would be both a time-consuming and unnecessary exercise. Much more desirable would be to list a canonical form (or small set of forms) for each verb in the dictionary, and then provide a 'recipe' for producing the full set of extant forms, and their associated features. *Computational morphology* provides mechanisms for implementing such recipes.

The topic of this chapter is morphological analysis as it applies, primarily, to written language. That is, when we speak of analyzing *words* we are speaking of analyzing words as they are written *in the standard orthography for the language in question*. The reason for this restriction is mainly historical, in that nearly all extant morphological analyzers work off written text, rather than, say, the phonetic transcriptions output by a speech recognizer. There is every reason to believe that the techniques that will be described here would apply equally well in the speech recognition domain, but as yet there have been no serious applications of this kind.

Section 2 gives some linguistic background on morphology. Secs. 3, 4, and 5 will be devoted to introducing basic techniques that have been proposed for analyzing morphology. Although many approaches to computational morphology have been proposed, there are only really two that have been at all widespread, and these can be broadly characterized as *finite-state* techniques, and *unification-based* techniques. In computational models of phonology or (orthographic) spelling changes, there has really been only one dominant type of model; namely, the finite-state models. In this chapter I shall focus on these widely used techniques. A broader overview that includes discussion of other techniques in computational morphology can be found Ref. 1.

Section 6 reviews probabilistic approaches to morphology, in particular methods that combine probabilistic information with categorical models. Although there has been relatively little work in this area, it is an area that shows a lot of potential.

The analysis of written words does not provide as tight a delimitation of the topic of this chapter as one might imagine. What is a *written word*? The standard definition that it is any sequence of characters surround by white space or punctuation will not always do. One reason it that there are many multiword entities—*in spite of, by and large, shoot the bull*, and other such—which from at least some points of view should be considered single lexical items. But an even more compelling reason is that there are languages—Chinese, Japanese, and Thai, among others—in which white space is never used to delimit words. As we shall see in Sec. 7, lexical analysis for these languages is inextricably linked with tokenization, the topic of Chap. 2.

Section 8 concludes the chapter with a summary, and a mention of some issues that relate to lexical analysis, but that do not strictly come under the rubric of computational morphology.

2. SOME LINGUISTIC BACKGROUND

Several different theoretical models of morphology have been developed over the years, each with rather specific sets of claims about the nature of morphology, and each with rather specific focuses in terms of the data that are covered by the theory. There is insufficient space here to review the various aspects of work on theoretical morphology that are relevant to computational models. (A fairly thorough review of some aspects of the linguistic underpinnings of computational morphology are given in Chap. 2 of Ref. 1.) However, it is worth presenting a high-level classification of morphological theories, to clarify the intellectual background tacitly assumed by most computational models of word-formation.

Linguistic models of morphology can be broadly classified into two main 'schools.' Following the terminology of Hockett [2], the first could be termed *item-and-arrangement* (IA), the second *item-and-process* (IP). Broadly speaking, IA models treat morphologically complex words as being composed of two or more atomic *morphemes*, each of which contributes some semantic, grammatical, and phonological information to the composite whole. Consider the German noun *Fähigkeit* 'ability,' which is derived from the adjective *fähig* 'able.' An IA analysis of *Fähigkeit* would countenance two morphemes, the base *fähig* and the affix *-keit*, which can be glossed roughly as *-ity*. Crucially, each of these elements—*Fähig* and *-keit*—would be entries in the lexicon, with the first being listed as belonging to the category A(djective) and the second belonging to a category that takes As as a base and forms N(oun)s.

IP models view things differently. While items belonging to major grammatical categories [Ns, As, V(erb)s] are listed in the lexicon, elements such as *-keit* have no existence separate from the morphological rules that introduce them. For example, in the IP-style analysis of Beard [3], there would be a morphological rule that forms abstract nouns out of adjectives. The morphosyntactic effect of the rule would be to change the category of the word from adjective to noun; the semantic effect would be to assign a meaning to the new word along the lines of 'state of being *X*,' where *X* is the meaning of the base; and the phonological effect (for certain classes of base) would be to add the affix *-keit*.

Even though IP models are arguably more in line with traditional models of grammar, IA models, on the whole, have been more popular in Generative Linguistics, largely because of the influence of the preceding American Structuralists, such as Bloomfield. IA models include Refs. 4–8. Nonetheless there have also been IP theories developed including Refs. 3, 9–12. However, although it is generally possible to classify a given morphological theory as falling essentially into one of the two camps, it should be borne in mind that there are probably no morphological theories that are 'pure' IA or IP.

Computational models have, on the whole, subscribed more to the IA view than the IP. As we shall see, it is fairly typical to view the problem of morphologically decomposing a word as one of parsing the word into its component morphemes, each of which has its own representation in the system's lexicon. This is not to say that there have not been computational models for which affixes are modeled as rules applying to bases [13], but such systems have definitely been in the minority.

3. COMPUTATIONAL PHONOLOGY

Any discussion of computational morphology must start with a discussion of computational phonology. The reason for this is simple. When morphemes are combined to form a word, often phonological processes apply that change the form of the morphemes involved. There are really only two ways to handle this kind of alternation. One, exemplified by Ref. 14 is to simply 'precompile' out the alternations and use *morphotactic* (see Sec. 4) constraints to ensure that correct forms of morphemes are selected in a given instance. The other, and more widely adopted approach, is to handle at least some of this alternation through the use of phonological models. Such models are the topic of this section.

Consider, for example, the formation of partitive nouns in Finnish. The partitive affix in Finnish has two forms, *-ta*, and *-tä*; the form chosen depends on the final harmony-inducing vowel of the base. Bases whose final harmony-inducing vowel is back take *-ta*; those whose final harmony-inducing vowel is front take *-tä*. The vowels *i* and *e* are not harmony inducing; they are transparent to harmony. Thus in a stem, such as *puhelin* 'telephone,' the last harmony-inducing vowel is *u*, so the form of the partitive affix is *-ta*.

Nominative	Partitive	Gloss
taivas	taivas + ta	'sky'
puhelin	puhelin + ta	'telephone'
lakeus	lakeut + ta	'plain'
syy	syy + tä	'reason'
lyhyt	lyhyt + tä	'short'
ystävällinen	ystävällinen + tä	'friendly'

This alternation is part of a much more general *vowel harmony* process in Finnish, one which affects a large number of affixes. Within the linguistics literature there have been two basic approaches to dealing with phonological processes such as harmony. One approach, which can be broadly characterized as rule-based ('procedural') posits that such processes are handled by rewrite rules. An example of a traditional string-rewriting rule to handle the Finnish case just described follows:[2]

a → ä/[ä, ö, y]C* ([i,e]C*)*___

This rule simply states that an *a* is changed into *ä* when preceded a vowel from the set *ä*, *ö*, *y*, with possible intervening *i*, *e*, and consonants. (This rule is a gross oversimplification of the true Finnish harmony alternation in several respects, but it will do for the present discussion.) A description of the form just given is quite outdated: more up-to-date rule-based analyses would posit a multitiered *autosegmental* representation where harmony features (in this case the feature [—back]) are spread from trigger vowels to the vowels in the affixes, either changing or filling in the features for those affixes found in the lexicon.[3] But the basic principle remains the same, namely that an input form is somehow modified by a rule to produce the correct output form.

[2] We assume here a basic familiarity with regular expression syntax.

[3] For a general introduction to such theories, see Ref. 15.

An alternative approach is declarative: one can think of the lexicon as generating all conceivable forms for an input item. So one would generate both, say, *lakeus + ta* and *lakeus + tä*. The function of the phonology is to select an appropriate form based on constraints. These constraints may be absolute ('hard') in the case of most work in the tradition of so-called *declarative phonology* [e.g., 16, 17]; or they may be ranked 'soft' constraints, as in Optimality Theory [e.g., 18], where the optimal analysis is chosen according to some ranking scheme.

Most work on computational phonology takes crucial advantage of a property of systems of phonological description as those systems are typically used: namely, that they are *regular* in the formal sense that they can be modeled as *regular relations* (if one is talking about rules) or *regular languages* (if one is talking about constraints).

The property of regularity is by no means one that is imposed by the linguistic formalisms themselves, nor is it necessarily imposed by most linguistic theories. That is, in principle, it is possible to write a rule of the form $\epsilon \rightarrow ab/a\_b$ which, if allowed to apply an unbounded number of times to its own output, will produce the set $a^n b^n$, $n > 0$, a context-free but nonregular language.[4] But unbounded applications of rules of this kind appear to be unnecessary in phonology. Similarly, one could write a rewrite rule that produces an exact copy of an arbitrarily long sequence. Such a rule would produce sets of strings of the form *ww*, and such a language is not regular. Rules of this second type may be appropriate for describing *reduplication* phenomena, though reduplication is typically bounded; thus, strictly speaking, can be handled by regular operations. But Culy reports on a case of seemingly unbounded reduplication in Bambara [19] (although this is presumably a morphological, rather than phonological rule, though, in turn, this is beside the point). So putting to one side the case of reduplication, and given that we never need to apply a rule an unbounded number of times to its own output, nearly all phonological operations can be modeled using regular languages or relations [20,21].

A. Rule-Based Approaches

Let us start with the case of phonological rewrite rules, and their implementation using finite-state transducers, as this is by far the most common approach used for phonological modeling in working computational morphology systems. Consider again the Finnish vowel harmony example described earlier. A simple finite-state transducer that implements the *a/ä* alternation is shown in Fig. 1.[5] In this transducer, state 0 is the initial state, and both states 0 and 1 are final states. The machine stays in state 0, transducing *a* to itself, until it hits one of the front vowels in the input *ä*, *ö*, *y*, in which case it goes to state 1, where it will transduce input *a* to *ä*.

Arguably the most significant work in computational phonology is the work of Kaplan and Kay at Xerox PARC [21], which presents a concrete method for the

[4] Here ϵ (conventionally) denotes the empty string.

[5] Readers unfamiliar with finite-state automata can consult standard texts such as Ref. 22. There is unfortunately much less accessible material on finite-state *transducers*: much of the material that has been written presumes a fair amount of mathematical background on the part of the reader. There is some background on transducers and their application to linguistic problems in Refs. 1, 21. One might also consult the slides from Ref. 23.

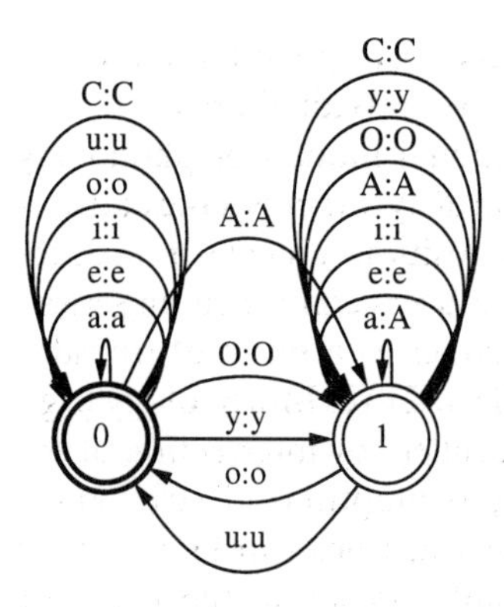

Fig. 1 A partial FST implementation of the Finnish vowel harmony rule. Here, *A* represents *ä*, *O* represents *ö* and *C* represents any consonant. The labels on the arcs represent the symbol-to-symbol transductions, with the input on the left and the output on the right of the colon.

compilation of rewrite rules into FSTs. Although only published in 1994, the work was started over a decade previously; in its early form, it was the inspiration for work on *two-level morphology* by Koskenniemi [24], as well as later work such as Refs. 25–27. The compilation algorithm presented by Kaplan and Kay is too complex to present in all its detail here; the reader is urged to consult their paper.[6] However the gist of the idea can be presented as follows. Imagine a phonologist applying the Finnish vowel harmony rule to a concrete example, say *lyhyt* + *tä*, and imagine that the phonologist was being more than usually careful in bookkeeping. The application of the rule can be notionally broken down into two phases: the identification of those parts of the input string that are going to be changed, and the actual changing of the parts. One can accomplish the first part by marking with brackets occurrences of potential contexts for the rule application.[7] The marked version of input *lyhyt* + *ta* would be *lyhyt* + *t*<*a*>. Note that the brackets have to be correctly restricted in that < should occur after a potential left context, and > before a potential right context—in this case null. The second part involves replacing a correctly bracketed input string with the appropriate output. Thus *lyhyt* + *t*<*a*> becomes *lyhyt* + *t*<*ä*>. The brackets may now be removed. The crucial insight of the Kaplan and Kay, and related algorithms is that all of these operations—bracket insertion, bracket restrictiotn, replacement, and bracket deletion—can be handled by finite-state transducers. Furthermore, FSTs can be composed together to produce a single FST that has the same input–output behavior as the sequential application of the individual machines. That is for any two transducers T_1 and T_2, one can produce the composition, $T_3 = T_1 \circ T_2$, which has the property that for any input I, $(I \circ T_1) \circ T_2$ is the same

[6] The algorithm in Ref. 27 is simpler, but is still relatively complex in all its details.

[7] This is not as simple as it sounds, as one application of the rule could *feed* a second application. This in fact happens in Finnish: in the example *pöydä* + *llä* + *kö* (table + ADESSIVE + QUESTION) 'on the table?', the first (boldface) *ä* generated by vowel harmony feeds the application of the rule to the second *ö*: compare the form of the *-ko* suffix in the word *kahvia* + *ko* (coffee + QUESTION). Thus the environments for the application will not in general be present in the initial string. This is handled in the rule compilation algorithms by a more involved placement of brackets than what is described here.

as $I \circ (T_1 \circ T_2) = I \circ T_3$. So, one can produce a single FST corresponding to the rule by composing together the FSTs that perform each of the individual operations.[8]

The main differences among rule-based computational models of phonology relate to the manner in which rules interact. Kaplan and Kay's model takes advantage of the observation that one can construct by composition a single transducer that has the same input–output behavior as a set of transducers connected 'in series.' Thus, one can model systems of ordered rewrite rules either as a series of transducers, or as a single transducer. Work in the tradition of Two-Level Morphology [e.g., 24, 26, 28, 29] applies sets of rules not in series but 'in parallel,' by intersection of the transducers modeling the rules.[9] One of the underlying assumptions of strict interpretations of two-level morphology is that all phonological rules can be stated as relations on lexical and surface representations, crucially obviating the need for intermediate representations. This may seem like a weakening of the power of rewrite rules, but two-level rules are more powerful than standard rewrite rules in one respect: they can refer to both surface and lexical representations, whereas standard rules can refer to only the input (lexical-side) level.[10]

From the point of view of the rule developer, the main difference is that in a two-level system one has to be slightly more careful about interrule interactions. Suppose that one has a pair of rules which in standard rewrite formalism would be written as follows:

$$\mathrm{w} \rightarrow \epsilon / \mathrm{V}\_\_\mathrm{V}$$
$$\mathrm{a} \rightarrow \mathrm{i} / \_\_\mathrm{i}$$

That is, an intervocalic *w* deletes, and an *a* becomes *i* before a following *i*. For an input *lawin* this would predict the output *liin*. In a two-level description the statement of the deletion rule would be formally equivalent, but the vowel coalescence rule must be stated such that it explicitly allows the possibility of an intervening deleted element (lexical *w*, surface ϵ). Despite these drawbacks two-level approaches have on the whole been more popular than ones based on ordered rules, and many systems and variations on the original model of Koskenniemi have been proposed; these include 'mixed' models combining both two-level morphology and serial composition [31], and multitape (as opposed to two-tape) models such as Ref. 32. Part of the reason for this popularity is undoubtedly due to an accident of history: when Kaplan and Kay first proposed modeling phonological rule systems in terms of serial composition of transducers, physical machine limitations made the construction of large rule sets by this approach slow and inefficient. It was partly to circumvent this ephemeral impracticality that Koskenniemi originally proposed the two-level

---

[8] This description does not accurately characterize all compilation methods. The two-level rule compilation method of Refs. 26 and 28, for example, depends on *partitions* of the input string into context and target regions. This partitioning serves the same function as brackets, although there are algorithmic differences.

[9] As Kaplan and Kay note, regular relations are closed under composition, they are not in general closed under intersection. Thus one cannot necessarily intersect two transducers. However, same-length relations and relations with only a bounded number of deletions or insertions are closed under intersection. As long as two-level rules obey these constraints, then it is possible to model systems of such rules using intersection.

[10] The formal relation between systems of two-level rules and systems of standard rewrite rules is not trivial. For some discussion see Ref. 1, pages 146–147, and Ref. 30.

approach which, because it was workable and because it was the first general phonological and morphological processing model, gained in popularity.

B. Declarative Approaches

The alternative to rule-based approaches is to state phonological regularities in terms of constraints. For the toy Finnish vowel harmony example such a model could be developed along the following lines. Imagine that the lexicon simply generates all variants for all elements as exemplified in Fig. 2. Then we could simply state a constraint disallowing *a* after front harmony-inducing vowels (FH), and *ä* after back harmony-inducing vowels (BH), with possibly intervening consonants and neutral vowels (NV):

$$\neg(FH(C \vee NV)^{*}\text{a}) \wedge \neg(BH(C \vee NV)^{*}\text{ä})$$

This constraint, implemented as finite-state acceptor and intersected with the acceptor in Fig. 2 will yield the single path *syytä.*

A computational model of phonology along these lines is Bird and Ellison's *one-level* phonology [33], in which phonological constraints are modeled computationally as finite-state acceptors, although of a state-labeled, rather than arc-labeled, variety of the kind we have been tacitly assuming here. Declarative approaches to phonology have been gaining ground in part because the more traditional rule-based approaches are (to some extent correctly) perceived as being overly powerful, and less explanatory. Still, although the linguistic reasons for choosing one approach over the other may be genuine, it is important to bear in mind that computational implementations of the two approaches—finite-state transducers versus finite-state acceptors—are only minimally different.

C. Summary

This rather lengthy digression on the topic of computational phonology was necessitated by the fact that word formation typically interacts in the phonology in rather complex ways. As a result, computational morphology systems must handle phonological alternations in some fashion; finite-state models are the most popular.

Although the topic of this section has been phonology, because the largest application of computational morphology is to written language, one is usually interested not so much in phonology as in orthographic spelling changes, such as the *y/i* alternation in English *try/tries.*[11] In some languages, such as Finnish, the orthography is sufficiently close to a phonemic transcription that this distinction is relatively unimportant. In other languages, English being a prime example, there is a

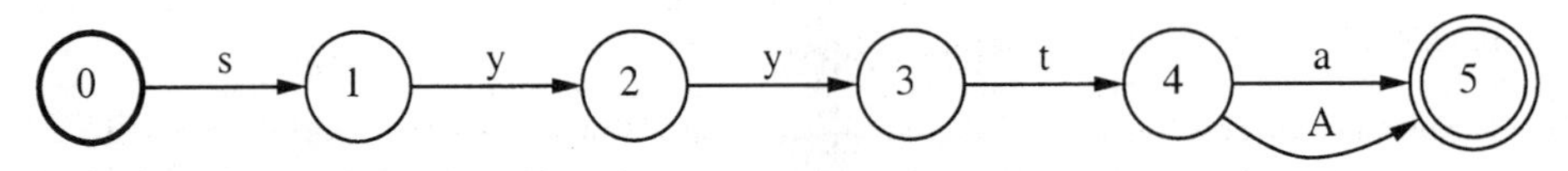

Fig. 2 An unpruned lattice for the Finnish partitive form *syy* + *tä* 'reason.' Again, *A* represents *ä*.

[11] There are applications, such as TTS, for which one needs to deal with both phonology and orthographic spelling changes.

large disparity between phonological and orthographic representations and alternations. But from the point of view of the techniques discussed here the distinction is not very material, as finite-state methods can be applied equally well to both.

4. FINITE-STATE TECHNIQUES IN MORPHOLOGY

Suppose one is building a syntactic parser for written Spanish. The lexical analysis component of such a system should be able to accept as input a written word form, and produce as output an annotation of that form in terms of grammatical features that are likely to be relevant for subsequent syntactic analysis. Such features would include information on the grammatical category of the word, but also other features, such as verbal person, number, tense, and mood features. For the purposes of this discussion, let us take a simple subset consisting of three verbs, *hablar* 'speak,' *cerrar* 'close,' and *cocer* 'cook,' conjugated in the present and preterite indicative forms:

Features	*hablar*	*cerrar*	*cocer*
ind pres 1 sg	hablo	**cie**rro	**cuez**o
ind pres 2 sg	hablas	**cie**rras	**cue**ces
ind pres 3 sg	habla	**cie**rra	**cue**ce
ind pres 1 pl	hablamos	cerramos	cocemos
ind pres 2 pl	habláis	cerráis	cocéis
ind pres 3 pl	hablan	**cie**rran	**cue**cen
ind pret 1 sg	hablé	cerré	cocí
ind pret 2 sg	hablaste	cerraste	cociste
ind pret 3 sg	habló	cerró	coció
ind pret 1 pl	hablamos	cerramos	cocimos
ind pret 2 pl	hablasteis	cerrasteis	cocisteis
ind pret 3 pl	hablaron	cerraron	cocieron

Represented here are two different conjugations, the *-ar* conjugation (*hablar* and *cerrar*) and the *-er* conjugation (*cocer*). Also represented are some vowel and consonant changes—the *c/z* alternation in *cocer*, and the diphthongization of the stem vowel in certain positions in *cerrar* and *cocer*. These stem changes are common, although by no means entirely predictable. In particular, whereas many verbs undergo the vowel alternations *o/ue*, *e/ie* there are also many verbs that do not (*comer* 'eat' does not for example). The correct generalization, for verbs that undergo the rule, is that the diphthong is found when the vowel is stressed. For the sake of simplicity in the present discussion we will assume that the alternation refers to a lexical feature **diph**, and is triggered by the presence of a single following unstressed syllable, a mostly reliable indicator that the current syllable is stressed.

Perhaps the simplest finite-state model that would generate all and only the correct forms for these verbs is an *arclist* model, borrowing a term from Ref. 14. Consider the arclist represented in the following table. On any line, the first column represents the start state of the transition, the second column the end state, the third the input, and the fourth the output; a line with a single column represents the final state of the arclist. Thus one can go from the START state to the state labeled *ar* (representing *-ar* verbs) by reading the string *habl* and outputting the string *hablar* **vb**. The final state of the whole arclist is labeled WORD. More generally, verb stems are represented as beginning in the initial (START) state, and going to a state that records their paradigm affiliation, *-ar* or *-er*. From these paradigm states, one can attain the final WORD state by means of appropriate endings for that paradigm. The verb stems for *cerrar* and *cocer* have lexical features **diph** and **c/z**, which trigger the application of spelling changes.

START	ar	habl	hablar **vb**
START	ar	cerr **diph**	cerrar **vb**
START	er	coc **diph c/z**	cocer **vb**
ar	WORD	+o#	+ **ind pres 1 sg**
ar	WORD	+as#	+ **ind pres 2 sg**
ar	WORD	+a#	+ **ind pres 3 sg**
ar	WORD	+amos#	+ **ind pres 1 pl**
ar	WORD	+'ais#	+ **ind pres 2 pl**
ar	WORD	+an#	+ **ind pres 3 pl**
ar	WORD	+'e#	+ **ind pret 1 sg**
ar	WORD	+aste#	+ **ind pret 2 sg**
ar	WORD	+'o#	+ **ind pret 3 sg**
ar	WORD	+amos#	+ **ind pret 1 pl**
ar	WORD	+asteis#	+ **ind pret 2 pl**
ar	WORD	+aron#	+ **ind pret 3 pl**
er	WORD	+o#	+ **ind pres 1 sg**
er	WORD	+es#	+ **ind pres 2 sg**
er	WORD	+e#	+ **ind pres 3 sg**
er	WORD	+emos#	+ **ind pres 1 pl**
er	WORD	+'eis#	+ **ind pres 2 pl**
er	WORD	+en#	+ **ind pres 3 pl**
er	WORD	+'i#	+ **ind pret 1 sg**
er	WORD	+iste#	+ **ind pret 2 sg**
er	WORD	+i'o#	+ **ind pret 3 sg**
er	WORD	+imos#	+ **ind pret 1 pl**
er	WORD	+isteis#	+ **ind pret 2 pl**
er	WORD	+ieron#	+ **ind pret 3 pl**
WORD			

This arclist can easily be represented as a finite-state transducer, which we will call D (Figures 3 and 4.)[12] The spelling changes necessary for this fragment involve rules to diphthongize vowels, change c into z, and delete grammatical features and the morpheme boundary symbol:

e → ie / ___ C* **feat*** **diph feat*** + V* C* #
o → ue / ___ C* **feat*** **diph feat*** + V* C* #
c → z / __ **feat*** **c/z** + [+back]
(feature ∨ **boundary)** → ϵ

Call the transducer representing these rules R. Then a transducer mapping between the surface forms of the verbs and their decomposition into morphs will be given by $(D \circ R)^{-1}$ (i.e., the *inverse* of $D \circ R$). An analysis of an input verb, represented by a finite-state acceptor S, is simply $I \circ (D \circ R)^{-1}$. Thus *cuezo* $\circ(D \circ R)^{-1}$ is *cocer* **vb** + **ind pres 1 sg**

Strictly finite-state models of morphology are in general only minor variants of the model just presented. For example, the original system of Koskenniemi [24] modeled the lexicon as a set of *tries*, which can be thought of as a special kind of finite automaton. Morphological complexity was handled by *continuation patterns*, which were annotations at the end of morpheme entries indicating which set of tries to continue the search in. But, again, this mechanism merely simulates an ϵ-labeled arc in a finite-state machine. Spelling changes were handled by two-level rules implemented as parallel (virtually, if not actually, intersected) finite-state transducers, as we have seen. Search on the input string involved matching the input string with the input side of the two-level transducers, while simultaneously matching the lexical tries with the lexical side of the transducers, a process formally equivalent to the composition $I \circ (D \circ R)^{-1}$. Variants of two-level morphology, such as the system presented in Ref. 31, are even closer to the model presented here. Tzoukermann and Liberman's work [14] takes a somewhat different approach. Rather than separately constructing D and R and then composing them together, the transducer $(D \circ R)^{-1}$ is constructed directly from a 'precompiled' arclist. For each dictionary entry, morphophonological stem alternants are produced beforehand, along with special tags that limit those alternants to appropriate sets of endings.

Other methods of constructing finite-state lexica present themselves beside the arclist construction just described. For example, one could represent the lexicon of stems (e.g., *habl-*) as a transducer from lexical classes (e.g., *ar-verb*) to the particular stems that belong to those classes. Call this transducer S. The inflectional paradigms can be represented as an automaton that simply lists the endings that are legal for each class: *ar-verb o* **ind pres 1 sg** would be one path of such an automaton, which we may call M. A lexicon automaton can then be constructed by taking the righthand projection of M composed with S concatenated with the universal machine $\Sigma^* : \pi_2[M \circ (S \cdot \Sigma^*)]$.[13] Yet another approach would be to represent the word gram-

[12] Actually, the transducer in Fig. 3 is the inverse of the transducer represented by the arclist.

[13] Following standard notational conventions, Σ denotes the alphabet of a machine and Σ^* the set of all strings (including the null string) over that alphabet. Projection is notated by π.

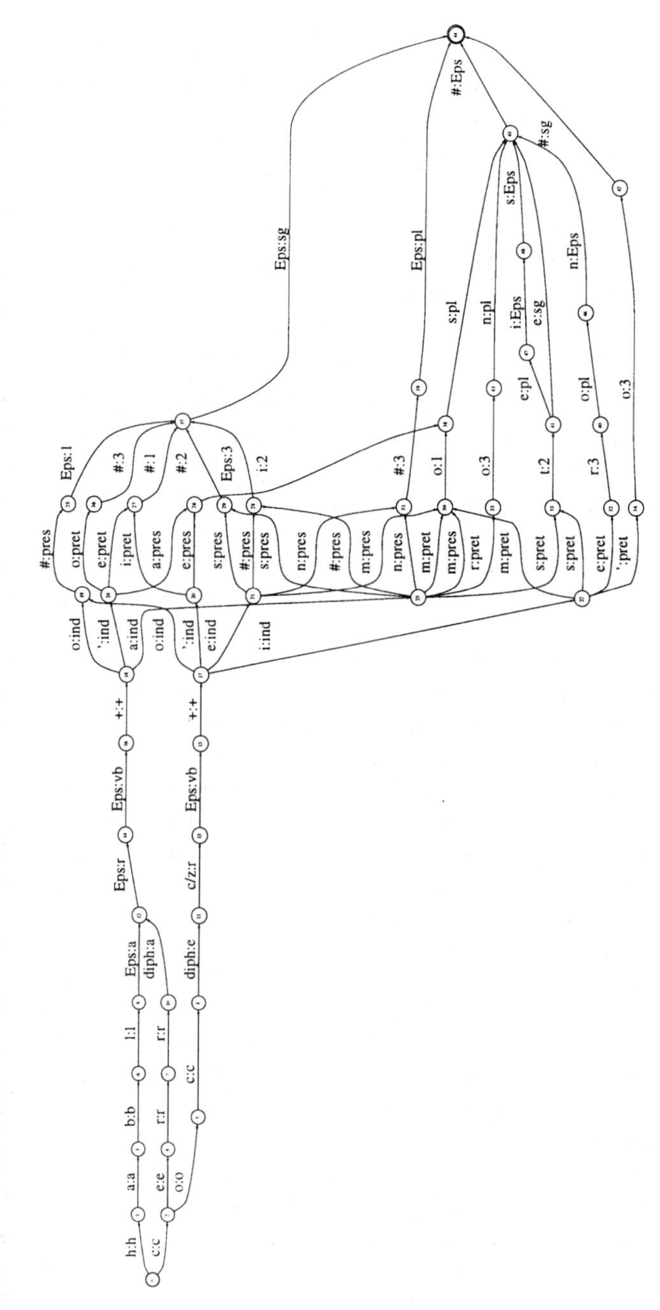

Fig. 3 A transducer for a small fragment of Spanish verbal morphology.

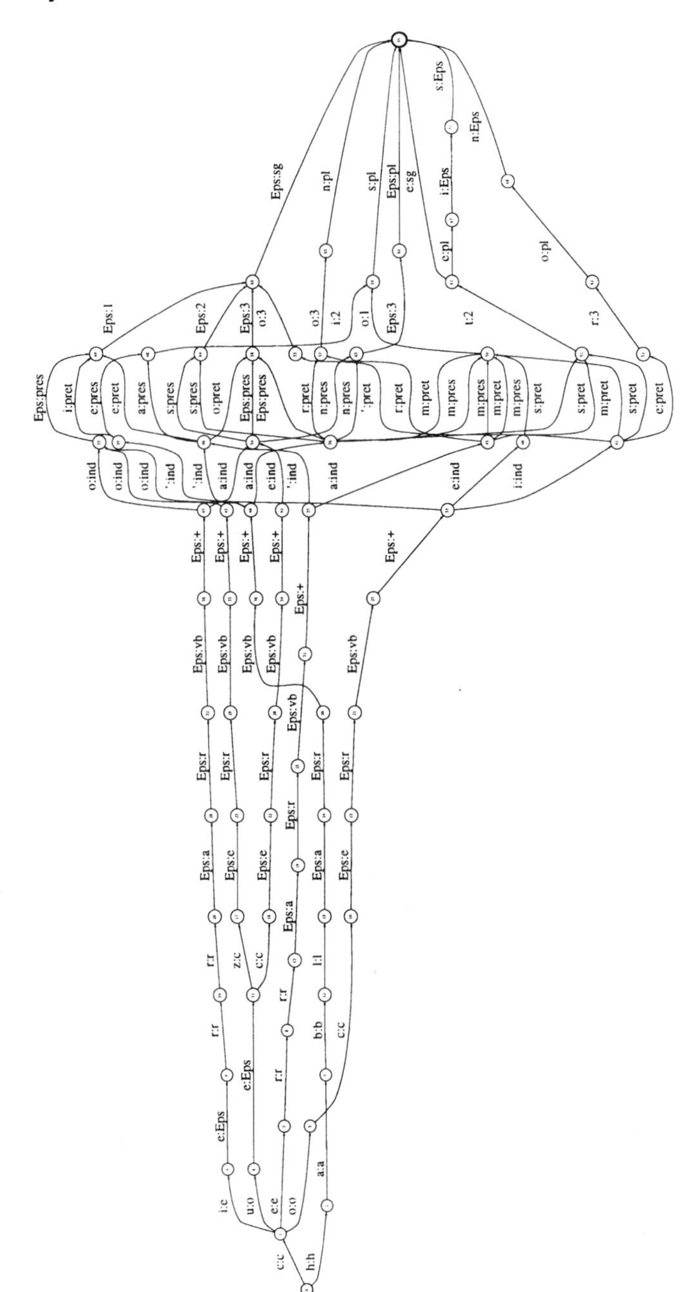

Fig. 4 A transducer that maps between surface and lexical forms for a small fragment of Spanish verbal morphology.

mar as a context-free grammar, and use a finite-state approximation algorithm [34] to produce a finite-state representation of the grammar.

The approach just sketched is sufficient for handling many morphological systems, but there are quite a few phenomena that are not so elegantly handled. One example from English involves cases for which the ability to attach a particular morpheme depends on the prior attachment of a morpheme that is not adjacent to it. The suffix *-able* productively attaches to verbs in English, not to other categories; the prefix *en-/em-* (unproductively) attaches to nouns or adjectives and makes verbs. As one might expect, *-able* can attach to verbs formed with *en-/em-*: *enrichable, empowerable, endangerable*. Crucially, *-able* either does not attach to the base nouns and adjectives, or else it has a very different meaning: **richable, powerable* (different meaning), **dangerable*. So it is only because of the prefix that *-able* is able to attach in these cases. The only way for a finite-state machine to model this is to duplicate a portion of the lexicon: there must be (at least) two entries for *rich, power,* and *danger,* one of which 'remembers' that *en-/em-* has been attached, the other of which remembers that it has not been attached. There are two possible solutions to this problem. One is simply to implement the morphology in a purely finite-state way anyway, but to hide the duplication of structure from the user by a clever set of grammar tools that allow the grammar developer to state the linguistic generalizations without worrying about how they are actually implemented as finite automata. The second solution is to construct higher-order computational models of morphotactics (the 'putting together' of morphemes to form a word), such as the unification-based methods described in Sec. 5.

The general class of models just described is particularly well suited to *concatenative* morphology, involving the stringing together of distinct affixes and roots. It is less well suited to handle so-called nonconcatenative morphology—phenomena such as reduplication, infixation, and the templatic morphology found in Semitic languages [see also Ref. 1, pp 159–170]. Some of these phenomena, such as infixation and reduplication, have never received a serious computational treatment, although it is certainly possible to handle them within strictly finite-state frameworks.

Others, such as Semitic template morphology, have proved more tractable. Various approaches to Semitic have been proposed [35–37]. An interesting recent approach is that of Kiraz [32], based in part on earlier work of Kay. Kiraz presents a multitape (rather than two-tape) transducer model, implementing regular n-relations, for $n > 2$. In this model one tape (typically) represents the input and the others represent different lexical tiers. To take a concrete example, consider the Arabic stem form *kattab*, the class II perfect active stem from the root *ktb*, which can be roughly translated as 'write.' According to standard autosegmental analyses [4], there are three clearly identifiable morphemes here, each on its own tier and each with their own predictable phonological, morphosyntactic, and (more or less) predictable semantic contributions to the form. These are the root itself; the vocalism, here *a*; and the template, which gives the general phonological shape to the word, and that for present purposes we will notate as *CVCCVC*. In Kiraz's model these three tiers are represented by three lexical tapes, which form part of a regular 4-relation with the surface tape. A typical representation would be as follows:

	a					*vocalism tape*
k		t			b	*root tape*
C	V	C	C	V	C	*pattern tape*
k	a	t	t	a	b	*surface tape*

The tapes are related by a set of *two-level* rules over *four* tapes, the particular formalism being an extension of that of Ref. 26 and previous work.

Finally, one issue that frequently comes up in discussions of finite-state morphology is efficiency. In principle, finite-state morphological transducers can become very large, and in the absence of efficient methods for representing them, and efficient algorithms for operations, such as composition, they can easily become unwieldy. Fortunately, efficient algorithms for these operations do exist [see e.g., 23, 38, 39].

5. UNIFICATION-BASED APPROACHES TO WORD STRUCTURE

Misgivings with the admittedly limited mechanisms for handling morpheme combinations within strictly finite-state morphology have led various researchers to augment the set of mechanisms available for expressing morphotactic restrictions. The most popular of these approaches involves *unification* [40]: the typical strategy is to use two-level morphology to model phonological alternations, with unification being used solely to handle morpheme combination possibilities.

An early example of this [41] uses a PATR-based unification scheme on acyclic finite automata, that effectively implements a chart parser. Consider the following example [41, p. 276], which models the affixation of *-ing* in English:

$$\begin{array}{ccccc} \textit{verb} & \rightarrow & \textit{verb} & + & \textit{ing} \\ 1 & & 2 & & 3 \end{array}$$

1. $< 2\ \textit{cat} >= \textit{verb}$
2. $< 3\ \textit{lex} >= \textit{-ing}$
3. $< 2\ \textit{form} >= \textit{inf}$
4. $< 1\ \textit{cat} >= \textit{verb}$
5. $< 1\ \textit{word} >=< 2 \textit{word} >$
6. $< 1\ \textit{form} >= [\textit{tense} : \textit{pres—part}]$

Say that the analyzer has already built an arc for an infinitive verb form, and is now analyzing a sequence *-ing*. The foregoing rule will license the construction of a new arc spanning the whole sequence, with the category feature *verb*, the tense features *pres—part*, and the lexical features inherited from the base verb. The equations license this construction because the *cat* and *form* of item 2 match the lexical features of the arc over the verb, and the *lex* of item 3 matches *-ing*. Rule 5 states that the derived word 1 is a form of the base word 2.

Restrictions necessary to handle long-distance dependencies of the kind involving *en-* and *-able* can easily be stated in such a framework. All one needs is a rule that states that *en-* attaches to nouns or adjectives and that the resulting word is a verb; another rule would allow *-able* to attach to verbs, forming adjectives. An example of this kind of discontinuous morphology in Arabic has been discussed

[37], which presents a two-level type model for Arabic morphology. Consider the paradigm for the active imperfect of the class I verbal form of *ktb* (stem *ktub*):

	Singular	Dual	Plural
1	**?a** + *ktub* + **u**	**na** + *ktub* + **u**	**na** + *ktub* + **u**
2/M	**ta** + *ktub* + **u**	**ta** + *ktub* + **aani**	**ta** + *ktub* + **uuna**
2/F	**ta** + *ktub* + **iina**	**ta** + *ktub* + **aani**	**ta** + *ktub* + **na**
3/M	**ya** + *ktub* + **u**	**ya** + *ktub* + **aani**	**ya** + *ktub* + **uuna**
3/F	**6a** + *ktub* + **u**	**ta** + *ktub* + **aani**	**ya** + *ktub* + **na**

Consider the prefix *ya-*, and the suffix *-u*, both of which occur in several positions in the paradigm. An examination of the slots in which they occur (and simplifying somewhat from Beesley's account), suggests that *ya-* and *-u* encode the following sets of features, respectively:

$$\begin{aligned} ya\text{-} &= (PERSON = 3 \wedge \\ &\quad ((GENDER = M \wedge (NUMBER = SG \vee NUMBER = DU)) \vee \\ &\quad (GENDER = unspecified \wedge NUMBER = PL)) \\ \text{-}u &= ((PERSON = 1 \wedge GENDER = unspecified \wedge NUMBER = unspecified) \\ &\quad \vee (PERSON = 2 \wedge GENDER = M \wedge NUMBER = SG) \\ &\quad \vee (PERSON = 3 \wedge GENDER = unspecified \wedge NUMBER = SG)) \end{aligned}$$

Even if the affixes are ambiguous by themselves, in a particular combination, such as *ya* + *ktub* + *u*, they are unambiguous. That is, the features for the combination *ya—u* are unambiguously ($PERSON = 3 \wedge GENDER = M \wedge NUMBER = SG$). This set of features can be arrived at by unifying the feature structures for the two affixes. Other work [32,42] includes both finite-state techniques for phonology and unification-based approaches for word structure.

A richer unification-based model [see, 43–45] that, in addition to unification, makes use of general *feature passing conventions* in the style of GPSG [46]. For example, they make use of a 'word-head convention' stated as follows [44, p. 296]:

> The WHead feature-values in the mother should be the same as the WHead feature-values of the right daughter.

In other words, for a predetermined set of WHead (word head) features, a complex word inherits the values for those features from the righthand daughter. For example it is assumed that the set of WHead features includes a plurality feature PLU. So, a word such as *dog* + *s* would inherit the feature specification ($PLU+$) from the affix *-s*, which is lexically specified with that feature specification.[14]

6. PROBABILISTIC METHODS

The methods that we have discussed so far have been categorical in that, although they may assign a set of analyses to a word, rather than just a single analysis, they

[14] One point that I have said nothing about is how to model interactions between morphosyntactic features and phonological rules. Various models for doing this have been proposed [32, 47].

provide no way of ranking those analyses. Now, the use of ad hoc weights to different types of analysis has a long tradition. For example in the DECOMP module of the MITalk TTS system [48], greater cost was assigned to compounds than to words derived by affixation, so that for *scarcity* one prefers an affixational analysis (*scare* + *ity*) over a compounding analysis *scar* + *city*. The use of real corpus-derived weights in models of computational morphology is rarer, however.

One exception is work of Heemskerk [49], who develops a probabilistic context-free model of morphology, where the probabilities of a particular morphological construction are derived from a corpus (in this case the CELEX database for Dutch). For a Dutch word, such as *beneveling* 'intoxication' (the literal gloss would be 'clouding up'), there are a large number of potential analyses that conform to regular Dutch spelling changes: *be* + *nevel* + *ing* (correct), *be* + *neef* + *eling*, *been* + *e* + *veel* + *ing*. ... Some of these can certainly be ruled out on purely morphotactic grounds, but not all. Of those that remain, Heemskerk's system ranks them according to the estimated probabilities of the context-free productions that would be necessary for their formation. Another probabilistic system is the finite-state Chinese word-segmentation system [50] and (see Sec. 7).

7. MORPHOLOGY AND TOKENISATION

The discussion in this chapter has been based on the premise that the input to lexical analysis has already been tokenised into words. But as we have seen in Chap. 2, there are languages such as Chinese, Japanese, and Thai, in which word boundaries are customarily not written and, therefore, tokenisation cannot be so cleanly separated from lexical analysis. Consider the case of Chinese, and as a concrete example, the approach based on weighted finite-state transducers [50]. Note that there have been many other publications on Chinese segmentation: a partial bibliography on this topic can be found in the paper cited here.

For the Mandarin sentence 日文章魚怎麼說 (*rì wén zhāng yú zěn mo shuō*) "How do you say octopus in Japanese?", we probably would want to say that it consists of four words, namely 日文 *rì-wén* 'Japanese', 章魚 *zhāng-yú* 'octopus', 怎麼 *zěn mo* 'how', and 說 *shuō* 'say'. The problem with this sentence is that 日 *rì* is also a word (e.g., a common abbreviation for Japan) as are 文章 *wén zhāng* 'essay,' and 魚 *yú* 'fish,' so there is not a unique segmentation. The task of segmenting Chinese text can be handled by the same kind of finite-state transduction techniques as have been used for morphology, with the addition of weights to model relative likelihoods of different analyses. So, one can represent a Chinese dictionary as a WFST D, with the input being strings of written characters and the output being lexical analyses. For instance, the word 章魚 'octopus' would be represented as the sequence of transductions 章:章/0.0 魚:魚/0.0 ϵ:noun/13.18, where 13.18 is the estimated cost (negative log-probability) of the word. Segmentation is then accomplished by finding the lowest weight string in $S \circ D^*$.[15]

[15] See Refs. 38, 50, and 51 for further information. For the purposes of this discussion we are assuming an interpretation of WFSTs where weights along a path are *summed* and for cases where there is more than one cost assigned to an analysis, the *minimal* cost is selected: this is the (*min*, +) interpretation. Other interpretations (formally: *semirings*) are possible, such as (+,·).

As with English, no Chinese dictionary covers all of the words that one will encounter in Chinese text. For example, many words that are derived by productive morphological processes are not generally found in the dictionary. One such case in Chinese involves words derived via the nominal plural affix 們-*men*. While some words in 們 will be found in the dictionary (e.g., 他們 *tā-men* 'they'; 人們 *rén-men* 'people'), many attested instances will not: for example, 小將們 *xiǎo-jiāng-men* 'little' (military) generals', 青蛙們 *qīng-wā-men* 'frogs.' To handle such cases, the dictionary is extended using standard finite state morphology techniques. For instance, we can represent the fact that 們 attaches to nouns by allowing ϵ-transitions from the final states of noun entries, to the initial state of a subtransducer containing 們. However, for the present purposes it is not sufficient merely to represent the morphological decomposition of (say) plural nouns, for one also wants to estimate the cost of the resulting words. For derived words that occur in the training corpus we can estimate these costs as we would the costs for an underived dictionary entry. For nonoccurring, but possible plural forms, we use the Good–Turing estimate [e.g., 52], whereby the aggregate probability of previously unseen members of a construction is estimated as N_1/N, where N is the total number of observed tokens and N_1 is the number of types observed only once; again, we arrange the automaton so that noun entries may transition to 們, and the cost of the whole (previously unseen) construction sums to the value derived from the Good–Turing esimate. Similar techniques are used to compute analyses and cost estimates for other classes of lexical construction, including personal names, and foreign names in transliteration.

Even in languages such as English or French, for which white space is used to delimit words, tokenisation cannot be done entirely without recourse to lexical information, because one find many multiword phrases—*in spite of, over and above, give up*—which semantically at least are clearly a single lexical entry. See Ref. 53 for relevant discussion.

8. FINAL THOUGHTS

Even with tokenization, we do not exhaust the set of lexical analysis problems presented by unrestricted natural language text. For applications such as TTS one also needs to find canonical forms for such things as abbreviations and numerals. Techniques from computational morphology have been applied in the case of numeral expansion. So, van Leeuwen [54] and Traber [55] model the expansion of expressions such as *123* as *one hundred twenty three* using rewrite rules, represented computationally as FSTs. In the Bell Laboratories multilingual TTS system [56], the problem is factored into two stages: the expansion of digit strings into sums of products of powers of ten ($1{\cdot}10^2 + 2{\cdot}10^1 + 3{\cdot}10^0$); and the mapping from that representation into number names (e.g., $2{\cdot}10^1 \rightarrow$ *twenty*). Both of these processes can be modeled using FSTs.

To summarize: the basic challenge for a lexical analysis system is to take (usually written) words as input and find lexical properties that are relevant for the particular application. The discussion in this chapter has been neutral on what these lexical properties are, and has focused rather on methods for obtaining lexical analyses from written representations of words. This is by design: much of the discussion in the literature on 'the lexicon' tacitly assumes that a particular kind of lexical information (usually semantic information) should be the output of a lexical analysis system. There is no basis for this bias. As outlined in the introduction, depending on the application one might be more interested in the semantic or pragmatic properties, syntactic properties, phonological properties or any combination of these.

ACKNOWLEDGMENTS

I thank George Kiraz and Evelyne Tzoukermann for comments on this chapter.

REFERENCES

1. R Sproat. Morphology and Computation. Cambridge, MA: MIT Press, 1992.
2. C Hockett. Two models of grammatical description. Word 10:210–231, 1954.
3. R Beard. Lexeme–Morpheme Base Morphology. Albany: SUNY, 1995.
4. J McCarthy. Formal problems in semitic morphology and phonology. PhD dissertation, The Massachusetts Institute of Technology, Cambridge, MA, 1979. Distributed by Indiana University Linguistics Club (1982).
5. A Marantz. On the Nature of Grammatical Relations. Cambridge, MA: MIT Press, 1984.
6. R Sproat. On deriving the lexicon. PhD dissertation, The Massachusetts Institute of Technology, Cambridge, MA, 1985. Distributed by MIT Working Papers in Linguistics.
7. M Baker. Incorporation: a Theory of Grammatical Function Changing. Chicago: University of Chicago Press, 1988.
8. R Lieber. Deconstructing Morphology: Word Formation in a Government-Binding Syntax. Chicago: University of Chicago Press, 1991.
9. P Matthews. Morphology. Cambridge, UK: Cambridge University Press, 1974.
10. M Aronoff. Word Formation in Generative Grammar. Cambridge, MA: MIT Press, 1976.
11. S Anderson. A-Morphous Morphology. Cambridge: Cambridge University Press, 1992.
12 WU Wurzel. Inflectional Morphology and Naturalness. Studies in Natural Language and Linguistic Theory. Dordrecht: Kluwer, 1989.
13. R Byrd, J Klavans, M Aronoff, F Anshen. Computer methods for morphological analysis. ACL Proceedings, 24th Annual Meeting, Morristown, NJ; Association for Computational Linguistics, 1986, pp. 120–127.
14. E. Tzoukermann and M Liberman. A finite-state morphological processor for Spanish. COLING-90, Vol 3. COLING, 1990, pp, 3:277–286.
15. J Goldsmith. Autosegmental and Lexical Phonology: An Introduction. Basil: Blackwell, 1989.
16. J Coleman. Phonological representations—Their names, forms and powers. PhD dissertation, University of York, 1992.
17. S Bird, E Klein. Phonological analysis in typed feature structures. Comput Ling 20:455–491, 1994.
18. A Prince, P Smolensky. Optimality theory. Tech Rep 2, Rutgers University, Piscataway, NJ, 1993.
19. C Culy. The complexity of the vocabulary of Bambara. Linguist Philos. 8:345–351, 1985.
20. CD Johnson. Formal Aspects of Phonological Description. The Hague: Mouton, 1972.
21. R Kaplan, M Kay. Regular models of phonological rule systems. Comput Linguist 20:331–378, 1994.
22. J Hopcroft, J Ullman. Introduction to Automata Theory, Languages and Computation. Reading, MA: Addison–Wesley, 1979.

23. M Mohri, M Riley, R Sproat. Algorithms for speech recognition and language processing. Tutorial presented at COLING-96, 1996. Available as xxx.lanl.gov/ps/cmp-1g/9608018.
24. K Koskenniemi. Two-level morphology: a general computational model for word-form recognition and production. PhD dissertation, University of Helsinki, Helsinki, 1983.
25. L Karttunen, K Beesley. Two-level rule compiler. Tech Rep P92-00149, Xerox Palo Alto Research Center, 1992.
26. E Grimley-Evans, G. Kiraz, S Pulman. Compiling a partition-based two-level formalism. Proceedings of COLING-96, Copenhagen, Denmark, COLING, 1996, pp 454–459.
27. M Mohri, R Sproat. An efficient compiler for weighted rewrite rules. In 34th Annual Meeting of the Association for Computational Linguistics. Morristown, NJ: Association for Computational Linguistics, 1996, pp 231–238.
28. GA Kiraz. SemHe: a generalized two-level system. In: 34th Annual Meeting of the Association for Computational Linguistics. Morristown, NJ: Association for Computational Linguistics, 1996, pp 159–166.
29. L Karttunen KIMMO: a general morphological processor. In L Karttunen, ed., Texas Linguistic Forum, 22. Austin, TX: University of Texas, 1983, pp 165–186.
30. L Karttunen. Directed replacement. 34th Annual Meeting of the Association for Computational Linguistics, Morristown, NJ: Association for Computational Linguistics, 1996, pp 108–115.
31. L Karttunen, R Kaplan, A Zaenen. Two-level morphology with composition. COLING-92. COLING, 1992, pp 141–148.
32. GA Kiraz. Computational nonlinear morphology with emphasis on semitic language. PhD dissertation, University of Cambridge, 1996. Forthcoming, Cambridge University Press.
33. S Bird, T. Ellison. One-level phonology: autosegmental representations and rules as finite automata. Comput Linguist 20:55–90, 1994.
34. FCN Pereira, RN Wright. Finite-state approximation of phrase–structure grammars. 29th Annual Meeting of the Association for Computational Linguistics. Morristown, NJ: Association for Computational Linguistics, 1991. pp, 246–255.
35. M Kay. Nonconcatenative finite-state morphology. ACL Proceedings, 3rd European Meeting. Morristown, NJ. Association for Computational Linguistics, 1987, pp. 2–10.
36. L Kataja, K Koskenniemi. Finite-state description of Semitic morphology: a case study of ancient Akkadian. COLING-88. Morristown, NJ: Association for Computational Linguistics, 1988, pp 313–315.
37. K Beesley. Computer analysis of Arabic morphology: a two-level approach with detours. Proceedings of the Third Annual Symposium on Arabic Linguistics, University of Utah, 1989.
38. F Pereira, M Riley. Speech recognition by composition of weighted finite automata. CMP-LG archive paper 9603001, 1996. www.lanl.gov/ps/cmp-1g/9603001.
39. M Mohri. Finite-state transducers in language and speech processing. Comput Linguist 23: 269–311, 1997.
40. S Shieber. An Introduction to Unification-Based Approaches to Grammar. Chicago: University of Chicago Press, 1986. Center for the Study of Language and Information.
41. J Bear. A morphological recognizer with syntactic and phonological rules. COLING-86. Morristown, NJ: Association for Computational Linguistics, 1986, pp 272–276.
42. H Trost. The application of two-level morphology to non-concatenative German morphology. COLING-90. Vol 2. Morristown, NJ: Association for Computational Linguistics, 1990, pp 371–376.
43. G Russel, S Pulman, G Ritchie, A Black, A dictionary and morphological analyser for English. COLING-86, Morristown, NJ: Association for Computational Linguistics, 1986, pp 277–279.

44. G Ritchie, S Pulman, A Black, G Russel. A computational framework for lexical description. Comput Linguist 13:290–307, 1987.
45. G Ritchie, G Russell, A Black, S Pulman. Computational Morphology: Practical Mechanisms for the English Lexicon. Cambridge, MA: MIT Press, 1992.
46. G Gazdar, E Klein, G Pullum, I Sag. Generalized Phrase Structure Grammar. Cambridge, MA: Harvard University Press, 1985.
47. J Bear. Morphology with two-level rules and negative rule features. COLING-88. Morristown, NJ: Association for Computational Linguistics, 1988, pp 28–31.
48. J Allen, MS Hunnicutt, D Klatt. From Text to Speech: the MITalk System. Cambridge: Cambridge University Press, 1987.
49. J Heemskerk. A probabilistic context-free grammar for disambiguation in morphological parsing. European ACL, Sixth European Conference. 1993, pp 183–192.
50. R Sproat, C Shih, W Gale, N Chang. A stochastic finite-state word-segmentation algorithm for Chinese. Comput Linguist 22: 1996.
51. F Pereira, M Riley, R Sproat. Weighted rational transductions and their application to human language processing. In: ARPA Workshop on Human Language Technology, Advanced Research Projects Agency, March 8–11, 1994, pp 249–254.
52. KW Church, W Gale. A comparison of the enhanced Good–Turing and deleted estimation methods for estimating probabilities of English bigrams. Comput Speech Lang 5: 19–54, 1991.
53. L Karttunen, J Chanod, G Grefenstette, A Schiller. Regular expressions for language engineering. J Nat Lang Eng 2:305–328, 1996.
54. H van Leeuwen. Too$_{L}$iP: a development tool for linguistic rules. PhD dissertation, Technical University Eindhoven, 1989.
55. C Traber. SVOX: the implementation of a text-to-speech system for German. Tech Rep 7, Swiss Federal Institute of Technology, Zurich, 1995.
56. R Sproat. Multilingual text analysis for text-to-speech synthesis. J Nat Lang Eng 2:369–380, 1996.

4

Parsing Techniques

CHRISTER SAMUELSSON

Xerox Research Centre Europe, Grenoble, France

MATS WIRÉN

Telia Research, Stockholm, Sweden

1. INTRODUCTION

This chapter gives an overview of rule-based natural language parsing and describes a set of techniques commonly used for handling this problem. By *parsing* we mean the process of analysing a sentence to determine its syntactic structure according to a formal grammar. This is usually not a goal in itself, but rather, an intermediary step for the purpose of further processing, such as the assignment of a meaning to the sentence. We shall thus take the output of parsing to be a hierarchical structure suitable for semantic interpretation (the topic of Chap. 5). As for the input to parsing, we shall assume this to be a string of words, thereby simplifying the problem in two respects (treated in Chaps. 2 and 3, respectively): First, our informal notion of a word disregards the problem of chunking, that is, appropriate segmentation of the input into parsable units. Second, the "words" will usually have been analysed in separate phases of lexical lookup and morphological analysis, which we hence exclude from parsing proper.

As pointed out by Steedman [101], there are two major ways in which processors for natural languages differ from processors for artificial languages, such as compilers. The first is in the power of the grammar formalisms used (the *generative capacity*): Whereas programming languages are mostly specified using carefully restricted subclasses of context-free grammar, Chomsky [20] argued that natural languages require more powerful devices (indeed, this was one of the driving ideas of Transformational Grammar.)[1] One of the main motivations for the need for

[1] For a background on formal grammars and formal-language theory, see Hopcroft and Ullman [38].

expressive power has been the presence of unbounded dependencies, as in English wh-questions:

Who did you give the book to _? {1}

Who do you think that you gave the book to _? {2}

Who do you think that he suspects that you gave the book to _? {3}

In Examples {1}–{3} it is held that the noun phrase "who" is displaced from its canonical position (indicated by "_") as indirect object of "give." Because arbitrary amounts of material may be embedded between the two ends, as suggested by {2} and {3}, these dependencies might hold at unbounded distance. Although it was later shown that phenomena such as this can be described using restricted formalisms [e.g., 2, 26, 30, 41], many natural language systems make use of formalisms that go far beyond context-free power.

The second difference concerns the extreme structural *ambiguity* of natural language. At any point in a pass through a sentence, there will typically be several grammar rules or parsing actions that might apply. Ambiguities that are strictly local will be resolved as the parser proceeds through the input, as illustrated by {4} and {5}:

Who has seen John? {4}

Who has John seen? {5}

If we assume left-to-right processing, the ambiguity between ""who" as a subject or direct object will disappear when the parser encounters the third word of either sentence ("seen" and "John", respectively). However, it is common that local ambiguities propagate in such a way that global ambiguities arise. A classic example is the following:

Put the block in the box on the table {6}

Assuming that "put" subcategorizes for two objects, there are two possible analyses of {6}:

Put the block [*in the box on the table*] {7}

Put [*the block in the box*] *on the table* {8}

If we add another prepositional phrase ("... in the kitchen"), we obtain five analyses; if we add yet another, we obtain fourteen, and so on. Other examples of the same phenomenon are nominal compounding and conjunctions. As discussed in detail by Church and Patil [21], "every-way ambiguous" constructions of this kind have a number of analyses that is superexponential in the number of added components. More specifically, the degree of ambiguity follows a combinatoric series called the Catalan Numbers, in which the nth number is given by

$$C_n = \binom{2n}{n}\frac{1}{n+1}.$$

In other words, even the process of just returning all the possible analyses would lead to a combinatorial explosion. Thus, much of the work on parsing—hence, much of the following exposition—deals somehow or other with ways in which the potentially enormous search spaces resulting from local and global ambiguity can be efficiently handled.

Finally, an additional aspect in which the problem of processing of natural languages differs from that of artificial languages is that any (manually written) grammar for a natural language will be incomplete—there will always be sentences that are completely reasonable, but that are outside of the coverage of the grammar. This problem is relevant to robust parsing, which will be discussed only briefly in this chapter (see Sec. 8.A).

2. GRAMMAR FORMALISMS

Although the topic of grammar formalisms falls outside of this chapter, we shall introduce some basic devices solely for the purpose of illustrating parsing; namely, context-free grammar and a simple form of constraint-based grammar. The latter, also called unification grammar, currently constitutes a widely adopted class of formalisms in computational linguistics (Fig. 1). In particular, the declarativeness of the constraint-based formalisms provides advantages from the perspective of grammar engineering and reusability, compared with earlier frameworks such as augmented transition networks (ATNs) (see Sec. 5).

A key characteristic of constraint-based formalisms is the use of feature terms (sets of attribute–value pairs) for the description of linguistic units, rather than atomic categories as in phrase-structure grammars.[2] Feature terms can be nested:

DCG Definite Clause Grammar (Pereira and Warren [73])

FUG Functional Unification Grammar (Kay [51, 52])

PATR (strictly speaking PATR-II; Shieber et al. [95], Shieber [91, 93])

CLE Core Language Engine (Alshawi [5])

TDL Type Definition Language (Krieger and Schäfer [58])

ALE Attribute Logic Engine (Carpenter and Penn [18])

TAG Tree-Adjoining Grammar (Joshi et al. [42, 41], Schabes et al. [86], Vijay-Shanker and Joshi [109])

LFG Lexical-Functional Grammar (Bresnan [14])

GPSG Generalized Phrase Structure Grammar (Gazdar et al. [30])

CUG Categorial Unification Grammar (Haddock et al. [34], Uszkoreit [108], Karttunen [47])

HPSG Head-driven Phrase Structure Grammar (Pollard and Sag [75])

Fig. 1 Examples of constraint-based linguistic formalisms (or tools).

[2] A phrase–structure grammar consists entirely of context-free or content-sensitive rewrite rules (with atomic categories).

their values can be either atomic symbols or feature terms. Furthermore, they are partial (underspecified) in the sense that new information may be added as long as it is compatible with old information. The operation for merging and checking compatibility of feature constraints is usually formalized as *unification*. Some formalisms, such as PATR, are restricted to simple unification (of conjunctive terms), whereas others allow disjunctive terms, sets, type hierarchies, or other extensions; for example, LFG and HPSG. In summary, feature terms have proved to be a versatile and powerful device for linguistic description. One example of this is unbounded dependency, as illustrated by Examples {1}–{3}, which can be handled entirely within the feature system by the technique of gap threading [46].

Several constraint-based formalisms are phrase-structure-based in the sense that each rule is factored in a phrase–structure backbone and a set of constraints that specify conditions on the feature terms associated with the rule (e.g., PATR, CLE, TDL, LFG, and TAG, although the latter uses tree substitution and tree adjunction rather than string rewritings). Analogously, when parsers for constraint-based formalisms are built, the starting-point is often a phrase–structure parser that is augmented to handle feature terms. This is also the approach we shall follow here.

We shall thus make use of a context-free grammar $G = \langle N, T, P, S \rangle$, where N and T are finite sets of nonterminal categories and terminal symbols (words), respectively, P is a finite set of production (rewrite) rules, and $S \in N$ is the start category. Each production of P is of the form $X \rightarrow \alpha$ such that $X \in N$ and $\alpha \in (N \cup T)^*$. If α is a word, the production corresponds to a lexical entry, and we call X a lexical or preterminal category.

Any realistic natural language grammar that licenses complete analyses will be *ambiguous*; that is, some sentence will give rise to more than one analysis. When a phrase-structure grammar is used, the analysis produced by the parser typically takes the form of one or several parse trees. Each nonleaf node of a parse tree then corresponds to a rule in the grammar: the label of the node corresponds to the left-hand side of the rule, and the labels of its children correspond to the right-hand-side categories of the rule.

Furthermore, we shall make use of a constraint-based formalism with a context-free backbone and restricted to simple unification, thus corresponding to PATR. A grammar rule in this formalism can be seen as an ordered pair of a production $X_0 \rightarrow X_1 \cdots X_n$ and a set of equational constraints over the feature terms of types $X_0, \cdots, X_n$. A simple example of a rule, encoding agreement between the determiner and the noun in a noun phrase, is the following:

$$
\begin{aligned}
&X_0 \rightarrow X_1 X_2 \\
&\quad \langle X_0\ category \rangle = NP \\
&\quad \langle X_1\ category \rangle = Det \\
&\quad \langle X_2\ category \rangle = N \\
&\quad \langle X_1\ agreement \rangle = \langle X_2\ agreement \rangle
\end{aligned}
$$

A rule may be represented as a feature term implicitly recording both the context-free production and the feature constraints. The following is a feature term corresponding to the previous rule (where $\boxed{1}$ indicates identity between the associated elements):

$$\begin{bmatrix} X_0 : & [\mathit{category}:\ NP\ ] \\ X_1 : & \begin{bmatrix} \mathit{category}: & \mathit{Det} \\ \mathit{agreement}: & \boxed{1}\ [\ ] \end{bmatrix} \\ X_2 : & \begin{bmatrix} \mathit{category}: & N \\ \mathit{agreement}: & \boxed{1} \end{bmatrix} \end{bmatrix}$$

When using constraint-based grammar, we will assume that the parser manipulates and outputs such feature terms.

3. BASIC CONCEPTS IN PARSING

A *recognizer* is a procedure that determines whether or not an input sentence is grammatical according to the grammar (including the lexicon). A *parser* is a recognizer that produces associated structural analyses according to the grammar (in our case, parse trees or feature terms).[3] A *robust parser* attempts to produce useful output, such as a partial analysis, even if the complete input is not covered by the grammar. More generally, it is also able to output a single analysis even if the input is highly ambiguous (see Sec. 8.A).

We can think of a grammar as inducing a search space consisting of a set of states representing stages of successive grammar-rule rewritings and a set of transitions between these states. When analysing a sentence, the parser (recognizer) must rewrite the grammar rules in some sequence. A sequence that connects the state S, the string consisting of just the start category of the grammar, and a state consisting of exactly the string of input words, is called a *derivation*. Each state in the sequence then consists of a string over $(N \cup T)^*$ and is called a *sentential form*. If such a sequence exists, the sentence is said to be grammatical according to the grammar. A derivation in which only the leftmost nonterminal of a sentential form is replaced at each step is called *leftmost*. Conversely, a *rightmost* derivation replaces the rightmost nonterminal at each step.

Parsers can be classified along several dimensions according to the ways in which they carry out derivations. One such dimension concerns rule invocation: In a *top-down* derivation, each sentential form is produced from its predecessor by replacing one nonterminal symbol X by a string of terminal or nonterminal symbols $X_1 \cdots X_n$, where $X \rightarrow X_1 \cdots X_n$ is a production of P. Conversely, in a *bottom-up* derivation, each sentential form is produced by replacing $X_1 \cdots X_n$ with X given the same production, thus successively applying rules in the reverse direction.

Another dimension concerns the way in which the parser deals with ambiguity, in particular, whether the process is *deterministic* or *nondeterministic*. In the former case, only a single, irrevocable choice may be made when the parser is faced with local ambiguity. This choice is typically based on some form of lookahead or systematic preference (compare Sec. 4.B).

---

[3] To avoid unnecessary detail, we will not say much about the actual assembling of analyses during parsing. Hence, what we will describe will resemble recognition more than parsing. However, turning a recognizer into a parser is usually not a difficult task, and is described in several of the references given.

A third dimension concerns the way in which the search space is traversed in the case of nondeterministic parsing: A *depth-first* derivation generates one successor of the initial state, then generates one of its successors, and then continues to extend this path in the same way until it terminates. Next, it uses backtracking to generate another successor of the last state generated on the current path, and so on. In contrast, a *breadth-first* derivation generates all the successors of the initial state, then generates all the successors at the second level, and so on. Mixed strategies are also possible.

A fourth dimension concerns whether parsing proceeds from left to right (strictly speaking front to back) through the input or in some other order, for example, inside-out from the heads.[4]

4. BASIC TOP-DOWN AND BOTTOM-UP PARSING

Parsers can be classified as working top-down or bottom-up. This section describes two such basic techniques, which can be said to underlie most of the methods to be presented in the rest of the chapter.

A. Basic Top-Down Parsing: Recursive Descent

A simple method for pure top-down parsing consists in taking a rule the left-hand side of which is the start symbol, rewrite this to its right-hand side symbols, then try to expand the leftmost nonterminal symbol out of these, and so on until the input words are generated. This method is called *recursive descent* [3] or *produce-shift* [97].

For simplicity and generality, we shall use a pseudodeductive format for specifying recursive-descent parsing (loosely modelled on the deductive system of Shieber et al. [94]; see Fig. 2). We thus assume that the parser manipulates a set of items, each of which makes an assertion about the parsing process up to a particular point. An item has the form $\langle \cdot \beta, j \rangle$, where β is a (possibly empty) string of grammar categories or words. Such an item asserts that the sentence up to and including word w_j followed by β is a sentential form according to the grammar. The dot thus marks the division between the part of the sentence that has been covered in the parsing process and the part that has not.[5] As the symbols of the string following the dot are recursively expanded, the item is pushed onto a stack.

To handle nondeterminism, the basic scheme in Fig. 2 needs to be augmented with a control strategy. Following Sec. 3, the two basic alternatives are depth-first search with backtracking or breadth-first search, maintaining several derivations in parallel.

Recursive descent in some form is the most natural approach for parsing with ATNs (see Sec. 5). The top-down, depth-first processing used with Definite Clause Grammars [73], as executed by Prolog, is also an instance of recursive descent. A problem with pure top-down parsing is that left-recursive rules (for example,

---

[4] A *head* is a designated, syntactically central category of the right-hand side of a grammar rule. For example, in the noun–phrase rule $NP \rightarrow Det\ A\ N$, the noun N is usually considered to be the syntactic head.

[5] Actually, the dot is superfluous and is used here only to indicate the relation with tabular approaches; compare Sec. 6.

Input: A string of words $w_1 \cdots w_n$.

Output: A set of items.

Method: If S is the start category, then generate an initial item $\langle \cdot S, 0 \rangle$. Do the following steps until the item $\langle \cdot, n \rangle$ has been generated or no more items can be generated.

> *Scan*: If there is an item $\langle \cdot w_{j+1}\beta, j \rangle$, then generate an item $\langle \cdot \beta, j+1 \rangle$. (In effect, consume the word w_{j+1}.)
>
> *Predict*: If there is an item $\langle \cdot Y\beta, j \rangle$ and a rule of the form $Y \rightarrow \gamma$, then generate an item $\langle \cdot \gamma\beta, j \rangle$. (In effect, substitute the element Y with the right-hand side γ of the rule.)

Fig. 2 A basic scheme for recursive–descent parsing.

NP → *NP PP*) lead to infinite recursion. This can be avoided by adopting some additional bookkeeping (compare Sec. 6.2B) or by eliminating left-recursion from the grammar [3]. More generally, however, the strong role of the lexical component in many constraint-based grammar frameworks makes top-down parsing in its pure form a less useful strategy.

B. Basic Bottom-Up Parsing: Shift–Reduce

The main data structure used in shift–reduce parsing is a parser stack, which holds constituents built up during analysis or words that are read from the input. Basically, the parser pushes (*shifts*) words onto the stack until the right-hand side of a production appears as one or several topmost elements of the stack. These elements may then be replaced by (*reduced* to) the left-hand-side category of the production. If successful, the process continues by successive shift and reduce steps until the stack has been reduced to the start category and the input is exhausted. [3].

Again, we shall use a pseudodeductive format for specifying shift–reduce parsing (loosely following Shieber et al. [94]). An item has the form $\langle \alpha \cdot, j \rangle$, where α is a (possibly empty) string of grammar categories or words. This kind of item asserts that α derives the sentence up to and including word w_j. The dot thus marks the division between the part of the sentence that has been covered in the parsing process and the part that has not.[6] Put differently, the string α corresponds to the parser stack, the right end of which is the top of the stack, and j denotes the word last shifted onto the stack.

Figure 3 provides a basic scheme for shift–reduce parsing. However, as it stands, it is impractical with any sizable grammar, since it has no means of (deterministically) choosing steps that will lead to success on a well-formed input. For example, if α appears on the top of the stack and $Y \rightarrow \alpha$ and $Y \rightarrow \alpha\beta$ are productions, it is not clear if the parser should shift or reduce. A *shift–reduce* (*parsing*) *table*, or *oracle*, can then be used for guiding the parser. The purpose of the parsing table is thus to endow the pure bottom-up processing with a certain amount of predictive ability. There are two situations that it must resolve: *shift–reduce conflicts*, in which

[6] Again, the dot is superfluous and is used only to indicate the relation with tabular approaches and LR parsing; compare Secs. 6 and 7.

Input: A string of words $w_1 \cdots w_n$.

Output: A set of items.

Method: If S is the start category, then generate an initial item $\langle \bullet, 0 \rangle$. Do the following steps until the item $\langle S \bullet, n \rangle$ has been generated or no more items can be generated.

> *Shift*: If there is an item of the form $\langle \alpha \bullet, j \rangle$, then generate an item $\langle \alpha w_{j+1} \bullet, j+1 \rangle$. (In effect, consume the word w_{j+1} and push it onto the stack.)
>
> *Reduce*: If there is an item of the form $\langle \alpha\gamma \bullet, j \rangle$ and a rule of the form $Y \rightarrow \gamma$, then generate an item $\langle \alpha Y \bullet, j \rangle$. (In effect, pop the stack element(s) γ and push the element Y onto the stack.)

Fig. 3 A basic scheme for shift–reduce parsing.

the parser can either shift a word onto the stack or reduce a set of elements on the stack to a new element (as seen in the foregoing), and *reduce–reduce conflicts*, in which it is possible to rewrite one or several elements on the stack by more than one grammar rule.[7]

Again, the nondeterminism can be handled by using breadth-first search or depth-first search with backtracking (in which case the parsing table might order the possible choices such that the more likely paths are processed first). Alternatively, one might try to keep to deterministic processing by ensuring that, as far as possible, a correct decision is made at each choice point. This approach was pioneered by Marcus [62] with his parser PARSIFAL. Marcus' aim was to model a particular aspect of hypothesized human linguistic performance; namely, that (almost all) globally unambiguous sentences appear to be parsed deterministically. With globally ambiguous sentences of the kind illustrated by example {6} PARSIFAL would deliver at most one of the legal analyses, possibly with a flag indicating ambiguity. (Alternatively, a semantic component might provide guidance to the parser.) More importantly, however, Marcus was interested in having the parser reject certain locally ambiguous sentences known as *garden-path sentences* [10]. An example is the following:

The boat floated down the river sank {9}

It has been claimed that the temporary ambiguity in {9} between "floated" as introducing a reduced relative clause and as a main verb forces the human sentence processor to back up when encountering the word "sank," thereby rendering the process nondeterministic (and increasing the effort). To allow processing of sentences such as those in Examples {4} and {5}, but not {9}, PARSIFAL depends on being able to delay its decisions a certain number of steps. It achieves this in two ways: First, PARSIFAL's shift–reduce machinery itself provides a way of delaying decisions until appropriate phrases can be closed by reductions. But PARSIFAL also makes use of a limited lookahead; namely, a three-cell buffer in which the next three words or NPs can be examined.[8]

[7] Sometimes preterminal categories, rather than actual words, are shifted to the stack; in this event, there may also be *shift–shift conflicts*.

[8] Under certain conditions, the lookahead may involve up to five buffer cells.

Although PARSIFAL can be seen as a shift–reduce parser, it has a richer set of operations than the basic scheme specified in Fig. 3, and makes use of a procedural rule format rather than phrase–structure rules. In subsequent work, Shieber [90] and Pereira [71] have shown how a similar deterministic behaviour can be obtained based on standard shift–reduce parsing. By adopting suitable parsing tables, they are thus able to account for garden-path phenomena as well as various kinds of parsing preferences discussed in the literature. Furthermore, in contrast with PARSIFAL, the grammars used by their parsers can be ambiguous.

Marcus' approach has been further developed by Hindle [36,37], who has developed an extremely efficient wide-coverage deterministic *partial* parser called FIDDITCH (compare Sec. 8). Furthermore, a semantically oriented account, in which garden-path effects are induced by the semantic context, rather than the sentence structure as such, is provided by Crain and Steedman [23]. For a discussion of Marcus' approach, see also Briscoe [15]. LR parsing, a more elaborate shift–reduce technique, in which the parsing table consists of a particular finite automaton, will be described in Sec. 7.

5. PARSING WITH TRANSITION NETWORKS

During the 1970s and early 1980s, Augmented Transition Networks (ATNs) constituted the dominating framework for natural language processing. To understand ATNs, it is useful to begin by looking at the simplest kind of transition network, which consists of a finite set of states (vertices) and a set of state transitions that occur on input symbols. This is called a *finite-state transition network* (FSTN; [31,111]). In automata theory, the equivalent abstract machine is called a *finite (state) automaton* [38]. Transition networks have a distinguished initial state and a set of distinguished final states. Each encountered state corresponds to a particular stage in the search of a constituent and corresponds to one or several dotted expressions (as used in Secs. 4, 6, and 7). Each arc of an FSTN is labelled with a terminal symbol; traversing the arc corresponds to having found some input matching the label. FSTNs can describe only the set of regular languages. Nevertheless, finite-state frameworks are sometimes adopted in natural language parsing on grounds of efficiency or modelling of human linguistic performance [22,27,57,77,80].

By introducing recursion, in other words, by allowing the label of an arc to be a nonterminal symbol referring to a subnetwork, a *recursive transition network* (RTN) is obtained [31,111]. In automata theory, the equivalent notion is a *pushdown automaton* [38]. An RTN thus consists of a set of subnetworks. Furthermore, it is equipped with a pushdown stack that enables the system to jump to a specified subnetwork and to return appropriately after having traversed it. An RTN can describe the set of context-free languages.

An ATN is an RTN in which each arc has been augmented with a *condition* and a sequence of *actions*. The conditions and actions may be arbitrary. The condition associated with an arc must be satisfied for the arc to be chosen, and the actions are executed as the arc is traversed. The purpose of the actions is to construct pieces of linguistic structure (typically, parts of a parse tree). This structure is stored in *registers*, which are passed along as the parsing proceeds. In addition, a global set of registers called the *hold list*, is commonly used for handling unbounded dependencies. Because of these very general mechanisms, the ATN framework is Turing-equivalent, allowing any recursively enumerable language to be described.

Mainly because of the way registers are typically used, the most straightforward processing regimen for ATN parsing is top-down, left-to-right (compare Sec. 4.A). Furthermore, depth-first search with backtracking is commonly used, but breadth-first is also possible. In practice, the efficiency of depth-first search can be increased by ordering the set of outgoing arcs according to the relative frequencies of the corresponding inputs.

The principal references to ATNs are Woods [113, 114] and Woods et al. [115]. For an excellent introduction, see Bates [8]. For a slightly different conception of ATNs (especially in terms of the structures generated), see Winograd [111, Chap. 5]. A comparison of ATNs with Definite Clause Grammars is made by Pereira and Warren [73], who show that the latter are able to capture everything essential from the former, with the added advantage of providing a declarative formalism.

6. TABULAR PARSING (CHART PARSING)

Because of the high degree of local ambiguity of natural language, the same subproblem often appears more than once during (nondeterministic) parsing of a sentence. One example of this is the analysis of the prepositional phrase "on the table" in Example {6}, which can be attached in two ways in the parse tree for the sentence. However, the prepositional-phrase analysis itself is independent of this attachment decision. It, therefore, can be useful to define equivalence classes of partial analyses, store these analyses in a table, and look them up whenever needed, rather than recomputing them each time as the nondeterministic algorithms in Secs. 4 and 5 would do. This technique, which is called *tabulation* (hence, *tabular parsing*), potentially allows exponential reductions of the search space, and thereby, provides a way of coping with ambiguity in parsing. An additional advantage of the tabular representation is that it is compatible with a high degree of flexibility relative to search and control strategies. The tabular framework in which this property has been most systematically explored is called *chart parsing*.[9]

A. Preliminaries

The table in which partial analyses are stored is often conceived of as a directed graph $C = \langle V, E \rangle$, the *chart*. V is then a finite, nonempty set of vertices, $E \subseteq V \times V \times R$ is a finite set of edges and R is a finite set of dotted context-free rules (see following). We assume that the graph is labelled such that each vertex and edge has a unique label, which we denote by a subscript (e.g., v_i or e_j).

The vertices $v_0, \cdots, v_n \in V$ correspond to the linear positions between each word of an n-word sentence $w_1 \cdots w_n$. An edge $e \in E$ between vertices v_i and v_j carries information about a (partially) analysed constituent between the correspond-

[9] To some extent, the use of the terms "tabular parsing" and "chart parsing" reflects different traditions: tabular parsing was originally developed within the field of compiler design [4,25,48,116], and chart parsing, more or less independently, within computational linguistics [44,49,50,53,103,104]. However, chart parsing often refers to a framework which, as a means to attain flexible control, makes use of *both* a table (in the form of a chart) and an agenda (see Sec. 6.D). More fundamentally, tabulation is a general technique that can be realized by other means than a chart (see Sec. 7.B).

ing positions. However, to avoid unnecessary detail, we have not included any element corresponding to the actual analysis in the edge tuple (but compare Sec. 6.E.) We write an edge as a triple

$$\langle v_s, v_t, X \rightarrow \alpha \cdot \beta \rangle$$

with starting vertex v_s, ending vertex v_t, and dotted context-free rule $X \rightarrow \alpha \cdot \beta$. The dotted rule asserts that X derives α followed by β, where α corresponds to the part of the constituent parsed so far. If β is empty, then the edge represents a completely analysed constituent and is called *inactive* (*passive*); otherwise it represents a partially analysed constituent and is called *active*. An active edge thus represents a state in the search for a constituent, including a hypothesis about the category of the remaining, needed input. If α is empty, then the edge is active and looping. It then represents the initial state, at which no input has been parsed in the search for the constituent.[10]

B. Chart Parsing with Context-Free Grammar

Figure 4 provides a scheme for chart parsing with a context-free grammar. It operates in a bottom-up fashion but makes use of top-down prediction, and thus corresponds to Earley's [4,25] algorithm.[11] Furthermore, it is equivalent to the top-down strategies of Kay [53, p. 60] and Thompson [103, p. 168].

The algorithm involves three repeated steps, which can be carried out in any order:

1. *Scan*: Add an inactive edge for each lexical category of the word.
2. *Predict (top-down)*: Add an edge according to each rule whose first left-hand-side category matches the needed category of the triggering active edge, unless an equivalent edge already exists in the chart.
3. *Complete*: Add an edge whenever the category of the first needed constituent of an active edge matches the category of an inactive edge.

In addition, a top-down initialization of the chart is needed. Note that, because a phrase–structure grammar (with atomic categories) is being used, the redundancy test in the prediction step corresponds to a simple equality test. The parsing scheme is specified in Fig. 4. For an illustration of a chart resulting from this scheme, using the example grammar in Fig. 5, see Fig. 6. The contents of the edges are further specified in Fig. 7. Typically, what is desired as output is not the chart as such, but the set of parse trees that cover the entire input and the top node of which is the start category of the grammar. Such a result can be extracted by a separate procedure, which operates on the set of S edges connecting the first and last vertex.

It is possible to obtain a pure bottom-up behaviour by replacing the top-down predictor with a corresponding bottom-up step:

[10] Chart parsing using both active and inactive edges is sometimes called *active chart parsing*. This is to distinguish it from the case when only inactive edges are used; that is, when the chart corresponds to a *well-formed substring table*.

[11] Edges correspond to *states* [25] or *items* [4, p. 320] in Earley's algorithm.

Input: A string of words $w_1 \cdots w_n$.

Output: A chart $C = \langle V, E \rangle$.

Method: Add an initial top-down prediction $\langle v_0, v_0, X_0 \rightarrow \cdot\, \alpha \rangle$ for each rule $X_0 \rightarrow \alpha$ such that $X_0 = S$, where S is the start category of the grammar. Repeat the following steps for each vertex v_j such that $j = 0, \ldots, n$ until no more edges can be added to the chart.

> *Scan*: If $w_j = a$, then for each lexical entry of the form $X_0 \rightarrow a$, add an edge $\langle v_{j-1}, v_j, X_0 \rightarrow a\, \cdot \rangle$.
>
> *Predict* (*top-down*): For each edge of the form $\langle v_i, v_j, X_0 \rightarrow \alpha \cdot X_m \beta \rangle$ and each rule of the form $Y_0 \rightarrow \gamma$ such that $Y_0 = X_m$, add an edge $\langle v_j, v_j, Y_0 \rightarrow \cdot\, \gamma \rangle$ unless it already exists.
>
> *Complete*: For each edge of the form $\langle v_i, v_j, X_0 \rightarrow \alpha \cdot X_m \beta \rangle$ and each edge of the form $\langle v_j, v_k, Y_0 \rightarrow \gamma \cdot \rangle$, add an edge $\langle v_i, v_k, X_0 \rightarrow \alpha X_m \cdot \beta \rangle$ if $X_m = Y_0$.

Fig. 4 A scheme for chart parsing with top-down prediction (Earley-style), using context-free grammar.

2. *Predict (bottom-up)*: Add an edge according to each rule whose first right-hand-side category matches the category of the triggering inactive edge, unless an equivalent edge already exists in the chart.

No initialization of the chart is needed for this. The algorithm is specified in Fig. 8, and an example chart is shown in Fig. 9.[12] It is often referred to as a *left-corner* algorithm, because it performs leftmost derivations (and the left corner of a rule is the leftmost category of its right-hand side). It is equivalent to the bottom-up strategies of Kay [53, p. 58] and Thompson [103, p. 168]. Furthermore, it is related to the Cocke–Kasami–Younger (CKY) algorithm [4, 48, 116], although, in particular, this matches categories from right to left in the right-hand side of a grammar rule and only makes use of inactive edges.

Because of the avoidance of redundant calculation, the problem of chart parsing with a context-free grammar is solvable in polynomial time. More specifically, the worst-case complexity of each of the two foregoing algorithms is $O(|G|^2 \cdot n^3)$, where n is the length of the input and $|G|$ is the size of the grammar expressed in terms of the number of rules and nonterminal categories (see, e.g.,

S	→	$NP\ VP$	Det	→	the
NP	→	$Det\ N$	N	→	old
NP	→	$Det\ A\ N$	N	→	man
VP	→	V	N	→	ships
VP	→	$V\ NP$	A	→	old
			V	→	man
			V	→	ships

Fig. 5 Sample grammar and lexicon.

[12] As formulated here, the algorithm does not work with empty productions.

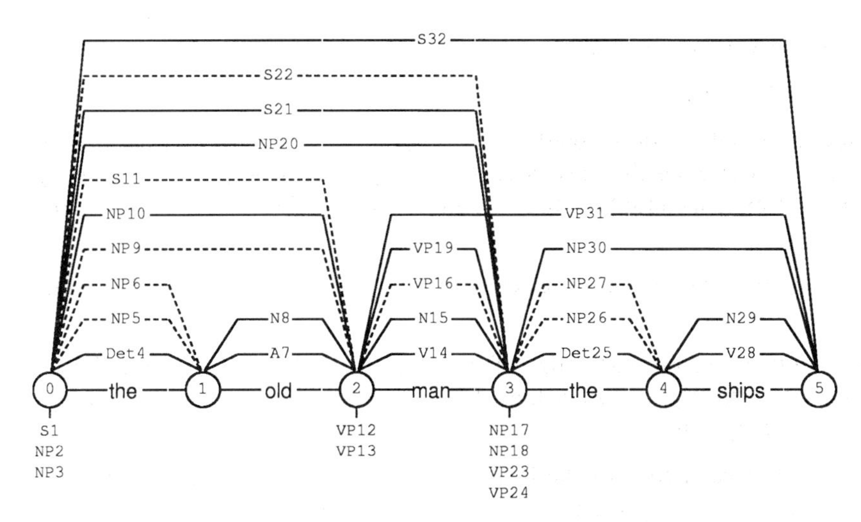

Fig. 6 Chart for the sentence "The old man the ships" using the Earley-style algorithm in Fig. 4 and the grammar given in Fig. 5. Inactive edges are drawn using continuous lines, active edges using dashed lines, and predicted (looping) edges are indicated below the vertices.

Earley [25], Aho and Ullman [4], or Barton et al. [7, Sec. 7.6].) For investigations of practical performance of chart parsing see, e.g., [11, 76, 88, 99, 106, 112].

C. Some Refinements

Various refinements of the foregoing two basic algorithms are possible. One option is filtering, which attempts to reduce the number of edges by making sure that only predictions that in some stronger sense are useful will be added to the chart. For example, pure top-down parsing has the problem that many predictions may be useless, because they are made before the input to which they are meant to apply has been scanned. Moreover, if empty categories are allowed (say, because the

S_1:	$\langle v_0, v_0, S \to \cdot NP\ VP \rangle$	NP_{17}:	$\langle v_3, v_3, NP \to \cdot Det\ N \rangle$
NP_2:	$\langle v_0, v_0, NP \to \cdot Det\ N \rangle$	NP_{18}:	$\langle v_3, v_3, NP \to \cdot Det\ A\ N \rangle$
NP_3:	$\langle v_0, v_0, NP \to \cdot Det\ A\ N \rangle$	VP_{19}:	$\langle v_2, v_3, VP \to V \cdot \rangle$
Det_4:	$\langle v_0, v_1, Det \to \text{the} \cdot \rangle$	NP_{20}:	$\langle v_0, v_3, NP \to Det\ A\ N \cdot \rangle$
NP_5:	$\langle v_0, v_1, NP \to Det \cdot N \rangle$	S_{21}:	$\langle v_0, v_3, S \to NP\ VP \cdot \rangle$
NP_6:	$\langle v_0, v_1, NP \to Det \cdot A\ N \rangle$	S_{22}:	$\langle v_0, v_3, S \to NP \cdot VP \rangle$
A_7:	$\langle v_1, v_2, A \to \text{old} \cdot \rangle$	VP_{23}:	$\langle v_3, v_3, VP \to \cdot V \rangle$
N_8:	$\langle v_1, v_2, N \to \text{old} \cdot \rangle$	VP_{24}:	$\langle v_3, v_3, VP \to \cdot NP \rangle$
NP_9:	$\langle v_0, v_2, NP \to Det\ A \cdot N \rangle$	Det_{25}:	$\langle v_3, v_4, Det \to \text{the} \cdot \rangle$
NP_{10}:	$\langle v_0, v_2, NP \to Det\ N \cdot \rangle$	NP_{26}:	$\langle v_3, v_4, NP \to Det \cdot N \rangle$
S_{11}:	$\langle v_0, v_2, S \to NP \cdot VP \rangle$	NP_{27}:	$\langle v_3, v_4, NP \to Det \cdot A\ N \rangle$
VP_{12}:	$\langle v_2, v_2, VP \to \cdot V \rangle$	V_{28}:	$\langle v_4, v_5, V \to \text{ships} \cdot \rangle$
VP_{13}:	$\langle v_2, v_2, VP \to \cdot V\ NP \rangle$	N_{29}:	$\langle v_4, v_5, N \to \text{ships} \cdot \rangle$
V_{14}:	$\langle v_2, v_3, V \to \text{man} \cdot \rangle$	NP_{30}:	$\langle v_3, v_5, NP \to Det\ N \cdot \rangle$
N_{15}:	$\langle v_2, v_3, N \to \text{man} \cdot \rangle$	VP_{31}:	$\langle v_2, v_5, VP \to V\ NP \cdot \rangle$
VP_{16}:	$\langle v_2, v_3, VP \to V \cdot NP \rangle$	S_{32}:	$\langle v_0, v_5, S \to NP\ VP \cdot \rangle$

Fig. 7 Contents of the chart edges obtained in Fig. 6.

Input: A string of words $w_1 \cdots w_n$.

Output: A chart $C = \langle V, E \rangle$.

Method: Repeat the following steps for each vertex v_j such that $j = 0, \ldots, n$ until no more edges can be added to the chart.

> *Scan*: As in Figure 4.
>
> *Predict* (*bottom-up*): For each edge of the form $\langle v_j, v_k, X_0 \rightarrow \alpha \cdot \rangle$ and each rule of the form $Y_0 \rightarrow Y_1\gamma$ such that $Y_1 = X_0$, add an edge $\langle v_j, v_j, Y_0 \rightarrow \cdot Y_1\beta \rangle$ unless it already exists.
>
> *Complete*: As in Figure 4.

Fig. 8 A scheme for chart parsing with bottom-up prediction, using context-free grammar.

grammar makes use of gaps), pure bottom-up parsing has the problem that every such category must be predicted at each point in the input string, because the parser cannot know in advance where the gaps occur.

To deal with problems such as these, a reachability relation $\mathcal{R}$ on the set of grammatical categories can be precompiled: $\mathcal{R}$ is defined as the set of ordered pairs $\langle X_0, X_1 \rangle$ such that there is a production $X_0 \rightarrow X_1 \cdots X_n$. If $\mathcal{R}^*$ is the reflexive and transitive closure of $\mathcal{R}$, then $\mathcal{R}^*(X, X')$ holds if and only if $X = X'$ or there exists some derivation from X to $X' \cdots$ such that X' is the leftmost category in a string of categories dominated by X.

Earley-style parsing with *bottom-up filtering* can then be obtained by augmenting the algorithm in Fig. 4 with a condition that checks if some category of a word w_j is reachable from the first needed category in α of an edge $\langle v_{j-1}, v_{j-1}, X \rightarrow \cdot\ \alpha \rangle$ about to be predicted; if not, this prediction is discarded. This corresponds to Kay's [53, p. 48] directed top-down strategy; it can also be seen as a one-symbol lookahead.

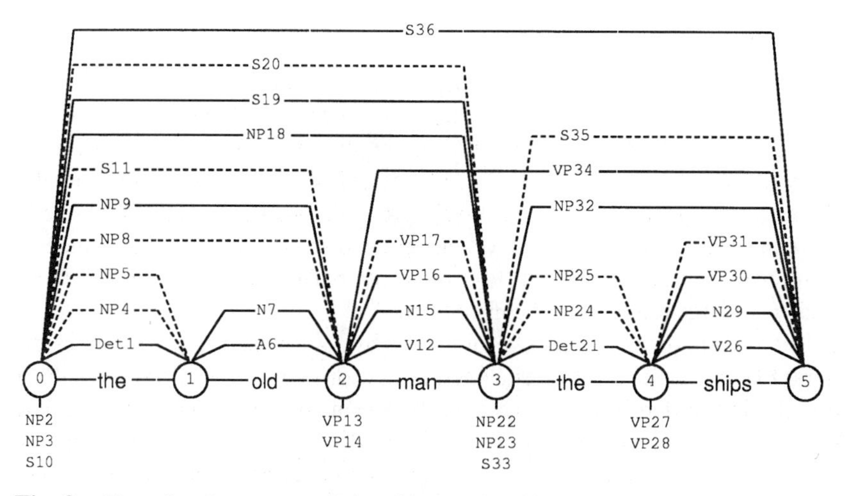

Fig. 9 Chart for the sentence "The old man the ships" using the pure bottom-up algorithm in Fig. 8 and the grammar in Fig. 5.

Analogously, bottom-up parsing with *top-down filtering* can be obtained by augmenting the algorithm in Fig. 8 with a condition that checks if the category of an edge $\langle v_k, v_k, X' \to \cdot\, \gamma \rangle$ about to be predicted is reachable from the first needed category in β of some incoming active edge $\langle v_i, v_k, X \to \alpha \cdot \beta \rangle$; if not, the edge about to be predicted is discarded. (This strategy needs an initial top-down prediction in order not to filter out predictions made at the first vertex.) This corresponds to the method described by Pratt [76, p. 424], who refers to the top-down filter as an oracle; it also corresponds to Kay's [53, p. 48] directed bottom-up parsing.

When several ambiguities are localized to the same part of a sentence, it is possible to use a more economic representation than the one above. For example, in the sentence

He put the block in the box on the table {10}

the verb phrase "put the block in the box on the table" has two readings in accordance with example {6}. Two inactive edges will thus be used to represent these readings. However, each of these edges will be treated in exactly the same way by any superior edge using it as a subedge. It is therefore possible to collapse the two edges into a single one in which two parse trees are shared. This technique is called (*local-ambiguity*) *packing* [106]; compare the end of Sec. 7.B.

Furthermore, other forms of rule invocation are possible. For example, in head-driven chart parsing the predictions are triggered from the heads instead of from the left corners, as in Fig. 8. For some work on this, see Bouma and van Noord [13], Kay [54], Satta and Stock [85] and Nederhof and Satta [68]. Another variant is bidirectional chart parsing [84], in which edges may be completed both to the left and to the right (in other words, the chart is here an undirected graph).

Chart (or tabular) parsing appears in several incarnations: For example, it can be seen as an instance of dynamic programming [9], and it has been formalized as deduction [74, 94], in which tabulation corresponds to maintaining a cache of lemmas. For general discussion of tabular parsing, see also Refs. 67, 89, 96.

D. Controlling Search

The algorithms in Figs. 4 and 8 do not specify any order in which to carry out the three parsing steps; in other words, they lack an explicit search strategy. To control search, an *agenda* can be used [53,104]. The idea is as follows: Every newly generated edge is added to the agenda. Edges are moved from the agenda to the chart one by one until the agenda is empty. Whenever an edge is moved to the chart, its consequences in terms of new (predicted or completed) edges are generated and themselves added to the agenda for later consideration.

Different search strategies can be obtained by removing agenda items in different orders: For example, treating the agenda as a (last-in-first-out stack results in a behaviour similar to depth-first search; treating it instead as a queue first-in-first-out stack) results in a behaviour similar to breadth-first search. Furthermore, by introducing a preference metric with respect to agenda elements, more fine-grained control is possible [28].

E. Chart Parsing with Constraint-Based Grammar

Basically, the chart parsers in Figs. 4 and 8 can be adapted to constraint-based grammar by keeping a feature term D in each edge, and by replacing the matching of atomic categories with unification. We thus take an edge to be a tuple

$$\langle v_s, v_t, X_0 \rightarrow \alpha \cdot \beta, D\rangle$$

starting from vertex v_s and ending at vertex v_t with dotted context-free rule $X_0 \rightarrow \alpha \cdot \beta$ and feature term D.

A problem in chart parsing with constraint-based grammar, as opposed to phrase-structure grammar, is that the edge-redundancy test involves comparing complex feature terms instead of testing for equality between atomic symbols. For this reason, we need to make sure that no previously added edge *subsumes* a new edge to be added [72, p. 199; 92]. One edge subsumes another edge if and only if the first three elements of the edges are identical and the fourth element of the first edge subsumes that of the second edge. Roughly, a feature term D subsumes another feature term D' if D contains a subset of the information in D'. The rationale for using this test is that we are interested only in adding edges that are less specific than the old ones, for everything we could do with a more specific edge, we could also do with a more general one. In general, the test must be carried out both in the prediction and completion step. (If an agenda is used, this needs analogous testing for the same reason. For efficiency it is also desirable that any existing edge that is subsumed by a new edge be removed, although this introduces some extra bookkeeping.)

Figure 10 provides a scheme for tabular parsing with a constraint-based grammar formalism (for its context-free counterpart, see Fig. 4). It corresponds to Shieber's [92] generalization of Earley's [25] algorithm. Here, for example, $D(\langle X_0$ *category*$\rangle)$ denotes the substructure at the end of the path $\langle X_0$ *category*$\rangle$ from the root of the feature structure D, and $D \sqcup E$ denotes the unification of the feature structures D and E.

As in the context-free case, alternative rule-invocation strategies can be obtained by changing the predictor. To generate more selective predictions, the feature terms of the left-hand-side category of the rule are unified with the needed category of the triggering active edge (rather than to just matching their *category* features, as would be the case if only the context-free backbone was used). Now, unification grammars may contain rules that lead to prediction of ever more specific feature terms that do not subsume each other, thereby resulting in infinite sequences of predictions. In logic programming, the occur check is used for circumventing a corresponding circularity problem. In constraint-based grammar, Shieber [92] introduced the notion of *restriction* for the same purpose. A restrictor removes those portions of a feature term that could potentially lead to nontermination. This is in general done by replacing those portions with free (newly instantiated) variables, which typically removes some coreference. The purpose of restriction is to ensure that terms to be predicted are only instantiated to a certain depth, so that terms will eventually subsume each other.

Input: A string of words $w_1 \cdots w_n$.

Output: A chart $C = \langle V, E \rangle$.

Method: Initialize the chart with an edge $\langle v_0, v_0, X_0 \to \cdot\ \alpha, D \rangle$ for each rule $\langle X_0 \to \alpha, D \rangle$ such that $D(\langle X_0\ category \rangle) = S$, where S is the start category of the grammar. Repeat the following steps for each vertex v_j such that $j = 0, \ldots, n$ until no more edges can be added to the chart.

> *Scan*: If $w_j = a$, then for each lexical entry of the form $\langle X_0 \to \cdot\, a, D \rangle$ add an edge $\langle v_{j-1}, v_j, X_0 \to a \cdot, D \rangle$.
>
> *Predict* (*top-down*): For each edge of the form $\langle v_i, v_j, X_0 \to \alpha \cdot X_m \beta, D \rangle$ and each rule of the form $\langle Y_0 \to \gamma, E \rangle$ such that the unification $E(\langle Y_0 \rangle) \sqcup D(\langle X_m \rangle)$ succeeds, add an edge $\langle v_j, v_j, Y_0 \to \cdot\ \gamma, E' \rangle$ unless it is subsumed by any existing edge. E' is the result on E of the restricted unification $E(\langle Y_0 \rangle) \sqcup D(\langle X_m \rangle)$.
>
> *Complete*: For each edge of the form $\langle v_i, v_j, X_0 \to \alpha \cdot X_m \beta, D \rangle$ and each edge of the form $\langle v_j, v_k, Y_0 \to \gamma \cdot, E \rangle$, add the edge $\langle v_i, v_k, X_0 \to \alpha X_m \cdot \beta, D' \rangle$ if the unification $E(\langle Y_0 \rangle) \sqcup D(\langle X_m \rangle)$ succeeds and this edge is not subsumed by any existing edge. D' is the result on D of the unification $E(\langle Y_0 \rangle) \sqcup D(\langle X_m \rangle)$.

Fig. 10 A scheme for chart parsing with top-down prediction (Earley-style), using constraint-based grammar.

7. LR PARSING

An LR parser is a type of shift–reduce parser that was originally devised for programming languages [56]. An LR parser constructs the rightmost derivation of a syntactic analysis in reverse. In fact, the "R" in "LR" stands for rightmost-derivation in reverse. The "L" stands for left-to-right scanning of the input string. The success of LR parsing lies in handling a number of production rules simultaneously by the use of prefix merging (compare, the end of the next section), rather than attempting to apply one rule at a time.

The use of LR parsing techniques for natural languages involves adapting them to high-coverage grammars that allow large amounts of syntactic ambiguity, and thus result in nondeterminism when parsing. The most important extensions to the original LR-parsing scheme addressing these issues are:

- Generalized LR (GLR) parsing, developed by Tomita [106], but whose theoretical foundation was provided by Lang [59], avoids exponential search by adopting a quasiparallel deterministic processing scheme, packing local ambiguity in the process. This is discussed in Sec. 7.B.
- Using constraint-based grammars employing complex-valued features to reduce the amount of (spurious) ambiguity [see, e.g., 66, 82]. This is discussed in Sec. 7.C.

Useful references to deterministic LR parsing and LR compilation are Sippu and Soisalon-Soininen [98] and Aho et al. [3; pp. 215–266]. We will here first discuss simple backtracking LR parsing.

A. Basic LR Parsing

An LR parser is basically a pushdown automaton, that is, it has a stack in addition to a finite set of internal states and a reader head for scanning the input string from left to right, one symbol at a time. The stack is used in a characteristic way: the items on the stack consist of alternating grammar symbols and states. The current state is simply the state on top of the stack. The most distinguishing feature of an LR parser, however, is the form of the transition relation—the action and goto tables. A nondeterministic LR parser can in each step perform one of four basic actions. In state S_i with lookahead symbol[13] X it can:

1. `accept`: Halt and signal success.
2. `error`: Fail and backtrack.
3. `shift` S_k: Consume the input symbol X, push it onto the stack, and transit to state S_k by pushing it onto the stack.
4. `reduce`: R_n: Pop off two items from the stack for each phrase in the RHS of grammar rule R_n, inspect the stack for the old state S_j now on top of the stack, push the LHS Y of the rule R_n onto the stack, and transit to state S_k determined by `goto` S_j Y S_k by pushing S_k onto the stack.

This is not a pushdown automaton according to some formal definitions found in the literature [e.g., 70, pp. 491–492], because the reduce actions do not consume any input symbols. From a theoretical point of view, this is not important, for it does not change the expressive power of the machine. From a practical point of view, this gives the parser the possibility of not halting by reducing empty (ϵ) productions and transiting back to the same state; care must be taken to avoid this. This is a variant of the well-known problem of bottom-up parsing in conjunction with empty productions.

Prefix merging is accomplished by having each internal state correspond to a set of dotted rules (as defined in Sec. 6), also called *items*[14]. For example, if the grammar contains the following three rules,

$$VP \rightarrow V \quad VP \rightarrow V\ NP \quad VP \rightarrow V\ NP\ NP$$

there will be a state containing the items

$$VP \rightarrow V \cdot \quad VP \rightarrow V \cdot NP \quad VP \rightarrow V \cdot NP\ NP$$

This state corresponds to just having found a verb (V). Which of the three rules to apply in the end will be determined by the rest of the input string; at this point no commitment has been made to any of them.

1. LR-Parsed Example

The example grammar of Fig. 11 will generate the internal states of Fig. 12. These in turn give rise to the parsing tables of Fig. 13. The entry `s2` in the action table, for example, should be interpreted as "shift the lookahead symbol onto the stack and transit to state S_2." The action entry `r7` should be interpreted as "reduce by rule R_7." The goto entries simply indicate what state to transit to once a phrase of that

[13] The lookahead symbol is the next symbol in the input string, that is, the symbol under the reader head.
[14] Not to be confused with stack items, or Earley items, compare the beginning of Sec. 6.B.

S	→	*NP VP*	(R_1)
VP	→	*V*	(R_2)
VP	→	*V NP*	(R_3)
VP	→	*V NP NP*	(R_4)
VP	→	*VP PP*	(R_5)
NP	→	*Det N*	(R_6)
NP	→	*Name*	(R_7)
NP	→	*NP PP*	(R_8)
PP	→	*Prep NP*	(R_9)

Fig. 11 Sample grammar.

type has been constructed. Note the two possibilities in states S_{11}, S_{12}, and S_{13} for lookahead symbol preposition, *Prep*, both of which must be tried: we can either shift it onto the stack or perform a reduction. This is a shift–reduce conflict, and it is the source of the nondeterminism when parsing sentence {11}.

Put the block in the box on the table {11}

The other source of nondeterminism is due to reduce–reduce conflicts, where it is possible to reduce by more than one rule. In this case, the RHS of one rule must be a suffix of the RHS of another one. For programming languages, these conflicts can be resolved by prescribing the language semantics, thereby specifying which alternative to choose, or the language syntax, to avoid thereby these situations altogether. The former strategy could also be applied when parsing natural languages to perform

S_0

$S' \rightarrow \cdot S$
$S \rightarrow \cdot NP\ VP$
$NP \rightarrow \cdot Det\ N$
$NP \rightarrow \cdot Name$
$NP \rightarrow \cdot NP\ PP$

S_1

$S \rightarrow NP \cdot VP$
$NP \rightarrow NP \cdot PP$
$VP \rightarrow \cdot V$
$VP \rightarrow \cdot V\ NP$
$VP \rightarrow \cdot V\ NP\ NP$
$VP \rightarrow \cdot VP\ PP$
$PP \rightarrow \cdot Prep\ NP$

S_2

$NP \rightarrow Det \cdot N$

S_3

$NP \rightarrow Name \cdot$

S_4

$S' \rightarrow S \cdot$

S_5

$S \rightarrow NP\ VP \cdot$
$VP \rightarrow VP \cdot PP$
$PP \rightarrow \cdot Prep\ NP$

S_6

$VP \rightarrow V \cdot$
$VP \rightarrow V \cdot NP$
$VP \rightarrow V \cdot NP\ NP$
$NP \rightarrow \cdot Det\ N$
$NP \rightarrow \cdot Name$
$NP \rightarrow \cdot NP\ PP$

S_7

$NP \rightarrow NP\ PP \cdot$

S_8

$PP \rightarrow Prep \cdot NP$
$NP \rightarrow \cdot Det\ N$
$NP \rightarrow \cdot Name$
$NP \rightarrow \cdot NP\ PP$

S_9

$VP \rightarrow VP\ PP \cdot$

S_{10}

$NP \rightarrow Det\ N \cdot$

S_{11}

$VP \rightarrow V\ NP \cdot$
$VP \rightarrow V\ NP \cdot NP$
$NP \rightarrow NP \cdot PP$
$NP \rightarrow \cdot Det\ N$
$NP \rightarrow \cdot Name$
$NP \rightarrow \cdot NP\ PP$
$PP \rightarrow \cdot Prep\ NP$

S_{12}

$VP \rightarrow V\ NP\ NP \cdot$
$NP \rightarrow NP \cdot PP$
$PP \rightarrow \cdot Prep\ NP$

S_{13}

$PP \rightarrow Prep\ NP \cdot$
$NP \rightarrow NP \cdot PP$
$PP \rightarrow \cdot Prep\ NP$

Fig. 12 The resulting internal states.

State	Action						Goto			
	Det	*N*	*Prep*	*Name*	*V*	*eos*	*NP*	*PP*	*S*	*VP*
S_0	s2			s3			S_1		S_4	
S_1			s8		s6			S_7		S_5
S_2		s10								
S_3	r7		r7	r7	r7	r7				
S_4						acc				
S_5			s8			r1		S_9		
S_6	s2		r2	s3		r2	S_{11}			
S_7	r8		r8	r8	r8	r8				
S_8	s2			s3			S_{13}			
S_9			r5			r5				
S_{10}	r6		r6	r6	r6	r6				
S_{11}	s2		s8/r3	s3		r3	S_{12}	S_7		
S_{12}			s8/r4			r4		S_7		
S_{13}	r9		s8/r9	r9	r9	r9		S_7		

Fig. 13 The corresponding LR parsing tables.

(some) disambiguation already at this stage, compare the discussion toward the end of Sec 4.B. This approach, however, is less popular today; more popular, instead, is letting the parser overgenerate and subsequently filter the set of analyses based on lexical collocation statistics. Lexical collocation statistics is often very important for structural disambiguation: for example, if we want to resolve the shift–reduce conflict when parsing the sentence pairs

John read a book on the train/topic {12}

we realize that the choice will depend crucially on which of the nouns "train" and "topic" is in the sentence. In the former case, John was on the train when he read the book, and the *PP* "on the train" attaches to the *VP* "read a book", whereas in the latter case the book was about some topic, and the *PP* "on the topic" attaches to the *NP* "a book".

Using these tables we can parse the sentence

John reads a book {13}

as follows:

Action	Stack	String
init:	$[S_0]$	*John reads a book*
s3:	$[S_0, Name, S_3]$	*reads a book*
r7:	$[S_0, NP, S_1]$	*reads a book*
s6:	$[S_0, NP, S_1, V, S_6]$	*a book*
s2:	$[S_0, NP, S_1, V, S_6, Det, S_2]$	*book*
s10:	$[S_0, NP, S_1, V, S_6, Det, S_2, N, S_{10}]$	ϵ
r6:	$[S_0, NP, S_1, V, S_6, NP, S_{11}]$	ϵ
r3:	$[S_0, NP, S_1, VP, S_5]$	ϵ
r1:	$[S_0, S, S_4]$	ϵ
acc:	$[S_0, S, S_4]$	ϵ

Initially state S_0 is pushed onto the empty stack. The name (*Name*) corresponding to the word "John" is shifted onto the stack and the parser transits to state S_3 by pushing it onto the stack. Then, *Name* on top of the stack is reduced to a noun phrase using rule R_7 ($NP \rightarrow Name$) by popping state S_3 and *Name* from the stack. The LHS noun phrase (*NP*) is pushed onto the stack and the parser transits to state S_1, determined by the entry for *NP* and state S_0 (on top of the stack after popping state S_3 and *Name*) in the goto table, by pushing S_1 onto the stack. Next, the verb (*V*) corresponding to the word "reads" is shifted onto the stack and the parser transits to state S_6 by pushing it onto the stack. Then the determiner (*Det*) corresponding to the word "a" is shifted onto the stack and the parser transits to state S_2. The noun (*N*) corresponding to the word "book" is shifted onto the stack and the parser transits to state S_{10}. At this point, the noun and the determiner on top of the stack are reduced to a noun phrase using rule R_6 ($NP \rightarrow Det\ N$) by popping state S_{10}, the noun, state S_2, and the determiner from the stack. The noun phrase is then pushed onto the stack, and the parser transits to state S_{11} by pushing it onto the stack. S_{11} is determined from the entry in the goto table for S_6 (on top of the stack after the pops) and the LHS symbol *NP*. Next, the noun phrase and the verb on top of the stack are reduced to a verb phrase (*VP*) using rule R_3 ($VP \rightarrow V\ NP$), which is pushed onto the stack, and the parser transits to state S_5. Then the verb phrase and the nown phrase on top of the stack are reduced to a sentence (*S*) using rule R_1 ($S \rightarrow NP\ VP$), which is pushed onto the stack, and the parser transits to state S_4. Finally, the input string is accepted.

2. *LR-Table Compilation*

Compiling LR-parsing tables consists of constructing the internal states (that is, sets of items) and from these deriving the shift, reduce, accept, and goto entries of the transition relation.

New states can be induced from previous ones; given a state S_i, another state S_j, reachable from it by `goto` $S_i\ X\ S_j$ (or `shift` $S_i\ X\ S_j$ if X is a terminal symbol), can be constructed as follows:

1. Select all items in state S_i where a particular symbol X follows immediately after the dot and move the dot to after this symbol. This yields the kernel items of state S_j.
2. Construct the nonkernel closure by repeatedly adding a so-called nonkernel item (with the dot at the beginning of the RHS) for each grammar rule whose LHS matches a symbol following the dot of some item in S_j.[15]

For example state S_1 of Fig. 12 can be constructed from state S_0 by advancing the dot of the items $S \rightarrow \cdot\ NP\ VP$ and $NP \rightarrow \cdot\ NP\ PP$ to form the items $S \rightarrow NP\ \cdot\ VP$ and $NP \rightarrow NP\ \cdot\ PP$ which constitute the kernel of state S_1. The remaining nonkernel items are generated by the grammar rules for *VP*s and *PP*s (the categories following the dot in the new kernel items); namely, rules R_2 R_3, R_4, R_5, and R_9. The resulting nonkernel items are:

---

[15] Note that this is a closure operation, as added nonkernel items may trigger adding further nonkernel items.

$$VP \rightarrow \cdot\ V \qquad VP \rightarrow \cdot\ VP\ PP$$
$$VP \rightarrow \cdot\ V\ NP \qquad PP \rightarrow \cdot\ Prep\ NP$$
$$VP \rightarrow \cdot\ V\ NP\ NP$$

By using this method, the set of all parsing states can be induced from an initial state the single kernel item of which has the top symbol of the grammar preceded by the dot as its RHS. In Fig. 12 this is the item $S' \rightarrow \cdot\ S$ of the state S_0.

The shift, goto and accept entries fall out automatically from this procedure. Any item where the dot is at the end of the RHS gives rise to a reduction by the corresponding grammar rule. Thus, it remains to determine the lookahead symbols of the reduce entries.

In *Simple LR (SLR)* the lookahead is any terminal symbol that can follow immediately after a symbol of the same type as the LHS of the rule. in *LookAhead LR (LALR)*, it is any terminal symbol that can immediately follow the LHS, given that it was constructed using this rule in this state. In general, LALR gives considerably fewer reduce entries than SLR, and thus results in faster parsing. LR(k) tables use k lookahead symbols, including these in the items. This gives a larger number of items and states.

The SLR lookaheads are straightforward to construct; we simply determine what grammar symbols can follow the LHS of the rule, and determine what (pre)-terminal symbol can start this symbol. The only two tricky points are that if we allow empty productions, some grammar symbols may have no realization in the input string, and we must take this into account by also looking at the symbols that follow these empty ones. The second tricky point is that the LHS symbol of the rule may occur at the end of some other grammar rule, and we must inspect the symbols that can follow this parent LHS symbol, and collect those terminal symbols that can start them.

For example, if we wish to find all terminal symbols that can follow an *NP*, to determine the lookahead symbols for reducing by any rule with *NP* as its LHS, we do as follows: from the grammar, we see that the three symbols that can follow *NP* are *VP* (rule R_1), *PP* (rule R_8), and *NP* (rule R_4). We further determine that only a *V* can start a *VP*, that only a *Prep* can start a *PP*, and that a *Det* or a *Name* can start an *NP*. We then note than an *NP* can be the last symbol of a *VP* in rules R_3 and R_4, and of a *PP* in the rule R_9, and that a *VP* can be followed by a *PP* or *eos* (rules R_5 and R_1, respectively), and that a *PP* can be followed by a *PP* by recursion through, for example, R_8, by a *VP* by transitivity through R_8 and R_1, by an *NP* through R_8 and R_4, or by *eos*, through, for example, R_5 and R_1. This adds *eos* as a lookahead symbol, as we have already found the ones that can start *VP*s, *PP*s, and *NP*s. As the grammar contains no empty productions, this completes the task.

Constructing the LALR lookahead symbols is somewhat more complicated, as is compiling LR(k) tables, and we refer the reader to Aho et al. [3; pp 240–244].

B. Generalized LR Parsing

Generalized LR (GLR) parsing addresses the issue of large amounts of ambiguity and nondeterminism by extending basic LR parsing with two concepts: a graph-structured stack (GSS) and a packed parse forest [107].[16] It also differs in being a

[16] Maintaining the GSS is a form of tabulation. For an approach to GLR parsing using a chart, see Nederhof and Setta [69].

breadth-first, accumulative search, synchronizing on shift actions, rather than a depth-first backtracking algorithm. Conceptually, a GLR parser works as follows:

1. Shift the next input symbol onto the stack(s). Do not keep the old stack(s), only the new one(s).
2. Perform all possible reduce actions on the stack(s). This will give rise to a new stack for each reduction. Keep the old stacks (before the reductions). Repeatedly perform all possible reduce actions on the new set of stacks, accumulating new stacks, until no further reduce actions are possible.
3. Goto 1.

This is a simple extension of the shift–reduce parsing algorithm given in Sec. 4.B. It is now guaranteed to find all analyses of any given input string, as it performs all possible combinations of reduce actions between the shift actions.

A graph-structured stack is used instead of a set of stacks[17]. This means that the new portions of the GSS constructed by shifting in step 1 are merged with the stack portions resulting from subsequent (repeated) reductions in step 2. After a shift action, two stack portions are merged if the new top states are equal. If we perform a reduction and the corresponding goto entry would imply creating a stack continuation that already exists, we simply use the existing one. The two partial parse trees corresponding to the two merged LHS symbols will then dominate the same substring, but be structurally different: As all surviving stack portions will have just shifted the word before the current string position (and possibly have been subjected to subsequent reductions), the dominated strings must end at the current string position. Because the same stack node is reached by the two reductions (after which the LHS symbol is pushed onto the GSS), the dominated strings must begin at the same string position. Because the action sequences producing the two derivations are different, the partial parse trees must also be different.

The partial parse trees associated with any nonterminal symbol will however not be recoverable from the GSS. For this reason, we will store the structure of each partial parse tree associated with each node in the GSS in a parse forest. By doing this in a clever way, we can avoid multiplying out the potentially exponentially many different parse trees that can be associated with the parsed word string: the LHS symbol of any reduction is constructed from a sequence of RHS symbols, each associated with a (set of) partial parse tree(s). The latter will be recorded in the packed parse forest, and the only information we need to specify for the node of the packed parse forest associated with the LHS of a reduction is the set of nodes in the packed parse forest associated with the RHS symbols. This means that the internal structure of each RHS symbol is encapsulated in its node in the packed parse forest. So, regardless of how many different ways these RHS symbols can, in turn, have been constructed, this ambiguity is contained locally at the corresponding nodes of the packed parse forest, and not multiplied out at the node corresponding to the current LHS of the production. As mentioned in Sec. 6.C, this is called local-ambiguity packing.

[17] Maintaining the GSS is a form of tabulation. For an approach to GLR parsing using a chart, see Nederhof and Satta [69].

The worst-case time complexity of the original GLR parsing algorithm is exponential, both in string length and grammar size. This is because there can theoretically be an exponential number of different internal states in the size of the grammar, and for certain grammars there may be inputs that force a parser to visit all states [40]. It can however be brought down to polynomial in string length by an optimization developed by Kipps [55]. It essentially involves avoiding a search of the graph-structured stack for the return state when reducing, instead employing a dynamically built table. Investigations of the practical performance of LR parsing are available [16, 19, 106].

C. LR Parsing and Constraint-Based Grammars

The problem of applying the LR-parsing scheme to constraint-based grammars, rather than context-free grammars for programming languages, for which it was originally intended, is that symbol matching no longer consists of checking atomic symbols for equality, but rather, comparing complex-valued feature terms; compare Sec. 6.E.

1. *Using a Context-Free Backbone Grammar*

The most common approach is to employ a context-free backbone grammar to compile the parsing tables and to parse input sentences. The full constraints of the underlying constraint-based grammar are then subsequently applied. The context-free backbone grammar is constructed by mapping the original unification-grammar feature bundles to atomic symbols.

A problem with this approach is that the predictive power of the constraint-based grammar is diluted when feature propagation is omitted, degrading parser performance, as the context-free backbone grammar will be less restrictive than the original constraint-based grammar. Another problem is that feature propagation is obstructed, which will further limit the restrictive power, and owing to empty productions, may lead to nontermination when parsing.

This problem can be relieved by using an appropriate mapping of feature terms to context-free symbols, minimizing the information lost, and methods for finding good mappings automatically have been proposed [16].

2. *Parsing with Constraint-Based Grammars*

It is thus desirable to use the full unification-grammar constraints when parsing and when constructing the parsing tables. There are three main problems to address here: The first one is how to define equivalence between linguistic objects, for these now consist of arbitrary complex-values feature bundles and not atomic symbols. In particular, the question is how to perform prefix merging.

The second question is how to maintain efficiency and avoid nontermination when compiling the parsing tables, in particular how to avoid adding more and more instantiated versions of the same item indefinitely when constructing the nonkernel closure. One might think that this could be done by adding only items that are not subsumed by any previous ones, but the grammar may allow generating an infinite sequence of items none of which subsumes the other. The problem can be solved by using restrictors, compare Sec. 6.E, to block out the feature propagation that leads to nontermination. Alternatively, as advocated in Nakazawa [66], we can add items without instantiations.

The third problem is how to incorporate the unification-grammar constraints when parsing, and in particular, how to incorporate feature propagation to reduce the search space and avoid nontermination.

Samuelsson [82] proposes a solution to these problems that achieves the same net effect as Nakazawa, by resorting to a context-free backbone grammar to define equivalence, but filtering on the unification-grammar constraints when compiling the parsing tables. In addition to this, the unification-grammar constraints are reintroduced into the parsing tables through linguistic generalization, see, e.g., [31; p. 232] over various objects; the grammar rule applied in a reduction is the linguistic generalization of the set of unification-grammar rules that map to the same context-free rule and could be applied at this point; each entry in the goto table is the generalization of the set of phrases that precede the dot in the items constructed from the original constraint-based grammar. The scheme is also viable for GLR parsing, using linguistic generalization for local-ambiguity packing.

Empty productions, for which the RHS of the grammar rule is empty, as discussed in Sec. 6.B, constitute a well-known problem for all bottom-up parsing schemes, as they have no realization in the input string, and can thus in principle be hypothesized everywhere in it. A common way of restricting their applicability is keeping track of moved phrases by gap threading, that is, by passing around a list of moved phrases to ensure that an empty production is only applicable if there is a moved phrase elsewhere in the sentence to license its use, see, e.g., [72, pp 125–129] for a more detailed description. As LR parsing is a bottom-up parsing strategy, it is necessary to limit the applicability of these empty productions by the use of top-down filtering. Block [12] and Samuelsson [82] offer methods for simulating limited gap threading in LR parsing.

8. DISCUSSION

When one attempts to implement a practically useful natural language system employing a general, linguistically based grammar, there is always a tension between this and trying to achieve fast and robust parsing. The last section briefly discusses some ways with which this problem has been dealt.[18]

A. Robustness

Ideally, the parsing of a sentence should result in a single analysis, or a limited number of analyses that can be adjudicated using semantic criteria. In practice, the result is often either zero or a very large number of analyses, possibly hundreds or thousands (at least with wide-coverage grammars). The first case (undergeneration) is the most difficult one and is caused by the unavoidable incompleteness of the grammar and lexicon or by errors in the input. To recover from this, some kind of robust fallback is required. One such approach works by *relaxing* constraints in such a way that a nongrammatical sentence obtains an analysis [39,64]. A potential problem with this is that the number of relaxation alternatives that are compatible with

[18] For further background on robust parsing, see Chap. 8; for related material, see also Chaps. 15 and 20.

analyses of the complete input may become extremely large. Another approach is *partial* parsing, that is, refraining from necessarily looking for a complete analysis and, instead, trying to recover as much structure as possible from grammatically coherent fragments of the sentence [1]. Some approaches along this line are Hindle [36, 37] within deterministic parsing, Lang [60] within chart parsing, Rekers and Koorn [79] within LR parsing, Joshi and Srinivas within Tree-Adjoining Grammar [43], Rayner and Carter [78] within a hybrid architecture, and Karlsson et al. [45] in the Constraint Grammar framework (for an exposition of the latter, see Chap. 15).

The second case (massive ambiguity) occurs because, by the very nature of its construction, a general grammar will allow a large number of analyses of almost any nontrivial sentence. However, in the context of a particular domain, most of these analyses will be extremely implausible. An obvious approach would then be to code a new, specialized grammar for each domain [17, 35, 87]. This may work well at least for small domains, but one loses the possibility of reusing grammatical descriptions across domains (and to a lesser extent languages), and of being able to import insights from theoretical linguistics. Hence, such an approach may be unsuitable, compare [99; p. 1].

An alternative approach is then to try to tune the parser or grammar, or both, for each new domain. One option is to select the most likely analysis from among a set of possible ones using a stochastic extension of the grammar at hand (compare Chap. 20). For example, Stolcke [102] describes how a chart parser can be adapted to use a probabilistic scoring function trained from a corpus, and Briscoe and Carroll [16] show how an LR parser, based on a general grammar, can make use of probabilities associated with parsing actions. Alternatively, some other type of numerical or discrete preference function can be used to select the highest-scoring analysis, see, e.g., [6] and [29] respectively.

A third approach is to take advantage of actual rule usage in a particular domain. For example, Grishman et al. [32] describe a procedure for extracting a reduced rule set from a general grammar, based on the frequency by which each rule is used when parsing a given corpus. Samuelsson and Rayner [83] make use of a further refined method known as grammar specialization. This is based on the observation that, in a given domain, certain groups of grammar rules tend to combine frequently in some ways, but not in others. On the basis of a sufficiently large corpus parsed by the original grammar, it is then possible to identify common combinations of rules of a (unification) grammar and to collapse them into single "macro" rules. The result is a specialized grammar that, compared with the original grammar, has a larger number of rules, but a simpler structure, allowing it to be parsed very fast using an LR parser (as described in Sec. 7.C.2).

B. Efficiency

Comparisons of the practical performance of parsers are complicated by differences in processor speed, implementation, grammar coverage, type of grammar, and quality of the structural descriptions assigned. Perhaps not surprisingly, relatively few controlled empirical comparisons have been reported, at least for wide-coverage grammars. Some investigations using phrase-structure grammars are Pratt [76], Slocum [99], Tomita [106], Wirén [112], and Billot and Lang [11]. Examples of investigations on constraint-based grammars are Shann [88], Bouma and van Noord [13], Nagata [65], Maxwell and Kaplan [63], and Carroll [19].

An important factor in parsing with constraint-based grammars is the amount of information represented by the phrase-structure backbone and the feature–term constraints, respectively. For lexicalist constraint-based grammars such as HPSG [75], the significance of the parsing algorithm is diminished, because unification may consume 85–90% of the parsing time [105]; however, compare [61]. In contrast, the parsing algorithm plays a more significant role in systems that make use of relatively substantial phrase-structure components, such as the CLE [5]. A related issue is the extent to which phrase-structure analysis and unification are interleaved during parsing [63].

Carroll [19] makes a thorough comparison of three unification-based parsers using a wide-coverage grammar of English, with a relatively substantial phrase–structure backbone. In particular, he found a rather moderate difference between an optimized left-corner parser and a nondeterministic LR parser in that the latter was roughly 35% faster and 30% more space efficient. Furthermore, his results are compatible with earlier findings [99; p. 5], which indicate that (phrase-structure-based) grammars often do not bring parsing algorithms anywhere near the worst-case complexity: Carroll found parsing times for exponential-time algorithms to be approximately quadratic in the length of the input for 1–30 words. For example, he achieved mean parsing times of close to 5–10 words per second for lengths from 20 down to 10 words. Carroll's results also indicate that sentence length may be a more significant factor than grammar size, in spite of the fact that the grammar-size constant usually dominates in calculations of worst-case time complexity. As he points out, these findings demonstrate that until more fine-grained methods are developed in complexity theory, empirical evaluations, in spite of their problems, seem more useful for studying and improving the practical performance of parsing algorithms.

Abney [1; p. 128] notes that the fastest parsers are deterministic, rule-based partial parsers, with speeds of on the order of thousands of words per second. Two examples are FIDDITCH [36,37] and ENGCG [110; p. 395]. In general, nondeterministic parsing licensing complete analyses lags far behind in terms of speed. However, implementational tuning may improve this [19; p. 288]; furthermore, optimization techniques, such as grammar specialization, have been reported to obtain speed-ups of at least an order of magnitude [83].

ACKNOWLEDGMENT

We gratefully acknowledge valuable comments by Giorgio Satta, Gregor Erbach, Eva Ejerhed, and Robert Dale on a draft version of this chapter. The work of the first author has been supported by Bell Laboratories and Xerox Research Centre Europe; the work of the second author has been supported by Telia Research.

REFERENCES

1. S Abney. Part-of-speech tagging and partial parsing. In: S Young, G Bloothooft, eds. Corpus-Based Methods in Language and Speech Processing. Dordrecht: Kluwer Academic, 1997, pp 118–136.

2. AE Ades, MJ Steedman. On the order of words. Linguist Philos 4:517–558, 1982.
3. AV Aho, R Sethi, JD Ullman. Compilers: Principles, Techniques, and Tools. Reading, MA: Addison–Wesley, 1986.
4. AV Aho, JD Ullman. The Theory of Parsing, Translation, and Compiling. Vol 1. Parsing. Englewood Cliffs, NJ: Prentice-Hall, 1972.
5. H Alshawi, ed. The Core Language Engine. Cambridge, MA: MIT Press, 1992.
6. H Alshawi, D Carter. Training and scaling preference functions for disambiguation. Comput Linguist 20(4):635–648, 1994.
7. GE Barton, RC Berwick, ES Ristad. Computational Complexity and Natural Language. Cambridge, MA: MIT Press, 1987.
8. M Bates. The theory and practice of augmented transition network grammars. In: L Bolc, ed. Natural Language Communication with Computers. Berlin: Springer-Verlag, 1978, pp. 191–259.
9. RE Bellman. Dynamic Programming. Princeton, NJ: Princeton University Press, 1957.
10. TG Bever. The cognitive basis for linguistic structures. In: JR Hayes, ed. Cognition and the Development of Language. New York: John Wiley & Sons, 1970.
11. S Billot, B Lang. The structure of shared forests in ambiguous parsing. Proc. 27th Annual Meeting of the Association for Computational Linguistics. Vancouver, BC, 1989, pp 143–151.
12. H–U Block. Compiling trace and unification grammar for parsing and generation. In: T Strzalkowski, ed. Reversible Grammar in Natural Language Processing. Norwell, MA: Kluwer Academic, 1993.
13. G Bouma, G van Noord. Head-driven parsing for lexicalist grammars: experimental results. Proc. Sixth Conference of the European Chapter of the Association for Computational Linguistics, Utrecht, 1993, pp 71–80.
14. J Bresnan, ed. The Mental Representation of Grammatical Relations. Cambridge, MA: MIT Press, 1982.
15. T Briscoe. Determinism and its implementation in PARSIFAL. In: Spark Jones and Wilks [100], pp 61–68.
16. T Briscoe, J Carroll. Generalized probabilistic LR parsing of natural language (corpora) with unification-based grammars. Comput Linguist 19(1):25–59, 1993.
17. RR Burton. Semantic grammar: an engineering technique for constructing natural language understanding systems. BBN Report 3453, Bolt, Beranek, and Newman, Inc., Cambridge, MA, 1976.
18. B Carpenter, G Penn. The attribute logic engine user's guide. User's guide Version 2.0.1. Pittsburgh, PA: Carnegie Mellon University, 1994.
19. J Carroll. Relating complexity to practical performance in parsing with wide-coverage unification grammars. Proc. 32nd Annual Meeting of the Association for Computational Linguistics, Las Cruces, NM, 1994, pp 287–294.
20. N Chomsky. Syntactic Structures. The Hague: Mouton, 1957.
21. KW Church, R Patil. Coping with syntactic ambiguity or how to put the block in the box on the table. Comput Linguist 8(3–4):139–149, 1982.
22. KW Church. On memory limitations in natural language processing. Report MIT/LCS/TM-216, Massachusetts Institute of Technology, Cambridge, MA, 1980.
23. S Crain, M Steedman. On not being led up the garden path: the use of context by the psychological syntax processor. In: DR Dowty et al., eds. Natural Language Parsing. Psychological, Computational, and Theoretical Perspectives. Cambridge, UK: Cambridge University Press, 1985, pp 320–358.
24. DR Dowty, L Karttunen, AM Zwicky, eds. Natural Language Parsing. Psychological, Computational, and Theoretical Perspectives. Cambridge, UK: Cambridge University Press, 1985.

25. J Earley. An efficient context-free parsing algorithm. Commun ACM 13(2):94–102, 1970. Also in Grosz et al, [33], pp 25–33.
26. E Ejerhed. The processing of unbounded dependencies in Swedish. In: E Engdahl, E Ejerhed, eds. Readings on Unbounded Dependencies in Scandinavian Languages. Stockholm: Almqvist & Wiksell, 1982, pp 99–149.
27. E Ejerhed. Finding clauses in unrestricted text by finitary and stochastic methods. In: Proc. Second Conference on Applied Natural Language Processing. Austin, TX, 1988, pp 219–227.
28. G Erbach. Bottom-up Earley deduction for preference-driven natural language processing. Dissertation, Universität des Saarlandes, Saarbrücken, Germany, 1997.
29. A Frank, TH King, J Kuhn, J Maxwell. Optimality theory style constraint ranking in large-scale LFG grammars. Proc. LFG 98 Conference, Brisbane, Australia, 1998.
30. G Gazdar, E Klein, G Pullum, I Sag. Generalized Phrase Structure Grammar. Oxford: Basil Blackwell, 1985.
31. G Gazdar, CS Mellish. Natural Language Processing in Prolog. Reading, MA: Addison–Wesley, 1989.
32. R Grishman, N Thanh Nhan, E Marsh, L Hirschman. Automated determination of sublanguage syntactic usage. Proc. 10th International Conference on Computational Linguistics/22nd Annual Meeting of the Association for Computational Linguistics, Stanford, CA 1984, pp 96–100.
33. B Grosz, K Sparck Jones, BL Webber, eds. Readings in Natural Language Processing. Los Altos, CA: Morgan Kaufmann, 1986.
34. N Haddock, E Klein, G Morrill, eds. Working Papers in Cognitive Science. Vol 1. Centre for Cognitive Science, University of Edinburgh, Scotland, 1987.
35. GG Hendrix, ED Sacerdoti, D Sagalowicz. Developing a natural language interface to complex data. ACM Trans Database Syst 3(2):105–147, 1978.
36. D Hindle. Acquiring disambiguation rules from text. Proc. 27th Annual Meeting of the Association for Computational Linguistics, Vancouver, BC, 1989, pp 118–125.
37. D Hindle. A parser for text corpora. In: A Zampolli, ed. Computational Approaches to the Lexicon. Oxford, UK: Oxford University Press, 1994.
38. JE Hopcroft, JD Ullman. Introduction to Automata Theory, Languages, and Computation. Reading, MA: Addison-Wesley, 1979.
39. K Jensen, GE Heidorn. The fitted parse: 100% parsing capability in a syntactic grammar of English. Proc. [First] Conference on Applied Natural Language Processing, Santa Monica, CA, 1983, pp 93–98.
40. M Johnson. The computational complexity of GLR Parsing. In: M Tomita, ed. Generalized LR Parsing. Norwell, MA: Kluwer Academic, 1991, 35–42.
41. AK Joshi. Tree adjoining grammars: how much context-sensitivity is required to provide reasonable structural descriptions? In: DR Dowty et al. eds, Natural Language Parsing. Psychological, Computational, and Theoretical Perspectives. Cambridge, UK: Cambridge University Press, 1995, pp 206–250.
42. AK Joshi, LS Levy, M Takahashi. Tree adjunct grammars. J of Comput Syst Sci 1975.
43. AK Joshi, B Srinivas. Disambiguation of super parts of speech (or supertags): almost parsing. Proc. 15th International Conference on Computational Linguistics. Vol. I, Kyoto, Japan, 1994, pp 154–160.
44. RM Kaplan. A general syntactic processor. In: Rustin [81], pp 193–241.
45. F Karlsson, A Voutilainen, J Heikkilä, A Anttila, eds. Constraint Grammar. A Language-Independent System for Parsing Unrestricted Text. The Hague: Mouton de Gruyter, 1995.

46. L Karttunen. D-PATR: a development environment for unification-based grammars. Proc. 11th International Conference on Computational Linguistics, Bonn, Germany, 1986, pp 74–80.
47. L Karttunen. Radical lexicalism. In: M Baltin, A Kroch, eds. Alternative conceptions of phrase structure. Chicago: University of Chicago Press, 1989.
48. T Kasami. An efficient recognition and syntax algorithm for context-free languages. Technical Report AF-CRL-65-758, Air Force Cambridge Research Laboratory, Bedford, MA, 1965.
49. M Kay. Experiments with a powerful parser. Proc. Second International Conference on Computational Linguistics [2ème conférence internationale sur le traitement automatique des langues], Grenoble, France, 1967.
50. M Kay. The MIND system. In: Rustin [81] pp 155–188.
51. M Kay. Functional unification grammar: a formalism for machine translation. Proc. 10th International Conference on Computational Linguistics/22nd Annual Meeting of the Association for Computational Linguistics, Stanford, CA, 1984, pp 75–78.
52. M Kay. Parsing in functional unification grammar. In: DR Dowty et al. eds. Natural Language Parsing. Psychological, Computational, and Theoretical Perspectives. Cambridge, UK: Cambridge University Press, 1985, pp 251–278.
53. M Kay. Algorithm schemata and data structures in syntactic processing. In: Grosz et al. [33] pp 35–70. Also Report CSL-80-12, Xerox PARC, Palo Alto, CA, 1980.
54. M Kay. Head driven parsing. Proc. [First] International Workshop on Parsing Technologies, Pittsburgh, PA, 1989, pp 52–62.
55. JR Kipps. GLR parsing in time $O(n^3)$. In: M Tomita, ed. Generalized LR Parsing. Nowell, MA: Kluwer Academic, 1991, 43–59.
56. DE Knuth. On the translation of languages from left to right. Inf Control 8(6):607–639, 1965.
57. K Koskenniemi. Finite-state parsing and disambiguation. Proc. 13th International Conference on Computational Linguistics, Vol 2, Helsinki, Finland, 1990, pp 229–232.
58. H–U Krieger, U Schäfer. TDL: a type description language for constraint-based grammars. Proc. 15th International Conference on Computational Linguistics, Vol 2, Kyoto, Japan, 1994, pp 893–899.
59. B Lang. Deterministic techniques for efficient non-deterministic parsers. In: J Loeckx, ed. Proc. Second Colloquium on Automata, Languages and Programming, Vol. 14, Saarbrücken, Germany, 1974, pp 255–269. Springer Lecture Notes in Computer Science, Heidelberg, Germany, 1974.
60. B Lang. Parsing incomplete sentences. Proc. 12th International Conference on Computational Linguistics, Budapest, Hungary, 1988, pp 365–371.
61. T Makino, M Yoshida, K Torisawa, J Tsujii. LiLFes—Towards a practical HPSG parser. Proc. 17th International Conference on Computational Linguistics/36th Annual Meeting of the Association for Computational Linguistics, Montreal, Quebec, Canada, 1998, pp. 807–811.
62. MP Marcus. A Theory of Syntactic Recognition for Natural Language. Cambridge, MA: MIT Press, 1980.
63. JT Maxwell III, RM Kaplan. The interface between phrasal and functional constraints. Comput Linguist, 19(4):571–590, 1993.
64. CS Mellish. Some chart-based techniques for parsing ill-formed input. Proc. 27th Annual Meeting of the Association for Computational Linguistics, Vancouver, BC, 1989, pp 102–109.
65. M Nagata. An empirical study on rule granularity and unification interleaving toward an efficient unification-based system. Proc. 14th International Conference on Computational Linguistics, Vol 1, Nantes, France, 1992, pp 177–183.

66. T Nakazawa. An extended LR parsing algorithm for grammars using feature-based syntactic categories. Proc. Fifth Conference of the European Chapter of the Association for Computational Linguistics, Berlin, 1991, pp 69–74.
67. M–J Nederhof. An optimal tabular parsing algorithm. Proc. 32nd Annual Meeting of the Association for Computational Linguistics, Las Cruces, NM, 1994, pp 117–124.
68. M–J Nederhof, G Satta. An extended theory of head-driven parsing. Proc. 32nd Annual Meeting of the Association for Computational Linguistics, Las Cruces, NM, 1994, pp 210–217.
69. M–J Nederhof, G Satta. Efficient tabular LR parsing. Proc. 34th Annual Meeting of the Association for Computational Linguistics, Santa Cruz, California, 1996, pp 239–246.
70. BH Partee, A ter Meulen, RE Wall. Mathematical Methods in Linguistics. Norwell, MA: Kluwer Academic, 1987.
71. FCN Pereira. A new characterization of attachment preferences. In: Dowty DR et al. eds. Natural Language Parsing. Psychological, Computational, and Theoretical Perspectives. Cambridge, UK: Cambridge University Press, 1985, pp 307–319.
72. FCN Pereira, SM Shieber. Prolog and Natural-Language Analysis. Number 10 in CSLI Lecture Notes. Chicago: University of Chicago Press, 1987.
73. FCN Pereira, DHD Warren. Definite clause grammars for language analysis—a survey of the formalism and a comparison with augmented transition networks. Artif Intell, 13(3):231–278, 1980.
74. FCN Pereira, DHD Warren. Parsing as deduction. Proc. 21st Annual Meeting of the Association for Computational Linguistics, Cambridge, MA, 1983, pp 137–144.
75. C Pollard, IA Sag. Head-Driven Phrase Structure Grammar. Chicago: University of Chicago Press, 1994.
76. VR Pratt. LINGOL—a progress report. Proc. Fourth International Joint Conference on Artificial Intelligence, Tbilisi, Georgia, USSR, 1975, pp 422–428.
77. SG Pulman. Grammars, parsers and memory limitations. Lang Cogn Proc 1(3):197–225, 1986.
78. M Rayner, D Carter. Fast parsing using pruning and grammar specialization. Proc. 34th Annual Meeting of the Association for Computational Linguistics, Santa Cruz, CA, 1996, pp 223–230.
79. J Rekers, W Koorn. Substring parsing for arbitrary context free grammars. Proc. Second International Workshop on Parsing Technologies, Cancun, Mexico, 1991, pp 218–224.
80. E Roche, Y Schabes, eds. Finite-State Language Processing. Cambridge, MA: MIT Press, 1997.
81. R Rustin, ed. Natural Language Processing. New York: Algorithmics Press, 1973.
82. C Samuelsson. Notes on LR parser design. Proc. 15th International Conference on Computational Linguistics, Vol 1, Kyoto, Japan, 1994, pp 386–390.
83. C Samuelsson, M Rayner. Quantitative evaluation of explanation-based learning as an optimization tool for a large-scale natural language system. Proc. 12th International Joint Conference on Artificial Intelligence, Sydney, Australia, 1991, pp 609–615.
84. G Satta, O Stock. Formal properties and implementation of bidirectional charts. Proc. Eleventh International Joint Conference on Artificial Intelligence, Detroit, MI, 1989, pp 1480–1485.
85. G Satta, O Stock. Head-driven bidirectional parsing: a tabular approach. Proc. [First] International Workshop on Parsing Technologies, Pittsburgh, PA, 1989, 43–51.
86. Y Schabes, A Abeillé, AK Joshi. Parsing strategies with 'lexicalized' grammars: application to tree adjoining grammars. Proc. 12th International Conference on Computational Linguistics, Budapest, Hungary, 1988, pp 270–274.
87. S Seneff. TINA: a natural language system for spoken language applications. Comput Linguist 18(1):61–86, 1992.

88. P Shann. The selection of a parsing strategy for an on-line machine translation system in a sublanguage domain. A new practical comparison. Proc. [First] International Workshop on Parsing Technologies, Pittsburgh, PA, 1989, pp 264–276.
89. BA Sheil. Observations on context free parsing. Stat Methods Linguist pp 71–109, 1976.
90. SM Shieber. Sentence disambiguation by a shift–reduce parsing technique. Proc. 21st Annual Meeting of the Association for Computational Linguistics, Cambridge, MA, 1983, pp 113–118.
91. SM Shieber. The design of a computer language for linguistic information. Proc. 10th International Conference on Computational Linguistics/22nd Annual Meeting of the Association for Computational Linguistics, Stanford, CA, 1984, pp 362–366.
92. SM Shieber. Using restriction to extend parsing algorithms for complex-feature-based formalisms. Proc. 23rd Annual Meeting of the Association for Computational Linguistics, Chicago, 1985, pp 145–152.
93. SM Shieber. An Introduction to Unification-Based Approaches to Grammar. Number 4 in CSLI Lecture Notes. Chicago: University of Chicago Press, 1986.
94. SM Shieber, Y Schabes, FCN Pereira. Principles and implementation of deductive parsing. J Logic Progr. 24(1–2):3–36, 1995.
95. SM Shieber, H Uszkoreit, FCN Pereira, JJ Robinson, M Tyson. The formalism and implementation of PATR-II. In: BJ Grosz, ME Stickel, eds. Research on Interactive Acquisition and Use of Knowledge, Final Report, SRI project number 1894. Menlo Park, CA: SRI International, 1983, pp 39–79.
96. K Sikkel. Parsing Schemata. A Framework for Specification and Analysis of Parsing Algorithms. Texts in Theoretical Computer Science. Berlin: Springer-Verlag, 1997.
97. S Sippu, E Soisalon-Soininen. Parsing Theory: Languages and Parsing, vol. 1. Berlin: Springer-Verlag, 1988.
98. S Sippu, E Soisalon-Soininen. Parsing Theory: LR(k) and LL(k) Parsing, vol. 2. Berlin: Springer-Verlag, 1990.
99. J Slocum. A practical comparison of parsing strategies. Proc. 18th Annual Meeting of the Association for Computational Linguistics, Philadelphia, PA, 1980, pp 1–6.
100. K Sparck Jones, Y Wilks, eds. Automatic Natural Language Parsing. Chichester, UK: Ellis Horwood, 1983.
101. MJ Steedman. Natural and unnatural language processing. In: Sparck Jones and Wilks [100], pp 132–140.
102. A Stolcke. An efficient probabilistic context-free parsing algorithm that computes prefix probabilities. Comput Linguist 21(2):165–202, 1995.
103. HS Thompson. Chart parsing and rule schemata in GPSG. Proc. 19th Annual Meeting of the Association for Computational Linguistics, Stanford, CA, 1981, pp 167–172.
104. HS Thompson. MCHART: a flexible, modular chart parsing system. Proc. Third National Conference on Artificial Intelligence, Washington, DC, 1983, pp 408–410.
105. H Tomabechi. Quasi-destructive graph unification. Proc. 29th Annual Meeting of the Association for Computational Linguistics, Berkeley, CA, 1991, pp 315–322.
106. M Tomita. Efficient Parsing for Natural Language. A Fast Algorithm for Practical Purposes. Number 8 in The Kluwer International Series in Engineering and Computer Science. Norwell, MA: Kluwer Academic, 1986.
107. M Tomita, ed. Generalized LR Parsing. Norwell, MA: Kluwer Academic Publishers, 1991.
108. H Uszkoreit. Categorial unification grammars. Proc. 11th International Conference on Computational Linguistics, Bonn, Germany, 1986.
109. K Vijay–Shanker, AK Joshi. Unification-based tree adjoining grammars. In: J Wedekind, ed. Unification-Based Grammars. Cambridge, MA: MIT Press, 1991.

110. A Voutilainen, P Tapanainen. Ambiguity resolution in a reductionistic parser. Proc. Sixth Conference of the European Chapter of the Association for Computational Linguistics, Utrecht, The Netherlands, 1993, pp 394–403.
111. T Winograd. Language as a Cognitive Process. Vol 1. Syntax. Reading, MA: Addison-Wesley, 1983.
112. M Wirén. A comparison of rule-invocation strategies in context-free chart parsing. Proc. Third Conference of the European Chapter of the Association for Computational Linguistics, Copenhagen, Denmark, 1987, pp 226–233.
113. WA Woods. Transition network grammars for natural language analysis. Commun ACM 13(10):591–606, 1970. Also in Grosz et al. [33], pp 71–87.
114. WA Woods. An experimental parsing system for transition network grammars. In: Rustin [81], pp 111–154.
115. WA Woods, RM Kaplan, B Nash–Webber. The lunar sciences natural language information system: final report. BBN Report 2378. Cambridge, MA: Bolt, Beranek, and Newman, Inc., 1972.
116. DH Younger. Recognition and parsing of context-free languages in time n^3. Inf Control 10(2):189–208, 1967.

5

Semantic Analysis

MASSIMO POESIO

University of Edinburgh, Edinburgh, Scotland

1. THE TASK OF SEMANTIC ANALYSIS

Parsing an utterance into its constituents is only one step in processing language. The ultimate goal, for humans as well as natural language-processing (NLP) systems, is to *understand* the utterance—which, depending on the circumstances, may mean incorporating the information provided by the utterance into one's own knowledge base or, more in general, performing some action in response to it. 'Understanding' an utterance is a complex process, that depends on the results of parsing, as well as on lexical information, context, and commonsense reasoning; and results in what we will call the SEMANTIC INTERPRETATION IN CONTEXT of the utterance, which is the basis for further action by the language-processing agent.

Designing a semantic interpreter involves facing some of the same problems that must be confronted when building a syntactic parser and, in particular, the problem of (semantic) AMBIGUITY: how to recognize the intended semantic interpretation of the utterance of a sentence in a specific context, among the many possible interpretations of that sentence. Semantic ambiguity may be the result of any of the following causes:

Lexical ambiguity: Many word-strings are associated with lexical entries of more than one syntactic category or with more than one sense. For example, according to the *Longman's Dictionary of Contemporary English* (LDOCE), the word *stock* can be used as both a noun and as a verb, and in each use it has more than one sense. The LDOCE lists 16 distinct senses for *stock* used as noun, including 'supply of something to use,' 'plan from which cuttings are grown,' 'group of animals used for breeding,' 'the money owned by a company and divided into shares,' 'a liquid made from the juices of meat and bones,' as well as special senses as in *taking stock*.

Referential ambiguity: Pronouns and other ANAPHORIC EXPRESSIONS typically have many possible interpretations in a context, as shown by the following example (from *The Guardian*), where the donation, the account, and the cheque all satisfy the syntactic and semantic constraints on the pronoun *it* in the third sentence:

... received a donation of pounds 75,000 on March 9, 1988. This amount was paid out of the Unipac NatWest Jersey account (67249566) by cheque number 1352. It was funded by a transfer of pounds 2,000,000 from PPI's account at NatWest Bishopsgate ... {1}

Scopal ambiguity: Ambiguity results when a sentence contains more than one quantifier, modal, and other expressions, such as negation, that from a logical point of view modify the parameters of interpretation. For example, the sentence *I can't find a piece of paper* can be used either to mean that there is a particular piece of paper that I can't find (cf., the preferred interpretation of *John didn't see a pedestrian crossing his path and almost hit him*) or to mean that I'm not able to find any piece of paper at all (cf., the preferred interpretation of *I don't own a car*). Scopal ambiguity is particularly interesting in that the number of possible interpretations grows factorially with the number of operators, so that Hobbs' reformulation of Lincoln's saying in {2} should have at least 6 readings, yet humans seem to have no problem processing this sentence.

In most democratic countries most politicians can fool most of the people on almost every issue most of the time. {2}

The task of designing a semantic interpreter is complicated because there is little agreement among researchers on what exactly the ultimate interpretation of an utterance should be, and whether this should be domain-dependent. Thus, whereas researchers often try to design general-purpose parsers, practical NLP systems tend to use semantic representations targeted for the particular application the user has in mind and semantic interpreters that take advantage of the features of the domain.

The organization of this chapter is as follows. We begin by reviewing some proposals concerning possible forms of semantic interpretation for natural language. We then discuss methods for deriving such a semantic interpretation out of natural language utterances; and finally, we consider one example of the role of context in semantic interpretation, anaphora resolution.

2. KNOWLEDGE REPRESENTATION FOR SEMANTIC INTERPRETATION

Research in Natural Language Processing (NLP) has identified two aspects of 'understanding' as particularly important for NLP systems. Understanding an utterance means, first of all, knowing what an appropriate response to that utterance is. For example, when we hear the instruction in {3} we know what action is requested, and when we hear {4} we know that it is a request for a verbal response giving information about the time:

Mix the flour with the water. {3}

What time is it? {4}

An understanding of an utterance also involves being able to draw conclusions from what we hear or read, and being able to relate this new information to what we already know. For example, when we semantically interpret {5} we acquire information that allows us to draw some conclusions about the speaker; if we know that Torino is an Italian town, we may also make some guesses about the native language of the speaker, and his preference for a certain type of coffee. When we hear {6}, we may conclude that John bought a ticket and that he is no longer in the city he started his journey from, among other things.

I was born in Torino. {5}

John went to Birmingham by train. {6}

The effect of semantic interpretation on a reader's intentions and actions is discussed in Chap. 7. In this chapter, we focus on the second aspect of understanding; that is, on the effect of semantic interpretation on an agent's knowledge base and on the inferences that that agent can make; and conversely, on how our prior knowledge affects the way we interpret utterances. These issues are a subpart of the concerns of the more general area of KNOWLEDGE REPRESENTATION, concerned with how knowledge is represented and used.

A great number of theories of knowledge representation for semantic interpretation have been proposed; these range from the extremely *ad hoc* (e.g., NLP systems for access to databases often convert natural language utterances into expressions of a database query language; but see Chap. 9 for a more general approach) to the very general (e.g., the theories of interpretation proposed in formal semantics). It is impossible to survey all forms of knowledge representation that have been proposed with a given application in mind; we will, therefore, concentrate here on general-purpose theories with a clear semantics, and in particular on logic-based theories of knowledge representation.

A. First-Order Logic

The best-known general-purpose theory of knowledge representation is FIRST-ORDER LOGIC. We will discuss it first because, notwithstanding its many limitations, it provides a good illustration of what a theory of knowledge representation is meant to do.

First-order logic, like all other theories of knowledge representation, gives us a way for expressing certain types of information: in the case of first-order logic, certain properties of sets of objects. This is done by providing a SYNTAX (i.e., a language for expressing the statements) and a SEMANTICS (i.e., a specification of what these statements mean relative to an intended domain); in the case of first-order logic, what kind of set-theoretic statements are expressed by the expressions of the language. Thus, in first-order logic, we can represent the information conveyed by natural language sentences stating that an object is a member of a certain set (say, *John is a sailor*) by means of a PREDICATE such as **sailor**, denoting a set of objects, and a TERM such as *j*, denoting John; the ATOMIC FORMULA **sailor**(*j*) expresses the statement. That formula is a syntactic way to express what we have called the semantic interpretation of the sentence. Using predicates of higher arity, we can also assign a semantic interpretation to sentences stating that certain objects stand in a certain relation: for example, *John likes Mary* could be translated by formulae such as **married**(j,m), where m is a term denoting Mary. The semantic interpretation of sentences asserting that a set is included in another (e.g., *Dogs are mammals*) can be expressed by means of the universal QUANTIFIER $\forall$, as in $\forall x \mathbf{dog}(x) \rightarrow \mathbf{mammal}(x)$, with the interpretation: the set of dogs is a subset of the set of mammals. The existential quantifier ($\exists$) can be used to capture the information that a certain set is not empty, as expressed by the sentence *I have a car*, which can be translated as $\exists x \mathbf{car}(x) \wedge \mathbf{own}(spkr,x)$. In first-order logic it is also possible to combine statements into more complex statements by means of so-called CONNECTIVES: for example, we can say that *John bought a refrigerator and a CD-player*, or that *John and Mary are happy*. In first-order logic, we can assign to these sentences the semantic interpretation expressed by formulas such as $\mathbf{happy}(j) \wedge \mathbf{happy}(m)$. Finally, in first-order logic, it is also possible to say that something it is *not* the case: for example, we can say that *John is not married*, $-\mathbf{married}(j)$.

For some applications, representing this information is all we need; for example, a simple railway enquiry system may just need to know that the user wants to go from city A to city B on a given day. More generally, however, we may want to draw some inferences from the information we have in our knowledge base. What makes first-order logic a logic is that it also includes a specification of the VALID conclusions that can be derived from this information (i.e., which sentences must be true given that some other sentences are true: This aspect of a logic is usually called PROOF THEORY.) Ideally, this set of inferences should be related to the semantics that we have given to the language: for example, if we know that statement {7a} ('All trains departing from Torino and arriving at Milano stop at Novara') is true, and that statement {7b} ('Train 531 departs from Torino and arrives at Milano') is also true, we can conclude that statement {7c}, 'train 531 stops at Novara' is also true. We want our system of inference to allow us to conclude this.

a. $\forall x(\mathbf{train}(x) \wedge \mathbf{depart}(x,torino) \wedge \mathbf{arrive}(x,milano)) \rightarrow \mathbf{stop}(x,novara)$
b. $\mathbf{train}(train531) \wedge \mathbf{depart}(train531,torino) \wedge \mathbf{arrive}(531,milano)$
c. $\mathbf{stop}(train531,novara)$

One of the reasons why first-order logic is so popular is that it has formalizations that are provably SOUND and COMPLETE (i.e., that allow us to deduce from a set of sentences in the language all and only the sentences that are consequences of those sentences according to the semantics specified in the foregoing). These formalizations are typically formulated in terms of INFERENCE RULES and a set of AXIOMS. An inference rule consists of a set of statements called PREMISES, and a statement called the CONCLUSION; the inference rule is a claim that if all premises are true, then the conclusion is true. An inference rule is VALID if this is indeed the case. An example of an inference rule is UNIVERSAL INSTANTIATION: if P is universally true then it is also true for a given object *a*. Another well-known example of a valid inference rule is MODUS PONENS: this inference rule states that if P is the case and P $\rightarrow$ Q is the case, then Q is the case. Together, these two inference rules allow us to conclude {7c} from the premises {7a} and {7b}. A well-known example of a sound and complete system of inference rules is NATURAL DEDUCTION (Gamut, 1991). Several theorem provers for full first-order logic exist, some of which are public domain: examples are Theorist, that uses linear resolution (Poole, et al., 1987) and lean-TAP, based on tableaux (Beckart and Posegga, 1996).

Because more or less everybody understands what a statement in first-order logic means, it acts as a sort of *lingua franca* among researchers in knowledge representation. On the other hand, first-order logic also has a number of problematic features when seen as a tool for capturing the inferences associated with understanding natural language utterances; many theories of knowledge representation have been proposed to address these problems.

One problem that is often mentioned is that inference with first-order logic is computationally expensive—indeed, in general, there is no guarantee that a given inference process is going to terminate. This suggests that it cannot be an appropriate characterization of the way humans do inferences, as humans can do at least some of them very quickly. Researchers have thus developed logics that are less powerful and, therefore, can lead to more efficient reasoning. Prolog is perhaps the best known example of a trade-off between efficiency and expressiveness. Many researchers think that the best way to find an efficient theory of knowledge representation for semantic interpretation is to look at the kind of inferences humans do efficiently. Work on SEMANTIC NETWORKS and FRAME-BASED REPRESENTATIONS has been inspired by these considerations.

For others, and especially for linguists, the problem with first-order logic is the opposite: it is not powerful enough. Formal semanticists, in particular, have argued for more powerful logics, either on the grounds that they provide more elegant tools for semantic composition, or that they can be used to model phenomena that cannot be formalized in first-order logic.

Finally, it has long been known that there is a fundamental difference between the proof theories proposed for first-order logic and the way humans reason: humans typically jump to conclusions on the basis of insufficient information, with the result that they make mistakes. Research on NONMONOTONIC REASONING (Brewka, 1991) and STATISTICAL THEORIES OF INFERENCE (e.g., Pearl, 1988) has aimed at developing theories of inference that capture this type of reasoning. We will not discuss these issues here.

B. Semantic Networks

Several psychological studies from the 1960s and the 1970s suggested that human knowledge is organized hierarchically. For example, studies by Quillian and others (1968) showed that it takes less time for humans to say whether sentence {8} is true than for {9}, and less time to verify {9} than {10}:

A canary is yellow. {8}

A canary has feathers. {9}

A canary eats food. {10}

These researchers concluded that human knowledge is organized in taxonomies called SEMANTIC NETWORKS, such as the one in Fig. 1. There are two basic kinds of objects in a semantic network: TYPES (also called GENERIC CONCEPTS) and INSTANCES (also known as TOKENS or INDIVIDUAL CONCEPTS). In Fig. 1, **THING**, **SOLID**, and

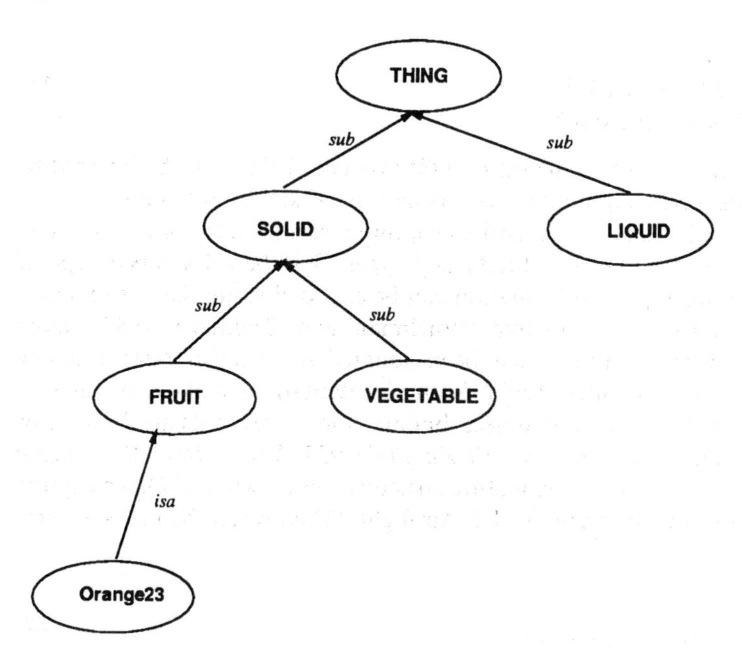

Fig. 1 A semantic network.

LIQUID, are types; **Orange23** is an instance. A semantic network such as the one in this example can encode two kinds of information: that an individual is an instance of a type, and that a type is a subtype of another. Both of these kinds of information are represented by arcs, although with different labels: an *isa* arc represents an instance relation, whereas a *sub* arc represents subtype information. The semantic network in Fig. 1, for example, encodes the information that **SOLID** and **LIQUID** are subtypes of **THING**, and that **Orange23** is an instance of **FRUIT**.

The essential form of inference supported by a semantic network of this type is *classification* based on arc traversal: that is, we can infer that **Orange23** is an instance of **THING** because there is a path consisting of a single instance of an *isa* link and zero or more **sub** links connecting **Orange23** to **THING**. A second form of inference that semantic networks are well-suited for is *vicinity*: it is possible to define how 'distant' two concepts are in terms of the number of arcs that it takes to go from one to the other. (Psychological studies have suggested that vicinity in this sense affects processes such as lexical disambiguation.)

Semantic networks can be interpreted in terms of set theory: types can be identified with sets, instances with elements, **sub** with the subset relation, and **isa** with the membership relation. In other words, semantic networks can be thought as a way of encoding a subset of the information that can be expressed in first-order logic, thus making certain kind of inferences—about set membership and set inclusion—efficient (Schubert, 1976; Hayes, 1979). For example, the information that **Orange23** is a **FRUIT**, that **FRUIT** is a kind of **SOLID**, and that **SOLID**s are a kind of **THING** could be expressed in first-order logic as follows:

a. **FRUIT**(*Orange*23)
b. ∀*X***FRUIT**(*X*)→**SOLID**(*X*) {11}
c. ∀*X***SOLID**(*X*)→**THING**(*X*)

That **Orange 23** is an element of the set of **THING**s could therefore be inferred by means of derivations involving modus ponens and universal instantiation.

Semantic networks can also be used to capture a second form of information: about the ATTRIBUTES or ROLES of objects and types. The best-developed type of formalism in which this type of information can be encoded is the class of semantic network theories derived from KL-ONE (Brachman and Schmolze, 1985); more recently, this type of representation has been adopted in so-called TYPED FEATURE STRUCTURE formalisms (Carpenter, 1992). Such representations make it possible to specify not only the type of a certain object, but also the values of its attributes. For example, by interpreting the sentence *US Air flight 923 departs from Washington National at 2:30 pm* in terms of typed feature structures as shown in {12}, we capture the information about the properties of US Air flight 923 conveyed by that sentence.

$$\begin{bmatrix} \textbf{flight} \\ \text{name} : US923 \\ \text{departure-airport} : \text{washington-national} \\ \text{departure-time} : \text{2:30pm} \end{bmatrix} \qquad \{12\}$$

Attribute–value representations such as that in {12}, are probably the most common form of knowledge representation used in NLP systems. In addition to information about a specific flight, these formalisms can also be used to specify, for example, that every flight has a departure time.

This augmented representation then supports, in addition to the classification inferences supported by all semantic networks, the type of inference called *role inheritance* (e.g., one can infer that flight *US*923 has an ARRIVAL-AIRPORT of type **airport** simply by following an **isa** link from *US*923 to the type **flight** and from there the **arrival-airport** link to an object of type **airport**). That humans can do these inferences is indicated by the use of so-called **bridging descriptions** (Clark, 1977): these are definite descriptions that refer to objects that have not been explicitly mentioned, but that are related to objects that have; for example, one can say *I wanted to take the USAir flight 923 the other day, but the departure time had changed.* Semantic networks provide a simple way of characterizing the inferences that take place when these definite descriptions are interpreted.

Just as in taxonomic information, the information about roles can be expressed in terms of first-order logic. Again, we can see semantic networks as specializations of first-order logic that specify particular inferences and include techniques for computing them efficiently.

C. Theories of Meaning Used in Formal Semantics

We cannot give here a satisfactory introduction to formal semantics and the issues it covers; we will simply provide a quick overview of some of the phenomena that motivate the search for more powerful formalisms. A good introduction to semantics is Chierchia and McConnell–Ginet (1990); the book by Gamut (1991) is a good introduction to logic and its use in semantics.

1. Modality and Intensionality

First-order logic is about what is true here and now. But in natural language we can do more than that: for example, we can entertain alternative possibilities, as in *You should have turned left when I told you so*, or about future events, as in *I will arrive in Paris tomorrow*. Conversion about plans such as those handled by the TRAINS-93 system (Allen et al., 1995; see also Chap. 14) are full of references to such alternative possibilities, expressed by modal statements such as *We should send an engine to Avon* or *We could use the boxcar at Bath instead.*

Researchers in philosophy and linguistics have proposed that such statements are about POSSIBLE WORLDS—states of affairs that differ from the 'actual' world. Technically, these proposals can be formalized by so-called MODAL LOGICS, which are logics in which the value of a statement depends not only on the universe of discourse and on a particular assignment of values to variables, but also on the world at which the statement is evaluated.

Other natural language expressions that make implicit reference to possible worlds are adverbials, such as *possibly* and *necessarily*, the modal auxiliaries *must, can, should,* and *would,* and so-called COUNTERFACTUAL CONDITIONALS (i.e., conditionals, such as *If my car were not broken I would gladly lend it to you*).

2. Adjectives

Adjectives are a second class of expressions whose meaning is difficult to model in first-order logic. For one thing, many adjectives are intrinsically VAGUE: what does it mean, for instance, to say that *John is tall*, or that *My bags are red*? In first-order logic, statements can only be true or false; but determining whether these statements are true or false is, at best, a highly subjective decision.

Adjectives also highlight another limitation at first-order logic: that it is best used to attribute INTERSECTIVE properties to objects. The statement *This is a red apple* can be interpreted by 'decomposing' the predicate in two: *X is red and X is an apple*. But the statement *This is the former principal* cannot be interpreted as *X is a principal and X is former*. One way to interpret this statement is to treat *former* as a PREDICATE MODIFIER: that is, as a function that, when applied to a property like *principal* denoting some set of objects here and now, returns a second set of objects—those individuals that were principals at some time in the past.

3. Quantifiers

All and *some* are only two of the many QUANTIFIERS of natural language. The full list includes expressions such as *most, many, few,* not to mention complex expressions such as *more than a half* or *hardly any*:

a. *Most books were rescued from the flood.* {13}
b. *Many books were rescued from the flood.*
c. *Few books were rescued from the flood.*
d. *Hardly any books were rescued from the flood.*

Barwise and Cooper (1981) proved that the meaning of quantifiers such as *most* cannot be expressed in terms of $\forall$ and $\exists$. According to the theory of generalized quantifiers (Barwise and Cooper, 1981; Keenan and Stavi, 1986), expressions such as *most* and *every* denote relations among sets. For example, *every* denotes the subset relation:

$$[\![\mathbf{every}(X,Y)]\!] = 1 \text{ iff } X \subseteq Y$$

Some denotes the relation that holds between two sets when their intersection is not empty, whereas *most* denotes the relation that holds when the intersection between a set X and a set Y is greater than the intersection between X and the complement of Y.

3. ASSIGNING A SEMANTIC INTERPRETATION TO UTTERANCES

The next question to be addressed when designing a semantic analyzer is: How are interpretations such as those discussed in the previous section assigned to utterances? In this chapter, we concentrate on theories of semantic composition, such as those developed in formal semantics. These theories are based on the assumption that the meaning of an utterance is constructed bottom-up, starting from the meaning of the lexical items. The process of meaning construction is driven by syntax, in the sense that the meaning of each phrase is derived from the meaning of its constitutents by applying determined semantic operations.

A warning: in real language, a parse tree covering the whole utterance cannot always be found; this is particularly true for spoken language. As a consequence, researchers attempting to build semantic interpreters capable of processing any kind of text in a robust fashion have developed methods to build semantic interpretations out of partial parses or that do without parsing altogether; this is often true in restricted domains, when the association of words with given concepts can be automatically acquired (see, e.g., Pieraccini et al., 1993). We will not discuss these other techniques; some possibilities are described in Allen (1995).

A. Syntax-Driven, Logic-Based Semantic Composition

We will begin by taking first-order logic as our target theory of knowledge representation, and by considering how the meaning of the utterance *John runs* is derived in a syntax-based approach. Let us assume the following simple grammar:

Phase Structure Rules:

- S → NP VP
- NP → PN
- VP → IV

Lexicon:

- PN → John
- IV → runs

Let us further assume that the semantic interpretation provided by the lexicon for *John* is simply its referent (i.e., the object in the world denoted by the constant *j*); and similarly, that the denotation of the intransitive verb *runs* is the unary predicate **runs**. We have then the situation in {14}:

```
        S
      /   \
   NP       VP
   |        |
   PN       IV
 "john"   "runs"
   j       runs
```
{14}

It should be easy to see how one can derive the intended interpretation for an utterance of *John runs*, **runs**(j): the semantic interpretation rules associated with the syntactic rules NP → PN and VP → IV ought to simply assign to the mothers the meaning of the daughters, whereas the semantic interpretation rule associated with the S → NP VP rule ought to construct a proposition by using the interpretation of the VP as a predicate and the interpretation of the NP as an argument. The process of semantic composition will then be as in {15}.

```
        S
     runs(j)
      /   \
   NP       VP
   j       runs
   |        |
   PN       IV
 "john"   "runs"
   j       runs
```
{15}

The questions that we have to address next are the following: what kind of semantic interpretation do we want to associate to the leaves, and what kind of operations can we perform to build the interpretation of higher constituents? We will use a feature–structure notation to specify the information associated with the nodes of a parse tree. Notice that each node in {15} is characterized by syntactic and semantic information at the same time: we will represent this by assuming that the feature structure characterizing the nodes of the parsing tree have the two attributes syn and sem. We will also use feature structures to represent lexical entries, as standardly done in HPSG (Pollard and Sag, 1994). The feature structures representing lexical entries will have an additional phon attribute. For example, the feature–structure representation at the top node of the parse tree in {15} will be as in {16}, whereas the representation of the lexical entry for *John* will be as in {17}.

$$\begin{bmatrix} \text{syn}: \begin{bmatrix}\text{cat}: \text{S}\end{bmatrix} \\ \text{sem}: \mathbf{runs}(j) \end{bmatrix} \qquad \{16\}$$

$$\begin{bmatrix} \text{phon}: \text{“John”} \\ \text{syn}: \begin{bmatrix}\text{cat}: \text{PN}\end{bmatrix} \\ \text{sem}: j \end{bmatrix} \qquad \{17\}$$

B. Semantic Interpretation with the Lambda Calculus

The approach to the formalization of semantic composition operations we are going to describe is one typically adopted in formal semantics. This approach is based on the hypothesis that the lexical meaning of words ought to specify their compositional properties (i.e., their role in determining the overall meaning of an utterance). The central idea is that meaning composition is similar to function application. The fact that a VP, such as *runs*, needs to be combined with an NP, such as *John*, to obtain a proposition is explained by hypothesizing that the lexical meaning of *runs* is a function that takes objects like *John* as arguments and returns propositions; in order to obtain the meaning of the sentence *John runs* we take the meaning of *runs* and apply it to the meaning of *John*.

The name often given to this approach derives from the λ notation for functions developed by Church (1940). Church noted that expressions such as $x+1$ were ambiguous, in that they were used both to denote a *function* (the one-increment function over integers that returns the value of it argument plus 1), as well as the *value* of that function for a given x. He, therefore, introduced the notation $\lambda x.x+1$ to denote functions, reserving the notation $x+1$ for their value. Expressions similar to $\lambda x.x+1$ are often called LAMBDA TERMS. The so-called LAMBDA CALCULUS is a system of inference that combines a language derived from Church's notation for describing complex terms denoting functions with inference rules for simplifying these terms. The logics obtained by adding lambda terms to first-order logic are usually known as TYPE THEORIES; the best known among these theories is the type theory used by Montague [see Gamut (1991) for details].

Let us see how this language can be used to assign meaning to natural language expressions. We assign as meaning to the verb *runs* the lambda term $\lambda x\mathbf{runs}(x)$, denoting a function from entities to truth values (these functions are usually characterized as objects of type $\langle$e,t$\rangle$). As before, the proper name *John* is translated as the term j. And the meaning for the whole sentence *John runs* is the lambda term obtained by applying the term for *runs* to the term for *John*, $[\lambda x\ \mathbf{runs}(x)](j)$, which, by the equivalence rules of the lambda calculus, is identical with the first-order formula $\mathbf{runs}(j)$.

We can now specify the lexical entries. The lexical entry for *John* is shown in {18}; the lexical entries for the verbs *runs* and *likes* are shown in {19} and {23}, respectively.

$$\text{"John"} \mapsto \begin{bmatrix} \text{phon} : \text{"John"} \\ \text{syn} : \begin{bmatrix} \text{cat} : \text{pn} \\ \text{agr} : \begin{bmatrix} \text{pers} : \text{sg} \\ \text{num} : 3 \end{bmatrix} \end{bmatrix} \\ \text{sem} : j \end{bmatrix} \qquad \{18\}$$

$$\text{"runs"} \mapsto \begin{bmatrix} \text{phon} : \text{"runs"} \\ \text{syn} : \begin{bmatrix} \text{cat} : \text{iv} \\ \text{agr} : \begin{bmatrix} \text{pers} : 3 \\ \text{num} : \text{sg} \end{bmatrix} \end{bmatrix} \\ \text{sem} : \lambda x\mathbf{runs}(x) \end{bmatrix} \qquad \{19\}$$

We will use a PATR-like notation to specify our syntactic rules.[1] In the notation we use, a rule is specified in the format

```
X ---> Y_1, ..., Y_n
       C_1
       ...,
       C_m
```

where X is the mother, Y_1, ..., Y_n are the daughters, and C_1, ... C_m are equational constraints on the feature structures associated with the mother and the daughters. An attribute path is represented as a sequence of attributes, as in <VP syn cat>. For example, the rule that specifies that a VP can consist of a single intransitive verb, and that the semantic interpretation of the VP is the same as the semantic interpretation of the IV, is specified as follows:

```
VP ---> IV
     <VP syn cat> = vp
     <IV syn cat> = iv                                {20}
     <VP syn agr> = <IV syn agr>
     <VP sem>     = <IV sem>
```

Most of this rule is concerned with syntactic constraints; the crucial bit for semantic interpretation is the last constraint, that specifies how the sem value percolates. The rule for specifying the semantic interpretation of an NP consisting of a single PN is similarly simple, and is shown in {21}.

```
NP ---> PN
     <NP syn cat> = np
     <PN syn cat> = pn                                {21}
     <NP syn agr> = <PN syn agr>
     <NP sem>     = <PN sem>
```

The rule for sentences is more interesting, because it involves application. Using the symbol * to represent application, this rule can be written as follows:

```
S ---> NP,VP
    <S syn cat>    = s
    <NP syn cat>   = np
    <VP syn cat>   = vp
    <NP syn agr>   = <VP syn agr>                     {22}
    <VP sem>       = Pred
    <NP sem>       = Arg
    <S  sem>       = Pred*Arg
```

It should be easy to see how the derivation in {15} is constructed.

Next, we can extend the grammar to transitive verbs. First of all, the lexical meaning of *likes* is as follows:

[1] For an introduction to unification-based formalisms and to the notations, see Shieber, (1986).

$$\text{"likes"} \mapsto \begin{bmatrix} \text{phon : "likes"} \\ \text{syn : } \begin{bmatrix} \text{cat : tv} \\ \text{agr : } \begin{bmatrix} \text{num : sg} \\ \text{pers : 3} \end{bmatrix} \end{bmatrix} \\ \text{sem : } \lambda y \lambda x \textbf{likes}(x, y) \end{bmatrix} \qquad \{23\}$$

The one interesting bit about this is that *likes* has not been assigned as an interpretation a two-argument function, as it might have been expected, but a function of type ⟨e,⟨e,t⟩⟩—that is, a function which maps entities into functions of type ⟨e,t⟩. With this kind of interpretation, we can derive the meaning of a sentence such as *Mary likes John* by first applying the meaning of *likes* to the meaning of *John*, thus deriving a meaning for the VP, which will be again a function; then applying the resulting meaning to the meaning of *Mary*, thus deriving the meaning for the sentence as a whole.[2] The phrase rule for transitive VPs is:

```
VP ---> TV,NP
     <VP syn cat> = vp
     <TV syn cat> = tv
     <NP syn cat> = np
     <VP syn agr> = <TV syn agr>                    {24}
     <V sem>      = Pred
     <NP sem>     = Arg
     <VP sem>     = Pred*Arg
```

The derivation of the meaning of the sentence *Mary likes John* can now be characterized as follows, where the numbered boxes indicate shared structure:

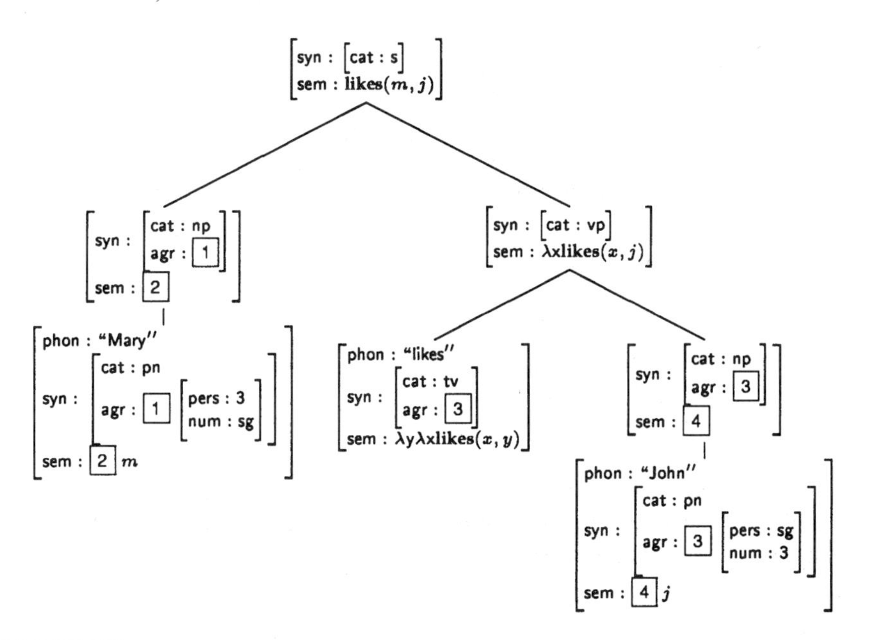

---

[2] It can be shown that these 'double one-argument functions' are equivalent to two-argument functions; this 'trick' (known as 'Currying') allows us to consider one-argument functions only, which makes the semantics simpler.

This might seem an overly complicated way of solving a rather simple problem; but the advantages of the lambda notation become apparent when trying to specify the semantic operations involved in the construction of more complex sentences. One such case are sentences contained *generalized conjunctions* (i.e., conjunctions of constituents other than sentences), such as NPS (as in *John and Mary run*) or VPS (as in *John ran and fainted*). Let us consider VP conjunction first. Intuitively, we would want to assign the sentence *John ran and fainted* the semantic interpretation in {25b}; at the same time, we would want to treat the VP *ran and fainted* as a constituent; that is, to assign to the sentence a syntactic analysis along the lines of {25c}. The problem is, how can be explain how {25b} is derived from {25c} without making too many changes to the rest of the grammar?

a. John ran and fainted {25}

b. $\mathbf{ran}(j) \wedge \mathbf{fainted}(j)$

c.

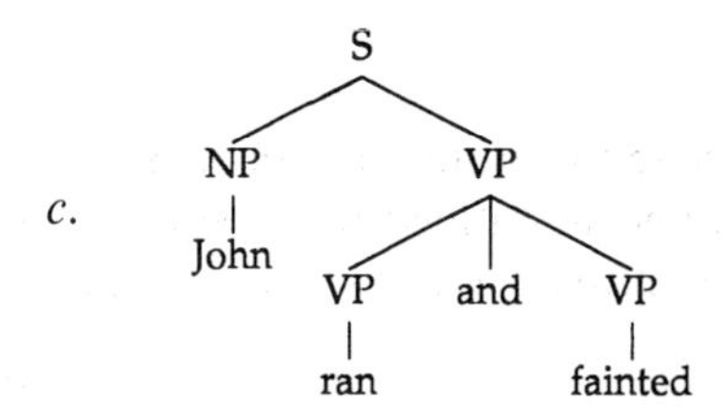

The problem is fairly easy to solve using the tools we have just introduced. One way of doing this is to introduce a rule for conjoined VPS that assigns to *ran and fainted* the meaning $\lambda x \mathbf{ran}(x) \wedge \mathbf{fainted}(x)$ by means of the following operation:

$$\lambda x VP_1(x) \wedge VP_2(x)$$

where VP_1 and VP_2 are the meanings of the two coordinated VPS. This operation takes two ⟨e,t⟩ predicates and returns a new ⟨e,t⟩ predicate by way of two applications and one abstraction. The syntactic rule can be expressed as in {26} where we have used the ^ infix notation for representing lambda expressions introduced by Pereira and Shieber (1987): in this notation, the lambda term λXP is represented as X^P. We have indicated conjunction with &.

```
VP ---> VP1,CONJ,VP2
      <VP1 syn cat>   = vp
      <VP2 syn cat>   = vp
      <CONJ syn cat>  = conj
      <VP1 syn agr>   = <VP2 syn agr>                    {26}
      <VP syn agr>    = <VP2 syn agr>
      <VP1 sem>       = Pred1
      <VP2 sem>       = Pred2
      <VP sem>        = X^((Pred1*X) & (Pred2*X))
```

The advantages of the lambda calculus can also be seen when we consider NP conjunction, as in *John and Mary ran*. Arguably, one would want to assign to this sentence the meaning in {27b} and the syntactic analysis in {27c}:

a. John and Mary ran
b. $\mathbf{ran}(j) \wedge \mathbf{ran}(m)$

c.  {27}

The analysis of these sentences most commonly found in the formal semantics literature depends on an idea about quantified NPS proposed in Montague (1973). The idea is that even though it is usual to look at verbs as predicates and at NPS as arguments, nothing prevents us from going about it the other way around (i.e., while still looking at VPS as properties, we could look at proper names as sets of properties)—the meaning of the NP *John* would be all the properties that are true of John (the object in the domain). This meaning could be expressed by a lambda term of the form {28}, denoting a function for which arguments are predicates that are then applied to the object John.

$$\lambda P_{\langle e,t \rangle} P(j) \quad \{28\}$$

Whether or not this kind of meaning is appropriate for proper names, it seems right for other kind of NPS, including conjoined NPS and quantified NPS (see later). This suggests that we could assign to the NP *John and Mary* the translation in {29}.

$$\lambda P_{\langle e,t \rangle} P(j) \wedge P(m) \quad \{29\}$$

This treatment of coordinated NPS allows us to keep the rest of the grammar unmodified, except that instead of using simple application in the sentence rule, we would need an operation of 'generalized application,' that would make the order of application dependent on the types of the constituents' meanings: in the case of a sentence such as *John runs*, where the meaning of the subject is of type e and the meaning of the VP of type ⟨e,t⟩, generalized application would involve applying the VP meaning to the subject meaning; in the case of *John and Mary ran*, instead, the meaning of the subject (of type ⟨⟨e, t⟩, t⟩) would be applied to the meaning of the VP.

C. Underspecification

Grammar does not always completely determine the interpretation of utterances; as we have seen at the beginning of this chapter, a word such as *stock* is typically going to have different meanings in different contexts, and the interpretation of anaphoric expressions is entirely determined by context. If we insist that semantic interpretation produce only completely disambiguated interpretations, the only interpretation strategy that we can adopt is for the grammar to generate all interpretations, if we want to be sure we generate the right one.

For lexical disambiguation, which is a fairly localized process, this is often the strategy used, typically using preferences acquired by statistical means to choose the most likely sense in the given context (Charniak, 1993). In the cases of anaphoric expressions and scope, generating all possible interpretations and filtering them later is implausible and expensive. The strategy developed in NLP to address these cases of incomplete determination of meaning is to have the grammar generate an UNDERSPECIFIED INTERPRETATION of the utterance—that is, an interpretation in which ambiguities are not completely resolved (van Deemter and Peters, 1996); the task of 'completing' this interpretation is left to contextual inference, which will presumably generate only one or a few preferred interpretations. Underspecified representations have been used most in connection with three types of context dependence:

1. Resolving underdetermined relations among predicates, such as those introduced by noun–noun modification or by prepositional modifications: for example, a *coffee cup* is a cup used to *contain* coffee, whereas a *paper cup* is a cup *made* of paper. Identifying the correct interpretation in each case depends on world knowledge.
2. Resolving anaphoric expressions, such as pronouns, definite descriptions (*the car*) or ellipsis: although syntax imposes some constraints on the possible solutions, identifying the referent of these expressions involves keeping track of which objects have been mentioned or are somehow 'around.'
3. Identifying the correct *scope* of quantifiers: as the following example from van Lehn (1978) shows, two sentences with the very same syntactic structure and consisting of almost exactly the same words can nevertheless have completely different interpretations:[3]

a. *Every researcher at IJCAI knows a dialect of Lisp.*
b. *Every reearcher at XEROX PARC knows a dialect of Lisp.* {30}

In all of these cases, the solution proposed in NLP was for the grammar not to attempt producing disambiguated interpretations in which these context dependencies were resolved, but to 'leave things as they are,' in the sense of producing an INTERMEDIATE LOGICAL FORM indicating, for example, that a noun–noun relation had to be identified, that an anaphoric expression had to be resolved, or that the relative scope of some quantifier still had to be determined (Woods, 1972; Schubert and Pelletier, 1982; Allen, 1995; Alshawi, 1992; Hobbs et al., 1992).

Many such 'logical forms' have been proposed; we will follow here the approach taken in Poesio (1994, 1996), in which semantic underspecification is seen as a special kind of partiality that can be characterized using the tools already introduced in feature-based formalisms to characterize syntactic partiality. Instead of introducing a special representation just to deal with ambiguity, we will use the syntactic structure itself (augmented with lexical information) as a logical form; underspecification in the semantic interpretation of a particular lexical element or phrase is indicated by leaving its `sem` slot empty. For example, the underspecified interpretation of the sentence *He runs* in the feature-based format we have introduced earlier would be as in {31}, where we have assigned an identifier to each sign (the *index* annotations), and we have assumed that each phrasal sign has a `head` and a `daughters` attribute, the latter of which is a list of signs.

[3] In these days of Common Lisp an explanation may be in order—back when Van Lehn constructed this example, every AI researcher was using a different dialect of Lisp, so the preferred interpretation of {30a} was very much the one in which different dialects of Lisp are involved. But XEROX PARC had developed their own dialect of Lisp, Interlisp; the much preferred interpretation of the second sentence was that there was a single dialect of Lisp that everybody at XEROX PARC knew.

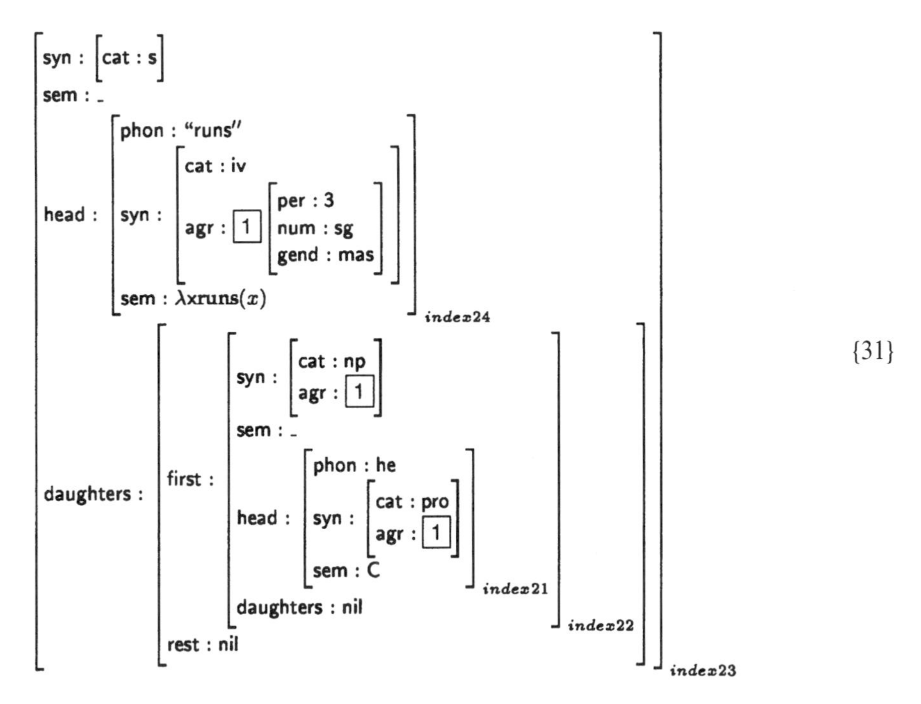

{31}

In 'logical forms' of this type, the initial position of quantifiers is automatically preserved, together with all other constituency information without introducing yet another level of representation. For example, the sentence *Every kid climbed a tree* may be given the underspecified representation in {32}, where the quantifiers are 'in situ' (we have removed most of the syntactic information to save space):

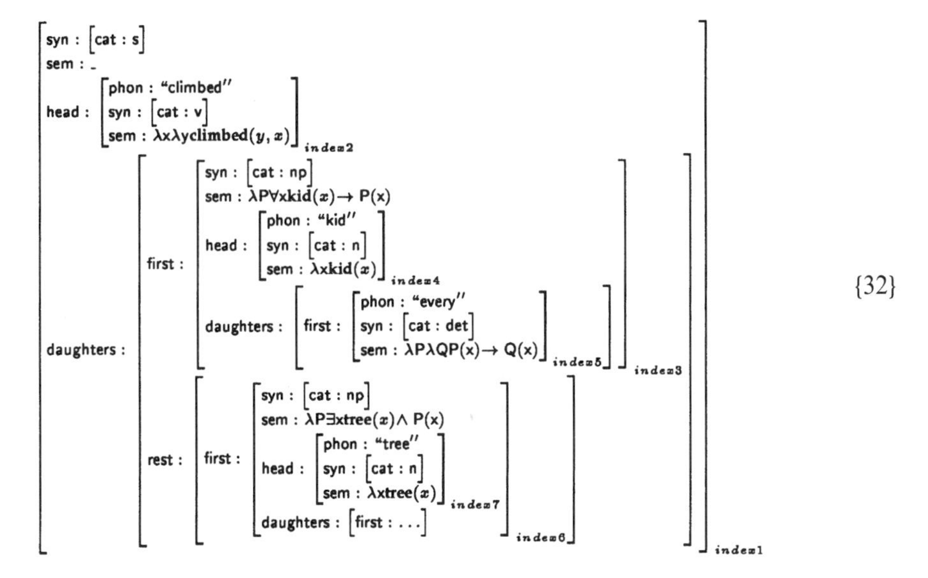

{32}

Besides preserving the position of quantifiers, this 'logical form' preserves other syntactic information which is useful for resolving anaphoric expressions, as we will see in the next section.

In many applications, we can consider the issue of ambiguity solved once we do not have to generate all completely disambiguated interpretations and we have efficient ways of generating the preferred interpretation; we will shortly see some of these techniques for the case of anaphoric expressions. From a semanticist's point of view, however, what we have said so far is at best a partial story: unless we say what expressions such as {31} mean, we are left at the end of semantic interpretation with an expression with no semantics. What is the meaning of these expressions and, most importantly, what can we do with them? (e.g., how do we know whether our algorithms for extracting disambiguated interpretations are correct?). Having a clear specification of the semantics of these expressions would also give us the opportunity to explore the question: are these interpretations really always solved? These questions have received much attention in recent years (see, e.g., Alshawi and Crouch, 1992; van Deemter, 1991; Poesio, 1991, 1996; Reyle, 1993) as well as the papers in van Deemter and Peters (1996).

4. THE ROLE OF CONTEXT IN SEMANTIC INTERPRETATION: INTERPRETING ANAPHORIC EXPRESSIONS

At the end of last section we said that, in general, grammar gives only a partial description of the meaning of an expression, to be completed by using context. In this section we review current techniques for one type of contextual resolution of underspecified interpretations: the resolution of anaphoric expressions.

A. Types of Anaphoric Expressions

The typical examples of anaphoric expression are pronouns, such as *he*, in a text similar to *John arrived. He looked tired.* In the preferred reading of this text, the pronoun *he is* interpreted as an 'abbreviated reference' to the individual John which is denoted by the expression *John.* Following the terminology introduced in Sidner (1979), we will say that the pronoun COSPECIFIES with the proper name *John*, and we will call John the ANTECEDENT of the pronoun.

Anaphoric expressions make a text more readable and emphasize the connection among its parts. There are several types of anaphoric expressions, differing in both their syntactic and semantic properties. Some of these types are illustrated in Fig. 2.

This text (a recipe posted to the newsgroup `rec.food.cooking`) contains at least three kinds of anaphoric expressions. Besides the pronouns *them* and *they*, the text includes DEFINITE DESCRIPTIONS such as *the dry yeast* and *the water*, and several examples of ELLIPSIS, as in *Stir—and let sit—*, where the arguments of *stir* and *let sit* have been omitted and have been recovered from context.

As this example shows, anaphoric expressions are exceedingly common in real texts. Other kinds of anaphoric expressions, in addition to those just discussed, include POSSESSIVE DESCRIPTIONS such as *her computer*, and ONE ANAPHORA, as in *He saw that the plugs were worn, and advised John to fit <u>some new ones</u>.* RESOLVING anaphoric expressions (i.e., identifying their antecedents so that, for example, we can understand what exactly should be put in the water in the example recipe) is one of the main problems confronting the designer of a natural language-processing system.

```
ingredients:
1/8 cup warm water
1 package dry yeast
1 1/4 cup hot water
1/3 packed brown sugar
4-5 cups white flour
lots of kosher salt
Tbs of backing soda

add THE DRY YEAST to THE WATER and let sit _ for a few minutes.
add THE REST OF THE WATER and sugar. Stir _ and let sit _.
slowly add THE FLOUR and stir. Add enough flour so you can
knead THE FLOUR in, but isn't sticky anymore. Take golf ball
size pieces of THE DOUGH and roll _ into pretzel shapes or
any shape you want. In a frying pan or pot - boil water with
backing soda (about 2 cups of water to a tablespoon of backing
soda).
Take THE PRETZELS and place THEM in THE WATER for 30 seconds
or until THEY float. put THEM on a generously greased and salted
cookie sheet. salt THEM generously on top as well. Back _ at 375
for 8-15 minutes.
```

Fig. 2 A recipe for pretzels.

B. Anaphoric Expressions and Underspecification

As discussed in the previous section, the task of resolving anaphoric expressions can be characterized in terms of the theory of underspecification we have assumed, as the problem of identifying the semantic value of expressions such as *he*. The lexicon specifies for these expressions an underspecified interpretation of the form in {33}.

$$he \Rightarrow \begin{bmatrix} \text{phon} : \text{“he”} \\ \text{syn} : \begin{bmatrix} \text{cat} : \text{pro} \\ \text{agr} : \begin{bmatrix} \text{gen} : \text{masc} \\ \text{num} : \text{sing} \\ \text{pers} : 3 \end{bmatrix} \end{bmatrix} \\ \text{sem} : \_ \end{bmatrix}_i \qquad \{33\}$$

'Resolving' this anaphoric expression involves filling in the missing information; that is, arriving at an interpretation such as, say, {34}.

$$he \Rightarrow \begin{bmatrix} \text{phon} : \text{“he”} \\ \text{syn} : \begin{bmatrix} \text{cat} : \text{pro} \\ \text{agr} : \begin{bmatrix} \text{gen} : \text{masc} \\ \ldots \end{bmatrix} \end{bmatrix} \\ \text{sem} : \text{j} \end{bmatrix}_i \qquad \{34\}$$

Resolving other cases of anaphoric expressions can be similarly seen as the problem of identifying such missing semantic interpretations.

C. Factors That Play a Role in Anaphora Interpretation

Interpreting anaphoric expressions involves, first of all, keeping track of what has been said, especially which possible antecedents have been introduced in the discourse (see following). The choice of an antecedent for an anaphoric expression among those available is affected by three major factors:

1. Syntactic information (e.g., about gender and number) and syntactic constraints: This information is especially important in the case of pronouns and ellipsis; for example, gender alone is sufficient to determine the interpretation of *he* and *she* in the following example:

   John met Mary the other day. He/She had just come off the bus. {35}

2. Common-sense knowledge: For example, both {36a} and {36b} are superficially unambiguous, but the pronoun *they* is most likely to refer to *the city council* in {36a} and to *the protesters* in {36b} (this example is from Winograd (1972)).

   a. *The city councilmen refused the protesters a permit because they advocated violence* {36}
   b. *The city councilmen refused the protesters a permit because they feared violence.*

3. Salience: All three occurrences of the pronoun *it* in the following first paragraph have the cheese as their antecedent. But in the second paragraph, *it* refers to the casserole dish, even though all we know about the referent of the pronoun *it* is that that object holds the heat and is round, both properties that could hold of cheese, as well as of casserole dishes. The hypothesis is that at any moment, certain discourse referents are more SALIENT than others, thus more likely to serve as antecedents.

   Classic Fondue (Fondue Neuchateloise) {37}

   The chief thing to remember about a cheese fondue is that the cheese must cook over very low heat or it will become stringy. It must also be kept hot, but over low heat, so that it will not heat too much and become tough.
   Equipment: In Switzerland a fondue is made in a round metal or earthenware pot, but a heavy earthenware or cast-iron casserole will serve as long as it holds the heat and is round in shape. The pot (caquelon) is put in the middle of the table . . .

A quantitative assessment of the effect of salience was carried out by Hobbs (1978), who found that between 90 and 95% of third-person singular pronouns had their antecedent in the same or the previous sentence.

D. Discourse Models

To resolve anaphoric expressions it is necessary to keep track of potential antecedents: this includes the entities mentioned in the text for discourse—old expressions—and entities in the 'immediate situation' for discourse—new expressions.

The antecedents introduced in a discourse are not necessarily objects with which the reader or the listener is acquainted: this is typically true, for example, with antecedents introduced by indefinite descriptions, as in *I met a guy at the bar yesterday*. Yet, this discourse can be continued with *He has a big beard* without the reader knowing the person the writer was talking about.

According to researchers in artificial intelligence, linguistics, and psychology, this is possible because readers, when processing a discourse, construct a DISCOURSE MODEL: a record of the entities mentioned in a discourse (usually called DISCOURSE REFERENTS or DISCOURSE MARKERS) together with the properties of these entitites that have been mentioned. Discourse-old anaphoric expressions are interpreted relative to the discourse model, rather than to a model of the world (Karttunen, 1976; Webber, 1979; Kamp, 1981; Heim, 1982).[4]

The most basic notion of discourse model is the HISTORY LIST: a list of all entities 'evoked' by a discourse, together with the properties that are relevant to reference resolution. These properties, minimally, include syntactic information, such as gender and number, and semantic information about the denotation of the discourse entity (i.e., the information contained in the sign encoding of the NP). In first approximation, that is, we can assume that the discourse model contains one discourse entity for each noun phrase occurring in the text, and that the information about each discourse entity simply consists of the sign representation of that noun phrase. For example, the information about the discourse entity introduced by a mention of the proper name *John* would be as follows:

$$\begin{bmatrix} \text{phon} : \text{“John”} \\ \text{syn} : \begin{bmatrix} \text{cat} : \text{pn} \\ \text{agr} : \begin{bmatrix} \text{gen} : \text{masc} \\ \text{num} : \text{sing} \\ \text{pers} : 3 \end{bmatrix} \end{bmatrix} \\ \text{sem} : j \end{bmatrix}_i \qquad \{38\}$$

In addition, the discourse model includes information about the semantic properties of the discourse entities (Grosz, 1977; Sidner, 1979). This information, minimally, includes the kind of information found in a semantic network, such as the type of the object (e.g., that j is a **person**) and its attributes.

This idea that sentences 'update' a context by adding new 'objects' to it is now well accepted in semantic theory. Although the interpretation of pronouns affects the truth-conditional interpretation of sentences, neither first-order logic nor the other formalisms for semantic interpretation previously discussed can explain how pronouns can cospecify with discourse referents introduced by indefinite descriptions. Consider for example the text in {39}:

There is an engine at Avon. <u>it</u> is hooked to a boxcar. {39}

[4] There is much less agreement over which sort of information is used to resolve discourse-new anaphoric expressions, and the relation of this information to the discourse model. One proposal is that listeners keep a record of the DISCOURSE SITUATION, which is very much like the record of the situation under discussion (Barwise and Perry, 1983). This proposal is developed in detail in Poesio (1994) and Poesio and Traum (1997).

The first-order logic representation of the first sentence in this discourse is shown in {40}: there is no way to construct an interpretation for the text in which the pronoun *it* in the second sentence is forced to denote the same object as the engine in the first sentence by simply conjoining the interpretation of the second sentence to the interpretation of the first sentence, as in {41}, because the second instance of the variable x would not be in the scope of the existential quantifier.

$$(\exists\, x\ \mathbf{engine}(x) \wedge \mathbf{at}(x,\ Avon)) \qquad \{40\}$$

$$(\exists\, x\ \mathbf{engine}(x) \wedge \mathbf{at}(x,\ Avon)) \wedge \exists\, y\ \mathbf{boxcar}(y) \wedge \mathbf{hooked\text{-}to}(x,y)) \qquad \{41\}$$

The solution proposed in Discourse Representation Theory (DRT) (Kamp, 1981; Heim, 1982; Kamp and Reyle, 1993) is inspired quite literally by the idea of a discourse model as introduced in the foregoing. According to DRT, semantic interpretation is a matter of incorporating the content of a sentence into the existing context. Contexts consist of DISCOURSE REPRESENTATION STRUCTURES (DRSs). A DRS is a pair consisting of a set of discourse referents together with a set of CONDITIONS expressing properties of these referents. Processing the first sentence of {39} results in the DRS in {42}.

$x\ w$
engine(x) **Avon**(w) **at**(x,w)

{42}

Note that this DRS includes a discourse referent for the engine (x) as well as for Avon (w). DRSs have a well-specified semantics; {42} is semantically equivalent to the FOL formula:

$$\exists x\ w\ \mathbf{engine}(x) \wedge \mathbf{Avon}(w) \wedge \mathbf{at}(x,w) \qquad \{43\}$$

Processing the second sentence involves, first of all, adding to the DRS in {42} a condition encoding the syntactic interpretation of this second sentence. Rewriting rules (the 'DRT construction algorithm') then apply to this second condition; the result of this process is the DRS in {44}, semantically equivalent to the FOL formula in {45}.

$x\ w\ y\ u$
engine(x) **Avon**(w) **at**(x,w) **boxcar**(y) **hooked-to**(u,y) u **is** x

{44}

$$(\exists x,\ w,\ y,\ u\ \mathbf{engine}(x) \wedge \mathbf{Avon}(w) \wedge \mathbf{at}(x,\ w) \wedge \mathbf{boxcar}(y)$$
$$\wedge\ \mathbf{hooked\text{-}to}(u,\ y) \wedge u = x) \qquad \{45\}$$

In the formulation of the theory just presented, discourse referents are made accessible beyond the clause in which they were introduced because the semantic interpretation procedure in DRT always begins by adding the syntactic interpretation of a new sentence to the existing DRS—thus, in effect, 'inserting' the new sentence in the scope of the top-most existential quantifiers. More sophisticated versions of the theory have been developed in recent years that achieve the same result without explicit logical form manipulations; in these theories, the interpretation of {39} is obtained by simply concatenating the interpretations of each sentence. Such theories allow us to contact a DRS on the basis of compositional semantic interpretation procedures, such as those discussed in Sec. 3.

For lack of space we will not be able to discuss these theories in any detail here; the general idea of one of them, Compositional DRT (Muskens, 1994), is to define DRSs within a type theory similar to those discussed in Sec. 3, ad well as a DRS concatenation operation ';' which makes it possible to merge two DRSs into one. Once these are defined, the assignment of meanings to expressions is modified so that (a) predicates become functions from individuals to DRSs instead of from individuals to propositions—for example, *runs* would translate as an abstraction over DRSs $\lambda x[|\mathbf{runs}(x)]$ (i.e., a DRS with no discourse referents and one condition, using the linear notation $[xy|\Phi]$ for DRSs) instead of simply as $\lambda x\mathbf{runs}(x)$—and (b) the meaning of quantifiers, in particular of indefinite descriptions such as *a boxcar*, is modified so that they introduce new discourse referents and they concatenate predicates, rather than simply modifying them: for example, the semantic translation of *a boxcar* is shown in {46}. Lexical entries for *runs* and *a* incorporating this semantics are shown in {47} and {48}.

$$\lambda P[y|\mathbf{boxcar}(y)];P(y) \qquad \{46\}$$

$$\text{"runs"} \mapsto \begin{bmatrix} \text{phon : "runs"} \\ \text{syn : } \begin{bmatrix} \text{cat : iv} \\ \text{agr : } \begin{bmatrix} \text{pers : 3} \\ \text{num : plur} \end{bmatrix} \end{bmatrix} \\ \text{sem : } \lambda x[|\mathbf{runs}(x)] \end{bmatrix} \qquad \{47\}$$

$$\text{"a"} \mapsto \begin{bmatrix} \text{phon : "a"} \\ \text{syn : } \begin{bmatrix} \text{cat : det} \\ \text{agr : } \begin{bmatrix} \text{num : sing} \end{bmatrix} \end{bmatrix} \\ \text{sem : } \lambda P \lambda Q[y|]; P(y); Q(y) \\ \text{do : y} \end{bmatrix} \qquad \{48\}$$

In {48} the discourse referent *y*, introduced by the indefinite article, is specified both by the `sem` and by the additional `do` feature, which makes it accessible for subsequent reference.

E. Pronoun Resolution

1. *The Task of Pronoun Disambiguation*

We have already presented our characterization of the task of pronoun resolution: filling in the missing semantic component of the underspecified interpretation of a pronoun (shown in {33}).

Now that we have introduced the notion of a discourse model, we can characterize more precisely the task of pronoun resolution as that of identifying the discourse referent evoked by the antecedent of the pronoun (rather than directly the referent of the antecedent): the case in which discourse referent x is identified as the antecedent of pronoun he_i can thus be represented as:

$$he \Rightarrow \begin{bmatrix} \text{phon} : \text{“he”} \\ \text{syn} : \begin{bmatrix} \text{cat} : \text{pro} \\ \text{agr} : \begin{bmatrix} \text{gen} : \text{masc} \end{bmatrix} \end{bmatrix} \\ \text{sem} : \text{x} \end{bmatrix}_i \qquad \{49\}$$

Resolving {33} involves searching the discourse model for a discourse referent with the appropriate syntactic and semantic characteristics: as discussed earlier, a candidate antecedent must match syntactic properties of the pronoun, such as gender and number, and must also satisfy semantic properties, such as selectional restrictions (e.g., a candidate antecedent for *it* in *Pour it in a pot* must be a liquid of some sort). These conditions are satisfied if we assume that the information associated with a discourse referent includes the information specified by the sign that evokes that referent; and that the semantic information about discourse referents includes at least the kind of information that is stored in a semantic network. In addition, a model of salience is needed; most research on pronoun resolution has concentrated on this problem. We consider this issue next.

2. *Centering Theory*

The simplest way of dealing with saliency is to adopt a recency based strategy (i.e., to keep a stack of discourse referents and look for the most recent discourse referent that satisfies the syntactic and semantic constraints on the pronoun). This strategy works reasonably well in simple cases; to handle more complex examples, theories of FOCUS have been developed. There are two types of focus theories:

a. *Activation-Oriented*

According to these theories, each discourse referent has a LEVEL OF ACTIVATION that increases or decreases depending on various factors. Discourse referents are discarded as possible antecedents when they go below some threshold level of activation. These kinds of models are very flexible and are often used in applications, but their success depends to a significant extent on the way in which the activation levels are determined; a lot of fine-tuning is required to achieve a satisfactory performance. Models of this type have been proposed by Kantor (1977) and, more recently, by Alshawi (1987) and Lappin and Leass (1994).

b. Discrete

The discrete theories make the claim that a fixed number of discourse entities are most salient at any given time, and provide rules of specifying how the data structures encoding this salience information is encoded. Being more restrictive, such theories are easier to verify than the activation-oriented ones; therefore, they tend to attract most of the attention of psycholinguists. The two main theories of this type are Sidner's focus theory (Sidner, 1979, 1983) and CENTERING THEORY (Grosz et al., 1995; Brennan et al., 1987) This latter is probably the most popular theory of local focus at the time of writing; therefore, we will consider it in some detail.

The hypothesis behind centering theory is that at any moment a text is "centered" around certain objects, and the more you stick to referring to these centers the more coherent your text is. This hypothesis is motivated by the contrast between {50} and {51}.

a. John went to his favorite music store to buy a piano. {50}
b. He had frequented the store for many years.
c. He was excited that he could finally buy a piano.
d. He arrived just as the store was closing for the day.

a. John went to his favorite music store to buy a piano. {51}
b. It was a store John had frequented for many years.
c. He was excited that he could finally buy a piano.
d. It was closing just as John arrived.

In {50}, every sentence is about John, and the pronoun *he* in subject position consistently refers back to John. In {51}, however, sentence (a) is about John, (b) is about the store, (c) is about John, and (d) is about the store again. Because the sentences emphasize different entities, the text is not felt to be very coherent.

Centering theory consists of a set of constraints on the FORWARD-LOOKING CENTERS (C_f) (the discourse referents introduced by an utterance) and the BACKWARD-LOOKING CENTER (C_b). This latter is the most salient discourse referent of utterance U_n that is realized in utterance U_{n+1}. The C_fs are ranked; most recent versions of centering theory assume that the ranking is determined by syntactic factors, so that, for example, the subject ranks more highly than the object. We illustrate the use of C_fs and C_b in the following example (Brennan et al., 1987). Each C_f is represented as a pair [SEM:PHON], where SEM is the denotation of the discourse referent and PHON is its phonetic realization. Note that all utterances have a C_b, except the first one.

a. Carl works at HP on the Natural Language Project. {52}
C_b: nil
C_f: ([POLLARD:Carl],[HP:HP],[NATLANG:Natural Language Project])

b. He manages Lyn.
C_b: [POLLARD:Carl]
C_f: ([POLLARD:A1],[FRIEDMAN:Lyn])

c. He promised to get her a raise.
C_b: [POLLARD:A1]
C_f: ([POLLARD:A2],[FRIEDMAN:A3],[RAISE:X1])

d. She doesn't believe him.
C_b: [POLLARD:A2]
C_f: ([FRIEDMAN:A4],[POLLARD:A5])

The first important claim of centering is that the C_b is, among the C_fs of the previous utterance, the most likely to be pronominalized:

RULE 1. If any element of $C_f(U_n)$ is realized by a pronoun in U_{n+1}, then $C_b(U_{n+1})$ is also realized as a pronoun.[5]

The second claim of centering theory is that the less the C_b changes, the more coherent a text is. Each utterance can be classified according to whether its C_b changes or stays the same, and according to whether the C_b of an utterance is also the most salient discourse referent of that utterance. This gives rise to the following classification of utterances:

Center Continuation: $C_b(U_n) = C_b(U_{n+1})$, and $C_b(U_{n+1})$ is the highest-ranked element of $C_f(U_{n+1})$.
Center Retaining: $C_b(U_n) = C_b(U_\mathrm{n} + 1)$, but $C_b(U_{n+1})$ is not the highest-ranked element of $C_f(U_{n+1})$. (For example, it is the object rather than the subject.)
Center Shifting: $C_b(U_n) \neq C_b(U_{n+1})$.

Center continuation is when the C_b remains the same; center retention is when the C_b remains the same, but that C_b is not the highest-ranked entity referred to in the current sentence. Center shifting is when we change to a new C_b (i.e., the sentence is about something else now). To these transitions, Kameyama (1986) added a **center establishment** transition: all second utterances of a text are going to be classified as transitions of this type. The second rule of centering is formulated as a preference over sequences of transitions:

RULE 2. Sequences of continuations are preferred over sequences of retaining; and sequences of retaining are preferred over sequences of shifting.

An algorithmic implementation of centering theory has been proposed in (Brennan et al., 1987). The algorithm consists of three steps: first all possible interpretations of an utterance are generated by enumerating all possible interpretations for pronouns and all possible choices of C_b; next, all those interpretations that violate syntactic constraints are eliminated; finally, the interpretations are ranked according to rules 1 and 2.

5. CONCLUSIONS

In this chapter, we have surveyed some of the most prominent approaches taken to semantic interpretation and representation within natural language processing, and we have looked at the use of these techniques relative to the particular problem of

[5] The rules and constraints of centering theory are not meant to be prescriptive; they should be interpreted as defaults, and the idea is that a text is going to be maximally coherent when these defaults are satisfied.

anaphoric reference resolution. The approaches described here begin to bring to semantic analysis the same level of sophistication as that found in current syntactic theories.

6. ANAPHORIC EXPRESSIONS

For space reasons we could discuss here only the resolution of pronominal references; however, it is important to bear in mind that there are many other forms of anaphora, as shown in Fig. 2. Two types of anaphoric expression that have also been extensively discussed in the literature are DEFINITE DESCRIPTIONS and ELLIPSIS.

- We have already touched on some of the issues that arise when considering definite descriptions. The initial interpretation of definite description can also be said to be underspecified in that the semantic value of the definite description is missing; the task of interpretation is to assign a discourse referent or object in the discourse situation or in the shared common ground as a value of the definite description. As in the case of pronouns, interpreting anaphoric definite descriptions involves maintaining a record of the discourse referents introduced in the text and searching these discourse referents for one that matches the properties of the definite description, keeping saliency into account. One important difference is that the notion of salience that affects definite description resolution appears to be different from the more local type that affects the interpretation of pronouns, and of which centering theory is a model. With definite descriptions, a notion of salience relative to the situation being described seems necessary. A second difference between resolving pronouns and resolving definite descriptions is that, whereas syntactic information (e.g., gender) plays a crucial role in determining whether a discourse referent is a potential antecedent for a pronoun, semantic information plays the most important role for definite descriptions. The classic work on definite description resolution is (Carter, 1987); for an example of recent work using corpus for evaluation, see (Porsio and Vieira, 1998).
- We have an ELLIPSIS when 'something is missing' from a clause, called the TARGET CLAUSE; the missing components have to be recovered from a so-called SOURCE CLAUSE. For example, VPs can be omitted from contexts in which two clauses are coordinated; the VPs are replaced the by expressions *did* or *did too*:

  a. *John read his paper before Bill did.* {53}
  b. *John sent Mary a gift, and Bill did too.*

  The problem of resolving the ellipsis in {53a} can be characterized again as that of resolving the underspecification in the target clause. There are two computing theories of how ellipsis is resolved:

  1. *The Syntactic View*: Ellipsis resolution involves reconstructing the missing syntactic structure of the target clause, or establishing a syntactic relation (such as binding) between this constituent and the source clause.

2. *The Semantic View*: A semantic procedure for interpreting elided constituents identifies a property (i.e., a predicate) which is applied directly to the denotation of an argument. The ellipsis is treated like a pronoun, and resolved by finding an antecedent in the source clause.

An example of a semantic approach to ellipsis resolution is HIGHER-ORDER UNIFICATION (Dalrymple et al., 1991). Robust algorithms for eclipses resolution are discussed in (Hardt, 1997).

Other than the forthcoming work by Poesio and Stevenson (1999) (which also includes an overview of psychological evidence), there is no recent review of work on anaphora; Hirst (1981) is a good introduction to work on anaphora up until 1981, and Carter (1987) a good update of work up until 1987 (with an additional discussion of the problems in integrating focusing with inference). Among the recently proposed algorithms for interpreting anaphora, that of Lappin and Leass (1994) has been thoroughly tested.

The models of discourse introduced in formal semantics are discussed in the *Handbook of Logic and Language* (van Benthem and ter Meulen, 1997), especially in the chapters by Muskens et al. and by Van Eijck and Kamp. A more extensive discussion of discourse structure and intention recognition can be found in Chap. 6 of this handbook.

REFERENCES

Allen JF. 1995. Natural Language Understanding, 2nd ed. Reading, MA: Addison–Wesley. 2nd ed.

Allen JF, LK Schubert, G Ferguson, P Heeman, CH Hwang, T Kato, M Light, N Martin, B Miller, M Poesio, DR Traum. 1995. The TRAINS project: a case study in building a conversational planning agent. *J Exp Theor AI* 7:7–48.

Alshawi H. 1987. Memory and Context for Language Interpretation. Cambridge, UK: Cambridge University Press.

Alshawi H, ed. 1992. The Core Language Engine. Cambridge, MA: MIT Press.

Alshawi H, R Crouch. 1992. Monotonic semantic interpretation. Proc. 30th. ACL, University of Delaware, Newark, pp 32–39.

Barwise J, Cooper. 1981. Generalized quantifiers and natural language. Linguist Philos 4:159–219.

Barwise J, J Perry. 1983. Situations and Attitudes. Cambridge, MA: MIT Press.

Beckart B, J Posegga. 1996. Lean TAP: Lean tableau-based deduction. Automated Reasoning.

Brachman RJ, HJ Levesque, eds. 1985. Readings in Knowledge Representation. San Mateo, CA: Morgan Kaufmann.

Brachman RJ, JG Schmolze. 1985. An overview of the KL-ONE knowledge representation system. Cognitive Sci 9(2):171–216, April–June.

Brennan SE, MW Friedman, and CJ Pollard. 1987. A centering approach to pronouns. Proc. ACL-87, June, pp 155–162.

Brewka G. 1991. Nonmonotonic Reasoning: Logical Foundations of Commonsense. Cambridge, UK: Cambridge, University Press.

Carpenter B. 1992. The Logic of Typed Feature Structures. Cambridge, UK: Cambridge University Press.

Carter DM. 1987. Interpreting Anaphors in Natural Language Texts. Chichester, UK: Ellis Horwood.

Charniak E. 1993. Statistical Language Learning. Cambridge, MA: MIT Press.

Chierchia G, S McConnell-Ginet. 1990. Meaning and Grammar: An Introduction to Semantics. Cambridge, MA: MIT Press.

Church A. 1940. A formulation of the simple theory of types. J Symbolic Logic 5:56–68.

Clark H H. 1977. Bridging. In: PN Johnson–Laird, PC Wason, eds. Thinking: Readings in Cognitive Science. New York: Cambridge University Press.

Dalrymple M, SM Shieber, FCN Pereira. 1991. Ellipsis and higher-order unification. Linguist Philos 14:399–452.

Gamut LTF. 1991. Logic, Language and Meaning. Chicago: University of Chicago Press.

Grosz BJ. 1977. The representation and use of focus in dialogue understanding. PhD dissertation, Stanford University, Palo Alto, CA.

Grosz BJ, AK Joshi, S Weinstein. 1986. Towards a computational theory of discourse interpretation. Unpublished ms.

Grosz BJ, AK Joshi, S Weinstein. 1995. Centering: a framework for modeling the local coherence of discourse. Comput Linguist 21:202–225. [Published version of Grosz et al., (1986).]

Hardt, D. 1997. An empirical approach to VP ellipsis Comput Linguist 23(4):525–541.

Hayes, PJ. 1979. The logic of frames. In: D Metzing, ed. Frame Conceptions and Text Understanding. Berlin: Walter de Gruyter pp 46–51. [Also published in Brachman and Levesque (1985).]

Heim I. 1982. The semantics of definite and indefinite noun phrases. PhD dissertation, University of Massachusetts at Amherst.

Hirst G. 1981. Anaphora in Natural Language Understanding: A Survey. Lecture Notes in Computer Science 119. Berlin: Springer-Verlag.

Hobbs JR 1978. Resolving pronoun references. Lingua 44:311–338.

Hobbs JR, Sticker M, Martin P, Edwards D. 1993. Interpretation as abduction. Artificial Intelligence 63:69–172.

Kameyama M. 1986. A property-sharing constraint in centering. Proc. ACL-86, pp 200–206.

Kamp H. 1981. A theory of truth and semantic representation. In: J Groenendijk, T Janssen, M Stokhof, eds. Formal Methods in the Study of Language. Amsterdam: Mathematical Centre.

Kamp H, U Reyle. 1993. From Discourse to Logic. Dordrecht: D Reidel.

Kantor RN. 1977. The management and comprehension of discourse connection by pronouns in English. PhD dissertation, Ohio State University, Department of Linguistics.

Karttunen, L. 1976. Discourse referents. In: J McCawley, ed. Syntax and Semantics 7—Notes from the Linguistic Underground. New York: Academic Press.

Keenan EL, J Stavi. 1986. Natural language determiners. Linguist Philos 9:253–326.

Lappin S, HJ Leass. 1994. An algorithm for pronominal anaphora resolution. Comput Linguist 20:535–562.

Montague R. 1973. The proper treatment of quantification in English. In: KJJ Hintikka, ed. Approaches to Natural Language. Dordrecht: D Reidel, pp 221–242. [Reprinted in Thomason (1974).]

Muskens R. 1994. A compositional discourse representation theory. In: P Dekker, M Stokhof, eds. Proceedings of the 9th Amsterdam Colloquium, pp 467–486.

Pearl J. 1988. Probabilistic Reasoning In Intelligent Systems: Networks of Plausible Inference. San Mateo, CA: Morgan Kaufmann.

Pereira FCN, SM Shieber. 1987. Prolog and Natural-Language Analysis. CSLI Lecture Notes. CSLI, Stanford.

Pieraccini R, E Levin, E Vidal. 1993. Learning how to understand language. Proceedings of Third Eurospeech, Berlin, pp 1407–1412.

Poesio M. 1991. Relational semantics and scope ambiguity. In: J Barwise, JM Gawron, G Plotkin, S Tutiya, eds. Situation Semantics and its Applications. Vol 2. Stanford, CA: CSLI, pp 469–497.

Poesio M. 1994. Discourse interpretation and the scope of operators. PhD dissertation, University of Rochester, Department of Computer Science, Rochester, NY.

Poesio M. 1996. Semantic ambiguity and perceived ambiguity. In: K van Deemter, S Peters, eds. Semantic Ambiguity and Underspecification. Stanford, CA: CSLI, pp 159–201.

Poesio M, R Stevenson. 2001. Salience: computational models and psychological evidence. Cambridge University Press.

Poesio M, D Traum. 1997. Conversational actions and discourse situations. Comput Intell 13:309–347.

Poesio M, R Vieira 1998. A corpus-based Investigation of definite description use. Computational Linguistics 21(2):183–216.

Pollard C, IA Sag. 1994. Head-Driven Phrase Structure Grammar. Chicago, University of Chicago Press.

Poole DL, R Goebel, R Aleliunas. 1987. Theorist: a logical reasoning system for defaults and diagnosis. In: NJ Cercone, G McCalla, eds. The Knowledge Frontier: Essays in the Representation of Knowledge. New York: Springer-Verlag, pp 331–352.

Quillian MR. 1968. Semantic memory. In: M Minsky, ed., Semantic Information Processing. Cambridge, MA: MIT Press, pp 227–270.

Reyle U. 1993. Dealing with ambiguities by underspecification: construction, representation and deduction. J Semant 10:123–179.

Schubert LK, FJ Pelletier. 1982. From English to logic: context-free computation of 'conventional' logical translations. Am J Comput Linguist 10:165–176.

Schubert LK. 1976. Extending the expressive power of semantic networks. Artif Intell 7(2):163–198.

Shieber S. 1986. An Introduction to Unifirication-Based Approaches to Grammar. Stanford, CA: CSLI.

Sidner CL. 1979. Towards a computational theory of definite anaphora comprehension in English discourse. PhD dissertation, MIT, Cambridge, MA.

Sidner CL. 1983. Focusing in the comprehension of definite anaphora. In: M Brady, R Berwick, eds. Computational Models of Discourse. Cambridge, MA: MIT Press.

Thomason RH, ed. 1974. Formal Philosophy: Selected Papers of Richard Montague. New York: Yale University Press.

van Benthem JFAK, A ter Meulen, eds. 1997. Handbook of Logic and Language. Amsterdam: Elsevier.

van Deemter K. 1991. On the Composition of Meaning. PhD dissertation, University of Amsterdam.

van Deemter K, S Peters, eds. 1996. Semantic Ambiguity and Underspecification. Stanford, CA: CSLI.

van Lehn KA. 1978. Determining the scope of English quantifiers. Technical Report AI-TR-483, Artificial Intelligence Laboratory, MIT, Cambridge, MA.

Webber BL. 1979. A Formal Approach to Discourse Anaphora. New York: Garland.

Winograd T. 1972. Understanding Natural Language, New York: Academic Press.

Woods WA, et al. 1972. The lunar sciences natural language information system: final report. Report 2378, BBN, Cambridge, MA.

6

Discourse Structure and Intention Recognition

KAREN E. LOCHBAUM

U S WEST Advanced Technologies, Boulder, Colorado

BARBARA J. GROSZ

Harvard University, Cambridge, Massachusetts

CANDACE L. SIDNER

Lotus Development Corporation, Cambridge, Massachusetts

1. INTRODUCTION

This chapter is concerned with two interrelated issues of discourse processing: the representation of discourse structure and the recognition of intentions. The *structure* of a discourse indicates how the utterances of the discourse group together into segments, as well as how those segments are related to each other. As we will show, *intentions* play a central role in determining this structure, as well as in determining the meaning of an utterance in context.

Representations of discourse structure are important for several language processing tasks. First, the structure of a discourse indicates the relevant context for interpreting and generating subsequent utterances of the discourse. Second, discourse structure plays a central role in accounting for a variety of linguistic phenomena including cue phrases and referring expressions. And third, theories of discourse meaning depend on theories of discourse structure.

Intention recognition involves determining the reasons why a speaker produced a particular utterance at a particular juncture in the discourse. The hearer's recognition of the speaker's intentions furthers the hearer's understanding of the utterance and thus aids the hearer in determining an appropriate response. Work on intention recognition has progressed from reasoning about single utterances, to reasoning about multiple utterances, to reasoning about discourse structure itself.

2. DISCOURSE STRUCTURE

Approaches to discourse structure may be divided into two types, which we will refer to as *informational* approaches and *intentional* approaches.[1] Informational approaches view discourse as structured according to the *content* of the discourse participants' utterances, whereas intentional approaches view discourse as structured according to the discourse participants' *intentions* in producing those utterances [2].

In this section, we survey a variety of approaches to discourse structure, highlighting two informational approaches, those of Hobbs [3,4] and of Mann and Thompson [5], and one intentional approach, that of Grosz and Sidner [6]. Taken together, these three approaches provide the foundation for all subsequent work in discourse structure.

A. Informational Approaches

Informational approaches to discourse structure are based primarily on reasoning about the information conveyed in the discourse participants' utterances; intentions are considered only implicitly if at all.[2] Informational approaches have been designed with a number of goals in mind. For example, Hobbs' work [3,4] is directed at the problems of discourse coherence, reference resolution, compound nominal interpretation, and the resolution of syntactic ambiguity and metonymy. Mann and Thompson [5] were concerned with providing a catalog of relations that an analyst could use to make plausible judgments about the structure of a monologue. Their work has had a significant influence within the natural language generation community. Each of these approaches will be discussed in more detail later.

Other informational approaches to discourse structure can be found in the literature; for reasons of space, we will discuss these only briefly here. A rule-based system to govern extended human–machine interactions was developed by Reichman–Adar [7]. This work involves a frame-based representation in which different types of conversational moves are represented according to their various functions, roles, and participants. Cohen [8] focused on formal arguments and developed a theory of discourse structure based on the possible evidence relations that can hold between the propositions expressed by a speaker's utterances. This structure can then be used by a system to reason about the validity of the argument and respond accordingly. Polanyi's work [9,10] differs from the others in being directed at discourse semantics, rather than pragmatics. Under Polanyi's approach, discourse structure is derived by incrementally building a discourse parse tree over the course of a discourse, according to rules specifying allowable attachments for new clauses. A semantic representation based on *discourse worlds* is created in lockstep with the parse tree.

[1] We follow Moore and Pollack [1] in our use of these terms.
[2] This is discussed in more detail in Secs. 2.A.2 and 2.C.

1. Hobbs' Approach

In Hobbs' model [3], discourse structure comprises text units and *coherence relations* among them. Consider the following sequence:

John can open Bill's safe. {S0}

He knows the combination. {S1}

Here, an Elaboration coherence relation is claimed to hold between the two sentences. Intuitively, this is because the second sentence expands on the first by specifying *how* John can open Bill's safe. More formally, Hobbs defines the Elaboration relation to hold between text units S0 and S1 "if a proposition P follows from the assertions of both S0 and S1 (but S1 contains a property of one of the elements of P that is not in S0)" [3].

In this example, the propositions expressed by the two sentences can be represented as follows:

can(John, open(Safe)) {S0′}

know(he, combination(Comb,y)) {S1′}

Here *John, Safe*, and *Comb* are literals and *he* and *y* are variables [3]. Then, by using domain axioms expressing general facts about bringing about states of affairs and more specific facts about the relation between safes and combinations, it is possible to reason from the first of these propositions to the following proposition:

know(John, cause(do(John, a), open(Safe))) {S0″}

"John knows that his doing some action a *opens the safe"*

Similarly, we can reason from the second proposition to the following:

know(he, cause(dial(z, Comb, y), open(y))) {S1″}

"he knows that z's *dialing the combination of* y *causes it to open."*

The requirements of the Elaboration relation will be satisfied by these two propositions if we recognize that dialing the combination of the safe is a specialization of doing some action that will open the safe. The second sentence thus elaborates the first by providing this additional property of how John can open the safe.

Hobbs [11] proposed a set of approximately twelve coherence relations, including Cause, Evaluation, Background, Parallel, and Elaboration. Although several linguists had previously proposed relations similar to those of Hobbs' (e.g., see the work of Grimes [12] and of Halliday and Hasan [13]), Hobbs' work extended these approaches by proposing an inference mechanism for reasoning about the relations. Hobbs' work also differed from previous approaches in incorporating reference resolution as a by-product of reasoning about coherence.[3] For example, to recognize that the Elaboration relation held in the foregoing example, we had to identify the

[3] Sidner [14] also addressed this problem, although in the context of a sentence completion task to test reading comprehension.

variables *he* and *z* with John and the variable *y* with the Safe. We thus discovered that the referent of "he" in the second sentence is John as a result of reasoning about coherence. Kehler [15] has shown how the recognition of coherence relations affects the interpretation of a variety of linguistic phenomena including verb phrase ellipsis, gapping, tense, and pronominal reference.

Hobbs et al.'s subsequent work [4] has expanded on his initial proposal by providing a more detailed mechanism for reasoning about coherence relations, as well as other discourse phenomena. In this subsequent work, interpreting an utterance involves providing the best explanation of why it would be true. Procedurally, this amounts to "proving" the logical form of the sentence using a technique called *weighted abduction*. The determination of the coherence relations that hold and the resolution of references, compound nominals, syntactic ambiguity, and metonymy are all by-products of this process.

Abduction involves reasoning according to the following schema:

$$\frac{\begin{array}{l}\forall x\ p(x) \rightarrow q(x)\\ q(A)\end{array}}{p(A)}$$

That is, based on the universal implication and the observed evidence $q(A)$, we infer that $p(A)$ is its cause or explanation. For example, suppose we have a rule stating that all instances or events of making spaghetti are instances of making pasta ($\forall e$ $MakeSpaghetti(e) \rightarrow MakePasta(e)$), and we see an instance of someone making pasta ($MakePasta(E1)$), then we might conclude that that person is making spaghetti ($MakeSpaghetti(E1)$).[4]

As this example illustrates, abduction is not valid inference; there may be many such axioms expressing possible explanations for the truth of $q(A)$. For example, the person we are watching might be making linguini or fettucini, rather than spaghetti. Weighted abduction uses a cost model to provide a means of choosing among the various possibilities to produce the "best" explanation for the observed phenomenon.

In the case of language processing, the phenomenon to be explained is the logical form of the utterance. The best explanation for an utterance is a proof of its logical form that uses the information conveyed in the previous discourse to account for the "given" information in the utterance while making minimal assumptions to account for the "new" information.

In summary: under Hobbs' approach, discourse structure comprises text units and coherence relations among them. These text units may range in size from single clauses to the entire discourse. Determining that a particular coherence relation holds requires reasoning about the propositions asserted by the text units, as well as general world knowledge.

2. *Mann and Thompson's Approach*

Mann and Thompson's model of discourse structure is known as Rhetorical Structure Theory or RST [5]. Similar to Hobbs, Mann and Thompson proposed a

[4] This example is from Kautz [16]

set of relations aimed at representing informational connections between utterances. Mann and Thompson's approach differs from Hobbs' in the types of relations used, as well as in the methods for reasoning about those relations. Mann and Thompson define a set of approximately twenty-five *rhetorical relations.* These include Elaboration, Summary, Contrast, Enablement, and Evidence. The discourse structure that results from these relations has been used to study a variety of linguistic issues, including discourse coherence [17].

Each rhetorical relation holds between two (or more) text spans, one of which is designated the Nucleus, the other the Satellite. The Nucleus conveys the central information of the two spans, while the Satellite conveys supporting information. As an example, the definition of the Purpose relation is given in Fig. 1.[5] Mann and Thompson [5] illustate its use with the text sequence below.

To see which Syncom diskette will replace the ones you're using now, {1}

send for your free "Flexi-Finder" selection guide and the name of the supplier nearest you. {2}

In this example, {2} is the Nucleus and {1} is the Satellite. Text span {2} describes an action the purpose of which is to realize the situation expressed in text span {1}.

The Motivation relation in Fig. 2 provides another example of a rhetorical relation. The text sequence below, taken from Moore and Pollack [1], illustrates its use.

Come home by 5:00 {3}

Then we can go to the hardware store before it closes. {4}

In this example, {3} is the Nucleus and {4} is the Satellite. Text span {4} describes a situation that provides motivation for performing the action described in text span {3}.

Mann and Thompson divided their relations into two types, known as subject matter and presentational, based on the effect they are intended to have on the reader. The intended effect of a subject matter relation is simply for the reader to recognize the relation in question. The Purpose relation in Fig. 1 is an example of such a relation. The intended effect of a presentational relation is to have some effect on the mental state of the reader, such as increasing the reader's belief in a proposi-

Relation Name: PURPOSE
Constraints on Nucleus (N): Presents an activity.
Constraints on Satellite (S): Presents a situation that is unrealized.
Constraints on the Nucleus and Satellite Combination: S presents a situation to be realized through the activity in N.
The Effect: The reader recognizes that the activity in N is initiated in order to realize S.
Locus of the Effect: N and S

Fig. 1 A definition of the Purpose relation. (From Ref. 5.)

[5] Mann and Thompson studied written texts, hence the use of the term "reader" in the definition.

Relation Name: MOTIVATION
Constraints on Nucleus (N): Presents an action in which the reader is the actor (including accepting an offer), unrealized with respect to the context of N.
Constraints on Satellite (S): None.
Constraints on the Nucleus and Satellite Combination: Comprehending S increases the reader's desire to perform the action presented in N.
The Effect: The reader's desire to perform the action presented in N is increased.
Locus of the Effect: N

Fig. 2 Definition of the Motivation Relation. (From Ref. 5.)

tion or his or her desire to act. The Motivation relation in Fig. 2 is an example of a presentational relation.

The distinction between subject matter and presentational relations corresponds approximately to the distinction we are drawing here between informational and intentional approaches to discourse structure [1]. Discourse structure, as construed by Mann and Thompson, thus comprises a mix of intentional and informational relations. We include Mann and Thompson's work in our discussion of informational approaches to discourse structure because the focus of their work has been purely on informational connections between utterances; their model does not explicitly reason about intentions. Mann and Thompson's rhetorical relations were designed to be used by an analyst to assign discourse structure based on plausible judgments about the writer and expected reader of a discourse. They were not intended to be used to model the process by which a reader interprets a writer's sentences. In fact, it has been a subject of some debate whether readers (or hearers) actually recognize such relations at all [6,18]. Nonetheless, RST relations have been used extensively for the problem of generating coherent discourse.

Hovy [19] was among the first to try to operationalize Mann and Thompson's rhetorical relations for the purposes of generation, although McKeown [20] had previously adapted a set of rhetorical predicates developed largely by Grimes [12].[6] Hovy formulated a subset of the RST relations as plan operators. For example, his plan operator, corresponding to the Purpose relation in Fig. 1, is shown in Fig. 3. A top-down planner [21] uses operators such as these to create a text plan satisfying a given communicative goal. The communicative goal is first matched against the Intended Effects field of some relation, whereupon backchaining occurs to satisfy the relation's constraints.[7] Further work in generation has gone on to extend and modify the use of RST relations in text planning. A summary of this work can be found in the next chapter of this book and elsewhere [22].

B. Intentional Approaches

In contrast to the informational approaches described in the foregoing section, Levy [24] and Grosz and Sidner [6] have argued that discourse is inherently intentional: people engage in discourse for a reason. They thus argue further that a theory of discourse structure cannot be based solely on the content of the discourse partici-

[6] McKeown's work is described in the next chapter.
[7] This type of reasoning is discussed in more detail in Sec. 3.A.

PURPOSE
Nucleus Constraints:
1. (BMB S H (ACTION ?act-1))
2. (BMB S H (ACTOR ?act-1 ?agt1))
Satellite Constraints:
1. (BMB S H (STATE ?state-1))
2. (BMB S H (GOAL ?agt-1 ?state-1))
3. (BMB S H (RESULT ?act-1 ?act-2))
4. (BMB S H (OBJ ?act-2 ?state-1))
Intended Effects:
1. (BMB S H (BEL ?agt-1 (RESULT ?act-1 ?state1)))
2. (BMB S H (PURPOSE ?act-1 ?state-1))

The expression (BMB S H P) represents that the speaker, S, believes that the speaker and hearer, H, mutually believe that P is true [23].

Fig. 3 A definition of the Purpose plan operator. (From Ref. 19.)

pants' utterances, but rather, must be based on the intentions that led to the production of those utterances.

1. *Grosz and Sidner's Approach*

Grosz and Sidner's theory of discourse structure [6] comprises three interrelated components: a linguistic structure, an intentional structure, and an attentional state. The linguistic structure is a structure that is imposed on the utterances themselves; it consists of discourse segments and embedding relations among them. The linguistic structure is thus similar to the informational discourse structures described in the foregoing section; text units, text spans, and discourse segments, all are simply groupings of utterances. The difference in the approaches lies in the types of relations that can hold between segments. In the linguistic structure, these relations are no more specific than an embedding relation.[8] Figure 4 contains a sample discourse in the domain of network management. The linguistic structure of the discourse is indicated by the bold rule grouping utterances into segments. Segments (2) and (3) are embedded relative to segment (1) in the discourse.

The intentional structure of discourse consists of the purposes of the discourse segments and their interrelations. Discourse segment purposes (henceforth, DSPs) are intentions that lead to the initiation of a discourse segment. In addition, they are intentions that are intended to be recognized, in the sense that they achieve their effect only if they are recognized.[9] There are two types of relations that can hold between DSPs, *dominance* and *satisfaction-precedence*. One DSP dominates another if the second provides part of the satisfaction of the first. That is, the establishment of the state of affairs represented by the second DSP contributes to the establishment of the state of affairs represented by the first. This relation is reflected by a corresponding embedding relation in the linguistic structure. One DSP satisfaction precedes another if the first must be satisfied before the second. This relation is reflected by a corresponding sibling relation in the linguistic structure.

[8] As will be discussed in more detail later, however, this embedding relation in the linguistic structure depends on information in the intentional structure.
[9] DSPs are thus similar to certain utterance-level intentions described by Grice [25].

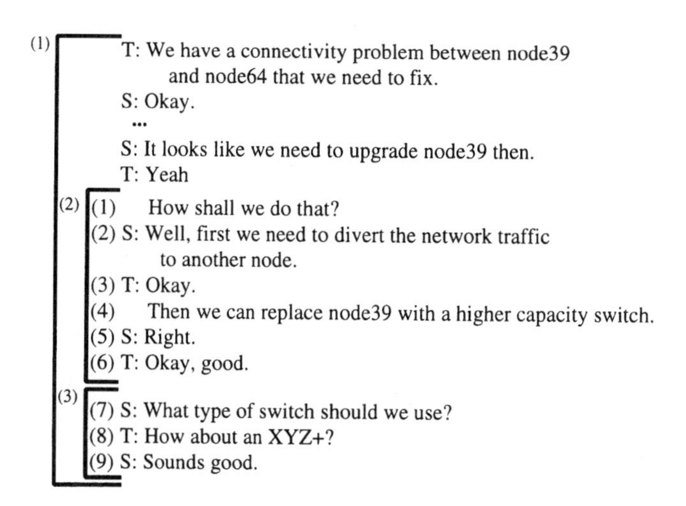

Fig. 4 A sample discourse demonstrating discourse structure.

Lochbaum [26,27] has developed a model of intentional structure based on the collaborative planning framework of SharedPlans [28,29]. SharedPlans specify mental state requirements for collaborative activities, such as discourse [28–30]; that is, they specify the beliefs and intentions that agents must hold to collaborate successfully. There are two types of SharedPlans, full SharedPlans (FSPs) and partial SharedPlans (PSPs). A set of agents have a full SharedPlan when all of the mental attitudes required for successful collaboration have been established. Until that point, the agents' plan will be only partial. Grosz and Kraus [29] introduce a new intention operator, *Int.Th*, as part of their SharedPlan framework. Int.Th represents an agent's *intention that* a proposition hold and is used by Lochbaum to model DSPs.

Figure 5 illustrates the role of SharedPlans in modeling intentional structure. Each segment of a discourse has an associated SharedPlan. The purpose of the segment is taken to be an intention that (Int.Th) the discourse participants form that plan. This intention is held by the agent who initiates the segment. Following Grosz and Sidner [6], we will refer to that agent as the ICP (the *initiating conversational participant*); the other participant is the OCP. Dominance relations between DSPs are modeled using *subsidiary* relations between SharedPlans. One plan is subsidiary to another if the completion of the first plan contributes to the completion of the second. That is, the completion of the first plan establishes one of the beliefs or intentions required for the agents to have the second plan. Satisfaction-precedence relations between DSPs correspond to temporal dependencies between SharedPlans.

In the dialogue of Fig. 4, two agents, the system (S) and the technician (T), are concerned with fixing a network connectivity problem. The purpose of the segment marked (1) in the dialogue can be glossed as follows:

> *"T intends that the agents collaborate to fix the connectivity problem between node39 and node64."*

This is represented in Lochbaum's model as follows:

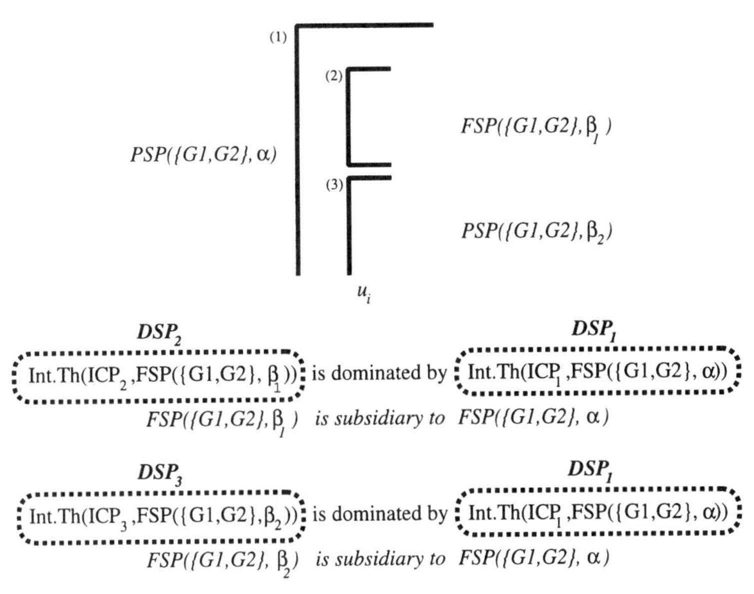

Fig. 5 Modeling intentional structure. (From Ref. 27.)

$$DSP_1 = Int.Th(t,FSP(\{s,t\},fix\_connectivity\ (node39,\ node64,\{s,t\}))).$$

The purpose of the segment marked (2) in Fig. 4 can be glossed as follows:

> *"T intends that the agents collaborate to determine a means of upgrading node39 to a different switch type."*

This is represented as follows:

$$\begin{aligned}DSP_2 = Int.Th(&t,\\ &FSP(\{s,t\},\\ &\quad Achieve(has.recipe(\{s,t\},\\ &\qquad upgrade(node39,SwitchType,\ \{s,t\}),R),\ \{s,t\}))).\end{aligned}$$

The SharedPlan used to model DSP_2 is subsidiary to that used to model DSP_1 because the completion of the first plan brings about a *knowledge precondition* [31–34] of the second. At the start of the segment marked (2) in Fig. 4, agents S and T have agreed to upgrade node39 in the network as part of fixing their connectivity problem. Unless the agents agree on a means of upgrading that node, they will not be able to complete their larger plan. Hence, the plan that is used to model DSP_2 is subsidiary to the plan that is used to model DSP_1. As a result, DSP_2 is dominated by DSP_1.

The third component of Grosz and Sidner's theory is attentional state. Attentional state serves as a record of those entities that are salient at any point in the discourse; it is modeled by a stack of focus spaces. A new focus space is pushed onto the stack with each new discourse segment. The objects, properties, and relations that become salient during the segment are entered into the focus space, as is the segment's DSP. Information from spaces lower down on the stack is accessible to those above, but not vice versa. Focus spaces are popped from the stack when their corresponding segments are completed. The primary function of attentional state is to constrain the range of information that must be considered in processing new utterances.

To further elucidate Grosz and Sidner's theory of discourse structure, we draw an analogy between the structure of a computer program and the structure of a discourse. Computer programs exhibit a block structure in which lines of code are grouped together into functions. Each line of code is part of a particular function because it contributes something to the purpose of that function. The functions can thus be compared to discourse segments. Each function also serves a purpose relative to the larger program of which it is a part. The purposes of the functions and their relations to each other, and to the program as a whole, may be compared with the intentional structure of a discourse. Programs also contains variables and constants that obey certain scoping rules. Variables may be used only in the functions in which they are defined. If variables of the same name exist in two different functions, they refer to different things. Attentional state may be compared with the use of scoping rules to determine the values of variables and constants.

The three components of Grosz and Sidner's theory—linguistic structure, intentional structure, and attentional state—provide a theory of discourse structure that simplifies and extends the treatment of a number of linguistic phenomena, including referring expressions, interruptions, and cue phrases [6]. The resolution of referring expressions depends on the attentional state of the discourse. In particular, the search for a referent is constrained to those entities represented in some space on the focus space stack; those entities in the topmost space are preferred over those below. Grosz [35] describes this process for the resolution of definite noun phrases. The resolution of pronouns requires an additional mechanism, such as *centering* [36,37]; see Chap. 5 for further discussion of the issues here. The treatment of interruptions is reflected in both the intentional structure and attentional state [38]. Interruptions have discourse segment purposes that are not related to those of the discourse in which they are contained. Although an interruption is linguistically embedded in a discourse, the intentional structure of the interruption is separate from that of the remaining discourse, as is its focus space stack. Cue phrases are realized in the linguistic structure, but can signal operations on the intentional structure or the attentional state or both. For example, phrases such as "but anyway" may indicate a pop of the focus space stack and the satisfaction of a DSP.

C. Discussion

The informational approaches described in Sec. 2.A rely primarily on discourse and domain information to determine the possible relations that can hold between segments of a discourse. Proponents of the intentional approach argue that, although such information is important for discourse processing, it cannot serve as the basis for determining discourse structure [6]. Discourse structure depends primarily on *why* the discourse participants are engaged in the discourse and its segments, and not on *what* they are engaged in.

There is some consensus now, however, that both types of information are necessary. Grosz and Sidner [6] have shown that reasoning about relations in the domain of the discourse can aid in reasoning about relations in the intentional structure and vice versa. For example, they describe a *supports* relation between propositions and a *generates* relation between actions that can be used to infer dominance relations between DSPs. Moore and Pollack [1] have shown that RST does not provide a sufficient model of discourse structure precisely because it does not allow both types of information to coexist. RST requires a single rhetorical relation, either informational (i.e., subject matter) or intentional (i.e., presentational), to hold between text spans [1]. And finally, Moore and Paris [39] have shown that generation systems that lack intentional information cannot respond effectively if a hearer does not understand or accept its utterances.

3. INTENTION RECOGNITION

As the foregoing discussion illustrates, intentions play an important role in representing discourse structure. Early work on intention recognition, however, was not based on recognizing or reasoning with discourse structure, but rather, on reasoning about single utterances. Subsequent work addressed the problems posed by discourses that comprise multiple utterances and segments. More recent work has begun to address the problem of recognizing the intentional structure of discourse itself. All of this work has been based on the idea that understanding an utterance involves determining why a speaker produced that utterance at the particular juncture of the discourse. It has also all been based on, and has subsequently contributed to, various artificial intelligence (AI) planning formalisms, and thus, is also known as plan recognition.

A. Foundations

The pioneering work of Cohen, Allen, and Perrault [40–42] on intention recognition brought together philosophical work on speech act theory [43,44] with AI work on planning systems [45]. In speech act theory, utterances are taken to be purposeful actions. People do things with their utterances; they request, warn, assert, inform, and promise, among other things. Utterances are not simply statements of truth or falsity, but rather, are "executed" to have some effect on the hearer. It thus follows that utterances may be reasoned about in the same way as other observable actions.

Cohen and Perrault [40] used AI formalisms that were originally designed to generate plans involving physical actions to generate plans involving communicative actions. In particular, they developed operators for the speech acts Request and Inform using STRIPS [45] operators. The basic STRIPS operators have evolved to model actions using (a) a header, (b) a precondition list, (c) a body,[10] and (d) an effects list. The header specifies the name of the action and any objects or parameters on which it operates; for example, *move*(*Box, FromLoc, ToLoc*) represents a moving action of a box from one location to another. The precondition list specifies the conditions that must be true in the world for the action to be performed; for example, *at*(*Box, FromLoc*) represents that the box must be at the first location. The body, if present, specifies the subactions that must be performed to accomplish the action in the header; for example, a body consisting of the action *push*(*Box, FromLoc, ToLoc*) represents that pushing the box from the first location to the second is one way of moving it. The effects list specifies the effect of the action

[10] The body component was introduced in NOAH [21].

on the world; for example, *at*(*Box,ToLoc*) represents that as a result of the action the box will be at the second location.

The preconditions and effects of speech acts are expressed not in terms of physical conditions, but rather, in terms of the beliefs and goals of the speaker and hearer. For example, Cohen and Perrault's formalization of the Inform speech act specifies that for a speaker to inform a hearer of some proposition, the speaker must believe the proposition and must want to inform the hearer of it. These are the operator's preconditions. The effect of the operator is that the hearer believes that the speaker believes the proposition. Cohen and Perrault used the speech act operators they developed in combination with other operators to show how the performance of utterances can lead to the performance of physical actions.

Whereas Cohen and Perrault addressed the problem of *generating* plans involving speech acts, Allen and Perrault [41] looked at the complementary problem of *recognizing* such plans. Here, the problem is to determine the beliefs and goals that led a speaker to produce a particular utterance. This is the problem of intention recognition. The recognition of the speaker's intentions allows the hearer to generate a more appropriate response to the speaker's utterance, one that addresses the information that is conveyed by the utterance both explicitly and implicitly. For example, in responding to the question "Do you have the time?", the hearer recognizes that the speaker's intention is to know the time and hence answers, "Yes, 3 PM," rather than "Yes, I do."

Allen [46] developed a model of plan recognition in which a speaker's larger goals or plans were recognized from his or her speech acts. Allen's model uses a series of rules to derive possible plans by reasoning both backward from the speech act representation of a speaker's utterance and forward from the hearer's expectations about the speaker's possible goals. Allen's rules were derived from the operators used to model actions. For example, the precondition–action rule states that if agent G wants to achieve proposition P and P is a precondition of act A, then G may want to perform A. This rule is for inferring plans. The corresponding plan construction rule states that if agent G wants to execute act A, then G may want to ensure that its precondition P is satisfied.

The plan recognition process begins from a set of partial plans. Each plan includes a representation of the speech act underlying the speaker's utterance paired with a possible goal of the speaker's based on the hearer's expectations. The plan inference rules operate on these partial plans to reason backward from the observed actions to the speaker's goal, whereas the plan construction rules are used to reason forward from the hearer's expectations. This process is guided by a set of heuristics that control the application of rules to favor those partial plans that are most likely to lead to the correct plan. The application of plan inference and plan construction rules continues according to the heuristics until the partial plan derived from reasoning backward can be merged with that derived from reasoning forward, and no other plan is highly enough rated as an alternative.

Figure 6 contains the plan that is derived from the utterance "When does the train to Windsor leave?" The speech act representation of the utterance is as follows:

REQUEST(A,S,INFORMREF(S,A,time1))

Here, A is the speaker, S is the system acting as the hearer, and time1 represents the departure time of the train to Windsor. A is requesting S to inform him of (the

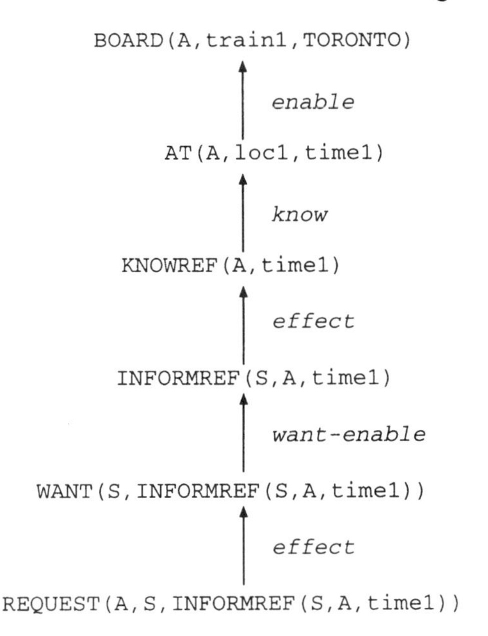

Fig. 6 Plan underlying the utterance "When does the train to Windsor leave?" (From Ref. 46.)

referent of) the time of the train to Windsor (from the Toronto train station). Each node in the plan graph in Fig. 6 represents an action that needs to be performed or a proposition that needs to be achieved. The labels next to the arrows in the graph indicate the relation between the actions and propositions, as derived from the plan operators. For example, working down from the top of the graph, the enable label indicates that A's being at the location of the train to Windsor at its departure time is a precondition of A's boarding the train at that time in Toronto. This link was derived using one of the action–precondition rules described previously. The know label indicates that A's knowing the referent of the time of the Windsor train is necessary for A's being at the location of the Windsor train at that time. The effect label indicates that the effect of S's informing A of the time of the Windsor train is that A knows the time. The want–enable link indicates that S's wanting to inform A of the time of the train is a precondition of actually informing him. And the second effect link indicates that the effect of A's requesting S to inform him of the time of the train is that S wants to inform him.

Once derived, this plan can then be used by the system to plan an appropriate response. The response process is modeled by first analyzing the recognized plan for any *obstacles*; obstacles are goals in the plan that the speaker cannot achieve. The system can then form a plan to address these obstacles, whether they were explicitly asked about or not. This plan forms the basis of the system's response. In the example shown in Fig. 6, one obstacle is readily apparent. Agent A needs to know the time of the train to Windsor. In addition, however, A will not be able to carry out this plan unless he also knows the location from which the train leaves. That is, he must also know the referent of *loc*1. On the basis of this reasoning, the system might thus respond to the speaker by saying "1600, gate 7."

Allen's work provides a model of plan recognition in the context of a single utterance exchange. His model can reason about indirect interpretations of utterances, such as "Can you tell me the time of the train to Windsor?" as well as utterance fragments such as, "The train to Windsor?" Allen and his colleagues' subsequent work on more general dialogue systems and speech act representations is described in Chap. 14.

More recent work in the area of single utterance reasoning includes that of Cohen and Levesque [47] and Perrault [48]. Their work provides a detailed mental state model of speech act processing. Cohen and Levesque develop a theory of speech acts within a theory of rational interaction. Perrault argues for the use of default logic to model the consequences of speech acts.

B. Extensions—Beyond Single Utterances

1. Accommodating Multiple Utterances

The next generation of work on intention recognition addressed the problem of recognizing a speaker's intentions across multiple utterances. In this case, the recognition algorithm must operate incrementally. The hearer infers as much information as possible from the speaker's first utterance, but may be able to identify only a set of candidate plans the speaker is pursuing, none of which is preferred over the other. The information contained in the speaker's subsequent utterances provides additional information that can be used to filter the initial candidate set of plans and also expand and instantiate them more completely.

Sidner's work on intention recognition [49–51], in addition to addressing the problem of incremental recognition, also takes a different philosophical view of the plan recognition problem from Allen's. Whereas Allen's model is based on the speech act work of Searle [44], Sidner's model is based on the work of the philosopher Grice [25,52]. The Gricean viewpoint adopted by Sidner is that in producing an utterance, a speaker intends to elicit a particular response in a hearer. Additionally, the speaker intends that the hearer recognize certain of the speaker's underlying plans in producing the utterance, and also intends that the hearer's recognition of those plans contributes in part to the hearer's recognition of the intended response. Whereas Allen's response mechanism is based on the general detection of obstacles, Sidner is able to distinguish responses that are intended from those that are helpful, but unintended.

In Sidner's model, a hearer's interpretation of an utterance is based on answering the question "Why did the speaker say that to me?" That is, the hearer needs to produce an explanation for the utterance in terms of her beliefs about the speaker's beliefs and plans, as well as her beliefs about the speaker's beliefs regarding her capacities to act. Sidner's plan recognition process begins not from the speech act representation of an utterance, but rather, from the surface level intentions of the utterance, as derived using Grice's theory of speaker meaning [25,52].

Sidner's adoption of the Gricean perspective also justifies the use of incremental recognition. If a speaker intends to elicit a particular response in a hearer, in part based on the hearer's recognition of his plans, then the speaker cannot expect the hearer to *guess* the correct plan when choice points are encountered in the plan recognition process. Rather, the speaker must supply enough information in his subsequent utterances to allow the hearer to recognize the intended response.

Carberry [53] also extended Allen's work to multiple utterances. Her work makes use of discourse principles to constrain the recognition process and also accounts for certain types of ill-formed input. Allen's heuristics for guiding recognition involve general action-based principles. For example, the rating of a partial plan is decreased in Allen's model if it contains an action the effects of which are already true at the time the action is to be performed. Carberry noted that in longer discourses, principles derived from attentional state may also be employed to rate plans. For example, if an agent is focused on the performance of a particular action, he will usually complete the discussion of that action before moving on to the discussion of another action. Otherwise, the agent will have to assume the overhead of reintroducing the first action later in the discourse. Carberry's recognition process is thus biased to account for new utterances based on the action currently in focus, rather than actions elsewhere in the plan space. Carberry developed a set of five focus-based heuristics to guide the recognition process. She also developed a set of heuristics to accommodate utterances in which an erroneous proposition is expressed about an attribute of an object or a relation.

2. *Accommodating Subdialogues*

Discourses containing multiple utterances often contain multiple segments as well. In the case of dialogues, these segments are known as subdialogues. Certain types of subdialogues, such as clarifications, corrections, and negotiations, pose an interesting challenge for plan recognition systems. The problem is that these subdialogues diverge from the expected flow of the discourse. In this section, and in Sec. 3.C, we examine various means of accommodating such subdialogues.

Litman and Allen [54] proposed the use of two types of plans to model clarification and correction subdialogues, *discourse plans* and *domain plans*. Domain plans represent knowledge about a task, whereas discourse plans represent conversational relations between utterances and plans. For example, an agent may use an utterance to *introduce*, *continue*, or *clarify* a plan.

In Litman and Allen's model, the process of understanding an utterance entails recognizing a discourse plan from the utterance and then relating that discourse plan to some domain plan. For example, in the dialogue of Fig. 7, utterance (3) is recognized as an instance of the CORRECT-PLAN discourse plan; with the utterance, the User is correcting a domain plan to add data to a KL-ONE [55] network.

Litman and Allen use a stack of plans to model attentional aspects of discourse. The plan stack after processing utterance (3) is shown in Fig. 8. The

(1) User: Show me the generic concept called "employee".
(2) System: OK. <system displays network>
(3) User: I can't fit a new ic below it.
(4) Can you move it up?
(5) System: Yes. <system displays network>
(6) User: OK, now make an individual employee concept whose first name is ...

Fig. 7 An example of a correction subdialogue. (From Ref. 50.)

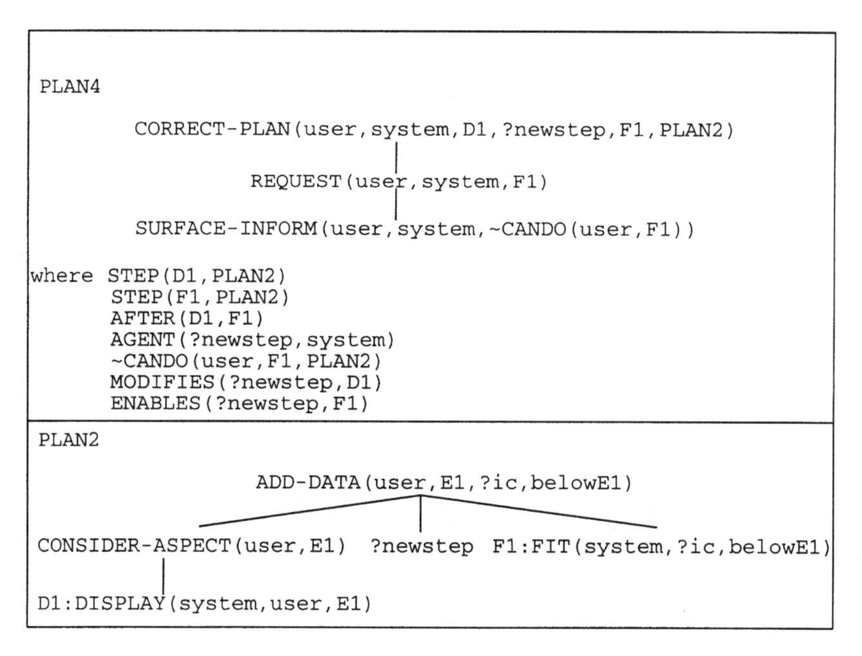

Fig. 8 Plan stack after processing utterance (3) of the dialogue in Fig. 7. (From Ref. 56.)

CORRECT-PLAN discourse plan on top of the stack indicates that the user and system are correcting a problem with the step labeled D1 in PLAN2 (the DISPLAY act of the ADD-DATA domain plan) by inserting a new step into PLAN2 (?newstep) before the step labeled F1 (the FIT step).

The plan stack after processing the User's subsequent utterance in (4) is shown in Fig. 9. The IDENTIFY-PARAMETER discourse plan indicates that utterance (4) is being used to identify the ?newstep parameter of the CORRECT-PLAN discourse plan.

Lambert and Carberry [57] have also introduced different types of plans to account for different types of subdialogues. They revised Litman and Allen's dichotomy of plans into a trichotomy of *discourse*, *problem-solving*, and *domain plans*. Their discourse plans represent means of achieving communicative goals, while their problem-solving plans represent means of constructing domain plans. Ramshaw [58] has augmented Litman and Allen's two types of plans with a different third type, *exploration plans*. This type of plan is added to distinguish those domain plans an agent has adopted from those it is simply considering adopting.

C. Extensions—Accommodating Differences in Belief and Intention

The approaches discussed in the previous sections adopt what has been called a "data-structure" view of plans [59–61]; they treat plans as collections of actions and model them using STRIPS-based operators. In contrast, the two approaches presented in this section adopt a mental-state perspective on plans. Pollack's work [60,62] was the first AI treatment of plans as mental phenomena. Lochbaum's work [26,27] builds on Pollack's model of plans, as well as on extensions to it for modeling

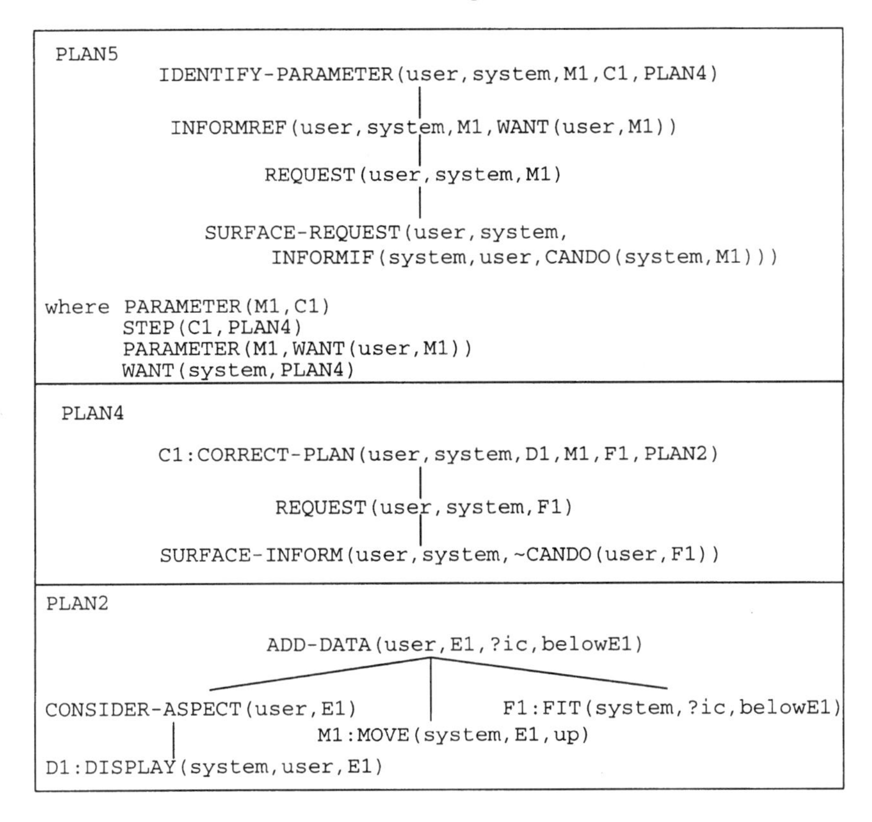

Fig. 9 Plan stack after processing utterance (4) of the dialogue in Fig. 7. (From Ref. 56.)

collaboration [28,29], to address the problem of recognizing and reasoning about the intentional structure of discourse.

Pollack [60,62] noted that previous approaches to plan recognition in discourse did not distinguish the beliefs of the speaker from those of the hearer. The implicit assumption in the previous work is that the operators are mutually known by both participants. These approaches thus fail to account systematically for discourses in which one participant holds beliefs that differ from those of the other. The discourse below, from Pollack [62], is such an example.

A: I want to talk to Kathy, so I need to find out the phone number of St. Eligius.
S: St. Eligius closed last month. Kathy was at Boston General, but she's already been discharged. You can call her at home. Her number is 555-1238.

To accommodate these types of discourses, Pollack draws a distinction between *recipes* and *plans.* Recipes are structured representations of information about actions; they represent what agents know when they know a way of doing something. Plans, on the other hand, are a collection of beliefs and intentions that are necessary to perform an action successfully.

Under Pollack's approach, the plan recognition process involves ascribing a set of beliefs and intentions to an agent on the basis of his utterances. A hearer recognizes an *invalid plan* when she ascribes a set of beliefs to the speaker that differ from those that she herself holds. Pollack showed that the appropriate response to the speaker in such cases depends on the particular type of discrepancy in belief. For example, in the foregoing discourse, S can infer that A's plan includes a belief that calling St. Eligius will contribute to his goal of talking to Kathy. S, however, believes that A cannot call St. Eligius and that even if he could it would not help him talk to Kathy. S's response thus conveys this information, as well as an alternative means for achieving A's goal, namely calling Kathy at home.

Pollack's approach involves a mental phenomenon view of single agent plans. The SharedPlan framework developed by Grosz et al. [28,29] extended Pollack's model to include the collaborative plans of multiple agents. SharedPlans were then used by Lochbaum [26,27] to create a model of intentional structure, as described in Sec. 2.B.1. Lochbaum also developed a model of intention recognition based on this structure.

In Lochbaum's model, the utterances of a discourse are understood in terms of their contribution to the SharedPlans associated with the segments of the discourse. Those segments that have been completed at the time of processing an utterance have a full SharedPlan (FSP) associated with them (e.g., see segment (2) in Fig. 5), whereas those that have not have a partial SharedPlan (PSP) (e.g., see segments (1) and (3) in Fig. 5).

For each utterance of a discourse, an agent must determine whether the utterance begins a new segment of the discourse, contributes to the current segment, or completes it [6]. For an utterance to begin a new segment of the discourse in Lochbaum's model, it must indicate the initiation of a subsidiary SharedPlan. For an utterance to contribute to the current segment, it must advance the partial SharedPlan associated with the segment toward completion. That is, it must establish one of the beliefs or intentions required for the discourse participants to have a full SharedPlan, but missing from their current partial SharedPlan. For an utterance to complete the current segment, it must indicate that the purpose of that segment has been satisfied. For that to be true, the SharedPlan associated with the segment must be complete. That is, all of the beliefs and intentions required of a full SharedPlan must have been established over the course of the segment.

The structure built by Lochbaum's model in processing the first subdialogue in Fig. 4 is shown in Fig. 10. Each box in the figure corresponds to a discourse segment

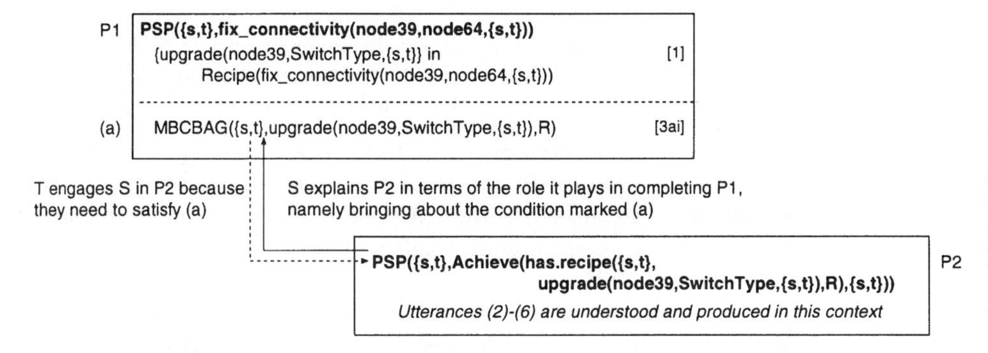

Fig. 10 Analysis of the dialogue in Fig. 4.

and contains the SharedPlan used to model the segment's purpose. Plan (P1) is used to model the purpose of the segment marked (1) in Fig. 4 and plan (P2) is used to model the purpose of the segment marked (2) (cf., DSP_1 and DSP_2 in Sec. 2.B.1). The information represented within each SharedPlan in Fig. 10 is separated into two parts. Those beliefs and intentions that have been established at the time of the analysis are shown above the dotted line, whereas those that remain to be established, but that are used in determining subsidiary relations, are shown below the line.

The solid arrow in the figure indicates a subsidiary relation between the SharedPlans, as explained by the text that adjoins it. As we noted in Sec. 2.B.1, plan (P2) is subsidiary to plan (P1) because the completion of (P2) brings about a knowledge precondition (as represented by the MBCBAG operator [29]) that is necessary for the completion of (P1). As a result, DSP_2 is dominated by DSP_1.

The process by which an agent recognizes subsidiary relations between plans can be characterized as one of explanation. Whereas at the utterance level, a hearer must explain why a speaker said what he did [49], at the discourse level, an OCP must explain why an ICP engages in a new discourse segment at a particular juncture in the discourse. The latter explanation depends on the relation of the new segment's DSP to the other DSPs underlying the discourse. In Lochbaum's model, this explanation process is modeled by reasoning about subsidiary relations between SharedPlans.

D. Discussion

Recognizing intentional structure requires recognizing intentions. As a result, Grosz and Sidner's theory of discourse structure provides a framework within which to integrate most of the work on intention recognition described in this chapter. However, to do so, it is essential to distinguish approaches that are based at the utterance level from those that are based at the discourse level.

This disparity between levels may be exemplified by comparing the structures that are built for the dialogue in Fig. 7 by Litman and Allen's model with those that are built by Lochbaum's model. The structure built by Litman and Allen's model was discussed in Sec. 3.B.2 and is shown in Fig. 9.

The structure built by Lochbaum's model is shown in Fig. 11. The plan labeled (P4) in Fig. 11 is used to model the *DSP* of the larger segment in Fig. 7. The *DSP* of this segment may be glossed as follows:

> *"U intends that the agents collaborate to add data to the network at some screen location."*

The plan labeled (P5) in the figure is used to model the *DSP* of the correction subdialogue, which may be glossed as follows:

> *"U intends that the agents collaborate to free up some space below the employee concept."*

Plan (P5) is subsidiary to plan (P4) because (P5) brings about a physical precondition (represented by the BCBA operator [29]) that is necessary for the completion of (P4): freeing up space on the screen satisfies a constraint of the recipe for adding data to the KL-ONE network, namely `freespace_for(Data, below(ge1))`.

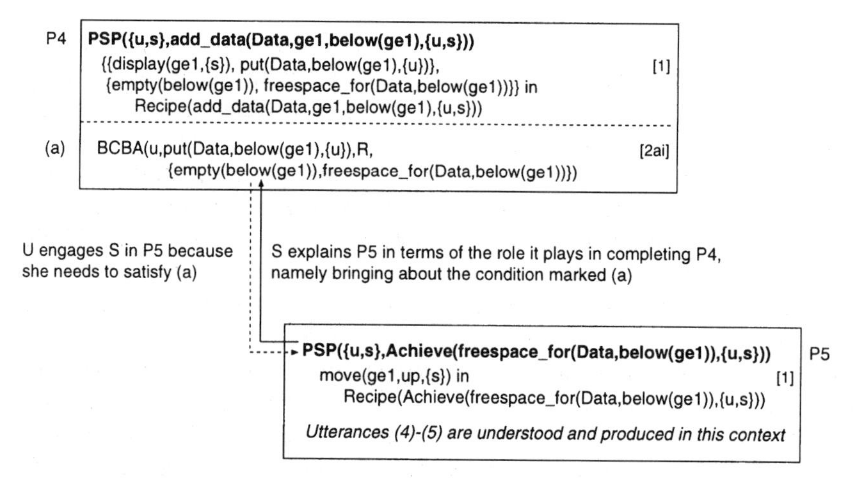

Fig. 11 Analysis of the dialogue in Fig. 7.

As indicated in Fig. 11, the utterances of the subdialogue in Fig. 7 are processed in the context of the plan in (P5). The move act described in utterance (4) of the discourse is thus understood as part of the recipe for freeing up space on the screen, rather than as part of the recipe for adding data to the network.

Intuitively, the entire dialogue (fragment) in Fig. 7 is concerned with adding data to a network, whereas the subdialogue is concerned with correcting a problem. Utterance (3) of the subdialogue identifies the problem, while utterance (4) suggests a method of correcting it. Lochbaum's representation in Fig. 11 accounts for this intuitive segmentation. In contrast, Litman and Allen's representation in Fig. 9 does not; each utterance introduces its own discourse, and possibly domain, plan. Utterance (3) of the subdialogue is thus understood as correcting a plan to add data to a network, whereas utterance (4) is understood as identifying a parameter of a discourse plan.

Litman and Allen's work addresses the problem of recognizing the propositional content of an utterance from its surface form, whereas Lochbaum's model begins from propositional content. However, Litman and Allen's approach provides only an utterance-to-utterance based analysis of discourse. As illustrated by the foregoing discussion, this type of analysis cannot capture the contribution of a subdialogue to the overall discourse in which it is embedded [26,27]. For a complete system, the speech act reasoning performed by approaches such as Litman and Allen's must be combined with the intentional structure reasoning performed by approaches such as Lochbaum's, to produce a model that recognizes both utterance-level and discourse-level intentions.

4. SUMMARY

Informational approaches to discourse structure focus on the ways in which the content of individual utterances relate to the utterances that precede (or follow) them. Relations among propositional content are taken not only to provide the coherence in the discourse, but also to form a basis for interpreting a range of context-dependent constructions (e.g., pronouns, compound nominals). Much of this work has focused on the texts produced by a single writer or speaker, and a major area of application has been the generation of multisentential texts, as discussed in other chapters of this handbook.

Intentional approaches to discourse structure postulate the centrality to discourse structure, and to discourse meaning, of determining the reasons (i.e., intentions) behind discourse segments and the utterances they comprise. Identifying utterance-level intentions is crucial to figuring out what an individual statement means, whereas identifying discourse segment purposes is essential to determining the structure of the discourse.

In short, both informational and intentional information is necessary to model discourse structure, and both utterance-level and discourse-level intentions must be recognized. Figure 12 provides a summary of the work discussed in this chapter. The starred entries indicate those papers containing algorithms, as well as theories.

Informational Approaches to Discourse Structure

- Cohen [8]
- Hobbs [3], [4]*
- Hovy [19]*
- Mann and Thompson [5, 17]
- Polanyi [9]*, [10]
- Reichman [7]*

Intentional Approaches to Discourse Structure

- Grosz and Sidner [6]
- Levy [24]
- Lochbaum [26, 27]*

Utterance-level Intention Recognition

- Allen [41, 46]*
- Carberry [53]*
- Cohen and Levesque [47]
- Cohen and Perrault [40]*
- Litman [54, 56]*
- Lambert and Carberry [57]*
- Perrault [48]
- Pollack [60, 62]
- Ramshaw [58]*
- Sidner [49, 50], [51]*

Discourse-level Intention Recognition

- Lochbaum [26, 27]*

Fig. 12 A summary of work on discourse structure and intention recognition.

REFERENCES

1. JD Moore, ME Pollack. A problem for RST: the need for multi-level discourse analysis. Comput Linguist, 18:537–544, 1992.
2. JR Hobbs. On the relation between the informational and intentional perspectives on discourse. In: EH Hovy, DR Scott, eds. Interdisciplinary Perspectives on Discourse. Heidelberg: Springer–Verlag, 1996.
3. JR Hobbs. Coherence and coreference. Cogn Sci 3:67–89, 1979.
4. JR Hobbs, ME Stickel, DE Appelt, P Martin. Interpretation as abduction. Artif Intell 63(1–2):69–142, 1993.
5. WC Mann, SA Thompson. Rhetorical structure theory: a theory of text organization. Technical Report RR–87–90, Information Sciences Institute (ISI), 1987.
6. BJ Grosz, CL Sidner. Attention, intentions, and the structure of discourse. Comput Linguist 12:175–204, 1986.
7. R Reichman–Adar. Extended person–machine interface. Artif Intell 22:157–218, 1984.
8. R Cohen. Analyzing the structure of argumentative discourse. Comput Linguist 13:11–24, 1987.
9. L Polanyi. A formal model of the structure of discourse. J Pragmat 12:601–638, 1988.
10. L Polanyi. The linguistic structure of discourse. Technical Report CSLI-96-200, Center for the Study of Language and Information, 1995.
11. JR Hobbs. On the coherence and structure of discourse. Technical Report CSLI-85–37, Center for the Study of Language and Information, 1985.
12. JE Grimes. The Thread of Discourse. The Hague: Mouton, 1975.
13. M Halliday, R Hasan. Cohesion in English. London: Longman, 1976.
14. CL Sidner [Bullwinkle]. Picnics, kittens and wigs: using scenarios for the sentence completion task. Proceedings of the 4th International Joint Conference on Artificial Intelligence (IJCAI-75), August 1975.
15. A Kehler. Interpreting cohesive forms in the context of discourse inference. PhD dissertation, Harvard University, Cambridge, MA, 1995. Available as Technical Report TR-11-95, Center for Research in Computing Technology, Division of Applied Sciences.
16. HA Kautz. A circumscriptive theory of plan recognition. In: PR Cohen, JL Morgan, ME Pollack, eds. Intentions in Communication. Cambridge, MA: MIT Press, 1990, pp 105–134.
17. WC Mann, SA Thompson. Relational propositions in discourse. Discourse Processes 9(1):57–90, 1986. Also available as USC Information Sciences Institute Research Report RR-83-115.
18. O Rambow, ed. Proceedings of the ACL-93 Workshop on Intentionality and Structure in Discourse Relations. Association of Computational Linguistics, Columbus, OH, June 1993.
19. EH Hovy. Planning coherent multisentential text. Proceedings of the 26th Annual Meeting of the ACL, Buffalo, NY, 1988, pp 163–169.
20. KR McKeown. Discourse strategies for generating natural language text. Artif Intell 27:1–42, 1985.
21. ED Sacerdoti. A Structure for Plans and Behavior. Amsterdam: North-Holland, 1977.
22. EH Hovy. Automated discourse generation using discourse structure relations. Artif Intell 63:341–385, 1993.
23. PR Cohen, HJ Levesque. Speech acts and rationality. Proceedings of the 23rd Annual Meeting of the ACL, Chicago, IL, 1985, pp 49–59.
24. DM Levy. Communicative goals and strategies: between discourse and syntax. In: T Givon, ed. Discourse and Syntax. New York: Academic Press, 1979; pp 183–210.
25. HP Grice. Utterer's meaning and intentions. Philos Rev 68:147–177, 1969.

26. KE Lochbaum. A collaborative planning model of intentional structure. Comput Linguist, 24:525–572, 1998.
27. KE Lochbaum. Using Collaborative Plans to Model the Intentional Structure of Discourse. PhD dissertation, Harvard University, Cambridge, MA, 1994. Available as Technical Report TR-25-94, Center for Research in Computing Technology, Division of Applied Sciences.
28. BJ Grosz, CL Sidner. Plans for discourse. In: PR Cohen, JL Morgan, ME Pollack, eds, Intentions in Communication. Cambridge, MA: MIT Press, 1990, pp 417–444.
29. BJ Grosz, S Kraus. Collaborative plans for complex group action. Artif Intell 86:269–357, 1996.
30. ME Bratman. Shared cooperative activity. Philos Rev 101:327–341, 1992.
31. J McCarthy, PJ Hayes. Some philosophical problems from the standpoint of artificial intelligence. In: B Meltzer, D Michie, eds. Machine Intelligence 4. Edinburgh: Edinburgh University Press, 1969, pp 463–502.
32. RC Moore. A formal theory of knowledge and action. In: JR Hobbs, RC Moore, eds. Formal Theories of the Commonsense World. Norwood, NJ: Ablex Publishing, 1985, pp. 319–358.
33. L Morgenstern. Knowledge preconditions for actions and plans. Proceedings of the 10th International Joint Conference on Artificial Intelligence (IJCAI-87), Milan, Italy, 1987, pp 867–874.
34. KE Lochbaum. The use of knowledge preconditions in language processing. Proceedings of the 14th International Joint Conference on Artificial Intelligence (IJCAI-95). Vol 2, Montreal, Canada, 1995, pp 1260–1266.
35. BJ Grosz. The representation and use of focus in a system for understanding dialogs. Proceedings of the Fifth International Joint Conference on Artificial Intelligence (IJCAI-77), Cambridge, MA, 1977, pp 67–76.
36. BJ Grosz, AK Joshi, S Weinstein. Centering: a framework for modelling the local coherence of discourse. Comput Linguist 21:203–225, 1995.
37. CL Sidner. Focusing in the comprehension of definite anaphora. In: M Brady, R Berwick, eds. Computational Models of Discourse. Cambridge, MA: MIT Press, 1983, pp 267–330.
38. BJ Grosz, CL Sidner. Discourse structure and the proper treatment of interruptions. Proceedings of the Ninth International Joint Conference on Artificial Intelligence (IJCAI-85), Los Angeles, CA, 1985, pp 832–839.
39. JD Moore, CL Paris. Planning text for advisory dialogues: capturing intentional and rhetorical information. Comput Linguist 19:651–694, 1993.
40. PR Cohen, CR Perrault. Elements of a plan-based theory of speech acts. Cogn Sci 3: 177–212, 1979.
41. JF Allen, CR Perrault. Analyzing intention in utterances. Artif Intell 15:143–178, 1980.
42. PR Cohen, CR Perrault, JF Allen. Beyond question-answering. In: W Lehnert, M Ringle, eds. Strategies for Natural Language Processing. Hillsdale, NJ: Lawrence Erlbaum Associates, 1982, pp 245–274.
43. JL Austin. How To Do Things with Words. London: Oxford University Press, 1962.
44. JR Searle. Speech Acts: An Essay in the Philosophy of Language. Cambridge, UK: Cambridge University Press, 1969.
45. RE Fikes, NJ Nilsson. STRIPS: a new approach to the application of theorem proving to problem solving. Artifi Intell 2:189–208, 1971.
46. JF Allen. Recognizing intentions from natural language utterances. In: M Brady, RC Berwick, eds. Computational Models of Discourse. Cambridge, MA: MIT Press, 1983, pp 107–166.

47. PR Cohen, HJ Levesque. Rational interaction as the basis for communication. In PR Cohen, JL Morgan, ME Pollack, eds. Intentions in Communication. Cambridge, MA: MIT Press, 1990, pp 221–255.
48. CR Perrault. An application of default logic to speech act theory. In: PR Cohen, JL Morgan, ME Pollack, eds. Intentions in Communication. Cambridge, MA: MIT Press, 1990, pp 161–186.
49. CL Sidner, DJ Israel. Recognizing intended meaning and speakers' plans. Proceedings of the 7th International Joint Conference on Artificial Intelligence (IJCAI-81), Vancouver, BC, Canada, 1981, pp 203–208.
50. CL Sidner. What the speaker means: the recognition of speakers' plans in discourse. Comput Math Appl 9:71–82, 1983.
51. CL Sidner. Plan parsing for intended response recognition in discourse. Comput Intell 1:1–10, 1985.
52. HP Grice. Meaning. Philos Rev 66:377–388, 1957.
53. S Carberry. Pragmatic modeling: toward a robust natural language interface. Comput Intell 3:117–136, 1987.
54. DJ Litman, JF Allen. A plan recognition model for subdialogues in conversations. Cogn Sci 11:163–200, 1987.
55. RJ Brachman, JG Schmolze. An overview of the KL-ONE knowledge representation system. Cogn Sci 9:171–216, 1985.
56. DJ Litman. Plan recognition and discourse analysis: an integrated approach for understanding dialogues. PhD dissertation, University of Rochester, 1985.
57. L Lambert, S Carberry. A tripartite plan-based model of dialogue. Proceedings of the 29th Annual Meeting of the ACL, Berkeley, CA, 1991, pp 47–54.
58. LA Ramshaw. A three-level model for plan exploration. Proceedings of the 29th Annual Meeting of the ACL, Berkeley, CA, 1991, pp 39–46.
59. ME Bratman. Taking plans seriously. Soc Theory Pract 9:271–287, 1983.
60. ME Pollack. A model of plan inference that distinguishes between the beliefs of actors and observers. Proceedings of the 24th Annual Meeting of the ACL, New York, 1986, pp 207–214.
61. ME Bratman. Intention, Plans, and Practical Reason. Cambridge, MA: Harvard University Press, 1987.
62. ME Pollack. Plans as complex mental attitudes. In: PR Cohen, JL Morgan, ME Pollack, eds. Intentions in Communication. Cambridge, MA: MIT Press, 1990, pp 77–103.

7

Natural Language Generation

DAVID D. McDONALD

Brandeis University, Arlington, Massachusetts

1. INTRODUCTION

Natural language generation (NLG) is the process by which thought is rendered into language. It has been studied by philosophers, neurologists, psycholinguists, child psychologists, and linguists. Here, we will examine what generation is to those who look at it from a computational perspective: people in the fields of artificial intelligence and computational linguistics.

From this viewpoint, the *generator*—the equivalent of a person with something to say—is a computer program. Its work begins with the initial intention to communicate, and then on to determining the content of what will be said, selecting the wording, and rhetorical organization, then fitting it to a grammar, through to formatting the words of a written text or establishing the prosody of speech. Today what a generator produces can range from a single word or phrase given in answer to a question, or as label on a diagram, through multi-sentence remarks and questions within a dialog, and even to full-page explanations and beyond, depending on the capacity and goals of the program it is working for—the machine "speaker" with something to say—and the demands and particulars of the context.

Modulo a number of caveats discussed later, the process of generation is usually divided into three parts, often implemented as three separate programs: (a) identifying the goals of the utterance, (b) planning how the goals may be achieved by evaluating the situation and available communicative resources, and (c) realizing the plans as a text.

Generation has been part of computational linguistics for as long as the field has existed, although it became a substantial subfield only in the 1980s. It appeared first in the 1950s as a minor aspect of machine translation. In the 1960s, random sentence generators were developed, often for use as a means of checking grammars. The 1970s saw the first cases of dynamically generating the motivated utterances of an artificial speaker: composing answers to questions put to database query programs and providing simple explanations for expert systems. That period also saw the first theoretically important generation systems. These systems reasoned, introspected, appreciated the conventions of discourse, and used sophisticated models of grammar. The texts they produced, although few, remain among the most fluent in the literature. By the beginning of the 1980s generation had emerged as a field of its own, with unique concerns and issues.

A. Generation Compared with Comprehension

To understand these issues, it will be useful to compare generation with its far more studied and sophisticated cousin, natural language comprehension. Generation is often misunderstood as a simple variation on comprehension—a tendency that should be dispelled. In particular, generation must be seen as a problem of construction and planning, rather than analysis.

As a process, generation has its own basis of organization, a fact that follows directly from the intrinsic differences in the information flow. The processing in language comprehension typically follows the traditional stages of a linguistic analysis: phonology, morphology, syntax, semantics, pragmatics and discourse; moving gradually from the text to the intentions behind it. In comprehension, the "known" is the wording of the text (and possibly its intonation). From the wording, the comprehension process constructs and deduces the propositional content conveyed by the text and the probable intentions of the speaker in producing it. The primary process involves scanning the words of the text in sequence, during which the form of the text gradually unfolds. The need to scan imposes a methodology based on the management of multiple hypotheses and predictions that feed a representation that must be expanded dynamically. Major problems are caused by ambiguity (one form can convey a range of alternative meanings), and by underspecification (the audience obtains more information from inferences based on the situation than is conveyed by the actual text). In addition, mismatches in the speaker's and audience's model of the situation (and especially of each other) lead to unintended inferences.

Generation has the opposite information flow: from intentions to text, content to form. What is already known and what must be discovered is quite different from the situation in comprehension, and this has many implications. The known is the generator's awareness of its speaker's intentions, its plans, and the structure of any text the generator has already produced. Coupled with a model of the audience, the situation, and the discourse, this information provides the basis for making choices among the alternative wordings and constructions that the language provides—the primary effort in deliberately constructing a text.

Most generation systems do produce texts sequentially from left to right, but only after having made decisions top-down for the content and form of the text as a whole. Ambiguity in a generator's knowledge is not possible (indeed one of the problems is to notice that an ambiguity has inadvertently been introduced into the text). Rather than underspecification, a generator's problem is to choose from an oversupply of possibilities on how to signal its intended inferences and what information to omit from mention in the text.

With its opposite flow of information, it would be reasonable to assume that the generation process can be organized similarly to the comprehension process, but with the stages in opposite order, and to a certain extent this is true: pragmatics (goal selection) typically precedes consideration of discourse structure and coherence, which usually precede semantic matters such as the fitting of concepts to words. In turn, the syntactic context of a word must be fixed before the precise morphological and suprasegmental form it should take can be known. However, we should avoid taking this as the driving force in a generator's design, because to emphasize the ordering of representational levels derived from theoretical linguistics would be to miss generation's special character; namely, generation is above all a planning process. Generation entails realizing goals in the presence of constraints and dealing with the implications of limitations on resources. (Examples of limited resources include the expressive capacity of the syntactic and lexical devices a given language happens to have, or the limited space available in a sentence or a figure title given the prose style that has been chosen.)

This being said, the consensus among persons who have studied both is that generation is the more difficult of the two. What a person needs to know to develop a computer program that produces fluent text is either trivial (the text is entered directly into the code, perhaps with some parameters, and produced as is—virtually every commercial program in wide use that produces text uses this "template" method), or else it is quite difficult because one has to work out a significant number of new techniques and facts about language that other areas of language research have never considered. It is probably no accident that for most of its history advances in NLG have come only through the work of graduate students on their PhD theses.

This also goes a long way toward explaining why so little work has been done on generation as compared with comprehension. At a general meeting, papers on parsing will outnumber those on generation by easily five to one or more. Most work on generation is reported at the international workshops on generation, which have been held nearly every year since 1983; many of those papers then appear in larger form in book collections. All of the edited books cited in the reference section of this chapter are the result of one of these workshops.

B. Computers Are Dumb

Two other difficulties with doing research on generation should be cited before moving on. One, just alluded to, is the relative stupidity of computer programs, and with it the lack of any practical need for natural language generation, as those in the field view it—templates will do just fine.

People who study generation tend to be more scientists than engineers and are trying to understand the human capacity to use language—with all its subtleties of nuance and the complexity, even arbitrariness, of its motivations. Computers, on the other hand, do not think very subtle thoughts. The authors of their programs, even artificial intelligence programs, inevitably leave out the rationales and goals behind the instructions for their behavior, and certainly no computer programs (with the exception of a few programs written specifically to drive generation systems) have any emotional or even rhetorical attitudes toward the people who are using them. Without the richness of information, perspective, and intentions that humans bring to what they say, computers have no basis for making the decisions that go into natural utterances; and it makes no sense to include a natural language generator in one's system if there is nothing for it to do.

This is a problem that will gradually go away. As data mining and other derivative information sources become common, the end-users of these systems will demand more attention to context and become dissatisfied with the rigid format and often awkward phrasing of texts produced by templates.

C. The Problem of the Source

The other difficulty is ultimately more serious, and it is in large part responsible for the relative lack of sophistication in the field as compared with other language processing disciplines. This is the problem of the source. We know virtually nothing about what a generation system should start from if it is to speak as well as people do.

In language comprehension the source is obvious; we all know what a written text or an acoustic signal is. In generation the source is a "state of mind" inside a speaker with "intentions" acting in a "situation"—all terms of art with very slippery meanings. Studying it from a computational perspective as we are here, we presume that this state of mind has a representation, but there are dozens of formal (consistently implementable) representations used within the artificial intelligence (AI) community that have (what we assume is) the necessary expressive power, with no a priori reason to expect one to be better than another as the mental source of an utterance. Worse yet is the absolute lack of consistency between research groups in their choice of primitive terms and relations—does the representation of a meal bottom out with "eat", or must that notion necessarily be expanded into a manner, a result, and a time period, with eat being just a runtime abstraction.

The lack of a consistent answer to the question of the generator's source has been at the heart of the problem of how to make research on generation intelligible and engaging for the rest of the computational linguistics community, and it has complicated efforts to evaluate alternative treatments even for persons in the field. Nevertheless, a source cannot be imposed by fiat. Differences in what information is assumed to be available, its relative decomposition when compared with the "packaging" available in the words or syntactic constructions of the language, what amounts and kinds of information are contained in the atomic units of the source and what sorts of compositions and other larger-scale organizations are possible—all of these have an influence on what architectures are plausible for generation and what efficiencies they can achieve. Advances in the field have often come precisely through insights into the representation of the course.

2. EXAMPLES OF GENERATED TEXTS: FROM COMPLEX TO SIMPLE AND BACK AGAIN

If we look at the development of natural language generation in terms of the sorts of texts different systems have produced, we encounter something of a paradox. As the field has advanced, the texts have become simpler. Only in the last few years have generation systems begun to produce texts with the sophistication and fluency that was present in the systems of the early 1970s.

A. Complex

One dramatic example developed during the earliest period is John Clippinger's program Erma (1977), which modeled an actual psychoanalytic patient talking to her therapist. It emulated one paragraph of speech by the patient excerpted from extensive transcripts of her conversations. The effort was joint work by Clippinger as his 1974 PhD thesis and Richard Brown in his Bachelors degree thesis (1974).

The paragraph was the result of a computationally complex model of the patient's thought processes: from the first identification of a goal, through planning, criticism, and replanning of how to express it, and finally linguistic realization. Clippinger and Brown's program had a multiprocessing capability—it could continue to think and plan while talking. This allowed them to develop a model of "restart" phenomena in generation including the motivation behind fillers like "uh" or dubitives like "you know." Text segments shown in parenthesis in the following are what Erma was planning to say before it cut itself off and restarted. In other respects this is an actual paragraph from a transcript of the patient reproduced in every detail.

> *You know for some reason I just thought about the bill and payment again. (You shouldn't give me a bill.) <Uh> I was thinking that I (shouldn't be given a bill) of asking you whether it wouldn't be all right for you not to give me a bill. That is, I usually by (the end of the month know the amount of the bill), well, I immediately thought of the objections to this, but my idea was that I would simply count up the number of hours and give you a check at the end of the month.*

There has yet to be another program described in the literature that can even begin to approach the human-like quality of this text. On the other hand, Erma only ever produced that one text, and some parameter-driven variations, and neither Brown's multilevel, resumable, interrupt-driven computational architecture, nor Clippinger's rich set of thinking, critiquing, and linguistic modules was ever followed up by other people.

B. Simple

By the end of the decade of the 1970s, generation began to be recognized as a field with shared assumptions and not just the work of scattered individuals. It also began to attract the attention of the research-funding community, a mixed blessing perhaps, because while the additional resources now allowed work on generation to be pursued by research teams instead of isolated graduate students, the need to conform to the expectations of other groups—particularly in the choice of source representation and conceptual vocabulary–substantially limited the creative options. Probably as a direct result, the focus of the work during the 1980s moved within the generator, and the representations and architecture of speaker became a black box behind an impenetrable wall.

Nevertheless, the greatly increased number of people working in the field led to many important developments. If the texts the various groups' systems produced were not of the highest quality, this was offset by increased systematicity in the techniques in use, and a markedly greater understanding of some of the specific issues in generation. Among these were the following:

- The implications of separating the processing of a generator into distinct modules and levels of representation, especially relative to what operations (lexical choice, linear ordering, and such), were positioned at which level
- The use of pronouns and other forms of subsequent reference
- The possibilities and techniques for "aggregating" minimal propositions to form syntactically complex texts
- The relation between how lexical choice is performed and the choice of representation used in the source

We will take up several of these issues toward the end of this chapter.

Here is an example of texts produced by systems developed in the late 1980s—a generator that is not at least this fluent today would be well behind the state of the art. This is from Meteer (1992), set in a military domain; the text shown here is an excerpt from a page-long generated text. Notice the use of simple formatting elements.

2. MISSION
10th Corps defend in assigned sector to defeat the 8th Combined Arms Army.
3. EXECUTION
a. 52d Mechanized Division
(1) Conduct covering force operations along avenues B and C to defeat the lead regiments of the first tactical echelon in the CFA in assigned sector.

A text such as this will never win any prizes for literature, but unlike its handcrafted predecessors of the 1970s, such texts can be produced mechanically from any comparable input without any human intervention or fine tuning.

The source for this text from the Spokesman system was a battle order data structure that was automatically constructed by a simulation system that fought virtual battles in detail against human troops in tank simulators—an excellent source of material for a generator to work with: for example, Meteer's system also realized English versions of the radio traffic among elements of the automated "opposing force."

A/74 TB has spotted a platoon of 3 vehicles at ES7000800.

A/74 TB's fuel level is 80%.

Given the rich architectures available in generation systems today, the production of detailed, if mundane, information derived directly from an application program has become almost a cookbook operation in the sense that practitioners of the art can readily engineer a system with these abilities in a relatively short time.

Much of what makes these modern generators effective is that they are applied to very specific domains, domains in which the corpus of text can be described as belonging to a "sublanguage." That is to say, they restrict themselves to a specialized area of discourse with a very focused audience and stipulated content, thereby reducing the options for word choice and syntactic style to a manageable set (see Chaps. 11 and 13 for further discussion of the role of sublanguages in natural language processing).

C. Today

Now, in the late 1990s, we have reached a point at which a well-designed and linguistically sophisticated system can achieve the fluency of the special purpose systems of the 1970s, but by operating on a better understood theoretical base. As an example of this, we will end this section by looking at Jacques Robin's Streak (1993, 1996). It too operates within a sublanguage, in this case the language of sports: it writes short summaries of basketball games. Similar to all news reporting, this genre is characterized by information-dense, syntactically rich summary texts, texts that remain challenging to the best of systems. With Streak, Robin has recognized the extensive references to historical information in these texts, and his experience has important implications for how to approach the production of summaries of all sorts.

Technically, Streak is a system based on revision. It begins by producing a representation of the simple facts that will provide anchors for later extensions. Here is an example of what it could start from (Robin 1996, p. 206):

> *Dallas, TX—Charles Barkley scored 42 points Sunday as the Phoenix Suns defeated the Dallas Mavericks 123-97.*

This initial text is then modified as salient historical or ancillary information about this game and the players' past records is considered. Here is the final form:

> *Dallas, TX—Charles Barkley tied a season high with 42 points and Danny Ainge came off the bench to add 21 Sunday as the Phoenix Suns handed the Dallas Mavericks their league worst 13th straight home defeat 123-97.*

Notice what has happened. Initial forms have been progressively replaced with phrases that carry more information: "scored N points" has become "tied a season high with N points." Syntactic formulations have been changed ["defeat X" has become "hand X (a) defeat"], with the new choice able to carry information that the original could not (the noun form of "defeat" can be modified by "their league worst" and "Nth straight home"). This is sophisticated linguistic reasoning that has been matched by only a few earlier systems.

3. THE COMPONENTS OF A GENERATOR

This section provides a high-level overview of the architectural elements that make up a natural language generation system. They will be fleshed out and exemplified in the sections that follow.

A. Speaker and Generator

To produce a text in the computational paradigm, there has to be a program with something to say—we can call this "the application" or "the speaker." And there must be a program with the competence to render the application's intentions into fluent prose appropriate to the situation—what we will call "the generator"—which is the natural language generation system proper.

Given that the task is to research or engineer the production of text or speech for a purpose—emulating what people do or making it available to machines—then both of these components, the speaker and the generator, are necessary. Studying the language side of the process without anchoring the work relative to the conceptual models and intentional structures of an application may be appropriate for theoretical linguistics or the study of grammar algorithms, but not for language generation.

As discussed in the foregoing, the very earliest work on sophisticated language production interleaved the functions of the speaker and the generator into a single system. Today, there will invariably be three or four components (if not a dozen) dividing the work among themselves according to a myriad of different criteria. We will discuss the philosophies governing these criteria later.

B. Components and Levels of Representation

Given the point of view we adopt in this chapter, we will say that generation starts in the mind of the speaker (the execution states of the computer program) as it acts on an intention to say something—to achieve some goal through the use of language.

Once the process is initiated, generation proper involves at least four interleaved tasks, regardless of the approach taken.

1. Information must be *selected* for inclusion in the utterance. Depending on how this information is reified into representational units (a property of the speaker's mental model), parts of the units may have to be omitted, other units added in by default, or perspectives taken on the units to reflect the speaker's attitude toward them.
2. The information must be given a *textual organization.* It must be ordered, both sequentially and in terms of linguistic relations such as modification or subordination. The coherence relationships among the units of the information must be reflected in this organization so that the reasons why the information was included will be apparent to the audience.
3. *Linguistic resources* must be chosen to support the information's realization. Ultimately these resources will come down to choices of particular words, idioms, syntactic constructions, productive morphological variations, and so on, but the form they take at the first moment that they are associated with the selected information will vary greatly between approaches. Note that to choose a resource is not ipso facto to simultaneously deploy it in its final form—a fact that is not always appreciated.
4. The selected and organized resources must be *realized* as an actual text or voice output. This stage can itself involve several levels of representation and interleaved processes.

These tasks are usually divided among three components as follows. The first two are often spoken of as deciding "what to say," the third deciding "how to say it."

1. The application program or "speaker." This does the thinking and maintains a model of the situation. Its goals are what initiate the process, and it is the representation of concepts and the world that this program provides which supply the source on which the other components operate.
2. A text planner. This selects (or receives) units from the application and organizes them to create a structure for the utterances as a text by employing some knowledge of rhetoric. It appreciates the conventions for signal-

ing information flow in a linguistic medium: what information is new to the interlocutors, what is old; what items are in focus; whether there has been a shift in topic. Now in the 1990s, this component is often broken out into two modules as discussed at the end of this section.

3. A linguistic component. This realizes the planner's output as an utterance. In its traditional form during the 1970s and early 1980s, it supplied all of the grammatical knowledge used in the generator. Today this knowledge is likely to be more evenly distributed throughout the system. This component's task is to adapt (and possibly to select) linguistic forms to fit their grammatical contexts and to orchestrate their composition. This process leads, possibly incrementally, to a surface structure for the utterance that is then read out to produce the grammatically and morphologically appropriate wording for the utterance.

How these roughly drawn components interact is a matter of considerable controversy and no little amount of confusion, as no two research groups are likely to agree on precisely what kinds of knowledge or processing appear in a given component or where its boundaries should lie.

One camp, making analogies to the apparent abilities of people, holds that the process is monotonic and indelible. A completely opposite camp extensively revises its (abstract) draft texts. Some groups organize the components as a pipeline; others use blackboards. Nothing conclusive about the relative merits of these alternatives can be said today. We are in a period during which the best advice is to let a thousand flowers bloom.

In addition to these components, there are representational levels. There are necessarily one or more intermediate levels between the source and the text simply because the production of an utterance is a serial process extended in time. Most decisions will influence several parts of the utterance at once, and consequently cannot possibly be acted on at the moment they are made. Without some representation of the results of these decisions there would be no mechanism for remembering them and utterances would be incoherent.

The consensus favors at least three representational levels, roughly the output of each of the components. In the first or "earliest" level, the information units of the application that are relevant to the text planner form a "message" level—the source from which the later components operate. Depending on the system, this level can consist of anything from an unorganized heap of propositions to an elaborate typed structure with annotations about the relevance and purposes of its parts.

All systems include one or more levels of surface syntactic structure. These encode the phrase structure of the text and the grammatical relations among its constituents. Morphological specialization of word stems and the introduction of punctuation or capitalization are typically done as this level is readout and the utterance uttered. Common formalisms at this level include systemic networks, Tree-Adjoining Grammar, and functional unification, though practically every linguistic theory of grammar that has ever been developed has been used for generation at one time or another. Nearly all of today's generation systems express their utterances as written texts—characters printed on a computer screen—rather than as speech. Consequently, generators rarely include an explicit level of phonological form or intonation.

Until relatively recently most systems went from the units that the text planner selected directly to surface structure. But it has been recognized that if the output is to be fluent while still keeping the architecture general and easily adapted to new tasks, then speaker-level to surface-level is too much to bridge in one step. This notion of a potential "generation gap" between the output of the planner and the needs and capacities of a sophisticated linguistic component was first articulated by Meteer (1991, 1992) as the problem of *expressibility*—does the generation system's architecture provide a principled means of ensuring that there will actually be a realization for the elements of its messages? If it does not, then the later stages may break or extensive, possibly nondeterministic, replanning may be required. By contrast, systems that ensure expressibility as a consequence of their design (as opposed to the diligence of their programmers) can operate deterministically at high efficiency.

This has focused attention on a middle level or levels of representation at which a system can reason about linguistic options without simultaneously being committed to syntactic details that are irrelevant to the problem at hand. At this level, abstract linguistic structures are combined with generalizations of the concepts in the speaker's domain-specific model and sophisticated concepts from lexical semantics. The level is variously called *text structure*, *deep syntax*, *abstract syntactic structure*, and the like. In some designs it will employ rhetorical categories, such as elaboration or temporal location. Alternatively, it may be based on abstract linguistic concepts, such as the matrix–adjunct distinction. It is usually organized as trees of constituents with a layout roughly parallel that of the final text. The leaves may be direct mappings of units from the application or may be semantic structures specific to that level.

It is useful in this context to consider a distinction put forward by the psycholinguist Willem Levelt (1989), between macro- and microplanning.

- *Macroplanning* refers to the process(es) that choose the speech acts, establish the content, determine how the situation dictates perspectives, and so on. It will be the subject of the next section in this chapter.
- *Microplanning* is a cover term for a group of phenomena: determining the detailed (sentence–internal) organization of the utterance, considering whether to use pronouns, looking at alternative ways to group information into phrases, noting the focus and information structure that must apply, and other such relatively fine-grained tasks. These, along with lexical choice, are precisely the set of tasks that fall into this nebulous middle ground that is motivating so much of today's work. We will put off discussion of these issues to the final section of this chapter.

4. APPROACHES TO TEXT PLANNING

Even though the classic conception of the division of labor in generation between a text planner and a linguistic component—where the latter is the sole repository of the generator's knowledge of language—was probably never really true in practice and is certainly not true today, it remains an effective expository device. In this section we will consider text planning in a relatively pure form, concentrating on the techniques for determining the content of the utterance and its large-scale (suprasentential) organization.

A. The Function of the Speaker

From the generator's perspective, the function of the application that it is working for is to set the scene. Because it takes no overtly linguistic actions beyond initiating the process, we are not inclined to think of the application program as a part of the generator proper. Nevertheless, the influence it wields in defining the situation and the semantic model from which the generator works is so strong that it must be designed in concert with the generator if high quality results are to be achieved. This is why we often speak of the application as the "speaker," emphasizing the linguistic influences on its design and its tight integration with the generator.

The speaker establishes what content is potentially relevant. It maintains an attitude toward its audience (as a tutor, reference guide, commentator, executive summarizer, copywriter, or whatever). It has a history of past transactions. It is the component with the model of the present state and its physical or conceptual context. The speaker deploys a representation of what it knows, and this implicitly determines the nature and the expressive potential of the "units" of speaker stuff that the generator must work from to produce the utterance (the source). We can collectively characterize all of this as the "situation" in which the generation of the utterance takes place, in the sense of Barwise and Perry (1983).

In the simplest case, the application consists of just a passive database of items and propositions. Here the situation is a selected subset of those propositions (the "relevant data") that has been selected through some means, often by following the thread of a set of identifiers chosen in response to a question from the user.

Occasionally, the situation is a body of raw data, and the job of speaker is to make sense of it in linguistically communicable terms before any significant work can be done by the other components. The literature includes several important systems of this sort. Probably the most thoroughly documented is the Ana system developed by Kukich (1986), in which the input is a set of timepoints giving the values of stock indexes and trading volumes during the course of a day; Chap. 11 discusses systems of this type in more detail.

When the speaker is a commentator, the situation can evolve from moment to moment in actual real time. One early system produced commentary for football (soccer) games that were being displayed on the user's screen. This led to some interesting problems on how large a chunk of information could reasonably be generated at a time: too small a chunk would fail to see the larger intentions behind a sequence of individual passes and interceptions, whereas too large a chunk would take so long to utter that the commentator would fall behind the action.

One of the crucial tasks that must often be performed at this juncture between the application and the generator is enriching the information that the application supplies so that it will use the concepts that a person would expect even if the application had not needed them. We can see an example of this in one of the earliest, and still among the most accomplished generation systems, Davey's Proteus (Davey, 1974).

Proteus played games of tic-tac-toe (noughts and crosses) and provided commentary on the results. Here is an example of what it produced:

> *The game started with my taking a corner, and you took an adjacent one. I threatened you by taking the middle of the edge opposite that and adjacent to the one which I had just taken but you blocked it and threatened me. I blocked your diagonal and forked you. If you had blocked mine, you would have forked me, but you took the middle of the edge opposite of the corner which I took first and the one which you had just taken and so I won by completing my diagonal.*

Proteus began with a list of the moves in the game it had just played. In this sample text the list was the following [Moves are notated against a numbered grid; square one is the upper left corner. Proteus (P) is playing its author (D).]

P:1 D:3 P:4 D:7 P:5 D:6 P:9

One is tempted to call this list of moves the "message" that Proteus's text planning component has been tasked by its application (the game player) to render into English—and it is what actually crosses the interface between them—but consider what this putative message leaves out when compared with the ultimate text: where are the concepts of move and countermove or of a fork? The game-playing program did not need to think in those terms to carry out its task and performed perfectly well without them, but if they were not in the text we would never for a moment think that the sequence described was a game of tic-tac-toe.

Davey was able to obtain texts of this complexity and naturalness only because he imbued Proteus with a rich conceptual model of the game and, consequently, could have it use terms like "block" or "threat" with assurance. As with most instances where exceptionally fluent texts have been produced, Davey was able to obtain this sort of performance from Proteus because he had the opportunity to develop the thinking part of the system as well as its strictly linguistic aspects and, consequently, could ensure that the speaker supplied rich perspectives and intentions with which the generator could work.

This, unfortunately, is a quite common state of affairs in the relation between a generator and its speaker. The speaker, as an application program carrying out a task, has a pragmatically complete, but conceptually impoverished, model of what it wants to relate to its audience. The concepts that must be made explicit in the text are implicit but unrepresented in the application's code and its remains to the generator (Proteus in this case) to make up the difference. Undoubtedly, the concepts were present in the mind of the application's human programmer, but leaving them out makes the task easier to program and rarely limits the application's abilities. The problem of most generators is, in effect, how to convert water to wine, compensating in the generator for the limitations of the application (McDonald Meteer, 1988).

B. Desiderata for Text Planning

The tasks of a text planner are many and varied. They include the following:

1. Construing the speaker's situation in realizable terms given the available vocabulary and syntactic resources, an especially important task when the source is raw data, e.g. precisely what points of the compass make the wind *easterly* (Bourbeau et al., 1990)
2. Determining the information to include in the utterance and whether it should be stated explicitly or left for inference

3. Distributing the information into sentences and giving it an organization that reflects the intended rhetorical force, coherence, and necessary cohesion given the prior discourse.

Because a text has both a literal and a rhetorical content, the determination of what the text is to say requires not only a specification of its propositions, statements, references, and so on, but also a specification of how these elements are to be related to each other as parts of a single text attempting to achieve a goal (what is evidence, what is a digression) and of how they are structured as a presentation to the audience to which the utterance is addressed. This presentation information establishes what is thematic, where there are shifts in perspective, how new text is to be understood to fit within the context established by the text that preceded it, and so on.

How to establish the simple, literal information content of the text is relatively well understood, and several different techniques have been extensively discussed in the literature. How to establish the rhetorical content of the text, however, is only beginning to be explored, and in the past was done implicitly or by rote through the programmer's direct coding. There have been some experiments in deliberate rhetorically planning, notably by Hovy (1988a) and DiMarco and Hirst (1993), and the work is usually placed under the heading of "prose style."

C. Pushing versus Pulling

To begin our examination of the major techniques in text planning, we need to consider how the text planner and speaker are connected. The interface between the two is based on one of two logical possibilities: "pushing" or "pulling."

The application can push units of content to the text planner, in effect telling the text planner what to say and leaving it the job of organizing the units into a text with the desired style and rhetorical effect. Alternatively, the application can be passive, taking no part in the generation process, and the text planner will pull units from it. In this scenario the speaker is assumed to have no intentionality and only the simplest motivation and ongoing state (often it is a database). Consequently, all of the work has to be done on the generator's side of the fence.

Text planners that pull content from the application establish the organization of the text hand in glove with its content, using models of possible texts and their rhetorical structure as the basis of its actions; their assessment of the situation determines which model they will use. Speakers that push content to the text planner typically use their own representation of the situation directly as the content source.

At the time of writing, the pull school of thought largely dominates new, theoretically interesting work in text planning, whereas virtually all practical systems are based on simple push applications or highly stylized, fixed "schema"-based pull planners. In what follows we will look at three general schools of text planning, one that pushes and two that pull.

D. Planning by Progressive Refinement of the Speaker's Message

This technique, often called *direct replacement*, is easy to design and implement, and it is by far the most mature approach of the three we will cover In its simplest form, it amounts to little more than is done by ordinary database report generators or mail–

merge programs when they make substitutions for variables in fixed strings of text. In its sophisticated forms, which invariably incorporate multiple levels of representation and complex abstractions, it has produced some of the most fluent and flexible texts in the field. Among the systems discussed in the introduction, three of the five did their text planning using progressive refinement: Proteus, Erma, and Spokesman.

Progressive refinement is a push technique. It starts with a data structure already present in the application's model of the world and its situation; and then it gradually transforms that data into a text. The semantic coherence of the final text follows from the underlying semantic coherence that is present in the data structure that the application presents to the generator as its message.

Progressive refinement, implemented literally as direct replacement, will produce reasonable texts, although their naturalness will depend strongly on how close to language the source is to begin with. Here, for example, is a rendering of a natural deduction proof of Russel's Barber Paradox. You can virtually read out the original lines of the logic from the sentences of this text (from unpublished work by D. Chester done in 1975 at the University of Texas at Austin).

> *Suppose that there is some barber such that for every person the barber shaves the person iff the person does not shave himself. Let A denote such a barber. Now he shaves himself iff he does not shave himself, therefore a contradiction follows. Therefore if there is some barber such that for every person the barber shaves himself iff the person does not shave himself then a contradiction follows. Therefore there is no barber such that for every person the barber shaves the person iff the person does not shave himself.*

At the expense of a considerably more complex system, the results do not have to be so stilted even when starting from the identical source. Here is the same proof as produced by the Mumble system in 1978. (Mumble is discussed later in the section on surface realization.) Here, a greater competence in the use of referring expressions, an elaborate system of lookahead (to determine the syntactic complexity of units before they are realized), and a certain liberty in the naming of logical constants combined to produce something close to what a person might say (McDonald 1983, p. 216).

> *Assume that there is some barber who shaves everyone who doesn't shave himself (and no one else). Call him Giuseppe. Now, anyone who doesn't shave himself would be shaved by Giuseppe. This would include Giuseppe himself. That is, he would shave himself, if and only if he did not shave himself, which is a contradiction. This means that the assumption leads to a contradiction. Therefore it is false, there is no such barber.*

Part of the refinement done by these planners is the deliberate grouping of message-level units according to semantic criteria that are natural to the domain (as opposed to the stylistic or syntactic criteria deployed in the "aggregation" techniques discusssed in the final section of this chapter). We can see a good example of this, again, in Davey's Proteus system.

The essence of the technique is to have the text planner layer additional information on top of the basic skeleton provided by the application, where in this case the skeleton is the sequence of moves. The ordering of the moves must still be respected in the final text, because Proteus is a commentator and the sequence of events described in a text is implicitly understood as reflecting sequence in the world. Proteus departs from the ordering only when it serves a useful rhetorical purpose, as in the example text in which it describes the alternative events that could have occurred if its opponent had made a different move early on.

On top of the skeleton, Proteus looks for opportunities to group moves into compound complex sentences by viewing the sequence of moves in terms of the concepts of tic-tac-toe. It looks for pairs of forced moves (i.e., a blocking move that counters a prior move that had set up two in a row). It looks for strategically important consequences (a move creating a fork). For each such semantically significant pattern that it knows how to recognize, Proteus has one or more patterns of textual organization that can express it. For example, the pattern "high-level action followed by literal statement of the move" yields *I threatened you by taking the middle of the edge opposite that.* Alternatively, Proteus could have used "literal move followed by its high-level consequence: *I took the middle of the opposite edge, threatening you.*

The choice of realization is left up to a "specialist," which takes into account as much information as the designer of the system knows how to bring to bear. Similarly, a specialist is employed to elaborate on the skeleton when larger-scale strategic phenomena occur. For a fork this prompts the additional rhetorical task of explaining what the other player might have done to avoid the fork.

Proteus' techniques are an example of the standard pattern for a progressive refinement text planner: a skeletal data structure that is a rough approximation of the final text's organization, with information provided by the speaker directly from its internal model of the situation. The structure then goes through some number of successive steps of processing and rerepresentation as its elements are incrementally transformed or mapped to structures that are closer and closer to a surface text, becoming progressively less domain oriented and more linguistic at each step.

Control is usually vested in the structure itself, using what is known as "data-directed control." Each element of the data is associated with a specialist or an instance of some standard mapping that takes charge of assembling the counterpart of the element within the next layer of representation. The whole process is often organized into a pipeline where processing can be going on at multiple representational levels simultaneously as the text is produced in its natural, "left-to-right," order as it would unfold if being spoken by a person.

A systematic problem with progressive refinement follows directly from its strengths; namely, its input data structure, the source of its content and control structure, is also a straightjacket. Although it provides a ready and effective organization for the text, the structure does not provide any vantage point from which to deviate from that organization, even if that would be more effective rhetorically. This remains a serious problem with the approach, and is part of the motivation behind the types of text planners we will look at next.

E. Planning Using Rhetorical Operators

The next text planning technique that we will look at can be loosely called *formal planning using rhetorical operators*. It is a pull technique that operates over a pool of relevant data that has been identified within the application. The chunks in the pool are typically full propositions—the equivalents of clauses if they were realized in isolation.

This technique, and the one that will be described next, assumes that there is no useful organization to the propositions in the pool or, alternatively, that such organization as there is, is orthogonal to the discourse purpose at hand and should be largely ignored. Instead, the mechanisms of the text planner look for matches between the items in the relevant data pool and the planner's abstract patterns, and select and organize the items accordingly.

Three design elements come together in the practice of operator-based text planning, all of which have their roots in work done in the later 1970s:

1. The use of formal means–ends reasoning techniques adapted from the robot–action-planning literature
2. A conception of how communication could be formalized that derives from speech–act theory and specific work done at the University of Toronto
3. Theories of the large-scale "grammar" of discourse structure.

Means-end analysis, especially as elaborated in the work by Saccerdoti (1977), is the backbone of the technique. It provides a control structure that does a top-down, hierarchical expansion of goals. Each goal is expanded through the application of a set of operators that describe a sequence of subgoals that will achieve it. This process of matching operators to goals terminates in propositions that can directly instantiate the actions dictated by terminal subgoals. These propositions become the leaves of a tree-structured text plan, with the goals as the nonterminals and the operators as the rules of derivation that give the tree its shape.

For a brief look at parts of this process in action, we can use the Penman system as described by Hovy (1990, 1991). Here is an example of its output; the domain is military manoeuvres at sea.

Knox, which is C4, is en route to Sasebo. It is at 18N 79E, heading SSW. It will arrive on 4/25. It will load for 4 days.

The source for this text was a pool of relatively raw propositions preselected from a Navy data base. Here is an excerpt (adapted from Hovy 1991, p. 87); notice how these propositions are linked through chains of reference to shared identifiers such as "a105":

```
(ship.employment a105)
(ship.r a105 Knox)
(ship.course.r a105 195)
(current.major.employment.r a105 e105)
(enroute e105)
(destination.r e105 Sasebo)
```

Penman's first step is not unlike what Davey's Proteus did. It applies a grouping expert to the data to clump it into appropriate (clause-sized) units and to identify aggregating lexical terms through the use of domain-specific rules. In this case, it knows that to become stationary after having been moving is to have "arrived," which is analogous to Proteus noticing that a particular pair of moves in tic-tac-toe constitutes a "threat and block." Here are some examples of some of these derived units:

```
((enroute e105)
 (ship.r e105 Knox)
 (destination.r e105 Sasebo)
 (heading.r e105 heading11416)
 (readiness.r e105 readiness11408)
 (next-action.r e105 arrive11400))

((arrive arrive11400)
 (ship.r arrive11400 Knox)
 (time.r arrive11400 870424)
 (next-action.r arrive11400 e107))

((position position11410)
 (ship.r position11410 Know)
 (longitute.r position11410 79)
 (latitude.r position11410 18))
```

Having constructed its input, the next step is to plan how to satisfy a specific communicative goal, as dictated by the person using the system. Here is the goal in this example; it is a representation of the question "Where is the Knox?".

```
(BMB Speaker Hearer (position-of e105 ?next))
```

The modal operator at the beginning of the goal is glossed as "believes, mutually believes" and comes from work at the University of Toronto that is part of a vast literature in speech–act theory and the fundamental logic of communication (Cohen and Perrault, 1979; Cohen and Levesque, 1985). As a whole, this goal amounts to a directive to cause it to happen that the Speaker (the generation system) and the Hearer (the person asking the question) both know where the Knox is. Because the Speaker is presumed to know the position, this means that the action it should take is to tell the Hearer what the position is.

That is the end to be achieved. The task of the text planner is to determine the means by which to achieve it, which comes down to searching its repertoire of planning operators. These operators amount to small chunks of knowledge about how goals can be achieved through the use of language. In this respect they are not dissimilar in function to the specialists employed by other kinds of text planners, but their form is highly schematic and uniform and their deployment is rigidly controlled by the planning formalism.

The plan operators have the following elements: a schematic description of the result they will achieve if they are used, the substeps that must be taken to do this (the expansion of the goal), stipulations of any preconditions that must be met before the operator is applicable, and any ancillary information that is needed by the particular grammar of discourse structure that is being used. In Penman, for this instance, the relevant planning operator is "sequence." Its elements are as follows (again adapted from Hovy, 1991 p. 89). Terms that are prefixed by a question mark are matching variables; notice that the result field of the operator will match the current goal.

Operator: Sequence
Result:
(BMB Speaker Hearer (position-of ?part ?next))
Nucleus + Satellite subgoals:
(BMB Speaker Hearer (next-action.r ?part ?next))
Nucleus subgoals:
(BMB Speaker Hearer ?part)
Satellite subgoals:
(BMB Speaker Hearer ?next)
Order: (nucleus satellite)
Relation-phrases: (" " "then" "next")

Application of the results pattern of this operator to the goal unit as it is instantiated in the datebase (recall that we are answering a question) will bind ?part to e105 (the nucleus) and ?next to arrive11400 (the satellite). These now become the next goals to be expanded. Furthermore, they have been given a fixed ordering within the text such that the "enroute" unit (e105) is to precede the "arrival" unit. Planning now proceeds by looking for operators for which the results fields (their "ends") match these two units, with further expansions ("substeps") as dictated by the subgoals of those operators (the "means").

The notions of nucleus and satellite are structural (syntactic) relations at the discourse level (i.e., describing the structure of paragraphs, rather than sentences) from a framework for discourse grammar known as Rhetorical Structure Theory (RST) (Mann and Thompson, 1987, 1988). Similar to virtually every other such system, RST characterizes the organization of extended, multisentence text as a tree. The nonterminal nodes are drawn from a vocabulary of the possible relations between different segments of a text as they contribute to the rhetorical effectiveness of the text as a whole. The precise set of rhetorical relations varies widely according to the analytical judgments of the people developing the system, but it will include notions, such as contrast, enabling condition, temporal sequence, motivation, and so on.

The process just described is the essence of this technique of formal planning using rhetorical operators. There are several interesting variations on this theme. Hovy's own treatment in Penman includes the notion of optional "growth points": locations within the discourse structure tree that are specified by the operators as available for information from the pool that has not already been incorporated into the text by the top-down hierarchical planning. This is how, for example, we go from the core text plan that would be realized as *Knox is enroute to Sasebo* to the elaborated plan *Knox, which is C4, is enroute to Sasebo. It is at 18X 79E, heading SSW.*

We have covered the architecture of the formal planning with rhetorical operators technique in some depth while just touching on the other elements that go into the design of these text planners. This is because it is the choice of architecture—the use of plan operators and hierarchical decomposition—that set this technique apart from the others. By contrast, the two other elements of the architecture—the idea that planning should be performed in terms of formal, axiomatized model of how a speaker comes to communicate with its audience, and the construal of discourse structure as consisting of a tree with rhetorical relations as its nodes—just provide the "content" of these planners: the information in the operators and the shape of the text plan. Other rationales for why a text is expected to be effective or other conceptions of what theoretical vocabulary should be used in organizing the structure of an extended discourse could equally well be used in an operator-driven, top-down, goal-expansion architecture (see, e.g., Marcu, 1997). That they have not is just a reflection of the accidents of history.

F. Text Schemas

The third important text-planning technique is the use of preconstructed, fixed networks that are referred to as *schemas* following the coinage of the person who first articulated this approach, McKeown (1985). Schemas are a pull technique. They make selections from a pool of relevant data provided by the application according to matches with patterns maintained by the system's planning knowledge—just like an operator-based planner. The difference is that the choice of (the equivalent of) operators is fixed, rather than actively planned. Means–end analysis-based systems assemble a sequence of operators dynamically as the planning is underway. A schema-based system comes to the problem with the entire sequence already in hand.

Given that characterization of schemas, it would be easy to see them as nothing more than compiled plans, and one can imagine how such a compiler might work if a means–ends planner were given feedback about the effectiveness of its plans and could choose to reify its particularly effective ones (though no one has ever done this). However, that would miss the important fact about system design that it is often simpler and just as effective to simply write down a plan by rote, rather than to attempt to develop a theory of the knowledge of context and communicative effectiveness that is deployed in the development of the plan and, then, from that attempt to construct the plan from first principles, which is essentially what the means–ends approach to text planning does. It is no accident that schema-based systems (and even more so progressive refinement systems) have historically produced longer and more interesting texts than means–end systems.

Schemas are usually implemented as transition networks, in which a unit of information is selected from the pool as each arc is traversed. The major arcs between nodes tend to correspond to chains of common object references between units: cause followed by effect, sequences of events traced step by step through time, and so on. Self-loops returning back to the same node dictate the addition of attributes to an object, side-effects of an action, and so on.

The choice of what schema to use is a function of the overall goal. McKeown's original system, for example, dispatched on a three-way choice between defining an object, describing it, or distinguishing it from another type of object. Once the goal is determined, the relevant knowledge pool is separated out from the other parts of the reference knowledge base and the selected schema is applied. Navigation through the schema's network is then a matter of what units or chains of units are actually present in the pool in combination with tests that the arcs apply.

Given a close fit between the design of the knowledge base and the details of the schema, the resulting texts can be quite good. Such faults as they have are largely the result of weakness in other parts of the generator and not in its content-selection criteria. Experience has shown that basic schemas can be readily abstracted and ported to other domains (McKeown et al., 1990). Schemas do have the weakness when compared with systems with explicit operators and dynamic planning that, when used in interactive dialogs, they do not naturally provide the kinds of information that is needed for recognizing the source of problems, which makes it difficult to revise any utterances that are initially not understood (Moore and Swartout, 1991; Paris, 1991; see also Chap. 6 for a more extensive discussion of related issues here). But, for most of the applications to which generation systems are put, schemas are a simple and easily elaborated technique that is probably the design of choice whenever the needs of the system or nature of the speaker's model make it unreasonable to use progressive refinement.

G. The Future: "Inferential Planners"

All of the techniques of text planning discusssed to this point, and indeed all of the planning done in any of the systems described so far, with the exception of Clippinger's Erma, are what we can call *literal* planners. All of the propositions in their input, push or pull, are realized virtually one-for-one in the final text. Each proposition appears either as a clause of its own, or as a simple reduction (aggregation) of a clause, such as when the proposition (color-of-house-1 red) is realized as an adjective in "the red house."

At the cutting edge of work on text planning are what we will call *inferential* planners. These systems take into account the inferences that the audience can be expected to make as they process an utterance in context. This allows them to have many of their input propositions realized implicitly in the mind of the audience as they fill in the information that they conclude must be implied by the literal content that they hear.

Take for example this example adapted from Di Eugenio (1994), in which the target text is

> "Could you go into the kitchen and get me the coffee urn?"

From the point of view of the (hypothetical) speaker that would produce this text, part of the information to be communicated is a proposition that identifies the location of the coffee urn—something that is only implicit in this text. The location of the urn is conveyed inferentially through the fact that the listener is directed to go into the kitchen, which is a location, and that the listener will carry this location forward and apply it to the action that they hear next.

Unfortunately, it is one thing to describe what is (probably) going on in an inferentially planned text, but it is another thing entirely to formalize the reasoning that must have gone on inside the (human) text planner to have appreciated that the option to convey the information inferentially was possible at all. There have been some attempts, but no one has yet developed a general, broadly applicable and well-demonstrated technique for inferential planning that is comparable with the well-understood push or pull techniques we have looked at.

5. THE LINGUISTIC COMPONENT

In this section we will look at the core issues in the most mature and well-defined of all the processes in natural language generation, the application of a grammar to produce a final text from the elements that were decided on by the earlier processing. This is the one area in the whole field in which we find true instances of what software engineers would call properly modular components: bodies of code and representations with well-defined interfaces that can be (and have been) shared between widely varying development groups, many using quite different approaches to macro- and micro- text planning.

A. Surface Realization Components

To reflect the narrow scope (but high proficiency) of these components, we will refer to them here as *surface realization components*: *surface* (as opposed to deep) because what they are charged with doing is producing the final syntactic and lexical structure of the text—what linguists in the Chomskian tradition would call a surface structure; and *realization* because what they do never involves planning or decision making: They are in effect carrying out the orders of the earlier components, rendering (realizing) their decisions into the shape that they must take to be proper texts in the target language.

The job of a surface realization component is to take the output of the text planner, render it into a form that can be conformed (in a theory-specific way) to a grammar, and then apply the grammar to arrive at the final text as a syntactically structured sequence of words that are read out to become the output of the generator as a whole. The relation between the units of the plan are mapped to syntactic relations. They are organized into constituents and given a linear ordering. The content words are given the grammatically appropriate morphological realizations. Function words ("to," "of," "has," and such) are added as the grammar dictates.

Note that this is a very utilitarian description of what a surface realization component is charged with doing. It leaves out any consideration of how texts are to achieve any significant degree of naturalness or fluency relative to what people do, or how a text might be cast into a particular style to fit its genre, or customized to accommodate its target audience. These are questions that are still without any good answers. Such fluency or stylistic capabilities as current systems have achieved is a matter of hand-crafting and idiosyncratic techniques. Furthermore, these abilities are likely not to be the province of realization components in the sense discussed here; in the past, all of the interesting decisions have always been made beforehand and incorporated into these components' inputs. Whether the boundary of this logically final component in the generation process will move outward to encompass these issues remains to be seen.

B. Producing Speech

The output of a generator is nearly always in the form of a written text. The text may be integrated with a graphic presentation (see, e.g., Feiner McKeown, 1991; Wahlster et al., 1993), or it may include formatting information, as for example in the dynamic construction of pages for the World Wide Web (Dale and Milosavljevic, 1996; Geldof, 1996), but only occasionally in the history of the field has the output of the generator been spoken. It is so rare that it is difficult to say at this point whether a sophisticated treatment would require its own processing component and representations or would be integrated into the surface realization component. Initial work in this area is described in Alter et al. (1997) and Teich et al. (1997)

C. Relation to Linguistic Theory

Practically without exception, every modern realization component is an implementation of one of the recognized grammatical formalisms of theoretical linguistics. It is also not an exaggeration to say that virtually every formalism in the alphabet soup of alternatives that is modern linguistics has been used as the basis of some realizer in some project somewhere.

The grammatical theories provide systems of rules, sets of principles, systems of constraints, and, especially, a rich set of representations, which, along with a lexicon (not a trivial part in today's theories), attempt to define the space of possible texts and text fragments in the target natural language. The designers of the realization components devise ways of interpreting these theoretical constructs and notations into an effective machinery for constructing texts that conform to these systems.

Importantly, all grammars are woefully incomplete when it comes to providing accounts (or even descriptions) of the actual range of texts that people produce. And no generator in the state of the art is going to produce a text that is not explicitly within the competence of the surface grammar it is using. Generation is in a better situation here than comprehension, however. As a constructive discipline, we at least have the capability of extending our grammars whenever we can determine a motive (by the text planner) and a description (in the terms of the grammar) for some new construction. As designers, we can also choose whether to use a construct or not, leaving out everything that is problematic. Comprehension systems, on the other hand, must attempt to read the texts with which they happen to be confronted, and so will inevitably be faced at almost every turn with constructs beyond the competence of their grammar.

D. Chunk Size

One of the side effects of adopting the grammatical formalisms of the theoretical linguistics community is that as a consequence every realization component that has been developed generates a complete sentence at a time (with a few notable exceptions such as Mumble, see later).

Furthermore, this choice of "chunk size" becomes an architectural necessity, not a freely chosen option. These realizers, as implementations of established theories of grammar, must adopt the same scope over linguistic properties as their parent theories do; anything larger or smaller would be undefined and inexpressible.

The requirement that the input to a surface realization component specify the content of an entire sentence at a time has a profound effect on the planners that must produce these specifications. Given a set of propositions to be communicated, the designer of a planner working in this paradigm is more likely to think in terms of a succession of sentences, rather than trying to interleave one proposition within the realization of another (although some of this may be accomplished by aggregation techniques, as described in the next section). Such lock-step treatments can be especially confining when higher-order propositions are to be communicated. For example, the natural expression of such information might be adding an "only" or a hedging phrase inside the sentence that realizes the argument proposition, yet the full-sentence-at-a-time paradigm makes this exceedingly difficult to notice as a possibility, let alone to carry it out.

E. Assembling Versus Navigating

Grammars, and with them the processing architectures of their realization components, fall into two camps.

1. The grammar provides a set of relatively small structural elements and constraints on their combination.
2. The grammar is a single, complex network or descriptive device that defines all the possible output texts in a single abstract structure (or in several structures, one for each major constituent type that it defines: clause, noun phrase, thematic organization, and so on).

When the grammar consists of a set of combinable elements, the task of the realization component is to select from this set and assemble them into a composite representation from which the text is then read out. When the grammar is a single structure, the task is to navigate through the structure, accumulating and refining the basis for the final text along the way and producing it all at once when the process has finished.

Assembly-style systems can produce their texts incrementally by selecting elements for the early parts of the text before the later and, thereby, can have a natural representation of "what has already been said," which is a valuable resource for making decisions about whether to use pronouns and other position-based judgments. Navigation-based systems, because they can see the whole text at once as it emerges, can allow constraints from what will be the later parts of the text to effect realization decisions in earlier parts, but they can find it difficult, even impossible, to make certain position-based judgments.

To get down to cases: among the small-element linguistic formalisms that have been used in generation we have conventional rewrite rule systems, Segment Grammar, and Tree Adjoining Grammar (TAG). Among the single-structure formalisms we have Systemic Grammar and any theory that uses feature structures: for example, HPSG and LFG. We will discuss just a very few of the surface realization components that have embodied these different approaches: Mumble (TAG) and production systems (generically) in the assembly camp; then Penman (Systemic Grammar) and FUF (feature structures) in the navigation camp (references are provided where the systems are discussed).

Before going into the particulars of grammar-based surface realization components, it is worthwhile to note that many realizers do not incorporate grammars at all, especially those written casually for expert systems. Instead, these use a combination of direct replacement and progressive refinement not unlike that one might expect to see in a good mail–merge program. They start with templates or schemas that are a mixture of words and embedded references to speaker–internal data structures or function calls to specialists. The templates are then interpreted, making the substitutions or executing the functions and replacing them with the resulting words, often recursively down to finer and finer levels of detail.

This is an effective technique, especially for a competent programmer, but it has sharp limitations. Experience has shown that the pragmatic limit in grammatical complexity is use of a subordinated clause; for example, the effort required to adapt a declarative sentence to be a relative clause, or to form "Harry wants to go home" from a composite internal expression such as wants [Harry, go-home(Harry)]. At that point the complexity is too much for realization components without proper grammars.

F. Mumble

Direct replacement, as simple as it is, is a good starting point for understanding how assembly-based realization components tend to work. In the original version of Mumble (McDonald, 1980, 1983), the text plan for an utterance would consist of a sequence of structures in the speaker's internal representation. As these were traversed top down and from left to right (relative to the eventual text), their elements were incrementally replaced with syntactic trees that had words and the next level of internal structures at their leaves; these syntactic trees, in turn, were traversed in the standard prenext order. Every time a structure in the speaker's representation was encountered, its set of possible realizations as given in its dictionary entry was consulted (occasionally, with the aid of some specialist-driven local text planning), a choice was made, and that bit of syntactic structure or wording substituted for the structure and the traversal continued. Words were emitted as they were encountered, and the utterance emerged in its natural order.

As the need to control the interaction between the planning of the discourse and its realization increased, Mumble incorporated more representational levels between the speaker and the realizer. The overall architecture remained the same, in the sense that the intermediate representations provided a stream for a traversing interpreter and the next level of representation was assembled from it by selecting from a set of alternatives according to the constraints of the context. Individual decisions became simpler because less was decided at each step.

The current version of Mumble, Mumble-86 (Meteer et al., 1987; McDonald and Pustejovsky, 1985), is reduced to an assembly and traversal machine. It assembles a surface structure of so-called elementary trees from its Tree Adjoining Grammar, for which the choice of trees is dictated by the representational level logically just above it. This level is a TAG derivation structure that indicates what trees to use and how they are to be attached to each other. Above that level is Meteer's Text Structure representation, which consists of a loose tree of carefully typed semantic constituents, each of which is guaranteed to have some realization in terms of a TAG that is grammatically consistent with its context in the Text Structure tree. The traversal of that tree is what constructs the derivation structure (referred to as a *linguistic specification*).

The assembly of the Text Structure is the job of a text planner such as Meteer's Spokesman (1991, 1992) or the planners being developed by Granville (1990) or McDonald (1992). These tend to work by instantiating a small, initial abstract semantic tree and then adding elements to it one by one, respecting the constraints that the various positions within the growing Text Structure tree impose. When stylistic criteria determine that an initial region of Text Structure should not be elaborated further, that region is traversed to assemble the derivation structure that, in turn, will drive Mumble-86.

Taken as a whole, this constitutes a system that selects from a repertoire of small elements at successive levels of representation, assembles a structure from them, and then traverses that structure to produce the next level of representation down. The work of surface realization per se is somewhat diluted across several levels, not just Mumble-86, since Mumble's surface structure, the succession of derivation trees that define it, and the Text Structure operate as a unified system.

G. Phrase Structure Rules

A substantial number of people work with assembly-based surface realization components that deploy conventional phrase structure grammar productions as their elements. These can profitably be viewed as hierarchical decomposition assembly systems in that they begin with an expression that describes the meaning of the whole utterance (invariably a sentence), and then incrementally decompose the expression (which is usually a conventional logical form) by selecting productions that will convey some portion of the input "semantics." Annotating the nonterminals on the righthand sides of each selected production is a pointer to what portion of the semantics of the whole production they are each responsible for and will be realized when that production is eventually rewritten. In this way the input is broken down into smaller and smaller elements as successive productions are selected and knit into the leaves of the growing derivation tree. The process terminates when all of the nonterminals have been rewritten as a consistent set of terminal words, which are then read out.

These systems truly are parsing algorithms run in reverse. They start with a proposition-based interpretation and search for a derivation tree (a parse) of rewrite rules that will take that interpretation to a grammatical sequence of words. Experiments with all of the standard natural language parsing algorithms can be found in the literature: top-down, bottom-up, Earley's algorithm, left-corner, and LL (see Chap. 4 for a discussion of these approaches in parsing; see Shieber et al., 1990; van Nord, 1990; Lager and Black, 1994; Harune et al., 1996; Wilcock 1998; and Becker, 1998 for their use in generation).

The greatest potential difficulty with realization components of this design is that they actually do have a hard boundary between realization and earlier macro- and microplanning activities, and they tend to strongly assume that all of the linguistic knowledge of the system will live inside the realizer. The results of the earlier processes must be summarized in the logical form that constitutes the realizer's input, and if this formula turns out to be inexpressible then there is no recourse.

It is no accident then that these systems are most popular in machine-translation applications (see Chap. 13). The depth of understanding of most of these systems is comparatively shallow, so the output of their parsing and comprehension processes can be comfortably expressed by a logical form. The fact that the source was a natural language text also helps with the potential problem of expressibility, because given a comprehensive transfer lexicon, the coherence of the source greatly increases the likelihood that the logical form will have a corresponding target text and that it is just a matter of searching through the space of possible derivations to find it.

H. Systemic Grammars

Understanding and representing the context into which the elements of an utterance are fit and the role of the context in their selection is a central part of the development of a grammar. It is especially important when the perspective that the grammarian takes is a functional, rather than a structural one—the viewpoint adopted in

Systemic Grammar. A structural perspective on language emphasizes the elements out of which language is built (constituents, lexemes, prosidics, and so on). A functional perspective turns this on its head and asks: What is the spectrum of alternative purposes that a text can serve (its "communicative potential")? Does it introduce a new object that will be the topic of the rest of the discourse? Is it reinforcing that object's prominence? Is it shifting the focus to somewhere else? Does it question? Enjoin? Persuade? The multitude of goals that a text and its elements can serve can provide the basis for a paradigmatic (alternative-based) rather than a structural (form-based) view of language.

The Systemic Functional Grammar (SFG) view of language originated in the early work of Halliday (1967, 1985) and, today, has a wide following. It has always been a natural choice for work in language generation (Davey's Proteus system was based on it) because much of what a generator must do is to choose among the alternative constructions that the language provides based on the context and the purpose they are to serve—something that a systemic grammar represents directly.

A systemic grammar is written as a specialized kind of decision tree: "if this choice is made, then this set of alternatives becomes relevant; if a different choice is made, those alternatives can be ignored, but this other set must now be addressed." Sets of (typically disjunctive) alternatives are grouped into systems (hence, *systemic grammar*), and connected by links from the prior choice(s) that made them relevant to the other systems that, in turn, they make relevant. These systems are described in a natural and compelling graphic notation of vertical bars listing each system and lines connecting them to other systems. (The Nigel systemic grammar, developed at ISI (Mann and Matthiessen 1983; Matthienssen 1983), required an entire office wall for its presentation using this notation.)

In a computational treatment of SFG for language generation, each system of alternative choices has an associated decision criteria. In the early stages of development, these criteria are often left to human intervention to exercise the grammar and test the range of constructions it can motivate (e.g., Fawcett, 1981). In the work at ISI, this evolved into what was called *inquiry semantics*, where each system had an associated set of predicates that would test the situation in the speaker's model and makes its choices accordingly. This makes it, in effect, a pull system for surface realization; something that in other publications has been called *grammar-driven* control as opposed to the *message-driven* approach of a system such as Mumble (McDonald et al 1987).

As the Nigel grammar grew into the Penman system (Penman Group, 1989), and gained a wide following in the late 1980s and early 1990s, the control of the decision-making and the data that fed it moved into the grammar's input specification and into the speaker's knowledge base. At the heart of the knowledge base—the taxonomic lattice that categorizes all of the types of objects that the speaker could talk about and defines their basic properties—a so-called Upper Model was developed (Bateman, 1997; Bateman et al., 1995). This set of categories and properties was defined in such a way that it was able to provide the answers needed to navigate through the system network. Objects in application knowledge bases built in terms of this Upper Model (by specializing its categories) are assured an interpretation in terms of the predicates that the systemic grammar needs because these are provided implicitly through the location of the objects in the taxonomy.

Mechanically, the process of generating a text using a systemic grammar consists of walking through the set of systems from the initial choice (which for a speech act might be whether it constitutes a statement, a question, or a command), through to its leaves, following several simultaneous paths through the system network until it has been completely traversed. Several parallel paths are followed because, in the analyses adopted by systemicists, the final shape of a text is dictated by three independent kinds of information: *experiential*, focusing on content; *interpersonal*, focusing on the interaction and stance toward the audience; and *textual*, focusing on form and stylistics.

As the network is traversed, a set of features that describe the text are accumulated. These may be used to "preselect" some of the options at lower strata in the accumulating text, as for example, when the structure of an embedded clause is determined by the traversal of the network that determines the functional organization of its parental clause. The features describing the subordinate's function are passed through to what will likely be a recursive instantiation of the network that was traversed to form the parent, and they serve to fix the selection in key systems, for example, dictating that the clause should appear without an actor (e.g., as a prepositionally marked gerund: "you blocked me by taking the corner opposite mine").

The actual text takes shape by projecting the lexical realizations of the elements of the input specification onto selected positions in a large grid of possible positions as dictated by the features selected from the network. The words may be given by the final stages of the system network (as expressed by the characterization of "lexis as most delicate grammar") or as part of the input specification.

I. Functional Unification Grammar

A functional, or purpose-oriented, perspective in a grammar is largely a matter of the grammar's content, not its architecture. What sets functional approaches to realization apart from structural approaches is the choice of terminology and of distinctions, the indirect relation to syntactic surface structure and, when embedded in a realization component, the nature of its interface to the earlier text planning components. Functional realizers ask questions about purpose, not about contents. Just as a functional perspective can be "implemented" in a system network, it can be implemented in an annotated TAG (Yang et al., 1991) or, in what we will turn to now, in a unification grammar.

A unification grammar is also traversed, but this is less obvious because the traversal is done by the built-in unification process and is not something that its developers actively consider. (Except for reasons of efficiency: the early systems were notoriously slow because nondeterminism led to a vast amount of backtracking; as machines have become faster and the algorithms have been improved, this is no longer a problem.)

The term *unification grammar* emphasizes the realization mechanism used in this technique, namely merging the component's input with the grammar to produce a fully specified, functionally annotated surface structure from which the words of the text are then read out. The merging is done using a particular form of unification; a thorough introduction can be found in McKeown (1985).

To be merged with the grammar, the input must be represented in the same terms. This means that it will already have a more linguistic character than, say, the input to Meteer's Text Structure, which is an example of a counterpart in the progressive refinement paradigm that operates at the same level of representation.

Unification is not the primary design element in these systems, however. It just happened to be the control paradigm that was in vogue when the innovative data structure of these grammars—feature structures—was introduced by linguists as a reaction against the pure-phrase structure approaches of the time (the late 1970s). Feature structures (FS) are much looser formalism than unadorned phrase structure; they consist of sets of attribute–value pairs. A typical FS will incorporate information from (at least) three levels simultaneously: meaning, (surface) form, and lexical identities. It allows general principles of linguistic structure to be stated more freely and with greater attention to the interaction between these levels than had been possible before.

The adaption of feature–structure-based grammars to generation was begun by Kay (1984), who developed the idea of focusing on functional relations in these systems—functional in the same sense as that employed in systemic grammar, with the same attendant appeal to persons working in generation who wanted to experiment with the feature–structure notation.

Kay's notion of a *functional* unification grammar (FUG) was first deployed by Appelt (1985), and then adopted by McKeown. McKeown's students, particularly Elhadad, made the greatest strides in making the formalism efficient. He developed the FUF system, which is now widely used (Elhadad, 1991; Elhadad and Robin, 1996). Elhadad also took the step of explicitly adopting the grammatical analysis and point of view of systemic grammarians, demonstrating quite effectively that grammars and the representations that embody them are separate aspects of system design.

6. CONCLUSIONS

This chapter has covered the basic issues and perspectives that have governed work on natural language generation. With the benefit of hindsight, it has tried to identify the axes that distinguish the different approaches that have been taken during the last thirty years—does the speaker intentionally "push" directives to the text planner, or does the planner "pull" data out of a passive database? Does surface realization consist of "assembling" a set of components or of "navigating" through one large structure?

Given this past, what can we say about the future? One thing we can be reasonably sure of is that there will be relatively little work done on surface realization. People working on speech or doing computational psycholinguistics may see the need for new architectures at this level, and the advent of a new style of linguistic theory may prompt someone to apply it to generation; but most groups will elect to see realization as a solved problem—a complete module that they can ftp from a collaborating site.

By that same token, the linguistic sophistication and ready availability of the mature realizers (Penman, FUF, and Mumble) will mean that the field will no longer sustain abstract work in text planning; all planners will have to actually produce text, preferably pages of it, and that text should be of high quality. Toy output that neglects to properly use pronouns or is redundant and awkward will no longer be acceptable.

The most important scientific achievement to look toward in the course of the next 10 years is the emergence of a coherent consensus architecture for the presently muddled "middle-ground" of microplanning. Between the point at which generation begins, where we have a strong working knowledge of how to fashion and deploy schemas, plan operators, and the like, to select what is to be said and give it a coarse organization, and the point at which generation ends, where we have sophisticated, off-the-shelf surface realization components that one can use with only minimal personal knowledge of linguistics, we presently have a grab bag of phenomena that no two projects deal with in the same way (if they handle them at all).

In this middle ground lies the problem of where to use pronouns and other such reduced types of "subsequent reference"; the problem of how to select the best words to use ("lexical choice"), and to pick among alternative paraphrases; and the problem of how to collapse the set of propositions that the planner selects, each of which might be its own sentence if generated individually, into fluent complex sentences that are free of redundancy and fit the system's stylistic goals (presently described as "aggregation").

At this point, about all that is held in common in the community is the names of these problems and what their effects are in the final texts. That at least provides a common ground comparing the proposals and systems that have emerged, but the actual alternatives in the literature tend to be so far apart in the particulars of their treatments that there are few possibilities for one group to build on the results of another.

To take one example, it is entirely possible that aggregation—the present term of art in generation for how to achieve what others call *cohesion* (Halliday and Hasan, 1976) or just *fluency*—is not a coherent notion. Consider that for aggregation to occur there must be separate, independent things to be aggregated. It might turn out that this is an artifact of the architecture of today's popular text planners and not at all a natural kind (i.e., something that is handled with the same procedures and at the same points in the processing for all the different instances of it that we see in real texts).

Whatever the outcome of such questions, we can be sure that they will be pursued vigorously by an ever burgeoning number of people. The special-interest group on generation (SIGGEN) has over 400 members, the largest of all the special interest groups under the umbrella of the Association for Computational Linguistics (ACL). The most recent international workshop on generation, held in August 1998, was also the largest with about 100 persons, most of whom were authors of papers given at the meeting. And the field is international in scope, with major research sites from Australia to Israel.

For someone new to the field, probably the single best place to begin to get a feel for what is going on today is to obtain the proceedings of the 1998 workshop, which is available from the ACL. To keep up with this field, one should look at SIGGEN's web site or join the mailing list; the web site can be reached via *www.acl-web.org*. This site includes pointers to many other NLG sites and resources.

The Turing Test for language generation is to produce a text that is indistinguishable from that which a person might produce in the same context. This has been achieved at least once for scientific papers (Gabriel, 1986), where three machine generated paragraphs describing a procedure were woven seamlessly into the rest of the paper. And there are many mundane reports and web pages today that are mixtures of generation from first principles (from a semantic model) and frozen templates selected by people that are hard to tell from the real thing.

Before this will become the norm for what a machine can produce, there is much challenging work remaining to be accomplished; this will keep those of us who work in the field engaged for years and decades to come. Whether the breakthroughs will come from traditional grant-funded research, or from the studios and basements of game makers and entrepreneurs, is impossible to say. Whatever the future, this is the best part of the natural language problem in which to work.

REFERENCES

Adorni G, M Zock, eds. Trends in Natural Language Generation, pp. 88–105.

Appelt D (1985). Planning English Sentences. Cambridge University Press.

Barwise J, J Perry (1983). Situations and Attitudes. MIT Press, Cambridge, MA.

Bateman JA (1997). Enabling technology for multilingual natural language: the KPML development environment. J Nat Language Eng 3(1), pp. 15–55.

Bateman JA, R Henschel, F Rinaldi (1995). Generalized upper model 2.1: documentation. Technical Report, GMD/Institute für Integrierte Publikations- und Informationssysteme, Darmstadt, Germany.

Becker J (1975). The phrasal lexicon, Proc. TINLAP-I, ACM, 1975, 60–64; also available as BBN Report 3081.

Becker T (1998). Fully lexicalized head-driven syntactic generation. In Proceedings of the 9th International Workshop on Natural Language Generation (available from ACL), pp. 208–217.

Bourbeau L, D Carcagno, E Goldberg, R Kittredge, A Polguère (1990). Bilingual generation of weather forcasts in an operations environment. COLING, Helsinki.

Brown R (1974). Use of multiple-body interrupts in discourse generation, Bachelor's dissertation, MIT, Cambridge, MA.

Clippinger J (1977). Meaning and Discourse: A Computer model of Psychoanalytic Speech and Cognition. Baltimore: Johns Hopkins Press.

Cohen PR, HJ Levesque (1985). Speech acts and rationality. Proceedings of the 23rd Conference of the ACL, Chicago, pp. 49–59.

Cohen PR, C Perrault (1979). Elements of a plan-based theory of speech acts. Cognitive Sci (3):177–212.

Conklin EJ (1983). Data-driven indelible planning of discourse generation using salience. PhD dissertation, University of Massachusetts at Amherst; available from the Department of Computer and Information Science as Technical Report 83–13.

Conklin EJ, D McDonald (1982). Salience: The key to selection in deep generation. **ACL-82**, June 16–18, University of Toronto, pp. 129–135.

Dale R (1988). Generating referring expressions in a domain of objects and processes. PhD dissertation, University of Edinburgh.

Dale R (1992). Generating referring expressions. MIT Press, Cambridge, MA.

Dale R (1995). Referring expression generation: problems introduced by one-anaphora. In: Hoeppner, Horacek, eds. Principles of Natural Language Generation, 1995, pp. 40–46.

Dale R, C Mellish, M Zock (1990). Current Research in Natural Language Generation. Boston, MA: Academic Press.

Dale R, E Hovy, D Rösner, O Stock, eds. (1992). Aspects of automated natural language generation. The 6th International Workshop of Natural Language Generation, Trento; Lecture Notes in Artif Intell 587, Springer–Verlag.

Dale R, M Milosavljevic (1996). Authoring on demand: natural language generation of hypermedia documents. Proceedings First Australian Document Computing Symposium (ADCS'96), Melbourne, Australia.

Dalianis H (1996). Concise natural language generation from formal specifications. PhD dissertation, Royal Institute of Technology, Sweden; Report Series No. 96-008, ISSN 1101–8526.

Dalianis H, E Hovy (1993). Aggregation in natural language generation. Proceedings of the 4th European Workshop on Natural Language Generation, Pisa, Italy, April 1993. Lecture Notes in Artif Intell 1036, Springer, 1996

Davey A (1978). Discourse Production. Edinburgh University Press.

DiMarco C, G Hirst (1993). A computational theory of goal-directed style in syntax. Comput Linguist 19:451–499.

Elhadad M (1991). FUF: the universal unifier user manual (v5). Technical Report CUCS-038-91, Department of Computer Science, Columbia University, New York.

Elhadad M, J Robin (1996). An overview of SURGE: a reusable comprehensive syntactic realization component. Technical Report 96-03, Department of Mathematics and Computer Science, Ben Gurion University, Beer Sheva, Israel.

Fawcett R (1981). Generating a sentence in systemic functional grammar. In Halliday and Martin, eds. Readings in Systemic Linguistics. London: Batsford.

Feiner S, K McKeown (1991). Automating the generation of coordinated multimedia explanations. IEEE Computer, 24(10):33–40.

Gabriel RP (1981). An organization of programs in fluid domains. PhD dissertation, Stanford University. Available as Stanford Artificial Intelligence Memo 342 (STAN-CA-81-856), 1981.

Gabriel RP (1986). Deliberate writing. In: McDonald D, Bolc L, eds. Natural Language Generating Systems. Berlin: Springer-Verlag, 1986, pp. 1–46.

Geldof S (1996). Hyper-text generation from databases on the Internet. Proceedings of the 2nd International Workshop on Applications of Natural Language to Information Systems, NLDB, 102–114, Amsterdam: IOS Press.

Green S, C DiMarco (1996). Stylistic decision-making in natural language generation. In: Trends in Natural Language Generation: An Artificial Intelligence Perspective. G Adorni, M Zock, eds. Springer-Verlag Lecture Notes in Artif Intell 1036.

Goldman N (1975). Conceptual generation. In: Schank, ed. Conceptual Information Processing. New York: Elsevier.

Granville R (1990). The role of underlying structure in text generation. Proceedings of the 5th International Workshop on Natural Language Generation, Dawson, Pennsylvania.

Halliday MAK (1967). Notes on transitivity and theme in English, Parts 1–3. J Linguist 3: 37–81; 199–244; 179–215.

Halliday MAK (1985). An Introduction to Functional Grammar. London: Edward Arnold.

Halliday MAK, R Hasan (1976). Cohesion in English. London: Longman.

Haruno M, Y Den, Y Matsumoto (1996). A chart-based semantic head drive generation algorithm. In: Adorni and Zock, 1996, pp. 300–313.

Hoeppner E, H Horacek, eds. (1995). Principles of Natural Language Generation, Papers from a Dagstuhl-Seminar. Bericht Nr. SI-12, Institute für Informatik, Gerhard-Mercator-Universitätp-GH, Duisburg, ISSN: 0942-4164.

Hovy E (1988a). Generating Natural Language Under Pragmatic Constraints. Hillsdale, NJ: Lawrence Erlbaum.

Hovy E (1988b). Generating language with a phrasal lexicon. In: McDonald Bolc, eds. 1998, pp. 353–384.

Hovy E (1990). Unsolved issues in paragraph planning. In: Dale et al., 1990, pp. 17–46.

Hovy E (1991). Approaches to the planning of coherent text. In: Paris et al., 1991, pp. 83–102.

Huang X, A Fiedler (1996). Paraphrasing and aggregating argumentative text using text structure. Proceedings of the 8th International Workshop on Natural Language Generation, Herstmonceux Castle, Sussex, June 12–15, 1996, pp. 21–30.

Lager T, WJ Black (1994). Bidirectional incremental generation and analysis with categorial grammar and indexed quasi-logical form. Proceedings of the 7th International Workshop on Natural Language Generation, Kennybunkport, Maine, pp. 225–228.

Kay M (1984). Functional unification grammar: A formalism for machine translation. Proceedings of Coling-84, pp. 75–78, ACL.

Kempen G (1987). Natural Language Generation. Dordrecht: Martinus Nijhoff.

Kukich K (1988). Fluency in natural language reports. In: McDonald and Bolc, 1988, pp. 280–312.

Levelt WJM (1989). Speaking. Cambridge, MA: MIT Press.

Mann WC, CMIM Matthiessen (1983). Nigel: a systemic grammar for text generation. Technical Report RR-83-105, Information Sciences Institute, Marina del Rey. Also in Benson Greaves, eds. You and Your Language. Pergamon Press 1984.

Mann WC, SA Thompston (1987). Rhetorical structure theory: a theory of text organization. In: L Polanyi, ed. The Structure of Discourse. Ablex.

Mann WC, SA Thompston (1988). Rhetorical structure theory: towards a functional theory of text organization. Text 8:243–281.

Marcu D (1997). From local to global coherence: a bottom-up approach to text planning. Proceedings of AAAI-97, pp. 629–635.

Matthiessen CMIM (1983). Systemic grammar in computation: the Nigel case. Proceedings of the First Annual Conference of the European Chapter of the Association for Computational Linguistics, Pisa.

McDonald DD (1983). Natural language generation as a computation problem: an introduction. In: Brady Berwick, eds. Computational Models of Discourse. Cambridge, MA: MIT Press, pp. 209–266.

McDonald DD (1992). Type-driven suppression of redundancy in the generation of inference-rich reports. In: Dale R, et al., eds. Aspects of Automated Natural Language Generation. Berlin: Springer-Verlag, 1992, pp. 73–88.

McDonald D, L Bolç, eds. (1988). Natural Language Generation Systems. Berlin: Springer-Verlag.

McDonald D, M Meteer (1988). From water to wine: generating natural language text from today's application programs. Proceedings Second Conference on Applied Natural Language Processing (ACL), February 9–12, Austin, Texas, pp. 41–48.

McDonald D, J Pustejovsky (1985). Description-directed natural language generation. Proceedings of IJCAI-85, UCLA, August, 18–23, pp. 799–805.

McDonald D, M (Meteer) Vaughan, J Pustejovsky (1987). Factors Contributing to Efficiency in Natural Language Generation. In Kempen, 1987, pp. 159–182.

McKeown KR (1995). Text Generation. Cambridge, UK: Cambridge University Press.

McKeown KR, M Elhadad, Y Fukumoto, J Lim, C Lombardi, J Robin, F Smadja (1990). Natural Language Generation in COMET. In: Dale R, et al., Current Research in Natural Language Generation. Boston, MA: Academic Press, 1990, pp. 103–140.

Meteer M (1991). Bridging the generation gap between text planning and linguistic realization. Comput Intell 7:296–304.

Meteer M (1992). Expressibility and the Problem of Efficient Text Planning. London: Pinter.

Meteer M, D McDonald, S Anderson, D Forster, L Gay, A Huettner, P Sibun (1987). Mumble-86: design and implementation. Technical Report 87–87, Department of Computer Information Science, University of Massachusetts at Amherst.

Moore JD (1995). Participating in Explanatory Dialogues. Cambridge, MA: MIT Press.

Moore JD, C Paris (1993). Planning texts for advisory dialogs: capturing intentional and rhetorical information. Computat Linguist 19:651–694.

Moore JD, WR Swartout (1991). A reactive approach to explanation: taking the user's feedback into account. In: C Paris, et al., Natural Language Generation in Artificial Intelligence and Computational Linguistics. Boston; Kluwer Academic, 1991, pp. 3–48.
Panaget F (1994). Using a textrual representational level component in the context of discourse or dialog generation. In: Proceedings of the 8th International Workshop on Natural Language Generation, Kennebunkport, Maine, June 21–24, 1994, pp. 127–136.
Paris CL (1991). Generation and explanation: building an explanation facility for the explainable expert systems framework. In: C Paris, et al., Natural Language Generation in Artificial Intelligence and Computational Linguistics. Boston: Kluwer Academic.
Paris CL (1993). User Modeling in Text Generation. Pinter: London.
Paris CL, WR Swartout, WC Mann, eds. (1991). Natural Language Generation in Artificial Intelligence and Computational Linguistics. Boston: Kluwer Academic.
Pattabhiraman T, N Cercone (1991). Special Issue: Natural Language Generation, Computat Intell [7] 4, Nov 1991.
Penman Natural Language Group (1989). The Penman documentation. USC Information Sciences Institute.
Reiter E (1990b). Generating descriptions that exploit a user's domain knowledge. In: RC Dale, et al., Current Research in Natural Language Generation. Boston: Academic Press, 1990, pp. 257–285.
Robin J (1993). A revision-based generation architecture for reporting facts in their historical context. In: Horacek and Zock, 1993, pp. 238–268.
Robin J (1996). Evaluating the portability of revision rules for incremental summary generation. Proceedings of the 34th Annual Meeting of the ACL, June 24–27, Santa Cruz, pp. 205–214.
Rubinoff R (1986). Adapting Mumble: experience with natural language generation. Technical Report, University of Pennsylvania, MS-CIS-86-32.
Sacerdoti E (1977). A Structure for Plans and Behavior. Amsterdam: North-Holland.
Scott P, M Steedman (1993). Generating contextually appropriate intonation. Proceedings, Sixth Conference of the European Chapter of the ACL, Utrecht, 1993, pp 332–340.
Shieber S, G van Nort, F Pereira, R Moore (1990). Semantic-head-driven generation. Computat Linguist 16(1):30–42, March.
Smadja F (1991). Retrieving collocational knowledge from textual corpora, an application: language generation. PhD dissertation, Columbia University, New York.
Smadja F, K McKeown (1991). Using collocations for language generation. In: T Pattabhiraman, N Circone, 1991, pp. 229–239.
Teich EEH, B Grote, J Bateman (1997). From communicative context to speech: integrating dialog processing, speech production and natural language generation. Speech Commun 21:73–99.
von Nord G (1990). An overview of head-driven bottom-up generation. In: Dale, Mellish, Zock, eds. Current Research in Natural Language Generation, Academic Press, Boston.
Wahlster W (1988). One word says more than a thousand pictures: on the automatic verbalization of the results of image sequences analysis systems. Bericht Nr. 25, DFKI, Saarbrüchen.
Wahlster W, E Andre, W Finkler, H-J Profitlich, T Rist (1993). Plan-based integration of natural language and graphics generation. Artific Intell 63:387–427.
Wilcock C (1998). Approaches to surface realization with HPSG. In: IWG-98, pp. 218–227.
Wilensky R (1976). Using plans to understand natural language. Proceedings of the Annual Meeting of the Association for Computing Machinery, Houston.
Winograd T (1972). Understanding Natural Language. New York: Academic Press.
Yang G, KF McCoy, K Vijay-Shanker (1991). From functional specification to syntactic structure: systemic grammar and tree-adjoining grammar, Comput Intell 7:207–219.

8

Intelligent Writing Assistance

GEORGE E. HEIDORN

Microsoft Research, Redmond, Washington

1. INTRODUCTION

This chapter discusses intelligent writing assistance, in particular that which is known as *text critiquing*. Text critiquing is done by a computer system that analyzes text for errors in grammar or style, and often such a system is referred to as a grammar checker. The term, *grammar checking*, narrowly defined means checking text for errors that a grammar book (e.g. [1]), would discuss, such as disagreement in number between a subject noun and its verb. *Style checking* refers to checking text for errors that a book on good writing style (e.g. [2]), would discuss, such as overuse of the passive voice. In practice, the term *grammar checker* is often used loosely to refer to systems that do grammar checking or style checking, or both. One could imagine other kinds of systems that give intelligent assistance to writers, such as for outline construction or plot construction, or even for automatically generating text. However, such systems are not readily available or in widespread use and, therefore, are beyond the scope of this chapter.

Grammar checkers, using the term in its widest sense, have been around for almost 20 years. Writer's Workbench [3] was probably the earliest to be widely used, on Unix systems. Then there were smaller systems with many similarities to Writer's Workbench that became available for the IBM PC in the early to mid-1980s. None of these systems did much sophisticated processing of the text above the level of words. Typically, errors were discovered by some sort of string matching. Then in the mid- to late 1980s, systems became available that could do some amount of parsing of the text, thereby doing a better job of real grammar checking. At IBM Research, the Epistle system [4] and later the Critique system [5] were well publicized with papers and demonstrations, and Critique even became part of a product for IBM mainframes in 1989 [6]. At about the same time, CorrecText Grammar Correction System [7] from Houghton–Mifflin (now from Inso Corporation) became a product for the PC, and the Grammatik Grammar-Checker [8] from Reference Software International (now part of Novell, Inc.) also became available. Not much has been published about the inner workings of the latter two, so it is difficult to know just how much parsing they do. However, from the kinds of critique that they present to the user, it is apparent that they do some amount of parsing. (In fact, Version 6 of Grammatik displays a form of sentence structure on request.) From the many published papers about Critique, it is well known that that system does a full syntactic analysis of every input sentence.

There is now a new kid on the grammar-checking block. Microsoft Word 97 [9], one of the most popular word processors for the PC, was released with its own built-in grammar checker that is based on work that was started by the natural language processing (NLP) group at Microsoft Research in 1992. This system has as its heart a full-blown, multipurpose natural language processor, which produces a syntactic parse structure and, when desired, a logical form, for each sentence. The information in these data structures can then be further analyzed by rules that check for violations of grammar and style. This grammar-checking system will be described in detail in this chapter as an example of a real-world NLP application that provides intelligent writing assistance.

This chapter discusses both the research aspects and the product development aspects of this work from the viewpoint of a research member of the team. The chapter begins with a description of the research NLP system that is behind the Microsoft grammar checker, followed by a section that describes how grammar checking is added to that NLP system. Then there are three sections discussing the productizing of this grammar checker for Microsoft Word 97. The first describes the features of this grammar checker in the product; the second discusses how our NLP system was hooked up to the word processor; and the third describes the techniques and tools used for testing the grammar checker. The concluding section makes some remarks about future work and about the value of product-oriented research for natural language processing.

2. THE MICROSOFT NLP SYSTEM

The NLP system that is behind the Microsoft grammar checker is a full-fledged natural language processing system that is also intended to be used for many other applications. It consists of a programming language and a runtime environment that are both specially tailored to the needs of an NLP system. The programming language, which is called G, has basically the syntactic appearance of the C language [10], but it gives special notational support to attribute–value data structures, called records, and provides an additional programming construct, called rules [11]. The runtime system, which is usually referred to as NLPWin, is a Microsoft Windows application that is written mostly in C and provides a grammar development environment and the functions needed to do natural language processing. That part of the processing that is written in G, such as the English analysis grammar, is translated into C by a program called Gtran, and then it is compiled and linked into the NLPWin executable.

Figure 1 gives an overall picture of the Microsoft NLP system. Actually, this is a rather generic picture for natural language processing, but it does serve to put a context on the rest of this discussion. As can be seen therein, the system is intended to do both analysis and generation of natural language text, to and from a meaning representation. At the current time, analysis work is being done for seven different languages (Chinese, English, French, German, Japanese, Korean, and Spanish), but generation is being done just for English. The generation work is nowhere near as far along as the analysis work, even for English, so no more will be said about it in this chapter.

Figure 2 gives a more detailed picture of the analysis portion of the Microsoft NLP system. The processing takes place in six stages. The first stage of processing is lexical, where the input text is segmented into individual tokens, which are primarily words and punctuation marks. The main processing at this stage does morphological analysis of words [12] and lookup of words in the dictionary. However, another important part of the processing at this stage deals with multiword items in the text. Some of these, such as multiword prepositions (e.g., "in back of"), are referred to as MWEs (Multi-Word Entries) and are stored as their own entries in the dictionary. Some others, such as dates and places, are called Factoids and are processed by rules written in G that are similar to, but simpler than, the main syntax rules. Additionally, there are Captoids, which are strings of mostly capitalized words, that are handled in an intelligent manner by a small set of Factoid-like rules. The data in the dictionary used by this system comes primarily from the online versions of the *Longman Dictionary of Contemporary English* [13] and *The American Heritage Dictionary* [14]. However, it is also augmented with additional information that has been entered manually or as the result of further processing [15]. The output from lexical processing is stored as a list of part-of-speech records, which are collections of attribute–value pairs that describe the words in detail.

The second stage of processing is called the syntactic sketch, and corresponds to what is typically called parsing. The input to this stage is the set of part-of-speech

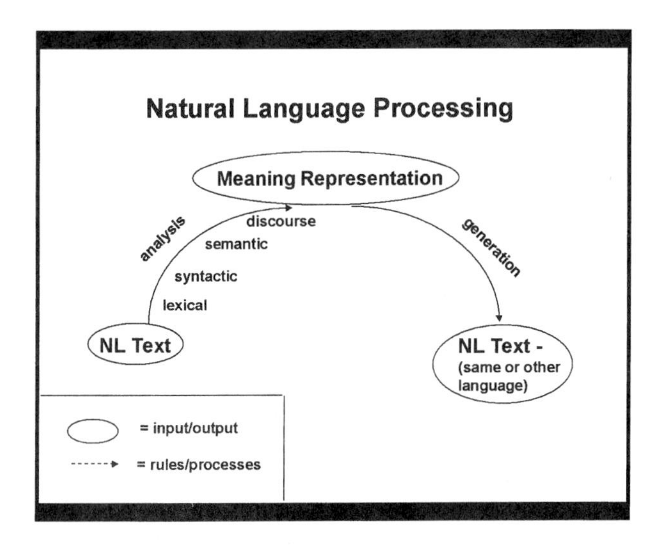

Fig. 1 The Microsoft NLP system.

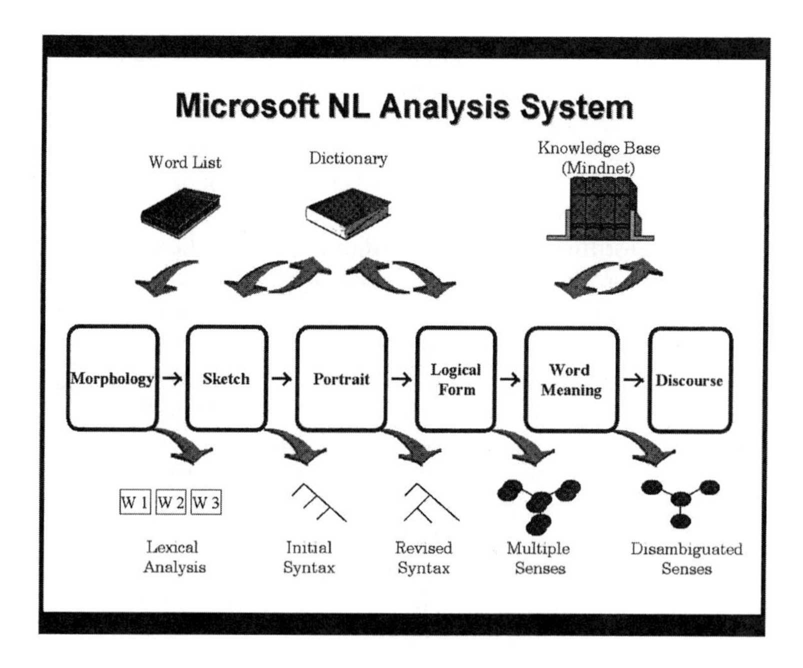

Fig. 2 The analysis portion of the Microsoft NLP system.

records produced by lexical processing, and the output is just an expanded set of records, sometimes called the chart in the NLP literature, which provide (primarily) syntactic descriptions of the various segments of the input text. These additional records are created as the result of applying augmented phrase structure grammar (APSG) rules [16] that have been written in the G programming language. There are approximately 125 rules, most of which are binary, with the remaining few being unary. In addition to implicitly producing a derivation tree based on the rules that are applied, they also produce an explicit computed tree that has a much flatter structure, with somewhat of a dependency flavor [17]. The processing at this stage is strictly syntactic and makes use of no information that would normally be considered semantic. However, the rules are quite complex, and in many places in these rules, particular word stems (which are called lemmas in this system) are referred to. The processing algorithm for this stage is basically just a bottom-up, multipath, chart-parsing algorithm, but it also makes use of both probabilistic information and a heuristic-scoring scheme to guide the process to a quicker and better termination point [18]. When the rules fail to produce a parse for an input text, a fitting procedure is used to try to put together the best pieces of the analysis that are in the chart to produce a fitted parse for the text. When the rules produce more than one parse for an input text, the alternatives are ordered according to the heuristic scores (called Pod scores in this system). The output of the syntactic sketch can be viewed as one or more trees, on which each node is actually a segment record consisting of attribute–value pairs to fully describe the text that it covers.

The third stage is called the syntactic portrait, because it is a refinement of the syntactic sketch. The purpose of this processing is to produce more reasonable attachments for some modifiers, such as prepositional phrases and relative clauses, that are merely attached to their closest possible modificand as a simplification in the sketch. We usually informally refer to this stage as reattachment, which we then further separate into syntactic reattachment and semantic reattachment. The first of these deals with those situations that can be handled using only syntactic information, such as moving an obvious prepositional phrase of time to clause level. The second handles those that require semantic information, such as knowing that it is more likely that a telescope is an instrument for seeing a bird, rather than being a part of the bird in the sentence, "I saw a bird with a telescope." Semantic reattachment makes use of semantic relations that are produced automatically by analyzing the text of definitions and example sentences in an online dictionary [19]. These semantic relations are stored in a rather large knowledge base that is now known as MindNet [20]. Fortunately, some applications of the Microsoft NLP system, such as the current grammar checker, can do an adequate job without doing semantic reattachment and, therefore, can be spared the added cost of that processing. Both types of reattachment processing are specified primarily in G rules, also, but these rules are applied by a different (simpler) algorithm than the G analysis rules. These rules may be thought of more along the lines of a production system. Just as with the sketch, the output of the syntactic portrait can also be viewed as one or more trees.

The fourth stage of processing shown in Fig. 2 produces logical forms, which are intended to make explicit the underlying semantics of the input text. This is done in five steps for each syntax tree. The first two steps apply transformations to the tree to handle some phenomena, such as extraposition and long-distance attachment. Then a skeleton logical form is created that corresponds roughly to the dependency structure implicit in the computed tree. The final two steps then massage this skeleton, fleshing it out to a complete logical form. Intrasentential pronoun anaphora and some verb phrase anaphora are handled here, also. Logical form processing is specified by sets of G rules of the same basic sort that are used to produce the portrait. In general, the output from this stage cannot be viewed as trees, but rather, as graphs with labeled and directed arcs, possibly forming cycles. As in earlier stages of processing, each node is actually a record with more detailed information than that shown in the graphs. The nodes are labeled with word lemmas, augmented by numbers to make them unique within a particular graph The displayed lemmas may be thought of as representing all senses of the corresponding words.

The fifth stage of processing deals with lexical disambiguation [i.e., determining the most appropriate sense (or senses) of each word in the input text]. Some of this comes about as a side effect of producing the syntactic portrait, such as selecting the sense of telescope that is "an instrument for seeing distant objects" when deciding that "with a telescope" modifies "saw," rather than "bird," in the sentence, "I saw a bird with a telescope." However, the bulk of the sense disambiguation is done in this separate stage by a set of G functions (not rules currently) that makes use of a wide variety of information, including both syntactic cues from the parse and the semantic relations that are in MindNet. The output from this stage is best viewed as an expanded logical form, with each node augmented by references to the relevant word senses in MindNet [21]. Because lexical disambiguation is not needed for the current form of the grammar checker, no more will be said about this stage of processing here.

The sixth, and final, stage of processing shown in Fig. 2 deals with discourse phenomena. The research and system-building that we are doing in this area is being done within the framework of Rhetorical Structure Theory (RST) and involves producing multisentence logical forms that also make explicit the discourse relations among clauses, using RST labels, such as Elaboration and Contrast [22]. Because this is currently the least-developed part of our analysis system and is not used by the current grammar checker, no more will be said about it here.

Figures 3 through 8 illustrate many of the points discussed in this section. Figure 3 shows two part-of-speech records of the sort that are produced by lexical processing (stage 1). These are two of the four records that are produced for the word "After" and are for the preposition and conjunction possibilities. (The other two would be for the adjective and the adverb uses.) Each record is displayed as a collection of attribute–value pairs, with the attribute names on the left and their corresponding values on the right.

Figure 4 shows the computed parse tree produced by the syntactic sketch (stage 2) for the sentence, "After running a mile he seemed tired." The top node of this tree is a DECL, which has at its immediate constituents a PP, an NP, a VERB, an AJP, and a CHAR. The asterisk attached to the VERB indicates that it is the head element at that level. The other constituents are either premodifiers or postmodifiers. (Every phrase at every level has a head and zero or more premodifiers and postmodifiers.) The numbers at the ends of node names have no special significance other than to make each node name unique for human reference purposes. Figure 5 is the derivation tree for the same sentence. It shows which rules were applied to which nodes to form higher level nodes. For example, applying the VPwPP1 rule to PP1 and VP4 produced VP1 and then applying the Sent rule to BEGIN1, VP1 and CHAR1

```
 -------------------------------
{Segtype    PREP
 Nodetype   PREP
 Nodename   PREP1
 Ft-Lt      1-1
 String     "After"
 CopyOf     REC40
 Lex        "After"
 Lemma      "after"
 Bits       TakesAn InitCap Tme
 Prob       1.00000 }
 -------------------------------
{Segtype    CONJ
 Nodetype   CONJ
 Nodename   CONJ1
 Ft-Lt      1-1
 String     "After"
 CopyOf     REC41
 Lex        "After"
 Lemma      "after"
 Bits       Subconj TakesAn
            InitCap Tme
 Prob       0.00119 }
 -------------------------------
```

Fig. 3 Two example lexical part-of-speech records.

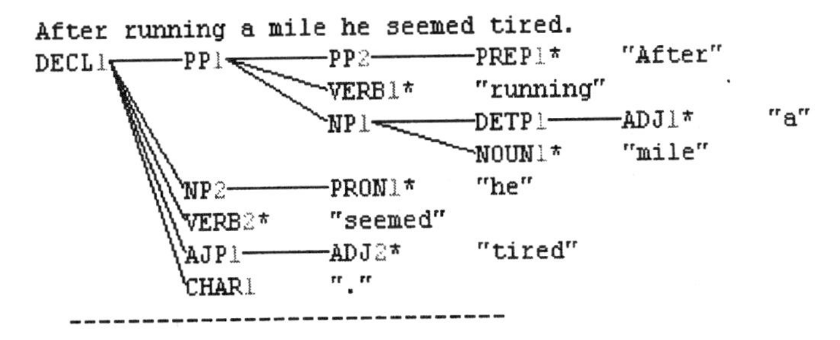

Fig. 4 An example parse tree.

produced DECL1. (The rule name VPwPP1 is short for VP with a PP to its left.) For this example sentence, the syntactic portrait processing (stage 3) does not modify the result of the syntactic sketch, at all.

Behind each node in these displayed trees is a record that gives a more detailed description of the corresponding segment of the input text. Figure 6 shows the segment record for DECL1, the node that covers the entire sentence. It is similar to the part-of-speech records shown in Fig. 3, but it has more attributes, and the values of many of these attributes are pointers to other (lower level) records. For example, the value of the Head attribute is a pointer to the record for the node VERB2. The display of the computed parse tree is produced by first displaying the

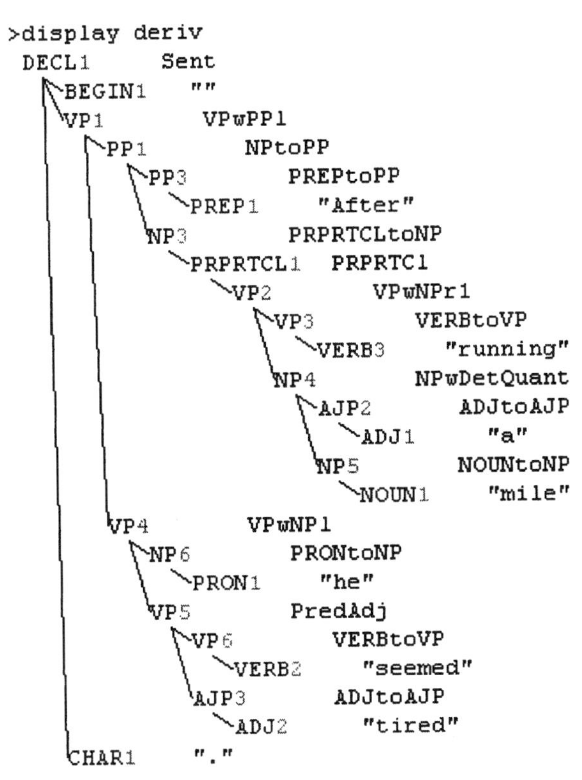

Fig. 5 An example derivation tree.

```
>display record DECL1
{Segtype     SENT
 Nodetype    DECL
 Nodename    DECL1
 Ft-Lt       0-8
 String      " After running a mile he seemed tired ."
 CopyOf      VP1
 Rules       (Sent VPwPP1 VPwNP1 PredAdj VERBtoVP)
 Constits    (BEGIN1 VP1 CHAR1)
 Lex         "seemed"
 Lemma       "seem"
 Bits        Pers3 Sing Past Closed
             L9 B0 Wv6 L1 L7 I3
             RtoSub F0 Wv8 Wv7 Wv4
             Wv6N I5 I6
 Prob        0.25645
 Prmods------PP1 "After running a mile"
             NP2 "he"
 Head--------VERB2 "seemed"
 Psmods------AJP1 "tired"
             CHAR1 "."
 Subject-----NP2 "he"
 Predadj-----AJP1 "tired"
 Props-------DECL1 " After running a mile he seemed tired ."
             PRPRTCL1 "running a mile"
 Pod         40
 Inverts-----PP1 "After running a mile"
 Nargs       1
 FrstV-------VERB2 "seemed"
 Vprp        (like)
 Predicat----VP5 "seemed tired"
 Topic-------NP2 "he"
 TopPPs------PP1 "After running a mile"
 Score       40.000000000 }
```

Fig. 6 An example segment record.

Nodename of a record and then recursively doing the same thing for each of the records in its Prmods (premodifiers), Head, and Psmods (postmodifiers) attributes. For the terminal nodes, which are always part-of-speech records, it also displays the word in the Lex attribute. In a similar manner, the display of the derivation tree is produced by using the Rules and Constits (constituents) attributes.

Figure 7 shows an abbreviated version of the VPwPP1 rule, which puts a PP and a VP together to form a larger VP during the syntactic sketch stage. In the derivation tree of Fig. 5, this rule could be seen putting PP1 ("After running a mile") together with VP4 ("he seemed tired") to form VP1 ("After running a mile he seemed tired"). (In this system, a VP is not just a verb and its auxiliaries, but may be any segment that has a verb as its head, such as a clause or even a whole sentence.) VPwPP1 is an APSG analysis rule written in the G programming language. The context-free grammar part of the rule (PP VP → VP) is fairly obvious in the figure, where the arrow separates the PP and VP on the left-hand side from the new VP to

```
VPwPP1:

    PP ( ^Comma(Prp) & ^Nappcomma(lastrec) & ^Precomma(lastrec) & ^SuspSUBCL &
         (forany(Prmods, [Comma]) -> Coords) &
         forall(firstrecs(PPobj), [Digits^=3 & Digits^=4]) &
         (forany(lastrecs, [Comma & ^Paren]) -> (Multcomma | Comma(lastphr))) &
         forall(lastrecs, [^Nomcomp | ^T5 | (Compl & Lemma(lasttokn)^="that")]) &
         (Gerund -> (^Rel(Postadv) | Postadv^=lastrec)) &
         Lemma(Prp) ^in? set{a an but x X} &
         forall(Coords, [Lemma(Prp) ^in? set{a an but x X}]) )

    VP ( ^Semiaux & ^Relpn & ^Paren &
         (forany(lastrecs(PP), [Nappcomma]) -> (^Pastpart | ^PPobj(first(Psmods)) |
            ^Comma(first(Psmods)))) &
         forall(lastrecs(PP), [Nappcomma -> (^Multcomma | Numbr ^agree? Numbr(VP))]) &
         (Nodetype(lastrec(PP))=="RELCL" -> (^Thatcomp(lasttokn(PP)) |
            Rel(first(Prmods(lastrec(PP)))))) &
         Nodetype(last(Psmods)) ^in? set{SREL TAG} &
         (Ord(Adj(Lex(lasttokn(PP)))) -> ^Num(Adj(Lex(firsttokn(first(Prmods)))))) &
         (Adv(Lex) -> (Prmods | Obj1 | (^Confus & Lemma ^in? set{no yes}))) &
         (Wh(Conj(Lex(PP))) -> (Prmods(PPobj(PP)) | YNQ)) &
         (Digits(first(Prmods)) -> (^Comma(first(Prmods)) | Prmods(first(Prmods)) |
            Nodetype(lasttokn(PP))^="NOUN")) &
         (Mnth(lasttokn(PP)) -> (^Ord(firsttokn) | ^Digits(firsttokn) |
            Digits(firsttokn)>2)) &
         ((Nom(Pron(Lex(lastrec(PP)))) & ^Obj(Pron(Lex(lastrec(PP))))) ->
            (Subject & Subject in? Prmods)) &
         (T5 -> (^Comma | (forall(Psmods, [^Oldsubcl]) &
            (^Nomcomp(Predcomp) | Compl(Predcomp) | ^Comma(lastphr(PP)))))) )

  --> VP { Prmods=PP++Prmods; Props=Props(PP)++Props; -SuspNREL;
           if (Subject(VP) ^in? Prmods(VP) & FortoPP(PP)) {Subject=PP; -VPInvert;}
           else if ((^Subject(VP) | VPInvert(VP)) & ^theresubj_test(VP)) MidPPs=PP++MidPPs;
           else {TopPPs=PP++TopPPs; Inverts=PP++Inverts;};
           Pod=Pod+Pod(PP);
           if (Lemma(lasttokn(PP))==";") Pod=Pod-4;
           if (^PPobj(PP) & Loc(Adv(Lex(PP)))) Pod=Pod-1;
           if (Subject in? Prmods(VP) | theresubj_test) Pod=Pod+1; }
```

Fig. 7 An example G analysis rule.

be formed on the right-hand side. In addition, each segment on the left-hand size of the rule is augmented with conditions on the corresponding record (and any record that can be reached from it), and the segment on the right-hand side is augmented with a procedure for creating the corresponding new record. It is not possible in the limited space of this chapter to go into any more detail about the G language, nor, unfortunately, are there currently any readily available publications describing it.

Figure 8 shows the logical form produced by the fourth stage of processing for the same sentence, "After running a mile he seemed tired." The nodes are considered to be concepts, and the relations between them are labeled (with fairly surfacey labels at this stage). For example, the Dsub (Deep subject) of seem1 is he1, and there is an "after" relation to run1. The information in parentheses with each node is the value of the Bits attribute in its underlying record. Even though the display makes the logical form look like a tree, it really is a more general graph. For example, the Dsub of run1 is he1, which is exactly the same node that is the Dsub of seem1. The node he1 appears twice in the display merely to simplify the drawing of lines.

Examples of the results of the fifth and sixth stages of processing (lexical disambiguation and discourse) are not shown in figures here, because they are not part of the processing done for the current grammar checker. However, for the example sentence being used here, the final result would appear similar to the logical form shown in Fig. 8, except that the appropriate word senses from MindNet would be added to the nodes (e.g., run101) and the appropriate discourse relations would be used (e.g., in an appropriate context, seem1 might become the Result of run1).

```
seem1 (+Past +L7)
 Dsub---he1 (+Masc +Pers3 +Sing +FindRef +Anim +Humn)
 Dadj---tired1 (+F0 +Psych)
 after---run1 (+T1 +Middle +Mov +Loc_sr)
           Dsub---he1
           Dobj---mile1 (+Indef +Pers3 +Sing +Conc +Count +Dst)
```

Fig. 8 An example logical form.

3. GRAMMAR CHECKING IN THE MICROSOFT NLP SYSTEM

The grammar checker, in its current form, does not use all six stages of processing. It uses all of the lexical stage (stage 1) and the entire syntactic sketch (stage 2), but just the syntactic reattachment portion of the syntactic portrait (stage 3). Also, it uses the logical form (stage 4) only under certain conditions, such as when the input text is passive or has a relative pronoun in it. The remaining stages of processing may come into play in later versions of the grammar checker, to diagnose errors of a more semantic sort and to detect problems that go beyond the sentence boundary.

We think of the grammar-checking component of this system not as a separate stage in the overall flow, but rather, as a branch off the mainline that occurs just after the syntactic portrait. This component is written in G and consists mostly of rules (not analysis rules, but the other kind that are used for most of the processing after the syntactic sketch) and also some supporting functions. There are several sets of rules, and each set deals with a particular type of syntactic constituent, such as a clause or a noun phrase. Each rule tests for a particular error situation and fires if it occurs. The overall flow of processing is to go through every node in the parse tree, from top to bottom, left to right, and for each node to consider each rule that is relevant for a node of that type. For example, if the node covers a clause, then the clause rules, such as the rule that handles subject-verb number disagreement, would be applied to that node. When the condition part of a rule (the left-hand side) is satisfied, then the action part of the rule (the right-hand side) is executed. The rules in the grammar-checking component are called descriptor rules, because they add descriptor records to the nodes in the tree to describe errors that are found.

A typical, but abbreviated, descriptor rule is shown in Fig. 9. This is the Desc_Comma5 rule, which diagnoses situations for which a comma should be added to an adverbial phrase (such as in the sentence, "After running a mile he seemed tired."). The conditions on the left-hand side of the rule (i.e., before the arrow) test a node in the parse tree (i.e., the record for that node and any records that can be reached from it) for those situations in which the rule should apply. The actions on the right-hand side (i.e., after the arrow) add information about that error to the record. This information is in the form of one or more descriptor records that are added to the Descrips attribute of the segment record in the tree by calling the function add_descrip.

Figure 10 shows the segment record for the PP1 node ("After running a mile") of the example parse tree in Fig. 4, after the Desc_Comma5 rule is applied to it. The Descrips attribute and the descriptor record that it has as its value can be seen near the bottom of the figure. The replacement string ("After running a mile,") shown there was created from the record (labeled PP4) that was constructed and passed to the function add_descrip in the actions on the right-hand side of the rule in Fig. 9.

```
Desc_Comma5:

  SYNREC (((Nodetype in? set{SUBCL AVP PRPRTCL AVPNP INFCL}) |
                (Nodetype=="PP" & PPobj)) &
           seg==first(Prmods(Parent)) &
           Nodetype(lasttokn) ^= "CHAR" &
           ^Theresubj &
           seg ^= Subject(Parent) &
           (Nodetype=="AVP" -> (^TheAVP & ^forany(Prmods,[TheAVP]))) &
           (Wh -> Lemma=="however") &
           ^forany(Coords,[Wh]) &
           (Nodetype(Head(Parent))=="VERB" | VPcoord(Parent)) &
           (Neg -> ^YNQ(Parent)) &
           ((Subject(Parent) &
              ((Ft(Subject(Parent))<Ft(FrstV(Parent)) & Ft(Subject(Parent))>Ft) |
               (VPcoord(Parent) & Ft(Subject)<Ft(FrstV(first(Coords(Parent)))))) |
             Nodetype(Parent)=="IMPR" |
             (Nodetype(Parent)=="QUES" & (YNQ(Parent) | WhQ(Parent)))))

 --> SYNREC { { segrec rec, commarec;
                commarec=segrec{Nodetype="CHAR"; Lemma=",";};
                rec=segrec{%%SYNREC; Psmods=Psmods++commarec;};
                add_descrip("Comma with Adverbials",0,rec); }; }
```

Fig. 9 An example G descriptor rule.

```
>display record PP1
{Segtype    PP
 Nodetype   PP
 Nodename   PP1
 Ft-Lt      1-4
 String     "After running a mile"
 CopyOf     NP3
 Rules      (TrLF_ControlatVP Desc_Comma5 NPtoPP PRPRTCLtoNP PRPRTC1 VPwNPr1 VERBtoVP)
 Constits   (PP1 PP1 PP3 NP3)
 Lex        "running"
 Lemma      "run"
 Bits       Pers3 Sing L9 X9 Wv6
            IO D1 T1 L1 L7 T5
            Asubj Loc_sr Unacc Mov
            Middle Wv4
 Prob       0.05363
 Prmods-----PP2 "After"
 Head-------VERB1 "running"
 Psmods-----NP1 "a mile"
 Gerund-----VERB1 "running"
 PPobj------NP3 "running a mile"
 Prp--------PP2 "After"
 Obj1-------NP1 "a mile"
 Props------PRPRTCL1 "running a mile"
 Pod        16
 Parent-----DECL1 "After running a mile he seemed tired ."
 Nargs      1
 FrstV------VERB3 "running"
 Object-----NP1 "a mile"
 Vptc       (along around away back down in off on out over through up across after)
 Vprp       (across after at from with over into on of through to against)
 Descrips
       {Ft-Lt      1-4
        Value      18
        DescType   "Comma with Adverbials"
        DescRepl---PP4 "After running a mile"
        DescReplStr "After running a mile," }
 SemNode----run1
 PrevCat    PP }
```

Fig. 10 An example segment record that includes a descriptor record.

```
>display desc

Comma with Adverbials:
        After running a mile    consider:  After running a mile,
```

Fig. 11 An example grammar error output in NLPWin.

Figure 11 shows the output produced by NLPWin when asked to display the descriptors in the example parse tree.

In the current system, there are about 200 descriptor rules. Approximately half of these rules are for diagnosing errors at the clause or phrase level, such as number disagreement, which depend very much on having a reasonable parse tree. The rest are for handling purely lexical problems, such as using the "n't" contraction in formal writing, which can be diagnosed with little or no syntactic context. The order of execution of the descriptor rules is not important. Each rule tends to be self-contained and does not depend on any information added to the tree by any previously executed descriptor rule. Where some ordering does come into account, however, is the order in which the critiques for a particular sentence are presented to the user. This is done by an algorithm that assigns a number to each descriptor such that critiques are presented in a primarily left-to-right order, but also taking into consideration the depth of each critiqued node within a phrase, with the deeper ones generally being presented first.

Most descriptor rules provide a suggested replacement string, but some provide only a message about the error. For example, in the simplest case, if "isn't" is found in formal writing, the replacement "is not" is suggested. Some of the more interesting cases of replacements include rewriting a passive clause in its active form. However, if there is no agentive by-phrase in a passive construction, then the passive verb is highlighted in a message, and no replacement is suggested. (For a while during its development, the system generated empty subjects, such as "someone," in the corresponding active form, but through usability testing, we determined that users did not want that.) In general, generating a replacement string involves considerations at all three of the clause, phrase, and word levels. For example, when rewriting a passive clause, the agentive by-phrase has to be moved to subject position (clause level), it has to be changed from a prepositional phrase into a noun phrase (phrase level), and its case has to be changed from objective to nominative (word level). Also, the subject becomes the direct object, the adverbial phrases have to maintain their proper positions in the clause, and the correct forms of the verbs in the verb phrase have to be generated. This can be seen in the following pair of sentences, the second of which was produced from the first by this grammar checker.

They were being carefully watched by me.

I was carefully watching them.

One class of critiques that is handled by a special mechanism that benefits greatly from the fact that this system is doing real parsing is what is often referred to as *confusable words*, the prototypical example of which is "its" vs. "it's." For example, both of these words, along with two others (all in italics) are used incorrectly in the first sentence in the following pair, which is corrected to the second sentence by this grammar checker (including preserving the italics), in four interactions with the user*. (Note that none of these words would be highlighted by a spelling checker, because they are all valid words. Note also that in this system these words are flagged not just because they are *potentially* confusable, but rather, because they appear to be *wrong* in this sentence when it is parsed.)

> *Its* an *alter* with a story *abut it's* construction.
>
> *It's* an *altar* with a story *about its* construction.

One part of the special mechanism for handling confusable words is an ini file that contains well over 100 pairs of confusable words. Some of them represent common grammar errors made by writers of English (such as "its" vs. "it's"); some of them represent common spelling errors (such as "alter" vs. "altar"); and some of them represent common typographical errors (such as "abut" for "about"). Roughly half of these specify that the confusion can go either way (such as "its" vs. "it's"), but the others are marked for only one direction (such as "abut" being typed for "about"—but not "about" for "abut," because that seems like an unlikely error). When a word that is in this file appears in text, the analysis system behaves as if the other word of the pair also appears at that same position in the text. Because of the bottom-up, multipath nature of the parsing algorithm, typically just one or the other word of the pair will result in a parse. If the word that results in a parse is not the word in the text, but rather the other one, then that other word is suggested to the user as a substitute for the word that was typed. For this mechanism to work based on just syntactic information, it is generally required that the words of a confusable pair do not have any parts of speech in common. Currently, some important pairs that violate this restriction, such as "affect" vs. "effect," are handled specially (and heuristically) by descriptor rules. If in later versions of the grammar checker some semantic processing is done, it should also be possible to handle these more difficult cases in a more general fashion.

For this system to be truly useful, it must be able to deal with unrestricted text. After all, the typical user of a word processor wants to write more than just linguistic textbook examples. On the other hand, the state of the art in natural language processing is such that there is much that any such system cannot properly handle at this time. NLPWin tries to parse any input string of up to about 100 words. For the grammar-checking application, it is often possible to produce useful critiques even if the parse is not exactly right (e.g., some modifying clauses and phrases might be attached to the wrong nodes). We sometimes refer to these as *approximate parses*. In addition, it is important to be able to deal with ill-formed input (e.g., producing a parse even where there is disagreement in number or person). We sometimes refer to this as *relaxed parsing*. When an input string is not parsed as a sentence, it may be because the current grammar simply cannot recognize it as a sentence, or it may be because the string is actually a fragment. The descriptor processing decides which it is, based on some characteristics of the *fitted parse* produced.

4. FEATURES OF THE MICROSOFT WORD 97 GRAMMAR CHECKER

The description of the grammar checker so far has been in the context of our general NLP system, NLPWin, which is a stand-alone Microsoft Windows application that

*At the time this chapter was written, all examples worked with Microsoft Word 97. However, some of the examples might behave differently in the more recent Microsoft Word 2000.

is used internally solely for research purposes. This section of the chapter begins the discussion of this grammar checker as it appears in Microsoft Word 97, a major Microsoft Windows application that is used worldwide by millions of people. (In addition to being shipped with the English versions of Microsoft Word 97, the English grammar checker was also shipped with all *non*-English versions of Microsoft Word 97.)

The productizing of our grammar checker was done using the well-established Microsoft product development process [23]. Detailed schedules were drawn up and followed over about a 20-month period. A working, but limited, version was produced early in the process, and then it was improved repeatedly until we had a shipping version. The usual product development team at Microsoft consists of three groups of people: program managers, who determine the features and specifications of the product; software development engineers, who implement these features according to the specifications; and testers, who test the features against the specifications. The work on this product followed the same paradigm, except that we also had researchers involved in all three parts, especially for the linguistic aspects. This section focuses on one aspect of the program management part of the process; namely, the features of the grammar checker as they appear to a user of Microsoft Word 97.

The use of the grammar checker in Word 97 can be quite easy, owing to a feature called *background grammar*, which was introduced with this version of Microsoft Word and is turned on by default. As the user enters text, or otherwise modifies the text of a document, Microsoft Word 97 regularly calls the grammar checker (using the functions described in the next section). As potential grammar errors are discovered, portions of the text on the screen are underlined with green squiggles to draw attention to them. The user can simply pay no attention to these or can right-click on a squiggle to obtain further information about the error at that position. Right-clicking brings up a context menu that typically offers one or two suggested replacements for the offending word or phrase, along with an option to ignore errors in this sentence (i.e., make the squiggle go away without changing anything) and another option to bring up the grammar dialogue box. An example of this context menu can be seen in Fig. 12, and an example of the corresponding grammar dialogue box can be seen in Fig. 13. In these figures, the grammar checker is suggesting that "mile" be replaced by "mile," (i.e., add a comma) in the sentence, "After running a mile he seemed tired."

Background *grammar* is a natural follow-on to background *spelling*, which had been introduced in Microsoft Word 95, two years earlier, in which words that are unrecognized by the spelling checker are underlined with *red* squiggles. This facility continues in Microsoft Word 97 and, in fact, the grammar checker and the spelling

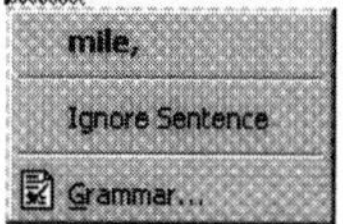

Fig. 12 An example of the grammar context menu in Microsoft Word 97.

After running a mile he seemed tired.

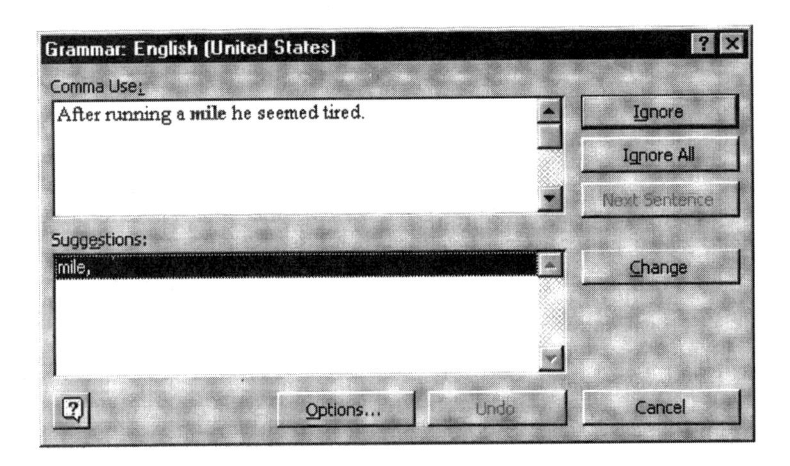

Fig. 13 An example of the grammar dialogue box in Microsoft Word 97.

checker present a consistent interface to the user. If a grammar user prefers not to do background checking, he or she may turn the feature off and do the checking later in a separate pass through the document, when errors are displayed one after another in dialogue boxes of the sort shown in Fig. 13.

For most of the 200 or so errors detected by this system, one or more suggested replacements are given, as described earlier. This varies all the way from simple word-level critiques, such as rewriting a contraction, to clause- or sentence-level critiques, such as rewriting a passive construction. However, in other cases the system is not able to suggest a replacement and instead simply gives a message about the error, such as when the deep subject of a passive construction is not readily available or when a sentence is too long. The user can then manually edit the text to correct the error or can simply choose to ignore it. In fact, the user additionally has those same two choices available for an error with a suggested replacement, too. If the user does not want to modify the text at all and does not want to continue to be told that there is an error there, the Ignore button can be selected.

Help information is available with each error diagnosis, and if requested by the user, is presented balloon-style by the Office Assistant, the on-screen character that is an integral part of Microsoft Office 97, of which Microsoft Word 97 is a part. This information is referred to as *brief explanations* and serves a very important function for the user, especially a new user. Each brief explanation has a user-friendly label that specifies the type of the error, followed by a few lines of text that describe that type of error, followed by a couple of examples to illustrate how to correct that type of error. The text of each brief explanation is specific for the *type* of error that has been diagnosed, but it is not tailored for the specific *instance* of the error. The total text of the brief explanations in the current system accounts for almost 15% of the sizes of the dll. Figure 14 shows the help information presented when the user clicked on the Help button (labeled with a question mark) in the grammar dialogue box of Fig. 13.

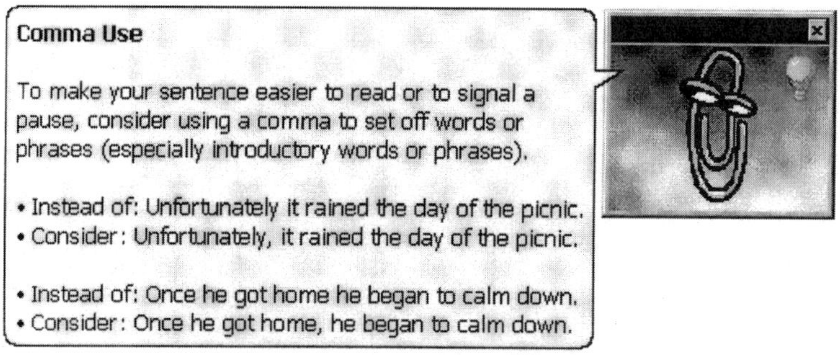

Comma Use

To make your sentence easier to read or to signal a pause, consider using a comma to set off words or phrases (especially introductory words or phrases).

- Instead of: Unfortunately it rained the day of the picnic.
- Consider: Unfortunately, it rained the day of the picnic.

- Instead of: Once he got home he began to calm down.
- Consider: Once he got home, he began to calm down.

Fig. 14 An example of the grammar help information in Microsoft Word 97.

The user of this grammar checker has some amount of control over the kinds of errors that are diagnosed. This control comes in two levels, known as *writing styles* and *options*. There are five writing styles available, with 21 options for each one. The writing styles vary from Casual to Formal, and they differ in which options are turned on. For example, only 5 of the options are turned on for Casual, but 20 are on for Formal. The options have names such as Capitalization, Punctuation, Passive Sentences, and Style—Wordiness. (Eight of the options have names that begin with Style, such as Style—Wordiness, to help make the distinction between grammar errors and style errors that was discussed in the opening section of this chapter.) The options that are on for each writing style are just the default settings, and it is easy for the interested user to modify those settings at any time. The default writing style is Standard, except when Microsoft Word 97 is being used as an e-mail editor, for which the default is Casual.

Figure 15 shows the spelling and grammar options dialogue box that appears when the user clicks on the Options button in the grammar dialogue box of Fig. 13. (The options dialogue box can also be reached from the Tools menu.) Here certain choices can be made, such as which writing style to use, and if the user clicks on the Settings button, the grammar settings dialogue box shown in Fig. 16 appears. Here the user can turn on and off the various options and also make a few other selections. Each of the 21 options is a collection of specific critiques. Some options, such as Negation, have only one critique, and some others, such as Misused words, have more than 30, with the average being about 10 critiques per option. The help information described earlier is specific to each critique, not merely to each option.

Figures 17 and 18 are intended to give the reader some idea of the range of critiques that this system makes. In the figures, which were produced using the Formal writing style, 19 different critiques are illustrated, ranging in difficulty from *Extra space between words* to *Passive construction*. The headings in these figures were put there manually for demonstration purposes and are not necessarily the official user-friendly labels for the errors shown there. Normal underlining is used in the figures to show the text that the grammar checker underlines with green squiggles. Following each sentence that has underlined text, there are one or more versions of the same sentence showing what it looks like after the user has accepted the replacements suggested by the grammar checker. There are also some examples of correct sentences that have no underlining, even though they bear much similarity

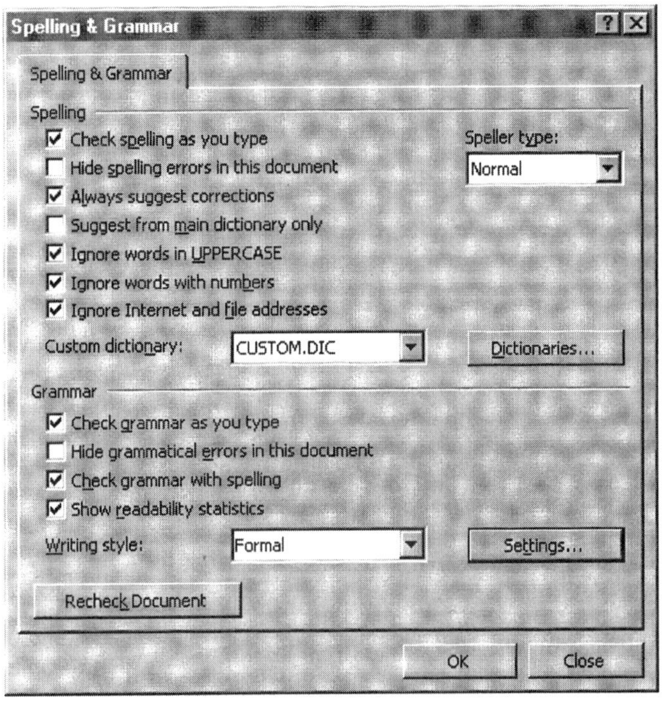

Fig. 15 An example of the grammar options dialogue box in Microsoft Word 97.

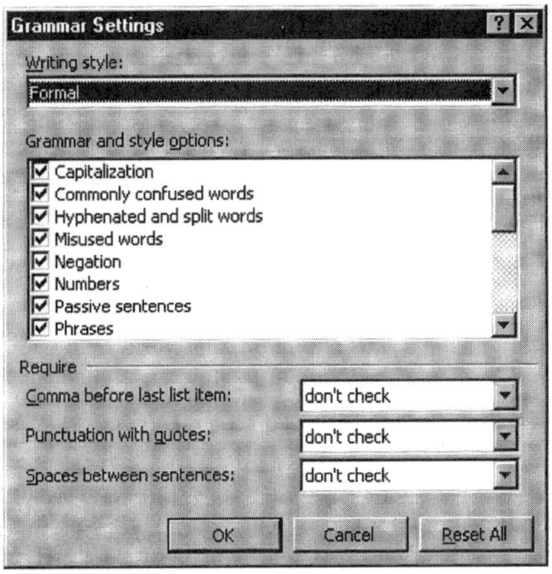

Fig. 16 An example of the grammar settings dialogue box in Microsoft Word 97.

Commas:
After running a mile he seemed tired.
After running a mile, he seemed tired.

After running a mile seemed long.
After running, a mile seemed long.

After running a mile seemed long he decided to retire.
After running a mile seemed long, he decided to retire.

Your/You're:
Your going to the movies with John is fine.

Your going to the movies with John.
You're going to the movies with John.
You are going to the movies with John.

Its/It's:
Its going to be a hot day.
It's going to be a hot day.
It is going to be a hot day.

The cat drank it's milk.
The cat drank its milk.

Their/There:
There cat went to sleep.
Their cat went to sleep.

Subject-Verb:
John are a big boy.
John is a big boy.

Chips using new technology was performing well.
Chips using new technology were performing well.
A chip using new technology was performing well.
Chips' using new technology was performing well.

A diagnostic chip for late model cars were performing well.
A diagnostic chip for late model cars was performing well.
Diagnostic chips for late model cars were performing well.

Passive:
The book on the Supernova was read by Chris.
Chris read the book on the Supernova.

The books on the Supernova were being read by him.
He was reading the books on the Supernova.

Many interesting papers on a variety of pertinent topics were presented by several researchers at this year's ANLP conference.
Several researchers at this year's ANLP conference presented many interesting papers on a variety of pertinent topics.

Fig. 17 Example critiques (with formal writing style).

to neighboring sentences that do have underlining. These were included in the figures as further illustrations of the power that comes from doing a full syntactic parse of each sentence when grammar checking, rather than trying to get by on simple string matching. Although the most interesting critiques are those that require a full syntactic parse, nevertheless, there are several useful critiques that this system provides that are based only on string matching, such as *Multiple spaces between words*, *Space before comma*, *Capitalize first word*, or *Sentence too long*. (The *Sentence too long* critique is triggered when a sentence exceeds either 40 or 50 words, depending on the writing style.)

Most of the sentences shown in Figs. 17 and 18 are rather short, just to focus on the error being diagnosed, but in real text, the sentences tend to be much longer, typically averaging close to 20 words, with many being in the 30 to 40-word range. The NLP system described here can achieve a complete parse for most long sentences, but sometimes the error analysis has to be done on a partial parse, in which

A/An:
John ate a apple.
John ate an apple.

Conjunction:
He is neither hungry and thirsty.
He is neither hungry nor thirsty.
He is both hungry and thirsty.

Split Words:
Everyone of the students passed the test.
Every one of the students passed the test.

Consistency:
Did you know either her cousin, aunt, or uncle?
Did you know her cousin, aunt, or uncle?

Between/And:
To find the driveway, look between the hedge or the mailboxes.
To find the driveway, look between the hedge and the mailboxes.

Extra Word:
Can you tell me where the child will be going to?
Can you tell me where the child will be going?

Capitalization:
Laura's Junior year in college exposed her to many new ideas.
Laura's junior year in college exposed her to many new ideas.

Hyphenation:
How can you tolerate his self righteous attitude?
How can you tolerate his self-righteous attitude?

Semicolon:
She did not make the cookies, she does not even have an oven.
She did not make the cookies; she does not even have an oven.

Mass/Count:
We needed to move a furniture store.

We needed to move a furniture.
We needed to move furniture.

Verbs:
He enjoys to talk to people.
He enjoys talking to people.

Possessive:
He had to wait for his managers return.
He had to wait for his manager's return.
He had to wait for his managers' return.

Spaces:
There is too much white space in this sentence.
There is too much white space in this sentence.

Fig. 18 More example critiques (with formal writing style).

case a more conservative stance is taken on reporting the possible errors found. On the other hand, the system also has to be able to deal with sentence fragments. In fact, one of the critiques is *Fragment*, which may be produced when a chunk of text separated by the sentence breaker can be parsed as a noun phrase, rather than as a sentence. Of course, noun phrases are fine as section headings or list items, so the grammar checker makes use of style format information that is passed to it from the word processor to try to do the right thing in these cases.

The grammar checker is also responsible for producing readability statistics if the user wants to see them. These appear in a simple box after the document has been processed and include counts of words, characters, paragraphs, and sentences. In addition, there are averages of sentences per paragraph, words per sentence, and characters per word, followed by the percentage of passive sentences, the Flesch reading ease score, and the Flesch–Kincaid grade level. Out of all of these, the percentage of passive sentences is the only statistic that requires the text to be parsed. The reading ease and grade level are based primarily on average words per sentence and syllables per word.

5. DEVELOPMENT OF THE MICROSOFT WORD 97 GRAMMAR CHECKER

This section discusses some of the work that had to be done to incorporate our grammar checker into Microsoft Word 97. Previous versions of Microsoft Word have had other grammar checkers incorporated into them, so there was already a defined grammar application program interface (API). The grammar API specififes many functions, but the main ones of interest to the discussion here are GramCheck and GramGetError. The first of these is called by the word processor to have one sentence checked. It passes a pointer to the first character of a sentence in a string of text, and expects to receive back a count of the number of characters in the sentence (thus, then also knowing where the next sentence begins). The second function is then called by the word processor, possibly multiple times, to obtain information about any errors found in that sentence so that they can be displayed to the user. The way that one writes a grammar checker for Microsoft Word is to write code for all of the functions specified in the grammar API and compile this code into a dynamic link library (DLL) file. In addition, any data needed at runtime, such as dictionary information, is put into an accompanying LEX file. For the grammar checker being described here, then, we wrote the set of functions for the grammar API, which we refer to as the API layer, primarily to make the necessary connections to functions in NLPWin.

One significant piece of code that was written for the API layer is the sentence breaker. Its purpose is to find the end of the sentence that starts at the text position passed in the call to GramCheck. This is done primarily by looking for end-of-sentence punctuation: namely, period, question mark, exclamation point, or carriage return; but it has to be done carefully, taking into account such things as abbreviations, spacing, uppercase, lowercase, ellipsis, quotation marks, and parentheses. So-called nonlinguistic critiques, such as improper spacing around punctuation, are also handled as part of this process, because they can be diagnosed without doing a linguistic analysis of the text. It was convenient for this first implementation of the grammar checker to do the sentence breaking and the nonlinguistic critiques separately from the main processing in NLPWin. However, this has resulted in some amount of duplicate code between the sentence breaker and lexical processing (i.e., tokenization is done in both places). In future versions these may be integrated, both to avoid the duplication and to possibly obtain the benefit of bringing linguistic analysis to bear on sentence breaking.

A large part of what goes on in the API layer is to communicate with the appropriate functions that already exist in NLPWin, to obtain the information required by the word processor when it calls into the grammar API. The most important such function is Analyze, which is called from GramCheck after it has isolated text that it considers a sentence. This function takes as input a text string that it then processes with all the machinery that has already been described in earlier sections of this chapter. The main result of this is to produce the information about any errors found in that string of text, to be able to respond to Microsoft Word's later calls on GramGetError. We wanted NLPWin to remain intact for its many other purposes, and we did not want to have to maintain multiple code bases. So, for instance, the Analyze function does not produce its output directly in the data structures required by the grammar API, but rather, uses an error encapsulation mechanism that we call a codestr (code string). This is just a variable length string of ASCII characters that is produced from the information in the descriptor records that were described in an earlier section. A codestr contains all the information about any errors found in a string of text that was isolated by the sentence breaker, such as how many errors, their identifications, their locations in terms of character positions, and any replacement strings. There are functions in the API layer that convert a codestr into the requisite grammar API data structure for passing back to Microsoft Word. Currently, the grammar checker keeps a recent cache of strings of text that it has processed, along with their corresponding codestrs, so that the strings need not be reprocessed when the user wants to see the details of the errors for a particular string.

Part of the challenge of constructing this grammar checker was to take a system that was developed primarily for research purposes and to make it small enough and fast enough to be part of a product. This was the focus of our development efforts over a period of about a year and a half. (At the same time, the linguistic aspects of the system were also dramatically improved.) At the beginning of this phase of the work, the NLPWin executable (exe) file was about 8M (megabytes) and it ran in a memory footprint of about 12M. The associated dictionary file was about 50M in its full form or about 7M in its reduced form that contained only syntactic information. Also, there were several other files that contained such data as word and rule probabilities and information about the various kinds of grammar errors.

Although there was a wide variety of work items that we undertook to make the program smaller and faster, the largest single factor in making it smaller was to design and implement a special *p-code*. It was mentioned earlier in this chapter that linguistic code (primarily APSG rules) that is written in the G language must be translated into corresponding code in C (by Gtran) to be executed as part of the NLPWin processing. With our special p-code, we introduce a level of interpretation, rather than direct execution. This p-code is a stack-oriented, byte-level code that can represent the kinds of entity–attribute–value processing done in this system in a very compact manner, especially for the conditions that are specified on the left-hand sides of rules. Most of the work that was done to bring about this p-code was done in the Gtran program, which was modified to translate rule conditions into p-code, instead of C if-statements (on an option setting). Gtran also generates the p-code interpreter, according to which instructions are actually used, and it produces the necessary calls to the interpreter from the C functions that it generates for the rules. This special p-code is approximately one-tenth the size of the equivalent C code and yet it only slows overall execution by about 20%. (A couple of the reasons for this good result are that the most often executed parts of the program are still in C; and also, making the memory footprint smaller, decreases the amount of time spent paging.)

During the course of this work, we also made many improvements to various data structures, which not only gave the system a smaller memory footprint, but also resulted in it running faster. We also made many algorithmic improvements to guide the parser to a good analysis more quickly. When the performance of the system doing grammar checking became acceptable, we then discovered that initialization was taking too long. We solved this problem by moving most of the initialization processing into Gtran, where the various ini files could be read to build the run-time arrays with their initial values specified by C initializers, thus eliminating the need for this processing at system initialization time. The size of the dictionary was first cut in half by doing a variety of specialized encodings of oft-repeated data, and then it was cut in half again by using standard data compression techniques.

As a result of this development work, the grammar checker that shipped with Microsoft Word 97 consists of two files: msgren32.dll, which is about 1.5 M, and msgr_en.lex, which is about 1.7M, for a total disk usage of about 3M. The incremental memory footprint is in the range of 2–3M. The speed of the grammar checker on a high-end PC (P6/200) is about eight sentences per second, where the average sentence length is close to 20 words. On an entry-level machine (486/50), the speed is about two sentences per second. It may be worth noting that all of these improvements in speed and size were done without disturbing the linguistic work. The linguists almost never had to change the way they were doing anything. All improvements were made *under the covers* by modifications to Gtran and also to various algorithms and their associated data structures.

6. TESTING THE MICROSOFT WORD 97 GRAMMAR CHECKER

One of the requirements for the production of a successful software product is to do thorough testing. Even though it is the responsibility of each developer and linguist to test his or her own code, the responsibility for making sure the overall product performs according to the specifications and in a manner that is acceptable to the users falls on the shoulders of the test team. This section discusses some of the testing that was done for the Microsoft Word 97 grammar checker.

Testing is an ongoing process that takes place at many levels. Certainly, when a developer or linguist adds a new feature or fixes a reported bug, he or she must test the new code to ascertain that it performs the desired function. However, at least as importantly, he or she must also do regression testing to make sure that the new code has not broken something that worked previously.

The lowest level of regression testing is sometimes called a smoke test. For us, this is a file of about 100 sentences originally chosen randomly from real documents. This file can be processed in a matter of seconds, as a coarse check on the state of the overall grammar-checking system. The next level is a set of about 3000 sentences of the sort that are found in a thorough grammar textbook. These can be run in just a few minutes on a four-processor server machine to test the syntactic sketch part of the system. Then there is a set of about 7000 sentences, some made up to exhibit certain grammar errors, but most from real text. These can be processed in 7 or 8 minutes to test all of the linguistic aspects of the grammar-checking system. The linguists who are working on the grammar rules and the descriptor rules typically run all three of these levels of regression testing many times each day. The linguists working on other parts, such as lexical processing and syntactic reattachment, also regularly run these same tests, plus others of their own.

The regression testing just described can be done on the linguist's own computer, but it is faster to run it on a four-processor server machine. In either event, it is just a matter of issuing a simple command and then looking at the files of differences produced, if any, where the differences are displayed in a convenient manner. The master files do not necessarily contain only correct results, but at least they contain the current expected results, and for any sentence can be easily updated with one mouse click. As time goes by and the system improves, the proportion of correct results in these master files tends to increase.

In addition to the basic regression testing just described, which is performed by the linguists themselves and is primarily testing only the basic linguistic functionality using NLPWin, there is the far-reaching product testing that is performed by the test team, which involves both system testing and additional linguistic testing. This process begins with something called the *nightly build.* Every night at 3 AM the *build machine* gathers up all the latest source code that has been checked into the source control repository by the developers and linguists, and then it builds the various system files, primarily executables and dictionaries. Build verification tests (BVTs) are run and, if successful, the system files are copied to a release point. Every few days, or when there is known to be some improvements of special interest, the tester in charge runs some further tests, partially manual and partially automated, and *blesses* the build. This then makes it fair game for the other testers to start looking for new problems or the resurgence of old ones. Additionally, real documents containing hundreds of thousands of sentences are run through it to observe differences in behavior from the previous time they were processed, using a variety of special driver programs written for this purpose.

The nightly build actually produces several different sets of files. For example, an NLPWin exe is built that corresponds to the linguists' development environment. In addition, two versions of the msgren32 dll are built, one being the *debug* version and the other the *retail* version. The debug version is larger and slower than the retail version because it includes additional output capabilities and is produced by the C compiler without doing any code optimization, but it can be more helpful for doing some kinds of debugging. The retail version is the one that eventually ships as the product. All of these executables are built from one set of source files so that everything is kept in sync. (#ifdefs are sprinkled liberally throughout the source files to accomplish this.) Various forms of the dictionary files are also built, differing primarily in the type and amount of data compression that is performed.

Given a new blessed build, the testers apply a range of manual and automated test methods to find and track down bugs. Whenever a bug is uncovered, an entry is made in the bug database for this project and assigned to the person responsible for fixing it. The developers and linguists constantly work on fixing the bugs assigned to them, and then assigning them back to the testers for closure. So, whenever there is a new blessed build, the testers first test for the bugs that have been fixed, so that they can be officially closed. Then they go on to find new ones by processing more text, some made up to test particular features, but mostly from real corpora. This work is guided by the project specifications that had been produced earlier in the project by the program management team. It is not unusual at some point in a project such as this to have hundreds of active bugs in the bug database.

We realized early on in the process that linguistic bugs (such as a bad parse or an incorrect critique) are different from system bugs (such as misplaced underlining or nonrepeatable results) and should be treated differently. Although it is possible to find and fix *almost all* system bugs by exercising all parts of the system, it is not possible at the current state of linguistic technology to similarly find and fix almost all linguistic bugs. There are an infinite number of different sentences, and each new one seen by the grammar checker might have some special combination of features that had not been properly dealt with before. The test team has devised a unified tool for finding, searching, and regressing linguistic bugs, to simplify the bookkeeping required and to try to more quickly make generalizations, rather than simply reporting individual incidents. An important aspect of this approach involves making an investment in the detailed manual tagging of many tens of thousands of sentences, to produce a database of reference analyses that can be used repeatedly for testing various versions of the system.

An important part of what the test team does is to summarize the bug data in interesting and useful ways. This provides the development team with important status information and can help them focus their efforts where they are most needed. For example, statistics were kept on how well each individual critique behaved (i.e., how many times it was correct, how many times it was false, and how many times it was missed). This information enabled the linguists to devote special attention to those critiques that were more troublesome. Then, when it came time to ship the product, this same information was used to determine which critiques should be left out of this first version because they were currently too unreliable. The system design provides a simple technique for disabling individual critiques. It is just a matter of deleting (or commenting out) lines from one of the ini files that are used by Gtran when building the system. The content of the writing styles and options also can be easily changed by modifying an ini file.

Another important part of what the test team does can be lumped under the term *measurement*. They are constantly measuring the *quality* of the product, by comparing it with the specifications and with competing products. They also analyze feedback from actual users, realizing that user perceptions can be just as important as numbers in judging quality. The testers regularly measure the *performance* of the product, reporting on its size and speed on various platforms, ranging from the minimum machine required to the most powerful one available. They also routinely use both manual and automated tests to exercise all of the features of the user interface to make sure they are functioning properly.

Still another important aspect of testing is *usability* testing, during which the product is used by real users. For example, focused usability testing was performed to determine if background grammar would be a convenient feature, and also to determine the best style for the brief explanations. All of these testing activities contribute substantially to producing a product of great robustness and flexibility.

Microsoft Word 97 went through two technical beta releases before the final product release, and the grammar checker was included in both of them. The number of user sites for the first beta was over a hundred and for the second one it was several hundred. Feedback from the beta users was analyzed and used to improve the system. The final product was reviewed in many trade publications, and the grammar checker fared quite well in most of them. We were pleased.

7. CONCLUSION

This chapter has presented a fairly detailed look at a system that provides a form of intelligent writing assistance, in particular the grammar checker that is part of

Microsoft Word 97, which shipped in early 1997. This system has at its heart a full-blown, multipurpose natural language processor called the Microsoft NLP system, which was described in Sec. 2. Other sections described how grammar checking was added to the basic NLP system and how that was added to Microsoft Word, along with discussions of the features of the grammar checker in Microsoft Word 97 and how it was tested. Significantly, it was pointed out that this work was done within the framework of the well-established Microsoft product development process.

Where does this work go from here? The Microsoft NLP system continues to be enhanced, especially in the direction of more semantic processing with MindNet and in the area of discourse processing. At the same time, grammar development is proceeding for other languages. The grammarians work closely together so that the commonality of the various languages can be exploited in such a way as to both simplify and improve the work on each one [24]. We expect that future versions of the Microsoft grammar checker will benefit from all of the work being done on the underlying NLP system. At some point, it will be possible to do a much better job of text critiquing by taking semantics and discourse phenomena into account, and it will be possible to critique text in languages other than English. In addition, many other applications of the underlying NLP system will eventually find their way into products.

Those of us who are working on this project feel strongly that research in NLP benefits enormously by having a product orientation. For example, it makes us face up to the hard practical problems in addition to the more fun theoretical ones. We cannot restrict ourselves to interesting linguistic examples, but have to deal with real text as written by real people. In addition, we have to be very concerned with issues of size and speed. Being associated with a successful and profitable product is nice from a resource standpoint, too. When we started the work on this grammar checker, there was just a handful of us involved. Then, as it became apparent that we were delivering what we had promised, the team grew. By the time the product shipped, there were at least 20 people directly involved. Now, between the NLP research group and the NLP development group in Microsoft Word, more than 50 people are working on some aspect of natural language processing for future Microsoft products.

Microsoft Word 97 is used by millions of people all over the world every day. Besides preparing the usual array of assorted documents, many people also use it as their e-mail editor. Because background grammar checking is on by default, there must be millions of sentences being analyzed by our NLP system every day. That gives us quite a sense of accomplishment. (On a more personal note, I am pleased to report that I found the grammar checker to be quite helpful in the preparation of this chapter.)

ACKNOWLEDGMENTS

The number of people intimately involved in this work is far too large for me to be able to adequately acknowledge each one individually. Therefore, let me just give blanket thanks to all of the people in the NLP research group and the NLG and Microsoft Word product groups, including supportive management, who contributed to the development of the Microsoft Word 97 grammar checker. It was, and continues to be, a great team effort.

REFERENCES

1. R Quirk, S Greenbaum, G Leech, J Svaartvik. A Grammar of Contemporary English. London: Longman Group, 1972.
2. W Strunk Jr, EB White. The Elements of Style. Boston: Allyn and Bacon, 1979.
3. NH Macdonald, LT Frase, P Gingrich, SA Keenan. The WRITER'S WORKBENCH: Computer aids for text analysis. IEEE Trans Commun 30:105, 1982.
4. GE Heidorn, K Jensen, LA Miller, RJ Byrd, MS Chodorow. The EPISTLE text-critiquing system. IBM Syst J 21:305, 1982.
5. SD Richardson, LC Braden-Harder. The experience of developing a large-scale natural language text processing system: CRITIQUE. Proceedings of the 2nd Conference on Applied NLP, Austin, Texas, 1988, pp. 195–202. Reprinted in Natural Language Processing: The PLNLP Approach. K Jensen, GE Heidorn, SD Richardson, eds. Boston: Kluwer Academic, 1993, p. 77.
6. IBM Critique Guide. IBM Publishing Systems ProcessMaster, Release 3, VM Edition (SC34-5139-00), Armonk, NY, 1989.
7. CorrecText Grammar Correction System. Inso Corporation, 1993.
8. Grammatik Grammar-Checker. Novell, 1996.
9. Microsoft Word 97 Word Processor. Microsoft Corporation, 1997.
10. BW Kernighan, DM Ritchie. The C Programming Language, 2nd ed. Englewood Cliffs, NJ: Prentice Hall, 1988.
11. GE Heidorn. Natural language inputs to a simulation programming system. PhD dissertation, Yale University, 1972. Also published as Technical Report NPS-55HD72101A, Naval Postgraduate School, Monterey, CA, 1972.
12. JP Pentheroudakis, L Vanderwende. Automatically identifying morphological relations in machine-readable dictionaries. Proceedings of the Ninth Annual Conference of the UW Centre for the New OED and Text Research, Oxford, England, 1993, pp. 114–131.
13. Longman Dictionary of Contemporary English. London: Longman Group, 1978.
14. The American Heritage Dictionary of the English Language, 3rd ed. Boston: Houghton Mifflin, 1992.
15. DA Coughlin. Deriving part-of-speech probabilities from a machine-readable dictionary. Proceedings of the Second International Conference on New Methods in Natural Language Processing, Ankara, Turkey, 1996, pp. 37–44.
16. GE Heidorn. Augmented phrase structure grammars. In: BL Webber, RC Schank, eds. Theoretical Issues in Natural Language Processing. Assoc. for Computational Linguistics, 1975, pp. 1–5.
17. K Jensen. Binary rules and non-binary trees: breaking down the concepet of phrase structure. In: A Manaster–Ramer, ed., Mathematics of Language. Amsterdam: John Benjamins Publishing, 1987, pp. 65–86.
18. SD Richardson. Bootstrapping statistical processing into a rule-based natural language parser. Proceedings of the Workshop on Combining Symbolic and Statistical Approaches to Language, Las Cruces, NM, 1994, pp. 96–103.
19. K Jensen, J-L Binot. Disambiguating prepositional phrase attachments by using online dictionary definitions. Comput Linguist 13:251, 1987.
20. WB Dolan, L Vanderwende, SD Richardson. Automatically deriving a structured knowledge base from on-line dictionaries. Proceedings of the Pacific Association for Computational Linguistics, Vancouver, BC, 1993, pp. 5–14.
21. W Dolan. Word sense ambiguation: clustering related senses. Proceedings of COLING94, Kyoto, Japan, 1994, pp. 712–716.
22. S Corston. Computing representations of the structure of written discourse. PhD dissertation, University of California at Santa Barbara, 1998.

23. MA Cusumano, RW Selby. How Microsoft builds software. Commun. ACM 40:53, June 1997.
24. M Gamon, C Lozano, J Pinkham, T Reutter. Practical experience with grammar sharing in multilingual NLP. In: J Burstein, C Leacock, eds. Proceedings of the Workshop on Making NLP Work in Practice, Madrid, Spain, 1997, pp. 49–56.

9

Database Interfaces

ION ANDROUTSOPOULOS

National Centre for Scientific Research "Demokritos," Athens, Greece

GRAEME RITCHIE

University of Edinburgh, Edinburgh, Scotland

1. INTRODUCTION

Natural language database interfaces (NLDBIS) are systems that allow users to access information stored in a database by formulating requests in natural language. For example, a NLDBI connected to the personnel database of a company would typically be able to answer questions such as the following. (We show user entries in *italics*, and system replies in **bold**.)

> *Who is the youngest employee in the sales department?*
> **John Smith**.
> *What is his salary?*
> **$25,000**.
> *Does any employee in the sales department earn less than John Smith?*
> **Yes, George Adams**.

NLDBIS have received particular attention within the natural language processing community [see Refs. 8, 28, 63 for previous reviews of the field[1]], and they constitute one of the first areas of natural language technology that have given rise to commercial applications.

[1]Some of the material in this chapter originates from Ref. 8.

This chapter is an introduction to key concepts, problems, and methodologies in the area of NLDBIS. It focuses on issues that are specific to NLDBIS, as opposed to general natural language processing (NLP) methods, but it also provides examples from NLDBIS in which the general techniques described in the earlier chapters of this book are used. As with the rest of the book, this chapter does not cover issues related to speech processing. This reflects the assumption in most current NLDBIS that the user's requests will be typed on a keyboard. As we comment briefly in Sec. 2, however, we believe that speech technology will play an important role in future NLDBIS.

The rest of this chapter is organised as follows: Sec. 2 provides a brief history of NLDBIS; Sec. 3 presents the typical architecture and the main components of modern NLDBIS; Sec. 4 discusses portability issues; Sec. 5 highlights some advanced issues: the "doctor-on-board" problem, database updates, and metaknowledge, modal, and temporal questions; Sec. 6 discusses NLDBIS with restricted input: namely, NLDBIS with controlled languages, and menu-based NLDBIS; Sec. 7 concludes by reflecting on the state of the art and the future of NLDBIS.

2. HISTORICAL BACKGROUND

The first NLDBIS appeared in the late 1960s, with the most well-known system of that period being LUNAR [83–86], a NLDBI to a database that contained chemical analyses of moon rocks. LUNAR demonstrated convincingly that usable NLDBIS could be built. It also introduced some innovative techniques (e.g., in the treatment of quantifiers) that had a significant influence on subsequent computational approaches to natural language.

Several other NLDBIS had appeared by the late 1970s (e.g., RENDEZVOUS [26], TORUS [61], PLANES [79], PHILIQA1 [68], and LADDER [45]).[2] Some of these early systems (e.g., PLANES, LADDER) used *semantic grammars*, an approach in which—roughly speaking—nonterminal symbols of the grammar reflect categories of world entities (e.g., *employee_name, question_about_manager*) instead of purely syntactic categories (e.g., *noun_phrase, sentence*; see Ref. 2 for more information). Semantic grammars allowed selectional restrictions (discussed in Sec. 3.E) to be encoded easily, and the resulting parse trees could be very close to logical formulae, eliminating the need to map from syntactic to semantic constructs. Semantic grammars, however, proved difficult to port to new knowledge domains (e.g., modifying a NLDBI to be used with a database about train schedules, rather than employees), and were gradually abandoned.

In the early 1980s, CHAT-80 was developed [80]. CHAT-80 incorporated some novel and ingenious techniques, and its implementation responded to queries very promptly. Its code was circulated widely, and it became a de facto standard demonstration of NLDBI capabilities. It also formed the basis of the experimental NLDBI that we will use as a source of illustrative examples here—MASQUE/SQL [5,7]. We used examples from MASQUE/SQL, mainly because we can say with authority what it would do, and partly because we consider its architecture and mechanisms as typifying a large class of NLDBIS.

---

[2]More information on the historical evolution of NLDBIS can be found in Ref. 11.

By the mid-1980s, NLDBIS had become a very popular research area, and numerous research prototypes were being implemented. Portability issues (see Sec. 4) dominated much of the NLDBI research of the time. TEAM [36,37,59], for example, was designed to be configured by database administrators with no background in linguistics, and ASK [75,76] allowed end-users to "teach" it new words at any point. ASK was actually an integrated information system, with its own built-in database, and the ability to interact with external applications (e.g., external databases and e-mail programs). The user could control any application connected to ASK by natural language requests. JANUS [22,46,81] had similar abilities to interface with multiple systems. DATALOG[3] [38,39], EUFID [71,72], LDC [13,14], TQA [29], and TELI [12], were also among the numerous research prototypes of the same period.

We note at this point that some natural language researchers use the term "database" to mean just "a lot of data." In this chapter, we mean much more than that. Most importantly, we assume that the data constitute a principled attempt to represent part of the world, and that they are structured according to a formally defined model. Database systems have evolved substantially over recent decades. The term *database systems* now denotes (at least in computer science) much more complex and principled systems than it denoted in the past. Many of the "database systems" of early NLDBIS would not deserve to be called database systems by today's systems.

Until the early 1980s, the standard way to interact with database systems was to use special database programming languages (e.g., SQL [60]), which are difficult for end-users to master. NLDBIS were seen as a promising way to make databases accessible to users with no programming expertise, and there was a widepsread optimism about their commercial prospects. In 1985, for example, Ovum Ltd. predicted that *By 1987 a natural language interface should be a standard option for users of database management system and 'Information Centre' type software, and there will be a reasonable choice of alternatives.* [49]

Since then, several commercial NLDBIS have appeared (e.g., Linguistic Technology's ENGLISH WIZARD, a descendant of INTELLECT[4] [43], which, in turn, was based on experience from ROBOT [40–42]; BBN'S PARLANCE, derived from the RUS [21] and IRUS [16] systems; IBM'S LANGUAGEACCESS [62]; Q&A from Symantec; NATURAL LANGUAGE; BIM'S LOQUI [20]; and ACCESS ELF).[5] Some of these systems are claimed to have been commercially successful. The use of NLDBIS, however, is much less widespread than it was once predicted, mainly because of the development of alternative graphic and form-based database interfaces (see, e.g., the discussion of Zloof's "query by example" technique in Ref. [77]). These alternative interfaces are arguably less natural to interact with, compared with NLDBIS. It has also been argued (e.g., [27,44]) that queries that involve quantification (e.g., *Which company supplies every department?*), negation (e.g., *Which department has no secretaries?*), or that require multiple database tables to be consulted are very difficult to formulate with graphic or form-based interfaces, whereas they can be expressed easily in natural language.

---

[3]This NLDBI has nothing to do with the subset of Prolog that is used as a database language [77].
[4]INTELLECT is currently owned by Platinum Technology.
[5]Few of these systems appear to be currently in the market. Consult `http://www.elf-software.com` and `http://www.englishwizard.com` for more information on two NLDBIS available at the time of writing.

Nevertheless, graphic and form-based interfaces have largely out-marketed NLDBIS, probably because their capabilities are often clearer to the users (see Sec. 3.C), and they are typically easier to configure (see Sec. 4). Experiments on the usability of NLDBIS, and comparisons with alternative database interfaces, are discussed in Refs. [18,23,32,48,69,82].

Perhaps as a result of their difficulties in the market, NLDBIS are no longer as fashionable a topic within academic research as they were in the 1980s. There is, however, a growing body of research on integrating speech recognition, robust interpretation, and dialogue-handling techniques, with the goal being to implement systems that engage users in spoken dialogues to help them perform certain tasks. We expect that this line of research will have a significant influence on future NLDBIS, giving rise to systems that will allow users to access databases by spoken dialogues, in situations for which graphic and form-based interfaces are difficult to use. This could lead, for example, to NLDBIS able to answer spoken queries over the phone [see, e.g. Ref. 1; also see Ref. 87, for information on the ATIS domain, where users make flight arrangements by interacting with a computer by spoken dialogues].

3. ARCHITECTURE AND MAIN COMPONENTS

This section discusses the typical architecture and the main modules of modern NLDBIS.

A. Architectural Overview

Ignoring some details, the architecture of most current NLDBIS is similar to that of Fig. 1. Roughly speaking, the part of the system that is labeled *linguistic front-end* translates the natural language input to an expression of some intermediate meaning representation language (MRL). The MRL expression is subsequently passed to the *database back-end.* This translates the MRL expression into a database language that is supported by the underlying database management system (DBMS; this is the part of the database system that is responsible for manipulating the information in the database). The resulting database language expression is then executed by the DBMS, to satisfy the user's request. For questions, the execution of the database language expression retrieves information from the database, which is reported back to the user.

This architecture has several portability advantages (discussed in Sec. 4). It also allows inferencing components to be added between the linguistic front-end and the database back-end to allow, for example, the NLDBI to deduce new facts from the contents of the database (we discuss this briefly in Sec. 5.A). In systems for which the natural language input is mapped directly to database language expressions, this inferencing would have to be carried out in the database language, which is particularly difficult because database languages are not designed to facilitate machine reasoning.

We discuss in the following the components of the architecture of Fig. 1 in more detail.

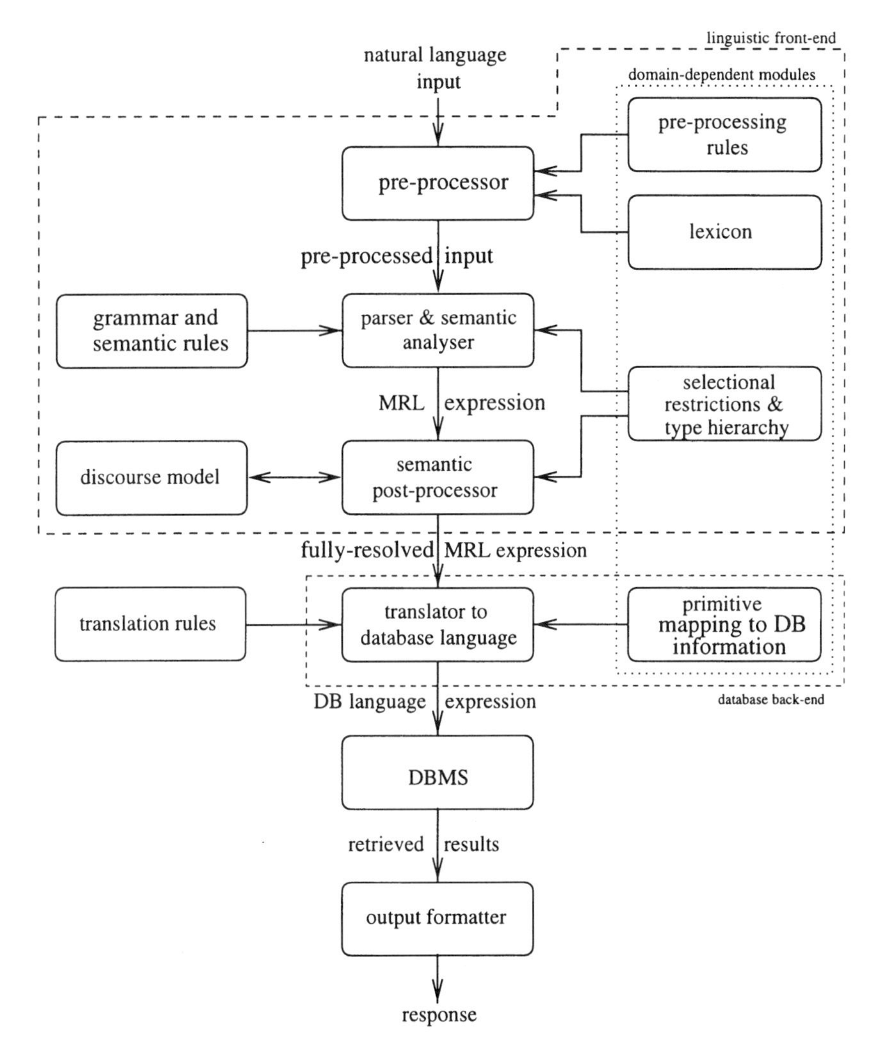

Fig. 1 Typical NLDBI architecture.

B. Preprocessing

The natural language input first undergoes a preprocessing phase. This tokenises the input, morphologically analyzes the words, and looks them up in a lexicon to retrieve their syntactic and semantic properties. The techniques of Chaps. 2–4 apply here. (We note, however, that most NLDBIS allow the user to type only single-sentence requests. Hence, sentence segmentation is usually not an issue. Punctuation is also typically ignored.)

Proper names (e.g., person names such as *Adams*, or flight numbers such as *BA742*) constitute a particular problem for NLDBIS. For example, the personnel database of a large company may contain information about thousands of employees. To be able to parse questions that contain employee names (e.g., *What is the salary of George Adams?*) the NLDBI must have entries in its lexicon for all these names (possibly separate entries for first names and surnames). Inserting manually an entry in the lexicon for each employee name is tedious, and it would also mean that the lexicon would have to be manually updated whenever new employees join the company.

One possible solution is to provide some mechanism that would automatically compute lexicon entries from proper names that appear in the database. (Here, the architecture of Fig. 1 has to be modified, to allow the preprocessor to access the database. The *preprocessing rules* of Fig. 1 would also have to be removed. They are used in an alternative preprocessing method, which is discussed later). In the personnel database case, whenever a word of the natural language input is not found in the lexicon, that word would be checked against a list of employee names held in the database). If the word is found in that list, an appropriate lexicon entry would be generated automatically using the database's information. For example, all the words that appear in the list of employee names could be assigned lexicon entries that classify the words as singular, proper-name, noun-phrases. If the database includes gender information, the lexicon entries could also show the gender of the denoted employee (this is useful for anaphora resolution; see Sec. 3.D).

This approach has the disadvantage that it introduces an additional database look-up during the preprocessing. In databases with a large number of proper names (e.g., databases for telephone directory enquiries) this additional look-up may be computationally expensive. This approach also runs into problems with questions that contain proper names not mentioned in the database. If, for example, the database does not contain the name *Johnson*, the NLDBI would fail to parse *Is Johnson in the database?* instead of responding negatively.

An alternative approach is to employ pattern-matching preprocessing rules and typographic conventions. In questions directed to a database about flights, for example, it may be safe to treat any word that consists of two or three letters followed by three or four digits (e.g., *BA742*) as a flight number. Any word that matches this pattern would be assigned a lexicon entry that classifies the word as singular, proper-name, noun-phrase. In the personnel database, person names could be identified by asking the user to type all words in lower-case letters, apart from person names whose first letter is to be capitalised, and by using a suitable pattern-matching rule.

This second approach does not run into problems with proper names not mentioned in the database. Without a database look-up, however, it may be difficult to obtain some information that we wish to be included in the generated lexicon entries for proper names. For example, including gender information in the entries for employee names is useful in anaphora resolution (see Sec. 3.D). It is generally difficult, however, to determine the genders of employees by applying pattern-matching rules on their names. It is also useful (e.g., for the mechanisms that will be discussed in Sec. 3.E) to include in the proper name entries information indicating the semantic types of the named entities (e.g., whether the proper name corresponds to a flight or a spare part). In some domains, proper names corresponding to different types of entities may be typographically very similar (e.g., there may be flight numbers such as *BA742* and spare part numbers such as *DX486*). In those cases, it may be difficult to determine the types of the named entities using pattern-matching rules, and one may be forced to introduce unnatural typographical notations.

C. Parsing and Semantic Analysis

The preprocessed input is subsequently parsed and analysed semantically using the techniques of Chaps. 4 and 5. This generates an expression in a meaning representation language (MRL; typically some form of logic) that is intended to capture formally what the NLDBI "understands" to be the semantics of the natural language input. MASQUE/SQL [5,7], for example, maps {1} (which is directed to a database containing geographic information) to {2}. (MASQUE/SQL's MRL is a subset of Prolog. Terms starting with capital letters are variables.)

What is the capital of each country bordering Greece? {1}

```
answer([Capital, Country]):-                                    {2}
  is_country(Country),
  borders(Country, greece),
  capital_of(Country, Capital).
```

Reply {2} shows that the user's question is a request to find all pairs [Capital, Country], such that Country is a country, Country borders Greece, and Capital is the capital of Country

In most NLDBIS (especially early ones), the parsing and semantic analysis are based on rather ad hoc grammars. There are, however, systems built on principled computational grammar theories (see, e.g., [6,24,65] for information on NLDBIS that are based on HPSG [64]. The grammars of these systems tend to be easier to comprehend and to extend. No matter how principled the grammar and how extensive the lexicon of the NLDBI are, there will always be user requests that the system fails to parse and analyze semantically in their entirety (e.g., sentences with unknown words, syntactic constructs not covered by the grammar, nongrammatical sentences). Natural language researchers are becoming increasingly interested in *robust-processing* techniques, that allow systems to recover from such failures. These techniques may, for example, combine the fragments of the natural language input that the system managed to process, to make reasonable guesses about the user's goals [70]. (Consult also Ref. 53 for an alternative method that can be used in NLDBIS.)

Most current NLDBIS, however, do not employ such mechanisms, generating simply an error message whenever they fail to process fully the user's input. If the failure was caused by an unknown word, the error message may report that word. In many other cases, however, the message may carry no information to help the users figure out which parts of their requests caused the failure. The situation is made worse by the fact that the linguistic coverage of most NLDBIS is not designed to be easily understandable by end-users. For example, MASQUE/SQL is able to answer *What are the capitals of the countries bordering the Baltic and bordering Sweden?*, which leads the user to assume that the system can handle all types of conjunctions. However, the system fails to process *What are the capitals of the countries bordering the Baltic and Sweden?* (MASQUE/SQL can handle conjunctions only if they involve progressive verb phrases.) The result is that the users are often forced to rephrase their requests, until a phrasing that falls within the linguistic coverage of the NLDBI is found. Because the users do not have a clear view of the NLDBI's linguistic coverage, several rephrasings may be needed. This is obviously very annoying for the users, and it may be one of the main reasons for the small market-share of NLDBIS. In contrast, the capabilities of graphic and form-based interfaces are usually clear from the options that are offered on the screen, and typically, any request that can be input can also be processed. We believe that it is crucial for NLDBIS to incorporate robust-parsing techniques, to overcome this problem (we return to this issue in Sec. 7). An alternative approach is to restrict severely, deliberately, and explicitly the natural language inputs that the users are allowed to enter, so that the linguistic capabilities of the system will be clearer to them. We discuss this approach in Sec. 6.

We should note, at this point, that there are also NLDBIS that appear to be using pattern-matching techniques instead of parsing and semantic analysis. [The authors of some of those NLDBIS seem unwilling to confirm this.] To illustrate a pattern-matching question-answering approach, let us assume that the database contains the following table that holds information about countries and their capitals. (In the examples of this chapter, we assume that the database is structured according to the *relational model* [25], in which information is stored in *relations*, intuitively tables consisting of rows and columns. Most of the techniques of this chapter apply to other database models as well.)

countries	
country	*capital*
Australia	Canberra
Greece	Athens
...	...

A simplistic pattern-matching system could use a rule such as the following, which would map *What is the capital of Greece?* to an appropriate MASQUE/SQL-like MRL expression:

If the input string matches the following pattern:
. . . "*capital*" . . . <*country_name*> . . .
then generate the MRL expression:

```
answer([Capital]):-
  capital_of(<country-name>, Capital).
```

The main advantage of the pattern-matching approach is its simplicity. The linguistic shallowness of this method, however, very often leads to irrelevant answers. In {1}, for example, the foregoing pattern-matching rule would cause the NLDBI to report the capital of Greece.

D. Semantic Postprocessing

The MRL expressions that are generated by the semantic analysis may be initially underspecified (i.e., they may leave the semantic contribution of some linguistic mechanisms unclear). For example, they may not specify the exact entities to which anaphoric expressions refer, or they may not specify the exact scopes of logical quantifiers (see the discussion in Chap. 5). There is usually one or more additional postprocessing phases (see Fig. 1) that resolve such underspecified elements of the logical expressions (e.g., they modify the logical expressions so that the referents of anaphoric expressions and the scopes of logical quantifiers are explicit; see Ref. 3 for a detailed discussion of how this is achieved in the CLE system).

Anaphoric expressions (e.g., *him*, *these*, *the project*) are supported by several NLDBIS. They are particularly useful because they allow follow-up questions to be short. For example, in the following interaction between the user and LOQUI [19], the user refers to the persons who report to E. Feron as *they*, instead of retyping the five names.

> *Who leads* TPI?
E. Feron
> *Who reports to him?*
C.Leonard, C.Willems, E.Bidonnet, P.Cayphas, J.P.Van Loo
> *What do they work on?*

project	worker
DOCDIS	C.Willems
	J.P.Van Loo
	P.Cayphas
EURS	C.Leonard
	C.Willems
	E.Bidonnet

> *Which of these are leaders?*
J.P.Van Loo

Anaphoric expressions are resolved by consulting a discourse model (see Chaps. 5 and 6). The discourse models of most NLDBIS are rather simplistic, as they are typically used only to resolve anaphoric expressions. They can be simply lists of entities that have been mentioned, along with information on the properties of these entities, and the locations in the discourse where the entities have been mentioned. For each anaphoric expression, the NLDBI searches this list until it finds an appropriate referent (e.g., for the pronoun *him*, a male singular entity has to be found). Various heuristics can be employed if an anaphoric expression has multiple possible referents in the list (e.g., the most recently mentioned possible referent may be preferred). (See Ref. 15 for information on the anaphora resolution module of a linguistic front-end. Reference 47 also discusses relatively simple methods for resolving pronoun anaphora that can be used in NLDBIS.)

More elaborate discourse models and reference resolution rules are needed to handle elliptical sentences, such as *How about* MIT? and *The smallest department?* in the following interaction (from Ref. 17) between the use and PARLANCE:

> *Does the highest paid female manager have any degrees from Harvard?*
Yes, 1.
> *How about* MIT?
No, none.
> *Who is the manager of the largest department?*

Name	Dept.	Count
Patterson	045	40

> *The smallest department?*

Name	Dept.	Count
Saavedra	011	2

(Consult Refs. 2 and 3 for a discussion of practical ellipsis resolution techniques.)

E. Type Hierarchy and Selectional Restrictions

During the semantic analysis and postprocessing, a type hierarchy of world entities, and a set of selectional restrictions are often employed. We discuss these in the following.

The type hierarchy shows the various types and subtypes of world entities that are modeled in the database. (The term "entity" should be interpreted with a broad meaning. It may include, for example, salaries or ages.) Figure 2 shows a fragment of a possible type hierarchy that could be used when interfacing with a company's database. (Similar hierarchies are used in TEAM [36,37], MASQUE/SQL [5,7], and the CLE ([3].) The hierarchy of Fig. 2 shows that, in the context of that particular database, a person can be either a client or an employee, that employees are divided into technicians, sales-persons, and managers, and so on. (The hierarchy does not need to be a tree. There could be, for example, a common subtype of *employee* and *client* for employees who are also clients of the company.)

The selectional restrictions typically specify the types of entities that the various predicate-arguments of the MRL may denote. To illustrate the use of the type hierarchy and the selectional restrictions, let us consider the hierarchy of Fig. 2, and let us assume that the database stores the salaries of all employees, but not the salaries of clients. (For simplicity, let us assume that an employee cannot also be a client.) Then, {3} will retrieve no information, because the database does not store salaries of clients. We would like the NLDBI to be able to detect that the question is conceptually problematic from the database's point of view.

What is the salary of each client? {3}

In MASQUE/SQL, {3} would receive an MRL expression such as {4}:

```
answer([Salary, Person]):-                    {4}
  is_client(Person),
  salary_of(Salary, Person).
```

The selectional restrictions could specify that (a) the argument of `is_client` must always denote an entity of type *client*, (b) that the first argument of `salary_of` must always denote a *salary*, and (c) that the second argument of `salary_of` must always denote an *employee*. Then, restriction (a) and the second line of expression {4} require the variable `Person` to denote a client. This, however, leads to a violation of restriction (c) in the third line of {4}: restriction (c) requires `Person` to

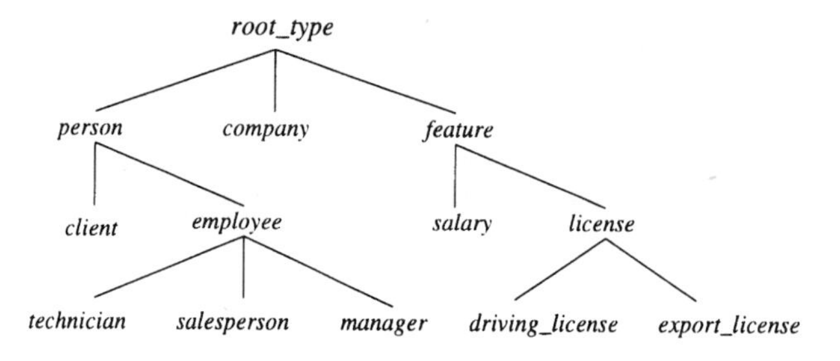

Fig. 2 A type hierarchy of world entities.

denote an employee; this is incompatible with the requirement that `Person` must denote a client, because neither *client* nor *employee* are subtypes of each other. The system would, thus, be able to detect that the question is conceptually anomalous. It is relatively easy to generate a message like "`A client does not have a salary.`" to report this anomaly to the user. This could be achieved, for example, by associating a message template of the form "`A` *X* `does not have a salary.`" with constraint (c), and by replacing *X* with the type-name of the second argument of `salary_of` whenever the constraint is violated.

In contrast, {5} would receive {6}. Assuming that the argument of `is_technician` must denote an entity of type *technician*, {6} does not violate any restriction, because *technician* is a subtype of *employee*.

What is the salary of each technician? {5}

```
answer([Salary, Person]):-                                          {6}
  is_technician(Person),
  salary_of(Salary, Person).
```

The type hierarchy and the selectional restrictions are also useful in disambiguation. For example, from a NLDBI's point of view, expression {7} is potentially ambiguous: *with a driving licence may refer either to employees* or to *company*. The two readings correspond to {8} and {9}, respectively.

List all employees in a company with a driving licence. {7}

```
answer([Employee]):-                                                {8}
  is_employee(Employee),
  is_company(Company),
  is_licence_for(driving, Licence),
  in(Employee, Company),
  has(Employee, Licence).

answer([Employee]):-                                                {9}
  is_employee(Employee),
  is_company(Company),
  is_licence_for(driving, Licence),
  in(Employee, Company),
  has(Company, Licence).
```

Of course, humans can figure out immediately that the intended reading is {8}, because typically persons and not companies have driving licences. (In *List all employees in a company with an export licence*, however, the situation would be reversed.) This knowledge can be encoded using selectional restrictions. One could specify that the arguments of `is_employee` and `is_company` must denote entities of types *employee* and *company*, respectively, that when the first argument of `is_licence_for` is `driving`, the second argument must denote a *driving_licence*, and that if the second argument of `has` denotes a *driving_licence*, the first argument must denote a *person*. This would rule out {9}, in which the second argument of `has` denotes a *driving_licence*, and the first one a *company*.

Anaphora resolution methods can also exploit the type hierarchy and the selectional restrictions. Knowing, for example, that only employees have salaries, limits the possible referents of *his* in *What is his salary?* to male employees, excluding any previously mentioned male clients.

Type hierarchies and selectional restrictions have been employed in many natural language processing areas. In systems that target broad knowledge domains (e.g., newspaper articles), type hierarchies, and selectional restrictions are often difficult (if at all possible) to construct, because of the very large number of involved entity types, and the enormous variety of possible relations between those types [see Ref. 2 for related discussion]. Databases, however, typically store information about very few types of entities; hence, type hierarchies and selectional restrictions become manageable. In fact, very often the entity types that are relevant to the database and many of their relations will already have been identified during the design of the database (e.g., in the form of "entity–relationship" diagrams [77]), and this information can be exploited when specifying the type hierarchy and the selectional restrictions of the NLDBI.

F. Paraphrasing the Input

Whenever the NLDBI "understands" a natural language request to be ambiguous, several MRL expressions are generated at the end of the postprocesing, each corresponding to what the NLDBI considers to be a possible reading of the input. (The input may be truly ambiguous, as in {10}. It may also be—as it often is—the case that the NLDBI has generated more readings than the input can actually have.)

Which disabled and female applicants speak German? {10}

Some NLDBIS employ heuristics in the form of preference measures to determine the most likely reading [see Ref. 3 for related discussion]. Another common technique is to generate an unambiguous paraphrase of each reading, asking the user to select the intended one. In {10}, for example, an NLDBI could ask the user to select one of the following:

Which applicants that are both disabled and female speak German? {11}

Which disabled and which female applicants speak German? {12}

For questions, if there are relatively few possible readings and the answers are short, the NLDBI may generate the answers to all the readings, showing which answer corresponds to which paraphrased reading, without asking the user. Paraphrasing the input is useful even when the NLDBI generates only one reading. In the following interaction [from Ref. 75] between the user and ASK, for example, ASK repeats the user's input with the pronoun replaced by what the system assumes to be its referent. This prevents the user from being misled, if the system has not chosen the referent the user had in mind.

> *Is there a ship whose destination is unknown?*

```
yes
```

> *What is it?*

```
What is [the ship whose destination is unknown]?
Saratoga
```

Repeating the user's input with anaphoric expressions replaced by their referents is relatively easy. (We can augment the list of previously mentioned entities of the discourse model to record the natural language expressions that introduce the entities; see Sec. 3.D.) In the general case, however, one has to use natural language generation techniques [see Chapter 7; Refs. 31 and 55 also discuss paraphrasing techniques in NLDBIS].

G. From Meaning Representation to Database Language

The MRL expressions that are generated at the end of the semantic postprocessing are subsequently translated into expressions of a database language (typically SQL [60]) that is supported by the underlying database management system (DBMS). (The DBMS is the part of the database system that is responsible for manipulating the information in the database.) Various methods to translate from forms of logic to and from database languages have been proposed [see, e.g., 2,5,33,56,58,67,77,78]. The full details of some of these proposals are highly technical; hence, they are beyond the scope of this chapter. Here, we only attempt to highlight some of the central ideas on which several of these proposals are based.

Let us return, for example, to the MRL expression of {2}, repeated here as {13}:

```
answer([Capital, Country]):-                                   {13}
  is_country(Country),
  borders(Country, greece),
  capital_of(Country, Capital).
```

Before MRL expressions such as {13} can be translated into database language, one has to link primitive MRL expressions, such as constants (e.g., `greece`) and predicates functors (e.g., `is_country`, `borders`), to database constructs. For MRL constants, one has to specify a systematic way to map them to database values (e.g., the country denoted by the MRL constant `greece` may be represented in the database as *Greece* or *GR*). In the case of predicate functors, one must link each functor that may appear in the MRL expressions to a database construct that shows for which arguments the predicates of the functor hold. (There are some cases for which this linking is impossible. We discuss this in Section 5.A.)

countries_info		
country	*capital*	*currency*
Italy	*Rome*	*Lira*
Australia	*Canberra*	*Dollar*
Greece	*Athens*	*Drachma*
...	...	...

borders_info	
borders1	*borders2*
Australia	*Pacific*
Greece	*Bulgaria*
Albania	*Greece*
Bulgaria	*Greece*
...	...

Assuming, for example, that the database contains the tables above, one can simply link the functor `borders` to the table *borders_info*, meaning that `borders(X, Y)` is true if and only if *borders_info* contains a row for X and Y. (We make a *closed-world assumption*, i.e., we assume that if a pair of two geographic areas is not included in *borders_info*, the two areas do not border each other.)

The database constructs to which functors are linked need not be stored directly in the database. They may be computed indirectly from the contents of the database. The functors `is_country` and `capital_of`, for example, could be linked to the tables *countries* and *capitals*. These are obtained by retaining only the first column or by dropping the third column of *countries_info* respectively. (In SQL, these derived tables can be defined as "views" [60].)

countries
country
Italy
Australia
Greece
...

capitals	
country	*capital*
Italy	*Rome*
Australia	*Canberra*
Greece	*Athens*
...	...

Once primitive MRL expressions have been linked to the information of the database, MRL formulas can be translated into database language using a set of translation rules. These are of two sorts: base (nonrecursive) rules that translate atomic MRL formulas [e.g., `borders(Country, greece)`], and recursive rules that translate nonatomic MRL formulas [e.g., {13}] by recursively calling other rules to translate their subformulas.

In {13}, a base rule could map `borders (Country, greece)` to {14}, which generates the table of {15}; {14} means that the resulting table should have the same contents as *borders_info*, except that only rows whose *borders*2 value is `Greece` should be retained. Intuitively, {15} shows all the argument values for which `borders(Country, greece)` holds. (In this case, the second argument is a constant; and hence, the second column of {15} contains the same value in all rows.)

```
SELECT *
FROM borders_info
WHERE borders2 = 'Greece'
```
{14}

(result of {14})	
*borders*1	*borders*2
Albania	*Greece*
Bulgaria	*Greece*
Turkey	*Greece*
...	...

{15}

The base rule that maps `borders(Country, greece)` to {14} could be formulated as follows: each atomic formula $\pi(\tau_1, \tau_2, \tau_3, \ldots, \tau_n)$ (where π is a predicate functor and $\tau_1, \ldots, \tau_n$ predicate-arguments) is to be mapped to the SQL expression of {16}, where *table* is the name of the database table to which the predicate functor is linked, $\kappa_1, \ldots, \kappa_n$ are all the MRL constants among $\tau_1, \ldots, \tau_n$, $h(\kappa_n)$ are the database values that correspond to these constants, and $\zeta, \ldots, \zeta_n$ are the names of the columns that correspond to the predicate-argument positions where $\kappa_1, \ldots, \kappa_n$ appear. (If there are no constants among $\tau_1, \ldots, \tau_n$, the WHERE clause is omitted.)

```
SELECT *                                                            {16}
FROM table
WHERE ζ1 = h(κ1) AND ζ2 = h(κ2) AND ζ3 = h(κ3) AND . . . AND ζn = h(κn)
```

Rule {16} would also map the `is_country(Country)` and `capital_of (Country, Capital)` of {13} to {17} and {18}, respectively.

```
SELECT *                                                            {17}
FROM countries
```

```
SELECT *                                                            {18}
FROM capitals
```

As an example of a recursive translation rule, an MRL conjunction of the form:

$$\pi_1(\tau_1^1, \ldots, \tau_{n_1}^1), \pi_2(\tau_1^2, \ldots, \tau_{n_2}^2), \ldots, \pi_k(\tau_1^k, \ldots, \tau_{n_k}^k)$$

could be mapped to {19}, where $trans(\pi_1(\tau_1^1, \ldots, \tau_{n_1}^1)), \ldots, trans(\pi_k(\tau_1^k, \ldots, \tau_{n_k}^k))$ are the SQL translations of $\pi_1(\tau_1^1, \ldots, \tau_{n_1}^1), \ldots, \pi_k(\tau_1^k, \ldots, \tau_{n_k}^k)$, and *variable_constraints* stands for an expression that requires columns corresponding to the same variable to have the same values. For example, {20} would be mapped to {21}, where "({17})," "({14})," "({18})" stand for the expressions of {17}, {14}, and {18}, respectively, {21} generates {22}, that intuitively shows all the combinations of predicate-argument values in {20} that make {20} true. (`r1.country` refers to the *country* column of the table generated by {17}, `r2.borders1` refers to the *borders*1 column of the table generated by {14}, and so on.) (Several improvements can be made to our translation rules, for example to remove identical columns from {22}.)

```
SELECT *                                                            {19}
FROM trans(π1(τ1^1, ..., τn1^1)), trans(π2(τ1^2, ..., τn2^2)), ..., trans(πk(τ1^k, ..., τnk^k))
WHERE variable_constraints
```

```
is_country(Country),                                                {20}
borders(Country, greece),
capital_of(Country, Capital)
```

```
SELECT *                                                            {21}
FROM ({17}) AS r1, ({14}) AS r2, ({18}) AS r3
WHERE r1.country = r2.borders1 AND r2.borders1 = r3.country
```

(result of (21)) {22}

r1.country	r2.borders1	r2.borders2	r3.country	r3.capital
Albania	Albania	Greece	Albania	Tirana
Bulgaria	Bulgaria	Greece	Bulgaria	Sofia
Turkey	Turkey	Greece	Turkey	Ankara
...	...	...	...	...

Another recursive rule would generate the SQL expression for the overall {13}, using {21} (the translation of {20}). In MASQUE/SQL, the resulting SQL expression would generate a table containing only the last column of {22}.

In some systems (e.g., [56]) the MRL expressions are translated first into more theoretical database languages (e.g., relational calculus [77]). Although these languages are not supported directly by the DBMSs, they are closer to logic-based MRLS, which simplifies the formulation of the translation rules and the proof of their correctness. In these cases, there is a separate mapping from the theoretical database language to the language that is actually supported by the DBMSs (e.g., SQL). This extra mapping is often trivial (e.g., part of SQL is essentially a notational variant of the tuple relational calculus).

At the end of the translation from MRL to database language, the generated database language expression is executed by the DBMS. This retrieves the appropriate information from the database (in the case of questions), or modifies the contents of the database (in the case of requests to update the database; we discuss these in Sec. 5.B).

H. Response Generation

So far, we have focused mainly on the interpretation of the natural language input. Generating an appropriate response to the user is also very important.

For questions, simply printing the database constructs that were retrieved by the execution of the database language expression (table rows in the case of relational databases) is not always satisfactory. For example, the retrieved database constructs may contain encoded information (e.g., department codes instead of department names). An output formatter (see Fig. 1) is then needed to convert the information to a more readable format. The retrieved information may also be easier to grasp if presented in graphic form (e.g., pie-charts). Commercial NLDBIS often provide such graphic output facilities.

In other cases, the NLDBIS may fail to "understand" the user's request. The cause of the failure (e.g., unknown word, syntax too complex, conceptually ill-formed input) should be explained to the user. (Some methods to detect conceptually ill-formed requests were discussed in Sec. 3.E.)

A more challenging task is to generate what is known as *cooperative responses*. These are needed when the user's requests contain false presuppositions, or do not express literally what the users want to know. These cases are illustrated in the following interaction with LOQUI (based on examples from {19} and {20}):

> > *Does every person that works on 20 projects work on* HOSCOM*?*
> `There is no such person.`
> > *Does David Sedlock work on something?*
> `Yes. BIM_LOQUI, MMI2, NLPAD.`

In the first question, the system has detected the false presupposition that there are people working on 20 projects, and it has generated a suitable message. In the second and third questions, the system did not generate a simple "yes" or "no"; it also reported additional information that the user probably wanted to know.

Occasionally, such cooperative responses can be generated using relatively simple mechanisms. Let us consider, for example, yes/no questions that contain expressions introducing existential quantifiers (e.g., *something, a flight*), similar to the second question of the foregoing dialogue. Whenever the answer is affirmative, one strategy is to print a "yes," followed by the answer to the question that results when the expressions that introduce existential quantifiers are replaced with suitable interrogatives (e.g., *what, which flight*; this is easier to achieve by operating on the MRL representation of the question). In the second question of the foregoing dialogue, this generates a *yes* followed by the answer to *On what does David Sedlock work?*

When the answer is negative, one can generate a *no*, followed by the answer to the question that results when a constraint of the original question is modified. In the following dialogue [based on an example from Ref. 49], the NLDBI has replaced the constraint that the flight must be an American flight, with the similar constraint that the flight must be a United flight.

> *Does American Airlines have a night flight to Dallas?*
`No, but United has one.`

One must be careful to limit the types of constraints that can be modified. In the previous dialogue, for example, the NLDBI must not be allowed to modify the constraint that the flight must be to Dallas, otherwise [as pointed out in Ref. 49] the dialogue could have been:

> *Does American Airlines have a night flight to Dallas?*
`No, but they have one to Miami.`

Kaplan [50–52] discusses relatively simple mechanisms for generating cooperative responses, including methods to detect false presuppositions. It is not always possible, however, to generate reasonable cooperative responses using simple mechanisms. Sometimes, to generate an appropriate cooperative response, the NLDBI must be able to reason about the user's intentions, as discussed in Chap. 6. This requires an inferencing component. We return to this issue in Sec. 5.A.

4. PORTABILITY

A significant part of NLDBI research has been devoted to portability issues. In the following, we discuss three kinds of NLDBI portability. Most of the discussion relates to the architecture of Fig. 1.

A. Knowledge-Domain Portability

Current NLDBIs can cope only with natural language requests that refer to a particular *knowledge domain* (e.g., requests only about the employees of a company, or only about train schedules). Before a NLDBI can be used in a new knowledge domain, it has to be *configured*. This typically involves modifying the preprocessing rules (see Sec. 3.B), the lexicon, the type hierarchy, and selectional restrictions (see Sec. 3.E), and the information that links primitive MRL expressions to database constructs (see Sec. 3.G). One of the advantages of the architecture of Fig. 1 is that it clearly separates these *domain-dependent* modules from the rest of the system. Different assumptions about the skills of the person who will configure the NLDBI are possible:

PROGRAMMER. In some systems a (preferably small and well-defined) part of the NLDBI's code has to be rewritten during the configuration. This requires the configurer to be a programmer.

KNOWLEDGE ENGINEER. Other systems provide tools that allow the configuration to be carried out without any programming. They still assume, however, that the configurer is familiar with knowledge representation techniques (e.g., logic), databases, and linguistic concepts. For example, MASQUE (an enhanced version of CHAT-80 from which MASQUE/SQL was developed) provides a "domain-editor" [10], that helps the configurer "teach" the system new words and concepts that are common in the new knowledge domain. The use of the domain-editor is illustrated in the following dialogue, in which the configurer teaches the system the verb *to exceed* (as in *Does the population of Germany exceed the population of Portugal?*):[6]

```
editor> add verb
what is your verb? exceed
what is its third singular present? exceeds
what is its past form? exceeded
what is its participle form? exceeding
to what set does the subject belong? numeric
is there a direct object? yes
to what set does it belong? numeric
is there an indirect object? no
is it linked to a complement? no
what is its predicate? greater_than
```

In the foregoing dialogue, apart from teaching the system the morphology of the various forms of the verb, the configurer has also declared that the verb introduces MRL predicates of the form `greater_than(X,Y)`, where `X` and `Y` denote entitites that belong to the *numeric* type of the type hierarchy (see Sec. 3.E).

Some NLDBIS (e.g., PARLANCE [17], CLE [3]) provide morphology modules that allow them to infer the various forms of the new words without asking the configurer. Other systems have large built-in lexicons that cover the most common words of typical database domains, minimising the need to elicit lexical information from the user. (According to the vendor of one current commercial NLDBI, their system comes with a built-in lexicon of 16,000 words.) Occasionally, it is also possible to construct automatically lexicon entries from the contents of the database (see e.g., the discussion on proper-names in Sec. 3.B).

DATABASE ADMINISTRATOR. In practice, NLDBIS may often be configured by the administrators of existing databases. In that case, it is reasonable to assume that the configurers will be familiar with database concepts and the targeted database, but not with MRLS or linguistics. This is the approach adopted by the designers of TEAM [36,37]. In the following dialogue [based on an example from Ref. 36], TEAM collects information about the database table ("`file`") `chip`. Notice that TEAM uses data-

[6] MASQUE's domain-editor does not actually shield completely the knowledge-engineer from the programming language. The knowledge engineer is still required to write some Prolog to relate MRL expressions to the database.

base terminology (e.g., "primary key"), but not terms from linguistics or logic (e.g., "participle" and "predicate" in the foregoing MASQUE dialogue).

```
file name: chip
fields: id maker width
subject: processor
synonyms for processor: chip
primary key: id
Can one say "Who are the processors?": no
Pronouns for subject (he, she, it, they): it
field: MAKER
type of field (symbolic, arithmetic, feature): symbolic
...
```

END-USER. Other NLDBI designers emphasise that the configuration can never be complete, and that the end-users should be allowed to teach the system new words and concepts using natural language. This approach is illustrated in the following interaction with PARLANCE [17], where the user defines the terms *yuppie* and *rich employee*:

> *define: a yuppie is a person under 30 with a graduate degree who earns over 5K per month*
> *define: a rich employee is an employee who earns over 100K per year*

Although large built-in lexicons and configuration tools have made NLDBIS much easier to port to new knowledge domains, configuring a NLDBI for a new knowledge domain is often still a complicated task, and it may still require at least some familiarity with basic natural language processing concepts. Alternative graphic and form-based database interfaces are usually easier to configure. As we noted in Sec. 2, this is probably one of the main reasons for the small market share of NLDBIS.

B. DBMS Portability

NLDBIS that adopt widely supported database languages (e.g., SQL [60]) can often be used with different DBMSS that support the same database language with only minor modifications. If the original database language is not supported by the new DBMS, the part of the NLDBI that is responsible for translating from MRL to database language (the *database back-end* of Fig. 1) has to be rewritten. In that case, one of the advantages of the architecture of Fig. 1 is that no modifications are needed in the linguistic front-end, provided that the knowledge domain remains the same.

The architecture of Fig. 1 also leads to the development of generic linguistic front-ends that, apart from NLDBIS, can be used as components of other natural language processing systems. (An example of such a system is CLE [3], which has been used as a component of both a NLDBI and a machine translation system.)

C. Natural Language Portability

Almost all of the existing NLDBIS require the user's requests to be formulated in English. Modifying an existing NLDBI to be used with a different natural language can be a formidable task, because the NLDBI may contain deeply embedded assumptions about the targetted natural language. Some information about an attempt to build a Portuguese version of CHAT-80 [80] has been provided [54]. Information about Swedish, French, and Spanish versions of the CLE can be found in Refs. [3,66].

5. ADVANCED ISSUES

Having introduced the basic components of the typical NLDBI architecture, we now move on to some more advanced issues, namely the "doctor-on-board" problem, database updates, metaknowledge, modal, and temporal questions.

A. The "Doctor-on-Board" Problem

The discussion of Sec. 3.G assumed that each predicate functor can be linked to a database construct (a table in the relational model) that shows the arguments for which the predicates of the functor hold. That database construct, it was assumed, is either stored directly in the database, or can be computed from the contents of the database. There are some cases, however, for which this assumption does not hold. The doctor-on-board problem [63] is a well-known example of such a case.

Let us imagine a database that contains only the following table, which holds information about the ships of a fleet. The third column shows if a doctor is on board the ship (*y*) or not (*n*).

ships_info		
ship	*crew*	*doctor*
Vincent	420	*y*
Invincible	514	*y*
Sparrow	14	*n*
...	...	...

Let us also consider question {23} which would receive the MASQUE/SQL-like MRL expression of {24} that shows that the answer should be affirmative if there is a `doctor D` that is on board Vincent.

Is there a doctor on the Vincent? {23}

```
answer([]):-
  is_doctor(D),
  on_board(D, vincent).
```
{24}

To apply the MRL to database language translation method of Sec. 3.G, one needs to link `is_doctor` to a database table that shows the entities that are doctors. This is impossible, however, because the database (which contains only *ships_info*) does not list the doctors. Similarly, `on_board` has to be linked to a table that shows which entities are on board which ships. Again, this table cannot be computed from the information in the database.

What is interesting is that {24} is equivalent to {25}, where `doctor_on_board (D,S)` is intended to be true if `is_doctor(D)` and `on_board(D,S)` are both true. Unlike {24}, {25} poses no problem for the translation method of Sec. 3.G, because `doctor_on_board` can be linked to a version of *ships_info* without the *crew* column.

```
answer{[]}:-                                          {25}
  doctor_on_board(D, vincent).
```

The problem is that {23} cannot be mapped directly to {25}. The MRL expression {24} is typical of what most natural language front-ends would generate, with predicates that are introduced by linguistic constituents [e.g., `is_doctor(X)` is introduced by the noun phrase *a doctor*, and `on_board(X, vincent)` by the prepositional phrase *on Vincent*]. What is needed is a mechanism to convert the MRL expressions that are generated by the linguistic front-end to equivalent MRL expressions that contain only predicate functors that can be linked to appropriate database tables. A mechanism that can, among other things, perform this conversion has been proposed [67]. Roughly speaking, the conversion is carried out by an inferencing component that employs rules, such as {26}.

$$(\texttt{doctor}(\xi_1), \texttt{on\_board}(\xi_1, \xi_2)) \equiv (\texttt{doctor\_on\_board}(\xi_2)) \qquad \{26\}$$

Expression {26} allows any MRL expression of the form $\texttt{doctor}(\xi_1)$, $\texttt{on\_board}(\xi_1, \xi_2)$ (where ξ_1 and ξ_2 stand for MRL terms) to be replaced by $\texttt{doctor\_on\_board}(\xi_2)$. Therefore {26} would license the conversion from {24} to {25}. Then, {25} would be translated into database language as discussed in Sec. 3.G.

Inferencing components are also useful in cases where the NLDBI has to reason about the user's intentions to generate cooperative responses (see Sec. 3.H). They can also be employed to allow the NLDBI to deduce new facts from the information in the database (e.g., with appropriate reasoning rules, a NLDBI connected to a hospital's database could answer *Which patients need attention overnight?* even if the database does not flag explicitly patients who need overnight attention). In some NLDBIS (e.g., INTELLECT), the inferencing component can be a full expert system, which acts as an intermediate layer between the linguistic front-end and the database back-end.

A detailed description of CLARE, a NLDBI with reasoning capabilities, has been provided [4]. CLARE is based on CLE [3], and includes the inferential mechanism of Ref. [67].

B. Database Updates

Apart from answering questions, some NLDBIS also allow the user to *update* the information in the database. This is illustrated in the following dialogue with ASK [75,76]:

> *What is the home port of the Tokyo Maru?*

`Yokohama`

> *Home port of the Tokyo Maru is Hong Kong.*

`Yokohama has been replaced by Hong Kong as home port of Tokyo Maru.`

ASK knows that a ship may have only one home port; hence, in the second request it replaced Yokohama by Tokyo Maru. In contrast, it knows that a ship may carry many types of cargo. Hence, the input *The Tokyo Maru carries coal.* would lead ASK to add coal to the previously known cargoes of the Tokyo Maru, instead of replacing the previous information.

Natural language database updates can be difficult to process because they may lead to unanticipated side effects. These side effects are due to database constraints that the user is not aware of. This is illustrated in the following dialogue (the example is borrowed from [30]).

> *List the employees and their managers*

```
employee  manager
Adams     Fisher
White     Baker
Brown     Jones
Smith     Jones
```

> *Change Brown's manager from Jones to Baker.*

```
Done.
```

> *List the employees and their managers.*

```
employee  manager

Adams     Fisher
White     Baker
Brown     Baker
Smith     Baker
```

Notice that Smith's manager also changed from Jones to Baker, although this was not requested. This happens because in the database (which contains the two following tables) employees are linked to managers indirectly, through their departments.

employees_table		
employee	*salary*	*department*
Adams	3000	*inventory*
White	3500	*marketing*
Brown	2500	*sales*
Smith	2500	*sales*

departments_table	
department	*manager*
sales	*Jones*
marketing	*Baker*
inventory	*Fisher*

To satisfy the user's request to change the manager of Brown, the system has changed the sales manager from Jones to Baker. This caused all other employees in the sales department (e.g., Smith) to receive Baker as their new manager. Users not aware of the database structure would find this behavior hard to explain. (NLDBI users are usually not informed about the structure of the database. Not having to be aware of the database structure is supposed to be an advantage of NLDBIS over graphic or form-based interfaces.)

The indirect link between employees and managers actually makes the update request in the foregoing dialogue ambiguous: instead of changing the sales manager from Jones to Baker, the NLDBI could have moved Smith from the sales department (which is managed by Jones) to the marketing department (which is managed by Baker). PIQUET [30] maintains a model of the user's conceptual view of the database. Whenever a user input reflects awareness or presupposition of a database object, link, or restriction, the model is modified accordingly. If the user enters an ambiguous request (in the foregoing sense), the NLDBI consults the model to select the reading that has the fewest side effects relative to the user's current view of the database. In the foregoing dialogue, after *List the employees and their managers.* has been answered, the user's view of the database corresponds to the following table:

employee	*manager*
Adams	*Fisher*
White	*Baker*
Brown	*Jones*
Smith	*Jones*

If *Change Brown's manager from Jones to Baker*, is interpreted as a request to change the sales manager from Jones to Baker, the answer causes the user's view to become as shown as follows on the left. A side effect (shown in bold) occurs. If, in contrast, the sentence is interpreted as a request to move Brown from the sales department to the marketing department, the user's view becomes as shown as follows on the right:

employee	*manager*
Adams	*Fisher*
White	*Baker*
Brown	*Baker*
Smith	**Baker**

employee	*manager*
Adams	*Fisher*
White	*Baker*
Brown	*Baker*
Smith	*Jones*

The second interpretation introduces no side effects to the user's view, and would have been preferred by PIQUET.

C. Metaknowledge, Modal, and Temporal Questions

This section discusses briefly some types of questions that present particular interest.

METAKNOWLEDGE QUESTIONS. Apart from questions about the entities that are represented in the database, the user may want to submit queries about the conceptual organisation of the database's knowledge (e.g., *What information does the database contain?, What are the properties of employees?*). These can be considered *metaknowledge questions*, as they refer to knowledge about the database's knowledge. The following example [75] shows how ASK reacts to a question of this kind.

> > *What is known about ships?*

```
Some are in the following classes: navy, freighter,
tanker
All have the following attributes: destination, home
port
. . .
```

Most NLDBIS do not support metaknowledge questions.

MODAL QUESTIONS. Database systems usually enforce a set of *integrity constraints* to guard against contradicting or incorrect information (e.g., an employee cannot have two dates of birth, and in practice, the age of an employee is never greater than 80), and to enforce corporate policies (e.g., that employees cannot earn more than their managers, or that all employees must be older than 20). A method has been presented [57] that exploits these integrity constraints to answer *modal questions*, questions that ask if something may or must be true (e.g., *Can an employee be 18 years old?, Is it the case that J.Adams must earn more than T.Smith?*).

TEMPORAL QUESTIONS. Most NLDBIS do not provide adequate support for temporal linguistic mechanisms For example, users are often allowed to use very few (if any) verb tenses, temporal adverbials (e.g., *for two hours, before 5:00pm*), or temporal subordinate clauses (e.g., *while J.Adams was personnel manager*).

Androutsopoulos and associates [6,9] discuss how some of these mechanisms can be supported when constructing NLDBIs for temporal databases (databases designed to handle time-dependent data).

6. RESTRICTED INPUT SYSTEMS

We noted in Sec. 3.C that one of the main problems of NLDBIs is that their linguistic coverage is usually not obvious to the user. This often has the annoying consequence that users are forced to rephrase their requests several times, until a phrasing that falls within the linguistic coverage of the system is found. To overcome this problem, most NLDBI developers attempt to expand the linguistic coverage of their systems, hoping that eventually it will be possible to process successfully most of the users' inputs. An alternative approach, which we discuss in this section, is to restrict drastically the linguistic coverage of the NLDBI, and to do this in a manner that allows the users to obtain a clearer view of the system's linguistic capabilities.

A. Controlled Language

One way to make the linguistic capabilities of NLDBIs clearer is to support only a natural language fragment for which syntax is made explicitly known to the user. (The term *controlled language* is often used to refer to such fragments.) The fragment must be selected carefully, to make its syntax easy to explain. A well-chosen fragment may also allow a direct mapping from natural language to database language.

In PRE [34], for example, the user was allowed to input only questions of the following pattern:

what is/are
conjoined noun phrases
  nested relative clauses
    conjoined relative clauses

The user could, for example, enter the following question [simplified example from Ref. 34]. Line 1 contains the conjoined noun phrases. Lines 2–4 contain the nested relative clauses (they are nested, because 3 refers to *schedules* in 2, and 4 refers to *appointments* in 3). Lines 5–7 contain the conjoined relative clauses (these are not nested; they all refer to the *orders* of 4).

what is/are	
the names, ids, and categories of the employees	(1)
who are assigned schedules	(2)
that include appointment	(3)
that are executions of orders	(4)
whose addresses contain 'maple' and	(5)
whose dates are later than 12/15/83 and	(6)
whose statuses are other than 'comp'	(7)

In contrast, the following question was not accepted by PRE. In this case, line 2 contains the only clause of the nested relative clauses part. Both 3 and 4 are conjoined relative clauses referring to the *schedules* of 2. However, 5 is nested relative to 4, because it refers to the *orders* of 4, not the *schedules* of 2. PRE's question pattern does not allow conjoined relative clauses to be followed by other nested relative clauses.

what are
the addresses of the appointments (1)
  that are included in schedules (2)
    whose call times are before 11:30 and (3)
    that are executions of orders (4)
      whose statuses are other than 'comp' (5)

Epstein claims that PRE's question pattern is easy to remember, although this is probably arguable.[7] PRE's question pattern is also chosen to simplify the retrieval of information from the database. PRE adopts an entity–relationship-like database model [77]. Figure 3, for example, shows the structure of the database to which the previous two examples refer.[8] Oval boxes correspond to entity types, rectangular boxes correspond to attributes of entities, and continuous lines correspond to relations between entities.

The conjoined noun phrases of PRE's question pattern correspond to a projection operator [77]. Projection operators specify which attributes (e.g., *name*, *address*) of an entity are to be reported. The nested relative clauses correspond to traversals of continuous line links (relations between entities). Finally, the conjoined relative clauses correspond to select operators. These pick individual entities of particular types according to certain criteria.

For example, to answer the first question of this section, PRE first transforms the conjoined relative clauses to a select operator. This selects all the *order* entities

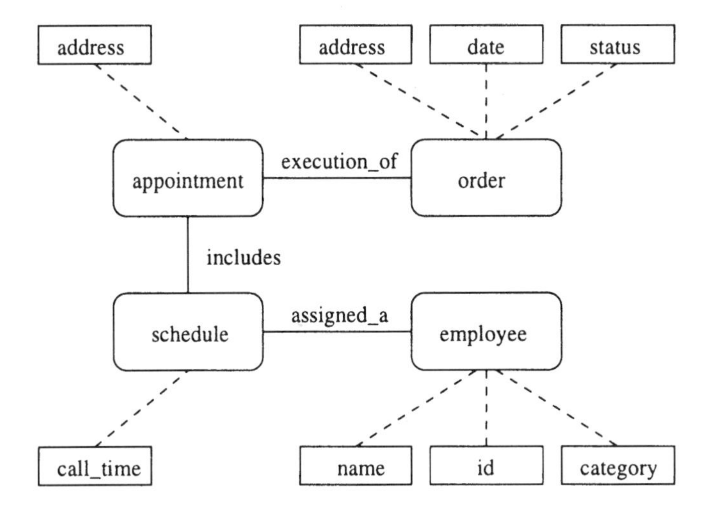

Fig. 3 The structure of a PRE-like database.

[7] APPEAL by Winformation Software adopts a similar controlled language approach, but with a more restricted and easier to remember syntax; see `http://wwwiz.com/home/mather`.

[8] [34] does not provide much information about PRE's database. Figure 3 may not reflect the exact form of PRE's database model.

that contain *maple* in their *address*es, and that have *date*s later than 12/15/83, and *status*es other than *comp*. Next, PRE transforms the nested relative clauses to a sequence of relations (continuous line) traversals. In our example, this connects the *orders* that were selected in the previous step to *appointment* by the *execution_of* relations, *appointment*s to *schedule* by the *includes* relation, and *schedule*s to *employees* by the *assigned_a* relation. Finally, PRE transforms the conjoined noun phrases to a projection operator. For each *employee* entity reached during the previous step, the system reports its *name*, *id*, and *category*.

The main disadvantage of NLDBIS with controlled languages is that the user has to be taught the syntax of the supported fragment. In applications for which the users need to be able to submit complex queries, it may not be easy to define a natural language fragment that is both rich enough to allow these queries and easy to understand.

B. Menu-Based Systems

Another approach is to require the users to form their requests by choosing words or phrases from menus displayed on the screen. Figure 4, for example, shows the initial screen of NLMENU [73,74]; the example is from Ref. 74. The highlighted border of the *commands* menu indicates that the first word of the user's request has to be chosen from that menu. (Symantec's Q&A offered a similar menu-based mode of interaction.)

NLMENU used a context-free semantic grammar (see Sec. 2). Whenever a new word or phrase was selected, NLMENU used the grammar to determine which words or

commands	***nouns***		***modifiers***
Find Delete	suppliers		whose part city is
Insert	parts		whose color is
attributes	shipments		whose part name is
weight	<specific suppliers>		whose part# is
quantity	<specific parts>		whose supplier city is
city	<specific shipments>		whose supplier name is
color	***comparisons***		whose supplier supplier# is
name	between		whose shipment part# is
part#	greater than		whose shipment supplier# is
supplier#	less than		whose supplier status is
status	greater than or equal to		whose part weight is
	less than or equal to		whose shipment quantity is
	equal to		which are shipments of
	connectors		which were shipped by
	the number of	of	who ship
	the average	and	who supply
	the total	or	which are supplied by
>			

Fig. 4 The initial NLMENU screen.

phrases could be attached to the string that the user had assembled up to that point, to form a request that the system could process. These words and phrases were then shown in the menus. Figure 5, for example, shows the NLMENU screen after entering *Find color and name of parts*. Notice how the *modifiers* menu has changed to contain only completions that the system can handle.

The main advantage of menu-based NLDBIS is that only requests that the systems can handle can be input, and the users can obtain an understanding of the systems' capabilities by browsing the menus. This approach works well in domains for which small dictionaries and short questions are adequate. With large dictionaries and longer questions, however, the menus proliferate, and become lengthy and difficult to use.

7. CONCLUSIONS

This chapter has attempted to serve two purposes: to introduce the reader to the field of NLDBIS by describing some of the central issues, and to highlight the state of the art in NLDBIS by outlining the facilities, methods, and problems of typical implemented systems. In the light of the discussion in the previous sections, the following observations seem valid to us.

The area of NLDBIS is mature enough, to the extent that usable systems can be constructed for real-world applications, and indeed several commercial NLDBIS have appeared.

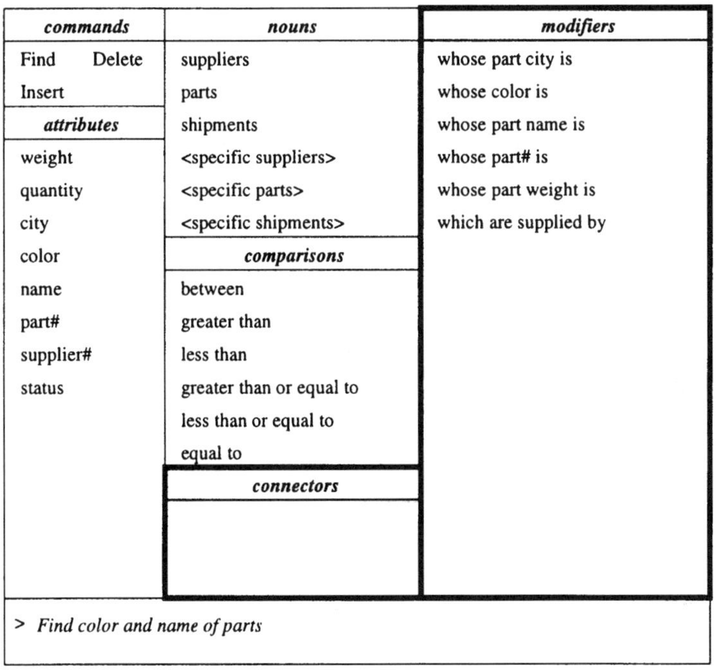

commands	***nouns***	***modifiers***
Find Delete	suppliers	whose part city is
Insert	parts	whose color is
attributes	shipments	whose part name is
weight	<specific suppliers>	whose part# is
quantity	<specific parts>	whose part weight is
city	<specific shipments>	which are supplied by
color	***comparisons***	
name	between	
part#	greater than	
supplier#	less than	
status	greater than or equal to	
	less than or equal to	
	equal to	
	connectors	

> *Find color and name of parts*

Fig. 5 NLMENU's screen at a later stage.

Although usable NLDBIS can and have been constructed, NLDBIS currently hold a rather small market share, mainly because of competition from graphic and form-based interfaces. Despite their limitations, these alternative interfaces are often more appealing, because their capabilities are usually clearer to the users, and they are easier to configure. To improve the position of their systems in the market, NLDBI developers need to address these two issues, and to explore applications for which the alternatives to NLDBIS are difficult to use.

One example of such an application might involve remote access to databases by telephone. There is a growing body of research [e.g., 1] integrating speech recognition, robust interpretation, and dialogue-handling techniques, with the goal of implementing systems that engage users in spoken dialogues to assist with certain tasks (see Chap. 14). This line of research is likely to have a significant effect on future NLDBIS, giving rise to systems that will allow users to access databases by speech, in situations where graphic and form-based interfaces are difficult to use (cf., the ATIS domain [87]).

In considering the obscurity of the capabilities of NLDBIS, we believe that this problem can be overcome by incorporating robust language interpretation methods and dialogue-handling techniques (see Chap. 14). These will allow NLDBIS to make reasonable guesses about the content of any user inputs that cannot be "understood" in their entirety. Also, such systems can engage in reasonably cooperative dialogues that will both help the users realise the capabilities of the NLDBI and also assist the system in determining the task to be performed.

The configuration difficulties of NLDBIS are more of an obstacle. Portability has been a primary research focus of NLDBI research for almost two decades. Although significant improvements have been made, NLDBIS still cannot compete with graphic or form-based interfaces in terms of ease of configuration. The situation may become worse if NLDBIS adopt robust interpretation and dialogue-handling techniques, as suggested in the foregoing, because many of these techniques rely heavily on domain-dependent information. There could be a possible niche for NLDBIS, nevertheless, in accessing a small number of very large, centrally held, widely accessed databases (e.g., telephone directories). The configuration of the NLDBIS for these databases could be assigned to NLDBI experts, with the understanding that it will be a lengthy and resource-consuming effort, the cost of which will be balanced by the subsequent wide use of the NLDBIS.

REFERENCES

1. D Albesano, P Baggia, M Danieli, R Gemello, E Gerbino, C Rullent. DIAL-OGOS: a robust system for human–machine spoken dialogue on the telephone. Proceedings of the International Conference on Acoustics, Speech, and Signal Processing, Munich, Germany, 1997.
2. JF Allen. Natural Language Understanding. Menlo Park, CA: Benjamin/Cummings, 1995.
3. H Alshawi, ed. The Core Language Engine. Cambridge, MA: MIT Press, 1992.
4. H Alshawi, D Carter, R Crouch, S Pulman, M Rayner, A Smith. CLARE—a contextual reasoning and cooperative response framework for the Core Language Engine. Final report, SRI International, Cambridge, UK, 1992.

5. I Androutsopoulos. Interfacing a natural language front-end to a relational database. Master's dissertation, University of Edinburgh, 1992.
6. I Androutsopoulos. A principled framework for constructing natural language interfaces to temporal databases. PhD dissertation, University of Edinburgh, 1996.
7. I Androutsopoulos, G Ritchie, P Thanisch. An efficient and portable natural language query interface for relational databases. In: PW Chung, G Lovegrove, M Ali, ed. Proceedings of the 6th International Conference on Industrial Engineering Applications of Artificial Intelligence and Expert Systems, Edinburgh. Langhorne, PA: Gordon & Breach Publishers, 1993, pp 327–330.
8. I Androutsopoulos, GD Ritchie, P Thanisch. Natural language interfaces to databases—an introduction. Nat Language Eng 1:29–81, 1995.
9. I Androutsopoulos, GD Ritchie, P Thanisch. Time, tense and aspect in natural language database interfaces. Nat Language Eng 4:229–276, 1998.
10. P Auxerre, R Inder. MASQUE modular answering system for queries in English—user's manual. Technical Report AIAI/SR/10, Artificial Intelligence Applications Institute, University of Edinburgh, 1986.
11. B Ballard, M Jones. Computational linguistics. In: SC Shapiro, ed. Encyclopedia of Artificial Intelligence. Vol. 1. New York: John Wiley & Sons, 1992, pp 203–224.
12. B Ballard, D Stumberger. Semantic acquisition in TELI. Proceedings of the 24th Annual Meeting of ACL, New York, 1986, pp 20–29.
13. BW Ballard. The syntax and semantics of user-defined modifiers in a transportable natural language processor: Proceedings of the 22nd Annual Meeting of ACL, Stanford, CA, 1984, pp 52–56.
14. BW Ballard, JC Lusth, NL Tinkham. LDC-1: a transportable, knowledge-based natural language processor for office environments. ACM Trans Office Inform Syst 2:1–25, 1984.
15. FA Barros, A De Roeck. Resolving anaphora in a portable natural language front end to a database. Proceedings of the 4th Conference on Applied Natural Language Processing, Stuttgart, Germany, 1994, pp 119–124.
16. M Bates, MG Moser, D Stallard. The IRUS transportable natural language database interface. In: L Kerschberg, ed. Expert Database Systems. Menlo Park, CA: Benjamin/Cummings, 1986, pp 617–630.
17. BBN Systems and Technologies. BBN Parlance Interface Software–System Overview. 1989.
18. JE Bell, LA Rowe. An exploratory study of ad hoc query languages to databases. Proceedings of the 8th International Conference on Data Engineering, Tempe, AZ, IEEE Computer Society Press, February 1992, pp 606–613.
19. BIM Information Technology. Loqui: An Open Natural Query System—General Description. 1991.
20. J-L Binot, L Debille, D Sedlock, B Vandecapelle. Natural language interfaces: a new philosophy. SunExpert Mag, pp 67–73, January 1991.
21. RJ Bobrow. The RUS system. In: Research in Natural Language Understanding, BBN Report 3878. Cambridge, MA: Bolt Beranek & Newman, 1978.
22. RJ Bobrow, P Resnik, RM Weischedel. Multiple underlying systems: translating user requests into programs to produce answers. Proceedings of the 28th Annual Meeting of ACL, Pittsburgh, PA, 1990, pp 227–234.
23. RA Capindale, RG Crawford. Using a natural language interface with casual users. Int J Man-Mach Stud 32:341–361, 1990.
24. N Cercone, P McFetridge, F Popowich, D Fass, C Groeneboer, G Hall. The systemX natural language interface: design, implementation, and evaluation. Technical Report CSS-IS TR 93-03, Centre for Systems Science, Simon Fraser University, Burnaby, BC, Canada, 1993.

25. EF Codd. A relational model for large shared data banks. Commun ACM 13:377–387, 1970.
26. EF Codd. Seven steps to RENDEZVOUS with the casual user. In: J Kimbie, K Koffeman, ed. Data Base Management. Amsterdam: North-Holland Publishers, 1974, pp 179–200.
27. PR Cohen. The role of natural language in a multimodal interface. Technical Note 514, Computer Dialogue Laboratory, SRI International, 1991.
28. A Copestake, K Sparck Jones. Natural language interfaces to databases. Knowledge Eng Rev 5:225–249, 1990.
29. F Damerau. Operating statistics for the transformational question answering system. Am J Comput Linguist 7:30–42, 1981.
30. J Davidson, SJ Kaplan. Natural language access to data bases: interpreting update requests. Comput Linguist 9:57–68, 1983.
31. AN De Roeck, BGT Lowden. Generating English paraphrases from formal relational calculus expressions. Proceedings of the 11th International Conference on Computational Linguistics, Bonn, Germany, 1986, pp 581–583.
32. SM Dekleva. Is natural language querying practical? Data Base pp 24–36, May 1994.
33. C Draxler. Accessing relational and higher databases through database set predicates in logic programming languages. PhD dissertation, University of Zurich, 1992.
34. SS Epstein. Transportable natural language processing through simplicity—the PRE system. ACM Trans Office Inform Syst 3(2):107–120, 1985.
35. B Grosz, K Sparck Jones, B Webber. Readings in Natural Language Processing. Los Altos, CA: Morgan Kaufmann, 1986.
36. BJ Grosz. TEAM: a transportable natural-language interface system. Proceedings of the 1st Conference on Applied Natural Language Processing, Santa Monica, CA. 1983, pp 39–45.
37. BJ Grosz, DE Appelt, PA Martin, FCN Pereira. TEAM: an experiment in the design of transportable natural-language interfaces. Artif Intell 32:173–243, 1987.
38. CD Hafner. Interaction of knowledge sources in a portable natural language interface. Proceedings of the 22nd Annual Meeting of ACL, Stanford, CA, 1984, pp 57–60.
39. CD Hafner, K Godden. Portability of syntax and semantics in datalog. ACM Trans Off Inform Syst 3:141–164, April 1985.
40. LR Harris. User-oriented data base query with the ROBOT natural language query system. Int J Man–Mach Stud 9:697–713, 1977.
41. LR Harris. The ROBOT system: natural language processing applied to data base query. Proceedings of the ACM'78 Annual Conference, 1978, pp 165–172.
42. LR Harris. Experience with ROBOT in 12 commercial natural language data base query applications. Proceedings of the 6th International Joint Conference on Artificial Intelligence, Tokyo, Japan, 1979, pp 365–368.
43. LR Harris. Experience with INTELLECT: artificial intelligence technology transfer. AI Mag 5(2):43–50, 1984.
44. LR Harris. Proliferating the data warehouse beyond the power user. White paper, Linguistic Technology Corporation, 1996.
45. G Hendrix, E Sacerdot, D Sagalowicz, J Slocum. Developing a natural language interface to complex data. ACM Trans Database Syst 3(2):105–147, 1978. Reprinted in Ref: 35, pp 563–584.
46. EW Hinrichs. Tense, quantifiers, and contexts. Comput Linguist 14(2):3–14, 1988.
47. CH Hwang, K Schubert. Resolving pronoun references. Lingua, 44:311–338, 1978. Reprinted in Ref. 35, pp 339–352.
48. M Jarke, JA Turner, EA Stohr, Y Vassiliou, NH White, K Michielsen. A field evaluation of natural language for data retrieval. IEEE Trans Software Eng, SE-11(1):97–113, 1985.

49. T Johnson. Natural Language Computing: The Commercial Applications. London: Ovum, 1985.
50. SJ Kaplan. Indirect responses to loaded questions. Theor Iss Nat Language Process 2:202–209, 1978.
51. SJ Kaplan. Cooperative responses from a portable natural language data base query system. Artif Intell 19:165–187, 1982.
52. SJ Kaplan. Cooperative responses from a portable natural language database query system. In: M Brady, RC Berwick, eds. Computational Models of Discourse. Cambridge, MA: MIT Press, 1983, pp 167–208.
53. MC Linebarger, LM Norton, DA Dahl. A portable approach to last resort parsing and interpretation. In: Human Language Technology–Proceedings of the ARPA Workshop, Princeton, NJ. Los Altos, CA: Morgan Kaufmann, 1993.
54. GP Lopes. Transforming English interfaces to other languages: an experiment with Portuguese. Proceedings of the 22nd Annual Meeting of ACL, Stanford, CA, 1984, pp 8–10.
55. BGT Lowden, AN De Roeck. REMIT: a natural language paraphraser for relational query expressions. ICL Tech J 5(1):32–45, 1986.
56. BGT Lowden, BR Walls, AN De Roeck, CJ Fox, R Turner. A formal approach to translating English into SQL. In: Jackson and Robinson, ed. Proceedings of the 9th British National Conference on Databases, 1991.
57. BGT Lowden, BR Walls, AN De Roeck, CJ Fox, R Turner. Modal reasoning in relational systems. Technical Report CSM-163, Dept. of Computer Science, University of Essex, 1991.
58. R Lucas. Database Applications Using Prolog. Halsted Press, 1988.
59. P Martin, D Appelt, F Pereira. Transportability and generality in a natural-language interface system. Proceedings of the 8th International Joint Conference on Artificial Intelligence, Karlsruhe, Germany, Los Altos, CA: Morgan Kaufmann, 1983, pp 573–581. Reprinted in Ref. 35, pp 585–593.
60. J Melton, AR Simon. Understanding the New SQL: A Complete Guide. San Mateo, CA: Morgan Kaufmann, 1993.
61. J Mylopoulos, A Borgida, P Cohen, N Roussopoulos, J Tsotsos, H Wong. TORUS: a step towards bridging the gap between data bases and the casual user. Inform Syst 2:49–64, 1976.
62. N Ott. Aspects of the automatic generation of SQL statements in a natural language query interface. Inform Syst 17:147–159, 1992.
63. CR Perrault, BJ Grosz. Natural language interfaces. In: HE Shrobe, ed. Exploring Artificial Intelligence. San Mateo, CA: Morgan Kaufmann, 1988, pp 133–172.
64. C Pollard, IA Sag. Head-Driven Phrase Structure Grammar. University of Chicago Press and CSLI Stanford, 1994.
65. F Popowich, P McFetridge, DC Fass, G Hall. Processing complex noun phrases in a natural language interface to a statistical database. Proceedings of the 15th International Conference on Computational Linguistics, Nantes, France, Vol 1. 1992, pp 47–52.
66. M Rayner, D Carter, P Bouillon. Adapting the Core Language Engine to French and Spanish. Proceedings of the Conference on Natural Language Processing and Industrial Applications, Moncton, Canada, 1996.
67. M Rayner. Abductive equivalential translation and its application to natural language database interfacing. PhD dissertation, Royal Institute of Technology, Stockholm, 1993.
68. RJH Scha. Philips question answering system PHILIQA1. In: SIGART Newslett no. 4 61. New York; ACM, 1977.
69. DW Small, LJ Weldon. An experimental comparison of natural and structured query languages. Hum Fact 25:253–263, 1983.

70. D Stallard, R Bobrow. The semantic linker—a new fragment combining method. In: Human Language Technology—Proceedings of the ARPA Workshop, Princeton, NJ. San Mateo, CA: Morgan Kaufmann, 1993, pp 37–42.
71. M Templeton. EUFID: a friendly and flexible frontend for data management systems. Proceedings of the 17th Annual Meeting of ACL, 1979, pp 91–93.
72. M Templeton, J Burger. Problems in natural language interface to DBMS with examples from EUFID. Proceedings of the 1st Conference on Applied Natural Language Processing, Santa Monica, CA, 1983, pp 3–16.
73. HR Tennant, KM Ross, M Saenz, CW Thompson, JR Miller. Menu-based natural language understanding. Proceedings of the 21st Annual Meeting of ACL, Cambridge, MA, 1983, pp 151–158.
74. R Tennant, KM Ross, CW Thompson. Usable natural language interfaces through menu-based natural language understanding. Proceedings of CHI'83, Conference on Human Factors in Computer Systems, Boston. ACM, 1983.
75. BH Thompson, FB Thompson. Introducing ASK, a simple knowledgeable system. Proceedings of the 1st Conference on Applied Natural Language Processing, Santa Monica, CA, 1983, pp 17–24.
76. BH Thompson, FB Thompson. ASK is transportable in half a dozen ways. ACM Trans Off Inform Syst, 3:185–203, 1985.
77. JD Ullman. Principles of Database and Knowledge-Base Systems. Vol 1. Rockville, MD: Computer Science Press, 1988.
78. A Van Gelder, RW Topor. Safety and translation of relational calculus queries. ACM Trans Database Syst 16:235–278, 1991.
79. DL Waltz. An English language question answering system for a large relational database. Commun ACM, 21:526–539, 1978.
80. D Warren, F Pereira. An efficient easily adaptable system for interpreting natural language queries. Comput Linguist 8:110–122, 1982.
81. R Weischedel. A hybrid approach to representation in the JANUS natural language processor. Proceedings of the 27th Annual Meeting of ACL, Vancouver, BC, 1989, pp 193–202.
82. S Whittaker, P Stenton. User studies and the design of natural language systems. Proceedings of the 4th Conference of the European Chapter of ACL, Manchester, England, April 1989, pp 116–123.
83. WA Woods. Procedural semantics for a question-answering machine. Proceedings of the Fall Joint Computer Conference. New York: AFIPS, 1968, pp 457–471.
84. WA Woods. Lunar rocks in natural English: explorations in natural language question answering. In: A Zampoli, ed., Linguistic Structures Processing. New York: Elsevier North-Holland, 1977, pp 521–569.
85. WA Woods. Semantics and quantification in natural language question answering. In: M Yovitz, ed. Advances in Computers. Vol 17. New York: Academic Press, 1978. Reprinted in Ref. 35, pp 205–248.
86. WA Woods, RM Kaplan, BN Webber. The lunar sciences natural language information system: final report. BBN Report 2378. Cambridge: MA: Bolt Beranek & Newman, 1972.
87. V Zue, S Seneff, J Polifroni, M Phillips, C Pao, D Goddeau, J Glass, E Brill. PEGASUS: a spoken language interface for on-line air travel planning. Proceedings of the ARPA Human Language Technology Workshop, Princeton, NJ, 1994.

10

Information Extraction

JIM COWIE

New Mexico State University, Las Cruces, New Mexico

YORICK WILKS

University of Sheffield, Sheffield, England

1. INTRODUCTION

Information extraction (IE) is the name given to any process that selectively structures and combines data that is found, explicitly stated or implied, in one or more texts. The final output of the extraction process varies; in every case; however, it can be transformed such that it populates some type of database. Information analysts working long term on specific tasks already carry out information extraction manually with the express goal of database creation.

One reason for interest in IE is its role in evaluating, and comparing, different natural language processing technologies. Unlike other NLP technologies, such as machine translation, the evaluation process is concrete and can be performed automatically. This, plus that a successful extraction system has immediate applications, has encouraged research funders to support both evaluations of and research into IE. It now seems that this funding will continue and will bring about the existence of working fielded systems; however, applications of IE are still scarce. A few well-known examples exist and other classified systems may also be in operation; but it is not the case that the level of the technology is such that it is easy to build systems for new tasks, or that the levels of performance are sufficiently high for use in fully automatic systems. The effect on long-term research on NLP is an issue of some debate: we return to consider this question in the final section of this chapter, in which we speculate on future directions in IE.

We begin our examination of IE by considering a specific example from the Fourth Message Understanding Conference (MUC-4; DARPA '92) evaluation. An examination of the prognosis for this relatively new, and as yet unproved, language technology follows, together with a brief history of how IE has evolved. The related problems of evaluation methods and task definition are examined. The current methods used for building IE extraction systems are then outlined; the term IE can be applied to a range of tasks, and we consider three generic applications.

2. AN EXAMPLE: THE MUC-4 TERRORISM TASK

The task given to participants in the MUC-4 evaluation (1991) was to extract specific information on terrorist incidents from newspaper and newswire texts relating to South America. Human analysts (here, the participants in the evaluation) prepared training and test data by manually performing extraction from a set of texts. The templates to be completed, either by humans or by computers, consisted of labelled slots, and rules were provided on how the slots were to be filled. For MUC-4 a flat record structure was used, with slots that had no associated information being left empty.

Without further commentary, we give a short example text and, in Fig. 1, its associated template:

> SANTIAGO, 10 JAN 90—[TEXT] POLICE ARE CARRYING OUT INTENSIVE OPERATIONS IN THE TOWN OF MOLINA IN THE SEVENTH REGION IN SEARCH OF A GANG OF ALLEGED EXTREMISTS WHO COULD BE LINKED TO A RECENTLY DISCOVERED ARSENAL. IT HAS BEEN REPORTED THAT CARABINEROS IN MOLINA RAIDED THE HOUSE OF 25-YEAR-OLD WORKER MARIO MUNOZ PARDO, WHERE THEY FOUND A FAL RIFLE, AMMUNITION CLIPS FOR VARIOUS WEAPONS, DETONATORS, AND MATERIAL FOR MAKING EXPLOSIVES.
>
> IT SHOULD BE RECALLED THAT A GROUP OF ARMED INDIVIDUALS WEARING SKI MASKS ROBBED A BUSINESSMAN ON A RURAL ROAD NEAR MOLINA ON 7 JANUARY. THE BUSINESSMAN, ENRIQUE ORMAZABAL ORMAZABAL, TRIED TO RESIST; THE MEN SHOT HIM AND LEFT HIM SERIOUSLY WOUNDED. HE WAS LATER HOSPITALIZED IN CURICO. CARABINEROS CARRIED OUT SEVERAL OPERATIONS, INCLUDING THE RAID ON MUNOZ' HOME. THE POLICE ARE CONTINUING TO PATROL THE AREA IN SEARCH OF THE ALLEGED TERRORIST COMMAND.

The template illustrates the two basic types of slot: strings from the text (such as "ENRIQUE ORMAZABAL ORMAZABAL") and "set fills," in which one of a set of predetermined categories must be selected (e.g., ROBBERY, GUN, and ACCOMPLISHED). On the surface the problem appears reasonably straightforward. The reader should bear in mind, however, that the definition of a template must be precise enough to allow human analysts to produce consistently filled templates (keys) and to give clear guidelines to the builders of automatic systems. We return to these problems in the following.

Template Slot ID	Fill Value
0. MESSAGE: ID	DEV-MUC3-0017 (NCCOSC)
1. MESSAGE: TEMPLATE	1
2. INCIDENT: DATE	07 JAN 90
3. INCIDENT: LOCATION	CHILE: MOLINA (CITY)
4. INCIDENT: TYPE	ROBBERY
5. INCIDENT: STAGE OF EXECUTION	ACCOMPLISHED
6. INCIDENT: INSTRUMENT ID	-
7. INCIDENT: INSTRUMENT TYPE	GUN: "-"
8. PERP: INCIDENT CATEGORY	TERRORIST ACT
9. PERP: INDIVIDUAL ID	"ARMED INDIVIDUALS" / "GROUP OF ARMED INDIVIDUALS WEARING SKI MASKS" / "MEN"
10. PERP: ORGANIZATION ID	-
11. PERP: ORGANIZATION CONFIDENCE -	-
12. PHYS TGT: ID	-
13. PHYS TGT: TYPE	-
14. PHYS TGT: NUMBER	-
15. PHYS TGT: FOREIGN NATION	-
16. PHYS TGT: EFFECT OF INCIDENT	-
17. PHYS TGT: TOTAL NUMBER	-
18. HUM TGT: NAME	"ENRIQUE ORMAZABAL ORMAZABAL"
19. HUM TGT: DESCRIPTION	"BUSINESSMAN": "ENRIQUE ORMAZABAL ORMAZABAL"
20. HUM TGT: TYPE	CIVILIAN: "ENRIQUE ORMAZABAL ORMAZABAL"
21. HUM TGT: NUMBER	1: "ENRIQUE ORMAZABAL ORMAZABAL"
22. HUM TGT: FOREIGN NATION	-
23. HUM TGT: EFFECT OF INCIDENT	INJURY: "ENRIQUE ORMAZABAL ORMAZABAL"
24. HUM TGT: TOTAL NUMBER	-

Fig. 1 Extracted terrorism template.

3. INFORMATION EXTRACTION: A CORE LANGUAGE TECHNOLOGY

Information extraction technology has not yet reached the market, but it could be of great significance to information end-user industries of all kinds, especially finance companies, banks, publishers, and governments. For instance, finance companies want to know facts, such as what company take-overs happened in a given time span on a large scale: they want widely scattered text information reduced to a simple database. Lloyds of London needs to know of daily ship sinkings throughout the world, and it pays large numbers of people to locate them in newspapers in a wide range of languages. These are potential uses for IE.

Computational linguistics techniques and theories are playing a strong role in this emerging technology, which should not be confused with the more mature technology of information retrieval (IR), which selects a relevant subset of documents from a larger set. IE extracts information from the actual text of documents. Any application of this technology is usually preceded by an IR phase, which selects a set of documents relevant to some query—normally a set of features or terms that appear in the documents. So, IE is interested in the structure of the texts, whereas one could say that, from an IR point of view, texts are just bags of words.

You can contrast these two ways of envisaging text-based information and its usefulness by thinking about finding, from the World Wide Web, what television programs you might want to watch in the next week. There is already a Web site in operation that provides textual descriptions of the programs showing on 25 or more British television channels; this is more text than most people can survey easily at a single session. On this Web site you can input the channels or genre (e.g., musicals or news) that interest you, and the periods when you are free to watch. You can also specify up to twelve words that can help locate programs for you (e.g., stars' or film directors' names). The Web site has a conventional IR engine behind it, and uses a standard boolean function of the words and genre or channel names you use to identify programs of potential interest. The results are already useful, although they treat the program descriptions as no more than bags of words.

Now suppose you also wanted to know what programs your favorite television critic liked: and suppose the Web site also had access to the texts of recent newspapers. An IR system cannot answer such a question because it requires searching review texts for films and seeing which ones are described in favorable terms. Such a task would require IE and some notion of text structure. In fact, such a search for program evaluations is not a best case for IE, and we mention it only because it is an example of the kind of leisure and entertainment application that will be so important in future informatics developments. To see that one only has to think of the contrast between the designed uses and the actual uses of the French Minitel system—designed for telephone number information, but actually used largely as an adult dating service.

An important insight is that what may seem to be wholly separate information technologies are really not so: MT and IE, for example, are just two ways of producing information to meet people's needs and can be combined in differing ways: for example, one could translate a document and then extract information from the result or vice versa, which would mean just translating the contents of the resulting templates. Which of these one chose to do might depend on the relative strengths of the translation systems available: a simpler one might be adequate only to translate the contents of templates, and so on. This last observation emphasizes that the product of an IE system—the filled templates—can be seen either as a compressed, or summarized, text itself, or as a form of database (with the fillers of the template slots corresponding to conventional database fields). One can then imagine new, learning-oriented, techniques such as data mining, being applied as a subsequent stage on the results of IE itself.

4. INFORMATION EXTRACTION: A RECENT ENTHUSIASM

Extracting information from text as a demonstration of "understanding" goes back to the early days of natural language processing (NLP). Early works by DeJong (1979) at Yale University focused on searching texts with a computer to fill predetermined slots in structures called scripts (Schank, 1977). These scripts are close to what would now more usually be called templates: structures with predetermined slots to be filled in specified ways, as a Film Director slot would be filled with a person's name. Film evaluations are not very script-like, but the scenario of ships sinking or the patterns of company takeovers are much more template–scenario-like and suitable for IE techniques.

Early commercially used systems, such as JASPER (from the Carnegie Group; Andersen et al., 1986), built for Reuters, depended on very complex handcrafted templates made up by analysts, and a very specific extraction task. However, the IE movement has grown by exploiting, and joining, the recent trend toward a more empirical and text-based computational linguistics; that is to say, by putting less emphasis on linguistic theory and trying to derive structures and various levels of linguistic generalization from the large volumes of text data that machines can now manipulate.

Information extraction, particularly in the context of automatic evaluation against human-produced results, is a relatively new phenomenon. The early Message Understanding Conferences, in 1987 and 1989, processed naval ship-to-shore messages. A move was then made to extract terrorism information from general newspaper texts. The task of developing the human-produced keys (template structures filled with data for specific texts) was shared among the MUC participants themselves. The combination of an evaluation methodology and a task that has definite applicability, and appears practicable, attracted the attention of various U.S. government agencies, who were prepared to pay for the development of large numbers of keys using professional information analysts. Progress in information extraction and the corresponding evaluation aspects were surveyed (Lehnert Cowie, 1996); broadly, one can say that the field grew very rapidly when ARPA, the U.S. defense agency, funded competing research groups to pursue IE, based initially on scenarios such as the MUC-4 terrorism events. To this task were added the domains of joint ventures and microelectronics fabrication development, with extraction systems for two languages, English and Japanese. All these tasks represent domains for which the funders want to replace the government analysts who read newspapers and then fill templates with recurring types of data: when and where a terrorist even took place, how many casualties occurred, and so on.

A fairly stable research and development (R&D) community has arisen around the Message Understanding Conferences. As well as the U.S. participants, a few groups from Europe, Canada, and Japan have also been involved. The idea of using a common task as a stimulus to research is useful, but it also has dangers. In particular, becoming so focused on performing well in the evaluation may actually force people to follow avenues that provide only short-term solutions. The other drawback is that the amounts of software development needed to produce the specific requirements of an extraction system are very large. A common plea at the MUC organizing committee is "let's not have the next evaluation next year; that way we'll get some time to do research." On the other hand, some usable technologies are appearing as a result of the focus on IE and the visibility of the evaluations to funders, both government and commercial. For example, recognizing and classifying names in text, not a task of particular interest to the NLP community, now proves to be possible at high levels of accuracy. However, the question of whether IE provides a good focus for NLP research is debatable; one key requirement for making IE a usable technology is the development of an ability to produce IE systems rapidly without using the full resources of an NLP research laboratory. However, in response to some of the concerns about the effect of the endeavour on basic research, the most recent MUCs have introduced a task, "coreference evaluation," with the goal of stimulating more fundamental NLP research.

The trend inside the ARPA Tipster Text Initiative, which provided funding for research on IE and IR, is an attempt to standardize NLP around a common architecture for annotating documents (Grishman, 1996). This has proved useful for building multicomponent NLP systems that share this common representation. NMSU's Computing Research Laboratory's Temple machine translation system (Zajac, 1996) and Oleada language-training system (Ogden, 1996) both use this representation system, as does the Sheffield GATE system described later in this chapter. This quest for some kind of standardization is now extending to specifying the kinds of information that drive IE systems. The idea is to have a common representation language that different developers can use to share pieces of an extraction system. Thus, for example, if someone has expended a great deal of effort developing techniques to recognize information on people in texts, this can be incorporated into someone else's system to recognize, for example, changes in holders of particular jobs. Reusable components of this type would certainly reduce the duplication of effort that is currently occurring in the MUC evaluations.

5. EVALUATION AND TEMPLATE DESIGN

Evaluation is carried out for IE by comparing the templates produced automatically by an extraction program with templates for the same texts produced by humans. The evaluation can be fully automatic. Thus, analysts produce a set of filled out templates or keys using a computer tool to ensure correct formatting and selection of fields; the automatic system produces its templates in the same form. A scoring program then produces sets of results for every slot.

Most of the MUC evaluations have been based on giving a one-point score for every slot correctly filled. Slots incorrectly filled are counted, as are spurious slots: these are slots that are generated and filled by the application, despite there being no corresponding information in the text. The total number of correct slots in a template (or key) is also known. These numbers allow two basic scores to be calculated; PRECISION, a measure of the percentage correctness of the information produced, and RECALL, a measure of the percentage of information available that is actually found.

These measures are adapted from work in information retrieval; it is not clear that they are quite so appropriate for IE. For example, in an object-style template, if a pointer to a person slot is filled, this counts as a correct fill; then, if the name is filled in the person object, this counts as a second correct fill. The method of counting correct slots can produce some paradoxical results for a system's scores. In MUC3, one single key that was filled with information about the killing of Jesuit priests was used as the extracted information for each of the test documents. This gave scores as accurate as many systems that were genuinely trying to extract information. Similarly, a set of keys with only pointers to objects and no strings in any other slots was submitted by Krupka in MUC-5 (ARPA 1993); this scored above the median of system performance. The point is that the details of how a score is achieved is important.

To give a flavor of what an IE system developer faces during an evaluation, we present in Fig. 2 a much reduced set of summary scores for the MUC-5 "microelec-

SLOT	POS	ACT	COR	PAR	INC	SPU	MIS	REC	PRE
<template>	100	100	100	0	0	0	0	100	100
content	123	134	94	0	2	38	27	76	70
subtotals	123	134	94	0	2	38	27	76	70
<entity>	121	131	91	0	9	31	21	75	69
name	121	131	77	3	20	31	21	65	60
location	58	47	25	4	4	14	25	46	57
nationality	36	19	14	0	4	1	18	39	74
type	121	131	91	0	9	31	21	75	69
subtotals	336	328	207	7	37	77	85	63	64
<micro-process>	124	134	94	0	2	38	28	76	70
process	124	134	84	0	12	38	28	68	63
developer	63	91	23	0	9	59	31	36	25
manufacturer	83	141	43	0	15	83	25	52	30
distributor	80	138	45	0	9	84	26	56	33
purchaser	25	36	13	0	1	22	11	52	36
subtotals	375	540	208	0	46	286	121	55	38
<layering>	44	57	36	0	1	20	7	82	63
type	44	57	32	2	3	20	7	75	58
film	13	2	0	0	1	1	12	0	0
temperature	5	5	2	0	0	3	3	40	40
device	13	9	6	0	0	3	7	46	67
equipment	39	57	20	0	13	24	6	51	35
subtotals	114	130	60	2	17	51	35	54	47
<lithography>	51	47	35	0	1	11	15	69	74
subtotals	161	165	90	5	12	58	54	57	56
<etching>	17	15	9	0	1	5	7	53	60
subtotals	39	37	16	2	5	14	16	44	46
<packaging>	12	15	10	0	0	5	2	83	67
subtotals	35	40	25	0	0	15	10	71	62

Fig. 2 A partial view of system summary scores–microelectronics template.

tronics" task. This presents total scores for a batch of documents; individual scores by document are also produced by the scoring program. The first column shows the names of the slots in the template objects. New objects are marked by delimiting angle brackets. The next columns are: the number of correct fills for the slot; the number of fills produced by the system; the number correct; the number partially matched (i.e., part, but not all, of a noun phrase is recognized); the number incorrect; the number generated that have no equivalent in the human-produced key (the earlier-mentioned spurious slots); the number missing; and finally, the recall and precision scores for this slot. At the end of the report are total scores for all slots, total scores for only slots in matched objects, and a line showing how many texts had templates correctly generated. Finally a score is given, the F measure, which combines retrieval and precision into one number. This can be weighted to favor recall or precision (typical combinations are P + R, 2P + R, and P + 2R).

A. The Effects of Evaluation

The aim of evaluation is to highlight differences between different NLP methods; to demonstrate improvement in the technology over time; and to push research in certain directions. These goals are, however, somewhat in conflict. One result of the whole evaluation process has been to push most of the successful groups into very similar methods based on finite state automata and partial parsing (Appelt, 1995; Grishman, 1996a). One key factor in the development process is to be able to score test runs rapidly and evaluate changes to the system. Slower, more complex systems are not well-suited for this rapid test development cycle. The demonstration of improvement over time implies that the same tasks be attempted repeatedly, year after year. This is an extremely boring prospect for system developers, so the MUC evaluations have moved to new tasks in every other evaluation. However, making a comparison of performance between old and new tasks is then extremely difficult.

The whole scoring system, coupled with the public evaluation process, can actually result in decisions being made in system development that are not ideal in terms of language processing, but that happen to give better scores.

One novel focus produced by IE is in what Donald Walker once called the "Ecology of Language." Before the MUC-based work, most NLP research was concerned with problems that were most easily tested with sentences containing no proper names. Why bother then with the idiosyncratic complexities of proper nouns? Walker observed that this "ecology" would have to be addressed before realistic text processing on general text could be undertaken. The IE evaluations have forced people to address this issue, and as a result highly accurate name recognition and classification systems have been developed. A separate Tipster evaluation was set up to find if accurate name recognition technology (better than 90% precision and recall) could be produced for languages other than English. The "Multilingual Named Entity Task" (MET; Merchant, 1996) was set up in a very short time and showed that scores of between 80 and 90% precision and recall were easily achievable for Spanish, Chinese, and Japanese.

B. Template Definition

The evaluation methodology depends on a detailed task specification. Without a clear specification, the training and test keys produced by the human analysts are likely to have low consistency. Often there is a cycle of discovery, with new areas of divergence between template designers and human template fillers regularly having to be resolved. This task involves complex decisions, which can have serious implications for the builders of extraction systems.

Defining templates is a difficult task involving the selection of the information elements required and the definition of their relations. This applied task has been further complicated in the evaluations by the attempt to define slots that provide

"NLP challenges" (such as determining if a contract is "being planned," "under execution," or "has terminated"). Often these slots became very low priority for the extraction system builders, as an attempt to fill them often had seriously prejudicial effects on the system score; often, the best approach is to simply select the most common option.

The actual structure of the templates used has varied from the flat record structure of MUC-4 to the more complex object-oriented definition used for Tipster and MUC-5 and MUC-6. This groups related information into a single object. For example, a person object might contain three strings (name, title, and age) and an employer slot, this being a pointer to an organization object. The information contained in both types of representation is equivalent; however, the newer object-style templates make it easier to handle multiple entities that share one slot, as it groups together the information related to each entity in the corresponding object. The readability of the key in printed form suffers as much of it consists of pointers.

The definition of a template consists of two parts: a syntactic description of the structure of the template, and a written description of how to determine whether a template should be filled, along with detailed instructions on how to determine the content of the slots. The description of the Tipster "joint venture" task extended to more than 40 pages. The simple task of name recognition, described in the following, required a description of seven pages in length. To see that this detail is necessary consider the following short extract from these instructions:

```
4.1 Expressions Involving Elision
Multi-name expressions containing conjoined modifiers
(with elision of the head of one conjunct) should be
marked up as separate expressions.
''North and South America'' <ENAMEX TYPE=''LOCATION''>
North</ENAMEX>
and <ENAMEX TYPE=''LOCATION''>South America</ENAMEX>
A similar case involving elision with number
expressions:
''10- and 20-dollar bills'' <NUMEX TYPE=''MONEY''>10</
NUMEX>- and <NUMEX TYPE=''MONEY''>20-dollar</NUMEX>
bills
In contrast, there is no elision in the case of single-
name expressions containing conjoined modifiers; such
expressions should be marked up as a single expression.
''U.S. Fish and Wildlife Service'' <ENAMEX TYPE=
''ORGANIZATION''>U.S. Fish and Wildlife Service</ENAMEX>
The subparts of range expressions should be marked up as
separate expressions.
''175 to 180 million Canadian dollars'' <NUMEX TYPE=
''MONEY''>175
</NUMEX> to <NUMEX TYPE=''MONEY''>180 million Canadian
dollars</NUMEX>
''the 1986-87 academic year'' the <TIMEX TYPE=''DATE''>
1986</TIMEX>-<TIMEX TYPE=''DATE'' ALT=''87''>87 academic
year</TIMEX>
```

A short sample of the syntactic description for the "microelectronics" template is given in the following. It should be noted that although the texts provided for this task include many on the packaging of microchips, they also included a few on the packaging of potato chips

```
<MICROELECTRONICS_CAPABILITY> :=
    PROCESS:            (<LAYERING> | <LITHOGRAPHY> | <ETCHING> |
                        <PACKAGING>)+
    DEVELOPER:                  <ENTITY>*
    MANUFACTURER:               <ENTITY>*
    DISTRIBUTOR:                <ENTITY>*
    PURCHASER_OR_USER:          <ENTITY>*
    COMMENT:            " "
<ENTITY> :=
    NAME:                       [ENTITY NAME]
    LOCATION:                 [LOCATION]*
    NATIONALITY:                [LOCATION_COUNTRY_ONLY]*
    TYPE:                       {COMPANY, PERSON, GOVERNMENT,
                                OTHER}
    COMMENT:            " "
<PACKAGING> :=
        TYPE:                   {{PACK_TYPE}}^
            PITCH:                          [NUMBER]
            PITCH UNITS:        {MIL, IN, MM}
            PACKAGE_MATERIAL:{CERAMIC, PLASTIC, EPOXY, GLASS,
            CERAMIC_GLASS, OTHER}*
            P_L_COUNT:                [NUMBER]*
            UNITS_PER_PACKAGE:[NUMBER]*
            BONDING                   {{BOND_TYPES}}*
            DEVICE:                           <DEVICE> *
            EQUIPMENT:                <EQUIPMENT> *
            COMMENT:            " "
```

C. New Types of Task

Three qualitatively different tasks are now being evaluated at the Message Understanding Conferences:

1. Name recognition and classification
2. Template element creation: simple structures linking information on one particular entity
3. Scenario template creation: more complex structures linking template elements

The first two tasks are intended to be domain-independent, and the third domain-specific. The degree of difficulty ranges from easy for the first through most difficult for the last. It was intended that each task would provide a base support for its successors, however, the non-decomposed output required for the name recognition and classification task may not provide sufficient information to support the template element creation task. The scenario template creation task is distinguished by the fact that a time constraint is placed on the system developers. The specifics of the task are announced only a month before the evaluation. As a result, groups possessing systems which can be rapidly adapted should do well at this task.

6. METHODS AND TOOLS

Practically every known NLP technique has been applied to the problem of information extraction. Currently, the most successful systems use a finite-state, automata-based approach, with patterns either being derived from training data and corpora, or being specified by computational linguists. The simplicity of this type of system design allows rapid testing of patterns using feedback from the scoring system. The linguistics experience of the system developers, and their experience in the development of IE systems remains, however, an important factor.

Systems relying solely on training templates and heuristic methods of combination have been attempted with some success for the microelectronics domain. In such cases, the system is using no NLP whatsoever. In the earlier MUC evaluations there was a definite bias against using the training data to help build the system. By MUC-3, groups were using the training data to find patterns in the text and to extract lists of organizations and locations. This approach, although successful, has the drawback that only in the early MUC evaluations were hundreds of training keys available.

Much work on learning and statistical methods has been applied to the IE task. This has given rise to several independent components that can be applied to the IE task. Part-of-speech taggers have been particularly successful. Recent research (Church, 1996) has shown that a number of quite independent modules of analysis of this kind can be built up independently from data, usually very large electronic texts, rather than being derived on the basis of either intuition or by virtue of some dependence on other parts of a linguistic theory.

These independent modules, each with reasonably high levels of performance in blind tests, include part-of-speech tagging; aligning texts sentence-by-sentence in different languages; syntactic analysis; and attaching word sense tags to words in texts to disambiguate them in context. That these tasks can be done relatively independently is very surprising to those who believed them all contextually dependent subtasks within a larger theory. These modules have been combined in various ways to perform tasks such as IE as well as more traditional NLP tasks such as machine translation.

The empirical movement, basing, as it does, linguistic claims on textual data, has another stream: the use in language processing of large language dictionaries (of single languages and bilingual forms) that became available about 10 years ago in electronic forms from publishers' typesetting tapes. These are not textual data in quite the foregoing sense, because they are large sets of intuitions about meaning set out by teams of lexicographers or dictionary makers. Sometimes they contain errors and inaccuracies, but they have, nevertheless, proved a useful resource for language processing by computer, and lexicons derived from them have played a role in actual working MT and IE systems (Cowie et al., 1993).

What such lexicons lack is a dynamic view of a language; they are inevitably fossilized intuitions. To use a well-known example: dictionaries of English normally tell you that the first, or main, sense of "television" is as a technology or a TV set, although now it is mainly used to mean the medium itself. Modern texts are thus out of step with even the most recent dictionaries. It is this kind of evidence that shows that, for tasks such as IE, lexicons must be adapted or "tuned" to the texts being analyzed. This has led to a new, more creative wave, in IE research: the need not just to use large textual and lexical resources, but to adapt them as automatically as possible, to enable systems to support new domains and corpora. This means both dealing with their obsolescent vocabulary and extending the lexicon with the specialized vocabulary of the new domain.

7. ASSEMBLING A GENERIC IE SYSTEM

Information extraction's brief history is tightly tied to the recent advances in empirical NLP, in particular to the development and evaluation of relatively independent modules for a range of linguistic tasks, many of which had been traditionally seen as inseparable, or achievable only within some general knowledge-based AI program. Hobbs (1995) has argued that most IE systems will draw their modules from a fairly predictable set; he has specified a "Generic IE System" that anyone can construct similar to a Tinkertoy from an inventory of the relevant modules, cascaded in an appropriate manner. The original purpose of this description was to allow very brief system presentations at the MUC conferences to highlight their differences from the generic system. Most systems contain most of the functionalities described in the following, but where exactly they occur and how they are linked together varies immensely. Many systems, at least in the early days, were fairly monolithic Lisp programs. External forces, such as a requirement for speed (which has meant reimplementation in C or C++), the necessary reuse of external components (such as Japanese segmentors), and the desire to have stand-alone modules for proper name recognition have imposed new modularity requirements on IE systems. We will retain Hobbs' division of the generic system for a brief exploration of the functionalities required for IE. Hobbs' system consists of 10 modules:

1. A Text Zoner, which turns a text into a set of segments.
2. A Preprocessor, which turns a text or text segment into a sequence of sentences, each of which being a sequence of lexical items.
3. A Filter, which turns a sequence of sentences into a smaller set of sentences by filtering out irrelevant ones.
4. A Preparser, which takes a sequence of lexical items and tries to identify reliably determinable small-scale structures.
5. A Parser, which takes a set of lexical items (words and phrases) and outputs a set of parse-tree fragments, which may or may not be complete.
6. A Fragment Combiner, which attempts to combine parse-tree or logical-form fragments into a structure of the same type for the whole sentence.
7. A Semantic Interpreter, which generates semantic structures or logical forms from parse-tree fragments.
8. A Lexical Disambiguator, which indexes lexical items to one and only one lexical sense; alternatively, this can be viewed as reducing the ambiguity of the predicates in the logical form fragments.
9. A Coreference Resolver, which identifies different descriptions of the same entity in different parts of a text.
10. A Template Generator which fills the IE templates from the semantic structures.

In the following we consider the functionality of each of these components in more detail.

A. The Text Zoner

The Text Zoner uses whatever format information is available from markup annotations and text layout to select those parts of the text that will be subjected to the remainder of the processes It isolates the rest of the system from the differences in possible text formats. Markup languages such as HTML and SGML (Goldfarb, 1990) provide the most explicit and well-defined source of such structural information. Most newswires also support some sort of convention for indicating fielded data in a text; special fields such as the dateline, giving the location and date of an article, can be recognized and stored in separate internal fields. Problematic portions of a text, such as headlines realised purely in uppercase, can be isolated for separate treatment. If paragraph boundaries or tables are also flagged explicitly, then the Text Zoner is the component of the system best paced to recognize them.

B. The Preprocessor

Sentences are not normally explicitly marked as such, even in SGML documents, so special techniques are required to recognize sentence boundaries; the techniques described in Chap. 2 can play a role here. For most languages, the main problem here is that of distinguishing the use of the full stop as a sentence terminator from its use as an abbreviation marker (as in "Dr," "Mr," "etc.," and so on) and other uses (as in the ellipsis ". . ."). Paradoxically, languages such as Japanese and Chinese, which appear to be more difficult to tokenise because they do not use spaces to separate lexical units, do not have this stop ambiguity problem and, therefore, can have sentences identified relatively easily.

Once the sentences are identified (or sometimes as a part of this process), it is necessary to identify lexical items and possibly convert them to an appropriate form for lexical lookup. Thus, we may convert each word in an English text to uppercase, while still retaining information about its original case usage. Tokenisation is, for the most part, relatively easy in most languages owing to the use of spaces (but see Chap. 2 for some of the more problematic issues that need to be addressed). Languages such as Japanese and Chinese provide particular problems here, and a special purpose tokenisation program is normally used; the Juman program produced at Kyoto University has been used by many sites as a preprocessor for Japanese. Juman also provides part-of-speech information and, typically, this type of lexical information is extracted at this stage of processing.

C. The Filter

Filtering plays a role within an IE system in two ways. Under the view that IE and IR are natural partners, we would normally assume that texts to be processed by the IE system have already been effectively filtered by an information retrieval system; therefore, the assumption is they do contain appropriate information. Many do, but a side effect of the retrieval process is to supply some bogus articles that, if passed through an IE system, would produce incorrect data. An oft-cited example from the microelectronics domain in Tipster was the inclusion of an article on the topic of "packaging potato chips." Thus, a filter may be used in an attempt to block those texts for which wrongful inclusion is an artifact of the IR process.

Within an IE system, however, the main objective of a filter is to reduce the load on the rest of the system. If relevant paragraphs can be identified, then the others can be abandoned. This is particularly important for systems that do extensive parsing on every sentence; clearly it is best to minimise the amount of text subjected to this kind of processing. However, if the filtering process makes mistakes, then there is the risk that paragraphs containing relevant information may be lost.

Normally, a filter process will rely on identifying either supporting vocabulary, or patterns, to support its operation. This may be in the form of simple word counting or by means of more elaborate statistical processing.

D. The Preparser

The preparser stage handles the "ecology of natural language" described earlier, and contains what is arguably the most successful of the results of IE so far; proper name identification and classification. Numbers (in text or numeric form), dates, and other regularly formed constructions are also typically recognized here. This may involve the use of case information, special lexicons, and context-free patterns, which can be processed rapidly. Often a second pass may be required to confirm shortened forms of names that cannot be reliably identified by the patterns, but that can be flagged more reliably once fuller forms of the names are identified. Truly accurate name classification may require some examination of context and usage. It is possible to provide many (relatively uncommon) instances for which simple methods will fail:

- "Tuesday Morning," the name of a chain of US stores
- "Ms Washington," which might be the name of a member of staff
- "nCube," a company, the name of which does not follow normal capitalization conventions
- China," a town in Mexico

A sophisticated system will pass all possible options to the next stage of processing, with a consequent increase in their complexity.

E. The Parser

Most systems perform some type of partial parsing. The necessity of processing actual newspaper sentences, which are often long and very complex, means the development of a complete grammar is impossible. Accurate identification of the structure of noun phrases, and subordinate clauses is, however, possible. This stage may be combined with the process of semantic interpretation described later.

F. The Fragment Combiner

The fragment combiner attempts to produce a complete structure for a sentence. It operates on the components identified in the parser and will use a variety of heuristics to identify possible relations between the fragments. Up to this point it can be argued that the processes used are relatively domain-independent; however, the subsequent steps rely more significantly on knowledge of the domain.

G. The Semantic Interpreter

A mapping from the syntactic structures to semantic structures related to the templates to be filled has to be carried out. The relation between the structures found in a sentence and the requirements of a specific template is carried out using semantic information. For example, semantic processing may use verb subcategorization information to check if appropriate types are found in the context around a verb or noun phrase. Simple techniques, such as identifying the semantic types of an apposition, may be used to produce certain structures. For example, "Jim Smith (human name), chairman (occupation) of XYZ Corp (Company name)" can produce two template objects; a person, employed by a company, and a company, which has an employee. The imposition of semantic restrictions also produces a disambiguation effect, because if appropriate fillers are found the template elements may not be produced.

At the end of this stage, structures will be available that contain fillers for some of the slots in a template.

H. The Lexical Disambiguator

The process of lexical disambiguation can occur either as a side effect of other processes (e.g., the semantic interpreter) or as a stand-alone stage inserted at a number of possible places within the stream of processing.

I. The Coreference Resolver

Coreference resolution is an important component in further combining fragments to produce fewer, but more completely filled, templates. It can be carried out in the early stages of processing, when pronouns and noun phrases can be linked to proper names using a variety of cues, both syntactic and semantic; it can also be delayed to the final stages of processing when semantic structures can be merged. Various kinds of coreference relations need to be handled by an IE system, including identity, meronymy (part-of relations), and event coreference. Reference to the original text, as well as to the semantic structures, may be required for successful processing. Strong merging may have the unfortunate effect of merging what should really be distinct events; just as unfortunate is the lack of merging two apparently distinct events when, in fact, only one occurred.

J. The Template Generator

Finally, the semantic structures have to be unwound into a structure that can be evaluated automatically or fed into a database system. This stage is fairly automatic, but may absorb a significant degree of effort to ensure that the correct formats are produced and that the strings from the original text are used.

K. The System as a Whole

Hobbs is surely right that most or all of these functions will be found somewhere in an IE system. The details of the processing carried out by the components and how they are put together architecturally, leave a great deal of scope for variation, however. As we noted, for example, Lexical Disambiguation is a process that can be performed early, on lexical items, or later, on semantic structures. Similarly, there is considerable dispute as to what should happen at the Parsing stage, because some systems use a form of syntactic parser, most now prefer some form of direct application of corpus-derived, finite-state patterns to the lexical sequences; this is a process that would once have been called "semantic parsing" (Cowie, 1993).

An important point to make is that IE is not, as some have suggested, wholly superficial and theory-free, with no consequences for broader NLP and CL. Many of the major modules encapsulate highly traditional CL/NLP tasks and preoccupations (e.g., syntactic parsing, word-sense disambiguation, and coreference resolution) and their optimization, individually or in combination, to very high levels of accuracy, by whatever heuristics, is a substantial success for the traditional concerns of the field.

8. THE SHEFFIELD GATE SYSTEM

The GATE system, designed at the University of Sheffield, has been evaluated in two MUCs and has performed particularly well at the named entity task. It incorporates aspects of the earlier DIDEROT TIPSTER system from NMSU (Cowie et al., 1993), and the POETIC system (Mellish et al., 1992) from the University of Sussex. There are two aspects to the Sheffield system. First, there is a core software environment called GATE (an acronym for General Architecture for Text Engineering; Cunningham, 1995) which attempts to meet the following objectives:

1. It should support information interchange between IE modules at the highest common level possible without prescribing a theoretical approach (although it should allow modules that share theoretical presuppositions to pass data in a mutually accepted common form).
2. It should support the integration of modules written in any source language, available either in source or binary form, and be available on any common platform.
3. It should support the evaluation and refinement of IE component modules, and of systems built from them, by a uniform, easy-to-use graphic interface that, in addition, should offer facilities for managing test corpora and ancillary linguistic resources.

GATE owes a great deal to collaboration with the TIPSTER architecture. The second aspect of the Sheffield system is VIE (for Vanilla Extraction System), an IE system built within GATE. One version of this system (LaSie) has entered two MUC evaluations (Gaizauskas, 1995).

A. GATE Design

GATE comprises three principal elements: GDM (the GATE Document Manager), based on the TIPSTER document manager; CREOLE (a Collection of REusable Objects for Language Engineering), this being a set of IE modules integrated with the system; and GGI (the GATE Graphical Interface), a development tool for language engineering R&D that provides integrated access to the services of the other components, and adds visualization and debugging tools.

Working with GATE the researcher will from the outset reuse existing components, and the common APIs of GDM and CREOLE mean only one integration mechanism must be learned. The intention is that, as CREOLE expands, more and more modules will be available from external sources.

B. VIE: An Application in GATE

Focusing on IE within the context of the ARPA Message Understanding Conference has meant fully implementing a system that

- Processes unrestricted "real-world" text containing large numbers of proper names, idiosyncratic punctuation, idioms, and so on
- Processes relatively large volumes of text in a reasonable time
- Needs to achieve only a relatively shallow level of understanding in a predefined domain area
- Can be ported to a new domain area relatively rapidly (a few weeks at most)

Given these features of the IE task, many developers of IE systems have opted for robust, shallow-processing approaches that do not employ a general framework for "knowledge representation," as that term is generally understood. That is, there may be no attempt to build a meaning representation of the overall text, nor to represent and use world and domain knowledge in a general way to help in resolving ambiguities of attachment, word sense, quantifier scope, coreference, and so on. Instead, shallow approaches typically rely on collecting large numbers of lexically triggered patterns for partially filling templates, as well as domain-specific heuristics for merging partially filled templates to yield a final, maximally filled template. This approach is exemplified in systems such as the SRI FASTUS system (Appelt, 1995) and the SRA and MITRE MUC-6 systems.

However, this is not the approach that we have taken in VIE. Although still not attempting "full" understanding (whatever that might mean), we do attempt to derive a richer meaning representation of the text than do many IE systems, a representation that goes beyond the template itself. Our approach is motivated by the belief (which may be controverted if shallower approaches prove consistently more successful) that high levels of precision in the IE task simply will not be achieved without attempting a deeper understanding of at least parts of the text. Such an understanding requires, given current theories of natural language understanding, both the translation of the individual sentences of the text into an initial, canonical meaning representation formalism and also the availability of general and domain-specific world knowledge together with a reasoning mechanism that allows this knowledge to be used to resolve ambiguities in the initial text representation and to derive information implicit in the text.

The key difference between the VIE approach and shallower approaches to IE is that the discourse model and intermediate representations used to derive it in VIE are less task- and template-specific than those used in other approaches. However, while committed to deriving richer representations than many IE systems, we are still attempting to achieve only limited, domain-dependent understanding, and hence, the representations and mechanisms adopted still miss much meaning. The approach we have adopted to KR does, nevertheless, allow us to address, in a general way, the problems of presupposition, coreference resolution, robust parsing; and inference-driven derivation of template fills. Results from the MUC-6 evaluation show that such an approach does no worse overall than shallower approaches, and we believe that, in the long run, its generality will lead to the significantly higher levels of precision that will be needed to make IE a genuinely usable NL technology.

Meanwhile, we are developing within GATE and using many LaSIe modules, a simpler finite state pattern matcher of the now classic type. We will then, within GATE, be able to compare the performance of the two sets of modules.

9. THE FUTURE

If we think along the lines sketched here we see that the first distinction drawn in this chapter, between traditional IR and the newer IE technologies, is not totally clear everywhere, but can itself become a question of degree. Suppose parsing systems that produce syntactic and logical representations were so good, as some now believe, that they could process huge corpora in an acceptably short time. One can then think of the traditional task of computer question answering in two quite different ways. The old way was to translate a question into a formalized language, such as SQL, and use it to retrieve information from a database, as in "Tell me all the IBM executives over 40 earning under $50K a year." But with a full parser of large corpora one could now imagine transforming the query to form an IE template; one could then search the entire text (not a database) for all examples of such employees. Both methods should produce exactly the same result starting from different information sources: a text versus a formalized database.

What we have called an IE template can now be seen as a kind of frozen query that one can reuse many times on a corpus and is, therefore, important only when one wants stereotypical, repetitive, information back rather than the answer to one-off questions.

"Tell me the height of Everest?" as a question addressed to a formalized text corpus is then neither IR nor IE, but a perfectly reasonable single request for an answer. "Tell me about fungi," addressed to a text corpus with an IR system, will produce a set of relevant documents, but no particular answer. "Tell me what films my favorite movie critic likes," addressed to the right text corpus, is undoubtedly IE as we saw, and will produce an answer also. The needs and the resources available determine the techniques that are relevant, and those, in turn, determine what it is to answer a question as opposed to providing information in a broader sense.

At Sheffield we are working on two applications of IE systems funded as European Commission LRE projects. One, AVENTINUS, is in the classic IE tradition, seeking information on individuals about security, drugs, and crime, and using classic templates. The other, ECRAN, a more research-oriented project, searches movie and financial databases and exploits the notion we mentioned earlier of tuning a lexicon such that one has the right contents, senses, and so on to deal with new domains and previously unseen relations.

In all this, and with the advent of speech research products and the multimedia associated with the Web, it is still important to keep in mind how much of our cultural, political, and business life is still bound up with texts, from manuals for machines, to entertainment news, to newspapers themselves. The text world is vast and growing exponentially: one should never be seduced by multimedia fun into thinking that text and the problem of how to deal with it and how to extract content from it are going to go away.

REFERENCES

Anderson PM, Hayes PJ, Heuttner AK, Schmandt LM, Nirenberg IB (1986). Automatic extraction. Proceedings of the Conference of the Association for Artificial Intelligence, Philadelphia, 1986, pp 1089–1093.

Aone C, Blejer H, Flank S, McKee D, Shinn S (1993). The Murasaki Project: multilingual natural language understanding. Proceedings of the DARPA Spoken and Written Language Workshop.

Appelt D, Bear J, Hobbs J, Israel D, Tyson M (1992). SRI International FASTUS system MUC-4 evaluation results. Proceedings of the Fourth Message Understanding Conference (MUC-4). San Mateo, CA: Morgan Kaufmann, 1992, pp 143–147.

Appelt D, Hobbs J, Bear J, Israel D, Kanneyama M, Tyson M (1993). SRI description of the JV-FASTUS system used for MUC-5. Proceedings of the Fifth Message Understanding Conference (MUC-5). San Mateo, CA: Morgan Kaufmann.

ARPA. The Tipster extraction corpus (available only to MUC participants at present). 1992. ARPA. Proceedings of the Fifth Message Understanding Conference (MUC-5), Baltimore, Maryland, 1993, Morgan Kaufman.

Basili R, Pazienza M, Velardi P (1993). Acquisition of selectional patterns in sub-languages. Machine Translation 8.

Communications of the ACM: Special Issue on Text Filtering 35(12), 1992.

Carlson LM, et al. (1993). Corpora and Data Preparation. Proceedings of the Fifth Message Understanding Conference (MUC-5). ARPA.

Church K, Young S, Bloothcroft G eds. (1996). Corpus-Based Methods in Language and Speech. Dordrecht: Kluwer Academic.

Ciravegna F, Campia P, Colognese A (1992). Knowledge extraction from texts by SINTESI. Proceedings of the 14th International Conference on Computational Linguistics (COLING92), Nantes, France, pp 1244–1248.

Cowie J, Wakao T, Guthrie L, Jin W, Pustejovsky J, Waterman S (1993). The Diderot information extraction system. Proceedings of the First Conference of the Pacific Association for Computational Linguistics (PACLING), Vancouver, BC.

Cowie J, Lehnert W (1996). Information extraction. In: Y Wilks, ed. Special NLP issue of Commun ACM.

Cunningham H, Gaizauskas R, Wilks Y (1995). GATE: a general architecture for text extraction. University of Sheffield, Computer Science Dept. Technical memorandum.

DARPA (1991). Proceedings of the Third Message Understanding Conference (MUC-3), San Diego, CA, 1991, Morgan Kaufmann.

DARPA (1992). Proceedings of the Fourth Message Understanding Conference (MUC-4), McLean, Virginia, 1992, Morgan Kaufmann.

DeJong GF (1979). Prediction and substantiation: a new approach to natural language processing. Cognitive Sci 3:251–273.

DeJong GF (1990). An overview of the FRUMP system. In: WG Lehnert, MH Ringle, eds. Strategies for Natural Language Processing. Hilldsale, NJ: Erlbaum 1982, pp 149–174.

Dorr B, Jones D (1996). The role of word-sense disambiguation in lexical acquisition: predicting semantics from syntactic cues. Proc. COLING96.

Gaizauskas R, Wakao T, Humphreys K, Cunnigham H, Wilks Y (1995). Description of the LaSIE System as Used for MUC-6. Proceedings of the Sixth Message Understanding Conference (MUC-6). DARPA.

Granger RR (1977). FOULUP: a program that figures out meanings of words from context. Proc. Fifth Joint International Conference on AI.

Grishman R, et al. (1996). Tipster text phase II architecture design. Proceedings of the Tipster Text Phase II Workshop, Vienna, Virginia, DARPA.

Goldfarb CF (1990). The SGML Handbook. Oxford: Clarendon Press.

Hobbs J, Appelt D, Tyson M, Bear J, Israel D (1992). SRI International: Description of the FASTUS system. Proceedings of the Fourth Message Understanding Conference (MUC-4), Morgan Kaufmann, pp 268–275.

Jacobs PS, Rau LF (1990). SCISOR: extracting information from on-line news. Commun ACM, 33(11):88–97.

Kilgarriff A (1993). Dictionary word-sense distinctions: an enquiry into their nature. Comput Hum.

Lehnert W, Cardie C, Fisher D, McCarthy J, Riloff E, Sonderland S (1992). University of Massachusetts: Description of the CIRCUS system. Proceedings of the Fourth Message Understanding Conference (MUC-4). Morgan Kaufmann, pp 282–288.

Lehnert W, Sundheim B (1991). A performance evaluation of text analysis technologies. AI Mag 12(3):81–94.

Levin B (1993). English Verb Classes and Alternations. Chicago, IL.

Mellish C, Allport A, Evans R, Cahill IJ, Gaizauskas R, Walker J (1992). The TIC message analyser. Technical Report CSRP 225, University of Sussex.

Merchant R, Okurowski ME, Chichor N (1996). The multi-lingual entity task (MET) overview. Proceedings of the Tipster Text Phase II Workshop, Vienna, Virginia, DARPA.

Ogden W, Bernick P (1996). OLEADA: user-centred TIPSTER technology for language instruction. Proceedings of the Tipster Text Phase II Workshop, Vienna, Virginia, DARPA.

Paik W, Liddy ED, Yu E, McKenna M (1993). Interpretation of proper nouns for information retrieval. Proceedings of the DARPA Spoken and Written Language Workshop.

Procter P, et al. (1994). The Cambridge language survey semantic tagger. Technical Report, Cambridge University Press.

Procter P, ed. (1978). Longman Dictionary of Contemporary English. Harlow: Longman.

Pustejovsky J, Anick P (1988). On the semantic interpretation of nominals. Proc. COLING88.

Rau L (1991). Extracting company names from text. Proceedings of the Seventh Conference on Artificial Intelligence Applications, Miami Beach, Florida, 1991.

Riloff E, Lehnert W (1993). Automated dictionary construction for information extraction from text. Proceedings of the Ninth IEEE Conference on Artificial Intelligence for Applications. IEEE Computer Society Press, pp 93–99.

Riloff E, Shoen J (1995). Automatically acquiring conceptual patterns without an annotated corpus. Proceedings Third Workshop on Very Large Corpora.

Sagers N (1981). Natural Language Information Processing: A Computer Grammar of English and Its Application. Reading, MA, Addison-Wesley.

Schank RC, Abelson RP (1977). Scripts, Plans, Goals and Understanding. Hillsdales, NJ: Lawrence Erlbaum Associates.

Sundheim BM, Chinchor NA (1993). Survey of the message understanding conferences. Proceedings of the DARPA Spoken and Written Language Workshop.

Wakao T, Gaizauskas R, Wilks Y (1996). Evaluation of an algorithm for the recognition and classification of proper names. Proc. COLING96.

Weischedel R (1991). Studies in the statistical analysis of text. Proceedings of the DARPA Spoken and Written Language Workshop. San Mateo, CA: Morgan Kaufmann, 1991, p 331.

Wilks Y (1978). Making preferences more active. Artif Intell 11.

Wilks Y, Slator B, Guthrie B, Guthrie L (1996). Electric words: Dictionaries, Computers and Meanings. Cambridge, MA: MIT Press.

Wilks Y. (1997) Senses and texts. Computers and the Humanities 34.

Zajac R, Vanni M (1996). The Temple Translator's Workstation Project. Proceedings of the Tipster Text Phase II Workshop, Vienna, Virginia, DARPA.

11

The Generation of Reports from Databases

RICHARD I. KITTREDGE* AND ALAIN POLGUÈRE

University of Montreal, Montreal, Quebec, Canada

1. INTRODUCTION

Report Generation (RG) means different things to different people. In the field of natural language processing (NLP), it has come to mean the production of professional-sounding natural language text, of one or more paragraphs in length, from some structured data input, such as one may find in a database or knowledge base.[1]

Some of the earliest applications of the field of Natural Language Generation (NLG) have been in automatically composing reports from structured data for tasks such as summarizing the daily activity on stock markets, forecasting the weather, and reporting the factual highlights of athletic events. Stereotypical texts in these domains are frequently seen in print media, heard on radio, or accessed through the Internet.

As will be seen, however, from the following discussion, the production of reports from data presupposes a certain amount of knowledge representation for the application domain. Given the effort required to build knowledge representations, it is no surprise that RG applications make most sense when the same knowledge representations needed to produce text can be reused (or are already available) for other purposes. Thus, a report in natural language may be just one of the output

[1]Another sense of the term is common in spreadsheets or other application programs, which can "generate" reports consisting of tables, graphics, and possibly canned text to summarize particular views of data.

**Also affiliated with:* CoGenTex, Inc., Ithaca, New York.

products of a broader knowledge-based application built on a common set of basic concepts.

Before looking at examples of RG systems, and examining approaches to planning and linguistically realizing these texts, we need a working definition of what we understand by the term *report*. Maybury characterizes reports as follows (1990):

> The most basic form of narration recounts events in their temporal order of occurrence. This occurs in a journal, record, account, or chronicle, collectively termed a report. Reports typically consist of the most important or salient events in some domain during one period of time (e.g., stock market report, weather report, news report, battle report). Sometimes reports focus on events and states involving one dimension of an agent as in a medical record, or a political record.

For many people a report can be any factual summary of events or situations. In this chapter, however, we start out by defining reports rather narrowly.

In Sec. 2, we identify four important properties of "true" reports, which we will then use to characterize the applications that have been developed to date. Section 3 describes the kinds of knowledge required to build RG systems. Section 4 then surveys the best-known RG systems in chronological order, summarizing basically *what* has been done to date. The next two sections compare and contrast *how* the systems have dealt with the problems of planning (see Sec. 5) and realization (see Sec. 6). In Sec. 7, we contrast RG systems with systems that generate texts that may appear similar to reports, but that lack one or more of our defining properties. In the conclusion, Sec. 8, we explain why so many of the RG systems have not yet been used for regular production of texts, and what the ingredients are for a useful RG application.

2. WHAT DO WE MEAN BY *REPORT*?

Our primary focus in this chapter is on written database reports in natural language.

A. Essential Conditions for a Report

We argue that our intuitive notion seems to involve four essential conditions on the situation and purpose of a report.

1. A recurrent situation of communication
2. A primary interest in objective data of fixed type
3. A temporal dimension in the data
4. Conceptual summarization over the data

Notice that these conditions involve the content (and situational context) of the report, rather than specific linguistic features of the report texts. In Sec. 2.C, we look at some of the linguistic correlates of these input conditions.

First, reports in our sense are produced and used in **recurrent situations**, where a single-view statement of fact is required. The factual statements are made within a stable domain, and from the same viewpoint each time the reporting situation occurs.

Second, reports must be based primarily on **objective data of fixed types**, where the data types and typical ranges of values for each type are basically unchanging over time. Most often, the objective data are quantifiable. Weather forecasts, for example, summarize data about the expected states of the atmosphere, using established parameters such as sky cover, wind speed and direction, precipitation type and amount, temperature, and visibility. When reports contain other information besides objective information, this typically plays a secondary role and occupies a back seat in the linguistic structure of the whole text.

Third, there is a **temporal dimension** in the data, so that the report deals with an evolving sequence of states or events. (In some domains, the events are simply significant changes between states.) Although the types of data remain the same, the values fluctuate over time. For example, a stock market report describes the changing values of company shares during the day, along with comparisons between closing prices, volume of trading, and such, for the day just ended and those values on previous days. But reports can also summarize predictions about future situations, and their evolution over time, the most noteworthy case being weather forecasts (often colloquially called weather "reports").

Fourth, reports typically involve **summarization**. The whole idea of a report is to give a concise snapshot view of an evolving situation, with priority given to its most salient features. The problem of generating concise summaries has both a conceptual and a linguistic dimension, as pointed out by McKeown et al. (1995). **Conceptual** summarization involves selection of salient information from a data source and its "packaging" at a conceptual level, whereas **linguistic** summarization concerns the choice of lexical and grammatical means to convey the chosen concepts in the most compact and effective form possible. It is **conceptual summarization** that we take to be a distinguishing property of true reports. This will be discussed in more detail in the following.

B. Reports Contrasted with Other Kinds of Generated Text

Given that true reports have the foregoing four defining properties, we can see that some types of texts, which have often been called reports, fall outside our intended scope. For example, a summary of a meeting, describing the agenda items, the discussion points, and the decisions made, is not a report in the sense just described because the information summarized is neither quantifiable nor (normally) of a few recurrent fixed types. The same applies to most "news reports."

Several NLG systems have generated texts from representations of complex structures, such as object model diagrams, thus carrying out a diagram explanation task. We reserve discussion of these systems to Sec. 7, and note in passing that they are distinct from RG systems in not having a temporal dimension in the input data, and usually in not carrying out conceptual summarization. Another important difference is that these latter systems do not take quantifiable data as their primary type of input.

An extreme case of generation from knowledge representations occurs when the sole function of the generator is to verbalize all the information in the representation. A recent example is Exclass (Caldwell and Korelsky, 1994; see Sec. 7.A), which generates bilingual job descriptions from conceptual representations. Exclass does no summarization of data, but just verbalizes as English (or French) text a sequence of conceptual frame structures representing various aspects of a public service job. There is also no temporal dimension to the output texts. This "conceptual readout"-type generator, lacking in both content determination and text structure planning, has a history going back at least to the work in the 1970s on story generation using scripts or frame structures.

C. Reports and Sublanguage

When experts in a particular field use language in a recurrent situation (i.e., the first condition for a true report), their specialized usage over time may give rise to a *sublanguage* (cf. Kittredge and Lehrberger, 1982). The notion of sublanguage applies to types of writing as broad and varied as research articles in a given scientific field, or as narrow and stereotyped as a stock market report. The writers who communicate under restrictions of a given domain and situation tend to evolve **distinctive patterns of word use, sentence structure, and text structure**, which can be described in a *sublanguage grammar and lexicon* (see also Chap. 13 for a discussion of sublanguage in the context of Machine Translation).

What characterizes all sublanguages is their having patterns of world class co-occurrence that reflect the semantics of the domain. For example, in stock market reports, specific verbs of movement, such as *rise*, *plunge*, or *creep up*, take nouns such as *stock*, *index*, or *issue*, as subject, but not human nouns such as *analyst*. The combinatorics of these word classes is a direct reflection of what is "sayable" in the domain.[2] Instead of writing a grammar with major lexical classes such as verb, noun, adjective, in a sublanguage it becomes useful to write a grammar using the domain subclasses. Moreover, the grammatically active lexical classes give empirical support for building a knowledge representation for the domain.

Reports in such narrow domains as weather forecasting, stock market, and sports, all represent sublanguages where the types of information being communicated is very limited. The number of sentence patterns tends to be just a small fraction of what would be possible in the language as a whole. Moreover, stereotypical sublanguage reports exhibit special **conventions on text structure**. These include both constraints on the global organization of information into sections and paragraphs, and on the organization of sentences within individual paragraphs. Given that reports in a particular sublanguage convey information of fixed types, the way information is presented within the text tends to become conventionalized.

Another typical linguistic feature of reports, which can be attributed to their frequent production and specialized situational usage, involves **conventions of visual formatting**. Within a given user community, text may be displayed on the page with particular typographical features, or with interspersed tabular or graphic information, or may use headers, labels, or such. For example, written weather forecasts in North America are traditionally given in upper-case type, using specialized punctuation. Each forecast begins with a header giving the date, time, place of origin, and period of time it covers.

---

[2] It is therefore "grammatical" to say "Stocks plunged" in the reporting sublanguage, but not to say "Analysts plunged."

3. KNOWLEDGE REQUIREMENTS FOR GENERATING REPORTS

Report generation lays strong emphasis on both the content and the professional style of output text. Many types of report are written for publication; hence, they are subject to stringent quality standards. Publicly disseminated reports often have legal or political implications, as in government-issued weather forecasts or economic summaries. In other cases, as in reports on the stock market or sporting events, the professional writing style plays an important role in maintaining reader interest.

Three types of knowledge are needed to generate high-quality reports: **domain concept definitions**, **domain linguistic knowledge**, and **domain communication knowledge**. We discuss each of these in turn in the following Secs. 3.A–3.C. The later sections of this chapter will then illustrate these types of knowledge in particular applications, and show how knowledge of each type is used operationally during the generation process.

There is one type of knowledge that is typically **lacking** in RG systems, and that is a detailed model of the user or interlocutor. Whereas advisory systems must maintain some representation of the user's knowledge and beliefs, RG systems have not required this. Some experiments have been conducted in configuring reports to suit varying needs and expertise levels of the user. But since reports are typically "one-shot" productions of text, there has been no need to track user feedback, maintain a user model dynamically, or otherwise represent the user in any other than a trivial way. Clearly, this difference is only one of emphasis, because one could easily imagine generating report in a dynamically changing context where such user modeling and dialogue tracking would become important.

A. Knowledge of Domain Concepts

Generators that produce data reports must begin by extracting from raw data the information needed to build a set of *messages* of interest to the user.[3] This requires that some basic domain concepts be defined in terms of the raw information categories. More complex concepts may then, as the application requires, be defined in terms of basic concepts. For example, in reporting on the labor market, the notion of a month-to-month "decrease" in employment can be defined directly in terms of comparison of numbers between consecutive months. The notion of a "slump" in employment, however, may be given a definition in terms of consecutive or predominating decreases over an extended period, or in terms of a more complex combination of trends in employment numbers, unemployment rate, averages across industries and regions, and so forth.

Generators that carry out conceptual summarization require, in addition to concept definitions, domain ontology information to plan succinct ways of summing up a number of related facts. However, very few report generators yet use such information in a general way.

[3]The term "message" has been used to refer to the prelinguistic form of a sentence, roughly its propositional content plus communicative information relating to emphasis, theme vs. rheme, and so on.

B. Domain Linguistic Knowledge

A second kind of knowledge of particular importance in RG applications involves linguistic patterns of usage at the sentence level. A fundamental part of this linguistic knowledge deals with word choices. In Canada, for example, marine winds with speed from 35 to 47 knots must be referred to as "gales." An important aspect of professional style involves word combinations (co-occurrences), such as that when winds decrease, they can be said to "diminish," whereas decreasing temperatures can be said to "lower," but never to "diminish." Word choices carry with them some requirements or preferences for using particular grammatic structures. The way in which words and word combinations are used in sentence structures is an important feature of the domain sublanguage, constituting an essential part of domain linguistic knowledge.

Domain linguistic knowledge can be extracted from a study of word distribution in a given sublanguage, based on a corpus of texts produced under natural conditions by domain experts. Because RG applications are held to a high standard of linguistic usage, detailed sublanguage study has tended to be a requirement for a credible application system.

Within an application system, domain linguistic knowledge resides primarily in the lexicon. Early experimental work in RG often made use of phrasal lexicons (Becker, 1975), in which strings of several words were used as lexical entries to simplify the linguistic rule base. Later work has tended to reject this approach, or combine it with one using more lexical knowledge, for phrasal lexicons tend to become unmanageable as the application size grows (for more on this point, see Secs. 4.A and 6.C).

C. Domain Communication Knowledge

The notion of domain communication knowledge (DCK) has been introduced to describe the priorities, ordering, and redundancy of information presentation that are appropriate for communication in a given domain, but that cannot be inferred easily from general principles (Rambow, 1990). DCK is particularly important in correctly generating reports of a stereotypical nature (Kittredge et al., 1991). For example, in Canadian marine forecasts, certain conventions exist about the order in which information is presented (winds, cloud cover, precipitation, fog and mist, visibility). Moreover, when the information to be communicated indicates hazardous weather conditions, certain warning statements must be inserted into the report in a way that indicates cross-serial dependencies between pieces of information. This requirement puts constraints on the method of text planning.

DCK involving text structure and information dependencies belongs neither to the domain conceptual knowledge required for reasoning and summarization, nor to the traditional sphere of linguistic knowledge. The existence of such knowledge, and the need to represent it explicitly has been fully appreciated only during attempts to build NLG applications (see Rambow and Korelsky, 1992).

4. REPORT GENERATION APPLICATIONS

This section provides an overview of several report generation systems developed since the early 1980s for specific applications. We give a brief statement of the application problem, characterize the input data, show an output text, and note the usage made of the system, if any, to date. Special knowledge requirements of each system are noted, as well as the difficulties and the lessons learned from the application. Only a general idea of the text-planning and realization approach is indicated here. Details of planning and realization used in RG systems are left for a comparative survey in Secs. 5 and 6, respectively.

It seems that very few RG systems have actually reached the stage of true application systems, enjoying a period of operational use. Only FoG (see Sec. 4.E) and PLANDOC (see Sec. 4.I) appear to have earned this distinction as of this writing. Certain other RG applications, such as LFS (see Sec. 4.F), have reached the "beta test" stage, in which significant testing has taken place by users at their own location. In some cases, as with Gossip (see Sec. 4.D), a system was built not just to test a theory or a new generation architecture, but with an attempt to show real functionality, even though users did not finally use or even seriously test the system. We have chosen to mention here all the systems we are aware of that produce reports in the sense outlined earlier, even when the project did not produce more than a prototype.[4] Future advances in NLG tools and cost-reduction techniques will undoubtedly make some of these past RG application "failures" come to fruition.

A. ANA and FRANA

To our knowledge, the first full-fledged report generator was the ANA system, built for a 1983 PhD thesis at the University of Pittsburgh (Kukich, 1983b, 1988). ANA provided very realistic summaries of daily stock market activity from a database of half-hourly quotes for the Dow Jones industrials and other indices on the New York exchange. Figure 1 shows a sample output text from ANA. The input data, taken from the Dow Jones news wire, includes end-of-day statistics for the reporting day

```
Thursday June 24, 1982

    wall street's securities markets meandered upward
through most of the morning, before being pushed
downhill late in the day yesterday.  the stock market
closed out the day with a small loss and turned in a
mixed showing in moderate trading.

    the Dow Jones average of 30 industrials declined
slightly, finishing the day at 810.41, off 2.76 points.
the transportation and utility indicators edged higher.

    volume on the big board was 55860000 shares
compared with 62710000 shares on Wednesday.  advances
were ahead by about 8 to 7 at the final bell.
```

Fig. 1 Stock market report generated by ANA.

[4]It is quite unlikely that our survey is truly exhaustive, although we believe it covers most systems reported in widely accessible NLG literature.

and preceding day. Note the variety of expressions used to convey nuances of upward and downward movement in stock prices.

ANA used four modules in sequence to generate reports on stock market events: (a) a fact generator, (b) a message generator, (c) a discourse organizer, and (d) a "text generator." The fact generator takes a stream of data (e.g., half-hourly price quotes for a stock index) and performs arithmetic operations to produce facts (i.e., fundamental concepts), for example, that the Dow Jones industrial average decreased in value by so many points over a half-hour interval. The message generator uses sets of facts to infer interesting messages (e.g., that the drop in the Dow Jones during the late afternoon period was very large). The fact generator and message generator together carry out the first stage of text planning. Next, the discourse organizer orders and groups the messages into incipient paragraphs according to topic, providing a second, higher-level text planning function. For example, the net daily changes for major Dow Jones indices will be grouped into the second paragraph of the report structure. The final module carries out the linguistic realization operations of picking words and phrases, fitting them into appropriate linguistic patterns.

An important part of ANA's realizer is a **phrasal lexicon**,[5] in which an expression such as *meandered upward through most of the morning* is treated as a unit. The lexical entry for such a verb phrasal expression allows for alternating tense forms of the head verb (e.g., *meandering, to meander*), but otherwise gives no information about the individual words or subphrases in the string. This technique permits a uniform and simple treatment of sets of phrases that have basically the same meaning in the sublanguage. It has the drawback that it is inefficient and hard to maintain when the number of phrases is large. To further improve the style of the output, ANA included some relatively advanced features, such as varying each sentence's length based on the length of the preceding sentence.

In 1985, Contant at the University of Montreal, built a French linguistic module to replace the final module (the text generator) of ANA (Contant, 1985). The resulting system, dubbed FRANA, produced French stock market reports from ANA's input data that were just as impressively fluent in French as ANA's reports were in English. FRANA showed that the three prelingustic operations of fact selection, message generation, and discourse organization could be made independently of the language (at least for languages as similar to English as French). However, some additional features were necessary in the text generation module to produce French, which requires more control over grammatical agreement.

Both ANA and FRANA built on previous sublanguage analysis of stock market reports at the University of Montreal during 1979–1981 (cf., Kittredge, 1982), which showed that naturally occurring reports can be mapped to **information formats**. These tabular arrangements of segmented text were first proposed by Harris (1954) as an analytical tool for scientific writing, and then developed at New York University (NYU) in the 1970s for information retrieval of medical reports (Sager, 1972, 1981). The information formats of stock market reports were equivalent to relational databases of market information, suggesting that databases were a feasible starting point for generating such reports (Kittredge, 1983).

[5]The use of this term in natural language generation literature is more simplistic than its original use by Becker (1975).

B. Medical Case Report Generation

The same kinds of medical reports analyzed at NYU in the 1970s also proved to be prime candidates for generation, because of their relatively fixed structure and content. A project at the Illinois Institute of Technology generated reports from hospital patient data in an (unnamed) prototype system, combining several earlier techniques in natural language processing (Li et al., 1986). Multiparagraph reports on stroke cases were produced by mapping patient data to information formats needed for the particular type of report. A text grammar for case reports was set up on the basis of an empirical study. The grammar characterized a report text on the level of content as a sequence of information formats. This was equivalent to providing a schema for the whole text structure. Individual sentences were produced by applying reverse transformations to the normalized sentences contained in the information formats, using a grammar provided by NYU's linguistic string parser. A relational lexicon was built for the stroke sublanguage, so that synonyms, hyperonyms, and other terms related to a given word could be used to avoid repetitions. Figure 2 shows a sample report fragment generated by this combination of techniques. The stroke reporter was beta-tested by physicians, but never entered regular service, as it was soon replaced by a simpler, but faster, report-authoring system (M. Evans, personal communication).

C. SEMTEX

SEMTEX grew out of a Japanese–German translation project (SEMSYN) at the University of Stuttgart that developed several demonstration applications using a semantic representation for individual sentences similar to a case frame (Roesner, 1987). Built as a prototype by 1986, SEMTEX produced paragraph-length texts on the German job market situation, including sentences reporting the reaction of political leaders. Its text planner created a list of frame structures, representing text content, from raw employment data for the current period and previous month and year. The realization component made a sentence from each case frame, being careful to avoid lexical repetition and introduce pronouns where required. One well-developed aspect of SEMTEX was its realization of temporal

```
Michael Reese Hospital Stroke Service Report

Patient 165 is a 39 year-old right handed white woman
admitted for a stroke with a moderate headache.  The
deficit came on when she got up in the middle of the
night.  It was maximal at onset.  At the onset of the
deficit, there was a moderate headache, a gradual onset of
obtundation, and vomiting within the first 12 hours, but
no seizure activity.

Past medical history revealed no stroke, TIA, or cardiac
disease.  There was no evidence that she had systemic
emboli or arteriosclerosis.
...
```

Fig. 2 Medical case report generated from information formats.

```
          Slight Reduction in the Unemployment Level

   NÜRNBERG/BONN (cpa) The number of unemployed in the
Federal Republic of Germany declined only very slightly
during October.  It fell from 2151600 to 2148800.  The
unemployment rate at the end of October stood at 8.6 per
cent.  For the comparable period of the previous year it
was also 8.6 per cent.  Government spokesman Ost saw the
decline as a positive development. ...
```

Fig. 3 Sample unemployment report from SEMTEX.[6]

references. The political comments at the end of each text also used several possible text schemata for positive and negative evaluations of the job market, to reflect comments coming from varying political constituencies (e.g., opposition parties, trade unions, and such). Although the texts were not long, the use of a phrasal lexicon combined with mixing of economic statistics and political comment made for relatively colorful texts (Fig. 3). SEMTEX was not developed beyond the prototype stage.

D. Gossip

The Gossip system (Generating Operating System Summaries in Prolog) was designed to provide computer operating system reports based on OS audit logs as input. A prototype was implemented for a large computer security research and development project by 1988 (Iordanskaja et al., 1991). Figure 4 shows a text produced by Gossip, preceded by a fragment of the input audit data used to generate the text. Each data record is in relational form, representing an elementary event (predicate ee) on the operating system. An elementary event has values for arguments such as user name; terminal ID; Unix command name; command parameters (e.g., file name); times for the start, stop, and duration of the event; CPU usage, and so on. The audit database is simply a chronological sequence of such events on the operating system, without selection for their importance. Gossip's job was to provide routine activity summaries in textual form. (Other types of text, such as reports on suspected violations of operating system security, were studied, but never implemented.)

A primary innovation of Gossip was the linguistic model implemented in its realizer. Gossip was the first text generator to be based on Meaning-Text theory

```
ee(martin,ttyp0,editor,[f1],8:30:00,9:10:32,0:40:32,240).
 ...

The system was used for 7 hours 32 minutes and 12 seconds.
The users of the system ran compilers and editors during
this time.  Compilers were run six times (the cpu-time
equal to 46% of the total cpu-time).  Editors were run
twelve times (the cpu-time equal to 53% of the total cpu-
time).  Two users, Martin and Jessie, logged on to the
system.  Jessie used the system for 63% of the time in use.
Jessie used the system for 40% of the time in use.
```

Fig. 4 Operating system event record (bold) and output report from Gossip.

[6] The original German output text is given here in its English translation (RK).

(Mel'čuk, 1981; Mel'čuk and Polguère, 1987). Moreover, it was the first implementation of this theory in natural language processing to make full use of Meaning-Text's semantic network representation of sentence meaning. In particular, the notion of theme/rheme distinction was implemented through all stages of linguistic processing to guide important aspects of text structuring, lexical selection, sentence planning, and word ordering (Iordanskaja et al., 1991, 1992).

One important difficulty in building Gossip turned out to be the lack of a suitable corpus of human-composed texts to use as a model of operating system reports. System administrators typically indicate significant system events using short canned text strings, or else use system audit expert systems that present a graphical summary of system activity and unusual events. (In other words, they do not have a well-defined written sublanguage.) To fill this gap, systems personnel were asked to compose activity summaries, or to evaluate simulated texts. Written documents on system security were also analyzed for vocabulary, word co-occurrence, and other features. But no clear consensus emerged among potential text users (e.g., managers) as to what text content and global structure would be most useful. The lack of a close-knit user community is one reason that Gossip never became operational, although parts of its design were reused in at least three later RG systems, FoG, LFS, and Project Reporter (see later discussion).

E. FoG

One of the few RG systems to enter regular use to date has been FoG, the **Fo**recast **G**enerator component of Environment Canada's Forecast Production Assistant (FPA) workstation (Kittredge et al., 1986,[7] Bourbeau et al., 1990; Goldberg et al., 1994). FoG was built between 1986 and 1990 and currently produces two distinct varieties of weather bulletins, *public* and *marine* forecasts, in both English and French, starting from forecast map data produced by an atmospheric-modeling program. The map data contain values for predicted air pressure, humidity, temperature, and other basic parameters at grid points on a North American map for several intervals throughout the 2- to 3-day forecast period. When displayed graphically, the map data are used to automatically draw a time sequence of weather maps with isobar lines and other annotations on the FPA workstation screen. Forecasters use interactive graphics tools to analyze the maps and add further information (e.g., about local area effects), which enhances the map data. After human-controlled graphical analysis and enhancement is complete, FoG operates automatically to interpret the enhanced map data to generate each forecast of a given type (public or marine) in both languages.

Figure 5 shows a fragment of a typical raw data set that serves as initial input to FoG. The data values given here are for predicted wind direction (in compass degrees), wind speed (in knots), precipitation type and air temperature (in degrees Celsius) at a particular grid point every hour (in Greenwich time) for the first eleven h starting after forecast issue time (1000Z).

Figure 6 shows the set of concepts that has been extracted from the full data set by the Space–Time Merger (STM) module of FoG. STM carries out two types of

[7]An early prototype for FoG was the RAREAS system, which used a simple phrasal grammar instead of the Meaning-Text language model.

```
1000Z    110    6 0    -22
1100Z    170    4 0    -22
1200Z    180    5 0    -21
1300Z    180    4 0    -21
1400Z    190    3 0    -21
1500Z    210    5 0    -20
1600Z    220    6 0    -20
1700Z    210    7 0    -20
1800Z    240   12 s    -19
1900Z    270   22 s    -19
2000Z    290   25 s    -20
2100Z    300   27 s    -21
```

Fig. 5 Raw input data to FoG's STM module.

operation on the raw data. First, for each weather parameter, it applies heuristic rules to the time series data in a process called time merging, to find one or more significant states that best characterize the parameter over the 2- to 3-day forecast period. For example, STM scans three days' worth of hourly wind speeds and directions (e.g., 300° at 27 knots) to find up to five consecutive wind concepts (e.g., northwest at 25–30 knots) for marine forecasts (where wind information is of primary importance). Some smoothing and averaging of values takes place if needed to reduce a large number of concepts to a sufficiently compact set to make a manageable forecast. The second operation on raw data is called space merging, and acts to compare the weather concepts in adjacent geographic areas to find cases where two or more contiguous areas can be given the same forecast. The concepts shown in Fig. 6 can be read as follows. The first line gives the names of the two geographic areas that have been grouped together by space merging (Belle Isle and Northeast Coast). The second through fifth lines give four wind concepts. The first wind concept states that between zero and 8 h after forecast issue time (i.e., 6 AM–2 PM) the direction will be variable (`. . . di a . . .`) and the speed will be between 5 and 5 knots (i.e., will be 5 knots). The second wind concept states that from 8 to 12 h the direction will be northwest, with speed between 25 and 30 knots, and so on. Following this are two precipitation concepts, stating that snow will occur between 8 and 17 h after issue time, with showers thereafter (until the 57th h). The following visibility concept says that during the same interval when snow occurs, the visibility is reduced to fair (3 nautical miles). The next three concepts, for freezing spray

```
group: (belle_isle northeast_coast);
winds:   (0 8 di a sp 5 5 );
winds:   (8 12 di nw sp 25 30 );
winds:   (12 35 di nw sp 15 15 );
winds:   (35 66 di a sp 5 5);
pcpn:    (0 s 8 17 );
pcpn:    (0 a 17 57 );
visib:   (8 17 3 3 fair fair s );
fr_spray: (0 14 e);
fr_spray: (14 28 ef);
fr_spray: (28 66 a);
temp:    (0 66 s -23 -19);
```

Fig. 6 Weather concepts sent from STM to the text planner.

```
BELLE ISLE
  NORTHEAST COAST.
  WINDS LIGHT INCREASING TO NORTHWEST 25 TO 30 KNOTS EARLY THIS
  AFTERNOON THEN DIMINISHING TO NORTHWEST 15 BY EVENING.
  WINDS DIMINISHING TO LIGHT WEDNESDAY AFTERNOON.
  SNOW BEGINNING EARLY THIS AFTERNOON THEN ENDING LATE THIS
  EVENING.  VISIBILITY FAIR IN SNOW.  MODERATE FREEZING SPRAY
  DEVELOPING NEAR DAWN THEN BECOMING SEVERE EARLY THIS
  EVENING.
  SEVERE FREEZING SPRAY ENDING WEDNESDAY MORNING.
  TEMPERATURES MINUS 23 TO MINUS 19.
  OUTLOOK FOR THURSDAY... LIGHT WINDS.
```

Fig. 7 English output text for concepts of Fig. 6.

conditions, state that during the first 14 h there will be freezing spray of normal intensity, followed by 14 h of intense freezing spray, which will then abate after the 28th h. During the entire 66-h forecast period, the temperature will range from −23 to −19°.

FoG's generator takes the STM concepts as input to the text planner. The output of the planner is an interlingual structure that serves as input to separate English and French realizers, which produces texts in the two languages (English shown in Fig. 7).

FoG has been used operationally since 1990 to produce marine and public forecasts on the FPA workstation at several locations in Eastern and Central Canada.

F. LFS/RTS/CPIS

From 1990 to 1993 the LFS project at CoGenTex-Montreal was devoted to building bilingual generators for three varieties of Canadian economic statistic summaries, dealing with (a) the labor force, (b) retail trade, and (c) the consumer price index. (The project took its name from Labor Force Statistics, the first of the three domains.) The Labor Force domain was very similar to the unemployment statistics domain treated by SEMTEX—more detailed in nature, but without the added political comments. The major aim was to generate the monthly reports that Statistics Canada publishes, using as input the data tables that accompany the English and French texts (Iordanskaja et al., 1992). A feasibility study had shown that the information in four such tables, appearing in Statistics Canada's official monthly publication, *The Labour Force*, is sufficient to determine the content of the accompanying published report. Figure 8 shows part of a data table (here in a form readable by a Prolog program) that appeared in this publication for November 1989.

An analysis of natural texts showed that their global structure is quite stereotyped in type and order of information conveyed. Figure 9 illustrates a typical text; here, part of the LFS system output for the data shown in Fig. 8.

LFS successfully generated reports as long as 3 pages, which approximated human-composed reports in complexity and style. The same approach was used for retail trade reports and for consumer price index reports (RTS and CPIS system variants). All three system variants were implemented with a graphical user interface

```
% ---------- "TABLE A"
% -----      Seasonally Adjusted Estimates by Age and Sex, Canada.

% ----------------------------------------------------------------------------------
%            Total Males   Fem 15-24y Males   Fem    25y+    Males   Fem
% ----------------------------------------------------------------------------------
[ labour_force  in thousands ,
  ( 11 , 1989 ) ,
           13600 , 7556 , 6044 , 2660 , 1408 ,  1252 ,  10940 ,   6148 ,  4792 ] .
[ labour_force  in thousands ,
  ( 10 , 1989 ) ,
           13538 , 7535 , 6003 , 2652 , 1399 ,  1253 ,  10886 ,   6136 ,  4750 ] .
[ labour_force  in thousands ,
  ( 09 , 1989 )43
           13528 , 7554 , 5974 , 2650 , 1407 ,  1252 ,  10878 ,   6147 ,  4731 ] .
% ----------------------------------------------------------------------------------
[ employment  in thousands ,
  ( 11 , 1989 ) ,
           12568 , 6989 , 5579 , 2352 , 1226 ,  1126 ,  10216 ,   5763 ,  4453 ] .
  ...
```

Fig. 8 Raw input employment data (part of Table 1) for LFS system.

to allow control over the display of data, the setting of linguistic parameters (e.g., length of sentences), and other output features. Users were also allowed to set certain thresholds of data salience. For example, the number of months used to define a trend in an index or a run (consecutive monthly increases or decreases in the same direction) was user-selectable. Likewise, the comparison of current levels with recent record high or low levels of an index could be controlled by selecting the number of previous months to include in the comparison. Figure 10 shows the CPIS system's

```
COMMENTARY

Overview

Estimates for November 1989 from Statistics Canada's Labour Force Survey
show that the seasonally adjusted level of employment rose by 32000 and
that the level of unemployment increased by 30000.  The unemployment rate
increased by 0.2 to 7.6

Employment

For the week ended November 3, 1989, the seasonally adjusted level of
employment was estimated at 12568000, up 32000 from October.  The increase
was concentrated among women aged 25 and over.   The employment /
population ratio remained virutally unchanged (62.1).

Employment among women aged 25 and over rose by 44000 and their employment
/ population ratio increased by 0.5 to 52.3.

Employment among men aged 25 and over fell by 12000 and their employment /
population ratio decreased by 0.3 to 72.5.

Part-time employment increased by 25000.  The increase was evenly
distributed between men and women.

Full-time employment remained virtually unchanged.  An increase among women
was offset by a decrease among men.

Employment fell by 10000 in agriculture, by 12000 in transportation,
communication and other utilities and by 12000 in primary industries other
than agriculture.  Employment rose by 68000 in services and by 20000 in
trade.  It remained virtually unchanged in the other sectors.

Employment rose by 11000 in Quebec, by 8000 in Alberta, by 6000 in British
Columbia and by 5000 in Ontario.  Employment decreased by 4000 in
Saskatchewan.  It remained virtually unchanged in the other provinces.
...
```

Fig. 9 LFS output text fragment for data shown in Fig. 8.

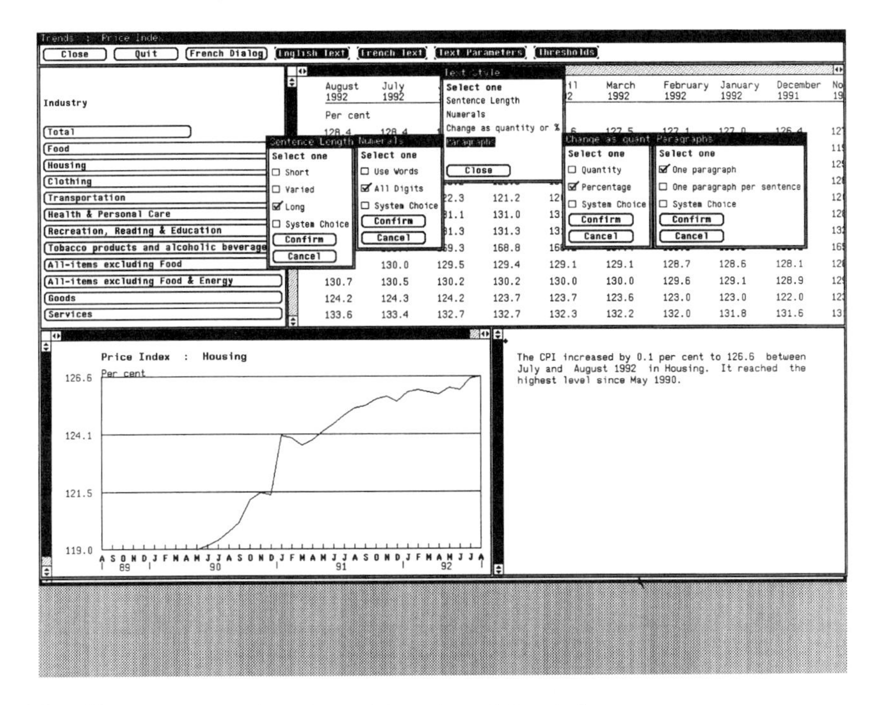

Fig. 10 RTS system interface with setting of linguistic parameters.

simultaneous display of numerical and graphical data, along with output text. The consumer price figures for the housing index up to August 1992 have been selected and displayed as a line graph. Output text has been generated in accordance with the menu choices displayed. For example, sentence length has been set to "long," which obliges the text planner to seek a compound sentence structure when several messages in the content can be integrated into a single sentence. A short output text is shown in the lower right window.

LFS was beta-tested by staff at Statistics Canada during 1993–1994. Although entirely satisfied with output text quality, analysts felt the need for a more complete integration of statistical knowledge with the generator, so that they could put more information into their reports. Unfortunately, the small annual volume of reports to be produced did not justify a significant investment in building up the level of statistical and domain knowledge available to the content selection component. An alternative use was examined for making short text summaries available to on-line browsers of statistical data, but this was considered premature, given the lack of a large community of data users with well-defined goals. As of this writing, no operational usage has been made of LFS, although some of its features have been reused in more recent work at CoGenTex.

G. Project Reporter

Large projects, which need to coordinate work on multiple tasks and subprojects, often maintain databases on expenditures of resources (labor hours and money) by task to allow project managers to track the progress of the project and compare it with the schedule and goals at the project outset. Ideally, project managers should be able to automatically generate a periodic textual report directly from the database. Project Reporter was first developed during 1991–1993 at CoGenTex in Ithaca, New York, USA, and designed to produce such reports in the context of large software-engineering environments (SEEs) that are dedicated to standardized software development processes (Korelsky et al., 1993).

Writing project management reports would seem to be an excellent candidate for automation. However, NLG can succeed here, as elsewhere, only to the extent that the database is rich in the relevant information, and the information is maintained up to date. Project reports often give not only information about expenditures of time and budget by task, but also about meetings, deliverables, changes in personnel, equipment and software, and unforeseen problems. Much of this information is not normally entered into a database.

Project management reports often require that linguistic semantic distinctions be made when referring to domain objects. For example, activities such as *coding an interface module* have duration, whereas events such as *the delivery of code* are considered points in time for the purposes of the reports. Durative tasks can be *completed*, but not point time tasks, which can only *occur* or *take place*, and so on. Relations between tasks must be represented, in particular by stating which tasks are subtasks of which other tasks, or by stating temporal dependencies between tasks, or their interaction through the production and use of project "artifacts," such as manuals. For example, one task might require the use of a manual that is produced by a (therefore) prior task. Moreover, some historical information about successive project states is required to report on project changes (e.g., modifications of schedule, budget, etc.).

The Project Reporter prototype implemented knowledge about a particular software life cycle support environment (SLCSE). A SLCSE-oriented report, which features some text about task overspending problems, is given in Fig. 11.

The initial development of Project Reporter was somewhat hampered by the fact that project report texts are not easy to collect into a corpus for study. There was no convenient public source for the texts, and no single large community of project report producers and users, so that the standards for content and style were idiosyncratic from one project to another. In other words, there was no established sublanguage to model. The first experiments with Project Reporter showed that (1) fairly sophisticated reports can be produced from properly constituted and maintained databases, but that (2) few software projects maintain rich databases in a standard form.

During 1998–99, Project Reporter was adapted to generate hypertext reports from the standardized databases of commercial project management software, and became a commercial product in early 1999 (for details, see www.cogentex.com/products/reporter). This application integrates textual status reports with tables and graphics in a form which is viewable through a standard web browser. The amount of text on each web page is limited to a few sentences, which serve to highlight important information that is not easily conveyed by the graphics (e.g., about a task that is behind schedule). The kinds of information presented are limited to the types that recur across most projects: costs, schedules, and personnel assignments for the whole project and for its component tasks. The importance of this common core of project information across project types has made such an application feasible.

```
Progress against milestones:
During September, software development of the Computer
Software Component (CSC) "Grammar" continued.  Coding and
testing was completed for its subordinate Computer Software
Units (CSUs) "Morphology Module" and "Surface Syntax Module".
The results were recorded in their Software Development Files
(SDFs).  Coding and testing continued for CSUs "Deep Syntax
Module" and "Semantics Module".  "Morphology Module" source
code was evaluated and incorporated into the Development
Configuration.  It was also placed under configuration
control.  Further test procedures for the CSC "Grammar" were
developed and recorded in its SDF.

Problem areas:
Technical personnel assigned to coding and testing of CSU
"Semantics Module" estimate that only 25% of this level 3
subtask was complete as of September 30, whereas 48% of the
original budget allotment has been spent.

Changes:
It was decided to increase the personnel allotment for coding
and testing of CSU "Semantics Module" by 160 hours.   The
allotment for integration and testing of the CSC "Grammar" was
decreased by the same amount.  ...
```

Fig. 11 Project management report fragment from Project Reporter (1993).

H. Streak

The Streak system was developed at Columbia during 1990–1994, mostly in the context of a PhD thesis project (cf., Robin, 1993).[8] The application domain was the production of basketball game summaries from box score data of the kind that can be obtained on-line. A major innovation of Streak was the incorporation of historical information about players, about teams, and their recent games into the summaries of specific games. This was achieved by a technique of **incremental revision**, which opportunistically adjoined the historical information to the framework of the basic game summary. A second innovation of Streak was its use of semantically rich words to convey more than one fact simultaneously. For example, the sentence *Denver outlasted Sacramento* conveys simultaneously the facts that (a) Denver beat Sacramento and that (b) the lead changed hands several times until the final minutes of the game. Figure 12 gives a sports summary generated by Streak from the game box score data plus a database of historical information. As with ANA, this domain favors the extensive use of synonymy, such as the use of a wide variety of verbs meaning “score” (e.g., *dish out, add, pump in*). Historical information phrases are indicated here in bold type face.

The historical information integrated into Streak's basketball game summaries is very similar in syntactic type to the background information found in many stock market reports. In stock market reports, information that does not belong to the trading data during the 10 AM–4 PM period tends to be syntactically subordinated. This can be seen as a distinction between core domain information, which is fore-

[8]See Chap. 7 for further discussion of STREAK.

```
Sacramento, Ca. - Michael Adams scored a career-high 44 points
Wednesday night, including seven 3-point baskets, to help the
short-handed Denver Nuggets end a five-game losing streak with
a 128-112 victory over the Sacramento Kings.

Adams, who was drafted and then discarded by the Kings four
seasons ago, made 17 of 26 field goals, including seven of 11
3-point attempts, and hit three of four free throws to break
his previous career high of 35 points. Adams also dished out
a game-high 10 assists and had five steals.
Rookie Chris Jackson added 22 points and center Blair
Rasmussen pumped in 21 and grabbed 12 rebounds for Denver,
which outscored Sacramento 19-4 during the final 5:01 of the
fourth quarter.
The Nuggets, who had only eight players available for the
game, improved to 2-12 on the road and 6-20 overall.
Rookie guard Travis Mays, playing his second game after
missing 11 with back spasms, scored a season-high 36 points
for the Kings, who lost their fourth straight. Lionel
Simmons added 24 points and 15 rebounds.
```

Fig. 12 Game summary from STREAK with historical information.

grounded, and secondary domain information, which in the case of stock market reports deals with economic news as well as historical information about the market in past trading sessions. For stock market reports, several specific syntactic devices, including nonrestrictive relative clauses, are used to "graft" information from the secondary domain(s) onto the framework of the core domain summary. Given sufficient linguistic analysis of the grammatical devices used, it might be possible to generate game summaries in one pass, without incremental revision, as an alternative.

I. PLANDOC

One of the most advanced applications of report generation, as well as one of the few systems to see actual productive use, has been PLANDOC (McKeown et al., 1994, 1995). Developed at Bellcore Research during 1992–94, PLANDOC generates summaries of telephone network planning activity, and has been used by regional telephone-operating companies to write the first draft of the documents that engineers must file to justify their decisions. It is noteworthy that users (i.e., engineers) were involved in every step of the design, implementation, and testing of PLANDOC.

Figure 13 gives the input data for PLANDOC's generator, followed by a handwritten explanation of the input. Figure 14 shows the corresponding text produced by PLANDOC. The data in Fig. 13 is the trace output from a software tool, called PLAN, which is used by Bellcore engineers to derive long-term telephone equipment plans based on growth forecasts and economic constraints.

The first module of PLANDOC is a fact generator that expands the explicit data in the PLAN trace into a more articulated knowledge-based form, including implicit information from the context of the application. This allows the following module, an ontologizer, to correctly add semantic information to each fact from the PLAN domain. These enriched facts are then sent to a discourse planner, which organizes the content and overall structure of the report. The output of the discourse planner is a sequence of semantic feature structures, one for each future sentence.

```
1.  RUNID fiberall FIBER 6/19/93 act yes
2. FA  1301  2 1995
3. FA  1201  2 1995
4. FA  1401  2 1995
5. FA  1501  2 1995
6. ANF  co  1103  2 1995 48
7. ANF  1201 1301  2 1995 24
8. ANF  1401 1501  2 1995 24
END. 856.0 670.2
```

Facts 2-5 are all fiber activations (FA). The second column for these facts indicates the site (CSA 1301 for line 2), the third indicates the quarter (quarter 2 for all of them) and the fourth column, the year. These three lines are all generated as the second sentence of the summary. Lines 6-8 are all cable placements (indicated by the code ANF). Column 2 indicates the starting section where the cable is placed (realized by "from section" in the summary), column 3 the ending section, column 4 the quarter, column 5 the year, and column 6 indicates the number of fibers in the cable. Facts 6-8 are realized as the third sentence of the summary.

Fig. 13 PLAN tracking information (input to PLANDOC fact generator), followed by an explanation of the data record structure.

```
Run-ID: FIBERALL

This saved fiber refinement included all DLC changes in RUNID
ALLDLC.  RUNID FIBERALL demanded that PLAN activate fiber for
CSAs 1201, 1301, 1401 and 1501 in 1995 Q2.  It requested the
placement of a 48-fiber cable from the CO to section 1103 and
the placement of 24-fiber cables from section 1201 to
section 1301 and from section 1401 to section 1501 in the
second quarter of 1995.  For this refinement, the resulting 20
year route PWE was $856.00K, a $64.11K savings over the BASE
plan and the resulting 5 year IFC was $670.20K, a $60.55K
savings over the BASE plan.
```

Fig. 14 PLANDOC output for input of Fig. 13.

These are then sent to the lexicalizer and surface generator for realization as the sequence of sentences in the output text.

The major problem faced by PLANDOC is in properly controlling conjunction and ellipsis (together called *aggregation*) of syntactically similar elementary sentences. The system uses a complex algorithm to cover a wide variety of types of aggregation.

This concludes our summary presentation of individual RG systems. In the following section we compare the techniques used to determine content and plan the global text structure in these systems.

5. TEXT PLANNING FOR REPORTS

By **text planning** we mean the computation of both the precise information content and the global structure of a text, before choosing the linguistic forms for the output sentences. Text planning acts to linearize and "package" the selected information into a sequence of sentence-sized chunks, sometimes referred to as **messages**,[9] each of which can then be passed to the realizer. Viewed as a "black box," text planning for report generation looks as follows:

Input: Structured data

⇓

Output: Sequence of sentence specifications

The two composite processes of text planning, sometimes called **content determination** and **text structuring**, may be carried out either sequentially or in some parallel or interleaved fashion. In any case, text planning typically builds a conceptual or semantic message structure that will trigger the production of an appropriate output sentence by the realizer.

Report generators differ in the level at which sentences are specified for development by the realizer. These and other questions in linguistic realization for RG systems will be dealt with in Sec. 6. The present section is divided into three parts. In Sec. 5.A, we identify the tasks that are typically associated with the text planners of RG systems. Next, in Sec. 5.B, we look at how report text planners have been implemented. Finally, in Sec. 5.C, we summarize some text-planning work aimed at configuring more flexible reports.

A. Tasks Performed by Text Planners

For the purposes of this section, we consider text planners at a "metalevel," disregarding idiosyncratic features of specific systems. We are interested here in the nature of the two quite distinct activities of content determination and text structuring. We examine each set of tasks in turn.

1. Content Determination for Reports

This first set of tasks can be quite heterogeneous: ranging from very "low-level" computation, such as character string manipulation, to sophisticated inferencing operations requiring detailed knowledge representation. Instead of trying to cover all cases here, we have identified four tasks that are almost necessarily performed by all RG systems.

a. *Analysis and Filtering of the Raw Input*

This includes the very basic task of reading the input and storing it in more convenient formats for subsequent processing. For instance, data organized in tabular form can be converted into sets of program objects, with associated values and parameters. A good illustration of this type of processing can be found in McKeown et al.'s (1995) presentation of Streak. The first task performed by this RG system is to transform tabular information appearing in the box score of a basketball match, such as:

[9]The term *message* was used by Kukich (1983b) as a semantic unit (roughly, proposition) destined to become a finished clause or sentence. Our use of the term assumes, in the general case, an additional communicative structure specification (e.g., theme/rheme), which constrains the way the semantic unit may be linearized as text.

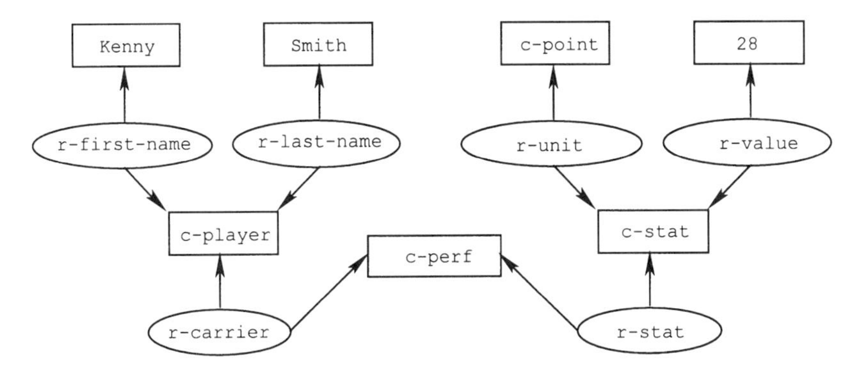

Fig. 15 Fact data structure built by STREAK'S planner. (From McKeown et al., 1995.)

```
Players        (...)        tp
               (...)
Kenny Smith    (...)        28
```

into complex data structures, of the type graphically represented in Fig. 15.

While performing tasks related to the "ingestion" of the input data, the planner can also filter out redundant or irrelevant information. For reports that can be tailored to given users of the system (e.g., marine forecasts for commercial mariners versus for recreational sailors), it is necessary to select which part of the input data is particularly relevant for the intended reader.

b. Using Quantitative Data to Find Concept Instances

It often happens that purely quantitative data has to be reinterpreted as instantiated concepts before its integration into sentence specifications. In some weather forecasts, for instance, forecasters may prefer the report generator to inform the user that *strong* or *light* winds are expected, rather than using actual numerical values (e.g., *winds 30 km/h*). Similarly, wind directions are calculated in compass degrees by the numerical forecast models that produce the input data, whereas actual weather forecasts must give this information using nonnumerical values; for instance, FoG's conceptualizer will instantiate the concept "south" from a wind direction value of 195° (see Goldberg et al., 1994).

Report generation typically involves multiple levels of concept instantiation. For example, from stock market data at half-hourly intervals ANA instantiates elementary concepts of increase or decrease in stock prices. From these, it can then instantiate a more complex concept such as that of "broadly-based decline in the market" from a predominance of decline instances across several market sectors and over several half-hour intervals.

c. Computing New Data from the Input

The concepts to be verbalized in a report may require that additional data be derived from the raw input data. For instance, the LFS planner computes the numerical change in employment level between two consecutive months to instantiate the "employment increase" concept that appears in the corresponding text:

Raw data	Corresponding text
[employment in thousands , (11 , 1989) , **12568** ...]. [employment in thousands , (10 , 1989) , **12536** ...] .	Estimates for November 1989 from Statistics Canada's Labour Force Survey show that the seasonally adjusted level of employment **rose by 32000** ...

The text "rose by 32000" in bold, taken from Fig. 9 in Sec. 4.F, corresponds to data derived by taking the arithmetic difference in the two monthly data values shown in bold to the left. Such inferences from raw data can take place at any stage of the planning process, depending on the type of input or output to be produced, the architecture of the planner, and so forth.

d. Computing New Data from Additional Information Sources

Reports are not necessarily generated from a single source of input. Manually generated reports are often based on two sets of data:

1. Actual input data, which is a database of basic facts that the reports must summarize
2. Additional databases, such as archives of past facts, domain knowledge bases, or other, that will be used to either interpret or complement the first type of data.

PLANDOC's "ontologizer," for instance, is a program module the task of which is to enrich the basic input facts given to the system with domain-specific knowledge. We have mentioned in Sec. 4.H that the Streak system uses a history of team or player statistics to enrich data found in the box score of a match. In FPA/FoG, previous weather conditions are used to compute proper weather warnings. The module of FPA that builds the conceptual representation of weather events to be expressed in the forecasts makes a follow-up of weather warnings, checking present conditions against preceding ones, to compute whether a gale warning, for instance, should be presented as new, continuing, or ending. Note that new "outside" data may either be found directly in the additional databases or computed from the information they contain.

A very interesting issue is the extent to which purely linguistic considerations can influence the process of content selection. Depending on the application domain, and existing constraints on the content and form of reports, the text planner may have to be guided by lexical and grammatical knowledge. For instance, one of the most important tasks FoG's planner has to perform is the grouping of regions with similar weather conditions, such that each group of regions will have its weather conditions described in one single paragraph (see the discussion of Space–Time Merging, in Sec. 4.E). For weather conditions to be considered as similar is often a function of the specific lexical items to be used to express them. For example, if three marine forecast regions A, B, and C are expecting wind speeds of 30, 35, and 45 knots respectively, regions B and C will be grouped together, despite the smaller arithmetical difference between regions A and B. This similarity is entirely based on the fact that winds from 35 to 47 knots are all conceptualized and linguistically referred to as *gales*. Because they belong to the same conceptual range, 35- and 45-knot winds are "closer" than 30- and 35-knot winds. This shows that linguistic distinctions can take precedence over physical ones when computing weather forecast content. What has been said here about weather forecasting applies equally to any technical field for which terms have a very precise and quantified interpretation. The output vocabulary predetermines to some extent what is considered significant in the input data.

Weather forecasts give us other interesting examples of content selection that does not strictly obey factual guidance. Sometimes, it is not even linguistic, but rather, psychological factors that will force the implementation of very specific planning procedures. For probabilities of precipitation, for instance, there is no way a forecast can boldly state a 50% chance of precipitation, even though 50% probability may be the numerical result of the weather model that feeds data to the text generator. The problem is that 50% essentially means "we can't tell," and this goes against what people expect from a weather forecast (or for that matter, from any kind of forecast). It may be necessary for the planner to "bend the truth" and predict, say, a 40 or 60% chance of precipitation (E. Goldberg, personal communication).

The facts expressed in reports do not all have the same status. Some may **have to** be communicated, whereas others may be expressed contingent on external factors (such as, for instance, the presence of other related facts). This is the case in Streak and PLANDOC, where a distinction is made between necessary and secondary (optional) information. In these systems, the selection of nonnecessary information is made after a first draft of the reports is generated. Optional information is introduced based on linguistic properties of elements present in the draft and, therefore, additional content selection is performed at the level of realization.

2. *Text Structuring*

Not all systems have a separate module to carry out text structuring (i.e., a module responsible for packaging information content into text structures). Often, the program component in charge of structuring the text into paragraphs and sentences is also in charge of the computing content at the level of paragraphs, sentences, or even clauses (propositions). In actual fact, when very strong predictions can be made about the structure of the text (i.e., when reports have very rigid structures), a *text grammar* model can be procedurally activated to look for relevant information in the input database.

Whether or not RG systems use separate modules for content selection and for text structuring, text structures are generally represented as trees of some sort. These can be actual tree-like hierarchies of text components, as in FoG and LFS, or text grammars that are formally equivalent to tree structures, as in the medical report generator discussed in Sec. 4.B. In fact, report text structures are often calculated by "trimming" a maximally complex tree structure down to one that expresses just the concepts instantiated from the data. This process of planning by tree trimming may also involve some local restructuring operations (see Carcagno and Iordanskaja, 1992).

Text planners are mainly in charge of ordering linguistic messages and grouping them into presentential chunks of the right size. In more sophisticated systems where the realizer possesses some sort of paraphrastic power, text planners are also in charge of computing the communicative organization of messages: that is, oppositions such as theme versus rheme (*Quebec's unemployment rate decreased by 0.5%* vs. *The unemployment rate decreased by 0.5% in Quebec*). This capacity may be important to achieve a smooth flow of text when sentences are long and complex. If the language of reports is simple and does not allow much syntactic variation, there may be no need for communicative structure specifications. In this case, text structuring can produce syntactic structures that can then be turned into finished sentences by a simple realizer.

To conclude this subsection, we need to mention a very important task that is often associated with text planners, namely: **lexical selection**. Although it is true that most systems perform some kind of lexical selection (or preselection) at the text-planning level, it is our belief that this problem logically belongs to the realization phase. It is possible to claim that all such selection performed at the level of planning is nothing but content or conceptual selection. It is only when actual linguistic structures are computed (i.e., at the realization level) that whatever appears in these structures can be considered as selected lexical items. Of course, for systems that do not choose between synonyms or otherwise play on lexical relations, a choice of concept often amounts to a choice of lexical item. But from a metalevel viewpoint, the selection of actual lexical items is, by definition, a linguistic problem, that properly belongs to realization. This is particularly clear in the context of multilingual RG systems (such as FoG and LFS), where actual linguistic (i.e., language-specific) lexical selection can be performed only at the realization level, from interlingual representations of sentence content.

B. Types of Implemented Text Planners

We see three main axes along which text planners can be distinguished: (a) the type of input they take, (b) the type of output they produce, and (c) the degree of modularity with which they process data. We will briefly examine each of these axes.

1. Type of Input

Most RG systems take input data mainly in the form of numerical tables. These tables can contain parameter values, such as wind speed at a given time, or employment rate in a given province of Canada, or trading value of a company's share, and so forth. In this case, events must be computed from the raw data, because the reports talk mainly about events of some sort (see Sec. 1.A).

The notion of *event* is not necessarily limited to a change of state. A state itself can very well be conceptualized as an event in the context of a report. Consider, for instance, the definition of *event* given in one of the user's guides produced for the FPA (Forecast Production Assistant, 1995):

> An *event* describes the physical value behind a segment of text. For example, "HIGH NEAR 12" is a segment of a forecast text reflecting a maximum temperature event of 12 degrees at a location for a specific time period.

Thus, for the FPA events include not only facts such as changes in wind direction or speed, beginning or ending of precipitation, but also states that are considered salient for the user.

For some RG systems the input database can itself be a record of a series of events. Such is true with Gossip (see Sec. 4.D), in which the input consists of collections of elementary events, such as logins taking place on a computer system. When the input is already packaged into events, there are usually higher-level concept instances built from the event data, such as total number of users and system resource usage reported by Gossip.

The foregoing observations show that it is not so much the event or nonevent nature of the input data that has to be considered if one wants to evaluate the type of work performed by the planner. What really matters is the gap that exists between the planner's input and the output in terms of **number** of concepts used, and the complexity of the calculations required to instantiate derived concepts.

2. *Type of Output*

The *output* of the planner is by (our) definition a formal specification of a text in terms of an ordered sequence of sentence specifications, or messages. There are two possibilities: (a) unilingual RG, in which the output specifications are language-specific, or (b) multilingual RG, in which the output is "interlingual." On top of the language-specific–interlingual distinction, planner output can be characterized according to the level of linguistic representation used. Most systems, such as FoG, have planner outputs expressed in some kind of syntactic representation.

A few RG systems, such as Gossip and LFS, have planned text using semantic representations (see Sec. 4.D and 4.F), typically when the output text is expected to contain long or complex sentences, or to require control of stylistic variation. The bilingual LFS system needs semantic representations, for English and French often differ in the syntactic structures used to realize equivalent sentences. In general, the richer the language required in the reports, in terms of linguistic complexity and variation, the deeper the linguistic specification coming from the planner needs to be.

3. *Degree of Modularity*

If one wants a system that can be maintained, improved, and adapted, modularity is essential. Modularity does not only mean that each piece of code that performs a given type of operation carries its own name and is clearly separated from the rest in a specific programming module. It also means that these pieces of code function in a relatively independent way (i.e., modifications in one module entail only minor, easily identified modifications in other modules).

C. Introducing Flexibility into Report Texts

Most of the RG systems built or designed to date have aimed at generating reports with relatively stereotyped (fixed) content and global structure (i.e., down to the paragraph level). In ANA and FoG, for example, the sequence of text paragraphs, and the order of information within paragraphs, is rigidly determined, even though the available data may not warrant the instantiation of a maximal text. This degree of conventionality has made preestablished schemas or tree structures the method of choice for most reported text planning. Working to instantiate concepts and build messages within preestablished structures for texts reduces the number of difficult decisions that might otherwise occur if planning on the basis of "first principles" were attempted. Plan tree structures may be trimmed to eliminate messages pertaining to data that is either unavailable or uninteresting in the current input. Moreover, plan trees also have the advantage of capturing many idiosyncracies of particular text types that would be hard to derive from general principles (see Sec. 3.C).

However, many varieties of report text, while following a rigid global structure, require some degree of flexibility in the planning process. For example, marine weather reports show significant variation in the number and complexity of sentences dealing with wind conditions, making it necessary for FoG to include aggregation rules that work on clusters of nodes in the instantiated plan tree. A different kind of flexibility is required in LFS, which orders sentences within some paragraphs on the basis of their relative salience. As the text plan is instantiated it is also adjusted to ensure that the most significant economic changes are discussed first.

Some recent RG systems are designed to "graft" information (in the form of pieces of linguistic specifications) onto a basic skeletal text plan. Streak, for example, produces a basic draft summary containing all necessary information. Then, based on the linguistic (lexical and grammatic) potential of this draft, additional nonessential information is added. PLANDOC uses syntactic constraints in each sentence plan to decide whether additional information will be added.

Other RG systems such as Project Reporter and LFS have given the user control over both text content and form through a graphic interface. Menu choices allow the suppression or expansion of areas of content, or as in LFS, the run-time setting of sensitivity thresholds for salient facts, and the greater or lesser aggregation of sentences to achieve stylistic variation. Note that when more control is given to the user to select text content and set priorities for salience and ordering of information presented, this can be seen as operating on the text plan, which is then procedurally invoked to request only the data needed from the database.

Another important way to introduce flexibility into report planning is to use rhetorical operators for top-down planning, as proposed by Hovy (1988b). Rhetorical, goal-directed text planning, which has been used for generating texts such as advisory dialogues, may not be appropriate for creating the macrostructure of most reports (see Sec. 3.C). Nevertheless, when reports require some reflection of reasoning about facts, there may be a role for rhetorical planning at the paragraph level.

6. APPROACHES TO THE IMPLEMENTATION OF LINGUISTIC REALIZATION

It has been claimed (e.g., by McDonald, 1993) that realization is a relatively trivial part of the generation process. Although RG systems usually require more effort on text planning than on realization, we would like to take issue with the idea that realization is in any sense a trivial operation, particularly when we are discussing **operational** RG systems, and when operations such as lexicalization and syntactic structuring are included in the realization process, as they have been in the four systems using the Meaning-Text approach (Gossip, FoG, LFS, Project Reporter).

Linguistic realizers perform **linguistic synthesis**, essentially taking sentence specifications (i.e., messages) as input and producing cohesive text as output. Unlike text planners, which may be implemented quite differently for reports than for other text types, RG realizers are hardly different from realizers for other types of NLG systems, except that

1. They implement rather strict stylistic conventions, both at the level of lexical choice (e.g., technical vocabularies) and grammatical structures.
2. They tend to specialize in lengthy texts, often extending to several paragraphs; thus, they may need to paraphrase the same content in multiple ways, to avoid repeating words and structures when possible.
3. They usually produce full sentences (possibly in a telegraphic style), and not the sentence fragments required in many dialogue systems.

We survey realization components of RG systems along three axes, each of which is covered in a separate subsection: use of a linguistic theory to structure linguistic descriptions (6.A), number and nature of the levels of linguistic representation (6.B), and types of lexicalization performed by the realizer (6.C).

A. Presence of a Linguistic Theory in the Background

Whether for RG or more general NLG, the main characteristic of realizers, which distinguishes them from text planners, is that they can be based on preexisting general theories: in this case, **linguistic** theories. There is no such thing as a global theory of "message computing," that could be used as a model for the implementation of planners. On the other hand, there **is** an abundance of formal linguistic theories, some of which are well-enough developed to help model the realization process.

One might expect that linguistic theories of the "generative" family would be favored by researchers and developers of RG systems. But such is not so. Most RG applications to date have been based on either Meaning-Text linguistics (Gossip, LFS, FoG, or other) or on systemic linguistics (Streak, PLANDOC, and such). We see at least two related reasons for this:

1. The best candidates for the implementation of realizers are linguistic approaches that put strong emphasis on representing meaning equivalence (paraphrase).
2. They must propose linguistic models that are formal devices for connecting message contents to linguistic forms.

These are salient properties of both Systemic Functional Grammar and Meaning-Text linguistics, that distinguish them from most generative approaches. In fact, the affinity between these theories and the generation process is so great that research projects in NLG are often used as means of performing "pure" linguistic research within these frameworks. (Such as for the Gossip project, for instance.)

Even though some linguistic theories seem better tailored as frameworks for implementing NLG realizers, it should be stressed that any linguistic approach, provided it is formal and covers all essential aspects of linguistic modeling, can be used for designing realizers. But are linguistic theories needed at all? For us, the answer is a clear "yes." Use of linguistic theories to structure realizers gives power and flexibility to RG systems, for mainly two reasons.

First, powerful realizers must (a) perform the synthesis of complex syntactic structures (active or passive sentences, subordinate and embedded clauses, coordination, and other); (b) model idiosyncratic behavior of lexical units (subcategorization, collocations, and such); (c) handle phenomena, such as pronominalization. In other words, they have to be based on the encoding of complex linguistic knowledge,

knowledge that is consistently modeled by means of a given linguistic theory. Linguistic theories impose on realizers both a conceptual structure (levels of representations, types of rules, structure of lexical descriptions) and, also, a terminology. This latter point may seem trivial, but without a theoretical framework, there is no unified terminology for language phenomena. Modeling linguistic knowledge within a coherent terminological and conceptual system allows persons involved in the design and maintenance of realizers to understand each other. There are cases where no existing linguistic theory describes the facts very well, or where the full power of a theory represents "overkill." This latter case occurs when the texts to be generated use a very small number of fixed sentence patterns, with trivial variations in them. Such is the case of "texts" such as:

```
Error : Variable myInt may not have been initialized.
```

But such semicanned texts are hardly full-fledged reports. They do not require true linguistic realization and, in fact, do not require any planning at all. In other words, the production of such texts does not really belong to the field of NLG. Reiter (1995) contrasts NLG with "template-based generation," (i.e., text production based on filling and concatenating sentence templates). Reiter discusses some of the pros and cons of each approach, and notes the need to base system design on a thorough cost–benefit analysis.

Second, the use of a coherent theory-based linguistic model permits a cleaner separation between the components of the realizer that capture declarative linguistic knowledge and the components that implement procedures for the activation of this knowledge. Linguistic theories generally make a clear distinction between linguistic knowledge proper (whether grammatical or lexical) and operations that are required to use this knowledge (when speaking or decoding natural language). Let us take a concrete example. The ancestor of the FoG system, RAREAS, was a prototype program developed for Environment Canada to demonstrate the feasibility of NLG for the automatic production of English weather forecasts. Although this program was successful as a prototype, it was literally hacked to meet the very tight time and budget constraints of the initial project. One main manifestation of this hacking was that the realizer of RAREAS did not make any clear separation between declarative linguistic knowledge and procedures to manipulate it. This realizer, like any realizer, was performing the translation of the conceptual–semantic representation of a linguistic message into a sentence. But the linguistic knowledge used to perform this translation was somehow embedded into the translation procedure itself. It is thus not unfair to say that RAREAS did not possess any explicit grammar and lexicon of English, or of the sub-English of weather forecasts. Although RAREAS seemed to do its job, as a demonstration system, things became more complicated when Environment Canada decided to move one step forward and asked for a bilingual version of RAREAS (English and French), to demonstrate the applicability of the approach to the Canadian bilingual context. The exercise of giving RAREAS bilingual capabilities made it clear to the authors of the system that the technology adopted was reaching its limits and that actual operational RG systems would never be developed without the implementation of clean and explicit linguistic models (Polguère et al., 1987).

To our knowledge, there is as yet no advanced RG system for which the realizer is **not** based on a linguistic theory, or on a hybrid of linguistic approaches—as with SURGE, which is claimed to be based on a mixture of systemic linguistics, Head-Driven Phrase Structure Grammar, the descriptive linguistics of Quirk et al. (1985), and even a pinch of Meaning-Text linguistics.

B. Levels of Linguistic Representation

Formal linguistic theories typically offer models of natural languages that are based on at least one level of representation of linguistic utterances. Some approaches can be "minimalist," whereas others can postulate an extreme stratification of linguistic modeling. This is directly reflected in the structure of RG realizers that are based on these models. In this subsection, we focus on the design of realizers of the Gossip/LFS/FoG family, that can be termed *(highly) stratificational.*

Meaning-Text linguistics postulates that a natural language is a functional device that maps the Semantic Representation (SemR) of a linguistic message onto the set of all Texts, actually the *Surface-Phonetic* representation of the texts, expressing this message. This mapping is performed through five intermediate levels of representations: Deep- and Surface–Syntactic Representations (D/SyntR), Deep- and Surface–Morphological Representations (D/SMorphR), and a Deep-Phonetic Representation (DPhonR). Meaning-Text realizers typically follow the successive levels of transition postulated by the theory, as indicated in Fig. 16, which summarizes the internal structure of LFS's realizer.[10]

Notice that, since LFS produces **written** texts, it does not implement any level of phonetic representation. Moreover, only one level of morphological representation is employed, because the morphological phenomena in these texts are quite simple. Figure 17 gives an example of two intermediate representations handled by the similar Gossip realizer (i.e., the Deep- and Surface–Syntactic Representations of the sentence *The users of the system ran compilers and editors during this time.* Kittredge and Polguère, 1991b).

When the reports to be generated are very simple, it may prove useful to "streamline" a full-fledged Meaning-Text realizer to eliminate unnecessary levels

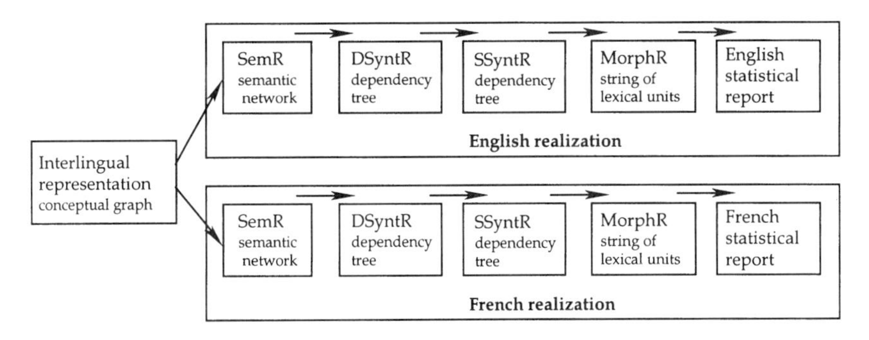

Fig. 16 Structure of LFS's Meaning-Text realizer.

[10]LFS's realizer has an internal structure that is very similar to that of Gossip's realizer. The main difference between the two is that Gossip is a monolingual system that performs only the production of reports in English.

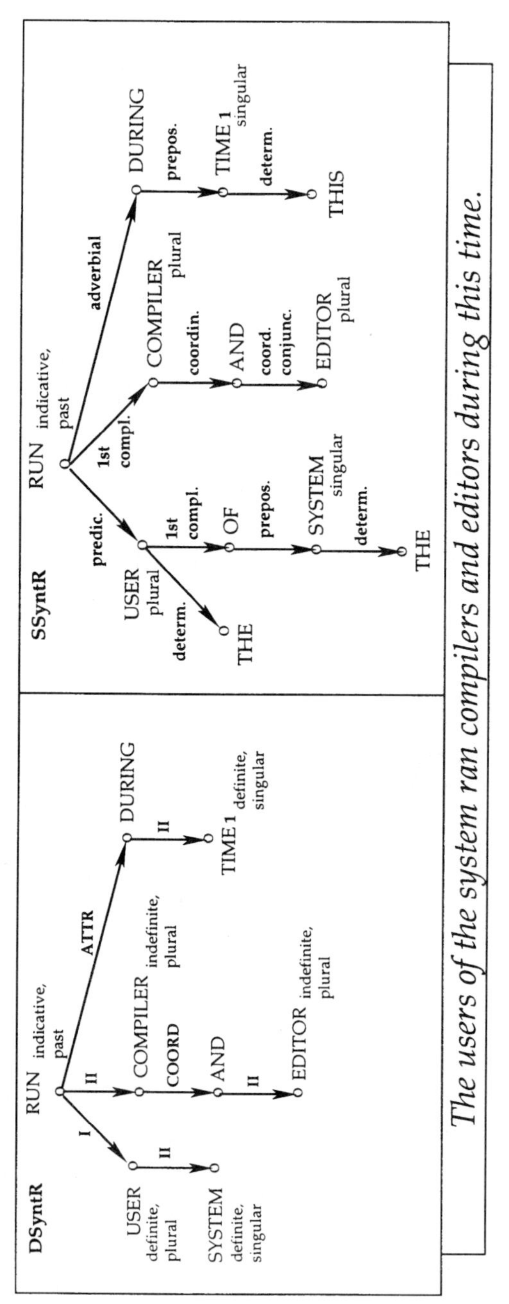

Fig. 17 Deep- and Surface-Syntactic trees produced by Gossip.

of representation. We have just examined the internal structure of LFS, which is almost maximal in terms of stratification. In contrast, FoG, which aims at the bilingual production of simple sublanguage texts, does not use a level of semantic representation, as can be seen in Fig. 18.

Not only does FoG lack a level of semantic representation, but its interlingual representations are not true conceptual graphs. They are tree-like structures, that resemble the (deep-)syntactic trees from which realization in each language begins (see Kittredge and Polguère, 1991a, for details). There are two main reasons for this:

1. Texts generated by FoG do not require much stylistic variation. In other words, in this type of weather forecast, a content to be expressed almost necessarily implies a given syntactic structure. Therefore, the realization model does not need to handle semantic networks, for which the main advantage, from a theoretical and practical point of view, is to serve as a basis for modeling equivalent syntactic formulations for any given sentence (i.e., paraphrases).
2. The English and French sublanguages of weather forecasts that FoG is designed to generate are very close to each other, especially for the type of syntactic structures involved. Therefore, it is possible to directly produce interlingual representations that are already almost syntactic in nature.

What are the justifications for using highly stratified realizers in LFS and FoG? Our experience with this type of system shows that, in spite of its apparent complexity, the resulting modularity offers advantages when it comes to maintenance and extension to other domains of application. Modifications will have very local or controlled effects; additionally, modularity makes it easier to diagnose any problems that may occur. FoG's realizer has the following two procedural characteristics:

1. It keeps track of all intermediate representations that are produced for the generation of each sentence.
2. It has a robust ("never crash") procedural design that keeps track of any local failures in the realization process.

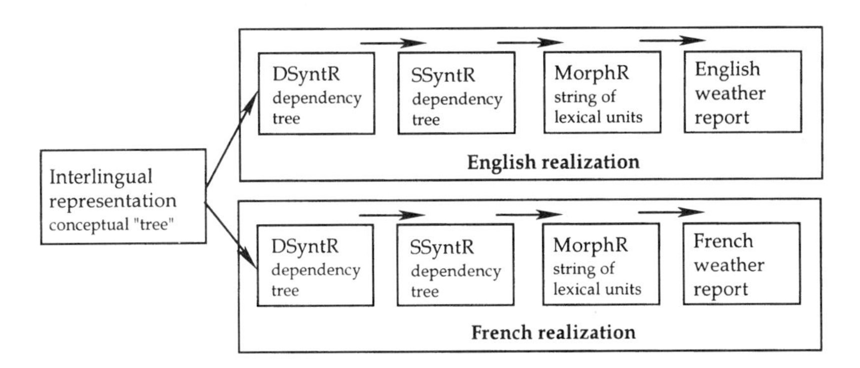

Fig. 18 Structure of FoG's Meaning-Text realizer.

These have proved very useful for realizer maintenance, to meet the changing requirements of different weather forecasting offices.

C. Types of Lexicalization

A very important issue for NLG in general, and RG in particular, is the problem of lexicalization, or lexical selection. In theory, nothing forbids RG systems from performing lexical selection at a very deep level, when planning report content. However, prerealization lexical choice has two main disadvantages.

First, it can hardly cope with multilingual generation, where reports of equivalent structures have to be generated in more than one language. In such a context, an interlingual representation of sentence content must be computed before the realization process can begin. This interlingual representation, by definition, cannot be based on preselected lexical units if it really has to be language-independent. In the multilingual realizers of LFS and FoG, shown in Figs. 16 and 18, lexical selection of full lexical units is performed, for each language (English and French), during the determination of Deep-Syntactic Representations. (The SemR level in LFS uses language-dependent semantic elements, which are not yet actual lexical units.) A given RG project may attempt only monolingual generation and, therefore, ignore this constraint. But performing lexical selection at the planning level will virtually **impose** a monolingual limitation to the system.

Second, lexicalization at the planning level is a "clean" option only when one deals with very constrained types of sublanguages, with very simple and fixed sentence patterns. In more "natural" languages, options are permitted—and often needed—for stylistic variations. These are options in the choice of lexical units as well as syntactic structures, and it is a well-known fact that there are very close relations between lexical units and the type of structures they can participate in. It is, therefore, very important, for flexible RG systems, to handle lexicalization and sentence structuring in a very related manner, thus performing lexical choice at a more superficial level than text planning. There are possible links between text planning and the type of lexical resources that are available in the targeted language(s). This is an aspect of the problem that has been well studied in the context of the Streak system (see Sec. 4.H). The revision process implemented in Streak allows *conflation* of the expression of facts through the use of semantically richer lexical units. Figure 19 presents an example of content conflation, taken from McKeown et al. (1995); this example shows that certain verbs such as "tie," may be introduced in a secondary process by which a draft text is relexicalized to efficiently accommodate some interesting but nonessential information (here, the fact that the number of points scored equals a personal record for the season).

Streak demonstrates clearly that "intelligent" lexicalization cannot be performed at a very deep level of processing and has to be tied to the whole of the realization process.

In addition to the problem of **when** lexicalization should take place, there is the interesting problem of the distinction between **different types** of lexicalization. One can distinguish between at least three main types of lexical units that have to be selected to produce informative and grammatical sentences:

1. **Full lexical units** (a.k.a., content words), directly introduced to express the message content itself

Fixed fact --i.e., necessary information introduced by the planner at an early stage, that triggers draft lexicalization:	'Karl Malone scored 39 points'
Floating fact --i.e., additional (less essential) information added during the revision process:	'These 39 points equal Karl Malone's season high'
Generated sentence with revised lexicalization (in bold):	*Karl Malone* ***tied*** *his season high with 39 points.*

Fig. 19 An example of revised lexicalization in STREAK. (From McKeown et al., 1995.)

2. **Collocational units**, which are meaningful, but for which choice is contingent on the choice of lexical units of the first type—for instance, the selection of *heavy* in *They suffered heavy losses* to express "intensification"
3. **Empty lexical units** (a.k.a., grammatical words), the introduction of which is controlled by either grammar rules (e.g., true auxiliary verbs, some determiners, and such, or the subcategorization frame of full lexical units (e.g., prepositions that are obligatory with a verb object)

Most RG systems make a distinction between lexical units of the first and third types, thus implementing two types of lexicalization, but ignore (or fail to encode in an explicit way) the problem of lexicalization of collocational lexical units. There are, to our knowledge, two noticeable exceptions.

First, ANA and a few other systems have made effective use of a **phrasal lexicon** i.e., one that describes not only independent lexical units, but also various kinds of phraseological phenomena. This approach was inspired by a proposal made in Becker (1975). Becker claimed that people speak by recycling chunks of already stored linguistic constructions and that word-by-word structuring of sentences plays a limited role in actual sentence production. For Becker, collocations (semi-idiomatic expressions, such as *suffer heavy losses*); plus "situational" utterances, such as *How can I repay you?*; proverbs; and the like, are overwhelmingly present in natural language texts. Clearly, such observations should be taken into consideration for an adequate implementation of lexicalization in NLG. ANA's phrasal lexicon was an easy way to give rich variety to generated stock market reports (but probably would not have been optimal for building a large operational system, because a large set of word strings would be hard to maintain and extend).

Second, at least one RG system based on Meaning-Text linguistics (namely, the LFS system discussed in Sec. 4.F) has implemented the notion of **lexical function** (see Wanner, 1996) to handle systematic collocations.[11] Lexical functions are formal devices used both in lexical entries, to encode collocations governed by the headword, and in the grammar itself, mainly to encode paraphrasing rules. To illustrate how this descriptive device can be used during the realization process, consider a LFS paraphrasing rule, shown in Figure 20, adapted from Iordanskaja et al. (1996).

[11]Lexical functions and their use in paraphrasing were studied in detail in the context of the Gossip system (see Sec. 4.1), but not systematically implemented in the generator (see Iordanskaja et al., 1991).

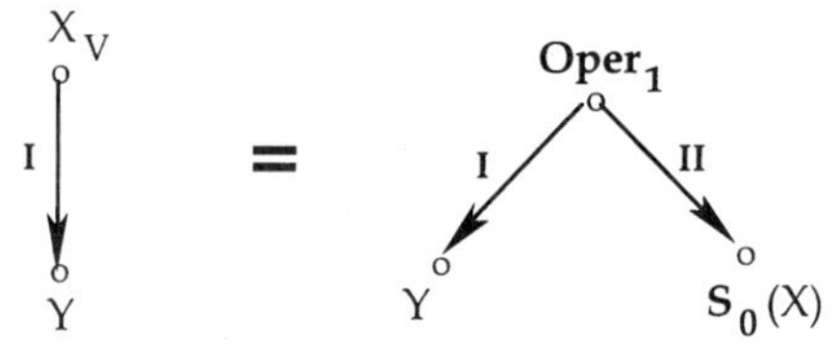

Employment[Y] *increased*[X] *in Quebec.* =
Employment[Y] *showed*[$\mathbf{Oper_1}$(*increase*$_N$)] *an increase*[$\mathbf{S_0}$(X)] *in Quebec.*

Fig. 20 A paraphrasing rule used in LFS. (From Iordanskaja et al., 1996.)

In this rule, two lexical functions are used:

1. $\mathbf{S_0}$ is the substantive (nominal) derivation of a verbal keyword
2. $\mathbf{Oper_1}$ is a semantically empty verb ("light" verb) that combines with a nominal keyword to produce a verbal structure.

According to Meaning-Text linguistics, most languages share the same kinds of lexical functions, and use such paraphrasing rules. Not all kinds of report texts, however, require a detailed account of collocations and paraphrasing.

D. Other Approaches to Realization—SURGE

Semantic Unification-Based Realization Grammar of English (SURGE) is a broad-coverage grammar of English that is the core component of the realizer of a significant number of NLG systems. A brief presentation of this realizer is particularly relevant here, as it is used in two RG systems mentioned earlier: PLANDOC and Streak.

That SURGE has been used in many different NLG systems is a direct consequence of its most interesting characteristic: it is a general-purpose grammar, not linked to a particular sublanguage or domain of application. In this sense, SURGE is in the same league as Nigel (the linguistic component of the Penman system). Similar to Nigel, SURGE is a grammar based on the general principles of Halliday's (1985) systemic linguistics. But in spite of its systemic "upbringing," SURGE also borrows from other linguistic approaches, notably from Head-Driven Phrase Structure Grammar (see Pollard and Sag, 1994) and from "traditional" grammar as formulated by Quirk et al. (1985).[12]

From a theoretical point of view, SURGE is, therefore, a linguistic melting-pot, which is not necessarily a weak point. At the implementation level, however, it is grounded in one uniform framework, the Functional Unification Formalism (FUF). FUF is more than just a formalism; it is a programming language that implements an extension of Kay's (1979) approach to the encoding of linguistic knowledge, Functional Unification Grammar (FUG). FUF extends the capabilities of standard FUG mainly in that it permits more powerful lexicalization procedures.

[12]Elhadad (1993) mentions that SURGE also borrows some ideas from Meaning-Text linguistics, although this appears to be limited to lexical organization in his ADVISOR II system.

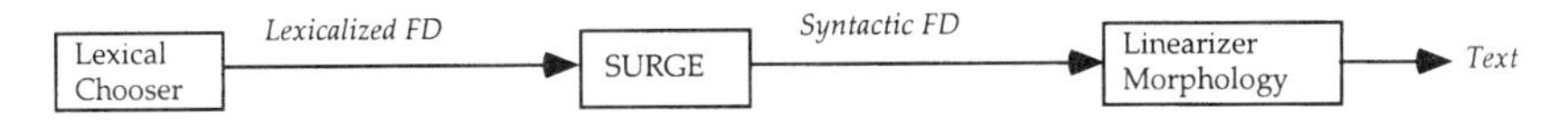

Fig. 21 The place of SURGE in the realization process. (From Elhadad, 1993.)

```
((cat clause)
 (process ((type material) (lex "work") (effective no)))
 (partic ((agent ((cat personal-pronoun)
               (person fist) (number singular)))))
 (tense present-progressive))
```

Fig. 22 Lexicalized FD for *I am working* (SURGE input). (From Elhadad, 1993.)

Information is encoded in FUF within complex structures that are called Functional Descriptions (FDs). These structures amount to sets of partially instantiated features that function as constraints for linguistic realization. It should be noted that FUF and the FDs it manipulates can be used at both the planning and realization levels, which offers a unified formalism as well as unified programming techniques at all levels of the NLG process. Moreover, as a computational formalism amounting to a programming language, FUF is not geared to any particular linguistic theory. In principle, it can be used to implement any linguistic framework, provided this framework can receive an interpretation in terms of the general principles underlying the unification-based approach. An extensive presentation of FUF can be found in Elhadad (1993).

SURGE's role in the generation process is to take *lexicalized* FDs and transform them into *syntactic* FDs, which are fully instantiated syntactic trees. Following SURGE there is additional postprocessing to (a) linearize the syntactic FDs and (b) apply morphological rules. The place of SURGE in the realization process is summed up in Fig. 21.

As indicated in this figure, lexical selection of open-class lexical units is performed before the execution of SURGE. For a concise presentation of lexicalization procedures in FUF-based NLG systems, see Elhadad et al. (1997). To give a more concrete idea of the task performed by SURGE, we present two figures. Figure 22 is a lexicalized FD that, given as input to SURGE, will give rise to the realization of the simple sentence *I am working*; Fig. 23 shows part of the resulting syntactic FD,

```
(process ((type material) (lex "work")
        (process-type material)
        (effective no) (agentive yes)
        (voice active)
        (subcat {oblique})
        (cat simple-verb-group) (generic-cat verb-group)
        (modality none) (tense {tense}) (polarity {polarity})
        (person {synt-roles subject person})
        (number {synt-roles subject number})
        (event ((cat verb) (lex {process lex})
              (ending present-participle)))
        (be-1 ((tense present) (person {process person})
             (number {process number}) (ending {process ending})
             (lex "be") (cat verb)))
        (pattern (dots be-1 dots event dots))))
```

Fig. 23 Part of the corresponding syntactic FD (SURGE output). (From Elhadad, 1993.)

corresponding to the instantiation of the **process** component of the lexicalized FD (in bold in the Fig. 22).

A quick examination of both figures shows that the task performed by SURGE amounts to mainly the filling in of incomplete structures, lexicalized and syntactic FDs being based on the same kind of formal structures.

7. GENERATION FROM "KNOWLEDGE BASES"

Until now, this chapter has focused on RG systems to give a unified characterization to this important class of application systems. Reports have been narrowly defined in terms of data summarization, for which the data is typically quantifiable and varies over time.

At this point it becomes interesting to compare and contrast RG systems with other types of application system that generate single-view sublanguage texts (not dialogues). In particular, we want to consider generation from knowledge representations that have more structure than just a relational view of time-varying data.

A. Conceptual Form Readout

One important kind of system, which has a long history in natural language processing, simply verbalizes a conceptual structure as text. Some early work on generating stories from script representations could fall under this label. A recent application of this kind, which will be revisited in more detail in the second part of this volume, is Exclass.

Exclass was built at CoGenTex-Montreal during 1991–1993 to verbalize conceptual representations of public sector jobs. The Exclass text generator (Caldwell and Korelsky, 1994) produces English and French descriptions of Canadian government jobs in the warehousing domain that have been built up and edited in conceptual frame-like forms. The conceptual forms were designed to meet the needs of computer-assisted job evaluation by a separate rule-based system. Content planning in Exclass is nonexistent, because human analysts select job content from standardized options using a menu-based interface. Text planning is a straightforward sequencing of several conceptual forms. Basically, Exclass just reads out the information in the conceptual frame, translating it into text. Aside from being bilingual, Exclass is noteworthy in using a categorial-based realization grammar, coupled with a lexicon that makes use of lexical functions from Meaning-Text linguistics. This allows virtually instantaneous generation, while preserving important linguistic differences between the two output languages. The relative efficiency of the realizer is also aided by a conceptual design that remains as close as possible to text surface structure.

B. Diagram Explanation

An important, but distinct, class of NLG system works not from database input, but rather from knowledge representations, such as representations of diagrams that may also have a graphical representation. Good examples of this kind of system are Joyce and ModelExplainer, described in the following. (Exclass works from job objects, which do not have diagrammatic representations.) The texts normally serve to describe or explain some aspect of the diagram, and, hence, lack a temporal dimension. Conceptual summarization is not a central or obligatory aspect of such object description systems, although one recent system, FlowDoc does summarize the features of a diagram.

We now briefly summarize a few systems that work from knowledge bases (KBs), and especially from representations of diagrams. As we will see, the applications are quite varied, but tend to lack the temporal dimension of data reports, and often lack a significant amount of summarization. (On the other hand, they tend to emphasize spatial, causal, or other high-level concepts.) The main purpose of their inclusion here is to give some greater perspective to the RG systems as a class.

JOYCE. The Joyce system (Rambow and Korelsky, 1992) was built at ORA-Ithaca during 1987–1989 as part of a software design environment called Ulysses, which included a graphical environment for designing secure distributed software systems. Joyce's output texts provided annotations of design diagrams, or summaries of secure information flow analysis (i.e., tracing the movement of information through the distributed system, using a narrative style). The text, by focusing on properties of components, generally complemented the information provided in the diagrams. One important aspect of Joyce was its explicit use of domain communication knowledge (see Sec. 3.B) in the text planner. DCK involves such things as the conventions for what to mention, and in what order, that are specific to a given domain and, hence, cannot be easily derived from general principles of text planning. Joyce made use of schemas to represent the DCK, but also added the ability of schemas to be executed as lists of instructions, permitting additional control and communication between schemas. For realization, Joyce reimplemented in Lisp the Prolog realizer developed in the Gossip system, but without using the semantic net representation level (see Sec. 4.D and 6.B).

FLOW DOC. FlowDoc (Passoneau et al., 1996) is a prototype text generator, developed at Bellcore, that summarizes information from work flow diagrams in the context of a business reengineering application. Input diagrams represent either current or possible future work flow representations that have been constructed using a GUI within a dedicated software environment. Verbalizing the work flows often requires conceptual summarizing of attributes of subtasks to reduce subflows to single nodes with appropriate labels. This calls for extensive ontological knowledge of the domain and metadomain, along with rhetorical relations and an ontology of message classes. Summarization also requires notions of salience for task participant, their decisions and actions. During sentence planning, the message class ontology is used to maintain the overall discourse organization during the aggregation of messages. FlowDoc was built with many of the components and features of PlanDoc and Streak, but undertakes more difficult summarization tasks than earlier systems. Its output texts include a variety of quantitative, structural and salience generalizations about the work flow. The overall goal of such systems is to help work flow analysts validate models, and find candidates for streamlining or reorganizing the workflow. By introducing a well-defined semantics of graph elements and domain ontology, FlowDoc also aims to introduce a more rigorous methodology into the application systems for drawing the work flow diagrams.

MODEX. ModelExplainer (or ModEx) (Lavoie et al., 1997) is an operational system for the description of object-oriented (OO) models used in software engineering. Built at CoGenTex-Ithaca during 1995–1997, it is being beta-tested by industrial software engineers. ModEx coordinates the output of tables, automatically generated text, and text entered freely by the user. Similar to FlowDoc, it aims to improve communication between domain experts, system analysts, and other personnel who collaborate to build OO software. Starting from the knowledge representation of the OO model as input, it produces paragraphs describing the relations between classes, along with paragraphs describing examples. Human-authored text adds information not deducible from the model, such as purpose relations between the classes. Text plans in ModEx are user-configurable at run time, and a hypertext interface allows users to browse through the model using standard Web tools. For realization, ModEx uses a generic realizer, RealPro (Lavoie and Rambow, 1997), a C+ + implementation of the Meaning-Text realizer used on earlier RG systems such as LFS. Unlike FlowDoc, ModEx does not use knowledge about the domain of the model, except for the model itself. This restriction limits the amount of conceptual summarization that it can carry out, but makes it easily portable to new domains.

Another related type of NLG system adds a kind of weak temporal order to a diagram to generate instructions for assembly, operation, or maintenance of a mechanical device or computer system. This kind of system, particularly when there is coordinated output of graphic image with text, can easilsy fall under the scope of multimedia systems, discussed in the next chapter of this volume, and is not discussed here.

8. CONCLUSIONS

We have presented a chronological overview of the most important RG systems and discussed the major approaches to the special problems of text planning (including content selection and text structuring) and realization used in generating reports. We have contrasted with RG systems a few of the best-known application systems which produce descriptive texts from knowledge representations such as OO model diagrams. Table 1 gives in tabular form a summary of the RG systems discussed in Sec. 4, as well as the KB explanation systems presented for comparison in Sec. 7. The column labeled T/CS refers to the presence (+) or absence (–) of a temporal dimension in the data (T) or use of conceptual summarization during text planning (CS).

RG applications are a relatively recent development in natural language processing compared with, for example, machine translation or database query systems. It is, therefore, not too surprising, relatively speaking, that not more systems have become operational. At the same time, given the accelerated pace of development in NLP in recent years, one might have expected more to result from the various RG projects to date. From our viewpoint, we can draw several conclusions.

First of all, several RG systems (e.g., ANA, FRANA, SEMTEX, and Streak) have grown out of academic thesis projects, with no pretention of immediate commercial application. The concerns of the possible end-users were, therefore, not factored into the design process. Moreover, the earlier systems (including also the RAREAS system that preceded FoG) were not constructed with sufficient modularity in the conceptual and linguistic components to allow these systems to be maintained and extended as serious applications. To cite one conceptual processing example, Kukich has noted (personal communication) that over time the ANA system began to describe the daily stock market in more dramatic terms, because the volume of the market had grown, and some rules were designed to use super-

Table 1 Summary Comparison of RG and Diagram Explanation Systems Mentioned in this Chapter*

System name	Place	Domain	Lang.	T/CS	Ling. model	Usage	Ref.
ANA (1983)	U. Pittsburgh	Stock market	English	+/+	Phrasal	Prototype	Kukich, 1983a,1983b
FRANA (1985)	U. Montreal	Stock market	French	+/+	Phrasal	Prototype	Contant, 1985
Medic.Reporter (1986)	Illinois IT	Stroke cases	English	+/?	String Gram.	beta Test	Li et al., 1986
SEMTEX (1986)	U. Stuttgart	Econ. Statistics	Ger./E	+/+	Phrasal?	Prototype	Roesner, 1987
RAREAS (1986)	U. Montreal	Weather f-cast	Eng./Fr	+/+	Phrasal	Prototype	Kittredge et al., 1986
GOSSIP (1989)	ORA-Montreal	OS security	English	+/+	Meaning-Txt	Prototype	Iordanskaja et al., 1988, 1991
FOG (1990)	CoGenTex-Mtl.	Weather f-cast	Eng./Fr	+/+	Meaning-Txt	**Operational**	Goldberg et al., 1994
LFS (1992)	CoGenTex-Mtl.	Econ. statistics	Eng./Fr	+/+	Meaning-Txt	beta Test	Iordanskaja et al., 1992
ProjectReporter (1993)	CoGenTex-Ith.	Project mgmt.	English	+/+	Meaning-Txt	beta Test	Korelsky et al., 1993
STREAK (1994)	U. Columbia	Sports events	English	+/+	Systemic	Prototype	McKeown et al., 1995
PLANDOC (1994)	Bellcore	Engineer.plans	English	?/+	Systemic	**Operational**	McKeown et al., 1994, 1995
Project Reporter (1999)	CoGenTex-Ith.	Project mgmt.	English	+/+	Phrasal	Commercial	www.cogentex.com/products/reporter
JOYCE (1989)	ORA-Ithaca	Design security	English	−/−	Meaning-Txt	Prototype	Rambow & Korelsky, 1992
EXCLASS (1993)	CoGenTex-Mtl.	Job descriptions	Eng./Fr	−/−	Categorial	Prototype	Caldwell & Korelsky, 1994
FLOWDOC (1996)	Bellcore	Work flow	English	−/+	Systemic	beta Test?	Passonneau et al., 1996
MODEX (1997)	CoGenTex-Ith.	OO modelling	English	−/?	Meaning-Txt	beta Test	Lavoie et al., 1997

*Additional information on these systems may be found in Sec. 4, 6, and 7 of this chapter.

latives based on absolute values of indices and trading volume. A more satisfactory application would have to constantly revise the notion of what constitutes a "normal" variation in price movements and trading volumes. ANA's phrasal lexicon and coarse-grained grammar would also pose problems for maintenance in any large-scale application. The limitations of early systems, however, are now easily overcome with approaches that have been tested in other domains.

Other obstacles have hindered the acceptance of report generation in certain application areas. For example, in the case of both Gossip and ProjectReporter, there was no true sublanguage that could be modeled. Either there was not a large group of users with a compelling need for natural language (i.e., no sublanguage in the case of Gossip), or there was insufficient uniformity in the way reports were written (i.e., a poorly defined sublanguage in the case of ProjectReporter). These conditions could change in the future as user groups become larger and their needs evolve.

The most important practical considerations are those of economic viability. Only in the case of weather forecasts has it been clear from the start that the volume of text needed justifies the investment in an RG system. There is a huge customer base for weather information, as witnessed by the many dozens of companies already serving this market, and much of that information is already delivered as text. Forecasters are under great pressure to communicate the volatile information, day and night, before it becomes useless. The growth of template-based approaches to weather forecast composition is another indication of the "market pull" in this application domain. it is not surprising that several new RG projects for weather forecasting (see Coch, 1997) are underway around the world today.

Closely related to the issue of economic viability is the question of whether or not the application domain has readily available knowledge representations from which a report generator can work. The FoG system has become operational, thanks to considerable work at Environment Canada with the FPA workstation, which helps build concepts from data to serve as a basis for a variety of graphical and nongraphic forecast products. It has taken several years for forecasters to formulate the relevant knowledge and test it for various operational requirements. The feasibility of report generation from the concepts has been a constant "technology-push" factor in this implementation, but it is doubtful whether the FPA would have been built to serve the needs of textual forecasting alone.

In closing, we should point to some recent trends in the area of report generation as covered in this chapter. First, we have seen that in high-payoff domains where texts are relatively simple, such as weather forecasting, there is strong competition from template-based approaches. The reason, it seems, is the desire on the part of user organizations to maintain control over their software, particularly when it contains domain knowledge. Weather agency programmers can build template systems, but do not yet understand NLG. For NLG to succeed, one of two things has to happen. Either NLG expertise needs to migrate to within the user agencies, or the tools for building and maintaining NLG (especially RG) systems need to become simpler to use, so that specialized training is reduced to a minimum. Presumably, both of these changes are in progress. In particular, we already see a movement in the latter area with the appearance of user configuration options for systems, such as ModEx, that have a specialized user community.

A second trend involves the rapid growth of document-authoring approaches to text production. Already in the 1993 LFS system, there was a possibility for user configuration of text content and style. Within the FoG system, users can modify concept values interactively when the generated text reveals that the input information is not valid. As seen in ModEx, it is now possible for generators to prepare a skeleton text that can then be enriched by added free text from human editors. Many kinds of RG system would, in fact, require this possibility to become truly useful, since many naturally occurring report texts occasionally include information that could not come from any database.

A third trend, which has only recently affected RG systems, is the availability of the data over the Internet, and the possible delivery of RG output through web servers. ModEx and the commercial version of Project Reporter are the clearest recent examples illustrating this trend, which is certain to grow in the near future.

ACKNOWLEDGMENTS

We gratefully acknowledge the support of the Natural Sciences and Engineering Research Council of Canada under grant OGPO200713, and the Quebec Ministry of Education under grant FCAR97-ER-2741.

REFERENCES

Becker, J. The phrasal lexicon. Theoretical Issues in Natural Language Processing—An Interdisciplinary Workshop in Computational Linguistics, Psychology, Linguistics, Artificial Intelligence. (B Nash-Webber, R Schank, eds.) Cambridge, MA: 1975, pp 60–63.

Bourbeau L, Carcagno D, Goldberg E, Kittredge R, Polguère A. Bilingual generation of weather forecasts in an operations environment. Proceedings of COLING-90, Helsinki, 1990, vol. 3, pp 318–320.

Caldwell D, Korelsky T. Bilingual generation of job descriptions from quasiconceptual forms. Proc of the Fourth Conference on Applied Natural Language Processing, Stuttgart, 1994, pp 1–6.

Carcagno D, Iordanskaja L. Content determination and text structuring: two interrelated processes. New Concepts in Natural Language Generation. (H Horacek, M Zock, eds). Pinter, 1993.

Contant C. Génération automatique de texte: application au sous-langage boursier français. MA dissertation, Linguistics Department, University of Montreal, Montreal, 1985.

DiMarco C, Hirst G, Wanner L, Wilkinson J. HealthDoc: customizing patient information and health information by medical condition and personal characteristics. Proceedings of AI in Patient Education. A Cawsey, ed. University of Glasgow, 1995.

Elhadad M. Using argumentation to control lexical choice: a unification-based implementation. PhD dissertation, Computer Science Department, Colombia University, 1993.

Elhadad M, McKeown K, Robin J. Floating constraints in lexical choice. Comput Linguist 23:195–239, 1997.

Environment Canada. Forecast Production Assistant. STM Programmer's Guide, 24 pages, 1995.

Goldberg E, Driedger N, Kittredge R. FoG: a new approach to the synthesis of weather forecast text. IEEE Expert 9(2):45–53, 1994.

Haliday MAK. An Introduction to Functional Grammar. London: Edward Arnold, 1985.

Harris Z. Discourse analysis. Language 28:1–30, 1952.

Hovy EH. Generating language with a phrasal lexicon. Natural Language Generation Systems. D McDonald, L Bolc, eds. New York: Springer-Verlag, 1988a.

Hovy EH. Planning coherent multisentential text. Proceedings of the 26th Annual Meeting of the ACL, Buffalo, NY, 1998b, pp 163–169.

Iordanskaja L. Communicative structure and its use during text generation. Int Forum Inform Document 17(2):15–27, 1992.

Iordanskaja L, Kim M, Kittredge R, Lavoie B. Integrated LFS/RTS/CPIS System, Final report to Communications Canada, CoGenTex-Montreal, August 1993.

Iordanskaja L, Kim M, Kittredge R, Lavoie B, Polguère. Generation of extended bilingual statistical reports. Proceedings of COLING-92, Nantes, 1992, pp 1019–1023.

Iordanskaja L, Kim M, Polguère A. Some procedural problems in the implementation of lexical functions for text generation. Lexical Functions in Lexicography and Natural Language Processing. L Wanner, ed. Amsterdam/Philadelpha: Benjamins, 1996, pp 279–297.

Iordanskaja L, Kittredge R, Polguère A. Implementing the meaning-text model for language generation. Addendum to Proceedings of COLING-88, Budapest, 1988.

Iordanskaja L, Kittredge R, Polguère A. Lexical selection and paraphrase in a meaning-text generation model. Natural Language Generation in Artificial Intelligence and Computational Linguistics. C Paris, W Swartout, W Mann, eds. Kluwer, 1991, pp 293–312.

Kay M. Functional unification grammar. Proceedings of the Berkeley Linguistic Society, 1979.

Kittredge R. Homogeneity and variation of sublanguages. R Kittredge, J Lehrberger, eds. 1982, pp 107–137.

Kittredge R. Semantic processing of texts in restricted sublanguages. Comput Math Appl 9(1):45–58, 1983.

Kittredge R, Korelsky T, Rambow O. On the need for domain communication knowledge. Comput Intell 7:305–314, 1991.

Kittredge R, Lehrberger J, eds. Sublanguage: Studies of Language in Restricted Semantic Domains. de Gruyter, 1982.

Kittredge R, Polguère A. Dependency grammars for bilingual text generation: inside FoG's stratificational model. Proceedings of the International Conference on Current Issues in Computational Linguistics. University Sains Malaysia, Penang, 1991a, pp 318–330.

Kittredge R, Polguère A. Generating extended bilingual texts from application knowledge bases. Proceedings of the International Workshop on Fundamental Research for the Future Generation of NLP, Kyoto, 1991b, pp 147–160.

Kittredge R, Polguère A., Goldberg E. Synthesizing weather forecasts from formatted data. Proceedings of COLING-86, Bonn, 1986, pp 563–565.

Korelsky T, McCullough D, Rambow O. Knowledge requirements for the automatic generation of project management reports. Proceedings of the Eighth Knowledge-Based Software Engineering Conference, IEEE Computer Society Press, 1993, pp 2–9.

Kukich K. The design of a knowledge-based report generator. Proceedings of the 21st Conf of the ACL, Cambridge, MA, 1983a, pp 145–150.

Kukich K. Knowledge-based report generation: a knowledge-engineering approach to natural language report generation. PhD dissertation, University of Pittsburgh, Pittsburgh, 1983b.

Lavoie B, Étude sur la planification de texte: application au sous-domaine de la population active. MS dissertation, Université de Montréal, Montreal, 1994.

Lavoie B, Rambow O. A fast and portable realizer for text generation systems. Proc of the Fifth Conference on Applied Natural Language Processing, Washington, DC, 1997, pp 265–268.

Lavoie B, Rambow O, Reiter E. Customizable descriptions of object-oriented models. Proceedings of the Fifth Conference on Applied Natural Language Processing, Washington, DC, 1997, pp 253–256.

Li P, Evens M, Hier D. Generating medical case reports with the linguistic string parser. Proceedings of AAAI-86, Philadelphia, 1986, pp 1069–1073.

Maybury M. Using discourse focus, temporal focus, and spatial focus to generate multisentential text. Proc of the Fifth International Workshop on Natural Language Generation, Dawson, PA, 1990, pp 70–78.

Maybury M. Generating summaries from event data. Inform Process Manage 31:735–751, 1995.

McDonald D. Issues in the choice of a source for natural language generation. Comput Linguist 19:191–197, 1993.

McKeown K. Text Generation: Using Discourse Strategies and Focus Constraints to Generate Natural Language Text. Cambridge: Cambridge University Press, 1985.

McKeown K, Kukich K, Shaw J. Practical issues in automatic documentation generation. Proceedings of the Fourth Conference on Applied Natural Language Processing, Stuttgart, 1994, pp 7–14.

McKeown K, Robin J, Kukich K. Generating concise natural language summaries. Inform Process Manage 31:703–733, 1995.

Mel'čuk I. Meaning-text models: a recent trend in Soviet linguistics. Annu Rev Anthropol 10:27–62, 1981.

Mel'čuk I. Dependency Syntax: Theory and Practice. Albany: State University of New York Press, 1988.

Mel'čuk I, Pertsov N. Surface Syntax of English. Philadelphia: John Benjamins, 1987.

Passonneau R, Kukich K, Robin J, Hatzivassiloglou V, Lefkowitz L, Jing H. Generating summaries of work flow diagrams. Natural Language Processing and Industrial Applications: Proc of the NLP+IA 96 Conference, Moncton, 1996, pp 204–210.

Polguère A. Structuration et mise en jeu procédurale d'un modèle linguistique déclaratif dans un cadre de génération de texte. PhD dissertation, University of Montreal, 1990.

Polguère A, Bourbeau, L, Kittredge R. RAREAS-2: bilingual synthesis of Arctic marine forecasts. Final report for Environment Canada, ORA Inc., Montreal, 1987.

Pollard C, Sag I. Head-Driven Phrase Structure Grammar, Chicago: University of Chicago Press, 1994.

Quirk R, Greenbaum S, Leech G, Svartvik J. A Comprehensive Grammar of the English Language. London: Longman: 1985.

Rambow O. Domain communication knowledge. Proceedings of the Fifth International Workshop on Natural Language Generation, Dawson, PA, 1990, pp 87–94.

Rambow O, Korelsky T. Applied text generation. Proceedings of the Third Conference on Applied Natural Language Processing, Trento, 1992, pp 40–47.

Reiter E, NLG vs. templates. Proceedings of the Fifth European Workshop on Text Generation, Leiden, 1995, pp 95–104.

Robin J. A revision-based generation architecture for reporting facts in their historical context. New Concepts in Natural Language Generation: Planning, Realization and Systems. H Horacek, M Zock, eds. London: Pinter, 1993.

Roesner D. The automated news agency: SEMTEX—a text generator for German. Natural Language Generation: New Results in Artificial Intelligence, Psychology and Linguistics. (G. Kempen, ed.) Dordrecht: Martinus Nijhoff, 1987, pp 133–148.

Roesner D, Stede M. Generating multilingual documents from a knowledge base: the TECHDOC project. Proceedings of COLING-94, Kyoto, 1994, pp 339–346.

Sager N. Syntactic formatting of science information. AFIPS Conference Proceedings 41, AFIPS Press, 1972, pp 791–800. (reprinted in R Kittredge and J Lehrberger, eds., 1982).

Sager N. Natural Language Information Processing. New York: Addison-Wesley, 1981.

Wanner L, ed. Lexical Functions in Lexicography and Natural Language Processing. Philadelphia: Benjamins, 1996.

White M, Caldwell DE. CogentHelp: NLG meets SE in a tool for authoring dynamically generated on-line help. Proceedings of the Fifth Conference on Applied Natural Language Processing, Washington, DC, 1997, pp 257–264.

12

The Generation of Multimedia Presentations

ELISABETH ANDRÉ

German Research Center for Artificial Intelligence, Saarbrücken, Germany

1. INTRODUCTION

Multimedia systems—systems that employ several media, such as text, graphics, animation, and sound for the presentation of information—have become widely available during the last decade. The acceptance and usability of such systems is, however, substantially affected by their limited ability to present information in a flexible manner. As the need for flexibility grows, the manual creation of multimedia presentations becomes less and less feasible. Although the automatic production of material for presentation is rarely addressed in the multimedia community, a considerable amount of research effort has been directed toward the automatic generation of natural language. The purpose of this chapter is to introduce techniques for the automatic production of multimedia presentations; these techniques draw on lessons learned during the development of Natural Language Generators.

Rapid progress in technology for the display, storage, processing, and creation of multimedia documents has opened completely new possibilities for providing and accessing information. Although the necessary infrastructure is already in place, we still need tools for making information available to users in a profitable way. Several authoring systems that support the human author in creating and changing multimedia documents are already commercially available. However, to satisfy the individual needs of a large variety of users, the human author would have to prepare an exponential number of presentations in advance. In the rapidly growing field of online presentation services, the situation is even worse. If live data has to be communicated, there is simply not enough time to manually create and continuously update presentations.

Intelligent multimedia presentation systems (IMMP systems) represent an attempt to automate the authoring process by exploiting techniques originating from Artificial Intelligence (AI), such as natural language processing, knowledge representation, constraint processing, and temporal reasoning. The intelligence of these systems lies in the fact that they base their design decisions on explicit representations of the application, and of presentation and contextual knowledge. Such systems go far beyond current multimedia systems, as they account for

- *Effectiveness* by coordinating different media in a consistent manner
- *Adaptivity* by generating multimedia presentations on the fly in a context-sensitive way
- *Reflectivity* by explicitly representing the syntax and semantics of a document

Because of these benefits, IMMP systems have gained widespread recognition as important building blocks for a large number of key applications, such as technical documentation, traffic management systems, educational software, and information kiosks (Fig. 1).

From a linguistic point of view, IMMP systems are interesting because communication by language is a specialized form of communication in general. Theories of natural language processing have reached a level of maturity whereby we can now investigate whether these theories can also be applied to other media, such as graphics or pointing gestures. Furthermore, the place of natural language as one of the most important means of communication makes natural language generation an indispensible component of a presentation system. Conversely, the integration of additional media may increase the acceptance of natural language components by avoiding communication problems resulting from the deficiencies of using just one medium.

The purpose of this chapter is to survey techniques for building IMMPSs, drawing on lessons learned during the development of natural language generators.

Application	**Sample Systems**
report generation	MAGIC [23], PostGraphe [29], SAGE [43], RoCCo [6]
instructions for operating technical devices	COMET [30], IDAS [60], PPP [10], Visual Repair [31] and WIP [5]
route directions	MOSES [48]
mission planning and situation monitoring	AIMI [51], CUBRICON [57], FLUIDS [36]
project management	EDWARD [17], IGING [26]
business forms	XTRA [3]
configuration of computer networks	MMI^2 [70]
education and training	PEA [54], MAGPIE [34], Herman the Bug [66], COSMO [45], Steve [62]
information kiosks	ALFRESCO [65], ILEX [44], PEBA-II [24], AiA [10]

Fig. 1 Applications for IMMP systems.

To facilitate the comparison of these systems, we will first present a generic reference model that reflects an implementation-independent view of the authoring tasks to be performed by an IMMP system. We then go on to present techniques for automating and coordinating these tasks.

2. A GENERIC REFERENCE MODEL FOR MULTIMEDIA PRESENTATION SYSTEMS

To enable the analysis and comparison of IMMP systems as well as the reuse of components, an international initiative has set up a proposal for a standard reference model for this class of systems (cf. [16]).

Besides a layered architecture, the proposal comprises a glossary of basic terms related to IMMPs. As the authors of the reference model point out, the terms *medium* and *modality* have been a source of confusion because they are used differently in different disciplines. Aiming at a pragmatic merger of a wide variety of approaches, the reference model uses the term *medium* to refer to different kinds of perceptible entities (e.g., visual, auditory, haptic, and olfactory), to different kinds of physical devices (e.g., screens, loudspeakers, and printers), and to information types (e.g., screens, loudspeakers, and printers), and to information types (e.g., graphics, text and video). The term *modality* is then used to refer to a particular means of encoding information (e.g., 2D and 3D graphics, written and spoken language).

Figure 2 outlines a simplified version of the reference model. It is composed of a knowledge server and a number of layers that stand for abstract locations for tasks, processes or system components:

Control Layer

The Control Layer embodies components that handle incoming presentation goals, presentation commands, and possibly further commands, such as stop or interrupt requests, that allow the user or external components to control the presentation process.

Content Layer

The Content Layer is responsible for high-level authoring tasks, such as selecting appropriate contents, content structuring, and media allocation. As a result, the

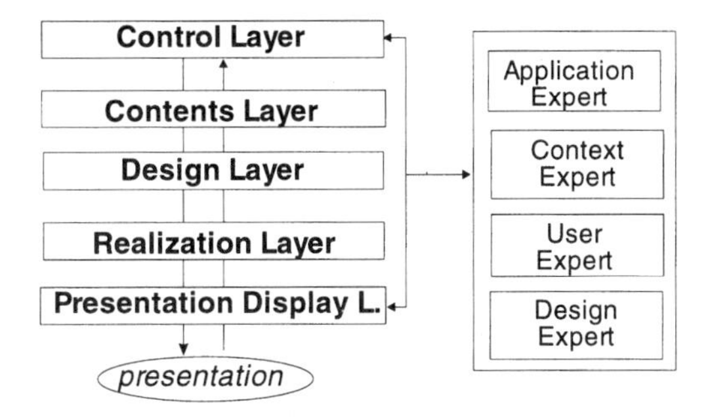

Fig. 2 A general reference model for IMMP systems.

Content Layer delivers a list of media- or modality-specific design tasks and a structural description that specifies how these tasks are related to each other.

Design Layer
The design layer embodies several media- or modality-specific design components. These design components can be seen as microplanners that transform the design tasks delivered by the content layer into a plan for the creation of media objects of a certain type. Furthermore, there is a layout design component that is responsible for setting up constraints for the spatial and temporal layout.

Realization Layer
The task of the realization layer is the media- or modality-specific encoding of information according to the design specifications that have been worked out in the superordinate Design Layer. For graphics, there may be a number of rendering components; for text, there may be components for grammatic encoding, linearization, and inflection. For the layout, realization includes the spatial arrangement of output and the design of a presentation schedule that takes account of the constraints delivered by the Design Layer.

Presentation Display Layer
The Presentation Display Layer describes the runtime environment for a presentation. It is responsible for dispatching media objects to suitable output devices such as a loudspeaker, a printer, or a computer screen (although the reference model abstracts from concrete devices). Furthermore, all coordination tasks between display devices are performed by this layer.

The *Knowledge Server* is a collection of knowledge sources that are shared among the layers. It consists of four expert modules, each of which represents knowledge of a particular aspect of the presentation process: application, user, context, and design.

In conventional multimedia systems, content selection and organization, media selection, and media encoding are usually performed by a human author. However, to classify a system as an IMMP system does not necessarily imply that all tasks sketched in the foregoing are automated. Usually, the system developers have concentrated on a specific subtask, whereas others are treated in a rather simplified way or neglected altogether. For example, XTRA and AIMI generate natural language and pointing gestures automatically, but rely on prestored tax forms and geographic maps, respectively. Other systems retrieve existing material from a database and adapt it to the user's current needs. For instance, Visual Repair [31] annotates preauthored video, whereas SAGE [21] modifies stored graphics by adding or deleting graphic objects.

3. TEXT-LINGUISTIC APPROACHES AS A METHODOLOGICAL BASIS

Encouraged by progress achieved in Natural Language Processing, several researchers have tried to generalize the underlying concepts and methods in a way such that they can be used in the broader context of multimedia generation. Although new questions arise (e.g., how to tailor text and graphics to complement each other) a number of tasks in multimedia generation, such as content selection and organization, bear considerable resemblance to problems faced in Natural Language Generation.

A. The Generation of Multimedia Presentations as a Goal-Directed Activity

Following a speech–act theoretical point of view, several researchers have considered the presentation of multimedia material as a goal-directed activity. Under this view, a presenter executes communicative acts, such as pointing to an object, commenting on an illustration, or playing back an animation sequence, to achieve certain goals. Communicative acts can be performed by creating and presenting multimedia material or by reusing existing document parts in another context (see also the distinction between the presentation display layer and the other four layers in the reference model). The following cases may be distinguished:

1. *The generation and the use of multimedia material are considered as a unit.* This case occurs, for example, when graphics are created and commented on while they are being viewed by the user.
2. *Multimedia material is created and used later by the same person.* This case occurs, for example, when someone prepares in advance the material to be used for a presentation.
3. *Multimedia material is created and used by different authors.* This case occurs, for example, when someone uses material retrieved from some other information source, such as the World Wide Web.

In the last two cases, the goals underlying the production of multimedia material may be quite different from the goals that are to be achieved by displaying it. For example, a graphic which has been generated to show the assembly of a technical device may be used on another occasion to show someone where he or she may find a certain component of this device. These examples illustrate that the relation between a multimedia document and its use is not trivial and cannot be described by a simple one-to-one mapping; instead, we have to clearly distinguish between the creation of material and its use.

Most multimedia systems are concerned only with the production of multimedia material. Recently, however, personalized user interfaces, in which life-like characters play the role of presenters explaining and commenting on multimedia documents, have become increasingly popular (see Sec. 6). In such applications, the need for a clear distinction between the design of material and its presentation becomes obvious and is also reflected by the system's architecture.

B. An Extended Notion of Coherence

Several text linguists have characterized coherence in terms of the coherence relations that hold between parts of a text [e.g., see 33,38]. Perhaps the most elaborate set is presented in Rhetorical Structure Theory (RST; cf. [50]), a theory of text coherence. Examples of RST relations are *Motivation*, *Elaboration*, *Enablement*, *Interpretation*, and *Summary*. Each RST relation consists of two parts: a *nucleus* that supports the kernel of a message, and a *satellite* that serves to support the nucleus.. RST relations may be combined into schemata that describe how a document is decomposed. Usually, a schema contains one nucleus and one or more

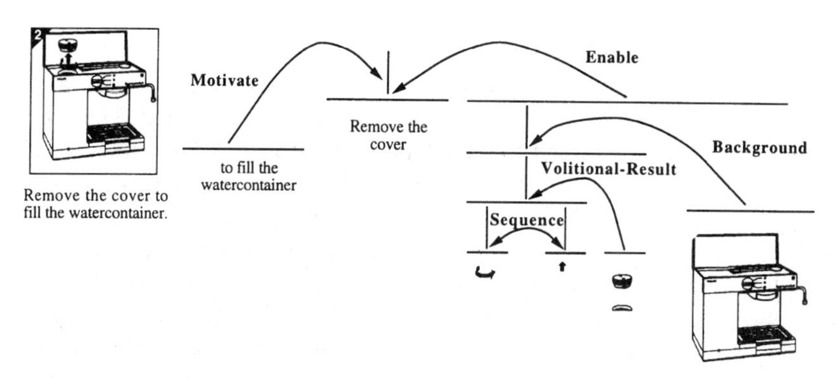

Fig. 3 Rhetorical structure of a sample document.

satellites related to the nucleus by an RST relation. For example, the *Request* schema consists of a nuclear request, and any number of satellites motivating or enabling the fulfillment of the request.

To generalize theories of text coherence to the broader context of multimedia, we have to analyze the relations between the component parts of a multimedia presentation. Earlier studies investigated only relations between pictures as a whole and text (i.e., they did not address the question of how a picture is organized [15,46]). An extensive study [4] of illustrated instructions has been carried out to find out which relations may occur between textual and pictorial document parts and how these relations are conveyed. It turns out that the structure of most instructions can be described by a slightly extended set of RST relations. New relations that have to be added include *illustration* and *label.*

Figure 3 shows the rhetorical structure of a document fragment.[1] We use the graphic conventions introduced by Mann and Thompson [50], representing relations between document parts by curved lines pointing from the nucleus to the satellite. The document is composed of a request, a motivating part, and a part that enables the user to carry out the action. Rhetorical relations can also be associated with individual picture parts. For example, the depiction of the espresso machine serves as a background for the rest of the picture:

C. Approaches to Content Selection and Content Organization

Because multimedia presentations follow structuring principles similar to those used in pure text, it seems reasonable to use text-planning methods for the organization of the overall presentation, as well as for structuring its textual parts. An essential advantage of a uniform-structuring approach is that not only relations within a single medium, but also relations between parts in different media, can be explicitly represented.

A number of IMMP systems make use of a notion of a schema based on that originally proposed by McKeown [52] for text generation. Schemata describe standard patterns of discourse by means of rhetorical predicates that reflect the relations

[1]The example is a slightly modified and translated version of instructions for the Philips espresso machine HD 5649.

between the parts of a presentation. Starting from a presentation goal (e.g., "the user should know how to operate a technical device"), a schema is selected. When traversing the schema, information from a given set of propositions is selected. The result of this selection process is forwarded to a media coordinator that determines which generator should encode the selected information. Examples of systems using a schema-based approach are COMET [30] and an earlier prototype of SAGE [64]. Whereas SAGE relies only on schemata to select the text contents, COMET employs schemata to determine the contents and the structure of the overall presentation.

As was shown by Paris [58], information concerning the effects of the individual parts of a schema are compiled out. If it turns out that a particular schema fails, the system may use a different schema, but it is impossible to extend or modify only one part of the schema. For text generation, this has been considered a major drawback and has led to the development of operator-based approaches [53] that enable more local revisions by explicitly representing the effects of each section of the presentation.

In the last few years, extensions of operator-based approaches have become increasingly popular for the generation of multimedia presentations, too. Examples include AIMI [51], FLUIDS [36], MAGIC [23], MAGPIE [34], PPP [9], WIP [5], and a recent extension of SAGE [43]. The main idea behind these systems is to generalize communicative acts to multimedia acts and to formalize them as operators of a planning system. The effect of a planning operator refers to a complex communicative goal, whereas the expressions in the body specify which communicative acts have to be executed to achieve this goal. Communicative acts include linguistic acts (e.g., inform), graphic acts (e.g., display), physical acts (e.g., gestures), and media-independent rhetorical acts (e.g., describe or identify). A detailed taxonomy of such acts has been proposed by Maybury [51]. Starting from a presentation goal, the planner looks for operators, the effect of which subsumes the goal. If such an operator is found, all expressions in the body of the operator will be set up as new subgoals. The planning process terminates if all subgoals have been expanded to elementary generation tasks that are forwarded to the medium-specific generators. During the decomposition, relevant knowledge units for achieving the goals are allocated and retrieved from the domain knowledge base, and decisions are taken concerning the medium or media combination required to convey the selected content.

An advantage of an operator-based approach is that additional information concerning media selection or the scheduling of a presentation can be easily incorporated and propagated during the content selection process. This method facilitates the handling of dependencies [8], as medium selection can take place during content selection and not only afterward, as in COMET.

Whereas WIP, COMET, and AIMI concentrate only on the creation of multimedia material, PPP, FLUIDS, and MAGIC also plan display acts and their temporal coordination (see Sec. 4.C). For instance, MAGIC synchronizes spoken references to visual material, with graphic highlighting when communicating information about a patient's postoperational status; PPP and FLUIDS synchronize speech with the display of graphic elements and the positioning of annotation labels to explain technical devices.

4. MEDIA COORDINATION

Multimedia presentation design involves more than just merging output in different media; it also requires a fine-grained coordination of different media. This includes distributing information onto different generators, tailoring the generation results to each other, and integrating them into a multimedia output.

A. Media Allocation

The media allocation problem can be characterized as follows: *Given a set of data and a set of media, find a media combination that conveys all data effectively in a given situation.* Essentially, media selection is influenced by the following factors:

Characteristics of the information to be conveyed
Characteristics of the media
The presenter's goals
User characteristics
The task to be performed by the user
Resource limitations

Earlier approaches rely on a classification of the input data, and map information types and communicative functions onto media classes by applying media allocation rules. For instance, WIP starts from ten communicative functions (*attract–attention*, *contrast*, *elaborate*, *enable*, *elucidate*, *label*, *motivate*, *evidence*, *background*, *summarize*) and seven information types (*concrete*, *abstract*, *spatial*, *covariant*, *temporal*, *quantification*, and *negation*) with up to ten subtypes. Examples of media allocation rules are as follows:

1. *Prefer graphics for concrete information (such as shape, color, and texture).*
2. *Prefer graphics over text for spatial information (e.g., location, orientation, composition) unless accuracy is preferred over speed, for which text is preferred.*
3. *Use text for quantitative information (such as most, some, any, exactly, and so on)*
4. *Present objects that are contrasted with each other in the same medium.*

This approach can be generalized by mapping features of input data to features of media (e.g., static–dynamic, arbitrary–nonarbitrary). An example of such a mapping rule is as follows [13]:

Data tuples, such as locations, are presented on planar media, such as graphs, tables, and maps.

However, because media allocation depends not only on data and media features, media allocation rules have to incorporate context information as well. Arens and his colleagues [13] proposed representing all knowledge relevant to the media allocation process in And-Or-Networks such as those used by Systemic Functional linguists to represent the grammars of various languages in a uniform formalism. Presentations are designed by traversing the networks and collecting at each node features that instruct the generation modules how to build sentences, construct diagrams, and so on.

Whereas only simple heuristics have been proposed for media selection, approaches relying on deeper inferences have been developed for the selection of graphic two-dimensional (2D)- techniques. For instance, APT [49] checks by means of formal criteria which information may be conveyed by a particular graphic technique (the *criterion of expressivity*) and how effectively such a technique may present the information to be communicated (the *criterion of effectiveness*). Casner [19] describes an approach in which media selection is influenced by perceptual factors. His system first analyses the user's task (e.g., the recognition of differences in temperature) and formulates a corresponding perceptual task (e.g., the recognition of differences in length) by replacing the logical operators in the task description with perceptual operators. Then, an illustration is designed that structures all data in such a way that all perceptual operators are supported and visual search is minimized.

B. Generation of Referring Expressions in Multimedia Environments

To ensure the consistency of a multimedia document, the media-specific generators have to tailor their results to each other. Referring expressions are an effective means for establishing coferential links between different media. In a multimedia discourse, the following types of referring expressions occur:

Multimedia-referring expressions refer to world objects by a combination of at least two media. Each medium conveys some discriminating attributes that, taken together, permit a proper identification of the intended object. Examples are natural language expressions that are accompanied by pointing gestures, and text–picture combinations, for which the picture provides information about the appearance of an object, whereas the text restricts the visual search space, as in "the switch on the frontside."

Cross–media-referring expressions do not refer to world objects, but to document parts in other presentation media [67]. Examples of cross-media referring expressions are "the upper left corner of the picture" or "Fig. x." In most cases, cross-media referring expressions are part of a complex multimedia referring expressions in which they serve to direct the reader's attention to parts of a document that also have to be examined to find the intended referent.

Anaphoric-referring expressions refer to world objects in an abbreviated form [37], presuming that they have already been introduced into the discourse, either explicitly or implicitly. The presentation part to which an anaphoric expression refers back is called the *antecedent* of the referring expression. In a multimedia discourse, we not only have to handle linguistic anaphora with linguistic antecedents, but also linguistic anaphora with pictorial antecedents, and pictorial anaphora with linguistic or pictorial antecedents (Fig. 4). Examples, such as "the hatched switch," show that the boundary between multimedia-referring expressions and anaphora is indistinct. Here, we have to consider whether the user is intended to employ all parts of a presentation for object disambiguation, or whether one wants him or her to infer anaphoric relations between them.

Usually, illustrations facilitate the generation of referring expressions by restricting the focus and providing additional means of discriminating objects from alternatives. A system can refer not only to features of an object in a scene, but also to features of the graphic model, their interpretation, and to the position of picture objects within the picture. Difficulties may arise, however, if it is unclear

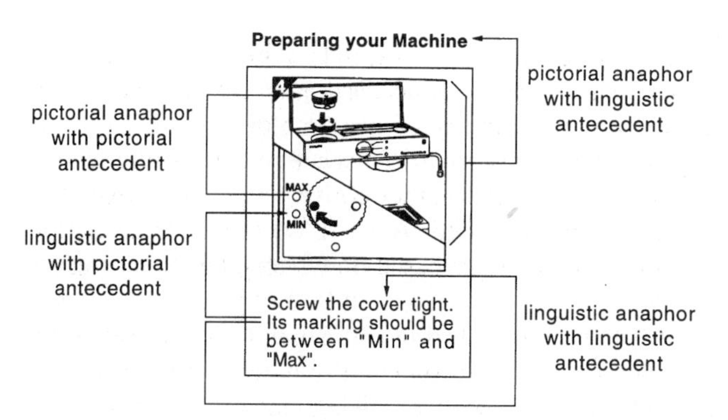

Fig. 4 Different types of anaphora occurring in a sample document.

whether a referring expression refers to an object's feature in the illustration or in the real world.

If a system refers to the interpretation of a graphic feature, it has to ensure that the user is able to interpret these encodings. For instance, the referring expression "the left resistor in the figure" is only understandable if the user is able to recognize certain images as resistor depictions. Furthermore, it must be clear whether a referring expression refers to an object's feature in the illustration or in the real world because these features may conflict with each other.

Often, spatial relations between images are used to discriminate objects from alternatives. Because a system cannot anticipate where images will be positioned within a picture or which layout will be chosen for text blocks and pictures, it does not make sense to compute such relations in advance. Some systems, such as COMET and WIP, rely on localization components to determine the position of images relative to an illustration (e.g., *the circle in the right section of the illustration*) or relative to other objects in the illustration (e.g., *the circle next to the triangle*). Furthermore, objects may be localized relative to parts of an illustration (e.g., the corners) or object groups [69].

An important prerequisite for the generation of referring expressions in a multimedia discourse is the explicit representation of the linguistic and pictorial context.

In the CUBRICON system [57], the linguistic context is represented by a focus list of entities and propositions to which the user of the system may refer through natural language or pointing gestures. The pictorial context represents which entities are visible, in which windows they are located, and which windows are visible on the screen. The XTRA system represents not only the linguistic context, but also maintains a data structure for graphics to which the user and the system may refer during the dialogue [61]. In XTRA's domain, the graphic context corresponds to a form hierarchy that contains the positions and the size of the individual fields as well as their geometric and logical relations. Furthermore, connections between parts of the form (e.g., *region437*) and the corresponding concepts in the knowledge base (e.g., *employer1*) are explicitly represented. Whereas XTRA and CUBRICON rely on different context models for the linguistic and pictorial context, EDWARD [22] uses a uniform model that considers context factors, such as the position in a sentence or visibility on the screen. Unlike these systems, WIP represents not only the semantics of individual images (e.g., *image-1* depicts *world-object-1*), but also the semantics of image attributes (e.g., the property of being colored red in a picture encodes the real-world property of being defective). To specify the semantic relation between information carriers and the information they convey, a relation tuple of the form (*Encodes carrier info context-space*) is used. A set of such encoding relations then forms the semantic description of an image based on which inferences processes may be performed [7].

Because image properties may change during the generation process, a presentation system has to ensure that the pictorial context is updated continuously. For instance, if the graphics generator chooses another viewing angle, some spatial relations will have to be updated as well.

C. Spatial and Temporal Coordination of the Output

Another coordination task is the integration of the individual generator results into a multimedia output. This includes the spatial arrangement of text blocks and graphics by means of a layout component. A purely geometric treatment of the layout task would, however, lead to unsatisfactory results. Rather, layout has to be considered as an important carrier of meaning. In particular, it may help indicate the intentions of the presenter, convey the rhetorical structure of a document, or draw the user's attention to relevant document parts. For example, two equally sized graphics can be contrasted by putting them beside one another, or one under the other.

Although there has been significant work on multimedia production, only a few approaches make use of the structural properties of the underlying multimedia information. The bulk of previous work on automatic layout has concentrated on single media types (e.g., in the context of graphics generation) and does not consider dependencies between layout design and properties of the raw content. Syntactic aspects of layout have been addressed, but not its communicative function. To handle these syntactic aspects, a large variety of techniques have been used, such as dynamic programming, graph-drawing algorithms, relational grammars, rule-based systems, genetic algorithms, constraint processing techniques, and efficient search and optimization mechanisms [see Ref. 40 for an overview].

The easiest way to enhance a natural language generator by formatting devices is to embed LaTeX or HTML markup annotations directly in the operators of a conventional planner [see also Ref. 39]. These annotations are then used by a rendering component (e.g., an HTML-browser) to produce the formatted document. The disadvantage of this method is that the system does not maintain any knowledge about the formatting commands and is thus not able to reason about their implications. Furthermore, it provides only limited control over the visual appearance of the final document.

To avoid these problems, WIP strictly distinguishes between the structural properties of the raw material and its visual realization [32]. Coherence relations between presentation parts (such as sequence or contrast) are mapped onto geometric and topologic constraints (e.g., horizontal and vertical layout, alignment, and symmetry) and a finite domain constraint solver is used to determine an arrangement that is consistent with the structure of the underlying information.

If information is presented over time, layout design also includes the temporal coordination of output units. The synchronization of media objects usually involves the following three phases:

1. *High-level specification of the temporal behavior of a presentation*
   During this phase, the temporal behavior of a presentation is specified by means of qualitative and metric constraints. Research in the development of multimedia authoring tools usually assumes that this task is carried out by a human author.
2. *Computation of a partial schedule which specifies as much temporal information as possible*
   From the temporal constraints specified in step 1, a partial schedule is computed which positions media objects along a time axis. Because the behavior of many events is not predictable, the schedule may still permit time to be stretched or shrunk between media events.
3. *Adaptation of the schedule at runtime*
   During this phrase, the preliminary schedule is refined by incorporating information about the temporal behavior of unpredictable events.

Most commercial multimedia systems are able to handle only those events with a predictable behavior, such as audio or video, and require the authors to completely specify the temporal behavior of all events by positioning them on a timeline. This means that the author has to carry out the foregoing first and third steps manually, and the second step can be left out because events with an unpredictable behavior are not considered.

More sophisticated authoring systems, such as FIREFLY [18] or CMIFED [35], allow the author to specify the temporal behavior of a presentation at a higher level of abstraction. However, the author still has to input the desired temporal constraints from which a consistent schedule is computed. In this case, the second and third steps are automatically performed by the system, whereas a human author is responsible for the first step.

Research in automatic presentation planning finally addresses the automatization of all three steps. In PPP, a complete temporal schedule is generated automatically starting from a complex presentation goal. Basically, PPP relies on the WIP approach for presentation planning. However, to enable both the creation of multimedia objects and the generation of scripts for presenting the material to the user, the following extensions have become necessary [9]:

1. *The specification of qualitative and quantitative temporal constraints in the presentation strategies*
   Qualitative constraints are represented in an Allen-style fashion [2], which permits the specification of 13 temporal relations between two named intervals, for example [*Speak1 (During) Point2*]. Quantitative constraints appear as metric (in)equalities, for example ($5 \leq$ *Duration Point2*).
2. *The development of a mechanism for building up presentation schedules*
   To temporally coordinate presentation acts, WIP's presentation planner has been combined with a temporal reasoner that is based on MATS (Metric/Allen Time System [42]). During the presentation-planning process, PPP determines the transitive closure over all qualitative constraints

and computes numeric ranges over interval endpoints and their differences. Then, a schedule is built up by resolving all disjunctions and computing a total temporal order.

A similar mechanism is used in MAGIC. However, MAGIC builds up separate constraint networks for textual and graphic output and temporally coordinates these media through a multistage negotiation process, whereas PPP handles all temporal constraints within a single constraint network, irrespective of which generator they come from.

5. THE INTEGRATION OF NATURAL LANGUAGE AND HYPERTEXT

If documents are presented online, it is quite straightforward to offer the user a hypermedia-style interface that allows him or her to jump from one document to another at the click of a mouse. World-Wide Web (WWW) browsers support the realization of such interfaces and make them available to large variety of users. In the ideal case, WWW documents should be customized to the individual user. Doing this manually, however, is not feasible because it would require anticipating the needs of all potential users and preparing documents for them. Even for a small number of users this may be a cumbersome task; it is simply not feasible for the WWW community of currently more than 100 million potential users.

The integration of natural language generation methods offers the possibility of creating hypermedia documents on demand, taking into account the user profile and his or her previous navigation behavior. For instance, if the user browses through the electronics pages of an online shop, this may be taken as evidence that he or she may also be interested in computer magazines. Therefore, a link may be added to the currently visited page that suggests also having a look at the computer magazine pages.

A benefit of hypermedia is that it provides an easy way of involving the user in the discourse planning process (see also [25]). Natural language generators often have to start from incomplete knowledge concerning the user's goals and interests. In such cases, it may be hard to determine which information to include in a presentation. The exploitation of hypermedia may alleviate this problem. Instead of overloading the user with information or risking to leave out relevant information, the user may determine whether or not to elaborate on particular presentation parts simply by selecting certain mouse-sensitive items.

To build a natural language generator that creates hypertexts, we have to modify the flow of control within the control–content layers of the architecture. Instead of building up a complete discourse structure in one shot, certain parts of it are only refined on demand. The basic idea is to treat subgoals that should be realized as hyperlinks analogously with the creation of clauses and forward them to design–realization components that generate corresponding mouse-sensitive items. If the user selects such an item when viewing the presentation, the control–content layer modules are started again with the goal corresponding to that item. The question of whether these modules should also be activated if the user clicks on an item a second time is a matter of debate in the hypermedia community. On the one hand, a dynamic hypermedia system should consider whether the user has already seen a page or not. On the other hand, most Web users today are acquainted only with static Web pages and might become confused if the contents and form of a Web page change each times they visit it.

There are various ways to integrate hypermedia facilities in a natural language generation system.

PEA [54] and IDAS [60] view hypertext as a means of realizing some simple form of dialogue and offer the user a hypertext-style interface that allows them to select mouse-sensitive parts of an explanation. On the basis of such mouse-clicks, the system generates a menu of follow-up questions that may be asked in the current context. In particular, the systems have to decide (a) where to include hyperlinks and (b) which follow-up questions to offer by consulting the user model and the dialogue history. For instance, PEA will not include a question in the menu if it assumes that the user already knows the answer to this question.

ALFRESCO [65] takes a different approach. This system exploits hypermedia as a browsing facility through an existing information space by generating text with entry points to an underlying preexisting hypermedia network. However, because only the initial text is automatically designed, the system has no influence on the links within the hypertext. Once a user enters the hyperspace, the system is no longer aware of his or her activities and has no possibility of adapting it to their needs.

More recent systems, such as ILEX [44] and PEBA-II [24], automatically compose hypertext from canned text and items from a knowledge base. To smoothly combine canned text with automatically generated text, the canned text may include various types of annotations. For instance, canned text in ILEX contains pointers to domain entities for which the system automatically creates referring expressions. The links between the single pages are automatically created based on the user profile and current situation.

The idea of dynamic hyperlinks can also be found in recent systems for assisted Web access, such as WebWatcher [41] and Letizia [47]. However, unlike ILEX and PEBA-II, these systems rely on handcrafted Web pages.

6. PERSONALIZED MULTIMEDIA PRESENTATION SYSTEMS

So far, we have addressed only the generation of multimedia material. Although this material may be coherent and even tailored to a user's specific needs, the presentation as a whole may fail because the generated material has not been presented in an appealing and intelligible way. This can often be observed when multimedia output is distributed across several windows, requiring the user to find out herself how to navigate through the presentation. To enhance the effectiveness of user interfaces, various research projects have focussed on the development of personalized user interfaces in which communication between user and computer is mediated by life-like agents (see [28] for an overview).

There are several reasons for using animated presentation agents in the interface. First, they allow for the emulation of presentation styles common in human–human communication. For example, they enable more natural referential acts that involve locomotive, gestural, and speech behaviors [45]. In virtual environments, animated agents may help users learn to perform procedural tasks by demonstrating their execution [62]. Furthermore, they can also serve as a guide through a presentation to release the user from orientation and navigation problems common in multi-window–multiscreen settings [11]. Last, but not least, there is the entertaining and emotional function of such animated characters. They may help lower the "getting started barrier" for novice users of computers or applications, and, as Adelson notes, ". . . interface agents can be valuable educational aids since they can engage students without distracting or distancing them from the learning experience" [1, p. 355].

To illustrate this, we use some examples taken from the PPP (**P**ersonalized **P**lan-based **P**resenter) system. The first application scenario deals with instructions for the maintenance and repair of technical devices, such as modems. Suppose the system is requested to explain the internal parts of a modem. One strategy is to generate a picture showing the modem's circuit board and to introduce the names of the depicted objects. Unlike conventional static graphics in which the naming is usually done by drawing text labels onto the graphics (often in combination with arrows pointing from the label to the object), the PPP Persona enables the emulation of referential acts that also occur in personal human–human communication. In the example in Fig. 5, it points to the transformer and utters "This is the transformer" (using a speech synthesizer). The example also demonstrates how facial displays and head movements help restrict the visual focus. By having the Persona look into the direction of the target object, the user's attention is directed to this object.

In the second example, the Persona advertises accommodation offers found on the WWW. Suppose the user is planning to take a vacation in Finland and, therefore, is looking for a lakeside cottage. To comply with the user's request, the system retrieves a matching offer from the Web and creates a presentation script for the PPP Persona which is then sent to the presentation viewer (e.g., Netscape Navigator with an inbuilt Java interpreter). When viewing the presentation, the PPP Persona highlights the fact that the cottage has a nice terrace by means of a *verbal annotation of a picture*; (i.e., Persona points to the picture during a verbal utterance; Fig. 6). When the graphic elements are generated automatically, as in the modem example, the presentation system can build up a reference table that stores the correspondences

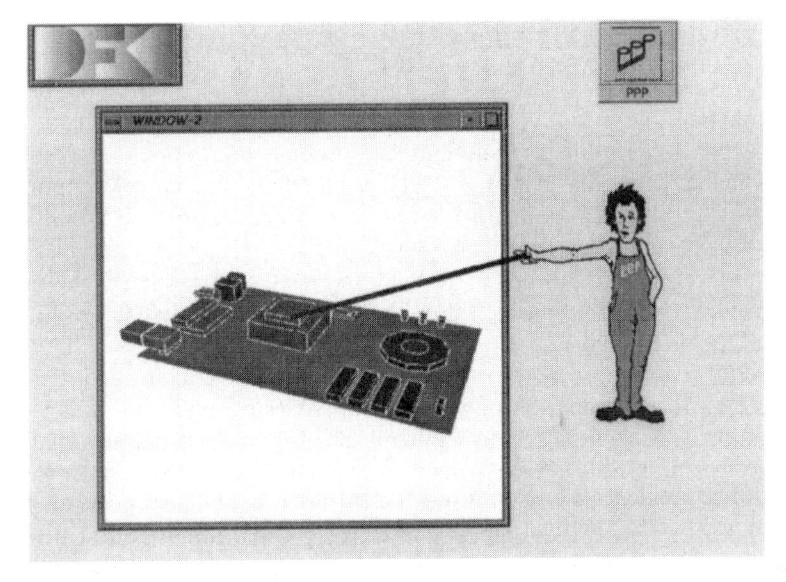

Fig. 5 The persona instructs the user in operating a technical device.

Fig. 6 PPP persona presents retrieval results from the Web using Netscape Navigator and Java.

between picture parts and domain concepts. Because scanned pictures are used in the traveling agent application, such a reference table has to be set up manually to enable pointing gestures to that material. However, frequently, the author of a Web page has already carried out the task of relating image regions to concepts. For example, many maps available on the Web are already mouse-sensitive; in these, the system just has to follow the links to find the concepts related to the mouse-sensitive regions.

According to its functional role in a presentation, an animated character must be conversant with a broad variety of presentation gestures and rhetorical body postures. It has to execute gestures that express emotions (e.g., approval or disapproval), convey the communicative function of a presentation act (e.g., warn, recommend, or dissuade), support referential acts (e.g., look at an object and point at it), regulate the interaction between the character and the user (e.g., establishing eye contact with the user during communication), and articulate what is being said.

From a technical point of view, it makes no difference whether we plan presentation scripts for the display of static and dynamic media, or presentation acts to be executed by life-like characters. Basically, we can rely on one of the temporal planners presented in Sec. 3.C if we extend the repertoire of presentation strategies by including strategies that control that Persona's presentation behavior. However, the behavior of a character is determined not only by the directives (i.e., presentation tasks) specified in the script. Rather, it follows the equation:

Persona behavior := directives + self-behavior

Such self-behaviors are indispensible to increase the Persona's vividness and believability [63]. They comprise idle-time actions, such as tapping with a foot; actions for indicating activity (e.g., turning over book pages); navigation acts, such as walking or jumping; and immediate reactions to external events, such as mouse gestures on the presented material.

Although it is possible to incorporate self-behaviors into the plan operators of a script generator, the distinction between task-specific directives on the one hand, and character-specific and situation-specific self-behaviors on the other, bears a number of advantages. From a conceptual point of view, it provides a clear borderline between a "*what to present*" *part* which is determined by the application, and a "*how to present*" *part* which, to a certain extent, depends on the particular presenter. From a practical perspective, this separation considerably facilitates the exchange of characters, and the reuse of characters for other applications.

Although planning techniques have proved useful for the generation of presentation scripts, the coordination of script directives and self-behaviors and their transformation into fine-grained animation sequences has to be done in real-time, and thus requires a method that is computationally less expensive. One solution is to precompile declarative behavior specifications into finite-state machines. Such an approach has been originally proposed for Microsoft's Peedy [14] and later been adapted for the PPP Persona. The basic idea is to compute for all possible situations beforehand which animation sequence to play. As a result, the system just has to follow the paths of the state machine when making a decision at runtime, instead of starting a complex planning process. Finite-state automata are also a suitable mechanism for synchronizing character behaviors. For instance, Cassell and colleagues [20] use so-called parallel transition networks (PaT-Nets) to encode facial and gestural coordination rules as simultaneous executing finite-state automata.

A character's face is one of its most important communication channels. Lip movements articulate what is being said, nose and eyebrow movements are effective means of conveying emotions, and eye movements may indicate interest or lack there of, just to mention a few examples. To describe possible facial motions performable on a face, most systems rely on the facial action coding system (FACS, [27]) or MPEG-4 facial animation parameters (FAPs). For instance, both Nagao and Takeuchi [56] and Cassell and colleagues [20] map FACS actions, such as Inner Brow Raiser, onto communicative functions, such as Punctuation, Question, Thinking, or Agreement. In both systems, speech controls the generation of facial displays: the speech synthesizer provides parameters, such as timing information, which form the input for the animation components.

The believability of a life-like character hinges on the quality of the output speech. Unfortunately, most life-like characters today only use the default intonation of a speech synthesizer. Here, the integration of natural language generation and speech synthesis technology offers great promise, because a natural language generator may provide the knowledge of an utterance's form and structure that a speech synthesizer needs to produce good output. Prevost [59] provides an example of such an approach: this work uses a Categorial Grammar to translate lexicalized logical forms into strings of words with intonational markings. Also noteworthy is the approach to speech synthesis adopted by Walker and colleagues [68], which also addresses the social background of the speaker and the affective impact of an utterance.

7. ARCHITECTURES FOR MULTIMEDIA PRESENTATION SYSTEMS

Whereas the reference model introduced in Sec. 2 abstracts away from a concrete implementation, the developer of an IMMP system has to commit to a particular architecture: he or she has to decide how to distribute the tasks corresponding to the single layers onto different components, and how to organize the information flow between these components.

Research on architectures for IMMP systems can be roughly divided into two camps. One group tries to find architectural designs that are able to handle interactions within and between layers. The other group focuses on the technically challenging task of integrating heterogeneous components that were not originally designed to work together.

Architectures proposed by the first group essentially differ in the organization of the processes to be performed by the content layer.

In the early SAGE prototype [64], relevant information is selected first and then organized by the text and graphics generators. Then, the generated structures are transformed into text an graphics. A disadvantage of this method is that text and graphics are built up independently of each other. Therefore, SAGE needs to revise generated textual and graphic document parts to tailor them to each other.

In COMET [30], a tree-like structure that reflects the organization of the presentation to be generated is built up first. This structure is passed to the media coordinator which annotates it with media information. Then, the tree is extended by the medium-specific generators in a monotonic manner: there is only a one-way exchange of information between the content layer and the design and realization layers. More attention has been devoted to dependencies between text and graphics generation. Examples include the coordination of sentence breaks and picture breaks and the generation of cross-references. To facilitate communication between text and graphics generation, all information to be communicated is represented in a uniform logical form representation language. Because both generators rely on the same formalism, they can provide information to each other concerning encoding decisions simply by annotating the logical forms.

Arens and colleagues [12] propose a strict separation of the planning and the medium selection processes. During the planning process, their system fully specifies the discourse structure, which is determined by the communicative goals of the presenter and the content to be communicated. Then, special rules are applied to select an appropriate medium combination. After medium selection, the discourse structure is traversed from bottom to top to transform the discourse structure into a presentation-oriented structure. A problem with this approach is that the presentation structure obviously has no influence on the discourse structure. The selection of a medium is influenced by the discourse structure, but the contents are determined independently of the medium.

MAGPIE [34] relies on an agent-based architecture for presentation design. As in COMET, a blackboard mechanism is used to handle the communication between the single agents. However, in MAGPIE, agents are hierarchically organized on the basis of task decomposition. That is agents may delegate tasks to other subordinated agents with whom they share a blackboard. For example, the group headed by the table generator agent includes the number agent, the icon agent, and the text agent.

Whereas the systems mentioned earlier rely on separate components for content selection or organization and media allocation, WIP uses a uniform-planning approach that accomplishes all these tasks concurrently. That is the content layer is realized just by one component, namely the presentation planner. The most important advantage of such an approach is that it facilitates the handling of dependencies among choices. As in COMET, the document plan generated by the presentation planner serves as the main data structure for exchanging information among the planner and the generators.

In the foregoing systems, the realization components forward media objects to the presentation display components, but do not specify how and when they should be presented to the user. Document parts are either shown to the user immediately after their production (incremental mode), or the systems wait until the production process is completed and then present all the material at once (batch mode). Consequently, these systems are not able to influence the order in which a user processes a document, or the speed of processing. Unlike these systems, MAGIC's and PPP's design and realization components also handle the temporal layout of presentations.

The second group is represented by Moran and colleagues [55] who focus on the combination and reuse of existing software components in new multimedia applications. To support the creation of applications from agents written in multiple languages and running on different platforms (e.g., a text generator running on a Sun and a graphics generator running on a Silicon Graphics), they propose an open agent-based architecture (OAA). The OAA agents communicate with each other in a high-level logical language called the Interagent Communication Language (ICL). A key role in OAA is played by the Facilitator Agent: this decomposes complex requests into elementary requests that are delegated to the individual agents. Thus, the Facilitator Agent may be compared with the content planner in the MAGIC system or the presentation planner in the PPP system. However, whereas the Facilitator has been conceived as a general-purpose communication agent, the MAGIC and PPP planners have been tailored to the coordination of presentation tasks in a multimedia environment. They also include mechanisms for media coordination and synchronization, whereas this task would be performed by a special agent in OAA.

8. CONCLUSION

The availability of new media opens up new ways of presenting information and leads to new research issues, such as the selection and coordination of media. On the one hand, the integration of multiple media adds complexity to the generation task because far more dependencies have to be handled. On the other hand, many theoretical concepts already developed in the context of natural language processing, such as speech acts and rhetorical relations, take on an extended meaning in multimedia discourse. A key observation of this chapter is that multimedia presentations follow structuring principles similar to those found in pure text. For this reason, text-planning methods can be generalized in such a way that they also become useful for the creation of multimedia presentations. Systems, such as MAGIC and PPP, show that the combination of text planners with a module for temporal reasoning even enables the generation of dynamic multimedia presentations. Indeed, the development of the first-generation of IMMP systems has been significantly influenced by research in natural language generation.

ACKNOWLEDGMENTS

This work has been supported by the BMBF under the contracts ITW 9400 7 and 9701 0. I would like to thank Robert Dale and Thomas Rist for their valuable comments.

REFERENCES

1. B Adelson. Evocative agents and multi-media interface design. Proceedings of the UIST'92 (ACM SIGGRAPH Symposium on User Interface Software and Technology), Monterey, CA, 1992, pp 351–356.
2. JF Allen. Maintaining knowledge about temporal intervals. Commun ACM 26:832–843, 1983.
3. J Allgayer, K Harbusch, A Kobsa, C Reddig, N Reithinger, D Schmauks. XTRA: a natural-language access system to expert systems. Int J Man–Mach Stud 31:161–195, 1989.
4. E André. Ein planbasierter Ansatz zur Generierung multimedialer Präsentationen. DISKI-108, INFIX-Verlag, Sankt Augustin, 1995.
5. E André, W Finkler, W Graf, T Rist, A Schauder, W Wahlster. WIP: the automatic synthesis of multimodal presentations. In: M Maybury, ed. Intelligent Multimedia Interfaces, AAAI Press, 1993, pp 75–93.
6. E André, G Herzog, T Rist. Generating multimedia presentations for RoboCup soccer games. In: H Kitano, ed. RoboCup-97: Robot Soccer World Cup I (Lect Notes Comput Sci), Springer, 1998, pp 200–215.
7. E André, T Rist. Referring to world objects with text and pictures. Proceedings of the 15th COLING, Vol 1, Kyoto, Japan, 1994, pp 530–534.
8. E André, T Rist. Generating coherent presentations employing textual and visual material. Artif Intell Re [Special Volume on the Integration of Natural Language and Vision Processing] 9(2–3):147–165, 1995.
9. E André, T Rist. Coping with temporal constraints in multimedia presentation planning. Proceedings AAAI-96, Vol 1. Portland, OR, 1996, pp 142–147.
10. E André, T Rist, J Müller. Employing AI methods to control the behavior of animated interface agents. Appl Artif Intell J 1999 13(4–5):415–448.

11. E André, T Rist, J Müller. WebPersona: a life-like presentation agent for the World-Wide Web. Knowledge-Based Syst, 1998 11(1):25–36.
12. Y Arens, E Hovy, S van Mulken. Structure and rules in automated multimedia presentation planning. Proceedings 13th IJCAI, Chambéry, France, 1993. Vol. 2, pp 1253–1259.
13. Y Arens, E Hovy, M Vossers. Describing the presentational knowledge underlying multimedia instruction manuals. In: M Maybury, ed. Intelligent Multimedia Interfaces, AAAI Press, 1993, pp 280–306.
14. G Ball, D Ling, D Kurlander, J Miller, D Pugh, T Skelly, A Stankosky, D Thiel, M van Dantzich, T Wax. Lifelike computer characters: the persona project at Microsoft. In: JM Bradshaw, ed. Software Agents, Menlo Park, CA: AAAI/MIT Press, 1997, pp 191–222.
15. S Bandyopadhyay. Towards an understanding of coherence in multimodal discourse. Technical Memo TM-90-01, Deutsches Forschungszentrum für Künstliche Intelligenz (DFKI), Saarbrücken, 1990.
16. M Bordegoni, G Faconti, S Feiner, MT Maybury, T Rist, S Ruggieri, P Trahanias, M Wilson. A standard reference model for intelligent multimedia presentation systems. Comput Stand Interfaces Int J Dev Appl Stand Comput Data Commun Interf 18(6–7):477–496, 1997.
17. E Bos, C Huls, W Claassen. Edward: full integration of language and action in a multimodal user interface. Int J Hum Comput Stud 40:473–495, 1994.
18. MC Buchanan, PT Zellweger. Automatically generating consistent schedules for multimedia documents. Multimedia Syst 1:55–67, 1993.
19. SM Casner. A task-analytic approach to the automated design of graphic presentations. ACM Trans Graph 10:111–151, 1991.
20. J Cassell, C Pelachaud, NI Badler, M Steedman, B Achorn, T Becket, B Douville, S Prevost, M Stone. Animated conversation: rule-based generation of facial expression, gesture and spoken intonation for multiple-conversational agents. Proceedings Siggraph'94, Orlando, 1994.
21. MC Chuah, SF Roth, J Kolojejchick, J Mattis, O Juarez. SageBook: searching data–graphics by content. Proceedings CHI-95, Denver, CO,, 1995, pp 338–345.
22. W Claassen. Generating referring expressions in a multimodal environment. In: R Dale, E Hovy, D Rösner, O Stock, ed. Aspects of Automated Natural Language Generation: Proceedings of the 6th International Workshop on Natural Language Generation, Berlin: Springer, pp 247–262, 1992.
23. M Dalal, S Feiner, K McKeown, S Pan, M Zhou, T Höllerer, J Shaw, Y Feng, J Fromer. Negotiation for automated generation of temporal multimedia presentations. ACM Multimedia 96, ACM Press, 1996, pp 55–64.
24. R Dale, M Milosavljevic. Authoring on demand: natural language generation in hypermedia documents. Proceedings of the First Australian Document Computing Symposium (ADCS'96), Melbourne, Australia, March 1996, pp 20–21.
25. R Dale, J Oberlander, M Milosavljevic, A Knott. Integrating natural language generation and hypertext to produce dynamic documents. Interact with Comput, 1998 11(2):109–135.
26. S Dilley, J Bateman, U Thiel, A Tissen. Integrating natural language components into graphical discourse. Proceedings of the Third Conference on Applied Natural Language Processing, Trento, Italy, 1992, pp 72–79.
27. P Ekman, WV Friesen. Facial Action Coding. Consulting Psychologists Press, 1978,
28. C Elliott, J Brzezinski. Autonomous agents as synthetic characters. AI Mag 19:13–30, 1998.
29. M Fasciano, G Lapalme. PostGraphe: a system for the generation of statistical graphics and text. Proceedings of the 8th International Workshop on Natural Language Generation, Sussex, 1996, pp 51–60.

30. SK Feiner, KR McKeown. Automating the generation of coordinated multimedia explanations. IEEE Comp 24(10):33–41, 1991.
31. BA Goodman. Multimedia explanations for intelligent training systems. In: M Maybury, ed. Intelligent Multimedia Interfaces. AAAI Press, 1993, pp 148–171.
32. W Graf. Constraint-based graphical layout of multimodal presentations. In: MF Costabile, T Catarci, S Levialdi, eds. Advanced Visual Interfaces (Proceedings of AVI '92, Rome, Italy), Singapore: World Scientific Press, 1992, 365–385.
33. JE Grimes. The Thread of Discourse. The Hague: Mouton, 1975.
34. Y Han, I Zuckermann. Constraint propagation in a cooperative approach for multimodal presentation planning. Proceedings 12th ECAI, Budapest, 1996, pp 256–260.
35. L Hardman, DCA Bulterman, G van Rossum. The Amsterdam hypermedia model: adding time and context to the Dexter model. Commun ACM 37(2):50–62, 1994.
36. G Herzog, E André, S Baldes, T Rist. Combining alternatives in the multimedia presentation of decision support information for real-time control. In: IFIP Working Group 13.2 Conference: Designing Effective and Usable Multimedia Systems, Stuttgart, Germany, 1998. A Sutcliffe, J. Ziegler, P. Johnson (eds.), p. 143–157, Kluwer.
37. G Hirst. Anaphora in Natural Language Understanding. Berlin: Springer, 1981.
38. J Hobbs. Why is a discourse coherent? Technical Report 176, SRI International, Menlo Park, CA, 1978.
39. EH Hovy, Y Arens. Automatic generation of formatted text. Proceedings AAAI-91, Anaheim, CA, 1991, pp 92–97.
40. W Hower, WH Graf. A bibliographical survey of constraint-based approaches to cad, graphics, layout, visualization, and related topics. Knowledge-Based Syst 9:449–464, 1996.
41. T Joachims, D Freitag, T Mitchell. Webwatcher: a tour guide for the World Wide Web. Proc 15th IJCAI, Nagoya, Japan, 1997, pp 770–775.
42. HA Kautz, PB Ladkin. Integrating metric and qualitative temporal reasoning. Proceedings AAAI-91, 1991, pp 241–246.
43. S Kerpedjiev, G Carenini, SF Roth, JD Moore. Integrating planning and task-based design for multimedia presentation. Proceedings 1997 International Conference on Intelligent User Interfaces, Orlando, FL, 1997, pp 145–152.
44. A Knott, C Mellish, J Oberlander, M O'Donnell. Sources of flexibility in dynamic hypertext generation. Proceedings 8th International Workshop on Natural Language Generation, Sussex, 1996.
45. J Lester, JL Voerman, SG Towns, CB Callaway. Deictic believability: coordinated gesture, locomotion, and speech in lifelike pedagogical agents. Appl Artif Intell J, 1999, 13(4–5):383–414.
46. JR Levin, GJ Anglin, RN Carney. On empirically validating functions of pictures in prose. In: DM Willows, HA Houghton, eds. The Psychology of Illustration, Basic Research. Vol 1. New York: Springer, 1987, pp 51–85.
47. H Lieberman. Letizia: an agent that assists web browsing. Proceedings 14th International Joint Conference on Artificial Intelligence, Montreal, August 1995, pp 924–929.
48. W Maaß. From vision to multimodal communication: incremental route descriptions. Artif Intell Rev (Special Volume on the Integration of Natural Language and Vision Processing] 8(2–3):159–174, 1994.
49. J Mackinlay. Automating the design of graphical presentations of relational information. ACM Trans Graphics 5(2):110–141, 1986.
50. WC Mann, SA Thompson. Rhetorical structure theory: a theory of text organization. Report ISI/RS-87-190, University of Southern California, Marina del Rey, CA, 1987.
51. MT Maybury. Planning multimedia explanations using communicative acts. Proceedings AAAI-91, Anaheim, CA, 1991, pp 61–66.
52. KR McKeown. Text Generation. Cambridge, MA: Cambridge University Press, 1985.

53. JD Moore, CL Paris. Planning text for advisory dialogues. Proceedings 27th ACL, Vancouver, BC, 1989, pp 203–211.
54. JD Moore, WR Swartout. Pointing: a way toward explanation dialogue. Proceedings AAAI-90, Boston, MA, 1990, pp 457–464.
55. DB Moran, AJ Cheyer, LE Julia, DL Martin, S Park. Multimodal user interfaces in the open agent architecture. Proceedings 1997 International Conference on Intelligent User Interfaces, Orlando, FL, 1997, pp 61–70.
56. K Nagao, A Takeuhi. Social interaction: multimodal conversation with social agents. Proceedings 32nd ACL, Vol. 1, 1994, pp 22–28.
57. JG Neal, SC Shapiro. Intelligent multi-media interface technology. In: JW Sullivan, SW Tyler, eds. Intell User Interfaces. New York: ACM Press, 1991, pp 11–43.
58. CL Paris. Generation and explanation: building an explanation facility for the explainable expert systems framework. In: CL Paris, WR Swartout, WC Mann, eds. Natural Language Generation in Artificial Intelligence and Computational Linguistics. Boston: Kluwer, 1991, pp 49–82.
59. S Prevost. An information structural approach to spoken language generation. Proceedings 34th ACL, Santa Cruz, CA, 1996, pp 294–301.
60. E Reiter, C Mellish, J Levine. Automatic generation of on-line documentation in the IDAS project. Proceedings of the Third Conference on Applied Natural Language Processing, Trento, Italy, 1992, pp 64–71.
61. N Reithinger. The performance of an incremental generation component for multimodal dialog contributions. In: R Dale, E Hovy, D Rösner, O Stock, eds. Aspects of Automated Natural Language Generation: Proceedings 6th International Workshop on Natural Language Generation. Berlin: Springer, 1992, pp 263–276.
62. J Rickel, WL Johnson. Animated agents for procedural training in virtual reality: perception, cognition, and motor control. Appl Artif Intell J 1999, 13(4–5):343–383.
63. T Rist, E André, J Müller. Adding animated presentation agents to the interface. In: J Moore, E Edmonds, A Puerta, eds. Proceedings 1997 International Conference on Intelligent User Interfaces, Orlando, Florida, 1997, pp 79–86.
64. SF Roth, J Mattis, X Mesnard. Graphics and natural language as components of automatic explanation. In: JW Sullivan, SW Tyler, eds. Intelligent User Interfaces. New York: ACM Press, 1991, pp 207–239.
65. O Stock. Natural language and exploration of an information space: the Al-Fresco Interactive System. Proceedings 12th IJCAI, Sidney, Australia, 1991, pp 972–978.
66. BA Stone, JC Lester. Dynamically sequencing an animated pedagogical agent. Proceedings AAAI-96, Vol. 1, Portland, Oregon, 1996, pp 424–31.
67. W Wahlster, E André, W Graf, T Rist. Designing illustrated texts: how language production is influenced by graphics generation. Proceedings 5th EACL, Berlin, Germany, 1991, pp 8–14.
68. M Walker, J Cahn, SJ Whittaker. Improving linguistic style: social and affective bases for agent personality. Proceedings of the First International Conference on Autonomous Agents. Marina del Rey: ACM Press, 1997, pp 96–105.
69. P Wazinski. Generating spatial descriptions for cross-modal references. Proceedings Third Conference on Applied Natural Language Processing, Trento, Italy, 1992, pp 56–63.
70. M Wilson, D Sedlock, J-L Binot, P Falzon. An architecture for multimodal dialogue. Proceedings Second Vencona Workshop for Multimodal Dialogue, Vencona, Italy, 1992.

13

Machine Translation

HAROLD SOMERS

UMIST, Manchester, England

1. INTRODUCTION

As an application of natural language processing (NLP), machine translation (MT) can claim to be one of the oldest fields of study. One of the first proposed non-numerical uses of computers, MT has recently celebrated its 50th birthday. That these celebrations were marked, among other things, by comments about how far MT still had to go to achieve its goals is an indication of the difficulty and depth of its problems. Characterized as the use of computers to perform translation of natural language texts, with or without human assistance, MT has to deal with almost every imaginable problem in computational linguistics, in at least two languages. Because of the necessity for this broad coverage, many of the problems (and indeed solutions) that we will mention in the following may also be found in other areas of NLP. MT has the special position of having to bring them all together.

In this chapter, we will approach MT first from a historical point of view, identifying the main trends both in research and development over the last 50 years. We will then focus on the particular aspects of translation that makes it especially difficult as an NLP task. We will next consider some of the design features that have been proposed in response to these problems. In this chapter we will focus on "traditional" symbolic approaches, whereas in Chap. 25, we will discuss some more recent alternative approaches that have a more empirical, data-driven basis. Alternative solutions to the problem of MT lie in the user's perspective, which we will consider in detail; and in the final section we will consider the way some of the problems can be addressed by different modes of use.

2. HISTORICAL OVERVIEW

The founding father of MT is generally agreed to be Warren Weaver, whose discussions with colleagues in the immediate postwar period led to him sending a memorandum in 1949 (reprinted in Locke and Booth, 1955) to some 200 of his colleagues. This is acknowledged as the launch of MT research in the United States (see Hutchins, 1986:24–30). Although Weaver's original idea to use some of the techniques of code-breaking proved unproductive, the general idea was seen as a worthwhile goal, and soon attracted government funding, at least in the United States. In 1951, Yehoshua Bar-Hillel became the first full-time MT researcher. A year later the Massachusetts Institute of Technology (MIT) hosted an international conference. A Russian–English system developed jointly by IBM and Georgetown University was demonstrated in 1954. A dedicated journal appeared, and work began in other countries (notably, the United Kingdom, the former Soviet Union, and Japan). The next 10 years saw worldwide activity, led by the United States, which invested 20 million dollars in MT research.

There were numerous groups conducting research in MT from the mid 1950s onward, and two basic approaches were adopted. Some groups took a "brute force" approach and based their work on the assumption that it was desirable, and indeed necessary, to come up with a working system as soon as possible, to take advantage of the enthusiasm that reigned at that time. They saw word-for-word translation systems as an adequate starting point that could be improved using feedback from the output. The basic need was for larger computer storage, and it was not felt necessary to solve all the foreseeable problems before trying to develop a system. Unfortunately, it was on the basis of this enthusiasm that hopes of a fully automatic system in the short-term grew rapidly. Other research groups preferred to take a slower, deeper, "perfectionist" approach, basing their research on linguistic issues, trying to take the human thought processes into account. They could be thought of as the early pioneers of Artificial Intelligence (AI). But their approach was equally hampered by the limitations of the computer technology of the time.

Despite all this enthusiasm, however, major problems began to appear, particularly because of the gap in knowledge between language specialists, on the one hand, and early computer scientists, on the other. Translators and linguists considered it impossible to define language in a manner rigid enough to produce good translation. The computer scientists, on the other hand, had no linguistic training and lacked an understanding of the intricacies of language. Early, overenthusiastic claims and promises about the prospects of "automatic translation" eventually led the U.S. government, in the early 1960s, to commission a report on the state of play: the now infamous ALPAC report (ALPAC, 1966) was hugely critical of MT as a whole, concluding that MT was more expensive, slower, and less accurate than human translation, and that there was no immediate or eventual prospect of usable MT. The ALPAC report may have convinced most people outside the field that MT had no hope of success, but many on the inside saw the report as short-sighted and narrow-minded. They had already realised many of the problems of automating the translation process (e.g., Bar-Hillel, 1951; Locke and Booth, 1955; Yngve, 1957) and did not accept that MT was forever doomed to failure. However, funds were no longer available and many MT research projects came to a halt.

Part of the problem was that these early MT systems were based on the idea that translation could be achieved basically by word-for-word substitution with some local adjustments. It soon became obivous that a more sophisticated, linguistically motivated, analysis of the text was necessary. The so-called second-generation of MT systems evolved this approach: the input text would be subject to a structural analysis, in which the relations between the words are discovered, and in this way differences in the way individual words should be translated could be resolved (see Sec. 3). This analysis was performed on the basis of linguistic rules, and many of the techniques described in Part I of this book—morphological analysis, parsing, semantic analysis—was developed as part of MT systems, or were quickly adapted for MT.

The basic approach of the second generation of MT systems, beginning in the later 1960s, but focussing on the 1970s, was to develop linguistic rule-writing formalisms that were implemented independently (in the classic manner of separating the general translation algorithms from the linguistic data that they used, the data in this case being grammars and lexicons). The analysis would result in a representation of the source text, and a major debating point throughout this period was the nature of this representation: in the "transfer" approach, the representation was of the syntactic structure of the source text. This would then undergo a transfer stage, which would result in a second representation more suitable for directly generating the target text. This contrasts with the "interlingua" approach, in which the meaning of the source text is somehow represented in a language-neutral form from which the target text can be directly generated without any intermediate transfer stage. This distinction will be discussed further in Sec. 4.

Since the late 1980s, the essentially syntax-oriented framework of transfer-based MT reearch has been replaced by lexicalist and knowledge-based approaches, with a lot of research focussing on domain-specific, sublanguage and controlled language approaches (see later discussion), including speech translation research. As we shall discuss in Chap. 25, statistical and corpus-based methods have also been developed. At the same time, there has been a great expansion of usage with the availability of cheaper PC-based software, online access to MT services, and in particular, the development of translation workstations with facilities for multilingual text editing, terminology databases, and "translation memories." There has also been great interest in the development of tools for multilingual authoring, text generation, database indexing and searching, and document abstracting, which incorporate automatic translation components.

At the beginning of the 1990s, there was a further change in the direction of MT research as many groups became interested in "empirical" approaches, based on large amounts of corpus data and essentially statistical, rather than linguistic, approaches. This strand of research will be dealt with in Chap. 25. Another feature of the most recent years of MT's history is the sudden influence of cheaper hardware and, above all, the Internet and World Wide Web. The former means that suddenly commercially viable systems are available to run on everyday PCs as found in many homes and offices, and with the availability of inexpensive software comes the possibility of ordinary users being able to use MT for their own purposes, which may differ quite substantially from the more formal requirements of corporate users. We will discuss this aspect more fully in Sec. 5, but one very obvious area in which this is true is the availability of Web-page and e-mail translation at the touch of a button. Advances in certain aspects of hardware development have also meant great strides in speech translation, an application for long thought to be too difficult to even contemplate. Because this book focusses on textual NLP, we will not dwell on this aspect too long, but mention only in passing that massive parallel processing along with more robust linguistic techniques have put speech MT firmly back on the agenda. Needless to say, translating speech presents a set of problems that barely overlap with those of text MT, on which we will concentrate here.

3. MT AS AN NLP PROBLEM

For most researchers, MT is essentially a linguistic problem with computational implications. It potentially encompasses or overlaps with all the problems that are dealt with in this book, with the added difficulty of the multilingual and contrastive element of translation.

The main difficulty of MT can be summed up in one word: **ambiguity**, although problems of style and interpretation should not be ignored.

A. Lexical Ambiguity

Similar to other NLP applications, ambiguity can arise at the lexical level, where a single word can have more than one interpretation and, hence, more than one translation. This is true from a monolingual perspective because morphology can be ambiguous (e.g., does *drier* mean 'more dry' or 'machine which dries'?), or because words can represent different parts of speech (*round* can be a noun, or a verb, or an adjective, or an adverb, or a preposition), or because words can have multiple meanings (e.g., *bank* 'financial institution' or 'side of a river'). All of the foregoing have an influence on translation because—unless the target language is closely related to the source language—it will be necessary to decide which meaning is intended to obtain the correct translation.

In addition, languages differ hugely in the way in which words correspond to concepts in the real world: for example, the English word *leg* corresponds to French *jambe*, *patte*, or *pied*, depending on whose leg it is. German has two words for *eat*, distinguishing human eating (*essen*) from animals performing the same activity (*fressen*). Spanish has only one word (*paloma*) for the bird distinguished in English as a *dove* or a *pigeon*. Examples such as these are extremely widespread and are not difficult to find.

These problems represent a particular difficulty for MT because computers often lack the necessary global understanding of the context to make the correct decision for their translation: although some of the ambiguities can be resolved by the kind of NLP techniques that have been presented in the earlier chapters of this book (for example, parsing can often identify the part of speech of an ambiguous word such as *round*), often only an understanding of what the text actually means can help to resolve the choice, and this understanding may come from the more or less immediate textual context, or may come from a general understanding of the real world.

For example, depending on the subject matter of the text, one or other reading of an ambiguous word may be more likely (e.g., *rail* as a type of bird or metal track that trains run on). Alternatively, neighbouring words may give a clue. Consider the word *ball*, which can indicate a spherical object or a dancing event. The correct interpretation is easy to obtain in sentences such as {1}, and {2} since the verb *roll* requires a spherical object as its grammatical subject, and *last* requires a durative event as its subject.

The ball rolled down the hill. {1}

The ball lasted until midnight. {2}

But consider the verb *hold* which means 'grasp' when its grammatical object is a physical object, but means 'organize' when its object indicates an event. So the two words *hold* and *ball* are both ambiguous in the sentence fragment {3}.

When you hold a ball ... {3}

B. Structural Ambiguity

Again, as in other NLP applications, ambiguity may arise from the different ways word sequences can be interpreted. **Structural ambiguity** arises when the combined lexical ambiguity of the words making up a sentence means that it has several interpretations. It is rather rare to find such examples in real life, as the context will usually determine whether, in a sentence such as {4}, we are talking about permission to partake of a hobby or work in a canning factory.

They can fish. {4}

However, for the computer, such problems can be quite common. More common are so-called **attachment ambiguities** such as sentence {5}, where some knowledge of the scientific domain is needed to know whether *thermal emission* is used for analysis or for investigation, a distinction that may have to be made explicit in the translation.

investigation of techniques of stress analysis by thermal emission {5}

Often-quoted further examples include {6a}, the potentially three-way ambiguous {6b}, and the perennial {6c}.

Visiting relatives can be boring. {6a}

Time flies like an arrow. {6b}

The man saw the girl with a telescope. {6c}

Although these are truly ambiguous for a computer without access to contextual knowledge, consider the very similar sentences in {7}, which are unambiguous to a human, but could present the same sort of ambiguity problem to a computer lacking the relevant linguistic knowledge.

Visiting hospitals can be boring. {7a}

Blow flies like a banana. {7b}

The man saw the girl with a hat. {7c}

If the MT system chooses the wrong interpretation, the resulting translation could be ludicrous, incomprehensible or, perhaps worse still, misleading.

C. Different Levels of Ambiguity

Structural ambiguities such as those illustrated in the foregoing can be further classified according to the "range" of the ambiguity they represent. **Local ambiguities** are the most superficial and, accordingly, often the easiest to deal with in translation (for example, because they persist in the target language). The attachment ambiguities already illustrated in {6a}, {6c} and {7c} are examples of this, where the ambiguity lies not in the words themselves, but in their juxtaposition. Further examples in {8} show this more clearly.

Did you read the story about the air crash in the jungle? {8a}

Did you read the story about the air crash in the paper? {8b}

John mentioned the book I sent to Susan. {8c}

This can be contrasted with **global ambiguity** caused by combinations of category ambiguities: different analyses involve different category choices and, hence, radically different translations, as in {9}.

He noticed her shaking hands. (i.e., *her hands were shaking* or *she was shaking hands with people*) {9a}

I like swimming. (i.e., to do or to watch) {9b}

A further categorization of ambiguity reflects the "depth" at which the ambiguity arises. For example, the sentences in {10} have a similar "surface structure," but at a deeper level differ in that *John* is the implied subject of *please* in {10a}, but its object in {10b}. Compare {10c}, which is truly ambiguous in this respect.

John is eager to please. {10a}

John is easy to please. {10b}

The chicken is ready to eat. {10c}

Example {11a} provides us with another subtle ambiguity of this type: to see that it is ambiguous, compare the sentences in {11b–e}, and consider a language that distinguishes the different interpretations (in case-marking, as in Finnish, or in the choice of the verb, as in German).

John painted the house. {11a}

John painted the window frame. {11b}

John painted the panorama. {11c}

John painted the canvas. {11d}

Da Vinci painted the Mona Lisa. {11e}

As a final example, consider the examples in {12} and {13}, which illustrate the problem of **anaphora resolution**, that is, identification of antecedents of pronouns. In translating into a language in which pronouns reflect grammatical differences, this is an important, but difficult, task for the computer. Consider what background information you, as a human, use to make the correct interpretation.

The monkey ate the banana because it was hungry. {12a}

The monkey ate the banana because it was ripe. {12b}

The monkey ate the banana because it was tea-time. {12c}

The soldiers shot at the women, and some of them fell. {13a}

The soldiers shot at the women, and some of them missed. {13b}

Chapter 5 discusses in detail approaches to these problems that have been developed in NLP research.

D. Solutions to the Problem of Ambiguity

Many of the problems described in the previous section were recognised very early on by the MT pioneers. Of the possible solutions suggested, some remain popular to this day, whereas others have been superseded.

The earliest approach involved using statistical methods to take account of the context. Lexical ambiguities would be interpreted one way or another depending on the surrounding words. This was typically implemented in a fairly crude way. For example, faced with a word that could either be a noun or a verb (of which there are many such words in English), one technique would be to look to see if the preceding word was an article (*a* or *the*). A similar technique might be evolved for translational ambiguities. For structural ambiguities, the program would look at longer sequences of parts of speech and identify which was the more probable. Although such techniques are often effective, in the sense that they deliver the right answer, they do not contribute to the notion of having the computer simulate understanding of the text in any sense. More sophisticated statistical techniques have been experimented with in recent years, as described in Chap. 25.

Because they recognised these limitations, researchers soon hit on the idea of involving a human user to help prepare the input for translation (preediting) or tidy up the output after the system has done its best (postediting). It was also recognised that limiting the scope and domain of the texts given to the system could result in better translations. All these ideas, which we shall look at in more detail later, and which are now commonly referred to by vendors of commercial MT software, were proposed way back in the early days of MT research.

Perhaps a more significant solution to these problems, however, was the incorporation of more sophisticated techniques of analysing the text to be translated, using methods proposed by linguists, and making use of linguistic theory coupled with more elaborate computer programs. Inasmuch as this approach distinguishes the early research in MT from much of the work done in the 1970s and 1980s, systems developed with these techniques are known as **second-generation** systems. We shall look in more detail at the important features of these systems in the next section.

4. BASIC DESIGN FEATURES

The incorporation of more sophisticated linguistic (and computational) techniques led in the 1970s and 1980s to the development of the so-called second-generation of MT design, characterized in particular by the **indirect** approach to translation, in which the source text is transformed into the corresponding target text via an intermediate, linguistically motivated, representation. This may purport to be a representation of the meaning of the text, or else, rather less ambitiously, a representation of the syntactic structure of the text. A further distinction is the transfer–interlingua issue. The difference between the two approaches is illustrated by the well-known "pyramid" diagram (Fig. 1), probably first used by Vauquois (1968): the deeper the analysis is, the less transfer is needed, the ideal case being the interlingua approach in which there is no transfer at all.

A. The Interlingua Approach

Historically prior, the **interlingua** approach represents a theoretically purer approach. The interlingual representation is an abstract representation of the meaning of the source text, capturing all and only the linguistic information necessary to generate an appropriate target text showing no undue influence from the original text. This turns out to be very difficult to achieve in practice, however. Even the very deepest of representations that linguists have come up with are still representations of text, not of meaning, and it seems inevitable that a translation system must be based on a mechanism that transforms the linguistic structures of one language into those of the other. This is unfortunate in a way, as there are many advantages of an interlingual approach, especially when one thinks of multilingual systems translating between many language pairs (consider that a system to translate between all 11 current official languages of the European Union would need to deal with 110 different language pairs), but the best attempts to implement this idea so far have

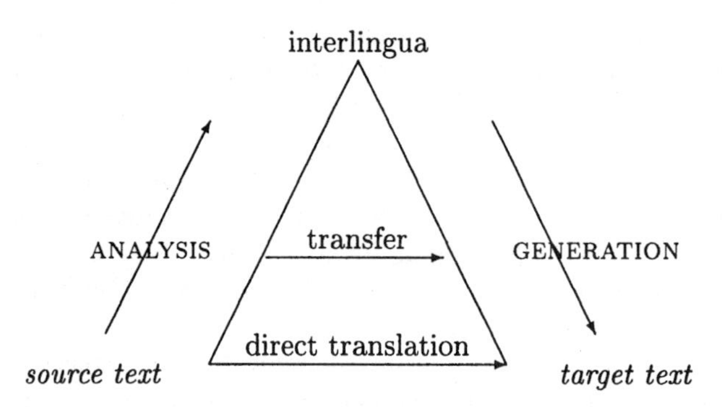

Fig. 1 The pyramid diagram.

had to be on a very small scale, and they have generally turned out translingual paraphrases, rather than translations.

B. Transfer

The more practical solution has been the **transfer** approach, which views translation as a three-stage process: (a) analysis of the input into a source-language syntactic structure representation, (b) transfer of that representation into the corresponding target-language structure, and (c) synthesis of the output from that structure. Although this approach has the disadvantage of requiring an additional stage of processing, there is a corresponding advantage in that the approach focusses on the *contrastive* element of translation, for it is precisely in the transfer stage that the differences between the languages can be brought out. Current systems working within this framework adopt an approach in which transfer is reduced as much as possible, ideally to the translation of the lexical items only, by the adoption of a "deep" intermediate representation that neutralises, as much as possible, the idiosyncrasies of the surface forms of the language, such as gender agreement, word order, tense, mood, and aspect in verbs and so on.

The issue of transfer vs. interlingua has been widely discussed in the literature: see particularly Arnold et al. (1994:80ff), Hutchins and Somers (1992:71ff), and further refererences cited there.

C. Computational Formalisms

Another theoretical issue in the design of an MT system is the computational one of how to compute and manipulate the representations described in the foregoing. Because there are broadly two problems here—what to do and how to do it, the former linguistic, the latter computational—attempts are often made to separate the two by providing computational formalisms or programming languages which a linguist can easily learn and work with. This usually means formalisms that are very similar to the grammar formalisms found in theoretical linguistics. But there is a further theoretical distinction between **declarative** and **procedural** formalisms. In a declarative formalism, the linguists have to think in terms of static relations and facts, and leave it to the computer to figure out how to combine them, whereas with a procedural formalism, they have to be more explicit about what to do and when. We can exemplify this quite easily: consider that *mice* is the plural of *mouse*. We might use this information in a number of different procedures; for example, in confirming that the subject of a sentence agrees with the verb, in determining how to translate *mice*, which is not in the bilingual dictionary, or when translating into English. So there is an obvious advantage in stating this piece of declarative information independently of the procedures that use it.

Early second-generation systems were typically more procedural in design. Often, the process of translation would be broken down into several subprocesses or modules, often motivated linguistically. One influential design was that of GETA's ARIANE system (Vauquois and Boitet, 1985; Boitet, 1987; Guilbaud 1987) discussed in detail in Hutchins and Somers (1992, Chap. 13) and in Whitelock and Kilby 1995, Chap. 9). In this system, the source text is first subjected to morphological analysis, then "multilevel syntactic analysis" consisting of a kind of parsing stage to recognise constituent structure (NPs, verb groups, and so on are recognised) followed by a kind of shallow semantic analysis stage during which syntactic and semantic functions are determined (subject, object, head, modifier, and so on). The resulting representation is then passed through a two-stage bilingual transfer process, the first "lexical," involving the choice of target-language words, the second "structural," in which the syntactic structure of the text is altered to suit the target language. The third stage, target-language generation, involves further syntactic restructuring followed by "morphological synthesis," which is when local agreement features, such as case, number, and gender, are set and the surface strings of the target text generated. This highly modular and stratificational design was much imitated in the 1970s, for example, in many of the Japanese systems, in early designs for the CEC's Eurotra project, and elsewhere.

An alternative, more declarative, design is exemplified by the **Shake and Bake** approach of Whitelock (1992, 1994) and Beaven (1992). This approach is based on a linguistic formalism called Categorial Unification Grammar (Wood, 1993), in which the basic idea is that instead of explicit rules about how words can be combined to form sentences, each word is associated with a set of features, and there are very general rules about how features can combine and co-occur. The analogy of Shake and Bake comes from the baking-for-idiots product where all the ingredients are put in a bag which is then shaken: the ingredients somehow "know" how they are meant to combine with each other and in what proportions. Then you put the mixture into the oven and it comes out as a cake, or a loaf of bread, or whatever. Similarly, in this system, the analysis and generation modules organize themselves depending on the basic ingredients (the words of the source text). Shake and Bake MT includes a bilingual lexicon in which the feature bundles that are derived for the source-language analysis are translated into corresponding feature-bundles for the target language, the bag is shaken up again, and out comes the correctly formed target text.

5. THE USER'S PERSPECTIVE

As mentioned in Sec. 3.D, there has been a shift in emphasis from fully automatic translation toward the reduced aim of **computer-assisted** (or aided) translation (CAT). We can distinguish broadly two types of CAT: in one, we have the use of partially automated computer tools by the human translator, which we can call **machine-aided human translation** (MAHT). The word processor is the most obvious tool now essential to almost all translators (assuming that the target language uses an alphabet for which word-processing tools have been developed). Texts can be received and delivered over the computer network or in machine-readable form on diskettes, and translations can be stored for future reference without filling endless filing cabinets with paper. On-line dictionaries, grammar and spelling checkers, an increasing number of terminological databanks, and other automated specialised information sources, have dramatically increased the speed and efficiency with which human translators can work.

The second type of CAT is when it is essentially the computer that performs the translation, although still under the control of the human user, who may or may not be a translator. This involves stages of contribution by the human to the process of translation, either before or after the machine has done what it can, or interactively during the translation process.

As soon as we start thinking about the relations between a system and its users, we must consider the different profiles of possible users of a CAT system. The obvious one is perhaps the translator looking for a way to speed up the translation process. But equally valid is the scenario in which the user may be the person who actually wants to make use of the translated document; for example, a scientist who has been sent a technical article in Russian or Japanese, and wants to know roughly what it is about, to see whether and which parts of it need translating "properly," or—a significantly widespread scenario nowadays—someone browsing the World Wide Web who comes across an interesting-looking Web page that happens to be in a language they cannot understand. Alternatively, the user might be the original author of a document, who wants to send a translation of it to someone in another country. Under certain circumstances, the document may be of a sufficiently stereotyped nature (for example, certain types of business letter, or a price list) that this work could be done by a system without the need for a human who knows the target language to check the output, as long as the recipient was warned that they were receiving "unrevised machine translation output."

In this latter case, we would want to be sure that the text that the system got to translate "unsupervised" was so straightforward that the system would not make any serious mistakes. There are two ways to do this: one involves "controlled language," the other "sublanguage." Both may involve the user in "preediting."

A. Controlled Language

Controlled language systems, sometimes also known as "restricted input," as the name suggests, operate on the understanding that the texts they are asked to translate will be constrained in some way, notably avoiding words or constructions that the system will not be able to translate. Computationally, it is relatively easy to check whether a text contains any "unknown" words, and not much more difficult to ascertain whether the sentences can be parsed by the system. If unknown words or "illegal" structures are detected, the user can then be asked to correct them, perhaps with reference to a style manual. Although, at first sight, there might be a negative reaction to this idea of a computer dictating what a writer can or cannot write, it should be pointed out that technical writers are quite used to the idea of pre- and proscribed vocabulary and style. The notion of "artificial languages" may be familiar to some readers (cf., Sager 1993:32ff)—although the term can be confusing because it is also sometimes used to describe formal codes such as programming languages. Many users of current (and earlier) commercial MT systems have reported tremendous success with the idea of restricted input (Elliston, 1979; Pym, 1990; Newton, 1992; and others). Notably, it has been reported that having texts written according to a style-sheet that permitted them to be machine translated actually led to an improvement in readability and style of the *original* text from the point of view of end-users (Lawson, 1979:81f). Certainly, the idea of a restricted subset of Basic English for consumption by nonnative readers predates MT (see Arnold et al. 1994:156ff for further discussion). In recent years, the controlled language movement has gained considerable momentum (see e.g., Adriaens, 1994; CLAW, 1996).

Closely related to the notion of restricted input is the so-called sublanguage approach. ***Sublanguage*** is a term that was introduced by researchers working in

Canada, notably on the Météo system which translates weather bulletins (Kittredge and Lehrberger, 1982). The notion will be familiar to some readers under different names such as "register," "LSP," and so on. The idea is that for certain text types and subject domains, the language used is *naturally* restricted in vocabulary and structures. It is fairly clear that the words used in a particular text will reflect the subject matter of the text: the term "jargon" is often used disparagingly, although it is also a useful technical term for linguists. If an MT system can recognise the subject matter of a text, then many possible lexical ambiguities may be resolved. Put another way, the lexicon for a sublanguage system can be greatly reduced in size, for many words, or alternative meanings and translations of words, can simply be omitted.

What is perhaps less obvious is that the same applies to the syntactic constructions found. For example, instruction manuals will typically contain many imperatives and simple declarative sentences. A legal contract may contain many conditional and hypothetical verb tenses. The often recurring and—of more interest—absent constructions for a given sublanguage mean that the grammars used by the system can be simplified, and again, potential ambiguities can be reduced or eliminated. So sublanguages are usually defined in terms of their specialised vocabulary, and the range of constructions used. This may or may not be done with reference to some notion of a "general" grammar of the language, because many sublanguages permit constructions that in another context would actually be considered as ungrammatical. For example, in medical reports, the verb *present* is used intransitively to describe symptoms as in {14a}: a construction that would sound odd under other circumstances. Similarly, sentences in weather bulletins often lack a tensed main verb, as in {14b}.

The patient presented with a sore throat. {14a}

Clear now, becoming cloudy later. {14b}

The sublanguage approach recognizes that certain sublanguages—for example the sublanguage of weather bulletins, the sublanguage of stock market reports, and the sublanguage of instruction manuals—have the natural lexical and syntactic restrictions that will make MT easier (and more reliable). It is also recognized that some sublanguages actually make MT harder. For example, some legal texts, such as patents, contain syntactic constructions that are in fact much *more* complex than everyday language (e.g., because helpful punctuation is generally omitted), which makes them *less* amenable to translation by machine. Telexes similarly, with their ellipses and generally "telegraphic" style introduce more ambiguities as compared with the "standard" language. A further point of interest, especially to researchers, is the extent to which corresponding sublanguages across languages share syntactic features (or not). It has been claimed (Kittredge, 1982) that the grammars of corresponding sublanguages have more in common across language pairs than with other sublanguages of the "parent" language: that is to say, notwithstanding the obvious lexical differences, the sublanguages of English and French instruction manuals are more similar than the (English) sublanguages of, say, stock market reports and holiday travel brochures. This point is somewhat debatable, and certainly some parallel sublanguages are quite strikingly different (cooking recipes, knitting instructions, and some kinds of job advertisments for example).

The sublanguage and controlled language approaches are very similar then, in that they rely on the fact that the potential complexity of the source texts is reduced. Perhaps the key difference between the two (and one which makes, for some observers, one approach acceptable where the other is not) is the *source* of the restrictions. In sublanguage systems, the restrictions occur naturally, because a *sublanguage* is defined as "semi-autonomous, complex semiotic systems, based on and derived from general language [whose] use presupposes special education and is restricted to communication among specialists in the same or closely related fields," (Sager et al., 1980), and it "reflects the usage of some 'community' of speakers, who are normally linked by some common knowledge about the domain which goes beyond the common knowledge of speakers of the standard language" (Kittredge, 1985). Sublanguage MT systems are designed to be able to handle the special structures found in the (already existing) sublanguages. We can contrast this with controlled language systems, for which the restrictions often arise from the perceived shortcomings of some existing MT system. That the resulting controlled language may turn out to be more appropriate to the community of end-users is a happy accident that should not hide the fact that it is the MT system that defines the sublanguage and not vice versa.

B. Preediting

Whatever the source of the restrictions on the system, it is usually necessary for there to be some check that the source text actually adheres to the restrictions. In the case of a sublanguage system, it may be hoped that this will happen naturally, although most sublanguage texts are "noisy" to some degree or other (e.g., when the weather report includes a friendly comment such as that in {15}.

··· *so you'd better take your umbrellas, folks.* {15}

Certainly in the case of artificially imposed restrictions, a phase of preprocessing is normally needed, during which the source text is made to conform to the expected input norms of the system.

This preprocessing is called **preediting**, and it may be carried out essentially manually (in much the same way that an editor prepares an article or book before publication), or the system itself may be able to assist, by pointing out possible deviations from the aforementioned agreed restrictions. This may be in a kind of "submit-and-return" system in which the text is given to the system and comes back with inappropriate vocabulary and constructions highlighted. Or the system may have an *interactive* preediting facility for which the user (or rather the preeditor) and the system work through the text together. The system may even be able to suggest corrections itself, on the basis of its "self-knowledge" about the vocabulary and grammar it "knows" how to handle.

C. Postediting

The other side of the human-assisted MT coin is **postediting**, during which a human user takes the results of the MT process and corrects or generally tidies them up ready for the end-user. Several points need to be made about this process. The first is one that is often ignored by critics of MT who fail to acknowledge that most human translation is subject to revision, so the need to revise MT output should not necessarily be seen as an added cost. Indeed, there is even evidence that revising MT output is in some senses easier than revising human translations, notably from the point of view of "peer sensitivity." Many observers report increased productivity (e.g., Magnusson-Murray, 1985); although there have been some dissenting voices (e.g., Wagner, 1985), these come mostly from translators who have been diverted from the task of translating to the (for them) more limiting task of postediting.

The amount of correction that needs to be done to the output from an MT system will crucially depend on the end use to which that text is to be put. One of the major advantages claimed for MT is that it can provide a rough and ready translation of a text that would otherwise have been completely inaccessible to the end users (e.g., because it is written in a foreign script), and from which they can identify whether the text needs to be translated "properly," and if so, which parts of it in particular. End users, as domain experts, may be able to understand a low-quality translation despite perhaps quite damaging translation errors (e.g., a misplaced negative) simply because they know the field well enough to "second guess" what the original author was probably trying to say. It has been reported (M Kay, personal communication) that early users of the Systran MT system which was used by the U.S. Air Force to translate documents from Russian into English got used to the kind of slavicised English that the system produced, claiming that it was "more accurate" than the results of postediting by translators who were not domain experts, a finding backed up by other evaluations of Systran (e.g., Habermann, 1986) and other systems [e.g., Sinaiko and Clare (1972) who found that mechanics' quality ratings of technical translations were consistently higher than translators'].

D. Interactive Systems

Whereas pre- and postediting involve a human working on the translation either before or after the MT system has done its work, another possibility is that the user collaborates with the system at run-time in what is generally called **interactive mode**. Interactive systems pause to ask the user to help them make linguistic decisions while the program is running. This might be the case where the source text has an unknown word (the user is asked to check its spelling and provide some grammatical information both about the source word and its translation), a lexical or syntactic ambiguity (the system cannot find sufficient criteria to make a choice, so consults the user), or a choice of target translation, again either lexical or syntactic.

Some modern commercial systems have a choice of interactive or default translation. In the latter, the system will make all the choices for itself, perhaps arbitrarily, or based on some "best guess" heuristics. For example, in attachment ambiguity, a good guess, in the absence of any other indications, based on general frequency in English, is to attach to the nearest noun (e.g. {16}, although sometimes this will give a false interpretation as in {17}).

percentage of children educated privately {16a}

decoration of rooms with no windows {16b}

total number of delegates broken down by sex {17}

One obvious problem with interactive systems is that the interactions slow down the translation process. In particular, there is evidence that users quickly tire of continued interactions requesting the most banal information or repeatedly asking the same question (e.g., Seal, 1992): "overinteraction" can lead to user alienation.

Another problem is the profile of the user that is implied by an interactive system: the program seeks help in understanding the source text, so it cannot be used by someone who knows only the target language (the end-user who wants to read the text); similarly, it asks questions about the style and syntax of the target text, so it cannot be used by someone who does not know the target language (e.g., someone writing a business letter to an overseas client); in fact, it also asks for decisions about the relation between the source and target language, how to translate an unknown word and so on, so the ideal user is probably a translator. And typically translators feel that they could probably do a better job more quickly without the intermediary of the system at all.

Two solutions to this problem offer themselves. The first is to give more control to the translator-as-user, inviting them to decide on a case-by-case basis whether to translate portions of the text manually, interactively, or automatically with postediting (cf., Somers et al., 1990). Certainly the latter two options will save the user a certain amount of typing, at the very least, and the idea is that as the user comes to know the system, they will learn how well it is likely to translate a given portion of text (which might be a whole paragraph, a sentence, or even a smaller stretch of text) and be able to decide what is the quickest way to deal with it.

The second solution is to allow the system to "learn" from its experience, and "remember" the corrections that the user suggested. Some systems have even been designed (e.g., Nishida and Takamatsu, 1990) so that they can learn new translation rules and patterns from user's interactions (or from a record of postediting changes). A more obvious implementation of this approach is in the now very widespread idea of Translation Memory, to which we return in the following.

E. Translator's Workbench

It has been suggested that all of the functions described in the foregoing could be combined into a flexible package for translators, often called a **Translator's Workbench**. This idea seems to have been first suggested by Kay in his now much-cited 1980 research manuscript; Melby (1982) describes a similar idea. Running on a personal computer, possibly linked by a network to a more powerful mainframe computer, the idea is that the translator can have access online to terminological resources, dictionaries, and other reference materials, as well as full or human-aided translation software. For output, spelling and grammar checkers, as well as desk-top publishing packages for (re-)formatting the finished work would all be available at the click of a button. For texts not already in machine-readable form, a scanner should be attached to the system.

One function that should be mentioned particularly is the idea of accessing previously translated texts on similar topics in what has come to be called a **Translation Memory**. The origin of this very simple idea is somewhat obscure: usually attributed to Kay (1980), several commentators seem to have made suggestions along these lines from the early 1970s onward (see Hutchins, 1998). The idea, in relation to MT, is that when translating a new piece of text (a sentence, for example) it is useful to have at one's fingertips examples of similar sentences and their translations. This allows the translator to ensure consistency of terminology and phraseology, and also, for an exact match, to turn translation into a simple cut-and-paste editing job. Recent research has looked at ways of making the matching process linguistically sophisticated, so that previously translated sentences of a similar content (as well as form) are retrieved. On the basis of large corpora of previously translated texts, for example the bilingual Canadian parliamentary proceedings, some researchers have provided quite sophisticated tools for translators (see Isabelle et al., 1993).

6. CONCLUSIONS

In this chapter we have discussed some of the main obstacles to be overcome in developing an MT system, and some of the solutions that have been proposed. We have concentrated on MT as a linguistic problem, while in a companion chapter elsewhere in this handbook (Chap. 25), we also look at a range of alternative approaches based on the availability of aligned bilingual corpora.

Whichever approach is taken (and indeed the different approaches are not necessarily incompatible, as the development of several hybrid systems can testify), it is certain that MT will continue to be a major goal of NLP researchers. In particular, the current climate of multilingual localisation and internationalisation of software, the spread of information through the World Wide Web and other resources, and the general recognition that speakers of languages other than English have an equal right to access this information explosion means that translation will be a major need in the next century, and more and more people will look toward computers to help them meet this need. In this chapter, we have outlined some of the difficulties that this need implies, and some of the techniques used to address the need. MT software is now readily available commercially at prices that even casual users can afford. The quality of the end-product is still limited, however, so MT research and development must follow multiple paths in the future: improving the basic quality of systems that are already available, but also spreading the technology ever wider, to encompass more and more of the world's languages. These are the joint challenges for the immediate future.

REFERENCES

Adriaens G (1994). Simplified English grammar and style correction in an MT framework: the LRE SECC project. In Translating and the Computer 16. London.

ALPAC (1966). Languages and Machines: Computers in Translation and Linguistics. A report by the Automatic Language Processing Advisory Committee, Division of Behavioral Sciences, National Research Council, National Academy of Sciences, Washington DC.

Arnold D, L Balkan, RL Humphreys, S Meijer, L Sadler (1994). Machine Translation: An Introductory Guide. Manchester: NEC Blackwell.

Bar-Hillel Y (1951). The state of machine translation in 1951. Am Document 2:229–237; reprinted in Y Bar-Hillel, Language and Information. Reading, MA: Addison-Wesley, 1964.

Beaven JL (1992). Shake-and-Bake machine translation. Proceedings of the fifteenth [sic] International Conference on Computational Linguistics, COLING-92, Nantes, pp 603–609.

Boitet C (1987). Research and development on MT and related techniques at Grenoble University. In: M King, ed. Machine Translation Today: The State of the Art, Edinburgh: Edinburgh University Press, pp 133–153.

CLAW (1996). Proceedings of the First International Workshop on Controlled Language Applications (CLAW 96). Leuven, Belgium: CCL.

Elliston JSG (1979). Computer aided translation: a business viewpoint. In: BM Snell, ed. Translating and the Computer. Amsterdam: North-Holland, pp 149–158.

Guilbaud J-P (1987). Principles and results of a German to French MT system at Grenoble University (GETA). In: M King, ed. Machine Translation Today: The State of the Art. Edinburgh: Edinburgh University Press, pp 278–318.

Habermann FWA (1986). Provision and use of raw machine translation. Terminologie et Traduction 1986/1:29–42.

Hutchins WJ (1986). Machine Translation: Past, Present, Future. Chichester: Ellis Horwood.

Hutchins WJ (1998). The origins of the translator's workstation. Mach Transl. 13:287–307.

Hutchins WJ, HL Somers (1992). An Introduction to Machine Translation. London: Academic Press.

Isabelle P, M Dymetman, G Foster, J-M Jutras, E Macklovitch, F Perrault, X Ren, M Simard (1993). Translation analysis and translation automation. Proceedings of the Fifth International Conference on Theoretical and Methodological Issues in Machine Translation TMI '93: MT in the Next Generation, Kyoto, Japan, pp 12–20.

Kay M (1980). The proper place of men and machines in language translation. Research Report CSL-80-11. Xerox Palo Alto Research Center, Palo Alto, CA October 1980; repr. in Mach Transl 12:3–23, 1997.

Kittredge RI (1982). Variation and homogeneity of sublanguages. In: RI Kittredge, J Lehrberger, eds. Sublanguage: Studies of Language in Restricted Semantic Domains. Berlin: de Gruyter, pp 107–137.

Kittredge RI (1985). The significance of sublanguage for automatic translation. In: S Nirenburg, ed. Machine Translation: Theoretical and Methodological Issues. Cambridge: Cambridge University Press, pp 59–67.

Kittredge RI, J Lehrberger, eds. (1982). Sublanguage: Studies of Language in Restricted Semantic Domains. Berlin: de Gruyter.

Lawson V (1979). Tigers and polar bears, or: translating and the computer. Incorp Linguist 18:81–85.

Locke WN, AD Booth, eds. (1955). Machine Translation of Languages. Cambridge, MA: MIT Press.

Magnusson-Murray U (1985). Operational experience of a machine translation service. In: V. Lawson, ed. Tools for the Trade: Computers and Translation 5. Amsterdam: North Holland, pp 171–180.

Melby AK (1982). Multi-level translation aids in a distributed system. In: J Horecký, ed. COLING 82: Proceedings of the Ninth International Conference on Computational Linguistics. Amsterdam: North-Holland, pp 215–220.

Newton J (1992). The Perkins experience. In: J. Newton, ed. Computers in Translation: A Practical Appraisal. London: Routledge, pp 46–57.

Nishida F, S Takamatsu (1990). Automated procedures for the improvement of a machine translation system by feedback from postediting. Mach Transl 5:223–246.

Pym PJ (1990). Pre-editing and the use of simplified writing for MT: an engineer's experience of operating an MT system. In: P Mayorcas, ed. Translating and the Computer 10: The Translation Environment 10 Years On. London: Aslib, pp 80–96.

Sager JC (1993). Language Engineering and Translation: Consequences of Automation. Amsterdam: John Benjamins.

Sager JC, D Dungworth, PF McDonald (1980). English Special Languages: Principles and Practice in Science and Technology. Wiesbaden: Brandstetter.

Seal T (1992) ALPNET and TSS: the commercial realities of using a computer-aided translation system. In: Translating and the Computer 13: The Theory and Practice of Machine Translation—A Marriage of Convenience? London: Aslib, pp 119–125.

Sinaiko HW, GR Klare (1972). Further experiments in language translation: readability of computer translations. ITL 15:1–29.

Somers HL, J McNaught, Y Zaharin (1990). A *user*-driven interactive machine translation system. Proceedings of SICONLP '90: Seoul International Conference on Natural Language Processing (Seoul), pp 140–143.

Vauquois B (1968). A survey of formal grammars and algorithms for recognition and transformation in machine translation. IFIP Congress-68 (Edinburgh), pp 254–260; reprinted in C Boitet, ed. Bernard Vauquois et la TAO: Vingt-cinq Ans de Traduction Automatique—Analectes. Grenoble: Association Champollin, 1988, pp 201–213.

Vauquois B, C Boitet (1985). Automated translation at Grenoble University. Comput Linguist 11:28–36.

Wagner E (1985). Rapid post-editing of Systran. In: V Lawson, ed. Tool for the Trade's Computers and Translation 5. Amsterdam: North Holland, pp 199–213.

Whitelock P (1992). Shake-and-bake translation. Proceedings of the fifteenth [sic] International Conference on Computational Linguistics, COLING-92, Nantes, pp 784–791.

Whitelock P (1994). Shake-and-bake translation. In: CJ Rupp, MA Rosner, RL Johnson, eds. Constraints, Language and Computation. London: Academic Press, pp 339–359.

Whitelock P, K Kilby (1995). Linguistic and Computational Techniques in Machine Translation System Design. London: UCL Press.

Wood MM (1993). Categorial Grammars. London: Routledge.

Yngve VH (1957). A framework for syntactic translation. Mechan Transl 4.3:59–65.

14

Dialogue Systems: From Theory to Practice in TRAINS-96

JAMES ALLEN, GEORGE FERGUSON, ERIC K. RINGGER*, and TERESA SIKORSKI ZOLLO

University of Rochester, Rochester, New York

BRADFORD W. MILLER

Cycorp, Austin, Texas

1. INTRODUCTION

In this chapter, we describe the TRAINS-96 system, an end-to-end spoken dialogue system that can interact with a user to solve simple planning problems. Because the goal of this work was to produce a working system, great emphasis was made on finding techniques that work in near real-time and that are robust against errors and lack of coverage. Robustness is achieved by a combination of statistical postcorrection of speech recognition errors, syntactically and semantically driven robust parsing, and extensive use of the dialogue context. Although full-working systems are difficult to construct, they do open the doors to end-to-end evaluation. We present an evaluation of the system using time-to-completion for a specified task and the quality of the final solution that suggests that most native speakers of English can use the system successfully with virtually no training.

The prime goal of the work reported here was to demonstrate that it is feasible to construct robust spoken natural dialogue systems. Although there has been much research on dialogue, there have been very few actual dialogue systems. Furthermore, most theories of dialogue abstract away many problems that must be faced to construct robust systems. Given well over 20 years of research in this area, if we cannot construct a robust system even in a simple domain, that bodes ill

* *Current affiliation:* Microsoft Research, Redmond, Washington.

for progress in the field. Without working systems, we are very limited in how we can evaluate the worth of various theories and models.

We chose a domain and task that was as simple as possible, yet could not be solved without collaboration between a human user and the system. In addition, there were three fundamental requirements we placed on the system:

1. It must run in near real-time.
2. The user should need minimal training and not be constrained in what can be said.
3. The dialogue should result in something that can be independently evaluated.

At the start of this experiment in November 1994, we had no idea whether our goal was achievable. Although researchers were reporting good accuracy (about 95%) for speech systems in simple question–answering tasks, our domain was considerably different and required a much more spontaneous form of interaction. We also knew that it would not be possible to directly use general models of plan recognition to aid in speech act interpretation (as proposed in Allen and Perrault, 1980; Litman and Allen, 1987; Carberry, 1990), as these models would not allow anything close to real-time processing. Similarly, it would not be feasible to use general planning models for the system back end or for planning its responses. We could not, on the other hand, completely abandon the ideas underlying the plan-based approach, as we knew of no other theory that could provide an account of the interactions in dialogue. Our approach was to try to retain the overall structure of plan-based systems, but to use domain-sepcific reasoning techniques to provide real-time performance.

Dialogue systems are notoriously hard to evaluate, as there is no well-defined "correct answer." So we cannot give accuracy measures, as is typically done, to measure the performance of speech recognition systems and parsing systems. This is especially true when evaluating dialogue robustness, which results from many different sources, including techniques for correcting speech recognition errors, using semantic knowledge to interpret fragments, and using dialogue strategies that keep the dialogue flowing efficiently, despite recognition and interpretation errors.

The approach we take is to use task-based evaluation. We measure how well the system does at helping the user solve a given problem. The two most telling measures are time-to-completion and the quality of the final solution. In the evaluation described later in this paper, we show that all our subjects were able to use TRAINS to solve problems with only minimal training. We also evaluated the overall effectiveness of our robust processing techniques by comparing spoken dialogues with keyboard dialogues by the same subjects. Even with a 30% word error rate, speech turned out to be considerably more efficient than keyboard.

2. THE TASK AND AN EXAMPLE SESSION

The domain in TRAINS is simple route planning. The user is given a map on a screen showing cities, routes, and the locations of a set of trains (Fig. 1), and a verbal specification of a set of destination cities where trains are needed. The task is to find the most efficient set of routes possible. The route planner used by the system is

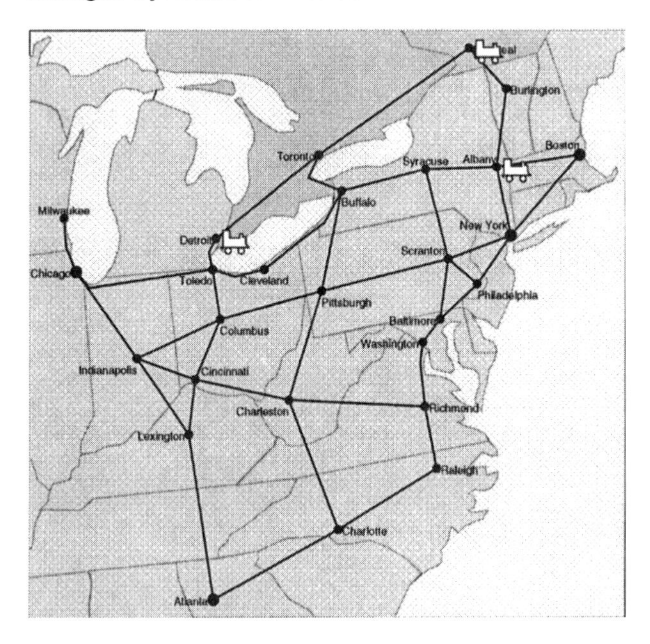

Fig. 1 The initial scenario.

deliberately weak so that interaction is needed to find good plans. Specifically, the planner cannot find routes longer than four hops without an intermediate city, and when it can generate a route, it randomly selects among the possibilities.

Typically, the examples in papers on dialogue systems are carefully hand-constructed and designed to show off specific system capabilities. Such examples, however, give little feeling for how robust the system is. Much more revealing is how the system responds to unplanned interactions with naive users, and how it interacts when things start to go wrong. As a result, we will not present any examples here in which everything went smoothly. Rather, we present a dialogue that stresses the system's capabilities, better illuminating what works and what areas require further improvement.

The following dialogue involved a person who had no expertise in programming, but had used the system a few times before. In the transcripts, we show the input as it was interpreted by the speech recognizer in uppercase letters, followed by what was actually said by the user. In cases where the recognition was perfect, we show only the speech recognition output. The dialogue contains a significant number of errors and has a 30% word error rate. The system display is also shown at key points in Figs. 2–4.[1] In this problem, the engines start at Detroit, Montreal, and Albany, and the goal is to get trains to Milwaukee, Lexington, and Washington.

S: Ok. I think I'm ready to start.
U: OKAY LET'S SEE CONTAIN FROM DETROIT TO WASHINGTON
Okay. Let's take a train from Detroit to Washington.

[1] A QuickTime movie can be found as TRAINS95-v1.3-Pia.qt.gz at http://www.cs.rochester.edu/research/trains/clips.

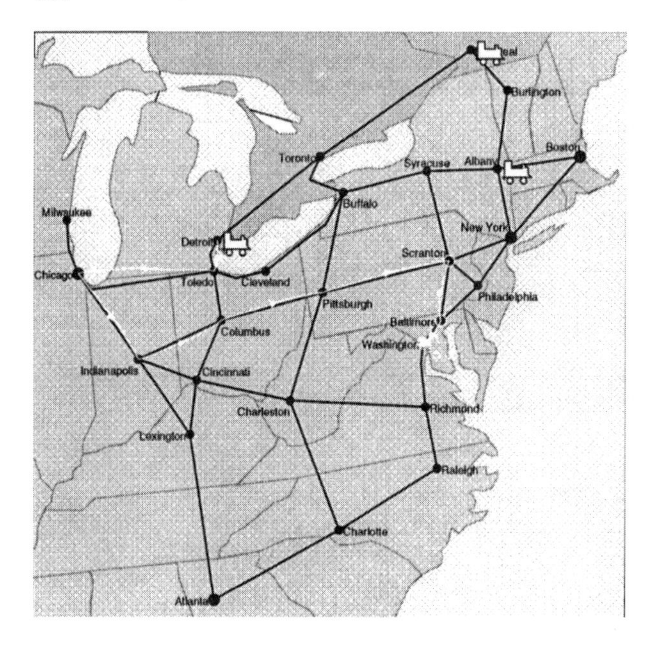

Fig. 2 The proposed route.

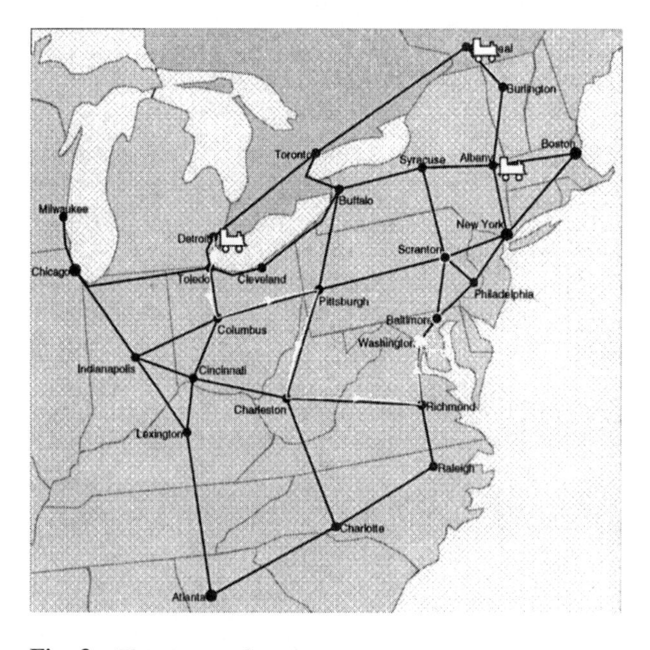

Fig. 3 The corrected route.

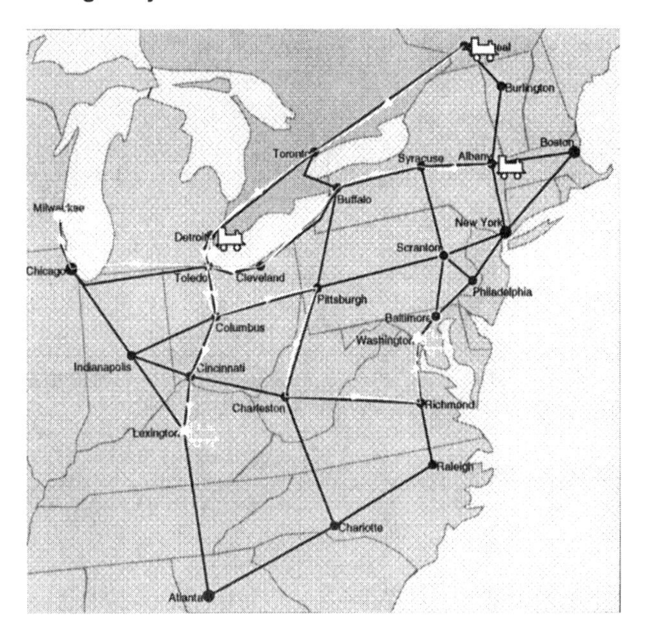

Fig. 4 The final routes.

S: What route would you like to get from Detroit to Washington?
U: LET'S GO VIA TOLEDO AND PITTSBURGH
S: (*displays Fig. 2*) The terminal at city Scranton is delaying traffic due to localized heavy winds. The terminal at city Baltimore is delaying traffic due to localized heavy winds. An additional 5 hours will be needed to travel through them due to decreased visibility.
U: NOW LET'S TAKE THE TRAIN FROM DETROIT TO WASHINGTON D S_X ANY
No. Let's take the train from Detroit to Washington via Cincinnati
S: (*displays Fig. 3*) Okay.
U: OKAY IT THAT'S OKAY NOW
Okay that that's okay now
S: Pardon me?
U: UP OKAY NOW LET'S TAKE THE TRAIN FROM MONTREAL TO LEXINGTON
Okay. Now let's take the train from Montreal to Lexington.
S: I need help choosing a route from Montreal to Lexington.
U: LET'S GO TO P_M TO TRY
Let's go via Detroit
S: (*displays a route to Detroit but not on to Lexington*) Is this OK?
U: YES NOW LET'S GO TO LEXINGTON
S: Yep.
U: OKAY NOW ADD STATE THE LAST TRAIN AND GO FROM ALBANY TO O_O'S
Okay Now let's take the last train and go from Albany to Milwaukee.

S: No. What city are you trying to have the engine at Montreal arrive at, if I caught your drift. I don't understand where we are sending the engine at Albany to.
U: ME A JET ADD ALBANY NEEDS TO GO TO MILWAUKEE
The engine at Albany needs to go to Milwaukee.
S: Tell me a route to use to get from Albany to Milwaukee, please.
U: I'D GO VIA BUFFALO
S: (*displays a route through Toronto and Buffalo*) *Yes.*
U: GO B_X SYRACUSE AT BUFFALO
Go via Syracuse and Buffalo.
S: (*displays Fig. 4*) *yeah.*
U: THAT'S COULD I CAN
That's great. I' done
S: Hey, its the programming.
U: I_NEED DONE
I'm done

3. BACKGROUND

The TRAINS system builds on a wide range of previous work that has investigated the use of plan-based models for natural language understanding and generation. This section explores some of the ideas that had the greatest influence on the system.

The fundamental assumptions underlying the approach were developed in the speech act tradition in philosophy, which focuses on the communicative actions (or **speech acts**) that are performed when a person speaks. Austin (1957) introduced many of the basic arguments for this viewpoint, and showed how utterances in language can be best analyzed as actions, such as requesting, promising, asserting, warning, congratulating, and so on. About the same time, Grice (1957) suggested the importance of intention recognition in communication, and argued that communication successfully occurs only when the speaker has an intention to communicate and intends the hearer to recognize these intentions as part of the understanding process. Grice (1975) later pointed out that to recognize these intentions, the conversants must assume they are cooperating to the extent that what is said is relevant and possible to interpret in the current context. Searle (1969, 1975) combined the ideas of Austin and Grice and developed an influential theory of speech acts based on the notion of intention recognition. Cohen and Perrault (1979) developed a computational model of speech acts as planning operators. Speech acts had preconditions and effects defined in terms of the intentions and beliefs of the speaker and hearer; for example, an INFORM is a statement uttered to get the hearer to believe that a certain fact is true. A key advantage was that this approach allowed models to connect nonlinguistic goals with linguistic actions. For instance, if Abe wants to open a safe but does not know the combination, and he believes Bea does know the combination, then he might reason (roughly) as follows: to open the safe I need to know the combination. I would know the combination if Bea told me. She could do this because she knows the combination, so I just have to get her to want to do the action, which I can do by asking her to do it. This quick summary avoids many subtleties, but gives a feel for the approach. The plan is summarized in Fig. 5. Note that there is an alternate way to get the safe open; that is, ask Bea to open it. With

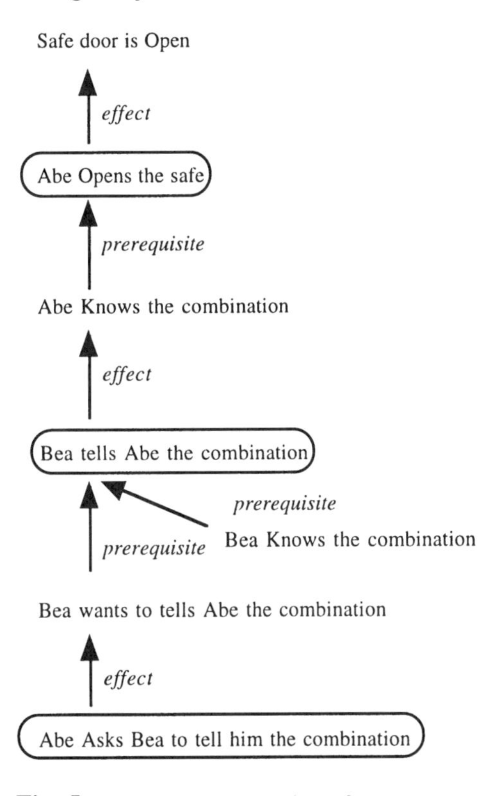

Fig. 5 Planning to open the safe.

this plan, the goal would still be achieved, but Abe would not end up knowing the combination. The plan-based model naturally accounts for such variations.

Cohen and Perrault's model accounts for how speech acts could be planned to accomplish the goals of an agent. Allen and Perrault (1981) then developed a model of language understanding based on intention recognition, for which they developed a formalism for plan inference (or plan recognition). Roughly speaking, this approach reversed the planning process: rather than working from a goal to the actions required to accomplish that goal, it worked from observed actions (the speech acts) and identified plausible goals that might have motivated the actions.

One particular difficulty for speech act interpretation is that speakers frequently speak indirectly. For instance, rather than explicitly asking someone for the time, persons typically ask indirectly by saying "Do you know the time?" or "Do you have a watch?" Perrault and Allen (1980) showed how the plan-based model could account for a wide range of indirect speech acts. Perhaps most important for our current work is that they proposed a model of a conversational agent: the agent observes surface speech acts (computed by the parser from utterances), uses plan recognition to identify underlying intentions, uses an "obstacle detection" process to identify problems in the user's plan, and selects goals to address, which then, leads to planning to achieve these goals, and ultimately leads to speech acts as a response. This model is significantly different from the approach taken in most natural language processing systems. For example, in a typical question-answering system, the response is based on the results of a query computed from the structure of the question. In the conversational agent model, the answer is determined from an analysis of the user's goals. This allows the system much more flexibility to recognize and address misconceptions, to provide additional information that was not explicitly requested, but that would be useful to the questioner, and to maintain the evolving context of the dialogue.

This work on speech acts provided a rich model for interpreting individual utterances, but did not account for the more global structure of dialogue, or discourse in general. Plan-based models have also been used to great advantage in modeling the global structure. This is especially true for the class of dialogues called task-oriented dialogues, in which the participants are communicating to accomplish some well-defined purpose. Although focusing on task-oriented dialogues might seem a limitation, it is worth noting that almost any application one can think of for human–computer dialogue falls into this category. This can be seen in the range of tasks that have been the focus of research, which includes planning routes as in the TRAINS system; selecting furniture for an apartment subject to budgetary and other constraints; planning student course schedules; assembling devices, such as water pumps; finding out information about flights or train schedules; and a wide range of other customer service agents (e.g., bank inquiries, catalogue ordering, automated telephone operators, and automated travel agents).

Grosz (1974, 1977) showed that the flow of topic in a task-oriented dialogue has a systematic relation with the structure of the task. It is not necessarily identical with the task, however, as not all aspects a task need be discussed, and the order of discussion does not necessarily have to follow exactly the order that steps would be executed. She argued that discourse is hierarchically organized, in which the flow of topics can be hierarchically organized in relation to the hierarchy in the task domain. Furthermore, a stack-like structure allows topics to be "suspended" and later "resumed." This organization suggests a significantly different organization for processes such as reference resolution that are often simply based on recency; that is, a definite noun phrase tends to refer to the object most recently mentioned that has the required characteristics. Grosz showed some counterexamples to the recency technique that could be easily explained within the hierarchical model.

Grosz and Sidner (1986) refined these ideas with a theory that separately modeled the **intentional state** and the **attentional state**. This work is described in detail in Chap. 6 of this handbook. The attentional state is a stack-like structure and is critical for defining the context for referential processing and disambiguation. For example, consider an example shown in Fig. 6, which is a simplified version of an actual keyboard dialogue collected in the COCONUT domain (Jordan and Moser, 1997). The sentences typed by A are marked as A.1, A.2, and so on, and those typed by B, with B.8, and so on. Note that in A.4, A suggests a chair. In B.11, B refers to the same chair to say that it seems expensive. But note that there was another chair more recently mentioned in A.6, but this does not seem to be considered. One reason for this is that A.5 through B.10 is a subdialogue that is addressing the interaction, rather than the task. Discussion of the task resumes in B.11, and B.11 and B.12 are interpreted relative to the context created by A.1 through A.4. This interpretation is further reinforced by considering what the discourse purposes are. A.2 through A.4 are clearly proposals for objects to purchase, and B.11 and B.12 are the responses to those proposals. The purpose underlying A.5 through B.10, on the other hand, are

A.1 I do not have a sofa for a better price
A.2 But I do have a floor-lamp, blue ($250)
A.3 I have a green table ($200) and
A.4 *a chair for 75 dollars.*

A.5 Sorry I am taking so much time.
A.6 I lost *a chair.*
A.7 Meghan is finding it.

B.8 Not a problem with the time
B.9 Sorry about the typo
B.10 My brain forgets that my fingers don't function as quick as it does

B.11 The lamp and table sound good
B.12 But *the chair* seems expensive

Fig. 6 A subdialogue.

more to do with maintaining a sense of collaboration and cooperation between the participants.

In the work described so far, plans have been used in two different roles. In speech act modeling and with discourse intentions, plans are used to model the communicative actions. In modeling topic flow, plans are used to model the task that the agents are performing. This suggests that multiple levels of plan reasoning may be required to fully model what is going on in a dialogue. Litman and Allen (1987, 1990) developed such a multilevel model that performed plan recognition simultaneously at two levels to determine an interpretation. One level was the **domain level**, in which the actual actions being discussed are modeled (e.g., in TRAINS, these are actions such as moving trains), and the other was called the **discourse level**, which included communicative acts, such as the speech acts and problem–solving-related acts, such as INTRODUCE–PLAN, IDENTIFY–PARAMETER, and so on. For example, consider a dialogue that starts as follows:

A.1 I want to get a train to Avon.
A.2. Let's use the engine at Bath.

Utterance A.1 would be recognized as an INTRODUCE–GOAL at the discourse level, and would cause the creation of a domain level plan with the goal of getting a train to Avon. Depending on the nature of the domain plan-reasoning algorithm, this might introduce an abstract action to move a train from somewhere to Avon. Utterance A.2 would then be recognized as an IDENTIFY–PARAMETER of the current domain plan. For this interpretation to make sense, the domain reasoner would need to find a way to fit the engine described into the plan. It could do this by instantiating the action of moving the engine at Bath to Avon. If this interpretation is accepted by the system, it would then execute the IDENTIFY–PARAMETER action, which would modify the domain plan. This is summarized in Fig. 7.

One of the key ideas in this work was that one can model discourse level clarifications and corrections by introducing another discourse level plan that expli-

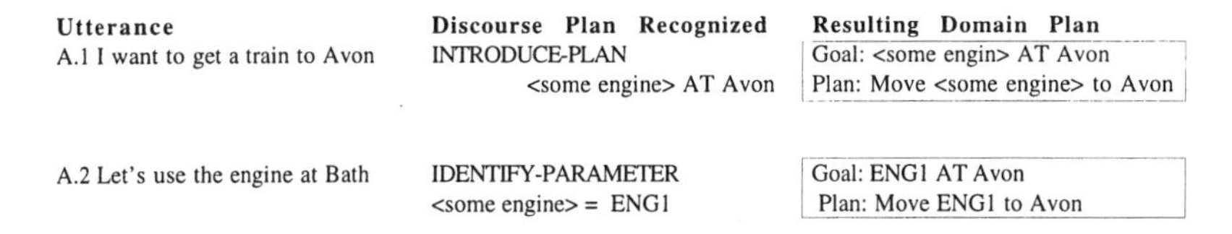

Utterance
A.1 I want to get a train to Avon
A.2 Let's use the engine at Bath
Discourse Plan Recognized
INTRODUCE-PLAN
<some engine> AT Avon
IDENTIFY-PARAMETER
<some engine> = ENG1
Resulting Domain Plan
Goal: <some engin> AT Avon
Plan: Move <some engine> to Avon
Goal: ENG1 AT Avon
Plan: Move ENG1 to Avon

Fig. 7 Multilevel plan interpretation.

citly considers and modifies the previous discourse level plan. Carberry (1990) and others have refined this model further, developing a modified version in which the actions are broken into three levels: the domain, the problem-solving, and the communication levels.

Despite the promise of all this work, when we started this project there still had not been a single robust dialogue system constructed using these principles. It is informative to consider why this was so, and then, how we should try to avoid these problems in the TRAINS system. There are at least four significant problems in using plan-based models in actual systems:

1. The knowledge representation problem: We have no fully adequate representational models of beliefs, intentions, and plans that account for all known problematic cases.
2. The knowledge engineering problem: Defining the information needed to cover the full range of situations and possible utterances in a dialogue for a realistic task is difficult and time-consuming.
3. The computational complexity problem: Known plan recognition and planning algorithms are too inefficient to run in close to real-time.
4. The noisy input problem: Speech recognition still involves a high word error rate on spontaneous speech, causing significant problems in parsing and interpretation.

For the most part, recent work in this area has focused on the knowledge representation problem, and has developed better representational models for intentions and speech acts (e.g., Cohen and Levesque 1990a,b) and for shared plans (Grosz and Sidner, 1990; Grosz and Kraus, 1996; see also Chap. 6 in this handbook).

We have addressed these problems from a different point of view. Rather than working to solve the full problem, we designed a limited domain and then focused only on the mechanisms that were required to produce a robust system in that domain. Thus, we are attacking the problems in a bottom-up, rather than a top-down, fashion. Critical to the success of this approach was the choice of a suitable domain. One too complex would make the task impossible, and one too easy would make the results not useful for subsequent work. Accordingly, we set the following criteria for the domain:

1. The domain should support a range of problems for people to solve, from trivial to challenging.
2. The domain should require as few representational constructs as possible to minimize the amount of explicit knowledge that needs to be encoded.
3. The domain should support a wide range of linguistic constructs to produce a rich set of dialogue behavior.

4. The domain should be intuitive so that a person can work in the domain and use the system with virtually no training.

The route-planning task in TRAINS fits these requirements quite well. It supports a wide range of problems from the trivial, such as finding a route for one train, to complex problems in scheduling many trains where the routes interact with each other. Even with the complex problems, however, the domain requires very few representational primitives to encode it. Basically, in TRAINS we need to represent locations (e.g., cities), connections (e.g., routes), simple objects (e.g., trains), and the location of all these objects over time. Furthermore, there are only two actions: GO from one location to another, and STAY at a location for some time period. Some simple experiments using two people solving problems in this domain verified that it was rich enough to generate a wide variety of linguistic constructions and dialogues, while still remaining considerably simpler than the full breadth of real language use. The domain naturally limits the speakers so that no training is required on how to speak to the system. Finally, the domain is quite intuitive, and people can learn the task after seeing or performing a single session.

The choice of the TRAINS domain goes a long way toward avoiding several of the foregoing problems, because the significant aspects of the entire TRAINS world can be encoded using only a few predicates. These are only the domain-level predicates, and we still need to represent the discourse level actions and mental states of the participants. We handle this problem by making some radical simplifying assumptions, which can be supported experimentally only by observing that these assumptions have not yet caused any problems in the system. The most radical is that we do not model the individual beliefs of the hearer. Rather, we have what can be thought of as a shared set of beliefs that includes all the information that has been conveyed in the dialogue (by language or by the graphic user interface), and then private information that only the system currently knows. Some of this is explicit knowledge (e.g., Avon is congested), whereas other information is discovered by the system's reasoning processes (e.g., there will be a delay because two trains are trying to use the same track at the same time).

As the plan is built, using the system's interpretations of suggestions from the user and the system's own suggestions based on its planning processes, all contributions end up as part of a single shared plan (which in TRAINS is always graphically displayed on the screen). Rather than trying to maintain separate versions of the plan (e.g., what S thinks the plan is, what S thinks the user thinks the plan is, and so on), a single plan is represented and we depend on the user or the system to offer corrections if the current plan is not to their liking. In many ways, this can be thought of as a primitive version of joint action models of discourse proposed by Clark (1996) and Heeman and Hirst (1995).

But having only a shared or joint plan does not eliminate the need for intention recognition. The system still needs to identify the user's intentions underlying an utterance to determine what changes the user is proposing (e.g., is "Go from Avon to Bath" a proposal of a new goal, or a correction of an existing route?). And, on the other side of the coin, we have not eliminated the need for the system to plan its own utterances. Thus, we still have the computational complexity problem: there are no efficient plan recognition or planning algorithms, even in domains as simple as TRAINS. We avoid this problem using a combination of techniques. At the discourse level, we effectively "compile out" the recognition and planning process and drive the interpretation and response planning by a set of decision trees that ask questions that might have been generated from a plan-based discourse model. This decision tree depends heavily on syntactic clues to an initial interpretation, which is then filtered and modified depending on how well the proposed operation fits in at the domain level.

Many decisions at the discourse level are reduced to questions about whether some operation makes sense at the domain level (e.g., whether it makes sense to modify the current route by avoiding Dansville). These questions are answered by a series of checks on constraints and preconditions of actions. For example, the system would reject a proposed modification of a route to avoid Dansville if that route did not currently go through Dansville. This uses a general principle that corrections implicitly define a condition on the current solution as well as identifying a property of the desired solution. For constraints of the form (AVOID Avon), the condition is that the current route goes through Avon. Note that this property does not hold for introducing or extending a route, and one can say "Go from Avon to Bath avoiding Corning" even though no route currently goes through Corning.

The system still eventually needs to do some planning at the domain level to find appropriate routes. This planning is performed by domain-specific reasoners. For instance, rather than planning routes from first principles using a general purpose planning algorithm, the system calls a fast route-finding procedure and then converts its results into a sequence of actions.

In summary, planning search is replaced by specialized-reasoning engines, and plan recognition is handled by decision trees that use constraint checking and plan reasoning (by the specialized-reasoning engines). The other significant factor in solving the complexity problem is simply the faster machines that are available today. Even our fast methods would have been impractical just 5 years ago.

The noisy input problem will be the focus of much of this chapter, as the presence of speech recognition errors greatly complicates all aspects of the system. Not only do we miss words that were said; the speech recognizer will insert erroneous words in their places. We address this problem by correcting recognizer errors that have occurred consistently in the past and using robust interpretation rules tempered by the validation of possible interpretations in context (i.e., the intention recognition process). Robust rules occur at all levels: the parser operates bottom-up so it can skip over uninterpretable input and build "islands of reliability"; semantic template matching is used to combine fragments into possible interpretations; speech acts are combined together at the discourse level to create better interpretations; intention recognition is used to attempt to identify the user's intent even when the content is not fully determined; information is combined across utterances to identify plausible interpretations; and finally, dialogue strategies use a variety of clarification, confirmation, and correction strategies to keep the dialogue running efficiently.

As a final comment on robustness, the strategies we used were developed bottom-up from direct experience with the system, rather than designed in advance. In particular, we built the initial prototype of the system in a 3-month period, and from then on ran sessions with users to identify what aspects of the system were actually causing dialogue failures. This allowed us to focus our activities. There were many things that we initially thought would be problematic, but that turned out to be handled by relatively simple techniques within the TRAINS domain.

4. THE SYSTEM

The TRAINS system is organized as shown in Fig. 8. At the top are the I/O facilities. The speech recognition system is the SPHINX-II system from CMU (Huang et al., 1993) with acoustic models trained from data in the Air Travel Information System (ATIS) domain. We have performed experiments using both the ATIS language model and a model constructed from TRAINS dialogue data. The speech synthesizer is a commercial product: the TRUETALK system from Entropic. The rest of the system was built at Rochester. The object-oriented Display Manager supports a communication language that allows other modules to control the contents of the display, and supports limited mouse-based interaction with the user. The speech recognition output is passed through the error-correcting postprocessor described in Sec. 4.A. The parser, described in Sec. 4.B, accepts input either from the post-processor (for speech) or the keyboard manager (for typed input), and produces a set of speech act interpretations that are passed to the dialogue manager, described in Sec. 4.D. The Display Manager also generates speech acts when the user uses the mouse. The Dialogue Manager (DM) consists of subcomponents-handling reference, speech act interpretation, and speech act planning (the verbal reasoner). It communicates with the Problem Solving Manager (PSM), described in Sec. 4.C, to obtain information about the problem-solving process and state of the plan so far, and with the domain reasoner for queries about the state of the world. When a speech act is planned for output, it is passed to the generator, which constructs a sentence to speak or a set of display commands to update the display.

The rest of this chapter describes the major modules in more detail, including the postprocessor, the parser, the dialogue manager, and the problem-solving man-

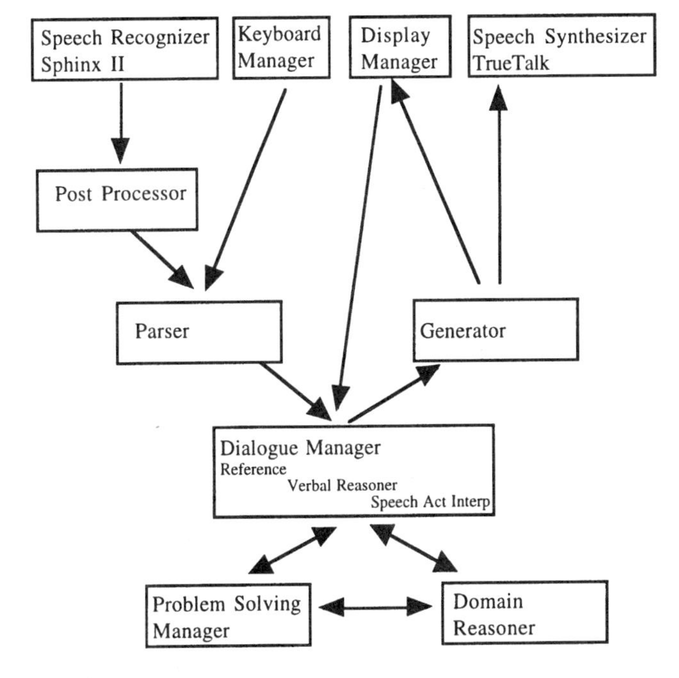

Fig. 8 The system architecture.

ager. The generator will not be discussed further, as it is currently a simple template-based system; other chapters in this handbook provide detailed expositions of issues in natural language generation and their solutions. Our generator uses templates associated with different speech act forms that are instantiated with descriptions of the particular objects involved. The form of these descriptions is defined individually for each class of objects in the domain. While primitive, it did not have a significant effect on the system's performance of the task.

A. Statistical Error Correction

The speech postprocessor (SPEECHPP) corrects speech recognition (SR) errors using a statistical model of the recognizer's past performance in the TRAINS domain. The following are examples of speech recognition errors that occurred in one particular dialogue; Appendix A contains the entire dialogue and the corrections attempted. In each example, the sequence of words tagged REF indicates what was actually said, while those tagged with SR indicate what the speech recognition system proposed, and PP indicates the output of S PEECHPP. Although the corrected transcriptions are not perfect, they are typically a better approximation of the actual utterance. As the first example shows, some recognition errors are simple word-for-word confusions:

SR: GO B_X SYRACUSE AT BUFFALO
PP: GO VIA SYRACUSE VIA BUFFALO
REF: GO VIA SYRACUSE AND BUFFALO

In the next example, a single word was replaced by more than one smaller word:

SR: LET'S GO P_M TO TRY
PP: LET'S GO P_M TO DETROIT
REF: LET'S GO VIA DETROIT

The postprocessor yields fewer errors by compensating for the mismatch between the language used to train the recognizer and the language actually used with the system. To achieve this, we adapted some techniques from statistical machine translation (such as Brown et al., 1990) to model the errors that Sphinx-II makes in our domain. Briefly, the model consists of two parts: a channel model, which accounts for errors made by the SR, and a language model that accounts for the likelihood of a sequence of words being uttered in the first place.

More precisely, given an observed word sequence O from the speech recognizer, S PEECHPP finds the most likely original word sequence S by finding the word sequence S that maximizes the expression $Prob(O|S)*Prob(S)$, where

$Prob(S)$ is the probability that the user would utter sequence S, and
$Prob(O|S)$ is the probability that the SR produces the sequence O when S was actually spoken.

For efficiency, it is necessary to estimate these distributions with relatively simple models by making independence assumptions. For $Prob(S)$, we train a word-bigram "back-off" language model (Katz, 1987) from hand-transcribed dialogues previously collected with the TRAINS system. For $Prob(O|S)$, we build a channel model that incorporates a model of one-for-one as well as one-for-two and two-for-one substitutions. We say that the channel is "fertile" because of the possibility that words multiply in the channel; for more details, see Ringger and Allen (1996). The channel model used in the example dialogue assumed only single word-for-word substitutions.

The channel model is trained by automatically aligning the hand transcriptions with the output of Sphinx-II on the utterances in the (S PEECHPP) training set and by tabulating the confusions that occurred.

We use a Viterbi beam-search to find the best correction *S* that maximizes the expression. This search is the heart of our postprocessing technique, but is widely known, so it is not described here (see Forney, 1973; Lowerre, 1986).

Having a relatively small number of TRAINS dialogues for training, we wanted to investigate how well the data could be employed in models for both the SR and the SPEECHPP. We ran several experiments to weigh our options. For a baseline, we built a class-based back-off language model for Sphinx-II using only transcriptions of ATIS spoken utterances. Using this model the performance of Sphinx-II alone was 58.7% on TRAINS-domain utterances. This is not necessarily an indictment of Sphinx-II, but simply reflects the mismatched language models.

From this baseline, we used varying amounts of training data exclusively for building models for the SPEECHPP; this scenario would be most relevant if the speech recognizer were a black-box and we did not know how to train its model(s). Second, we used varying amounts of the training data exclusively for augmenting the ATIS data to build a language model for Sphinx-II. Third, we combined the methods, using the training data both to extend the language model in Sphinx-II and to then train SPEECHPP on the newly trained system.

The results of the first experiment are shown by the second curve from the bottom of Fig. 9, which indicates the performance of the S PEECHPP over the baseline

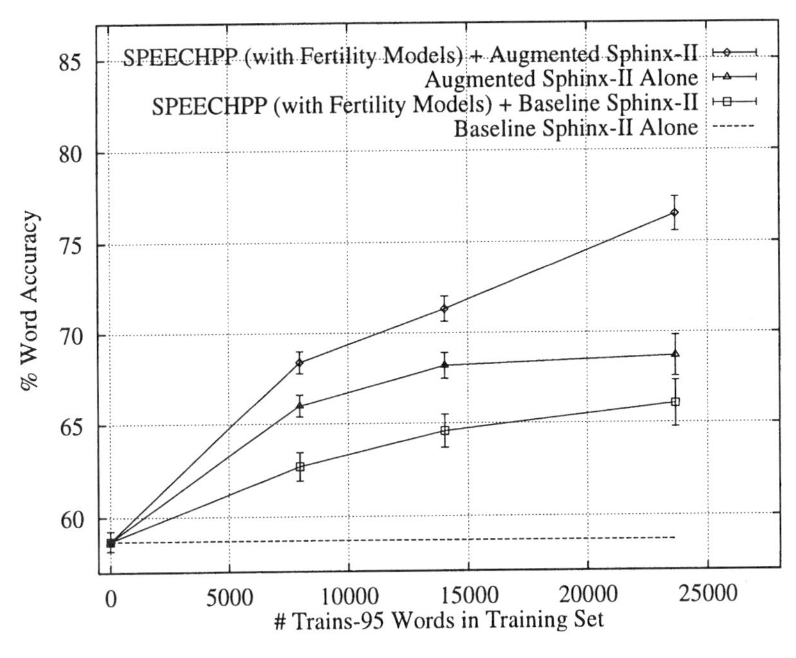

Fig. 9 Postprocessing evaluation.

Sphinx-II. The first point comes from using approximately 25% of the available training data in the S PEECHPP models. The second and third points come from using approximately 50 and 75%, respectively, of the available training data. The curve clearly indicates that the S PEECHPP does a reasonable job of boosting our word recognition rates over baseline Sphinx-II. We did not use all of our available data, because the remainder was used to determine the test results by repeated leave-one-out cross-validation.

Similarly, the results of the second experiment are shown by the middle curve. The points reflect the performance of Sphinx-II (without S PEECHPP) when using 25, 50, and 75% of the available training data in its language model. These results indicate that equivalent amounts of training data can be used with greater effect in the language model of the SR than in the postprocessor.

Finally, the outcome of the third experiment is reflected in the uppermost curve. Each point indicates the performance of the SPEECHPP using a set of models trained on the behavior of Sphinx-II for the corresponding point from the second experiment. The results from this experiment indicate that even if the language model of the SR can be modified, then the postprocessor trained on the same new data can still significantly improve word recognition accuracy. Hence, whether the SR's models are tunable or not, the postprocessor is in neither case redundant.

B. Robust Parsing

Given that errors are inevitable, robust parsing techniques are essential. We use a pure bottom-up parser (using the algorithm described in Allen, 1995; see also Chap. 4 in this handbook) to identify the possible constituents at any point in the utterance based on syntactic and semantic restrictions. Every constituent in each grammar rule specifies both a syntactic category and a semantic category, plus other features to encode cooccurrence restrictions, as found in many grammars. The semantic features encode selectional restrictions, most of which are domain-independent. For example, in effect there is no general rule for PP attachment in the grammar; rather, there are rules for temporal adverbial modification (e.g., *at eight o'clock*), locational modification (e.g., *in Chicago*), and so on.

The end result of parsing is a sequence of speech acts, rather than a syntactic structure. Viewing the output as a sequence of speech acts has a significant influence on the form and style of the grammar. It forces an emphasis on encoding semantic and pragmatic features in the grammar. There are, for instance, numerous rules that encode specific conventional speech acts (e.g., *That's good* is a CONFIRM, *Okay* is a CONFIRM/ACKNOWLEDGE, *Let's go to Chicago* is a SUGGEST, and so on). Simply classifying such utterances as sentences would miss the point. Thus, the parser computes a set of plausible speech act interpretations based on the surface form, similar to the model described in Hinkelman and Allen (1989).

We use a hierarchy of speech acts that encodes different levels of vagueness, with the root being a speech act that indicates content with an unspecified illocutionary force. This allows us to always have an illocutionary force identified, which can be refined as more of the utterance is processed. The final interpretation of an utterance is the sequence of speech acts that provides the "minimal covering" of the input (i.e., the shortest sequence that accounts for the input). If an utterance was completely uninterpretable, the parser would still produce an output: a speech act with no specified illocutionary force or content. The speech act hierarchy in TRAINS is shown in Fig. 10. Figure 11 gives the logical form computed by the parser for the sentence "Okay now go from Avon to Bath please." It includes a list of all objects mentioned in the utterances, the paths, as well as other information, such as politeness cues, a reliability measure, and some syntactic information. A fairly traditional logical form could be obtained by starting at the value of the SEMANTICS slot and chasing down the pointers through the objects in paths. The advantage of this distributed representation is that we obtain similar representations from fully parsed utterances and from utterances that consist of only fragments. This uniformity is crucial for subsequent robust processing.

Consider an example of how an utterance from the sample dialogue is parsed, showing how speech recognition errors are handled. The input, after postprocessing, is "Okay now I take the last train in go from Albany to is." The best sequence of speech acts to cover this input consists of three acts:

a CONFIRM ("Okay")

a TELL, with content to take the last train ("Now I take the last train")

a REQUEST to go from Albany ("go from Albany")

Note that the "to is" at the end of the utterance is simply ignored as it is uninterpretable. Although not present in the output, the presence of unaccounted words will lower the parser's confidence score that it assigns to the interpretation.

The actual utterance was "Okay now let's take the last train and go from Albany to Milwaukee." Note that while the parser is not able to reconstruct the complete intentions of the user, it has extracted enough to continue the dialogue in a

Speech Act Hierarchy	Use	Example
SPEECH-ACT	uninterpretable utterances	"Us the go Avon"
TELL	simple declaratives and fragments	"Avon is congested"
ID-GOAL	utterance explicitly introduces a goal	"I want to take a train to Bath"
EVALUATION	declarative that evaluates an option	"That's good"
REQUEST	generic directive act for imperatives	"Go to Corning"
SUGGEST-ACTION	suggestions of actions	"Let's go to Corning"
YN-QUESTION	yes-no questions	"Is the train at Avon?"
WH-QUESTION	wh question	"Where are the trains?"
AGREEMENT		
ACCEPT/CONFIRM	accepting a proposal	"OK", "yes"
REJECT	rejecting a proposal	"No, that won't work"
CONVENTIONAL		
CLOSE	conventional closing	"good-bye"
GREET	conventional greeting	"Hi"
NOLO-COMPRENDEZ	signal non-comprehension	"huh?"
APOLOGIZE	apologizing	"I'm sorry"
NOLO-PROBLEMO	responding to an apology	"no problem"
ACKNOWLEDGE	acknowledgement of understanding	"OK", "uh-huh"

Fig. 10 The speech act hierarchy in TRAINS.

```
(CONFIRM
        :SEMANTICS OKAY
        :RELIABILITY 100
        :MODE :DISPLAY
        :INPUT (OKAY ))
(REQUEST
        :OBJECTS       ((DESCRIPTION (STATUS NAME) (VAR V2242) (CLASS CITY)
                               (LEX AVON) (SORT INDIVIDUAL))
                        (DESCRIPTION (STATUS NAME) (VAR V2242) (CLASS CITY)
                               (LEX BATH) (SORT INDIVIDUAL)))
        :PATHS         ((PATH (VAR V2238) (CONSTRAINT (AND (FROM V2238 V2242)
                                                           (TO V2238 V2253)))))
        :SEMANTICS     (PROP (VAR V2231) (CLASS GO-BY-PATH)
                               (CONSTRAINT (AND (LSUBJ V2231 *YOU*) (LOBJ V2231 V2238)))
        :SOCIAL-CONTEXT  (POLITENESS PLEASE)
        :RELIABILITY 100
        :MODE TEXT
        :SYNTAX        ((SUBJECT . *YOU*) (OBJECT))
        :INPUT  (GO FROM AVON TO BATH PLEASE))
```

Fig. 11 The logical form for "Okay Go from Avon to Bath please".

reasonable fashion by invoking a clarification subdialogue. Specifically, it has correctly recognized the confirmation of the previous exchange (act 1), and recognized a request to move a train from Albany (act 3). Act 2 is an incorrect analysis, and results in the system generating a clarification question that the user ends up ignoring. Thus, as far as furthering the dialogue is concerned, the system has done reasonably well.

C. The Problem Solver and Domain Reasoner

The next phase of processing of an utterance is the interpretation by the dialogue manager. Before we consider this, however, we first examine the system's problem-solving capabilities, as the dialogue manager depends heavily on the problem-solving manager when interpreting utterances. The problem solver is called to evaluate all speech acts that appear to constrain, extend, or change the current plan and then to perform the requested modification to the plan once an interpretation has been determined. It maintains a hierarchical representation of the plan developed so far. This is used to help focus in the interpretation process, as well as to organize the information needed to invoke the domain reasoner to reason about the plans under construction (in this case, routes).

1. *Goals, Solutions, and Plans*

In traditional-planning systems (e.g., Fikes and Nilsson, 1971; Weld, 1994), goals are often simply the unsatisfied preconditions or undeveloped actions in the hierarchy. Once a plan is fully developed, it is hard to distinguish between actions that are present because they are the goals and actions that were introduced because they seemed a reasonable solution to the goal. Losing this information makes many operations needed to support interactive planning very difficult. Consider plan revision, a critical part of interactive planning. If we cannot tell what the motivating goals were, it is difficult to determine what aspects of the plan should be kept and what can be changed. As a very simple example, consider two situations with identical solutions in the TRAINS domain: in the first case, the goal is to go from Toledo to Albany, and the solution is to take the route from Toledo, through Columbus, Pittsburgh, and Scranton. In the second case, the goal is to go from Toledo to Albany avoiding Buffalo, with the same route. Say the user wants to change the route in each case, and indicates this by a simple rejection. In the first case, the planner might suggest the route through Cleveland, Buffalo, and Syracuse, because it seems shorter than the original. In the second plan, though, this would be an inappropriate suggestion, as it does not meet the goal, which requires that the route avoid Buffalo.

In TRAINS, the goal hierarchy captures the goals and subgoals that have been explicitly developed through the interaction with the user. Particular solutions to the goals, on the other hand, are not in the hierarchy explicitly, although they are associated with the goals that they address. The goals provide an organization of the solution that is useful for many purposes—for revising parts of the plan, for summarizing the plan, and for identifying the focus of the plan for a particular interaction. For example, consider the problem solving state shown in Fig. 12.

There are two subplans, which in this particular case are independent of each other as they involve different trains, cities, and tracks. The planned actions for ENG1 satisfy the stated goal of going from Chicago to Albany avoiding Buffalo, whereas the planned actions for ENG2 satisfy the goal of going from Boston to Atlanta. If we needed an alternative solution to the first subgoal, the new solution should still avoid Buffalo. With the second goal, the only part that is crucial to the goal is the origin and destination.

The goal hierarchy plays a crucial role in managing the focus of attention during the interaction. In general, a user will move up and down the goal hierarchy in fairly predictable ways as they work on one problem and then move to the next. The routes are constructed by a specialized domain reasoner, whose results are then converted into a solution by the problem-solving manager.

2. *The Problem-Solving Operations*

From studying human–human interactions in joint problem solving (Heeman and Allen, 1995), and from our experience in building the TRAINS system, we have identified an initial set of problem-solving interactions that characterize the operations that can be performed by the user through speech acts. These operations define an abstract domain-independent model of a problem-solving system that can sup-

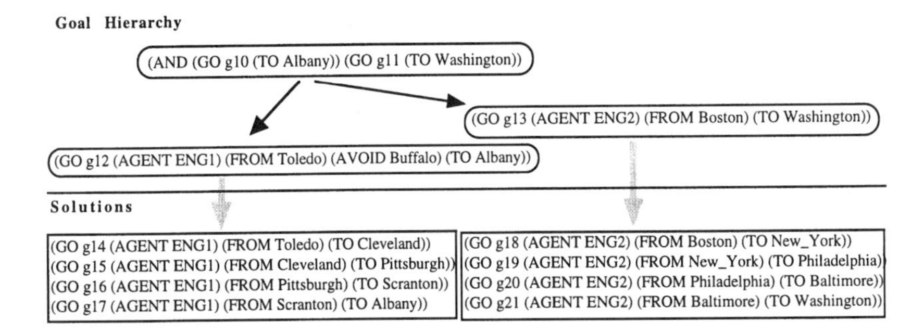

Fig. 12 A simple problem-solving state.

port mixed initiative planning. Each operation is associated with a *focus node*, which indicates the subgoal currently under consideration as well as its associated solution, if there is one. Sometimes an operation changes the goal, whereas other times it changes only the solution, leaving the goals unchanged. The operations supported by the problem-solving manager in TRAINS are specified in Fig. 13.

The TRAINS problem solver implements the strategy of automatically looking for a solution whenever a new goal is introduced or modified (see Ferguson et al., 1996). Thus, to the user it is not always apparent whether an interaction changes the goals or just changes the solution. Recognizing which of these cases is appropriate, however, is crucial for the system, especially in determining how to interpret future interactions and revisions.

Figure 14 identifies the correct problem-solving action and the problem-solving action intended for each of the user's utterances in the sample dialogue. Note that this shows what the user actually said and what operation was intended, rather than what the system actually did. In several cases, speech recognition errors prevented the system from identifying the correct intention. But each utterance is considered given the prior context in the dialogue, including the system responses based on errors up to that point.

The problem-solving manager provides a wide range of reasoning capabilities that both allow the dialogue manager to determine what speech acts make sense at any particular time int he dialogue, as well as providing the interface to the domain reasoners that then actually do the work required. We now turn to the dialogue manager, which determines the overall system's behavior.

D. Robust Discourse Processing

The dialogue manager is responsible for interpreting the speech acts in context, invoking the problem-solving actions, formulating responses, and maintaining the system's idea of the state of the discourse. It maintains a discourse stack similar to the attentional state of Grosz and Sidner (1986). Each element of the discourse stack captures a context for interpreting utterances that consists of

The domain or discourse goal motivating the segment
The focus node in the problem-solving hierarchy
The object focus and history list for the segment

NEW-SUBPLAN	Introduces a new goal (or subgoal) to be accomplished
REFINE	Refines a goal specification into a more specific goal
EXTEND	Extends a plan by adding an additional action or goal
MODIFY	Modifies an existing plan, removing some previous part of the plan
DO-WHAT-YOU-CAN	Attempts to interpret an action as a REFINE, EXTEND, or MODIFY operation
CANCEL	Removes a goal, subgoal, or object specified in the plan
CONFIRM	Verifies the structure and content of an interpretation
REJECT-SOLUTION	Rejects a current solution for a goal (and the system produces another)
SELECT-SOLUTION	Selects a specific solution out of a range of options
NEW-SCENARIO	Defines a new set of background assumptions for a new planning task
DELETE-PLAN	Deletes a specific plan structure from the plan hierarchy
UNDO	Backs up to a previous problem solving state

Fig. 13 The problem-solving operations in TRAINS.

Utterance as Actually Said	Intended Problem Solving Operation
Okay let's send a train from Detroit to Washington	NEW-SUBPLAN with goal <Go from Detroit to Washington>
Let's go via Toledo and Pittsburgh	REFINE producing goal <GO from Detroit to Washington via Toledo and Pittsburgh>
No Let's take the train from Detroit to Washington via Cincinnati	REJECT, MODIFY to produce goal <GO from Detroit to Washington via Cincinnati Avoiding Toledo and Pittsburgh>
Okay That's Okay now	CONFIRM goal <GO from Detroit to Washington via Cincinnati Avoiding Toledo and Pittsburgh>
Okay Now let's take the train from Montreal to Lexington	NEW-SUBPLAN with goal <GO from Montreal to Lexington>
Let's go via Detroit	REFINE producing goal <GO from Montreal to Lexington via Detroit>
Yes Now let's go to Lexington	EXTEND goal <GO from Montreal to Detroit> to <GO from Montreal to Lexington via Detroit>
Okay Now let's take the last train and go from Albany to Milwaukee	NEW-SUBPLAN with goal <GO from Albany to Milwaukee>

Fig. 14 The intended problem-solving operations behind each utterance.

Information on the status of problem-solving activity (e.g., has the goal been achieved yet or not).

Figure 15 shows a simple discourse stack consisting of two contexts, each pointing to a node in the problem-solving hierarchy. Such a situation could arise if the participants were discussing how to get from Atlanta to Toronto, and then one suggested that they work on getting from Atlanta to Pittsburgh first. The subgoal now becomes the domain-level motivation in the dialogue, and the original goal remains on the stack indicating that it may be resumed at a later stage. The system always attempts to interpret an utterance relative to the context that is at the top of the discourse stack. If no good interpretation is possible, it tries contexts lower in the stack, popping off the higher contexts. If no context produces a good interpretation, it calls the problem-solving manager to allow it to attempt to construct a relevant context based on the goal hierarchy. Note that the problem-solving hierarchy does not correspond to the intentional state in Grosz and Sidner's model because the hierarchy captures the domain-level goals, not the discourse segment purposes. Grosz and Sidner's intentional states most closely correspond to the problem-solving

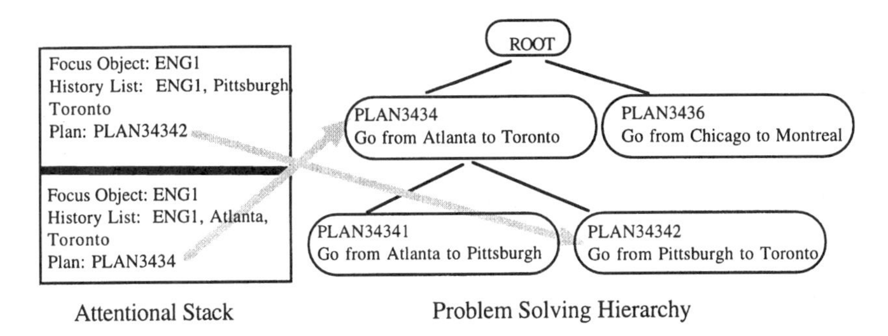

Fig. 15 The attentional stack and its relation to the problem-solving state.

acts that were discussed in the previous section. These acts are not explicitly represented in the discourse state, although the discourse state information is critical to allow the problem solver to identify appropriate information.

A fundamental principle in the design of TRAINS was a decision that, when faced with ambiguity, it is better to choose a specific interpretation and run the risk of making a mistake than to generate a clarification subdialogue. This strategy is used at several different levels. At the domain level, when a goal is specified, there are typically many different possible solutions. Rather than making the user specify enough information to uniquely select a solution, the system chooses one and presents it to the user for comment (Ferguson et al., 1996). At the discourse level, the system is often faced with uncertainty over the intended operation, often because of recognition errors. In these cases, it often selects one of these interpretations and continues the dialogue, rather than entering a clarification subdialogue. The success of this strategy depends on the system's ability to recognize and interpret subsequent corrections if they arise. Significant effort was made in the system to detect and handle a wide range of corrections, both in the grammar, the discourse processing, and the domain reasoning. In later systems, we plan to specifically evaluate whether this decision was the correct one.

The verbal reasoner is organized as a set of prioritized rules that match patterns in the input speech acts and the discourse state. These rules allow robust processing in the face of partial or ill-formed input as they match at varying levels of specificity, including rules that interpret fragments that have no identified illocutionary force. For instance, one rule would allow a fragment such as "to Avon" to be interpreted as a suggestion to extend a route, or an identification of a new goal. The prioritized rules are used, in turn, until an acceptable result is obtained. The discourse interpretation rules may reinterpret the parser's assignment of illocutionary force if it has additional information to draw on.

As an example of the discourse processing, consider how the system handles the user's first utterance in the dialogue, "Okay let's send contain from Detroit to Washington." From the parser we get three acts:

A CONFIRM/ACKNOWLEDGE (*Okay*)

A TELL involving some uninterpretable words (*Let's send contain*)

A TELL act that mentions a route (*from Detroit to Washington*)

The dialogue manager sets up its initial conversational state and passes the act to reference for identification of particular objects, and then hands the acts to the verbal reasoner. Because there is nothing on the discourse stack, the initial confirm has no effect. (Had there been something on the stack, e.g., if the system had just asked a question, then the initial act might have been taken as an answer to the question, or if the system had just suggested a route, it would be taken as a confirmation). The following empty TELL is uninterpretable and, hence, ignored. Although it is possible to claim the "send" could be used to indicate the illocutionary force of the following fragment, and that a "container" might even be involved, the fact that the parser separated out the speech act indicates there may have been other fragments lost. The last speech act could be a suggestion of a new goal to move from Detroit to Washington. After checking that there is an engine at Detroit, this interpretation is accepted. In the version of the system used for this dialogue, the planner was limited to finding routes involving fewer than four hops, thus it was unable to generate a path between these cities. It returns two items:

1. An identification of the speech act as a suggestion of a goal to take a train from Detroit to Washington
2. A signal that it could not find a solution to satisfy the goal.

The discourse context is updated and the verbal reasoner generates a response to clarify the route desired, which is realized in the system's response: "What route would you like to get from Detroit to Washington?"

As another example of robust processing, consider an interaction later in the dialogue in which the user's response "no" is misheard as "now". "Now let's take the train from Detroit to Washington do S_X Albany" (instead of "No let's take the train from Detroit to Washington via Cincinnati"). Since no explicit rejection is identified because of the recognition error, this utterance looks like a confirm and continuation on with the plan. Thus, the verbal reasoner calls the problem solver to extend the path with the currently focused engine (say *engine1*) from Detroit to Washington.

The problem solver realizes that *engine1* is not currently in Detroit, so this cannot be a route extension. In addition, there is no other engine at Detroit, so this is not plausible as a focus shift to a different engine. Since *engine1* originated in Detroit, the verbal reasoner then decides to reinterpret the utterance as a correction. Because the utterance adds no new constraints, but there are the cities that were just mentioned as having delays, it presumes the user is attempting to avoid them, and invokes the domain reasoner to plan a new route avoiding the congested cities. The new path is returned and presented to the user.

Although the response does not address the user's intention to go through Cincinnati owing to the speech recognition errors, it is a reasonable response to the problem the user is trying to solve. In fact, the user decides to accept the proposed route and forget about going through Cincinnati. In other cases, the user might persevere and continue with another correction, such as *No, through Cincinnati*. Robustness arises in the example because the system uses its knowledge of the domain to produce a reasonable response. The primary goal of the discourse system, namely to contribute to the overall goal of helping the user solve a problem with a minimum number of interactions, is best served by the "strong commitment" model; typically it is easier to correct a poor plan, than having to keep trying to specify a perfect one, particularly in the face of recognition problems.

5. EVALUATING THE SYSTEM

Although examples can be illuminating, they do not really address the issue of how well the system works overall. That can be demonstrated only by controlled experimentation. To see how this can be done, consider an evaluation that we performed on TRAINS-95, an earlier version of the system described here. To explore how well the system robustly handles spoken dialogue, we designed an experiment to contrast speech input with keyboard input. The experiment evaluated the effect of input medium on the ability to perform the task and explored user input medium preferences. Task performance was evaluated in terms of two metrics: the amount of time taken to arrive at a solution, and the quality of the solution. Solution quality for our domain is determined by the amount of time needed to travel the planned routes.

Sixteen subjects for the experiment were recruited from undergraduate computer science courses. None of the subjects had ever used the system before. The procedure was as follows:

1. The subject viewed an online tutorial lasting 2.4 min that explained the mechanics of using the system, but did not explain the capabilities of the system nor train the user on how to interact with it.
2. The subject was then allowed a few minutes to practice speech in order to allow them to adapt to the speech recognizer.
3. All subjects were given identical sets of five tasks to perform, in the same order. Half of the subjects were asked to use speech first, keyboard second, speech third, and keyboard fourth. The other half used keyboard first and then alternated. All subjects were given a choice of whether to use speech or keyboard input to accomplish the final task.
4. After performing the final task, the subject completed a questionnaire.

The tasks were of similar complexity to the task solved in the example dialogue. Each involved three trains, and in each situation there was at least one delay affecting what looks like a reasonable route. An analysis of the experiment results shows that the plans generated when speech input was used are of similar quality to those generated when keyboard input was used. However, the time needed to develop plans was significantly lower when speech input was used. In fact, overall, problems were solved using speech in 68% of the time needed to solve them using the keyboard. Figure 16 shows the task completion time results, and Fig. 17 gives the solution quality results, each broken out by task.

Of the 16 subjects, 12 selected speech as the input mode for the final task and 4 selected keyboard input. Three of the 4 selecting keyboard input had actually experienced better or similar performance using keyboard input during the first four tasks. The fourth subject indicated on his questionnaire that he believed he could solve the problem more quickly using the keyboard; however, that subject had solved the two tasks using speech input 19% faster than the two tasks he solved using keyboard input.

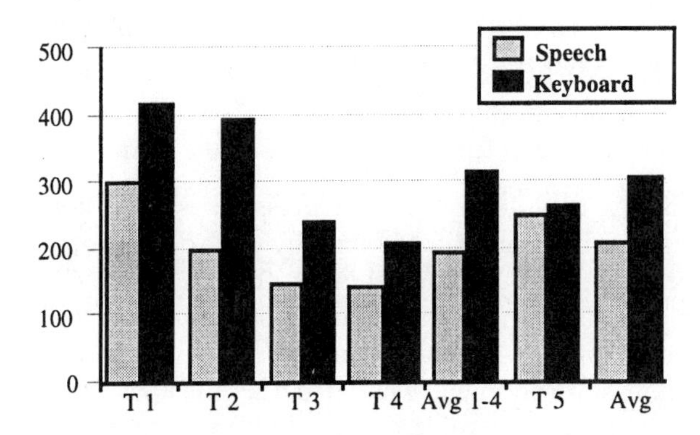

Fig. 16 Time to completion by task.

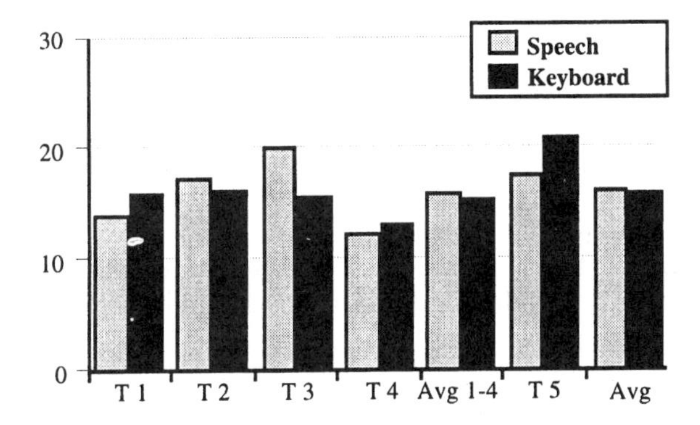

Fig. 17 Length of solution by task.

Of the 80 tasks attempted, there were 7 in which the stated goals were not met. In each unsuccessful attempt, the subject was using speech input. There was no particular task that was troublesome and no particular subject that had difficulty. Seven different subjects had a task for which the goals were not met, and each of the 5 tasks was left unaccomplished at least once.

A review of the transcripts for the unsuccessful attempts revealed that in three of the seven failure cases, the subject misinterpreted the system's actions, and ended the dialogue believing the goals were met. Each of the other four unsuccessful attempts resulted from a common sequence of events. After the system proposed an inefficient route, word recognition errors caused the system to misinterpret rejection of the proposed route as acceptance. The subsequent subdialogues intended to improve the route were interpreted to be extensions to the route, causing the route to "overshoot" the intended destination. This analysis led us to redesign the problem-solving component of the system, to the version described in this paper, so that the system could better identify when the user returned to a previously completed route to modify it.

6. DISCUSSION

We know of no previous dialogue system that can be successfully used by untrained users to perform concrete tasks as complex as those dealt with in the TRAINS system. Several speech systems, such as PEGASUS (Zue et al., 1994), WHEELS (Teng et al., 1996), and RAILTEL (Bennacef et al., 1996), support some dialogue-style interactions, but they do not involve explicit reasoning about the domain to perform their tasks (i.e., the task being performed can be hard-coded in advance and never changes during the dialogue). The AGS demonstrator (Sadek et al., 1996) is notable in that it uses a general model of a dialogue agent to drive the dialogue, but similar to the other foregoing systems does not require any explicit domain reasoning to perform the task. One system that does perform explicit domain reasoning is the Duke system (Smith et al., 1994, 1995), which has a sophisticated domain reasoner and uses an integrated problem-solving component to drive the dialogue. However, the system uses a highly restricted subset of English and requires significant training to use. Because of the scalability of the domain, TRAINS provides a unique test bed for evaluating the research in spoken dialogue systems.

Our research strategy was bottom-up with inspiration from the "top-down" theories in previous work. We have attempted to build a fully functional system in the simplest domain possible, and now plan to increase the complexity of the domain and focus on the problems that most significantly degrade overall performance. This leaves us open to the criticism that we are not using the most sophisticated models available. In an extreme case, consider our generation strategy. Template-based generation is not new and is clearly inadequate for more complex generation tasks. In fact, when starting the project we thought generation would be a major problem. However, the problems we expected have not arisen. Although we could clearly improve the output of the system even in this small domain, the current generator does not appear to drag the system's performance down. We tried the simplest approaches first and only then generalized those algorithms for which inadequacies clearly degrades the performance of the system.

Likewise, we view the evaluation as only a very preliminary first step. Although our evaluation might appear that we are addressing an HCI issue of whether speech or keyboard is a more effective interface, this comparison was not our goal. Rather, we used the modality switch as a way of manipulating the error rate and the degree of spontaneous speech. Although keyboard performance is not perfect because of typing errors (we had a 5% word error rate on keyboard), it is considerably less error-prone than speech. All we conclude from this experiment is that our robust-processing techniques are sufficiently good that speech is a viable interface in such tasks, even with high word error rates. In fact, it appears to be more efficient than the keyboard equivalent. Furthermore, subjects preferred the speech interface when given the choice (cf. Rudnicky, 1993).

Despite the limitations of the current evaluation, we are encouraged by the first step. It seems obvious to us that progress in dialogue systems is intimately tied to finding suitable evaluation measures. And task-based evaluation seems one of the most promising candidates, for it measures the effect of a proposed technique on the effectiveness of the interaction, rather than an abstract accuracy figure that may or may not have any effect on the overall functionality and effectiveness of the system.

Another area for which we are open to criticism is that we used algorithms specific to the domain to produce effective intention recognition, disambiguation, and domain planning. Thus, the success of the system may be a result of the domain and say little about the plan-based approach to dialogue. For the current system, it is hard to defend ourselves against this criticism. However, this is a first step in what we see as a long, ongoing process. To look at it another way: if we were not able to successfully build a system by employing whatever we could, then there is little hope for finding more general effective solutions. More generally, however, we believe that domain-specific knowledge will always be necessary to efficiently solve realistically complex problems. Our work on the problem solver in TRAINS is directed toward finding principled ways of incorporating such knowledge.

Even though there are clearly many places in which our system requires further work, it does set a new standard for spoken dialogue systems. More importantly, it will allow us to address a larger number of research issues in the future in a much more systematic way, supported by empirical evaluation.

ACKNOWLEDGMENTS

This work was supported in part by ONR/ARPA grants N0004-92-J-1512 and N00014-95-1-1088, and NSF grant IRI-9503312. Many thanks to Alex Rudnicky, Ronald Rosenfeld, and Sunil Issar at CMU for providing the Sphinx-II system and related tools.

APPENDIX A: THE COMPLETE DIALOGUE SHOWING POSTPROCESSING

Here is a complete transcript of postcorrection repairs for the sample dialogue.

REF: OKAY LET'S TAKE THE TRAIN FROM DETROIT TO WASHINGTON
SR: OKAY LET'S SEE CONTAIN FROM DETROIT TO WASHINGTON
PP: OKAY LET'S SEND CONTAIN FROM DETROIT TO WASHINGTON
REF: LET'S GO VIA TOLEDO AND PITTSBURGH
SR: LET'S GO VIA TOLEDO AND PITTSBURGH
PP: LET'S GO VIA TOLEDO AND PITTSBURGH
REF: NO LET'S TAKE THE TRAIN FROM DETROIT TO WASHINGTON VIA CINCINNATI
SR: NOW LET'S TAKE THE TRAIN FROM DETROIT TO WASHINGTON D S_X ANY
PP: NOW LET'S TAKE THE TRAIN FROM DETROIT TO WASHINGTON DO S_X ALBANY)
REF: OKAY THAT—THAT'S OKAY NOW
SR: OKAY IT THAT'S OKAY NOW
PP: OKAY IT THAT'S OKAY NOW
REF: OKAY NOW LET'S TAKE THE TRAIN FROM MONTREAL TO LEXINGTON
SR: UP OKAY NOW LET'S TAKE THE TRAIN FROM MONTREAL TO LEXINGTON
PP: UP OKAY NOW LET'S TAKE THE TRAIN FROM MONTREAL TO LEXINGTON
REF: LET'S GO VIA DETROIT
SR: LET'S GO P_M TO TRY
PP: LET'S GO P_M TO DETROIT
REF: YES NOW LET'S GO TO LEXINGTON
SR: YES NOW LET'S GO TO LEXINGTON
PP: YES NOW LET'S GO TO LEXINGTON
REF: OKAY NOW LET'S TAKE THE LAST TRAIN AND GO FROM ALBANY TO MILWAUKEE
SR: OKAY NOW ADD STATE THE LAST TRAIN AND GO FROM ALBANY TO O_O'S
PP: OKAY NOW I TAKE THE LAST TRAIN IN GO FROM ALBANY TO IS
REF: THE ENGINE AT ALBANY NEEDS TO GO TO MILWAUKEE
SR: ME A JET ADD ALBANY NEEDS TO GO TO MILWAUKEE
PP: ME THE JET AT ALBANY INSTEAD TO GO TO MILWAUKEE

REF: UH GO VIA BUFFALO
SR: I'D GO VIA BUFFALO
PP: UH GO VIA BUFFALO
REF: GO VIA SYRACUSE AND BUFFALO
SR: GO B_X SYRACUSE AT BUFFALO
PP: GO VIA SYRACUSE VIA BUFFALO
REF: THAT'S GOOD I'M DONE
SR: THAT'S COULD I CAN
PP: THAT'S GREAT UH CAN
REF: I'M DONE
SR: I_NEED DONE
PP: I'M DONE

APPENDIX B: A SAMPLE KEYBOARD DIALOGUE

U: Let's take the train in Charlotte to Lexington
S: Yeah
U: Can we go through Atlanta instead?
S: Yes
U: much better
S: Oh?
U: Let's move from Toronto to Washington
S: What route would you like to get from Toronto to Washington
U: through Montreal.
S: Yes
Is this ok?
U: good, now can we go down through Albany
S: Huh?
U: from Montreal, let's take the train south through Burlington and Albany.
S: OK
Is this ok?
U: Yes, now let's move the train in Boston to Washington.
S: OK
City New York is congested due to unusually heavy traffic.
An additional 5 hours will be needed to travel through there.
U: let's go around New York instead.
S: OK
U: How about going through Albany.
S: Yep
U: I'm sorry, can we move the train in Boston to Philadelphia instead.
S: It would take a far more powerful and expensive computer than me to figure the answer to that one out.
U: I'm sorry, can we move the train from Boston to Philadelphia.
S: OK
S: But, Your routes cross at City Baltimore.
Trains will take an additional 5 hours to move through the crossed cities.
U: Let's redo the route for Boston.
S: What was that?

U: Let's clear Boston's current route.
S: Yeah
U: Let's clear Boston's current route.
S: Yes
U: Let's move from Boston to Philadelphia
S: OK
U: I think we are done.

REFERENCES

Allen JF. 1995. Natural Language Understanding. 2nd ed. Redwood City, CA: Benjamin-Cummings.

Allen JF, G Ferguson, B Miller, E Ringger 1995. Spoken dialogue and interactive planning. Proceedings of the ARPA SLST Workshop, Morgan Kaufmann. CA: San Meteo.

Allen JF, CR Perrault 1980. Analyzing intention in utterances. Artif Intell 15:143–178.

Austin JL. 1962. How to Do Things with Words. New York: Oxford University Press.

Bennacef S, L Devillers, S Rost, L Lamel 1996. Dialog in the RAILTEL telephone-based system. Proceedings of the International Symposium on Dialogue, ISSD 6, Philadelphia.

Brown, PF, J Cocke, SA Della Pietra, VJ Della Pietra, F Jelinek, JD Lafferty, RL Mercer, PS Roossin 1990. A statistical approach to machine translation. Comput Linguist 16(2):79–85.

Carberry S. 1990. Plan Recognition in Natural Language Dialogue. Cambridge, MA: MIT Press.

Clark, H. 1996. Using Language. London: Cambridge University Press.

Cohen PR, HJ Levesque 1990a. Intention is choice with commitment. Artif Intell 42:3.

Cohen PR, HJ Levesque 1990b. Rational interaction as the basis for communication. In: Cohen et al. eds, Intentions in Communications, Cambridge MA: MIT Press, 1990, pp 221–256.

Cohen PR, J Morgan, M Pollack 1990. Intentions in Communication. Cambridge, MA: MIT Press.

Cohen PR, CR Perrault 1979. Elements of a plan-based theory of speech acts. Cogn Sci 3: 177–212.

Ferguson GM, JF Allen, BW Miller 1996. TRAINS-95: towards a mixed-initiative planning assistant. Proceedings Third Conference on Artificial Intelligent Planning Systems (AIPS-96).

Fikes RE, NJ Nilsson 1971. STRIPS: a new approach to the application of theorem proving to problem solving. Artif Intell 2(3/4):189–208.

Forney GE Jr 1973. The Viterbi algorithm. Proc. IEEE 61:266–278.

Grice HP 1957. Meaning. Philos Rev 66:377–388. Reprinted in D Stenburg, L Jakobovits, eds. Semantics. New York: Cambridge University Press, 1971.

Grice HP 1975. Logic and conversation. In: P Cole, J Morgan, eds. Syntax and Semantics. Vol. 3: Speech Acts. New York: Academic Press, pp 41–58.

Grosz B, 1974. The structure of task oriented dialog. IEEE Symp. Speech Recognition. Reprinted in L Polanyi, ed. The Structure of Discourse. Norwood, NJ: Ablex, 1986.

Grosz B, 1977. The representation and use of focus in a system for understanding dialogs. Proc IJCAI, pp 67–76.

Grosz B and C Sidner 1986. Attention, intention and the structure of discourse. Comput Linguist 12(3).

Grosz B and C Sidner 1990. Plans for discourse. In: Cohen et al. Intentions in Communications. Cambridge, MA: MIT press, pp 417–444.

Grosz B and S. Kraus, 1996. Collaborative plans for complex group action. In: Artif Intell 86(2): 269–357.

Heeman P, JF Allen 1995. The TRAINS 93 Dialogues, TRAINS Technical Note 94-2, Department of Computer Science, University of Rochester. Speech data is available on CD-ROM from the Linguistics Data Consortium at the University of Pennsylvania.

Heeman P, G Hirst 1995. Collaborating on referring expressions. Comput Linguist 21(3).

Hinkelman E, JF Allen 1989. Two constraints on speech act ambiguity, Proc ACL.

Huang XD, F Alleva, HW Hon, MY Hwang, KF Lee, R Rosenfeld 1993. The Sphinx-II speech recognition system: an overview. Computer, Speech Language.

Jordan PW, M Moser 1997. Multi-level coordination in a computer-mediated conversation. Dissertation, University of Pittsburgh.

Katz SM 1987. Estimation of probabilities from sparse data for the language model component of a speech recognizer. IEEE Trans Acoustics, Speech, Signal Proc pp. 400–401.

Litman D, JF. Allen 1979. A plan recognition model for subdialogues in conversation. Cogn Sci 11(2):163–200.

Litman DJ, JF Allen 1990. Discourse processing and commonsense plans. In: Cohen et al., eds. Intentions in Communication. Cambridge, MA: MIT Press, 1990, pp 365–388.

Lochbaum KE, 1994. Using Collaborative Plans to Model the Intentional Structure of Discourse. PhD Dissertation, Harvard University, Cambridge, MA.

Lowerre B, R Reddy 1986. The harpy speech understanding system. In: Trends in Speech Recognition. Apple Valley, MI: Speech Science Publications. Reprinted in Waibel, 1990, pp 576–586.

Perrault CR, JF Allen 1980. A plan-based analysis of indirect speech acts. Am J Comput Linguist 6(3–4):167–182.

Ringger EK, JF Allen 1996. A fertility channel model for post-correction of continuous speech recognition. Fourth International Conference on Spoken Language Processing (ICSLP Ô96). Philadelphia, PA.

Runicky A 1993. Mode preference in a simple data-retrieval task. *Proceedings of the ARPA Workshop on Human Language Technology*, San Mateo: Morgan-Kaufmann.

Sadek MD, A Ferrieux, A Cozannet, P Bretier, F Panaget, J Simonin 1996. Effective human–computer cooperative spoken dialogue: the AGS demonstrator. Proceedings of the International Symposium on Dialogue, ISSD 6, Philadelphia.

Searle J. 1969. Speech Acts. London: Cambridge University Press.

Searle J. 1975. Indirect speech acts. In: P Cole, J Morgan, eds. Syntax and Semantics. vol 3: Speech Acts. New York: Academic Press, pp 59–82.

Seneff S, V Zue, J Polifroni, C Pao, L Hethington, D Goddeau, J Glass 1995. The preliminary development of a displayless PEGASUS system. Proceedings Spoken Language Systems Technology Workshop, Jan 1995. San Mateo, CA: Morgan Kaufmann.

Smith R and RD Hipp 1994. Spoken natural Language dialog Systems: A Practical Approach. London: Oxford University Press.

Smith R, R Hipp, A Biermann 1995. An architecture for voice dialogue systems based on prolog-style theorem proving. Comput Linguist 21(3).

Teng H, S Busayapongchai, J Glass, D Goddeau, L Hetherington, E Hurley, C Pao, J Polifroni, S Seneff, V Zue 1996. WHEELS: a conversational system in the automobile classifieds domain. Proceedings International Symposium on Dialogue, ISSD 6, Philadelphia.

Waibel A, K-F Lee, eds 1990. Readings in Speech Recognition. San Mateo, CA: Morgan Kaufmann.

Weld D. 1994. An introduction to least commitment planning. AI Mag, winter.

15

Empirical Approaches to Natural Language Processing

HAROLD SOMERS
UMIST, Manchester, England

1. INTRODUCTION

A major development in the last 10 or 15 years in natural language processing (NLP) has been the emergence of so-called empirical approaches. Largely enabled by the increasing power and sophistication of computers, these empirical approaches focus on the use of large amounts of data, together with procedures involving statistical manipulations. Because of this orientation toward data (as opposed to theory) and statistics (as opposed to linguistics), the empirical approaches have sometimes been seen as being in some kind of opposition to the more traditional "rationalist" approaches, as covered in Part I of this book. In fact, very few of the researchers using these methods would see themselves as "antilinguistics," and frequently, the empirical methods are used alongside traditional methods in many different ways.

The main unifying thread in the empirical approaches, is the use of large amounts of data, specifically **corpus** data. The term "corpus" is used to refer to any collection of language material (although see later); for our purposes we will concentrate on textual corpora, even though a collection of recordings of speech would also constitute a corpus and, indeed, there are many collections of transcriptions of spoken material.

The field of **corpus linguistics** has a long history, predating by decades the emergence of "empirical NLP" (see, e.g., McEnery and Wilson, 1996; Kennedy, 1998). Even before the invention of the computer, exhaustive (and no doubt exhausting) work on large collections of linguistic data was reported; for example, in lexicography (all the great dictionaries since Samuel Johnson's time have been based on examination of real texts), language acquisition (the diary studies of the 19th century), dialect studies, spelling reform, and above all in literary studies (e.g., authorship attribution, see later). With the appearance of the computer, the field of Corpus Linguistics could have expected to be energised, except that at about this time Linguistics in general came under the huge influence of Chomsky, who steered the field very firmly away from the empirical approaches. He was particularly negative about the use of corpus data, famously dismissing all corpora as "skewed," and urging linguists to concentrate on "competence," rather than "performance."

Nevertheless, work on corpora continued, with the publication in 1964 of the Brown corpus (Francis and Kučera, 1964) a notable landmark. Since that date, many other important and significant corpora have become available, including the Lancaster–Oslo–Bergen (LOB) corpus (Johansson et al., 1978), the British National Corpus (BNC; Burnard, 1995), as well as several collections of machine-readable texts in various languages [e.g., the Association for Computational Linguistics' Data Collection Initiative (ACL, 1994) and the jointly organized Text Encoding Initiative (TEI; Ide and Sperberg-McQueen, 1990], and including some collections of multilingual "parallel" corpora such as the Canadian Hansard parliamentary proceedings in English and French, the proceedings of the Hong Kong legislative council in Cantonese and English, various documents from official Swiss sources in French, German, Italian, and English, and data from the European Union, often in up to 11 languages.

It is appropriate, therefore, that the first chapter in Part 2 addresses issues of corpus collection and annotation, both of which will be important for any of the techniques and applications described here.

2. CHAPTER 16: CORPUS ANNOTATION

We have made the distinction in the foregoing between "corpora" and "collections of machine-readable texts," a distinction often made by corpus linguists, and crucially identifying two important aspects of the text collection. One aspect may be termed "planning," and concerns the element of deliberate selection and filtering of samples, so that, for example, a random collection of text samples would not constitute a corpus per se. The other important aspect is corpus **annotation**, whereby the texts are not merely collected in a machine-readable form, but subjected to some kind of analysis, which is reflected in the way the texts are stored.

As a potential user of a corpus, you may be fortunate in that an existing annotated corpus may be appropriate for your planned use; but, rather, it may be that you need to create and annotate your own corpus. Thompson's Chap. 16 "Corpus Creation for Data-Intensive Linguistics" will be of help here. Thompson leads the reader through the process of finding suitable data, legally acquiring the data, preparing it for encoding, deciding what to encode and then sharing this newly created resource with other potential users.

3. CHAPTER 17: PART-OF-SPEECH TAGGING

One of the most common types of annotation of corpus material is to indicate the part of speech (**POS**) or grammatical category of each word. The process is known as "POS tagging," or often simply **tagging**, and the annotations attached to the words in the corpus are known as **tags**. This is a task that is closely related to the task of parsing described in Chap. 4, and the POS tags could in theory be determined using the techniques described earlier. However, it is more usual, perhaps because of the large volume of data to be tagged, to use techniques based not on grammars, but on statistical techniques, in particular probabilities. One difference between parsing and tagging is that parsing normally aims to analyze the overall syntactic structure of the text, whereas tagging has the less ambitious goal of identifying only the grammatical categories of the individual words (although often, because of lexical ambiguity, the correct determination of the POS tag for one word is dependent on the POSs of neighbouring words). Because parsing and tagging, therefore, have different goals and also use different techniques it is usual to use the terms "parsing" and "tagging" in this distinctive way.

One of the most widely used tagging algorithms was developed by Brill, who is also the author of our overview Chap. 17 on tagging. Because tagging does not make use of a grammar to predict sequences of tags, the usual technique is to use statistical probabilities of tag sequences to calculate the most likely tag for a given word based on the surrounding context, especially the preceding context. Brill's chapter explains and clarifies the different approaches to this problem. One distinction is between "supervised" and "unsupervised" tagging. In the former, the system "learns" the probabilities for the various sequences from a manually tagged **training** corpus. Markov models are usually used for this purpose. In unsupervised tagging, the training is done automatically using the Forward–Backward or Baum–Welch algorithm. Brill then goes on to explain some other methods for training and learning.

4. CHAPTER 18: ALIGNMENT

An important corpus-based resource is the bilingual **parallel** corpus, containing a text and its translation (or, in some cases, corresponding texts in two languages, neither of which is necessarily the "source" text, but they are nevertheless translations of each other). Such a resource can be very useful for automatically extracting bilingual contrastive information, notably word and term equivalences, but also grammatical knowledge. A prior requirement for such a corpus, however, is that it be **aligned** (i.e., that the corresponding portions of text in the two languages be identified). This alignment can be at a very coarse level—section or paragraph—or at the finer level of sentence or even phrase and word.

For many texts, alignment at the paragraph level is quite straightforward, especially if the texts are close translations of each other. Occasionally, extraneous factors, such as misplaced footnotes or figure captions, can cause problems, but often alignment at this level is quite straightforward. Much more difficult is alignment at any finer level. Even at the sentence level, a text and its translation can often differ in that one language will use two sentences, whereas the other uses one, or in the order that the sentences appear. Even the very notion of "sentence" can be quite different from one language to another (in Chinese, for example, a single sentence, as identified by a "period," usually corresponds to three or four English sentences). For this reason, as Wu explains, the very term *alignment* has come to mean something special and different in the domain of bilingual corpus linguistics.

The problem of alignment, then, and its solution can be very complex. Wu first explains the essential concepts needed to understand the various algorithms that he presents. He then describes briefly the problem of alignment at the coarsest level (document structure) before turning to the more complex sentence alignment problem. A variety of techniques are presented, ranging from comparison of relative lengths of sentences (in characters or words); comparison of similar lexical content; use of physically similar strings ("cognates"), such as proper names, numbers, dates, and so on; ending with techniques that combine several of these. Wu then concentrates on the application of these techniques to "noisy" bilingual corpora; that is, corpora for which the sentences are known not to be neatly aligned, owing to omissions, reversals, layout differences, or simply linguistic differences between the languages, including writing system (e.g., English and Chinese or Japanese), as well as lexical and grammatical differences (e.g., where one of the languages has a very rich system of inflection, or allows new compound words to be coined as in German, and so on). Wu then considers alignment at the word level, followed by a section on alignment of structurally analyzed texts. Finally, he mentions bilingual lexicon acquisition techniques, for which the aim is to extract bilingual lexical correspondences automatically.

5. CHAPTERS 19 AND 20: COMPUTING SIMILARITY

For many applications in NLP, it is necessary to compute the similarity or, conversely, the distance, between two objects. These may be individual words, sequences of words, or linguistic structures, such as trees or graphs.

When applied to words, a similarity measure usually relates to meaning, broadly defined. It is useful in Information Retrieval to expand the scope of a query: if, for example, a user asks for documents about *aeroplanes*, the system might usefully also search for documents mentioning *airplanes*, and *planes*, these being synonyms of the word originally specified. If the user is interested in documents about *cars*, they may also be interested in documents about specific types of cars (*sports cars*, *station wagons*, *campers*), or other motor vehicles (*buses*, *trucks*). In the analysis of textual data, as in automatic summarization or classification, families of similar words can help automatically pinpoint the topic of a text. Another application of word similarity is in Machine Translation, specifically in the translator's aid known as "Translation Memory": this is a piece of software that searches a database of previous translations for sentences that resemble the current sentence to be translated, which the translator can then use as a model. Obviously, a sentence that differs from the current one only in the matter of one or two words, which are syntactically or semantically similar, might be a good candidate for the purpose.

The similarity between two words can be measured with reference to large corpora by looking at the **distribution** of the two words: if they occur in the context of a similar set of words, then they may be judged to be similar. Similarity of distribution can be measured by several different statistical techniques, as explained by Dagan in Chap. 19, "Contextual Word Similarity." In Lebart and Rajman Chap. 20, more detail is given about specific applications of word similarity measures. They also touch briefly on some measures of similarity beyond the word level.

6. CHAPTER 21: COLLOCATIONS

Another technique based on the distribution of words in corpora is the identification of "collocations." Two words are collocated if the occurrence of one is closely related to the occurrence of another. An extreme case of this is the idiom or fixed phrase, for example, under normal circumstances the word *helter* invariably occurs immediately next to the word *skelter* (and vice versa). (Notice that this general rule is, in fact, broken by the preceding sentence, however!) Of more interest is the recognition that certain families of words occur together (not necessarily in the same sentence, but in the same text) significantly more often than chance would predict. If these collocations can be identified, this provides a useful source of information, that can help to disambiguate homographs, aid translation, determine the choice of near synonyms in text generation, help identify the general topic of a piece of text, and so on. McKeown and Radev's Chap. 21 exemplifies a number of uses of collocation, and discusses the statistical measures employed to identify them.

7. CHAPTER 22: STATISTICAL PARSING

One technique with obvious parallels with those in the first part of this collection is discussed in Carroll's Chap. 22 on statistical parsing. "Parsing" can be contrasted with tagging (see foregoing) in that the aim is to arrive at a structural representation of a text. Chapter 4 discussed more conventional or traditional approaches to parsing; here we are concerned with the use of statistical and probabilistic approaches to the same problem.

Carroll begins with a number of "grammar-driven" approaches to parsing, starting with the **probabilistic** variant of the standard context-free grammar, in which each rule has associated with it a probability score (the scores for rules expanding the same left-hand side must add up to 1). Although the individual probabilities can be set manually, to seed a grammar, more usually they are derived from a corpus or a tree-bank. He then discusses context-dependent probabilistic models, in which the probabilities associated with the rules are not entirely independent, reflecting the reality in language that some combinations of categories are more or less likely according to other, contextual, factors. Several approaches to this problem are described. The section ends with a discussion of lexicalised probabilistic models, where statistical information associated with the individual words, rather than the more abstract categories, is also incorporated into the technique.

The chapter then focuses on "data-driven" approaches that do not involve a formal grammar, but instead, rely on similarity computations relative to a corpus of parsed text, a **tree-bank**. There are a number of techniques for arriving at the most plausible parse tree for a given input text, among them the rather crude "hill-climbing" method, and the more sophisticated "simulated-annealing" technique, adapted from chemical engineering. Carroll also discusses the highly practical "data-oriented parsing" approach, and the technique of parsing by tree-transformation rules.

The final section briefly discusses a technique that, unlike the others discussed here, has as its aim to propose representations in the form of logical structures, rather than syntax trees.

8. CHAPTER 23: AUTHORSHIP IDENTIFICATION AND COMPUTATIONAL STYLOMETRY

The remaining chapters in this part describe applications in NLP that rely on or make use of large amounts of data. The first of these, described by McEnery and

Oates (see Chap. 23), is an application that predates the availability of corpora, and even the availability of computers for that matter. "Stylometry" is the attempt to quantify aspects of writing style, and is a part of the more general task of authorship identification. This task, most familiar in the form of **literary studies** ("Who wrote the Bible?," "Is this a previously unknown Shakespeare play?," and such), also has important applications in social history (resolving cases of disputed authorship), and even in a **forensic** setting (e.g., identifying forgeries or revealing whether a confession has been coerced).

The authors discuss some of the problems in authorship identification and stylometry, and then describe a variety of measures and techniques from the simplest word counts to the more complex statistical analyses, in each case illustrating the technique with references to the literature.

9. CHAPTER 24: LEXICAL KNOWLEDGE ACQUISITION

In Chap. 24, Matsumoto and Utsuro discuss the extraction from large-scale corpora of lexical information. In other words, a corpus of text is seen as a collection of words (or, more exactly, word forms) in context, from which information about their use can be extracted automatically.

Among the applications discussed in this chapter are extraction of terminology (i.e, technical domain-specific vocabulary, often consisting of a multiword terms); and the inference of grammatical information about words, especially their cooccurrence properties (what other words they occur with, or structures they occur in). Identification of different **word senses** is also a useful goal: intuitively we can recognize homonymy and polysemy, but this chapter shows that we can use statistical measures of word distributions to identify variety of meaning and collocation. Different word senses can be identified by looking at the vocabulary used in the surrounding text, or by recognizing different grammatical constructions associated with different meanings: for example, *count* meaning 'enumerate' takes a direct object, whereas *count* meaning 'rely' often occurs with the preposition *on*.

Matsumoto and Utsuro then turn their attention to the acquisition of bilingual information from parallel corpora. Here they deal with the possibility of extracting comparative linguistic knowledge, especially **lexical translation equivalents**, by studying the relative distribution of word forms in parallel texts. There are many problems here, especially because of the different characteristics of individual languages, but a number of statistical techniques are described in this chapter, and their relative success is compared.

10. CHAPTER 25: EXAMPLE-BASED MACHINE TRANSLATION

In Part I, Somers described the history and state of the art in Machine Translation, concentrating on techniques involving traditional NLP tools. In Chap. 25 he continues the story, describing the latest approaches to the problem of translation using corpus-based and statistical methods. The main method is called **example-based** MT (EBMT), and involves the use of a bilingual parallel corpus of previous translations, which are used as a model for the new translation. Somers discusses the translator's aid known as **Translation Memory**, where relevant previous translations are presented to the human translator to use as model or not, at their own discretion. In

EBMT the system itself has to figure out how to reuse the example to produce an appropriate translation. As Somers points out, the technology has many problems, and is most often used in conjunction with more traditional methods, in a number of different combinations.

11. CHAPTER 26: WORD-SENSE DISAMBIGUATION

Hinted at in some of the earlier chapters, an important application of statistical methods with large corpora is the identification of alternative **word senses.** Yarowsky's Chap. 26 identifies various aspects of the problem, including different sources of word-sense ambiguity as well as different reasons for needing to disambiguate, such as Information Retrieval, Message Understanding, and Machine Translation. He then reviews early approaches to the problem before concentrating on corpus-based techniques. Similar to tagging (see foregoing), a distinction can be made between "supervised" and "unsupervised" techniques, the former requiring some kind of priming using already analysed corpora. Supervised methods include decision-tree methods and use of Bayesian classifiers, among others. Unsupervised approaches to word-sense disambiguation include clustering algorithms based on vector spaces, iterative bootstrapping algorithms, and various class models.

12. FURTHER READING

Each of the chapters is accompanied by a thorough bibliography. In the introductory chapter to Part 1 we were able to refer the reader to several general textbooks and introductions for symbolic approaches to NLP. Because empirical approaches are much more recent, it is more difficult to identify similar reading matter in this area. On the topic of corpus linguistics, the books by McEnery and Wilson (1996) and Kennedy (1998), already cited, can be recommended. Barnbrook (1996) and Thomas and Short (1996) also cover some of the topics we have chosen to include here. Readers with adequate French would find Habert (1995) a useful reference. Collections of research articles in the field can be recommended, although obviously these tend to be more technical, rather than introductory. One such collection is Jones and Somers (1997). Other places to look out for relevant research reports are in journals, such as *Computational Linguistics*, and in proceedings of conferences of the Association for Computational Linguistics, especially its Special Interest Group SIGDAT, which organizes conference series such as "Workshop on Very Large Corpora" and "Conference on Empirical Methods in Natural Language Processing."

REFERENCES

ACL 1994. Association for Computational Linguistics European Corpus Initiative: Multilingual Corpus 1. CD-ROM produced by HCRC, University of Edinburgh and ISSCO, University of Geneva, published by the Association for Computational Linguistics.

Barnbrook G 1996. Language and Computers: A Practical Introduction to the Computer Analysis of Language. Edinburgh: Edinburgh University Press.

Burnard L, ed. 1995. User's Reference Guide for the British National Corpus, Version 1.0. Oxford: Oxford University Computing Services.
Francis WN, H Kučera 1964. Manual of Information to Accompany "A Standard Sample of Present-Day Edited American English, for Use with Digital Computers." Providence, RI: Department of Linguistics, Brown University.
Habert B 1995. Traitement probabilistes et corpus. Special issue of Traitement Automatique des Langues. 36 (1–2).
Ide N, CM Sperberg-McQueen 1990. Outline of a standard for encoding literary and linguistic data. In: Y Choueka, ed. Computers in Literary and Linguistic Research: Proceedings of the 15th ALLC Conference, 1988. Geneva: Slatkine, pp 215–232.
Johansson S, G Leech, H Goodluck 1978. Manual of Information to Accompany the Lancaster–Oslo–Bergen Corpus of British English, for Use with Digital Computers. Oslo: Department of English, Oslo University.
Jones D, H Somers 1997. New Methods in Language Processing. London: UCL Press.
Kennedy G 1998 An Introduction to Corpus Linguistics. London: Longman.
McEnery T, A Wilson 1997. Corpus Linguistics. Edinburgh: Edinburgh University Press.
Thomas J, M Short 1996. Using Corpora for Language Research. London: Longman.

16

Corpus Creation for Data-Intensive Linguistics

HENRY S. THOMPSON

University of Edinburgh, Edinburgh, Scotland

1. INTRODUCTION

The success of the whole data-intensive or corpus-based approach to (computational) linguistics research rests, not surprisingly, on the quality of the corpora themselves. This chapter addresses this issue at its foundation, at the level of methodology and procedures for the collection, preparation, and distribution of corpora.

This chapter starts from the assumption that corpus design and component selection are separated from the issues addressed in this chapter. It also considers textual material only—spoken material raises many altogether different issues.

Where possible,[1] issues will be addressed in general terms, but it is in the nature of the enterprise that the majority of the problems and difficulties lie in the details. Accordingly, the approach adopted here is an example-guided one, drawing on experience in the production of several large multilingual CD-ROM-based corpora.[2]

2. COLLECTION

Material suitable for the creation of corpora does not grow on trees, and even if it did, the orchard owner would have a say in what you could do with it. It is all too

[1] A note on terminology: When writing on this subject, one is immediately confronted with the necessity of choosing a word or phrase to refer to text stored in some digital medium, as opposed to text in general or printed or written text. I am unhappy with all the phrases I can think of, and so will use "eText" throughout for this purpose.

[2] All trademarks are hereby acknowledged.

easy to embark on a corpus-based research or development effort without allocating anywhere near enough resources to simply getting your hands on the raw material, with the legal right to use it for your intended purposes. This section covers various issues that may arise in this phase of a corpus creation project, both technical and legal.

A. Finding the Data

There is a lot of eText out there, more every day, but actually locating eText appropriate to your needs can be quite difficult. Stipulate that your intended application, your corpus design or simply your inclination, have identified financial newspaper data, or legal opinions, or popular fiction, as your target text type. How do you go about locating suitable sources of eText of this type?

It is best to tackle this from two directions at once. On the one hand, virtually any text type has some kind of professional association connected to its production: the association of newspaper editors, or legal publishers, or magazine publishers. In this day and age, such a group, in turn, almost certainly has a subgroup or a working party concerned with eText (although they are unlikely to call it that). The association concerned will almost certainly be willing to make the membership of such a group available on request, provided it is clear the motivation involves research and is not part of a marketing exercise.

The key to success once you have a list of possible contacts is **personal contact**. Letters and telephone calls are *not* the right medium through which to obtain what you want. Usually, the persons involved in such activities will respond positively to a polite request for a face-to-face meeting, if you make it clear you are interested in exploring with them the possibility of making use of their eText for research purposes. A letter, followed up by a telephone call, requesting a meeting, has always worked in my experience. The simple fact of your willingness to travel to their establishment, together with the implied complement your interest represents, appears to be sufficient to gain initial access.

The other way in is through your own professional contacts. The Internet is a wonderful resource, and your colleagues out there are in general a friendly and helpful group. A message or news item directed to the appropriate fora (the `linguist` and `corpora` mailing lists, and the `comp.ai.nat-lang` newsgroup, see Sec. 5 for details) asking for pointers to existing corpora or known sources of eText of the appropriate text type may well yield fruit.

By far the best bet is to combine both approaches. If you can obtain a personal introduction from a colleague to a member of an appropriate professional association, you have a much greater chance of success. We got off to a flying start in one corpus collection project by being introduced to a newspaper editor by a colleague, and then using *his* agreement to participate as a form of leverage when we went on to visit other editors to whom we did not have an introduction.

So what do you say once you are in someone's office, you have introduced yourself, and you actually have to start selling? Here is a checklist that has worked for me:

- Enthusiastic description of the science
- No commercial impact
- Respect for copyright

- Visibility
- Technology payback
- The "bandwagon" effect

The first of these is easy to forget, but my experience underlines its importance. Usually people, even quite senior people, are flattered and excited by the thought that their eText has a potential part to play in a respectable scientific enterprise. Do not be afraid of this naive enthusiasm—it does no harm to encourage people's speculative ideas about your project, however uninformed they may be. Indeed sometimes you will be surprised and learn something useful.

But once this phase of the conversation is slowing down, you have to move quickly to reassure them that this scientific contribution will not impinge on their business. The line to take here is "Form, not content," [i.e., "we want your data because it is (a particular type of) English (or French or Thai)"] not because we want to exploit its content. An offer to use older material, or every other [pick an appropriate unit: issue, article, entry, $\cdots$], can help make this point clear.

Closely related to the previous point, you cannot emphasize too early your commitment to respect and enforce (insofar as you can) their **copyright** of the material involved. You cannot honourably negotiate with major eText owners unless you are prepared to take copyright seriously, and you will need to convince them that this is true. There is no quicker way to find yourself headed for the door than to suggest that once text is on the Internet, no-one owns it anymore. If you are genuinely of that opinion yourself, you should think twice before getting into the corpus business at all, or at any rate look exclusively to materials already clearly in the public domain.

There are also some concrete benefits to the owner of the eText from contributing, which you should point out. The story of the *Wall Street Journal* is often useful here: for many years after the publication of the 1994 ACL/ECI CD-ROM the *Wall Street Journal was* "English for computational purposes," establishing the brand name in a preeminent position, for free, in a situation with huge potential downstream impact. You do not want to overpromise the potential influence of your work, but it is perfectly reasonable to say something to the effect that if they give you the data, and your work is successful, their name will be at the heart of an area of technology development (financial management systems, medical informatics, or other) which, in turn, may develop into a market opportunity for them.

This leads on to the next concrete payback. Some domains in general, or some potential providers in particular, are just on the verge of recognising their eText as a real business resource. An offer to hand back not only the normalised data at the end of the day, but also to make the normalising process available, may in such circumstances be quite appealing.

Finally, if this is not the first visit you have made, letting them know that you already have agreement from so-and-so can be a good idea. It is a judgment call whether this is a good idea if so-and-so is a business competitor, but if they are from another country or work in another language, it is almost certainly a good idea.

B. Legally Acquiring the Data

Once you have convinced the owners of some eText that providing you with data is something they would like to do, you still have some hurdles to jump before you can start negotiating with the technical people. With luck, you should not have to pay for data, but if you do, a rough guideline of what I am aware has been paid in the past is around $100 (U.S.) per million words of raw data, in other words not very much.

Whether you pay for it or not, you need more than just informal agreement to give you the data—you need a **licence** that allows you to do what you intend to do with it. This is where the point above about respect for copyright comes home. You either have to undertake not to distribute the material at all, that is to be sure it never gets beyond your own laboratory, or preferably, you have to negotiate a licence that allows further distribution. To obtain this, you will almost certainly have to agree the terms of the licence agreement *you* will require from anyone to whom you redistribute the data. You will also need to agree just what uses you (or other eventual recipients) can make of the material. Someone giving you data for free is entitled to restrict you to noncommercial uses, but you should be careful to obtain the right to modify the material, since up-translation of character set and markup and addition of further markup is likely to be on your agenda.

The best thing to do if you are seriously intending to distribute corpus materials is to look at some sample licence agreements that one of the large corpus providers, such as the LDC (Linguistic Data Consortium) or ELRA (European Language Resources Association), use (see Sec. 5 for contact details). Indeed if you are intending any form of publication for your corpus, you may find that involving one of those nonprofit establishments in the negotiations, so that the licence is actually between them and the supplier, may be to your advantage.

C. Physically Acquiring the Data

Not the least because what you are likely to get for free is not the most recent, the physical medium on which eText is delivered to you may not be under your control. If it is from archives of some sort, it may be available *only* in one physical form, and even if it is on-line in some sense, technical staff may be unwilling (especially if you are getting the data for free) or unable to deliver it in a form you would prefer. Mark Liberman (personal communication) described the first delivery of material for the ACL European Corpus Initiative (ECI) as follows: "Someone opened the back door to the warehouse, grunted, and tossed a cardboard box of nine-track mag tapes out into the parking lot."

If you are lucky, you will obtain the material in a form compatible with your local computer system. If not, your first stop is whatever form of central computing support your institution or organisation can provide. Without this, or if they are unable to help, there are commercial firms that specialise in this sort of thing, at a price. Searching for *data conversion* with a World Wide Web search engine is a reasonable starting point if you are forced down this road.

3. PREPARATION

In the wonderful world of the future, all electronic documents will be encoded using Unicode (ISO 10646) and tagged with SGML (Standard Generalized Markup Language) (ISO 8879), but until that unlikely state of bliss is achieved, once you have your hands on the raw material for your corpus, the work has only just begun—indeed you will not know for a while yet just how *much* work there is.

For the foreseeable future, particularly because of the rate at which technology penetrates various commercial, governmental, and industrial sectors, as well as issues of commercial value and timeliness, much of the material available for corpus creation will be encoded and structured in out-of-date, obscure, or proprietary ways. This section covers the kind of detective work that may be involved in discovering what the facts are about your material in its raw state, the desired outcome, and plausible routes for getting from one to the other. The discussion is separated into two largely independent parts: **character sets and encoding** and **document structure**.

A. Character Set Issues

What do you want to end up with, what do you have to start with, and how do you get from one to the other? These questions are not as simple as you might think (or hope).

1. *Choosing a Delivery Character Set and Encoding*

Without getting into the religious wars that lurk in this area, let us say that characters are abstract objects corresponding to some (set of) graphic objects called glyphs, and that a character set consists inter alia of a mapping from a subsequence of the integers (called **code-points**) to a set of **characters**. An **encoding**, in turn, is a mapping from a computer-representable byte- or word-stream to a sequence of code-points. ASCII, Unicode, JIS, and ISO-Latin-1 (ISO 8859-1) are examples of character sets. UTF-8 is an encoding from eight-bit byte-streams to ASCII, ISO-Latin-1, and Unicode. Shift-JIS is an encoding from eight-bit byte-streams to JIS. The best place for a detailed introduction to all these issues is Lunde (1993).

So which should you use to deliver your corpus? There is a lot to be said for international (ISO) standards, but if *all* your data and your users are Japanese, or Arabic, then national or regional de facto standards may make sense, even if they limit your potential clientele. As an overbroad generalisation, use ASCII or regional standards if you are concerned with a large existing customer base with conservative habits, use the identity encoding of ISO-Latin-n if your material will fit and your customer base is relatively up-to-date, but for preference use UTF-8-encoded Unicode. This will ensure the widest possible audience and the longest lifetime, at the expense of some short-term frustration as tool distribution catches up. Note that you pay *no* storage penalty for using UTF-8 for files that contain only ASCII code-points.

2. *Recoding*

The hardest thing about dealing with large amounts of arbitrary non-SGML eText is that you cannot trust anything anyone tells you about it. This applies particularly at the level of character sets and encoding. My experience suggests that the *only* safe thing to do is to construct a **character histogram** for the entire collection as the very first step in processing. A character histogram is a tabulation of how many times each code point in the file occurs. For example, if you think you have a file using an 8-bit encoding, you simply need to count the number of times each byte between 0 and 255 occurs, and display the results in tabular form. For 16-bit encodings, things can get a bit more complex: at this point in the evolution of eText, it is probably best to start with an 8-bit histogram no matter what. Note that if you are not sure about whether an 8- or 16-bit encoding has been used, some preliminary detective work with a **byte-level viewer** is probably required. For substantial amounts of data, it is worth writing a small C program to build the histogram—alas it is not something for which I am aware that a public domain source exists. Once you have obtained that, you need to look *very* carefully at the results. To understand what you are looking at, you will also need a good set of character-set and encoding **tabulations**. The most comprehensive tabulation of 8-bit encodings I know of is "RFC 1345," which can be found on the World Wide Web at a number of places (see Sec. 5 for details). For Chinese, Japanese, and Korean character sets and encodings, the work of Lunde (1993) is invaluable.

A preliminary indication of the character set and encoding will usually be given by its **origin**. The first thing to do is to confirm that you are in the right ballpark by checking the code-points for the core alphabetic characters. I usually work with a multicolumn tabulation, illustrated in Table 1, showing ASCII/ISO-8859-1 glyph or name (since that is what is most likely to appear on your screen for want of something better) code-point in octal and hex, "count" (the number of occurrences of the character in the eText), RFC 1345 code, and name in currently hypothesised character set, code-point in target character set, or encoding if translation is required, and any comment.

The example in Table 1 shows extracts from a tabulation for a German eText that appears to be using something close to IBM Code Page 437. Note the difference between code points 5d (*]*) and 81 (*x*): the first is recoded to a point *other* than what the hypothesised encoding would suggest, whereas the latter, as indicated by the comment, is recoded consistently with that hypothesis.

The only way to deal with material about which you have even the *slightest* doubt is to examine every row of the histogram for things out of the ordinary. Basically, any count that stands out on either the high or the low side needs to be investigated. For instance, in the example in Table 1, the count for 09*^[* is suspiciously high, and indeed turns out to be playing a special role, whereas the counts for

Table 1 Extract from Tabulation of German eText

Char	Octal	Hex	Count	RFC 1345	Hypothesised name	Recode	Comment
^[	\011	09	26062	HT	Character tabulation (ht)		Page ref if before number
!	\041	21	6283	!	Exclamation mark		
"	\042	22	25	"	Quotation mark	ô	
#	\043	23	0	Nb	Number sign		
Y	\131	59	1136	Y	Latin capital letter *y*		
Z	\132	5a	51403	Z	Latin capital letter *z*		
[	\133	5b	30	<(	Left square bracket	Ä	
\	\134	5c	14	//	Reverse solidus	Ö	
]	\135	5d	79	)>	Right square bracket	Ü	
^	\136	5e	0	'>	Circumflex accent		
x	\201	81	257074	u:	Latin small letter *u* with diaeresis	ü	yes

22 *"* and 5b-d *[, \,]* are suspiciously *low*, so they in turn were examined and context used to identify their recodings.

Note that it is not always necessary to be (or have access to someone who is) familiar with the language in question to do this work. In the example in Table 1, the suspicious code-points were not the only ones used for the target character. For instance, not only was 5b used for Latin capital A with diaeresis, but also 8e, the correct code-point for this character in the IBM 437 encoding. This was detected by finding examples of the use of the 5b code-point (using a simple search tool such as `grep`), and then looking for the result with some *other* character in the crucial place. For instance if we search for 5b (left square bracket *[*) in the source eText, we find inter alia the word *[gide*. If we look for words of the form `.gide`, where the full-stop matches any character, we find in addition to *[gide* also *\216gide*, and *no others*. When *[ra* and *[nderungsantrag* yield the same pattern, it is clear, even without knowing any German, that we have hit the right recoding, and a quick check with a German dictionary confirms this, given that octal 216 is hex 8e.

Even if you do have access to someone who knows the language, this kind of preliminary hypothesis testing is a good idea, because not only will it confirm that the problem code-point is at least being deployed consistently, but it will also reduce the demands you make on those you ask for help.

Once you have identified the mechanical recoding you need to do, how you go about doing it depends on the consistency of the recoding. If your eText makes consistent and correct use of only one encoding, then the `recode` software from the Free Software Foundation can be used to translate efficiently between a wide range of encodings. If, as in the foregoing example, a mixture of encodings has been used, then you will have to write a program to do the job. Almost any scripting language will suffice to do the work if your dataset is small, but if you are working with megabytes of eText you probably will want to use C. If you do have, or have access to, C expertise, `recode` is a good place to start, as you can use it to produce the declaration of a mapping array for the dominant encoding to get you started. I also use `recode` to produce the listing of names and code-points for hypothesised source encodings, as illustrated in the foregoing.

3. *Word-Boundary Issues*

Once you have done the basic recoding, you still may have another tedious task ahead of you, if line-breaks have been introduced into your eText in a nonreversable way. Various code-points may be functioning as some sort of cross between end-of-line and "soft" hyphen (i.e., hyphenation introduced by some word-processing package), but you cannot assume that deleting them will always be the right thing to do. One trick that may be helpful is to use the text itself to produce a word list for the language involved. You can then use that as a filter to help you judge what to do with unclear cases.

For example, if preliminary investigations suggest that *X* is such a character, and you find the sequence *abcXdef* in your eText, if you have *abcdef* as a word elsewhere in the text, but no occurrences of either *abc* or *def*, then you can pretty confidently elide the *X*.

As in the simpler-recoding cases, you may well be left with a residuum of unclear cases for investigation by hand or referral to a native speaker, but doing as much as you can automatically is obviously to everyone's advantage when dealing with eText of any size.

Finally, you must take care to construct a histogram of your recoded eText as a final check that you have not slipped up anywhere. Constructing a word list can also be a useful sanity check at this point. You should view it sorted both by size (bugs often show up as very short or very long "words") and by frequency and size (all frequent words are worth a look, and frequent long words or infrequent short words are *ipso facto* suspicious).

B. Document Structure Issues

Unless your eText is extremely unusual, it will have at least *some* structure. Almost every eText has words organised into sentences and larger units: paragraphs, quotations, captions, headings, and so on. The original may also include more explicitly presentational information, such as font size and face changes, indentation, and the like, some of which may reflect structural or semantic distinctions you wish to preserve.

Two primary questions arise in response to the structure:

1. How will you notate the structural aspects of your eText?
2. How will you process your originals to yield versions marked up with the desired notation?

1. Deciding What to Notate Explicitly

The matter of when to leave things as they are and when to delimit structure explicitly is by no means obvious. Below a certain level you will undoubtedly stick with the orthographic conventions of your source language (e.g., words are notated with their letters adjacent to one another, in left-to-right order, with spaces in between). Above that level you will want to use *some* conventions to identify structural units, but it is less clear whether orthographic conventions, even if they are available, are appropriate. Two factors will determine the point at which orthographic convention is no longer appropriate.

1. If you need to include annotation systematically for elements of structure at a given level, then it is best not to use a (likely to be idiosyncratic) orthographic convention, but rather, to make use of one of several more formal approaches. In some cases this may even operate at the level of words or subword units, if for example, every word in your eText is annotated with its part of speech.
2. At some point the orthographic convention becomes too hard to interpret mechanically with any reliability. Human readers can easily distinguish figure captions, section titles, and single-line paragraphs when they are set off from surrounding text by horizontal or vertical white space, but mechanical processing is unlikely to do as well.

The rule of thumb is that whenever possible, difficult programming jobs should be done once, by you, the corpus creator, rather than being left to the many users of your corpus to do many times, *provided* they can be done *reliably*.

a. *A Note on the Vexed Matter of Sentences*

Sentences are a difficult matter. In all but the most straightforward of eText, reliably and uncontroversially determining where sentences start and end is impossible. Only if you are very sure that you will not be fooled by abbreviations, quotations, tables, or simple errors in the source text should you attempt to provide mechanically explicit formal indication of sentence structure. If you must have a go, at least do not replace the orthographic signals that accompany sentence boundaries in your eText, but *add* sentence boundary information in such a way that your users can easily ignore or remove altogether, and then be left to make their own mistakes.

2. *How is Explicit Structural Information Recorded?*

Once you have decided what aspects of your eText to delimit and label, the question of how to do this in the most user-friendly and reusable way arises. There are three main alternatives, with some subsidiary choices.

1. Design your own idiosyncratic annotation syntax.
2. Use a database:
   a. Use a widely available spreadsheet.
   b. Use a widely available nonspecialised database.
   c. Use a database designed to store eText.
3. Use a standard markup language such as SGML or XML (see later):
   a. Use a public document-type definition (DTD) such as TEI or CES (see later).
   b. Design your own DTD.

Although (1) was the option of choice until fairly recently, there is no longer any justification for the fragility and nonportability that this approach engenders, and it will not be considered further here.

The database approach has some things to recommend it. The simplest version, in which the lowest level of structure occupies a column of single spreadsheet cells, has the advantage that the necessary software (e.g., Microsoft Excel, Lotus Spreadsheet) is very widely available. This approach also makes the calculation of simple statistics over the corpus very straightforward, and most spreadsheet packages have built-in visualisation tools as well. This approach does not scale particularly well, nor admit to batch processing or easy transfer to other formats, but may be a good starting point for small eText.

I know of only one example, MARSEC, in their case spoken language transcripts, that attempted to use an ordinary relational database for corpus storage (Knowles, 1995), but for a small corpus this is also at least, in principle, possible, although once again portability and scalability are likely to be problems.

The third database option has recently been widely adopted within that portion of the text-processing community in the United States who are funded by ARPA, in particular those involved in the TREC (Text Retrieval Conference) and MUC (Message Understanding Conference) programmes. A database architecture for eText storage and structural annotation called the **Tipster architecture** (see Sec. 5) has been specified and implemented. This architecture is also at the heart of a framework for supporting the modular development of text-processing systems called

GATE (General Architecture for Text Engineering) developed at the University of Sheffield (England).

SGML is the Standard Generalized Markup Language (ISO 8879; cf. Goldfarb, 1990; van Herwijnen, 1990). **XML** is the eXtensible Markup Language, a simplified version of SGML originally targetted at providing flexible document markup for the World Wide Web. They both providing a low-level grammar of annotation, which specifies how markup is to be distinguished from text. They also provide for the definition of the structure of families of related documents, or document **types**. The former says, for example, that to annotate a region of text it should be bracketted with **start and end tags**, in turn delimited with angle brackets, as in example {1}.

```
<quotation>To be or not to be</quotation>                {1}
```

Qualifications may be added to tags using **attributes**, as in {2}.

```
<quotation author='WS'>. . . </quotation>                {2}
```

Document-type definitions (**DTDs**) are essentially context-free grammars of allowed tag structures. They also determine which attributes are allowed on which tags. Software is available to assist in designing DTDs, creating documents using a given DTD, and validating that a document conforms to a DTD.

This is not the place for a detailed introduction to SGML or the Tipster architecture, nor would it be appropriate to enter in to a detailed comparison of the relative strengths and weaknesses of the Tipster architecture as compared with XML for example, for local representation and manipulation of eText, which is a contentious matter already addressed in various scholarly papers. There is, however, little if any debate that for **interchange**, SGML or XML are the appropriate representation: both the LDC and ELRA now use SGML for all their new distributions. Furthermore, the Tipster architecture mandates support for export and import of SGML-annotated material. Accordingly, and in keeping with our own practice, in the remainder of this chapter we will assume the use of SGML or XML.

What DTD should you use? Is your eText so special that you should define your own grammar for it, with all the attendent design and documentation work this entails? Or can you use one of the existing DTDs that are already available? These fall into two broad categories: DTDs designed for those authoring electronic documents [e.g., **RAINBOW** (somewhat dated now) or **DocBook**] and those designed for marking up existing text to produce generally useful eText for research purposes [e.g., the **Text Encoding Initiative** (TEI) and the **Corpus Encoding Standard** (CES); see Sec. 5 for details].

TEI has published a rich modular set of DTDs for marking up text for scholarly purposes. All the tags and attributes are extensively documented, making it easier for both producer and consumer of eText. In addition to the core DTD, there are additional components for drama, verse, dictionaries, and many other genres. CES is an extension of the TEI intended specifically for use in marking up large existing eText, as opposed to producing eText versions of preexisting, nonelectronic documents.

Should you use SGML or XML? In the short-term the TEI and CES DTDs are available only in SGML form, so using them means using SGML. But an XML version of the TEI DTD is expected to be available soon (watch the TEI mailing list) and it costs almost nothing to make the individual documents that make up your eText be XML-conformant. Because all valid XML documents are valid SGML documents, you will still be able to validate your eText against the TEI DTD, while still being well placed to use the rapidly growing inventory of free XML software. If you are designing your own DTD, you probably know enough to make the decision unaided, but in the absence of a strong reason for using full SGML—for example, you really need SUBDOC (a facility that allows the "black box" embedding of documents with different DTDs) or CONREF (a facility for having an attribute–value *or* element content, but not both), or content model exceptions (limited non–context-free modifications to the grammar of your documents' markup)—I would recommend sticking with XML. In my view the apparent benefit of most of SGML's minimisation features, which are missing from XML, is illusory in the domain of eText: omitted (or short) tags, `shortref` and `datatag` are almost always more trouble than they are worth, particularly when markup is largely both produced and exploited (semi)-automatically.

3. *Up-translation*

At the beginning of this section, two issues were introduced: what means to use to annotate structure, and how to actually get that annotation added, using the markup system you have chosen, to your eText. We turn now to the second issue, called **up-translation** in the SGML context.

The basic principle of formal markup is that all structure which is unambiguously annotated should be annotated consistently using the documented formal mechanism. This means any preexisting, more or less ad hoc markup, whether formatting codes, white space, or whatever has to be replaced with, say, XML tags and attributes. Depending on the nature and quality of the preexisting markup, this may be straightforward, or it may be extremely difficult.

In the case of small texts, this process can be done by hand, but with any substantial body of material, it will have to be automated. Proprietary tools designed for this task are available at a price (e.g., DynaTag or OmniMark). The no-cost alternative is to use a text-oriented scripting language: `sed`, `awk`, and `perl` have historically been the tools of choice, and these are now available for WIN32 machines as well as under UNIX.

The basic paradigm is one of stepwise refinement, writing small scripts to translate some regularity in the existing markup into tags and attributes. As in the case of character-set normalisation, discussed earlier, heuristics and by-hand post-processing may be necessary if the source material is inconsistent or corrupt.

What is absolutely crucial is to record every step that is taken, so that the up-translation sequence can be repeated whenever necessary. This is necessary so that when, for example, at stage 8 you discover that a side effect of stage 3 was to destroy a distinction that only now emerges as significant, you can go back, change the script used for stage 3 and run the process forward from there again without much wasted effort. This, in turn, means that when by-hand postprocessing is done, if possible the results should be recorded in the form of **context-sensitive patches** (the output of `diff -c`), which can be reapplied if the up-translation has to be redone. It is impossible to overestimate the value of sticking to this disciplined approach: in processing substantial amounts of eText you are bound to make mistakes or simply change your mind, and only if the overhead in starting over from the beginning is kept as low as possible will you do it when you should. Note also that because it is the infrequent cases that tend to be the sources of difficulty, in large eText you will usually find that problems are independent of one another, so your patches and scripts that are not directly involved with the problem are likely to continue to work.

Another tactic that often proves helpful is to be as profligate in your use of intermediate versions as the file storage limits you are working with allow. Using shell scripts to process all files in one directory through the script for a given stage in the up-translation, saving the results in another directory, and so on for each stage, functions again to reduce the cost of backing up and redoing the process from an intermediate position.

At the risk of descending into "hackery," do not forget that, although shell scripts, as well as most simple text-processing tools, are line-oriented, `perl` is not, and by judicious setting of `$/` you can sequence through your data in more appropriate chunks.

Another parallel with character-set normalisation is that validating the intermediate forms of your eText as you progressively up-translate it is a valuable diagnostic tool.

To sum up: Up-translation is a software development task, *not* a text-editing task, and all the paraphernalia of software engineering can and should be brought to bear on it.

4. DISTRIBUTION

After having put all that work in, you really ought to make the results of your work available to the wider community. This section briefly introduces the main issues that you will need to confront: what medium to use for the distribution, what tools if any should accompany the data, what documentation should you provide, how should you manage the legal issues and exactly what should you include in the distribution.

A. Media Issues

For the last few years there has been only one plausible option for the distribution of large eText, namely CD-ROM. A CD holds over 600 Mb of data, which is large enough for fairly substantial collections of text. Small amounts of textual material (say under 25 Mb) are probably better **compressed** (see following) and made available through the Internet (but see 4.D. concering licences). The drawback of CDs is that the economics and preproduction hassle vary considerably with the number you need to distribute.

In large quantities (i.e., hundreds) having CDs pressed by a pressing plant can be quite inexpensive, as the marginal cost per disc is very small (a few dollars) and the cost of a master disc from which the copies are pressed is falling, as the market is quite competitive. Exactly how much mastering costs depends on how much work you do ahead of time: if you send a premastered CD image and all the artwork needed for label and insert (if any), it can be as low as 1500 dollars (US), but this can rise quickly the more work you leave for the producer. If you think your eText is of sufficient interest to warrant a large press run, you may well be better off negotiating with the LDC or ELRA to handle preproduction and pressing for you.

In small numbers you can use write-once CDs and do your own production using a relatively inexpensive Worm drive, but this becomes both expensive and tedious for more than a dozen or so discs.

Finally, note, that although the standard for CD-ROMs (ISO-9660, sometimes referred to as **High Sierra**) is compatible with all three major platforms (UNIX, WIN32, and Macintosh), it is by nature a lowest–common-denominator format, so that, for example, symbolic links and long file names are best avoided.

In a number of ways CDs are not particularly satisfactory: they are too big for modest eText, too small for large eText, particularly when accompanied by digitised speech (a single CD will only hold about 2 h of uncompressed high-quality stereo digitised speech) and slow to access. It may be that new technology such as Zip and SuperDrive will provide an alternative for modest eText, and DVD (digital video disk) *may* be an improvement for some larger eText, but the situation for very large datasets does not look likely to improve soon.

B. Tools and Platforms

There was once a time when distributing eText meant either distributing the tools to access it along with the eText itself, or condemning users to build their own. Fortunately, the last few years have seen a rapid increase in the availability of free software for processing SGML, and most recently XML, for both WIN32 and UNIX platforms, so the use of one of these markup languages pretty much obviates the necessity of distributing tools. Special-purpose tools that take advantage of, say, XML markup can add considerably to the value of eText, so if you have built them you should certainly consider including them. For example, along with the SGML-marked-up transcripts and the digitised audio files, the CD-ROMs for our own HCRC Map Task Corpus (Andersen et al., 1991) included software that links the two, so that while browsing the transcripts the associated audio is available at the click of a button.

It must be said that the situation is not as favorable on the Macintosh. Large-scale processing of eText is not well-supported, although free XML-processing tools are beginning to be ported to the Macintosh, including our own **LT XML** tool suite and API (see Sec. 5).

1. (De)compression

One of the more vexed questions associated with distribution is whether or not to compress your eText, and if so how. Particularly if the size of the uncompressed eText is just above the limit for some candidate distribution medium, the question is sure to arise. My inclination is not to compress, unless it makes a big difference in ease of distribution; for example, by reducing your total size to what will fit on a single CD-ROM:

- Transparency: If you do not compress (and you do use XML or SGML), then any tool that can process or display text files will work with your eText as distributed.
- Portability: Although free tools and libraries for enabling the decompression of several forms of efficient compression (most notably `zip` and `gzip` formats), they are not universal, and there is some correlation between platform type and preferred compression format.

- Speed (not): Not long ago, CD-ROM readers were slow enough that it was actually faster to access compressed eText than uncompressed from CD-ROM: the time to decompress on an average-speed processor was considerably less than the time saved by having to transfer on average 60% fewer bytes. But in the last few years CD-ROM speeds have increased even faster than processor speeds, so this is no longer true, at least for the time being.

2. *Validation*

If we assume you are using XML or SGML to annotate your eText, you should certainly use a DTD, and you should certainly confirm that your eText as valid for that DTD. For SGML the state-of-the-art, up-to-date validator is free: Clark's **SP**, available for WIN32 or UNIX (see Sec. 5). For XML the list is larger and still changing rapidly, so the best bet is to check one of the tools summary pages.

3. *Search and Retrieval*

What is the use of structure if you cannot exploit it? (Actually, it is worth using rigorously defined markup even if you cannot or will not exploit it, just for the consistency it guarantees and the documentation it provides.) Several free tools now exist to allow structure to figure in simple searches of XML or SGML annotated eText (e.g., **LT XML**, **sgrep**; see Sec. 5). There is clearly scope for structure sensitivity in concordancing and other more advanced text–statistical tools, but as yet such tools have not emerged from the experimental stage.

C. Documentation

Another benefit to using the TEI DTD (or the TEI-based CES DTD) is that it requires a carefully thought-out and detailed header, which records not only basic bibliographic information, but also relevant information about encoding, markup, and up-translation procedures used, if any. Any eText consumer needs to know all this, as well as the meaning of any markup for which the semantics is not already published (i.e., in the TEI or CES specifications). Note also that bibliographic information falls into two categories: information about the original **source** of the eText, and information about the eText itself.

D. License Agreements

If you have had to sign license agreements with the copyright owners of any of the sources of your eText, you may well be obliged thereby in turn to require agreement to license terms from anyone to whom you distribute that eText. Even if you have not, you must make clear to anyone who comes across your eText the terms under which they may make use of it, if any. The contribution you made in producing a body of eText is copyright by you the minute you publish it in any way, so even if you wish anyone to be free to use it in any way they choose (and it is within *your* rights to grant this) you need to grant them license to do so explicitly.

So at the very least you need to incorporate a copyright notice in your eText, either granting a license, or stating clearly that unlicensed use is not allowed, and the means, if any, whereby a license can be obtained. In either case, most people also include some form of disclaimer of responsibility for errors and such, although the legal status of such notices is not entirely clear. Here is one we have used:

> <YOUR NAME HERE> makes no representations about the suitability of this data for any purpose. It is provided "as is" without express or implied warranty. <YOUR NAME HERE> disclaims all warranties relative to this data, including all implied warranties of merchantability and fitness, in no event shall <YOUR NAME HERE> be liable for any special, indirect, or consequential damages or any damages whatsoever, action of contract, negligence, or other tortious action, arising out of or in connection with the use of this data.

When you actually need a license agreement, which a user acknowledges as governing their access to the eText, you should at least look to existing precedent from the experts, the LDC or ELRA and, if necessary, and you have the patience, talk to a lawyer.

For eText of major significance, particularly if you have a legal obligation as a result of license terms *you* have agreed to, you may well choose to require physical signature of a paper license agreement or payment before releasing copies. For eText where this would be overkill, and which is small enough to be distributed on the Internet, we have been content with an interactive licensing process, either with or without an e-mail step to obtain positive identification. Most intermediate-level introductions to Web site construction contain enough information and examples to put one of these together.

E. What to Distribute

This may seem obvious: you distribute the eText, right? Not quite, or rather, not only. Along with the eText itself, its documentation, and copyright notice, your distribution package should, if at all possible, include the raw original sources, in as close as possible a form to that which you yourself started with (but note that here is the obvious place that compression may be both necessary and appropriate). You should never be so sure of your recoding and up-translation, or as confident that your markup is what others need and want, to restrict them to your version of the sources alone. Replication of results is not a paradigm often observed in this branch of science, but the advent of data-intensive processing and theorising make it much more appropriate, and in my view it should apply even at the humble level we have been discussing in this chapter.

5. RESOURCE POINTERS

Here is a list of pointeres to the resources referred in the foregoing. In keeping with the nature of the business, they are almost all pointers to World Wide Web pages, and as such subject to change without notice. With luck, there is enough identifying text so that a search engine will be able to locate their current incarnation by the time you read this.

CES (Corpus Encoding Standard) WWW home page at `http://www.cs.vassar.edu/CES/`

`comp.ai.nat-lang` Internet newsgroup

`corpora` subscribe to this mailing list by sending an e-mail with the message "sub corpora firstname lastname" to corpora@hd.uib.no There is also a file archive available by anonymous ftp at nora.hd.uib.no, or http://www.hd.uib.no/corpora/archive.html

DocBook document-type description language maintained by the Davenport Group, http://www.sil.org/sgml/gen-apps.html#davenport

ELRA (European Language Resources Association), WWW home page at http://www.icp.grenet.fr/ELRA/home.html

GATE architecture developed at the Department of Computer Science, University of Sheffield, England.
http://www.dcs.shef.ac.uk/research/groups/nlp/gate/

HCRC Language Technology Group http://ww.ltg.ed.ac.uk/software/xml/

HCRC Map Task Corpus CD-ROM available from Human Communication Research Centre, University of Edinburgh, Scotland.

LDC (Linguistic Data Consortium)

linguist moderated Internet bulletin board at http://www.linguistlist.org

LT XML toolkit for XML documents http://www.ltg.ed.ac.uk/software/xml/

recode software for character set conversion: any Gnu software archive, for example, ftp.warwick.ac.uk/pub/gnu/recode-3.4.tar.gz

RAINBOW ftp://ftp.ifi.uio.no/pub/SGML/Rainbow

RFC 1345 Any RFC archive, e.g. http://andrew2.andrew.cmu.edu/rfc/rfc1345.html

SGML http://www.sil.org/sgml/sgml.html

SGML/XML tools http://www.sil.org/sgml/publicSW.html, http://www.infotek.no/sgmtool/guide.htm

sgrep tool for searching structured text files, developed at the Department of Computer Science, Helsinki University: http://www.cs.helsinki.fi/~jjaakkol/sgrep.html

SP (object-oriented toolkit for SGML parsing) http://www.jclark.com/sp/

TEI (Text Encoding Initiative) WWW home page at http://www-tei.uic.edu/orgs/tei/

Tipster architecture developed at the Department of Computer Science, New York University.
http://www.cs.nyu.edu/cd/faculty/grishman/tipster.html

XML (eXtensible Markup Language) http://www.ucc.ie/xml/

REFERENCES

ACL/ECI (1994). Association for Computational Linguistics European Corpus Initiative: Multilingual Corpus 1. CD-ROM produced by HCRC, University of Edinburgh, Scotland, and ISSCO, University of Geneva, Switzerland, published by the Association for Computational Linguistics.

Andersen AH, M Bader, EG Bard, E Boyle, G Doherty, S Garrod, S Isard, J. Kowto, JM McAllister, J Miller, CF Sotillo, HS Thompson, R Weinert (1991). The HCRC Map Task Corpus. Lang Speech 34:351–366.

Goldfarb C (1990). The SGML Handbook. Oxford: Oxford University Press.

Lunde K (1993). Understanding Japanese Information Processing. Sebastopol, CA: O'Reilly & Associates, Japanese version published in 1995 by Softbank Corporation, Tokyo.

Kowles G (1995). Converting a corpus into a relational database: SEC becomes MARSEC. In: G Leech, G Myers, J Thomas, eds. Spoken English on Computer: Transcription, Mark-up and Application. London: Longman, pp 208–219.
van Herwijnen E (1990). Practical SGML. Dordrecht: Kluwer Academic.

17

Part-of-Speech Tagging

ERIC BRILL

Microsoft Research, Redmond, Washington

1. INTRODUCTION TO TAGGING

Determination of the part of speech (POS) of a word is an important precursor to many natural language processing (NLP) tasks. POS-tagged text is often used as a reasonable compromise between accuracy and utility. Tagged text provides a great deal of information beyond simply an unanalyzed sequence of words. Although less informative than a full syntactic parse that additionally provides higher-level structural information and relations between words, POS tagging can be done much more accurately and quickly than parsing. Also, good taggers can typically be developed for a domain much more rapidly than good parsers can.

Some of the tasks for which tagging has proved useful include machine translation, information extraction, information retrieval, and higher-level syntactic processing. In machine translation, the probability of a word in the source language translating into a word in the target language is highly dependent on the POS of the source word. For example, the word *record* in English can be translated as either *disque* or *enregistrer* in French. However, once we know that *record* is a noun, then the *disque* translation is almost certain.

In information extraction, patterns used for extracting information from text frequently make reference to POSs. To extract information of the form: `acquire (X,Y)` from the *Wall Street Journal*, one might use a template such as {1}.

(DET)? (PROPER NOUN) + (*acquired|bought*) (DET)?(PROPER NOUN) + {1}

`X` `acquire` `Y`

This is a regular expression that matches any string containing an optional determiner (DET) followed by one or more proper nouns, followed by either the word *acquired* or *bought*, followed by an optional determiner and one or more proper nouns. For instance, from {2a} would be extracted {2b}.

International Business Machines bought Violet Corporation. {2a}

`acquire (International Business Machines, Violet Corporation)` {2b}

One can go quite far in information extraction by building and extraction patterns referring to words and POSs (Cardie, 1997).

POS tagging can also be used for information retrieval (IR). Most current IR systems treat a query as a bag of words. When the person uses a word, they are using a particular instance of that word, with a particular POS and word sense. By augmenting the query a person gives to an IR system with POS information (as well as adding tags to the documents potentially being retrieved), more refined retrieval is possible.

When higher-level information is necessary, such as phrases or relation between phrases, tagging often serves as a precursor to these higher-level syntactic-processing systems. For instance, several programs have been written that do "noun phrase chunking," finding the noun phrases in each sentence. Almost all these programs use a combination of word and POS information to learn either regular expressions for noun phrases in a sentence that are likely to indicate the beginning or ending of a phrase (Church, 1988; Ramshaw and Marcus, 1994). Tagging can also be important for speech synthesis. For instance, the word *produce* is pronounced differently depending on whether it is a noun or a verb.

There are two things that make tagging a difficult problem: (a) many words have multiple POSs, and (b) new words appear all the time. Because words can have multiple POSs, one cannot simply look up the POS of a word and be done, but rather, one must use contextual cues to determine which of the possible POSs is correct for the word in any particular context. The word *can* can be a noun, a verb, or an auxiliary verb. In {3a}, which can be glossed as {3b}, three usages of the word *can* appear.

We can can the can. {3a}

We are able to put the can into a can. {3b}

It is not difficult for a person, based on the context in which each word appears, to determine that the first *can* is an auxiliary, the second is a verb and the third is a noun. So one of the difficult tasks in tagging is to provide the machine with the knowledge also necessary to be able to disambiguate from the set of allowable tags based on context.

As words are constantly being invented (e.g., Shakespeare introduced approximately 1500 words into the English language), as well as falling in and out of favor, no matter how complete our word list is, we will always encounter words not on the list, as well as novel word usages. Therefore, a tagger will need a method not just for choosing between possible tags for a word, but also for determining the tags of new words. This can be done based on both contextual information and information about the word itself, such as affixes. For example, in {4}, it is easy to determine that the made-up word *goblanesque* is an adjective, based upon the environment in which it appears and the suffix *-esque*.

His goblanesque writing was a refreshing treat. {4}

2. FIRST ATTEMPTS AT AUTOMATIC TAGGING

Work on building programs to tag text automatically originated in the 1950s and 1960s. One of the earliest programs (Greene and Rubin, 1971) was created as a means to tag semiautomatically the Brown Corpus (Francis and Kŭcera, 1982), a 1-million-word corpus of English containing text from many different writing genres. This program worked by determining word–environment combinations that, with very high certainty, indicated a particular tag should be chosen. These high-certainty environments were discovered manually. They were then applied to tag reliable regions in the corpus, after which the corpus was tagged and corrected manually. This method did not achieve high rates of accuracy, compared with modern systems, most likely because they did not have a manually tagged corpus to develop and test their manually created rules empirically.

Klein and Simmons (1963), constructed a totally automatic tagger. Their tagger assumed, first, that a lexicon was available that listed the allowable parts of speech for every word that would be encountered. The first step in tagging was to assign all allowable tags to each word. So for instance, sentence {5a} might appear as {5b}.

We can race the race. {5a}

We/Pron can/N_V _Aux race/N_V the/Det race/N_V {5b}

A sequence of handwritten rules, each in turn, then deletes certain tags as possibilities in certain environments. For example, a rule might state: "If a word is tagged with both an N tag and a V tag, and it occurs immediately after a Det, then remove the V tag." As rules are applied, tags are removed. It was possible that after all rules applied there would still be cases of ambiguity. It is impossible to compare their tagger with modern taggers, because they ran experiments on a corpus to which no other taggers have been applied.

However, there is a recently developed tagger that is based on the same principle as the tagger of Klein and Simmons, but it was developed over a many-year period and uses much more linguistically sophisticated rules than those used previously. It is based on Constraint Grammars (Karlsson et al., 1995), a grammar formalism specified as a list of linguistic constraints, and has been applied to POS tagging with success. Although a manually developed tagger can often capture sophisticated constraints that could not readily be learned by the current machine-learning algorithms, tagger development is quite slow and the resulting system cannot easily be ported to a new tag set, domain, or language.

3. THE FIRST AUTOMATICALLY TRAINED TAGGERS: MARKOV MODELS

A. Supervised Tagging

Since the time of the original manually built POS taggers, several **corpora** have been built containing millions of words that have been carefully tagged by hand. For American English, the two most commonly used corpora are the Brown Corpus, mentioned earlier, and the Penn Treebank Corpus (Marcus et al., 1993). The Penn Treebank contains the Brown Corpus retagged with a different set of tags, over a million words of the *Wall Street Journal*, and over a million words of phone conversation transcripts. For British English, there is the British National Corpus, which is similar to, although much bigger than, the Brown Corpus, but contains British texts. There are currently tagged corpora in many different languages, and plans are being made to create tagged texts for even more languages in the near future.

With the availability of such corpora, as well as relatively inexpensive computer processing and storage, it has become possible to explore automatic tagger training as an alternative to laborious manual construction. The first attempts at automatic training were based on **Markov models** (Derouault and Jelinek, 1984; Church, 1988). In addition to having a great influence on persons interested in developing taggers, this result was one of the first clear demonstrations that linguistic knowledge can effectively be extracted automatically from online corpora, serving as a catalyst to the movement toward empirical, machine-learning based natural language processing. Tagging was done using Markov model technology, which had previously proved extremely effective for building robust speech recognizers. A Markov model is a finite-state machine. From a state, the model emits a symbol and then moves to another state. For every state there are associated two probability distributions: the probability of emitting a particular symbol and the probability of moving to a particular state.

First, we will give the mathematics behind the model, then present the model itself. The task of tagging is to find the sequence of tags $T = \{t_1, t_2, t_3, \cdots t_n\}$ that is optimal for a word sequence $W = \{w_1, w_2, w_3, \cdots w_n\}$. Probabilistically, this can be framed as the problem of finding the most probable tag sequence for a word sequence [i.e., $\text{argmax}_T \mathrm{P}(T|W)$]. We can apply Bayes' law {6}.

$$\mathrm{P}(T|W) = \mathrm{P}(T)^*\mathrm{P}(W|T)/\mathrm{P}(W) \qquad \{6\}$$

Since $\mathrm{P}(W)$, the probability of the word sequence, will be the same regardless of the tag sequence chosen, we can then search for $\text{argmax}_T \mathrm{P}(T)^*\mathrm{P}(W|T)$.

Looking at the first term, we have Eq. {7}.

$$\begin{aligned} \mathrm{P}(T) &= \mathrm{P}(t_1, t_2, t_3 \cdots t_n) \\ &= \mathrm{P}(t_1)^*\mathrm{P}(t_2|t_1)^*\mathrm{P}(t_3|t_1, t_2)^* \cdots ^* \mathrm{P}(t_n|t_1, t_2, t_3 \cdots t_{n-1}) \end{aligned} \qquad \{7\}$$

It would be impossible to obtain reliable estimates for many of these probabilities. To make this tenable, a simplifying assumption is made. In particular, it is assumed that the probability of a tag is dependent only on a small, fixed number of previous tags. This is called a **Markov assumption**. The Markov assumption clearly does not hold for English, as was shown by Chomsky (1956). We can construct sentences in which two words arbitrarily far apart influence each other. For example, in the context {8a}, the probability of the singular verb *eats*, is much more likely than the plural verb *eat*, because the word *dog*. But this is still true of {8b}.

The dog... {8a}

The dog on the hill by the brook in the park owned by the sailor with a wooden leg ... {8b}

Although the Markov assumption does not hold in actuality, it is a simplification that makes estimating P(*T*) feasible and, as will be seen, good tagging results can still be achieved despite the fact that the assumption is false. If we assume the probability of a tag depends only on one previous tag (this is called a **bigram model**), then we have Eq. {9}.

$$P(T) = P(t_1)^* P(t_2|t_1)^* P(t_3|t_2)^* \cdots ^* P(t_n|t_{n-1}) \quad \{9\}$$

Given a manually tagged corpus, it is easy to estimate these probabilities: $P(t_i|t_j)$ is simply the number of times we see tag t_i immediately following t_j in the corpus, divided by the number of times we see tag t_j. Next, by making the simplifying assumption that the relation between a word and a tag is independent of context, we can simplify $P(W|T)$ to be as in Eq. {10}, involving probabilities that can again be easily derived from a tagged training corpus.

$$P(W|T) \approx P(w_1|t_1)^* P(w_2|t_2)^* \cdots ^* P(w_n|t_n) \quad \{10\}$$

We can relate this to a Markov model as follows: For the bigram model, in which a tag depends on only one previous tag, each state in the model corresponds to a POS tag. Then, the state transition probabilities are simply the $P(t_i|t_j)$ probabilities, the probability of moving from one POS state to another, and the symbol emission probabilities are simply the $P(w_i|t_j)$, the probability of a word being emitted from a particular tag state. The tagged sentence {11} would have been generated by first going to the Det state and emitting *The*, then moving to the Noun state and emitting *race*, then moving to the Verb state and emitting *ended*, as shown in Fig. 1.

The/Det race/Noun ended/Verb {11}

Given this model, P(Det Noun Verb | *The race ended*) is estimated as in Eq. {12}.

P(Det | START) * P(Noun | Det) * P(Verb | Noun) * P(*The* | Det)* P(*race* | Noun) * P(*ended* | Verb). {12}

We now know how to derive the probabilities needed for the Markov model, and how to calculate $P(T|W)$ for any particular T, W pair. But what we really need is to be able to find the most likely T for a particular W. This can be seen as the problem of being given the output of a Markov model and then having to determine what state sequence would have the highest probability of having generated that output.

One simple solution would be to iterate over all tag sequences that have the same length as W, compute $P(T|W)$ for each, and pick the best one. The problem with this is that if we have a tag set of size S, the run-time of this approach is $\mathbf{O}(S^{|W|})$. Fortunately, there is an algorithm that allows us to find the best T in the linear time, the **Viterbi algorithm** (AJ Viterbi. Error bounds for convolutional codes and an asymptotically optimum decoding algorithm. IEEE Transactions on Information

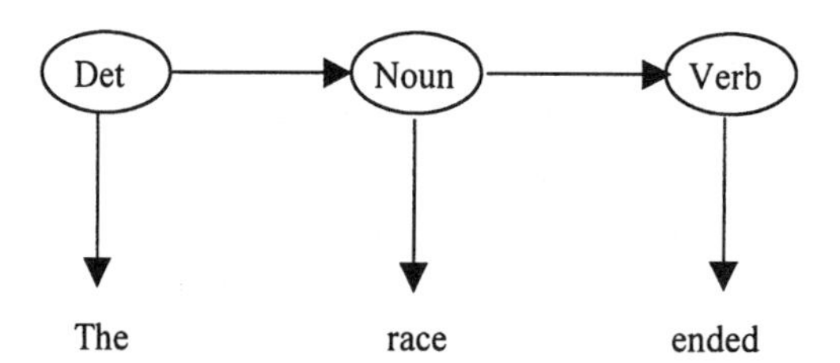

Fig. 1 Markov model for sentence {11}.

Theory. IT-13(2), 260–269, 1967). The idea behind the algorithm is that of all the state sequences where the ith state is X, only the most probable of these sequences need be considered. In particular:

```
For t=1 to Number_of_Words_In_Sentence
  {For each state b ∈ S
     {For each state a ∈ S
        For the best state sequence ending in state a at
           time t-1, Compute the probability of that
           state sequence going to state b at time t.}
     Determine the most-probable state sequence ending in
        state b at time t.}
```

So if every word can have S possible tags, then the Viterbi algorithm that runs in $\mathbf{O}(S^2 * |W|)$ time, or linear time with respect to the length of the sentence. In fact, the true run-time is actually considerably better than the S^2 constant, for the average number of tags per word is significantly lower than the total number of allowable tags.

Several papers have presented results using Markov models for tagging, as described in the foregoing (Church, 1988; Weischedel et al., 1993). If there are no words in the test set that did not appear in the training set (the **closed vocabulary** assumption), accuracies of close to 96–97% are typically achieved in English when training on about 10^6 words of text. There are methods within this framework for dealing with unknown words, but they will not be discusssed here. Briefly, for words not seen in the training set, $P(w_i|t_j)$, is estimated based on features of the unknown word, such as whether the word contains a particular suffix.

B. Unsupervised Tagging

In the foregoing discussion, it was assumed that a large manually tagged corpus was available from which we could obtain the probabilities needed by the Markov model. Unfortunately, we are often not fortunate enough to have access to such a resource. This is obviously true if we wish to perform POS tagging in a language for which such a corpus does not currently exist. But it is even true if we are tagging a new domain and need relatively high tagging accuracy. Weischedel et al. (1993) showed that when training and testing on two similar corpora (e.g., training on the 1988 *Wall Street Journal* and testing on the 1989 corpus of the same newspaper), very good results can be obtained; but if the training and test genres are not a good fit (e.g., training on the *Wall Street Journal* and testing on a corpus of air travel reservation transcripts), accuracy degrades significantly.

Fortunately, there is a technique for training a Markov model that does not require a manually annotated corpus. The technique is called the **forward–backward algorithm**, also referred to as the **Baum–Welch algorithm** (Baum, 1972), and is a special case of the more general class of **Estimation Maximization** (EM) algorithms. The algorithm works as follows:

```
1. Assign initial probabilities to all parameters
2. Repeat:
   Adjust the probabilities of the parameters to increase the
   probability the model assigns to the training set.
3. Until training converges
```

Step 1 can be performed a number of different ways. If no information is known a priori about the model, all parameters can be made equally probable or parameters can be assigned random probabilities. If there is domain knowledge available, an informed initial guess can be made for the parameter values. Step 2 is a bit more complicated. The forward–backward algorithm gives a method for performing 2, such that training is guaranteed to converge on a local maximum for parameter values; in other words, in each training iteration we are guaranteed that the parameter values will be adjusted such that the probability of the training set will not decrease. The idea is that for each training iteration, we estimate probabilities by counting just as we did when an annotated corpus was available. But instead of basing our counts on the "true tags" provided by the human annotators, we base them on the tags and corresponding probabilities that the current model assigns. For example, when a tagged training corpus is available, on seeing ...*saw*/Verb... we would increment by 1 our count for the number of verbs seen and the number of times *saw* and the tag Verb occurred together. If instead our model assigns Verb to *saw* in this context with probability 0.75 and Noun with probability 0.25, then we would increment by 0.75 our count for the number of verbs seen and the number of times *saw* and Verb occurred together, and would increment by 0.25 our count for the number of nouns seen and the number of times *saw* and Noun occurred together. Training typically is stopped when the increases in the probability of the training set between iterations drops below some small value.

Although this training algorithm is indeed guaranteed to adjust the parameter values in a way such that it finds a locally best set of values from the initial parameter settings, the converged on model is not necessarily globally optimal. The forward–backward algorithm has been applied with mixed results as a method for training a tagger without using a manually tagged corpus. In Kupiec (1992), results were presented showing that a tagger resulting from forward–backward training was able to achieve accuracies comparable with those achieved by the best taggers trained on manually annotated text. However, Merialdo (1994) obtained much less favorable results. There he demonstrated that the algorithm frequently overtrained; although each training iteration improved the probability assigned to the training set, it actually decreased the accuracy of tagging the test set. Merialdo's experiments showed that training straight from a very small amount of manually tagged text gave better results than using the forward–backward algorithm on a large amount of untagged text. In addition, his results showed that given a tagger trained from manually tagged text, and using the parameter settings of this tagger as the initial settings for running the forward–backward algorithm, at times the forward–backward algorithm was able to improve the tagger accuracy, but other times, the accuracy decayed.

C. Other Algorithms for Automatic Training

Markov model-based taggers are the most commonly used taggers. However, there has recently been a great deal of progress in developing alternative methods for automatic-tagger building, mostly stemming from the realm of Machine Learning. Successful taggers have been built based on many standard machine-learning algorithms, such as **neural networks** (Schmid, 1994), **constraint relaxation and nearest neighbor** (Daelemans et al., 1996), and **decision trees** (Schmid, 1997). We will discuss here two recent developments in Machine Learning for POS tagging: transformation-based tagging and maximum entropy tagging. All of the trainable taggers yet developed are based on the same insight: that a local environment of one to three words and POS tags around a word are often sufficient for determining the proper tag for that word. But although all taggers are using more or less the same cues for making tagging decisions, each approach uses this information in a different way.

1. *Transformation-Based Learning*

Although the Markov model approach to tagging has an advantage over the earlier rule-based approaches in that it can be trained automatically, the clarity of simple rules in these early systems is abandoned for large tables of opaque statistics. Transformation-based learning is an attempt to preserve the benefit of having linguistic knowledge in a readable form while extracting linguistic information automatically from corpora, as opposed to composing these rules by hand (Brill, 1995). The output of the learner is an ordered sequence of transformations of the form shown in Eq. {13}.

$$\text{Tag}_i \rightarrow \text{Tag}_j \text{ in context } C \qquad \{13\}$$

There are three components to a transformation-based learner:

1. An initial state annotator
2. A set of allowable transformations
3. An objective function

Learning works as follows (Fig. 2). First, the training text is assigned some initial guess for the proper tags by the **initial state annotator**. This initial state annotator can be any program that assigns a tag to every word, ranging in sophistication from randomly assigning tags to being a full-blown Markov model tagger. The most commonly used initial state annotator assigns every word its most likely tag, as indicated in the training corpus. So every occurrence of *can* in the foregoing example sentence {3a} would initially be labeled as an auxiliary verb.

The learner is given a set of allowable transformation types: {14} shows some examples of transformation types.

Change a tag from X to Y if: {14}

a. The previous word is W.
b. The previous tag is t_i and the following tag is t_j.
c. The tag two before is t_i and the following word is W.

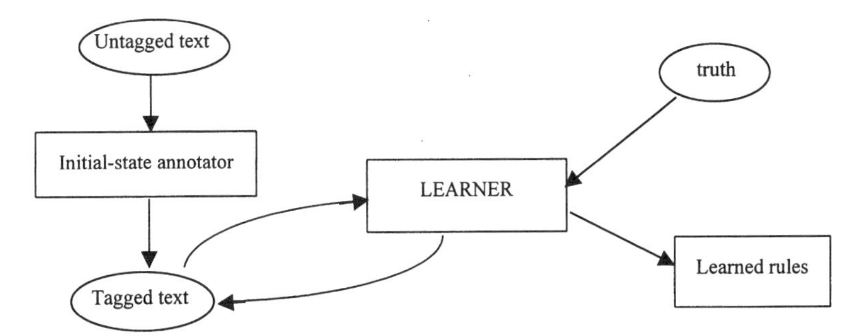

Fig. 2 Transformation-based learning.

In the transformation-based tagger described in Brill (1995), there are a total of 18 different transformation templates, each making reference to some combination of words and tags within a window of three of the word being tagged. In each iteration, the learner searches for the transformation for which the application to the training corpus results in the best improvement according to the objective function. Because we want to maximize tagging accuracy, we simply use accuracy as our objective function, meaning that in each learning iteration the learner finds that transformation for which the application results in the greatest error reduction on the training set. (Note that this is different from the objective function used in Markov model tagging: There the objective function is to maximize the probability the model assigns to the training corpus.) This transformation is then appended to the end of the transformation list and applied to the training corpus. Once training is completed, the tagger is run by first tagging fresh text with the initial-state annotator, then applying each transformation, in order, wherever it can apply.

2. *Learning Without a Tagged Text*

Learning can also be done without a tagged text (Brill and Pop, in press), with only some minor changes to the transformation-based learner described in the foregoing. Instead of tagging each word with its most likely tag, each word is tagged with a list of all possible tags (as was done in Klein and Simmons 1963). Instead of changing a tag, transformations delete tags from the list of possible tags for a particular word in a particular context. The main difference is that a new objective function is needed. Because we no longer have access to a manually tagged corpus in learning, we cannot use accuracy as our criterion for choosing a transformation. Rather than give the specific measure used, we will give the intuition behind the measure. If we look through a corpus and notice that after the word *the* there are many words that can be nouns and verbs, and many words that can only be nouns, but there are few or no words in that context that can only be verbs, this is very strong evidence for a rule such as {15}.

Noun_or_Verb → Noun if the previous word is *the* {15}

In experiments reported in Brill and Pop (in press), this method of training without a manually tagged text was seen to be less susceptible to overtraining than the forward–backward algorithm.

The transformation-based tagger has several advantages over the Markov model-based tagger. First, a small set of very simple rules is learned. For example, one learned rule is as in {16}.

Change a tag from Verb to Noun if the previous word is a Determiner {16}

The rules are easy to understand, making development and debugging much easier, and allowing easy interlacing of machine-learned and human-generated rules. Second, the transformation list can be compiled into a finite-state machine (Roche and Schabes, 1995), resulting in a very fast tagger: roughly ten times faster than the fastest Markov- model tagger. This is done by first making a finite-state machine for every learned rule. These finite-state machines are then composed together, and the resulting machine is determinised. This results in a tagger for which tagging is done by simply making a transition from one state to another in a deterministic finite-state transducer, a very inexpensive and fast operation. Although the operations for building this transducer could theoretically result in a very large transducer with an exponential explosion in states, for the experiments presented in Roche and Schabes (1995) a very manageable-sized transducer results. Another advantage of the transformation-based learner is that it is less rigid in what cues it uses to disambiguate a particular word. It can choose the most appropriate cues, rather than using the n preceding POS tags for disambiguating all words.

The Markov-model tagger has one big advantage over the transformation-based tagger: it not only outputs tags, but it can also provide tag probabilities. This allows the tagger output to be readily combined with other probabilistic components. For instance, a Machine Translation system could be more refined in its reasoning if it was told that a certain word was a noun with probability 0.6 than if it was just told that the tagger thinks the correct tag for that word is a noun.

3. *Maximum Entropy Tagging*

The final method for automatic training combines some of the good properties of both transformation-based tagging and Markov model tagging: **Maximum entropy tagging** (Ratnaparkhi, 1996). Maximum entropy tagging allows great flexibility in the cues that are used to disambiguate words, while outputting probabilities along with tags.

The maximum entropy framework, similar to the transformation-based approach, readily allows one to train a tagger using a diverse set of features, choosing the appropriate features for each particular context. The maximum entropy framework does this by finding a single probability model consistent with all the constraints that are specified. The idea behind this framework is to derive a probability model which is (a) consistent with the training data, and (b) maximally agnostic beyond what the training data indicates.

In other words, the goal is to find a model with maximum entropy (i.e., maximum randomness or minimum additional structure) that still satisfies the constraints provided from the training data. The probability model is taken over a space $H \times T$, where H is the set of environments a word appears in and T is the set of possible POS tags. Just as environments were defined to which a transformation can make reference (such as previous two words, previous word and next tag, and so on), the maximum entropy model specifies a set of features from the environment in which a word occurs that will be useful for tag prediction. For example (taken from Ratnaparkhi, 1996), for tagging one could specify the environment to be as in Eq. {17}.

$$h_i = \{w_i, w_{i+1}, w_{i+2}, w_{i-1}, w_{i-2}, t_{i-1}, t_{i-2}\} \qquad \{17\}$$

In other words, the environment for word i is the surrounding two words and the previous two tags. Given such a history, a set of binary features is then defined. For instance {18} is the jth feature, and is "on" or "off" based on certain properties of the environment.

$$f_j(h_i, t_i) = \begin{cases} 1 \text{ if suffix}(w_i) = \text{"ing" and } t_i = \text{PastPartVerb} \\ 0 \text{ otherwise} \end{cases} \qquad \{18\}$$

Features are generated from feature templates. The template from which {18} was derived is {19}.

$$X \text{ is a suffix of } w_i, |X| < 5 \text{ AND } t_i = T \qquad \{19\}$$

where X and T are variables. First a set of features and their observed probabilities are extracted from a training set. Through a method known as **Generalized Iterative Scaling**, a probability model is then generated that is the maximum entropy model consistent with the observed feature probabilities.

Once the model has been trained, application is similar to Markov model tagging, in that the goal is to find the most probable tag sequence according to the probability model. This is actually carried out using a **beam search**, always keeping the n most likely tag sequences up to the word being tagged. This approach allows us to build a probability model to be used for POS tagging. But unlike the Markov model approach, there is a great deal of flexibility in what contextual cues can be used. In fact, any binary-valued feature can be incorporated into the model.

In experiments presented in: Eric Brill and Jun Wu. Classifier combination for improved lexical disambiguation. Proceedings of the 17th International Conference on Computational Linguistics, 1998, a maximum entropy tagger, Markov model-based tagger and a transformation-based tagger all achieved approximately the same level of accuracy. This could be attributed to several possible reasons:

1. We are past the point where differences in accuracy can be measured using the common test corpora currently being used. Every corpus will have a degree of noise, owing to human annotator errors and genuine linguistic ambiguity, which will prevent any system from achieving perfect tagging accuracy as measured by comparing tagger output with the corpus. It may be that taggers have exceeded that level of accuracy.
2. The "80/20 rule" of engineering might apply to tagging. The 80/20 rule states that for any problem, one can quickly and easily solve 80% of that problem (or some such number), and the remaining 20% will be extremely difficult. So with tagging perhaps all the current taggers are solving the 80%, and we need a richer understanding of linguistics and more poweful machine-learning techniques before we can crack the remaining 20%.

As the number of publicly available tagged corpora increases, and more sophisticated learning algorithms are developed for tagging, we should soon achieve a better understanding of what accounts for the current perceived bottleneck in progress, and we hope will gain insight into what can be done to overcome it.

REFERENCES

Baum L 1972. An inequality and associated maximization technique in statistical estimation for probabilistic functions of a Markov process. Inequalities 3:1–8.

Brill E 1995. Transformation-based error-driven learning and natural language processing: a case study in part-of-speech tagging. Comput Linguist 21:543–565.

Brill E, M Pop (in press). Unsupervised learning of disambiguation rules for part of speech tagging. In: K Church, ed. Natural Language Processing Using Very Large Corpora. Dordrecht: Kluwer Academic Press.

Cardie C 1997. Emprical methods in information extraction. AI Mag 18(4):65–80.

Chomsky N 1956. Three models for the description of language. IRE Trans Inform Theory IT-2:113–124.

Church KW 1988. A stochastic parts program and noun phrase parser for unrestricted text. Second Conference on Applied Natural Language Processing, Austin, TX, pp 136–143.

Daelemans W, J Zavrel, P Berck, S Gillis 1996. MBT: a memory-based part of speech tagger-generator. Proceedings of the Fourth Workshop on Very Large Corpora, Copenhagen, pp 14–27.

Derouault A, F Jelinek 1984. Modèle probabiliste d'un langage en reconnaissance de la parole. Ann Telecommun.

Francis WN, H Kučera 1982. Frequency Analysis of English Usage: Lexicon and Grammar. Boston: Houghton Mifflin.

Greene BB, G Rubin 1971. Automated Grammatical Tagging of English. Providence, RI: Department of Linguistics, Brown University.

Karlsson F, A Voutilainen, A Anttila 1995. Constraint Grammar. Berlin: Mouton de Gruyter, 1995.

Klein S, R Simmons 1963. A computational approach to grammatical coding of English words. J ACM 10:334–347.

Kupiec J 1992 Robust part of speech tagging using a hidden Markov model. Comput Speech Lang 6.

Marcus MP, B Santorini, MA Marcinkiewicz 1993. Building a large annotated corpus of English: the Penn Treebank. Comput Linguist 19:313–330.

Merialdo B 1994. Tagging English text with a probabilistic model. Comput Linguist 20:155–172.

Ramshaw L, M Marcus 1995. Exploring the nature of transformation-based learning. In: J Klavans, P. Resnik eds. The Balancing Act: Combining Symbolic and Statistical Approaches to Language. Cambridge, MA, MIT Press, pp 135–156.

Ratnaparkhi A 1996. A maximum entropy model for part-of-speech tagging. Proceedings of the Conference on Empirical Methods in Natural Language Processing. Philadelphia, PA, pp 133–142.

Roche E, Y Schabes 1995. Deterministic part-of-speech tagging with finite state transducers. Comput Linguist 21:227–253.

Schmid H 1994. Part-of-speech tagging with neural networks. COLING 94: The 15th International Conference on Computational Linguistics, Kyoto, Japan, pp 172–176.

Schmid H 1997. Probabilistic part-of-speech tagging using decision trees. In: DB Jones, HL Somers, eds. New Methods in Language Processing. London: UCL Press, pp 154–164.

Weischedel R, M Meteer, R Schwartz, L Ramshaw, J Palmucci 1993. Coping with ambiguity and unknown words through probabilistic models. Comput Linguist 19: 359–382.

18

Alignment

DEKAI WU

The Hong Kong University of Science and Technology, Kowloon, Hong Kong

1. INTRODUCTION

In this chapter we discuss the work done on automatic alignment of parallel texts for various purposes. Fundamentally, an alignment algorithm accepts as input a **bitext** (i.e., a bilingual parallel text), and produces as output a **map** that identifies corresponding passages between the texts. A rapidly growing body of research on bitext alignment, beginning about 1990, attests to the importance of alignment to translators, bilingual lexicographers, adaptive machine translation systems, and even ordinary readers. A wide variety of techniques now exist, ranging from the most simple (counting characters or words), to the more sophisticated, sometimes involving linguistic data (lexicons) that may or may not have been automatically induced themselves. Techniques have been developed for aligning passages of various granularities: paragraphs, sentences, constituents, collocations, and words. Some techniques make use of apparent morphological features. Others rely on cognates and loan words. Of particular interest is work done on languages that do not have a common writing system. The robustness and generality of different techniques has generated much discussion.

BILINGUAL CONCORDANCES. Historically, the first application for bitext alignment algorithms was to automate production of the cross-indexing for bilingual concordances (Warwick and Russell, 1990; Karlgren et al., 1994; Church and Hovy, 1993). Such concordances are consulted by human translators to find the previous contexts in which a term, idiom, or phrase was translated, thereby helping the translator maintain consistency with preexisting translations, which is important in government and legal documents. An additional benefit of bilingual concordances is a large increase in navigation ease in bitexts.

BITEXT FOR READERS. Aligned bitexts, in addition to their use in translators' concordances, are also useful for bilingual readers and language learners.

LINKED BILINGUAL CONCORDANCES AND BILINGUAL LEXICONS. An aligned bitext can be automatically linked with a bilingual lexicon, providing a more effective interface for lexicographers, corpus annotators, as well as human translators. This was implemented in the BICORD system (Klavans and Tzoukermann, 1990).

TRANSLATION VALIDATION. A word-level bitext alignment system can be used within a translation-checking tool that attempts to automatically flag possible errors, in the same way that spelling and style checkers operate (Macklovitch, 1994). The alignment system in this case is primed to search for deceptive cognates (*faux amis*) such as *library*/*librarie* 'bookshop,' in English and French.

AUTOMATIC RESOURCE ACQUISITION. These days there is as much interest in the side effects of alignment as in the aligned text itself. Many algorithms offer the possibility of extracting knowledge resources: bilingual lexicons listing word or collocation translations, or translation examples at the sentence, constituent, or phrase level. Such examples constitute a database that may be used by example-based machine translation systems (Nagao, 1984), or they may be taken as training data for machine-learning of transfer patterns. Alignment has also been employed to infer sentence bracketing or constituent structure as a side effect.

In the following sections, we first discuss the general concepts underlying alignment techniques. Each of the major categories of alignment techniques is considered, in turn, in subsequent sections: document-structure alignment, sentence alignment, alignment for noisy bitexts, word alignment, constituent and tree alignment, and bilingual-parsing alignment.

We attempt to use a consistent notation and conceptual orientation throughout. For this reason, some techniques we discuss may appear superficially quite unlike the original works from which our descriptions were derived.

2. DEFINITIONS AND CONCEPTS

A. Alignment

The problem of aligning a parallel text is characterized more precisely as follows:

INPUT. A bitext $(\mathbf{C},\mathbf{D})$. Assume that $\mathbf{C}$ contains C passages and $\mathcal{C}$ is the set $\{1,2, \ldots, C\}$, and similarly for $\mathbf{D}$.[1]

OUTPUT. A set $\mathcal{A}$ of pairs (i, j), where (i, j) is a *coupling* designating a passage $\mathbf{C}_i$ in one text that corresponds to a passage $\mathbf{D}_j$ in the other, and $1 \leq i \leq C$ and $1 \leq j \leq D$. (In the general case, $\mathcal{A}$ can represent a nontotal many-to-many map from $\mathcal{C}$ to $\mathcal{D}$.)

The term "alignment" has become something of a misnomer in computational linguistics. Technically, in an alignment the coupled passages must occur in the same order in both texts, that is, with no crossings. Many alignment techniques have their roots in speech recognition applications, for which acoustic waveforms need to be aligned with transcriptions that are in the same order. In bitext research, the term alignment originally described the reasonable approximating assumption that para-

[1]We use the following notational conventions: bold letters ($\mathbf{X}$) are vectors, calligraphic letters ($\mathcal{X}$) are sets, and capital (nonbold) letters (X) are constants.

graph and sentence translations always preserve the original order. However, "word alignment" was subsequently coopted to mean coupling of words within sentences, even when word-coupling models do not assume that order is preserved across translation. To avoid confusion, we will use the term *monotonic alignment* whenever we mean "alignment" in its proper sense.

In a *bijective alignment*, every passage in each text is coupled with exactly one passage in the other text. A bijective alignment is *total*, meaning that no passage remains uncoupled. In practice, bijective alignments are almost never achievable except at the chapter or section granularity, or perhaps at the paragraph granularity for extremely tight translations. Ordinarily, we aim for a *partial alignment*, in which some passages remain uncoupled *singletons*; or we aim for a *many-to-many alignment*, in which passages may be coupled to multiple passages. Another often-useful approximating assumption is that the alignment is a (partial) function from a language-2 position to a language-1 position (but not necessarily vice versa). Such a function is an *injective* map and is written $i = \mathbf{a}(j)$. A bijective alignment is injective in both directions. For convenience, we will sometimes freely switch between the set notation $\mathcal{A}$ and the function notation $\mathbf{a}(\cdot)$ to refer to the same injective alignment.

A common generalization on the alignment problem is *hierarchical alignment*. The simplest case is to align sentences within paragraphs that are themselves aligned. A more complex case is to couple nested constituents in sentences.

Unless we specify otherwise, *alignments* are nonmonotonic, many-to-many, partial, and nonhierarchical.

B. Constraints and Correlations

Every alignment algorithm inputs a bitext and outputs a set of couplings. The techniques are nearly all statistical, because of the need for robustness in the face of imperfect translations. Much of the variation between techniques lies in the other kinds of information—constraints and correlations—that play a role in alignment. Some alignment techniques require one or more of these inputs, and bring them to bear on the alignment hypothesis space. Others derive or learn such kinds of information as by-products. In some cases the by-products are of sufficient quality that they can be taken as outputs in their own right, as mentioned earlier. Important kinds of information include the following.

BIJECTIVITY CONSTRAINT. Bijectivity is the assumption that the coupling between passages is 1-to-1 (usually in the sense of partial bijective maps, which allow some passages to remain uncoupled). This assumption is inapplicable at coarser granularities than the sentence level. However, it is sometimes useful for word level alignment, despite being clearly inaccurate. For example, consider when the words within a sentence pair are being aligned (Melamed, 1997b). If only one word in the language-1 sentence remains uncoupled and similarly for the language-2 sentence, then the bijectivity assumption implies a preference for coupling those two words, by process of elimination. Such benefits can easily outweigh errors caused by the inaccuracies of the assumption.

MONOTONICITY CONSTRAINT. This assumption reduces the problem of coupling bitext passages to a properly monotonic alignment problem: coupled passages occur in the same order in both sides of the bitext.

KNOWN ANCHORS. An anchor is a definitely coupled pair of passages or boundaries, and is a hard constraint. As shown in Fig. 1, anchor passages can be thought of as confirmed positions within the matrix that represents the alignment candidate space. The points (0,0) and (C,D) are the *origin* and *terminus*, respectively.

Taken together with the monotonicity constraint, a set of anchors divides the problem into a set of smaller, independent alignment problems. Between any two anchors are passages the alignment of which is still undetermined, but for which couplings must remain inside the region bounded by the anchors. As shown in Fig. 2, the correct alignment could take any (monotonic) path through the rectangular region delimited by the anchors. We call the candidate subspace between adjacent anchors a *slack range*.

There are two common cases of anchors:

1. *End constraints.* Most techniques make the assumption that the origin and terminus are anchor boundaries. Some techniques also assume a coupling between the first and last *passages*.
2. *Incremental constraints.* A previous processing stage may produce an alignment of a larger passage size. If we are willing to commit to the coarser alignment, we obtain a set of anchor boundaries.

The latter kind of anchor occurs in *iterative refinement* schemes, which progressively lay down anchors in a series of passes that gradually restrict the alignment candidate space. At the outset, candidate couplings may lie anywhere in the entire rectangular matrix; eventually, they are restricted to many small rectangular regions. Each region between adjacent anchors is a smaller alignment subproblem that can be

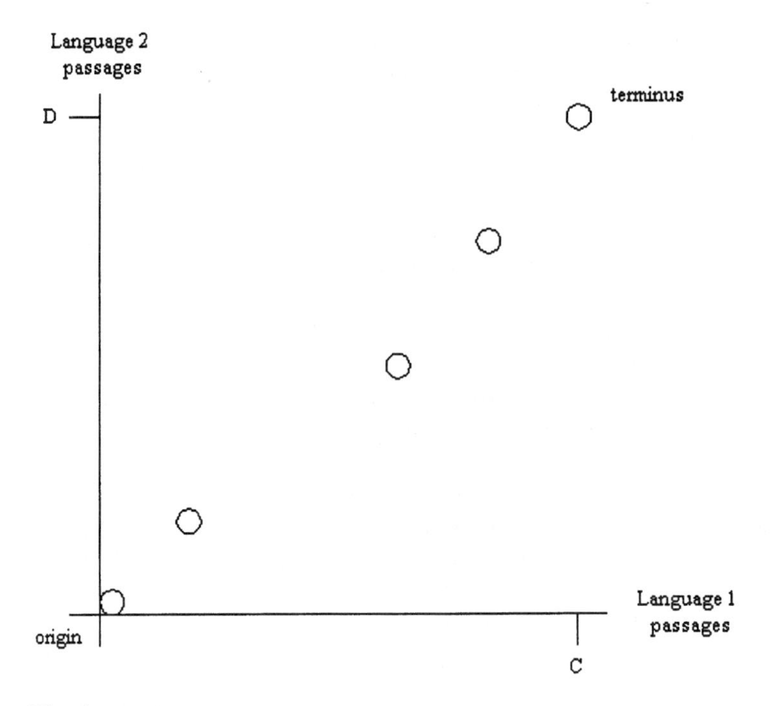

Fig. 1 Anchors.

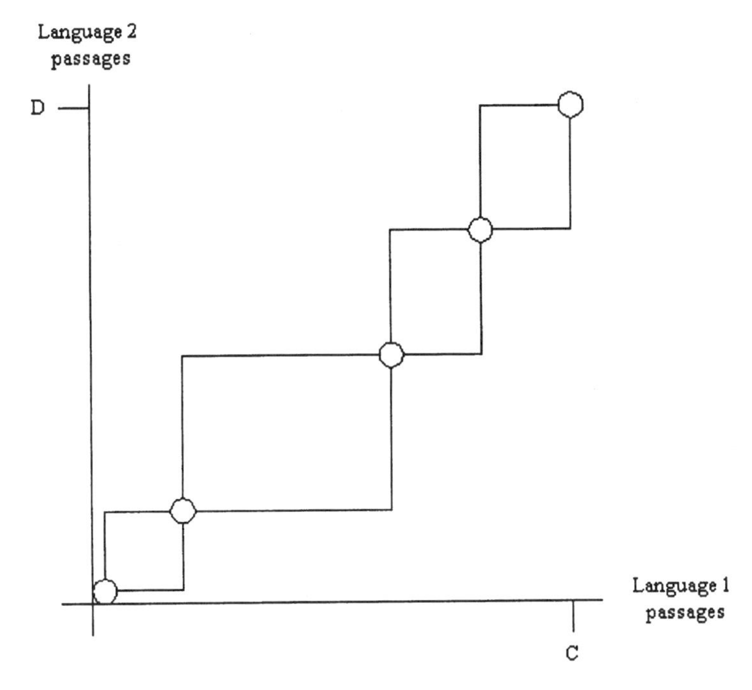

Fig. 2 Slack ranges between anchors.

processed independently. A variation on this schems is *hierarchical iterative refinement*, which produces a hierarchical alignment by using one pass at each granularity in the hierarchical structure. The first pass aligns passages at the coarsest level (commonly, sections or paragraphs); this level is chosen so that the total numbers of passages (C and D) are small. Committing to the output of this stage yields a set of anchor boundaries. The next pass aligns passages at the next finer level (commonly, paragraphs or sentences). This approach is taken so that the alignment subproblem corresponding to any slack range has small C and D values. The alignment is refined on each pass until the sentence level granularity is reached; at each pass, the quadratic cost is kept in check by the small number of passages between adjacent anchors.

BANDS. A common heuristic constraint is to narrow the rectangular slack ranges so that they more closely resemble bands, as shown in Fig. 3. We call this *banding*. Banding relies on the assumption that the correct couplings will not be displaced too far from the average, which is the diagonal between adjacent anchors. Different narrowing heuristics are possible.

One method is to model the variance assuming the displacement for each passage is independently and identically distributed. This means the standard deviation at the midpoint of a $C{:}D$ bitext is $O(\sqrt{C})$ for the language-1 axis and $O(\sqrt{D})$ for the language-2 axis. Kay and Röscheisen (1993) approximate this by using a banding function such that the maximum width of a band at its midpoint is $O(\sqrt{C})$.[2]

[2]They do not give the precise function.

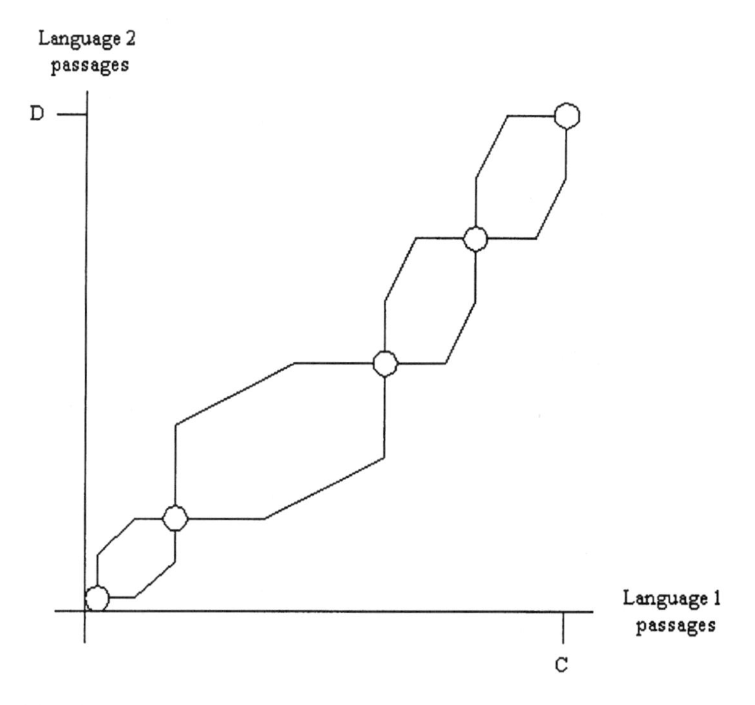

Fig. 3 Banding the slack ranges using variance.

Another method, from Simard and Plamondon (1996), is to prune areas of the slack range that are further from the rectangle's diagonal than some threshold, where the threshold is proportional to the distance between the anchors. This results in bands of the shape in Fig. 4.

Banding has the danger of overlooking correct couplings if there are large differences in the translations. Kay and Röscheisen (1993) report a maximum displacement of ten sentences from the diagonal, in a bitext of 255:300 sentences. However, most bitext, especially in larger collections, contains significantly more noise. The width of the band can be increased, but at significant computational expense. Section 5.B describes an effective counterstrategy that makes no initial banding assumptions.

GUIDES. A heuristic constraint we call *guiding*, from Dagan et al. (1993), is applicable when a rough *guide* alignment already exists as the result of some earlier heuristic estimate. The preexisting alignment can be used as a guide to seek a more accurate alignment. If we assume the preexisting alignment is described by the mapping function $\mathbf{a}^0(j)$, a useful alignment range constraint is to define an allowable deviation from $\mathbf{a}^0(j)$ in terms of some distance d: a language-1 position i is a possible candiate to couple with a language-2 position j iff

$$\mathbf{a}^0(j) - d \leq i \leq \mathbf{a}^0(j) + d \qquad \{1\}$$

This is depicted in Fig. 5. We denote the set of (i, j) couplings that meet all alignment range constraints as $\mathcal{B}$.

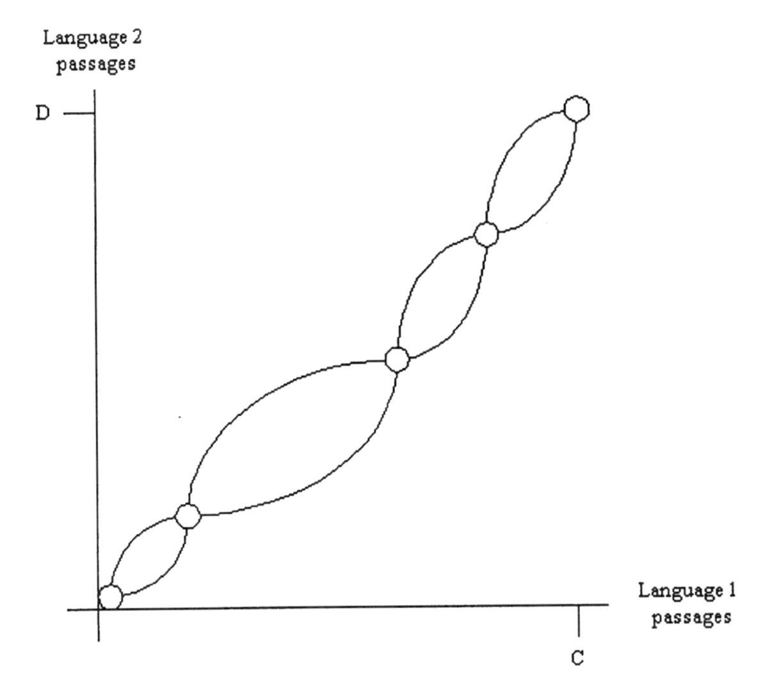

Fig. 4 Banding the slack ranges using width thresholds.

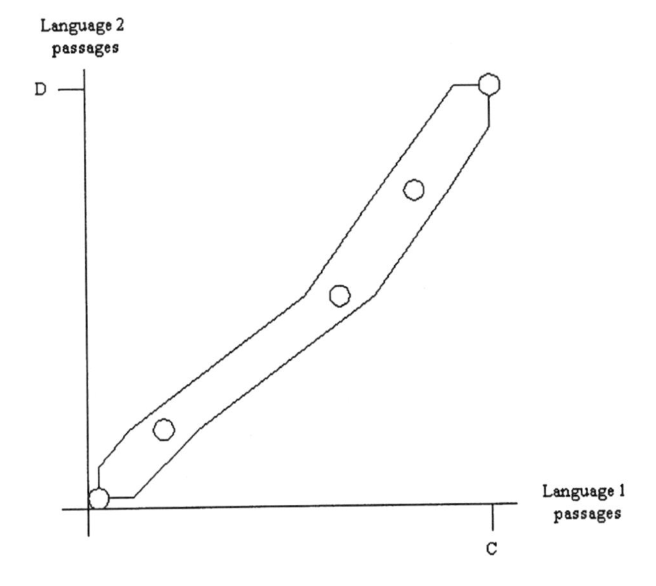

Fig. 5 Guiding based on a previous rough alignment.

ALIGNMENT RANGE CONSTRAINTS. Monotonicity, anchors, banding, and guiding are all special cases of alignment range constraints. There are many other possible ways to formulate restrictions on the space of allowable alignments. In general, alignment range constraints play an important role. Some alignment techniques are actually computationally infeasible without strong a priori alignment range constraints. Other techniques employ iterative modification of alignment range constraints, similar to iterative refinement with anchors.

BILINGUAL LEXICON CONSTRAINTS. A bilingual lexicon holds known translation pairs. Bilingual lexicons usually list single words, but may also contain multiword collocations, particularly for nonalphabetic languages in which the difference is not so clear.

The strongest word-coupling constraints come from words (or tokens, including punctuation) that have 1-to-1 deterministic translations. These in effect provide anchors if monotonicity is assumed. However, the rarity of such cases limits their usefulness.

A bilingual lexicon usually supplies word translations that are 1-to-many or many-to-many. The set of known word translations can be used to cut down on the space of candidate couplings. Clearly, whatever pruning method is used must still permit unknown word translations, for translation is not always word-by-word, and bilingual lexicons have imperfect coverage (especially in methods that learn a bilexicon as they perform the alignment).

A *graded bilingual lexicon* stores additional information about the *degree* of correlation or association in word pairs. This can be particularly clean in probabilistic alignment techniques. Various other statistical or ad hoc scores can also be employed.

COGNATES. Cognates are word pairs with common etymological roots; for example, the word pair *financed:financié* or *government:gouvernement* in English and French. For alphabetic languages that share the same (or directly mappable) alphabets, it is possible to construct heuristic functions that compare the spelling of two words or passages. In the simplest case, the function returns true or false (a decision function), acting similar to a low-accuracy, easily constructed bilingual lexicon with low memory requirements. Alternatively, a function that returns a score acts as a graded bilingual lexicon. In either case, a cognate funtion can be used either in place of or in addition to an alignment bilingual lexicon.

PASSAGE LENGTHS. The lengths of passages above the sentence granularity can be strong features for determining couplings. Most reported experiments indicate that the correlation is relatively strong even for unrelated languages, if the translation is tight. This means that the usefulness of this feature is probably more dependent on the genre than the language pair. Length-based methods are discussed in Sec. 4.A.

LANGUAGE UNIVERSALS. Outside of the relatively superficial features just discussed, relatively little attempt has been made to bring to bear constraints from theories about language-universal grammatical properties. One exception is discussed in Sec. 9. Language-universal constraints apply to all (or a large class) of language pairs, without making language-specific assumptions, and they apply at the sentence and word granularities.

3. DOCUMENT–STRUCTURE ALIGNMENT TECHNIQUES

Various corpus-specific heuristics may be used to align passages at the document–structure granularity. Usually, these are merely the preprocessing for a finer alignment.

At the coarsest level, bitexts may be aligned at the file level. For example, transcripts such as the Canadian Hansards or United Nations corpora are usually stored one file per session; in this case the date is sufficient to align the sessions.

Depending on the domain, the text may contain certain labels that are easily identified and can be coupled with high reliability. Section headings are a common example of this, particularly if they are numbered consistently across the languages. In government transcripts, dialogues usually contain labels to identify the speakers. Explicit markup, for either font style or document structure, is sometimes available. By precoupling such labels, a set of anchors can be obtained.

Because document–structure techniques are by their nature corpus-specific, we will not discuss them further, except to point out their usefulness for the first pass(es) when alignment is performed using iterative refinement.

4. SENTENCE ALIGNMENT TECHNIQUES

Techniques for sentence alignment are generally applicable to larger units as well, such as paragraphs and sections. By taking advantage of this, hierarchical iterative refinement usually works rather well, by first aligning paragraphs, yielding paragraph pairs for which internal sentences can subsequently be aligned.

Broadly speaking, sentence alignment techniques rely on sentence lengths, on lexical constraints and correlations, or on cognates. Other features could no doubt be used, but these approaches appear to perform well enough.

A. Length-Based Sentence Alignment

The length-based approach examines the lengths of the sentences. It is the most easily implemented technique, and performs nearly as well as lexical techniques for tightly translated corpora, such as government transcripts. The overall idea is to use dynamic programming to find a minimum cost (maximum probability) alignment, assuming a simple hidden generative model that emits sentences of varying lengths. Purely length-based techniques do not examine word identities at all, and consider the bitext as nothing more than a sequence of sentences the lengths of which are the only observable feature, as depicted in Fig. 6. The length-based approach to sentence alignment was first introduced by Gale and Church (1991) and Brown et al. (1991), who describe essentially similar techniques. The most thorough evaluation results are found in an updated article (Gale and Church, 1993).

The generative model that underlies the technique is a one-state hidden Markov model, as shown in Fig. 7, for which S is both the start and final state. The model emits bitext as a series of “parallel clumps” or “beads,” each bead being a short clump of E sentences in language-1 coupled with a short clump of F sentences in language-2. For example, Fig. 8 shows one sequence of beads that could have generated the bitext of Fig. 6.

Lengths of sentences in the bitext

Language 1:	10	30	15	12		
Language 2:	12	14	15	12	11	2

Fig. 6 Example sentence lengths in an input bitext.

The output distribution contains two factors: the bead type and the bead's clump lengths. For the bead type, the usual practice is to allow at least the following *E*:*F* configurations: 0:1, 1:0, 1:1, 1:2, and 2:1. Allowing 2:2 beads in addition is reasonable. It can be convenient to think of these bead types as operations for translating language-1 passages into language-2, respectively: insertion, deletion, substitution, expansion, contraction, and merger. Beads with three or more sentences in one language are not generally used, despite the fact that 1:3, 2:3, 3:3:, 3:2, and 3:1 sentence translations are found in bitext once in a while. This is because alignment of such passages using only the sentence length feature is too inaccurate to make the extra parameters and computation worthwhile. A simple multinomial distribution over the bead types can be estimated from a small hand-aligned corpus. Alternatively, EM can be used to estimate this distribution simultaneously with the clump length distributions.

Given the bead type, the length l_1 of the language-1 clump is determined next. The length of each sentence in the language-1 clump is assumed to be independent, and to follow some distribution. This distribution is implicitly Poisson in the case of Gale and Church (1991). In Brown et al. (1991), relative frequencies are used to estimate probabilities for short sentence lengths (up to approximately 80 words), and the distribution for longer sentences is fit to the tail of a Poisson. Because the empirical distribution for short sentences is fairly Poisson-like (Fig. 9), these variations do not appear to have a significant effect.

Finally, the length l_2 of the language-2 clump is determined, by assuming that its difference from the length of the language-1 clump follows some distribution,

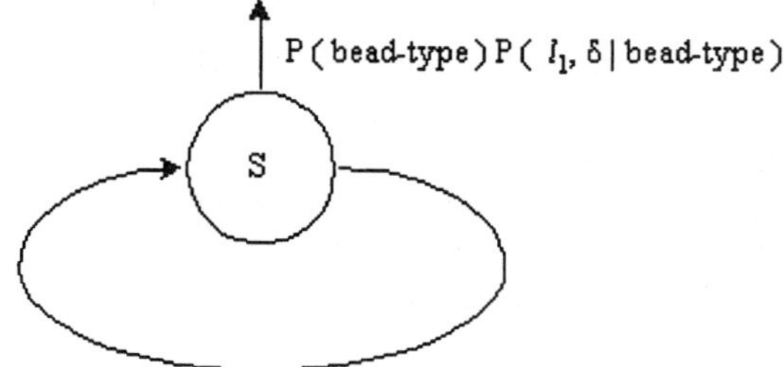

Fig. 7 Markov model interpretation for the bead generation process.

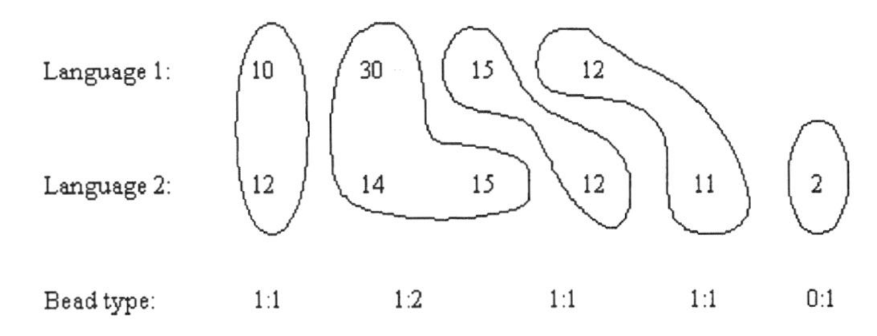

Fig. 8 Beads.

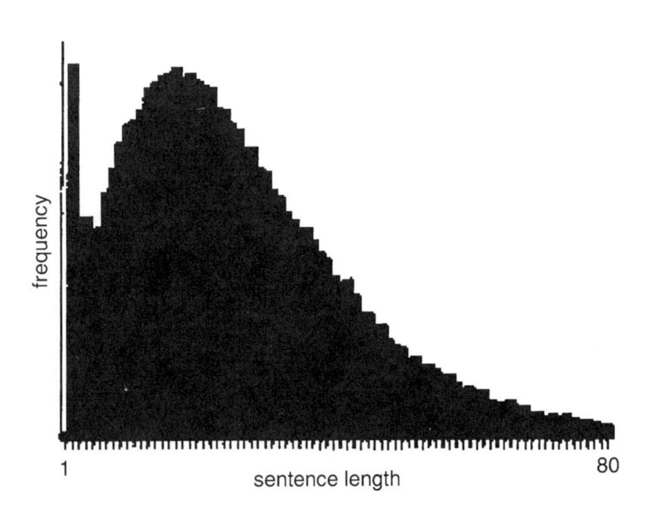

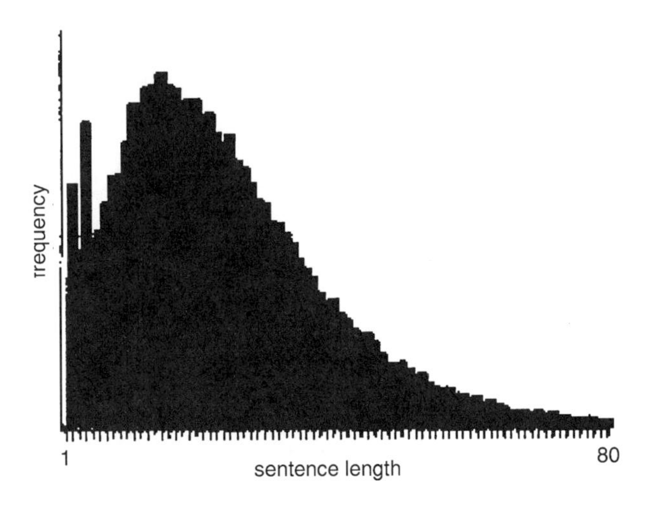

Fig. 9 Empirical distribution on sentence lengths. (From Brown et al., 1991)

usually a normal distribution. The difference functions used by Gale and Church (1991) Eq. {2} and Brown et al. (1993) Eq. {3} are as follows:

$$\delta(l_1, l_2) = \frac{(l_2 - l_1 c)}{\sqrt{l_1 s^2}} \qquad \{2\}$$

$$\delta(l_1, l_2) = \log \frac{l_2}{l_1} \qquad \{3\}$$

Aside from normalizing the mean and variance to one's arbitrary preference, the only decision is whether to take the logarithm of the sentence length difference, which sometimes produces a better fit to the empirical distribution.

These assumptions are sufficient to compute the probability of any bead. It is convenient to write a candidate bead as (*bead-type, i,j*) where the language-1 clump begins with the *i*th passage in **C** and the language-2 clump begins with the *j*th passage in **D**. For instance, a candidate 1:2 bead that hypothesizes coupling $\mathbf{C}_{31}$ with $\mathbf{D}_{36}$, $\mathbf{D}_{37}$ has the estimated probability $\hat{P}(1{:}2,31,36)$.

Given this generative model, the actual alignment algorithm relies on dynamic programming (Bellman, 1957) to find the maximum probability alignment. The recurrence has the structure of dynamic time-warping (DTW) models and is based on the fact that the probability of any sequence of beads can be computed by multiplying the probability of last bead with the total probability of all the beads that precede it. As usual, it is convenient to maintain probabilities in the log domain, so we define the maximum log probability up to passages (i, j) in $(\mathbf{C}, \mathbf{D})$ to be

$$\delta(i, j) = \log P^* \, (\mathbf{C}_{0..i}, \mathbf{D}_{0..j}) \qquad \{4\}$$

The recurrence chooses the best configuration over the possible types of the last bead.

$$\textit{Basis} \qquad \delta(0, 0) = 0 \qquad \{5\}$$

$$\textit{Recursion} \qquad \delta(i, j) = \min \left\{ \begin{array}{c} \delta(i, j-1) - \log \hat{P}(0{:}1, i, j-1) \\ \delta(i-1, j) - \log \hat{P}(1{:}0, i-1, j) \\ \delta(i-1, j-1) - \log \hat{P}(1{:}1, i-1, j-1) \\ \delta(i-1, j-2) - \log \hat{P}(1{:}2, i-1, j-2) \\ \delta(i-2, j-1) - \log \hat{P}(2{:}1, i-2, j-1) \\ \delta(-2i, j-2) - \log \hat{P}(2{:}2, i-2, j-2) \end{array} \right\} \qquad \{6\}$$

Note that the form of the recurrence imposes a set of slope constraints on the time warping.

The most significant difference between the methods is that Gale and Church's (1991) method measures sentence lengths in terms of number of characters, whereas Brown et al. (1991) use number of words (as determined by heuristic segmentation). Gale and Church (1993) report that using characters instead of words, holding all other factors constant, yields higher accuracy (in their experiment, an error rate of

4.2% for characters as compared with 6.5% for words). The reasons are not immediately obvious, although Gale and Church (1993) use variance measures to argue that there is less uncertainty because the number of characters is larger (117 characters per sentence, as opposed to 17 words). However, these experiments were conducted only on English, French, and German, for which the large number of cognates improve the character length correlation.

Wu (1994) showed that for a large English and Chinese government transcription bitext, where the cognate effect does not exist, Gale and Church's method is somewhat less accurate than for English, French, and German, although it is still effective. Church et al. (1993) show that sentence lengths are well correlated for the English and Japanese AWK manuals, but do not actually align the sentences. We know of no experimental results comparing character and word length methods on noncognate languages; the lack of such experiments is partly because of the well-known difficulties in deciding word boundaries in languages such as Chinese (Chiang et al., 1992; Lin et al., 1992; Chang and Chen, 1993; Lin et al., 1993; Wu and Tseng, 1993; Sproat et al., 1994; Wu and Fung, 1994).

Bitexts in some language pairs may exhibit highly dissimilar sentence and clause structures, leading to very different groupings than the simple beads we have been considering. For example, although the sentence length correlations are strong in the English–Chinese government transcriptions used by Wu (1994), Xu and Tan (1996) report that the CNS English–Chinese news articles they use have very different clause and sentence groupings and, therefore, suggest generalizing the beads to allow many-clause–many-clause couplings.

In general, length-based techniques perform well for tightly translated bitexts. However, they are susceptible to misalignment if the bitext contains long stretches of sentences with roughly equal length, as for example, in dialogues consisting of very short utterances, or in itemized lists. One possible solution is discussed in the section on lexical techniques.

The basic algorithm is $O(CD)$ in both space and time, that is, approximately quadratic in the number of passages in the bitext. This is prohibitive for reasonably large bitexts. Three basic methods for circumventing this problem are banding, the hierarchical variant of iterative refinement, and thresholding.

Banding is often a feasible approach because the true alignment paths in many kinds of bitexts lie close enough to the diagonal that they fall within reasonable bands. However, banding is not favored when the other two approaches can be used, because they make fewer assumptions about the alignment path.

Hierarchical iterative refinement appears to work well for tightly translated bitexts, such as government transcripts. Generally, only a few passes are needed: session or section (optional), paragraph, speaker (optional), and sentence.

Thresholding techniques prune the δ matrix so that some alignment prefixes are abandoned during the dynamic-programming loop. Many variants are possible. Relative thresholding is more appropriate than absolute thresholding; the effect is to prune any alignment prefix for which the probability is excessively lower than the most-probable alignment prefix of the same length. Beam search approaches are similar, but limit the number of "live" alignment prefixes for any given prefix length. It is also possible to define the beam in terms of a upper–lower bounded range (j^-, j^+), so that the loop iteration that computes the ith column of $\delta(i, j)$ only considers $j^- \leq j \leq j^+$. This approach is similar to banding, except that the center of the band is dynamically adjusted during the dynamic-programming loop.

Thresholding can cause large warps (deletions, insertions) to be missed. Chen (1993) suggests a resynchronization method to improve the robustness of relative thresholding schemes against large warps, based on monitoring the size of the "live" prefix set during the dynamic-programming loop. If this set reaches a predetermined size, indicating uncertainty over the correct alignment prefix, the presence of a large warp is hypothesized and the alignment program switches to a resynchronization mode. In this mode, both sides of the bitext are linearly scanned forward from the current point, seeking rare words. When corresponding rare words are found in both sides, a resynchronization point is hypothesized. After collecting a set of possible resynchronization points, the best one is selected by attempting sentence alignment for some significant number of sentences following the candidate resynchronization point, taking the resulting probability as an indication of the goodness of this resynchronization point. Resynchronization can improve error rates significantly when the bitext contains large warps.

B. Lexical Sentence Alignment

The prototypical lexically based sentence alignment technique was proposed by Kay and Röscheisen (1988) in the first paper to introduce an iterative refinement solution to the bitext alignment problem.[3] The method employs banding, and has the structure shown in the algorithm in Fig. 10.[4]

Although this algorithm can easily be implemented in a straightforward form, its cost relative to lexicon size and bitext length is prohibitive unless the data structures and loops are optimized. In general, efficient implementations of the length-based methods are easier to build and yield comparable accuracy, so this lexical algorithm is not as commonly used.

A notable characteristic of this algorithm is that it constructs a bilingual lexicon as a by-product. Various criterial functions for accepting a candidate word pairing are possible, rating either the word pair's degree of correlation or its statistical significance. For the correlation between a candidate word pair (w, x), Kay and Röscheisen (1988) employ Dice's coefficient (van Rijsbergen, 1979) Eq. {8}, whereas Haruno and Yamazaki (1996) employ mutual information (Cover and Thomas, 1991) Eq. {8}. For the statistical significance, Kay and Röscheisen simply use frequency, and Haruno and Yamazaki use t-score Eq. {9}.

$$\text{Dice} = \frac{2N(w, x)}{N(w) + N(x)} \qquad \{7\}$$

[3]A simplified version of the method was subsequently applied to a significantly larger corpus by Catizone et al. (1989).

[4]For the sake of clarifying the common conceptual underpinnings of different alignment techniques, our description uses terms different from Kay and Röscheisen (1988). Roughly, our $\mathcal{A}$ is the Sentence Alignment Table, our candidate space is the Alignable Sentence Table, and our bilexicon is the Word Alignment Table.

1. Initialize the set of anchor (sentence) alignments $\mathcal{A}$

2. **repeat**

3. Compute the (sentence alignment) candidate space by banding between each adjacent pair of anchors

4. Collect word-similarity statistics from the candidate space

5. Build a bilexicon containing sufficiently similar words

6. Collect sentence-similarity statistics from the candidate space, with respect to the bilexicon

7. Add sufficiently confident candidates to the set of alignments $\mathcal{A}$

8. **until** no new candidates were added to $\mathcal{A}$

9. **return** $\mathcal{A}$

Fig. 10 Lexical sentence alignment algorithm.

$$\mathrm{MI} = \log \frac{NN(w, x)}{N(w)N(x)} \tag{8}$$

$$\mathrm{t} = \frac{P(w, x) - P(w)P(x)}{\sqrt{P(w, x)/N}} \tag{9}$$

In either case, a candidate word pair must exceed thresholds on both correlation and significance scores to be accepted in the bilingual lexicon.

The bilingual lexicon can be initialized with as many preexisting entries as desired; this may improve alignment performance significantly, depending on how accurate the initial anchors are. Even when a good preexisting bilingual lexicon is available, accuracy is improved by continuing to add entries statistically, rather than "freezing" the bilingual lexicon. Haruno and Yamazaki (1996) found that combining an initial bilingual lexicon (40,000 entries from a CD-ROM) with statistical augmentation significantly outperformed versions of the method that either did not employ the CD-ROM bilingual lexicon or froze the bilingual lexicon to the CD-ROM entries only. Results vary greatly depending on the bitext, but show consistent significant improvement across the board. Precision and recall are consistently in the mid-90% range, up from figures in the 60 to 90% range.

With respect to applicability to non–Indo-European languages, Haruno and Yamazaki's experiments show that the algorithm is usable for Japanese–English bitexts, as long as the Japanese side is presegmented and tagged. Aside from the variations already discussed, two other modified strategies are employed. To improve the discriminatory ability of the lexical features, an English–Japanese word coupling is allowed to contribute to a sentence-similarity score only when the English word occurs in only one of the candidate sentences to align to a Japanese sentence. In addition, two sets of thresholds are used for mutual information and *t*-score, to divide the candidate word couplings into high-confidence and low-confidence classes. The high-confidence couplings are weighted three times more heavily. This strategy attempts to limit damage from the many false translations that may be statistically acquired, but still allow the low-confidence bilingual lexicon entries to influence the later iterations when no more discriminative leverage is avaliable from the high-confidence entries.

It is possible to construct lexical sentence alignment techniques based on underlying generative models that probabilistically emit beads, similar to those used in length-based techniques. Chen (1993) describes a formulation in which each bead consists of a multiset of word pairs. Instead of distributions that govern bead lengths, we have distributions governing the generation of word pairs. Word order is ignored, for this would be unlikely to improve accuracy significantly, despite greatly worsening the computational complexity. Even with this concession, the cost of alignment would be prohibitive without a number of additional heuristics. The basic method employs the same dynamic-programming recurrence as Eq. {6}, with suitable modifications to the probability terms; the quadratic cost would be unacceptable, especially because the number of word-pair candidates per head candidate is exponential in the bead length. Chen employs the relative thresholding heuristic, with resynchronization. The probability parameters, which are essentially the same as for the length-based methods except for the additional word-pair probabilities, are estimated using the Viterbi approximation to EM.

Similar to Kay and Röscheisen's method, this method induces a bilingual lexicon as a side effect. For the method to be able to bootstrap this bilexicon, it is important to provide a seed bilingual lexicon initially. This can be accomplished manually, or by a rough estimation of word-pair counts over a small manually aligned bitext. The quality of bilingual lexicons induced in this manner has not been investigated, although related EM-based methods have been used for this purpose (see later discussion).

The alignment precision of this method on Canadian Hansards bitext may be marginally higher than the length-based methods. Chen (1993) reports 0.4% error as compared with 0.6% for Brown et al.'s method (but this includes the effect of resynchronization, which was not employed by Brown et al.). Alignment recall improves by roughly 0.3%. Since the accuracy and coverage of the induced bilingual lexicon has not been investigated, it is unclear how many distinct word pairs are actually needed to obtain this level of performance improvement. The running time of this method is "tens of times" slower than length-based methods, despite the many heuristic search approximations.

To help reduce the running time requirements, it is possible to use a less expensive method to impose an initial set of constraints on the coupling matrix. Simard and Plamondon (1996) employ a cognate-based method within a framework similar to SIMR (discussed later) to generate a set of anchors, and then apply banding.

The *word_align* method (Dagan et al., 1993) may also be used for lexical sentence alignment, as discussed in a later section.

C. Cognate-Based Sentence Alignment

For some language pairs, such as English and French, the relatively high proportion of cognates makes it possible to use cognates as the key feature for sentence alignment. Simard et al. (1992), who first proposed using cognates, manually analyzed approximately 100 English–French sentence pairs from the Canadian Hansards, and found that roughly 21% of the words in such sentence pairs were cognates. In contrast, only 6% of the words in randomly chosen nontranslation sentence pairs were cognates.

Cognate-based alignment, in fact, can be seen as a coarse approximation to lexical alignment, in which the word pairs are based on a heuristic operational definition of cognates, rather than an explicitly listed lexicon. The cognate identification heuristic used by Simard et al. considers two words w, x to be cognates if:

1. w, x are identical punctuation characters.
2. w, x are identical sequences of letters and digits, with at least one digit.
3. w, x are sequences of letters with the same four-character prefix.

This heuristic is clearly inaccurate, but still discriminates true from false sentence pairs fairly well: it considers 30% of the words in bilingual sentence pairs to be cognates, versus only 9% in nontranslation sentence pairs.

Given such a definition, any of the lexical alignment methods could be used, simply by substituting the heuristic for lexical lookup. The techniques that require probabilities on word pairs would require that a distribution of suitable form be imposed. Simard et al. use the dynamic-programming method; however, for the cost term, they use a scoring function based on a log-likelihood ratio, rather than a generative model. The score of a bead containing c cognates and average sentence length n is:

$$\text{Score} = -\log\left(\frac{P(c|n,t)}{P(c|n,\neg t)} \times P(bead - type)\right) \qquad \{10\}$$

The pure cognate-based method does not perform as well as the length-based methods for the Canadian Hansards. In tests on the same bitext sample, Simard et al. obtain a 2.4% error rate, whereas Gale and Church's (1991) method yields a 1.8% error rate. The reason appears to be that even though the mean number of cognates differs greatly between coupled and noncoupled sentence pairs, the variance in the number of cognates is too large to separate these categories cleanly. Many true sentence pairs share no cognates, whereas many nontranslation sentence pairs share several cognates by accident.

Cognate-based methods are inapplicable to language pairs, such as English and Chinese, for which no alphabetic matching is possible.

D. Multifeature Sentence Alignment Techniques

It is possible to combine sentence-length, lexical, and cognate features in a single model. Because a length-based alignment algorithm requires fewer comparison tests than a lexical alignment algorithm, all other things being equal, it is preferable to employ length-based techniques when comparable accuracy can be obtained. However, length-based techniques can misalign when the bitext contains long stretches of sentences with roughly equal length. In this case, it is possible to augment the length features with others. One method of combining features is by incorporating them into a single generative model. Alternatively, the less expensive length-based feature can be used in a first pass; afterward, the uncertain regions can be realigned using the more expensive lexical or cognate features. Uncertain regions can be identified either using low log probabilities in the dynamic-programming table as indicators of uncertain regions, or using the size of the "live" prefix set during the dynamic-programming loop.

The method of Wu (1994) combines sentence-length with lexical word-pair features using a single generative model. A limited word-pair vocabulary is chosen before alignment; to be effective, each word pair should occur often in the bitext and should be highly discriminative (i.e., the word translation represented by the word pair should be as close to deterministic as possible). The total set of chosen word pairs should be small; otherwise, the model degenerates into full lexical alignment with all the computational cost. The model assumes that all words in a bead, except for those that belong to the set of chosen word pairs, are generated according to the same length distribution as in the basic length-based model. The dynamic programming algorithm is generalized to find the maximum probability alignment under this hybrid model. The degree of performance improvement with this method is highly dependent on the bitext and the word-pair vocabulary that is chosen.

Simard et al. (1992) combine sentence length with cognate features using a two-pass approach. The first pass is Gale and Church's (1991) method, which yields a 1.8% error rate. The second pass employs the cognate-based method, improving the error rate to 1.6%.

E. Comments on Sentence Alignment Techniques

Sentence alignment can be performed fairly accurately regardless of the corpus length and method used. Perhaps surprisingly, this holds even for lexicon-learning alignment methods, apparently because the decreased lexicon accuracy is offset by the increased constraints on alignable sentences.

The advantages of the length-based method are its ease of implementation and speed. Although it is sometimes argued that the asymptotic time complexity of efficiently implemented lexical methods is the same as length-based methods, even then the constant factor for them would be significantly costlier than the length-based methods.

The advantages of the lexical methods are greater robustness, and the side effect of automatically extracting a lexicon.

Multifeature methods appear to offer the best combination of running time, space, accuracy, and ease of implementation for aligning sentences in tightly translated bitexts.

5. ALIGNMENT TECHNIQUES FOR NOISY BITEXTS

Noisy bitexts present difficulties that require a different set of approaches than the sentence alignment techniques discussed in the previous section. By "noisy" we mean that the bitext contains significant amounts of material that is not tightly translated on even sentence boundaries, so that the assumptions of the bead model break down. Noise can arise from many sources, for example:

- Nonliteral translation
- Out-of-order translation
- Omitted sections

- Floating passages: footnotes, figures, headers, and such
- OCR errors
- Sentence segmentation errors

Beyond a certain point, even the resynchronization methods are not worthwhile. For such cases, a more robust group of techniques have been developed, taking a less-structured view of bitext.

Methods for noisy bitext alignment usually output a set of anchors, rather than a total mapping of sentences. There are two reasons for this. The presence of noise in the bitext inherently means that we can expect only a partial alignment. In addition, given the typical sources of noisy bitext, it is more difficult to accurately identify structural units, such as paragraphs and sentences (due to missing blank lines, floating passages, missing punctuation).

Given the set of anchors, one can always interpolate linearly between the anchors to provide an alignment up to whatever resolution is desired—paragraph, sentence, word, character, or byte. The monotonicity assumption is not really valid at finer granularities than the sentence level, so any such alignment will contain many coupling errors. For many concordancing applications, such errors may be acceptable as long as the alignment is within a sentence or so of the true coupling.

A. Cognate-Based Alignment for Noisy Bitexts

The *char_align* method (Church, 1993) aligns passages at the character level; it views bitext as merely a pair of character sequences. Clearly, the two sides of a bitext will not have the same character order (or even word order), so the alignment assumption is simply an approximation that makes it convenient to find the rough vicinity of the true coupling, for example, to arrive within a sentence or two of the corresponding passages. This resolution is sufficient for some applications in translators' aids, or as an early pass in a more complex alignment scheme.

The method is cognate-based, and as such possesses similar strengths and weaknesses to cognate-based sentence alignment. The same character encoding must be used in both sides of the bitext. The cognate identification heuristic is as follows: the two characters $\mathbf{C}_s$ and $\mathbf{D}_t$ at positions s and t are considered possible cognates when the (character) 4-grams beginning at s and t are identical, that is

$$(\mathbf{C}_s \ldots \mathbf{C}_{s+3}) = (\mathbf{D}_t \ldots \mathbf{D}_{t+3}) \qquad \{11\}$$

Memory considerations are even more important than for sentence alignment, because C and D are in terms of characters instead of sentences. The method deals with this through banding, and by initially approximating the $C \times D$ alignment matrix at lower resolutions. Although the method is iterative, it does not use anchors, and employs only a single, constant-width band between the endpoints of the corpus, that is gradually narrowed. The iteration has the structure shown in the algorithm in Fig. 11.

The "dotplot" cognate density matrix is a low-resolution discretization of the alignment space, that (a) can fit into memory, and (b) captures the density of possible cognates through the alignment space. Church (1993) uses low-pass filtering and thresholding to down-sample the matrix, but it is also possible to employ a simpler approach of partitioning both the C and D axes into uniformly spaced bins. A dotplot is graphically depicted in Fig. 12.

1. b ? maximum width of band that fits in memory
2. **repeat**
3. Compute the "dotplot" cognate density matrix
4. Search for the best alignment path A through the dotplot
5. b ? maximum deviation of any point in A from the diagonal
6. **until** b does not decrease
7. Linearly interpolate a path through A up to desired resolution (e.g., individual characters)

Fig. 11 *Char _align* algorithm.

A dotplot constitutes a (nonprobabilistic) score function on which we can perform the standard dynamic-programming search for a best path (at the resolution of the dotplot). Subsequently, the band can be narrowed by taking the maximum deviation of this path from the diagonal as the new width of the band. The memory saved by banding can be used to increase the resolution and reiterate the process. Several iterations eventually yield the highest-resolution band that fits in memory yet still contains the alignment path. Finally, the resulting best path can be interpolated to give an alignment at any desired resolution, up to the level of individual characters.[5]

The performance of *char_align* is highly dependent on the language pair and the bitext. Church (1993) reports good performance on English and French, with accuracy usually within the average length of a sentence.

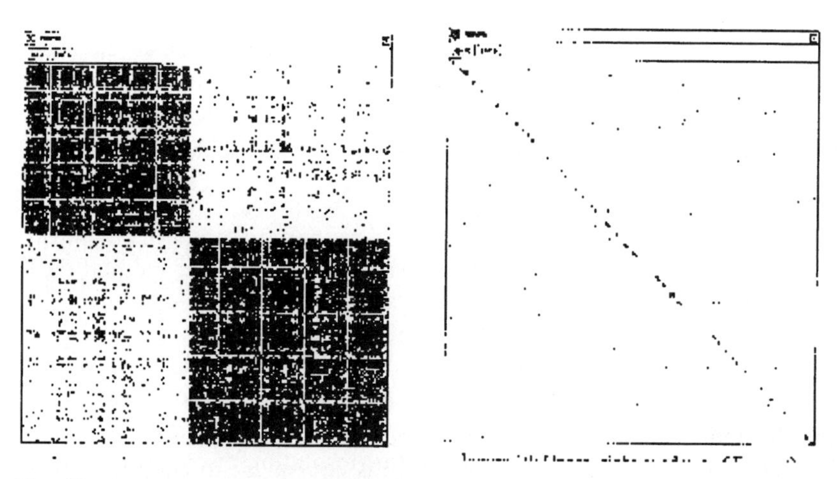

Fig. 12 Example of a dotplot: The right-hand picture shows the upper-right quadrant of the main plot enhanced by signal processing.

[5]For implementational convenience, Church (1993) uses a version of the alignment space that is rotated 45° to make the band horizontal instead of diagonal.

In general, cognate-based methods do not work well for languages that do not share the same alphabet, or nonalphabetic languages (particularly the CJK languages). One exception is for technical material such as the AWK manual, where there are enough English terms scattered through the Japanese translation to provide anchors to some extent (Church et al., 1993).

B. Lexical Alignment for Noisy Bitexts

The "DK-vec method" (Fung and McKeown, 1994, 1997; Fung, 1995) employs lexical features in a manner well-suited to noisy bitexts. The key representational insight is to use the *recency* between repeat occurrences of a word. *Recency* is defined as the number of bytes since the previous occurrence of the same word. Recency appears to be a feature that can remain reliable even in a noisy bitext. (Note, that although sentence length is also a feature that measures distances, sentence length is not robust in noisy bitexts.)

The method gains robustness against large insertions and deletions by avoiding initial banding assumptions. Instead, it uses many *individual* word pairs to produce many separate crude alignments. Afterward, all the crude alignments are analyzed together, to produce a better composite alignment.

Alignment proceeds in several phases as in the algorithm shown in Fig. 13. The method begins by hypothesizing many candidate word pairs, choosing the words from the midfrequency range. High-frequency words are usually function words for which the usage pattern is too language-specific to help anchor an alignment. On the other hand, low-frequency words do not provide enough anchor points over the bitext. Each candidate word pair (w, x) is then scored. Each individual score computation can actually be thought of as a miniature stand-alone instance of the dynamic-programming alignment discussed earlier, where:

- *Passage* is taken to mean a stretch of the text beginning with w (or x), and ending just before the next occurrence of w (or x). In sentence alignment, a passage was a sentence, but here, sentence segmentation is never performed.
- *Bead* is taken to mean a 0:1, 1:1, or 1:0 coupling of passages.
- *Bead cost* is taken to mean the difference between the byte length of its language-1 and language-2 clumps (i.e., $|l_1 - l_2|$).
- *Score* is taken to mean the *normalized* cost of the alignment path. *Normalized* means to divide the accumulated cost of a path by the number of beads on the path, giving the average cost *per bead.* Normalization is important to be able to compare scores between *different* candidate word pairs, which can break the same bitext into completely different numbers of passages.[6]

Interestingly, this miniature alignment step can be thought of as being both length-based *and* lexical, because coupling passages here is equivalent to coupling occurrences of w and x and the lengths of the passages are dictated by w and x.

[6]In principle, sentence alignment should perhaps also have used normalization, but the practical effect is much less significant because the number of sentences is much larger.

1. /* *Build a set of candidate word pairs.* */}

2. **for** each language-1 word w in the mid-frequency range

3. **for** each language-2 word x in the mid-frequency range

4. *CandidatePairs* ? $(w,x) \cup$ *CandidatePairs*

5. /* *Extract alignment lexicon and candidate anchor set.* */

6. **for** each word pair $(w, x) \in$ *CandidatePairs*

7. *TempBilexicon* $\leftarrow \{(w, x)\}$

8. Apply lexical alignment to the entire bitext using *TempBilexicon*

9. *Score*$(w, x) \leftarrow$ the alignment path's normalized score

10. *WordAnchors*$(w, x) \leftarrow$ the set of anchor points

11. **if** *Score*$(w, x) >$ *threshold*

12. *AlignmentBilexicon* $\leftarrow (w,x) \cup$ *AlignmentBilexicon*

13. *CandidateAnchors* $\leftarrow$ *WordAnchors*$(w,x) \cup$ *CandidateAnchors*

14. /* *Select anchors from candidate anchor set.* */

15. Use a greedy heuristic search to prune the candidates (see text) *Anchors*$(w, x) \leftarrow$ the remaining anchors

Fig. 13 The DK-vec algorithm.

Any candidate word pair (w, x) for which the score exceeds a preset threshold is accepted as a member of the alignment lexicon.[7] In other words, we take all the word pairs that align the bitext reasonably well, where "reasonably well' means that the alignment is fairly close to being diagonal. Note that it is possible to adjust the threshold dynamically if not enough candidates are found, under the suspicion that the bitext may be extremely warped, so the alignment should be allowed to deviate farther from the diagonal.

The final set of anchors is obtained by choosing a path through the candidate anchor set. Because the candidate anchors already all lie relatively close to the diagonal, Fung and McKeown (1994) suggest a method that employs a simple greedy heuristic. Let the candidate anchor set be recorded in terms of byte positions, sorted in order of the position in the language-1 text, that is, $(i_0, j_0), (i_1, j_1), \ldots, (i_{A-1}, j_{A-1})$ such that $i_0 < i_1 \ldots < i_A$. We proceed through the list in order, accepting a candidate (i, j) if it meets the slope constraint in Eq. {12}.

$$\text{Lower threshold} < \frac{j - j'}{i - i'} < \text{upper threshold} \qquad \{12\}$$

[7]The alignment lexicon is called a "primary lexicon" in the original papers.

where (i', j') is the immediately preceding candidate anchor to be accepted. Alternatively, another dynamic programming pass can be performed on the candidate anchor set, especially when the alignment is suspected to be extremely warped.[8]

A simple, efficient implementation of the alignment is possible, taking advantage of the fact that the bilingual lexicon contains only a single word pair and that the passages being aligned are delimited by occurrences of those words. For each candidate word (in both language-1 and language-2), all the recency values are stored in a vector known as a "DK-vec" (Fung and McKeown, 1994). For each candidate word pair, dynamic time warping between the two words' respective DK-vecs yields the score.

The recency feature remains fairly reliable, even in a noisy bitext, as long as the words that are chosen are relatively monosemous and occur often enough in the corpus. The method's lack of assumptions about sentence segmentations, along with its efficiency at pursuing alignments that deviate far from the diagonal, contribute to its ability to handle noisy bitexts.

The DK-vec method is well suited for nonalphabetic languages. In fact, development of the DK-vec method was partly motivated by such problems, with the work of Fung and McKeown (1997) applied primarily to English–Japanese and English–Chinese. Extensive evaluations on other languages are not yet available. Somers (1998) reports an evaluation of the front-end method for comparing DK-vecs of candidate word pairs, on a number of word pairs involving English, French, German, Dutch, Spanish, and Japanese, using the ACL/ECI ESPRIT and ITU parallel corpora. (Note this is *not* an evaluation of alignment accuracy, but only the scoring metric for constructing the alignment lexicon.) Precision of the proposed word pairs in the alignment lexicon varied greatly, ranging from 0% (English–French) to 34% (English–Japanese) to 75% (French–Spanish). Many factors apparently contribute to the variance: the bitext size, the human translators, the thresholds for selecting midfrequency words, and so on. For purposes of bitext alignment, good alignments can be obtained using a bilingual lexicon without high accuracy or coverage (just as with lexical sentence alignment).

6. MULTIFEATURE ALIGNMENT FOR NOISY BITEXTS

The "Smooth Injective Map Recognizer" or SIMR algorithm (Melamed, 1997a) also operates at the word granularity, but generates candidate couplings by combining the correlation scores from cognates, bilingual lexicon, or any other criteria that can be scored. The criteria are combined as a series of filters. Thresholds for the filters are set by trial and error.

Several other interesting ideas are suggested by Melamed, as follows:

Instead of dynamic programming, a greedy search strategy that finds one chain at a time can be used. The search algorithm proceeds monotonically through the candidate matrix from the origin toward the terminus, finding successive *chains*, each

[8] The method was actually developed primarily for the purpose of extracting term translation bilingual lexicons. Alignment constitutes just the first part of the original method, which goes on to use the alignment to extract a much larger bilingual lexicon.

chain being a set of six to nine couplings. Afterwards, the union of all the chains gives the complete output alignment $\mathcal{A}$. The main loop of the search algorithm greedily accepts each chain. The last coupling of each chain (i.e., the coupling closest to the terminus) is accepted as an anchor, which constrains the search for the next chain. Each iteration of the search loop only needs to consider the slack range bounded by the anchor from the previous chain and the terminus. We search for the best chain that begins at the anchor.

Instead of searching through the entire slack range, Melamed suggests an expanding search heuristic. We search in a smaller rectangular area of the same C: D proportion as the entire matrix, also lower-bounded at the anchor, but upper-bounded some small distance Δ away. If no acceptable chain can be found, Δ is increaed; this repeats until some acceptable chain is found.

The search for an acceptable chain can be implemented efficiently if "acceptable" is defined in terms of displacement from the diagonal. All candidate couplings within the rectangle are sorted by their displacement from the rectangle's diagonal (note that this diagonal is parallel to the main diagonal between the origin and terminus). The chain with the minimum RMS deviation from the diagonal must be a contiguous subsequence of the sorted candidates. An acceptable chain is one for which deviation from the diagonal is below a preset threshold. Finding the chain takes only $O(kc + c \log c)$ time, where c is the number of candidate couplings and $6 \leq k \leq 9$.

7. WORD ALIGNMENT TECHNIQUES

For true coupling of words (as opposed to simple linear interpolation up to the word granularity), we must drop the monotonicity constraint. Motivations to move to a more accurate lexical coupling model include the following:

1. *Alignment Accuracy.* In principle, lexical coupling without false monotonicity assumptions leads to higher accuracy.
2. *Sparse Data.* Smaller bitexts should be alignable, with bilingual lexicons automatically extracted in the process. The lexical alignment models discussed earlier require large bitexts to ensure that the counts for good word pairs are large enough to stand out in the noisy word-coupling hypothesis space.
3. *Translation Modeling.* Accurate lexical coupling is necessary to bootstrap learning of structural translation patterns.

Complexity of coupling at the word level rises as a result of relaxing the monotonicity constraint. Partly for this reason, the assumption is usually made that the word coupling cannot be many-to-many. On the other hand, models that permit one-to-many couplings are feasible. The TM-Align (Macklovitch and Hannan, 1996) employs such a model, namely "Model 2" of the IBM series of statistical translation models (Brown et al., 1988, 1990, 1993). This model incorporates a *fertility* distribution that governs the probability on the number of language-2 words generated by a language-1 word. Macklovitch and Hannan report that 68% of the words were correctly aligned. Broken down more precisely,

78% of the content words are correctly aligned, compared with only 57% of the function words.[9]

The *word_align* method of Dagan et al. (1993) employs a dynamic-programming formulation, similar to those already discussed, but has slope constraints that allow the coupling order to move slightly backward as well as forward. The method does not involve any coarser (section, paragraph, sentence) segmentation of the bitexts. The output of the basic algorithm is a partial word alignment.

The method requires as input a set of alignment range constraints, to restrict the search window to a feasible size inside the dynamic-programming loop. Dagan et al. employ *char_align* as a preprocessing stage to produce a rough alignment $\mathbf{a}^0\cdot$, and then use guiding to restrict the alignments considered by *word_align*.

An underlying stochastic channel model is assumed, similar to that of the IBM translation model (Brown et al., 1990, 1993). We assume each language-2 word is generated by a language-1 word. Insertions and deletions are permitted, as in the sentence alignment models. However, there are no strings of beads, because the monotonicity assumption has been dropped: the language-2 words do not necessarily map to language-1 words in order. To model this, a new set of *offset probabilities* $o(k)$ are introduced. The offset k of a pair of coupled word positions i, j is defined as the deviation of the position i of the language-1 word from where it "should have been" given where j is:

$$k(i,j) = i - \mathbf{a}'(j) \tag{13}$$

where $\mathbf{a}'(j)$ is where the language-1 word "should have been." We determine this by linear extrapolation along the slope of the diagonal, from the language-1 position $\mathbf{a}(j^-)$ corresponding to the previous language-2 word:

$$\mathbf{a}'(j) = \mathbf{a}(j^-) + (j - j^-)\frac{C}{D} \tag{14}$$

where j^- is the position of the most recent language-2 word to have been coupled to a language-1 word (as opposed to being an insertion). Given a language-1 string $\mathbf{w}$, the channel model generates the language-2 translation $\mathbf{x}$ with the probability Eq. {15}.

$$P(\mathbf{x}|\mathbf{w}) = \sum_{\mathbf{a}} K \prod_{j=0}^{D-1} P(x_j|w_{a(j)}) \cdot o[\mathbf{a}(j) - \mathbf{a}'(j)] \tag{15}$$

In theory this is computed over all possible alignments $\mathbf{a}$, but we approximate by considering only those within the alignment range constraints in Eq. {16}.[10]

$$P(\mathbf{x}|\mathbf{w}) - \sum_{\mathcal{A}\subset\mathcal{B}} K \prod_{(i,j)\in\mathcal{A}} P(x_j|w_i) \cdot o[i - \mathbf{a}'(j)] \tag{16}$$

Assuming we have estimated distributions $\hat{P}(x|w)$ and $\hat{o}(k)$, the alignment algorithm searches for the most probable alignment $\mathcal{A}^*$ Eq. {17}.

---

[9] Although the IBM statistical translation models are all based on word alignment models, they themselves have not reported accuracy rates for word alignment.

[10] Remember that $\mathcal{A}$ and $\mathbf{a}$ refer to the same alignment.

$$\mathcal{A}^* = \arg\max_{\mathcal{A} \subset \mathcal{B}} \prod_{(i,j) \in \mathcal{A}} P(x_j | w_i) \cdot o[i - \mathbf{a}'(j)] \tag{17}$$

The dynamic-programming search is similar to the sentence alignment cases, except on the word granularity and with different slope constraints. It considers all values of j proceeding left-to-right through the words of the language-2 side, maintaining a hypothesis for each possible corresponding value of i. The recurrence is as follows. Let $\delta(i, j)$ denote the minimum cost partial alignment up to the jth word of the language-2 side, such that the jth word is coupled to the ith position of the language-1 side.

$$\begin{aligned}
\delta(i,j) &= \min \begin{cases} \min_{i^- : (i^-, j^-) \subset \mathcal{B} \wedge i^- - d \leq i \leq i^- + d} \delta(i^-, j-1) - \log \hat{P}(x_j | w)_i) - \log \hat{o}(i - \mathbf{a}'(j)), \\ \delta(i, j-1) - \log \hat{P}(x_j | \varepsilon), \\ \delta(i-1, j) - \log \hat{P}(\varepsilon | w)_i) \end{cases} \\
&= \min \begin{cases} \min_{i^- : (i^-, j^-) \subset \mathcal{B} \wedge i^- - d \leq i \leq i + d} \delta(i^-, j-1) - \log \hat{P}(x_j | w_i) - \log \hat{o}\left(i - \mathbf{i}^- + (j - j^-)\frac{C}{D}\right), \\ \delta(i, j-1) - \log \hat{P}(x_j | \varepsilon), \\ \delta(i-1, j) - \log \hat{P}(\varepsilon | w_i) \end{cases}
\end{aligned} \tag{18}$$

The i^- values that need to be checked are restricted to those that meet the alignment range constraints, and lie within d of i. In practice, the probabilities of word insertion and deletion, $\hat{P}(x_j | \varepsilon)$ and | $\hat{P}(\varepsilon | w_i)$, can be approximated with small flooring constants.

The offset and word-pair probability estimates, $\hat{o}(k)$ and $\hat{P}(x|w)$, are estimated using EM on the bitext before aligning it. The expectations can be accumulated with a dynamic-programming loop of the same structure as that for alignment. An approximation that Dagan et al. (1993) make during the EM training is to use the initial rough alignment $\mathbf{a}^0(j)$ instead of $\mathbf{a}'(j)$.

An experiment performed by Dagan et al. (1993) on a small 160,000-word excerpt of the Canadian Hansards produced an alignment in which about 55% of the words were correctly aligned, 73% were within one word of the correct position, and 84% were within three words of the correct position.

Quantitative performance of *word_align* on nonalphabetic languages has not been extensively evaluated, but Church et al. (1993) align the English and Japanese AWK manuals using the technique. The Japanese text must first be segmented into words, which can lead to an incorrectly segmented text.

It is possible to use *word_align* as a lexical sentence alignment scheme, simply by postprocessing its output down to the coarser sentence granularity. The output set of word couplings can be interpreted as a set of candidate anchors, of higher credibility than the usual set of possible word couplings. Using dynamic programming again, the postprocessor may choose the set of sentence beads that maximizes the (probabilistically weighted) coverage of the candidate anchor set. However, it is not clear that the additional word-order model would significantly alter performance over a model such as that of Chen (1993).

Wu (1995a) describes a word alignment method that employs a structural approach to constraining the offsets, using an underlying generative model based on a stochastic transduction grammar. This is described later in the Sec. 9 on "biparsing."

8. STRUCTURE AND TREE ALIGNMENT TECHNIQUES

A *structure alignment* algorithm produces an alignment between constituents (or sentence substructures) within sentence pairs of a bitext. The passages to be aligned are the nodes in constituent analyses of the sentences; the output set $\mathcal{A}$ contains pairs of coupled nodes. All existing techniques process the bitext one sentence-pair at a time, so sentence alignment always precedes structure alignment. Coupled constituents may be useful in translators' concordances, but they are usually sought for use as examples in example-based machine translation systems (Nagao, 1984; see also Chap. 25). Alternatively, the examples constitute training data for machine learning of transfer patterns.

Tree alignment is the special case of structure alignment in which the output $\mathcal{A}$ must be a strictly hierarchical alignment. (The same constraint is applicable to dependency tree models.) This means that tree alignment obeys the following:

> **Crossing Constraint.** Suppose two nodes in language-1 p_1 and p_2 correspond to two nodes in language-2 q_1 and q_2, respectively, and p_1 dominates q_1. Then q_1 must dominate q_2.

In other words, couplings between subtrees cannot cross each another, unless the subtrees' immediate parent nodes are also coupled to each other. Most of the time this simplifying assumption is accurate, and it greatly reduces the space of legal alignments and, thereby, the search complexity. An example is shown in Fig. 14, in which both *Security Bureau* and *police station* are potential lexical couplings to the first three Chinese characters in the figure, but the crossing constraint rules out the dashed-line couplings because of the solid-line couplings. The crossing constraint reflects an underlying cross-linguistic hypothesis that the core arguments of frames tend to stay together over different languages. A special case of the crossing constraint is that a constituent will not be coupled to two disjoint constituents in the other languages, although it may be coupled to multiple levels within a single constituent subtree.

Structure alignment is usually performed using a *parse–parse–match* strategy. Tree alignment methods require as input the constituent analysis of each side of the bitext (with the exception of the biparsing methods described later). Unfortunately, it is rarely possible to obtain bitext in which the constituent structures of both sides have been marked. However, if suitable monolingual grammars for each of the languages are available, each side can (independently) be parsed automatically, yielding a low-accuracy analysis of each sides, before the tree alignment begins. A variant on this is to supply alternative parses for each sentence, either explicitly in a list (Grishman, 1994) or implicitly in a well-formed substring table (Kaji et al., 1992). Note that the parsed bitext is not parallel in the sense that corresponding sentences do not necessarily share a parallel constituent structure. It is difficult, if not impossible, to give an interpretation based on some underlying generative model.

The kind of structures that are to be coupled clearly depend on the linguistic theory under which the sides are parsed. The simplest approach is to use surface structure, which can be represented by bracketing each side of the bitext (Sadler and Vendelmans, 1990; Kaji et al., 1992). Another approach was described by Matsumoto et al. (1993), who use LFG-like (Bresnan, 1982) unification grammars to parse an English–Japanese bitext. For each side, this yields a set of candidate

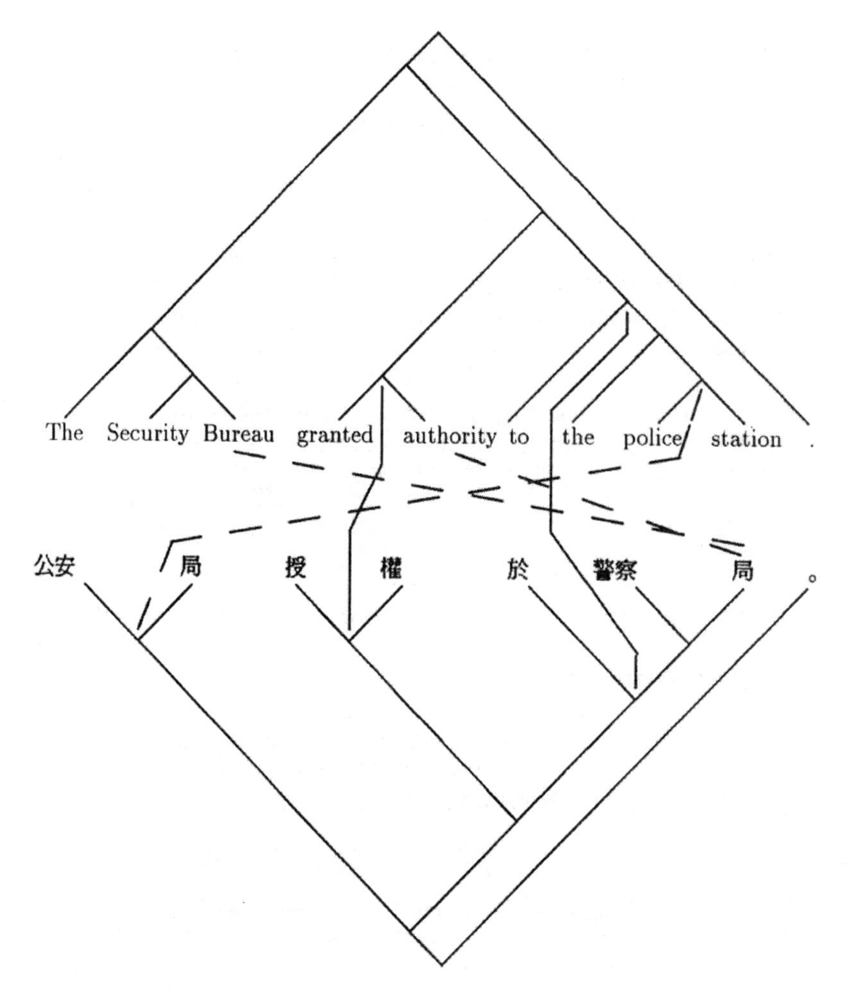

Fig. 14 The crossing constraint.

feature-structures corresponding to the constituent structures. The feature-structures are simplified to dependency trees. Structural alignment is then performed on the dependency trees, rather than the original constituent trees. (Alignment of dependency trees can be performed in essentially the same way as alignment of constituent trees.)

A. Cost Functions

Various cost functions may be employed. Some of the qualitative desiderata, along with exemplar cost functions, are as follows:

Couple leaf nodes (words) that are lexical translations. This is simply word alignment, recast here as the special case of structure alignment at the leaf level. A bilingual lexicon is assumed. Filtering heuristics may be employed, for example, to ignore any words that are not open-class or content words. The simplest cost function of this type is Eq. {19}.

$$Cost(p, q) = \begin{cases} -1 & \text{if } (w, x) \in \text{Bilexicon} \\ & \quad \text{where } w = Word(p), x = Word(q) \\ 0 & \text{otherwise} \end{cases} \qquad \{19\}$$

where the pair of leaf nodes p, q holds the words $Word(p)$ and $Word(q)$.

Alternatively, if a probabilistic bilingual lexicon is available, a soft version of the cost function may be used, in which the negative log probability of a word pair is taken as its cost (instead of -1).

Couple leaf nodes (words) that are similar. To overcome the spotty coverage of most bilingual lexicons, thesauri may be employed. One approach (Matsumoto et al., 1993) employs a single language-1 thesaurus to estimate the cost (dissimilarity) between a word pair w, x. All possible translations of the language-2 word x are looked up in the language-1 thesaurus, and the length l of the shortest path from any one of them to w is taken as the cost. Matsumoto et al. use an additional biasing heuristic that always subtracts a constant d from the path length to obtain l (if the path length is less than the d, then l is assumed to be zero) [Eq. {20}].

$$Cost(p, q) = \begin{cases} -6 & \text{if } (w, x) \in \text{Bilexicon} \\ & \quad \text{where } w = Word(p), x = Word(q) \\ l & \text{otherwise} \end{cases} \qquad \{20\}$$

Couple internal nodes that share coupled leaf nodes (words). Let $\mathcal{A}_{words}$ denote the subset of $\mathcal{A}$ that deals with coupling of leaf nodes (usually this subset is precomputed in an earlier stage, according to one of the foregoing criteria and possibly with additional constraints). Let $Words(p)$ be the set of leaves of p that are deemed important, for example, all the content words. The simplest cost function of this type is Eq. {21}.

$$Cost(p, q) = \begin{cases} -1 & \text{if } (w, x) \in \mathcal{A}_{word} \\ & \quad \text{for all } w \in Words(p), x \in Words(q) \\ 0 & \text{otherwise} \end{cases} \qquad \{21\}$$

This cost function is what Kaji et al. (1992) in effect use, permitting nodes to be matched only if they share exactly the same set of coupled content words.

A softer approach is to maximize the number of shared coupled leaf nodes [Eq. {22}].

$$Cost(p, q) = \sum_{w \in Words(p), x \in Words(q)} Cost(w, x) \qquad \{22\}$$

Again, probabilistic versions of these cost functions are straightforward, if a probabilistic bilingual lexicon is available.

Couple nodes that share as many coupled children or descendants as possible. Similar ad hoc cost functions to those in the foregoing can be formulated. The idea is to maximize structural isomorphism. Through the recursion, this also attempts to share as many coupled leaf nodes as possible.

B. Algorithms

Although authors vary greatly on notations and descriptions, nearly all structure alignment algorithms can, in fact, be formulated as search algorithms that attempt to minimize the total cost of $\mathcal{A}\$$ the entire set of couplings [Eq. {23}].

$$\sum_{(p,q)\in\mathcal{A}} Cost(p, q) \qquad \{23\}$$

Search techniques can be heuristic or exhaustive, and can employ greedy, backtracking, beam, or exhaustive strategies.

In general, constituent alignment has lower complexity than tree alignment, as there is no need to enforce the crossing constraint. Kaji et al. (1992) describe a simple bottom-up greedy strategy. Suppose that a bracketed sentence pair contains P language-1 nodes and Q language-2 nodes. The algorithm simply considers all $P \times Q\$$ possible couplings between node pairs p, q, starting with the smallest spans. Any pair for which the cost is less than a preset threshold is accepted. Thus, as many pairs as possible are output, so long as they meet the threshold.

For tree alignment as opposed to constituent alignment, bottom-up greedy search is still possible, but performance is likely to suffer owing to the interaction of the coupling hypothesis. Crossing constraints deriving from early miscouplings can easily preclude later coupling of large constituents, even when the later coupling would be correct. For this reason, more sophisticated strategies are usually employed.

Matsumoto et al. (1993) employ a branch-and-bound search algorithm. The algorithm proceeds top-down, depth-first. It hypothesizes constituent couplings at the highest level first. For each coupling that is tried, a lower bound on its cost is estimated. The algorithm always backtracks to expand the hypothesis with the lowest expected cost. The hypothesis is expanded by further hypothesizing couplings involving the constituent's immediate children (subconstituents).

Grishman (1994) employs a beam search. Search proceeds bottom-up, hypothesizing couplings of individual words first. Only hypotheses for which the cost is less than a preset threshold are retained. Each step in the search loop considers all couplings involving the next larger constituents consistent with the current set of smaller hypotheses. The costs of the larger hypotheses depend on the previously computed costs of the smaller hypotheses.

Dynamic-programming search procedures can be constructed. However, the worst-case complexity is exponential because it grows with the number of permutations of constituents at any level. The procedure of Kaji et al. (1992) employs two separate well-formed substring tables as input, one for each sentence of the input sentence pair, that are computed by separate dynamic-programming chart parsing processes. However, the actual coupling procedure is greedy, as described in the foregoing. A true dynamic-programming approach (Wu, 1995c) is described later, in Sec. 9 (in which permutation constraints guarantee polynomial time complexity).

C. Strengths and Weaknesses of Structure Alignment Techniques

The most appropriate application of constituent alignment approaches appears to be for example-based machine translation models. Given the same bitext, constituent alignment approaches can produce many more examples than strict tree alignment approaches (driving up recall), although many of the examples are incorrect (driving down precision). However, the nearest-neighbor tactics of example-based machine translation models have the effect of ignoring incorrect examples most of the time, so recall is more important than precision.

Strict tree alignment approaches tend to have higher precision, since the effects of local disambiguation decisions are propagated to the larger contexts, and vice versa. For this reason, these methods appear to be more suitable for machine learning of transfer patterns.

Constitutent alignment has three main weaknesses stemming from the parse–parse–match approach:

1. *Appropriate, robust, monolingual grammars may not be available.* This condition is particularly relevant for many non-Western–European languages, such as Chinese. A grammar for this purpose must be robust because it must still identify constituents for the subsequent coupling process even for unanticipated or ill-formed input sentences.
2. *The grammars may be incompatible across languages.* The best-matching constituent types between the two languages may not include the same core arguments. While grammatical differences can make this problem unavoidable, there is often a degree of arbitrariness in a grammar's chosen set of syntactic categories, particularly if the grammar is designed to be robust. The mismatch can be exacerbated when the monolingual grammars are designed independently, or under different theoretical considerations.
3. *Selection between multiple possible constituent couplings may be arbitrary.* A constituent in one sentence may have several potential couplings to the other, and the coupling heuristic may be unable to discriminate between the options.

9. BIPARSING ALIGNMENT TECHNIQUES

Bilingual parsing ("biparsing") approaches use the same grammar to simultaneously parse both sides of a bitext (Wu, 1995a). In contrast with the parse–parse–match approaches, discussed in the preceding section, biparsing approaches readily admit interpretations based on underlying generative *transduction grammar* models. A single transduction grammar governs the production of sentence pairs that are mutual translations. The most useful biparsers are the probabilistic versions, which work with stochastic transduction grammars that, in fact, constitute *bilingual language models*.

Biparsing approaches unify many of the concepts discussed in the foregoing. The simplest version accepts sentence-aligned, unparsed bitext as input. The result of biparsing includes parses for both sides of the bitext, plus the alignment of the constituents.

There are several major variants of biparsing. It is possible to perform biparsing alignment *without* a grammar of either language, using the BTG technique (see

Sec. 9.D). On the other hand, biparsing can make use of a monolingual grammar if one exists. If any a priori brackets on the input bitext are available, biparsing can accept them as constraints.

A. Transduction Grammars

A transduction grammar describes a structurally correlated pair of languages. It generates sentence pairs, rather than sentences. The language-1 sentence is (intended to be) a translation of the language-2 sentence.

Many kinds of transduction grammars exist, but we concentrate here on the context-free varieties. The class of transduction grammars based on context-free productions are sometimes formally known as *syntax-directed transduction grammars* or SDTGs (Aho and Ullman, 1969a,b, 1972). The motivation for the term "syntax-directed" stems from the roots roots of SDTGs in compiler theory. The input–output connotation of the term is not very appropriate in the context of biparsing.[11]

The narrowest class of context-free SDTGs, appropriately known as *simple transduction grammars* (Lewis and Stearns, 1968), are too inflexible for the present purpose. Simple transduction grammars are the same as ordinary CFGs, except that instead of terminal symbols, they have terminal symbol pairs that represent word translation pairs. Simple transduction grammars are inadequate because they require exactly the same word order for both languages.

In general SDTGs, word-order variation between languages is accommodated by letting the symbols on each production's right-hand side appear in different order for language-1 and language-2. Simple transduction grammars turn out just to be the subclass of SDTGs for which the right-hand side symbols are required to have the same order for both languages. In contrast, general SDTGs allow any permutation of the right-hand side symbols.

Formally, we denote a transduction grammar by $G = (\mathcal{N}, \mathcal{W}_1, \mathcal{W}_2, \mathcal{R}, S)$ where $\mathcal{N}$ is a finite set of nonterminals, $\mathcal{W}_1$ is a finite set of words (terminals) of language 1, $\mathcal{W}_2$ is a finite set of words (terminals) of language 2, $\mathcal{R}$ is a finite set of rewrite rules (productions), and $S \in \mathcal{N}$ is the start symbol. The space of word pairs (terminal pairs) [Eq. {24}]

$$\mathcal{X} = (\mathcal{W}_1 \cup \{\varepsilon\}) \times (\mathcal{W}_2 \cup \{\varepsilon\}) \qquad \{23\}$$

contains lexical translations denoted x/y and singletons denoted x/ε or ε/y, where $x \in \mathcal{W}_1$ and $y \in \mathcal{W}_2$.

Any SDTG trivially implements the crossing constraint. This is because in the generative model, at any time the rewrite process substitutes for only a single constituent, which necessarily corresponds to contiguous spans in (both of) the sentences.

The main issue for alignment models based on transduction grammars is how much flexibility to allow. We have already seen that simple transduction grammars are insufficiently expressive. However, computational complexity for general SDTGs is excessively expensive. No polynomial-time biparsing algorithm is known for the

[11] Finite-state transducers correspond to the subclass of SDTGs that have regular grammar backbones.

general class, as growth in number of possible alignments follows the number of permutations of right-hand side symbols, which is exponential.

A subclass of SDTGs called *inversion transduction grammars* or ITGs, offer a balance of expressiveness and computational complexity (Wu, 1995a,c). Expressiveness appears adequate from both theoretical and empirical analyses (Wu, 1995b, forthcoming), excluding so-called free word-order languages and "'second-order" phenomena such as raising and topicalization. At the same time, a polynomial-time biparsing algorithm exists.

There are three equivalent ways to define the class of ITGs, as follows:

1. *ITGs are the subclass of SDTGs with only straight or inverted productions.* This means the order of right-hand side symbols for language-2 is either the same as language-1 (*straight* orientation) or exactly the reverse (*inverted* orientation). A straight production is written $A \rightarrow [a_1 a_2 \ldots a_r]$, and an inverted production is written $A \rightarrow < a_1 a_2 \ldots a_r >$, where $a_i \in \mathcal{N} \cup \mathcal{X}$ and r is the rank of the production. Any ITG can be converted to a normal form, where all productions are either lexical productions or binary-fanout productions (Wu, 1995c), as in Theorem 1:
   Theorem 1. For any inversion transduction grammar G, there exists an equivalent inversion transduction grammar G' in which every production takes one of the following forms:

   $$\begin{array}{lll} S \rightarrow \varepsilon/\varepsilon & A \rightarrow x/\varepsilon & A \rightarrow [BC] \\ A \rightarrow x/y & A \rightarrow \varepsilon/y & A \rightarrow \langle BC \rangle \end{array}$$

   The theorem leads directly to the second characterization of ITGs.
2. *ITGs are the subclass of SDTGs with productions of rank* ≤ 2. That is, any SDTG whose productions are all binary-branching is an ITG. The equivalence follows trivially from the fact that the only two possible permutations of a rank-2 right-hand side are straight and inverted.
3. *ITGs are the subclass of SDTGs with productions of rank* ≤ 3. It can be shown that all six possible permutations of a rank-3 right-hand side can be generated using only straight and inverted productions in combination.

The expressiveness of ITGs relative to word order in different natural languages is not straightforward to characterize formally. Some light is shed by Fig. 15, which enumerates how ITGs can deal with transposition of four adjacent constituents. This is important because the number of core arguments of a frame is normally fewer than four, in nearly all linguistic theories. There are $4! = 24$ possible permutations of four adjacent constituents, of which 22 can be produced by combining straight and inverted productions. The remaining two permutations are highly distorted "inside-out" alignments, which are extremely rare in (correctly translated) bitext.[12] For more than four adjacent constituents, many permutations cannot be generated and do not appear necessary.

---

[12]In fact, we know of no actual examples in any parallel corpus for languages that do not have free word order.

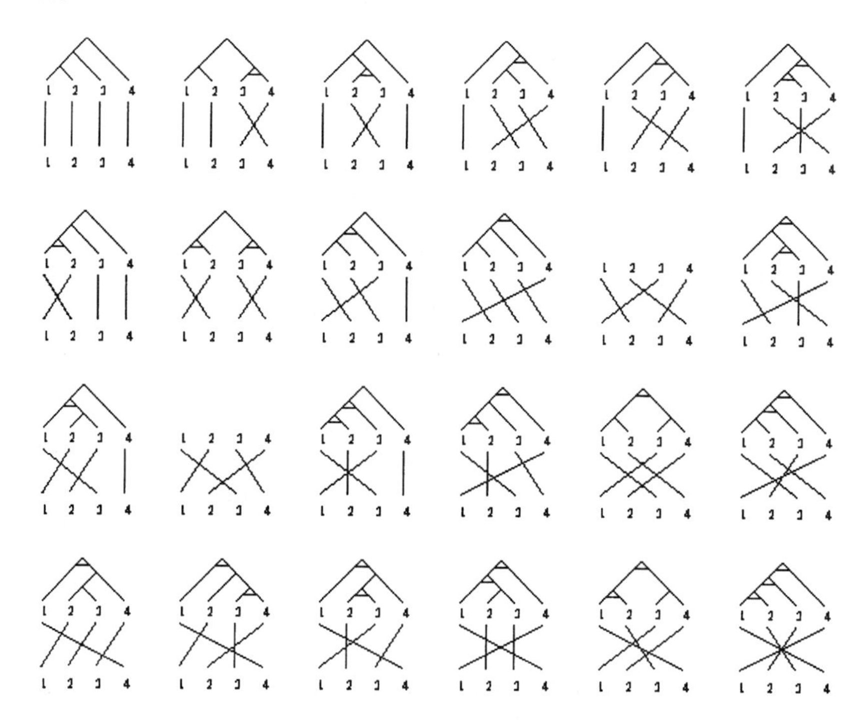

Fig. 15 The 24 complete alignments of length 4, with ITG parses for 22. All nonterminal and terminal labels are omitted. A horizontal bar under a parse tree node indicates an inverted production.

The fundamental assumption of ITGs is a language universals hypothesis, that the core arguments of frames, which exhibit great ordering variation between languages, are relatively few and surface in syntactic proximity. This assumption over-simplistically blends syntactic and semantic notions. That semantic frames for different languages share common core arguments is more plausible than syntactic frames, but ITGs depend on the tendency of syntactic arguments to correlate closely with semantics. If in particular cases this assumption does not hold, the biparsing algorithm can attempt to contain the damage by dropping some word couplings (as few as possible). More detailed analysis is given in Wu (forthcoming).

B. Cost Functions

A cost function can be derived naturally from the generative model. First a stochastic version of the ITG is created by associating a probability with each production. Just as for ordinary monolingual parsing, probabilizing the grammar permits ambiguities to be resolved by choosing the maximum likelihood parse. For example, the probability of the rule NN $\xrightarrow{0.4}$ [A N] is $a_{\mathrm{NN}\to[\mathrm{A\ N}]} = 0.4$. The probability of a lexical rule A $\xrightarrow{0.001}$ x/y is $b_{\mathrm{A}}(x, y) = 0.001$. Let W_1, W_2 be the vocabulary sizes of the two languages, and $\mathcal{N} = A_1, \ldots, A_N$ be the set of nonterminals with indices $1, \ldots, N$. (For conciseness, we sometimes abuse the notation by writing an index when we mean the corresponding nonterminal symbol, as long as this introduces no confusion.) As with stochastic CFGs, the probabilities for a given left-hand side symbol must sum to unity [Eq. {24}].

$$\sum_{1 \leq j,k \leq N} (a_{i \rightarrow [jk] + ai \rightarrow \langle jk \rangle}) + \sum_{\substack{1 \leq x \leq W \\ 1 \leq y \leq W_{21}}} b_i(x, y) = 1 \qquad \{24\}$$

The same constituent structure applies to both sides of the sentence pair, unlike the earlier structure–tree alignment cases. This means the alignment $\mathcal{A}$ is more naturally thought of in terms of a single shared parse tree, rather than a set of couplings between constituents. We denote the parse tree by a set $\mathcal{Q}$ of nodes $q_0, \ldots, q_{Q-1}$. Note that $\mathcal{Q}$ can always be transformed back into a set of couplings $\mathcal{A}$.

The natural cost function to be minimized is the entropy (negative log probability) of the parse tree $\mathcal{Q}$ [Eq. {25}].

$$Cost(\mathcal{Q}) - \sum_{q \in \mathcal{Q}} \begin{cases} a_{\Pr\ oduction(q)} & \text{if } q \text{ is an internal node} \\ b_{\Pr\ oduction(q)} & \text{otherwise} \end{cases} \qquad \{25\}$$

C. Algorithms

The biparsing algorithm searches for the minimum cost parse tree on the input sentence pair. The probabilistic cost function optimizes the overlap between the structural analysis of the two sentences. The algorithm resembles the recognition algorithm for HMMs (Viterbi, 1967) and CKY parsing (Kasami, 1965; Younger, 1967).

Let the input language-1 sentence be $\mathbf{e}_1, \ldots, \mathbf{e}_T$ and the corresponding input language-2 sentence be $\mathbf{c}_1, \ldots, \mathbf{c}_V$. As an abbreviation we write $\mathbf{e}_{s..t}$ for the sequence of words $\mathbf{e}_s, \mathbf{e}_{s+1}, \ldots, \mathbf{e}_t$ and similarly for $\mathbf{c}_{u..v}$; also, $\mathbf{e}_{s..s} = \varepsilon$ is the empty string. It is convenient to use a 4-tuple of the form $q = (s, t, u, v)$ to identify each node of the parse tree, where the substrings $\mathbf{e}_{s..t}$ and $\mathbf{c}_{u.v.}$ both derive from the node q. Denote the nonterminal label on q by $\ell(q)$. Then for any node $q = (s, t, u, v)$, define Eq. {26}.

$$\delta_q(i) = \delta_{stuv}(i) = \max_{\text{subtrees of } q} P[\text{subtree of } q, \ell(q) = i, i \Rightarrow \mathbf{e}_{s..t}/\mathbf{c}_{u..v}] \qquad \{26\}$$

as the maximum probability of any derivation from i that successfully parses both $\mathbf{e}_{s..t}$ and $\mathbf{c}_{u..v}$. Then the best parse of the sentence pair has probability $\delta_{0,T,0,V}(S)$.

The algorith computes $\delta_{0,T,0,V}(S)$ using the following recurrences. The argmax notation is generalized to the case where maximization ranges over multiple indices, by making the argument vector-valued. Note that [ ] and ⟨ ⟩ are just constants. The conditions $(S - s)(t - S) + (U - u)(v - U) \neq 0$ is a way to specify that the substring in one but not both languages may be split into an empty string ε and the substring itself; this ensures that the recursion terminates, but permits words that have no match in the other language to map to an ε instead.

1. **Initialization**

$$\delta_{t-1,t,v-1,v}(i) = b_i(\mathbf{e}_t/\mathbf{c}_v),\ 1 \leq t \leq T,\ 1 \leq v \leq V \qquad \{27\}$$

$$\delta_{t-1,t,v,v}(i) = b_i(\mathbf{e}_t/\varepsilon),\ 1 \leq t \leq T,\ 0 \leq v \leq V \qquad \{28\}$$

$$\delta_{t,t,v-1,v}(i) = b_i(\varepsilon/\mathbf{c}_v), 0 \le t \le T, 1 \le v \le V \qquad \{29\}$$

2. **Recursion**

$$\text{For all } i, s, t, u, v \text{ such that} \begin{cases} 1 \le i \le N \\ 0 \le s < t \le T \\ 0 \le u < v \le V \\ t - 2 + v - u > 2 \end{cases}$$

$$\delta_{stuv}(i) = \max[\delta^{[]}_{stuv}(i), \delta^{\langle\rangle}_{stuv}(i)] \qquad \{30\}$$

$$\theta_{stuv}(i) = \begin{cases} [] \text{ if } \delta^{[]}_{stuv}(i) \le \delta^{\langle\rangle}_{stuv}(i) \\ \langle\rangle \text{ otherwise} \end{cases} \qquad \{31\}$$

where

$$\delta^{[]}_{stuv}(i) = \max_{\substack{1\le j\le N \\ 1\le k\le N \\ s\le S\le t \\ u\le U\le v \\ (S-s)(t-S)+(U-u)(v-U)\ne 0}} a_{i\to[jk]}\delta_{sSuU}(j)\delta_{StUv}(k) \qquad \{32\}$$

$$\begin{bmatrix} \mathbf{l}^{[]}_{stuvv}(i) \\ \boldsymbol{\kappa}^{[]}_{stuv}(i) \\ \sigma^{[]}_{stuv}(i) \\ \upsilon^{[]}_{tuv}(i) \end{bmatrix} = \operatorname*{arg\,max}_{\substack{1\le j\le N \\ 1\le k\le N \\ s\le S\le t \\ u\le U\le v \\ (S-s)(t-S)+(U-u)(v-U)\ne 0}} a_{i\to[jk]}\delta_{sSuU}(j)\delta_{StUv}(k) \qquad \{33\}$$

$$\delta^{\langle\rangle}_{stuv}(i) = \max_{\substack{1\le j\le N \\ 1\le k\le N \\ s\le S\le t \\ u\le U\le v \\ (S-s)(t-S)+(U-u)(v-U)\ne 0}} a_{i\to\langle jk\rangle}\delta_{sSuU}(j)\delta_{StUv}(k) \qquad \{34\}$$

$$\begin{bmatrix} \mathbf{l}^{\langle\rangle}_{stuvv}(i) \\ \boldsymbol{\kappa}^{\langle\rangle}_{stuv}(i) \\ \sigma^{\langle\rangle}_{stuv}(i) \\ \upsilon^{\langle\rangle}_{tuv}(i) \end{bmatrix} = \operatorname*{arg\,max}_{\substack{1\le j\le N \\ 1\le k\le N \\ s\le S\le t \\ u\le U\le v \\ (S-s)(t-S)+(U-u)(v-U)\ne 0}} a_{i\to\langle jk\rangle}\delta_{sSuU}(j)\delta_{StUv}(k) \qquad \{35\}$$

3. **Reconstruction**

Initialize by setting the root of the parse tree to $q_1 = (0, T, 0, V)$ and its nonterminal label to $\ell(q_1) = \mathrm{S}$. The remaining descendants in the optimal parse tree are then given recursively for any $q = (s, t, u, v)$ by:

$$\text{LEFT}(q) = \begin{cases} \text{NIL} & \text{if } t - s + v - u \le 2 \\ (s, \sigma_q^{[]}(\ell(q)), u, v_q^{[]}(\ell(q))) & \text{if } \theta_q(\ell(q)) = [\,] \text{ and } t - s + v - u > 2 \\ (s, \sigma_q^{\langle\rangle}(\ell(q)), u, v_q^{\langle\rangle}(\ell(q)), v) & \text{if } \theta_q(\ell(q)) = \langle\,\rangle \text{ and } t - s + v - u > 2 \end{cases} \quad \{36\}$$

$$\text{RIGHT}(q) = \begin{cases} \text{NIL} & \text{if } t - s + v - u \le 2 \\ (\sigma_q^{[]}(\ell(q)), t, v_q^{[]}(\ell(q)), v) & \text{if } \theta_q(\ell(q)) = [\,] \text{ and } t - s + v - u > 2 \\ (\sigma_q^{\langle\rangle}(\ell(q)), t, u, v_q^{\langle\rangle}(\ell(q))) & \text{if } \theta_q(\ell(q)) = \langle\,\rangle \text{ and } t - s + v - u > 2 \end{cases} \quad \{37\}$$

$$\ell(\text{LEFT}(q)) = t_q^{\theta_q(\ell(q))}(\ell(q)) \quad \{38\}$$

$$\ell(\text{RIGHT}(q)) = \kappa_q^{\theta_q(\ell(q))}(\ell(q)) \quad \{39\}$$

The time complexity of this algorithm in the general case is $\Theta(N^3 T^3 V^3)$ where N is the number of distinct nonterminals and T and V are the lengths of the two sentences. (Compare this with monolingual parsing, which is faster by a factor of V^3.) The complexity is acceptable for corpus analysis that does not require real-time parsing.

D. Grammars for Biparsing

Biparsing techniques may be used without a specific grammar, with a coarse grammar, or with detailed grammars.

No language-specific grammar (the BTG technique). In the minimal case, no language-specific grammar is used (this is particularly useful when grammars do not exist for both languages). Instead, a generic *bracketing transduction grammar* or BTG (40) is used:

$$\begin{aligned} &A \xrightarrow{a[]} [A\ A] \\ &A \xrightarrow{a\langle\rangle} \langle A\ A \rangle \\ &A \xrightarrow{b_{ij}} u_i/v_j \text{ for all } i, j \text{ lexical translation pairs} \\ &A \xrightarrow{b_{i\varepsilon}} u_i/\varepsilon \text{ for all } i \text{ language-1 vocabulary} \\ &A \xrightarrow{b_{\varepsilon j}} \varepsilon/v_j \text{ for all } j \text{ language-2 vocabulary} \end{aligned} \quad \{40\}$$

All probability parameters may be estimated by EM (Wu, 1995d). However, in practice, alignment performance is not very sensitive to the exact probabilities, and rough estimates are adequate. The b_{ij} distribution can be estimated through simpler EM-based bilingual lexicon learning methods, as discussed later. For the two singleton productions, which permit any word in either sentence to be unmatched, a small ε-constant can be chosen for the probabilities $b_{i\varepsilon}$ and $b_{\varepsilon j}$, so that the optimal bracketing resorts to these productions only when it is otherwise impossible to match the singletons. Similarly, the parameters $a_{[]}$ and $a_{\langle\rangle}$ can be chosen to be very small relative to the b_{ij} probabilities of lexical translation pairs. The result is that the maximum-likelihood parser selects the parse tree that best meets the combined lexical translation preferences, as expressed by the b_{ij} probabilities.

BTG biparsing can be seen as being similar in spirit to *word_align*, but without positional offsets. The maximum probability word alignment is chosen, but with little or no penalty for crossed couplings, as long as they are consistent with constituent structure (even if the coupled words have large positional offsets). The assumption is that the language-universal core arguments hypothesis modeled by ITGs is a good constraint on the space of alignments allowed. The *word_align* bias toward preferring the same word order in both languages can be initiated, to a large extent, by choosing $a_{\langle\rangle}$ to be slightly smaller than $a_{[]}$ and thereby giving preferring the straight orientation.

Empirically, the BTG technique is fairly sensitive to the bilingual lexicon accuracy and coverage. This is due to the fact that a missed word coupling can adversely affect many couplings of its dominating constituents. Also, in practice, additional heuristics are useful to compensate for ambiguities that arise from the underconstrained nature of a BTG. (The extreme case is where both sides of a sentence pair have the same word order; in this case there is no evidence for any bracketing.) A number of heuristics are discussed in Wu (forthcoming).

Coarse grammar. Performance over the BTG technique may often be improved by introducing a small number of productions to capture frequent constituent patterns. This can be done by writing a very small grammar either for one of the languages (Wu, 1995d) or for both languages simultaneously. The generic bracketing productions should still be retained, to handle all words that do not fit the constituent patterns.

Detailed grammar. A detailed monolingual grammar for one of the languages can be converted into the ITG, by doubling each production into straight and inverted versions. This biases the constituents that will be aligned to fit the selected language, at the expense of degraded parsing of the other language.

E. Strengths and Weaknesses of Biparsing Alignment Techniques

Biparsing models have a somewhat stronger theoretical basis for selecting alignments, as they are clearly formulated relative to a generative bilingual language model. This permits probabilistic trading-off between the amount of information in the monolingual parses versus the lexical correspondences.

Biparsing techniques may be used with prebracketed bitext, just as with parse–parse–match tree alignment techniques, by including the brackets as a priori constraints in the dynamic-programming search (Wu, forthcoming). The performance is similar, except that the ordering constraints are slightly stronger with ITGs. Otherwise, the exhaustive dynamic-programming search is more reliable than the heuristic search used by tree alignment methods.

The BTG technique can be used to produce a rough bracketing of bitext when no other parsers or grammars are available. It is the only approach under circumstances for which no grammar is available for one or both of the languages (thereby, precluding the possibility of preparsing the sides of the bitext). Clearly, it is the weakest of all the structure–tree-biparsing alignment methods in terms of parsing accuracy. The flip side is that the BTG technique infers new bracketing hypotheses, which can be used for grammar induction.

Similarly, if a grammar is available for only one of the languages, a biparsing approach can use just the single grammar, whereas parse–parse–match techniques do not apply. The biparsing technique is also useful if two grammars are available, but use very different constituent structures (as mentioned earlier, structure or tree alignment methods may be unable to come up with good constituent couplings under these circumstances).

The major weakness of (non-BTG) biparsing techniques is that it is difficult and time-consuming to write a single grammar that parses both sides well. It is relatively easy to parse one side well, by using a grammar that fits one of the languages. However, the more language-specific details in the grammar, the more nonterminals it requires to keep the rules in sync with both languages. For this reason, it is important to retain backup generic productions for robustness.

Neither biparsing, nor structure–tree alignment techniques that work on surface constituency structures, can accommodate the general case of "second-order transformations," such as raising, topicalization, *wh*-movement, and gapping. Such transformations can cause the surface constituency structure to lose its isomorphism with the "deep" structure's frames and core arguments. No consistent set of couplings is then possible between the surface tree structures of the two languages. For these phenomena, working on the feature-structure tree instead of the constituency tree may hold more hope.

For languages without explicit word boundaries, particularly Asian languages, the biparsing techniques are particularly well suited. With a trivial extension to the biparsing algorithm, the segmentation of the text can be chosen in tandem with choosing the bracketing and couplings, thereby, selecting a segmentation that optimally fits the alignment (Wu, 1995c). The separate well-formed substring tables approach (Kaji et al., 1992) could be extended similarly for structure or tree alignment methods.

10. AUTOMATIC ACQUISITION OF BILINGUAL LEXICONS

Many alignment algorithms produce potentially useful translation lexicon entries as a side effect of alignment. Because bilingual lexicon acquisition is a significant area in its own right, a few comments relative to the alignment algorithms are in order.

Lexical sentence alignment methods are obviously one way to extract lexicons. However, as we have seen, the lexicons obtained in this way may not have high precision or recall. Thus, these methods are usually used to obtain first-pass alignments, which are followed by stronger methods for extracting higher-accuracy word translation candidates.

Word alignment methods can be applied to a bitext, and a candidate bilingual lexicon can be extracted straight from the bitext. This is done simply by counting word cooccurrences from the couplings, yielding relative frequency estimates of the word translation probabilities. Similar approaches can straightforwardly be taken with the output of structure or tree and biparsing alignment.

A more accurate approach is to iterate the word-alignment and counting stages until convergence, and to consider different alignments weighted by their probabilities. This is the basis of EM methods such as the IBM models. Word co-occurrences are counted with aligned sentencees (Wu and Xia, 1994). Alternatively, since Dagan et al. (1993) and Resnik and Melamed (1997) do not have aligned sentences, a different definition of word cooccurrence is used. A constant-width region is defined by using the alignment as a guide. The count for a candidate word pair is incremented whenever an occurrence of the word pair falls within the region. EM training (or approximate versions thereof) are then applicable. Wu and Xia (1995) and Melamed (1996) describe filtering procedures to prune the candidate word translations after training, based on their probabilities or other criteria.

A different approach developed by Fung and McKeown (1997) employs a nonlinear variant of the "K-vec" method (Fung and Church, 1994) to extract n-best term translation lists, given a bitext with a set of anchors.[13]

The anchors divide the bitext into $*A*$ segments. (That is, the anchors carve A slack ranges out of the coupling matrix.) The K-vec methods are based on the assumption that two correlated words should appear in the same segment (slack range). A binary *K-vec* vector of dimension A is created for each word. The ith bit of the K-vec is true if that word appears in the ith segment at least once. Candidate word pairs are scored by using the *weighted mutual information* between the words' K-vecs.

Many of the alignment techniques can produce bilingual lexicon entries with high precision rates, in the 90% range. However, much lower rates are to be expected for recall. Recall rates are rarely reported, in large part because it is difficult to evaluate recall, for this would require a complete list of all bilingual lexicon entries that "could" be learned from a bitext. Note that it is possible to obtain high-alignment accuracies at the sentence level, even if the acquired bilingual lexicon has very low coverage, because we may not need many anchors to obtain a correct sentence-level alignment.

The methods we have described are not restricted to single words. For example, Kupiec (1993) employs the EM-based lexical method on a version of the Canadian Hansards in which the noun phrases of each sentence have already been segmented by a preprocessor, to extract a bilingual lexicon containing translations for entire noun phrases. Similarly, for English and Chinese, Wu and Xia (1994) deal with the absence of word boundaries in Chinese text by applying EM to a bitext in which the Chinese side is preprocessed by a heuristic word segmenter.

[13]Fung and McKeown use the anchors produced by the DK-vec method, but any method for obtaining a large number of highly reliable anchors is suitable.

Accurate learning of bilexicons requires a reasonably long corpus. For example, Kay and Röscheisen (1993) report roughly 50% precision for corpora of 20–60 sentences, whereas precision rises to nearly 100% for corpora longer than 150 sentences.

Semiautomatic bilingual lexicon acquisition is another promising approach, in translators' aids. Dagan et al. (1993) report that translators using their system (Dagan and Church, 1994) were producing bilingual terminology lexicons at a rate of 60–100 terms per hour, compared with about 30 terms per hour using *char_align* output, and "extremely lower rates" before the alignment tools. The system builds aligned bilingual concordances using *word_align*. It counts the alternative translations for each term, listing them by their frequency for the translators' consideration. The translators are also shown the passages from the bilingual concordances, to help them make judgments. Candidates can be accepted or rejected with a minimum of keystrokes.

11. CONCLUSION

Alignment of matching passages can be performed at various granularities: document-structure, sentence, word, or constituent. A variety of techniques are available for each granularity, with trade-offs in speed and memory requirements, accuracy, robustness to noisy bitext, ease of implementation, and suitability for unrelated and nonalphabetic languages. Sources of leverage include lexical translations, cognates, end constraints, passage lengths, and assumptions about monotonicity, approximate monotonicity, and word order.

Alignment techniques support various applications, ranging from translators' and lexicographers' workstations to statistical and example-based machine translation systems. Increasingly, alignment algorithms that are faster, more accurate, and more robust are being used for automatic and semiautomatic resource acquisition, especially extraction of bilingual lexicons and phrasal translation examples.

REFERENCES

Aho AV, JD Ullman 1969a. Syntax directed translations and the pushdown assembler. J Comput Syst Sci 3:37–56.

Aho AV, JD Ullman 1969b. Properties of syntax directed translations. J Comput Syst Sci 3:319–334.

Aho AV, JD Ullman 1972. The Theory of Parsing, Translation, and Compiling. Englewood Cliffs, NJ: Prentice-Hall.

Bellman R. 1957 Dynamic Programming. Princeton, NJ: Princeton University Press.

Bresnan J, ed. 1982. The Mental Representation of Grammatical Relations. Cambridge, MA: MIT Press.

Brown PF, J Cocke, SA Della Pietra, VJ Della Pietra, F Jelinek, RL Mercer, P Roossin 1988. A statistical approach to language translation. COLING, International Conference on Computational Linguistics, Budapest, pp 71–76.

Brown PF, J Cocke, SA Della Pietra, VJ Della Pietra, F Jelinek JD Lafferty, RL Mercer, PS Roossin 1990. A statistical approach to machine translation. Comput Linguist 16:29–85.

Brown, PF SA Della Pietra, VJ Della Pietra, RL Mercer 1993. The mathematics of statistical machine translation: parameter estimation. Comput Linguist 19:263–311.

Brown PF, JC Lai, RL Mercer 1991. Aligning sentences in parallel corpora. 29th Annual Meeting of the Association for Computational Linguistics, Berkeley, CA., pp 169–176.

Catizone R, G Russell S Warwick 1989. Deriving translation data from bilingual texts. Proceedings of the First Lexical Acquisition Workshop, Detroit.

Chang C-H, C-D Chen 1993. HMM-based part-of-speech tagging for Chinese corpora. Proceedings of the Workshop on Very Large Corpora, Columbus, OH, pp 40–47.

Chen SF 1993. Aligning sentences in bilingual corpora using lexical information. 31st Annual Meeting of the Association for Computational Linguistics, Columbus, OH, pp 9–16.

Chiang T-H, J-S Chang, M-Y Lin, K-Y Su 1992. Statistical models for word segmentation and unknown resolution. Proceedings of ROCLING-92, Taiwan, pp 121–146.

Church KW 1993. Char_align: a program for aligning parallel texts at the character level. 31st Annual Meeting of the Association for Computational Linguistics, Columbus, OH: pp 1–8.

Church KW, I Dagan, W Gale, P Fung, J Helfman, B Satish 1993. Aligning parallel texts: do methods developed for English–French generalize to Asian languages? Proceedings of the Pacific Asia Conference on Formal and Computational Linguistics.

Church KW, EH Hovy 1993. Good applications for crummy machine translation. Mach Transl 8:239–258.

Cover TM, JA Thomas 1991. Elements of Information Theory. New York: Wiley.

Dagan I, KW Church 1994. Termight: identifying and translating technical terminology. Fourth Conference on Applied Natural Language Processing, Stuttgart, pp 34–40.

Dagan I, KW Church, WA Gale 1993. Robust bilingual word alignment for machine aided translation. Proceedings of the Workshop on Very Large Corpora, Columbus, OH, pp 1–8.

Fung P 1995. A pattern matching method for finding noun and proper translations from noisy parallel corpora. 33rd Annual Meeting of the Association for Computational Linguistics, Cambridge, MA, pp 236–243.

Fung P, KW Church 1994. K-vec: a new approach for aligning parallel texts. COLING-94 Fifteenth International Conference on Computational Linguistics, Kyoto, pp 1096–1102.

Fung P, K McKeown 1994. Aligning noisy parallel corpora across language groups: word pair feature matching by dynamic time warping. Technology Partnerships for Crossing the Language Barrier, Proceedings of the First Conference of the Association for Machine Translation in the Americas, Columbia, MD, pp 81–88.

Fung P, K McKeown 1997. A technical word- and term-translation aid using noisy parallel corpora across language groups. Mach Transl 12:53–87.

Gale WA, KW Church 1991a. A program for aligning sentences in bilingual corpora. 29th Annual Meeting of the Association for Computational Linguistics, Berkeley, CA, pp 177–184.

Gale WA, KW Church 1991b. A program for aligning sentences in bilingual corpora. Technical Report 94, AT&T Bell Laboratories, Statistical Research.

Gale WA, KW Church 1993. A program for aligning sentences in bilingual corpora. Comput Linguist 19:75–102.

Grishman R 1994. Iterative alignment of syntactic structures for a bilingual corpus. Proceedings of the Second Annual Workshop on Very Large Corpora, Kyoto, pp 57–68.

Haruno M., T Yamazaki 1996. High-performance bilingual text alignment using statistical and dictionary information. 34th Annual Meeting of the Association for Computational Linguistics, Santa Cruz, CA, pp 131–138.

Kaji H, Y Kida, Y Morimoto 1992. Learning translation templates from bilingual text. Proceedings of the fifteenth International Conference on Computational Linguistics, COLING-92, Nantes, pp 672–678.

Karlgren H, J Karlgren, M Nordström, P Pettersson, B Wahrolén 1994. Dilemma—an instant lexicographer. Proceedings of the Fifteenth International Conference on Computational Linguistics, Kyoto, pp 82–84.

Kasami T 1965. An efficient recognition and syntax analysis algorithm for context-free languages. Technical Report AFCRL-65-758, Air Force Cambridge Research Laboratory, Bedford, MA.

Kay, M, M Röscheisen 1988. Text-translation alignment. Technical Report P90–00143, Xerox Palo Alto Research Center, Palo Alto, CA.

Kay M, M Röscheisen 1993. Text-translation alignment. Comput Linguist 19:121–142.

Klavans J, E Tzoukermann 1990. The BICORD system. COLING 90: Proceedings of the Thirteenth International Conference on Computational Linguistics Helsinki, vol. 3, pp 174–179.

Kupiec J 1993. An algorithm for finding noun phrase correspondences in bilingual corpora. 31st Annual Meeting of the Association for Computational Linguistics, Columbus, OH, pp 17–22.

Lewis PM, RE Stearns 1968. Syntax-directed transduction. J Asoc Comput Mach 15:465–488.

Lin M-Y, T-H Chiang, K-Y Su 1993. A preliminary study on unknown word problem in Chinese word segmentation. Proceedings of ROCLING-93, Taiwan pp 119–141.

Lin Yi-C, T-H Chiang, K-Y Su 1992. Discrimination oriented probabilistic tagging. Proceedings of ROCLING-92, Taiwan, pp 85–96.

Macklovitch E 1994. Using bi-textual alignment for translation validation: the transcheck system. Technology Partnerships for Crossing the Language Barrier, Proceedings of the First Conference of the Association for Machine Translation in the Americas, Columbia, MD, pp 157–168.

Macklovitch E, M-L Hannan 1996. Line 'em up: advances in alignment technology and their impact on translation support tools. In Expanding MT Horizons, Second Conference of the Association for Machine Translation in the Americas, Montreal, pp 145–156.

Matsumoto Y, H Ishimoto, T Utsuro 1993. Structural matching of parallel texts. 31st Annual Meeting of the Association for Computational Linguistics, Columbus, OH, pp 23–30.

Melamed ID 1996. Automatic construction of clean broad-coverage translation lexicons. Expanding MT Horizons, Second Conference of the Association for Machine Translation in the Americas, Montreal, pp 125–134.

Melamed ID 1997a. A portable algorithm for mapping bitext correspondence. 35th Annual Meeting of the Association for Computational Linguistics and 8th Conference of the European Chapter of the Association for Computational Linguistics, Madrid, pp 305–312.

Melamed ID 1997b. A word-to-word model of translational equivalence. 35th Annual Meeting of the Association for Computational Linguistics and 8th Conference of the European Chapter of the Association for Computational Linguistics, Madrid, pp 490–497.

Nagao M 1984. A framework of a mechanical translation between Japanese and English by analogy principle. In: A Elithorn, R Banerji, eds. Artificial and Human Intelligence. Amsterdam: North-Holland, pp 173–180.

Resnik P, ID Melamed 1997. Semi-automatic acquisition of domain-specific translation lexicons. Fifth Conference on Applied Natural Language Processing, Washington, DC, pp 340–347.

Sadler V, R Vendelmans 1990. Pilot implementation of a bilingual knowledge bank. COLING 90: Proceedings of the Thirteenth International Conference on Computational Linguistics, Helsinki, vol 3, pp 449–451.

Simard M, GF Foster, P Isabelle 1992. Using cognates to align sentences in bilingual corpora. TMI-92, Proceedings of the Fourth International Conference on Theoretical and Methodological Issues in Machine Translation, Montreal, pp 67–81.

Simard M, P Plamondon 1996. Bilingual sentence alignment: balancing robustness and accuracy. Expanding MT Horizons, Second Conference of the Association for Machine Translation in the Americas, Montreal, pp 135–144.

Somers HL 1998. Further experiments in bilingual text alignment. Int J Corpus Linguist 3:115–150.

Sproat R, C Shih, W Gale, N Chang 1994. A stochastic word segmentation algorithm for a Mandarin text-to-speech system. 32nd Annual Meeting of the Association for Computational Linguistics, Las Cruces, NM, pp 66–72.

van Rijsbergen, CJ 1979. Information Retrieval, 2nd edn. London: Butterworths.

Viterbi AJ 1967. Error bounds for convolutional codes and an asymptotically optimal decoding algorithm. IEEE Trans Inform Theory 13:260–269.

Warwick-Armstrong S, G Russell 1990. Bilingual concordancing and bilingual lexicography. Proceedings of EURALEX '90, Malaga, Spain.

Wu D 1994. Aligning a parallel English–Chinese corpus statistically with lexical criteria. 32nd Annual Meeting of the Association for Computational Linguistics, Las Cruces, NM, pp 80–87.

Wu D 1995a. An algorithm for simultaneously bracketing parallel texts by aligning words. 33rd Annual Meeting of the Association for Computational Linguistics, Cambridge, MA pp 244–251.

Wu D 1995b. Grammarless extraction of phrasal translation examples from parallel texts. TMI-95, Proceedings of the Sixth International Conference on Theoretical and Methodological Issues in Machine Translation, Leuven, Belgium, pp 354–372.

Wu D 1995c. Stochastic inversion transduction grammars, with application to segmentation, bracketing, and alignment of parallel corpora. Proceedings of IJCAI-95, Fourteenth International Joint Conference on Artificial Intelligence, Montreal, pp 1328–1334.

Wu D 1995d. Trainable coarse bilingual grammars for parallel text bracketing. Proceedings of the Third Annual Workshop on Very Large Corpora, Cambridge, MA pp 69–81.

Wu D (in press). Stochastic inversion transduction grammars and bilingual parsing of parallel corpora. Comput Linguist.

Wu D, P Fung 1994. Improving Chinese tokenization with linguistic filters on statistical lexical acquisition. Fourth Conference on Applied Natural Language Processing, Stuttgart, pp 180–181.

Wu D, X Xia 1994. Learnng an English–Chinese lexicon from a parallel corpus. Technology Partnerships for Crossing the Language Barrier, Proceedings of the First Conference of the Association for Machine Translation in the Americas, Columbia, MD, pp 206–213.

Wu D, X Xia 1995. Large-scale automatic extraction of an English–Chinese lexicon. Mach Transl 9:285–313.

Wu Z, G. Tseng 1993. Chinese text segmentation for text retrieval: achievements and problems. J Am Soc Inform Sci 44:532–542.

Xu D, CL Tan 1996. Automatic alignment of English–Chinese bilingual texts of CNS news. ICCC-96, International Conference on Chinese Computing, Singapore, pp 90–97.

Younger DH 1967. Recognition and parsing of context-free languages in time n^3. Inform Control 10:189–208.

19

Contextual Word Similarity

IDO DAGAN

Bar-Ilan University, Ramat Gan, Israel

1. INTRODUCTION

Identifying different types of similarities between words has been an important goal in Natural Language Processing (NLP). This chapter describes the basic statistical approach for computing the degree of similarity between words. In this approach a word is represented by a **word co-occurrence vector** in which each entry corresponds to another word in the lexicon. The value of an entry specifies the frequency of joint occurrence of the two words in the corpus; that is, the frequency in which they co-occur within some particular relations in the text. The degree of similarity between a pair of words is then computed by some similarity or distance measure that is applied to the corresponding pairs of vectors.

This chapter describes in detail different types of lexical relations that can be used for constructing word co-occurrence vectors. It then defines a schematic form for vector-based similarity measures and describes concrete measures that correspond to the general form. We also give examples for word similarities identified by corpus-based measures, along with "explanations" of the major data elements that entailed the observed similarity.

This chapter describes in detail the basic vector-based approach for computing word similarity. Similarities may be computed between different lexical units, such a word strings, word lemmas, and multiword terms or phrases. We shall use the term "word" to denote a lexical unit, but the discussion applies to other units as well.

A. Applications of Word Similarity

The concept of word similarity was traditionally captured within thesauri. A *thesaurus* is a lexicographic resource that specifies semantic relations between words, listing for each word related words such as synonyms, hyponyms, and hypernyms. Thesauri have been used to assist writers in selecting appropriate words and terms and in enriching the vocabulary of a text. To this end, modern word processors provide a thesaurus as a built-in tool.

The area of information retrieval (IR) has provided a new application for word similarity in the framework of **query expansion**. Good free-text retrieval queries are difficult to formulate because the same concept may be denoted in the text by different words and terms. Query expansion is a technique in which a query is expanded with terms that are related to the original terms that were given by the user, in order to improve the quality of the query. Various query expansion methods have been implemented, both by researchers and in commercial systems, that rely on manually crafted thesauri or on statistical measures for word similarity (Frakes, 1992).

Word similarity may also be useful for disambiguation and language modeling in the area of NLP and speech processing. Many disambiguation methods and language models rely on word co-occurrence statistics that are used to estimate the likelihood of alternative interpretations of a natural language utterance (in speech or text). Because of data sparseness, though, the likelihood of many word co-occurrences cannot be estimated reliably from a corpus, in which case statistics about similar words may be helpful.

Consider for example the utterances in {1}, which may be confused by a speech recognizer.

The bear ran away. {1a}

The pear ran away. {1b}

A typical language model may prefer the first utterance if the word co-occurrence *bear ran* was encountered in a training corpus whereas the alternative co-occurrence *pear ran* was not. However, owing to data sparseness it is quite likely that neither of the two alternative interpretations was encountered in the training corpus. In such cases information about word similarity may be helpful. Knowing that *bear* is similar to other animals may help us collect statistics to support the hypothesis that animal names can precede the verb *ran*. On the other hand, the names of other fruits, which are known to be similar to the word *pear*, are not likely to precede this verb in any training corpus. This type of reasoning was attempted in various disambiguation methods, where the source of word similarity was either statistical (Grishman et al., 1986; Schütze, 1992, 1993; Essen and Steinbiss, 1992; Grishman and Sterling, 1993; Dagan et al., 1993, 1995; Karov and Edelman, 1996; Lin, 1997) or a manually crafted thesaurus (Resnik, 1992, 1995; Jiang and Conrath, 1997).

Although all the foregoing applications are based on some notion of "word similarity" the appropriate type of similarity relations might vary. A thesaurus intended for writing assistance should identify words that resemble each other in their meaning, such as *aircraft* and *airplane*, which may be subtituted for each other. For query expansion, on the other hand, it is also useful to identify contextually related words, such as *aircraft* and *airline*, that may both appear in relevant target documents. Finally, co–occurrence-based disambiguation methods would benefit from identifying words that have similar co-occurrence patterns. These might be words that resemble each other in their meaning, but may also have opposite meanings, such as *increase* and *decrease*.

B. The Corpus-Based Approach

Traditionally, thesauri have been constructed manually by lexicographers. Similar to most lexicographic tasks, manual thesaurus construction is very tedious and time-consuming. As a consequence, comprehensive thesauri are available only for some languages, and for the vocabularies of very few professional domains, such as the medical domain. Furthermore, although vocabularies and their usage patterns change rapidly, the process of updating and distributing lexicographic resources is slow and lags behind the evolution of language. These problems may be remedied, at least to some extent, by automatic and semiautomatic thesaurus construction procedures that are based on corpus statistics.

The common corpus-based approach for computing word similarity is based on representing a word (or term) by the set of its **word co-occurrence** statistics. It relies on the assumption that the meaning of words is related to their patterns of co-occurrence with other words in the text. This assumption was proposed in early linguistic work, as expressed in Harris' **distributional hypothesis**: "... the meaning of entities, and the meaning of grammatical relations among them, is related to the restriction of combinations of these entities relative to other entities" (Harris, 1968, p 12). The famous statement "You shall know a word by the company it keeps!" (Firth, 1957, p 11) is another expression of this assumption.

Given the distributional hypothesis, we can expect that words that resemble each other in their meaning will have similar co-occurrence patterns with other words. For example, both nouns *senate* and *committee* co-occur frequently with verbs such as *vote, reject, approve, pass,* and *decide*. To capture this similarity, each word is represented by a **word co-occurrence vector**, which represents the statistics of its co-occurrence with all other words in the lexicon. The similarity of two words is then computed by applying some vector similarity measure to the two corresponding co-occurrence vectors. Section 2 describes how different types of word co-occurrence vectors can be constructed by considering different types of co-occurrence relations (see Sec. 2.C relates to the altenative representation of **document occurrence vector**, which is commonly used in information retrieval). Section 3 defines a schematic form for vector-based similarity measures, whereas Sec. 4 describes concrete measures and discusses their correspondence to the schematic form. Section 5 lists examples for word similarities identified by corpus-based measures, along with "explanations" of common co-occurrences patterns that entailed the observed similarity.

2. CO-OCCURRENCE RELATIONS AND CO-OCCURRENCE VECTORS

In the corpus-based framework, a word is represented by data about its joint co-occurrence with other words in the corpus. To construct representations, we should first decide what counts as a co-occurrence of two words and specify informative types of relations between adjacent occurrences of words. Different types of co-occurrence relations have been examined in the literature, for computing word similarity as well as for other applications. These relations may be classified into two general types: grammatical relations, which refer to the co-occurrence of words within specified syntactic relations, and nongrammatical relations, which refer to the co-occurrence of words within a certain distance (window) in the text. As will be discussed in the following, the types of relations used in a particular word similarity system will affect the types of similarity that will be identified.

A. Grammatical Relations

Lexical co-occurrence within syntactic relations, such as subject–verb, verb–object, and adjective–noun, provide an informative representation for linguistic information. Statistical data on co-occurrence within syntactic relations can be viewed as a statistical alternative to traditional notions of selectional constraints and semantic preferences. Accordingly, this type of data was used successfully for various broad-coverage disambiguation tasks.

The set of co-occurrences of a word within syntactic relations provides a strong reflection of its semantic properties. It is usually the meaning of the word that restricts the identity of other words that can co-occur with it within specific syntactic relations. For example, *edible* items can be the direct object of verbs, such as *eat, cook,* and *serve,* but not of other verbs, such as *drive*. When used for assessing word similarity, grammatical co-occurrence relations are likely to reveal similarities between words that share semantic properties.

To extract syntactically based lexical relations, it is necessary to have a syntactic parser. Although the accuracy of currently available parsers is limited, limited accuracy is sufficient for acquiring reliable statistical data, with a rather limited amount of noise that can be tolerated by the statistical similarity methods. Yet, the use of a robust parser may be considered a practical disadvantage in some situations, because such parsers are not yet widely available and may not be sufficiently efficient or accurate for certain applications.

B. Nongrammatical Relations

Nongrammatical co-occurrence relations refer to the joint occurrence of words within a certain distance (window) in the text. This broad definition captures several subtypes of co-occurrence relations such as n-grams, directional and nondirectional co-occurrence within small windows, and co-occurrence within large windows or within a document.

1. *N-Grams*

An *n*-gram is a sequence of *n* words that appear consecutively in the text. These *n*-gram models are used extensively in language modeling for automatic speech recognition systems [see Jelinek et al. (1992) for a thorough presentation of this topic], as well as in other recognition and disambiguation tasks. In an *n*-gram model, the probability of an occurrence of a word in a sentence is approximated by its probability of occurrence within a short sequence of *n* words. Typically, sequences of two or three words (bigrams or trigrams) are used, and their probabilities are estimated from a large corpus. These probabilities are combined to estimate the a priori probability of alternative acoustic interpretations of the utterance to select the most probable interpretation.

The information captured by *n*-grams is largely just an indirect reflection of lexical, syntactic, and semantic relations in the language. This is because the production of consecutive sequences of words is a result of more complex linguistic structures. However, *n*-grams have practical advantages for several reasons: it is easy to formulate probabilistic models for them, they are very easy to extract from a corpus and, above all, they have proved to provide useful probability estimations for alternative readings of the input.

Word similarity methods that are based on bigram relations were tried for addressing the data sparseness problem in *n*-gram language modeling (Essen and Steinbiss, 1992; Dagan et al., 1994). Word similarities that are obtained by *n*-gram data may reflect a mixture of syntactic, semantic, and contextual similarities, as these are the types of relations represented by *n*-grams. Such similarities are suitable for improving an *n*-gram language model which, by itself, mixes these types of information.

2. *Co-occurrence Within a Large Window*

A co-occurrence of words within a relatively large window in the text suggests that both words are related to the general topic discussed in the text. This hypothesis will usually hold for frequent co-occurrences, that is, for pairs of words that often co-occur in the same text. A special case for this type of relations is co-occurrence within the entire document, which corresponds to a maximal window size.

Co-occurrence within large windows was used in the work of Gale et al. (1993) on word-sense disambiguation. In this work co-occurrence within a maximal distance of 50 words in each direction was considered. A window of this size captures context words that identify the topic of discourse. Word co-occurrence within a wide context was also used for language modeling in speech recognition in the work of Lau et al. (1993), when the occurrence of a word affects the probability of other words in the larger context. In the context of computing word similarity, co-occurrence within a large window may yield topical similarities between words that tend to appear in similar contexts.

3. *Co-occurrence Within a Small Window*

Co-occurrence of words within a small window captures a mixture of grammatical relations and topical co-occurrences. Typically, only co-occurrence of content words is considered because these words carry most semantic information.

Smadja (1993) used co-occurrence within a small window as an approximation for identifying significant grammatical relations without using a parser. His proposal relies on an earlier observation that 98% of the occurrences of syntactic relations relate words that are separated by at most five words within a single sentence (Martin, 1983). Smadja used this fact to extract lexical collocations, and applied the extracted data to language generation and information retrieval. Dagan et al. (1993, 1995) use this type of data as a practical approximation for extracting syntactic relations. To improve the quality of the approximation, the direction of co-occurrence is considered, distinguishing between co-occurrences with words that appear to the left or to the right of the given word. The extracted data is used to compute word similarities, which capture both semantic similarities, as when using grammatical relations, but also some topical similarities, as when using co-occurrence within a larger context.

Another variant of co-occurrence within a small window appears in the work of Brown et al. (1991). They use a part-of-speech tagger to identify relations such as "the first verb to the right" or "the first noun to the left," and then use these relations for word-sense disambiguation in machine translation. This type of relation provides a better approximation for syntactically motivated relations while relying only on a part of speech tagger, which is a simpler resource compared with syntactic parsers.

C. Word Co-occurrence Vectors

The first step in designing a word similarity system is to select the type of co-occurrence relations to be used, and then to choose a specific set of relations. For example, we may decide to represent a word by the set of its co-occurrences with other words within grammatical relations, as identified by a syntactic parser, and then select a specific set of relations to be used (subject–verb, verb–object, adjective–noun, and so on). Alternatively, we can use co-occurrences within a small window of three words and distinguish between words that occur to the left or to the right of the given word. Notice that in both cases there is more than one type of relation to be used. For example, when representing the word *boy*, and considering its co-occurrences with the verb *see*, the representation should distinguish between cases in which *boy* is either the subject or the object of *see*.

In formal terms, a word u is represented by a set of **attributes** and their frequencies. An attribute, $att = <w, rel>$, denotes the co-occurrence of u with another word w within a specific relation *rel*. For example, when representing the word *boy*, its occurrences as the subject and the object of the verb *see* will be denoted by the two attributes $<see, subj>$ and $< see, obj >$. The word u is represented by a **word co-occurrence vector** that has an entry for each possible attribute of u. The value of each entry is the frequency in which u co-occurs with the attribute, denoted by *freq(u,att)*. The co-occurrence vector of u thus summarizes the statistics about the co-occurrences of u with other words, while representing the particular types of relations that are distinguished by the system. In the special case in which only one type of relation is used, such as the relation "the word preceding u" (bigram), the relation *rel* can be omitted from the attribute, which will consist only of the word w. In this case, the length of the co-occurrence vector is equal to the number of words in the lexicon.

Co-occurrence vectors are typically sparse because a word typically co-occurs only with a rather small subset of the lexicon; hence, with a small set of attributes. Accordingly, the co-occurrence vector of u is represented in memory as a sparse vector that specifies only **active** attributes for u, that is, attributes that actually co-occur with it in the corpus. A sparse co-occurrence vector may also be considered as a set of attribute–value pairs, where the value of an attribute is its co-occurrence frequency with u (notice that the order of entries in the co-occurrence vector is not important as long as it is the same for all vectors). This chapter uses the notation of vectors, rather than attribute–value pairs, which is the common notation in the literature of IR and statistical NLP.

The similarity and distance measures described in the following sections take as input two co-occurrence vectors, which represent two words, and measure the degree of similarity (or dissimilarity) between them. The vector representation is general and enables us to separate the two major variables in a word similarity system: the types of co-occurrence relations for the attributes of the vectors and the type of similarity measure that is used to compare pairs of vectors. These two variables are orthogonal to each other: any type of co-occurrence relation can be combined with any type of similarity measure. The set of co-occurrence relations that is chosen affects the nature of similarities that will be identified, such as semantic similarity versus topical similarity. The concrete similarity measure that is chosen affects the resulting similarity values between different word pairs. For example, the most similar word to u may be v_1 according to one similarity measure and v_2 according to another measure, where both values are based on the same vector representation.

The IR literature suggests another type of vector representation for computing word similarities (Frakes, 1992). In this framework, a word u is represented by a **document occurrence vector**, which contains an entry for each document in the corpus. The value of an entry denotes the frequency of u in the corresponding document. Word similarity is then measured by the degree of similarity between such vectors, as computed by some similarity or distance measure. Two words are considered as similar by this method if there is a large overlap in the two sets of documents in which they occur.

Notice that a document occurrence vector is substantially different from a word co-occurrence vector, in which each entry corresponds to a word in the lexicon, rather than to a document. Notice also that a word co-occurrence vector that is constructed from the relation "joint occurrence in a document" represents exactly the same statistical information as a document occurrence vectors. However, the information is represented differently in the two types of vector, which yields different similarity values even if the same similarity metric is used. This chapter focuses on the word co-occurrence representation, although the similarity metrics of the following sections are relevant also for document occurrence vectors.

3. A schematic structure of similarity measures

Given representations of words as co-occurrence vectors, we need a mathematical measure for the degree of similarity between pairs of vectors. Either a similarity measure or a distance measure can be used: the value given by a similarity measure is proportional to the degree of similarity between the vectors (higher values for higher similarity), whereas the value of a distance measure is proportional to the degree of dissimilarity between the vectors (higher values for lower similarity).[1] In the rest of this chapter we refer to the two types of measures interchangeably, making the distinction only when necessary. This section describes, in a schematic way, the major statistical factors that are addressed by similarity and distance measures. The next section presents several concrete measures and relates their structure to the general scheme.

Similarity measures are functions that compute a similarity value for a pair of words, based on their co-occurrence vectors and some additional statistics. These functions typically consist of several components that quantify different aspects of the statistical data. The following subsections specify the major statistical aspects, or factors, that are addressed by similarity measures.

---

[1]Our use of the term **distance measure** is not restricted to the mathematical notion of a **distance metric**; it is rather, used in general sense of denoting a measure of dissimilarity.

A. Word-Attribute Association

The first aspect to be quantified is the degree of **association** between a word u and each of its attributes, denoted by $assoc(u, att)$. The simplest way to quantify the degree of association between u and att is to set $assoc(u, att) = freq(u, att)$, which is the original value in the vector entry corresponding to att. This measure is too crude, though, because it overemphasizes the effect of words and attributes that are frequent in the corpus and, therefore, are a priori more likely to co-occur together. Most similarity measures use more complex definitions for $assoc(u, att)$ which normalize for word frequencies and possibly scale the association value according to individual attribute weights.[2]

Mathematical definitions for word association were developed also for other applications, such as extracting collocations and technical terms (see Chap. 7). In these tasks, obtaining a good measurement for the degree of association is the goal of the system. In the context of word similarity, association measurement is only an intermediate step that appropriately scales the values of vector entries to facilitate their comparison with other vectors.

B. Joint Association

The degree of similarity between two words u and v is determined by comparing their associations with each of the attributes in the two co-occurrence vectors. The two words will be considered similar if they tend to have similar degrees of association (strong or weak) with the same attributes. To capture such correspondence, similarity measures include a component that combines two corresponding association values, $assoc(u, att)$ and $assoc(v, att)$, into a single value. We call this component the **joint association** of u and v relative to the attribute att, and denote it by $joint(assoc(u, att), assoc(v, att))$. The joint operation is intended to compare the two values $assoc(u, att)$ and $assoc(v, att)$, producing a high value when the two individual association values are similar (this is so for similarity measures; for distance measures the joint association value is low when the two values are similar).

In some measures the comparison between the two association values is by the joint operation, such as when defining the joint operation as the ratio or difference between the two values. In other measures the joint operation is defined by a different type of mathematical operations, such as multiplication, that do not directly express a comparison. In these cases the comparison is expressed in a global manner, so that the normalized sum (see following section) of all joint associations will be high if, on average, corresponding association values in the two vectors are similar. The joint association operation may also include a weighting factor for each attribute, which captures its "importance" relative to similarity calculations.

[2]When using a concrete similarity measure one might think of the values in the co-occurrence vector of u as denoting the value $assoc(u, att)$, for each attribute att. We prefer to consider $freq(u, att)$ as the original value of a vector entry and $assoc(u, att)$ as being part of the similarity formula. This view yields a general vector representation that summarizes the raw statistics of the word and may be fed into different similarity measures.

C. Normalized Sum of Joint Associations

The joint association of u and v relative to the attribute *att* can be considered as the contribution of the individual attribute to the similarity between u and v. The overall similarity between the two words is computed by summing the joint association values for all attributes. This sum should be normalized by a normalization factor, denoted by *norm*, whenever there is a variance in the "length" of co-occurrence vectors of different words. That is, different words may have a different number of active attributes (attributes with which they co-occur), as well as different association values with their attributes. Normalization of the sum is required to avoid scaling problems when comparing different word pairs. In some measures the normalization is implicit, because the *assoc* or *joint* operations already produce normalized values. In these cases the factor *norm* is omitted.

By using the three foregoing factors, we now define a schematic form of similarity measures:

$$sim(u, v) = \frac{1}{norm} \sum_{att} joint[assoc(u, att), assoc(v, att)] \quad \{2\}$$

The schematic form illustrates the statistical "reasoning" captured by similarity measures. Distance measures have the same schematic form, where the *joint* operation produces higher values when the two association values are dissimilar. Many concrete similarity measures follow this schematic form closely. Although some variation in the form of a similarity formula and its components may be possible, the statistical reasoning behind the formula still corresponds pretty much to the general schema.

4. CONCRETE SIMILARITY AND DISTANCE MEASURES

Word-attribute associations, joint associations, and normalization may be captured by different mathematical formulas. This section presents several concrete measures that have appeared in the literature, analyzes their structure, and illustrates their correspondence to the general schema of Eq. {2}. Although quite a few alternative measures appear in the literature, somewhat little work has been done to compare them empirically and to analyze the effects of their components. Therefore, further research is necessary to obtain a better understanding of the desired form of word similarity measures. The measures described here give a range of different types of alternatives, although the list is not comprehensive.

A. "Min/Max" Measures

By the name "Min/Max" measures we refer to a family of measures that have the following general form {3}:

$$sim(u, v) = \frac{\sum_{att} \min[assoc(u, att), assoc(v, att)]}{\sum_{att} \max[assoc(u, att), assoc(v, att)]} \quad \{3\}$$

This form may be considered as a weighted version of the Tanimoto measure, also known as the Jaccard measure (see, e.g., in Ruge, 1992). The Tanimoto measure assigns *assoc* values that are either 1, if the attribute is active, or 0, if it is inactive, whereas the Min/Max measure allows for any *assoc* value.

Relative to the schema {2}, we can interpret the "min" operation as playing the role of the *joint* operation, whereas the sum of "max" values is the normalizing factor *norm*. This selection of *joint* and *norm* may be motivated by the following interpretation. Consider {4} as being composed of two components.

$$joint[assoc(u, att), assoc(v, att)] = \min[assoc(u, att), assoc(v, att)] \quad \{4\}$$

$$\frac{\min[assoc(u, att), assoc(v, att)]}{max[assoc(u, att), assoc(v, att)]} \quad \{5\}$$

Equation {5} is the ratio that compares the two association values. This ratio ranges between 0 and 1, obtaining higher values when the two association values are closer to each other. The ratio is multiplied by a weighting factor, $\max[assoc(u, att), assoc(v, att)]$, which means that the effect of an attribute on the comparison of two words is determined by its maximal association value with one of the words. Thus, the *joint* operation is composed of the product of two components: the first measures the closeness of the two association values and the second weighs the importance of the attribute relative to the given pair of words. Finally, to obtain the complete formula, the normalization factor *norm* is simply the sum of all attribute weights.

1. Log-Frequency and Global Entropy Weight

Given the general "Min/Max" form it is necessary to define the *assoc* operation. Grefenstette (1992, 1994) defines *assoc* as in Eq. {6}, which was applied to co-occurrence vectors that are based on grammatical co-occurrence relations.

$$assoc(u, att) = \log[freq(u, att) + 1] \cdot Gew(att) \quad \{6\}$$

where $freq(u, att)$ is the co-occurrence frequency of the word u with the attribute att. *Gew* stands for **global-entropy-weight** for the attribute, which measures the general "importance" of the attribute in contributing to the similarity of pairs of words Eq. {7}.

$$Gew(att) = 1 \frac{1}{\log nrels} \sum_{v} -P(v|att) \cdot \log(P(v|att)) \quad \{7\}$$

where v ranges over all words in the corpus, *nrels* is the total number of co-occurrence relations that were extracted from the corpus, and the basis of the logarithm is 2. The sum $\sum_v -P(v|att) \cdot \log(P(v|att))$ is the entropy of the empirical probability distribution (as estimated from corpus data) $P(v|att)$, which is the probability of finding the word v in an arbitrary co-occurrence relation of the attribute att. The normalization value log *nrels* is an upper bound for this entropy, making the normalized entropy value $1/\log nrels \sum_v -P(v/|att) \cdot \log(P(v|att))$ range between 0 and 1. This value, in proportion to the entropy itself, is high if *att* occurs with many different words and the distribution $P(v|att)$ is quite uniform, which means that the occurrence of *att* with a particular word v is not very informative. *Gew* is inversely proportional to the normalized entropy value, obtaining high values for attributes that are relatively "selective" about the words with which they occur.

In summary, Grefenstette's association value is determined as the product of two factors: the absolute co-occurrence frequency of u and att, the effect of which is moderated by the log function, and a global weight for the importance of the attribute. Note that the combination of these two factors balances (to some extent) the effect of attribute frequency. A frequent attribute in the corpus (corresponding to a frequent word w) is likely to occur more frequently with any other word, leading to relatively high values for $freq(u, att)$. This may overamplify the effect of co-occurrence with frequent attributes. On the other hand, since a frequent attribute is likely to occur with many different words, it is also likely to have a relatively high entropy value for $P(v|att)$, yielding a relatively low *Gew* value. This effect of *Gew* may be compared with the commonly used IDF (Inverse Document Frequency) measure for attribute weight in IR, which is inversely proportional to the frequency of the attribute. The *Gew* measure was borrowed from the IR literature (Dumais, 1991).

2. Mutual Information

Dagan et al. (1993, 1995) have also used the "Min/Max" scheme for their similarity measure, which was applied to co-occurrence relations within a small window. They adopted a definition for *assoc* [Eq. {8}] that is based on the definition of mutual information in information theory (Cover and Thomas, 1991) and was proposed earlier for discovering associations between words (Church and Hanks, 1990; Hindle, 1990; see also Chap. 6).

$$assoc(u, att) = \log\frac{P(u, att)}{P(u)P(att)} = \log\frac{P(att|u)}{P(att)} = \log\frac{P(u|att)}{P(u)} \qquad \{8\}$$

Mutual information normalizes for attribute frequency, as can be seen in the conditional form of the formula. Its empirical estimation is sensitive to low-frequency events: because of data sparseness, the estimated value for $P(u, att)$ may be too high for many word–attribute pairs that co-occur once or twice in the corpus, yielding an inflated association for these pairs. This is a well-known problem when using mutual information to discover collocations or technical terminology, when identifying specific associations is the primary goal of the system. The problem is less severe when the measure is used to quantify many associations simultaneously as the basis for comparing two co-occurrence vectors.

B. Probabilistic Measures

The measures described in this subsection are comparative probability distributions. In the context of computing word similarity, each word is represented by a probability distribution constructed by defining $assoc(u, att) = P(att|u)$. That is, the word u is represented by the conditional probability distribution that specifies the probability for the occurrence of each attribute given the occurrence of u. The similarity (or distance) between two words u and v is then measured as the similarity (or distance) between the two conditional distributions $P(att|u)$ and $P(att|v)$.

1. KL Divergence

The **Kullback–Leibler (KL) divergence**, also called **relative entropy** (Cover and Thomas, 1991), is a standard information theoretic measure of the dissimilarity between two probability distributions. It was used for distributional word clustering

(Pereira et al., 1993) and for similarity-based estimation in Dagan et al. (1994). When applied to word-attribute co-occurrence it takes the following form [Eq. {9}].

$$D(u\|v) = \sum_{att} P(att|u) \cdot \log \frac{P(att|u)}{P(att|v)} \cdot \log \frac{P(att|u)}{P(att|v)} \qquad \{9\}$$

The KL divergence is a nonsymmetrical measure, as in general $D(u\|v) \neq D(v\|u)$. Its value ranges between 0 and infinity, and is 0 only if the two distributions are identical. Relative to the schema {2}, the *joint* operation is composed of two components: a log-scaled ratio that compares the two association values $P(att|u)$ and $P(att|v)$, and a weighting factor $P(att|u)$ for the contribution of *att* to $D(u\|v)$.

$D(u\|v)$ is defined only if $P(att|v)$ is greater than 0 whenever $P(att|u)$ is. This condition does not hold in general when using the Maximum Likelihood Estimator (MLE), for which the estimate for $P(att|v)$ is 0 when $freq(att, v) = 0$. This forces using a smoothed estimator that assigns nonzero probabilities for all $P(att|)$ even when $freq(att, v) = 0$. However, having zero association values for many word–attributes pairs gives a big computational advantage, which cannot be exploited when using the KL divergence as a dissimilarity measure. Furthermore, the need to use a smoothing method both complicates the implementation of the word similarity method and may introduce an unnecessary level of noise into the data.The problem is remedied by the symmetrical measure of **total divergence to the average**, which is described next. KL-divergence was used for word similarity by Dagan et al. (1994), who later adopted the total divergence to the average.

2. *Total Divergence to the Average*

The total divergence to the average, also known as the **Jensen–Shannon divergence** (Rao, 1982; Lin, 1991), was used for measuring word similarity by Lee (1997) and by Dagan et al. (1997) and is defined as Eq. {10}.

$$A(u, v) = D\left(u\| \frac{u+v}{2}\right) + D\left(v\| \frac{u+v}{2}\right) \qquad \{10\}$$

where $u + v/2$ is a shorthand [Eq. {11}].

$$\frac{u+v}{2} = \frac{1}{2}(P(att|u) + P(att|v)) \qquad \{11\}$$

$A(u, v)$ is thus the sum of the two KL-divergence values between each of the distributions $P(att|u)$ and $P(att|v)$ and their average $u + v/2$. It can be shown (Lee, 1997; Dagan et al., 1997) that $A(u, v)$ ranges between 0 and 2 log 2.

By definition, $u + v/2$ is greater than 0 whenever either $P(att|u)$ or $P(att|v)$ is. Therefore, the total divergence to the average, unlike KL-divergence, does not impose any constraints on the input data. MLE estimates, including all zero estimates, can be used directly.

3. *L_1 Norm (Taxi-Cab Distance)*

The **L_1 norm**, also known as the **Manhattan distance**, is a measure of dissimilarity between probability distributions [Eq. {12}].

$$L_1(u, v) = \sum_{att} |[LP(att|u) - P(att|v)]| \qquad \{12\}$$

This measure was used for measuring word distance by Lee (1997) and Dagan et al. (1997). It was shown that $A(u, v)$ ranges between 0 and 2. Relative to schema {2}, the *joint* operation compares the two corresponding association values $P(att|u)$ and $P(att|v)$ by measuring the absolute value of their difference. The use of a difference (rather than a ratio) implicitly gives higher importance to attributes when at least one of the two association values is high.

Empirical work, summarized in Dagan et al. (1997) and detailed in Lee (1997), compared the performance of total divergence to the average and L_1 norm [as well as **confusion probability** (Essen and Steinbiss, 1992)] on a pseudo–word-sense disambiguation task. Their method relies on similarity-based estimation of probabilities of previously unseen word co-occurrences. The results showed similar performance of the disambiguation method when based on the two dissimilarity measures, with slight advantage for the total divergence to the average measure.

C. The Cosine Measure

The **cosine measure** has been used extensively for IR within the framework of the vector space model (Frakes, 1992, Chap. 14). It was applied also for computing word similarity, for example, by Ruge (1992), who used co-occurrence vectors that are based on grammatical relations. The cosine measure is defined as [Eq. {13}].

$$\cos(u, v) = \frac{\sum_{att} assoc(u, att) \cdot assoc(v, att)}{\sqrt{\sum_{att} assoc(w_1, att)^2} \cdot \sqrt{\sum_{att} assoc(w_2, att)^2}} \qquad \{13\}$$

This is the scalar product of the normalized, unit-length, vectors produced by the *assoc* values. For schema {2}, the *joint* operation can be interpreted as the product of the two corresponding association values whereas the *norm* factor is the product of the lengths of the two *assoc* vectors. Combined together, these two operations entail a high cosine value if corresponding *assoc* values are overall similar (obtaining the effect of a comparison), giving a higher impact for larger *assoc* values.

Given the general cosine form, it is necessary to define the *assoc* operation. Ruge (1992) found the definition $assoc(u, att) = \ln(freq(u, att))$ to be more effective than either setting $assoc(u, att) = freq(u, att)$ or $assoc(u, att) = 1$ *or* 0 according to whether the attribute is active or inactive for u. In analogy to other measures, one might try adding a global weight for the attribute (like *Gew*) or some normalization for its frequency. Ruge tried a somewhat different attribute weight, which obtains a low value when attribute frequency is either very high or very low. However, this global attribute weighting did not improve the quality of the similarity measure in her experiments.

5. EXAMPLES

This section gives illustrative examples for word similarities identified by the methods described in this chapter. The statistical data was collected from a sample of about 10,000 Reuters articles, which were parsed by a shallow parser. The co-occurrence vectors were based on statistics of grammatical co-occurrence relations, as described in Sec. 2.A.

Table 1 lists the most similar words for a sample of words, sorted by decreasing similarity score, as computed by four different similarity measures. Table 2 provides data about common context words that contributed mostly to the similarity scores of given pairs of words. For example, the first line of Table 2 specifies that the words *last*, *next*, and so on, co-occurred, as modifying adjectives, with both nouns *month* and *week*.

As can be seen from the tables, many of the similarities identified indeed correspond to meaningful semantic similarities, yet some degree of noise exists (the examples chosen are relatively "“clean,” other examples may include a higher level of noise). These results indicate that there is still a lot of room for future research that will improve the quality of statistical similarity measures and reduce the amount of noise in their output.

6. CONCLUSIONS

Automatic computation of various forms of word similarity has been recognized as an important practical goal. Word similarities are useful for several applications for which the statistical computation may be used either in a fully automatic manner or in a semiautomatic way, letting humans correct the output of the statistical measure. Substantial research work in this area has resulted in several common principles and

Table 1 Sample of Most Similar Words by Four Different Similarity Measures: Total Divergence to the Average, Cosine Measure with $assoc(u, att) = \ln(freq(u, att))$, Min/Max with Log-Frequency and Global Entropy Weight

Word	Measure	Most similar words
month	Average-KL	year week day quarter night December thanks
	Cosine	year week day quarter period dlrs January February
	Minmax-Gew	week year day quarter period February March sale
	Minmax-MI	week year day January February quarter March
shipment	Average-KL	delivery sale contract output price export registration
	Cosine	delivery contrasale output price
	Minmax-Gew	delivery output contract surplus period registration
	MI	delivery output contract registration surplus election
copper	Average-KL	rubber nickel LPG Opec composite sorghum issue
	Cosine	rubber composite LPG truck issue Opec sorghum
	Minmax-Gew	nickel zinc cathode gold coffee Jeep silver spot
	Minmax-MI	nickel zinc cathode coffee Jeep coal semiconductor
boost	Average-KL	accommodate curb increase reduce depress keep
	Cosine	accommodate depress curb increase cut translate reduce
	Minmax-Gew	reduce keep curb cut push increase limit exceed hit
	Minmax-MI	curb keep reduce hit push increase exceed cut limit
accept	Average-KL	reject submit formulate disseminate approve consider
	Cosine	disseminate formulate reject submit unrestricted
	Minmax-Gew	reject approve submit consider threaten adjust extend
	Minmax-MI	submit threaten approve reject reaffirm consider pass

Table 2 Common Context Words, Listed by Syntactic Relation, Which Were Major Contributors to the Similarity Score of Word-1 and Word-2

Word-1	Word-2	Rel	Common context words
month	week	a-n	last next few previous past several
		n-n	bill period
		n-prep-n	end sale export
	year	a-n	last next first previous past few recent several
		n-n	period
		n-prep-n	end dlrs export
shipment	delivery	n-n	april June March January
		n-prep-n	cent dlrs June April January year
		n-prep-n	tonne corn
	export	n-n	coffee corn June January grain
		n-prep-n	tonne year January Europe
		s-v	rise fall
		n-prep-n	tonne tariff sugar
agency	news	n-n	official Tass Tanjug Association APS SPK MTI
	ministry	s-v	say announce add
		n-n	statement source official
	department	s-v	say report note assign point announce
		n-n	statement plan official
law	regulation	n-n	banking
		a-n	federal new German current
		n-prep-n	accordance change
		s-v	allow
		v-o	change
	legislation	n-n	trade reform preference
		a-n	special new recent
		n-prep-n	change year
		s-v	require call

Key: **n-n**, noun–noun (noun compound); **a-n**, adjective–noun; **n-prep-n**, noun–preposition–noun; **s-v**, subject–verb; **v-o** verb–object. The underscored position in the relation corresponds to the position of each of the two similar words (Word-1/2) whereas the other position corresponds to the common context word.

a variety of concrete methods. Yet, methods of this type are very rarely integrated in operational systems. Further research is required to establish "common practices" that are known to be sufficiently reliable and useful for different applications.

ACKNOWLEDGMENTS

The writing of this chapter benefited from the knowledge and results acquired in joint work with colleagues on statistical word similarity. The measure of Sec. 4.A.2 was developed in joint work with Shaul Marcus and Shaul Markovitch. The probabilistic measures of Sec. 4.B were proposed for the purpose of measuring word similarity in joint work with Lillian Lee and Fernando Pereira. The scheme of Sec. 3 and the examples of Sec. 5 were obtained in joint work with Erez Lotan at Bar Ilan University.

REFERENCES

Brown P, SA Della Pietra, VJ Della Pietra, R Mercer 1991. Word sense disambiguation using statistical methods. 29th Annual Meeting of the Association for Computational Linguistics, Berkeley, CA, pp 264–270.

Church KW, P Hanks 1990. Word association norms, mutual information, and lexicography. Comput Linguist 16:22–29.

Cover TM, JA Thomas 1991. Elements of Information Theory. New York: John Wiley and Sons.

Dagan I, L Lee, F Pereira 1997. Similarity-based methods for word sense disambiguation. 35th Annual Meeting of the Association for Computational Linguistics and 8th Conference of the European Chapter of the Association for Computational Linguistics, Madrid, pp 56–63.

Dagan I, S Marcus, S Markovitch 1993. Contextual word similarity and estimation from sparse data. In 31st Annual Mmeeting of the Association for Computational Linguistics, Columbus, Ohio, pp 164–171. Comput Speech Lang 9:123–152.

Dagan, I, S. Marcus, S Markovitch, 1995. Contextual word similarity and estimation from sparse data.

Dagan I, F Pereira, L Lee 1994. Similarity-based estimation of word cooccurrence probabilities. 32nd Annual Meeting of the Association for Computational Linguistics, Las Cruces, NM, pp 272–278.

Dumais ST 1991. Improving the retrieval of information from external sources. Behav Res Methods Instr Comput 23:229–236.

Essen U, Steinbiss 1992. Co-occurrence smoothing for stochastic language modeling. Proceedings of ICASSP, vol 1, pp 161–164.

Firth JR 1957. A synopsis of linguistic theory 1930–1955. In: Philological Society, ed., Studies in Linguistic Analysis. Oxford: Blackwell, pp 1–32. Reprinted in F Palmer, ed. Selected Papers of J. R. Firth. Longman, 1968.

Gale W, K Church, D Yarowsky 1993. A method for disambiguating word senses in a large corpus. Comput Human 26:415–439.

Grefenstette G 1992. Use of syntactic context to produce term association lists for text retrieval. Proceedings of the Fifteenth Annual International ACM SIGIR Conference on Research and Development in Information Retrieval, Copenhagen, pp 89–97.

Grefenstette G 1994. Exploration in Automatic Thesaurus Discovery. Dordrecht: Kluwer Academic.

Grishman R, L Hirschman, NT Nhan 1986. Discovery procedures for sublanguage selectional patterns—initial experiments. Comput Linguist 12:205–214.

Grishman R, J Sterling 1993. Smoothing of automatically generated selectional constraints. Proceedings of DARPA Conference on Human Language Technology, San Francisco, CA, pp 254–259.

Harris ZS 1968. Mathematical Structures of Language. New York: Wiley, 1968.

Hindle D 1990. Noun classification from predicate-argument structures. 28th Annual Meeting of the Association for Computational Linguistics, Pittsburgh, PA, pp 268–275.

Jelinek F, RL Mercer, S Roukos 1992. Principles of lexical language modeling for speech recognition. In: S Furui, MM Sondhi, eds. Advances in Speech Signal Processing. New York: Marcel Dekker, pp 651–699.

Jiang JJ, DW Conrath 1997. Semantic similarity based on corpus statistics and lexical taxonomy. Proceedings of International Conference Research on Computational Linguistics (ROCLING). Taiwan.

Karov Y, S Edelman 1996. Learning similarity-based word sense disambiguation from sparse data. Proceedings of the Fourth Workshop on Very Large Corpora, Copenhagen, pp 42–55.

Lee L 1997. Similarity-based approaches to natural language processing. PhD dissertation, Harvard University, Cambridge, MA.

Lin D 1997. Using syntactic dependency as local context to resolve word sense ambiguity. 35th Annual Meeting of the Association for Computational Linguistics and 8th Conference of the European Chapter of the Association for Computational Linguistics, Madrid, pp 64–71.

Lin J 1991. Divergence measures based on the Shannon entropy. IEEE Trans Inform Theory 37:145–151.

Martin WJR, BPF Al, PJG van Sterkenburg 1983. On the processing of text corpus: from textual data to lexicographical information. In RRK Hartman, ed. Lexicography: Principles and Practice. Academic Press, London.

Pereira F, N Tishby, L Lee 1993. Distributional clustering of English words. 31st Annual Meeting of the Association for Computational Linguistics, Columbus, OH, pp 183–190.

Rao CR 1982. Diversity: its measurement, decomposition, apportionment and analysis. Sankyha: Indian J Stat 44(A): 1–22.

Resnik P 1992. Wordnet and distributional analysis: a class-based approach to lexical discovery. AAAI Workshop on Statistically-Based Natural Language Processing Techniques, Menlo Park, CA, pp 56–64.

Resnik P 1995. Disambiguating noun groupings with respect to WordNet senses. Proceedings of the Third Workshop on Very Large Corpora, Cambridge, MA, pp 54–68.

Ruge G 1992. Experiments on linguistically-based term associations. Inform Process Manage 28:317–332.

Schütze H 1992. Dimensions of meaning. Proceedings of Supercomputing '92, Minneapolis, MN, pp 787–796.

Schütze H 1993. Word space. In: SJ Hanson, JD Cowan, CL Giles, eds. Advances in Neural Information Processing Systems 5. San Mateo, CA: Morgan Kaufman, pp 895–902.

Smadja F 1993. Retrieving collocations from text: Xtract. Comput Linguist 19:143–177.

20

Computing Similarity

LUDOVIC LEBART

Ecole Nationale Supérieure des Télécommunications, Paris, France

MARTIN RAJMAN

Swiss Federal Institute of Technology, Lausanne, Switzerland

1. INTRODUCTION

The analysis of similarities between textual entities in high-dimensional vector spaces is an active research field. This chapter describes specific techniques of statistical multivariate analysis and multidimensional scaling adapted to large sparse matrices (singular value decomposition, latent semantic analysis, clustering algorithms, and others). Examples of textual data visualization and classification concern the domains of textual data analysis and information retrieval.

Computing similarities between textual entities is a central issue in several domains, such as Textual Data Analysis, Information Retrieval, and Natural Language Processing. In each of these domains, proximities between textual data are used to perform a large variety of tasks:

- In Textual Data Analysis (TDA), similarities will serve as a basis for the description and exploration of textual data, for the identification of hidden structural traits and for prediction.
- In Information Retrieval (IR), the computation of similarities between texts and queries is used to identify the documents that satisfy the information need expressed by a query.
- In Natural Language Processing (NLP), similarities are often very useful to resolve (or at least reduce) ambiguities at various processing levels (morphosyntactic, syntactic, or even semantic).
- In Text Mining (TM), similarities are used to produce synthetic representations of sets of documents to enable Knowledge Extraction out of textual data.

The purpose of this chapter is to describe the different approaches, methods, and techniques that are used in some of the foregoing domains to deal with similarities between textual entities.

Section 2 (Lexical Units and Similarity Measures) is devoted to a discussion of the linguistic units that can be used to represent texts for similarity computations. Some of the measures that are frequently used to assess similarities (or dissimilarities) are briefly presented, and some of their important properties will be analyzed.

Section 3 (Singular Value Decomposition and Clustering Techniques) presents the basic techniques for exploratory statistical analysis of multivariate data. Particular attention is given to the application of such techniques to visualization (principal axes methods) and organization (cluster analysis) of textual data (Lebart et al., 1998).

Section 4 (Characteristic Textual Units, Modal Sentences or Texts) presents so-called characteristic words or characteristic element calculations that are used to identify units, the frequencies of which are atypical in each of the parts of a corpus. An automatic selection of modal documents or modal texts makes it possible to restore words into their context, and to possibly characterize generally voluminous text parts with smaller pieces of text (sentences, paragraphs, documents, survey responses).

Section 5 (Text Categorization) studies the discriminatory power of texts in the statistical sense. How can a text be attributed to an author (or to a time period)? Is it possible to assign an individual to a group on the basis of his or her response to an open question? How can a document in a textual database be classified (i.e., assigned to preexisting categories)? The corresponding methods, often referred to as Text Categorization in IR, are closely related to those currently used in the domains of Pattern Recognition and Artificial Neural Networks.

Section 6 (Enhancing Similarities) examines the possible enhancements of similarities and distances when further information is taken into account, such as external corpora and semantic networks.

Finally, Sec. 7 (Validation Techniques) is devoted to the assessment and validation of results (visualization, categorization) that can be obtained on the basis of computed similarities.

2. LEXICAL UNITS AND SIMILARITY MEASURES

The techniques used to compute similarities in each of specific domains such as TDA, IR, or NLP, are quite different, but they all fit in the same general two-step approach:

1. The textual entities are associated with some specific structures that will serve as a basis for the computation of the similarities; the precise nature of the associated structures is dependent on the domain of application that is considered. In TDA, lexical profiles are often used, whereas in IR and TM (possibly weighted) distributions of indexing keywords are taken into consideration. In NLP, contextual co-occurrence vectors are sometimes considered. It is important to notice that the associated structures are always represented in the form of elements of a high-dimensional vector space that we will call the **representation space**.

2. A mathematical tool is chosen to measure, in the reprsentation space, the proximities that will be representative for the similarities between the textual entities. In TDA, the χ^2 distance (see later) is a frequent choice. In IR, similarities based on the cosine measure are routinely used, whereas in TM and NLP, information theory-based measures are often preferred (such as the Kullback–Leibler distance, based on relative entropy, or associations, based on mutual information).

A. Lexical Units for the Representation of Texts

To produce the structures that will represent the texts for the computation of similarities, texts have to be decomposed in simpler lexical units. Several choices can be made, as different units have different degrees of relevance in each particular field of research. They also have different advantages (and disadvantages) as far as practical implementation is concerned.

For instance, a researcher exploring a set of articles assembled from a database in the field of chemistry might require that the noun *acid* and its plural *acids* be grouped into the same unit, to be able to query all of the texts at once on the presence or absence of one or other word. On the other hand, in the field of political text analysis, researchers have observed that singular and plural forms of a noun are often related to different, sometimes opposite, concepts (e.g., the opposition in recent text of *defending freedom* vs. *defending freedoms*, referring to distinct political currents). In this latter case, it could often be preferable to code the two types of elements separately and to include both in the analysis.

1. Decompositions Based on Graphic Forms (or Words)

A classic simple way of defining textual units in a corpus of texts is to use graphic forms (or words, or types). To obtain an automatic segmentation of a text into occurrences of graphic forms, a subset of characters must be specified as delimiting characters (all other characters contained in the text are then considered as nondelimiting characters) (see also Chap. 4).

A series of nondelimiters the bounds of which at both ends are delimiters is an **occurrence**. Two identical series of nondelimiters constitute two occurrences **(tokens)** of the same word **(type)**. The entire set of words in a text is its **vocabulary**. In the context of this type of segmentation, a **text** is a series of occurrences that are separated from one another by several delimiters. The total number of occurrences contained in a text is its **size** or its **length**. This approach assumes that each character in a text has only one status; that is, that the text has been cleaned of any coding ambiguities.

2. Decompositions Involving More Sophisticated Linguistic Knowledge

The graphic elements resulting from automatic segmentation may be further processed to integrate more sophisticated linguistic knowledge into the representations. Morphosyntactic tagging (i.e., automatic association of parts of speech with the word occurrences) or lemmatization (i.e., automatic reduction of inflected forms to some canonical representation, such as infinitive for the verbs, singular for the nouns, and such) are today standard preprocessing steps for textual data sets.

Notice, however, that such linguistic preprocessing requires the availability of adequate NLP tools (see Chap. 11, Sec. 3.B).

3. *Repeated Segments*

Even after setting aside words with a purely grammatical role, the meaning of words is linked to how they appear in compound words, or in phrases and expressions that can either inflect or completely change their meanings. For example, expressions such as *social security*, *living standard*, and the like, have a meaning of their own that cannot be simply derived from the meaning of their constituents. It is thus useful to count larger units consisting of several words. Thse elements can be analyzed in the same ways as words. But decisions concerning how to group occurrences of words cannot be made without bringing subjectivity into the analytical process. Units such as repeated segments (see Lafon and Salem, 1983; Salem, 1984) can be automatically derived from the text. In subsequent sections we shall demonstrate how these formal objects are used with different statistical methods. Note that numerous dictionaries of phrases and compound words are available in the domain of corpus linguistics [e.g., for English Fagan (1987); for French Silberztein (1993) and Gross and Perrin (1989); for Italian Elia (1995)]. The repeated-segments approach is significantly different: it deals with the detection of repeated sequences of words within a given corpus, whether or not they constitute frozen phrases or expressions.

4. *Quasi-Segments*

The methods for which the objective is to find co-occurrences of two or more words in a text have been developed particularly in information retrieval. Procedures that select repeated segments in a corpus are not yet able to find repetitions that are slightly altered by the introduction of an adjective or by a minor lexical modification of one of the components. Becue and Peiro (1993) proposed an algorithm that is used to find (repeated) quasi-segments. This algorithm can be used, for example, to assemble into one element (*play*, *sport*) the sequences *play a sport* and *play some sports*. However, quasi-segments are even more numerous than segments, and counting them gives rise to problems in their selection and display.

B. Some Examples of Proximity Measures

1. *Similarities for Information Retrieval*

To illustrate the use of textual similarity measures in the domain of Information Retrieval, we will consider two of the similarities used in the SMART information retrieval system (Salton and Buckley, 1988, 1990). SMART is one of the reference systems implementing the vector-space retrieval model, which currently corresponds to one of the most widely used models for information retrieval systems.

In SMART, each textual entity (be it a document or a query) is represented by a vector of weights, each weight measuring the importance of a specific indexing term (word, word phrase, or thesaurus term) assigned to the textual entity. If m is the size of the indexing term set, the vector $[D]$ associated with a given textual entity D is therefore of the form of Eq. {1}.

$$[D] = [w_1, w_2, \ldots, w_m] \qquad \{1\}$$

where w_i is the weight assigned to the ith element in the indexing term set.

Retrieval is then achieved by searching for the documents that maximize the similarity between the vector associated with a query and the vectors associated with the documents present in the base. Several similarity measures can be considered.

SMART ATC SIMILARITY: For any two documents D_i and D_j, the atc similarity measure is defined by Eq. {2}, with D defined as in Eq. {3}, and w_{ik} as in Eq. {4},

$$\text{atc}_{\text{sim}}(D_i, D_j) = \cos(D_i, D_j) \qquad \{2\}$$

$$D_i = w_{ik} \qquad \{3\}$$

$$w_{ik} = \begin{cases} \frac{1}{2}\left(\frac{1+p_{ik}}{\max_l p_{il}}\right)\log\frac{N}{n_k} & \text{if pik}=0 \\ 0 & \text{otherwise} \end{cases} \qquad \{4\}$$

where w_{ik} is the weight of term T_k in document D_i, p_{ik} is the relative document frequency of T_k in D_i given by Eq. {5}, N represents the total number of documents in the collection, n_k the number of documents with term T_k,

$$p_{ik} = \frac{f_{ik}}{\sum_k f_{ik}} \qquad \{5\}$$

where f_{ik} is the document frequency of T_k in D_i.

SMART ATN SIMILARITY Defined as in Eq. {6},

$$\text{atn}_{\text{sim}}(D_i, D_j) = [D_i] \cdot [D_j] \qquad \{6\}$$

where · (dot) denotes the inner product of vectors and the vectors $[D_i]$ and $[D_j]$ are defined in same way as for the atc similarity.

We now discuss some of the properties of the similarity measures. Let us denote

$|D|$, the norm (i.e., the length) of the vector $[D]$ so that $|D|^2 = [D] \cdot [D]$)

$[D_1 \backslash D_2]$, the restriction of the vector $[D_1]$ to only those coordinates that are null in the vector $[D_2]$

$[D_1|D_2]$, the restriction of the vector $[D_1]$ to only those coordinates that are not null in the vector $[D_2]$; in informal terms, $[D_1|D_2]$ is the "intersection" of $[D_1]$ with $[D_2]$ or the projection of $[D_1]$ onto $[D_2]$

max $[D]$, the maximal coordinate of the vector $[D]$

Then, Eq {7} is true for the IR similarities.

If $\max[D_1] = \max[D_1|D_2]$ and $\max[D_2] = \max[D_2|D_1]$,
then $\text{sim}_{\text{atn}}(D_1, D_2) = \text{sim}_{\text{atn}}(D_1|D_2, D_2|D_1)$ {7}

Property {7} means that, under the conditions on the maxima, the atn similarity is sensitive to only the common parts of the textual items compared. The atn similarity measure, therefore, is well suited to computing similarities in the cases where similarities between excerpts are sufficient to entail similarities between the whole documents. In informal terms, the atn similarity is sensitive to the "number" of matching words in the documents.

Note, however, that this property is true only if $\max[D_i \backslash D_j]$ (i.e., the maximal coordinate value **outside the intersection**) is smaller than $\max[D_i/D_j]$ (i.e., the maximal coordinate value **inside the intersection**). This means that the similarity actually does have a sensitivity to the outside-intersection coordinates, but limited to the integration of the maximal value.

In Information Retrieval, the atn similarity can be used to search for information within documents or for short text excerpts, such as single sentences, for which text similarity that depends on the proportion of matching terms is not indicative of coincidence in text meaning.

$$\text{atc}_{\text{sim}}(D_1, D_2) = \text{atn}_{\text{sim}}(D_1, D_2)/(|D_1| \cdot |D_2|) \qquad \{8\}$$

In informal terms, property {8} says that the atc similarity is sensitive to the "proportions" of matching words in the documents.

2. *TDA Distances: The χ^2 Distance*

The χ^2 distance measure is defined as in Eq. {8a} (for a more detailed presentation see later Sec. 3.B),

$$\mathrm{d}\chi^2(D_i, D_j) = |Di - Dj| \qquad \{8a\}$$

with the weights for the vectors associated with the document defined as in Eq. {9}.

$$w_{ik} = \frac{p_{ik}}{\sqrt{n_k}} \qquad \{9\}$$

The χ^2 distance measure has the property shown in Eq. {10}.

$$d\chi^2(D_1, D_2)^2 = d\chi^2(D_1|D_2, D_2|D_1)^2 + |D_1 \backslash D_2|^2 + |D_2 \backslash D_1|^2 \qquad \{10\}$$

Property {10} shows that the χ^2 distance is a proximity measure that is particularly sensitive to the outside-intersection differences between the textual items. The sensitivity to the differences is not surprising because any distance is, by definition, a dissimilarity and, therefore, a function that increases when differences between the compared items increase. What is, however, specific for the χ^2 distance (and will not be for the KL-distance (see Sec. 2.B.3)), is that the outside intersection differences play an important role for the computation of the dissimilarity. The consequence of this is that the χ^2 distance is a priori not well suited for the cases where the sizes of the textual items are very different (and in particular for the evaluation of proximities between short queries and larger documents).

3. *TM Distances: The Kullback–Leibler Distance*

The Kullback–Leibler (KL) distance between two documents D_i and D_j can be defined as the symmetrical Kullback–Leibler distance [Eq. {11}] between the probability distributions $[p_1, p_2, \ldots, p_m]$, where p is defined as in Eq. {12}, which can be derived from the frequency vectors $[f_1, f_2, \ldots, f_m]$ associated with the documents.

$$\mathrm{d}_{\mathrm{KL}}(D_i, D_j) = \sum_{k|p_{ik}*p_{jk}\neq 0} p_{ik} \log\frac{p_{ik}}{p_{jk}} + p_{jk} \log\frac{p_{jk}}{p_{ik}} \qquad \{11\}$$

$$p_i = \frac{f_i}{\sum_k f_k} \qquad \{12\}$$

An alternative, and probably more understandable expression for the KL distance is Eq. {13}.

$$\mathrm{d_{KL}}(D_i, D_j) = \sum_{k|p_{ik}*p_{jk}\neq 0} (p_{ik} - p_{jk})(\log p_{ik} - \log p_{jk}) \qquad \{13\}$$

Note, that here we use the symmetrical expression for the KL distance. Notice also that the term "distance" is not strictly correct because the KL distance does not verify the triangle inequality. Therefore, strictly speaking, the symmetrical version of the KL distance is a dissimilarity.

The KL distance has the following property. As, by definition, the KL distance is a sum restricted to the pairs of coordinates in the lexical profiles that are simultaneously strictly positive, it corresponds to a dissimilarity exclusively sensitive to **inside-intersection** differences. Similar to the atn similarity, it can, therefore, be used for documents of significantly different sizes.

4. *Some General Properties of the Various Proximity Measures*

A first important property of all the proximity measures mentioned in the foregoing is that they are all defined in terms of relative document frequencies and not in terms of absolute document frequencies. A consequence of this is that the computed similarities will be sensitive to only the "profiles" of the documents and not to the absolute values of some frequencies. This corresponds to a quite desirable property that guarantees that the composite document obtained by the concatenation of an arbitrary number of duplicates of the same elementary document will be, as far as the similarities are concerned, strictly identical with the elementary document itself.

A second property that is characteristic of the different proximity measures described here is that, apart from the KL distance, they all use a weighting scheme for the dimension of the vectorial representation space [the $\log(N/n_k)$ factor for the IR similarities and the $1/\sqrt{n_k}$ factor for the TDA distance]). Such weighting schemes can be seen as normalization procedures that tend to integrate within the proximity computation the notation of **discriminating power** associated with the dimensionality of the representation space. The underlying idea is that a dimension that "selects" a large number of textual items (i.e., a dimension for which a large number of items have strictly positive coordinates) also provides less discrimination in the whole item set and, therefore, should be given less weight in the proximity evaluation. This explains the use of weighting factors based on the inverse document frequency $1/n_k$.

5. *Association Measures for NLP*

Generally, the search for preferential associations is an important factor in applications involving NLP. In this way some ambiguities encountered in optical scanning of characters can be removed (at least in terms of probability) by considering neighboring words that have already been recognized, if the probabilities of association are known. Disambiguation in the course of a morphosyntactic analysis can be achieved under the same conditions. The work of Church and Hanks (1990) belongs to the same general framework. These authors suggest using as a measure of association between two words x and y the **mutual information** index $I(x, y)$ resulting from the Information Theory of Shannon [see Eq. {14}].

$$I(x, y) = \log_2\left(\frac{P(x, y)}{P(x)P(y)}\right) \qquad \{14\}$$

where $P(x)$ and $P(y)$ are the probabilities of words x and y in a corpus, and $P(x, y)$ is the probability of neighboring occurrences of these two words; x precedes y (thus there is no symmetry for x and y), and closeness is defined by a distance counted as a number of words. Thus, for English-language texts, these authors recommend considering as neighbors two words separated by fewer than five words. This procedure seems efficient for identifying some phrasal verbs. In a corpus of 44 million occurrences, the authors show that the verb *set* is followed first by *up* [$I(set, up) = 7.3$] and then by *off* [$I(set, off) = 6.2$], then, by *out, on, in* and so on.

3. SINGULAR VALUE DECOMPOSITION AND CLUSTERING TECHNIQUES

The main purpose of multidimensional descriptive data analysis methods is to describe tables such as tables of measures, classification tables, incidence tables, contingency tables, or presence–absence tables. These methods provide analysis with global panoramas by themes, tools that test the coherence of the information and procedures for selecting the most relevant patterns. Visualization tools (principal axis planes) and clustering techniques (methods that group individuals by simultaneously taking into account several basic characteristics) make it possible for the analyst to gain a macroscopic view of the raw data. The lexical tables to which we plan to apply these techniques are special contingency tables. The statistical individual being counted in each cell of the table is the occurrence (or token) of a textual unit: word, lemma, repeated segment, or other. The rows of the table, for example, are words, the frequency of which in the data is greater than a given threshold, and the columns are groupings of text: speakers, categories of speakers, authors, documents, and so forth. In general, cell (i, j) contains the number of occurrences of word i in text group j. This next paragraph briefly reviews the basic principles and practicalities of using these methods.

A. Basic Principles of Principal Axes Methods

The common principle to all multidimensional statistical methods is the following: each one of the two dimensions of a rectangular table of numerical data can be used to define distances (or proximities) among the elements of the other dimension. Thus, the table's columns (which might be variables, attributes, and such) are used to define the distances among the table's rows (which might be individuals, observations, and the like) with the help of appropriate formulae. Similarly, the table's rows are used to calculate distances among the table's columns. In this fashion tables of distances are obtained. These tables are the source of complex geometric representations that describe the similarities among the rows and among the columns of the rectangular tables being analyzed. The problem is to make these visualizations as understandable and as intuitive as possible, while keeping at a minimum the loss of information that results from summarizing the data.

Briefly, there are two broad types of methods for performing these data reductions—**principal axes** methods, largely based on linear algebra, and singular value decomposition (Eckart and Young, 1936)—produce graphic representations on which the geometric proximities among row-points and among column-points translate statistical associations among rows and among columns. Principal components analysis and correspondence analysis, belong to this family of methods.

Clustering or classification methods that create groupings of rows or of columns into clusters (or into families of hierarchical clusters) are presented in the following section. These two families of methods can be used on the same data matrix, and they complement one another very effectively. However, it is obvious that the rules of interpretation for the representations obtained through these reduction techniques are not as simple as those of elementary descriptive statistics.

B. Correspondence Analysis

Correspondence analysis is a technique for describing contingency tables (or cross-tabulations) and certain binary tables (also known as **presence–absence** tables). This description essentially takes the form of a graphic representation of associations among rows and among columns. Correspondence analysis was first presented in a systematic way as a flexible technique for exploratory analysis of multidimensional data by Benzécri et al. (1969). A variety of independent sources are either precursors to correspondence analysis or involved in further research. Among them we mention Guttman (1941) and Hayashi (1956), both of whom mention quantification methods, Nishisato (1980) who mentions dual scaling, and Gifi (1990), who discusses homogeneity analysis to designate multiple correspondence analysis. Presentations and examples are found in Greenacre (1984), Lebart et al. (1984), Benzécri (1992), and Gower and Hand (1996).

1. *Notation for Profiles*

Let us denote by f_{ij} the general term of the frequency table the elements of which have been divided by the grand total k. This table has n rows and p columns. According to the usual notation, $f_{i.}$ designates the sum of the elements of row i and $f_{.j}$ is the sum of the elements of column j of this table. The profile of row i is the set of p values [Eq. {15a}]. The profile of column j is the set of n values [Eq. {15b}].

$$\frac{f_{ij}}{f_{i.}}, j = 1, ..., p \qquad \{15a\}$$

$$\frac{f_{ij}}{f_{.j}}, i = 1, ..., n \qquad \{15b\}$$

2. *Reduction of Dimensionality*

Because higher-dimensional spaces are difficult to inspect, it becomes necessary to reduce the dimensionality of the points, which cannot be obtained without a certain loss of information. Principal axes methods (principal component analysis, factor analysis, correspondence analysis) aim to find the lower-dimensional subspace that most accurately approximates the original distributions of points. Principal components analysis and correspondence analysis are used under different circumstances: **principal components analysis** (which can be viewed as a purely descriptive variant of factor analysis) is used for tables consisting of continuous measurements. **Correspondence analysis** is best adapted to contingency tables (cross-tabulations). Most of these methods provide the user with a sequence of nested subspaces. That means that the best one-dimensional subspace (a straight line) is included in the best two-dimensional subspace (a plane) which, in turn, is included in the best three-dimensional subspace, and so on. In such a series of nested subspaces, the two-dimensional one is a very special case because it is compatible with our most usual communication devices, such as sheets of paper or video screens. The term **biplot**, coined by Gabriel (1971), is sometimes used to designate various methods leading to two-dimensional displays of multivariate data, as exemplified by Gower and Hand (1996).

3. *What Is the Meaning of the Proximities?*

We now define what is meant by "close." To do so, we explain how the distances, the planar approximations of which are shown in graphs, are calculated. The profiles, which are series of n or p numbers, depending on whether we are dealing with rows or columns, are used to define points in spaces of p or n dimensions (more exactly, $p-1$ and $n-1$ dimensions, because the sum of the components of each profile is equal to 1). Thus, the distances between points are defined in these high-dimensional spaces. The analysis phase consists in reducing these dimensions to obtain a visual representation that distorts the distances as little as possible.

The distance between two row-points i and i' is given by Eq. {16}.

$$d^2(i,i') = \sum_{j=1}^{p} \frac{1}{f_{.j}} \left(\frac{f_{ij}}{f_{i.}} = \frac{f_{i'j}}{f_{i'.}} \right)^2 \qquad \{16\}$$

Similarly, the distance between two column-points j and j' is given by Eq. {17}.

$$d^2(j,j') = \sum_{i=1}^{n} \frac{1}{f_{i.}} \left(\frac{f_{ij}}{f_{.j}} - \frac{f_{ij'}}{f_{.j'}} \right)^2 \qquad \{17\}$$

This distance, which is called the χ^2 **distance**, is very much like the Euclidean distance (the sum of squares of differences between components of the profiles), but a weighting factor is also involved. This weight is the reciprocal of the frequency corresponding to each term: $1/f_{.j}$ for each term in the sum that defines $d^2(i,i')$, and $1/f_{i.}$ for each term in the sum that defines $d^2(j,j')$. The χ^2 distance has a remarkable property that is called **distributional equivalence**. Briefly, this property means that the distances among rows (or columns) remains the same when we merge two columns (or rows) that have identical profiles. This invariance property induces a certain stability in the results: thus two texts that have the same lexical profile can be considered either separately or as one, without affecting any of the other distances between rows or between columns.

4. *Principal Axes Methods in Information Retrieval*

Today, Automatic IR has become an autonomous discipline with its own publications, conferences, software, terminology, and concepts (see Salton and McGill, 1983; Salton, 1988). This specialization owes its existence to the size and context of the problems encountered. Generally, very large and sparse matrices are being dealt with. These matrices come from specialized vocabulary and key-word counts, depending on the subject area, within bodies of work that include several hundred thousand documents that are often short and sometimes stereotyped. This operations is often very pragmatic. The idea is to have a tool that works, with established failure rates and predefined systems of costs and constraints. In this context, there are several outside sources of information that can be called on to resolve classification problems: syntactic analyzers, preliminary steps toward gaining an understanding of the search, dictionaries or semantic networks to lemmatize and eliminate ambiguities within the search, and possibly artificial corpora that experts resort to (see the pioneering work of Palermo and Jenkins, 1964). However, many of these techniques, and among them the most efficient ones—according to their authors—use multivariate tools that are very similar to those advocated by Benzécri (1977) or by Lebart (1982) in the case of large, sparse matrices. For instance Furnas et al. (1988) and Deerwester et al. (1990) suggest, under the name of **Latent Semantic Indexing**, an approach for which the basic assumption is that the term–document relations implicitly represented in the term-by-document matrix obtained from the document collection are, in fact, obscured by the lexical variability in the word choice (see also Bartell et al., 1992; Berry, 1996). A **Singular Value Decomposition** (SVD) is therefore realized on the original term-by-document matrix and the usual lexical profiles are replaced by the coordinates of the documents in the subspace spanned by the k first-principal vectors produced by SVD. This new representation encodes the associative relations between word and documents in a way that no longer uniquely relies on the words. Two documents can be close together in the k-dimensional subspace without sharing any terms. Such a method is very similar to discriminant analysis (see later) performed on the first-principal axes of a correspondence analysis. In fact these authors use singular value decomposition, which is the basis of both correspondence analysis and principal components analysis. Other authors emphasize the complementarity between models and descriptive methods in IR (see Fuhr and Pfeifer, 1991) as well as the importance of visualizations that provide the best possible overviews (Fowler et al., 1991). A case can be made for stating that the tendency is toward a gradual lessening of barriers between the various disciplines, and that "hard-nosed decisionists" are recognizing more and more the importance of the descriptive and exploratory phases of analysis.

C. Cluster Analysis of Words and Texts

Clustering techniques constitute a second family of data analysis techniques in addition to principal axes methods. Clustering techniques are divided into two major families:

1. Methods of hierarchical clustering, which allow us to obtain a hierarchy of groups partially nested in one another, starting with a set of elements that

are characterized by variables (or elements for which pairwise distances are known)

2. partitioning methods, or direct clustering methods, that produce simple segments or partitions of the population under analysis without the intermediate step of a hierarchical cluster analysis

These methods are better adapted to very large data sets. The methods of both families can be combined into a mixed approach. In practice the results obtained with clustering methods have revealed themselves to be essential complements to the results provided by singular value decomposition and correspondence analysis, described in the previous section. In analyzing lexical tables, correspondence analysis uncovers major structural traits that apply simultaneously to the two sets (rows and columns) being analyzed. However, when the number of elements is large, the displays are harder to read. The same holds true if the text being analyzed is long, even if the number of text parts is not that large. The number of terms increases rapidly even when the analysis is restricted to units occurring with a frequency greater than a given threshold. A second reason that motivates a combined use of principal axes methods and clustering has to do with enriching the representations from a multidimensional viewpoint. The groupings obtained are based on distances calculated in the whole space, and not only with the first-principal coordinates. Therefore, they are able to correct some of the distortions that are inherent in projections into a lower-dimensional space. Numerous contributions in the field of IR make use of clustering techniques together with other visualization tools (Chalmers and Chitson, 1992; Dubin, 1995; Iwayama and Tokunaga, 1995; Hearst and Pedersen, 1996; Hull et al., 1996).

1. *A Brief Review of Hierarchical Clustering*

The first book dealing with hierarchical cluster analysis is probably Sokal and Sneath (1963). Other general manuals include Hartigan (1975), Gordon (1981), and Kaufman and Rousseeuw (1990). Interesting critical reviews are given by Cormack (1971), Gordon (1987), and Bock (1994).

The basic principle of hierarchical agglomeration is very simple. We start with a set of n elements (which we shall refer to here as **basic elements** or **terminal elements**), each of which has a weight, and among which distances have been calculated (so there are $n(n-1)/2$ distances among the possible pairs). First the two elements that are closest to one another are agglomerated. This agglomerated pair now constitutes a new element, for which both a weight and distances to each of the elements still to be clustered, can be calculated. In practice there are several ways of proceeding that fit this definition, which explains the large variety of clustering methods. After this step, there remain $n-1$ elements to be clustered. Once again the two closest elements are agglomerated, and this process is repeated $n-1$ times until the set of elements is completely used up. The last, $(n-1)$th operation groups together the whole set of elements into one large cluster, which coincides with the original data set. Each of the groupings created at each step by following this method is called a **node**. The set of terminal elements corresponding to a node is called a **cluster**. The principle of this method is **agglomerative clustering**, the most prevalent family of methods.

2. *The Dendrogram*

The cluster analysis thus obtained can be represented in several different ways. The representation as a hierarchical tree or **dendrogram** is probably the most revealing one. The grouping obtained at each step of the hierarchical clustering algorithm brings together elements that are the closest to one another. They constitute a node of the hierarchy, and their distance is the index attached to the node. The farther the grouping procedure progresses (the closer we come to the top of the tree), the greater the number of already agglomerated elements, and the greater the minimal distance between the clusters that remain to be agglomerated. An example is shown in Fig. 1.

3. *Cluster Analysis of Terms*

Even though cluster analyses performed on a set of texts and those done on a set of terms appear to be linked from a formal viewpoint, the two types of analyses meet different needs, leading to different utilizations of the method. When texts are being analyzed (literary, political, historical texts), the number of words is generally far greater than the number of text parts. The dendrogram has so many terminal elements that any global summarization becomes rather complicated. In practice, the interpretation of such a large cluster analysis is performed by considering in order of priority the associations that appear at the two extremities of the dendrogram:

1. Clusters at the lower levels of the hierarchy, comprising agglomerations of words with a very small index (i.e., clusters that become agglomerated at the outset of the clustering process: these associations group together sets of words the distribution profiles of which are very similar among text.
2. Higher-level clusters, often comprising many words, that are analyzed as entities. At higher levels of clusters, the main observed oppositions of the first-principal plane (spanned by the first two principal axes) are generally found.

Hierarchical cluster methods can be used to find textual co-occurrences within sentences (Fig. 1), as well as fixed length text segments and paragraphs.

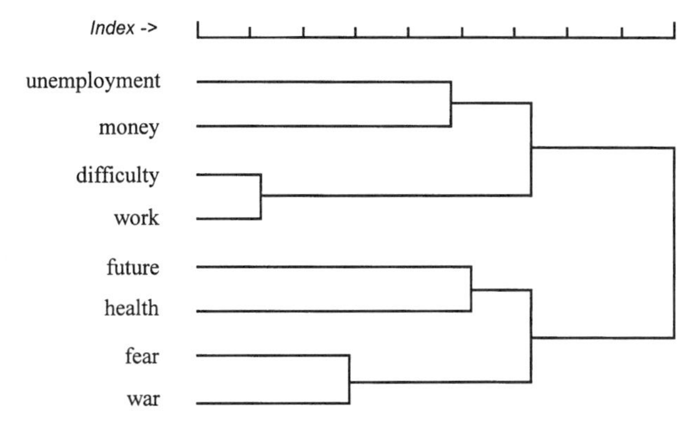

Fig. 1 Sketch of a dendrogram showing associations between terms within the responses to an open question in a sample survey.

4. CHARACTERISTIC TEXTUAL UNITS, MODAL SENTENCES OR TEXTS

The visualization tools of the preceding sections produce global panoramas of lexical tables. These panoramas can be generated whether or not the data are aggregated, whether they are natural partitions of a body of work (chapters, works of one author, articles, discourses, documents, and such), or whether they result from artificial groupings (open-ended responses organized by categories, documents organized by subject matter, and so forth). The spatial representations provided by correspondence analysis can be enhanced through the use of parameters called **characteristic elements** or **characteristic textual units**. They are used to describe groups of texts by displaying textual units that are present in a given group either a great deal more or a great deal less than in the overall corpus. These elements can be words, lemmas, or segments. In a more generalized and, therefore, more complex fashion, it is also possible to exhibit, for any part of a corpus, characteristic associations consisting of collections of elements. Thus, in the case of open-ended questions, response groupings may be characterized by a few responses that are designated as **modal responses** (also called characteristic responses). Briefly, modal responses generally contain a large number of characteristic terms within a particular text. In the case of automatic IR, modal responses might be modal documents (documents that characterize a particular theme). Thus the term "modal response" is used whether we are dealing with actual responses to open-ended questions or with documents. For longer documents, context elements of variable length (e.g., phrases or paragraphs) can be displayed.

A. Characteristic Elements

The method presented in the following is an adaptation of classic statistical tests to textual material. The goal is to build a simple tool that can detect, within each of the parts of a corpus, which elements are used frequently, as well as which elements tend to be used rarely. The hypergeometric model has been chosen because of its relative simplicity. Note, that this model is not particularly analogous to text composition by a human being, a process that is far more complex: in this process, it is not possible to assume even the simplest of the hypotheses required within a hypergeometric model for performing statistical tests (e.g., independence of random samples). Consequently, the computed indices do not by any means represent directly interpretable probabilities. Rather, they are of a purely descriptive nature. We shall use these indices to select characteristic terms within each of the parts.

1. *Computation of Characteristic Elements*

We denote the following quantities (Fig. 2), expressed as occurrences of simple terms (terms can be either types or lemmas):

k_{ij} subfrequency of word i in part j of corpus
$k_{i.}$ frequency of word i in entire corpus
$k_{.j}$ size (number of tokens) of part number j
$k_{..}$ size (number of tokens) of corpus (or, simply k)

We start by imagining a population of objects with a total count k. Among all of these objects, we suppose that $k_{i.}$ objects are "marked," that is, that they possess a feature that differentiates them from the others: for example, objects corresponding

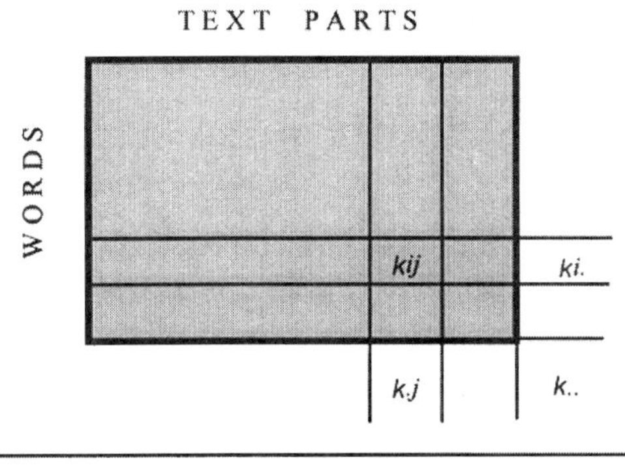

k..	size of corpus
ki.	frequency of word in corpus
kij	frequency of word in text part
k.j	size of text part

Fig. 2 The four parameters for calculating characteristic elements.

to tokens of a same type, the total frequency of which is $k_{i.}$. The remaining objects of our population are all mixed together into a subtotal and considered as "unmarked." The number of unmarked objects, therefore, is equal to $k - k_{i.}$. Now, using random sampling without replacement, we draw a sample that contains exactly $k_{i.}$ objects in our population. After this drawing, the number of marked objects contained in the sample is denoted as k_{ij}. The numbers $k, k_{i.}, k_{.j}$, just defined constitute the parameters of the model. The probability distribution for a sampling without replacement under the hypothesis of independence is the hypergeometric distribution. This distribution is similar to the multinomial distribution when the samples are relatively small compared with the population (then samplings with and without replacement are virtually the same). We denote as *Prob*$(k, k_{i.}, k_{.j}, n)$ the calculated probability of exactly n marked objects appearing in a sample without replacement of a sample of size $k_{.j}$ among a population of total count k, knowing that the entire population has $k_{i.}$ marked objects. The classic hypergeometric distribution formula is Eq. {18}.

$$Prob(k, k_{i.}, k_{.j}, n) = \frac{\binom{k_{i.}}{n}\binom{k - k_{i.}}{k_{.j} - n}}{\binom{k_{.}}{k_{.j}}} \qquad \{18\}$$

Now we can use the probability distribution thus built on the parameters $k, k_{.j}$, and $k_{i.}$ to assess the absolute frequency n_0 observed when drawing our sample—still under the assumption that the hypotheses of independence mentioned earlier hold true. The probability Psup(n_0) is equal to the sum of all the probabilities for values of n that are greater than or equal to n_0. It is thus the sum of the probabilities *Prob*$(k, k_{i.}, k_{.j}, n)$ for values of n between n_0 and $k_{i.}$. If the probability Psup(n_0) is very small for this sample, we conclude that the sample has an "unusually" large number of occurrences of the term (word, segment, lemma) under study. Evidently, "unusually" small numbers of occurrences can be detected in a similar fashion.

2. *List of Characteristic Units*

Diagnostics thus obtained by the method of characteristic elements in the table under analysis can then be reordered into more readable reports. Such a report contains an ordering of the characteristic elements within each of the parts of the corpus. This report identifies the units (words or segments) that are overrepresented as well as the units that are underrepresented within each part. The characteristic textual unit diagnostics are sorted in order of increasing probability, that is from the most characteristic to the most banal.

3. *The Problem of Multiple Comparisons*

The simultaneous calculation of several characteristic element diagnostics runs into the obstacle of multiple comparisons, a problem familiar to statisticians. Suppose that the text parts are perfectly homogeneous and thus that the hypothesis of independence between the words and the parts is true. Under these conditions, on the average, out of 100 calculated characteristic element diagnostics, 5 are significant relative to the probability threshold 5%. In fact, this 5% threshold makes sense only for a single test, and not for multiple tests. In other words, the unsuspecting user will always find "something significant" at the 5% level. A practical way to solve this difficulty is to choose a stricter threshold. In the context of analysis of variance, several procedures have been devised to overcome this difficulty. As an example, the **Bonferroni method**, briefly, recommends dividing the probability threshold by the number of tests (number of comparisons in the case of the design of experiments). This reduction of the probability threshold is generally considered as too severe. Classic overviews and discussions about multiple comparisons are found in O'Neill and Wetherhill (1971) and Mead (1988). However, in the present context, we are dealing only with orders of magnitude for the probability thresholds.

B. Modal Sentences or Texts

A simple, straightforward technique addresses several objections that have been raised concerning the fragmentation inherent in any analysis that is limited to isolated words without placing them in their immediate context. This technique consists of the automatic selection of **modal texts** (also called **characteristic documents** or **modal responses** in other types of applications). Modal texts are not manufactured texts that depict each text grouping, but actual texts or documents that are chosen because they are representative of a given category of individuals. There are two ways in which modal texts are chosen: the first is based on calculations that make use of characteristic elements. The second is based on distance computations according to simple geometric criteria (the χ^2 distance).

1. *Selection of Modal Texts Using Characteristic Elements*

A first way of selecting modal texts is the following: a modal text of a grouping is a text that contains, as much as possible, the most characteristic words of this grouping. To each word i belonging to grouping j can be associated a test-value $t(i, j)$ that measures the deviation between the relative frequency of the word within group j

relative to its global frequency calculated on the entire set of texts. This deviation is normalized to be able to be considered as a standardized normal variable under the hypothesis of random distribution of the word under consideration in the groups.

2. Practical Computation of Test Values

Within this framework, and using the notation defined in Sec. 4.A, estimates of the expectation $m(i,j)$ and of the variance $s^2(i,j)$ of the hypergeometric variable k_{ij} observed in the cell (i,j), under the hypothesis of independence, are given in Eqs. {19} and {20}.

$$m(i,j) = \frac{k_{ij}}{k} \qquad \{19\}$$

$$s^2(i,j) = k_{i.}\left(1 - \frac{k_{i.}}{k}\right)\left(\frac{k - k_{.j}}{k - 1}\right) \qquad \{20\}$$

Thus, the quantity $t(i,j)$ such that

$$t(i,j) = \frac{k_{ij} - m(i,j)}{s(i,j)} \qquad \{21\}$$

is approximately a standardized normal variable under the hypothesis of random distribution of the word under consideration in the groups. Under this hypothesis, the test value $t(i,j)$ lies between the values -1.96 and $+1.96$, with a probability of 0.95. However, as this calculation depends on a normal approximation of the hypergeometric distribution, it is only used when the counts within a grouping are not too small relative to the total count.

3. The Selection of Modal Texts from Characteristic Words

A simple empirical formula involves associating with each text the mean rank of the words it contains: if this mean rank is small, it means that the text contains only words that are very characteristic of the grouping. Even better than ranks, test values can be used (the smallest mean rank then corresponds to the highest mean test value). It is rather obvious that some choices have to be made: for instance, between a short text made up of a few very characteristic words, and a longer text, that necessarily contains some less characteristic words. This calculation mode (mean test value) has the property of favoring short texts, whereas the χ^2 criterion tends to favor lengthy texts.

4. Selection of Modal Texts Using χ^2 Distances

The principle of this selection is as follows: once the texts (k documents or open-ended responses) have been numerically coded, they give rise to a rectangular table **T** with k rows (documents or texts) and as many columns (v) as there are selected words. A grouping of document or texts is a set of row vectors, and the mean lexical profile of this grouping is obtained by calculating the mean of the row vectors of this set. If this grouping is based on the categories of a nominal variable coded in a binary table **Z**, the aggregated lexical table **C** is calculated with Eq. {22}.

$$\mathbf{C} = \mathbf{T}'\mathbf{Z} \qquad \{22\}$$

Thus, we can calculate distances between documents and groupings of the documents. Documents (rows of **T**) and groupings of documents (columns of **C**) are both represented by vectors in the *v-dimensional* space of terms. These distances express the deviation between the profile of a document and the mean profile of the group to which the document belongs. The distance that is used between these profiles of frequencies is the χ^2 distance, thanks to its distributional properties stressed in Sec. 3. The distance between a row-point i of **T** and a column-point m of **C** is then given by the Eq. {23}.

$$d^2(i, m) = \sum_{j=1}^{v} \frac{t_{..}}{t_{.j}} \left(\frac{t_{ij}}{t_{.j}} - \frac{c_{jm}}{c_{.m}} \right)^2 \qquad \{23\}$$

with the following notation:

$t_{..}$ designates the overall sum of the elements of table **T**; that is, the total number of occurrences.
$t_{.j}$ designates the sum of the elements of column j of **T** (number of occurrences of term j).
$t_{i.}$ designates the sum of the elements of row i of **T** (length of text I).
c_{jm} is the jth element of column m of **C** (number of occurrences of the word j in the grouping m).
$c_{.m}$ is the sum of the elements of column m of **C** (total number of occurrences of grouping m).

For each grouping, these distances can be sorted in increasing order. Thus, the most representative texts relative to the lexical profile (i.e., those who distances are the smallest) can be chosen.

5. TEXT DISCRIMINATION

The exploratory statistical methods discussed in the previous sections complement a broad spectrum of techniques devoted to a type of research often called **decision research**, even though they are actually decision aids. In textual statistics, decision aids are used for attributing a text to an author or a period, choosing a document within a database in response to a query, and coding information expressed in natural language. The statistical procedures that can be used for these attributions, choices, or coding are based on textual discriminant analysis. Their aim is to predict the assignment of an "individual" to a group or a category (among several possible groups or categories) on the basis of variables measured on this individual. Early discriminant analyses were carried out on biometric and anthropometric measures by the statisticians Fisher (1936) and Mahalanobis (1936), who were attempting to predict membership in an ethnic group on the basis of measurements of the skeleton. They were the first to use the technique that is sometimes known as **linear discriminant analysis**: it is one of the oldest methods, and it is also one of the methods that is most commonly used today. The prediction is made possible thanks to a learning phase carried out on a set of individuals for whom both variables and groups are known (learning or training sample).

Among the most common applications of textual analysis techniques related to discriminant analysis, there are two major areas of concern in such analysis, as follows:

1. Applications in Information Retrieval, automatic coding, and analysis of open-ended responses, essentially deal with content, the meaning, and the substance of texts. The way in which a document is formulated is less important than that it belongs to a group of documents that are relatively homogeneous in their content. This first family of applications is referred to as text categorization.
2. At the other extreme, applications to literary corpora (for example, attribution to authors or dating) attempt to go beyond a text's content to capture the characteristics of its form (often equivalent to its style) on the basis of statistical distributions of vocabulary, and indices and ratios. The idea is to capture the "invariant" aspects of an author or of an era, that may be hidden to the ordinary reader (see Chap. 23).

A. Text Categorization

In the analyses that follow, which are used in IR or in analyses or coding of open-ended responses in surveys, the intent of discriminant analysis is to take into account both form and content, or in some cases, content alone.

1. General Principles

The basic principles common to the most widespread methods of discriminant analysis are as follows:

1. There are n observations characterized by p variables; in other words, there are n points classified into K groups, labeled 1 to K.
2. There are also m points in the same p-dimensional space as the previous n points (that is, they are characterized by the same variables or the same profiles), but these points are not classified into groups. The purpose is to assign each of these m points to the most-probable group among a set of predefined groups.

A straightforward method consists of calculating in each case the K mean profiles in the p-dimensional space and then assigning each new point to its closest profile. The shape of each of the K subconfigurations of points can also be taken into account. It can be seen that the concept of predicting assignment to a group can be complex. Many technical variants exist that make it possible to interpret calculated distances in terms of probabilities. The **local Mahalanobis distance** of point X to group k, which is used in quadratic discriminant analysis, is

$$d_k(X) = (X - m_k)' S_k^{-1} (X - m_k) \qquad \{24\}$$

where S_k is the internal covariance matrix of group k with mean point (center of gravity) m_k (see Anderson, 1984; McLachlan, 1992). The major discriminant analysis methods that are suited to large matrices of qualitative data are:

- Discriminant analysis under conditional independence (see Goldstein and Dillon, 1978).

- Discriminant analysis by direct density estimation (Habbema et al., 1974; Aitchison and Aitken, 1976).
- Discriminant analysis by the method of nearest neighbors (see e.g., Hand, 1981).
- Discriminant analysis on principal coordinates (which is used to apply to qualitative variables the same methods used for quantitative variables: factorial discriminant analysis, quadratic discriminant analysis, and so on). We will discuss this method later (Wold, 1976; Benzécri, 1977).

A thorough review of most of these methods is found in McLachlan (1992), but they can be applied only after the basic statistical units have been determined, a crucial step in textual data analysis.

2. *Discriminant Analysis and Modal Texts or Documents*

The numbering phase and the preliminary transformation steps result in a matrix **T** of order (n, v), the rows of which are the n documents or texts and the columns are the v different terms (words, segments, lemmas, or other). The K groups define a partition of the documents or texts. They can be described by a binary matrix **Z** of order (n, K). We saw earlier (see Sec. 4) that Eq. {21} can be used to calculate the aggregate lexical table **C** for any categorical variable whose categories are coded in a binary table **Z**. Thus, lexical profiles can be compared within various groups of the population. Several tools can be used to help read the aggregated lexical tables: correspondence analysis, associated clustering techniques (see Sec. 3), lists of characteristic words, and lists of modal responses (see Sec. 4). Each of these tools can serve in itself as a technique of discrimination. In particular, the process of calculating modal texts is very close to the most common forms of discriminant analysis.

3. *Modal Documents or Responses That Use Characteristic Words*

Recall that characteristic elements or words (see Sec. 4) are words that appear with "abnormal" frequency in a part of the corpus or in the responses of a group of individuals. A modal text for a given group has been defined as a text belonging to that group, and containing the words that characterize the group as a whole. Thus, a text that happens to contain many characteristic words and very few contracharacteristic words is a very strong characteristic for that group. Thus, to each text i of group g can be associated the sum $s_i(g)$ of the test values relative to the words that constitute the text (the contracharacteristic words are assigned negative test values). The modal texts are associated with the largest values of $s_i(g)$. It can be considered that $s_i(g)$ defines a discriminant analysis rule, because for a given individual i, the value of g for which $s_i(g)$ is maximum can define the group to which i is assigned with the highest probability.

4. *Modal Texts According to a Distance Criterion*

A text (or document) is a row-point of **T**, or a vector with v components. We learned how to calculate distances between documents and groupings of these documents, for documents (rows of **T**) and groupings of documents (columns of **C**) are all represented by vectors in the same space. Selecting characteristic documents translates into finding for a given group g the document i that minimizes the distance

$d^2(i, g)$. Conversely, the categorization problem consists in finding, for a given document i, the group g that minimizes $d^2(i, g)$.

B. Discriminant Analysis Regularized Through Preliminary Singular Value Decomposition

Correspondence analyses can be performed to describe the **C** tables that are contingency tables (in which the counts concern occurrences of words) or sparse tables **T**. They provide a visualization of the associations among words and groups or categories. They also make it possible to replace qualitative variables (simple presence or frequency of a word) with numeric variables (the values of principal coordinates), and thus to apply the classic methods of discriminant analysis (linear or quadratic) after calculating the coordinates. Correspondence analysis thus serves as a "bridge" between textual data (that are qualitative, and often sparse) and the usual methods of discriminant analysis. But most important, a sort of filtering of information is accomplished (see Wold, 1976; Benzécri, 1977) by dropping the last principal coordinates. This process strengthens the predictive power of the procedure as a whole. These properties are applied in IR (see Furnas et al., 1988; Deerwester et al., 1990; Yang, 1995).

This is the meaning of the concept of **regularizing discriminant analysis**. Regularization techniques (see Friedman, 1989) strive to make discriminant analysis possible in cases that statisticians deem to be "poorly posed" (hardly more individuals than variables) or "ill posed" (fewer individuals than variables). Such cases arise when samples of individuals are small, or when the number of variables is large, which is often the situation in textual statistics. The opportunity to work on a small number of principal coordinates becomes a definite advantage.

6. ENHANCING SIMILARITIES

The starting point is to consider each text as described by its lexical profile (i.e., by a vector that contains the frequency of all the selected units in the text). The units could be words (or lemmas, or types), segments (sequences of words appearing with a certain frequency) or quasi-segments (segments allowing noncontiguous units). The corpus is represented by an (n, p) matrix **T** the ith row (i.e., ith document) of which contains a vector for which p components are the frequencies of units (words). However, a typical classification algorithm applied to the rows of **T** could lead to disappointing or misleading results:

1. The matrix **T** could be very sparse, many rows could have no element at all in common.
2. A wealth of metadata is available and needs to be utilized (syntactic relations, semantic networks, external corpus, and lexicons, or others).

To make more meaningful the distances between the lexical profiles of the texts we can add new variables obtained from a morphosyntactic analyzer; it is then possible to tag the text units depending on their category. It is then possible to complement the p-vector associated to a response or a text with new components. It is also useful to take into account the available semantic information about units, information that can be stored in a **semantic weighted graph**. From now on, we will focus on this latter issue and discuss the different ways of deriving such a graph from the data themselves.

A. Semantic Graph and Contiguity Analysis

There is no universal rule to establish that two words are semantically equivalent, despite the existence of synonym dictionaries and thesauri. In particular, in some sociological studies, it could be simplistic to consider that two different words (or expressions) have the same meaning for different categories of respondents. But it is clear that some units having a common extratextual reference are used to designate the same "object." Whereas the syntactic metainformation can provide the user with new variables, the semantic information defined over the pairs of terms is described by a graph that can lead to a specific metric structure (Fig. 3).

1. The semantic graph can be constructed from an external source of information (a dictionary of synonyms, a thesaurus, for instance). In such a case, a preliminary lemmatization of the text must be performed.
2. It can be built up according to the associations observed in a separate (external) corpus.
3. Possibly, the semantic graph can also be extracted from the corpus itself. In this case, the similarity between two terms is derived from the proximity between their distributions (lexical profiles) within the corpus.

The p vertices of this weighted undirected graph are the distinct units (words) $j, (j = 1, \ldots, p_1)$. The edge (j, j') exists if there is some nonzero similarity $s(j, j')$ between j and j'. The weighted associated matrix $\mathbf{M} = (m_{ij})$, of order (p, p) associated with this graph, contains in the line j and column j' the weight $s(j, j')$ of the edge (j, j'), or the value 0 if there is no edge between j and j'.

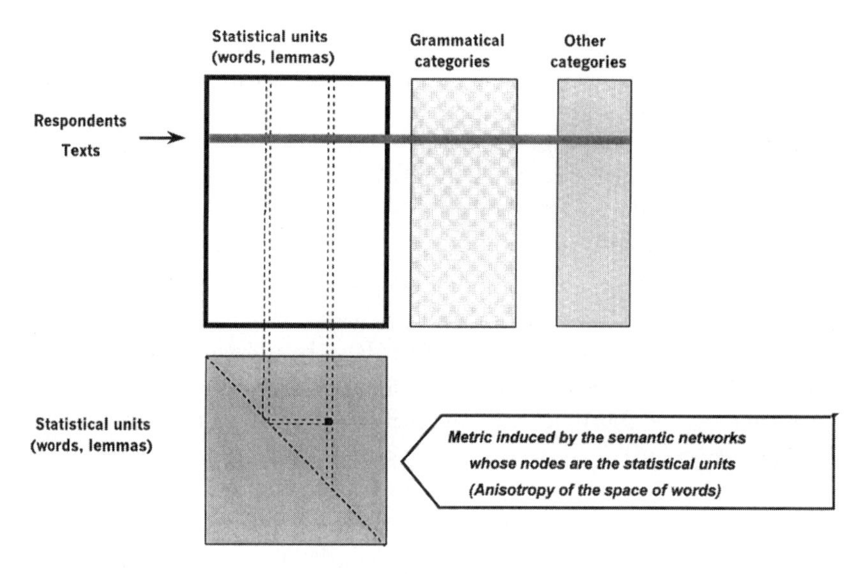

Fig. 3 How to use metadata in text classification: metric or new variables?

The repeated presence of a pair of words within a same sentence of a text is a relevant feature of a corpus. The words can help to disambiguate each other (see e.g., Lewis and Croft, 1992). To take into account co-occurrence relations allows one to use words in their most frequent contexts. In particular, we can also describe the most relevant co-occurrences by using a weighted undirected complete graph linking lexical units. Each pair of units is joined by an edge weighted by a co-occurrence intensity index.

At this stage, we find ourselves within the scope of a series of descriptive approaches working simultaneously with units and pairs of units (see e.g., Art et al., 1982), including contiguity analysis or local analysis. These visualization techniques are designed to modify the classic methods based on Singular Value Decomposition by taking into account a graph structure over the entries (row or column) of the data table. The classification can then be performed either

1. By using as input data the principal coordinates issued from these contiguity analyses, or
2. By computing a new similarity index between texts. This new index is built from generalized lexical profiles [i.e., original profiles complemented with weighted units that are neighbors (contiguous) in the semantic graph].

B. Different Types of Associated Graphs

1. *The Case of an External Semantic Graph*

A simple way to take into account the semantic neighbors leads to the transformation in Eq. {25}

$$\mathbf{Y} = \mathbf{T}(\mathbf{I} + \alpha\mathbf{M}) \qquad \{25\}$$

where $\mathbf{M}$ is the matrix associated with the graph defined previously, and α a scalar allowing the calibration of the importance given to the semantic neighborhood (Becue and Lebart, 1998). This is equivalent to providing the p-dimensional space of words with a metric defined by the matrix [Eq. {26}].

$$\mathbf{Q} = (\mathbf{I} + \alpha\mathbf{M})^2 \qquad \{26\}$$

This leads immediately to a new similarity index that can be used for the classification of the rows. Owing to the size of the data tables involved, the classification is often performed using distances computed with the principal coordinates.

2. *The Case of an Internal Co-occurrence Graph: Self-Learning*

A possible matrix $\mathbf{M}$ could be Eq. {27},

$$\mathbf{M} = \mathbf{C} - \mathbf{I} \qquad \{27\}$$

where $\mathbf{C}$ is the correlation matrix between words (allowing negative weights, which correspond to a negative co-occurrence intensity measure); let the columns of $\mathbf{C}$ have their variances equal to 1 {28}.

$$\mathbf{C} = \frac{1}{n}\mathbf{T}^T\left(\mathbf{I} - \frac{1}{n}\mathbf{U}\right)\mathbf{T} \qquad \{28\}$$

where $\mathbf{I}$ is an identity matrix, with $u_{ij} = 1$ for all i and j). Since Eq. {29},

$$\mathbf{Y} = \mathbf{T}(\mathbf{I} + \alpha(\mathbf{C} - \mathbf{I})) \qquad \{29\}$$

the matrix **S** to be diagonalized when performing a Principal Components Analysis of **Y**, reads as Eq. {30}.

$$\mathbf{S} = \frac{1}{n}\mathbf{Y}^T\left(\mathbf{I} - \frac{1}{n}\mathbf{U}\right)\mathbf{Y} = (1-\alpha)^2\mathbf{C} + 2\alpha(1-\alpha)\mathbf{C}^2 + \alpha^2\mathbf{C}^3 \qquad \{30\}$$

Therefore, the eigenvectors of **S** are the same as the eigenvectors of **C**. However, to an eigenvalue μ of **C** corresponds the eigenvalue λ of **S** such that Eq. {31} holds.

$$\mu = (1 = \alpha)^2\lambda + 2\alpha(1-\alpha)\lambda^2 + \alpha^2\lambda^3 \qquad \{31\}$$

The effect of the new metric is simply to reweight the principal coordinates when recomputing the distances to perform the classification. If $\alpha = 1$, for instance, we get μ^3 instead of μ, thus the relative importance of the first eigen values is strongly increased. Such properties contribute to the illumination of the prominent role played by the first principal axes, particularly in the techniques of Latent Semantic Indexing used in automatic IR. Different approaches using co-occurrences have been used in the domain of IR. For example, techniques based on average co-occurrence vectors, such as in the Distributional Semantics approach proposed in Rungsawang and Rajman (1995), Rajman and Rungsawang (1995), Rungsawang (1997) following Harris (1954), or based on context vectors such as in Schütze (1992a,b).

Another way of deriving a matrix **M** from the data itself is to perform a hierarchical classification of words, and to cut the dendrogram at a low level of the index. It can either provide a graph associated with a partition, or a more general weighted graph if the nested set of partitions (corresponding to the lower values of the index) is taken into account. However, in our context, it would be awkward to confine ourselves to semantic graphs having a partition structure because the relation of semantic similarity is not transitive (consider, for example, the semantic chain: fact–feature–aspect–appearance–illusion).

It is more appropriate to assign to each pair of words (i, j) a value [Eq. {32}],

$$m_{ij} = \begin{cases} f(d(i,j)) & \text{for } d(i,j) < \mathbf{T} \\ 0 & \text{for } d(i,j) > \mathbf{T} \end{cases} \qquad \{32\}$$

where $f(t)$ is a decreasing function of t, and $d(i,j)$ is the index corresponding to the smallest set containing both elements i and j.

7. VALIDATION TECHNIQUES

A. How to Assess Similarities

As similarities are never used for themselves, but are always a means for some specific task or application (visualization, clustering, retrieval, and such), it is meaningless to try to define "bad" or "good" similarities. In other words, a similarity can only be more or less adequate for the task or application at hand, and its quality will be evaluated on the basis of the performance of some associated system. Depending on the domains, several adequacy measures have been defined. In IR, for instance, the notions of **precision** and **recall** are routinely used to measure the effectiveness of retrieval systems. Briefly, these notions are defined as follows. For a given collection of texts, and a given query q, the set of all the documents within the collection that are considered as relevant (i.e., as satisfying the information need expressed by the query) Pert(q) is determined, for example, by an expert in the domain. For a given query q, the set of items retrieved by the system will be denoted by Ret(q). The precision P(q) of the retrieval system for the query q is then defined as the proportion of pertinent documents in Ret(q), whereas the recall R(q) is the ratio of pertinent documents in Ret(q) to the total number of pertinent documents as defined by Pert(q). The precision is thus the proportion of documents that are relevant among all those that are retrieved, whereas the recall is the proportion of documents that are retrieved among all those that are relevant. The precision (recall) of a retrieval system will then be defined as the average of P(q) (R(q)) over all queries q for which Pert(q) is available. Because the performance of a retrieval system is the result of a complex process in which the similarity is only one component, it provides only a rather indirect assessment of the quality of similarity. Its advantage, however, is that it corresponds to objective (yet empirical) evaluation procedures that can be realized in well-defined evaluation campaigns, such as the TREC (Text REtrieval Conference) campaigns regularly organized in the IR field.

B. Assessment of Visualizations

To assess the patterns observed among the visual representations of similarities between words or between documents, calculations of **stability** and **sensibility** are probably the most convincing validation procedures. These calculations consist of verifying the stability of the configurations obtained after modifying the initial term-by-document matrix (or any lexical contingency table) in various ways.

Certain specific simulations can give information concerning sampling fluctuation. The **Bootstrap resampling technique** (Efron, 1982) will allow us to decide whether the observed patterns are significant in a statistical sense, as opposed to being the results of random noise (see e.g., Gifi, 1990; Greenacre, 1984). The basic idea is rather simple: several "replicates" of the initial data table are built through a sampling with replacement among the statistical units (terms or documents). After repeating the exercise a large number of times, replicate points are positioned in the initial visualization space (they are projected afterward onto the subspace spanned by the first principal axes). The convex hull enclosing the replicates corresponding to a specific element provides us with an informal confidence region for the location of this element. The dissimilarities characterized by nonoverlapping (or slightly overlapping) regions are worth interpreting. Evidently, more adapted simulation schemes could be used, taking advantage of specific available information about the data table. The general principle of stability calculation remains the same.

C. Validation of a Text Categorization (Discriminant Analysis)

The many options available to analysts (choosing and building statistical units, choosing a discriminant analysis technique) should not puzzle or discourage them. The quality of a discriminant analysis can be evaluated through precise criteria that are simpler than those used in exploratory methods. To validate a discriminant analysis, it is necessary to perform tests on part of the sample (the **test sample**) that is not used to calculate the discriminant function. The discriminant function is calculated on a portion of the initial sample that is called the **learning sample** (or **training sample**). One of the simplest criteria for validity is either recall or the precision, which are calculated for the learning sample (giving an overly optimistic view of the quality of the discriminant analysis) and for the test sample (giving a more exact view of this quality).

This distinction among subsamples is fundamental, and actually rather intuitive if the learning phase is described in the same terms as the learning procedures in artificial intelligence. By arriving at a maximum percentage of correctly classified cases on the learning sample, the discriminant analysis procedure uses all of the distinctive aspects of that sample, and thus includes random noise to attain its ends. In some cases, if the number of parameters is large, the percentage of correctly classified individuals can reach 100% in the learning sample, and the percentage can be very small on any other sample: this is the phenomenon of learning by heart. One can say that the procedure has learned without understanding, that is without capturing the general structural traits, without making a distinction between the structural information and the ancillary information in the learning sample.

The true value of a discriminant analysis is measured on only one or, even better, a series of test samples. McLachlan (1992) performed an interesting review of the methods of error calculation in discriminant analysis. Techniques such as the Jackknife and the Bootstrap [see Efron (1982) for a review of these two families of methods] make it possible to estimate the degree of confidence one can have in a discriminant analysis procedure. The same holds true for their variants, Leaving-one-out (or cross-validation) techniques (see Lachenbruch and Mickey, 1968; Stone, 1974; Geisser, 1975) that consist in building n samples by leaving out one individual each time and applying the discriminant function to this individual (these techniques are costly and can be applied to only small samples).

REFERENCES

Aitchison J, CGG Aitken (1976). Multivariate binary discrimination by the kernel method. Biometrika 63:413–420.

Anderson TW (1984). An Introduction to Multivariate Statistical Analysis (2nd ed.). New York: John Wiley & Sons.

Art D, R Gnanadesikan, JR Kettenring (1982). Data based metrics for cluster analysis. Utilitas Mathematica. 21A:75–99.

Bartell BT, GW Cottrell RK Belew (1992). Latent semantic indexing is an optimal special case of multidimensional scaling. In: N Belkin et al., eds. Proceedings of the 15th International ACM-SIGIR Conference on Research and Development in Information Retrieval. New York: ACM Press, pp 161–167.

Becue M, L Lebart (1998). Clustering of texts using semantic graphs. Application to open-ended questions in surveys. In: C Hayashi et al., eds. Data Science, Classification and Related Methods. Tokyo: Springer, pp 480–487.

Becue M, R Peiro (1993). Les quasi-segments pour une classification automatique des réponses ouvertes. Actes des 2ndes Journées Internationales d'Analyse des Données Textuelles (Montpellier). Paris: ENST, pp 310–325.

Benzécri J-P (1977). Analyse discriminante et analyse factorielle. Les Cahiers de l'Analyse des Données 2:369–406.

Benzécri J-P (1992). Correspondence Analysis Handbook (Trans TK Gopalan). New York: Marcel Dekker.

Benzécri J-P, et al. (1969). Statistical analysis as a tool to make patterns emerge from clouds. In: S Watanabe, ed. Methodology of Pattern Recognition. London: Academic Press, pp 35–74.

Berry MW (1996). Low-rank orthogonal decompositions for information retrieval applications. Numeric Linear Algebra Appl 1:1–27.

Bock HH (1994). Classification and clustering: problems for the future. In: E Diday et al., eds. New Approaches in Classification and Data Analysis. Berlin: Springer Verlag, pp 3–24.

Chalmers M, P Chitson (1992). Bead: exploration in information visualization. ACM/SIGIR '92, Copenhagen, Denmark, pp 330–337.

Church KW, P Hanks (1990). Word association norms, mutual information and lexicography. Comput Linguist 16:22–29.

Cormack RM (1971). A review of classification. J R Statist Soc A 134:321–367.

Cover TM, JA Thomas (1991). Elements of Information Theory. New York: John Wiley & Sons.

Deerwester S, ST Dumais, GW Furnas, TK Landauer, R Harshman (1990). Indexing by latent semantic analysis. J Am Soc Inform Sci 41:391–407.

Dubin D (1995). Document analysis for visualization. Proceedings of the 18th Annual International Conference on Research and Development in Information Retrieval (ACM-SIGIR'95), Seattle, WA, pp 199–204.

Eckart C, G Young (1936). The approximation of one matrix by another of lower rank. Psychometrika 1:211–218.

Efron B (1982). The Jackknife, the Bootstrap and Other Resampling Plans. Philadelphia: SIAM.

Elia A (1995). Per una disambiguazione semi-automatica di sintagmi composti: i dizionari electronici lessico-grammaticali. In: R Cipriano, S Bolasco, eds. Ricerca Qualitativa e Computer. Milano: Franco Angeli.

Fagan J (1987). Experiments in automatic phrase indexing for document retrieval; an examination of syntactic and non syntactic methods. Proceedings of the 10th Annual International Conference on Research and Development in Information Retrieval (ACM-SIGIR'87), New Orleans, pp 91–101.

Fisher RA (1936). The use of multiple measurements in taxonomic problems. Ann Eugen 7:179–188.

Fowler RH, WAL Fowler, BA Wilson (1991). Integrating query, thesaurus, and documents through a common visual representation. Proceedings of the 14th International ACM Conference on Research and Development in Information Retrieval, Chicago, pp 142–151.

Friedman JH (1989). Regularized discriminant analysis. J Am Statist Assoc 84:165–175.

Fuhr N, U Pfeifer (1991). Combining model-oriented and description-oriented approaches for probabilistic indexing. Proceedings of the 14th International ACM Conference on Research and Development in Information Retrieval, Chicago, pp 46–56.

Furnas GW, S Deerwester, ST Dumais, TK Landauer, RA Harshman, LA Streeter, KE Lochbaum (1988). Information retrieval using a singular value decomposition model of latent semantic structure. Proceedings 14th International ACM Conference on Research and Development in Information Retrieval, Chicago, pp 465–480.

Gabriel KR (1971). The biplot graphical displays of matrices with application to principal component analysis. Biometrika 58:453–467.

Geisser S (1975). The predictive sample reuse method with applications. J Am Statist Assoc 70:320–328.

Gifi A (1990). Non-linear Multivariate Analysis. Chichester: Wiley.

Goldstein M, WR Dillon (1978). Discrete Discriminant Analysis, Chichester: Wiley.

Gordon AD (1981). Classification: Methods for the Exploratory Analysis of Multivariate Data London: Chapman & Hall.

Gordon AD (1987). A review of hierarchical classification. J R Statist Soc A 150:119–137.
Gower JC, DJ Hand (1996). Biplots. London: Chapman & Hall.
Greenacre M (1984). Theory and Applications of Correspondence Analysis. London: Academic Press.
Gross M, D Perrin (1989). Electronic Dictionaries and Automata in Computational Linguistics. Berlin: Springer Verlag.
Guttman L (1941). The quantification of a class of attributes: a theory and method of a scale construction. In: P Horst, ed. The Prediction of Personal Adjustment. New York: SSCR, pp 251–264.
Habbema DF, J Hermans, K van den Broek (1974). A stepwise discriminant analysis program using density estimation. In: G Bruckman, ed. COMPSTAT 1974, Proceedings in Computational Statistics. Vienna: Physica Verlag.
Hand DJ (1981). Discrimination and Classification. New York: Wiley.
Harris ZS (1954). Distributional structure. Word 2:146–162.
Hartigan JA (1975). Clustering Algorithms. New York: John Wiley & Sons.
Hayashi C (1956). Theory and examples of quantification (II). Proc Inst Statist Math 4:19–30.
Hearst MA, JO Pedersen (1996). Reexamining the cluster hypothesis: Scatter/Gather on retrieval results. Proceedings of the 19th Annual International Conference on Research and Development in Information Retrieval, ACM/SIGIR'96, Zurich, pp 76–84.
Hull DA, JO Pedersen, H Schütze (1996). Method combination for document filtering. Proceedings of the 19th Annual International Conference on Research and Development in Information Retrieval, ACM/SIGIR'96, Zurich, pp 279–287.
Iwayama M, T Tokunaga (1995). Cluster-based text categorization: a comparison of category search strategies. Proceedings 18th Annual International Conference on Research and Development in Information Retrieval, ACM/SIGIR'95, Seattle, WA, pp 273–280.
Kaufman L, PJ Rousseeuw (1990). Finding Groups in Data. New York: John Wiley & Sons.
Lachenbruch PA, MR Mickey (1968). Estimation of error rate in discriminant analysis, Technometrics 10:1–11.
Lafon P (1981). Dépouillements et statistiques en lexicometrie. Paris: Slatkine-Champion, 1984.
Lafon P, A Salem (1993). L'inventaire des segments répétés d'un texte. Mots 6:161–177.
Lebart L (1982). Exploratory analysis of large sparse matrices, with application to textual data, COMPSTAT. Vienna: Physica Verlag, pp. 67–76.
Lebart L, A Morineau, K Warwick (1984). Multivariate Descriptive Statistical Analysis. New York: Wiley.
Lebart L, A. Salem, E. Berry (1998). Exploring Textual Data. Dordretch: Kluwer.
Lewis DD, WB Croft (1990). Term clustering of syntactic phrases. Proceedings 13th International ACM Conference on Research and Development in Information Retrieval, New York, pp 385–395.
Mahalanobis PC (1936). On the generalized distance in statistics. Proc Natl Inst Sci India 12:49–55.
McLachlan GJ (1992). Discriminant Analysis and Statistical Pattern Recognition. New York: Wiley.
Mead R (1988). The Design of Experiments: Statistical Principles for Practical Applications. Cambridge: Cambridge University Press.
Nishisato S (1980). Analysis of Categorical Data: Dual Scaling and its Application. Toronto: University of Toronto Press.
O'Neill R, GG Wetherhill (1971). The present state of multiple comparison methods. J R Statist Soc B 33:218–241.
Palermo D, J Jenkins (1984). Word Association Norm. Minneapolis: University of Minnesota Press.

Rajman M, A Rungsawang (1995). A new approach for textual information retrieval. Proceedings 18th Annual International ACM/SIGIR Conference on Research and Development in Information Retrieval, poster session, Seattle, WA.

Rungsawang A (1977). Information Retrieval Based on Distributional Semantics (recherche documentaire à base de sémantique distributionnelle). PhD dissertation ENST 97E041, Ecole Nationale Supérieure des Télécommunications, Paris.

Rungsawang A, M Rajman (1995). Textual information retrieval based on the concept of distributional semantics, JADT 1995: III Gionarti Internazionale di Analisi Statisticha dei Dati Testuali, Rome, vol 1, pp 237–244.

Salem A (1984). La typologie des segments répétés dans un corpus, fondée sur l'analyse d'un tableau croisant mots et textes. Les Cahiers de l'Analyse des Données 9:489–500.

Salton G (1988). Automatic Text Processing: The Transformation, Analysis and Retrieval of Information by Computer. New York: Addison–Wesley.

Salton G, C Buckley (1990). Flexible text matching in information retrieval. Technical Report 90-1158, Department of Computer Science, Cornell University, Ithaca, NY.

Salton G, C Buckley (1988). Term weighting approaches. Autom Text Retriev Inform Processing Manage 24:513–523.

Salton G, MJ McGill (1983). Introduction to Modern Information Retrieval. New York: McGraw-Hill.

Schütze H (1992a). Context space. Working Notes of the AAAI Fall Symposium on Probabilistic Approaches to Natural Language, Cambridge, MA, pp 113–120.

Schütze H 1992b. Dimensions of meaning. IEEE Proceedings of Supercomputing '92. Minneapolis, MN, pp 787–796.

Silberztein, M (1993). Dictionnaires electroniques et analyse automatique de textes. Le Système INTEX. Paris: Masson.

Sokal RR, PHA Sneath (1963). Principles of Numerical Taxonomy. San Francisco: Freeman & Co.

Stone CJ (1974). Cross-validatory choice and assessment of statistical predictions. J R Statist Soc B: 36:111–147.

Wold S (1976). Pattern recognition by means of disjoint principal component models. Pattern Recogn 8:127–139.

Wong MA (1982). A hybrid clustering method for identifying high density clusters. J Am Statist Assoc 77:841–847.

Yang Y (1995). Noise reduction in a statistical approach to text categorization. Proceedings 18th Annual International ACM/SIGIR Conference on Research and Development in Information Retrieval, Seattle, WA, pp 256–263.

21

Collocations

KATHLEEN R. McKEOWN

Columbia University, New York, New York

DRAGOMIR R. RADEV

University of Michigan, Ann Arbor, Michigan

1. INTRODUCTION

This chapter describes a class of word groups that lies between idioms and free-word combinations. Idiomatic expressions are those in which the semantics of the whole cannot be deduced from the meanings of the individual constituents. Free-word combinations have the properties that each of the words can be replaced by another without seriously modifying the overall meaning of the composite unit, and if one of the words is omitted, a reader cannot easily infer it from the remaining ones. Unlike free-word combinations, a collocation is a group of words that occur together more often than by chance. On the other hand, unlike idioms, individual words in a collocation can contribute to the overall semantics of the compound. We present some definitions and examples of collocations, as well as methods for their extraction and classification. The use of collocations for word sense disambiguation, text generation, and machine translation is also part of this chapter.

Although most computational lexicons represent properties of individual words, in general most people would agree that one knows a word from the "company that it keeps" (Firth, 1935). Collocations are a lexical phenomenon that has linguistic and lexicographic status as well as utility for statistical natural language paradigms. Briefly put, they cover word pairs and phrases that are commonly used in language, but for which no general syntactic or semantic rules apply. Because of their widespread use, a speaker of the language cannot achieve fluency without incorporating them in speech. On the other hand, because they escape characterization, they have long been the object of linguistic and lexicographic study in an effort to both define them and include them in dictionaries of the language.

It is precisely because they are observable in language that they have been featured in many statistical approaches to natural language processing (NLP). Since they occur repeatedly in language, specific collocations can be acquired by identifying words that frequently occur together in a relatively large sample of language; thus, **collocation acquisition** falls within the general class of corpus-based approaches to language. By applying the same algorithm to different domain-specific corpora, collocations specific to a particular **sublanguage** can be identified and represented.

Once acquired, collocations are useful in a variety of different applications. They can be used for **disambiguation**, including both word-sense and structural disambiguation. This task is based on the principle that a word in a particular sense tends to co-occur with a different set of words than when it is used in another sense. Thus, *bank* might co-occur with *river* in one sense and *savings* and *loan* when used in its financial sense. A second important application is **translation**: because collocations cannot be characterized on the basis of syntactic and semantic regularities, they cannot be translated on a word-by-word basis. Instead, computational linguists use statistical techniques applied to aligned, parallel, bilingual corpora to identify collocation translations and semiautomatically construct a **bilingual collocation** lexicon. Such a lexicon can then be used as part of a machine translation system. Finally, collocations have been extensively used as part of language **generation** systems. Generation systems are able to achieve a level of fluency, otherwise not possible, by using a lexicon of collocations and word phrases during the process of word selection.

In this chapter, we first overview the linguistic and lexicographic literature on collocations, providing a partial answer to the question "What is a collocation?" We then turn to algorithms that have been used for acquiring collocations, including word pairs that co-occur in flexible variations, compounds that may consist of two or more words that are more rigidly used in sequence, and multiword phrases. After discussing both acquisition and representation of collocations, we discuss their use in the tasks of disambiguation, translation, and language generation. We will limit our discussion to these topics; however, we would like to mention that some work has been done recently (Ballesteros and Croft, 1996) in using collocational phrases in cross-lingual information retrieval.

2. LINGUISTIC AND LEXICOGRAPHIC VIEWS OF COLLOCATIONS

Collocations are not easily defined. In the linguistic and lexicographic literature, they are often discussed in contrast with free-word combinations at one extreme and idiomatic expressions at the other, collocations occurring somewhere in the middle of this spectrum. A **free-word combination** can be described using general rules; that is, in terms of semantic constraints on the words that appear in a certain syntactic relation with a given headword (Cowie, 1981). An **idiom**, on the other hand, is a rigid word combination to which no generalities apply; neither can its meaning be determined from the meaning of its parts, nor can it participate in the usual word-order variations. Collocations fall between these extremes, and it can be difficult to draw the line between categories. A word combination fails to be classified as free and is termed a collocation when the number of words that can occur in a syntactic relation with a given headword decreases to the point that it is not possible to describe the set using semantic regularities.

Thus, examples of free-word combinations include *put* + [object] or *run*+[object] (i.e., 'manage') when the words that can occur as object are virtually open-ended. In the case of *put*, the semantic constraint on the object is relatively open-ended (any physical object can be mentioned) and thus the range of words that can occur is relatively unrestricted. In the case of *run* (in the sense of 'manage' or 'direct') the semantic restrictions on the object are tighter, but still follow a semantic generality: an institution or organization can be managed (e.g., *business*, *ice cream parlor*). In contrast to these free-word combinations, a phrase such as *explode a myth* is a collocation. In its figurative sense, *explode* illustrates a much more restricted collocational range. Possible objects are limited to words, such as *belief*, *idea*, or *theory*. At the other extreme, phrases such as *foot the bill* or *fill the bill* function as composites, where no words can be interchanged and variation in usage is not generally allowed. This distinction between free-word combinations and collocations can be found with almost any pair of syntactic categories. Thus, *excellent/good/ useful/useless dictionary* are examples of free-word adjective + noun combinations, while *abridged/bilingual/combinatorial dictionary* are all collocations. More examples of the distinction between free-word combinations and collocations are shown in Table 1.

Because collocations fall somewhere along a continuum between free-word combinations and idioms, lexicographers have faced a problem in deciding when and how to illustrate collocations as part of a dictionary. Thus, major themes in lexicographic papers address the identification of criteria that can be used to determine when a phrase is a collocation, characteristics of collocations, and representation of collocations in dictionaries. Given that collocations are lexical, they have been studied primarily by lexicographers and by relatively fewer linguists, although early linguistic paradigms that place emphasis on the lexicon are exceptions (e.g., Halliday and Hasan, 1976; Mel'čuk and Pertsov, 1987). In this section, we first describe properties of collocations that surface repeatedly across the literature. Next we present linguistic paradigms that cover collocations. We close the section with a presentation of the types of characteristics studied by lexicographers and proposals for how to represent collocations in different kinds of dictionaries.

A. Properties of Collocations

Collocations are typically characterized as arbitrary, language- (and dialect)-specific, recurrent in context, and common in technical language (see overview by Smadja, 1993). The notion of **arbitrariness** captures the fact that substituting a synonym for one of the words in a collocational word pair may result in an infelicitous lexical

Table 1 Colocations vs. Idioms and Free-Word Combinations

Idioms	Collocations	Free-word combinations
to kick the bucket	to trade actively	to take the bus
dead end	table of contents	the end of the road
to catch up	orthogonal projection	to buy a house

combination. Thus, for example, a phrase such as *make an effort* is acceptable, but *make an exertion* is not; similarly, *a running commentary*, *commit treason*, *warm greetings*, all are true collocations, but *a running discussion*, *commit treachery*, and *hot greetings* are not acceptable lexical combinations (Benson, 1989).

This arbitrary nature of collocations persists across languages and dialects. Thus, in French, the phrase *régler la circulation* is used to refer to a policeman who *directs traffic*, the English collocation. In Russian, German, and Serbo-Croatian, the direct translation of *regulate* is used; only in English is *direct* used in place of *regulate*. Similarly, American and British English exhibit arbitrary differences in similar phrases. Thus, in American English one says *set the table* and *make a decision*, whereas in British English, the corresponding phrases are *lay the table* and *take a decision*. In fact, in a series of experiments, Benson (1989) presented nonnative English speakers and later, a mix of American English and British English speakers, with a set of 25 sentences containing a variety of American and British collocations. He asked them to mark them as either American English, British English, World English, or unacceptable. The nonnative speakers got only 22% of them correct, while the American and British speakers got only 24% correct.

While these properties indicate the difficulties in determining what is an acceptable collocation, on the positive side it is clear that collocations occur frequently in similar contexts (Benson, 1989; Cowie, 1981; Halliday and Hasan, 1976). Thus, although it may be difficult to define collocations, it is possible to **observe** collocations in samples of the language. Generally, collocations are those word pairs that occur frequently together in the same environment, but do not include lexical items that have a high overall frequency in language (Halliday and Hasan, 1976). The latter include words such as *go*, *know*, and so forth, which can combine with just about any other word (i.e., are **free-word combinations**) and thus, are used more frequently than other words. This property, as we shall see, has been exploited by researchers in natural language processing to identify collocations automatically. In addition, researchers take advantage of the fact that collocations are often domain-specific; words that do not participate in a collocation in everyday language often do form part of a collocation in technical language. Thus, *file* collocates with verbs such as *create*, *delete*, *save*, when discussing computers, but not in other sublanguages.

Stubbs (1996) points out some other interesting properties of collocations. For example, he indicates that the word *cause* typically collocates with words expressing negative concepts, such as *accident*, *damage*, *death*, or *concern*. Conversely, *provide* occurs more often with positive words such as *care*, *shelter*, and *food*.

B. Halliday and Mel'čuk

Many lexicographers point back to early linguistic paradigms which, as part of their focus on the lexicon, do address the role of collocations in language (Mel'čuk and Pertsov, 1987; Halliday and Hasan, 1976). Collocations are discussed as one of five means for achieving lexical cohesion in Halliday and Hasan's work. Repeated use of collocations, among other devices, such as repetition and reference, is one way to produce a more cohesive text. Perhaps because they are among the earliest to discuss collocations, Halliday and Hasan present a more inclusive view of collocations and are less precise in their definition of collocations than others. For them, collocations include any set of words, the members of which participate in a semantic relatiaon.

Their point is that a marked cohesive effect in text occurs when two semantically related words occur in close proximity in a text, even though it may be difficult to systematically classify the semantic relations that can occur. They suggest examples of possible relations, such as complementarity (e.g., *boy*, *girl*), synonyms, and near-synonyms, members of ordered sets (e.g., *Monday*, *Tuesday*; *dollars*, *cents*), part–whole (e.g., *car*, *brake*), as well as relations between words of different parts of speech (e.g., *laugh*, *joke*: *blade*, *sharp*; *garden*; *dig*). They point to the need for further analysis and interpretation of collocations in future work; for their purpose, they simply lump together as collocations all lexical relations that cannot be called referential identity or repetition.

In later years, Mel'čuk provided a more restricted view of collocations. In the meaning–text model, collocations are positioned within the framework of **lexical functions** (LFs). An LF is a semantico-syntactic relation that connects a word or phrase with a set of words or phrases. LFs formalize the fact that in language there are words, or phrases, the usage of which is bound by another word in the language. There are roughly 50 different simple LFs in the meaning–text model, some of which capture semantic relations (e.g., the LF **anti** posits a relation between antonyms), some of which capture syntactic relations (e.g., $\mathbf{A_0}$ represents nouns and derived adjectivals such as *sun–solar*), whereas others capture the notion of restricted lexical co-occurrence. The LF **magn** is one example of this, representing the words that can be used to magnify the intensity of a given word. Thus, **magn**(*need*) has as its value the set of words {*great*, *urgent*, *bad*}, whereas **magn**(*settled*) has the value {*thickly*}, and **magn**(*belief*) the value {*staunch*}. $\mathbf{Oper_1}$ is another LF that represents the semantically empty verb that collocates with a given object. Thus, the $\mathbf{Oper_1}$ of *analysis* is {*perform*}.

C. Types of Collocations

In an effort of characterize collocations, lexicographers and linguists present a wide variety of individual collocations, attempting to categorize them as part of a general scheme (Allerton, 1984; Benson, 1989; Cowie, 1981). By examining a wide variety of collocates of the same syntactic category, researchers identify similarities and differences in their behavior; in the process, they come a step closer to providing a definition. Distinctions are made between grammatical collocations and semantic collocations. **Grammatical collocations** often contain prepositions, including paired syntactic categories, such as verb + preposition (e.g., *come to*, *put on*), adjective + preposition (e.g., *afraid that*, *fond of*), and noun + preposition (e.g., *by accident*, *witness to*). In these cases, the open-class word is called the **base** and determines the words it can collocate with, the **collocators**. Often, computational linguists restrict the type of collocations they acquire or use to a subset of these different types (e.g., Church and Hanks, 1989). **Semantic collocations** are lexically restricted word pairs, for which only a subset of the synonyms of the collocator can be used in the same lexical context. Examples in this category have already been presented.

Another distinction is made between **compounds** and **flexible word pairs**. Compounds include word pairs that occur consecutively in language and typically are immutable in function. Noun + noun pairs are one such example, which not only occur consecutively but also function as a constituent. Cowie (1981) notes that compounds form a bridge between collocations and idioms because, like collocations, they are quite invariable, but they are not necessarily semantically opaque. Since collocations are recursive, collocational phrases including more than just two words can occur. For example, a collocation such as *by chance*, in turn, collocates with verbs such as *find*, *discover*, *notice*. Flexible word pairs include collocations between subject and verb, or verb and object; any number of intervening words may occur between the words of the collocation.

D. Collocations and Dictionaries

A final, major recurring theme of lexicographers is where to place collocations in dictionaries. Placement of collocations is determined by which word functions as the base and which functions as the collocator. The base bears most of the meaning of the collocation and triggers the use of the collocator. This distinction is best illustrated by collocations that include "support" verbs: in the collocation *take a bath*, *bath* is the base and the support verb *take*, a semantically empty word in this context, the collocator. In dictionaries designed to help users encode language (e.g., generate text), lexicographers argue that the collocation should be located at the base (Haussmann, 1985). Given that the base bears most of the meaning, it is generally easier for a writer to think of the base first. This is especially true for persons learning a language. When dictionaries are intended to help users decode language, then it is more appropriate to place the collocation at the entry for the collocator. The base–collocator pairs listed in Table 2 illustrate why this is so.

3. EXTRACTING COLLOCATIONS FROM TEXT CORPORA

Early work on collocation acquisition was carried out by Choueka et al. (1983). They used **frequency** as a measure to identify a particular type of collocation, a sequence of adjacent words. In their approach, they retrieved a sequence of words that occurs more frequently than a given threshold. Although they were theoretically interested in sequences of any length, their implementation is restricted to sequences of two to six words. They tested their approach on an 11-million–word corpus from the *New York Times* archives, yielding several thousand collocations. Some examples of retrieved collocations include *home run*, *fried chicken*, and *Magic Johnson*. This work was notably one of the first to use large corpora and predates many of the more mainstream corpus-based approaches in computational linguistics. Their metric, however, was less sophisticated than later approaches; because it was based on frequency alone, it is sensitive to corpus size.

Table 2 Base–Collocator Pairs

Base	Collocator	Example
noun	verb	*set the table*
noun	adjective	*warm greetings*
verb	adverb	*struggle desperately*
adjective	adverb	*sound asleep*
verb	preposition	*put on*

Church and Hanks (1989) and Church et al. (1991) used a **correlation-based metric** to retrieve collocations; in their work, a *collocation* was defined as a pair of words that appear together more than would be expected by chance. To estimate correlation between word pairs, they used **mutual information** as defined in Information Theory (Shannon, 1948; Fano, 1961).

If two points (words) x and y have probabilities $P(x)$ and $P(y)$, then their mutual information $I(x, y)$ is defined as in Eq. {1} (Church et al., 1989),

$$I(x; y) = \log_2 \frac{P(x, y)}{P(x)P(y)} \tag{1}$$

where $P(x, y)$ is the probability of seeing the two words x and y within a certain window.

Whenever $(x, y) = P(x)P(y)$, the value of $I(x; y)$ becomes 0, which is an indication that the two words x and y are not members of a collocation pair. If $I(x; y) < 0$, then the two words are in complementary distribution (Church et al., 1989). Other metrics for computing strength of collocations are discussed in Oakes (1998).

Church et al.'s approach was an improvement over that of Choueka et al., in that they were able to retrieve interrupted word pairs, such as subject + verb or verb + object collocations. However, unlike Choueka et al., they were restricted to retrieving collocations containing only two words. In addition, the retrieved collocations included words that are semantically related (e.g., *doctor–nurse*, *doctor–dentist*) in addition to true lexical collocations.

Smadja and McKeown (Smadja, 1991, 1993; Smadja and McKeown, 1990) addressed acquisition of a wider variety of collocations than either of the two other approaches. This work featured the use of several **filters** based on linguistic properties, the use of several **stages** to retrieve word pairs along with compounds and phrases, and an **evaluation** of retrieved collocations by a lexicographer to estimate the number of true lexical collocations retrieved.

Their system **Xtract** began by retrieving word pairs using a frequency-based metric. The metric computed the **z-score** of a pair by first computing the average frequency of the words occurring within a ten-word radius of a given word and then determining the number of standard deviations above the average frequency for each word pair. Only word pairs with a z-score above a certain threshold were retained. In contrast to Choueka et al.'s metric, this metric ensured that the pairs retrieved were not sensitive to corpus size. This step is analogous to the method used by both Choueka et al. and Church et al., but it differs in the details of the metric.

In addition to the metric, however, Xtract used three additional filters based on linguistic properties. These filters were used to ensure an increase in the accuracy of the retrieved collocations by removing any that were not true lexical collocates. First, Xtract removed any collocations of a given word for which the collocate can occur equally well in any of the ten positions around the given word. This filter removed semantically related pairs, such as *doctor–nurse*, where one word can simply occur anywhere in the context of the other; in contrast, lexically constrained collocations will tend to be used more often in similar positions (e.g., an adjective + noun collocation would more often occur with the adjective several words before the noun). A second filter noted patterns of interest, identifying whether a word pair was always used rigidly with the same distance between words or whether there is more than one position. Finally, Xtract used syntax to remove collocations where a given word did not occur significantly often with words of the same syntactic function. Thus, verb + noun pairs were filtered to remove those that did not consistently present the same syntactic relation. For example, a verb + noun pair that occurred equally often in subject + verb and verb + object relations would be filtered out.

After retrieving word pairs, Xtract used a second stage to identify words that co-occurred significantly often with identified collocations. This way it accounted for the recursive property of collocations noted by Cowie (1981). In this stage, Xtract produced all instances of appearance of the two words (i.e., concordances) and analyzed the distributions of words and parts of speech in the surrounding positions, retaining only those words around the collocation that occurred with probability greater than a given threshold. This stage produced **rigid compounds** (i.e., adjacent sequences of words that typically occurred as a constituent such as noun compounds) as well as **phrasal templates** (i.e., idiomatic strings of words possibly including slots that may be replaced by other words). An example of a compound is {2}, while an example of a phrasal template is {3}.

the Dow Jones industrial average {2}

*The NYSE's composite index of all its listed common stocks fell *NUMBER* to *NUMBER*.* {3}

Xtract's output was evaluated by a lexicographer to identify precision and recall: 4000 collocations produced by the first two stages of Xtract, excluding the syntactic filter, were evaluated in this manner. Of these, 40% were identified as good collocations. After further passing these through the syntactic filter, 80% were identified as good. This evaluation dramatically illustrated the importance of combining linguistic information with syntactic analysis. Recall was measured only for the syntactic filter. It was noted that of the good collocations identified by the lexicographer in the first step of output, the syntactic filter retained 94% of them.

4. USING COLLOCATIONS FOR DISAMBIGUATION

One of the more common approaches to word-sense disambiguation involves the application of additional constraints on the words for which sense is to be determined. Collocations can be used to specify such constraints. Two major types of constraints have been investigated. The first uses the general idea that the presence of certain words near the ambiguous one will be a good indicator of its most likely sense. The second type of constraint can be obtained when pairs of translations of the word in an aligned bilingual corpus are considered.

Research performed at IBM in the early 1990s (Brown et al., 1991) applied a statistical method using as parameters the context in which the ambiguous word appeared. Seven factors were considered: the words immediately to the left or to the right of the ambiguous word, the first noun and the first verb both to the left and to the right, as well as the tense of the word if it was a verb or the first verb to the left of the word otherwise. The system developed indicated that the use of collocational information results in a 13% increase in performance over the conventional trigram model in which sequences of three words are considered.

Work by Dagan and Itai (1994) extended the ideas set forth in Brown et al.'s work. They augmented the use of statistical translation techniques with linguistic information, such as syntactic relations between words. By using a bilingual lexicon and a monolingual corpus of one language, they were able to avoid the manual tagging of text and the use of aligned corpora.

Other statistical methods for word-sense disambiguation are discussed in Chap. 26.

Church et al. (1989) have suggested a method for word disambiguation in the context of Optical Character Recognition (OCR). They suggest that collocational knowledge helps choose between two words in a given context. For example, the system may have to choose between *farm* and *form* when the context is either {3} or {4}.

federal ... credit {4a}

some ... of {4b}

In {4a}, the frequency of *federal* followed by *farm* is 0.50, whereas the frequency of *federal* followed by *form* is 0.039. Similarly in {4b}, the frequencies of *credit* following either *farm* or *form* are 0.13 and 0.026, respectively. One can, therefore, approximate the probabilities for the trigram *federal farm credit*, which is $(0.5 \times 10^{-6}) \times (0.13 \times 10^{-6}) = 0.065 \times 10^{-12}$ and for *federal form credit*, which is $(0.039 \times 10^{-6}) \times (0.026 \times 10^{-6}) = 0.0010 \times 10^{-12}$. Because the first of these probabilities is 65 times as large as the second one, the OCR system can safely pick *farm* over *form* in {4a}. Similarly, *form* is 273 times more likely than *farm* in {4b}. Church et al. also note that syntactic knowledge alone would not help in such cases, as both *farm* and *form* are often used as nouns.

5. USING COLLOCATIONS FOR GENERATION

One of the most straightforward applications of collocational knowledge is in natural language generation. There are two typical approaches applied in such systems: the use of phrasal templates in the form of canned phrases and the use of automatically extracted collocations for unification-based generation. We will describe some of the existing projects using both of these approaches. At the end of this section we will also mention some other uses of collocations in generation.

A. Text Generation Using Phrasal Templates

Several early text-generation systems used **canned phrases** as sources of collocational information to generate phrases. One of them was Unix consultant (UC), developed at Berkeley by Jacobs (1985). The system responded to user questions related to the Unix operating system and used text generation to convey the answers. Another such system was Ana, developed by Kukich (1983) at the University of Pittsburgh, which generated reports of activity on the stock market. The underlying paradigm behind generation of collocations in these two systems was related to the reuse of canned phrases, such as the following from Kukich (1983): *opened strongly, picked up momentum early in trading, got off to a strong start.*

On the one hand, Kukich's approach was computationally tractable, as there was no processing involved in the generation of the phrases, whereas on the other hand, it did not permit the flexibility that a text generation system generally requires. For example, Ana needed to have separate entries in its grammar for two quite similar phrases: *opened strongly* and *opened weakly*.

Another system that made extensive use of phrasal collocations was FOG (Bourbeau et al., 1990). This was a highly successful system that generated bilingual (French and English) weather reports that contained a multitude of canned phrases such as {5}.

Temperatures indicate previous day's high and overnight low to 8 a.m. {5}

In general, canned phrases fall into the category of **phrasal templates**. They are usually highly cohesive; and the algorithms that can generate them from their constituent words are expensive and sophisticated.

B. Text Generation Using Automatically Acquired Collocational Knowledge

Smadja and McKeown (1991) have discussed the use of (automatically retrieved) collocations in text generation.

The Xtract system mentioned earlier (see Sec. 3) used statistical techniques to extract collocations from free text. The output of Xtract was then fed to a separate program, Cook (Smadja and McKeown, 1991), which used a functional unification paradigm FUF (Kay, 1979; Elhadad, 1993) to represent collocational knowledge, and more specifically, constraints on text generation imposed by the collocations and their interaction with constraints caused by other components of the text generation system. Cook could be used to represent both compound collocations (such as *the Down Jones average of 30 Industrial Stocks*) and predicative collocations (such as *post–gain* or *indexes–surge*).

Cook represented collocations using attribute–value pairs, such as **Synt-R** (the actual word or phrase in the entry), **SV-collocates** (verbal collocates with which the entry is used as the subject), **NJ-collocates** (adjectival collocates that can modify the noun), and others. For example, if **Synt-R** contained the noun phrase *stock prices*, some possible values for th **SV-collocates** would be *reach*, *chalk up*, and *drift*. Using such representations, Cook was able to generate a sentence such as{6}.

X chalked up strong gains in the morning session. {6}

Cook's lexicalization algorithm consisted of six steps:

1. Lexicalize topic
2. Propagate collocational constraints
3. Lexicalize subtopics
4. Propagate collocational constraints
5. Select a verb
6. Verify expressiveness

A comparison between the representation of collocations in Ana and Cook will show some of the major differences in the two approaches: whereas Ana kept full phrases with slots that could be filled by words obeying certain constraints,

Cook kept only the words in the collocation and thus avoided a combinatorial explosion when several constraints (of collocational or other nature) needed to be combined.

Another text generation system that makes use of a specific type of collocations is SUMMONS (Radev and McKeown, 1997). In this system, the authors have tried to capture the collocational information linking an entity (person, place, or organization) with its description (premodifier, apposition, or relative clause) and to use it for generation of referring expressions. For eample, if the system discovers that the name *Ahmed Abdel-Rahman* is collocated with *secretary-general of the Palestinian authority*, a new entry is created in the lexicon (also using FUF as the grammar for representation) linking the name and its description for later use in the generation of references to that person.

C. Other Techniques

An interesting technique used by Iordanskaja et al. (1990) in the GOSSiP system involved the modification of the structure of a semantic network before generation to choose a more fluent wording of a concept. Their work makes of Meaning-Text Theory (Mel'čuk and Pertsov, 1987), using lexical functions to represent collocations in the generation lexicon. Lexical functions allow very general rules to perform certain kinds of paraphrases. For example, this approach allows for the generation of {7a} as a paraphrase of {7b}, where the support verb *made* and the noun collocate replace the verb *use*. Other generation systems have also explored the use of Meaning-Text Theory to handle generation of collocations (Nirenburg, 1988; McCardell Doerr, 1995).

Paul made frequent use of Emacs Friday. {7a}

Paul used Emacs frequently Friday. {7b}

6. TRANSLATING COLLOCATIONS

Because collocations are often language-specific and frequently cannot be translated compositionally, researchers have expressed interest in statistical methods that can be used to extract bilingual pairs of collocations for parallel and nonparallel corpora.

Note that one cannot assume that a concept expressed by way of a collocation in one language will use a collocation in another language. Let us consider the English collocation *to brush up a lesson*, which is translated into French as *repasser une leçon* or the English collocation *to bring about*, for which the Russian translation is the single-word verb *osushtestvljat'*. Using only a traditional (noncollocational) dictionary, it is hard-to-impossible to find the correct translation of such expressions. Existing phraseological dictionaries contain certain collocations, but are by no means sufficiently exhaustive.

Luckily for natural language researchers, there exist a large number of bilingual and multilingual aligned corpora (see Chap. 25). Such bodies of text are an invaluable resource in machine translation in general, and in the translation of collocations and technical terms in particular.

Smadja et al. (1996) have created a system called Champollion,[1] which is based on Smadja's collocation extractor, Xtract. Champollion uses a statistical method to translate both flexible and rigid collocations between French and English using the Canadian Hansard corpus.[2] The Hansard corpus is prealigned, but it contains a number of sentences in one of the languages that do not have a direct equivalent in the other. Champollion's approach includes three stages:

1. Identify syntactically or semantically meaningful units in the source language.
2. Decide whether the units represent constituents or flexible word pairs.
3. Find matches in the target languages and rank them, assuming that the highest-ranked match for a given source-language collocation is its translation in the target language.

Champollion's output is a bilingual list of collocations ready to use in a machine translation system. Smadja et al. indicate that 78% of the French translations of valid English collocations were judged to be good by the three evaluations by experts.

Kupiec (1993) describes an algorithm for the translation of a specific kind of collocation; namely, noun phrases. He also made use of the Canadian Hansard corpus. The algorithm involved three steps:

1. Tag sentences in the (aligned) corpus.
2. Use finite-state recognizers to find noun phrases in both languages.
3. Use iterative reestimation to establish correspondences between noun phrases.

Some examples retrieved are shown in Table 3. An evaluation of his algorithm has shown that 90 of the 100 highest ranking correspondences are correct.

A tool for semiautomatic translation of collocations, *Termight*, is described in Dagan and Church (1994). It is used to aid translators in finding technical term correspondences in bilingual corpora. The method proposed by Dagan and Church used extraction of noun phrases in English and word alignment to align the head and tail words of the noun phrases to the words in the other language. The word sequence between the words corresponding to the head and tail is produced as the translation. *Termight* was implemented within a full-editor environment that

Table 3 Translations Extracted from the Canadian Hansard Corpus

English collocation	French collocation
late spring	*fin du printemps*
Atlantic Canada Opportunities Agency	*Agence de promotion économique du Canada atlantique*

[1]The French egyptologist Jean-François Champollion (1790–1832) was the first to decipher the ancient Egyptian hieroglyphs using parallel texts in Egyptian, Demotic, and Greek found on the Rosetta stone.
[2]The Canadian Hansard corpus contains bilingual reports of debates and proceedings of the Canadian parliament.

allows practical use of its results as an aid to translators. Because it did not rely on statistical correlation metrics to identify the words of the translation, it allowed the identification of infrequent terms that would otherwise be missed owing to their low statistical significance.

The reader should not remain with the impression that French and English are the only two languages for which research in translation of collocations has been done. Language pairs involving other languages, such as Japanese, Chinese, Dutch, and German have also been investigated. Fung (1995) used a pattern-matching algorithm to compile a lexicon of nouns and noun phrases between English and Chinese. The algorithm has been applied on the Hong Kong government bilingual corpus (English and Cantonese). Wu and Xia (1994) also compute a bilingual Chinese–English lexicon, although they have less of a focus on the inclusion of terms. They use Estimation–Maximization (Dempster et al., 1977) to produce word alignment across parallel corpora and then apply various linguistic filtering techniques to improve the results.

Researchers have investigated translation of Japanese phrases using both parallel and nonparallel corpora. Fung (1997) uses morphosyntactic information in addition to alignment techniques to find correspondences between English and Japanese terms. Her algorithm involves tagging both the English and Japanese texts of a parallel corpus, extracting English NPs from the English side and aligning a subset of Japanese translations with the English side manually for training. Frequency information and learning through consideration of unknown words are used to produce the full lexicon. Exploratory work in the use of nonparallel corpora for translation, a potentially much larger resource than parallel corpora, exploits the use of correlations between words to postulate that correlations between words in one text are also likely to appear in the translations. Tanaka and Iwasaki (1996) demonstrate how to use nonparallel corpora to choose the best translation among a small set of candidates, whereas Fung (1997) uses similarities in collocates of a given word to find its translation in the other language.

In other work, van der Eijk (1993) has compared several methods for automatic extraction and translation of technical terminology in Dutch and English. He achieves best results under the assumption that technical terms are always NPs; therefore, candidate terms can be pinpointed using a combination of a pattern matcher and a part of speech tagger. Some examples of the terms retrieved by his system are shown in Table 4.

7. RESOURCES RELATED TO COLLOCATIONS

Two classes of resources might be of interest to researchers interested in the extraction or translation of collocations. Several dictionaries of collocations exist either on

Table 4 Dutch–English Technical Term Pairs

Dutch term	English term
hardnekkige weerzin	persisting aversion
vroegtijdige standaardisatie	early standardization
wisselwerking ... produkten	inter-working ... products

paper or in a CD-ROM format. We would like to note four such dictionaries: the *Collins Cobuild Dictionary*, the BBI *Combinatory Dictionary of English*, NTC's *Dictionary of Phrasal Verbs and Other Idiomatic Verbal Phrases*, and the *Dictionary of Two-Word Verbs for Students of English*.

Cobuild (Sinclair, 1987) is the largest collocational dictionary, for which the CD-ROM version gives access to 140,000 English collocations and 2.6 million examples of how these collocations are used. The collocations and examples are extracted from the 200-million word Bank of English corpus (see `http://titania.cobuild.collins.co.uk/boe_info.html`. *Cobuild* provides an on-line service, Cobuild*Direct* that provides access to both concordances and collocations from its corpus. Cobuild*Direct* is available from `http://titania.cobuild.collins.co.uk/direct_info.html`.

The BBI dictionary (Benson et al., 1986) is geared toward learners of English and focuses on lexical and grammatical collocations, including 19 verb patterns. Because the goal of the BBI is to make it easy for a learner to find collocations, collocations are placed within the entry for the base (see Sec. 2.D). The BBI was evaluated for ease of use through two experiments. In the first, nonnative speakers were asked to fill in a missing word in a set of sentences, in which each contained a blank. Their performance consistently improved when they used the BBI in the task (from 32% acuracy to 93% accuracy). In the second task, Russian speakers were given a list of Russian collocations along with the associated English collocations, each of which was missing a word. Again, accuracy improved with the use of the BBI (from 40–45 to 100%).

NTC's dictionary (Spears, 1986) covers 2796 verbs and 13,870 definitions or paraphrases of their collocational usage with different prepositions. Even though the focus of this dictionary is primarily on idiomatic collocations (the meaning of which cannot be extracted from the meanings of the constituent words), because its primary audience includes persons learning English as a second language, it also includes a large number of commonly used collocations.

The *Dictionary of Two-Word Verbs for Students of English* (Hall, 1982) specializes in phrasal verbs, such as *add up*, *keep on*, and *pile up*. The dictionary includes some 700 such collocations along with examples and transitivity information.

In addition to dictionaries, researchers can use the software package TACT (Text Analysis Computing Tools) that consists of 16 programs for text retrieval and analysis of literary texts; it also contains a component, *Usebase*, which allows retrieval of collocations from a corpus. TACT is accessible from `http://-www.chass.utoronto.ca/cch/tact.html`.

8. SUMMARY

Although definitions of collocations have varied across research projects, that they are observable in large samples of language has led to successes in their use in various statistical applications. Sophisticated approaches to collocation acquisition for representation in a lexicon now exist. These semiautomatically developed phrasal lexicons have been used for the tasks of language generation, machine translation, and to some extent, information retrieval. In addition, identification and integration of collocations and lexical context have also played a central role in tasks such as statistical approaches to word sense disambiguation.

In addition to the original research works cited in this chapter, we would like to bring to the attention of readers two overviews of collocations in Stubbs (1996) and Oakes (1998).

REFERENCES

Allerton DJ 1984. Three or four levels of co-occurrence relations. Lingua 63:17–40.

Ballesteros L, WB Croft 1996. Dictionary-based methods for cross-lingual information retrieval. Proceedings 7th International DEXA Conference on Database and Expert Systems Applications, pp 791–801.

Benson M 1989. The structure of the collocational dictionary. Int J Lexicogr 2:1–14.

Benson M, E Benson, R Ilson 1986. The BBI Combinatory Dictionary of English: A Guide to Word Combinations. Philadelphia: John Benjamins.

Bourbeau L, D Carcagno, E Goldberg, R Kittredge A Polguère 1990. Bilingual generation of weather forecasts in an operations environment. In: COLING-90: Papers Presented to the 13th International Conference on Computational Linguistics, vol 1, pp 90–92.

Brown PF, SA Della Pietra, VJ Della Pietra, RL Mercer 1991. Word-sense disambiguation using statistical methods. 29th Annual Meeting of the Association for Computational Linguistics, Berkeley, CA, pp 264–270.

Choueka Y, T Klein E Neuwitz 1983. Automatic retrieval of frequent idiomatic and collocational expressions in a large corpus. J Liter Linguist Comput 4:34–38.

Church K, P Hanks 1989. Word association norms, mutual information, and lexicography. 27th Meeting of the Association for Computational Linguistics, Vancouver, BC, pp 76–83.

Church K, W Gale, P Hanks, D Hindle 1989. Parsing, word associations and typical predicate–argument relations. DARPA Speech and Natural Language Workshop, Harwich Port, MA.

Church K, W Gale, P Hanks, D Hindle 1991. Using statistics in lexical analysis. In: U Zernik, ed. Lexical Acquisition: Exploiting On-Line Resources to Build a Lexicon. Hillsdale, NJ: Laurence Erlbaum, pp 116–164.

Cowie AP 1981. The treatment of collocations and idioms in learner's dictionaries. Appl Linguist 2:223–235.

Dagan I, K Church 1994. TERMIGHT: identifying and translating technical terminology. 4th Conference on Applied Natural Language Processing, Stuttgart, Germany, pp 34–40.

Dagan I, A Itai, 1994. Word sense disambiguation using a second language monolingual corpus. Comput Linguist 20:563–596.

Dempster AP, NM Laird, DB Rubin 1977. Maximum likelihood from incomplete data via the EM algorithm [with discussion]. J R Statist Soc B 39:1–38.

Elhadad M 1993. Using argumentation to control lexical choice: a unification-based implementation. PhD dissertation, Columbia University, New York, 1993.

Fano R 1961. Transmission of Information: A Statistical Theory of Information. Cambridge, MA: MIT Press.

Firth JR 1935. The technique of semantics. Trans Philol Soc pp 36–72.

Fung P 1995. A pattern matching method for finding noun and proper noun translations from noisy parallel corpora. 33rd Annual Conference of the Association for Computational Linguistics, Cambridge, MA, 236–233.

Fung P 1997. Using word signature features for terminology translation from large corpora. PhD dissertation, Columbia University, New York.

Hall J 1982. Dictionary of Two-Word Verbs for Students of English. New York: Minerva Books.

Halliday MAK, R Hasan 1976. Cohesion in English. London: Longman.

Haussmann F 1985. Kollokationen im deutschen Wörterbuch: ein Beitrag zur Theorie des lexicographischen Beispiels. In: H Bergenholtz, J Mugdon, eds. Lexikographie und Grammatik. Tübingen: Niemeyer.

Iordsanskaja LN, R Kittredge, A Polguère 1990. Lexical selection and paraphrase in a meaning-text generation model. In: CL Paris, WR Swartout, WC Mann, eds. Natural Language Generation in Artificial Intelligence and Computational Linguistics. Dordrecht: Kluwer Academic.

Jacobs PS 1985. A knowledge-based approach to language production. PhD dissertation, University of California, Berkeley.

Kay M 1979. Functional grammar. Proceedings of the 5th Annual Meeting of the Berkeley Linguistic Society.

Kukich K 1983. Knowledge-based report generation: a knowledge engineering approach to natural language report generation. PhD dissertation, University of Pittsburgh, 1983.

Kupiec J 1993. An algorithm for finding noun phrase correspondences in bilingual corpora. 31st Annual Meeting of the Association for Computational Linguistics, Columbus, OH, pp 17–22.

McCardell Doerr, R 1995. A lexical-semantic and statistical approach to lexical collocation extraction for natural language generation. PhD dissertation, University of Maryland, Baltimore, MD.

Mel'čuk, IA, NV Pertsov 1987. Surface-Syntax of English, a Formal Model in the Meaning-Text Theory. Philadelphia: Benjamins.

Nirenburg S 1988. Lexicon building in natural language processing. Program and Abstracts of the 15th International Conference of the Association for Literary and Linguistic Computing, Jerusalem, Israel.

Oakes M 1998. Statistics for Corpus Linguistics. Edinburgh: Edinburgh University Press.

Radev DR, KR McKeown 1997. Building a generation knowledge source using Internet-accessible newswire. 5th Conference on Applied Natural Language Processing, Washington, DC, pp 221–228.

Shannon CE 1948. A mathematical theory of communication. Bell Syst Technol 27:379–423, 623–656.

Sinclair JM, ed. 1987. Collins COBUILD English Language Dictionary. London: Collins, http://titania.cobuild.collins.co.uk/.

Smadja F 1991. Retrieving collocational knowledge from textual corpora. An application: language generation. PhD dissertation, Columbia University, New York.

Smadja F 1993. Retrieving collocations from text: Xtract. Comput Linguist 19:143–177.

Smadja F, K McKeown 1990. Automatically extracting and representing collocations for language generation. 28th Annual Meeting of the Association for Computational Linguistics, Pittsburgh, PA, pp 252–259.

Smadja F, K McKeown 1991. Using collocations for language generation. Comput Intell 7.

Smadja F, KR McKeown, V Hatzivassiloglou 1996. Translation collocations for bilingual lexicons: a statistical approach. Comput Linguist 22:1–38.

Spears R 1996. NTC's Dictionary of Phrasal Verbs and Other Idiomatic Expressions. Lincolnwood, IL: NTC Publishing Group.

Stubbs M 1996. Text and Corpus Analysis. Oxford: Blackwell Publishers.

Tanaka K, H Iwasaki 1996. Extraction of lexical translations from nonaligned corpora. COLING-96: The 16th International Conference on Computational Linguistics, Copenhagen, Denmark, pp 580–585.

van der Eijk P 1993. Automating the acquisition of bilingual terminology. Sixth Conference of the European Chapter of the Association for Computational Linguistics, Utrecht, Netherlands, pp 113–119.

Wu D, X Xia 1994. Learning an English–Chinese lexicon from a parallel corpus. Technology Partnerships for Crossing the Language Barrier: Proceedings of the First Conference of the Association for Machine Translation in the Americas, Columbia, MD, pp 206–213.

22

Statistical Parsing

JOHN A. CARROLL

University of Sussex, Brighton, England

1. INTRODUCTION

Practical approaches to natural language (NL) parsing must address two major problems: (a) how to resolve the (lexical, structural, or other) ambiguities that are inherent in real-world natural language text, and (b) how to constrain the form of analyses assigned to sentences, while still being able to return "reasonable" analyses for as wide a range of sentences as possible.

Parsing systems within the generative linguistics tradition (Part I of this handbook; see also, e.g., Gazdar & Mellish, 1989) employ hand-built grammars in conjunction with parsing algorithms that either (a) return all possible syntactic analyses, which would then be passed on to detailed, domain-dependent semantic and pragmatic processing subsystems for disambiguation; or (b) use special purpose, heuristic parsing algorithms that are tuned to the grammar and that invoke hand-coded linguistic or domain-specific heuristics and/or grammar relaxation techniques to perform disambiguation and cope with extragrammatical input, respectively. The major drawbacks to these approaches to parsing are that

- Computing the full set of syntactic analyses for a sentence of even moderate length with a wide-coverage grammar is often intractable.
- If this is possible, there is still the problem of how to apply semantic processing and disambiguation efficiently to a representation of a (possibly very) large set of competing syntactic analyses.
- Although linguistic theories are often used as devices for explaining interesting facts about a language, actual text in nontrivial domains contains a wide range of poorly understood and idiosyncratic phenomena, forcing any

grammar to be used in a practical system to go beyond established results and requiring much effort in filling gaps in coverage.

- Hand-coding of heuristics is a labour-intensive task that is prone to mistakes and omissions, and makes system maintenance and enhancement more complicated and expensive.
- Using domain-specific hand-coded knowledge hinders the porting of a system to other domains or sublanguages.

In part motivated by some of these weaknesses, and drawing on work in the area of **corpus linguistics** (see, e.g., McEnery & Wilson, 1996) which had collected large bodies of machine-readable (and in some cases linguistically annotated) textual data, in the mid-1980s researchers started to investigate the application of statistical approaches to parsing. Two broad types of technique are evident in work in this area, even from the outset: the **grammar-driven approach**, and the **data-driven approach**.

In the grammar-driven approach a generative grammar is used to define—more or less broadly—the language that can be parsed and the class of analyses that will be returned, and distributional biases in the language (as observed in training data) are encoded in associated statistical data. When parsing, in cases of ambiguity, the statistical information is used either to rank multiple competing parses or to select a single, preferred parse. Within the generative linguistics tradition, grammars include constraints purely intended to prevent acceptance of ungrammatical inputs or generation of spurious incorrect analyses. However, if a disambiguation mechanism exists the grammar can afford to impose fewer constraints on the language accepted because analyses that are less plausible for a given sentence would, in principle, be assigned lower likelihoods.

The data-driven approach dispenses with a generative grammar in the usual sense, instead deriving grammatical knowledge through the learning of analysis schemas or structural biases from manually parsed training data (**treebanks**). When parsing a sentence, the goal is for the most predominant structures in the training data that are applicable in the context of the sentence to be the ones that appear in the analysis produced. Of course, this approach trades manual effort in constructing a grammar with the requirement for a substantial, representative treebank. Some grammar-driven approaches require parsed data, but in general data-driven approaches make much more intensive use of it. The treebanks that have been the most widely used to date are:

- The Penn Treebank (Marcus et al., 1993), with 1 million words drawn from the *Wall Street Journal*, and a small sample of ATIS-3 data. Other material that has also been published as part of the treebank includes parsed versions of the Brown Corpus (Francis & Kucera, 1982), IBM computer manuals, and transcripts of radio broadcasts.
- The IBM/Lancaster Treebank (Black et al., 1993), containing 800,000 words extracted from IBM computer manuals.
- SUSANNE (Sampson, 1996), consisting of a 150,000-word representative sample of the Brown Corpus.

These resources all contain only (American) English. There are still no substantial treebanks for other languages, although this situation is changing.

For reasons of robustness, statistical parsers are sometimes designed to be applied to a partially or fully disambiguated sequence of lexical syntactic categories, rather than the original words in a sentence. Simplifying the input to the parser in this way circumvents many problems of lexical coverage suffered by systems that require detailed syntactic subcategorisation information encoding, for example, verb valencies (Jensen, 1991; Briscoe & Carroll, 1993), as well as taking advantage of the robustness and relatively high accuracy of part-of-speech tagging (e.g., Garside et al., 1987; DeRose, 1988).

2. GRAMMAR-DRIVEN APPROACHES

A. Probabilistic Context-Free Grammar

One of the most investigated and basic grammar-driven statistical frameworks is **Probabilistic Context-Free Grammar** (PCFG). PCFG is a variant of context-free grammar (CFG) in which a probability is associated with each production in such a way that the probabilities of all productions with the same left-hand-side, non-terminal add up to 1. An example of a very simple PCFG with just eight productions is the following:

S → NP VP	(1.0)	{1}
VP → Vt NP	(0.4)	{2}
VP → Vi	(0.6)	{3}
NP → ProNP	(0.4)	{4}
NP → Det N	(0.3)	{5}
NP → NP PP	(0.3)	{6}
N → N N	(0.3)	{7}
PP → P NP	(1.0)	{8}

Each derivation of a sentence is assigned a probability that is the product of probabilities of all the rules utilised. Thus the probability of tree (a) in Fig. 1 is $1.0 \times 0.4 \times 0.4 \times 0.3 = 0.048$ (one occurrence each of rules 1, 4, 2, and 5). The probability of a sentence is defined as the sum of the probabilities of all derivations of that sentence, and it is the case that for any "reasonable grammar" (Booth and Thompson, 1973) the sum of the probabilities of all the sentences accepted by the grammar is 1.

The individual rule probabilities can be estimated using Hidden Markov Model (HMM) techniques. The Baum–Welch algorithm (Baum, 1972) is an efficient technique for performing iterative reestimation of the parameters of a (hidden) stochastic **regular grammar**; Baker (1979) extends Baum–Welch reestimation to CFGs in Chomsky Normal Form (CNF) giving the **Inside–Outside algorithm** which automatically assigns a probability to each CF production on the basis of

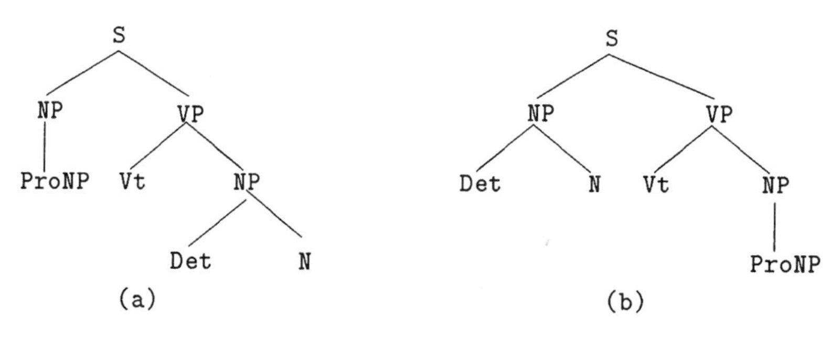

Fig. 1 Context-free grammar derivations.

its frequency of occurrence in all possible analyses of each sentence in the training corpus. Kupiec (1991) further extends the reestimation technique to arbitrary (non-CNF) CFGs. Starting with an initial set of production probabilities, the algorithm iterates over the training corpus, modifying them in a way such as to increase the sum of probabilities of the sentences in the training corpus. The algorithm is guaranteed to converge to a local maximum (for a more detailed exposition of this and other statistical material in this chapter, see Manning and Schütze, 1999). The running time of the algorithm is proportional to the cubes of the sentence length and number of nonterminals in the grammar, so it becomes expensive for very long sentences and grammars with large category sets. Also, considering all possible analyses of each sentence of the training corpus means that this technique will produce good results only if the rule probabilities estimated over all possible analyses in the training data correlate with the frequency of rules in correct analyses, and if the "noise" in the training data created by incorrect parses is effectively factored out. Whether this is so depends on several factors including the number of "false-positive" sentences with only incorrect analyses, and the amount of heterogeneity in the training corpus.

Fujisaki et al. (1989), Sharman et al. (1990), and Pereira & Schabes (1992; see also Schabes et al., 1993) describe corpus analysis experiments using PCFGs of various sizes and formats, but in each case trained using a version of the Inside–Outside algorithm. Fujisaki et al. used a predefined nonprobabilistic grammar of 7550 CF productions, and derived rule probabilities by training on a corpus of 4206 sentences of plain text (with an average sentence length of approximately 11 words). Sharman et al. used a grammar containing 100 terminals and 16 nonterminals in ID/LP format (Gazdar et al., 1985; Sharman, 1989).[1] Pereira & Schabes used a CNF grammar with a (small) category set containing 15 nonterminals and 48 terminals, corresponding to the Penn Treebank lexical tag set (Santorini, 1990). The training data consisted of 770 sentences, represented as tag sequences, drawn from the treebank. Each group used the Viterbi (1967) algorithm when parsing to select the most probably analysis. Fujisaki et al.

[1]ID/LP grammars separate the two types of information encoded in CF rules—immediate dominance and linear precedence—into two rule types which together define a CFG. This allows probabilities concerning dominance, associated with ID rules, to be factored out from those concerning precedence, associated with LP rules.

reparsed 84 sentences from their training corpus, and in 72 cases the most probable analysis was correct giving a "success" rate (per sentence) of 85%. Sharman et al. parsed 42 sentences of 30 words or less drawn from their training corpus, and found that for 88% of them the most probable analysis was identical or "similar" to the correct analysis. Pereira & Schabes found that for unseen test sentences the most probable analysis was compatible with the manually assigned analysis in up to 78% of cases.

Black et al (1993) give a detailed description of the development of a treebank, together with a statistical grammar and a 3000-word lexicon for parsing computer manual text. The grammar was developed manually in a "frequency-based" manner based entirely on corpus data. Large-scale grammar development and enhancement was made possible by implementing the grammar in terms of feature-based rule "templates," each of which encoded a (possibly large) set of related CF productions. Before training, the templates (700 in all) were expanded by replacing each feature-based category with one of more of a set of more specific manually specified "mnemonic" categories. Then this version of the grammar was trained using the Inside–Outside algorithm, constrained to take into account only grammar analyses consistent with the treebank, and at each rule application checking for the local compatibility of partially specified categories. On unseen IBM computer manual sentences of length 7–17 words, in 76% of cases parsing gave a bracketing that was consistent (but not necessarily identical) with the correct one.

1. *Acquisition of PCFGs from Corpora*

Fujisaki et al. and Black et al.'s grammars were hand-crafted, whereas Sharman et al. and Pereira & Schabes' grammars were inferred from the training material that was also used for estimation of the rule probabilities. Sharman et al. constructed a grammar and initial probabilities based on the frequency of ID and LP relations in a manually parsed corpus (treebank) of about one million words of text. Pereira & Schabes used the Inside–Outside algorithm to infer the grammar as well as the rule probabilities, with an initial grammar containing all possible CNF rules over the given category set. The system was trained in an unsupervised mode on plain text, and also in a "semisupervised" mode in which the manually parsed version of the corpus was used to constrain the set of analyses used during reestimation. (In supervised training, analyses were accepted if they produced bracketings consistent with those assigned manually.) Pereira & Schabes found that, in the unsupervised mode, the accuracy was degraded from 78 to 35%; supervised training also converged faster to a stable set of probabilities. These results underline the importance of supervised training when the grammar itself is being inferred as well as the probability of individual rules: accuracy is much improved, and also the grammar is guaranteed to converge to one that is linguistically plausible. Moreover, with supervised training the computational cost of reestimation is reduced, from proportional to the cube of the sentence lengths, to linear. Briscoe & Waegner (1993) obtain similar results—while still training on plain text—by imposing linguistic constraints on the initial grammar (such as each rule must have a head) and biasing initial probabilities to favour a set of hand-coded, linguistically motivated "core" rules. Nevertheless, their computational expense limits such techniques to simple grammars with category sets of a dozen or so nonterminals, or to training on manually parsed data.

A CFG can be created by directly reading productions of the analyses in a treebank (forming a **treebank grammar**) in such a way that each single-level subtree gives rise to a production in which the mother of the tree is the left-hand side nonterminal and the children in the tree are the rule daughters. Charniak (1996) describes an experiment in which productions were extracted in this way from a corpus of 300,000 words taken from the Penn Treebank, resulting in a grammar of 10,605 rules (of which only 3,943 occurred more than once). Rule probabilities were estimated directly from the treebank, computed simply by normalising frequency counts to 1 over each set of rules with the same mother category; the probability of a parse was defined as the product of rule probabilities, as usual, but with a heuristic bias applied to favour right-branching structures (since syntactic structure in English tends to be right-branching). When parsing word-tag sequences, Charniak reports superior results to comparable systems. In the first instance this result seems surprising, but it is partly due to the fact that, although rules are missing from the grammar, it overgenerates so much that coverage is not a problem, and also partly to the right-branching correction which counteracts an observed tendency for CF parsers to return more centre-embedding structures than is warranted.

2. *Practical Parsing with Large PCFGs*

A CFG induced from a treebank with a relatively small nonterminal alphabet will overgenerate massively. Such grammars may place "almost no constraints on what part of speech might occur at any point in the sentence" (Charniak, 1996:8). This extreme ambiguity places large demands on a parser. Caraballo & Charniak (1996) use a bottom-up, best-first parser and investigate various scoring measures for subanalyses to be able to prune those that are unlikely to form part of the most probable global analysis. Sekine & Grishman (1995) also use a bottom-up strategy to parse with a large context-free probabilistic treebank grammar, and report a crucial improvement in speed from compiling the right-hand sides of productions into a finite-state automaton, thus factoring common prefixes and reducing the number of partial hypotheses that have to be stored. Magerman & Weir (1992) describe *Picky* which uses a three-phase bidirectional, heuristic approach to parsing to significantly reduce the number of edges produced by a CKY-like chart-parsing algorithm.

Another approach to practical parsing with wide-coverage CFGs is to use a parsing algorithm that incorporates top-down prediction and look-ahead to avoid pursuing analytical paths that could not lead to a global analysis. The most powerful predictive technique for unambiguous CFGs is LR (Knuth, 1965), and an LR parse table derived from a PCFG can be used to drive a nondeterministic LR parser (Tomita, 1987). Wright & Wrigley (1989), Wright (1990), and Ng & Tomita (1991) describe techniques for distributing PCFG production probabilities around the corresponding parse table in such a way that the probabilities assigned to analyses are identical with those defined by the original grammar. The results obtained with this approach will be exactly equivalent to parsing with the original PCFG directly, but parser throughput may be better.

B. Context-Dependent Probabilistic Models

In PCFG, the association of a probability with each production means, crudely, that more common grammatical constructions would tend to be preferred over less common ones; however, defining the probability of a derivation as the product of context-free rule probabilities relies on the assumption that rule applications are statistically independent. This, however, is not true in actual language: for instance a noun phrase (NP) is more likely to be realized as a pronoun in subject position than elsewhere (Magerman & Marcus, 1991). Thus the tree configuration (b) in Fig. 1 is, in reality, less likely than (a), so unless a PCFG multiplied out the grammar to split NPs into subject and nonsubject NPs, it must necessarily assign both configurations the same probability because the same rules are used the same number of times and only one global probability can be associated with the relevant production. For similar reasons, for NPs like {9}, the grammar assigns the same probability (0.00243) to both of the two derivations it licenses (one in which *with the telescope* modifies *the man in the park* and the other in which it modifies just *the park*).

the man with the telescope in the park {9}

Ideally, a language model should be able to distinguish alternative structural configurations of this kind in case they occur with different frequencies in the particular type of text under consideration.

HMM reestimation techniques are used widely in speech recognition, where the derivation of a given input is not of interest; rather what is sought is the best (most likely) model of the input (i.e., the string of words realized by the input speech signal). Language models trained using HMM reestimation maximise the sum of the probabilities of the sentences in a corpus given a grammar. However, it is debatable whether this property is what is required when the task is to select a single, plausible derivation as the analysis for each sentence, because there is no clear connection between the probability of a derivation and that of a sentence. For parsing, it would be more appropriate to maximise the probability of the correct derivation given a sentence in the training corpus. Thus, it is unclear that such language models are suitable for the task of parse selection. Chiang et al. (1992) remark on this problem and propose an **adaptive learning** procedure which improves accuracy by adjusting probabilistic model parameters directly during training to increase the difference in probabilistic scores assigned to correct and to incorrect analyses.

PCFG continues to be widely investigated in speech recognition work as a word prediction technique that integrates well with lower-level HMM processing. However, most research in NLP has moved on to investigating approaches in which the probabilistic model is sensitive to details of parse context, and is also trained to maximise the probability of the *correct* derivation. For example, Chitrao & Grishman (1990) describe a context-dependent probabilistic model in which the probability of each possible expanding production for each nonterminal appearing on the right-hand side of a rule is recorded separately, and report a 23% decrease in in incorrect analyses over a previous (standard) PCFG.[2] In Pearl (Magerman & Marcus, 1991) the probability of a derivation is estimated as the geometric mean

[2]Charniak & Carroll (1994) extend PCFG in a similar way by associating a set of probabilities with each production, one for each possible nonterminal one level higher in the parse tree. However, their motivation for doing this is to construct an improved language model for English (reflecting the true distribution of words in sentences), rather than a more accurate parser.

of the product of the conditional probabilities of each CFG rule appearing in the derivation given the (nonterminal trigram) context of its dominating rule and the part-of-speech trigram at that point. In contrast to PCFG, these approaches can distinguish (to a limited extent) the probability of different derivations in which the same rules are applied but in different structural configurations—for example, in compound nouns such as *railway station buffet* which could be analysed as in {10a} or {10b}.

railway [station buffet] {10a}

[railway station] buffet. {10b}

More discriminating models of syntactic context have been obtained by associating probabilistic information with intermediate parser configurations as defined by an LR parse table (Knuth, 1965). An **LR parse state** encodes information about the left context at any point in a parse, and a limited amount of right context can be captured with (usually) one symbol of look-ahead. Thus, probabilities are associated directly with an LR parse table, rather than simply with rules of the grammar. Deriving probabilities relative to the parse context can allow a probabilistic parser to distinguish situations in which identical rules reapply in different ways across different derivations (cf Fig. 1) or apply with differing probabilities in different contexts (cf, the distribution of pronouns in subject position and elsewhere). Su et al. (1991; see also Chang et al., 1992) define a general probabilistic method of parse ranking for LR parsing with CFG, distinguishing syntactic contextual factors by associating scores with **shift** actions in the parse table. Briscoe & Carroll (1993) describe a system in which a score is associated with each transition in an LALR(1) parse table constructed from the context-free backbone of a wide-coverage unification-based grammar of English. Unification of the "residue" of features not incorporated into the backbone is performed at parse time when **reduce** actions are applied. Unification failure results in the associated derivation being assigned a probability of zero. Probabilities are assigned to transitions in the parse table by a process of supervised training based on computing the frequency with which transitions would be used when parsing a training corpus. Carroll & Briscoe (1996) describe a more robust version of this system that uses a grammar of punctuation and part-of-speech labels to assign "phrasal" analyses to arbitrary corpus text.

History-based grammar (**HBG**) (Black et al., 1991, 1993) is a probabilistic disambiguation technique that provides a method for estimating the parameters of a disambiguation model combining any arbitrary set of contextual factors. From a corpus of bracketed sentences, HBG builds a **decision tree** encoding the relevant factors in an analysis that determine the correct parse of a sentence. Parsing works bottom-up, left-to-right, constructing a left-most derivation, and at each stage the model can use information located anywhere in the partial derivation tree to determine the probabilities of applying different rules at that point. As in PCFG, the probability of a complete parse is the product of the rule application probabilities. Black et al. report experiments using the IBM/Lancaster Treebank of sentences from computer manuals in conjunction with an existing broad-coverage phrase-structure grammar developed for that domain. An HBG decision tree was grown for the grammar from a training set of about 9000 trees, at each node conditioning rule probabilities on the node's syntactic and semantic labels, the rule applied at the node's parent, the constituent's index as a child of this parent, and the parent's lexical head(s). The grammar was trained using the Inside–Outside algorithm on 15,000 trees, producing a PCFG. Testing on unseen treebank sentences of 7–17 words, the HBG model returned a labelled analysis consistent with the correct one for 75% of the sentences as opposed to 60% for the PCFG, a reduction in error rate of 37%.

C. Lexicalised Probabilistic Models

In traditional approaches to parsing, the input is in the form of (ambiguous) sequences of lexical syntactic categories, rather than the words they were derived from. Probabilistic versions of parsers within this paradigm can, therefore, attempt to disambiguate only on the basis of training data indicating common and uncommon structural configurations in syntactic analyses. However, it is evident from several separate research efforts into the resolution of prepositional phrase (PP) attachment ambiguities—where a PP can attach either to a preceding verb or to its direct object (e.g., in {11} whether the PP *in the park* is more closely associated with *saw* or *the man*)—that adding word-level information can aid disambiguation.

I saw the man in the park. {11}

Hindle & Rooth (1993) demonstrate that an approach to resolving these types of ambiguity based purely on structural preferences will have an error rate of 33–67%. They achieve an error rate of only 20% by finding instances in a training corpus where the attachment is unambiguous (e.g., when the NP object is a pronoun) and accumulate counts of how frequently a particular preposition is attached to a particular verb, and how often to a particular (head) noun. These statistics are then used to determine which decision to make in cases where the attachment is ambiguous. Collins & Brooks (1995) extend this approach using a more principled probabilistic model that also takes into account the head of the NP following the preposition, and report an accuracy of 85% (on different data).

Empirical work into the wider problem of assigning complete syntactic analyses to sentences in general text has also demonstrated the value of lexical probabilistic information. Charniak (1995) reports an experiment in which he compared the accuracy of a PCFG to that of the same grammar using a modified probabilistic model for which the probability of a constituent in a parse tree depended not only on the production used at that point, but also on the lexical head of the constituent and those of its parent and grandparent. On a set of sentences containing words for which there was ample lexical data, the lexicalised model returned a parse that was structurally consistent with the correct analysis in significantly more cases than the standard PCFG, in fact, giving a reduction in the error rate of 41.1%. This improvement is an upper bound, though, because word-level data is typically highly sparse; a practical system in a nontrivial domain would need to be able to "back off" to non-lexical statistics in the frequent cases where the lexical data was inadequate. On a randomly constructed test corpus without any artificial lexical bias, Briscoe & Carroll (1997) observed more modest gains in accuracy when they took a purely structural context-dependent probabilistic model and augmented it with probabilistic information for individual verbs. They computed phrasal parses for 380,000 words of unrestricted text and used these to hypothesize verbal complementation frames and associated probabilities for 356 verb lemmas occurring in 250 test sentences. The lexicalized version of the parser exhibited a 7% reduction in error rate on the test corpus.

An alternative approach to incorporating lexical probabilistic information into a parsing system is to start with a **lexicalised grammar** and associate statistical information directly with this. Indeed, Fujisaki et al.'s (1989) system described in Sec. 2.A used a lexicalised CFG in Greibach Normal Form (in which each production may contain only a terminal category as its left-most daughter). Thus, for example, the probability that the verb *believe* is followed by an NP object can be differentiated from the probability that it is followed by a sentential complement (see {12a} v. {12b}).

I believe him. {12a}

I believe he will go. {12b}

This might explain the relatively good reported accuracy when reparsing sentences from the original training corpus. Accuracy would be expected to be degraded, though, on unseen sentences, given that a relatively small training corpus (of only 4206 sentences) was used and lexical data is by nature sparse. Fujisaki et al. also propose a technique for integrating lexical and context-dependent probabilistic information within PCFG: to differentiate different noun–noun compound structures they create 5582 instances of four morphosyntactically identical rules for classes of word forms with distinct bracketing behaviour. However, this would greatly enlarge the grammar and, hence, the size of training corpus required.

Building on existing (nonprobabilistic) work on lexicalised grammar formalisms, Schabes (1992) and Resnik (1992) define stochastic versions of Lexical Tree Adjoining Grammar (LTAG; Schabes et al., 1988). In LTAG each lexical item is associated with one or more elementary tree structures in a way that allows properties related to that lexical item (such as subcategorization and agreement information) to be expressed wholly within the elementary structure. Although presented differently and arrived at independently, Schabes' and Resnik's versions are essentially equivalent (so only the former is described here). Schabes formalises his approach using a stochastic version of Linear Indexed Grammar (LIG), a context-free-like grammar formalism. An LTAG is compiled to a LIG by decomposing the elementary trees into single-level trees and introducing additional productions that explicitly represent every possible tree combination possibility. Each LIG production is assigned a probability (which is 1 for the decomposition productions, for they are deterministic). The additional LIG productions encode the probability of combining a particular elementary tree at a specific place in another elementary tree. This means that the frequency information associated with them is, to some extent, dependent on context. Furthermore, as both Schabes and Resnik point out, by taking advantage of LTAG's enlarged domain of locality, probabilities can model both lexical and structural co-occurrence preferences. Schabes also describes a technique based on the Inside–Outside algorithm for estimating the LIG probabilities, but does not present any experimental results.

3. DATA-DRIVEN APPROACHES

The probabilistic parsing techniques described in Sec. 2 use generative grammars that are either (a) hand-constructed or (b) inferred from training data. In the first case, the grammars are labour-intensive to develop and inevitably suffer from undergeneration. For example, Briscoe & Carroll's (1995) hand-written "phrasal" grammar designed to parse unrestricted English fails to analyse 21% of sentences in a large, balanced corpus; Black et al.'s (1993) hand-written grammar developed to parse computer manual text containing a restricted vocabulary of only 3000 words is unable to generate the correct analysis structure for 4% of sentences of length 7–17 words. For some applications, though, a certain amount of undergeneration might be acceptable. In the second case, when grammars are derived automatically, the training data must be preparsed or the form of the grammar tightly constrained, otherwise the learning process is computationally intractable.

In contrast, data-driven approaches dispense with a generative grammar, returning analyses that are patterned on those in a treebank. Rich models of context are often used to make up for the lack of linguistic constraints that would—in a grammar-driven approach—have been supplied by a formal grammar. However, as a side effect, the (combinations of) syntactic phenomena that can be parsed are necessarily limited to those present in the training material: being able to deal properly with new texts would normally entail further substantial treebanking efforts and possibly also major improvements in the efficiency of storage and deployment of derived syntactic knowledge. A further disadvantage of not using a hand-constructed grammar is that the form of the analyses assigned to sentences is governed entirely by the representations specified in the treebank, and so is not fully determinate because there is no underlying formal grammar. The analyses would, therefore, not support principled methods of semantic interpretation.

A. Parsing Using Lexical–Structural Relation Data

Perhaps the first application of corpus-based techniques to automatic parsing was Sampson's (1986) simulated annealing parser, APRIL. During the training phase APRIL collects statistics from manually parsed corpus material on the relative prevalence of different tree configurations, and uses these to construct a metric that for a given tree returns its "plausibility." Parsing a sentence is a process of searching for the most plausible labelled tree that conforms to the input. The plausibility of a tree in this system is defined as a combination of the plausibilities of all of its "local" subtrees (i.e., trees of depth 1); the plausibility of a local tree is defined by transition probabilities between nodes in a finite-state transition network labelled with the categories of the tree's parent and children. The search for the most plausible tree is based on a **hill-climbing** technique: the search starts at random point in the space of possible solutions (i.e., analysis trees), examines the point's immediate neighbours (defined in terms of a random sequence of tree transformation operations), and moves to the best neighbour (i.e., the most plausible); this process is repeated until the current solution, or tree, is better (more plausible) than all of its neighbours.

A simple hill-climbing search works well if the evaluation (plausibility) function over the solution space takes the form of a convex surface; for example, the upper half of a sphere. If it does not, however, the search is likely to become "trapped" at a **local optimum** and not be able to move away to a higher-scoring region elsewhere in the solution space. For natural language, the solution space, as defined by the evaluation function and the permissible set of tree transformations, is highly irregular so APRIL uses **simulated annealing**, a more sophisticated type of search. When parsing, initially almost every transformation is permitted, so the solution space is traversed at a coarse level of detail; after successive preset numbers of transitions, decreases in the plausibility measure are gradually limited with the objective of narrowing in on a globally high-scoring region. The process thus becomes closer and closer to pure hill-climbing over a localised area as an optimum is reached (Fig. 2). In experiments with parsing word-tag sequences, a version of the APRIL system achieved correctness of rule application in the mid- to high 80%.

The SPATTER parser (Magerman, 1995) combines lexical and structural relation data using the history-based grammar technique (Black et al., 1991) described in Sec. 2.B. However, SPATTER dispenses with Black et al.'s generative grammar component, instead recording in an "extension" feature in a parse tree node structural information specifying whether that node is either the root of the tree; the first, middle, or last child of a constituent; or the only child of a unary constituent. During training on treebanked sentences the correct parse tree is constructed in a bottom-up left-to-right fashion by adding nodes with appropriate category labels, extension values, and lexical heads (inherited from the head child as specified in a node label/children lookup table). Each addition is used as a training example for the construction of the decision tree that associates probabilities with contextual factors. Factors that can be encoded consist of information about the values of the label, extension, or head word or tag (a) at the current node, (b) at the sibling nodes one or two distant to right or left, and (c) at child nodes. Parsing is a search for the highest probability tree, defined as the product of the node addition probabilities. The lack of a grammar means that SPATTER's search space is extremely large, for there is no mechanism constraining the structure of analyses hypothesised: they are determined solely by the decision tree probabilities. The search procedure works in two phases. The first is depth-first to find a reasonably probable complete parse for the sentence. Then unexplored partial analyses left over from the first phase are extended in a breadth-first manner to see if they can produce a higher-probability parse. With training and test data taken from the *Wall Street Journal* portion of the Penn Treebank, SPATTER identifies labelled constituents with over 84% accuracy on sentences of 4–40 words in length.

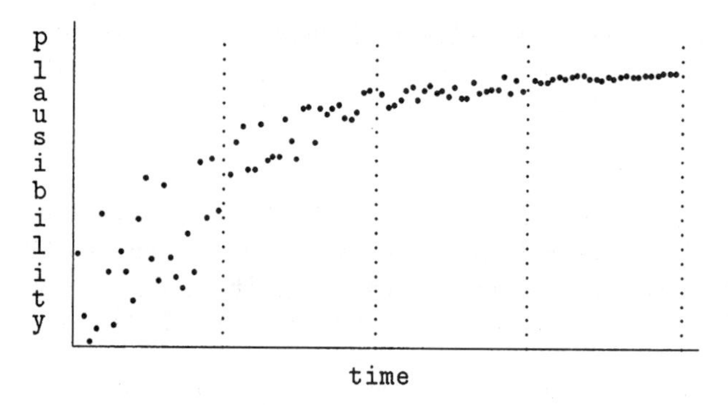

Fig. 2 Search using simulated annealing. *Vertical lines indicate imposition of stricter limits on decreases in the plausibility measure.*

Collins (1996) presents a conceptually simpler lexical probabilistic model that incorporates information only on head–modifier relations between pairs of words. The words considered are the heads of "BaseNPs," minimal NPs, which themselves do not contain an NP. Each BaseNP head in a sentence—with the exception of the head of the sentence itself—is defined to modify exactly one other, and the relation between them is encoded as a triple of nonterminal labels consisting of the modifier nonterminal, the category of their common parent node, and the label of the modified word. Training estimates the probability of a relation, given the word/part-of-speech tag tuples of the pairs of words involved and a measure based on their linear order and distance apart in the sentence (on the premise that surface order is significant and more distant relations are likely to be weaker). Where there is insufficient data, probability estimates are backed-off to replace words by tags. The parser takes a part-of-speech–tagged sentence, computes possible BaseNPs, and searches for the parse that maximises the product of the BaseNP and head–modifier relation probabilities. Because there is no grammar, the parser effectively has to search through the space of all nonterminal triples that have been observed in the training data. The search can be optimised, however, because if there are two similar constituents analysing the same words, the lower-probability one can be discarded. To further improve parsing efficiency, all constituents for which the probability is below a certain threshold are discarded, but if no parse is found the threshold is lowered and parsing attempted again. A heuristic beam search strategy is also used that prunes constituents the probability of which is below a given factor of the best over the same words. Collins' system is much faster than SPATTER and achieves comparable or better accuracy on the same *Wall Street Journal* data.

B. Data-Oriented Parsing

Bod (1995, 1999) observes that there can be many significant dependencies between linguistic units, for example, words or phrases, that go beyond what a linguistic model of a language (i.e., a generative grammar) would normally attempt to capture. Similarly, some dependencies cannot be encoded in probabilistic models of the form used by Magerman and Collins (see foregoing) that use a fixed set of lexical/structural relations, each of which is assumed to be independent of the others. For example, in the Air Travel Information System (ATIS) corpus (Hemphill et al., 1990) the generic NP *flights from X to Y* (as in sentences such as {13}) occurs very frequently. In this domain the dependencies between the words in the NP—but without *X* and *Y* filled in—are so strong that in ambiguity resolution it should arguably form a single statistical unit.

Show me flights from Dallas to Atlanta. {13}

This cannot be achieved in a traditional linguistic grammar formalism except by analysing each dependency of this kind individually as a special case, which would then tie the grammar in strongly to the particular corpus.[3] To capture these types of dependency within a general-purpose parsing system, Bod describes a Data-Oriented

[3] A general-purpose grammar could be automatically "tuned" to a particular corpus, though; for example, Rayner & Carter (1996) describe a "grammar specialisation" technique that prunes and restructures a grammar relative to a corpus to decrease ambiguity and improve parser throughput.

Parsing (DOP) framework which dispenses with a grammar in the traditional sense and constructs analyses of sentences purely out of segments of parse trees from a training corpus. In the DOP models he investigates in detail, analyses are built from subtrees of arbitrary depth, and the probability of a particular derivation is the product of the probabilities of the subtrees used. Frequently occurring subtrees would tend to encode significant dependencies between their constituent parts, and so where applicable these would be more likely to appear in an analysis than less frequent alternatives.

However, when subtrees are allowed to have depths of more than 1, a single analysis of a given sentence can generally be derived in many different ways, using different combinations of subtrees of different depths that cover the input text in different ways (Fig. 3). For this reason the probability of an analysis is defined as the sum of the probabilities of all of its possible derivations. The computation of the most-probable analysis cannot use a Viterbi-like search as does PCFG because the objective is to maximise the *sum* of derivation probabilities for an analysis, rather than just maximising the probability of a single derivation. Bod, therefore, uses a Monte-Carlo algorithm, in which random portions of the search space are sampled to find an analysis for which the probability can be guaranteed to be within an arbitrarily small threshold of the most probable. When parsing unseen text drawn from the ATIS corpus DOP is very accurate, but is also extremely slow (taking several minutes of CPU time on average to parse each sentence). It is thus unlikely that using current techniques and computer hardware it would be computationally feasible for DOP to be trained on larger treebanks or to parse text from less-restricted domains.

C. Parsing by Tree Transformation Rules

Transformation-based, error-driven techniques have been successfully used in assigning part-of-speech labels to words in text (Brill, 1992), among other tasks.

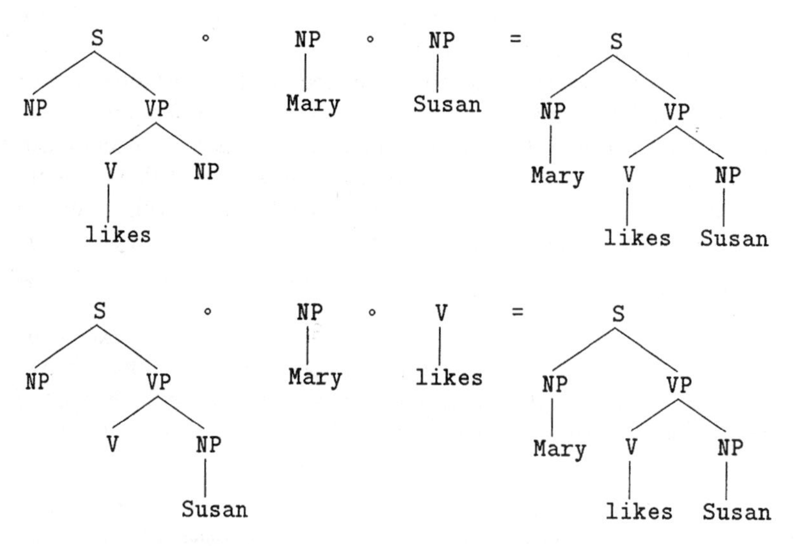

Fig. 3 Different derivations of the same analysis in DOP.

The approach is not probabilistic, but nevertheless is still statistical because it is driven by the distribution of relevant information in an annotated training corpus. Brill (1993) applies a variant of the same techniques to parsing, and shows that a robust and relatively accurate parser can be learned from only a small corpus of bracketed, binary-branching, part-of-speech–annotated training text. The technique starts by assuming that all sentences are uniformly right-branching, but with final punctuation attached high—for example, the annotated sentence {14a} would lead to the initial structure {14b} being built.

The/Det dog/N barked/V ./. {14a}

((The/Det (dog/N barked/V)) ./.) {14b}

Then transformation rules are automatically learned to transform the sentence structures into ones that better match the types of structure that are present in the training corpus. A rule consists of a change to be applied in a given triggering environment. Brill describes a repertoire of twelve rule types:

- *Add* | *delete* a *left* | *right* parenthesis to the *left* | *right* of part-of-speech tag *X* (8 rule types)
- *Add* | *delete* a *left* | *right* parenthesis between tags *X* and *Y* (4 rule types)

After a transformation, further adjustments are made to the structure to ensure it remains balanced and binary-branching. An example of a rule that might be learned is {15}.

delete a left parenthesis to the right of part-of-speech tag `Det` (determiner) {15}

This would result in the transformation illustrated in Fig. 4. Training consists of making the initial right-branching assumption for each sentence in the corpus, followed by a "greedy" search, repeatedly testing each possible rule instantiation against the whole corpus and applying to the corpus the rule that makes the bracketings as a whole conform most closely to the correct ones. To parse a new sentence, the right-branching structure is imposed, and the transformation rules are applied one by one in order.

Rule-driven parsing is completely deterministic and can easily be implemented very efficiently, but the training process is potentially intractable: Brill reports training times of the order of weeks. Vilain & Palmer (1996) make Brill's approach practical by devising fast algorithms that prune the search space, reducing training times to hours. They also systematically explore variants of the approach and

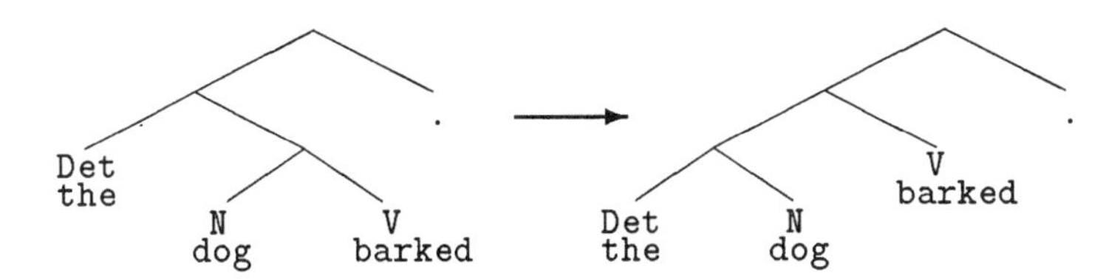

Fig. 4 Application of a tree transformation rule.

conclude that the procedure is robust, in the sense that variations in the training conditions make little difference to the accuracy of the resulting parser, and that relatively small amounts of training material are needed to attain reasonable accuracy.

4. PROBABILISTIC PREDICATE–ARGUMENT STRUCTURE

The techniques described so far in this chapter associate probabilistic information with syntactic rules, parser configurations, or portions of syntactic structure to produce a ranked set of syntactic analyses or a single most-highly preferred analysis. Whether or not the results of syntactic processing are to be subject to semantic interpretation, the techniques rest on the assumption that the probabilistic information will model semantic and pragmatic constraints in the particular corpus being analysed. An alternative, more direct approach would be to define a probabilistic model at the logical form (LF) level, because correct disambiguation depends ultimately on sentence meaning. The advantage of this approach is that since LFs abstract away from syntax, relevant relations between words in a sentence can be extracted uniformly without regard for syntactic transformations (such as passivisation, topicalisation, extraposition); the particular syntactic construct linking them (verbal complement, nominal modifier); or idiosyncratic predicates (e.g., control verbs, or verbs taking a dummy subject).

The CLARE system (Alshawi et al., 1992) assigns preference measures to (syntactic) structural properties, semantic head collocations, particular resolutions of anaphora, and scoping permutations in LF representations during intermediate stages of semantic processing. Alshawi & Carter (1994) describe a hill-climbing technique for optimising the weightings associated with each type of preference. Chang et al (1992) present an alternative framework that integrates lexical, syntactic, and semantic preferences into a single-evaluation function operating on a syntax tree containing syntactic and semantic annotations. Major categories have a semantic feature that contains a set of semantic tags (such as "stative" for verbs, "animate" for nouns), which are passed up the parse tree from heads to parent nodes. The probability of a given derivation is computed while parsing as the product of the probabilities of each set of annotations at the frontier of the current (partial) tree, given the previous set. Miller et al. (1994) apply HMM techniques to natural language understanding, producing tree-structured meaning representations directly from text in the ATIS domain. The leaves of the representations are the words in the sentence, and the internal nodes, abstract concepts such as "flight" and "flight origin".

A major disadvantage of disambiguation using semantic information, and a reason for the comparatively small number of research efforts in this area, is that it is difficult to construct or obtain sufficient quantities of suitable, semantically annotated training material. Collecting such data for this type of approach is an open-ended, and usually subjective and labour-intensive manual process, although it may be possible to derive limited types of relevant data (semi)automatically, for example from machine-readable dictionaries (e.g., Alshawi, 1989; Copestake, 1990), or as a result of further processing of syntactically annotated text.

REFERENCES

Alshawi H 1989. Analysing the dictionary definitions. In: B Boguraev, E Briscoe, eds. Computational Lexicography for Natural Language Processing. Harlow, UK: Longman, pp 153–169.

Alshawi H, D Carter 1994. Training and scaling preference functions for disambiguation. Comput Linguist 20:635–648.

Alshawi H, D Carter, R Crouch, S Pulman, M Rayner, A Smith 1992. CLARE: A Contextual Reasoning and Cooperative Response Framework for the Core Language Engine. Cambridge, UK: SRI International.

Baker J 1979. Trainable grammars for speech recognition. In: D Klatt, J Wolf, eds. Speech Communication Papers for the 97th Meeting of the Acoustical Society of America. Cambridge, MA: MIT, pp 547–550.

Baum L 1972. An inequality and associated maximization technique in statistical estimation for probabilistic functions of Markov processes. Inequalities 3:1–8.

Black E, R Garside, G Leech, eds. 1993. Statistically-Driven Computer Grammars of English: the IBM/Lancaster Approach. Amsterdam: Rodopi.

Black E, S Abney, D Flickenger, C Gdaniec, R Grisham, P Harrison, D Hindle, R Ingria, F Jelinek, J Klavans, M Liberman, M Marcus, S Roukos, B Santorini, T Strzalkowski 1991. A procedure for quantitatively comparing the syntactic coverage of English grammars. Proceedings of the 4th DARPA Speech and Natural Language Workshop, Pacific Grove, CA, pp 306–311.

Bod R 1995. Enriching linguistics with statistics: performance models of natural language. PhD dissertation, University of Amsterdam, ILLC dissertation series 95–14.

Bod R 1999. Beyond Grammar. Stanford, CA: CSLI Press.

Booth T, R Thompson 1973. Applying probability measures to abstract languages. IEEE Trans Comput C-22.442–450.

Brill E 1992. A simple rule-based part of speech tagger. Third Conference on Applied Natural Language Processing, Trento, Italy, pp 152–155.

Brill E 1993. Automatic grammar induction and parsing free text: a transformation-based approach. 31st Annual Meeting of the Association for Computational Linguistics, Columbus, OH, pp 259–265.

Briscoe E, J Carroll 1993. Generalised probabilistic LR parsing for unification-based grammars. Comput Linguist 19:25–60.

Briscoe E, J Carroll 1995. Developing and evaluating a probabilistic LR parser of part-of-speech and punctuation labels. 4th ACL/SIGPARSE International Workshop on Parsing Technologies, Prague, pp 48–58.

Briscoe E, J Carroll 1997. Automatic extraction of subcategorization from corpora. Fifth Conference on Applied Natural Language Processing, Washington, DC, pp 356–363.

Briscoe E, N Waegner 1992. Robust stochastic parsing using the inside–outside algorithm. AAAI Workshop on Statistically-based NLP Techniques, San Jose, CA, pp 39–53.

Caraballo S, E Charniak 1996. Figures of merit for best-first probabilistic chart parsing. Proceedings of the Conference on Empirical Methods in Natural Language Processing, Philadelphia, PA, pp 127–132.

Carroll J, E Briscoe 1996. Apportioning development effort in a probabilistic LR parsing system through evaluation. ACL SIGDAT Conference on Empirical Methods in Natural Language Processing, Philadelphia, PA, pp 92–100.

Chang J-S, Y-F Luo, K-Y Su 1992. GPSM: a generalized probabilistic semantic model for ambiguity resolution. 30th Annual Meeting of the Association for Computational Linguistics, Newark, DE, pp 177–184.

Charniak E 1995. Parsing with context-free grammars and word statistics. Technical report CS-95-28, Department of Computer Science, Brown University, Providence, RI.

Charniak E 1996. Tree-bank grammars. Technical Report CS-96-02, Department of Computer Science, Brown University, Providence, RI.

Charniak E, G Carroll 1994. Context-sensitive statistics for improved grammatical language models. Proceedings of the 12th National Conference on Artificial Intelligence (AAAI '94), Seattle, WA, pp 728–733.

Chiang T-H, Y-C Lin, K-Y Su 1992. Syntactic ambiguity resolution using a discrimination and robustness oriented adaptive learning algorithm. Proceedings of the fifteenth [sic] International Conference on Computational Linguistics: COLING-'92, Nantes, Fr, pp 352–359.

Chitrao M, R Grishman 1990. Statistical parsing of messages. Workshop on Speech and Natural Language Processing, Asilomar, CA, pp 263–266.

Collins M 1996. A new statistical parser based on bigram lexical dependencies. 34th Annual Meeting of the Association for Computational Linguistics, Santa Cruz, CA, pp 184–191.

Collins M, J Brooks 1995. Prepositional phrase attachment through a backed-off model. Proceedings of the 3rd Workshop on Very Large Corpora, Cambridge, MA, pp 27–38.

Copestake A 1990. An approach to building the hierarchical element of a lexical knowledge base from a machine-readable dictionary. Workshop on Inheritance in Natural Language Processing, Tilburg, Netherlands, pp 19–29.

DeRose S. 1988. Grammatical category disambiguation by statistical optimization. Comput Linguist 14:31–39.

Francis W, H Kučera, with the assistance of A Mackie, 1982. Frequency Analysis of English Usage: Lexicon and Grammar. Boston, MA: Houghton Mifflin.

Fujisaki T, F Jelinek, J Cocke, E Black, T Nishino 1989. A probabilistic method for sentence disambiguation. 1st International Workshop on Parsing Technologies, Pittsburgh, PA, pp 105–114.

Garside R, G Leech, G Sampson 1987. The Computational Analysis of English: A Corpus-based Approach. London: Longman.

Gazdar G, E Klein, G Pullum, I Sag 1985. Generalized Phrase Structure Grammar. Oxford: Blackwell.

Gazdar G, C Mellish 1989. Natural Language Processing in Lisp. London: Addison-Wesley.

Hemphill C, J Godfrey, G Doddington 1990. The ATIS spoken language systems pilot corpus. Workshop on Speech and Natural Language Processing. Hidden Valley, PA, pp 96–101.

Hindle D, Rooth M 1991. Structural ambiguity and lexical relations. 29th Annual Meeting of the Association for Computational Linguistics, Berkeley, CA, pp 229–236.

Jensen, K 1991. A broad-coverage natural language analysis system. In: M Tomita, ed. Current Issues in Parsing Technology. Dordrecht: Kluwer, pp 261–276.

Knuth D 1965. On the translation of languages from left to right. Inform Control 8:607–639.

Kupiec J 1991. A trellis-based algorithm for estimating the parameters of a hidden stochastic context-free grammar. Workshop on Speech and Natural Language. Pacific Grove, CA, pp 241–246.

Magerman DM 1995. Statistical decision-tree models for parsing. 33rd Annual Meeting of the Association for Computational Linguistics, Boston, MA, pp 276–283.

Magerman DM, M Marcus 1991. Pearl: a probabilistic chart parser. 2nd International Workshop on Parsing Technologies, Cancún, Mexico, pp 193–199.

Magerman DM, C Weir 1992. Efficiency, robustness and accuracy in picky chart parsing. 30th Annual Meeting of the Association for Computational Linguistics, Newark, DE, pp 40–47.

Manning C, H Schütze 1999. Foundations of Statistical Natural Language Processing. Boston, MA: MIT Press.

Marcus M, B Santorini, M Marcinkiewicz 1993. Building a large annotated corpus for English: the Penn Treebank. Comput Linguist 19:313–330.

McEnery T, A Wilson A 1996. Corpus Linguistics. Edinburgh. Edinburgh University Press.

Miller S, R Bobrow, R Ingria, R Schwartz 1994. Hidden understanding models of natural language. 32nd Annual Meeting of the Association for Computational Linguistics, Las Cruces, NM, pp 25–32.

Ng S-K, M Tomita 1991. Probabilistic parsing for general context-free grammars. 2nd International Workshop on Parsing Technologies, Cancún, Mexico, pp 154–163.

Pereira F, Y Schabes 1992. Inside–outside re-estimation for partially bracketed corpora. 30th Annual Meeting of the Association for Computational Linguistics, Newark, DE, pp 128–135.

Rayner M, D Carter 1996. Fast parsing using pruning and grammar specialization. 34th Annual Meeting of the Association for Computational Linguistics, pp 223–230.

Resnik P 1992. Probabilistic tree-adjoining grammar as a framework for statistical natural language processing. Proceedings of the fifteenth [sic] International Conference on Computational Linguistics: COLING-'92, Nantes, France, pp 418–424.

Sampson G 1995. English for the Computer. Oxford: Oxford University Press.

Sampson G 1996. Evolutionary Language Understanding. London: Cassell.

Santorini B 1990. Penn Treebank Tagging and Parsing Manual. CIS Department, University of Pennsylvania, Philadelphia.

Schabes Y 1992. Stochastic lexicalized tree adjoining grammars. Proceedings of the fifteenth [sic] International Conference on Computational Linguistics: COLING-'92, Nantes, France, pp 426–432.

Schabes Y, A Abeillé, A Joshi 1988. Parsing strategies with "lexicalized" grammars: application to tree adjoining grammars. Proceedings of the 12th Conference on Computational Linguistics, Budapest, Hungary, pp 578–583.

Schabes Y, M Roth, R Osborne 19 . Parsing of the *Wall Street Journal* with the inside–outside algorithm. Sixth Conference of the European Charter of the Association for Computational Linguistics, Utrecht, Netherlands, pp 341–347.

Sekine S, R Grisham 1995. A corpus-based probabilistic grammar with only two non-terminals. Proceedings of the 4th International Workshop on Parsing Technologies, Prague, Czech Republic, pp 216–223.

Sharman R 1989. Probabilistic ID/LP grammars for English. Report 217, IBM UK Scientific Centre, Winchester, England.

Sharman R, F Jelinek, R Mercer 1990. Generating a grammar for statistical training. Workshop on Speech and Natural Language, Hidden Valley, PA, pp 267–274.

Su K-Y, J-N Wang, M-H Su, J-S Chang 1991. GLR parsing with scoring. In: M Tomita, ed. Generalized LR Parsing. Dordrecht: Kluwer, pp 93–112.

Tomita M 1987. An efficient augmented-context-free parsing algorithm. Comput Linguist 13:31–46.

Vilain M, D Palmer 1996. Transformation-based bracketing: fast algorithms and experimental results. Proceedings of the Workshop on Robust Parsing at ESSLI '96, Brighton, England, pp 93–102.

Viterbi A 1967. Error bounds for convolutional codes and an asymptotically optimum decoding algorithm. IEEE Trans Inform Theory IT-13:260–269.

Wright J 1990. LR parsing of probabilistic grammars with input uncertainty for speech recognition. Comput Speech Lang 4:297–323.

Wright J, E Wrigley 1989. Probabilistic LR parsing for speech recognition. 1st International Workshop on Parsing Technologies, Pittsburgh, PA, pp 193–202.

23

Authorship Identification and Computational Stylometry

TONY McENERY AND MICHAEL OAKES

Lancaster University, Lancaster, England

1. INTRODUCTION

The identification of authors has been an ongoing topic of research since at least the middle of the 19th century.[1] Often the impetus for this research is the identification of the authorship of disputed works. As we will see, however, identifying anonymous authors is also of interest.

Human curiosity aside, the identification of an author or set of authors for a document is often of importance. To give an example from natural language processing (NLP), it would be of benefit for systems that "train" themselves to particular users to be able to identify an author automatically, rather than to undertake some user identification dialogue. Away from NLP the need is more acute and instantly demonstrable:

- Scholars studying texts, say for instance those of Shakespeare, want to be able to attribute texts accurately.
- It may be possible that the application of techniques for the automated identification of authors to software authoring may help track down the writers of computer viruses (Sallis et al., 1996).

[1]The authors would like to acknowledge input received from Harald Baayen, Andrew Gray, Fiona Tweedie, and Andrew Wilson in the production of this chapter.

- In political writing, as we will see, the identification of anonymous authors has recurred as an issue over the past three centuries at least.[2] In the 1996 American Presidential campaign, author identification methods were used to identify the author of an anonymous work of "fiction" in the United States (*Primary Colors*), which angered Democratic Party members by its seeming bias against them. This last example is an interesting one, because it shows the potential efficacy of author identification techniques. By building up a large sample of e-mails from potential suspects and doing some linguistic detective work, the Democrats got their man.
- Determining whether a confession is real or not in court can literally be a matter of life and death. Authorship attribution techniques have been used, for instance in the UK (Morgan, 1991; Campbell, 1992), to examine forged confessions.

2. HOW DOES AUTHOR IDENTIFICATION WORK?

Although we have given an example of successful authorship identification, it is not possible to leap from this example to the conclusion that automated author identification is mature as an area of study, and that no work remains to be done. In the *Primary Colors* case, for example, a great deal of human intuition was used. Similarly, most computer-based work in forensic linguistics could best be described as machine-aided, rather than automatic (see, e.g., Butler et al., 1991). When it comes to automatic author identification, as we will see, it is the case that only now are we pushing beyond the use of lexical statistics, and much work remains to be done.

Before we proceed with a discussion of stylometry, to consider what potentially automatic author identification could take place, it is useful to consider how author identification has been achieved to date. Author identification techniques can be classified into one of four varieties[3]: namely, historical research, cipher-based decryption, applied or forensic linguistics, and various stylometric techniques.

We will deal with each of these in turn, although the bulk of the chapter will be devoted to the fourth technique, as it is there that the potential for credible automated author identification may be realised.

Authorship has often been ascribed in the past, and still is today, on the basis of **historical evidence**. For example, an argument against Christopher Marlowe having written any of *Edward III*, is that it was published in 1596, 3 years after his death. However, it has been pointed out that there are strong illusions to the Spanish Armada in the battle scenes in lines 1038–1316. As the Armada was defeated in

[2]We will only cast our net back to the disputed authorship of the Swift-Arbuthnot letters. Disputes concerning political writings stretch back much farther than this. For example, the authorship of Caesar's political tracts, the *Anticatones*, is disputed, with the second book thought to be the work of Paulus Hirtius. This shows the depth of interest in authorship that grips the human mind. The *Anticatones* no longer exist, yet their authorship is questioned.

[3]In producing these four categories, we are expressly excluding a fifth: physical evidence. Some attributions can be called into doubt by an analysis of handwriting and carbon dating of so called original manuscripts. The *Hitler Diaries* is a famous case that springs to mind here. As this variety of author identification is of minimal interest to Language Engineering, we are expressly excluding it from our categorization.

1588, it could well be, historically speaking, that the play was written at any point up to Marlowe's death in 1593, and published posthumously. Obviously, this form of author identification could be described euphemistically as "knowledge intensive," or more honestly, as beyond the realms of possibility given the state of the art in computational linguistics. The curious reader may, however, be interested in Leech and Falon (1992) which presents a study that did, on lexical grounds, detect differences in culture (between modern British English and American English). Although not concerned with authorship studies at all, such a technique may allow a rough approximation to some type of "automated carbon-dating" of language if we look at cultural change through time.

Although of dubious worth, there is a body of work in the examination of Renaissance literature and the Bible that claims to be able to identify authors on the basis of **ciphers** which, it is claimed, the authors systematically included in their works. The claim is that various authors encoded their name throughout a work using a special secret code. If one is able to decipher the code, one can tell who wrote, say *Coriolanus*, as the author's name will literally be written all over it. In Biblical pseudoscholarship, cryptography is invoked to provide evidence that God had a hand in the writing of the Bible.[4] The cryptographic debate in Renaissance literature is most often used to try to prove that Francis Bacon wrote Shakespeare's work. It was largely debunked as a theory by Friedman and Friedman (1957), but resurfaced in Penn (1987), backed up by computer analysis, which claimed to have found Bacon's name written all over everything Shakespeare supposedly wrote.

The work has been quite roundly trounced by a superb essay on the World Wide Web by Terry Ross,[5] in which, by applying the code-breaking rules laid down by Penn, Ross tested the theory on a range of texts clearly not written by Bacon (Spenser's *Faerie Queen*, the Bible, Caesar's *Gallic Wars*, *Hiawatha*, *Moby Dick*, and *The Federalist Papers* [see later among others]). Needless to say, it appeared Bacon had written them all,[6] as his name appeared, encoded, throughout the texts. Indeed, they appeared even more frequently at times than in the works of Shakespeare, in which he was supposed to have left a deliberate message. Profound doubt must hang over the use of this technique in author identification.

Yet another form of author identification comes from the application of wide-ranging manual linguistic analyses of text, applying the techniques of **forensic linguistics** (Svartvik, 1968; Coulthard, 1993). This has a close affinity with stylometry, which is the main focus of this chapter, but we will set it apart here as the focus of a forensic linguistic analysis of a text need not be quantitative. Forensic linguistics also employs such a wide range of linguistic evidence as to make the adoption of forensic linguistic techniques wholesale in computational linguistics infeasible given the current state of NLP. Although it is possible to imagine a machine undertaking lexical and rudimentary syntactic analysis online, it is simply not yet possible for a machine to undertake, say, an automated conversation analysis or detailed syntactic analysis

[4]See `http://www.av1611.org/genesis.html` for a bewildering account of this research from "Dial-the-truth" ministries.

[5]See `http://www.bcpl.lib.md.us:80/~tross/ws/will.html#5b`.

[6]If you want to try to find Bacon in your works, get the Perl and UNIX scripts that carry out the decryption from `http://www.bcpl.lib.md.us:80/~tross/ws/bacipher.html`.

of a text with sufficient accuracy to make the results of forensic linguistics available immediately.

3. STYLOMETRY

Stylometry is the fourth type of authorship attribution technique we have identified, and as noted, forms the focus of this chapter because it provides us with the best chance, given the current state of the art of NLP, of developing online authorship identification systems.

Stylometry is an attempt to capture the essence of the **style** of a particular author by reference to a variety of quantitative criteria, usually lexical in their nature, called **discriminators**. Studies undertaken since the early 1960s have been surprisingly successful in distinguishing texts by different authors and adding weight to scholarly attributions of text authorship. So we may assume that the stylometric approach is sufficiently successful to be worth close examination. Before we proceed, however, it is useful to note three things. First, authorship attribution is only one of the aims of stylometry. Stylometry is also used to study changes in an author's style over time to establish a chronology of authorship. This is a point that will be examined further in Sec. 4. Second, stylometry has been widely and continually criticised, for example by Sams (1995). This is not to say that we should not consider stylometric results; rather we should do so aware of their shortcomings. Finally, stylometrists are at pains to point out that their techniques are not foolproof, for reasons we will examine shortly.

As will be seen in the following sections, the series of features used as discriminators in stylometric studies are quite diverse. With the notable exceptions of Kjell (1994) and Baayen et al. (1996), however, the discriminators are overwhelmingly lexical in nature. Indeed Holmes (1994) even goes so far as to say "to date no stylometrist has managed to establish a methodology which is better able to capture the style of a text than that based on lexical items" (p. 87). As we will see later, there is now reason to doubt what Holmes says, but let us for the moment assume that we have a set of lexical criteria that we believe distinguish a certain author. What measures would we want to use on those features to determine whether a given document was written by that author? Since we would want our measure to be reliable, it would be preferable if the technique worked for a range of authors and we should also want a measure that was theoretically well founded (i.e., had known properties and features).

Stylometry has not always provided this. In this survey, we will restrict ourselves to an examination of techniques that seem to meet these three criteria.

Stylometry is really the area in which computationally tractable approaches to author identification have been developed. Note that we are not saying that **software systems** have been developed. In nearly all of the cases we study in this chapter, the production of a piece of author identification software has not been the goal of the study. Most of these studies have developed analyses of texts by hand, or, by using simple feature frequency-counting programs, have extracted the relevant data and processed that data either using statistical software packages or by manual statistical analysis. A little lateral thinking is needed here. We are looking at work that seems primarily interested in the examination of a statistical technique (e.g., Mosteller and Wallace, 1964) or the attribution of authorship to specific texts (e.g., Merriam, 1993,

1996). In the latter, the scholars concerned have generally been interested only in distinguishing texts that are of concern to their main scholarly pursuit (e.g., Shakespearean studies). The application of their work to anonymous Internet authors will most likely have been of no concern to them. It is hardly surprising, therefore, that reusable authorship identification software packages have not been forthcoming. However, when the survey of existing work is complete, it will be apparent to any reader that the production of a suite of programs for authorship identification is quite possible on the basis of work in stylometry undertaken to date.

4. PROBLEMS WITH THE ATTRIBUTION OF AUTHORSHIP STYLE

In the previous section, several problems with stylometry were alluded to. Before we actually examine stylometric techniques, these problems must be examined more fully.

A. Homogeneity of Authorship

A clear underlying assumption in stylometry is that although authors may consciously influence their own style, there will always be a subconscious exercise of a consistent style throughout their work (spelled out by Robinson, 1992:381). It must be pointed out that there is no clear and indisputable evidence of the existence of such features. Indeed, to be credible, this assumption has to be instantly reduced in scope. It is clear that this is true of some authors more than others, and that it is more true for the same author at some times than others. As a case in point, let us consider the disputed works of Goldsmith, as reported by Dixon and Mannion (1993). The 18th century Anglo–Irish author Oliver Goldsmith wrote a series of anonymous essays for magazines, such as *The Bee* and *The Busy Body*. Over a hundred essays have been attributed to him since his death. Yet part of the problem in determining which are actually his comes from the versatility of his writing. Although stylometrists may find it quite possible in most cases to distinguish authors of modern fiction (Baayen et al., 1996) the Augustan prose of Goldsmith is another matter. Goldsmith changes the manner of his writing when attributing speech, his style for narrative and reported speech being quite different, with the style for reported speech changing to reflect the supposed speaker.[7] The very versatility of a writer is a problem. Also, there is strong evidence to identify Goldsmith as part of a "school" of writing: in the Dixon and Mannion study, some of his works were indistinguishable from those of two other Anglo–Irish writers. Such identification with a school of writing may, it appears, occasionally mask any personal tendencies. So whereas homogeneity of authorship seems to hold sometimes, it does not hold all of the time.[8]

---

[7]Do not think that this is a peculiarity of Georgian writers alone; Kenny (1986:100) also finds attributions that may in fact be "the work of a single, extremely versatile author."

[8]See Clayman (1992:387) for a further discussion of the influence of intertextuality on authorship attribution.

B. Heterogeneity of Authorship Over Time

Paradoxically, stylometrists have also used quantitative features to study the changes of style of a single author over time. Why paradoxically? Well, as Clayman (1992:387) points out, if you accept that authorship attribution can proceed on the basis of unaltering subconscious behaviour, you cannot simultaneously accept that this behaviour changes over time. As this chapter is concerned with authorship attribution and not chronology of authorship, we can comfortably bypass this point. It could reemerge as a point of significance, however, if authorship systems do become implemented as online systems. What we may see is that an author, over time, becomes different, and identification of that author fails as their subconscious style shifts.

C. Authorship and Genre

The final point is an important one for which there is firm experimental evidence. Differences in **genre** seem to be more pronounced than differences in author. The type of work that Biber (1995) has been doing is a little like stylometry, except that he has been using a range of quantitative features to distinguish genres of text. It seems that in doing so he has been following stronger distinctions than between authoring styles, because where genre changes, differences in authorship often become suppressed, because the shift in style between genres is so pronounced that the subtle differences in authorship become blocked out by the gross differences in genre style. Baayen et al. (1996) found in an experimental examination of authorship styles and genre that "differences in register can be much stronger than differences within a register between texts of different authors" (p. 122). So, much as variations of style within a text can confound authorship identification, so too can comparison of authorship across genre boundaries.

5. BASIC METHODOLOGY

The basic methodology of stylometry is to discover a range of features that successfully discriminate among authors. These discriminants often vary between pairs of authors: discriminants that can distinguish Marlowe and Shakespeare will not necessarily distinguish Goldsmith and Johnson. However, there is at least one technique that is author-independent. Kjell (1994) used **frequencies of letter pairs** within texts to discriminate between authors (he worked with *The Federalist Papers*). However, his results were poorer than other studies on the same texts with discriminators developed in a more text-specific, labour-intensive way.

More typically, discriminants, as noted, are **lexical** in their nature The number of discriminants in any given study varies. In the following, however, is a list of typical discriminants that may be used either as a whole, or in some subset:

A. Word or Sentence Length

Early attempts at stylometry have used measures based on average word length, or the distribution of different word lengths, but this measure seems to be more an indicator of genre than individual authorship and has been discredited. The average sentence length has been used as a discriminant by some authors. It has been effective for certain Greek and English authors (Barr, 1994), but is by no means a universal discriminant, again being too easily under the conscious control of the author. On the other hand, the **distribution** of syllables and their **relative frequencies** (i.e., the lengths of the gaps between words of the same syllabicity) has been seen to be a fairly promising avenue by Brainerd (1974), especially when subjected to statistical analysis.

B. Vocabulary Richness

Various measures to test vocabulary richness are used (i.e., the diversity or repetitiveness of the words an author uses). By far the most familiar is the **type/token ratio** (i.e., the ratio of the number of different words [types] to the total number of words [tokens] in a text collection). For example, if in a 100,000-word text collection, author A is found to use 10,000 different words. The type/token ratio = 10,000:100,000 = 0.10. This ratio has the disadvantage of being unstable relative to variation in sample size: whereas the number tokens may increase without limit, the possible maximum number of types is more or less finite (i.e., the total vocabulary of the language). Valid type/token ratio comparisons, therefore, must be made with comparable sample sizes.

To compensate for this, various measurers have been proposed that are independent of sample size. **Simpson's Index** (Simpson, 1949) measures the chance that two words arbitrarily chosen from a text sample will be the same [Eq. {1}],

$$D = \frac{\sum_{i} r(r-1)V_r}{N(N-1)} \qquad \{1\}$$

where V_r is the number of types that occur r times ($r = 1, 2, 3, ..., i$), and N is the number of tokens in the sample. Another measure devised by Yule (1944), called Yule's Characteristic or **Yule's *K***, is based on the idea that the occurrence of a given word is a chance occurrence that can be modelled as a Poisson distribution [Eq. {2}].

$$K = 10^4 \frac{\sum_{r} r^2 V_r - N}{N^2} \qquad \{2\}$$

The familiar notion of **entropy** can also be seen as a measure of vocabulary richness, that is the degree of internal consistency of a text: the lower the value, the more uniform the sample. The formula for entropy (H) involves the probability p_i of each type i, which is easily calculated (by dividing V_i by N), and is normalised by taking the value of N into account [Eq. {3}].

$$H = -100 \frac{\sum_{i} p_i \log p_i}{\log N} \qquad \{3\}$$

A simple measure of vocabulary richness, proposed by Brunet (1978), is shown in Eq. {4},

$$W = N^{v-\alpha} \qquad \{4\}$$

where α is a constant in the range 0.165–0.172.

Finally, let us mention Honoré's R, which takes account of the particular class of words that occur just once each in the sample (V_1), so-called *hapax legomena*, and the ratio of these to the total number of different types, V, as in Eq. {5}.

$$R = \frac{100 \log N}{1 - (V_1/V)} \qquad \{5\}$$

C. Word Ratios

Honoré's R is one example of a measure based on word ratios, in this case the ratio of V_1s to V. Other studies have looked at the relative distribution of function words, or have compared specific high-frequency words (which are not necessarily closed-class vocabulary items, e.g., a verb such as *said* may be highly frequent in a text with a lot of reported speech) with a standard list of frequent words such as Taylor's (1987) list of ten function words.[9] Mosteller and Wallace's classic study of *The Federalist Papers* was based on the relative frequency of "marker words": synonym pairs such as *on* and *upon*, *while* and *whilst*, and so on, which were distinctive for the two candidate authors. By the same token, a text sample may be distinctive in its lack of use of words which are common elsewhere with the author in question.

D. Letter Based

Letter-pair frequencies were mentioned in the foregoing. Other measures focussing on letters of the alphabet have been used; for example, in authorship attribution in Greek texts. The Greek alphabet contains several rarely used characters. The frequency of the occurrence of these characters may be a useful discriminant (Ledger, 1995:86). Letter discrimination may also specify a position in a word for the letter to occur, especially at the beginning or end of a word. Some letters may be grouped in an obvious way (e.g., vowels vs. consonants) so that "proportion of words beginning with a vowel" could be a useful measure.

Letter-based measures are of interest, in particular, because they focus on an aspect of language which is much less under the control of the author and, hence, it is less susceptible to fluctuation due to genre. Having said that, it must also be said that there is no obvious reason why such measures should turn out to be distinctive on an individual basis. Studies making this assumption have had mixed results, and it is more usual to use measures such as these in combination, for example, in multivariate methodologies (see later discussion).

E. Part of Speech

Just as the incidence of particular individual words can provide a measure, so too can relative measures of particular parts of speech. However, these measures are

[9]These ten important function words are *but*, *by*, *for*, *no*, *not*, *so*, *that*, *the*, *to*, and *with*.

rather more susceptible to genre differences than author differences [e.g., Antosch (1969) suggested that the adjective/verb ratio was a good discriminator of genre, folk tales having a high ratio, scientific writing a low ratio, for obvious reasons]. The repetition of certain function words and sequences of function words in variable patterns is a commonly used discriminant.[10]

6. EXPERIMENTAL METHODOLOGY

The experimental methodology used to test authorship attribution algorithms is quite straightforward. Let us assume that we have some texts the authorship of which is in question, with two possible candidates, authors A and B. Initially, a corpus of works where there is certain attribution for A and B is collected, as well as a third corpus of texts the authorship of which is uncertain, but thought to be either A or B. Care is taken to ensure that the works collected are of the same genre, for reasons outlined earlier in this chapter. In systems that need human selection of discriminants for the author pair (which is the overwhelming majority of cases) these corpora are then used as the basis of discriminant selection. The efficacy of the discriminants and associated measure can then be tested on the corpus of texts with certain attribution: for any measure and set of discriminants to be credible, it must be able to distinguish texts for which the author is clearly known. This stage of the process may be cyclic, with discriminants, and even measures, changing until the system operates satisfactorily for this particular author pair. The adaptations undertaken may be **manual**, with human intervention to change the rule set, or may be the product of some **machine learning** algorithm (see Sec. 7.E and 7.F). When the technique works on these texts, it can then be run over texts where the authorship is uncertain between the two authors. At this stage, the hope is that the system's ability to discriminate between texts that have certain authorship translates readily into being able to discriminate between texts with disputed authorship.

New measures, as will be seen, are often first tested on well-known difficult cases. With their efficacy proved on a difficult case, it is then assumed that their efficacy on an easier case is strengthened all the more. One such "difficult case," that has been mentioned already and will be seen over and over again in the next section, is that of *The Federalist Papers*. It is, therefore, worth considering *The Federalist Papers* in more detail, as their use as a test-bed for authorship studies is widespread, and readers not familiar with the brief but lively history of American political thought may be unaware of them. The so-called Federalist Papers were a series of anonymous essays written, under the pen name Publius, for publication in New York newspapers in the late 18th century. The aim of these essays was to persuade New Yorkers to vote for the ratification of the then new U.S. Constitution.[11] It later

[10]See previous note. Function words have proved time and again to be a fruitful source of discriminants. However, see Sec. 7.G for an explanation of why their importance may in part be illusory.

[11]They failed in their immediate goal. However, history was on their side, and the United States became a federal state. The Anti-Federalist papers, a series of anonymous articles that persuaded New Yorkers to vote *against* the constitution, receive very little attention today, and we are not aware of any attempts at attribution of authorship to these papers. There is a moral here. If you would like to look at *The Federalist Papers* they are available on the web at `http://lcweb2.loc.gov:80/const/fedquery.html`. To view the Anti-Federalist Papers try `http://colossus.net:80/wepinsto/rot/afp/`.

emerged that the papers had been written by Alexander Hamilton, John Jay, and James Madison. Authorship problems arise because both Hamilton and Madison claimed 12 of the papers as their own. The papers are viewed as a difficult case for authorship attribution, as it appears that the styles of Hamilton and Madison are surprisingly similar. Furthermore, as the papers have been used in so many studies now, they provide a useful litmus test: the discriminants that are effective are quite well known, so it is possible to try out new techniques and compare results. The accuracy figures produced by these experiments can then be compared informally to establish some idea of how effective a range of measures are.

7. MEASURES

In the following sections we will consider a series of statistical measures used upon discriminants in an attempt to attribute authorship. In each section specific methodologies will be covered only in so far as they are novel. Where a statistical technique that is well known is involved, bibliographic references but no detailed description will be given. In listing discriminants, only a general indication of lexical discriminants (e.g., function words, content words) will be given, not, as a rule, an exhaustive list of each lexical discriminant.

A. Bayesian Probability

Study	Text	Measures	Discriminants
Mosteller and Wallace (1964, 1984)	*The Federalist Papers*	Classic Bayesian inference and non–Bayesian linear discriminant function	30 "marker words," the best of which were *upon, whilst, there, on, while, vigor, by, consequently, would, voice*

This is the classic text in modern stylometry (for earlier work see Yule, 1944). Mosteller and Wallace took the notorious Federalist Papers and determined a very credible attribution of authorship on the basis of a range of discriminants and a series of studies. They used Bayesian analysis in doing so, and a brief glance at their articles shows that they were at least as interested in Bayes' theorem as in authorship attribution, if not more.

One may reasonably ask why there are no other, more recent, studies listed here if this was such a trail-blazing piece of research. Although it was indeed trail-blazing, Mosteller and Wallace also, within the same book, demonstrate that results that are every bit as effective can be acquired by shunning Bayes' theorem and, instead, dealing with a non-Bayesian linear discriminant function. The difficulty of wading through their main Bayesian analysis would be enough to set even the most mathematically minded a serious challenge. As similar results are available through a more familiar route, it is hardly surprising that Bayesian inferencing has been quietly dropped from the repertoire of stylometry.

Although we as researchers can neither condemn nor condone the abandonment of Bayesian reasoning in stylometry, surely it at least remains to be said that a thorough, ongoing comparison of the efficacy of a Bayesian approach with other contemporary approaches in stylometry is needed.

B. Univariate Analyses

Study	Text	Measures	Discriminants
Merriam (1993)	Shakespeare's plays, Marlowe's *Tamburlaine* plus the disputed *Edward III*	z-Scores	Taylor's list of ten function words, proportional pairs

Univariate analyses are worthy of mention only from a historical viewpoint. Multivariate analyses (as will be shown later) are now in the ascendant over univariate analyses. However, the Messiam study is listed here as it is of particular interest, having been replicated, as a multivariate study, in 1996 producing the same results (see later). So although the lack of sophistication of univariate statistics may make them unappealing, they can still mirror the results of multivariate statistics, at least in some cases.

C. Cumulative Sum Charts

Study	Text	Measures	Discriminants
Hilton and Holmes (1993)	A range of authors of fiction, and *The Federalist Papers*	*t*-Test	Length of sentence and occurrence of various linguistic "habits"

The cumulative sum chart test is worthy of mention, only so that it may be dismissed. Developed over a series of papers from the early 1970s (e.g., Bee, 1971; Morton, 1978; Farringdon, 1996), cumulative sum ("cusum") charts would fall outside of our remit if it had not been for the work of Hilton and Holmes (1993). Cusum charts are based on the usual stylometric assumption of authorial "fingerprints" at some level of style. Cusums determine the value of a specified feature throughout a text (say, the number of two-letter words) typically on a per sentence basis. These values are then mapped to show the real values plotted against a curve showing the mean value as it develops throughout the text. So, if we had the values 2, 3, and 5 for some discriminant, these would be compared against a developing mean of 2, 2.5, and 3.3. The result is a graph for a given feature showing the deviation of individual values from the mean of that value to that point, which is supposed to be distinctive for an individual author. In the work of Morton and Michaelson, interpretation of the graphs is purely impressionistic, and prone to distorted interpretation (e.g., based on the scale used in the plotting graph: see Hilton and Holmes, 1993:75). However, Hilton and Holmes (1993) have developed a version of the test, "weighted cusums" that has a better statistical foundation, giving results equivalent to a *t*-test (see also Bissell, 1995). Although the test is attractive, requiring only small portions of text, it remains of marginal interest to us, as its reliability has been quite poor (Hilton and Holmes, 1995:80).

D. Multivariate Analyses

Study	Text	Measures	Discriminants
Holmes (1991, 1992)	Mormon scriptures	Cluster analysis	Vocabulary richness (five measures)
Burrows (19920	Various	Principal components	Common high-frequency words
Dixon & Mannion (1993)	Anonymous essays of Oliver Goldsmith	Correspondence, principal components, and cluster analysis	22 discriminants, including sentence length; mainly function worlds
Ledger (1995)	Letters of St. Paul	Cluster analysis	Specific letter frequencies, relative frequency of words with a specific final letter, type/token ratio
Holmes & Forsyth (1995)	*The Federalist Papers*	Cluster analysis	Vocabulary richness (6 measures)
Mealand (1995)	Gospel of Luke	Correspondence analysis	Nine parts of speech, function words, word length in letters, relative frequency of certain letters in word, initial position
Greenwood (1995)	Gospel of Luke and *Acts*	Nonlinear mapping and cluster analysis	Common word frequencies
Merriam (1996)	Shakespeare's plays, Marlowe's *Tamburlaine*, plus the disputed *Edward III*	Principal components analysis	Function word combinations, many in combination with Taylor's list

The techniques of multivariate analysis are well known, and do not need repeating here. The interested reader should consult a good general text, such as Oakes (1997), Krzanowski (1988), or Mardia et al. (1979), for a description of the mathematics behind multivariate analysis. We will restrict ourselves here to a consideration of what is so attractive in the use of multivariate analysis, and of why each of the foregoing principal techniques has appeal for studies of authorship attribution.

1. Correspondence Analysis

This is a technique that looks at a two-way contingency table (an $n \times n$ matrix of differences in n variables, which have a multivariate normal distribution,[12] measured between each author, A and B), with similarity and dissimilarity being measured by the χ^2 formula. Popularised by the works of Greenacre (1984, 1993), the technique has a fairly wide application within authorship studies, as can be seen in the fore-

[12]See Krzanowski (1988:214) for a further discussion of this point.

going list of studies. It has the advantage over principal component and factor analysis of displaying data in row points, but also in column points. The resulting charts are easily amenable to visual interpretation, as similar documents will cluster within the same space in the chart. If all of the documents in one cluster are known works of author A plus two unattributed documents, the conclusion that can be drawn is obvious. The visual nature of the results obviates the need, in most cases, for human analysts to consult numeric details related to each point. This contrasts sharply with factor analysis, for which factor loadings must be looked up to interpret the dimensions represented in the data. Note that this advantage is strictly marginal if the aim of a system designer is to develop automated author identification. Indeed, from the point of view of computational processing of authorship attribution tests, the visual aspect of correspondence analysis is no benefit at all.

2. *Cluster Analysis*

The idea behind clustering is that, given a series of measures related to a series of documents, it should be possible to estimate which objects are more similar to each other on the basis of the values of the measurements observed. To take our authorship example, if we have a series of discriminants, these are our measures. On the basis of the values of our measures, we may find that two documents for which the authorship is unknown look like—cluster toward—the work of one author, rather than those of another. Clustering may go through a variety of cycles, with each cycle reducing the number of clusters in the model by progressively reclustering the clusters.

A wide range of clustering techniques is available, for example, **average linkage** (where items in the same cluster have the values of their measurements averaged out before any further clustering takes place). A general disadvantage of clustering as opposed to correspondence analysis is that the latter is generally held to be more reliable (Gordon, 1981:136).

3. *Principal Components Analysis*

The idea behind principal components analysis is attractive for authorship attribution. It seeks measures that vary a great deal between samples in preference to measures that remain relatively stable. The idea clearly links to authorship studies: we want to identify the measures that show us the differences not the similarities between authors. Unlike correspondence analysis, the discriminants in a principal components analysis need not possess a multivariate normal distribution. There is some evidence (Baayen et al., 1996:129) that a use of standardised scores (analysing the correlation structure) produces better results in author identification than an analysis of the covariance structure.

E. Neural Networks

Of the foregoing studies, the results achieved by Kjell are of particular interest. His study is, in some ways, the most linguistically naïve of all the studies examined in this chapter. Yet although his results do not match up to the work undertaken with lexical discriminators and beyond, the results are fair. His work also has the great advantage that there is no need to select manually a set of discriminators on a "by author pair" basis. Given sufficient computing power, it is possible to calculate on the basis of |*alphabet*| × |*alphabet*| sized sets of discriminants.[13] As such, Kjell's approach, of all of those listed, is clearly the one most amenable to rapid implementation as an online system.

Study	Text	Measures	Discriminants
Matthews and Merriam (1993, 1994)	Works of Shakespeare, Fletcher, and Marlowe	Back propagation algorithm running on a multilayer perceptron	Five ratios of various word pairs
Kjell (1994)	*The Federalist Papers*	Back propagation algorithm running on a multilayer perceptron	Letter pair frequencies

F. Genetic Algorithms

Study	Text	Measure	Discriminants
Holmes and Forsyth (1995)	*The Federalist Papers*	Evolutionary training algorithm	30 marker words

Holmes and Forsyth used a Darwinian training algorithm to develop a set of discriminating rules using a system called BEAGLE. The system started with a random set of rules, then trained on a set of *The Federalist Papers* for which the authorship was known. The training algorithm allowed successful discriminants to survive, whereas those that were unsuccessful died out. The initial rules were initially based upon the 49 "marker" words (high-frequency function words) and accurately attributed 10 of the disputed papers to Madison. When the test was rerun with a smaller group of 30 marker words (those developed by Mosteller and Wallace, 1984), results improved. In improving, the genetic algorithm whittled the 30 marker words down to 8, which survived in its rule base. This set of rules successfully attributed all 12 of the unknown Federalist papers to Madison.

Holmes and Forsyth (1995:125) list a variety of supposed benefits and disadvantages for their system. From the point of view of online automated authorship attribution systems, however, the principal benefit of this approach must be its ability to pare away rules. It will run with the most parsimonious rule set available if trained properly. This is a useful feature, for redundant rules, which may possibly even degrade performance, are discarded by the system.

[13]Although it should be noted that Kjell did have limited computing power, and followed the traditional experimental route of selecting a set of discriminants from the full potential set of 26 × 26 discriminants. Kjell does outline a means of identifying the "significant" letter pairs automatically, however, so one need not be profligate with computing power.

G. Corpus Annotation and Author Identification

Study	Text	Measure	Discriminants
Baayen et al. (1996)	Modern crime fiction	Principal components analysis	Syntactic rewrite rules, vocabulary richness, and frequent words

This study has been listed separately from the others, not because the measure it used is different—the study employed a variety of multivariate analysis—but rather, because this is the first study of authorship to date to make use of a syntactically annotated corpus.[14] The use of syntactically annotated corpus data was of importance: for the first time authorship studies could reliably use features related to constituent structure within sentences. The work of two authors within the genre of crime writing was used. The authors' work constituted part of the Nijmegen corpus, a corpus developed at the University of Nijmegen by a team led by Jan Aarts. The authorship of the text was known to van Halteren, but he passed on the texts anonymised to Baayen and Tweedie, who had to develop techniques to distinguish them. For their study, Baayen and Tweedie used rewrite rules (similar to phrase-structure rules) as a discriminant. They looked at two potential varieties of syntactic discriminants: the 50 most frequent rewrite rules, and the 50 most infrequent rewrite rules. In parallel, they evaluated these new discriminants against more traditional discriminants; namely, vocabulary richness and frequent words.

There were a series of advantages to using syntactic discriminants. First, the syntactic discriminants seemed less prone to being "fooled" by the variations of author style, which we have discussed previously. Second, the analyses that relied on syntactic discriminants were the most successful. Third, of the syntactic criteria, the one that looked at the lowest frequency rewrite rules is the most robust in terms of identifying authors across genre boundaries. All in all, it would appear that it could well be that this aspect of authorial style is less prone to conscious variation than the lexical level: it could be that at this level of analysis lies the key to authorship identification.

In some ways the findings, especially the third one, are not at all surprising. Although, as we have noted, Holmes (1994) has asserted the primacy of lexical discriminants, others have not been so sure. Ledger (1989:1) seems to strike the right note when he says of stylometry:

> If one were restricted to mention a single weak point which undermines so many enquiries in the field, it must surely be that at the level of word frequency so much variation is possible that reliable predictions cannot be made. (p. 1)

Holmes is both right and wrong when he asserts the primacy of lexical determinants. Although the definition of authorial style may include lexical discriminants, it must also include a wide range of other levels of linguistic description (see Leech and

[14]It is not the first to make use of an annotated corpus, however. A morphosyntactically annotated version of the Greek New Testament was used in the Mealand (1995) study. This had been produced manually by Friberg and Friberg (1981) using an interactive concordance package.

Short, 1981. So in a sense, Holmes is quite wrong, as if we move beyond stylometry to stylistics and forensic linguistics it is plain to see that discriminants other than lexical ones must be of supreme importance in the attribution of authorship. However, as already noted, the feasibility of wide-scale annotation of various levels of linguistic description is limited. Although a wide range of annotation schemes[15] has been developed, which could conceivably be applied to stylometry, the work involved in introducing linguistic annotations manually to a text (which is how most types of annotations must now be made) is such that the dominance of lexical approaches is assured until NLP can broaden this bottleneck in authorship attribution studies.

So in fairness to Holmes, when we consider the feasibility of using anything but lexical and part-of-speech–based discriminants, Holmes, to this point in time, is quite right in his assessment of lexical discriminants.

The discussion of the importance of nonlexical discriminants allows us to return to a point we raised earlier: function words as discriminants may have a somewhat illusory purpose. Could it be, as Baayen et al. speculate (1996:121), that function-word counts are nothing but crude approximations to syntactic discriminants? This certainly sounds plausible, and explains why function words have been so effective as discriminants. The true discriminant appears to be syntactic, and because of the close relation between function words and syntax, function words appear as a somewhat imperfect mirror image of one of the features we really want to look at, the syntax of the author. Until, however, reliable parsing algorithms become available, the power of this technique can only be tested on manually or semimanually annotated treebanks.[16]

8. CONCLUSION

This chapter has considered a very wide range of authorship attribution studies. However, for readers interested in implementing a system that allows "quick-and-dirty" authorship identification, the work of Kjell seems to be the likeliest route. However, this chapter has also outlined several lines of experimentation that could be followed up, for example a reevaluation of Boolean measures, or the harnessing of a constraint grammar tagger to exploit syntactic discriminants in an online system.

It is doubtless that work in authorship attribution will continue, and that new measures and discriminants will be applied as time goes on. However, it is also certain that as NLP products, such as parsers, become more robust and reliable, their influence on authorship attribution studies may be such that a synthesis of work from the two fields may lead to effective online authorship identification systems in the future.

[15]See Garside et al. (1997) for a comprehensive list of corpus annotation schemes.

[16]Although another unexploited possibility would be to use the constraint tagger approach (Karlson et al., 1995). This would allow a reliable partial, automated syntactic analysis on line.

REFERENCES

Antosch F (1969). The diagnosis of literary style with the verb–adjective ratio. In: L Dolezel, RW Bailey, eds. Statistics and Style. New York: American Elsevier.

Baayen H, H van Halteren, F Tweedie (1996). Outside the cave of shadows: using syntactic annotation to enhance authorship attribution, Literary Linguist Comput 11:121–132.

Barr GK (1994). Scale in literature with reference to the *New Testament* and other texts in English and Greek. PhD dissertation, University of Edinburgh.

Bee RE (1971). Statistical methods in the study of the masoretic text of the *Old Testament*. J R Statist Soc A 134:611–622.

Biber D (1995). Dimensions of Register Variation. Cambridge: Cambridge University Press.

Bissell AF (1995). Weighted cumulative sums for text analysis using word counts. J R Stat Soc A 158:525–545.

Brainerd B (1974). Weighing Evidence in Language and Literature: A Statistical Approach. Toronto: University of Toronto Press.

Brunet E (1978). Vocabulaire de Jean Girandoux: Structure et Évolution. Paris: Slatkine.

Burrows JF (1992). Not unless you ask nicely: the interpretative nexus between analysis and information. Literary Linguist Comput 7:91–109.

Butler J, D Glasgow, AM McEnery 1991. Child testimony: the potential of forensic linguistics and computational analysis for assessing the credibility of evidence. J Fam Law 21:65–74.

Campbell D (1992). Writings on the wall. The Guardian, Wed. October 7, 1992, p. 25.

Clayman DL (1992). Trends and issues in quantitative stylistics. Trans Am Philol Assoc 122:385–390.

Coulthard RM (1983). Beginning the study of forensic texts: corpus, concordance, collocation. In: MP Hoey, ed. Data, Description, Discourse. London: Harper Collins, pp 86–97.

Dixon P, D Mannion (1993). Goldsmith's periodical essays—a statistical analysis of eleven doubtful cases, Literary Linguist Comput 8:1–19.

Farringdon JM (1996). Analysing for Authorship: A Guide to the Cusum Technique. Cardiff: The University of Wales Press.

Friburg B, T Friburg (1981). Morphologically Tagged Electronic Text of the Greek *New Testament*. CATTS.

Friedman WF, ES Friedman (1957). The Shakespeare Ciphers Examined. London: Cambridge University Press.

Garside R, G Leech, T McEnery (1997). Corpus Annotation. London: Addison–Wesley–Longman.

Gordon AD (1981). Classification: Methods for the Exploratory Analysis of Multivariate Data. London: Chapman & Hall.

Greenacre MJ (1984). Theory and Application of Correspondence Analysis. London: Academic Press.

Greenacre MJ (1993). Correspondence Analysis in Practice. London: Academic Press.

Greenwood HH (1995). Common word frequencies and authorship in Luke's Gospel and Acts. Literary Linguist Comput 10:183–188.

Hilton ML, DI Holmes (1993). An assessment of cumulative sum charts for authorship attribution. Literary Linguist Comput 8:73–80.

Holmes DI (1991). Vocabulary Richness and the Prophetic Voice. Literary Linguist Comput 6:259–268.

Holmes DI (1992). A stylometric analysis of Mormon scripture and related texts. J R Statist Soc A 155:91–120.

Holmes DI (1994). Authorship attribution. Comput Humanities 28:87–106.

Karlsson F, A Voutilainen, J Heikkala, A Anttila, eds. (1995). Constraint Grammar: A Language Independent System for Parsing Unrestricted Text. Berlin: Mouton de Gruyter.

Kenny AJ (1986). A Stylometric Study of the *New Testament*. Oxford: Oxford University Press.

Kjell B (1994). Authorship determination using letter pair frequency features with neural network classifiers. Literary Linguist Comput 9:119–124.

Krzanowski WJ (1988). Principles of Multivariate Analysis. Oxford: Clarendon Press.

Leary TP (1990). The Second Cryptographic Shakespeare. Omaha, NE: Westchester House Publishers.

Ledger GR (1989). Re-counting Plato—A Computer Analysis of Plato's Style. Oxford: Clarendon Press.

Ledger GR (1995). An exploration of differences in the Pauline epistles using multivariate statistical analysis. Literary Linguist Comput 10:85–98.

Leech G, R Fallon (1992). Computer corpora: what do they tell us about culture? ICAME J 16:1–22.

Leech GN, MH Short (1981). Language and Fiction: A Linguistic Introduction to English Fictional Prose. London: Longman.

Mardia KV, JT Kent, JM Bibby (1979). Multivariate Analysis. London: Academic Press.

Matthews R, T Merriam (1993). Neural computation in stylometry I: An application to the works of Shakespeare and Fletcher. Literary Linguist Comput 8:203–210.

Mealand DL (1995). Correspondence analysis of Luke. Literary Linguist Comput 10:171–182.

Merriam TVN (1993). Marlowe's hand in *Edward III*. Literary Linguist Comput 8:59–72.

Merriam TVN (1996). Marlowe's hand in *Edward III* revisited. Literary Linguist Comput 11:19–22.

Merriam TVN, RAJ Matthews (1994). Neural computation in stylometry II: an application to the works of Shakespeare and Marlowe. Literary Linguist Comput 9:1–6.

Morgan B (1991). Authorship test used to detect faked evidence. Times Higher Educational Supplement, 9th August, 1991.

Morton AQ (1978). Literary Detection. East Grinstead: Bowker Publishing.

Mosteller F, DL Wallace (1984). Applied Bayesian and Classical Inference: The Case of the Federalist Papers. Reading, MA: Addison–Wesley.

Oakes MP (1998). Statistics for Corpus Linguistics. Edinburgh: Edinburgh University Press.

Robinson J (1992). Ora avni, the resistance of reference: linguistics, philosophy and the literary text. J Aesthet Art Crit 50:258.

Sallis PJ, SG McDonnell, A Aakjaer (1996). Software forensics: old methods for a new science. Proceedings of Software Engineering: Education and Practice (SE:E&P 96), Dunedin, NZ, pp 481–485.

Sams E (1995). The Real Shakespeare: Retrieving the Early Years, 1564–1594. New Haven: Yale University Press.

Simpson EH (1949). Measurement of diversity. Nature 163:168.

Svartvik J (1968). The Evans statements: a case for forensic linguistics. English Dept. of the University of Gothenborg.

Taylor G (1987). The canon and chronology of Shakespeare's plays. In: S Wells, G Taylor, eds. William Shakespeare: A Textual Companion. Oxford: Clarendon Press, pp 69–145.

Yule GU (1944). The Statistical Study of Literary Vocabulary. London: Cambridge University Press.

24

Lexical Knowledge Acquisition

YUJI MATSUMOTO AND TAKEHITO UTSURO

Nara Institute of Science and Technology, Ikoma, Nara, Japan

1. INTRODUCTION

The advent of large-scale online linguistic resources makes it possible to acquire linguistic knowledge in some automatic way. A number of different types of linguistic knowledge are obtainable from online resources, which is the main topic of this chapter. Corpus-based statistical part-of-speech taggers and parsers are successful examples of linguistic knowledge acquisition for language analysis. Disambiguation at syntactic and semantic levels is another successful application of corpora to tackle the notoriously hard problem. Although most of the techniques are using more lexical-level information, such as lexical dependency and lexical association, compilation of a general-purpose lexicon is still difficult for various reasons.

LEXICAL REPRESENTATION. There is virtually no consensus on the representation of the lexicon. Subcategorization frames for predicates are one of the few examples for which a common representation has, been agreed to some extent. Still there are various different ways of defining subcategorization frames, from surface case frames to semantically deeper argument structures, or even lexical conceptual structures. Some other devices for lexical representation include semantic classes in a form of flat or hierarchical classifications. Several knowledge-representation formalisms have been proposed in the artificial intelligence (AI) community based on Frame theory, and KL-ONE (Brachman and Schmolze, 1985), or its derivatives, are often used for natural language tasks. Still, the vocabulary of most such systems is very small and has rarely increased to a practical level.

A lexically centered view of grammars is becoming popular, as is seen in recent lexicalized grammar theories such as Head-Driven Phrase Structure Grammar (Pollard and Sag, 1994) and lexicalized Tree Adjoining Grammar (Joshi and

Schabes, 1992). In many such grammar theories, subcategorization frames serve as the basic lexical representation of predicates. Acquisition of subcategorization frames is one of the major topics in this chapter.

Although few, there have been proposals for lexical representation. The "Generative Lexicon" (Pustejovsky, 1995) exploits lexical construction in detail for generative use of lexicon. Pustejovsky et al. (1993) also propose a preliminary method for obtaining lexical knowledge along with the Generative Lexicon by means of corpus analysis. See also Saint-Dizier and Viegas (1995) and Pustejovsky and Boguraev (1996) for issues on lexical semantics.

APPLICATION ORIENTEDNESS. The depth and the sort of lexical knowledge required in the lexicon depends on the application for which it is used. Especially in Machine Translation, real understanding may not be of the foremost importance, as understanding does not necessarily guarantee high-quality translation. Rather, what is required is a direct correspondence of words or expressions between the source and target languages. The second major topic in this chapter is the automatic construction of a bilingual lexicon.

DOMAIN DEPENDENCY. Word usage drastically changes according to the domain, and words may have nonstandard usage or uncommon meanings in a specific sublanguage. For example, *abduction* in computer science always means a reasoning process of inferring a premise from an observation, and never kidnapping.

QUANTITATIVE ISSUE. A lexicon for practical application requires a vocabulary of at least several tens of thousands of entries. Manual compilation of a lexicon requires a tremendous human labor. It is also difficult to keep consistency between lexical entries.

There have been several attempts to construct lexicons or lexical knowledge bases, among which the EDR dictionaries (EDR, 1995) and the CYC knowledge base (Lenat, 1995) are representative general-purpose lexicons for language processing and for commonsense reasoning. Although they are undoubtedly valuable resources, they are not fully exploited in wide applications, because the depth and breadth of lexical knowledge vary according to the domain and the application, and some manipulation is necessary to adapt such existing knowledge sources to the intended domain and purposes. Corpora are valuable sources of domain-specific information and are useful for acquiring lexical knowledge as well as for enhancing existing lexicons.

This chapter gives an overview of techniques for automatic acquisition of lexical knowledge from online resources, especially for corpora. After giving an overview of lexical knowledge acquisition at various levels in next section, we focus on two active areas of automatic construction of the lexicon: subcategorization-frame acquisition for verbs, and acquisition of translation lexicons from bilingual corpora.

2. LEXICAL KNOWLEDGE ACQUISITION FOR LANGUAGE ANALYSIS: AN OVERVIEW

Lexical knowledge is classifiable in many ways. This section briefly overviews acquisition of various lexical information from linguistic resources (corpora, machine-readable dictionaries).

A. Lexical Collocation and Association

Lexical collocation and association provide important and useful information for lexicography, and a considerable amount of interesting research into the acquisition of such information from corpora has been done in last decade, including Church and Hanks (1990) and Smadja (1993). These topics are covered in Chap. 21.

B. Disambiguation

Ambiguity is one of the major obstacles to natural language processing and arises at almost every level of processing. Disambiguation in part-of-speech tagging is effectively resolved by corpus-based statistical techniques. Up-to-date supervised learning algorithms achieve tagging accuracy higher than 97% (see Chap. 17). Probabilistic grammars are typically used to achieve syntactic ambiguity resolution by learning of probabilistic parameters. Lexical dependency is crucial for high-accuracy parsing (see Magerman, 1995; Hogenhout and Matsumoto, 1996; Collins, 1996; Charniak, 1997; and also Chap. 22 for details). Corpora are quite useful for acquiring rules for resolving specific forms of syntactic ambiguities such as those caused by PP-attachment (Hindle and Rooth, 1993; Collins and Brooks, 1995; Kayaalp, et al., 1997). Identification of the proper sense of polysemous words is another important issue, which is discussed in Chap. 26.

C. Lexical Similarity and Thesaurus Construction

A thesaurus is a hierarchical representation of a lexicon, in which words are arranged according to their semantic similarity. Several hand-crafted thesauri are available and are widely used in natural language and information retrieval systems. However, domain-specific knowledge should be integrated into the thesaurus to make better use of it. Automatic construction or automatic augmentation of a thesaurus by means of corpus-based semantic similarity is the topic of Chap. 19.

D. Semantic Hierarchy Extraction from a Machine-Readable Dictionary

Machine-Readable Dictionaries (MRDs) are no doubt another useful resource from which to acquire lexical knowledge. Because common descriptions of terms in a dictionary consist of a semantically general term (*genus* term) and complementary explanation to distinguish the term from the other-related concepts (*differentia*), they seem especially useful for extracting taxonomical (hyponymy–hypernymy) relations between senses.

Early experimental work was reported by Amsler (1981). He reported a feasibility study on the extraction of taxonomical information of nouns and verbs using the *Marrian Webster Pocket Dictionary*. Identification of the genus terms in the definition sentences and sense disambiguation were done manually. Through an experimental simulation, the following observations were made:

- Since definition sentences are described in natural language and sense disambiguation of genus terms is not easy, only a tangled hierarchy is obtainable.

- General genus terms appearing on higher levels of the taxonomy (such as, *cause*, *thing*, and *class*) tend to form loops.
- Some specific words appearing before *of*, for example, *a type of*, are not the genus terms, and the words following such an expression are the genus terms. By classifying those expressions containing *of*, some other relations like "part-of" can be identified.

Chodorow et al. (1985) gave heuristic rules to extract genus terms from *Webster's 7th New Collegiate Dictionary*. Genus terms of verbs are obtainable simply by extracting a verb following *to* in the definition sentence with an accuracy of almost 100%. They also reported that 98% of correct genus terms for nouns are extracted by specifying pattern-matching rules for finding the head noun and by describing some specific expressions containing *of*.

Even if the genus terms are extracted, their sense ambiguity must be resolved to obtain a clean taxonomy. Guthrie et al. (1990) utilized the "box codes" (semantic class codes) and the subject codes (domain codes) for sense disambiguation of genus terms.

Lexical knowledge acquisition from MRDs has not reached a practical level for the foregoing reasons. However, automatic acquisition of taxonomical information from corpora is far more difficult. Ide and Véronis (1993) give a comprehensive survey of this issue. They report how the use of multiple MRDs improves the result of the extraction and also suggest the complementary use of MRDs and corpora. The BICORD system (Klavans and Tzoukermann, 1990) is an example of such an attempt at augmenting a bilingual dictionary by the information obtained from bilingual corpora.

E. Terminology Extraction

A lexicon is not a collection of bare words. Especially in technical domains, multiword technical terms that act as lexical entities should be identified and included in the lexicon. Recent work shows that very simple criteria are quite effective in extracting multiword technical terms from documents. Justeson and Katz (1995) propose an algorithm for technical-term identification that uses structural patterns of noun phrases (NPs) and a simple frequency count. The only constraint they place on technical terms is that they must be NPs consisting of only adjectives, nouns, and optionally prepositions (which may be restricted to *of*), and they must appear twice or more in the text. Similar results are reported by Daille (1996), in which candidate French (and English) technical terms are defined by simple NP structures. Daille tries to figure out what sort of statistical scores may serve as an additional filter to choose the "good" multiword technical terms from the candidates, and reports that the simple frequency count turns out to be a very good filter. Daille also claims that log-likelihood ratio (see Sec. 4.B.4) is effective to achieve high accuracy. Daille et al. 1994 and Dagan and Church (1997) extend this work on monolingual terminology extraction to the task of bilingual terminology extraction.

F. Automatic Compilation of the Lexicon

All of the aforementioned topics are partial tasks of compiling lexicons. Acquisition of lexical knowledge for automatic lexicon compilation is not an easy job. What is worse, we do not know precisely what to learn to construct a lexicon. Subcategorization frames are the most fundamental piece of lexical knowledge of verbs, and automatic acquisition of subcategorization frames is the most intensively studied problem in this field. The next section gives a detailed survey of the techniques for subcategorization-frame acquisition from corpora. Although most of the work aims at extracting frames of *surface cases*, several attempts have been made to acquire *deep case frames*, or *thematic frames*, which is semantic, rather than syntactic, knowledge (Liu and Su, 1993; Dorr et al., 1994; Dorr and Jones, 1996; Oishi and Matsumoto, 1995). The basic idea of acquiring semantic knowedge from the surface-level syntactic information is to apply rules mapping syntactic clues to thematic frames. For example, Dorr et al. (1994) and Dorr and Jones (1996) automatically extract mapping rules from the surface-level subcategorization frames for Levin's (1993) verb classes and then assign a thematic frame to each of the verb classes by hand. Oishi and Matsumoto (1995, 1997) propose a mapping rule from the alternation patterns of surface-level subcategorization frames to thematic roles, which is to be applied to thematic role acquisition. Furthermore, there exist several works on acquiring other types of semantic knowledge of verbs, such as aspectual information (Brent, 1991b; Dorr, 1992; Oishi and Matsumoto, 1997a,b).

The application-oriented lexicon is another important target of automatic lexicon acquisition. As stated in the previous section, the information needed in a lexicon depends greatly on the application. Among others, Information Retrieval or Extraction and Machine Translation are important practical application areas of natural language processing. The former usually does not require an in-depth language analysis, but rather, a superficial analysis, such as lemmatization or shallow parsing. The important semantic information in the lexicon is the semantic classes or the semantic similarity between words. Acquisition of a thesaurus specialized for the application domain is the important topic, which is discussed in Chap. 19. For the latter, a high quality and natural translation requires a translation lexicon specialized for each target domain. Most domain-specific technical terms do not appear in ordinary dictionaries. Since translation errors in technical terminology are likely to lead to misunderstanding, having a domain-specific bilingual dictionary is indispensable for good translation. Besides technical terminology, there are several expressions in a language that cannot be translated on a word-for-word basis, and the range of expressions varies depending on the language pair. Some English expressions may be translated more or less literally into French, but not into Japanese. Automatic techniques for finding correspondences between words or phrases in parallel corpora are now intensively studied, and this is the second main topic of this chapter.

3. ACQUISITION OF SUBCATEGORIZATION FRAMES

A. Overview

The **subcategorization frame** has been considered as one of the most fundamental and important elements of lexical knowledge in lexical semantics as well as in natural language processing (NLP). Basically, representing a subcategorization frame involves at least the following four aspects:

1. Case patterns or case dependencies

2. Sense restriction of argument or adjunct nouns
3. Classification of polysemous senses
4. Subcategorization preference

First, in a subcategorization frame, it is imperative to indicate what types of syntactic categories *can* or *must* co-occur with the verb as its arguments or adjuncts. For example, in English, it is necessary to indicate whether the verb takes an object or a prepositional phrase headed by a certain preposition, and so on. Usually, the types of syntactic categories of arguments and adjuncts are called **case slots** of the verb, and the patterns of case slots in a subcategorization frame are sometimes referred to as **case patterns** or **case dependencies**. Second, it is also informative to indicate what types of words can appear in each case slot as the argument of the verb. Usually, types of words are represented by some semantic categories or semantic classes constituting a conceptual hierarchy, such as a thesaurus. This restriction on the **semantic type** of each case slot is often referred to as **sense restriction** or **selectional restriction**. Third, because a verb may have several senses, and syntactic behavior or usages of the verb may also vary according to its senses, it is necessary to classify those polysemous senses of the verb, especially from the viewpoint of lexical semantics. Finally, in the context of utilizing lexical knowledge in syntactic analysis, subcategorization frames are useful for resolving syntactic ambiguities in parsing. This use of subcategorization frames is often referred to as **subcategorization preference**, and the way to determine subcategorization preference is also an important part of the knowledge about subcategorization frames.

Although there exist several hand-compiled subcategorization-frame lexicons that are publicly available, manual compilation suffers because it requires a huge amount of manual labor, hand-compiled subcategorization lexicons are hard to extend, and are also unstable, in that the criteria may be inadvertently changed or may receive different interpretations during compilation, leading to inconsistencies. On the other hand, in recent years, since large-scale linguistic resources such as raw, part-of-speech (POS)-tagged or parsed corpora have become available, attempts to extract subcategorization frames automatically from those corpora have been made and several techniques have been invented. Although these techniques are still somewhat immature or at the research stage, and there exists no reliable subcategorization frame lexicon automatically compiled from corpora, the results of these attempts are encouraging and promising.

The following subsections describe works on extracting subcategorization frames from various kinds of corpora topic by topic. First, Sec. 3.B surveys works on extracting case patterns from unparsed corpus. These works concentrate on inventing techniques for identifying appropriate subcategorization frames through syntactic analysis of raw or POS-tagged texts. Next, Sects. 3.C through 3.D.2 describe works on extracting subcategorization frames from parsed corpora, including extracting sense restrictions on argument nouns (see Sec. 3.C), learning dependencies of case slots using some statistical measures (see Sec. 3.D), classifying polysemous verb senses (see Sec. 3.D.1), and learning models of subcategorization preference (see Sec. 3.D.2). Section 3.D.3 summarizes works on extracting subcategorization frames from parsed corpora, and discusses their similarities and differences. Finally, Sec. 3.E summarizes this section.

B. Extraction of Case Patterns from an Unparsed Corpus

In this section we discuss the extraction of case patterns from corpora that are not parsed or even, in most of the research reported, not tagged: in other words, it is a raw corpus. As summarized in Table 1, almost all the works study extraction of English subcategorization frames from various kinds of corpora including the *Wall Street Journal* (WSJ) and *New York Times* (NYT), with the exception of de Lima (1997), who deals with German.

In general, the process of subcategorization-frame (SF) extraction consists of the following three steps:

1. POS tagging and syntactic analysis
2. SF (subcategorization frame) detection by pattern matching
3. Statistical filtering

1. POS Tagging and Syntactic Analysis

First, each sentence in a raw corpus is POS-tagged and then syntactically analyzed using some (finite-state) parser. For example, in Ushioda et al. (1993), only NPs in each sentence are parsed by some shallow NP parser (and then SF candidates are detected using patterns written in regular expressions). Others also use some syntactic analyzer such as a finite-state parser (Manning, 1993), a shallow parser (de Lima, 1997), or a probabilistic LR parser (Briscoe and Carroll, 1997). These approaches differ in the treatment of syntactic ambiguities that arise during parsing. In Manning, syntactically ambiguous sentences are skipped and ignored. In Briscoe and Carroll, the most probable parsing result produced by the probabilistic LR parser is selected. In de Lima, syntactically ambiguous phrases are left unattached and those ambiguities are resolved during the statistical filtering phase.

Brent (1991, 1993) takes an approach that differs from the others in that he utilizes only a small number of local morphosyntactic cues as well as a small dictionary of closed-class words, such as pronouns, determiners, auxiliary verbs, and conjunctives, and does not use a POS tagger or a parser. Brent considers only the six types of subcategorization frames listed in Table 2a, none of which has prepositional phrases as arguments. Brent's system uses only local morphosyntactic cues that are reliable enough to identify those six types of subcategorization frames.

2. SF Detection by Pattern Matching

Second, from the parsing results, candidate subcategorization frames are detected by pattern matching. Table 1 also lists the numbers of SF types considered in each work. Brent (1991, 1993) and Ushioda et al. (1993) consider six SF types listed in Table 2a. Manning considers 19 SF types (some parameterized for a preposition), some of which are listed in Table 2b, and Briscoe and Carroll consider 160 SF types.[1] De Lima considers six SF types, each of which is parameterized for a preposition. Each of the patterns for SF detection is designed to identify one of those SF types considered.

[1]Those 160 SF types are obtained by manually merging the SF types exemplified in the Alvey NL Tools (ANLT) dictionary (Boguraev et al., 1987) and the COMLEX Syntax dictionary (Grishman et al., 1994), and then adding about 30 SF types found by manual inspection of unclassified patterns.

Table 1 Summary of Extraction of Subcategorization Frames from Untagged Corpora

Work	Extraction method	Corpus	SF extracted	Evaluation[a]
Brent, 1991	SF detection by pattern matching; utilizes a small dictionary and morphosyntactic cues; no POS tagging or parsing	Untagged WSJ, 2.6 M words	6 SF types 2258 verbs	Token recall: 3–40% Token precision: 97–99%
Brent, 1993	Brent (1991) plus statistical filtering	Untagged Brown corpus	6 SF types 87 verbs	(Against corpus) Type recall: 60% Type precision: 96%
Ushioda et al., 1993	NP parser plus regular expressions	POS-tagged WSJ, 300 K words	6 SF types 33 verbs	Token precision: 83%
Manning, 1993	POS tagger plus finite-state parser; statistical filtering	Untagged NYT, 4.1 M words	19(+ prep) SF types 3103 verbs	Token recall: 82% (against dictionary) Type recall: 43% Type precision: 90%
Briscoe and Carroll, 1997	POS tagger plus probabilistic LR parser, pattern-matching, statistical-filtering	Untagged, Susanne/SEC/LOB total 1.2 M words	160 SF types 14 verbs	Token recall: 81% (against corpus) Type recall: 43% Type precision: 77% (against dictionary) Type recall: 36% Type precision: 66%
de Lima, 1997	Shallow parser, pattern-matching, statistical-filtering	German newspaper corpus; 36 M words	6(+ pre) SF types 15,178 verbs	(against dictionary) Type recall: 30% Type precision: 28%

[a]Token recall/precision: recall/precision averaged for a verb occurrence in corpus compared with corpus annotation or hand analysis.
Type recall/precision: recall/precision averaged for SF type.
Against corpus: compared with corpus annotation or hand analysis.
Against dictionary: compared with some existing dictionary.

Table 2 Examples of SF Types

SF type	Description
(a) Six SF types considered in Brent (1991, 1993) and Ushioda et al. (1993)	
NP only	Transitive verbs
Tensed clause	Finite complement
Infinitive	Infinitive clause complement
NP + clause	Direct object and finite complement
NP + infinitive	Direct object and infinitive complement
NP + NP	Ditransitive verbs
(b) Selection of 19 SF types considered in Manning (1993)	
IV	Intransitive verbs
NP only	Transitive verbs
That clause	Finite *that* complement
Infinitive	Infinitive clause complement
ING	Participial VP complement
P(`prep`)	Prepositional phrase headed by `prep`
NP + *that* clause	Direct object and *that* complement
NP + infinitive	Direct object and infinitive complement
NP + NP	Ditransitive verbs
NP + P(`prep`)	Direct object and prepositional phrase headed by `prep`

3. *Statistical Filtering*

It can happen that SF candidates detected in the previous SF detection step contain adjuncts, or they are simply wrong owing to errors of the POS tagger or the parser. The statistical-filtering step attempts to determine whether one can be highly confident that the detected SF candidates are actually subcategorization frames of the verb in question. Brent (1993) proposes applying the technique of hypothesis testing on binomial distribution. Manning and Briscoe and Carroll also take Brent's approach of hypothesis testing. In the case of de Lima, the SF detection step allows ambiguities and outputs a set of candidate subcategorization frames. Then, the EM algorithm (Dempster et al., 1977) is applied to assign probabilities to each frame of the set of candidate subcategorization frames and the most-probable frames are selected.

4. *Evaluation*

Extracted subcategorization frames are evaluated in terms of **recall and precision**. Usually, at least one of the following three types of recall and precision measures is used.

1. **Token** recall–precision: recall–precision in terms of how many occurrences of verbs in the corpus are assigned correct subcategorization frames, compared with the correct corpus annotation or hand analysis, and calculated as an average for a verb occurrence in the corpus.
2. **Type** recall–precision: recall–precision in terms of how many extracted SF types are correct, and calculated as an average for a SF type.

Against corpus: compared with the correct corpus annotation or hand analysis.

Against dictionary: compared with subcategorization frame definitions in some existing dictionary.

For example, to calculate dictionary recall–precision, Manning compares the extracted subcategorization frames with the entries listed in the *Oxford Advanced Learner's Dictionary of Current English* (OALD) (Hornby, 1989), and Briscoe and Carroll compare their frames with the entries obtained by merging those of the ANLT dictionary and the COMLEX Syntax dictionary. They also evaluate the extracted subcategorization frames in terms of parsing recall–precision.

Generally speaking, type recall–precision against the corpus is greater than against a dictionary, because there are SF types listed in existing dictionaries with no occurrence of instances in the corpus. In the case of Brent, precision is almost 100%, whereas recall is much smaller. This is because Brent's system uses reliable cues only. In other works, major sources of errors are mistakes of the POS tagger or the parser. For example, adjuncts are incorrectly judged to be arguments because of the mistakes during parsing.

C. Extraction of Sense Restriction

In work on extracting or learning sense restrictions on the arguments in subcategorization frames, first, collocations of verbs and argument nouns are collected, and then, those instances of collocation are generalized and sense restrictions on nouns are represented by conceptual classes in an existing thesaurus.

1. Methods

Resnik (1992) first pointed out the limitations of the approach of extracting lexical association at the level of word pairs (Church and Hanks, 1990). The limitations are that often either the data is insufficient to provide reliable word-to-word correspondences, or the task requires more abstraction than word-to-word correspondences permit. Resnik proposes a useful measure of word–class association by generalizing an information-theoretic measure of word–word association. The proposed measure addresses the limitations of the word level lexical association by facilitating statistical discovery of facts involving word *classes* rather than individual words.

Resnik's measure of word-class association can be illustrated by the problem of finding the prototypical object classes for verbs. Let $\mathcal{V}$ and $\mathcal{N}$ be the sets of all verbs and nouns, respectively. Given a verb $v \in \mathcal{V}$ and a noun class $c \subseteq \mathcal{N}$, the joint probability of v and c is estimated as in Eq. {1}.

$$P(v, c) \approx \frac{\sum_{n \in c} count(v, n)}{\sum_{v' \in \mathcal{V}} \sum_{n' \in \mathcal{N}} count(v', n')} \qquad \{1\}$$

The **association score** $A(v, c)$ of a verb v and a noun class c is defined as in Eqs. {2} and {3}.

$$A(v, c) = P(c|v) \log \frac{P(v, c)}{P(v)P(c)} \qquad \{2\}$$

$$= P(c|v) I(v; c) \qquad \{3\}$$

The association score takes the mutual information (MI) between the verb and a noun class, and scales it according to the likelihood that a member of the class will actually appear as the object of the verb. The first term of the conditional probability measures the generality of the association, whereas the second term of the MI measures the specificity of the co-occurrence of the association.

Resnik applies this method to the collocations extracted from the POS-tagged Brown corpus (Francis and Kučera, 1982) included as part of the Penn Treebank (Marcus et al., 1993). First, POS-tagged sentences of the Brown corpus are syntactically analyzed by the Fidditch parser (Hindle, 1983), and then, the collocations of the verbs and the object nouns are extracted. Sense restriction is represented by the classes in the WordNet thesaurus (Miller, 1995).[2] For a selection of 15 verbs chosen randomly from the sample, the highest-scoring object classes are searched for, which are listed in Table 3.

Li and Abe (1995) and Abe and Li (1996) formalize the problem of generalizing case slots as that of estimating a model of probability distribution over some portion of words, and propose generalization methods based on the Minimum Description Length (MDL) principle (Rissanen, 1984): a well-motivated and theoretically sound principle of statistical estimation in information theory. In Li and Abe's works, it is assumed that a **thesaurus** is a tree in which each leaf node stands for a noun, while each internal node represents a noun class, and domination stands for set inclusion,

Table 3 Prototypical Object Classes Based on Resnik's Association Score

$A(v, c)$	Verb v	Object noun class c	Object nouns
0.16	call	⟨someone, [person]⟩	(name, man, ...)
0.30	catch	⟨sooking_at, [look]⟩	(eye)
2.39	climb	⟨stair, [step]⟩	(step, stair)
1.15	close	⟨movable_barrier, [...]⟩	(door)
3.64	cook	⟨meal, [repast]⟩	(meal, supper, dinner)
0.27	draw	⟨cord, [cord]⟩	(line, thread, yarn)
1.76	eat	⟨nutrient, [food]⟩	(egg, cereal, meal, mussel, celery, chicken, ...)
0.45	forget	⟨conception, [concept]⟩	(possibility, name, word, rule, order)
0.81	ignore	⟨question, [problem]⟩	(problem, question)
2.29	mix	⟨intoxicant, [alcohol]⟩	(martini, liquor)
0.26	move	⟨article_of_commerce, [...]⟩	(comb, arm, driver, switch, bit, furniture, ...)
0.39	need	⟨helping, [aid]⟩	(assistance, help, support, service, ...)
0.66	stop	⟨vehicle, [vehicle]⟩	(engine, wheel, car, tractor, machine, bus)
0.34	take	⟨spatial_property, [...]⟩	(position, attitude, place, shape, form, ...)
0.45	work	⟨change_of_place, [...]⟩	(way, shift)

[2]WordNet is an online lexical reference system the design of which is inspired by current psycholinguistic theories of human lexical memory. English nouns, verbs, adjectives, and adverbs are organized into synonym sets, each representing one underlying lexical concept. Different relations link the synonym sets. WordNet contains more than 118,000 different word forms and more than 90,000 different word senses.

as in the example in Fig. 1. Then they restrict the possible partitions of the thesaurus to those that exist in the form of a **cut** in the tree. A cut in a tree is any set of nodes in the tree that defines a partition of the leaf nodes, viewing each node as representing the set of all the leaf nodes it dominates. For example, in the thesaurus of Fig. 1, there are five cuts: [ANIMAL], [BIRD, INSECT], [BIRD, bug, bee, insect], [swallow, crow, eagle, bird, INSECT], and [swallow, crow, eagle, bird, bug, bee, insect]. As the model of probability distribution, Li and Abe (1995) introduces the class of *tree cut model* of a fixed thesaurus tree as a probability distribution over a tree cut. Formally, a tree cut model is represented by a pair consisting of a tree cut, and a probability parameter vector of the tree cut. For example in Fig. 1, given the frequency counts below each leaf noun, $M = ([\text{BIRD, bug, bee, insect}], [0.8,0,0.2,0])$ is a tree cut model. Figure 1 also gives probability parameters for the other tree cut models. In practice, the tree cut model is used to represent the conditional probability distribution $P(n|v, s)$ of noun n over a tree cut, which appears at the slot s of the verb v.

In Li and Abe (1995), the description length for a model is calculated in the following way, and a model with the MDL is selected. Given a tree cut model M and data $S_{v,s}$ of nouns appearing at the slot s of the verb v, its total description length $L(M)$ is computed as the sum of the model description length $L_{mod}(M) + L_{par}(M)$, and the data description length $L_{dat}(M)$, Eq. {4}.

$$L(M) = L_{mod}(M) + L_{par}(M) + L_{dat}(M) \tag{4}$$

L_{mod} and L_{par} are calculated by Eq. {5} and Eq. {6}, respectively

$$L_{mod}(M) = \log|\mathcal{M}| \tag{5}$$

$$L_{par}(M) = \frac{\mathcal{K}}{2} \times \log|S_{v,s}| \tag{6}$$

where $\mathcal{M}$ denotes the set of the cuts in the tree, and $\mathcal{K}$ denotes the number of (free) parameters in the tree cut model. $L_{dat}(M)$ is calculated using the probability distribution $\hat{P}_M(n|v, s)$ with the parameters estimated by the MLE (Maximum Likelihood Estimate), given the tree cut Γ of M Eqs. {7} through {9}

$$L_{dat}(M) = -\sum_{n \in S_{v,s}} \log \hat{P}_M(n|v, s) \tag{7}$$

$$\forall n \in C, \hat{P}_M(n|v, s) = \frac{1}{|C|} \times \hat{P}_M(C|v, s) \tag{8}$$

$$\forall C \in \Gamma, \hat{P}_M(C|v, s) = \frac{f(C|v, s)}{|S_{v,s}|} \tag{9}$$

where $f(C|v, s)$ denotes the total frequency of nouns in class C in sample $S_{v,s}$. Finally, a model $\hat{M}$ with the MDL is selected Eq. {10}.

$$\hat{M} = \underset{M}{\operatorname{argmin}}\ L(M) \tag{10}$$

The description lengths for the tree cut models in Fig. 1 are shown in Table 4. [Note that it is only necessary to calculate and compare $L_{par}(M) + L_{dat}(M)$, as every cut has an equal $L_{mod}(M)$.] These figures indicate that the tree cut [BIRD, INSECT]

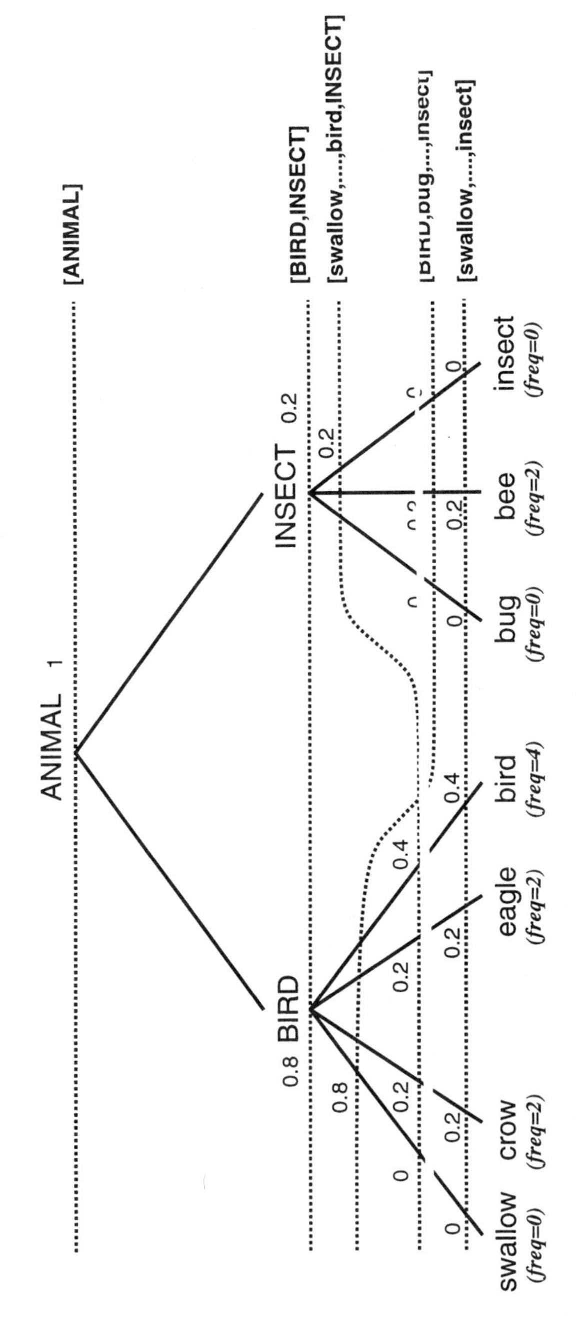

Fig. 1 An example of the thesaurus and the tree cut models.

Table 4 Description Lengths of the Tree Cut Models

Tree cut model	L_{par}	L_{dat}	$L_{par} + L_{dat}$
[ANIMAL]	0.00	28.07	28.07
[BIRD, INSECT]	1.66	26.39	28.05
[BIRD, bug, bee, insect]	4.98	23.22	28.20
[swallow, crow, eagle, bird, INSECT]	6.64	22.39	29.03
[swallow, crow, eagle, bird, bug, bee, insect]	9.97	19.22	29.19

gives the best model according to the MDL principle. After an optimal model M is selected, they use the conditional probability $\hat{P}_M(n|v, s)$ as a measure of the collocation of the verb v and the noun n appearing at the slot s of v.

Compared with Li and Abe (1995), Abe and Li (1996) employ the **association norm** $A(n, v)$ as a measure of co-occurrence of a verb v and a noun n (assuming that n appears at a given slot s of v) Eq. {11}.

$$A(n, v) = \frac{P(n, v)}{P(n)P(v)} \qquad \{11\}$$

where $P(n, v)$ denotes the joint distribution over the nouns and the verbs (over $N \times V$), and $P(n)$ and $P(v)$ the marginal distributions over N and V induced by $P(n, v)$, respectively. Abe and Li formulate the problem of optimizing this association norm as a subtask of learning the conditional distribution $P(n|v)$, by exploiting the identity Eq. {12}.

$$P(n|v) = A(n, v) \cdot P(n) \qquad \{12\}$$

As in Li and Abe (1995), the probability distributions $P(n)$ and $P(n|v)$ are represented as tree-cut models, and the following two-step estimation method is proposed based on the MDL principle. The method first estimates $P(n)$ as $\hat{P}(n)$ using MDL, and then estimates $P(n|v)$ for a fixed v by applying MDL on the hypothesis class of $\{A \cdot \hat{P}(n) | A \in \mathcal{A}\}$ for some given class $\mathcal{A}$ of representations for association norm. Formally, the class $\mathcal{A}$ of representations for association norm is defined as a pair (τ, p) of a tree cut τ and a function p from τ to $\mathcal{R}$ (the set of real numbers), and called the **association tree-cut model** (ATCM). The estimation of A is therefore obtained as a side effect of a near optimal estimation of $P(n|v)$.

For the experiments of Li and Abe, the training data are obtained from the texts of the tagged WSJ corpus, which contains 126,084 sentences. From the texts, triples of the form (`head, slot_name, slot_value`) are extracted. As the thesaurus, the WordNet thesaurus is transformed into an appropriate tree structure and then used for representing noun classes. For the direct object slot of the verb *buy*, Fig. 2 shows selected parts of the ATCM of Abe and Li (1996), as well as the tree cut model of Li and Abe (1995).

Abe and Li (1996) list the following general tendencies that can be observed in these results. First, many of the nodes that are assigned high A values by the ATCM are not present in the tree cut model, because they have negligible absolute frequencies. Some examples of these nodes are ⟨*property*, *belonging*, $\cdots$, ⟩, ⟨*right*⟩, ⟨*ownership*⟩, and ⟨*part*, $\cdots$⟩. Abe and Li (1996) claim that those nodes do represent suitable direct objects of *buy*, and the fact that they are picked up by the ATCM

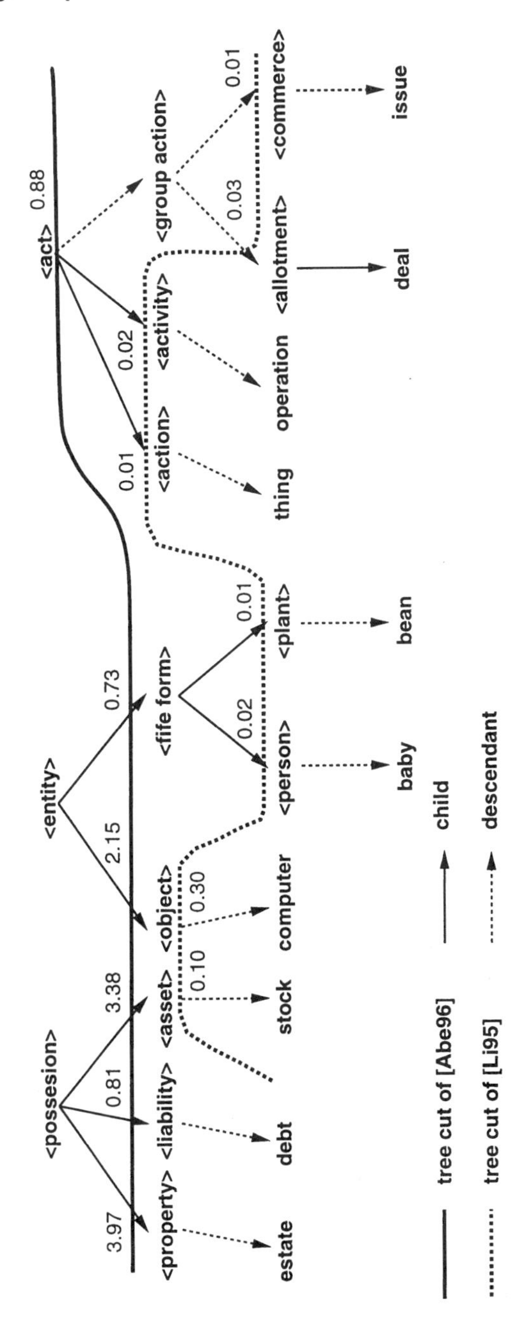

Fig. 2 An example of WordNet tree structure and the generalization results.

despite their low absolute frequencies seems to confirm the advantage of the ATCM over the tree cut model. Second, the cut in the ATCM is always "above" that of the tree cut model. For example, the four nodes ⟨*action*⟩, ⟨*activity*⟩, ⟨*allotment*⟩, and ⟨*commerce*⟩ in the tree cut model are all generalized as one node ⟨*act*⟩ in the ATCM, reflecting the judgment that despite their varying absolute frequencies, their association norms with "*buy*" do not significantly deviate from one another. In contrast, the nodes ⟨*property*⟩, ⟨*asset*⟩, and ⟨*liability*⟩ are kept separate in the ATCM, for the first two have high A values, whereas ⟨*liability*⟩ has a low A value, which is consistent with the intuition that one does not want to buy debt.

2. Evaluation

Resnik (1993), Li and Abe (1995), and Abe and Li (1996) evaluate the results of generalizing sense restriction of subcategorization frames in terms of the accuracy of prepositional-phrase attachment (PP-attachment) task.[3] The method used in Resnik (1993) is a variant of the association score proposed in Resnik (1992), which is called **selectional association** $SA(v, c)$. For the sake of comparison, Abe and Li (1996) apply the three methods of Resnik and Li and Abe to the same training and test data and report their coverage and accuracy. From the bracketed corpus of the Penn Treebank (WSJ corpus) were selected 19,739 triples of the forms (a) `(verb, preposition, noun)` and (b) $(\texttt{noun}_1, \texttt{preposition}, \texttt{noun}_2)$, and 820 quadruples of the form (c) $(\texttt{verb}, \texttt{noun}_1, \texttt{preposition}, \texttt{noun}_2)$. From the training triples of form (a), the generalized classes of the slot values of noun are obtained according to the three methods. Also from the training triples of form (c), the generalized classes of the slot values of $\texttt{noun}_2$ are obtained according to the three methods. Disambiguation of PP-attachment of the test quadruples using the selectional association of Resnik (1993) is done by comparing the values $SA_{\texttt{prep}}(\texttt{verb}, \texttt{noun}_2)$ and $SA_{\texttt{prep}}(\texttt{noun}_1, \texttt{noun}_2)$. Disambiguation using the tree-cut model of Li and Abe (1995) is done by comparing the condition probabilities $\hat{P}_{\texttt{prep}}(\texttt{noun}_2 | \texttt{verb})$ $\hat{P}_{\texttt{prep}}(\texttt{noun}_2 | \texttt{noun}_1)$. Disambiguation using the ATCM of Abe and Li (1996) is done by comparing the values $\hat{A}_{\texttt{prep}}(\texttt{noun}_2 | \texttt{verb})$ and $\hat{A}_{\texttt{prep}}(\texttt{noun}_2 | \texttt{noun}_1)$. Table 5 shows the coverage and accuracy of these three methods. In the table, "Default" refers to the method of always attaching $(\texttt{preposition}, \texttt{noun}_2)$ to $\texttt{noun}_1$. In terms of accuracy, the differences between the three methods are not significant, whereas, in terms of coverage, the ATCM of Abe and Li significantly out-performs both Resnik's selectional association and Li and Abe's tree cut model.

Abe and Li claim that the fact that their ATCM appears to do better than the tree cut model confirms the intuition that the association norm $A(n, v)$ is better suited for the purpose of disambiguation than the conditional probability. They also claim that the fact that their ATCM out-performs Resnik's selectional association confirms that the estimation method for the association norm based on MDL is not only theoretically sound but excels in practice, since Resnik's selec-

[3] Ribas (1995) also evaluates the association score of Resnik (1992) in terms of the accuracy of the task of word–sense disambiguation.

Table 5 Results of PP-Attachment Disambiguation Using Sense Restriction Information

Method	Coverage (%)	Accuracy (%)
Default	100.0	70.2
Selectional association (Resnik, 1993)	63.7	94.3
Tree cut model (Li and Abe 1995)	73.3	94.6
ATCM (Abe and Li 1996)	80.0	95.2

tional association is a heuristic method based on essentially the same notion of association norm.[4]

D. Learning Case Dependencies from Parsed Corpora

In Sec. 3.B, we listed and compared works on extracting case patterns from unparsed corpus. In those works, one of the most important processes is to identify unambiguous collocations of verbs and argument nouns. Unlike those works, this section describes approaches to learning dependencies of case slots from syntactically analyzed corpora, using some measures for identifying statistical dependencies among case slots.

Li and Abe (1996) assume that the dependencies of case slots can be represented as "dendroid distributions" (i.e., each variable directly depends on, at most, one other variable). They employ Suzuki's (1993) algorithm for learning an optimal dendroid distribution based on the MDL principle and apply it to the task of learning dependencies of case slots. According to Suzuki, any pairs of variables (X_i, X_j) are judged as dependent if their mutual information $\mathrm{I}(X_i, X_j)$ satisfies the threshold condition in Eq. {13}

$$\mathrm{I}(X_i, X_j) > (k_i - 1)(k_j - 1)\frac{\log N}{2N} \qquad \{13\}$$

where k_i and k_j denote the number of possible values assumed by the variables X_i and X_j, respectively; N is the input data size. From the bracketed corpus of the Penn Treebank, 181,250 collocations of verbs and argument nouns are extracted as the training data. Among 357 verbs appearing more than 50 times in the training data, 59 verbs are judged as having case slots dependent on each other. Table 6 lists selected examples of those verbs as well as the dependent case slots.

Li and Abe (1996) also evaluate the case dependencies extracted in terms of the accuracy of the PP-attachment task. 93 test examples of the form `(verb, noun`$_1$`, preposition, noun`$_2$`)` are extracted, where `verb` is among those verbs for which object and `preposition` case slots are judged as dependent. Out of those 93 test examples, PP-attachment is correct for 90 examples, and this achieves 97% accuracy. This result is compared with that of applying the disambiguation method based on the assumption of case independence proposed by Li and Abe (1995), which gives only 79 (85%) correct attachments. Li and Abe (1996) conclude that the PP-attach-

[4]However, note that the methods of Resnik (1993) and Abe and Li (1996), which are based on the association norm, are computationally expensive because they require the computation of the marginal distribution $P(n)$ of the noun n or $P(c)$ of the noun class c over the whole training corpus.

Table 6 Verbs and Their Dependent Case Slots Extracted by the Li and Abe's (1996) Method

Verb	Dependent case slots		Verb	Dependent case slots	
add	object	to	acquire	from	for
blame	object	for	apply	for	to
buy	object	for	boost	from	to
climb	object	from	climb	from	to
compare	object	with	fall	from	to
convert	object	to	grow	from	to
defend	object	against	improve	from	to
explain	object	to	raise	from	to
file	object	against	sell	to	for
focus	object	on	think	of	as

ment accuracy is significantly improved using the information on case dependencies. They also try to identify case dependencies at the WordNet thesaurus class levels. First, an optimal tree-cut model is learned by Li and Abe's (1995) generalization method, and then all the pairs of variables at this optimal tree-cut level are checked to see whether or not the inequality of Eq. {13} holds. They report that no two case slots are determined as dependent, and claim that this is because the number of parameters in this optimal tree-cut model is very large compared with the size of the available data.

As can be seen from the foregoing description, Li and Abe (1996) formulate a method that directly detects dependencies between case slots. Compared with Li and Abe (1996), Haruno (1995a,b) propose to measure the usefulness of a whole subcategorization frame that may include more than one case slot as well as sense restriction of case slots represented by thesaurus classes. Haruno (1995a) extends Resnik's association score to the cases of whole subcategorization frames that may include more than one case slot. Haruno (1995b) applies an information-theoretic data-compression technique to the task of learning subcategorization frames, which may include more than one case slot as well as sense restriction of case slots represented by thesaurus classes. Haruno's method in the (1995b) paper is briefly described in the following.

The underlying assumption of the method is that the more compactly a surface case frame (which is equivalent to "a subcategorization frame"; for simplicity, the term "case frame" is used by Haruno) compresses example collocations, the more plausible it is. This assumption is a form of Occam's razor (Rissanen, 1989). The compactness of a case frame F is measured by the **case frame utility** Eq. {14}.

$$Utility(F) = explicit\text{-}bits(F) - (CL(F) + PL(F)) \text{ (bits)} \qquad \{14\}$$

where $explicit\text{-}bits(F)$ is the encoding length of explicitly enumerating all the examples covered by F, $CL(F)$ the encoding length of the case frame F itself, and $PL(F)$ the encoding length of a proof delineating how examples are reproduced using the case frame F. The greater this difference between $explicit\text{-}bits(F)$ and the sum of $CL(F)$ and $PL(F)$, the more compact is the case frame F. First, the encoding length

$CL(F)$ of the case frame F is defined as the following: if there exist P case slots in the target language and Cat noun classes in the thesaurus, to encode a case frame F that has N case slots,

1. $\text{Log } N + \log\binom{P}{N}$ bits to encode case particle used, and
2. $N\log(Cat)$ bits to encode the noun classes of N case slots are required. $CL(F)$ is defined as the sum of these lengths Eq. {15}.

$$CL(F) = \log N + \log\binom{P}{N} + N\log(Cat) \quad \text{(bits)} \tag{15}$$

Explicit-bits(F) is then defined as the sum of the encoding length $CL(E)$ over all the examples E which is covered by F, Eq. {16}.

$$Explicit\text{-}bits(F) = \sum_{E \text{ covered by } F} CL(E) \quad \text{(bits)} \tag{16}$$

Finally, let an example E be covered by a case frame F, E has a case slot s, and a node v exists between the thesaurus class node of the slot s of F and the leaf class of the same slot s of E. Then, supposing that $Branch(v)$ be the number of branches at the node v, $PL(F)$ is calculated by summing up the entropy values $\log(Branch(v))$ of every node v Eq. {17}.

$$PL(F) = \sum_{E} \sum_{\text{slot } s \text{ of } E} \sum_{v} \log[Branch(v)] \tag{17}$$

To avoid exhaustive search, Haruno employs the following bottom-up covering strategy. By generalizing collocation examples, a sequence of case-frame candidates is generated, out of which the most plausible case frame is selected. Iterating this procedure generates several case frames for a polysemous verb, each of which corresponds to a single sense or usage of the verb. From one year's Japanese newspaper articles containing about 75 million words, the training collocation examples are extracted by hand. Haruno uses a thesaurus of 12 hierarchy levels that contains about 2800 classes as well as 400,000 words (Ikehara et al. 1993). Table 7 shows the first ten case frames for a Japanese polysemous verb *noru* ('ride,' 'advise,' 'be on the right track,' or other) learned by this learning strategy. The first, the fourth, the sixth, and the seventh case frames correspond to the sense of "ride," whereas the others correspond to more or less idiomatic usages of the verb *noru* (e.g., the second "be on the right track," the third "advise"). The seventh is a more general representation of the first one, and this is because the first is judged as more compact than the seventh according to the **case-frame utility** measure. Haruno also claims that these results reflect the co-occurrence existing in the corpus, although the lexical association is not considered explicitly, and that only obligatory case slots are included in the extracted case frames without any explicit facilities aiming at dividing obligatory or optional case slots.

1. *Classification of Polysemous Senses*

For polysemous verbs, it is necessary to classify polysemous senses of those verbs and extract a subcategorization frame for each sense or usage. As described in the previous section, Haruno (1995a,b) applies a bottom-up covering algorithm to the

Table 7 Subcategorization Frames of *noru* Learned by Haruno's (1995b) Method

Rank	Utility (bits)	Subcategorization frame[a,b]
1	8561	[*ni*('on'):VEHICLE, *ga*(nom):MAN AND WOMAN]
2	5442	[*ga*(nom:ABSTRACT, *ni*('on'):*track*]
3	824	[*ga*(nom):HUMAN, *ni*('on'):*consultation*]
4	410	[*ni*('on'):CONCRETE]
5	306	[*ga*(nom):HUMAN, *ni*('on'):*wind*]
6	254	[*de*('at'):ADMINISTRATIVE REGION, *ni*('on'):vehicle]
7	233	[*ni*('on'):VEHICLE, *ga*(nom):HUMAN]
8	194	[*ni*('on'):SPIRIT]
9	2662	[*ga*(nom):HUMAN, *ni*('on'):SITUATION]
10	1590	[*ni*('on'):*boom*]

[a]UPPER CASE: Japanese nonleaf classes (e.g., VEHICLE, HUMAN).
[b] *Italicized*: Japanese leaf classes (e.g., *track, wind*).

task of extracting subcategorization frames of polysemous verbs, which generates several frames for each verb. Subcategorization frames are generated in the order of compactness which is measured in terms of the case-frame utility. Table 8 gives the results of extracting subcategorization frames for 11 typical Japanese polysemous verbs by Haruno's method and also compares them with the entries in the hand-crafted IPAL (1987) subcategorization-frame lexicon. The column "Recall" indicates how many IPAL case frames are actually learned from the newspaper corpus. Recall rates for highly polysemous verbs that have ten or more IPAL case frames, such as *kakeru* and *utsu*, are relatively low. Haruno claims that this is mainly due to the characteristics of the newspaper corpus. For those highly polysemous verbs, it often happens that the newspaper corpus contains instances of only some portion of IPAL case-frame entries. Haruno also points out that the case frames learned from the corpus tend to be finer than IPAL case frames, because the semantic hierarchy used by Haruno contains many more semantic classes than that used in the IPAL case-frame lexicon.

Utsuro et al. (1993) and Utsuro (1996) propose methods of classifying senses of polysemous verbs using a parallel corpus and apply the methods to the task of classifying senses of Japanese polysemous verbs. In this approach, semantic classes of English predicates extracted from a Japanese–English parallel corpus are quite useful for sense classification of Japanese polysemous verbs. In the method, first, parallel sentences are structurally matched by the method of Matsumoto et al. (1993)[5] and then a matching result is obtained. This matching result resembles ordinary training instances for subcategorization–frame-learning programs except that it contains a feature for the semantic class of the English predicate in the parallel sentences. Using information about the semantic classes of English predicates, it becomes much easier to classify polysemous senses of the verb. For example,

[5]Another advantage of using a parallel corpus is that syntactic ambiguities of both monolingual sentences can be resolved in the course of structural matching of parallel sentences, because English and Japanese sentences usually have quite different syntactic structures as well as different patterns of syntactic ambiguities.

Table 8 Evaluation of Haruno's (1995b) Sense Classification of Japanese Polysemous Verbs

Verb	English equivalents	No. of examples	No. of subcat. Frames extracted	IPAL	Recall (%)
kakeru	hang, spend	4455	32	15	46.7
noru	ride, advise	1582	17	6	83.3
tatsu	stand, leave	1523	21	6	83.3
nozomu	face, desire	650	9	3	100.0
kaku	write	604	5	2	100.0
nomu	drink, accept	459	6	4	75.0
utsu	hit, impress	448	14	10	50.0
tsunoru	collect, become serious	400	7	—	—
koeru	exceed, cross	283	3	3	100.0
tomonau	accompany, entail	184	4	4	100.0
ukabu	float, occur	145	8	4	75.0

although the Japanese verb *kakeru* is a typical Japanese polyseme and corresponds to English verbs such as *hang*, *spend*, *play some music*, and so on, the verb *hang* in the English part helps to classify senses of the Japanese verb *kakeru*.

Training instances are collected for each Japanese polysemous verb, and are classified to maximize the MI of the English predicate class and the Japanese part of the training instance. In this classification process, pairs of an English predicate class and the Japanese part of the training instance that have highly positive correlations are selected. Table 9 shows examples of pairs with highly positive correlations for four Japanese polysemous verbs *haru*, *hiku*, *hiraku*, and *kanau*. These four verbs are typical Japanese polysemes, each of which has both intransitive and transitive senses; Table 9 lists examples classifying the intransitive or transitive senses.

The training instances are extracted from a corpus of about 40,000 translation examples included in a machine-readable Japanese–English dictionary (Shimizu and Narita, 1979). *Roget's Thesaurus*[6] is used as the English thesaurus, and *Bunrui Goi Hyō* (BGH) (NLRI, 1993)[7] as the Japanese thesaurus. The Japanese part f_J (see "Jap. Case-Frame" in Table 9) does not have to contain all the case slots of the training instances, but could have only some part of them. Actually, in the pairs listed in Table 9, every Japanese case frame f_J consists of only one case slot that has the highest positive correlation with the English predicate class.

For nine Japanese polysemous verbs, Utsuro evaluates the results of sense classification by comparing them with a manual classification that is based mainly on the classification of IPAL subcategorization frame lexicon. Table 10 lists the evaluation results.

[6] *Roget's Thesaurus* has a seven-layered abstraction hierarchy and over 100,000 words are allocated at the leaves.

[7] BGH has a six-layered abstraction hierarchy and more than 60,000 Japanese words are assigned at the leaves and its nominal part contains about 45,000 words. The classes of BGH are represented as numerical codes, in which each digit denotes the choice of the branch in the thesaurus. BGH's nominal part consists of the classes starting with 11–15, which are subordinate to "abstract relations," "agents of human activities," "human activities," "products," and "natural objects and natural phenomena."

Table 9 Examples of Intransitive or Transitive Distinction in Japanese Polysemous Verbs

Japanese verb	Eng. pred. class (level,[a] example)	Jap. case frame (level, example)	Examples	MI score
haru	*expensive* Leaf	*ga*(nom):*ne(price)* Leaf	3	0.299
	Special sensation −3, *freeze*	*ga*(nom):15130-11-10 Leaf, *kōri* ('ice')	3	0.237
	Acts −2, *persist, stick to*	*wo*(acc):13040 5, *gōjō* ('obstinacy')	7	0.459
hiku	Decrease −1, *subside*	*ga*(nom):151 3, *kōzui* ('floods')	2	0.109
	Results of Reasoning −2, *catch, have*	*wo*(acc):15860-11 6, *kaze* ('cold')	26	0.421
hiraku	Intellect 1, *open*	*ga*(nom):14(Products) 2, *to* ('door')	12	0.339
	hold Leaf	*wo*(acc):13510-1 6, *kaigi* ('meeting')	3	0.114
kanau	Completion	*ga*(nom):1304	8	0.460
	Leaf−1,realize	4, *negai* ('desire')		
	Quantity −3, *equal*	*ni*(dat):12000-3-10 Leaf, *kare* ('he')	8	0.504

[a]"Level" means level in the thesaurus; initial capital: English nonleaf classes (e.g., Acts, Intellect); numerical codes: Japanese classes (e.g., 13040); *italicized*: English and Japanese leaf classes (e.g., *persist*, *kouzui*).
Source: Utsuro, 1996.

"Cl." and "Ex." indicate the numbers of clusters and examples, respectively. The columns headed "One-sense" show the totals where each cluster contains examples of only one hand-classified sense; "Cl." and "Ex." list the number of such clusters and the total number of examples contained in such clusters, respectively. "Acc." indicates the accuracy, measured as the ratio of the number of examples contained in "one-sense" clusters to the total number of examples. This is 100% for four of the nine verbs, and 93.3% on average. The column "Hand-class" indicates the total number of senses according to the manual classification, and the "Ratio" is the total number of clusters divided by the total of hand-classified senses, corresponding to the average number of clusters into which one hand-classified sense is divided. Utsuro claims that the sense classification method has achieved almost *pure* classification in terms of the rate of One-sense cluster with the help of English predicate classes, regardless of the relatively small training data. The result seems a little finer than hand-classified, mainly because clusters that correspond to the same

Table 10 Evaluation of Sense Classification of Japanese Polysemous Verbs by the Utsuro Method (1996)

Verb	English	Slots[a]	Total[a]		One-sense[a]			Hand-class	Ratio
			Cl.	Ex.	Cl.	Ex.	Acc. (%)		
agaru	rise	nom	41	74	39	69	93.2	17	2.41
ageru	raise	acc	54	107	52	93	86.9	18	3.00
aku	open (itr)	nom	12	29	12	29	100.0	8	1.50
haru	spread (tr/itr)	nom/acc	19	36	17	30	83.3	11	1.73
hiku	subside, pull	nom/acc	40	105	40	105	100.0	23	1.74
hiraku	open (tr/itr)	nom/acc	15	54	13	50	92.6	10	1.50
kakeru	hang	acc	45	103	42	86	83.5	25	1.80
kanau	realize, conform	nom/dat	14	31	14	31	100.0	3	4.67
kau	buy	acc	15	75	15	75	100.0	4	3.75
Average							93.3		2.46

[a]Cl., cluster; Ex., example; itr, intransitive verb; tr, transitive verb; nom, nominative *ga*; acc, accusative *wo*; dat, dative *ni*.

hand-classified sense are separately located in the human-made thesaurus, and it is not easy to find exactly one representative class in the thesaurus.

2. *Learning Subcategorization Preference Models*

For the evaluation of subcategorization frames extracted from corpora, the performance in syntactic ambiguity resolution based on subcategorization preference is quite important. Actually, in syntactic analysis, subcategorization frames are used as one of the most important items of lexical knowledge for resolving syntactic ambiguities. Among those works on learning subcategorization frames from parsed corpora which we have described so far, Resnik (1993), Li and Abe (1995), and Abe and Li (1996) provide measures of the strength of subcategorization preference, and Li and Abe (1996) also evaluate the estimated case dependencies in terms of the performance in PP-attachment task. However, these works are limited in that these measures cannot be considered as total models of subcategorization preference, and consider the sense restriction of *only one case slot* in the first three works cited, or consider case dependencies with no sense-restriction generalization in the case of Li and Abe (1996). Compared with those works, this section describes work on directly learning total models of subcategorization preference from a parsed corpus.

a. Methods

Utsuro and Matsumoto (1997) extend the association score of Resnik (1992) to a measure of subcategorization preference that considers dependencies of more than one case slot. They also propose an iterative procedure for learning parameters of the subcategorization preference function and apply it to learning subcategorization preferences of Japanese verbs. Utsuro et al. (1997) apply the **maximum entropy model** learning technique (Berger et al., 1996; Della Pietra et al., 1997) to the task of probabilistic model learning of subcategorization preference. Miyata et al. (1997) employ the Bayesian network, and propose a method of learning according to the

MDL principle. Among these works, the following briefly describes the method of Utsuro et al. (1997).

In the maximum entropy-modeling approach, generalization of sense restriction (see Sec. 3.C) and case dependencies (see Sec. 3.D) are represented as **features** in a uniform way. Especially, features are allowed to overlap and this is quite advantageous when considering generalization of sense restriction and case dependencies in parameter estimation. The set $\mathcal{S}$ of active features is constructed from the set $\mathcal{F}$ of candidate features.[8] Candidate features are generated by considering all the possible patterns of case dependencies as well as sense restriction generalization of argument nouns. Then, the conditional probability distribution $p_{\mathcal{S}}(e_p|v)$ of the argument nouns e_p given the verb v is defined according to the maximum entropy-modeling framework. In this probability distribution, a **parameter** is assigned to each feature. The values of the parameter are estimated using the **Improved Iterative Scaling** (IIS) algorithm.

In Utsuro et al. (1997), the training instances are extracted from the EDR (1995) Japanese bracketed corpus, which contains about 210,000 sentences collected from newspaper and magazine articles. Because verbs are independent of each other in this model, the training data set is constructed for each verb, and its size varies from 200 to 500. BGH is again used as the Japanese thesaurus.

b. Evaluation

Utsuro et al. (1997) evaluate the performance of the model in the task of subcategorization preference of a verb-final Japanese sentence with a subordinate clause. Suppose that the word sequence in Eq. {18} represents a verb-final Japanese sentence with a subordinate clause, where *N*s are nouns, *p*s are case-marking postpositional particles, and *v*s are verbs, and the first verb v_1 is the head verb of the subordinate clause.

$$\underline{N_x - p_x}\;\; N_{11} - p_{11} \cdots N_{1l} - p_{1l}\;\; v_1\;\; N_{21} - p_{21} \cdots N_{2k} - p_{2k}\; v_2 \qquad \{18\}$$

Japanese sentences of this type have the subcategorization ambiguity of the postpositional phrase $N_x - p_x$ (i.e., whether $N_x - p_x$ is subcategorized for by v_1 or v_2). Example {19} illustrates the case where the postpositional phrase *terebi de* has the ambiguity of subcategorization.

<u>*terebi-de*</u> *mōketa shōnin-wo mita.* {19}

BY/ON-TV EARNED-MONEY MERCHANT-ACC SAW

(a) '(Somebody) saw a merchant who earned money by (selling) television.'

(b) 'On television, (somebody) saw a merchant who earned money.'

If the phrase *terebi-de* 'by/on TV' modifies the verb *mōketa* 'earned money,' interpretation (a) is meant. On the other hand, if that phrase modifies the verb *mita* 'saw,' interpretation (b) is taken. Test instances are artifically generated from collocation instances of the verbs v_1 and v_2 (such as e_1^+ and e_2^+ in Fig. 3) which are not used in the training. Erroneous test instances like e_1^- and e_2^- in Fig. 3 are generated by choosing

---

[8]More specifically, the set $\mathcal{S}$ of active features is built up by the feature selection procedure, in which a feature that produces the greatest increase in log-likelihood of the training sample is successively added to $\mathcal{S}$.

a case element $p_x : N_x$ at random and moving it from v_1 to v_2. Then, the precision of subcategorization preference is measured as the rate of test instances for which the correct collocation pair has a greater probability value than the erroneous pair [i.e., the inequality in Eq. (20) holds].

$$p_{S_1}(e^+_{p1}|v_1)p_{S_{21}}(e^+_{p2}|v_2) > p_{S_1}(e^-_{p1}|v_1)p_{S_2}(e^-_{p2}|v_2) \qquad \{20\}$$

From collocation instances of seven Japanese verbs, *agaru* 'increase' or 'elevate,' *ch igau* 'differ,' *kau* 'buy,' *motoduku* 'base,' *ōjiru* 'respond,' *sumu* 'live,' and *tsunagaru* 'connect,' 3360 test instances are generated. Coverage of the subcategorization preference model (i.e., the rate that for each case slot of the correct collocation instances e_1^+ and e_2^+, there exists at least one active feature that corresponds to the case slot, is 60.8%). For those test instances that are covered by the subcategorization preference model, the precision is 83.1%. The conditions on subcategorization preference can be relaxed and it can be determined with 100% coverage if the following heuristics are used: first, the collocations for which case slots are all covered are preferred, and second, the probability value of $p_S(e_p|v)$ is used for all the collocation instances even if some of their case slots are not covered by the model. With these heuristics, the precision becomes 79.0%.[9,10] In an experiment of relatively smaller scale, they compare this model with the one that assumes independence of case slots in the same task of subcategorization preference. Although the test set coverage of the model with case-slot independence is about 70–80%, its precision (that of covered test instances only, and also that with the heuristics above) is 20–30% lower than with case-slot dependencies. This result confirms that the case dependencies detected during the feature selection procedure are effective in the task of subcategorization preference.[11]

$$\boxed{\text{Correct}} \qquad\qquad\qquad\qquad \boxed{\text{Error}}$$

$$e_1^+ = \begin{bmatrix} pred : v_1 \\ p_{11} : N_{11} \\ \vdots \\ p_{1l} : N_{1l} \\ \underline{p_x : N_x} \end{bmatrix} \quad e_2^+ = \begin{bmatrix} pred : v_2 \\ p_{21} : N_{21} \\ \vdots \\ p_{2l} : N_{2k} \end{bmatrix} \iff e_1^- = \begin{bmatrix} pred : v_1 \\ p_{11} : N_{11} \\ \vdots \\ p_{1l} : N_{1l} \end{bmatrix} \quad e_2^- = \begin{bmatrix} pred : v_2 \\ p_{21} : N_{21} \\ \vdots \\ p_{2k} : N_{2k} \\ \underline{p_x : N_x} \end{bmatrix}$$

Fig. 3 Artificially generated test instances.

---

[9]These precision and coverage values are at an optimal point of the features selection procedure (i.e., a point that is judged as having [in total] the highest precision and coverage in the subcategorization preference task.

[10]In this experiment, the noun class generalization level of each feature is limited to above the level 5 from the root node in the thesaurus. Utsuro et al. (1997) claims that precision will be improved by incorporating all the BGH thesaurus classes including leaf classes. Actually, the method of Utsuro and Matsumoto (1997) incorporates all the BGH thesaurus classes including leaf classes and achieves 74–84% coverage as well as more than 96% precision.

[11]Utsuro et al. (1997) point out that the feature selection procedure presented in Berger et al. (1996) and Della Pietra et al. (1997) is not suitable for the tasks that consider class generalization, and claim that low test-set coverage will be improved by employing another feature selection strategy that starts from general and independent features and gradually selects more specific and dependent features.

3. Dimension

Table 11 lists works on extracting subcategorization frames from parsed corpora that are described in Sec. 3C–3.D.2, and classifies them according to five aspects:

1. Extraction of sense restriction
2. Learning case dependencies
3. Classification of polysemous senses of a verb
4. Learning models of subcategorization preference
5. Target languages

In the table, for each of the issues 1–4, a check-mark "$\surd$" indicates that this study can be regarded as considering the issue, a "?" indicates that it considers the issue only partially, and a cross "×" otherwise.

Although Li and Abe (1996) evaluate the extracted case dependencies in terms of the subcategorization preference as a heuristic in the PP-attachment task, they do not provide any measure of subcategorization preference. So, we put "?" in the "Subcategorization preference" column (4). The methods of Utsuro et al. (1993) and Utsuro (1996) put no restriction on case dependencies, and thus, in principle, are capable of detecting case dependencies. However, the primary goal of these methods is to discover strong dependencies between the English class information and the Japanese frame. Therefore, for the purpose of classifying polysemous senses of a verb, it becomes unnecessary to detect case dependencies within the Japanese frame. Actually, in the results of Table 10, all the Japanese case frames have only one case slot. So we put "?" in the "Case dependencies" column (2).

Most of the works listed in Table 11 can be considered as employing similar methodologies in that they assume some probabilistic models of subcategorization frames and select an optimal model according to some information–theoretic criteria, such as the MDL principle. This approach has been quite effective in selecting appropriate generalization levels of sense restriction as well as in detecting dependencies among case slots. This approach also has another advantage in

Table 11 Extraction of Subcategorization Frames from Parsed Corpora: Summary

	(1)	(2)	(3)	(4)	(5)
Resnik (1992, 1993), Li and Abe (1995), Abe and Li (1996)	$\surd$	×	×	$\surd$	English
Li and Abe (1996)	×	$\surd$	×	?	English
Haruno (1995a,b)	$\surd$	$\surd$	$\surd$	×	Japanese
Utsuro et al. (1993), Utsuro (1996)	$\surd$	?	$\surd$	×	Japanese
Utsuro and Matsumoto (1997), Utsuro et al. (1997), Miyata et al. (1997	$\surd$	$\surd$	×	$\surd$	Japanese

that it is theoretically sound. This contrasts with the methods of Li and Abe (1995, 1996) and Abe and Li (1996) which consider the two issues of sense restriction generalization and case dependencies separately, Haruno (1995), Utsuro and Matsumoto (1997), and Miyata et al. (1997) successfully show that it is quite straightforward to consider the two issues in a uniform way without any significant change in the methodology.

In terms of subcategorization preference, those works listed in Table 11 can be roughly divided into two groups according to the measures of subcategorization preference employed: one group (Li and Abe, 1995); Utsuro et al., 1997; Miyata et al., 1997) employs the conditional probability of the argument noun(s) given the verb, and the other (Resnik, 1993; Abe and Li, 1996; Utsuro and Matsumoto, 1997) employs some sort of association norm between the verb and the argument noun(s). As described in Sec. 3.C, Abe and Li (1996) claim that the association norm is better suited for the purpose of disambiguation than the conditional probability, although its computation is somehow more expensive.

Another characteristic of Table 11 is that no work considers both "Classification of Polysemous Senses" (3) and "Subcategorization Preference" (4). This is because it is not necessary to classify polysemous senses for the purpose of measuring subcategorization preference. It is quite possible to consider the former when measuring the latter, although it might be difficult to prove that this is really effective in subcategorization preference. Even from the viewpoint of lexical semantics, it is not easy to evaluate objectively the results of classifying polysemous senses of a verb. As described in Sec. 3.D.1, sense classifications existing in the hand-compiled subcategorization lexicons, such as IPAL, are useful as standards of sense classification. However, those classifications based only on human intuition sometimes suffer from lack of consistent criteria for sense classification.

E. Summary

This section summarizes those works on extracting subcategorization frames from unparsed/parsed corpora. As Tables 1 and 10 show, there is a correlation between the methodologies employed and the language to which they are applied. In those works on extracting case patterns from *unparsed* corpora, it is quite common to predetermine the types of subcategorization frames considered in the extraction. In addition, most works consider the extraction of English subcategorization frames. On the other hand, among those works on extracting subcategorization frames from *parsed* corpora, those which consider extraction of English subcategorization frames are limited in that they consider the extraction of sense restrictions on only one argument noun (Resnik, 1992, 1993; Li and Abe, 1995; Abe and Li, 1996) or the dependencies of two case slots without sense restriction (Li and Abe, 1996). All other works that consider the extraction of the whole subcategorization frame describe the extraction of Japanese subcategorization frames (Haruno, 1995a,b; Utsuro et al., 1997; Utsuro and Matsumoto 1997; Miyata et al., 1997), and the types of subcategorization frames are not predetermined in these works.

It seems quite straightforward to apply the methodology of extracting the whole subcategorization frame from *parsed* corpora to the extraction of English subcategorization frames. Therefore, it is to be hoped that there will be work on extracting English subcategorization frames from an existing English parsed corpus, such as the bracketed corpus of the Penn Treebank. On the other hand, in the case of extracting subcategorization frames from *unparsed* corpora, existing techniques seems to be restricted to the languages in which the types of subcategorization frames can be easily predetermined, such as English and German. In the languages such as Japanese, there are too many possible combinations of case-marking postpositional particles, and it is not practical to predetermine the set of possible subcategorization frames. Therefore, in languages such as Japanese, it is not easy to detect subcategorization frames by pattern-matching techniques. Existing methods are also limited in that they consider only extraction of case patterns. More work needs to be done to extract subcategorization frames with sufficient information including (a) case dependencies, (b) sense restriction of argument nouns, (c) classification of polysemous senses of a verb, and (d) subcategorization preference.

Finally, as we mentioned in Sec. 3.A, the methods described in this section are still somehow immature or at the research stage, and there exists no method reliable enough to compile a subcategorization frame lexicon automatically from some corpus. More efforts toward making these methods practicable are necessary.

4. ACQUISITION OF A BILINGUAL LEXICON

A. Overview

Machine Translation is one of the most important application areas of natural language processing and has been studied most intensively. The recent increased availability of bilingual texts has made it possible to use translation correspondences found in such texts for the automatic compilation of a bilingual lexicon. This section overviews the background and various techniques for this task.

Because the prerequisites and background to the research in bilingual lexicon acquisition are diverse, we will first give a summary of several important elements of the research in this field.

B. Background to Bilingual Lexicon Acquisition Research

1. Source Texts

Every research assumes online bilingual texts as the basic resource for bilingual lexicon acquisition. The most well-known resource is the Canadian Hansards, the proceedings of Canadian Parliament, recorded in both English and French. A number of other language pairs have also been studied.

Bilingual texts exist in various forms. They may be **parallel** or **nonparallel**. Parallel texts are ones for which the bilingual texts are mutual translations from one language to the other, keeping most of the essential information and the order of description. Some parallel texts may not preserve all the information between the two texts; that is, the target text may lack some information or may hold some extra information in comparison with the source text. Such texts are called **noisy** parallel texts.

It is not always true that one sentence is translated into exactly one sentence in the other language. The alignment of a parallel texts is, in general, a many-to-many mapping between sentences. Automatic alignment of parallel texts has been studied intensively, and quite accurate alignment is possible (see Chap. 18 for details). We use the term **alignment** here as an order-preserving mapping between sentences in parallel texts. The mapping between lexical items or between expressions or phrases in parallel texts that will feature in the bilingual lexicon is called the **correspondence**, which may not be order-preserving in the texts. This section concerns the techniques to find such correspondences in bilingual texts.

2. *Preprocessing of Source Texts*

Although some works directly use the bilingual texts intact, most of the works assume some preprocessing of the texts. There are various levels of possible preprocessing as follows:

LEMMATIZATION/SEGMENTATION. In **lemmatization** (or **stemming**), changed forms of words, such as inflected forms, are restored to the original base forms. There is a trade-off between doing and not doing lemmatization because some collocation may require some of its elements in a specifically changed form. For languages such as Japanese, Chinese, and Korean, segmentation of sentences into lexical items should be performed at the same time because in these languages words in sentences are not separated by white spaces.

PART-OF-SPEECH (POS) TAGGING. POS information is in some cases useful to define affinity between types of expressions of phrases in the two languages. Currently, large-scale POS-tagged corpora are available. Furthermore, up-to-date POS-taggers, most of which learn statistical parameters or rules from tagged corpora, achieve considerably high and stable accuracy (see Chap. 22). Several studies rely on existing POS-taggers.

SHALLOW-PARSING/NP RECOGNIZER. Noun phrases are important elements in sentences, and convey the conceptual information of entities that compose most of the domain-specific technical terms as described in Sec. 2. Fortunately, in English and some other languages, the forms of NPs without postmodifiers, such as relative clauses or prepositional phrases, have rather few variations and are easily defined in simple regular expressions. An efficient NP-recognizer is programmable as a finite-state automaton or as a pattern matcher. A number of authors implement an NP-recognizer as the preprocessor for the texts.

PARSING/BRACKETING. To analyze real-world texts syntactically is a very hard task, and not so many works rely on accurate parsing programs. Still, currently parsing systems give quite good results with relatively short sentences. When the structural correspondences are targeted for extraction of translation patterns, parsing or bracketing of sentences is performed to some extent.

3. *Unit of Correspondence*

Most of the early works in bilingual lexicon acquisition seek single-word correspondences. To acquire a bilingual lexicon suitable for translation of technical documents, units of translation should not necessarily be restricted to one word to one word. As described in Justeson and Katz (1995), most technical terms are noun phrases, so that at least an NP-recognizer or some other technique to handle sequences of multiple words is necessary. Moreover, some domain-specific collocations involve stereotyped or conventional expressions that must be learned as they are. For example, Smadja's (1993) XTRACT system finds long and interesting collocations from the WSJ corpus such as {21}.

The Dow Jones average of 30 industrials fell *NUMBER* points to *NUMBER*. {21}

Such collocations must be incorporated into the bilingual lexicon to realize natural and comprehensive translation. Quite different types of constituents are taken as the units of bilingual correspondences:

WORDS. Only word-to-word correspondences are taken into account.

NOUN PHRASES. An NP-recognizer or a pattern matcher is used to extract NPs in two languages. Those NPs together with single words are taken as the unit. Note that ambiguities arise in the determination of a unit in that a word may be taken as a unit per se or as just a part of another unit, and this may alternate according to the usage.

COLLOCATIONS. Word sequences of virtually unrestricted length and grammatical construction are possible collocations, and they must be acquired in bilingual lexicon. When a collocation consists of a fixed sequence of words, it is called a **rigid** collocation, and when word order is variable or some optional elements can be inserted it is called a **flexible** collocation.

TREE/DEPENDENCY STRUCTURES. When some syntactic analysis is possible, more accurate correspondences may be specifiable between sentences. Syntactic structure may be represented in several forms. A **parse tree** is a phrase-structure tree representation of a sentence, in which each subtree has a grammatical category name (a label). A **bracketed** sentence is a label-free parse tree. A **dependency tree** is a tree consisting of lexical items as the nodes, where an edge connects a word with its direct modifier. The modified word is called the **head** and the modifier is called the **dependant**.

It is not easy to find bilingual correspondences at the tree-structure level. However, such information is no doubt useful not only for acquisition of the bilingual lexicon, but also for the comparative study of languages. Collection of bilingual tree banks with tree-level correspondences is a hard but important goal for future work on annotated corpora. Sadler and Vendelmans (1990) propose tools and an environment for achieving such a task. Bilingual corpora so constructed may directly serve as the source knowledge for example-based machine translation (see Chap. 25). The activities for automatic construction of bilingual corpora with structural correspondences are briefly summarized at the end of Sec. 4.C.

4. Similarity Measures

To identify the bilingual correspondences, a measure of similarity needs be defined between words or expressions over two languages. This section gives an overview of similarity measures proposed in the literature. When the similarity is defined by means of occurrence frequencies of words or expressions in bilingual texts, it can be explained based on the **contingency table** (Table 12).

In this table, x and y stand for the occurrences of those expressions in the corpus, while $\neg x$ and $\neg y$ indicate nonoccurrences of them in the corpus. Frequency is counted as the number of aligned sentences (or regions). For example, $freq(x, y)$ is the number of aligned sentences of which the former and the latter contain x and y respectively. Similarly, $freq(x, \neg y)$ is the number of aligned sentences that contain x, but not y. The number of sentences that contain x is $a + b$. The total number of aligned sentences, N, is $a + b + c + d$. Similarity measures that are often used are

Table 12 Contingency Table

	y	$\neg y$
x	$freq(x, y) = a$	$freq(x, \neg y) = b$
$\neg x$	$freq(\neg x, y) = c$	$freq(\neg x, \neg y) = d$

mutual information (MI) Eq. {22}, the ϕ^2 **statistic** (Gale and Church, 1991) Eq. {23}, the **Dice coefficient** (or "Dice's coefficient") (Salton and McGill, 1983) Eq. {24}, and the **log-likelihood ratio** (Dunning, 1993) Eq. {25}.

$$
\begin{aligned}
I(x; y) &= \log_2 \frac{prob(x, y)}{prob(x)prob(y)} \\
&= \log_2 \frac{\frac{freq(x,y)}{N}}{\frac{freq(x)}{N} \ \frac{freq(y)}{N}} \\
&= \log_2 \frac{freq(x, y)\ N}{freq(x) \ \ freq(y)} \\
&= \log_2 \frac{aN}{(a+b)(a+c)}
\end{aligned}
$$

$$\phi^2(x, y) = \frac{(ad - bc)^2}{(a+b)(a+c)(b+d)(c+d)} \qquad \{23\}$$

$$
\begin{aligned}
Dice(x, y) &= \frac{2prob(x, y)}{prob(x) + prob(y)} \qquad \{24\} \\
&= \frac{2 \ \frac{freq(x,y)}{N}}{\frac{freq(x)}{N} + \frac{freq(y)}{N}} \\
&= \frac{2 \ \ freq(x, y)}{freq(x) + freq(y)} \\
&= \frac{2a}{2a + b + c}
\end{aligned}
$$

$$
\begin{aligned}
\text{Log-like} &= f(a) + f(b) + f(c) + f(d) - f(a+b) \qquad \{25\} \\
&\quad - f(a+c) - f(b+d) - f(c+d) + f(a+b+c+d)
\end{aligned}
$$

where $f(x) = x \log x$

In addition, Brown et al. (1988, 1990, 1993) use **translation probability** $t(f|e)$, the probability that an occurrence of the source word e is translated into the target word f. This value is estimated through the expectation maximization (EM) algorithm so that the sentences in the training bilingual corpus achieve the highest overall translation probability.

Statistics for the contingency table (see Table 12) cannot be defined if clean parallel corpora are not available. However, when large-scale bilingual texts in the same or similar domain are available, one can expect that lexical items in two distinct languages that amount to translation equivalents share a number of characteristics that are defined by their local contexts. This means that semantically similar words display similar behavior in both languages, and this may become the key to finding the correspondence of lexical items between two languages. Research based on this assumption of **contextual similarity** is introduced in a later section.

The precise definition of MI in information theory is the value calculated in Eq. {22} averaged over all x and y, and the value in Eq. {22} is known as **specific MI**. In this field of research, the simpler term is generally used however. Since MI is defined as the logarithm of the fraction of joint probability of x and y and the product of marginal probabilities of x and y, it takes a large positive value if mutual occurrence is higher than a mere chance, and takes a negative value if x and y occur less frequently than by chance. MI becomes zero if x and y are independent.

Mutual information is a widely used similarity measure and gives good statistics for relatively frequent events (Church and Hanks, 1990). However, it picks up too much for infrequent events. It is, therefore, used with another condition, such as making *freq*(x, y) as a threshold, or estimating statistical significance based on a *t-test* (Church et al., 1991). The t-test estimates the difference between the estimated means of two events by the proportional divergence over the standard distribution. When the standard deviation of the events are known as σ_1 and σ_2 and their estimated means are p_1 and p_2, t is defined as in Eq. {26}.

$$t = \frac{p_1 - p_2}{\sqrt{\frac{\sigma_1^2}{n_1} + \frac{\sigma_2^2}{n_2}}} \qquad \{26\}$$

Note, that if $prob(x|y)$ and $prob(x|\neg y)$ are compared over the unconditioned distribution, where the variance is estimated as $\sigma^2 = prob(x)(1 - prob(x))$, then t is given as Eq. {27}. This is actually equal to the square root of the χ^2-statistic.

$$t = \frac{prob(x|y) - prob(x|\neg y)}{\sqrt{prob(x)[1 - prob(x)]\left(\frac{1}{a+b} + \frac{1}{c+d}\right)}} \qquad \{27\}$$

$$= \sqrt{(a+b+c+d)(ad-bc)^2/(a+b)(a+c)(b+d)(c+d)}$$

Both the Dice coefficient and the ϕ^2-statistic take a value between 0 and 1. Dice produces 0 when $a = 0$, and ϕ^2 produces 0 when there is no correlation between a, b and c, d (equivalently between a, c and b, d). They both give the value 1 when $b = c = 0$.

Let us regard the corpus data as Beroulli trials of observing the occurrence of word x. When we regard the probability density function of the observation number k out of n trials as the function of various probability parameter θ with fixed n and k, it is called a **likelihood function** of the parameter θ and is written $L(\theta|n, k)$. The value of the function is called the likelihood. The likelihood ratio is the ratio of the maximum value of the likelihood under the null hypothesis to the maximum value of the likelihood under the alternative hypotheses. When the null hypothesis assumes that there is only one parameter of the occurrence probability for x [$prob(x)$] and the composite hypothesis assumes that there are two parameters conditioned over the occurrence and nonoccurrence of y, [$prob(x|y)$ and $prob(x|\neg y)$], the formula in Eq. {25} is derived by taking the logarithm of the inverse likelihood ratio (Dunning, 1993).

Table 13 shows examples of various combinations of the entities of the contingency table (by fixing the total number of aligned regions at 200,000) along with

Table 13 Comparison of Similarity Measures

	a	b	c	d	MI	Dice	ϕ^2	Log-like	wMI	wDice
1	7,120	2,675	1,068	189,137	4.1502	0.7919	0.6171	43,682.68	14.7746	10.1339
2	1,108	8,687	98	190,107	4.2295	0.2014	0.0985	6,137.61	2.3432	2.0373
3	142	3,275	14	196,569	5.7355	0.0795	0.0371	1,067.89	0.4072	0.5683
4	50	3,367	6	196,577	5.7076	0.0288	0.0128	369.75	0.1427	0.1625
5	2,636	780	510	196,074	5.6164	0.8034	0.6413	21,605.05	7.4024	9.1301
6	131	695	1,105	198,069	4.6816	0.1271	0.0157	632.21	0.3066	0.8937
7	2	0	0	199,998	16.6096	1.0000	1.0000	50.05	0.0166	1.0000
8	2	1	1	199,996	15.4397	0.6667	0.4444	42.41	0.0154	0.6667
9	200	100	100	199,600	8.7959	0.6667	0.4438	2,339.19	0.8796	5.0959
10	200	1	1	199,798	9.9514	0.9950	0.9901	3,137.70	0.9951	7.6058
11	10	2,737	5,037	192,216	−2.7933	0.0026	0.0003	81.94	−0.0140	0.0085

their corresponding similarity values. Since MI picks up too much for mutually rare events (see row 8), some authors use a weighted MI Eq. {28} (Fung, 1997).

$$prob(x, y) * MI \frac{a}{N} \log_2 \frac{aN}{(a+b)(a+c)} \tag{28}$$

Some authors use a weighted Dice coefficient Eq. {29} to signify higher co-occurring events (see rows 8 and 9) (Kitamura and Matsumoto, 1996).

$$log_2 freq(x, y)) * Dice = log_2 a \frac{2a}{2a+b+c} \tag{29}$$

5. *Types of Algorithm*

Once the similarity measure is defined, the algorithms to identify bilingual correspondences can be categorized into the following two types. There are exceptions, and the algorithms in the same category vary in their details. In any case, the objective of the algorithms is almost the same in that they intend to maximize the overall value of similarity.

EM-like iterative reestimation: Brown et al. (1993) define the translation probabilities and directly use the EM algorithm for their estimation. Even when probabilities are not used, it is an ordinary method that the similarity values of correspondence are determined gradually by an iterative process, using the similarity value in the preceding step to reestimate the new values. The final results are given as the limit values as the algorithm converges.

Greedy determination: The corresponding pairs are determined step by step according to the similarity values. A pair will never be reconsidered once it is fixed as a corresponding pair. Iteration may be incorporated such that only the correspondences with high confidence are fixed, and the remaining undergo the next iteration of recalculation of the similarity values.

Besides the form of the algorithm, a number of heuristics are used to improve the precision and the efficiency. Because the number of combinations that should be taken into account is usually huge, reduction of the number of combinations is quite important for these purposes. The following is a list of typical heuristics:

MRD (Machine Readable Dictionary) heuristics: When some translation pairs are already known to be very stable at least in the domain, such pairs need not be learned. Moreover, when an aligned sentence-pair contains the word x and y in a fixed translation pair (x, y), there is no need to consider the possibility of correspondences including x or y.

Cognate heuristics: Languages in the same family contain many words that have the same origin so that their spellings are quite similar. Such words are called **cognates**. Although the definition of cognates differs according to different authors, this heuristic works the same as the MRD heuristic.

POS (part-of-speech) heuristics: Between some language pairs, words of some POSs are always translated into words of the same POS. Even between language pairs in quite different families there are some constraints within POS correspondences. This implies that putting a preference on some POS combinations is useful to restrict the search space.

Positional heuristics: There must be some correlation between the positions of a word and its translation in the aligned sentences. For example, it is likely that the first word in an English sentence has its translation at the beginning part in the corresponding French sentence. An English verb appearing near the beginning of a sentence often has its translation at the end of the corresponding Japanese sentence.

6. *Evaluation*

Similar to other corpus-based techniques, the acquired bilingual lexicon is evaluated by **precision** and **recall**, and they may be computed either **by token** or **by type**.

- Precision is the ratio of the correct corresponding pairs to all acquired pairs. The correctness of corresponding pairs may be difficult to define and is in many cases subjective.
- Recall is the ratio of the source language vocabulary in the acquired bilingual lexicon to what is supposed to be in the lexicon.

When precision or recall is evaluated by token, all corresponding links in a test bilingual corpus (this may be same as the training corpus) are taken into account. When precision or recall is evaluated by type, each entry in the acquired bilingual lexicon is judged. The judgment is usually difficult because a word or an expression in the source language can have more than one translation, and the correctness of translation depends on the context.

Brown et al. (1993) used **perplexity** to evaluate their acquired lexicon. Perplexity is roughly the average number of expected words in a language model. As their model is a translation model from English to French, the perplexity of French text is computed on each reestimation step of EM algorithm, in which a steady decrease in perplexity is observed.

Another more direct automated evaluation method is proposed by Melamed (1995). His method, called bitext-based lexicon evaluation (BiBLE), places links between the words in a test bilingual text in accordance with the acquired bilingual lexicon. The number of links for a word type is compared with the number of all occurrences of the word type in the corpus to compute its **hit rate**. Hit rates for all word types in the lexicon are averaged, and this average hit rate is used for the estimation of the precision of the lexicon. The idea behind this is that a better bilingual lexicon will contain better translation pairs and will yield more links in the test corpus.

When each word in the bilingual lexicon has more than one target word and they are ranked in an order, the hit rate of a word type can be calculated for each kth rank, in which higher ranked target is privileged in placing links. By accumulating the ranked hit rate up to some k, k-best accumulative hit rates are easily calculated.

Although a conflict may arise when more than one source word selects the same target word as its translation, and it is not straightforward to extend this algorithm to a bilingual lexicon that has larger units than words, this kind of automatic evaluation method is very important.

C. Bilingual Lexicon Acquisition from Parallel Corpora

This section gives a brief overview of research on bilingual lexicon acquisition, especially from parallel corpora. Some assume that a sentence-level alignment has been done on the corpora.

The work of Brown et al. (1988, 1990, 1993) was the pioneering work of statistical machine translation, and proposed various models for estimation of translation probabilities of word-to-word correspondences between English and French. They used the Canadian Hansard aligned corpus. Although their proposal is a complete translation system based on statistics, we just look at the definition of translation probabilities. Suppose that a French sentence $\mathbf{f} = f_1, f_2, \cdots, f_m$ is a translation of English sentence $\mathbf{e} = e_1, e_2, \cdots, e_l$, where, f_j is the jth word in $\mathbf{f}$ and e_i is the ith word in $\mathbf{e}$. By $t(f|e)$, we mean the probability that the English word e is translated into the French word f, and estimate it by the corpus. In the following model (model 1 of Brown et al., 1993), the initial probability of any $t(f|e)$ does not affect the result. The expected number of times that e translates into f in aligned sentences $\mathbf{e}$ and $\mathbf{f}$, the count of f given e, is defined as in Eq. {30}.

$$c(f|e; \mathbf{f}, \mathbf{e}) = \frac{t(f|e)}{t(f|e_0) + t(f|e_l) + \cdots + t(f|e_l)} \sum_{j=1}^{m} \delta(f, f_j) \sum_{i=1}^{l} \delta(e, e_i) \qquad \{30\}$$

Here, e_0 denotes an imaginary null word so that $t(f|e_0)$ means that f is not a translation of any word in e. δ is Kronecker's delta function, where the $\delta(f, f_j) = 1$, if $f = f_j$, and 0 otherwise. The two summations of δs are for counting the number of possible correspondences of f and e in the sentence pair $\mathbf{f}$ and $\mathbf{e}$. They show that the maximum likelihood estimator for $t(f|e)$ is defined as in Eq. {31}.

$$t(f|e) = \frac{\sum_{(\mathbf{f},\mathbf{e}) \in Corpus} c(f|e; \mathbf{f}, \mathbf{e})}{\sum_{f} \sum_{(\mathbf{f},\mathbf{e}) \in Corpus} c(f|e; \mathbf{f}, \mathbf{e})} \qquad \{31\}$$

The EM-algorithm is applied to reestimate $t(f|e)$ using the foregoing eqations. They extend this model in various ways. Because a source word and the corresponding target word may have a tendency to appear in relatively similar positions in both sentences, especially with closely related language pairs, in model 2 they introduce

alignment probability denoted by $a(i|j, m, l)$, which is the probability that the jth word in a French sentence of length m is the translation of the ith word in an English sentence of length l. This is a way to implement positional heuristics. They further enhanced their model to models 3, 4, and 5, for which we do not go into the detail here. We only mention that model 3 enhances model 2 by adding the **fertility probability** that specifies how many French words each English word is translated into. Whereas the alignment probabilities are still defined at each position independently in model 3, model 4 is equipped with the parameters to take account of relative positions of words in a phrase if the fertility is greater than 1. Model 5 removes a deficiency that may arise in the previous models; that is, the French sentence produced may contain vacant positions or duplicate positions so that it may not form a straight string of words.

Because the number of parameters increases along with the sophistication of the models, and furthermore, model 1 is proved to give the only one true maximum, the EM-algorithm is run first with model 1 and the model is gradually replaced by better models in the training iteration. Table 14 shows examples of acquired translation and fertility probabilities [ϕ denotes the fertility number and $n(\phi|e)$ is its probability].

Kupiec (1993) employs a similar method for finding correspondences between French and English NPs. HMM-based POS taggers, and finite-state NP recognizers are used to extract NPs from sentences in both languages beforehand, so that only the correspondences between NPs are taken into consideration. An EM-like iteration is adopted to reestimate the probabilities of correspondences. Besides this preprocessing, the difference between this and Brown et al.'s model 1 is that the new counts are estimated not by Eq. {30} obtained from maximum likelihood estimation, but directly by the previous probability Eq. {32}. The precision is evaluated only with the top 100 correspondences and is estimated to be about 90% correct.

$$c(f|e; \mathbf{f}, \mathbf{e}) = t(f|e) \sum_{j=1}^{m} \delta(f, f_j) \sum_{i=1}^{l} \delta(e, e_i) \qquad \{32\}$$

Dagan et al. (1993) apply some modification to Brown et al.'s model 2 for finding word-level alignment in nonaligned parallel corpora. Church's (1993) character-based alignment algorithm *char_align* is first used to obtain a rough alignment

Table 14 Examples of Translation Probabilities for *national* and *the*

Probabilities for *national*				Probabilities for *the*			
f	$t(f\|e)$	ϕ	$n(\phi\|e)$	f	$t(f\|e)$	ϕ	$n(\phi\|e)$
nationale	0.469	1	0.905	le	0.497	1	0.746
national	0.418	0	0.094	la	0.207	0	0.254
nationaux	0.054			les	0.155		
nationales	0.029			l'	0.086		
				ce	0.018		
				cette	0.011		

Source: Brown et al., 1993.

at word level. Because sentence-level alignment is not assumed, he defines the alignment probability by the *offset* probability $o(k)$, which is the probability that the corresponding target word appears at k words away from the position to which the source word is connected. The offset is defined within the window size of 20, i.e., $-20 \leq k \leq 20$. He reports that his program *word_align* significantly improves the result of *char_align*. As *word_align* produces an alignment for all words, keeping the ordering 100% correctness is not possible. He evaluated the precision in accordance with the offset. When applied to the Canadian Hansard corpus, except for very high-frequency function words and infrequent words (fewer than three occurrences), the precision is 55%. However, 84% of correct correspondences are found within the offset of 3, which shows the very good performance of the algorithm. Similar precisions are achieved even with much more noisy data produced by OCR scanning.

Gale and Church (1991) propose a method of word correspondence that uses the ϕ^2-statistic for the similarity measure. Their algorithm is a greedy iterative method, in which they use smaller data in early iterations and gradually increase the size of the training data. At the end of each iteration, the program selects the pairs (x, y) that give a much higher ϕ^2 value than any other pairs of the form (x, z) or (w, y). Then, the size of the training corpus is increased, and all the selected pairs are removed and ϕ^2 values are recalculated for the next iteration. They also propose a kind of cognate heuristics that says that a pair is selected as a translation pair if it appears much more often than by chance and both words in the pair have the same first five characters with the words in an already selected pair. They report that more than 6000 pairs are acquired from a corpus of 220,000 corresponding regions and the cognate heuristics doubled the number of acquired pairs. In both cases, the precision is about 98%.

A similarity measure is used by van der Eijk (1993) that is very close to MI, in which positional heuristic is used to weight the cooccurrence frequencies of terms and a frequency-based heuristic is used to filter the candidates. The parallel corpora are aligned at sentence level, then taggers and NP recognizers are used to extract NPs. Correspondences between words and noun phrases are tested. The similarity measure is equivalent to Eq. {33}.

$$\frac{prob(y|x)}{prob(y)} = \frac{a/(a+b)}{(a+c)/N} = \frac{aN}{(a+b)(a+c)} \qquad \{33\}$$

One difference from MI is that the cooccurrence frequency a is actually replaced by another value a' that takes into account the positional difference in that "the score is decreased proportionally to the distance from the expected position." Similar to MI, the foregoing measure tends to overestimate very low-frequency terms. A frequency-based filter is employed that uses a threshold function defined by Eq. {34}.

$$\frac{prob(y|x)}{prob(x)} = \frac{aN}{(a+b)^2} \qquad \{34\}$$

The experiments were performed with Dutch–English parallel corpora and show that the positional heuristic is very effective in raising precision. By moving the threshold, the trade-off between recall and precision is easily controlled. Comparisons with other similarity measures were also performed and similar results were obtained both for MI plus filtering and for the ϕ^2-statistic when positional heuristics were employed.

Kumano and Hirakawa (1994) propose a method to find NP correspondences using linguistic evidence, as well as statistical evidence. They use a nonaligned Japanese–English corpus and a Japanese–English MRD. They first align the corpora using the MRD to obtain rough-corresponding units (a unit may be a phrase, a sentence, or a sequence of sentences). An NP recognizer is run on just the Japanese part of the corpus and for each Japanese NP J_i any English string E_j of length from 1 to $2 \times |J_i|$ is taken as the candidate translation. The statistical measure they use is Eq. {35}.

$$TSL \stackrel{\text{def}}{=} \frac{freq(J_i, E_j)}{freq(J_i)} = \frac{a}{a+b} \tag{35}$$

For the linguistic evidence, they use two criteria: the difference of length of Japanese and English terms, and word-for-word translation correspondences found in the MRD. They define a formula *TLL* to measure the similarity. The overall similarity of a Japanese NP and an English string is defined by a linear combination, $(TSL + \beta TLL)/(1 + \beta)$, in which they propose that the value of β should alter according to the value of $freq(J_i)$. Experiments are performed on a fairly small Japanese–English corpus and achieve over 70% precision for the top candidate and over 80% for top three candidates.

Melamed (1997) proposed a method for finding word-level correspondences with two kinds of interleaving procedures. The first procedure is based on a greedy iterative algorithm to identify word-to-word correspondences (links) using the likelihood ratio that is defined by the second procedure. In the second procedure, the overall word-to-word–linking probability is maximized by means of the maximum likelihood estimation, assuming two probabilistic parameters λ^+ and λ^-. The former is the probability that correct translation pairs of words are actually linked, and the latter that incorrect pairs are linked. The values of the parameters that maximize the total joint probabilities of link frequency on observed links are then estimated.

Suppose $B(k|n, p)$ is the probability that k pairs are linked out of n cooccurrences for which k has a binomial distribution with parameters n and the probability p. Then, the probability $P(u, v)$ of the link frequency of a pair (u, v) is defined by a linear combination of two probabilities [Eq. {36}].

$$P(u, v) = \tau B(k_{(u,v)}|n_{(u,v)}, \lambda^+) + (1 - \tau)B(k_{(u,v)}|n_{(u,v)}, \lambda^-) \tag{36}$$

He shows that τ can be thought of as a function of λ^+ and λ^- and estimates the values of λ^+ and λ^- that maximize the total joint link probability. An example using the Hansard corpus shows a clear global maximum, and a hill-climbing algorithm is used to find the best λ^+ and λ^- combination. The likelihood ratio is then defined by the ratio of true positive cases to false positive cases as in Eq. {37}.

$$L(u, v) = \frac{B(k_{(u,v)}|n_{(u,v)}, \lambda^+)}{B(k_{(u,v)}|n_{(u,v)}, \lambda^-)} \tag{37}$$

This likelihood ratio is again used for the next iteration of the first procedure, and the processes are repeated until they converge.

The evaluation is done by types, and the correctness is defined in two ways. A word pair (u, v) is *correct* if u and v co-occur at least once as a direct translation, and is *incomplete* if they are a part of a direct translation. The precisions are plotted by the recall rate, and are as high as 90% even if the incomplete pairs are considered incorrect. When the incomplete pairs are considered correct, it achieves 92.8% precision with 90% recall and 99% precision close to 40% recall.

So far, we have considered correspondence between words or NPs. Several attempts have been made at extracting more complicated structures that are of great importance for bilingual lexicons that specialize in domain-specific applications. Although most of the technical terms are NPs, there are a number of domain-specific collocations other than NPs.

As introduced in Sec. 4.B.3, Smadja's XTRACT is a system to extract such collocations. One approach to bilingual collocations is first to extract collocations from the source-language part of the bilingual corpus and then to find the corresponding expressions from the target-language part. Kumano and Hirakawa's work on acquiring NP correspondences is in this line of research. Smadja et al.'s (1996) *Champollion* pursues this line using XTRACT to acquire English collocations (both rigid and flexible). Dice threshold and absolute frequency threshold (set at 5) are used to identify the French translation for each English collocation. To avoid combinatorial explosion, the algorithm first selects French expressions of length 1 (i.e., single words) that pass the two thresholds, then the passed words are extended by increasing the length in the next iteration. By gradually increasing the length of expressions, those exceeding the thresholds are saved and then the expression with the highest Dice value is selected as the translation candidate. During this procedure, French expressions are treated as just bags of words. So, finally, the word order is determined by collecting all sentences containing the selected bag of words. A number of ideas to reduce computational overheads are reported in the paper.

Haruno et al. (1996) propose a similar method for Japanese–English collocations. Japanese and English rigid collocations are first extracted based on N-gram statistics. Then, for each language, the acquired rigid expressions are combined gradually by means of MI to construct a tree structure for each sentence. Finally, Japanese and English trees are related at each level using the graph-matching method proposed in Matsumoto et al. (1993). Because the trees are not necessarily constructed from consecutive expressions, translation expressions with flexible collocations are automatically acquired.

Kitamura and Matsumoto (1996) propose a method for acquiring rigid collocations of unrestricted length from Japanese–English-aligned corpora. They used a weighted Dice coefficient [Eq. {38}] to measure the similarity between expressions. Original corpora are morphologically analyzed and strings of content words (nouns, adjectives, verbs, and adverbs) are taken into consideration. A greedy iterative algorithm is employed. With each Japanese–English-aligned corpus with 10,000 sentence pairs of three different domains, they achieve over 90% precision.

$$\log_2 freq(x, y) Dice = \log_2 a \frac{2a}{2a + b + c} \qquad \{38\}$$

Some other methods directly try to find syntax-level correspondences. All of the research in this direction defines a similarity measure between phrases and puts some structural constraint that should be satisfied in the matched trees. Either of two kinds of tree structures is used: phrase structure trees or dependency trees.

Kaji et al. (1992) propose a method of finding tree-to-tree correspondences between Japanese and English aligned sentences. CKY parsers are applied to both

Japanese and English sentences to obtain possible phrase structures. The correspondences between phrases in the two languages are estimated by the common pairs of words that appear in a bilingual MRD. The phrase structure level correspondences are searched for in a bottom-up manner. Syntactic ambiguities are resolved during the matching process. They also propose an acquisition of translation templates from the results.

Matsumoto et al. (1993) use disjunctive dependency trees of Japanese- and English-aligned sentences to express syntactic ambiguity in a single structure. Similarity between substructures (subtrees) is defined as a function of the pairs of words that appear in a bilingual MRD or as estimated by the bilingual corpus using the Dice measure (Kitamura and Matsumoto, 1995; Matsumoto and Kitamura, 1997). Corresponding Japanese and English disjunctive trees are matched top-down to find structurally isomorphic trees that maximize the total similarity. The ambiguity (disjunction) is resolved during the matching process. The authors use this method to acquire fertile translation patterns with selectional constraints.

Meyers et al. (1996) describe a bottom-up algorithm for matching structure-sharing dependency trees. A bilingual MRD is used as a word-to-word similarity measure. A constraint is placed that the lowest common ancestors of two corresponding pairs of nodes between the matched trees should be preserved across the trees.

Wu (1995, 1997) defines inversion transduction grammars (ITGs) that directly express parallel phrase structures of two languages. An ITG production rule defines a dominance relation of phrases simultaneously between two languages. The ordering of daughter phrases in a rule may be switched so that the word order difference can be absorbed.

D. Bilingual Lexicon Acquisition from Nonparallel Corpora

Although parallel corpora are a valuable resource for bilingual lexicon acquisition, the availability of large-scale parallel corpora is quite limited. On the other hand, the availability of monolingual corpora is much higher and increasing very rapidly. Moreover, as is pointed out in Fung (1995), acquisition of a bilingual lexicon from parallel corpora is "at best a reverse engineering of the lexicon" that translators already know. Some results have been reported using nonparallel corpora in related domains for acquisition of bilingual lexicon. All research in this direction relies on the fact that semantically similar terms appear in similar contexts.

Fung (1995) defines the similarity between a pair of words by their **context heterogeneity**. The context heterogeneity of a word w is a two-dimensional vector (x, y), where x is the proportion of different types of tokens appearing to the left of w; that is, if a is the number of different types appearing immediately to the left of w, and c is the number of occurrences of w, $x = a/c$. The value of y is defined in the same way to the right of w. The similarity between a pair of Chinese and English words is defined by the Euclidean distance of their context heterogeneity vectors. Fung (1997) extends the idea by giving a more informative vector to each word in nonparallel corpora. A bilingual MRD is used, from which known translation word pairs called **seed words** are decided beforehand. These seed words are used to define vectors that characterize candidate words. The similarity between the candidate word and each seed word is defined by the weighted MI, shown in Eq. {28}, and becomes a coordinate in the vector. Note that the similarity is calculated within a monolingual corpus. The co-occurrence in a monolingual corpus is defined either within a clause, a sentence, or paragraphs, depending on the size and quality of the corpus. Because the seed words have one-to-one correspondence between the languages, the vectors for words in both languages have the same dimension. The similarity of a pair of words in different languages is defined either by Euclidean distance or by the Cosine Measure. Two experiments have been reported. The first interestingly uses two distinct corpora in same language, different parts of English WSJ, and tests how accurately the same words are identified. Using 307 seed words and about 600 test words, the precision of the top candidate was 29% and that of top 100 candidates was 58%. The second experiments were performed on Japanese and English corpora. Although the number of test words is small (19 Japanese words and 402 English words), the results achieved a precision of 20–50%. As the author describes, this could be quite usable as a translator aid.

Kaji and Aizono (1996) report a similar technique to acquire translation pairs from bilingual corpora. Taggers and NP recognizers are run on the corpora to identify content words and noun compounds for the translation units. A context-based similarity measure between term pairs is defined by means of a bilingual MRD. Correlation of two words in a monolingual corpus is defined by co-occurrenes within sentences. They do not define any anchoring word, such as seed words, but take into account all content words appearing in the MRD and occurring with the candidate. Each candidate is associated with a vector composed of the frequency counts of the co-occurring words. The similarity of a Japanese–English term pair is then defined by the proportion of the number of word counts in the vector that can be connected to another word in the opposite word vector by the bilingual MRD, in which the MRD is considered to be defining a many-to-many mapping, and a word is connected to another word if at least one mapping between them is defined in the MRD. Several filtering heuristics are proposed, of which we omit the details. Although this method is applicable to nonparallel corpora as well as to parallel ones, experiments are performed with fairly small-sized parallel corpora. The recall is calculated as the proportion of the extracted correct pairs over the number of terms in the Japanese corpus. The results show recall of about 30% and precision of about 88%.

Rapp (1995) gives another idea for using context-based similarity defined across nonparallel corpora. The background intuition is the same as those already presented. If two words in a language are related to each other, their translation equivalents must be related in the other language. The similarity values between all the pairs of n words in a language can be represented in a form of an $n \times n$ matrix, in which the element a_{ij} is the similarity between the words w_i and w_j. Suppose we pick up n words from one language and their translations in another language. If the foregoing intuition is true, the difference between two $n \times n$ matrices constructed for those languages becomes minimal when their word orders correspond. Co-occurrence is defined by the appearance within the window size of 13 words, and the three kinds of similarity measures [Eq. {39}–{41}] are tested.

$$a_{ij} = \frac{freq(w_i, w_j)^2}{freq(w_i)freq(w_j)} \qquad \{39\}$$

$$a_{ij} = freq(w_i, w_j) \qquad \{40\}$$

$$a_{ij} = \frac{freq(w_i, w_j)}{freq(w_i)freq(w_j)} \qquad \{41\}$$

Large-scale English and German nonparallel corpora are used and the difference between the matrices are measured by Eq. {42}

$$s = \sum_{i=1}^{N} \sum_{j=1}^{N} |e_{ij} - g_{ij}| \qquad \{42\}$$

where e_{ij} is the i, j element of the English matrix and g_{ij} the German. N is the size of the vocabulary of both languages. The two matrices are normalized beforehand so that all the elements in a matrix sum up to the same value. Because it is difficult to apply this idea to a real setting, a feasibility study is reported. First, German words are selected that are the translations of English vocabulary. Then the word order of the German matrix is randomly permuted and the difference is computed; that is, the experiments are performed the other way round. The results show that the more the number of noncorresponding positions increases, the larger the difference between the matrices. The first formula of the three gives the steepest increase.

Tanaka and Iwasaki (1996) extend the idea further. Because one-to-one correspondence between translation equivalents does not necessarily hold in a word-based model, they enhance the model by introducing the translation matrix T, of which the element t_{ij} defines the probability that ith word in the source language has the jth word in the target language as its translation. Then the co-occurrence similarity a_{uv} in language A is translated into the other language B by the formula in Eq. {43}.

$$s = \sum_{u,v} t_{ku} a_{uv} t_{lv} \qquad \{43\}$$

If the uth word and vth word in A correspond to kth word and lth word in B, this formula gives a value close to b_{kl}. Then, the original problem is replaced to find the translation matrix T that gives the smallest value for $|T^t AT - B|$.

5. CONCLUSIONS

This chapter has presented an overview of techniques for automatic acquisition of lexical knowledge from online resources, especially from corpora. After considering lexical knowledge acquisition at various levels, we have focused on two active areas of automatic construction of lexicon: subcategorization frame acquisition of verbs, and acquisition of a translation lexicon from bilingual corpora.

Research on subcategorization frame acquisition of verbs was classified into two categories: (a) extraction of case patterns from unparsed corpora; and (b) acquisition of subcategorization frames from *parsed* corpora. The techniques employed in those two categories are quite different, and the types of information included in the acquisition results also differ. However, it is quite desirable that some common techniques are devised in the future so that they are applicable to subcategorization frame acquisition from both unparsed and parsed corpora.

For the acquisition of a bilingual lexicon, we first considered six issues on bilingual lexicon acquisition: (a) source texts, (b) preprocessing of source texts, (c) unit of correspondence, (d) similarity measures, (e) types of algorithm, and (f) evaluation. Then, we classified existing works into two categories: bilingual lexicon acquisition from parallel or nonparallel corpora. So far, most of the attempts at bilingual lexicon acquisition concentrate on the correspondence between words or NPs, and significant progress has been made in the acquisition of word/NP-level translation lexicons. On the other hand, the number of works on acquiring translation knowledge involving more complicated structures, such as *flexible* collocations and syntax-level translation patterns, is quite limited. It is very important to develop robust techniques for acquiring such higher-level translation knowledge from bilingual corpora. This requires parallel corpora of much larger size than for word/NP-level translation knowedge. Furthermore, those parallel corpora need to be aligned at the sentence or syntactic structure levels. Therefore, it is also necessary to apply existing techniques of aligning parallel texts to the (semi-)automatic construction of (aligned) parallel corpora that are large enough for acquiring that kind of higher-level translation knowledge.

REFERENCES

Abe N, H Li (1996). Learning word association norms using tree cut pair models. Proceedings of the 13th International Conference on Machine Learning, Bari, Italy, pp 3–11.

Amsler RA (1981). A taxonomy for English nouns and verbs. 19th Annual Meeting of the Association for Computational Linguistics, Stanford, CA, pp 133–138.

Berger AL, SA Della Pietra, VJ Della Pietra (1996). A maximum entropy approach to natural language processing. Comput Linguist 22:39–71.

Boguraev B, T Briscoe, J Carroll, D Carter, C Grover (1987). The derivation of a grammatically indexed lexicon from the Longman Dictionary of Contemporary English. 25th Annual Meeting of the Association for Computational Linguistics, Stanford, CA, pp 193–200.

Brachman RJ, JG Schmolze (1985). An overview of the KL-ONE knowledge representation system, Cognit Sci 9:171–216.

Brent MR (1991). Automatic acquisition of subcategorization frames from untagged text. 29th Annual Meeting of the Association for Computational Linguistics, Berkeley, CA, pp 209–214.

Brent MR (1991). Automatic semantic classification of verbs from their syntactic contexts: an implemented classifier for stativity. Fifth Conference of the European Chapter of the Association for Computational Linguistics, Berlin, Germany, pp 222–226.

Brent MR (1993). From grammar to lexicon: unsupervised learning of lexical syntax. Comput Linguist 19:243–262.

Briscoe T, J Carroll (1997). Automatic extraction of subcategorization from corpora. Fifth Conference on Applied Natural Language Processing, Washington, DC, pp 356–363.

Brown PF, J Cocke, S Della Pietra, V Della Pietra, F Jelinek, RL Mercer, PS Roosin (1988). A statistical approach to language translation. Proceedings of the 12th International Conference on Computational Linguistics (COLING), Budapest, Hungary, pp 71–76.

Brown PF, J Cocke, S Della Pietra, V Della Pietra, F Jelinek, JD Lafferty, RL Mercer, PS Roosin (1990). A statistical approach to machine translation. Comput Linguist 16:79–85.

Brown PF, S Della Pietra, V Della Pietra, RL Mercer (1993). The mathematics of statistical machine translation: parameter estimation. Comput Linguist 19:263–311.

Charniak E (1997). Statistical parsing with a context-free grammar and word statistics. Proceedings of the 14th National Conference on Artificial Intelligence, Providence, RI, pp 598–603.

Chodorow M, R Byrd, G Heidorn (1985). Extracting semantic hierarchies from a large on-line dictionary. 23rd Annual Meeting of the Association for Computational Linguistics, Chicago, IL, pp 299–304.

Church KW, P Hanks (1990). Word association norms, mutual information, and lexicography. Comput Linguist 16:22–29.

Church KW, W Gale, P Hanks, D Hindle (1991). Using statistics in lexical analysis. In U Zernik, ed. Lexical Acquisition: Exploiting On-line Resources to Build a Lexicon, Hillsdale, NJ: Lawrence Erlbaum Associates, pp 115–164.

Church KW (1993). Char_align: a program for aligning parallel texts at the character level. 31st Annual Meeting of the Association for Computational Linguistics, Columbus, OH, pp 1–8.

Collins M, J Brooks (1995). Prepositional phrase attachment through a backed-off model. Proceedings of the Third Workshop on Very Large Corpora, Cambridge, MA, pp 27–38.

Collins M. (1996) A new statistical parser based on bigram lexical dependencies. 34th Annual Meeting of the Associatioan for Computational Linguistics, Santa Cruz, CA, pp 184–191.

Dagan I, KW Church, WA Gale (1993). Robust bilingual word alignment for machine aided translation. Proceedings of the Workshop on Very Large Corpora: Academic and Industrial Perspectives, Columbus, OH, pp 1–8.

Dagan I, K Church (1997). Termight: coordinating humans and machines in bilingual terminology acquisition. Mach Transl 12:89–107.

Daille B (1996). Study and implementation of combined techniques for automatic extraction of terminology. In JL Klavans, P Resnik, eds. Balancing Act: Combining Symbolic and Statistical Approaches to Language. Cambridge, MA: MIT Press, pp 49–66.

Daille B, E Gaussier, JM Lange (1994). Towards automatic extraction of monolingual and bilingual terminology. COLING 94. 15th International Conference on Computational Linguistics, Kyoto, Jpn, pp 515–521.

de Lima EF (1997). Acquiring German prepositional subcategorization frames from corpora. Proceedings of the 5th Workshop on Very Large corpora, Beijing and Hong Kong, pp 153–167.

Della Pietra S, V Della Pietra, J Lafferty (1997). Inducing features of random fields. IEEE Trans on Pattern Anal Mach Intell 19:380–393.

Dempster AP, NM Laird, DB Rubin (1977). Maximum likelihood from incomplete data via the EM algorithm. J R Stat Soc 39:1–38.

Dorr BJ (1992). A parameterized approach to integrating aspect with lexical–semantic for machine translation. 30th Annual Meeting of the Association for Computational Linguistics, Newark, DE, pp 257–264.

Dorr BJ, J Garman, A Weinberg (1994). From syntactic encodings to thematic roles: building lexical entries for interlingual MT. Mach Transl 9:221–250.

Dorr BJ, D Jones (1996). Role of word sense disambiguation in lexical acquisition: predicting semantics from syntactic cues. COLING-96: 16th International Conference on Computational Linguistics, Copenhagen, Den, pp 322–327.

Dunning T (1993). Acurate methods for statistics of surprise and coincidence. Comput Linguist 19:61–74.

EDR (Electronic Dictionary Research Institute, Ltd., Japan) (1995). Electronic Dictionary Technical Guide. Tokyo: EDR.

Francis W, H Kŭcera (1982). Frequency Analysis of English Usage, Lexicon and Grammar. Boston, MA; Houghton Mifflin.

Fung P (1995). Compiling bilingual lexicon entries from a non-parallel English–Chinese corpus. Proceedings of 3rd Workshop on Very Large Corpora, Cambridge, MA, pp 173–183.

Fung P (1997). Finding terminology translation from non-parallel corpora. Proceedings of 5th Workshop on Very Large Corpora, Beijing and Hong Kong, pp 192–202.

Gale W, K Church (1991). Identifying word correspondences in parallel texts. Proceedings of Speech and Natural Language Workshop, Orange Grove, CA, pp 152–157.

Grishman R, C Macleod, A Meyers (1994). Complex syntax: building a computational lexicon. COLING 94: 15th International Conference on Computational Linguistics, Kyoto, Jpn, pp 268–272.

Guthrie L, BM Slator, Y Wilks, R Bruce (1990). Is there content in empty heads? COLING–90: Papers presented to the 13th International Conference on Computational Linguistics, Helsinki, Finland, vol 3, pp 138–143.

Haruno M (1995a). A case frame learning method for Japanese polysemous verbs. Proceedings of the AAAI Spring Symposium: Representation and Acquisition of Lexical Knowledge: Polysemy, Ambiguity, and Generativity, Stanford, CA, pp 45–50.

Haruno M (1995b). Verbal case frame acquisition as data compression. Proceedings of the 5th International Workshop on Natural Language Understanding and Logic Programming, Lisbon, Portugal, pp 97–105.

Haruno, M, S Ikehara, T Yamazaki (1996). Learning bilingual collocations by word-level sorting. COLING-96: 16th International Conference on Computational Linguistics, Copenhagen, Denmark, pp 525–530.

Hindle D (1983). Deterministic parsing of syntactic non-fluencies. 21st Annual Meeting of the Association for Computational Linguistics, Cambridge, MA, pp 123–128.

Hindle D, M Rooth (1993). Structural ambiguity and lexical relations. Comput Linguist 19:103–120.

Hogenhout WR, Y Matsumoto (1996). Training stochastical grammars on semantical categories. In: S. Wermter, E Riloff, G Scheler, eds. Connectionist, Statistical and Symbolic Approaches to Learning for Natural Language Processing. Berlin:Springer-Verlag, pp 160–172.

Hornby AS (1989). Oxford Advanced Learner's Dictionary of Current English, 4th ed. Oxford: Oxford University Press.

Ide N, J Véronis (1993). Extracting knowledge bases from machine-readable dictionaries: have we wasted our time? Proceedings of International Conference on Knowledge Building and Knowledge Sharing, Tokyo, Japan, pp 257–266.

Ikehara S, M Miyazaki, A Yokoo (1993). Nichiei kikaihon-yaku no tameno imi kaiseki yō no chishiki to sono bunkainō [Classification of language knowledge for meaning analysis in machine translation]. Trans Inform Process Soc Jpn 34:1962–1704.

IPA (Information–Technology Promotion Agency, Japan) (1987). Keisanki Yō Nihongo Kihon Dōshi Jisho IPAL (Basic Verbs) [IPA Lexicon of the Japanese Language for Computers IPAL (Basic Verbs)]. Tokyo: Information–Technology Promotion Agency.

Joshi AK, Y Schabes (1992). Tree adjoining grammars and lexicalized grammars. In: M Nivat, A Podelski, eds. Tree Automata and Languages. Amsterdam: Elsevier, pp 409–431.

Justeson JS, SM Katz (1995). Technical terminology: some linguistic properties and an algorithm for identification in text. Nat Lang Eng 1:9–27.

Kaji H, Y Kida, Y Morimoto (1992). Learning translation templates from bilingual text. Proceedings 15th International Conference on Computational Linguistics, Nantes, France, pp 672–678.

Kaji H, T Aizono (1996). Extracting word correspondences from bilingual corpora based on word co-occurrence information. COLING–96. 16th International Conference on Computational Linguistics, Copenhagen, Denmark, pp 23–28.

Kayaalp M, T Pedersen, R Bruce (1997). A statistical design making method: a case study on prepositional phrase attachment. Proceedings of the Workshop on Computational Natural Language Learning, Madrid, Spain, pp 33–42.

Kitamura M, Y Matsumoto (1995). A machine translation system based on translation rules acquired from parallel corpora. International Conference on Recent Advances in Natural Language Processing RANLP '95, Tzigov Chark, Bulgaria, pp 27–36.

Kitamura M, Y Matsumoto (1996). Automatic extraction of word sequence correspondences in parallel corpora. Proceedings of the 4th Workshop on Very Large Corpora, Copenhagen, Denmark, pp 79–87.

Klavans J, E Tzoukermann (1990). The BICORD system: combining lexical information from bilingual corpora and machine readable dictionaries. COLING–90: 13th International Conference on Computational Linguistics, Helsinki, Finland, vol 3, pp 174–179.

Kumano A, H Hirakawa (1994). Building an MT dictionary from parallel texts based on linguistic and statistical information. COLING 94: 15th International Conference on Computational Linguistics, Kyoto, Japan, pp 76–81.

Kupiec J (1993). An algorithm for finding noun phrase correspondences in bilingual corpora. 31st Annual Meeting of the Association for Computational Linguistics, Columbus, OH, pp 17–22.

Lenat DB (1995). CYC: a large-scale investment in knowledge infrastructure. Commun ACM 38(11):33–88.

Levin B (1993). English Verb Classes and Alternations. Chicago, IL: University of Chicago Press.

Li H, N Abe (1995). Generalizing case frames using a thesaurus and the MDL principle. International Conference on Recent Advances in Natural Language Processing, Tzigov Chark, Bulgaria, pp 239–248.

Li H, N Abe (1996). Learning dependencies between case frame slots. COLING–96: 16th International Conference on Computational Linguistics, Copenhagen, Denmark, pp 10–15.

Liu R-L, VW Soo (1993). An empirical study on thematic knowledge acquisition based on syntactic clues and heuristics. 31st Annual Meeting of the Association for Computational Linguistics, Columbus, OH, pp 243–250.

Magerman DM (1995). Statistical decision-tree models for parsing. 33rd Annual Meeting of the Association for Computational Linguistics, Cambridge, MA, pp 276–283.

Manning CD (1993). Automatic acquisition of a large subcategorization dictionary from corpora. 31st Annual Meeting of the Association for Computational Linguistics, Columbus, OH, pp 235–242.

Marcus MP, B Santorini, MA Marcinkiewicz (1993). Building a large annotated corpus of English: the Penn Treebank. Comput Linguist 19:313–330.

Matsumoto Y, H Ishimoto, T Utsuro (1993). Structural matching of parallel texts. 31st Annual Meeting of the Association for Computational Linguistics, Columbus, OH, pp 23–30.

Matsumoto Y, M Kitamura (1997). Acquisition of translation rules from parallel corpora. In: R Mitkov, N Nicolov, eds. Recent Advances in Natural Language Processing: Selected Papers from RANLP'95, Amsterdam: John Benjamins, pp 405–416.

Melamed ID (1995). Automatic evaluation and uniform filter cascades for inducing N-best translation lexicons. Proceedings of the 3rd Workshop on Very Large Corpora, Cambridge, MA, pp 184–198.

Melamed ID (1997). A word-to-word model of translation equivalence. 35th Annual Meeting of the Association for Computational Linguistics and 8th Conference of the European Chapter of the Association for Computational Linguistics, Madrid, Spain, pp 490–497.

Meyers A, R Yangarber, R Grishman (1996). Alignment of shared forests for bilingual corpora, COLING–96: 16th International Conference on Computational Linguistics, Copenhagen, Denmark, pp 460–465.
Miller GA (1995). WordNet: a lexical database for English. Commun ACM 38:39–41.
Miyata T, T Utsuro, Y Matsumoto (1997). Bayesian network models of subcategorization and their MDL-based learning from corpus. Proceedings of the 4th Natural Language Processing Pacific Rim Symposium, Phuket, Thailand, pp 321–326.
NLRI (National Language Research Institute) (1993). Bunrui Goi Hyō [Word List by Semantic Principles]. Tokyo, Shuei Shuppan.
Oishi A, Y Matsumoto (1995). A method for deep case acquisition based on surface case pattern analysis. Proceedings of the 3rd Natural Language Processing Pacific Rim Symposium, Seoul, Korea, pp 678–684.
Oishi A, Y Matsumoto (1997a). Automatic extraction of aspectual information from a monolingual corpus. 35th Annual Meeting of the Association for Computational Linguistics and 8th Conference of the European Chapter of the Association for Computational Linguistics, Madrid, Spain, pp 352–359.
Oishi A, Y Matsumoto (1997b). Detecting the organization of semantic subclasses of Japanese verbs. Int J Corpus Linguist 2:65–89.
Pollard C, IA Sag (1994). Head-Driven Phrase Structure Grammar. Chicago, IL. University of Chicago Press.
Pustejovsky J, S Bergler, P Anick (1993). Lexical semantic techniques for corpus analysis, Comput Linguist 19:331–358.
Pustejovsky J (1995). The Generative Lexicon. Cambridge, MA: MIT Press.
Pustejovsky J, B Boguraev (1996). Lexical Semantics: The Problem of Polysemy. Oxford: Oxford University Press.
Rapp R (1995). Identifying word translations in non-parallel texts. 33rd Annual Meeting of the Association for Computational Linguistics, Cambridge, MA, pp 320–322.
Resnik P (1992). WordNet and distributional analysis: a class-based approach to lexical discovery. Proceedings of the AAAI-92 Workshop on Statistically-Based Natural Language Programming Techniques, Stanford, CA, pp 48–56.
Resnik P (1993). Semantic classes and syntactic ambiguity. Proceedings of the Human Language Technology Workshop, Princeton, NJ, pp 278–283.
Ribas F (1995). On learning more appropriate selectional restrictions. Seventh Conference of the European Chapter of the Association for Computational Linguistics, Dublin, Ireland, pp 112–118.
Rissanen (1984). Universal coding, information, prediction, and estimation. IEEE Trans Inform Theory IT-30:629–636.
Rissanen J (1989). Stochastic Complexity in Statistical Inquiry. Singapore: World Scientific.
Sadler V, Vendelmans (1990). Pilot implementation of a bilingual knowledge bank. COLING–90: 13th International Conference on Computational Linguistics, Helsinki, Finland, vol 3, pp 449–451.
Saint–Dizier, P, Viegas eds (1995). Computational Lexical Semantics. Cambridge: Cambridge University Press.
Salton G, MJ McGill (1983). Introduction to Modern Information Retrieval. New York: McGraw-Hill.
Shimizu M, S Narita. eds (1979). Japanese–English Dictionary. Tokyo: Kodansha Gakujutsu Bunko.
Smadja F (1993). Retrieving collocations from text: Xtract. Comput Linguist 19:143–177.
Smadja F, KR McKeown, V Hatzivassiloglou (1996). Translating collocations for bilingual lexicons: a statistical approach. Comput Linguist 22:1–38.

Suzuki J (1993). A construction of Bayesian networks from databases based on an MDL principle. Proceedings of the 9th Conference on Unertainty in Artificial Intelligence, Washington, DC, pp 66–273.

Tanaka K, H Iwasaki (1996). Extraction of lexical translations from non-aligned corpora. COLING–96: 16th International Conference on Computational Linguistics, Copenhagen, Denmark, pp 580–585.

Ushioda A, DA Evans, T Gibson, A Waibel (1993). The automatic acquisition of frequencies of verb subcategorization frames from tagged corpora. Proceedings of Workshop: Acquisition of Lexical Knowledge from Text, Columbus, OH, pp 95–106.

Utsuro T, Y Matsumoto, M Nagao (1993). Verbal case frame acquisition from bilingual corpora. 13th International Joint Conference on Artificial Intelligence, Chambéry, France, pp 1150–1156.

Utsuro T (1996). Sense classification of verbal polysemy based on bilingual class/class association. COLING–96: 16th International Conference on Computational Linguistics, Copenhagen, Denmark, pp 968–973.

Utsuro T, Y Matsumoto (1997). Learning probabilistic subcategorization preference by identifying case dependencies and optimal noun class generalization level. Fifth Conference on Applied Natural Language Processing, Washington, DC, pp 364–371.

Utsuro T, T Miyata, Y Matsumoto (1997). Maximum entropy model learnng of subcategorization preference. Proceedings of the 5th Workshop on Very Large Corpora, Beijing and Hong Kong, pp 246–260.

van der Eijk P (1993). Automating the acquisition of bilingual terminology. Sixth Conference of the European Chapter of the Association for Computational Linguistics, Utrecht, Netherlands, pp 113–119.

Wu D (1995). Stochastic inversion transduction grammars, with application to segmentation, bracketing, and alignment of parallel corpora. 14th International Joint Conference on Artificial Intelligence, Montréal, Québec, pp 1328–1335.

Wu D (1997). Stochastic inversion transduction grammars and bilingual parsing of parallel corpora. Comput Linguist 23:377–403.

25

Example-Based Machine Translation

HAROLD SOMERS

UMIST, Manchester, England

1. INTRODUCTION

This chapter follows on from Chap. 13 in describing some corpus-based and empirical approaches to machine translation (MT). It is concerned primarily with variously named approaches to MT that we can generally characterise as "example-based MT" (EBMT), although we shall also consider the purely statistical approach of the IBM group.

In 1988, at the second TMI conference at Carnegie Mellon University, IBM's Peter Brown shocked the audience by presenting an approach to MT that was quite unlike anything that most of the audience had ever seen or even dreamed of before. IBM's "purely statistical" approach flew in the face of all the received wisdom about how to do MT at that time, eschewing the rationalist linguistic approach in favour of an empirical corpus-based one.

There followed something of a flood of "new" approaches to MT, few as overtly statistical as the IBM approach, but all having in common the use of a corpus of translation examples, rather than linguistic rules, as a significant component. This apparent difference was often seen as a confrontation, especially, for example, at the 1992 TMI conference in Montreal, which had the explicit theme "Empiricist vs. Rationalist Methods in MT" (Isabelle, 1992), although already by that date most researchers were developing hybrid solutions using both corpus-based and theory-based techniques.

In this chapter, we will review the achievements of a range of approaches to corpus-based MT, which we will consider variants of EBMT, although individual authors have used alternative names, perhaps wanting to bring out some key difference that distinguishes their own approach: "analogy-based," "memory-based," and "case-based" are all terms that have been used. These approaches all share the use of a corpus or database of already translated examples, and involve a process of matching a new input against this database to extract suitable examples that are then recombined in an analogical manner to determine the correct translation.

Two variants of the corpus-based approach stand somewhat apart from the scenario suggested here. One, which we will not discuss at all in this chapter, is the Connectionist or Neural Network approach: some MT research within this paradigm is discussed in Chap. 27.

The other major "new paradigm" is the purely statistical approach already mentioned, and usually identified with the IBM group's Candide system (Brown et al., 1990, 1993), although the approach has also been taken up by several other researchers. The statistical approach is clearly example-based, in that it depends on a bilingual corpus, but the matching and recombination stages that characterise EBMT are implemented in a quite different way in these approaches; more significant is that the important issues for the statistical approach are somewhat different, focusing, as one might expect, on the mathematical aspects of estimation of statistical parameters for the language models. We will describe these approaches in our overview.

2. EBMT AND TRANSLATION MEMORY

EBMT is sometimes confused with the related technique of "translation memory" (TM) (see Chap. 13, Sec. 5.E). This problem is exacerbated because the two gained wide publicity at roughly the same time, and also by the (thankfully, short-lived) use of the term "memory-based translation" as a synonym for EBMT. Although they have in common the idea of reuse of examples of already existing translations, they differ in that TM is an interactive tool for the human translator, whereas EBMT is an essentially automatic translation technique or methodology. They share the common problems of storing and accessing a large corpus of examples, and of matching an input phrase or sentence against this corpus; but having located a (set of) relevant example(s), the TM leaves it to the human to decide what, if anything, to do next, whereas for EBMT the hard work has only just begun.

One other thing that EBMT and TM have in common is the long time period that elapsed between the first mention of the underlying idea and the development of systems exploiting the ideas. The original idea for EBMT can be found in a paper presented by Nagao, at a 1981 conference, although not published until 3 years later (Nagao, 1984). The essence of EBMT, called "machine translation by example-guided inference, or machine translation by the analogy principle" by Nagao, is succinctly captured by his much quoted statement:

> Man does not translate a simple sentence by doing deep linguistic analysis, rather, Man does translation, first, by properly decomposing an input sentence into certain fragmental phrases [...], then by translating these phrases into other language phrases, and finally by properly composing these fragmental translations into one long sentence. The translation of each fragmental phrase will be done by the analogy translation principle with proper examples as its reference (Nagao, 1984:178f).

Nagao correctly identified the three main components of EBMT: matching fragments against a database of real examples, identifying the corresponding translation fragments, and then recombining these to give the target text. Clearly, EBMT involves two important and difficult steps beyond the matching task that it shares with TM.

Mention should also be made, at this point, of the work of the DLT group in Utrecht, often ignored in discussions of EBMT, but dating from about the same time as (and probably without knowledge of) Nagao's work. The matching technique suggested by Nagao involves measuring the semantic proximity of the words, using a thesaurus. A similar idea is found in DLT's "Linguistic Knowledge Bank" of example phrases described in Pappegaaij et al. (1986) and Schubert (1986).

3. UNDERLYING PROBLEMS

In this section we will review some of the general problems underlying example-based approaches to MT. Starting with the need for a database of examples (i.e., parallel corpora), we then discuss how to choose appropriate examples for the database, how they should be stored, various methods for matching new inputs against this database, what to do with the examples once they have been selected, and finally, some general computational problems on speed and efficiency.

A. Parallel Corpora

Because EBMT is corpus-based MT, the first thing that is needed is a parallel aligned corpus.[1] Machine-readable parallel corpora are, in this sense, quite easy to come by: EBMT systems are often felt to be best suited to a sublanguage approach, and an existing corpus of translations can often serve to define implicitly the sublanguage that the system can handle. Researchers may build up their own parallel corpus or may locate such corpora in the public domain. The Canadian and Hong Kong Parliaments both provide huge bilingual corpora in the form of their parliamentary proceedings, the European Union is a good source of multilingual documents, and many World Wide Web pages are available in two or more languages. However, not all these resources necessarily meet the sublanguage criterion.

The alignment problem can be circumvented by building the example database manually, as is sometimes done for TMs, when sentences and their translations are added to the memory as they are typed in by the translator.

B. Suitability of Examples

The assumption that an aligned parallel corpus can serve as an example database is not universally made. Several EBMT systems work from a manually constructed database of examples, or from a carefully filtered set of "real" examples.

---

[1] By *parallel* we mean a text together with its translation. By *aligned*, we mean that the two texts have been analysed into corresponding segments; the size of these segments may vary, but typically corresponds to sentences (see Chap. 18). Interestingly for some corpus linguists, the term "translation corpus" is used to indicate that the texts are mutual translations, whereas "parallel corpus" refers to any collection of multilingual texts of a similar genre. Other researchers prefer the term "comparable corpus" (see McEnery and Wilson, 1996:60n).

There are several reasons for this. A large corpus of naturally occurring text will contain overlapping examples of two sorts: some examples will mutually reinforce eaach other, either by being identical, or by exemplifying the same translation phenomenon. However, other examples will be in conflict: the same or similar phrase in one language may have two different translations for no reason other than inconsistency.

If the examples reinforce each other, this may or may not be useful. Some systems involve a similarity metric (see following, and Chap. 20) which is sensitive to frequency, so that a large number of similar examples will increase the score given to certain matches. But if no such weighting is used, then multiple similar or identical examples are just extra baggage and, in the worst-case, may present the system with a choice—a kind of "ambiguity"—which is simply irrelevant: in such systems, the examples can be seen as surrogate "rules," so that, just as in a traditional rule-based MT system, having multiple examples (rules) covering the same phenomenon leads to overgeneration.

Nomiyama (1992) introduces the notion of "exceptional examples," while Watanabe (1994) goes further in proposing an algorithm for identifying examples such as the sentences in {1} and {2}.[2]

Watashi wa kompyuutaa o kyooyoosuru. {1a}
I (subj) COMPUTER (obj) SHARE-USE.
I share the use of a computer.

Watashi wa kuruma o tsukau. {1b}
I (subj) CAR (obj) USE.
I use a car.

Watashi wa dentaku o shiyoosuru. {2}
I (subj) CALCULATOR (obj) USE.

I share the use of a calculator. {2a}
I use a calculator. {2b}

Given the input in {2}, the system might choose {2a} as the translation because of the closer similarity of *calculator* to *computer* than to *car* (the three words for *use* being considered synonyms). So {1a} is an exceptional example because it introduces the unrepresentative element of *share*. The situation can be rectified by removing example {1a} or by supplementing it with an unexceptional example.

Distinguishing exceptional and general examples is one of a number of means by which the example-based approach is made to behave more like the traditional rule-based approach. Although it means that "example interference" can be minimised, EBMT purists might object that this undermines the empirical nature of the example-based method.

[2]I have adapted Watanabe's transcription, and corrected an obvious misprint in {2a}.

C. How Are Examples Stored?

EBMT systems differ quite widely in how the translation examples themselves are actually stored. Obviously, the storage issue is closely related to the problem of searching for matches.

In the simplest case, the examples may be stored as pairs of strings, with no additional information associated with them. Sometimes, indexing techniques borrowed from information retrieval (IR) can be used: this is often necessary when the example database is very large, but there is an added advantage that it may be possible to make use of a wider context in judging the suitability of an example. Imagine, for instance, an example-based dialogue translation system, wishing to translate the simple utterance *OK*. The Japanese translation for this might be *wakarimashita* 'I understand,' *iidesu yo* 'I agree,' or *ijoo desu* 'let's change the subject,' depending on the context.[3] It may be necessary to consider the immediately preceding utterance both in the input and in the example database. So the system could broaden the context of its search until it found enough evidence to make the decision about the correct translation.

If this kind of information was expected to be relevant on a regular basis, the examples might actually be stored with some kind of contextual marker already attached. This was the approach taken in the proposed MEG system (Somers and Jones, 1992).

Early attempts at EBMT—in which the technique was often integrated into a more conventional rule-based system—stored the examples as fully annotated tree structures with explicit links (e.g., Sato and Nagao, 1990). Figure 1 shows how the Japanese example in {3} and its English translation is represented.

Kanojo wa kami ga nagai. {3}
SHE (topic) HAIR (subj) IS-LONG.
She has long hair.

More recent systems have adapted this somewhat inefficient approach, and annotate the examples more superficially. In Jones (1996), the examples are POS-tagged, carry a Functional Grammar predicate frame and an indication of the sample's rhetorical function. In Collins and Cunningham's (1995) system, the examples

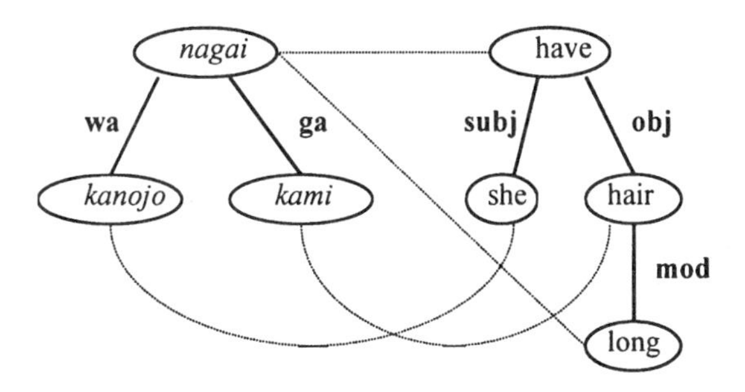

Fig. 1 Representation scheme for statement {3}. (From Watanbe, 1992.)

[3]Examples are from Somers et al. (1990:274).

are tagged, and carry information about syntactic function. Somers et al. (1994) use only tags.

In Furuse and Iida's (1992) proposal, examples are stored in one of three ways: (a) literal examples; (b) "pattern examples," with variables instead of words; and (c) "grammar examples," expressed as context-sensitive rewrite rules, using semantic features. Each type is exemplified in {4–6} respectively.

Sochira ni okeru → We will send it to you {4}
Sochira wa jimukyoku desu → This is the office

X o onegai shimasu → may I speak to X′ {5}
(X = *jimukyoku* 'office')

X o onegai shimasu → please give me the X′
(X = *bangoo* 'number')

N1 N2 N3 → N3′ of N1′ (N1 = *kaigi* 'meeting,' N2 = *kaisai* 'opening,' N3 = *kikan* 'time') {6}
N1 N2 N3 → N2′ N3′ for N1′ (N1 = *sanka* 'participation,' N2 = *mooshikomi* 'application,' N3 = *kyooshi* 'form')

We will see how these different example types are matched in the next section; but what is clear is the hybrid nature of this approach, where the type (a) examples are pure strings, type (c) are effectively "transfer rules" of the traditional kind, with type (b) half-way between the two.

1. *The Translation Model in Statistical Translation*

At this point we might also mention the way examples are "stored" in the *statistical* approaches. In fact, in these systems, the examples are not stored at all, except inasmuch as they occur in the corpus on which the system is based. What *is* stored is the precomputed statistical parameters that give the probabilities for bilingual word pairings, the "translation model." The "language model" (see following), which gives the probabilities of target word strings being well-formed, is also precomputed, and the translation process consists of a search for the target-language string that optimises the product of the two sets of probabilities, given the source-language string. If **e** is the target-language string, and **f** the source-language string, then Eq. {7} is the "Fundamental Equation of Machine Translation," as Brown et al. (1993) put it, where $\hat{\mathbf{e}}$ is the largest-language string for which $\Pr(\mathbf{e}|\mathbf{f})$ is the greatest.

$$\hat{\mathbf{e}} = \operatorname{argmax}_{\mathbf{e}} \Pr(\mathbf{e}) \Pr(\mathbf{f}|\mathbf{e}) \qquad \{7\}$$

Equation {7} is related to the probability $\Pr(\mathbf{e}|\mathbf{f})$ by Bayes' theorem [Eq. {8}]; since $\Pr(\mathbf{f})$ remains constant for all candidate **e**s, it can be excluded from Eq. {7}.

$$\Pr(\mathbf{e}|\mathbf{f}) = \frac{\Pr(\mathbf{e})\Pr(\mathbf{f}|\mathbf{e})}{\Pr(\mathbf{f})} \qquad \{8\}$$

The translation model, $\Pr(\mathbf{f}|\mathbf{e})$, is described in Brown et al. (1990) as depending on "alignment probabilities" for the words in **e** and **f**, these being complicated by the variable "fertility" of the source-language words relative to the target-language words. The alignment probability is an estimation of the likelihood that a given source-word is "responsible" for the appearance of a given target-word in the translation. The fertility aspect reflects that source- and target-language words may not be in one-to-one mappings: for example, the English word *not* might be aligned with the two French words *ne* and *pas*. Brown et al. allow a wide range of fertility values from 0 to 25, although in practise values higher than 2 are rarely supported by the statistical evidence. In addition to the alignment probabilities for the individual words, the translation model has to take sequencing into account. Brown et al. characterise as "distortion" the extent to which the order of the target-language words differs from that of the source-language words.

In Brown et al. (1993), the authors go into considerable detail concering their translation model, and offer five variants. In the first, naive, model, the probability of alignment depends only on the probabilities assigned to its constituent bilingual word pairings. In model 2 the probabilities are enhanced by probabilities relating to relative position: alignments that minimise distortion are favoured. Model 3 incorporates the idea of fertility mentioned earlier, whereas models 4 and 5 depend on the number and identity of the word pairings.

D. Matching

The first task in an EBMT system is to take the source-language string to be translated and to find the example (or set of examples) that most closely match it. This is also the essential task facing a TM system. This search problem depends on the way the examples are stored. In the statistical approach, the problem is an essentially mathematical one of maximising a huge number of statistical probabilities. In more conventional EBMT systems the matching process may be more or less linguistically motivated.

All matching processes necessarily involve a distance or similarity measure. In the most simple case, where the examples are stored as strings, the measure may be a traditional character-based, pattern-matching one. In the earliest TM systems (ALPS' "Repetitions Processing"; see Weaver, 1988), only exact matches, *modulo* alphanumeric strings, were possible: {9a} would be matched with {9b}, but the match in {10} would be missed.

This is shown as A in the diagram. {9a}

This is shown as B in the diagram. {9b}

The large paper tray holds up to 200 sheets of A3 paper. {10a}

The small paper tray holds up to 400 sheets of A4 paper. {10b}

In the case of Japanese–English translation, which many EBMT systems focus on, the notion of character-matching can be modified to take account of the fact that certain "characters" (in the orthographic sense: each Japanese character is represented by two bytes) are more discriminatory than others (e.g., Sato, 1991). This introduces a simple linguistic dimension to the matching process, and is akin to the well-known device in IR, for which only keywords are considered.

Perhaps the "classic" similarity measure, suggested by Nagao (1984) and used in many EBMT systems, is the use of a thesaurus. Here, matches are permitted when words in the input string are replaced by near synonyms (as measured by relative distance in a hierarchically structured vocabulary) in the example sentences. This measure is particularly effective in choosing between competing examples, as in Nagao's examples, where, given {11a,b} as models, we choose the correct translation of *eat* {12a,b} as *taberu* 'eat (food)' or *okasu* 'erode,' on the basis of the relative distance from *he* to *man* and *acid*, and from *potatoes* to *vegetables* and *metal*.

A man eats vegetables. *Hito wa yasai o taberu.* {11a}

Acid eats metal. *San wa kinzoku o okasu.* {11b}

He eats potatoes. *Kare wa jagaimo o taberu.* {12a}

Sulphuric acid eats iron. *Ryuusan wa tetsu o okasu.* {12b}

In a little-known research report, Carroll (1990) suggests a trigonometric similarity measure based on both the relative length and relative contents of the strings to be matched: the relevance of particular mismatches is reflected as a "penalty," and the measure can be adjusted to take account of linguistic generalisations (e.g., a missing comma may incur a lesser penalty than a missing adjective or noun). Carroll hoped that his system would be able, given {13} as input, to offer both {14a and b} as suitable matches.

When the paper tray is empty, remove it and refill it with appropriate size paper. {13}

When the bulb remains unlit, remove it and replace it with a new bulb. {14a}

If the tray is empty, refill it with paper. {14b}

The availability to the similarity measure of information about syntactic classes implies some sort of analysis of both the input and the examples. Cranias et al. (1994) describe a measure that takes function words into account, and makes use of POS tags. Furuse and Iida's (1994) "constituent boundary parsing" idea is not dissimilar. Veale and Way (1997) use sets of closed-class words to segment the examples.

Earlier proposals for EBMT, and proposals where EBMT is integrated within a more traditional approach, assumed that the examples would be stored as tree structures, so the process involves a rather more complex tree-matching (e.g., Watanabe, 1995; Matsumoto et al., 1993).

In the multiengine Pangloss system (see later discussion), the matching process successively "relaxes" its requirements, until a match is found (Nirenburg et al., 1993, 1994): the process begins by looking for exact matches, then allows some deletions or insertions, then word-order differences, then morphological variants, and finally POS-tag differences, each relaxation incurring an ever-increasing penalty.

E. Adaptability and Recombination

After having matched and retrieved a set of examples, with associated translations, the next step is to extract from the translations the appropriate fragments, and to combine these to produce a grammatical target output. This is arguably the most difficult step in the EBMT process: its difficulty can be gauged by imagining a source-language monolingual trying to use a TM system to compose a target text. The problem is twofold: (a) identifying which portion of the associated translation corresponds to the matched portions of the source text, and (b) recombining these portions in an appropriate manner. Compared with the other issues in EBMT, this one has received considerably less attention.

Sato's approach, as detailed in his 1995 paper, takes advantage of the fact that the examples are stored as tree structures, with the correspondences between the fragments explicitly labelled. So problem (a) effectively disappears. The recombination stage is a kind of tree unification, familiar in computational linguistics. Watanabe (1992, 1995) adapts a process called "gluing" from Graph Grammars.

The problem is further eased, in the case of languages such as Japanese and English, by the fact that there is little or no grammatical inflection to indicate syntactic function. So, for example, the translation associated with *the handsome boy* extracted, say, from {15}, is equally reusable in either of the sentences in {16}. This, however, is not the case for a language such as German (and many others), where the form of the determiner, adjective, and noun, all can carry inflections to indicate grammatical case, as in {17}.

The handsome boy entered the room. {15}

The handsome boy ate his breakfast. {16a}

I saw the handsome boy. {16b}

Der schöne Junge aß seinen Frühstück. {17a}

Ich sah den schönen Jungen. {17b}

Collins and Cunningham (1997) stress this question of whether all examples are equally reusable with their notion of "adaptability." Their example retrieval process includes a measure of adaptability that indicates the similarity of the example not only in its internal structure, but also in its external context.

In Somers et al. (1994), the recombination process considers the left and right context of each fragment in the original corpus, which gives as "hooks" tagged words that can then be compared with the context into which the system proposes to fit the fragment. As a further measure, the system attempts to compare the target texts composed by the recombination process with the target-language side of the original corpus, reusing the matching algorithm as if the proposed output were, in fact, an input to be translated: the ease with which the generated text can be matched against the corpus is a measure of the verisimilitude of the constructed sentence.

One other approach to recombination is that taken in the purely statistical system: similar to the matching problem, recombination is expressed as a statistical modelling problem, the parameters having been precomputed. This time, it is the "language model" that is invoked, with which the system tries to maximise the product of the word-sequence probabilities. This model is computed in terms of *n*-grams. If a given sentence **e** is considered to be a string of words $(w_1, w_2, w_3, \ldots, w_n)$ then Pr(**e**) can be expressed as the product of the probabilities of each word given the preceding string [Eq. {18}].

$$\Pr(w_1, w_2, w_3 \ldots, w_n) = \Pr(w_1)\Pr(w_2|w_1)\Pr(w_3|w_1, w_2)\ldots\Pr(w_n|w_1, w_2, \ldots, w_{n-1}) \quad \{18\}$$

However, a formulation such as Eq. {18} involves the estimation of too many parameters, for many of which there will be little or no evidence in the corpus, or for which the value will be very small. As a compromise, it has proved sufficient to take the probabilities of shorter sequences of *n* words ("*n*-grams"), where *n* typically takes the value 2 or 3. Thus, Eq. {18} can be reexpressed as Eq. {19} for the "bigram" model.

$$\Pr(w_1, w_2, w_3 \ldots, w_n) = \Pr(w_1)\Pr(w_2|w_1)\Pr(w_3|w_2)\ldots\Pr(w_n|w_{n-1}) \quad \{19\}$$

F. Computational Problems

All the approaches mentioned so far have to be implemented as computer programs, and significant computational factors influence many of them. One criticism to be made of the approaches such as Sato and Nagao (1990), Watanabe (1992), and even Jones (1996), which store the examples as fully annotated structures, is the huge computational cost in terms of creation, storage, and complex retrieval algorithms. Sumita and Iida (1995) is one of the few papers to address this issue explicitly, turning to parallel processing for help, a solution also adopted by Kitano (1994) and Sato (1995).

One important computational issue is speed, especially for those of the EBMT systems that are used for real-time speech translation. The size of the example database will obviously affect this, and it is thus understandable that some researchers are looking at ways of maximising the effect of the examples by identifying and making explicit significant generalisations. In this way the hybrid system has emerged, assuming the advantages of both the example-based and rule-based approaches.

4. FLAVOURS OF EBMT

So far we have looked at various solutions to the individual problems that make up EBMT. In this section, we prefer to take a wider view, to consider the various different contexts in which EBMT has been proposed. In many cases, EBMT is used as a component in an MT system which also has more traditional elements: EBMT may be used in parallel with these other "engines," or just for certain classes of problems, or when some other component cannot deliver a result. Also, EBMT methods may be better suited to some kinds of applications than others. And finally, it may not be obvious any more what exactly the dividing line is between EBMT and so-called traditional rule-based approaches. As the opening paragraphs of this chapter suggest, EBMT was once seen as a bitter rival to the existing paradigm, but there now seems to be a much more comfortable coexistence.

A. Suitable Translation Problems

Let us consider first the range of translation problems for which EBMT is best suited. Certainly, EBMT is closely allied to *sublanguage* translation, not least because of EBMT's reliance on a real corpus of real examples: at least implicitly, a corpus can go a long way toward defining a sublanguage. On the other hand, nearly all research nowadays in MT is focused on a specific domain or task, so perhaps all MT is sublanguage MT.

More significant is that EBMT is often proposed as an antidote to the problem of "structure-preserving translation as first choice" (see Somers, 1987:84) inherent in MT systems which proceed on the basis of structural analysis. Because many EBMT systems do not compute structure, the source-language structure cannot, by definition, be imposed on the target language. Indeed, some of the early systems in which EBMT is integrated into a more traditional approach explicitly use EBMT for such cases:

> When one of the following conditions holds true for a linguistic phenomenon, RBMT [rule-based MT] is less suitable than EBMT.
>
> (a) Translation rule formation is difficult.
> (b) The general rule cannot accurately describe [the] phenomen[on] because it represents a special case.
> (c) Translation cannot be made in a compositional way from target words (Sumita and Iida, 1991:186).

B. Pure EBMT

Very few research efforts have taken an explicitly "purist" approach to EBMT. One exception is Somers et al. (1994), in which the authors wanted to push to the limits a "purely non-symbolic approach" in the face of what they felt was a premature acceptance that hybrids were the best solution. Not incorporating any linguistic information that could not be derived automatically from the corpus became a kind of dogma.

The other nonlinguistic approach is the purely statistical one of Brown et al. (1990, 1993). In fact, their aspirations were much less dogmatic, and in the face of mediocre results, they were soon resorting to linguistic knowledge (Brown et al., 1992); not long afterward the group broke up, although other groups have taken up the mantle of statistics-based MT (Vogel et al., 1996; Wang and Waibel, 1997).

Other approaches, as we have seen in the foregoing, although remaining more or less true to the case-based (rather than theory-based) approach of EBMT, accept the necessity to incorporate linguistic knowledge either in the representation of the examples, or in the matching and recombination processes. This represents one kind of hybridity of approach; but in this section we will look at hybrids in another dimension, in which the EBMT approach is integrated into a more conventional system.

C. EBMT for Special Cases

As the foregoing quotation from Sumita and Iida shows, one of the first uses envisaged for the EBMT approach was when the rule-based approach was too difficult. The classic case of this, as described in one of the earliest EBMT papers (Sumita et al., 1990; Sumita and Iida, 1991), was the translation of Japanese adnominal particle constructions (*A no B*), where the default or structure-preserving translation (*B of A*) is wrong 80% of the time. In Sumita and Iida's traditional rule-based system, the EBMT module was invoked just for this kind of example (and a number of other similarly difficult cases). In a similar way, Katoh and Aizawa (1994) describe how only "parameterizable fixed phrases" in economics news stories are translated on the basis of examples, in a way very reminiscent of TM systems.

D. Example-Based Transfer

Because their examples are stored as tree structures, one can describe the systems of Sato and Nagao (1990) and Sato (1991) as "example-based transfer": source-language inputs are analysed into dependency representations in a conventional manner, only transfer is on the basis of examples, rather than rules, and then generation of the target-language output is again performed in a traditional way.

E. Deriving Transfer Rules from Examples

Some researchers take this scenario a step further, using EBMT as a research technique to build the rule base, rather than a translation technique per se. We can see this in the case of Furuse and Iida's (1992) distinction of three types of "example" {4}–{6} in Sec. 3.C: They refer to "string-level," "pattern-level," and "grammar-level" transfer knowledge, and it seems that the more abstract representations are *derived* from examples by a process of generalisation.

Kaji et al. (1992) describe their "two-phase" EBMT methodology, the first phase involving "learning" of templates (i.e., transfer rules) from a corpus. Each template is a "bilingual pair of pseudo-sentences" (i.e., example sentences containing variables). The translation templates are generated from the corpus first by parsing the translation pairs and then aligning the syntactic units with the help of a bilingual dictionary, resulting in a translation template as in Fig. 2. This can then be generalised by replacing the coupled units with variables marked for syntactic category, also shown in Fig. 2. Kaji et al. do not make explicit the criteria for choosing the units, although they do discuss the need to refine templates that give rise to a conflict, as in {20 and 21}.

play baseball → *yakyu o suru* {20a}

play tennis → *tenisu o suru* {20b}

play X[NP] → X[NP] *o suru* {20c}

play the piano → *piano o hiku* {21a}

play the violin → *baiorin o hiku* {21b}

play X[NP] → X[NP] *o hiku* {21c}

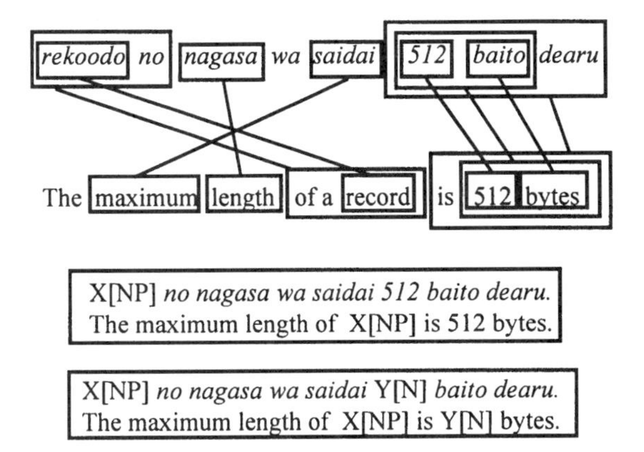

Fig. 2 Generation of translation templates from aligned example. (From Kaji et al., 1992.)

Nomiyama (1992) similarly describes how examples ("cases") can be generalised into rules by combining them when similar segments occur in similar environments, this similarity being based on semantic proximity as given by a hierarchical thesaurus.

Almuallim et al. (1994) and Akiba et al. (1995) report much the same idea, although they are more formal in their description of how the process is implemented, citing the use of two algorithms from Machine Learning. Interestingly, these authors make no claim that their system is therefore "example-based." Also, many of the examples that they use to induce the transfer rules are artificially constructed.

F. EBMT as One of a Multiengine System

One other scenario for EBMT is exemplified by the Pangloss sytem, in which EBMT operates in parallel with two other techniques: knowledge-based MT and a simpler lexical transfer engine (Frederking and Nirenburg, 1994; Frederking et al., 1994). Nirenburg et al. (1994) and Brown (1996) describe the EBMT aspect of this work in most detail. What is most interesting is the extent to which the different approaches often mutually confirm each other's proposed translations, and the comparative evidence that the multiengine approach offers.

5. CONCLUSIONS

In this chapter, we have seen a range of applications all of which might claim to "be" EBMT systems. So one outstanding question might be: What counts as EBMT? Certainly, the use of a bilingual corpus is part of the definition, but this is not sufficient. Almost all research on MT nowadays makes use at least of a "reference" corpus to help define the range of vocabulary and structures that the system will cover. It must be something more, then.

EBMT means that the main knowledge-base stems from examples. However, as we have seen, examples may be used as a device to shortcut the knowledge-acquisition bottleneck in rule-based MT, the aim being to generalise the examples as much as possible. So part of the criterion might be whether the examples are used at run-time or not: but by this measure, the statistical approach would be ruled out;

although the examples are not used to derive rules in the traditional sense, still at run-time there is no consultation of the database of examples.

The original idea for EBMT seems to have been couched firmly in the rule-based paradigm: examples were to be stored as tree structures, so rules must be used to analyse them: only transfer was to be done on the basis of examples, and then only for special, difficult cases. After the comparative success of this approach, and also as a reaction to the apparent stagnation in research in the conventional paradigm, the idea grew that EBMT might be a "new" paradigm altogether, even in competition with the old. This confrontational aspect has quickly died away, and in particular, EBMT has been integrated into more traditional approaches (and vice versa, one could say) in many different ways.

We will end this chapter by mentioning, for the first time, some of the advantages that have been claimed for EBMT. Not all the advantages that were claimed in the early days of polemic are obviously true. But it seems that at least the following do hold, inasmuch as the system design is primarily example-based (e.g., the examples may be "generalised", but corpus data is still the main source of linguistic knowledge):

1. Examples are real language data, so their use leads to systems that cover the constructions that really do occur, and ignore the ones that do not, so overgeneration is reduced.
2. The linguistic knowledge of the system can be more easily enriched, simply by adding more examples.
3. EBMT systems are data-driven, rather than theory-driven: therefore, because there are no complex grammars devised by a team of individual linguists, the problem of rule conflict and the need to have an overview of the "theory," and how the rules interact, is lessened. (On the other hand, as we have seen, there is the opposite problem of conflicting examples.)
4. The example-based approach seems to offer some relief from the constraints of "structure-preserving" translation.

EBMT is now seen not as a rival to rule-based methods, but as an alternative, available to enhance and, sometimes, replace it. Nor is research in the purely rule-based paradigm finished. As mentioned in Somers (1997:116), the problem of scaling-up remains, as do a large number of interesting translation problems, especially with new uses for MT (e.g., web-page and e-mail translation) emerge.

REFERENCES

Akiba Y, M Ishii, H Almuallim, S Kaneda (1995). Learning English verb selection rules from hand-made rules and translation examples. Proceedings of the Sixth International Conference on Theoretical and Methodological Issues in Machine Translation, Leuven, Belgium, pp 206–220.

Almuallim H, Y Akiba, T Yamazaki, A Yokoo, S Kaneda (1994). Two methods for learning ALT-J/E translation rules from examples and a semantic hierarchy. COLING 94, 15th International Conference on Computational Linguistics, Kyoto, pp 57–63.

Brown PF, J Cocke, SA Della Pietra, VJ Della Pietra, F Jelinek, JD Lafferty, RL Mercer, PS Roossin (1990). A statistical approach to machine translation. Comput Linguist 16:79–85.

Brown PF, SA Della Pietra, VJ Della Pietra, JD Lafferty, RL Mercer (1992). Analysis, statistical transfer, and synthesis in machine translation. In: Isabelle P, ed. TMI-92, Montreal: CCRIT-CWARC, 1992 pp 83–100.

Brown PF, SA Della Pietra, VJ Della Pietra, RL Mercer (1993). The mathematics of statistical machine translation: parameter estimation. Comput Linguist 19:263–311.

Brown RD (1996). Example-based machine translation in the Pangloss system. COLING–96: 16th International Conference on Computational Linguistics, Copenhagen, pp 169–174.

Carroll JJ (1990). Repetitions processing using a metric space and the angle of similarity. Report No. 90/3, Centre for Computational Linguistics, UMIST, Manchester.

Collins B, P Cunningham (1995). A methodology for example based machine translation. CSNLP 1995: 4th Conference on the Cognitive Science of Natural Language Processing, Dublin.

Collins B, P Cunningham (1997). Adaptation guided retrieval: approaching EBMT with caution. Proceedings of the 7th International Conference on Theoretical and Methodological Issues in Machine Translation, Santa Fe, NM, pp 119–126.

Cranias L, Papageorgiou, S Piperidis (1994). A matching technique in example-based machine translation. COLING 94, 15th International Conference on Computational Linguistics, Kyoto, pp 100–104.

Frederking R, S Nirenburg (1994). Three heads are better than one. 4th Conference on Applied Natural Language Processing, Stuttgart, pp 95–100.

Frederking R, S Nirenburg, D Farwell, S Helmreich, E Hovy, K Knight, S Beale, C Domashnev, D Attardo, D Grannes, R Brown (1994). Integrating translations from multiple sources within the Pangloss Mark III machine translation system. Technology Partnerships for Crossing the Language Barrier: Proceedings of the First Conference of the Association for Machine Translation in the Americas, Columbia, MD, pp 73–80.

Furuse O, H Iida (1992). An example-based method for transfer-driven machine translation. In: Isabelle P, ed. TMI-92, Montreal, pp 139–150.

Furuse O, H Iida (1994). Constituent boundary parsing for example-based machine translation. COLING 94, 15th International Conference on Computational Linguistics, Kyoto, pp 105–111.

Isabelle P, ed (1992). Quatrième colloque international sur les aspects théoriques et méthodologiques de la traduction automatique; Fourth International Conference on Theoretical and Methodological Issues in Machine Translation. Méthodes empiristes versus méthodes rationalistes en TA; Empiricist vs. Rationalist Methods in MT. TMI-92. Montréal: CCRIT-CWARC.

Jones D (1996). Analogical Natural Language Processing. London: UCL Press.

Kaji H, Y Kida, Y Morimoto (1992). Learning translation templates from bilingual text. Proceedings of the fifteenth [sic] International Conference on Computational Linguistics, COLING–92, Nantes, pp 672–678.

Katoh N, T Aizawa (1994). Machine translation of sentences with fixed expressions. 4th Conference on Applied Natural Language Processing, Stuttgart, pp 28–33.

Kitano H (1994). Speech-to-Speech Translation: A Massively Parallel Memory-Based Approach. Boston: Kluwer.

Matsumoto Y, H Ishimoto, T Utsuro (1993). Structural matching of parallel texts. 31st Annual Meeting of the Association for Computational Linguistics, Columbus, OH, pp 23–30.

McEnery T, A Wilson (1996). Corpus Linguistics. Edinburgh: Edinburgh University Press.

Nagao M (1984). A framework of a mechanical translation between Japanese and English by analogy principle. In: A Elithorn, R Banerji, eds. Artificial and Human Intelligence. Amsterdam: North-Holland, pp 173–180.

Nirenburg S, S Beale, C Domashnev (1994). A full-text experiment in example-based machine translation. International Conference on New Methods in Language Processing (NeMLaP), Manchester, pp 78–87.

Nirenburg S, C Domashnev, DJ Grannes (1993). Two approaches to matching in example-based machine translation. Proceedings of the Fifth International Conference on Theoretical and Methodological Issues in Machine Translation: MT in the Next Generation, Kyoto, pp 47–57.

Nomiyama H (1992). Machine translation by case generalization. Proceedings of the fifteenth [sic] International Conference on Computational Linguistics, COLING-92, Nantes, pp 714–720.

Pappegaaij BC, V Sadler, APM Witkam, eds. (1986). Word Expert Semantics: An Interlingual Knowledge-Based Approach. Dordrecht: Reidel.

Sato S (1991). Example-based translation approach. Proceedings of International Workshop on Fundamental Research for the Future Generation of Natural Language Processing (FGNLP), Kyoto, pp 1–16.

Sato S (1995). MBT2: a method for combining fragments of examples in example-based machine translation. Artif Intell 75:31–49.

Sato S, M Nagao (1990). Toward memory-based translation. COLING–90, Papers Presented to the 13th International Conference on Computational Linguistics, Helsinki, vol 3, pp 247–252.

Schubert K (1986). Linguistic and extra-linguistic knowledge: a catalogue of language-related rules and their computational application in machine translation. Comput Transl 1:125–152.

Somers H (1987). Some thoughts on interface structure(s). In: W Wilss, K-D Schmitz, eds. Maschinelle Übersetzung—Methoden und Werkzeuge. Tübingen: Niemeyer, pp 81–99.

Somers HL (1997). The current state of machine translation. MT Summit VI: Machine Translation Past Present Future, San Diego, CA, pp 115–124.

Somers HL, D Jones (1992). Machine translation seen as interactive multilingual text generation. Translating and the Computer 13: The Theory and Practice of Machine Translation—A Marriage of Convenience? London: Aslib, pp 153–165.

Somers H, I McLean, D Jones (1994). Experiments in multilingual example-based generation. CSNLP 1994: 3rd conference on the Cognitive Science of Natural Language Processing, Dublin.

Somers HL, J Tsujii, D Jones (1990). Machine translation without a source text. COLING–90, Papers presented to the 13th International Conference on Computational Linguistics, Helsinki, vol 3, pp 271–276.

Sumita E, H Iida (1991). Experiments and prospects of example-based machine translation. 29th Annual Meeting of the Association for Computational Linguistics, Berkeley, CA, pp 185–192.

Sumita E, H Iida (1995). Heterogeneous computing for example-based translation of spoken language. Proceedings of the Sixth International Conference on Theoretical and Methodological Issues in Machine Translation, Leuven, Belgium, pp 273–286.

Sumita E, H Iida, H Kohyama (1990). Translating with examples: a new approach to machine translation. The Third International Conference on Theoretical and Methodological Issues in Machine Translation of Natural Language, Austin, TX, pp 203–212.

Veale T, A Way (1997). *Gaijin*: a bootstrapping approach to example-based machine translation. International Conference, Recent Advances in Natural Language Processing, Tzigov Chark, Bulgaria, pp 239–244.

Vogel S, H Ney, C Tillmann (1996). HMM-based word alignment in statistical translation. COLING–96: 16th International Conference on Computational Linguistics, Copenhagen, pp 836–841.

Wang Y-Y, A Waibel (1997). Decoding algorithm in statistical machine translation. 35th Annual Meeting of the Association for Computational Linguistics and 8th Conference of the European Chapter of the Association for Computational Linguistics, Madrid, pp 366–372.

Watanabe H (1992). A similarity-driven transfer system. Proceedings of the fifteenth [sic] International Conference on Computational Linguistics, COLING–92, Nantes, pp 770–776.

Watanabe H (1994). A method for distinguishing exceptional and general examples in example-based transfer systems. COLING 94, 15th International Conference on Computational Linguistics, Kyoto, pp 39–44.

Watanabe H (1995). A model of a bi-directional transfer mechanism using rule combinations. Mach Transl 10:269–291.

Weaver A (1988). Two aspects of interactive machine translation. In: M Vasconcellos, ed. Technology as Translation Strategy. State University of New York at Binghamton (SUNY), pp 116–123.

26

Word-Sense Disambiguation

DAVID YAROWSKY

Johns Hopkins University, Baltimore, Maryland

1. INTRODUCTION

Word-sense disambiguation (WSD) is essentially a classification problem. Given a word such as *sentence* and an inventory of possible semantic tags for that word, which tag is appropriate for each individual instance of that word in context? In many implementations these labels are major sense numbers from an online dictionary, but they may also correspond to topic or subject codes, nodes in a semantic hierarchy, a set of possible foreign language translations, or even assignment to an automatically induced sense partition. The nature of this given sense inventory substantially determines the nature and complexity of the sense disambiguation task.

Table 1 illustrates the task of sense disambiguation for three separate sense inventories: (a) The dictionary sense number in *Collins COBUILD English Dictionary* (Sinclair et al., 1987); (b) a label corresponding to an appropriate translation into Spanish; and (c) a general topic, domain, or subject-class label. Typically, only one inventory of labels would be used at a time, and in the case of Table 1 each of the three inventories has roughly equivalent discriminating power. Sense disambiguation constitutes the assignment of the most appropriate tag from one of these inventories corresponding to the semantic meaning of the word in context. Section 2.B discusses the implications of the sense inventory choice on this task.

The words in context surrounding each instance of *sentence* in Table 1 constitute the **evidence sources** with each classification can be made. Words immediately adjacent to the target word typically exhibit the most predicative power. Other words in the same sentence, paragraph, and even entire document typically contribute weaker evidence, with predictive power decreasing roughly proportional to their

Table 1 **Sense Tags** for the Word Sentence from Different Sense Inventories

COBUILD dictionary	Spanish translation	Subject class	Instance of the target word in context
Noun–2	*sentencia*	LEGAL	... for a maximum *sentence* for a young ...
Noun–2	*sentencia*	LEGAL	... of the minimum *sentence* of seven years ...
Noun–2	*sentencia*	LEGAL	... were under the *sentence* of death ...
Noun–2	*sentencia*	LEGAL	... criticize a *sentence* handed down ...
Noun–1	*frase*	LING	... in the next *sentence*, they say ...
Noun–1	*frase*	LING	... read the second *sentence* because ...
Noun–1	*frase*	LING	... as the next *sentence* is an important ...
Noun–1	*frase*	LING	... is the second *sentence* which I ...
Noun–1	*frase*	LING	... listen to this *sentence* uttered by a ...

distance from the target word. The nature of the syntactic relations between potential evidence resources is also important. Section 4 will discuss the extraction of these contextual evidence sources and their use in supervised learning algorithms for word-sense classification. Section 5 will discuss unsupervised and minimally supervised methods for sense classification when costly hand-tagged training data is unavailable or is not available in sufficient quantities for supervised learning. As a motivating precursor to these algorithm-focused sections, Sec. 3 will provide a survey of applications for WSD and Sec. 6 will conclude with a discussion of current research priorities in sense disambiguation.

2. WORD-SENSE INVENTORIES AND PROBLEM CHARACTERISTICS

One of the key defining properties of the sense disambiguation task is the nature of the sense inventory used: its source, granularity, hierarchical structure, and treatment of part-of-speech (POS) differences.

A. Treatment of Part-of-Speech

Although sense ambiguity spans part of speech [e.g., 1—river *bank* (noun); 2—financial *bank* (noun); and 3—to *bank* an airplane (verb)], the large majority of sense-disambiguation systems treat the resolution of part-of-speech distinctions as an initial and entirely separate **parsing** process (see Chap. 5) or **tagging** process (see Chap. 11). The motivation for this approach is that part-of-speech ambiguity is best resolved by a class of algorithms driven by local syntactic sequence optimization having a very different character from the primarily semantic word associations that resolve within-part-of-speech ambiguities.

The remainder of this chapter will follow this convention, assuming that a POS tagger has been run over the text first and focusing on remaining sense ambiguities within the same part of speech. In many cases, the POS tags for surrounding words will also be used as additional evidence sources for the within-part-of-speech sense classifications.

B. Sources of Sense Inventories

The nature of the sense disambiguation task depends largely on the source of the **sense inventory** and its characteristics.

- **Dictionary-based inventories**: Much of the earliest work in sense disambiguation (e.g., Lesk, 1986; Walker and Amsler, 1986) involved the labeling of words in context with sense numbers extracted from machine-readable dictionaries. Use of such a reference standard provides the automatic benefit of the "free" classification information and example sentences in the numbered definitions, making it possible to do away with hand-tagged training data altogether. Dictionary-based sense inventories tend to encourage hierarchical classification methods and support relatively fine levels of sense granularity.
- **Concept hierarchies (e.g., WordNet)**: One of the most popular standard sense inventories in recent corpus-based work, especially on verbs, is the WordNet **semantic concept hierarchy** (Miller, 1990). Each "sense number" corresponds to a node in this hierarchy, with the BIRD sense of *crane* embedded in a concept path from HERON-LIKE-BIRDS through BIRD to the concept LIVING-THING and PHYSICAL-ENTITY. This inventory supports extensive use of **class-based inheritance** and **selectional restriction** (e.g., Resnik, 1993, 1997).
- **Domain tags/subject codes (e.g., LDOCE)**: The online version of the *Longman Dictionary of Contemporary English* (*LDOCE*) (Procter, 1978), assigns general domain or subject codes (such as *EC* for economic/financial usages, and *EG* for engineering usages) to many, but not all, word senses. In the cases for which sense differences correspond to domain differences, *LDOCE* subject codes can serve as sense labels (e.g., Guthrie et al., 1991; Cowie et al., 1992), although coverage is limited for non–domain-specific senses. Subject codes from hierarchically organized thesauri such as *Roget's 4th International* (Chapman, 1977) can also serve as sense labels (as in Yarowsky, 1992).
- **Multilingual translation distinctions**: Sense distinctions often correspond to translation differences in a foreign language and, as shown in the example of *sentence* in Table 1, these translations (such as the Spanish *frase* and *sentencia*) can be used as effective sense tags. Parallel polysemy across related languages may reduce the discriminating power of such sense labels (as discussed in Sec. 3.C), but this problem is reduced by using translation labels from a more distantly related language family. The advantages of such a sense inventory are that (a) it supports relatively direct application to Machine Translation (MT), and (b) sense-tagged training data can be automatically extracted for such a sense inventory from parallel bilingual corpora (as in Gale et al., 1992a,c).
- **Ad hoc and specialized inventories:** In many experimental studies with a small example set of polysemous words, the sense inventories are often defined by hand to reflect the sense ambiguity present in the data. In other cases, the sense inventory may be chosen to support a particular application (such as a specialized meaning resolution in message-understanding systems).

- **Artificial sense ambiguities ("pseudo-words"): Pseudo-words**, proposed by Gale et al. (1992d), are artificial ambiguities created by replacing all occurrences of two monosemous words in a corpus (such as *guerilla* and *reptile*) with one joint word (e.g., *guerilla–reptile*). The task of deciding which original word was intended for each occurrence of the joint word is largely equivalent to determining which "sense" was intended for each occurrence of a polysemous word. The problem is not entirely unnatural, as there could well exist a language in which the concepts *guerilla* and *reptile* are indeed represented by the same word owing to some historical–linguistic phenomenon. Selecting between these two meanings would naturally constitute WSD in that language. This approach offers the important benefit that potentially unlimited training and test data are available and that sense ambiguities of varying degrees of subtlety can be created on demand by using word pairs of the desired degree of semantic similarity, topic distribution, and frequency.
- **Automatically induced sense inventories:** Finally, as discussed in Sec. 5.A, recent work in unsupervised sense disambiguation has utilized automatically induced semantic clusters as effective sense labels (e.g., Schütze, 1992). Although these clusters may be aligned with more traditional inventories such as dictionary sense numbers, they can also function without such a mapping, especially if they are used for secondary applications such as information retrieval for which the effective sense partition (rather than the choice of label) is most important.

C. Granularity of Sense Partitions

Some disambiguation can be performed at various levels of subtlety. Major meaning differences called **homographs** often correspond to different historical derivations converging on the same orthographic representation. For example, the homographs I–III for *bank* in Fig. 1 entered English through the French *banque*, Anglo–Saxon *benc*, and French *banc*, respectively. More subtle distinctions, such as between the (I.1) financial *bank* and (I.2) general repository sense of *bank*, typically evolved through later usage, and often correspond to quite clearly distinct meanings that are likely translated into different words in a foreign language. Still more subtle distinctions, such as between the (I.1a) general institution and (I.1b) physical building senses of financial *bank* are often difficult for human judges to resolve through context (e.g., *He owns the bank on the corner*), and often exhibit parallel **polysemy** in other languages.

The necessary level of granularity clearly depends on the application. Frequently, the target granularity comes directly from the sense inventory (e.g., whatever level of distinction is represented in the system's online dictionary). In other cases, the chosen level of granularity derives from the needs of the target application: those meaning distinctions that correspond to translation differences are appropriate for MT, whereas only homograph distinctions that result in pronunciation differences (e.g., *bass*/bæs/ vs. /beɪs/) may be of relevance to a text-to-speech synthesis application.

Such granularity issues often arise in the problem of evaluating sense-disambiguation systems, and how much penalty one assigns to errors of varying

I *Bank* - REPOSITORY

I.1 Financial Bank

I.1a - an institution

I.1b - a building

I.2 General Supply/Reserve/Inventory

II *Bank* - GEOGRAPHICAL

II.1 Shoreline

II.2 Ridge/Embankment

III *Bank* - ARRAY/GROUP/ROW

Fig. 1 Different meanings of bank.

subtlety. One reasonable approach is to generate a **penalty matrix** for misclassification sensitive to the functional semantic distance between any two senses/subsenses of a word. Such a matrix could be derived automatically from hierarchical distance in a sense tree, as shown in Table 2.

Such a penalty matrix could also be based on **confusability** or **functional distance** within an application (e.g., in a speech-synthesis application, only those sense-distinction errors corresponding to pronunciation differences would be penalized). Such distances could also be based on **psycholinguistic data**, such as experimentally derived estimates of similarity or confusability (Miller and Charles, 1991; Resnik, 1995).

In this framework, rather than computing system accuracy with a boolean match/no-match weighting of classification errors between subsenses (however subtle the difference), a more sensitive weighted accuracy measure capturing the relative seriousness of misclassification errors could be defined as in Eq. {1}.

$$\text{Weighted accuracy} = \frac{1}{N}\sum_{i=1}^{N} d(c_{s_i}, a_{s_i}) \qquad \{1\}$$

Table 2 Example Pairwise Semantic Distance Between the Word Senses of *bank*

	I.1a	I.1b	I.2	II.1	II.2	III
I.1a	0	1	2	4	4	4
I.1b	1	0	2	4	4	4
I.2	2	2	0	4	4	4
II.1	4	4	4	0	1	4
II.2	4	4	4	1	0	4
III	4	4	4	4	4	0

where the distance $d(c_i, a_j)$ is the normalized pairwise penalty or cost of misclassification between an assigned sense (a_j) and correct sense (c_j) over all N test examples.

If the sense-disambiguation system assigns a probability distribution to the different sense/subsense options, rather than a hard boolean assignment, the weighted accuracy could be defined as in Eq. {2})

$$\text{Weighted accuracy} = \sum_{i=1}^{N} \sum_{j=1}^{s_1} d(c_{s_i}, a_{s_i}) \times P(s_j | w_i, context_i) \qquad \{2\}$$

where for any test example i of word w_i having senses s_i, the probability mass assigned by the classifier to incorrect senses is weighted by the communicative distance or cost of that misclassification. Similar cross–entropy-based measures could be used as well.

D. Hierarchical Versus Flat Sense Partitions

Another issue in sense disambiguation is that many sense inventories represent only a flat partition of senses, with no representation of relative semantic similarity through hierarchical structure. Similar **cross–entropy**-based measures could be used as well. Furthermore, flat partitions offer no natural label for underspecification or generalization for use when full subsense resolution cannot be made. When available, such hierarchical sense/subsense inventories can support top-down hierarchical sense classifiers such as in Secs. 4.C and 5.D, and can contribute to evaluation of partial correctness in evaluation.

E. Idioms and Specialized Collocational Meanings

A special case of fine-granularity sense inventories is the need to handle **idiomatic** usages or cases in which a specialized sense of a word derives almost exclusively from a single collocation. *Think tank* and *tank top* (an article of clothing) are examples. Although these can in most cases be traced historically to one of the major senses (e.g., CONTAINER *tank* in the two foregoing examples), these are often inadequate labels and the inclusion of these idomatic examples in training data for the major sense can impede learning. Thus inclusion of specialized, collocation-specific senses in the inventory is often well justified.

F. Regular Polysemy

The term **regular polysemy** refers to standard, relatively subtle variations of usage or aspect that apply systematically to classes of words, such as physical objects. For example, the word *room* can refer to a physical entity {3} or the space it encloses {4}. The nouns *cup* and *box* exhibit similar ambiguities. The rich complexity of regular polysemy and specialized approaches to its resolution are covered in detail in Chap.

The room was painted red. {3}

A strong odor filled the room. {4}

3. APPLICATIONS OF WSD

Sense disambiguation tends not to be considered a primary application in its own right, but rather, is an intermediate annotation step that is utilized in several end-user applications.

A. Applications in Information Retrieval

The application of WSD to Information Retrieval (IR) has had mixed success. One of the goals in IR is to map the words in a document or in a query to a set of **terms** that capture the semantic content of the text. When multiple morphological variants of a word carry similar semantic content (e.g., *computing*, *computer*), **stemming** is used to map these words to a single term (e.g., *comput-*). However, when a single word conveys two or more possible meanings (e.g., *tank*), it may be useful to map that word into separate distinct terms (e.g., *tank*-1 'military vehicle' and *tank*-2 'container') based on context.

The actual effectiveness of WSD on bottom-line IR performance is unclear, however. Krovetz and Croft (1992) and Krovetz (1997) argue that word-sense disambiguation *does* contribute to the effective separation of relevant and nonrelevant documents, and even a small domain-specific document collection exhibits a significant degree of lexical ambiguity (over 40% of the query words in one collection).

In contrast, Sanderson (1994) and Voorhees (1993) present a more pessimistic perspective on the helpfulness of WSD to IR. Their experiments indicate that in full IR applications, it offers very limited additional improvement in performance, and much of this is due to resolving part-of-speech distinctions (*sink* the verb vs. *sink*, a bathroom object). Although Schütze and Pedersen (1995) concur that dictionary-based sense labels have limited contribution to IR, they found that automatically induced sense clusters (see Sec. 5.A) are useful, as the clusters directly characterize different contextual distributions.

A reasonable explanation of the foregoing results is that the similar disambiguating clues used for sense tagging (e.g., {*Panzer* and *infantry* with *tank*} selecting for the military sense of *tank*), are also used directly by IR algorithms (e.g., {*Panzer*, *tank*, and *infantry*} together indicate relevance for military queries). The additional knowledge that *tank* is in sense 1 is to a large extent simply echoing the same contextual information already available to the IR system in the remainder of the sentence. Thus sense tagging should be more productive for IR in the cases of ambiguities resolved through a single collocation rather than the full sentence context (e.g., *think tank* ≠ 'container'), and for added discriminating power in short queries (e.g., *tank*-1 *procurement policy* vs. just *tank procurement policy*).

B. Applications in Message or Text Understanding

Understanding word meanings is clearly central to the ultimate task of text understanding. But even in more restricted (sub)tasks such as extracting information about specific entities in a text, being able to classify such entities appropriately (e.g., does an instance of the word *drug* in context refer to a medicine or narcotic) is a significant challenge. In previous DARPA-sponsored Message Understanding Conference (MUC) tasks, the range of sense ambiguity has been relatively limited because of the restricted subject domains. However, interest is growing in "glass-box"-style evaluations that specifically test whether the salient lexical ambiguities that are present have been correctly resolved. Also, one of the most important current message-understanding tasks, **named-entity classification**, is basically sense disambiguation for proper names. An example is the classification of *Madison* as a person, place, company, or event. The training-data formats, evidence sources and sense-disambiguation algorithms presented in Secs. 4 and 5 are directly applicable to this classification task.

C. Applications in MT

It should be clear from Sec. 2 that lexical translation choice in MT is similar to word-sense tagging. There are substantial divergences, however.

In some cases (such as when all four major senses of *interest* ('desire to know,' 'advantage,' 'stake or share,' 'financial growth') translate into French as *intérêt*), the target language exhibits parallel ambiguities with the source and full-sense resolution is not necessary for appropriate translation choice. In other cases, a given sense of a word in English may correspond to multiple similar words in the target language that mean essentially the same thing, but have different preferred or licensed collocational contexts in the target language. For example, *sentencia* and *condena* are both viable translations into Spanish for the (noun) sense of the English word *sentence* in the legal context. However, *condena* rather than *sentencia* would be preferred when associated with a duration (e.g., *life sentence*). Selection between such variants is largely an optimization problem in the target language.

Nevertheless, monolingual sense-disambiguation algorithms may be utilized productively in MT systems once the mapping between source-language word senses and corresponding target-language translations has been established. This is clearly true in interlingual MT systems, where source-language sense-disambiguation algorithms can help serve as the lexical semantic component in the analysis phase. Brown et al. (1991) have also utilized monolingual sense disambiguation in their statistical transfer-based MT approach, estimating a probability distribution across corresponding translation variants and using monolingual language models to select the optimal target word sequence given these weighted options.

D. Other Applications

Sense-disambiguation procedures also may have commercial applications as intelligent dictionaries, thesauri, and grammar checkers. Students looking for definitions of or synonyms for unfamiliar words are often confused by or misuse the definitions/synonyms for contextually inappropriate senses. Once the correct sense has been identified for the currently highlighted word in context, an intelligent dictionary/thesaurus would list only the definition(s) and synonym(s) appropriate for the actual document context. A somewhat indirect application is that the algorithms developed for classic sense disambiguation may also be productively applied to related lexical ambiguity resolution problems exhibiting similar problem characteristics. One such closely related application is accent and diacritic restoration in Spanish and French, studied using a supervised sense-tagging algorithm in Yarowsky (1994).

4. EARLY APPROACHES TO SENSE DISAMBIGUATION

WSD is one of the oldest problems in natural language processing (NLP). It was recognized as a distinct task as early as 1955, in the work of Yngve (1955) and later Bar-Hillel (1960). The target application for this work was MT, which was of strong interest at the time.

A. Bar-Hillel: An Early Perspective on Sense Disambiguation

To appreciate some of the complexity and potential of the sense-disambiguation task, it is instructive to consider Bar-Hillel's early assessment of the problem. Bar-Hillel felt that sense disambiguation was a key bottleneck for progress in MT, one that ultimately led him and others to conclude that the problem of general MT was intractable given current, and even foreseeable, computational resources. He used the now famous example {5} of the polysemous word *pen* as motivation for this conclusion.

> Little John was looking for his toy box. Finally he found it. *The box was in the **pen***. John was very happy. {5}

In his analysis of the feasibility of MT, Bar-Hillel (1960) argued that even this relatively simple sense ambiguity could not be resolved by electronic computer, either current or imaginable:

> Assume, for simplicity's sake, that *pen* in English has only the following two meanings: (1) a certain writing utensil, (2) an enclosure where small children can play. I now claim that no existing or imaginable program will enable an electronic computer to determine that the word *pen* in the given sentence within the given context has the second of the above meanings, whereas every reader with a sufficient knowledge of English will do this "automatically." (Bar-Hillel, 1960).

Such sentiments helped cause Bar-Hillel to abandon the NLP field. Although one can appreciate Bar-Hillel's arguments, given their historical context, the following counterobservations are warranted:

Bar-Hillel's example was chosen to illustrate where selectional restrictions fail to disambiguate: both an enclosure *pen* and a writing *pen* have internal space and hence admit the use of the preposition *in*. Apparently more complex analysis regarding the relative size of toy boxes and writing pens is necessary to rule out the second interpretation.

What Bar-Hillel did not seem to appreciate at the time was the power of associational proclivities, rather than hard selectional constraints. One almost never refers to what is "in" a writing pen (except in the case of *ink*, which is a nearly unambiguous indicator of writing pens by itself), whereas it is very common to refer to what is "in" an enclosure pen. Although the trigram *in the pen* does not *categorically* rule out either interpretation, *probabilistically* it is very strongly indicative of the enclosure sense and would be very effective in disambiguating this example even without additional supporting evidence.

Thus, although this example does illustrate the limitations of selectional constraints and the infeasible complexity of full pragmatic inference, it actually represents a reasonably good example for where simple collocational patterns in a probabilistic framework may be successful.

B. Early AI Systems: Word Experts

After a lull in NLP research following the 1966 ALPAC report, semantic analysis closely paralleled the development of artificial intelligence (AI) techniques and tended to be embedded in larger systems such as Winograd's SHRDLU (1973) and Woods' LUNAR (1972). Word-sense ambiguity was not generally considered as a separate problem, and indeed did not arise very frequently given the general monosemy of words in restricted domains.

Wilks (1973, 1975) was the one of the first to focus extensively on the discrete problem of sense disambiguation. His model of **Preference Semantics** was based primarily on selectional restrictions in a Schankian framework, and was targeted at the task of MT. Wilks developed frame-based semantic templates of the form exemplified in Fig. 2, which were used to analyze sentences such as {6} by finding the maximally consistent combination of templates.

The policeman interrogates the crook. {6}

Small (1980) and Small and Rieger (1982) proposed a radically lexicalized form of language processing using the complex interaction of **word experts** for parsing and semantic analysis. These "experts" included both selectional constraints and hand-tailored procedural rules, and were focused on multiply ambiguous sentences such as {7}.

The man eating peaches throws out a pit. {7}

Hirst (1987) followed a more general word–expert-based approach, with rules based primarily on selectional constraints with back-off to more general templates for increased coverage. Hirst's approach also focused on the dynamic interaction of these experts in a marker-passing mechanism called **Polaroid words.**

Cottrell (1989) addressed similar concerns regarding multiply conflicting ambiguities as in {8} in a **connectionist** framework, addressing the psycholinguistic correlates of his system's convergence behavior.

Bob threw a ball for charity. {8}

C. Dictionary-Based Methods

To overcome the daunting task of generating hand-built rules for the entire lexicon, many researchers have turned to information extracted from existing dictionaries. This work became practical in the late 1980s with the availability of several large-scale dictionaries in machine-readable format.

```
policeman --> ((folk sour)((((notgood man)obje)pick)(subj man)))

interrogates --> ((man subj)((man obje)(tell force)))

crook --> ((((notgood act) obje)do)(subj man))

crook --> ((((((this beast)obje)force)(subj man))poss)(line thing))
```

Fig. 2 Examples of sense definitions from Wilks' *Preference Semantics.*

Lesk (1986) was one of the first to implement such an approach, using overlap between definitions in *Oxford's Advanced Learner's Dictionary of Current English* to resolve word senses. The word *cone* in *pine cone* was identified as a 'fruit of certain evergreen trees' (sense 3), by overlap of both the words *evergreen* and *tree* in one of the definitions of *pine*. Such models of strict overlap clearly suffer from **sparse data problems**, as dictionary definitions tend to be brief; without augmentation or class-based generalizations they do not capture nearly the range of collocational information necessary for broad coverage.

Another fertile line of dictionary-based work used the subject codes in the online version of Longman's LDOCE (see Sec. 2.B). These codes, such as *EC* for economic/financial usages and *AU* for automotive usages, label specialized, domain-specific senses of words. Walker and Amsler (1986) estimated the most appropriate subject code for words such as *bank* having multiple specialized domains, by summing up dominant presence of subject codes for other words in context. Guthrie et al. (1991) and Cowie et al. (1992) enriched this model by searching for the globally optimum classifications in the cases of multiple ambiguities, using **simulated annealing** to facilitate search. Veronis and Ide (1990) pursued a **connectionist** approach using cooccurrences of specialized subject codes from *Collins' English Dictionary*.

D. Kelly and Stone: An Early Corpus-Based Approach

Interestingly, perhaps the earliest corpus-based approach to WSD emerged in the 1975 work of Kelly and Stone, nearly 15 years before data-driven methods for WSD became popular in the 1990s. For each member of a target vocabulary of 1815 words, Kelly and Stone developed a flowchart of simple rules based on a potential set of patterns in the target context. These included the morphology of the polysemous word and collocations within a plus or minus four-word window, either for exact word matches, parts-of-speech, or 1 of 16 hand-labeled semantic categories found in context.

Kelly and Stone's work was particularly remarkable for 1975 in that they based their disambiguation procedures on empirical evidence derived from a 500,000-word text corpus rather than their own intuitions. Although they did not use this corpus for automatic rule induction, their hand-built rule sets were clearly sensitive to and directly inspired by patterns observed in sorted key word in context (KWIC) concordances. As an engineering approach, this data-driven but hand-tailored method has much to recommend it even today.

5. SUPERVISED APPROACHES TO SENSE DISAMBIGUATION

Corpus-based sense disambiguation algorithms can be productively divided into two general classes: those based on **supervised** and **unsupervised** learning. Although there is considerable debate over where to draw the boundary between the two, a useful heuristic is that supervised WSD algorithms derive their classification rules or statistical models directly or predominantly from sense-labeled training examples of polysemous words in context. Often hundreds of labeled training examples per word sense are necessary for adequate classifier learning, and shortages of training data are a primary bottleneck for supervised approaches. In contrast, unsupervised algorithms do not require this direct sense-tagged training data, and in their purest form induce sense partitions from strictly untagged training examples. Many such approaches do make use of a secondary knowledge source, such as the WordNet semantic concept hierarchy to help bootstrap structure from raw data. Such methods can arguably be considered unsupervised, as they are based on existing independent knowledge sources with no direct supervision of the phenomenon to be learned. This distinction warrants further discussion, however, and the relatively conservative term **minimally supervised** will be used here to refer to this class of algorithms.

A. Training Data for Supervised WSD Algorithms

Several collections of hand-annotated data sets have been created with polysemous words in context labeled with the appropriate sense for each occurrence, instance or example. The most comprehensive of these is the WordNet **SEMCOR semantic concordance** (Miller et al., 1994), where a growing percentage of the 1-million–word Brown corpus has been annotated with the appropriate concept node in WordNet. Vocabulary coverage is wide and balanced, while the number of examples per polysemous word is somewhat limited. A small but widely cited data set is the ***line* corpus** (Leacock et al., 1993a), with 2094 examples of the word *line* tagged with one of its six major senses. Yarowsky (1992, 1995) and Bruce (1995; Bruce and Wiebe, 1994) have each developed sense-tagged data sets for 12 different polysemous words, with between 300 and 11,000 tagged examples for each word. Ng and Lee (1996) are currently building a sense-tagged corpus of the 191 most frequent polysemous words in English, using SEMCOR tags, with approximately 1000 examples of each. These data sets are publically available from the Linguistic Data Consortium (`ldc@unagi.cis.upenn.edu`).

It is useful to visualize sense-tagged data as a table of tagged words in context (typically the surrounding sentence or ±50 words), from which specialized classification features can be extracted. Table 3 shows the example of the polysemous word *plant* exhibiting a manufacturing plant and living plant sense.

Relevant features include, but are not limited to, the surrounding raw words, lemmas (word roots), and part-of-speech (POS) tags, often itemized by a relative position or syntactic relation, but in some models represented as a position-independent **bag of words**. An example of such feature extraction from the foregoing data is shown in Table 4.

Once these different features are extracted from the data, it is possible to compute the **frequency distribution** of the sense tags for each feature pattern. Table 5 illustrates this for several different feature types, with $f(M)$ indicating the fre-

Table 3 Example of Sense-Tagged Words in Context

Sense tag	Instance of polysemous word in context
MANUFACT	... from the Toshiba *plant* located in ...
MANUFACT	... union threatened *plant* closures...
MANUFACT	... chloride monomer *plant*, which is ...
LIVING	... with animal and *plant* tissues can be ...
LIVING	... Golgi apparatus of *plant* and animal cell ...
LIVING	... the molecules in *plant* tissue from the ...

Table 4 Example of Feature Extraction

	Word -2			Word -1			Word 0	Word+1
Sense tag	Class	Word	POS	Class	Lemma	POS	Word	Word
MANUFACT		the	DET	CORP	Toshiba	NP	Plant	Located
MANUFACT	BUSN	union	NN		threaten	VBD	Plant	Closures
MANUFACT	CHEM	chloride	NN	CHEM	monomer	NN	Plant	,
LIVING	ZOOL	animal	NN		and	CON	Plant	Tissues
LIVING		apparatus	NN		of	PREP	Plant	And
LIVING	CHEM	molecules	NNS		in	PREP	Plant	Tissue

quency of the feature pattern as the manufacturing sense of *plant*, and $f(L)$ giving the frequency of the living sense of *plant* for this feature pattern. These raw statistics will drive almost all of the classification algorithms discussed in the following.

Note that word order and syntactic relations can be of crucial importance for the predicative power of word associations. The word *open* occurs within $\pm k$ words of *plant* with almost equal likelihood of both senses, but when *plant* is the direct object of *open* it exclusively means the manufacturing sense. The word *pesticide* immediately to the left of *plant* indicates the manufacturing sense, but in any other position the distribution in the data is 6 to 0 in favor of the living sense. This would suggest that there are strong advantages for algorithms that model collocations and syntactic relations carefully, rather than treating contexts strictly

Table 5 Frequency Distribution of Various Features Used to Distinguish Two Senses of *plant*

Feature type	Feature pattern	$f(M)$	$f(L)$	Majority sense
WORD +1	*plant growth*	0	244	LIVING
WORD +1	*plant height*	0	183	LIVING
LEMMA +1	*plant size*/N	7	32	LIVING
LEMMA +1	*plant closure*/N	27	0	MANUFACT
WORD −1	*assembly plant*	161	0	MANUFACT
WORD −1	*nuclear plant*	144	0	MANUFACT
WORD −1	*pesticide plant*	9	0	MANUFACT
WORD −1	*tropical plant*	0	6	LIVING
POS +1	*plant* <NOUN>	561	2491	LIVING
POS +1	<NOUN> *plant*	896	419	MANUFACT
WORD $\pm k$	*car* within $\pm k$ words	86	0	MANUFACT
WORD $\pm k$	*union* within $\pm k$ words	87	0	MANUFACT
WORD $\pm k$	*job* within $\pm k$ words	47	0	MANUFACT
WORD $\pm k$	*pesticide* within $\pm k$ words	9	6	MANUFACT
WORD $\pm k$	*open* within $\pm k$ words	20	21	LIVING
WORD $\pm k$	*flower* within $\pm k$ words	0	42	LIVING
Verb/Obj	*close*/V, Obj = *plant*	45	0	MANUFACT
Verb/Obj	*open*/V, Obj = *plant*	10	0	MANUFACT
Verb/Obj	*water*/V, Obj = *plant*	0	7	LIVING

as unordered bags of words. An empirical study of this observation and its implications for algorithm design is presented in Yarowksy (1993, 1994).

The following sections will contrast several supervised learning algorithms that can be applied to the foregoing data, including decision trees, Bayesian classifiers, decision lists, and a comparative survey of other classification algorithms. The choice of features, mechanisms for generalization in the face of sparse data and the chosen decision architecture are the major sources of variation among these supervised algorithms.

B. Decision Tree Methods

Decision trees are hierarchically nested series of (typically binary) questions, that for any instance of example data yields a unique path to a leaf node, and outputs a single member of the classification inventory (e.g., word senses) most probable at that node. Decision tree implementations for WSD are surprisingly rare, given their ubiquitous presence in other fields of machine learning. One likely reason is their difficulty handling the data fragmentation resulting from the very high feature dimensionality of typical WSD data.

In one of the earliest empirical machine-learning approaches to sense disambiguation, Black (1988) trained decision trees on three types of features: *LDOCE* subject codes (DG), a set of words and word attributes extracted from the concordance line of the ambiguous word (DS1), and a combination of method DS1 and manually developed subject categories (DS2). Mean classification accuracy was similar between method DS1 (72%) and DS2 (75%), with *LDOCE* subject codes alone (method DG) performing the weakest (47% accuracy). Baseline accuracy for Black's data (random sense selection) was 37%.

Brown et al. (1991) presented work that forms the basis of subsequent Decision Tree methods in WSD, aimed at the application of French-to-English MT. They utilized the **Flip–Flop algorithm** (Nadas et al., 1991) for generating a single question (for use in the root node of a decision tree) that maximally partitions French examples based on their English translation. In one of their three instances, examples of the French word *prendre* are partitioned into two sets, one of which is most likely translated as *to_take* and the other as *to_make* or *to_speak*. The distinguishing feature used is exclusively the noun to the right, with the French words *mesure* and *note* indicative of the *to_take* sense, while *décision* and *parole* are indicative of the *to_make* or *to_speak* sense. The Flip–Flop algorithm begins with a random partition of features, and incrementally identifies better partitions that increase the **mutual information** between the partitioned instances and their English translation. As the mutual information after a given repartition is at least as great as before (i.e., is monotonically increasing) and because the upper bound is fixed at 1 bit (for the binary split), the algorithm will eventually converge on a French vocabulary partition reasonably predictive of the English translation. These tests can be repeated on different possible features (such as verb tense) and the best partitioning question is selected as the root node of the decision tree. Although not explored and evaluated in the paper, this node-partitioning algorithm may be recursively applied to child nodes in a decision tree. Brown et al.'s algorithm was not evaluated independently as a classification technique for WSD. However, their preliminary results showed that the use of the basic single partition induction procedure improved performance of the

IBM French–English translation system by 22% (from 37 acceptable translations to 45 acceptable translations on a set of 100 test examples).

C. Bayesian Classifiers

Bayesian classifiers are one of the oldest learning algorithms, developed in depth on the task of authorship identification by Mostellar and Wallace (1964). Their general theoretical framework is based on Bayes' rule [Eq. {9}].

$$P(class_i|E) = \frac{P(class_i) \times P(E|class_i)}{P(E)} \tag{9}$$

where (in the case of sense disambiguation) $class_i$ represents one of the k possible senses of a target ambiguous word and where E is a combination of potential evidence on which the classification may be based, typically the words in surrounding context. $P(class_i|E)$ is the probability of $class_i$ given in particular that the target ambiguity has $class = i$, $P(E|class_i)$ is the probability of a given configuration of evidence E when the target ambiguity has $class = i$, and $P(E)$ is the underlying probability of a given configuration of evidence E.

When resolving a binary sense ambiguity ($k = 2$ possible clauses), it is productive to look at the ratio $P(class_1|E)$ over $P(class_2|E)$, which using Bayes' rule expands to Eq. {10}.

$$\frac{P(class_1|E)}{P(class_2|E)} = \frac{P(class_1)}{P(class_2)} \times \frac{P(E|class_1)}{P(E|class_2)} \tag{10}$$

One can simplify the computation of $P(E|class_i)$ by making the assumption that the conditional probabilities of the individual pieces of evidence e_j that compose E are independent of one another, as in Eq. {11},

$$P(E|class_i) = \prod_{E_j \in E} P(e_j|class_i) \tag{11}$$

and, therefore, Eq. {12}.

$$\frac{P(class_1|E)}{P(class_2|E)} = \frac{P(class_1)}{P(class_2)} \times \prod_{e_j \in E} \frac{P(e_j|class_1)}{P(e_j|class_2)} \tag{12}$$

Gale et al. (1992a) applied this model to the task of WSD, treating E as the immediately surrounding context (e.g., ±50 words) around a target polysemous word having one or two senses ($class_1 = sense_1$, $class_2 = sense_2$) as in Eq. {13},

$$\frac{P(sense_1|Context)}{P(sense_2|Context)} = \frac{P(sense_1)}{P(sense_2)} \times \prod_{e_j \in E} \frac{P(w_j|sense_1)}{P(w_j|sense_2)} \tag{13}$$

where w_j is a word in the surrounding context ($Context$) of a polysemous target word having senses $sense_1$ and $sense_2$. In addition, they investigated methods for **interpolated smoothing** in the face of sparse data from O(100,000) unique collocational features.

In previous WSD work there has always been a shortage of sense-tagged training data on which to estimate the probabilities such as $P(w_j|sense_i)$. Gale et al. addressed this problem by using a parallel-aligned bilingual corpus (the Canadian

Hansard parliamentary debates in French and English) and used the French translations of the ambiguous word as sense tags. Polysemous words in their study included *duty* (translated as *droit* 'tax' or *devoir* 'obligation') and *drug* (translated as *médicament* 'medicine' or *drogue* 'illicit drug').

This approach had the advantages of both (a) providing automatically acquired sense tags and (b) relevance to the task of MT. It suffered, however, from the very incomplete coverage and skewed sense distributions of the Hansard corpus and also from the problem that bilingual translation distinctions do not always correspond directly with monolingual sense distinctions as discussed in Sec. 3.C. Overall, their classification performance averaged 90% on a training set of 17,138 tagged examples of six polysemous words (*duty*, *drug*, *land*, *language*, *position*, and *sentence*).

Bruce and Wiebe (1994), Bruce (1995), and Pedersen and Bruce (1997) proposed the use of Bayesian networks and the general class of decomposable models for WSD. They noted that the assumption of independence in the foregoing technical termed "naive" Bayes algorithm does not generally hold, and attempted to model the correlations between features that would help compensate for this. They handled the combinatorial complexity of the possible sets of feature correlations by identifying the most robustly estimated feature associations, and used a G^2 **goodness of fit test** for model evaluation.

The outcome of this approach is that the few correlations that are modeled are modeled very accurately, but the algorithm tends to yield small models and models with very limited feature sets exploited for a given word (typically fewer than 11 features). For example, the optimal model selected for the ambiguous noun *interest* includes the three lexical features *percent, in*, and *rate*, the parts-of-speech of the words at offsets ± 2 (*r1pos*, *r2pos*, etc.), and the ending of the ambiguous word (*interest* or *interests*) yielding a network decomposition as in Eq. {14}.

$$\begin{aligned} &P(\text{percent}, \text{in}, \text{rate}, r1pos, r2pos, l1pos, l2pos, ending, tag) \\ &\quad = P(l1pos, l2pos|tag) \times P(r1pos|tag) \times P(r2pos|tag) \times P(\text{in}|tag) \\ &\quad \times P(\text{rate}|tag) \times P(\text{percent}|tag) \times P(ending|tag) \times P(tag) \end{aligned} \qquad \{14\}$$

In Bruce (1995), a study of 12 polysemous nouns, verbs, and adjectives achieved mean accuracy of 84% on data with a baseline (majority sense) performance of 72%.

It is interesting that the network models selected by the Bruce (1995) algorithm, including the foregoing *interest* example, rarely captured dependencies among features; all chosen configurations were nearly or exactly equivalent to the minimal Bayesian model assuming independence among all features. Bruce explained this outcome as the result of relatively small training sample sizes and a model-selection process driven more by recall than precision. Nevertheless, her results indicate that relatively simple network topologies may yield comparable or superior performance, especially if the use of a full Bayesian network requires a greatly reduced feature space for computational tractability, as in the Bruce study. This result is consistent with the conclusions of Mooney (1996) in following Sec. 5.D.

D. Other Supervised Methods

Decision lists (Rivest, 1987) are linearly ordered sequences of pattern–action rules, in which only the first (most confident) matching rule is utilized. Yarowsky (1994, 1995) has applied them to a variety of disambiguation tasks, and proposed an interpolated learning strategy to help overcome sparse data problems. In addition to high perspicuity and efficiency, decision lists are especially well suited to data sets with highly nonindependent features that often have very low entropy, a good characterization of the complex, but often repetitive, collocational features for WSD described in Sec. 4.A. They are less well suited for unordered bag-of-words representations of context, as observed in Mooney (1996) and discussed in Sec. 5.D.

Content vector models such as those used in IR (Salton and McGill, 1983) have been applied by Schütze (1992, 1995) and others using unordered bags of words to model the surrounding 25–100 words of context for a word just as would be used to model a full document for IR purposes. They are quite similar in practice to naive Bayes algorithms for this application, differing primarily in vector similarity measures.

Nearest neighbor algorithms have been pursued by Ng (1997) and Ng and Lee (1996) using PEBLS (Cost and Salzberg, 1993). The general framework of this work is that the sense tag that a new vector is assigned is based on a weighted vote of the tags of the *k* most similar vectors. Ng's results are quite similar to the performance of naive Bayes.

E. Comparative Studies

Several surveys have systematically compared the efficacy of different learning procedures for WSD on uniform feature sets and identical training and test data.

Leacock et al. (1993b) contrasted a **neural network**, **Bayesian classifier**, and **IR-vector model**. Their experiments were conducted using the *line* corpus described in Sec. 5.A. Each of the three systems used a vector of the unordered bag of words in context as input. In the neural network this was represented as a binary bit vector, whereas the other techniques included term weighting that effectively down-weighted frequent terms. In the neural network and vector model, terms were stemmed first; the Bayesian model used raw, unstemmed words, a potentially significant source of difference in the results. Several network topologies were explored for the neural network. Curiously, the network **without a hidden layer** of units (a perceptron) performed as well as or better than those **with hidden layers**. Comparative results showed the neural net performing slightly better (76%) than the other two methods (71% Bayesian and 72% vector model), but with the differences not statistically significant, and quite possibly owing to differences such as the presence or absence of stemming.

A similar comparative study of Bayesian, content vector, and neural net models was conducted by Lehman (1994), also using the Leacock et al. *line* data sets. Results were very similar, with the neural net model (76% recall and 75% precision) outperforming the Bayesian model (71% recall and 73% precision) and the content vector model (72% recall and 72% precision).

A more recent and more extensive comparative study was performed by Mooney (1996), also using the *line* data. The **Naive Bayes** model (Bayesian classifier assuming independence) and **simple perceptron** achieved significantly higher accuracy than other classifiers across all different training set sizes tested. Lower ranked classifiers were, in order of performance, a **PFOIL Decision List** (Mooney, 1995), **C4.5 Decision Tree**, **3 Nearest Neighbors**, and **PFOIL DNF** and **CNF** formulae. The PFOIL DNF, CNF, and Decision List were significantly slower to train, and the Naive Bayes and 3 Nearest Neighbors were slowest to test on new data. These results suggest that algorithms such as Bayesian classifiers and perceptrons that use a weighted combination of all features are more effective in high-dimensional WSD feature spaces than algorithms, such as decision trees that exploit only a small set of the most informative features.

However, all three of these comparative studies used the relatively impoverished data representation of an unordered bag of words. The most predictive evidence sources, adjacent collocations and local syntactic relations, were not identifiable in this representation; hence, they were not exploitable. One would expect decision trees and decision lists to perform much better when the more predictive and refined collocational and syntactic features are included in the data.

6. UNSUPERVISED APPROACHES TO WSD

Although there have been recent improvements in the availability of sense-tagged training data, insufficient quantities exist for many words and finer-grained sense distinctions. Almost no sense-tagged training data exists for languages other than English. A wide range of techniques in unsupervised and minimally supervised learning offer the potential to overcome the knowledge acquisition bottleneck slowing progress in this field.

A. Vector–Space-Based Sense Clustering Algorithms

Two of the earliest unsupervised WSD algorithms were introduced by Zernik (1991) and Schütze (1992), utilizing the framework of IR-style content vectors noted in Sec. 5.C. Just as documents can be represented as vectors $< W_1, W_2, W_3, W_4, \ldots, W_V >$ in V-dimensional space (where $V =$ vocabulary size, and W_i is a weighted function of word frequency in that document), the context surrounding a polysemous word can be similarly represented as if it were a separate document. Given a **vector similarity function**, context vectors may be clustered hierarchically into partitions of the vector set of high relative within-cluster similarity. The goal of unsupervised learning in this case is that the induced clusters will also correspond to word contexts exhibiting the same sense. To the extent that similar word distributions in surrounding context correlates with similar word sense, such a partitioning algorithm is likely to yield clusters of vectors with primarily the same sense. For example, Fig. 3 shows four clusters that have been identified in the text (indicated by the smaller boxes). The true sense labels associated with each context vector are identified by labels (A, B, C, D). Although these labels were not known at the time of clustering, in this particular example the induced partitions exhibit relatively high-sense purity within clusters.

Many different clustering algorithms and similarity functions may be used in this general framework. Zernik (1991) utilized **bottom-up hierarchical agglomerative clustering** based on **cosine similarity**. Schütze (1992) utilized a top-down Bayesian algorithm **Autoclass** (Cheesman et al., 1988) and a partially randomized algorithm **Buckshot** (Cutting et al., 1992). Schütze also used **singular value decomposition**

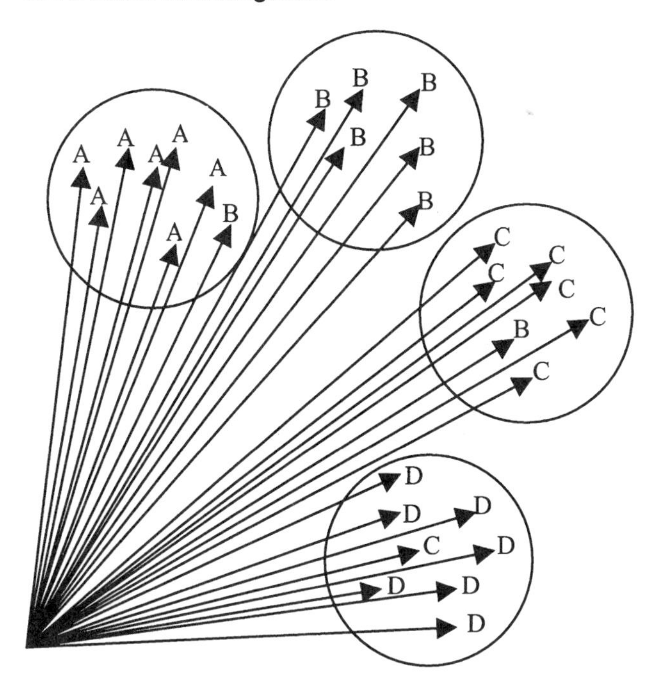

Fig. 3 Four vector–space sense clusters (letters corresponding to true sense of IO's).

(Deerwester et al., 1990) to reduce the dimensionality of the vector space, using **canonical discriminant analysis** to optimize the weighting of dimensions in the space for maximal cluster separation. This reduced dimensionality, called **Word Space** by Schütze, helped overcome some of the sparse data problems associated with similarity measures such as cosine distance.

One significant problem with this approach is that without at least some sense-labeled context vectors, it is difficult to map the induced clusters to a sense number in an established sense inventory that would be useful for a secondary task (such as MT). Both Zernik and Schütze used manual post hoc alignment of clusters to word senses in their WSD experiments. Because top-level cluster partitions based purely on distributional information do not necessarily align with standard sense distinctions, Zernik and Schütze generated partitions of up to ten separate sense clusters and manually assigned each to a fixed sense label (based on the hand-inspection of 10–20 sentences per clusters). Semiautomated post hoc alignments are possible, if at least some information on the reference senses such as a dictionary definition is available. Without such a mapping, one cannot evaluate sense-tagging accuracy relative to an inventory, only cluster purity. Section 3.A describes an application, IR, for which such alignment is not necessary (only the unlabeled sense partitions are needed). It is also possible to treat this task as unsupervised **sense induction**, and use a list of distinctive words in each cluster to label the cluster. Nevertheless for most applications sense tagging is crucially a disambiguation task relative to an established sense inventory tied to the application.

B. Iterative Bootstrapping Algorithms

Assuming that some information, such as a dictionary definition, is available to do post hoc alignment of automatically induced clusters and established word senses, why not use this same information at the *beginning* of the algorithm when it can serve as a seed and search–space-minimizing guide?

Hearst (1991) proposed an early application of **bootstrapping** based on relatively small hand-tagged data sets used in supervised sense tagging. Similar to Schütze and Zernik, Hearst used a term-vector representation of contexts and cosine similarity measure to assign new test contexts to the most similar vector centroid computed from her tagged training examples. She later augmented the existing centroids with additional untagged vectors, assigned to their closest labeled centroid, with the goal of alleviating some of the sparse data problems in the original tagged training data. Classifiers using these augmented training sets performed nearly the same as those using the purely hand-tagged data, and in some cases, the presence of the additional sentences hurt performance.

Yarowsky (1995) developed an algorithm for multipass iterative bootstrapping from different types of small seed data for each target word sense. These starting configurations included 'minimal seed' information as basic as just the two sense labels (e.g., MANUFACTURING and LIVING for the word *plant*), as well as other experiments using dictionary definitions and a small set of hand-labeled collocations as minimal seed data. A key component of this work was the process by which this very small seed data was reliably and robustly expanded eventually to tag all instances of the word in a corpus. The driving engines for this procedure were two empirically studied properties: (a) polysemous words strongly tend to exhibit only one sense per collocation (Yarowsky, 1993) and (b) polysemous words strongly tend to exhibit only one sense within a given discourse (Gale et al., 1992b).

In these studies, *collocation* was defined as an arbitrary association or juxtaposition of words, with the effects measured sensitive to syntactic relations and distance between the two words, as well as factors such as the part of speech of the words. Yarowsky observed that for collocations with a polysemous word (such as *plant*) when in adjacent collocation with another noun (e.g., *plant closure*), on average 98% of the instances of a given collocation refer to the same word sense. As collocational distance increases to 20 words (e.g., *flower* occurring within 20 words of *plant*), over 80% of the instances of a given collocation refer to the same sense. Verbs and adjectives tend to have more localized collocational predictive reliability, while noun collocations even 100 words distant still strongly tend to select the same sense.

In a separate study (Yarowsky, 1995) it was observed that for a sample set of 12 polysemous nouns and verbs, on average 50.1% of the instances of each polysemous word occurred in a discourse with another instance of the same word. Of those discourses with two or more occurrences of the polysemous word, the probability that any randomly chosen instance of the polysemous word would have the same sense as another randomly chosen instance of the same word was measured at 99.8% for the sample set of words. Clearly, the one-sense-per-discourse tendency holds with high accuracy, and can be exploited for approximately half of the instances of polysemous words in the corpus.

Together, the strong one-sense-per-discourse and one-sense-per-collocation tendencies can be jointly exploited in an iterative bootstrapping framework. This

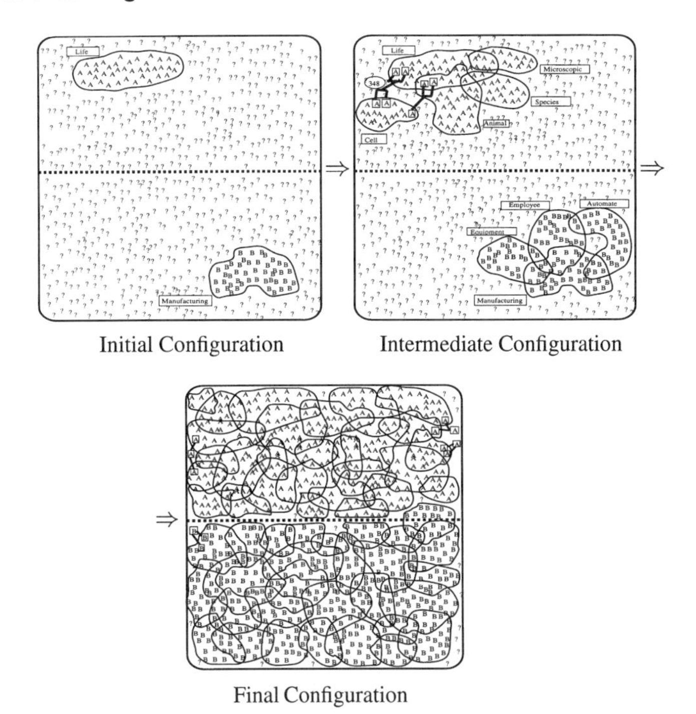

Fig. 4 Iterative bootstrapping framework (from the initial seed words).

is illustrated graphically in Fig. 4. The algorithm begins with as few as two seed collocations that are indicative of the target senses. Examples of the word *plant* that contain the seed collocations are initially labeled as A = *life* and B = *manufacturing* for the two major senses of the polysemous word *plant*). The remainder of the examples of the polysemous word are initially untagged, represented graphically in Fig. 4 as **?**. A supervised classifier (based on decision lists in this case) is then trained on the seed sets, and applied to the residual untagged data. New collocations indicative of either sense are learned, and those classifications with confidence above a threshold are added to the next round's training set. Simulated annealing is applied to this confidence threshold to help avoid overtraining. The one-sense-per-discourse tendency is also exploited as both a bridge to add new collocational examples to the training model and as effective filter or to decrement the probability of initially misclassified examples given strong evidence for the competing sense from other examples of the polysemous word in the discourse. After multiple iterations of the bootstrapping steps, the third configuration in the figure indicates final convergence at a partition of the example space corresponding closely to the true partition (dotted line). Mean accuracy of this algorithm starting from only two seed words was 90.6%. Starting from words in an online dictionary definition when fully using the one-sense-per-discourse model achieves 96.5%. This slightly outperforms a baseline (decision list) supervised-learning algorithm *not* using discourse co-occurrence information (96.1% accuracy), indicating the power of the one-sense-per-discourse tendency to effectively guide and constrain incremental bootstrapping.

Table 6 Contextual Word-Class Model Exemplified with *crane*

Example	The engine of the *crane* was damaged	Flocks of *cranes* nested in the swamp
P(MACHINE|context)	0.650	0.002
P(BIRD|context)	0.007	0.370
P(MINERAL|context)	0.005	0.004
Best sense/class label	MACHINE	BIRD

C. Topic-Driven Class Models

Another productive unsupervised approach exploits the fact that most nouns in a word list are monosemous, and thus it is possible to build robust detectors for large noun classes that can then be used to classify polysemous words that are members of multiple classes. Consider the example of the word *crane* which can be a member of the class BIRD and MACHINE. Lists of other birds (*heron*, *grebe*, *pelican*, *hawk*, *seagull*, and so on) and machines (*jackhammer*, *drill*, *tractor*, *bulldozer*, and so on) can be extracted from online resources, such as *Roget's Thesaurus*. It is then possible to build a Bayesian classifier or context-vector classifier for these two classes, by taking an aggregate profile of the typical context of all members of the class. For example, *egg*, *nest*, *feeding*, and *migration*, are all contextual clues learned for the BIRD class. Note that some members of the word classes (e.g., *drill*, *hawk*) are polysemous, but the noise introduced by including them in the training data is tolerable. Although nonbird examples, such as congressional *hawks* and golfing *eagles* are part of the training data for the aggregate-context models, (a) their numbers are modest, (b) secondary senses tend to be distributed uniformly across many different topic areas, while (c) the common contextual properties of the class (e.g., eggs, nests, migration, and such) are focused and exhibit adequate signal-to-noise ratio. Thus, these contextual **word-class** models trained on undisambiguated data can be applied to their polysemous members such as *crane* to decide which class membership (e.g., BIRD or MACHINE) is most probable for each individual context, as exemplified in Table 6.

Because word-class labels can serve as word-sense labels, and because word-class detectors can be trained effectively on somewhat noisy polysemous data, as long as a class inventory is available in which class members are primarily monosemous and share common properties, accurate WSD can be achieved without sense-tagged training data. This algorithm, developed in Yarowsky (1992), achieved 92% mean accuracy on a test set of 12 polysemous words using *Roget's Thesaurus* as the primary sense inventory and topic-sensitive Bayesian classifiers as the primary class detection algorithm.

D. Hierarchical Class Models Using Selectional Restriction

Similar class-based models can be applied to the problem of resolving noun and verb senses through selectional restriction and selectional preference. Resnik (1993, 1997) has defined **selection preference** as a probabilistic distribution over a concept hierarchy, such as WordNet. Consider the the parent (hypernym) classes for the words in WordNet illustrated in Fig. 5.

```
gin -> {BEVERAGE,DEVICE,GAME}

vodka -> {BEVERAGE}

coffee -> {BEVERAGE,TREE,SEED,COLOR}

pint -> {VOLUME_UNIT}

cup -> {VOLUME_UNIT,VESSEL,TROPHY}
```

Fig. 5 Hypernym classes for five words, from WordNet.

Resnik has extracted a large set of verb–object pairs from a parsed corpus (e.g., *drink*(*vodka*), *drink*(*gin*), *drink*(*pint*), or *play*(*gin*)). He then computed statistics on the common objects of the verb *drink*, the WordNet classes of these objects, their parent classes, and so forth. At any level of generality, he can express the likelihood of *P*(WORDNET_CLASS|*drink*), where *P*(BEVERAGE|*drink*) is based on the frequency of members of the class BEVERAGE as objects of the word *drink*. In the case of polysemous words, such as *gin*, partial credit is uniformly assigned to each of the class options (e.g., *P*(BEVERAGE|*drink*), *P*(DEVICE|*drink*), and *P*(GAME|*drink*)). As in the case of Yarowsky's class models in Sec. 6.C, the noise introduced by this spurious polysemy will be tolerable, and *P*(BEVERAGE|*drink*) and *P*(VOLUME_UNIT|*drink*) will be the two most likely object classes observed in a corpus, even under the partial weighting scheme for spurious secondary senses. Thus, as in Sec. 6.C, the class label of any individual verb–object pair [e.g., *drink*(*gin*)] can be resolved as the most likely of *gin*'s WordNet class to be drunk, given the probabilistic model of selectional preference. Similarly, as *P*(VOLUME_UNIT|*drink*) is higher than *P*(TROPHY|*drink*), the correct sense of *drink*(*cup*) is also selected.

Similarly, verb senses in Resnik's framework can be defined as the most likely WordNet class of the verb's object. The pair *drink*(*cup*) is also correctly assigned the verb sense corresponding to *drink*(VOLUME_UNIT) rather than *drink*(TROPHY). Collectively, Resnik's work clearly demonstrates the ability for raw corpus cooccurrence statistics combined with an existing hierarchical class-based lexicon to identify the underlying semantic "signal" in the noise of polysemy.

7. CONCLUSION

Corpus-based, statistical algorithms for WSD have demonstrated the ability to achieve respectable accuracy resolving major sense distinctions when adequate training data is available. Given the high cost and relatively limited supply of such data, current research efforts are focusing on the problems of:

- Exploiting other potential sources of automatically sense-tagged training data for supervised learning, such as parallel bilingual corpora
- Improving the speed and efficiency of human annotation through active learning algorithms that dynamically guide interactive-tagging sessions based on currently unsatisfied information need

- Developing algorithms that can better bootstrap from the lexical and ontological knowledge present in existing references sources such as online dictionaries, WordNet, thesauri, or other driving minimally supervised algorithms on unannotated corpora
- Using unsupervised clustering and sense induction for applications (e.g., IR) that do not require alignment of the induced sense partitions to an existing sense inventory

Another major open challenge is the broad-coverage resolution of finer-grained sense and subsense distinctions that have been underaddressed in past research. Richer feature representations that capture more refined lexical, syntactic, pragmatic, and discourse-level dependencies may be necessary, demanding improved algorithms for extracting such information from corpora and other available knowledge sources. Thus future progress in sense disambiguation depends heavily on parallel progress in the other text-analysis tasks described in this book.

REFERENCES

Bar-Hillel Y (1960). Automatic translation of languages. In: D Booth, RE Meagher, eds. Advances in Computers. New York: Academic Press.

Black E (1988). An experiment in computational discrimination of English word senses. *IBM J Res Dev* 232:185–194.

Brown P, S Della Pietra, V Della Pietra, R Mercer (1991). Word sense disambiguation using statistical methods. 29th Annual Meeting of the Association for Computational Linguistics, Berkeley, CA, pp 264–270.

Bruce R (1995). A statistical method for word-sense disambiguation. PhD dissertation, New Mexico State University.

Bruce R, J Wiebe (1994). Word-sense disambiguation using decomposable models. 32nd Annual Meeting of the Association for Computational Linguistics, Las Cruces, NM, pp 139–146.

Chapman R (1977). Roget's International Thesaurus, 4th ed. New York: Harper & Row.

Cottrell G (1989). A Connectionist Approach to Word Sense Disambiguation. London: Pitman.

Cowie J, J Guthrie, L Guthrie (1992). Lexical disambiguation using simulated annealing. Proceedings, DARPA Speech and Natural Language Workshop, Harriman, NY.

Cutting D, D Carger, J Pedersen, J Tukey (1992). Scatter-gather: a cluster-based approach to browsing large document collections. Proceedings, SIGIR'92, Copenhagen.

Deerwester S, S Dumais, G Furnas, T Landauer, R Harshman (1990). Indexing by latent semantic analysis. J Am Soc Inform Sci 41:391–407.

Gale W, K Church, D Yarowsky (1992a). A method for disambiguating word senses in a large corpus. Comput Hum 26:415–439.

Gale W, K Church, D Yarowsky (1992b). One sense per discourse. Proceedings of the 4th DARPA Speech and Natural Language Workshop, Harriman, NY, pp 233–237.

Gale W, K Church, D Yarowsky (1992c). Using bilingual materials to develop word sense disambiguation methods. Fourth International Conference on Theoretical and Methodological Issues in Machine Translation, Montréal, pp 101–112.

Gale W, K Church, D Yarowsky (1992d). On evaluation of word–sense disambiguation systems. 30th Annual Meeting of the Association for Computational Linguistics, Columbus, OH, pp 249–256.

Gale W, K Church, D Yarowsky (1994). Discrimination decisions for 100,000-dimensional spaces. In: A Zampoli, N Calzolari, M Palmer, eds. Current Issues in Computational Linguistics: In Honour of Don Walker. Dordrecht: Kluwer Academic, pp 429–450.

Guthrie J, L Guthrie, Y Wilks, H Aidinejad (1991). Subject dependent co-occurrence and word sense disambiguation. 29th Annual Meeting of the Association for Computational Linguistics, Berkeley, CA, pp 146–152.

Hearst M (1991). Noun homograph disambiguation using local context in large text corpora. In: Using Corpora, Waterloo, Ontario: University of Waterloo.

Hearst M (1994). Context and structure in automated full-text information access. PhD dissertation, University of California at Berkeley.

Hirst G (1987). Semantic Interpretation and the Resolution of Ambiguity. Cambridge: Cambridge University Press.

Kelly E, P Stone (1975). Computer Recognition of English Word Senses. Amsterdam: North-Holland.

Krovetz R (1997). Lexical acquisition and information retrieval. In: PS Jacobs, ed. Text-Based Intelligent Systems: Current Research in Text Analysis, Information Extraction and Retrieval. Schenectady, NY: GE Research and Development Center, pp 45–64.

Krovetz R, W Croft (1992). Word sense disambiguation using machine-readable dictionaries. Proceedings of the Twelfth Annual International ACM SIGIR Conference on Research and Development in Information Retrieval, pp 127–136.

Leacock C, G Towell, E Voorhees (1993a). Corpus-based statistical sense resolution. Proceedings, ARPA Human Language Technology Workshop, Plainsboro, NJ, pp 260–265.

Leacock C, G Towell, E Voorhees (1993b). Towards building contextual representations of word senses using statistical models. Proceedings, SIGLEX Workshop on Acquisition of Lexical Knowledge from Text, Columbus, OH.

Lehman J (1994). Toward the essential nature of statistical knowledge in sense resolution. Proceedings of the Twelfth National Conference on Artificial Intelligence, pp 734–471.

Lesk M (1986). Automatic sense disambiguation: how to tell a pine cone from an ice cream cone. Proceedings of the 1986 SIGDOC Conference, New York.

Miller G (1990). WordNet: an on-line lexical database. Int J Lexicogr 3.

Mosteller F, and D Wallace (1964). Inference and Disputed Authorship: The Federalist. Reading, MA: Addison–Wesley.

Nadas A, D Nahamoo, M Picheny, J Powell (1991). An iterative "flip-flop" approximation of the most informative split in the construction of decision trees. Proceedings of the IEEE International Conference on Acoustics, Speech and Signal Processing, Toronto, Canada.

Ng H, H Lee (1996). Integrating multiple knowledge sources to disambiguate word sense: an exemplar-based approach. 34th Annual Meeting of the Association for Computational Linguistics, Santa Cruz, CA, pp 40–47.

Procter P (1978). Longman Dictionary of Contemporary English. Longman Group: Harlow, UK.

Resnik P (1993). Selection and information: a class-based approach to lexical relationships. PhD dissertation, University of Pennsylvania, Philadelphia, PA.

Rivest RL (1987). Learning decision lists. Mach Learn 2:229–246.

Salton G, M. McGill (19–). Introduction to Modern Information Retrieval. San Francisco: McGraw-Hill.

Schütze H (1992). Dimensions of meaning. Proceedings of Supercomputing '92.

Schütze H (1995). Ambiguity in language learning: computational and cognitive models. PhD dissertation. Stanford University, Stanford, CA.

Sinclair J, P Hanks, G Fox, R Moon, P Stock, eds. (1987). Collins COBUILD English Language Dictionary. London and Glasgow: Collins.

Small S, C Rieger (1982). Parsing and comprehending with word experts (a theory and its realization). In: W Lehnert, M Ringle, eds. Strategies for Natural Language Processing. Hillsdale, NJ: Lawrence Erlbaum.

Veronis J, N Ide (1990). Word sense disambiguation with very large neural networks extracted from machine readable dictionaries. Proceedings of the 13th International Conference on Computational Linguistics, COLING–90, Helsinki, pp 389–394.

Voorhees E (1993). Using WordNet to disambiguate word senses for text retrieval. Proceedings of SIGIR'93, Pittsburgh, PA, pp 171–180.

Walker D, R Amsler (1986). The use of machine-readable dictionaries in sublanguage analysis. In: R Grishman, R Kittredge, eds. Analyzing Language in Restricted Domains: Sublanguage Description and Processing. Hilsdale, NJ: Lawrence Erlbaum, pp 69–84.

Wilks Y (1975). A preferential, pattern-seeking semantics for natural language inference. Artif Intell 6:53–74.

Yarowsky D (1992). Word-sense disambiguation using statistical models of Roget's categories trained on large corpora. Proceedings of the Fifteenth International Conference on Computational Linguistics, COLING–92, Nantes, France, pp 454–460.

Yarowsky D (1993). One sense per collocation. Proceedings, ARPA Human Language Technology Workshop, Princeton, NJ pp 266–274.

Yarowsky D (1994a). Decision lists for lexical ambiguity resolution: application to accent restoration in Spanish and French. 32nd Annual Meeting of the Association for Computational Linguistics, Las Cruces, NM, pp 88–95.

Yarowsky D (1994b). A comparison of corpus-based techniques for restoring accents in Spanish and French text. Proceedings, 2nd Annual Workshop on Very Large Corpora, Kyoto, Japan, pp 19–32.

Yarowsky D (1995). Unsupervised word sense disambiguation rivaling supervised methods. 33rd Annual Meeting of the Association for Computational Linguistics, Cambridge, MA, pp 189–196.

Yngve V (1995). Syntax and the problem of multiple meaning. In: W Locke, D Booth, eds. Machine Translation of Languages, New York: Wiley.

Zernik U (1991). Train1 vs. train2: tagging word senses in a corpus. In: U Zernik, ed. Lexical Acquisition: Exploiting On-Line Resources to Build a Lexicon. Hillsdale, NJ: Lawrence Erlbaum, pp 91–112.

27

NLP Based on Artificial Neural Networks: Introduction

HERMANN MOISL

University of Newcastle, Newcastle-upon-Tyne, England

1. INTRODUCTION

This introduction aims to provide a context for the chapters that comprise Part 3 of the handbook. It is in two main sections. The first section introduces artificial neural network (ANN) technology, and the second gives an overview of NLP-specific issues.

The quantity of research into the theory and application of ANNs has grown very rapidly over the past two decades or so, and there is no prospect of being able to deal exhaustively either with ANN technology or with ANN-based NLP in a relatively few pages. The discussion is therefore selective. Specifically,

1. The introduction to ANN technology makes no attempt to survey the current state of knowledge about ANNs, but confines itself to presentation of fundamental ANN concepts and mathematical tools.
2. The overview of ANN-based NLP surveys the historical development of the subject and identifies important current issues. It does not, however, go into detail on specific techniques or systems; such detail is provided by the chapters that follow. What is important in any research area is to some extent subjective, but the topics chosen for inclusion are well represented in the literature.

Citation of the relevant research literature is also necessarily selective, because it is, by now, very extensive. In general, the aim has been to provide a representative sample of references on any given topic.

2. ANN TECHNOLOGY

There are numerous general textbooks on ANNs [Refs. 7,23,26,36,114,124,160,164, 278,281 are a representative selection]. The appearance of the *Parallel Distributed Processing* volumes [281] sparked off the current wave of interest in ANNs, and they are still an excellent introduction to the subject; there is also a clear and very accessible account [[60]; see Chap. 3]. Further information on what follows is available from these and other textbooks.

The quintessential neural network is the biological brain. As the name indicates, an artificial neural network is a man-made device that emulates the physical structure and dynamics of biological brains to some degree of approximation. The closeness of the approximation generally reflects the designer's motivations—a computational neuroscientist aiming to model some aspect of brain structure or behaviour will, for example, strive for greater biological fidelity than an engineer looking for maximum computational efficiency—but all ANNs have at least this much in common:

- Structurally, they consist of more or less numerous interconnected artificial neurons, also called "nodes" or "units."
- They communicate with an environment by means of designated input and output units.
- Signals received at the input units are propagated to the remaining units through the interconnections and reappear, usually transformed in some way, at the output units.

Figure 1 shows a generic ANN; here the circles represent the units, and the arrows directional connections between units. The generic ANN is rarely if ever used in practice, but numerous variations on it have been developed; more is said about this later in the chapter.

Many researchers argue that, although biological brain structure need not, and in practice usually does not, constrain ANN architecture except in the most general terms just listed, what is known of brain structure should inform development and application of ANN architectures, for two main reasons. Firstly, it can be productive of new ideas for ANN design, and secondly, because the human brain is the only device known to be capable of implementing cognitive functions, it is sensible to emulate it as closely as possible [1E 9, 218]. The case for this is unanswerable, but in view of the space limitations of this handbook there is simply no room for anything beyond a brief synopsis of ANN technology. (The reader who wants an accessible path into the (voluminous) literature on brain science is referred to [9] and [60].)

A. Architecture

Here, ANN architecture is taken to mean the combination of its topology, that is, the physical layout and behavioural characteristics of its components, and its learning mechanism. Topology and learning are discussed separately.

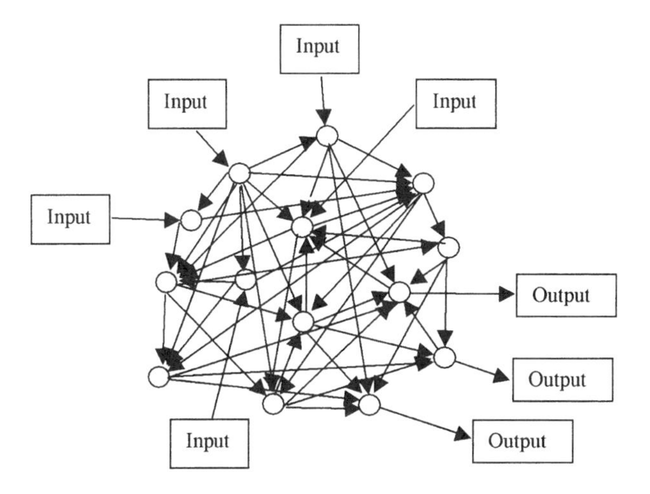

Fig. 1 A generic ANN.

1. Topology

This subsection develops the generic net mentioned earlier. It first of all looks more closely at the generic net, and then considers restrictions on its topology and the effects of these restrictions on network behaviour.

a. Generic ANN Characteristics

- The generic net is a physical device.
- It consists of units and interconnections among units.
- The units are partitioned into three types: input units that receive signals from an environment, output units that make signals available to an environment, and "hidden" units that are internal to the net and not directly accessible from outside.
- Each unit has at least one and typically many connections through which it receives signals. The aggregate of these signals at any time *t* elicits a response from the unit that, in the case of input and hidden units, is propagated along all outgoing connections to other units in the net, and in the case of output units is made available to the environment.
- Units can be of different types in the sense that they may respond differently to their inputs.
- Any given connection between units may be more or less efficient in transmitting signals; this relative efficiency is referred to as "connection strength." There is typically a significant variation in strength among the connections in a net.
- Signals applied at the input units are propagated throughout the net via connections and emerge at the output units, usually transformed in some way. The transformation of the input signals, and therefore the response of the net to environmental signals in general, is conditioned by the nature of the constituent units' response to incoming signals, the pattern of interconnection among units together with the strengths of those connections, and

the processing dynamics; more is said of processing dynamics later in the chapter.

b. Restrictions on the Generic Net

As noted, the generic net is never used in practice. Instead, restrictions are imposed on aspects of network topology, and these restrictions crucially affect network behaviour:

UNIT RESPONSE. A single unit in the generic net has some number of incoming connections. Designate the unit as n, the number of incoming connections as k, the strength of any given incoming connection i as w_i (where $w =$ "weight"), and the unit output as o. Signals s_i are applied to each of the w_i at some time t. Then the total input to n is $\sum_1^k s_i w_i$; that is, each signal is multiplied by the corresponding weight to determine how strongly it will be propagated to n, and n sums the signals thus weighted. This sum is then transformed into an output signal o as some function of the inputs: $o = f(\sum_1^k s_i w_i)$, as shown in Fig. 2. The output function f may be linear, so that $\sum_1^k s_i w_i$ is increased or decreased by some constant amount or simply output unaltered; a net consisting of units with such a linear output function can implement only a restricted range of input–output behaviours. Alternatively, f may be nonlinear. There is an arbitrary number of possible nonlinear functions that might be applied, but in practice the nonlinearity is restricted to the binary step function and approximations of it. The binary step function says that if $\sum_1^k s_i w_i$ exceeds some specified threshold the unit outputs a signal, usually represented as 1, and if it does not exceed the threshold it fails to respond, where failure is typically represented as 0. Continuous approximations of the binary step function such as the frequently used logistic sigmoid $f(x) = 1/1 + e^{-x}$ (where $x = \sum_1^k s_i w_i$) are S-shaped. Here, the unit response to $\sum_1^k s_i w_i$ is not restricted to binary values, but can take real values in some interal, most often 0..1 or −1..1.

NUMBER OF UNITS. The number of input and output units of an ANN is determined by the dimensionality of the input and output data: n-dimensional input data requires n input units, one for each dimension, and so also for outputs, where $n = 1, 2, \ldots$ There is no limit in principle on n, but there are practical limits, the most important of which is that, as the dimensionality grows, network training becomes

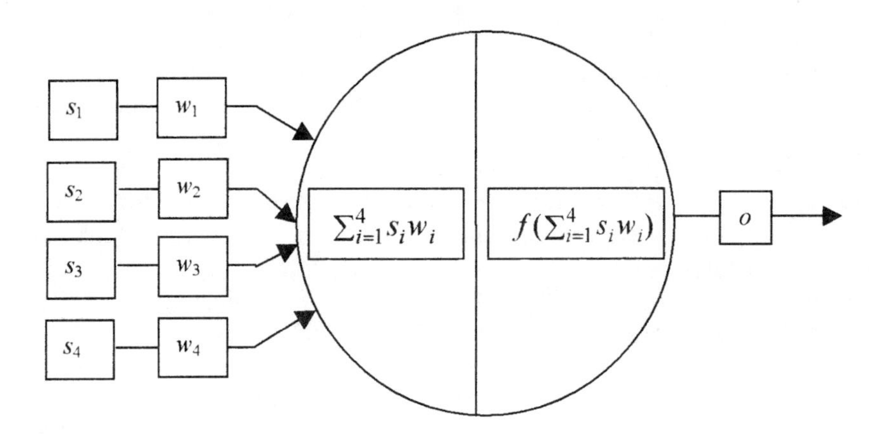

Fig. 2 A single ANN unit.

more and more time-consuming and, eventually, intractable for commonly used ANN architectures. It is, therefore, important to keep the dimensionality as small as possible consistent with the need to preserve essential application-specific informtaion; various ways of preprocessing data to achieve information-preserving data dimensionality reduction have been developed (see below).

The number of hidden units required for a net to learn a given data set and then generalize acceptably well is one of the most researched issues in the theory and application of ANNs. Theoretical and empirical results have consistently shown that, as the size of the data set that the ANN is required to learn grows, so the number of hidden units must increase. However, there is still no reliable way to determine with any precision how many hidden units are required in a particular application. The traditional approach has been heuristic search: try a variety of hidden unit sizes, and use the one that works best. Several more principled and efficient approaches to determining the optimum number of hidden units have since been developed. Again, more is said about this later.

CONNECTIVITY. In the generic net the units are arbitrarily interconnected. Such a topology is rarely if ever used. Instead, some systematic restriction is typically imposed on connectivity. Some of the connections in the generic net can be removed, as in Fig. 3. The units can then be arranged so that all the input units form a group, all the units connected to the input units form another group, all the units connected to the second group form a third group, and so on, as shown in Fig. 4. A layered structure has emerged, based on connectivity. Such layering is a fundamental concept in ANN design, and it raises important issues.

How many layers should a net have?

In the literature, some authors count layers of units and others count layers of connections. The latter is in the ascendant, for good reason, and we adopt that convention here.

A net can have any number of layers, but for economy of design and processing efficiency the aim is to have as few as possible. The single-layer net is maximally

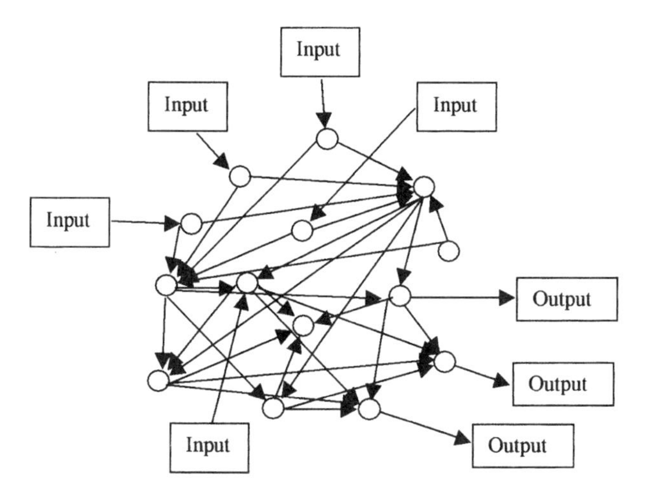

Fig. 3 Generic ANN with connections selectively removed.

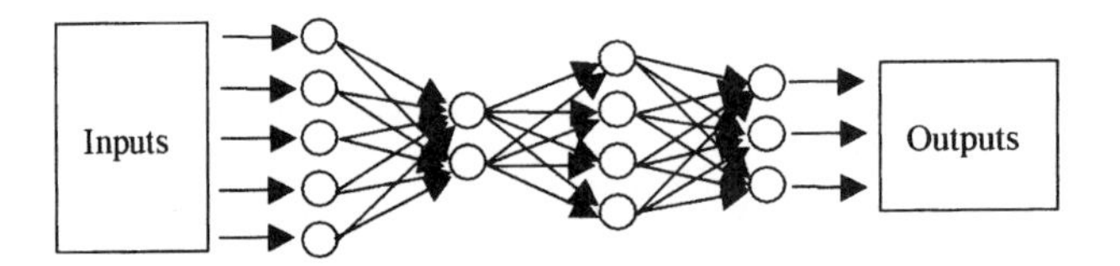

Fig. 4 Rearrangement of Fig. 3 into a layered structure.

simple. Early ANNs were single-layered, and they are still used in certain applications, but there are input–output behaviours that no single layer net can implement [229]. Two-layer nets can overcome this restriction; in fact, for any input–output behaviour that a net with more than two layers can implement, there is always a two-layer net than implements the same behaviour. In theory, therefore, more than two layers are never required, but in practice it may be convenient or even necessary to use more than two.

What should the pattern of connectivity between layers be?

In the foregoing layered modification of the generic net, all the connections point in the same direction, so that input signals are propagated from the inputs units through intermediate layers of units to the output units. ANNs with this arrangement of connections are called feedforward nets. Feedback connections are also possible, however. Assume a two-layer net, that is, one with three layers of units labelled i, j, k, as depicted in Fig. 5. Here, in addition to the connections between layers i, j and j, k, there are connections from layer j back to layer i, so that outputs from the units in layer j are propagated not just to layer k, but also become input to layer i. An ANN with one or more feedback connections is called a recurrent net.

Feedforward and recurrent nets differ in how they process inputs. To see the difference, assume an encoding scheme in which each alphabetic character and each digit 0..9 is assigned a unique binary representation. Assume also a net that outputs the representation of 1 if an input string consists entirely of alphabetic characters, and the representation of 0 otherwise. Now input (the encoding of) the string *abc* to a feedforward net, as in Fig. 6. The signal is applied to the input units, and the corresponding signal emerges at the output units. In a physical net there would necessarily be a small propagation delay, but this is routinely disregarded in practice: the input–output behaviour of a feedforward net is taken to be instantaneous. A recurrent net, on the other hand, processes input sequentially over continuous time

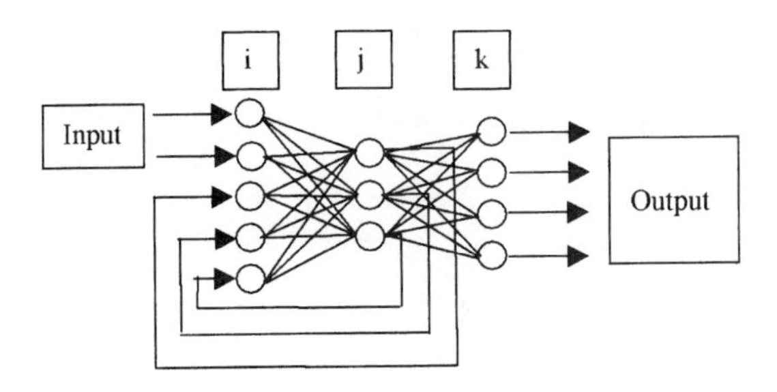

Fig. 5 A layered ANN with feedback connections.

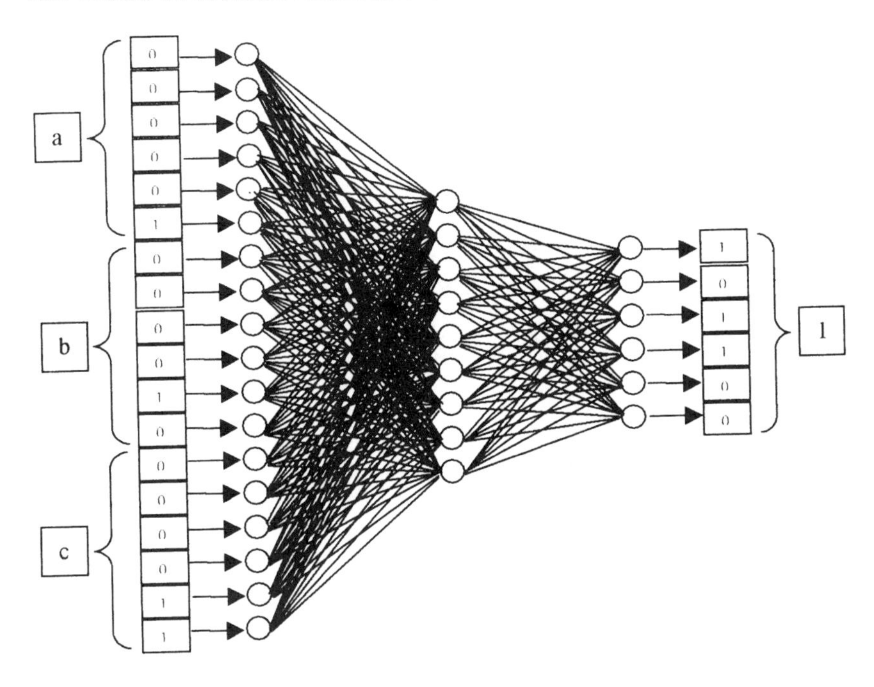

Fig. 6 A feedforward ANN for string processing.

or in discrete time steps, as in Fig. 7. Assuming discrete time at t_0 input to the net is the representation of a together with the initialized outputs of the layer j units fed back by the recurrent connections. This generates output in layer j which, together with b, is input at time t_1, and so on to the end of the string. Because speech and NL text are inherently sequential, recurrent nets are important in ANN-based NLP. Figure 7 shows a frequently used feedback topology, but various others are possible. These varieties, and issues such as higher-order feedback connections, are discussed in the cited textbooks.

The two ANN topographies most frequently used in NLP work are the two-layer feedforward net, with sigmoid hidden unit activation function known as the multilayer perceptron (MLP), and the MLP with feedback connections from the hidden units back to the input, known as the simple recurrent network or SRN. Both of these are exemplified in the foregoing. Various other topologies, such as Kohonen and radial basis function nets, are also extensively used, however [see e.g., 114].

2. *Learning*

There is no necessary connection between ANNs and learning. It is possible to configure the connections in an ANN manually in such a way as to give it some desired behaviour [i.e., 121,122,246]. Learning capability is, however, probably the single most attractive aspect of ANNs for reserachers, and the overwhelming majority of ANN-based work uses it.

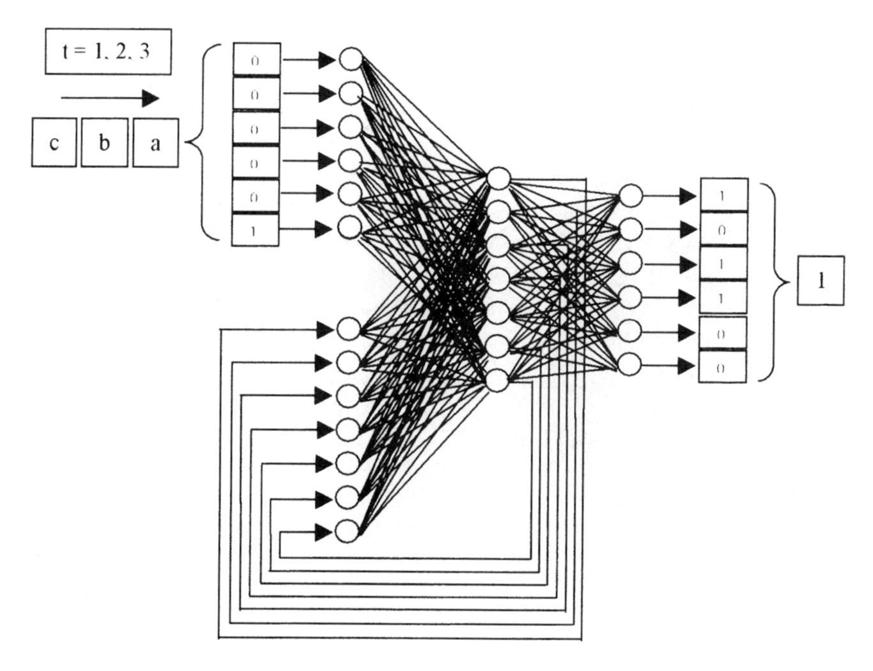

Fig. 7 A recurrent ANN for string processing.

Given a sufficiently complex net, that is, one with enough hidden units, the aim of ANN learning is to find a set of connection weights that will allow the net to implement some desired behaviour. An ANN learns by adjusting its connection strengths in response to signals from an environment such that, once training is deemed to be complete by the designer, the net responds in predictable ways to signals that it encountered in the course of training, and in reasonable ways to signals that it did not encounter during training, where the designer is the judge of what is reasonable. Such learning assumes a broadly stationary environment. For example, a net that is to function as a speech processor is trained on speech signals for which the probability distribution does not change significantly over time. Once trained, the net is expected to behave reasonably in response to novel speech input from the same distribution, but not to, say, visual signals, which bear no systematic relation to speech.

ANN learning mechanisms are standardly categorized into supervised and unsupervised algorithms:

1. A *supervised learning algorithm* requires predefinition of a training set of (input, target output) signal pairs: for each input, the net must learn to output the target signal. In other words, the net is taught a prespecified input–output behaviour. The essence of supervised learning is as follows. The input component of a training pair is presented at the input units of the ANN. The signal is propagated through the net, and the response that emerges at the output units is compared with the target output component of the training pair. If the target and actual outputs correspond within tolerances, no change is made to the network weights. Otherwise, the

weights are adjusted in a way such that they decrease the difference between the actual response and the target output the next time the input in question is presented. This adjustment process continues incrementally until actual and target outputs coincide. How exactly the appropriate weight adjustments are made is the province of the various available ANN learning algorithms. By far the most popular of these is back-propagation, which was introduced in its present form by Rumelhart and McClelland in 1986 [281], since which time numerous modifications and extensions have been proposed. Back-propagation is described in numerous places, including the textbooks cited at the outset of this section; a good introductory account is given [108].

2. An *unsupervised learning algorithm* does not use target outputs, and the net thus does not learn a prespecified input–output behaviour. Rather, an ANN is presented with inputs and expected to self-organize in response to regularities in the training set without external guidance, such that the learned response is in some sense interesting or useful. A Kohonen net, for example, is a single layer feedforward topology the input units of which are connected to a two-dimensional grid of output units, as shown in Fig. 8. Given a training set of n-dimensional vectors, the Kohonen learning algorithm adjusts the connections so that the distance among the training vectors in n-dimensional space is reflected in unit activation in two-dimensional output space: once a Kohonen net is trained, each input vector is associated with a particular region of output unit activation, and the distance among regions is proportional to the distance among inputs in vector space. The result is called a topographic map. No target output is involved: the Kohonen learning algorithm allows the net to self-organize.

B. Mathematical Modelling

The foregoing discussion has tried to give an account of ANNs as physical devices that can be trained to display interesting physical behaviours. An intuitive understanding of ANNs at this level is useful for most people, but it is not sufficient. The limitations become apparent as one works with ANNs and has to confront a range of design and analysis issues. Which ANN architecture is appropriate for a given application? How much data is required? How large does the net have to be? How

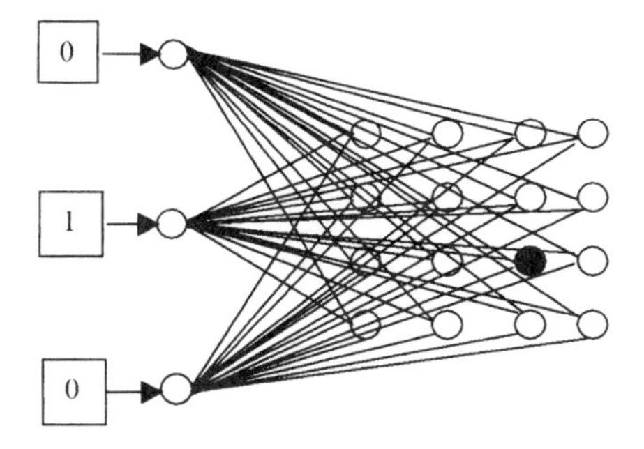

Fig. 8 A Kohonen net.

long will training take? Once trained, how can the net's behaviour be understood? One can approach these and other issues experimentally, but that is at best inefficient. The alternative is to model ANNs mathematically, and then to apply established mathematical methods to the models in addressing matters of design and analysis [103,317]. The latter approach is now standard; some essentials are outlined in what follows.

To undertake mathematical modelling, physical nets have to be represented as mathematical objects. Linear algebra is fundamental in this:

- Assemblies of input, hidden, and output units are represented as vectors the components of which take numerical values, so that the ith unit activation in an n-unit assembly corresponds to the ith element in a vector of length n.
- Connections between an m-unit and an n-unit assembly are represented as a two-dimensional $m \times n$ matrix, and numerical matrix values represent connection strengths: the value in cell $m_i n_j$ represents the connection strength between the ith unit in assembly m and the jth unit in assembly n.
- In the evaluation of a unit output $o = f(\sum_1^k s_i w_i)$, the calculation of the sum in brackets is represented by the inner product of the input vector s and the corresponding row of the weight matrix w.

There are currently three main approaches to mathematical modelling of ANNs: as dynamical systems, as computational systems, and as statistical algorithms [317]. We look briefly at each.

1. *ANNs as Dynamical Systems*

A physical system that changes over some time span is a physical dynamical system [9,38,88,140,165,166,243,244,310,317,346,361]. Dynamical systems theory is the branch of mathematics that studies the long-term behaviour of physical dynamical systems; this is done by constructing mathematical models of physical systems, and then analyzing the properties of the models.

Analysis of ANNs as dynamical systems generally uses state–space modelling, the key concepts of which are:

a. *State Space*

A state space is an n-dimensional euclidean or noneuclidean space. Assuming the former, the dimensionality of the space is determined by the number of variables in the model: n variables determine an n-dimensional space, where each orthogonal axis measures a different variable. Let $x_1, x_2, \ldots x_n$ be variables of some model M. Then the vector of variable values $x_1(t), x_2(t), \ldots, x_n(t)$ is the state of M at time t. Now assume an initial state at time t_0 and plot the state vector in state space as the system changes for $t > 0$. The result is a trajectory in state space that represents the state evolution of the model. If this is repeated for a large number of initial states, the result is a collection of state trajectories called a phase portrait.

b. *Invariant Manifolds*

A manifold is a k-dimensional region in n-dimensional space, where k is strictly smaller than n. In a dynamical system model, the n-dimensional volume defined by the vector of initial values collapses onto a smaller-dimensional volume or manifold as the system evolves over time. The manifold to which the system evolves from a given starting condition is called an invariant manifold or attractor. Typically, more than one initial vector will collapse onto the same manifold; the set of vectors that evolve onto a given manifold is called the manifold's basin of attraction. Invariant manifolds thereby define subspaces of state space to which trajectories evolve, and within which they remain, for specific initial conditions.

There are various types of manifold. The simplest is the fixed-point attractor, a 0-dimensional manifold that represents a physical system that evolves to, and in the absence of disturbance remains in, a fixed and inchanging state; the classic example is a pendulum, which eventually comes to rest and remains so unless moved. Limit cycles are trajectories that loop back on themselves, and represent oscillatory motion in the physical system being modelled. In addition, more complex attractors are possible, including fractal and chaotic ones.

Invariant manifolds may or may not be stable when the system has settled down and is subsequently disturbed. If the system settles back down to the predisturbance attractor, the manifold is stable. The state may, however, move away from the attractor as a result of the disturbance and go to another attractor in the state space.

A general dynamical system model is a triple $M = (S, T, f)$, where S is a state space, T is a temporal domain, and $f = S \times T \rightarrow S$ is a state transition function that describes how the state of the model evolves over time. Depending on how S, T, and f are defined, dynamical models can be subcategorized in various ways. S can, for example, be a continuous or discrete state space; T can be continuous or discrete time; f can be linear or nonlinear. In addition, the model can be autonomous or nonautonomous. For the last of these, dynamical systems theory draws a distinction between modelling a physical system's intrinsic behaviour, and modelling the effect which an external environment has on the system; such environmental influences are called inputs. In an autonomous system the effects of inputs are not modelled, whereas they are for nonautonomous systems. An autonomous system is started with known initial state values and then allowed to evolve over time without interaction with an environment, that is, without input. In many applications, however, nonautonomous systems are more interesting because the physical systems being modelled are generally subject to environmental influences.

There are two sorts of ANN dynamical systems to consider: activation dynamics and weight dynamics. For activation dynamics, assume a net containing n units. These n units are the state variables of the activation dynamical system, and the state of the system at time t is the vector of unit activation values $u_1, u_2, \ldots u_n$. If the weights and initial conditions or input are held constant, the evolution of the unit activation vector is the activation dynamics of the net. The state space of the weight dynamics, on the other hand, is the space of weight matrices, the dimension of which is that of the number of trainable weights in the net. Given some initial weight matrix W, the weight dynamics is the evolution of W as the net is trained using a learning algorithm. The key issue in weight dynamics is convergence to a point attractor, that is, for the weights to evolve so that they stop changing, as this indicates that learning is complete. The activation dynamics of various types of ANN exhibit the range of long-term behaviours characteristic of dynamic systems, from fixed-point to chaotic. The aim in most applications is to have the activation dynamics converge to point attractors, but not invariably; for example, there is some work that exploits fractal dynamics for NLP.

2. *ANNs as Computational Systems*

Given arbitrary sets A and B, a function f is a subset S of the Cartesian product A $\times$ B such that, for each pair $(a, b) \in$ S, $a \in$ A is a uniquely associated with $b \in$ B. An algorithm is said to compute f if, given the first component a of a pair $(a, b) \in$ S, it returns the second component b, that is, if it generates the set of pairs that constitutes f. Now, if the physical inputs and outputs of a net N are interpreted as A and B, and if N's physical input–output behaviour is consistent with S under that interpretation, then N can be seen as an algorithm for computing f, that is, as a computational system [for discussions of computation see Ref. 60, Chap. 3, and the journal *Minds and Machines* 1994; 4:377–488].

a. Computability

Computability in standard automata theory studies the classes of function that can be computed by various types of automata. The corresponding study of ANN computability does the same for various types of ANN topology [120,278,317]. The key results thus far for present purposes are as follows:

1. Under unboundedness assumptions analogous to those made in automata theory, certain ANN topologies have been shown to be Turing-equivalent, that is, for any function that a Turing Machine computes, there is an ANN with one of these topologies that computes the same function, and vice versa. These topologies include some of the most frequently used types of net [see survey in Ref. 285]: multilayer perceptrons with any one of a broad class of nonlinear hidden-layer activation functions, including the more or less standard sigmoid [372,373, reprinted in Ref. 374], radial basis function networks [137], and recurrent networks [308; but see Ref. 321]. Single-layer feedforward nets, on the other hand, are known not to be Turing-equivalent: there are some functions that no such net can compute [i.e., 26].
2. ANNs that are bounded in terms of the number of units or the precision of connection strengths are computationally equivalent to finite-state automata. Because all physically implemented ANNs are necessarily bounded, they are all limited to finite state computational power. This does not compromise ANNs relative to automata; however, because the same applies to the latter—every implemented Turing Machine is equivalent to a finite-state machine. In real-world applications automata are given sufficient memory resources to compute the finite subset of the required function in practice, whereas ANNs are given more units or higher-resolution connections to achieve the same end.

These results say only that, for some function f, there exists an ANN topology and a set of connections that compute f. There is no implication that the requisite set of connections can be learned with a net of tractable size, or within a reasonable time; see further discussion in what follows.

b. Computational Complexity of Learning

Every computation has time and space resource requirements, and these vary from algorithm to algorithm. Computational complexity theory [353] studies both the relative resource requirements of different classes of algorithm for a given function, and also the rate at which resource requirements increase as the size of particular instances of the function being computed grows, where "size" can be informally understood as the amount of input data needed to describe any particular instance of a function, or, more formally, as the number of n of bits used to encode the input data. As n increases, so do the resource requirements. The question is how quickly they increase relative to n—linearly, say, or polynomially. If the increase is too rapid, the function becomes intractable to compute within reasonable time or space bounds.

Computational complexity theory has a direct and important application to ANN learning [5,31,36,81,187,188,278,317]: the problem of scaling. Given some ANN topography, the aim of learning is to find a set of connection weights that will allow the net to compute some function of interest; the size of the problem is the number of weights or free parameters in the net. The number of weights in a net is a function of the number of input, output, and hidden units, which, as we have seen, are determined by the data-encoding scheme adopted in any given application, and the size of the data set that the net is required to learn. As the number of parameters used to encode the data or the number of data items increases, therefore, so does the number of connections. Now, empirical results have repeatedly shown that the time required to train an ANN increases rapidly with the number of network connections, or network complexity, and fairly quickly becomes impractically long. The question of how ANN learning scales with network complexity, therefore is, crucial to the development of large nets for real-world applications. For several commonly used classes of ANN, the answer is that learning is an NP-complete or intractable problem—*We cannot hope to build connectionist networks that will reliably learn simple supervised learnng tasks* [187].

This result appears to bode ill for prospects of scaling ANNs to large, real-world applications. It is all very well to know that, in theory, ANNs with suitable topologies and sufficient complexity can implement any computable function, but this is of little help if it takes an impractically long time to train them. The situation is not nearly as bad as it seems, however. The intractability result is maximally general in that it holds for all data sets and all ANN topologies. However theoretically interesting, such generality is unnecessary in practice [19], and a variety of measures exist that constrain the learning problem such that intractability is either delayed or circumvented; these include the following [23,26,271]:

- Restricting the range ANN topologies used [187,278].
- Developing mechanisms for determining the optimal network complexity for any given learning problem, such as network growing and pruning algorithms [26,364].
- Biasing the net toward the data to be learned. Empirical evidence has shown that biasing network topology in this way can substantially speed up learning, or even permit data to be loaded into a fixed-size net which, without biasing, could not load the data at all [33,106]. Such biasing involves incorporation of prior knowledge about the problem domain into the net in various ways [80,122,183,245,345].
- Explicit compilation of knowledge into initial network weights [121,134,246,248,249].
- Preprocessing of inputs by feature extraction, where the features extracted reflect the designer's knowledge of their importance relative to the problem.

Feature extraction also normally involves reduction of the number of variables used to represent the data and, thereby, the number of input units, with consequent reduction in network complexity [23,26,114].

- Incremental training, during which the net is trained with data that is simple initially, and that increases in complexity as learning proceeds [64,65,106].
- Transfer: the weights of a net N_1 to be trained are initialized using weights from another net N_2 that has already been successfully trained on a related problem [128,269,302,337–340].

c. *Automata Induction*

From a computational point of view, the activation dynamics of a feedforward net are of no interest because, for a given initial condition or input, it always converges to the same point attractor. The dynamics of feedforward nets, therefore, are disregarded for computational purposes, and the input–output mapping is taken to be instantaneous. The dynamics of recurrent ANNs (RANNs) are on the other hand of considerable computational interest, because the state space trajectory of a RANN in response to a sequence of input signals is computationally interpretable as the state transitions of an automaton in response to an input symbol string, and as such the dynamics of a RANN in response to a set of input signal sequences is interpretable as an automaton processing a language L. If the dynamics are learned from input–output data rather than explicitly compiled into the net, moreover, the RANN, from a computational point of view, can be taken to have approximated an automaton that defines L or, equivalently, to have inferred the corresponding grammar. RANNs are therefore amenable to analysis in terms of a well-developed formal language and automata theory [39,318,342,354].

Since the mid-1980s there has been a good deal of work on the use of RANNs for grammatical inference [4,130–136,142,185,221,246–248,266,297,298,362,379,380]. The training of a Simple Recurrent Network (SRN) as a finite-state acceptor is paradigmatic: given a language L and a finite set T of pairs (a, b), where a is a symbol string and b is a boolean that is true if $a \in$ L and false otherwise, train an SRN to approximate a finite-state acceptor for L from a proper subset T$'$ of T. The SRN, like various other RANN topologies used for grammatical inference, is a discrete-time, continuous-space dynamical system. To extract discrete computational states, the continuous ANN state space is partitioned into equivalence classes using, for example, statistical clustering algorithms based on vector distance, and each cluster is interpreted as a single computational state [39,132,142,250,341,362; see also Ref. 81]. Any finite state machine extracted in this way is a possible computational interpretation of the RANN, but it is not unique, because the number of states extracted depends on the granularity of the continuous space partitioning and on the partitioning algorithm used [94,95].

Grammatical inference using RANNs is not without its problems. The most important of these are the following:

1. A RANN can be interpreted as a finite-state automaton. Nevertheless, it remains a dynamical system that approximates, but does not implement, the automaton in the strict sense of implementation. For short strings the approximation is usually quite close, but as string length increases,

instabilities in the network dynamics can cause the net's behaviour to diverge from that of the abstracted machine [39,142,193–195].

2. There can be systematic dependencies between symbols that occur at different sequential positions in strings; subject–verb number agreement in English is an example. A problem with using RANNs for grammatical inference has been that, as the sequential distance between lexical items in a dependency pair increases, so does the difficulty of learning the dependency [22,23]. An SRN processes an input string $a_1, a_2 \ldots, a_n$ sequentially, and represents the lexical processing history up to any given symbol a_i in state S_i; that is, in the configuration of hidden layer units when a_i is the current input. Assume now that there is a dependeny between a_i and a_j later in the string. When processing arrives at a_j, the farther back in the sequence a_i is, the weaker its representation in S_j, until the resolution in the hidden units is insufficient to retain a memory of the net's having seen a_i and the dependency is lost. A variety of solutions have been proposed [102,168,214,248,254,291,360].
3. Theoretically, grammatical inference is an intractable problem. Gold [138] showed that even the simplext class of languages, the regular languages, cannot be learned in reasonable time from a finite set of positive examples. On this basis one might assume that, whatever its successes relative to particular and usually quite small or simple languages, general RANN-based grammatical inference is hopeless. That would be an overinterpretation, however. In practice, there are measures that render grammatical inference tractable, such as provision of negative as well as positive examples, or incorporation of prior knowledge of the grammar to be inferred into the learning mechanisms [52,108,215].

Work on grammatical inference is not confined to finite-state acceptors or indeed to more general finite-state machines. Pushdown automata and Turing machines have also been inferred [78–80,327,376,380]. Chapter 29 provides a detailed discussion of grammatical inference and automata induction.

3. *ANNs as Statistical Algorithms*

Inferential statistics and ANN learning both discover regularities in data and model the processes that generate that data. It is not, therefore, surprising that ANNs are now seen as one approach to statistical modelling [6,26–28,48,186,277,278,284,317, 358,375]; ANNs *are in essence statistical devices for inductive inference* [364]. For present purposes, the most important aspect of ANN-based statistical modelling is regression analysis. Given a set of independent variables $x_1, x_2, \ldots, x_n$, for example the various factors affecting computer network loading, and a dependent variable y, for example, the average length of time it takes for any given user to obtain a response at some level of loading, regression analysis aims to infer the relation between independent and dependent variables from a sample data set of (independent variable, dependent variable) value pairs. The inferred relation is intended as a model for the system that generated the data: it is taken to describe the relation between independent and dependent variable values not only for the sample data set, but also for the population as a whole. For example, assuming the case in which there is only one independent and one dependent variable, data plotted on x, y axes might look like (a) or (b) in Fig. 9. For (a), it is clear that the relation between independent and dependent variables is basically linear, and as such, the general form for a suitable model would be a linear polynomial $y = \alpha_0 + \alpha_1 x^1$, where the α_i are constants chosen so that the line that the resulting expression describes best fits the data, and where "best" is usually defined as the line for which the sum of the squared distances from itself to each data point is a minimum. This is the regression line. The line does not pass through all or even many of the data points; there is a random scatter on either side. The relation between independent and dependent variables that the regression line and the corresponding polynomial model is thus probabilistic: the value the model predicts for y given some x not in the data set is the mean of a random scatter to either side of y, as in Fig. 9. Similarly, the shape of (b) suggests a quadratic polynomial of the general form polynomial $y = \alpha_0 + \alpha_1 x^1 + \alpha_2 x^2$ as a suitable model, where, again, the values chosen for the parameters are such that the resulting expression describes the line of best fit through the data.

To see how ANNs can be used as regression models, we concentrate on the best-studied topology: the two-layer feedforward MLP with sigmoid activation hidden units and real-valued output. The input layer corresponds to the independent variables, the output layer to the dependent variable(s), the network weights to the parameters α_i, and the training set of (input, target output) pairs to the (independent variable, dependent variable) data set. ANN learning then becomes an algorithm for determining the parameters in the expression for the regression line through the data or, in other words, a statistical method for function approximation. The resulting model takes the form not of a multivariate polynomial, but of an artificial neural network.

As noted, a conventional (i.e., non-ANN) statistical model must represent not only the relations between independent and dependent variables for the data sample on which it is based, but also for the population from which the sample was drawn. The same is true of ANN-based regression models: loading is not enough. For loading, all that is required is that a net learn a given set T of data pairs, such that, when the first element of a pair $(a, b) \in$ D is input to the loaded net, the corresponding second element is output. But, in almost all applications, the net is expected to generalize. Given a population D of (input, target output) pairs and a proper subset T of D as a training set, then if T is successfully loaded, the expectation is that for every pair $(a_j, b_j) \in$ D and $\notin$ T, a_j input to the net will output b_j within

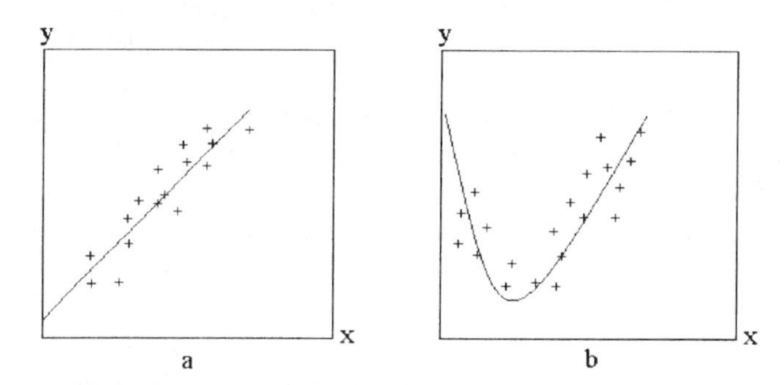

Fig. 9 Linear and quadratic relations between independent and dependent variables.

required tolerances. The generalization property of ANNs is much cited in the literature, but it is by no means automatic. To design a net that generalizes well, the relation between network complexity—the number of connections in the net—and the size of the training data set has to be understood.

Given a training data set, how complex does the net that learns that data need to be for adequate generalization? A net that is either too complex or not complex enough will fail to generalize adequately. To see why this is so, we consider once again a univariate data set and fit three different regression lines to it, as in Fig. 10.

The first regression line corresponds to some linear polynomial $y = \alpha_0 + \alpha_1 x^1$, the second to some cubic polynomial $y = \alpha_0 + \alpha_1 x^1 + \alpha_2 x^2 + \alpha_3 x^3$ and the third to some higher-order polynomial $y = \alpha_0 + \alpha_1 x^1 + \alpha_2 x^2 + \ldots + \alpha_n x^n$. The linear curve fits the data poorly, the cubic one fits the general trend of the data, but actually passes through only a few points, and the higher-order one gives the best fit in that it passes through each point exactly. The linear expression is clearly not a good model for the data. Surprisingly, however, neither is the higher-order one; the best of them is the cubic curve. Consider what happens for some value of x not in the sample. The cubic polynomial returns the corresponding y value on the regression curve, and it is consistent with the random scatter of data points around the curve; that is, the model has predicted a y value in response to x that is in the same distribution as the sample data. Under identical circumstances, the higher-order polynomial may return a y value that is relatively far from the sample data points, and thus does not predict a y value in the same distribution as the sample. The problem is that the higher-order polynomial has fit the sample data too exactly and thus failed to model the population well. In short, given a data set, there is an optimal number of terms in the polynomial for regression modelling. In ANN terms, this corresponds to network complexity: too complex a net overfits the data, whereas a net that is not complex enough undefits it, similar to the foregoing linear polynomial. It also goes without saying that it is not only the size of the data set that is important, but also the distribution of the samples. Even a large number of data points will not result in good generalization if, for example, they are all clustered in one isolated region of the population distribution. Training data must be chosen in such a way that it represents the population data distribution well.

The results of theoretical work on the interrelation of network complexity, data set size, and generalization capability [19,26,173,175,375] have, in practical terms,

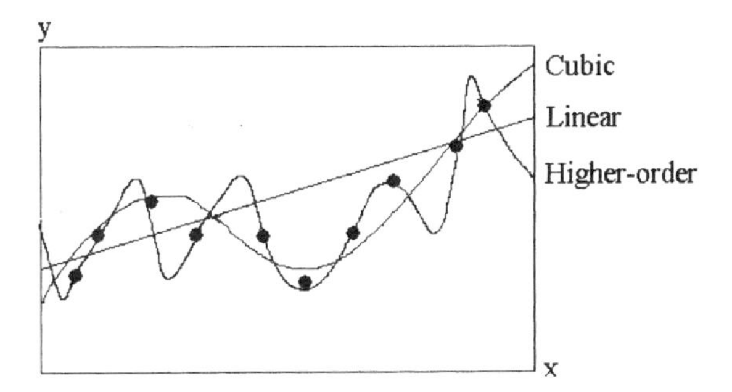

Fig. 10 Three different fits to a data distribution.

yielded a rule of thumb: good generalization is attained when the number of weights in the net is small compared with the number of data items in the training set. Experimental results, however, have shown that good generalization can be achieved with more complex nets or, equivalently, fewer data than this rule of thumb would indicate; clearly, additional factors, such as the nature of the function being approximated and the characteristics of the learning algorithm, are also involved [208,209].

In practical terms, determination of network complexity for a given data set has historically been a matter of trial and error, and to a large extent remains so: try nets of varying complexity, and use the one that generalizes best. More principled approaches have been developed, however, such as algorithms that increase or decrease the number of connections in the course of network training, with the aim of thereby arriving at an optimum topology; these and other methods are reviewed [26,271].

3. ANN-BASED NLP

This section is in two main parts. The first outlines the motivation for using ANN technology in NLP, and the second gives an overview of the history and current state of ANN-based NLP.

A. Motivation

In general, a new technology is adopted by a research community when it offers substantial advantages over what is currently available, and any associated drawbacks do not outweigh those advantages. NLP has, from the outset, been dominated by a technology based on explicit design of algorithms for computing functions of interest, henceforth called "symbolic NLP" for reasons that will emerge, and implementation of those algorithms using serial computers. It is only since the early 1980s that alternative technologies have become available, one of which is ANNs. In what follows we look at the main advantages and disadvantages of ANNs relative to symbolic NLP.

1. *Advantages*

Various advantages of ANNs are cited in the literature [14,23,35,93,124,160,251]. The three discussed in what follows are the most frequently cited and, in the case of the first and second, arguably the most important.

a. Function Approximation

We have seen that ANNs can approximate any computable function as closely as desired. The function f that a given ANN approximates is determined by the parameter values, or weights, associated with the ANN, and these parameter values are learned from a data set $D \subset f$. In principle, therefore, NLP functions can be approximated from data using ANNs, thereby bypassing the explicit design of algorithms.

What should one want to dispense with explicit design? Looking back on several decades' work on symbolic artificial intelligence (AI), some researchers have come to feel that a variety of problems, including some NLP ones, are too difficult to be solved by explicit algorithm design, given the current state of software technology [93,124,218,305], and have instead turned to function approximation techniques, such as ANNs; Arbib [9], referring to the class of adaptive systems to which ANNs belong, writes: *The key motivation for using learning networks is that it may be too hard to program explicitly the behavior that one sees in a black box, but one may be able to drive a network by the actual input/output behavior of that box...to cause it to adapt itself into a network which approximates that given behavior* (p.20). A, and probably the, chief advantage of ANN technology for NLP, therefore, is that it offers an alternative way of implementing NLP functions that have thus far proved difficult to implement using explicit algorithm design.

It should be noted that function approximation from data is not exclusive to ANNs. There is renewed interest in non-ANN machine learning in the AI community [146,169,170,307,355,371]. The claim here is not that ANNs are the only or even the best approach to function approximation in the AI/NLP domain [206,210], but rather that they offer one possible technology for it.

b. Noise Tolerance

Practical NLP systems must operate in real-world environments, and real-world environments are characterized by noise, which can for present purposes be taken as the presence of probabilistic errors in the data—spelling or syntax errors, for example. A frequent criticism of symbolic AI/NLP systems is that they are brittle in the sense that noisy input which the designer has not taken into account can cause a degree of malfunction out of all proportion to the severity of the input corruption [i.e., 218]. The standard claim is that ANNs are far less brittle, so that the performance of an ANN-based system will "degrade gracefully" in some reasonable proportion to the degree of corruption.

Noise tolerance is a by-product of ANN function approximation by nonlinear regression. We have seen that ANNs approximate a function from data by fitting a regression curve to data points, and that the best approximation—the one that generalizes best—is not the curve that passes through the data points, but the one that captures the general shape of the data distribution. Most of the data points in a noisy environment will be at some distance from the regression curve; if the input corruption is not too severe, the regression model will place the corresponding output in or near the training data distribution.

Symbolic NLP systems have historically been heavily dependent on generative linguistic theory. This was and is reasonable: linguistic theory has been intensively developed for several decades, and it constitutes a substantial body of knowledge about how natural language works. But mainstream generative linguistic theory concerns itself with what has traditionally been called competence [3,63] models of the human language faculty, disregarding, as a matter of principle, the myriad contingencies of real-world linguistic usage. To the extent that it is based on generative linguistic theory, therefore, a symbolic NLP system is an intrinsically competence design that has to be supplemented with mechnisms that allow it to operate with an acceptable level of resilience in response to noisy real-world input. There is no theoretical obstacle to this: the process of adapting a competence virtual machine is a matter of software development. In practice, though, the often-cited brittleness of symbolic systems indicates that this has so far proved problematical. ANN function approximation does away with the need for this adaptation process. There is no a priori competence model, only performance data from which the net learns the required mapping. What the net learns is not an implementation of a competence virtual machine, but rather an implementation of some (nonexplicitly specified) performance virtual machine [52].

c. ANNs and Biological Brains

A frequently cited advantage of ANNs is that they are brain-like to greater or lesser degrees and, therefore, more suitable for difficult AI problems such as vision, speech, and natural language understanding than are conventional AI methods. The argument goes like this. We know that the biological brain implements functions such as vision, speech, and so on. For artificial implementation of such functions it makes practical sense to work with a processing architecture that is as close as possible to that of the brain. ANNs are closer to brains in terms of processing architectures than serial computers. Consequently, ANNs are to be preferred for AI applications. There is no theoretical justification for this, because (certain types of) ANN and the class of Turing Machines are computationally equivalent. The argument is a pragmatic one that will, presumably, be decided on empirical grounds one day. For the moment, its validity is a matter of personal judgment.

A final note. ANNs had a resurgence of popularity in the mid-1980s, and from some overenthusiastic claims made at the time one might have thought that they were a philosopher's stone for the problems that beset a wide range of computationally oriented disciplines. Since then, it has become increasingly clear that they are not. The observation that the class of ANNs can implement the computable functions directly parallels the one that Turing Machines are capable of the same thing: it is reassuring that the solution to any given (computable) problem exists, but that does not help one find it. For Turing Machine-based computation, this is a matter of identifying an appropriate algorithm. For ANNs it is a matter of training. But, as we saw earlier, successful function approximation and consequent noise tolerance involves appropriate choice of data and network architecture. The function approximation and noise tolerance here cited as technological advantages are not automatic properties of all ANNs in all applications. Rather, they must be attained by theoretically informed design and experiment.

2. Disadvantages

Two of the main problems associated with ANNs, scaling and generalization, have already been discussed, together with possible solutions. To these must be added:

a. Inscrutability

A network that has loaded training data and generalizes well realizes a desired behaviour, but it is not immediately obvious how it does so; the set of connection strengths determine the behaviour, but direct examination of them does not tell one very much [16]. Because of this, ANNs acquired a reputation as "black box" solutions soon after their resurgence in the early–mid-1980s, and have consequently been viewed with some suspicion, particularly in critical application areas such as medical diagnosis expert systems, where unpredictable behaviour, or even the possibility of unpredictable behaviour, is unacceptable. ANNs are, however, no longer the black boxes they used to be. We have looked at mathematical tools for understanding them, and various analytical techniques have been developed [16,37,64,77,144]. Inscrutability has, in short, become less of a disadvantage in the application of ANNs.

b. Representation of Structure

A foundational assumption of current mainstream thinking about natural language is that sentences have syntactic and semantic hierarchical structure. It is easy to represent such structure using symbolic AI/NLP techniques, but is problematic in ANNs. In fact, the ability of ANNs to represent structure has been a major research issue, of which more is said later in this chapter.

3. Discussion

It needs to be stressed that, in the foregoing comparison of symbolic and ANN technology, the intention was not to argue that one is necessarily "better" than the other in a partisan sense. They are alternative technologies, each with its strengths and weaknesses and, in an NLP context, can be used pragmatically in line with one's aims [99,141,169,213,333]. The current position relative to NLP is that ANN-based systems, *while becoming ever more powerful and sophisticated, have not yet been able to provide equivalent (let alone alternative superior) capabilities to those exhibited by symbolic systems* [101, p 391].

B. History and Current State of ANN-Based NLP

In the 1930s and 1940s, mathematical logicians formalized the intuitive notion of an effective procedure as a way of determining the class of functions that can be computed algorithmically. A variety of formalisms were proposed–recursive functions, lambda calculus, rewrite systems, automata, artificial neural networks—all of them equivalent in terms of the class of functions they can compute [126]. Automata theory was to predominate in the sense that, on the one hand, it provided the theoretical basis for the architecture of most current computer technology and, on the other, it is the standard computational formalism in numerous science and engineering disciplines. The predominance was not immediate, however, and artificial neural networks in particular continued to be developed throughout the 1950s and 1960s [124,160,169,222,264,335]. Indeed, the perceptron, introduced in the late 1950s, caused considerable scientific and even popular excitement because it could learn from an environment, rather than having to be explicitly configured. But, in 1969, Minsky and Papert [229] showed that there were computable functions that perceptrons could not compute [see discussion in Ref. 261], as a consequence of which ANN-based research activity diminished significantly, and throughout the 1970s automata theory became the dominant computational formalism. Some researchers persevered with ANNs, however, and by the early 1980s interest in them had begun to revive [111]. In 1986 Rumelhart and McClelland published their now-classic *Parallel Distributed Processing* [281] volumes. Among other things, these proposed the backpropagation learning algorithm, which made it possible to train multilayer nets and thereby to overcome the computational limitations that Minsky and Papert had demonstrated for perceptrons. The effect was immediate. An explosion of interest both in the theory and application of ANNs ensued, and that interest continues today.

One of the ANN application areas has been NLP, and the purpose of this section is to survey its development. Unfortunately, the application of ANN technology to NLP is not as straightforward to document as one might wish. The Preface noted that the development of NLP is historically intertwined with that of several other language-oriented disciplines—cognitive science, AI, generative linguistics, computational linguistics [on this see Ref. 9, pp. 11–17 and 31–34]. In general, the interaction of these disciplines has been and continues to be important, because insights in any one of them can and often do benefit the others. It remains, however, that each discipline has its own research agenda and methodology, and it is possible to waste a good deal of time engaging with issues that are simply irrelevant to NLP. Now, it happens that ANN-based research into natural language has historically been strongly cognitive science-oriented, and the cognitive science agenda has driven work on NL to a large extent. For present purposes it is important to be clear about the significance of this for NLP. This section, therefore, is in two parts. The first considers the significance of the cognitive science bias for ANN-based NLP, and the second then gives an overview of work in the field.

1. ANN-based NLP and Cognitive Science

The formalisms invented in the 1930s and 1940s to define computable functions were soon applied to modelling of aspects of human intelligence, including language [17,159; see also discussion in *Minds and Machines* 1994; 4:377–490]. There have been two main strands of development. One is based on automata and formal language theory, and has come to be known as the "symbolic" paradigm. The other is based on ANNs and is known as the "connectionist" or "subsymbolic" paradigm. Until fairly recently, the symbolic paradigm dominated thinking about natural language in linguistics, cognitive science, and AI. It reached its apotheosis in the late 1970s, when Newell and Simon proposed the Physical Symbol System Hypothesis (PSSH), in which "physical symbol system" is understood as a physical implementation of a mathematically stated effective procedure, the prime example of which is the Turine Machine [see discussion in Ref. 334]:

> The necessary and sufficient condition for a system to exhibit general intelligent action is that it be a physical symbol system. Necessary means that any physical system that exhibits general intelligence will be an instance of a physical symbol system. Sufficient means that any physical symbol system can be organized further to exhibit general intelligence [240].

The PSSH was based on existing results in linguistics, cognitive science, and AI, and was intended both as an agenda for future work in those disciplines—it *sets the terms in which we search for a scientific theory of mind* [240]—and was also widely accepted as such. Thus, by 1980, these disciplines were all concerned with physical symbol systems: in essence, the first two proposed cognitive virtual architectures, and the third implemented them. At this time, however, interest in ANNs was being revived by cognitive scientists who saw them as an alternative to the dominant symbolic paradigm in cognitive modelling, and a debate soon arose on the relative merits of the PSSH- and ANN-based approaches to cognitive modelling. The PSSH case was put in 1988 by Fodor and Pylyshyn (FP) [116], who labelled what we are here calling the symbolic as the "classical" position, and Smolensky (SM) [310] argued the ANN-based case; between them they set the parameters for subsequent discussion. It is important to be as clear as possible about the issues, so a summary is presented here.

FP and SM agreed on the following (all quotations in what follows are from [116] and [310]):

- For any object of scientific study, there are levels of description, all potentially true of that object. The level of description chosen depends on the sort of explanation required.
- The level of interest in the debate between what will henceforth be referred to as the symbolists and the subsymbolists is the cognitive level, and the aim is to specify models of (aspects of) human cognition.
- The cognitive level is defined by the postulation of representational mental states, which are states of the mind that encode states of the world; mental states have semantics. Discussions at the cognitive level, therefore, must address mental architectures based on representational states.
- There is a fundamental disagreement between symbolists and subsymbolists about the nature of mental representations and the processes that operate on them.

Thereafter, they differed. We consider their positions separately.

The symbolist position which FP articulated descends directly from the PSS hypothesis, and thus proposes cognitive architectures that compute by algorithmic manipulation of symbol structures. On this view of cognitive modelling, the mind is taken to be a symbol-manipulation machine. Specifically:

- There are representational primitives: symbols. FP refer to these as atomic representations.
- Being representational, symbols have semantic content; that is, each symbol denotes some aspect of the world.
- A representational state consists of one or more symbols, each with an associated semantics, in which (i) there is a distinction between structurally atomic and structurally molecular representations, (ii) structurally molecular representations have syntactic constituents that are themselves either structurally molecular or structurally atomic, and (iii) the semantic content of a representation is a function of the semantic contents of its syntactic parts, together with the syntactic structure.

- Input–output mappings and the transformation of mental states *are defined over the structural properties of mental representations. Because these have combinatorial structure, mental processes apply to them by virtue of their form.*

Together, the foregoing features define cognitive architectures that are intended to be taken literally in the sense that they constrain their physical realization. *In particular, the symbol structures in a classical model are assumed to correspond to real physical structures in the brain, and the combinatorial structure of a representation is supposed to have a counterpart in the structural relations among physical properties of the brain. This is why Newell (1980) speaks of computational systems such as brains as physical symbol systems. This bears emphasis because the classic theory is committed to there being not only a system of physically instantiated symbols, but also the claim that the physical properties onto which the structure of the symbols is mapped are the very properties that cause the system to behave as it does. A system which has symbolic expressions, but whose operations does not depend on the structure of these expressions does not qualify as a classical machine.*

Therefore, *if in principle syntactic relations can be made to parallel semantic relations, and if in principle you have a mechanism whose operation on expressions are sensitive to syntax, then it is in principle possible to construct a syntactically driven machine whose state transitions satisfy semantic criteria of coherence. The idea that the brain is such a machine is the foundational hypothesis of classical cognitive science.*

SM distinguished symbolic and subsymbolic paradigms in cognitive science. The symbolic paradigm corresponds directly to the classic position just outlined, whereas the subsymbolic paradigm is the one that he himself proposed. The subsymbolic paradigm defines models that are *massively parallel computational systems that are a kind of dynamical system*. Specifically:

- The representational primitives are called "subsymbols." They are like classic symbols in being representational, but unlike them in being finer-grained: they *correspond to constituents of the symbols used in the symbolic paradigm... Entities that are typically represented in the symbolic paradigm as symbols are typically represented in the subsymbolic paradigm as a large number of subsymbols*. A subsymbol in a subsymbolic ANN model corresponds directly to a single processing unit.
- Being representational, subsymbols have a semantic content, that is, each subsymbol denotes some aspect of the world. The difference between symbolic and subsymbolic models lies in the nature of the semantic content. SM distinguishes two semantic levels, the conceptual and the subconceptual: *The conceptual level is populated by consciously accessible concepts, whereas the subconceptual one is comprised of finer-grained entities beneath the level of conscious concepts*. In classic models, symbols typically have conceptual semantics, that is, semantics that correspond directly to the concepts that the modeller uses to analyze the task domain, whereas subsymbols in subsymbolic models have subconceptual semantics; the semantic content of a subsymbol in a subsymbolic ANN model corresponds directly to the activity level of a single processing unit.
- In the symbolic paradigm, as noted, input–output mappings and the transformation of mental states *are defined over the structural properties of mental representations. Because these have combinatorial structure, mental processes can apply to them by virtue of their form*. This is not so in the subsymbolic case. Subsymbolic representations are not operated on by processes that manipulate symbol structures in a way that is sensitive to their combinatorial form because subsymbolic representations do not have combinatorial form. Instead, they are operated on by numeric computation. Specifically, a subsymbolic ANN model is a dynamical system the state of which is a numerical vector of the activation values of the units comprising the net at any instant t. The evolution of the state vector is determined by the interaction of (a) the current input, (b) the current state of the system at t, and (c) a set of numerical parameters corresponding to the relative strengths of the connections among units.

How do the symbolic and subsymbolic paradigms relate to one another as cognitive models? FP dismissed the subsymbolic paradigm as inadequate for cognitive modelling. SM was more accommodating.

FP argued as follows. Symbolists posit constituency relations among representational primitives, that is, among symbols, which allows for combinatorial representations. For subsymbolists the representational primitives are units or aggregates of units, but only one primitive relation is defined among units: causal connectedness. In the absence of constituency relations, subsymbolic ANN models cannot have representational states with combinatorial syntactic and semantic structure. Because subsymbolic representations lack combinatorial structure, mental processes cannot operate on them in the structure-sensitive way characteristic of symbolic models. *To summarize, classical and connectionist* [=subsymbolic] *theories disagree about the nature of mental representations. For the former but not the latter, representations characteristically exhibit combinatorial constituent structure and combinatorial semantics. Classical and connectionist theories also disagree about the nature of mental processes: for the former but not the latter, mental processes are characteristically sensitive to the combinatorial structure of the representations on which they operate. These two issues define the dispute about the nature of cognitive architecture.* Now, any adequate cognitive model must explain the productivity and systematicity of cognitive capacities (on systematicity see [143,241,242,304]). Symbolic models appeal to the combinatorial structure of mental representations to do this, but subsymbolists cannot: *Because it acknowledges neither syntactic nor semantic structure in mental representations, it treats cognitive states not as a generated set but as a list*, and among other things lists lack explanatory utility. Because they cannot explain cognitive productivity and systematicity, subsymbolic models are inadequate as cognitive models. Subsymbolic models may be useful as implementations of symbolically defined cognitive architectures, but this has no implications for cognitive science.

SM proposed his view of the relation between symbolic and subsymbolic paradigms in the following context: Given a physical system S and two computational descriptions of its behaviour—a "lower level" description μ, say an assembly language program, and a "higher-level" description M, say a Pascal program, what possible relations hold betwen μ and M? Three possibilities are proposed:

1. Implementation: Both μ and M are *complte, formal, and precise* accounts *of the computation performed by S*. Here μ can be said to implement M in that there is nothing in the lower-level account that is not also in the higher-level one.
2. Elimination: μ is a *complete, formal, and precise* account of S, and M bears no systematic relation to it. Here, M has no descriptive validity, and μ eliminates M.
3. Refinement: μ is a *complete, formal, and precise* account of S, and M is not. There are, however, systematic relations between μ and M, so that M can be said to approximately describe S. Here μ is said to refine M.

SM's proposal was that subsymbolic models refine symbolic ones, rather than, as FP had suggested, implementing them. In a now-famous analogy, he likened the relation between symbolic and subsymbolic paradigms to that which obtains between the macrophysics of Newtonian mechanics and the microphysics of quantum theory. Newtonian mechanics is not literally instantiated in the world according to the microtheory, because fundamental elements in the ontology of the macrotheory, such as rigid bodies, cannot literally exist according to the microtheory. In short, *in a strictly literal sense, if the microtheory is right, then the macrotheory is wrong.* This does not, however, mean that Newtonian mechanics has to be eliminated, for it has an explanatory capacity that is crucial in a range of sciences and branches of engineering, and that a (strictly correct) quantum mechanical account lacks; such explanatory capacity is crucial in SM's view. Thus, cognitive systems *are explained in the symbolic paradigm as approximate higher-level regularities that emerge from quantitative laws operating on a more fundamental level*—the subconceptual—*with different semantics.* Or, put another way, symbolic models are competence models that idealize aspects of physical system behaviour, whereas subsymbolic models are performance models that attempt to describe physical systems as accurately as possible.

As noted, these two positions set the parameters of a debate that was to continue to the present day. It began with a long series of peer commentaries appended to Smolensky's article, and both Smolensky and Fodor subsequently expanded on their positions [117–119,311,312,314]. In addition, numerous other researchers joined the discussion; a representative sample is [21,30,40,42,59,60–62,64,66,67,89,92,93,99,108,169,171,172,177,241,267,304,347,350–352,355–357,365]; see also the discussion in *Minds and Machines* 1994. There is no way we can follow the debate any further here, or presume to judge the complex issues it has raised. We do, however, need to be clear about its implications for ANN-based NLP.

Firstly, the debate has forced a reexamination of fundamental ideas in cognitive science and AI [i.e., 64,67,93], and its results are directly relevant to NLP. It has, moreover, already been noted that much of the ANN-based research on NL is done within a cognitive science framework. ANN-based NLP cannot, therefore, afford to ignore developments in the corresponding cognitive science work. This is not bland ecumenism, but a simple fact of life.

Secondly, notwithstanding what has just been said, it remains that the cognitive science focus of the debate can easily mislead the NLP researcher who is considering ANNs as a possible technology and wants to assess their suitability. The debate centres on the nature of cognitive theories and on the appropriateness of symbolist and subsymbolist paradigms for articulation of such theories. These issues, however intrinsically interesting, are orthogonal to the concerns of NLP as this handbook construes them. Cognitive science is concerned with scientific explanation of human cognition (on explanation in cognitive science see [63,64] and *Minds and Machines* 1998, 8), including the language faculty, whereas NLP construed as language engineering has no commitment to explanation of any aspect of human cognition, and NLP systems have no necessary interpretation as cognitive models. The symbolist argument that the ANN paradigm is inadequate in principle for framing cognitive theories is, therefore, irrelevant to NLP as understood by this handbook, as are criticisms of particular ANN language-processing architectures in the literature on the grounds that they are "cognitively implausible," or fail to "capture generalizations," or do not accord with psycholinguistic data.

Thirdly, once the need for cognitive explanation is factored out, the debate reduces to a comparison of standard automata theory and ANNs as computational technologies [169,171]. So construed, the relation is straightforward. We have taken the aim of NLP to be design and construction of physical devices that have specific behaviours in response to text input. For design purposes, the stimulus–response behaviour of any required device can be described as a mathematical function, that is, as a mapping from an input to an output set. Because, moreover, the stimulus–response behaviour of any physical device is necessarily finite, the corresponding input–output sets can also be finite, with the consequence that every NLP function is computable; in fact, the sizes of the I/O sets are specifiable by the designer and, therefore can be defined in such a way as to make the function not only finite and thereby theoretically computable, but also computationally tractable. As such, for any NLP mapping, there will be a Turing Machine—a PSS—that computes it. But we have seen that certain classes of ANN are Turing equivalent, so there is no theoretical computability basis for a choice between the two technologies. The choice hinges, rather, on practical considerations such as ease of applicability to the problem in hand, processing efficiency, noise and damage tolerance, and so on. A useful illustrative analogy is selection of a programming language to implement a virtual machine that computes some function. All standard-programming languages are equivalent in terms of the functions they can implement, but some are more suitable for a given problem domain than others in terms of such things as expressiveness or execution speed: assembler would be much harder to use in coding some complex AI function than, say LISP, but once done it would almost certainly run faster [169,171].

And finally, the debate has set the agenda for ANN-based language-oriented research in two major respects: the paradigm within the research is conducted, and the ability of ANNs to represent compositional structure. Both these issues are discussed in what follows.

2. *ANN-Based NLP: An Overview*

ANN-based NL research [46,55,101,276,296,299,301] began, fairly slowly, in the early 1980s with papers on implementing semantic networks in ANNs [167], visual word recognition [139,216,279], word–sense disambiguation [72–74], anaphora resolution [275], and syntactic parsing [109,294,295,309]. In 1986, Lehnert published a paper [212] on the implications of ANN technology for NLP, an indication that this early work had by then attracted the attention of mainstream work in the field. Also, 1986 was the year in which the *Parallel Distributed Processing* volumes [281] appeared, and these contained several chapters on language: McClelland and Kawamoto on case role assignment, McClelland on word recognition, and Rumelhart and McClelland on English past-tense acquisition. All of these were to be influential, but the last-named had an effect out of all proportion to the intrinsic importance of the linguistic issue it dealt with. Rumelhart and McClelland [280] presented an ANN that learned English past-tense morphology from a training set of (past-tense, present-tense) verb form pairs, including both regular ("-ed") and irregular formations. They considered their net as a cognitive model of past-tense morphology acquisition on the grounds that its learning dynamics were in close agreement with psycholinguistic data on past-tense acquisition in children and, because it was able to generalize the regular-tense formation to previously unseen present-tense forms after training, that it had learned an aspect of English morphology. Crucially, though, the net did this without reference to any explicit or implicit PSS architecture. This was quickly perceived as a challenge by symbolist cognitive scientists, and it became a test case in the symbolist vs. subsymbolist debate outlined earlier. Pinker and Prince [252] made a long and detailed critique, in response to which the Rumelhart and McClelland model was refined by a succession of researchers [64,82,83,108,219,258,259]; see also 220 and the discussion in Chap. 31].

From an NLP point of view, the chief importance of Rumelhart and McClelland's work and its successive refinements is not in its validity as a cognitive model, but in the impetus that it gave to ANN-based NL research. It made 1986 a watershed year, in the sense that the number of language-oriented papers has increased dramatically since then. Disregarding speech and phonology because of this handbook's focus on text processing, there has been further work on a wide variety of topics, a representative selection of which follows:

Morphology [64,82,83,108,128,129,147–149,219,220,234,258,259,268]
Lexical access [257,293]
Lexical category learning [104–106,115]
Noun phrase analysis [370]
Anaphora resolution [1]
Prepositional phrase attachment [71,343]
Grammaticality judgment [3,205,207,210,378]
Syntax acquisition [45,53,54,56–58, 104–106,108,144,362,377]
Parsing [47,73,145,161–163,174,178–180,198–200,227,231,270,320,324,334, 359,363]
Case role assignment [217,224,228,324,325]
Lexical semantics [11,12,69,90,110,113,235,238,239,272–274,286–290,319, 323,366]
Lexical disambiguation [34,190]
String semantics [3,71,144,228,233,323,325,326,367]
Metaphor interpretation [236,369]
Reasoning [201–203,306,328–331]
Full text understanding [35,223,225,226]
Language generation [76,85–87,127,150,197,286]

A range of practical NLP applications have also been developed. Chapters 33–37 give a representative sample.

The extent of NLP-related work since 1986 precludes any attempt at a useful summary here. Instead, the rest of this chapter discusses several important general issues in ANN-based NLP, leaving detailed consideration of specific topics and research results to the chapters that follow.

a. Research Paradigms

The symbolist–subsymbolist debate has resulted in a trifurcation of ANN-based natural language-oriented research, based on the perceived relation between PSSH- and ANN-based cognitive science and AI:

- The symbolic paradigm accepts FP's view of the position of ANNs relative to cognitive science. It considers ANNs as an implementation technology for explicitly specified PSS virtual machines, and it studies ways in which such implementation can be accomplished. NL-oriented work in this paradigm is described in Chap. 30.
- The subsymbolic paradigm subdivides into what is sometimes called "radical connectionism," which assumes no prior PSSH analysis of the problem

domain, but relies on inference of the appropriate processing dynamics from data, and a position that, in essence, considers prior PSSH analysis of the problem domain as a guide to system design or as an approximate or competence description of the behaviour of the implemented system. References [91–97,356] exemplify the radical position, and Smolensky, in a manner of speaking the father of subsymbolism, has in various of his writings [310,314–316] taken the second. The subsymbolic paradigm is described in Chap. 31.

- The hybrid paradigm, as its name indicates, is a combination of the symbolic and the subsymbolic. It uses symbolic and subsymbolic modules as components in systems opportunistically, according to what works best for any given purpose. A subsymbolic module might, for example, be used as a preprocessor able to respond resiliently to noisy input, whereas the data structures and control processes are conventional PSS designs [99,169,171,218,332,333,367,368]. The hybrid paradigm is described in Chap. 32.

Interest in the hybrid paradigm has grown rapidly in recent years and, to judge by relative volumes of research literature, it is now the most often used of the foregoing three alternatives in engineering-oriented applications such as NLP. It is not hard to see why this should be so. The hybrid paradigm makes full use of theoretical results and practical techniques developed over several decades of PSSH-based AI and NLP work, and supplements it with the function-approximation and noise-resistance advantages of ANNs when appropriate. By contrast, the symbolic and subsymbolic paradigms are in competition with established PSSH-based theory or methodology. On the one hand, the symbolic paradigm has yet to demonstrate that it will ever be superior to conventional computer technology as an implementation medium for PSS virtual machines [i.e., 99]. On the other hand, the subsymbolic paradigm essentially disregards existing PSSH-based NLP theory and practice, and starts afresh. Of the three paradigms, therefore, it is the least likely to generate commercially exploitable systems in the near future, although it is the most intriguing in pure research terms.

b. Representation

The most fundamental requirement of any NLP system is that it represent the ontology of the problem domain [64,114,124,300,303,304]; see also Chap. 28]. One might, for example, want to map words to meanings, or strings to structural descriptions: words, meanings, strings, and structures have to be reprsented in such a way that the system can operate on them so as to implement the required mapping. Most ANN-based NL work has been directly or indirectly concerned with this issue, and this section deals with it in outline.

Before proceeding, a particular aspect of the representation issue has to be addressed. As we have seen, the symbolist–subsymbolist debate made representation of compositional structure a major topic in ANN-based cognitive science. FP claimed that ANNs were incapable of doing so and, accordingly, dismissed them as inadequate for cognitive modelling. In response, adherents of ANN-based cognitive science have developed a variety of structuring mechanisms. Now, FP insists on representation of compositional structure in cognitive modelling on explanatory grounds: it captures the productivity and systematicity of cognitive functions such as language. But NLP is not interested in cognitive explanation. The aim is to design and implement devices having some specified I/O behaviour. Whatever its importance for cognitive science, therefore, is representation of compositional structure is an issue for ANN-based NLP?

Generative linguistic theory provides a well-developed and, by now, natural way of thinking about NL, a foundational assumption of which is that sentences are compositionally structured objects. Symbolist AI and NLP have shared this assumption, and designed systems in which compositional data structures are manipulated by structure-sensitive processes. This is one way of approaching the design and implementation of NLP devices, but not the only way [64]. System identification theory poses the black box problem, which asks: Given a physical device with an observable input–output behaviour, but for which internal mechanism is not open to inspection, what is its internal mechanism? The answer, in Arbib's words, is this: *Even if we know completely the function, or behavior, of a device, we cannot deduce from this a unique structural description ... The process of going from the behavior of a system to its structural description is then not to be thought of as actually identifying the particular state variable form of the system under study, but rather that of identifying a state variable description of a system that will yield the observed behavior, even though the mechanism for generating that behavior may be different from that of the observed system* [8, pp. 38–39 for discussion see Refs. 42 and 43]. Now, we have taken the aim of NLP to be design and construction of physical devices with a specified I/O behaviour, so the identification problem applies directly: given a desired I/O behaviour, what mechanism should be used to produce it? One answer is a Turing Machine, as discussed earlier, but it is not the only answer [196]. It is possible to use automata of complexity classes lower than that of Turing Machines, and even finite state machines to compute NLP functions [51–55, 230]: we have already seen that NLP functions are necessarily finite, and any finite function can be computed by a finite-state machine. But finite-state machines can represent only trivial compositionality that reduces to simple sequentiality; it was because of their inability to represent nontrivial structure—that is, simultaneous left- and right-branching dependency—that Chomsky originally rejected them as models for NL sentence structure 49, p.24]. Indeed, one could even use a continuous space dynamical system, for which compositional discrete symbol structures are undefined. All three mechanisms are theoretically capable of generating the required observable behaviour and, assuming they are appropriately configured, they are equivalent for NLP purposes.

In principle, therefore, compositional structure is not necessary for NLP. It may, however, be useful in practice. A discrete-time, continuous-space dynamical system, such as a two-layer feedforward ANN, with sigmoid activation function, may theoretically be capable of implementing any computable function, but, for some particular function, is finding the required weight parameters computationally tractable, and will network complexity have reasonable space requirements? It may well turn out to be that compositional structure makes implementation of certain NLP functions easier or indeed tractable; the need for compositional structure in ANN-based NLP is an empirical matter and, consequently, researchers need to be aware of the structuring mechanisms developed by ANN-based cognitive science.

There are two fundamentally different approaches to ANN-based representation [60,64,108,336,348,349]:

1. Local representation: Given some set E of objects to be represented and a set N of network units available for representational use, a local representation scheme (or "local scheme" for short) allocates a unique unit or group of units $\in N$ for each object $e \in E$.
2. Distributed representation: Given the same sets E and N, a distributed representational scheme uses all the units $n \in N$ to represent each $e \in E$ [348,349].

The difference is exemplified in the pair of representational schemes for the integers 0..7 shown in Fig. 11. In the local scheme each bit represents a different integer, whereas in the distributed one, all the bits are used to represent each integer, with a different pattern for each. Because, in the local scheme, each bit stands for one and only one integer, it can be appropriately labelled, but in the distributed scheme, no bit stands for anything on its own; no bit can be individually labelled, and each is interpretable only in relation to all the others.

Local and distributed schemes both have advantages [60,99,108,124,300,305, 336]. Much of the earlier work used localist representation, and although the balance has now shifted to the distributed approach, significant localist activity remains [15,35,101,110,133,201,203,274,305,306] (see also the discussion in Chap. 32). In what follows, local and distributed approaches to representation of primitive objects and of compositional structure in ANNs are discussed separately.

LOCAL REPRESENTATION

Representation of primitives

Local representation of primitive objects is identical with that in the PSSH approach: in a PSS each object to be represented is assigned a symbol, and in local ANN representation each object is assigned a unit in the net.

Representation of structure

Local representation of primitive objects is straightforward, but representation of compositional structure is not. The difficulty emerges from the following example [24]. Assume a standard AI blocks world consisting of red and blue triangles and squares. A localist ANN has to represent the possible combinations.

	Local	Distrib
1	0000001	0000001
2	0000010	0000010
3	0000100	0000011
4	0001000	0000100
5	0010000	0000101
6	0100000	0000110
7	1000000	0000111

Fig. 11 Local and distributed representations of the integers 1–7.

One unit in the net is allocated to represent "red", another for "blue," another for "triangle," and yet another for "square." If one wants to represent "red triangle and blue square," the obvious solution is to activate all four units. Now represent "blue triangle and red square" using the same procedure. How can the two representations be distinguished? The answer is that they cannot, because there is no way to represent the different colour-to-shape bindings in the two cases. In other words, there is no way to represent constituency in a net such as this. Localists have developed a variety of binding mechanisms to overcome this problem [24,102,305,306,329].

DISTRIBUTED REPRESENTATION

Representation of primitives

In distributed ANNs, each primitive object is represented as a pattern of activity over some fixed-size group g of n units, or, abstractly, as a vector v of length n in which the value in any vector element v_i represents the activation of unit g_i, for $1 \leq i \leq n$. Such representation has two properties that localist schemes lack.

Firstly, the relation between a representation and what is represents can be nonarbitrary in a distributed scheme. In a localist scheme, the relation is arbitrary: each node (or node group) in a localist net represents a primitive object, and it does not matter which node is chosen for which object. Such arbitrariness extends to certain kinds of distributed representation as well. ASCII encoding can, for example, be taken over directly into a text-based NLP system to represent alphanumeric characters, such that the 256 eight-bit codes are distributively represented over eight network units. This gives advantages of efficiency and damage resistance, but the representation is arbitrary in the sense that each code assignment depends only on binary number notation and the order in which the characters happen to have been arranged. If one wanted to be perverse, the code-to-character assignment could be altered without affecting the usefulness of the scheme, because there is no reason to prefer any particular code for any particular character. Distributed representation can, however, be made nonarbitrary by using feature vectors. Compare, for example, ASCII and feature–vector representations for alphanumeric characters in Fig. 12.

In the feature–vector scheme, each element of the grid is a pixel; the rows are concatenated from the upper right, assigning value-1 for dark and 0 otherwise, and the resulting vector represents the physical shape of the corresponding letter. As such, the assignment of representation to what it represents is nonarbitrary.

Secondly, a nonarbitrary distributed scheme can represent similarities among the primitive objects. In any distributed scheme, arbitrary or nonarbitrary, the dimensionality n of the vectors defines an n dimensional space within which the representations have a similarity structure in the sense that some vectors are closer than others in that space. In a nonarbitrary scheme, however, the similarity structure systematically reflects similarities among represented objects. Referring to Fig. 12, F and T are more visually similar than either O and F or O and T. In the ASCII scheme there is no systematic relation between vector distance among codes and visual similarity among bitmaps: the ASCII codes for O and F are most similar, those for O and T are farthest apart. Distance among feature–vectors, however, corresponds directly to visual similarity.

	ASCII	Bitmap	Bitmap vector
F	01000100	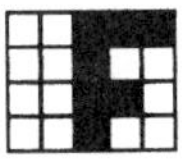	00111001000011000100
O	01001101		01110100011000101110
T	01010010	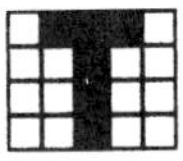	01110001000010000100

Fig. 12 ASCII and feature–vector representations of alphanumeric characters.

Until fairly recently, feature–vector representations were handcrafted, and were thus conditioned by individual designers' analyses of what is significant in task domains. There has been a move away from such explicitly designed distributed representational schemes to learned ones. For example, a feedforward MLP can be used for this purpose by training it to autoassociate vectors, that is, by making each vector v_i in some data set both input and target output. If the hidden layer is smaller than the input–output layers, then v_i at the input of a trained net will generate a "compressed representation" of itself on the hidden units, which can be used as a representation of v_i in subsequent processing. Examples of learned distributed representations in NLP applications [101] are FGREP [223,225] (described in Chap. 37) and xRAAM [211].

Representation of structure

Using a distributed ANN to represent compositional structure is difficult because arbitrarily complex structures have to be represented with a fixed-size resource, that is, over some specific group of units. To see this, assume that the primitive objects in a given domain are represented as feature vectors. An ANN that uses distributed representations by definition uses all the available units for each vector. There would be no difficulty about individually representing "man," for example, or "horse." But how would the net represent "man" and "horse" at the same time? Even more difficult is representation of relations, such as "man on horse." The problem, therefore, has been to find ways of overcoming this difficulty.

The crucial insight came from van Gelder in 1990 [348; discussed in Ref. 64]. He argued that distributed representations can be compositional, but not necessarily in the sense intended within the PSSH paradigm. Van Gelder's paper has become very influential in ANN-based cognitive science generally; its importance for present purposes lies in the distributed compositional representation that it proposes, and that also underlies several distributed ANN-based representational mechanisms. Because of its importance, a reasonably detailed account of it is given here.

Van Gelder states a set of "minimal abstract conditions" that any representational scheme must satisfy for it to be compositional:

There is a set of primitive types (symbols, words, and such) P_i. For each type there is available an unbounded number of tokens.

There is a set of expression types R_i. For each type there is available an unbounded number of tokens.

There is a set of constituency relations over these primitive and expression types.

He then focusses on the distinction between the types and tokens in these conditions, and makes that distinction the basis of his proposal for distributed compositional representation. The argument goes like this. Specification of a particular representational scheme is bipartite. On the one hand, the primitive and expression types together with the abstract constituency relations among them have to be stated. On the other hand, it is necessary to state what the tokens of the primitive and expression types look like physically, and how primitive and expression tokens can be combined to generate new expression tokens. The physical specification is necessary because it provides a notation in terms of which the abstract specification can be stated and applied; without a notation, there is no way to talk about the abstract representational scheme. The standard way of specifying compositional representational schemes conflates the abstract and physical specifications. Thus, the specification, *If A and B are wff, then (A&B) is a wff*, uses the physical symbols *A*, *B* and *(A&B)* to specify two primitive types and their combination is an expression type without making the distinction between abstract and physical explicit. This is necessarily the case, because notation exists to permit formal statement of abstractions. It does, however, obscure the distinction, maintenance of which is crucial to the argument for distributed compositional representation. In particular, given any set of primitive types, primitive expressions, and constituency relations, and given also a set of primitive tokens, there is more than one way of physically instantiating the abstract constituency relations by combining primitive tokens into expression tokens. Two sorts of such "modes of combination" are cited:

1. Concatenative combination. The tokens of an expression's constituents are physically present in the expression token. Thus, the logical expression [*(P&O)&R*] contains the expression token *(P&Q)* and the primitive token *R*, and *(P&Q)* itself contains the primitive tokens *P* and *Q*. This concatenative compositionality is spatial: primitive tokens are placed alongside one another in a physical sequence, as in [*(P&Q)&R*].

   A concatenative representational scheme says two things: *When one is describing a representation as having a concatenative structure, one is making more than just the grammatical point that it stands in certain abstract constituency relations. One also says that it will have a formal structure of a certain kind, that is, a structure such that the abstract constituency relations among expression types find direct, concrete instantiation in the physical structure of the corresponding tokens. This is called syntactic structure. Thus, the syntactic structure of the representation is the kind of formal structure that results when a concatenative mode of combination is used.*
2. Nonconcatenative combination. The familiar compositional schemes—natural languages, programming languages, mathematical and logical languages—are all concatenative. This makes it easy to lose sight of the possibility that there might be alternatives to concatenative combination

> of tokens for representation of abstract constituency relations. To represent abstract constituency relations, it is sufficient to specify general, effective, and reliable procedures for generating expression tokens from constituent tokens, and for decomposing expressions into their constituents. One way of doing this is to specify processes operating on concatenative representations, but there is no reason, in principle, why tokens of constituents should be literally present in token expressions. If general, effective, and reliable procedures for generation and decomposition of nonconcatenative expression tokens can be specified, then the scheme can legitimately be said to represent the corresponding abstract constituency relations. Any scheme that specifies such procedures is said to be functionally compositional. Van Gelder cites Goedel numbering for formal languages as an example, the details of which would take us too far afield; the important features for present purposes are that it provides the required general, effective, and reliable procedures for constructing expression tokens from constituent tokens and for decomposing expression tokens into constituent tokens, and that the generated expression tokens do not literally contain the constituent tokens, as they do in concatenative schemes.

Concatenation cannot work with an ANN that uses distributed representation because the size of an expression token must increase with the complexity of the abstract constituency that it represents, but the ANN representational resource is limited to the size of a single primitive token. The importance of nonconcatenative representation is that it breaks the link between abstract complexity and spatial representation size: because it does not require constituent tokens to be physically present in an expression token, it becomes possible, in principle, to represent abstract constituency relations over a fixed-size resource. What is needed in ANN terms are "general, effective, and reliable procedures" to compose constituent tokens into and to decompose them from expression tokens represented over the representational units of the net. Several such nonconcatenative mechanisms have been proposed [101,242,348], chief among them tensor products [313], recursive autoassociative memories (RAAM) [262,263; see also Ref. 41], and holographic-reduced descriptions [254–256]. These are discussed in Chaps. 30 and 31.

c. Sequential Processing

Text processing is inherently sequential in the sense that word tokens arrive at the processor over time. ANN-based work on NL has addressed this sequentiality using three main types of network architecture [23]:

MULTILAYER PERCEPTRONS. MLPs are a feedforward architecture, and time is not a parameter in feedforward nets: they map inputs to outputs instantaneously. Consequently, MLPs appear to be inappropriate for sequential processing. Nevertheless, the earlier ANN-based NL work used them for this purpose by, in effect, spatializing time. Given a set of symbol strings to be processed, the MLP is given an input layer large enough to accommodate the longest string in the set, as in Fig. 13. This was unwieldy both because, depending on the input encoding scheme, it could result in large nets that take a long time to train, and because of the inherent variability in the length of NL strings. It is now rarely used.

TIME-DELAY NEURAL NETWORKS. A time-delay neural network (TDNN) is an MLP for which the input layer is a buffer k elements wide, as shown in Fig. 14. It processes input dynamically over time $t_0, t_1 \ldots t_n$ by updating the input buffer at each t_i and propagating the current input values through the net to generate an output. The problem here is buffer size. For example, any dependencies in a string for which the lexical distance is greater than the buffer size will be lost. In the limiting case, a buffer size equal to the input string length reduces to an MLP. TDNNs have been successfully used for finite-state machine induction [70] and in NLP applications [23,32,33].

RECURRENT NETWORKS. Recurrent networks (RANN) use a fixed-size input layer to process strings dynamically; RANNs used for NL work are discrete-time, and input successive symbols in a string at time steps $t_0, t_1 \ldots t_n$, as in Fig. 15. The net's memory of sequential symbol ordering at any point in the input string is maintained in the current state, which is fed back at each time step.

NL research using RANNs has proceeded in concert with the work on automata induction described earlier [53,54,56–58,68,104–108,185,204,205,207,210, 297,298]. Because RANNs are dynamical systems, they can display the range of behaviours that characterize such systems. ANN-based NL work has so far used mainly fixed-point dynamics, but there have been some who exploit more complex dynamics [13,29,177,194,265] (see also discussion in [23,108] and Chap. 31).

Other approaches to dynamic processing of symbol sequences have been developed [232], such as sequential Kohonen nets [44,181], but most existing ANN-based NL work has used the foregoing three varieties.

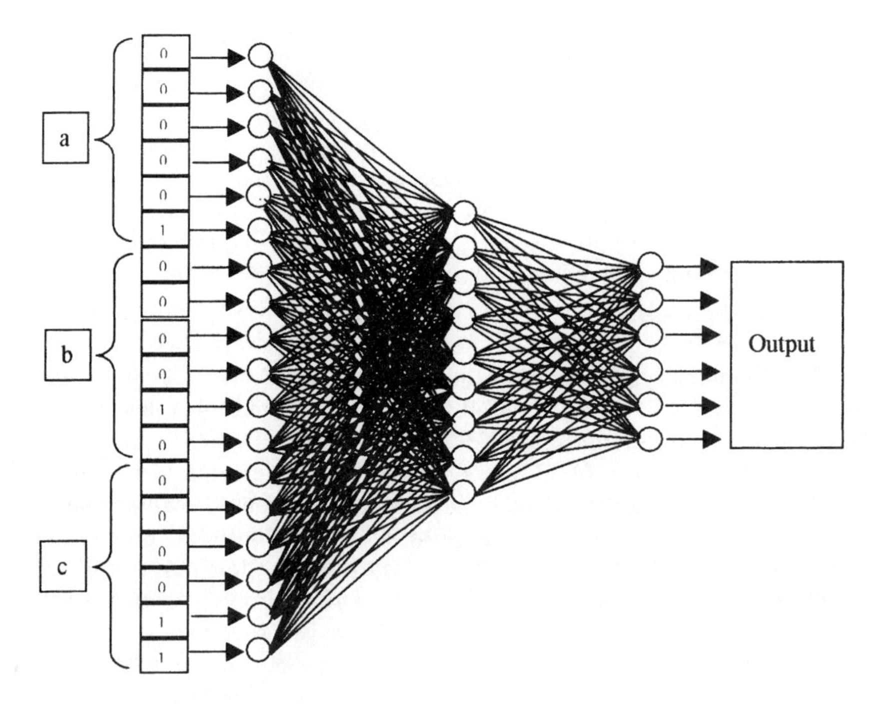

Fig. 13 An MLP for string processing.

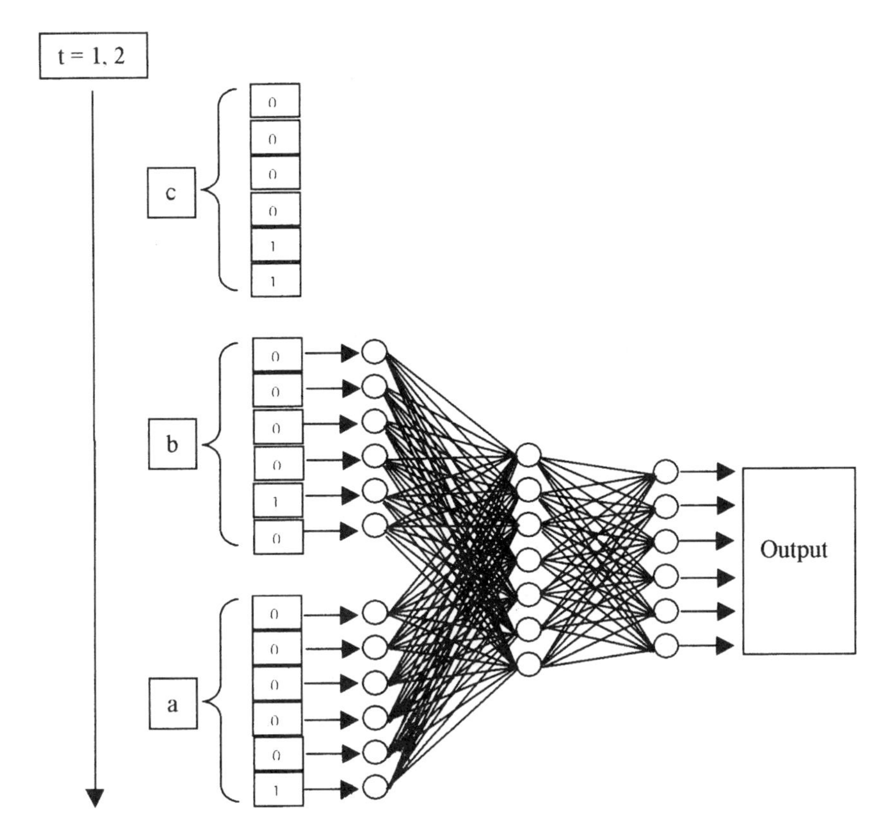

Fig. 14 A TDNN for sequential string processing.

Dynamic processing of language is usefully understood in the context of work on the dynamical systems approach to cognitive modelling [20,66,67,107,108,191,267,351].

d. Tabula Rasa Learning

The overwhelming majority of ANN-based language work involves learning, and is thus describable as language acquisition research in a broad sense. There is a large generative linguistics and cognitive science literature on language acquisition, the aim of which is to explain how children learn the linguistic knowledge characteristic of adult native speakers. Such explanation is typically stated in terms of a generative linguistic theoretical framework, and based on the "poverty of the stimulus" argument: the examples of language usage available to children is insufficiently rich to allow adult linguistic competence to be inferred without some innate set of constraints that make the acquisition process tractable; this set of constraints is known as Universal Grammar [50,253]. ANN-oriented linguists and cognitive scientists on the other hand argue that *rich linguistic representations can emerge from the interaction of a relatively simple learning device and a structured linguistic environment* [260; see also 3,64,108], although most now agree that some form of innate constraint is necessary [17,18,108,182,302]; on the nature of such constraints; however,

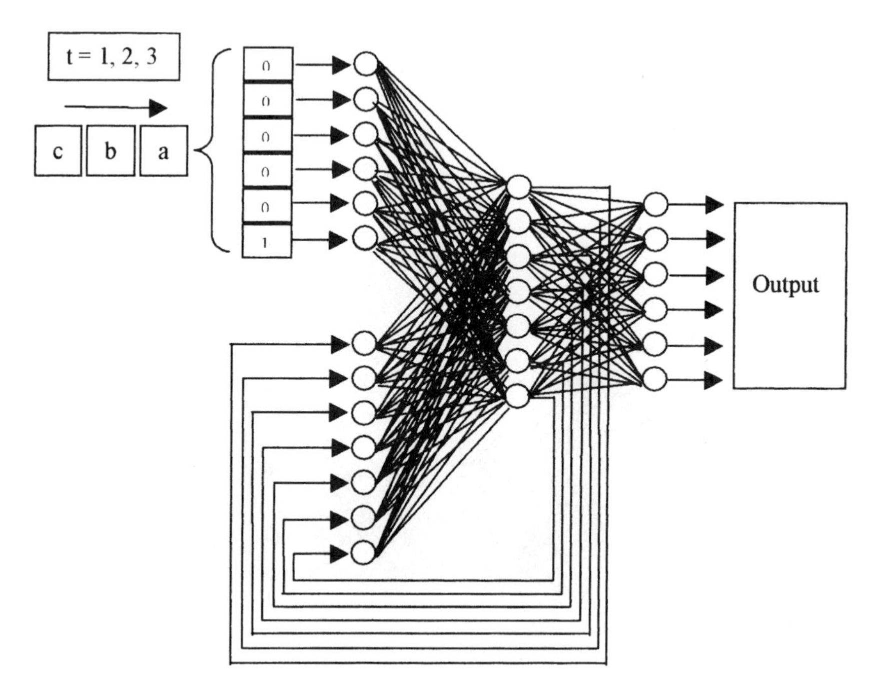

Fig. 15 A RANN for sequential string processing.

see other discussions [52,53,57,64,108]. The difference between them and generativists is not absolute, in the sense that they support *tabula rasa* learning as a general strategy for language acquisition, but rather one of degree: they differ from generativists on the relative importance of innate constraints and environmental factors, tending to emphasize the latter. On language acquisition see also Chap. 29.

In the discussion of computational complexity of ANN learning we looked briefly at ways of initializing nets with knowledge of the problem domain, thereby constraining the learning process and rendering it more tractable. These are general techniques, and are thus applicable to NL language learning. In the NLP literature, however, two main emphases have emerged [64]:

INCREMENTAL TRAINING. As we have seen, domain knowledge can be incorporated into a net by appropriate initialization of weights. NL-oriented work has, however, preferred to achieve this by incremental training, whereby domain knowledge is not explicitly compiled into network weights, but acquired from data. For some language-learning task, the net is initially trained on short, syntactically simple strings, and as these are learned, increasingly longer and more complex strings are introduced into the training set. Elman [106] found that, for an NL set S containing strings that varied in structural complexity from monoclausal to multiple layers of embedding, a given RANN with randomly initialized weights—that is, a *tabula rasa* net lacking any systematic prior knowledge—could not learn the S if the whole of S was used for training, but could if the strings were presented such that the monoclausal ones were presented first, then those with one level of embedding, and so on. In the first phase of training, all 10,000 strings of the training set were monoclausal. Once the net had learned these, the training set was redefined to include 7,500 monoclausal strings and 2,500 with embedded clauses, then 5,000 monoclausal and 5,000 complex, and finally 10,000 complex. Loading of the final 10,000 complex strings was successful, whereas the attempt to train the same *tabula rasa* net on the final training set of 10,000 complex strings had been unsuccessful [see also Refs. 64,65,135].

TOPOLOGICAL STRUCTURING. Theoretically, a single MLP with a sufficient number of hidden units can implement any computable function. Therefore, it should be possible to implement any required NLP function using a sufficiently large MLP. Early work in fact did use single MLPs to solve small language-processing problems, but there is now a widespread recognition that this approach will not work for large, real-world applications: *We cannot feed 20 years of raw sensory input to a 3-layer, feed-forward, back-error propagation network and then expect a college graduate to result* [199, p. 46; 101,110]. As NLP systems have grown in size and complexity, they have increasingly used modular architectures, in which modules for specific tasks are interconnected in accordance with the preanalyzed requirements of the problem domain, and thereby compute the required function globally [84,125,184]. Miikkulainen's language-understanding system, described in Chap. 37, is an example [see also 225,226,228]; the L_0/NTL project is another [110,113]. Hybrid architectures are a special case of modular ones in which ANN and PSS-based modules interact.

Why are modular systems expected to work where large single-ANN ones are not? Essentially because the structuring of the modules relative to one another reflects prior knowledge of the task domain, and is thus a way of incorporating prior constraints into a learning system to aid tractability, as discussed earlier.

An alternative approach to topological structuring focuses on the amount of memory available to the learning system. In the same series of experiments referred to in the preceding subsection on incremental training, Elman [106] presented the entire training set to the net in one batch, but varied the amount of feedback memory, starting small and gradually increasing it in the course of training. The result was that the net was able to learn the training set, but only in stages: initially, when memory was most restricted, it learned only the simplest strings in the training set and, as memory was gradually increased, it was able to learn strings of increasing syntactic complexity [see also 64].

e. Meaning

There are some NLP applications, such as document search, for which the meaning of the text being processed is not an issue. In others, semantic interpretation of text is necessary, but reasonably straightforward; an example would be an NL command interpreter for a database front end, where both the syntax of input strings and the semantic interpretations to which they are mapped are both well defined and severely restricted relative to normal linguistic usage. When, however, one moves to AI-oriented applications such as (more or less) unrestricted NL-understanding systems, semantic interpretation becomes a difficult and still largely unresolved problem. ANNs do not provide an easy solution, but they do offer a promising alternative to existing PSSH-based approaches.

Meaning is variously understood by different disciplines and by researchers within them. It does, however, seem noncontroversial to say that meaning of NL linguistic expressions has to do with denotation of states of the world, and that semantic interpretation is a mapping from strings to denotations. That, in any case, is what is assumed here. PSSH-based AI and NLP systems have implemented the mapping by constructing system-internal representations of some aspect of the world—the "domain of discourse"—and then relating input strings to the representation [25,192]. How best to represent the word has become of the research discipline in its own right—knowledge representation—and numerous formalisms exist. Systems that use logic formalisms, for example, transform input strings into logical propositions, and these are then related to the domain representation using a deductive inference mechanism to arrive at a semantic interpretation. Some ANN-based work on semantic interpretation continues in the PSSH tradition, in the sense that they use explicitly designed domain representations. Other work takes a radically different approach, however: input strings are mapped not to explicitly designed representations of the world which, inevitably, reflect a designer's analysis of what is significant in the task domain, but to representations that are learned from the world through transducers without designer intervention. At its most ambitious, this line of research aims to embed NLP systems in robotic agents that not only receive inputs from an environment by, say, visual, acoustic, and tactile transducers, but also interact with and change the environment by means of effectors. The aim is for such agents to develop internal world representations by integrating inputs and internal states through self-organization based on adaptive interaction with the environment: *Concepts are thus the "system's own," and their meaning is no longer parasitic on the concepts of others (the system designer)* [97]. In particular, agents would learn to represent the meanings of words and expressions from their use in specific environment-interactive situations. Work on this is proceeding[11,12,75, 92,93,96–98,100,101,112,113,176,237–239,251,272–274,283,288,319,322,356], although it must be said that, to keep experimental simulations tractable, the goal of real-world interaction is often reduced to explicitly designed microworlds reminiscent of ones like the famous SHRLDU in the PSSH tradition. On these matters see also Chap. 31.

The work just cited is symptomatic of developments in cognitive science and AI that emphasize the role of the real-world environment in the explanation of cognition and in the design and constriction of artificially intelligent machines [66,67,93,151–158].

REFERENCES

The references cited in the text and listed in the following appear in conventionally published books, journals, conference proceedings, and technical reports. As in virtually all other areas of academic study, however, a major new source of research information has developed in recent years: the Internet. Subject-specific Web sites with a wealth of information and resources now exist, and most research groups and individual researchers maintain their own sites, where the status of current work is reported, and from which it is typically possible to download existing research papers and preprints of work to appear in conventional paper format. Because this information source is now effectively indispensible, this list of references is prefaced with a list of Web URLs that are particularly useful for ANN-based NLP. As all Web users know, URLs can be volatile, and those given are correct at the time of writing, but where there has been a change it is usually possible to find the site using a search engine.

A. Web Resources

1. *NLP-Specific Sites*

- R Cole, J Mariani, H Uszkoreit, A Zaenen, V Zue, eds. Survey of the State of the Art in Human Language Technology. Center for Spoken Language Understanding, Oregon Graduate Institute of Science and Technology: http://www.cse.ogi.edu/CSLU/HLTsurvey/
- Association for Computational Linguistics: http://www.cs.columbia.edu/~radev/newacl

2. *Directories*

- Galaxy: http://galaxy.einet.net/galaxy/Engineering-and-Technology/Computer-Technology.html
- Pacific Northwest National Laboratory http://www.emsl.pnl.gov:2080/proj/neuron/neural
- Yahoo. Neural Networks: http://www.yahoo.co.uk/Science/Engineering/Electrical_Engineering/Neural_Networks/
- Yahoo: Natural Language Processing http://www.yahoo.co.uk/Science/Computer_Science/Artificial_Intelligence/Natural_Language_Processing/

3. *Coordinating Organizations*

- European Network in Language and Speech (ELSNET): http://www.elsnet.org
- IEEE Neural Network Council: http://www.ewh.ieee.org/tc/nnc/index.html
- International Neural Network Society: http://www.inns.org

4. *Research Groups*

Most research groups maintain Web sites. Space precludes any attempt at an exhaustive listing. What follows is a selection of those that have proven particularly useful; all contain links to other sites. Individual researchers' pages are not included. These can be found using a Web search engine.

- The Neural Theory of Language Project (formerly the L_0 project): http://www.icsi.berkeley.edu/NTL
- UTCS Neural Networks Research Group: http://net.cs.utexas.edu/users/nn
- Austrian Research Institute for Artificial Intelligence (ÖFAI): http://www.ai.univie.ac.at/oefai/nn/nngroup.html
- Artificial Intelligence Research Laboratory, Department of Computer Science, Iowa State University: http://www.cs.iastate.edu/~honavar/aigroup.html
- Brain and Cognitive Sciences, University of Rochester: http://www.bcs.rochester.edu/bcs

- Language and Cognitive Neuroscience Lab, University of Southern California:
  http://siva.usc.edu/coglab
- NEC Research Institute:
  http://external.nj.nec.com/homepages/giles/
- Center for the Neural Basis of Cognition, Carnegie Mellon University/ University of Pittsburgh
  http://www.cnbc.cmu.edu/
- Interactive Systems Lab, Carnegie Mellon University/University of Karlsruhe:
  http://www.is.cs.cmu.edu/ISL.neuralnets.html
- Neural Network Theory at Los Alamos, Los Alamos National Laboratory:
  http://www-xdiv.lanl.gov/XCM/neural/neural_theory.html
- Neural Networks Research Centre, Helsinki University of Technology:
  http://www.cis.hut.fi/research
- Neural Computing Research Group (NCRG), Aston University
  http://www.ncrg.aston.ac.uk
- Natural Language Processing Group, University of Sheffield:
  http://www.dcs.shef.ac.uk/research/groups/nlp/
- Centre for Neural Networks (CNN), King's College, London:
  http://www.mth.kcl.ac.uk/cnn/

5. *Publication Archives*

The main electronic print archive is Neuroprose at ftp://archive.cis.ohio-state.edu/pub/neuroprose/; a directory of others is at NCRG, http://neural-server.aston.ac.uk/NN/archives.html

6. *Discussion Groups*

- Neuron Digest:
  Contact address: neuron-request@psych.upenn.edu
- Connectionists:
  Contact address: connectionists-request@cs.cmu.edu

B. Cited References

1. R Allen. Several studies on natural language and back-propagation. Proceedings of the First Annual International Conference on Neural Networks, 1987.
2. R Allen. Sequential connectionist networks for answering simple questions about microworlds. Proceedings of the Tenth Annual Conference of the Cognitive Science Society, 1988.
3. J Allen, M Seidenberg. The emergence of grammaticality in connectionist networks. In: B Macwhinney, ed. Emergentist Approaches to Language: Proceedings of the 28th Carnegie Symposium on Cognition, Hillsdale, NJ: Lawrence Erlbaum.
4. N Alon, A Dewdney, T Ott. Efficient simulation of finite automata by neural nets. J ACM 38:495–514, 1991.
5. M Anthony. Probabilistic analysis of learning in artificial neural networks: the PAC model and its variants. Neural Comput Surv 1:1–47, 1997.
6. S Amari. Learning and statistical inference. In: M Arbib, ed. The Handbook of Brain Theory and Neural Networks. Cambridge, MA: MIT Press, 1995.

7. J Anderson. An Introduction to Neural Networks. Bradford Books/MIT Press, 1995.
8. M Arbib. Brains, Machines, and Mathematics. 2nd ed. New York: Springer, 1987.
9. M Arbib, ed. The Handbook of Brain Theory and Neural Networks, Cambridge, MA: MIT Press, 1995.
10. M Aretoulaki, G Scheler, W Brauer. Connectionist modelling of human event memorization process with application to automatic text summarization. AAAI Spring Symposium on Intelligent Text Summarization, Stanford, CA, 1998.
11. D Bailey, N Chang, J Feldman, S Narayanan. Extending embodied lexical development. Proceedings of CogSci98, 1998, Hillsdale, NJ: Lawrence Erlbaum.
12. D Bailey, J Feldman, S Narayanan, G Lakoff. Embodied lexical development. Proceedings of the 19th Annual Meeting of the Cognitive Science Society, Stanford University Press, 1997.
13. B Baird, T Troyer, F Eeckman. Synchronization and grammatical inference in an oscillating Elman net. In: S Hanson, J Cowan, C Giles, eds. Advances in Neural Information Processing Systems 5, Morgan Kaufman, 1993.
14. J Barnden. Artificial intelligence and neural networks. In: M Arbib, ed. The Handbook of Brain Theory and Neural Networks. Cambridge, MA: MIT Press, 1995.
15. J Barnden Complex symbol-processing in Conposit, a transiently localist connectionist architecture. In: R Sun, L Bookman, eds. Computational Architectures Integrating Neural and Symbolic Processes. Dordrecht: Kluwer, 1995.
16. R Baron. Knowledge extraction from neural networks: a survey. Neurocolt Technical Report Series NC-TR-94-040, 1995.
17. E Bates, J Elman. Connectionism and the study of change. In: M Johnson, ed. Brain Development and Cognition. Oxford: Blackwell, 1993.
18. E Bates, D Thal, V Marchman. Symbols and syntax: a Darwinian approach to language development. In: N Krasnegor, ed. The Biological and Behavioural Determinants of Language Development. Hillsdale, NJ: Lawrence Erlbaum, 1991.
19. E Baum. When are *k*-nearest neighbor and backpropagation accurate for feasible-sized set of examples? In: Hanson, G Drastal, R Rivest, eds. Computational Learning Theory and Natural Learning Systems. Cambridge, MA: MIT Press, 1994.
20. W Bechtel, A Abrahamsen. Connectionism and the mind. Oxford: Blackwell, 1991.
21. W Bechtel. Representations and cognitive explanations: assessing the dynamicist's challenge in cognitive science. Cogn Sci 22:295–318, 1998.
22. Y Bengio, P Simard, P Frasconi. Learning long-term dependencies with gradient descent is difficult. IEEE Trans Neural Networks 5:157–166, 1994.
23. Y Bengio. Neural Networks for Speech and Sequence Recognition. International Thomson Computer Press, 1996.
24. E Bienenstock, S Geman. Compositionality in neural systems. In: M Arbib, ed. The Handbook of Brain Theory and Neural Networks, Cambridge, MA: MIT Press, 1995.
25. L Birnbaum. Rigor mortis: a response to Nilsson's "Logic and artificial intelligence," Artif Intell 47:57–77, 1991.
26. C Bishop. Neural Networks for Pattern Recognition. Clarendon Press, 1995.
27. C Bishop. Classification and regression. In: E Fiesler, R Beale, eds. Handbook of Neural Computation. Oxford: Oxford University Press, 1997.
28. C Bishop. Generalization. In: E Fiesler, R Beck, eds. Handbook of Neural Computation. Oxford: Oxford University Press, 1997.
29. A Blair, J Pollack. Analysis of dynamical recognizers. Neural Comput 9:1127–1242, 1997.
30. D Blank, L Meeden, J Marshall. Exploring the symbolic/subsymbolic continuum: a case study of RAAM. In: J Dinsmore, ed. The Symbolic and Connectionist Paradigms: Closing the Gap. Hillsdale, NJ: Lawrence Erlbaum, 1992, pp 113–148.

31. A Blum, R Rivest. Training a 3-node neural network is NP-complete. Proceedings of the 1988 IEEE Conference on Neural Information Processing. Morgan Kaufmann, 1988.
32. U Bodenhausen, S Manke. A connectionist recognizer for on-line cursive handwriting recognition. Proceedings of the International Conference on Acoustics, Speech, and Signal Processing, 1994.
33. U Bodenhausen, P Geutner, A Waibel. Flexibility through incremental learning: neural networks for text categorization. Proceedings of the World Congress on Neural Networks (WCNN), 1993.
34. L Bookman. A microfeature based scheme for modelling semantics. Proceedings of the International Joint Conference on Artificial Intelligence, 1987.
35. L Bookman. Trajectories through Knowledge Space. A Dynamic Framework for Machine Comprehension. Dordrecht: Kluwer, 1994.
36. N Bose, P Liang. Neural Network Fundamentals with Graphs, Algorithms, and Applications. New York: McGraw-Hill, 1996.
37. A Browne, ed. Neural Network Analysis, Architectures, and Applications. Institute of Physics Publishing, 1997.
38. T Burton. Introduction to Dynamic Systems Analysis. New York: McGraw-Hill, 1994.
39. M Casey. The dynamics of discrete-time conputation, with application to recurrent neural networks and finite state machine extraction. Neural Comput 8:1135–1178, 1996.
40. D Chalmers. Subsymbolic computation and the Chinese room. In: J Dunsmore, ed. The Symbolic and Connectionist Paradigms: Closing the Gap. Hillsdale, NJ: Lawrence Erlbaum, 1992.
41. D Chalmers. Syntactic transformations on distributed representations. Connect Sci 2: 53–62, 1990.
42. D Chalmers. A computational foundation for the study of cognition, TR-94-03, Philosophy/Neuroscience/Psychology Program, Washington University, 1994.
43. B Chandrasekaran, S Josephson. Architecture of intelligence: the problems and current approaches to solutions. In: V Honavar, L Uhr, eds. Artificial Intelligence and Neural Networks: Steps Toward Principled Integration, Academic Press, 1994.
44. G Chappel, J Taylor. The temporal Kohonen map. Neural Networks 6:441–445, 1993.
45. N Chater, P Conkey. Finding linguistic structure with recurrent neural networks. Proceedings of the Fourteenth Annual Meeting of the Cognitive Science Society. Hillsdale, NJ: Lawrence Erlbaum, 1992.
46. N Chater, M Christiansen. Connectionism and natural language processing. In: S Garrod, M Pickering, eds. Language Processing. Psychology Press, 1999.
47. C Chen, V Honavar. A neural network architecture for syntax analysis. ISU CS-TR 95-18, Department of Computer Science, Iowa State University, 1995.
48. B Cheng, D Titterington. Neural networks: a review from a statistical perspective. Stat Sci 9:2–54, 1994.
49. N Chomsky. Syntactic Structures. The Hague: Mounton & Co., 1957.
50. N Chomsky. Knowledge of Language. Praeger, 1986.
51. M Christiansen. The (non)necessity of recursion in natural language processing. Proceedings of the Fourteenth Annual Conference of the Cognitive Science Society. Hillsdale, NJ: Lawrence Erlbaum, 1992.
52. M Christiansen. Language learning in the full, or why the stimulus might not be so poor after all. TR-94-12, Philosophy/Neuroscience/Psychology Program, Washington University, 1994.
53. M Christiansen, N Chater. Generalization and connectionist language learning. Mind & Lang 9:273–287, 1994.

54. M Christiansen, N Chater. Natural language recursion and recurrent neural networks, TR-94-13, Philosophy/Neuroscience/Psychology Program, Washington University, 1994.
55. M Christiansen, N Chater. Connectionist natural language processing: the state of the art. Special issue of Cogn Sci on Connectionist Models of Human Language Processing: Progress and Prospects. Cognitive Science 23, 1999, 417–437.
56. M Christiansen, N Chater. Toward a connectionist model of recursion in human linguistic performance. Cognitive Science 23, 1999, 157–205.
57. M Christiansen, J Devlin. Recursive inconsistencies are hard to learn: a connectionist perspective on universal word order correlations. Proceedings of the 19th Annual Cognitive Science Society Conference. Lawrence Erlbaum, 1997.
58. M Christiansen, M MacDonald. Processing of recursive sentence structure: testing predictions from a connectionist model. work in progress.
59. P Churchland. On the nature of explanation: a PDP approach. Physica D 42:281–292, 1990.
60. P Churchland, T Sejnowski. The Computational Brain. Cambridge, MA: MIT Press, 1992.
61. A Clark. Beyond eliminativism. Mind & Lang 4:251–279, 1989.
62. A Clark. Microcognition: Philosophy, Cognitive Science, and Parallel Distributed Processing. Cambridge, MA: MIT Press, 1989.
63. A Clark. Connectionism, competence, and explanation. J Philos Sci 41:195–222, 1990.
64. A Clark. Associative Engines. Connectionism, Concepts, and Representational Change. Cambridge, MA: MIT Press, 1993.
65. A Clark. Representational trajectories in connectionist learning. Mind & Mach 4:317–332, 1994.
66. A Clark. The dynamical challenge. Cogn Sci 21:461–481, 1997.
67. A Clark. Being There. Putting Brain, Body, and World Together Again. Cambridge, MA: MIT Press, 1997.
68. A Cleeremans, S Servan–Schreiber, J McClelland. Finite state automata and simple recurrent networks. Neural Comput 1:372–381, 1989.
69. D Clouse, G Cottrell. Regularities in a random mapping from orthography to semantics. Proceedings of the Twentieth Annual Cognitive Science Conference. Lawrence Erlbaum, 1998.
70. D Clouse, C Giles, B Horne, G Cottrell. Time-day neural networks: representation and induction of finite-state machines. IEEE Trans Neural Networks 8:1065–1070, 1997.
71. C Cosic, P Munro. Learning to represent and understand locative prepositional phrases. Proceedings of the Tenth Annual Meeting of the Cognitive Science Society. Lawrence Erlbaum, 1998.
72. G Cottrell, S Small. A connectionist scheme for modelling word sense disambiguation. Cogn & Brain Theory 6: 1983.
73. G Cottrell. Connectionist parsing. Proceedings of the Seventh Annual Conference of the Cognitive Science Society, 1985, pp 201–212.
74. G Cottrell. A Connectionist Approach to Word Sense Disambiguation. Pitman, 1989.
75. G Cottrell, B Bartell, C Haupt. Grounding meaning in perception. Proceedings of the German Workshop on Artificial Intelligence, 1990.
76. G Cottrell, K Plunkett. Acquiring the mapping from meaning to sounds. Connect Sci 6:379–412, 1995.
77. M Craven, J Shavlik. Extracting tree-structured representations of trained networks. In: Advances in Neural Information Processing Systems 8. Cambridge, MA: MIT Press, 1996.
78. S Das, C Giles, G Sun. Capabilities and limitations of a recurrent neural network with an external stack memory in learning context-free grammars. Proceedings of the

International Joint Conference on Neural Networks. China: Publishing House of Electronics Industry, 1992, vol 3.
79. S Das, C Giles, G Sun. Learning context-free grammars: capabilities and limitations of a recurrent neural network with an external stack memory. Proceedings of the Fourteenth Annual Conference of the Cognitive Science Society. Lawrence Erlbaum, 1992.
80. S Das, C Giles, G Sun. Using prior knowledge in an NNPDA to learn context-free languages. In: C Giles, S Hanson, J Cowan, eds. Advances in Neural Information Processing Systems 5. Morgan Kaufmann, 1993.
81. S Das, M Mozer. Dynamic on-line clustering and state extraction: an approach to symbolic learning. Neural Networks 11:53–64, 1998.
81a. B DasGupta, H Siegelmann, E Sontag. On the complexity of training neural networks with continuous activation functions. IEEE Trans Neural Networks 6:1490–1504, 1995.
82. K Daugherty, M Seidenberg. Rules or connections? The past tense revisited. Proceedings of the Fourteenth Annual Meeting of the Cognitive Science Society. Lawrence Erlbaum, 1992.
83. K Daugherty, M Hare. What's in a rule? The past tense by some other name might be called a connectionist net. Proceedings of the 1993 Connectionist Models Summer School. Lawrence Erlbaum, 1993.
84. M de Francesco. Modular topologies. In: E Fiesler, R Beck, eds. Handbook of Neural Computation. Oxford: Oxford University Press, 1997.
85. G Dell. A spreading activation theory of retrieval in language production. Psychol Rev 93:283–321, 1986.
86. G Dell. The retrieval of phonological forms in production: tests of predictions from a connectionist model, J Mem Lang 27:124–142, 1988.
87. G Dell, L Burger, W Svec. Language production and serial order: a functional analysis and a model. Psychol Rev 104:123–147, 1997.
88. R Devaney. An Introduction to Chaotic Dynamical Systems. 2nd ed. Addison–Wesley, 1989.
89. J Dinsmore, ed. The Symbolic and Connectionist Paradigms: Closing the Gap. Lawrence Erlbaum, 1992.
90. G Dorffner. Taxonomies and part-whole hierarchies in the acquisition of word meaning—a connectionist model. Proceedings of the 14th Annual Conference of the Cognitive Science Society. Lawrence Erlbaum, 1992.
91. G Dorffner. Radical connectionism for natural language processing. TR-91-07, Oesterreichisches Forschungsinstitut fuer Artificial Intelligence, Wien, 1991.
92. G Dorffner. A step toward sub-symbolic language models without linguistic representations. In: R Reilly, N Sharkey, eds.. Connection Approaches to Natural Language Processing. Hillsdale, NJ: Lawrence Erlbaum, 1992.
93. G Dorffner. Neural Networks and a New Artificial Intelligence. Thomson Computer Press, 1997.
94. G Dorffner. Radical connectionism—a neural bottom-up approach to AI. In: Neural Networks and a New Artificial Intelligence. Thomson Computer Press, 1997.
95. G Dorffner. A radical view on connectionist language modelling. In G Dorffner, ed. Konnektionismus in Artificial Intelligence und Kognitionsforschung. Berlin: Springer, 1990.
96. G Dorffner, E Prem, H Trost. Words, symbols, and symbol grounding. TR-93-30, Oesterreichisches Forschunginstitut fuer Artificial Intelligence, Wien, 1993.
97. G Dorffner, E Prem. Connectionism, symbol grounding, and autonomous agents. Proceedings of the 15th Annual Conference of the Cognitive Science Society, 1993.
98. M Dyer, V Nenov. Language learning via perceptual/motor experiences. Proceedings of the 15th Annual Conference of the Cognitive Science Society. Lawrence Erlbaum, 1993.

99. M Dyer. Symbolic neuroengineering for natural language processing: a multilevel research approach. In: J Barnden, J Pollack, eds. Advances in Connectionist and Neural Computation Theory 1, Ablex, 1991.
100. M. Dyer. Grounding language in perception. In: V Honavar, L Uhr, eds. Artificial Intelligence and Neural Networks: Steps Toward Principle Integration. New York: Academic Press, 1994.
101. M Dyer. Connectionist natural language processing: a status report. In: R Sun, L Bookman, eds. Computational Architectures Integrating Neural and Symbolic Processes. Dordrecht: Kluwer, 1995.
102. S El Hihi, Y Bengio. Hierarchical recurrent neural networks for long-term dependencies. In: M Mozer, DS Touretzky, M Perrone, eds. Advances in Neural Information Processing Systems 8. Cambridge, MA: MIT Press, 1996.
103. S Ellacott, D Bose. Neural Networks: Deterministic Methods of Analysis. International Thomson Computer Press, 1996.
104. J Elman. Finding structure in time. Cogn Sci 14:172–211, 1990.
105. J Elman. Distributed representation, simple recurrent networks, and grammatical structure. Mach Learn 7:195–225, 1991.
106. J Elman. Learning and development in neural networks: the importance of starting small. Cognition 48:71–99, 1993.
107. J Elman, Language as a dynamical system. In: R Port, T van Gelder, eds. Mind as Motion: Explorations in the Dynamics of Cognition, MIT Press, 1995.
108. J Elman, E Bates, M Johnson, A Karmiloff–Smith, D Parisi, K Plunkett. Rethinking Innateness. A Connectionist Perspective on Development. Cambridge, MA: MIT Press, 1996.
109. M Fanty. Context-free parsing in connectionist networks. TR174, Department of Computer Science, University of Rochester, 1985.
110. J Feldman. Structured connectionist models and language learning. Artif Intell Rev 7:301–312, 1993.
111. J Feldman, D Ballard. Connectionist models and their properties. Cogn Sci 6:205–254, 1982.
112. J Feldman, G Lakoff, A Stolcke, W Hollbach, S Weber. Miniature language acquisition: a touchstone for cognitive science. TR-90-009, ICSI, Berkeley, 1990.
113. J. Feldman, G Lakoff, D Bailey, S Narayanan. T Regier, A Stolcke. L_0—the first five years of an automated language acquisition project. Artif Intell Rev 10: 103–129, 1996.
114. E Fiesler, R Beale. Handbook of Neural Computation. Oxford University Press, 1997.
115. S Finch, N Chater. Learning syntactic categories: a statistical approach. In: M Oaksford, G Brown, eds. Neurodynamics and Psychology. New York: Academic Press, 1993.
116. J Fodor, Z Pylyshyn. Connectionism and cognitive architecture: a critical analysis. Cognition 28:3–71, 1988.
117. J Fodor, B McLaughlin. Connectionism and the problem of systematicity: why Smolensky's solution doesn't work. In: T Horgan, J Tienson, eds. Connectionism and the Philosophy of Mind. Kluwer, 1991.
118. J Fodor, B McLaughlin, Connectionism and the problem of systematicity: why Smolensky's solution won't work. Cognition 35:183–204, 1990.
119. J Fodor. Connectionism and the problem of systematicity (continued): why Smolensky's solution still doesn't work. Cognition 62:109–119, 1997.
120. S Franklin, M Garzon. Computation in discrete neural nets. In: P Smolensky, et al., eds. Mathematical Perspectives on Neural Networks. Hillsdale, NJ: Lawerence Erlbaum, 1996.

121. P Frasconi, M Gori, G Soda. Injecting nondeterministic finite state automata into recurrent networks. Technical Report, Dipartimento di Sistemi e Informatica, Firenze, 1993.
122. P Frasconi, M Gori, G Soda. Recurrent neural networks and prior knowledge for sequence processing: a constrained nondeterministic approach. Know Based Syst 8: 313–322, 1995.
123. P Frasconi, M Gori, M Maggini, G Soda. Representation of finite state automata in recurrent radial basis function networks. Mach Learn 23:5–32, 1996.
124. S Gallant. Neural Network Learning and Expert Systems. Cambridge, MA: MIT Press, 1993.
125. P Gallinari. Modular neural net systems, training of. In: M Arbib, ed. The Handbook of Brain Theory and Neural Networks. Cambridge, MA: MIT Press, 1995.
126. A Galton. The Church–Turing thesis: its nature and status. AISB Q 74:9–19, 1990.
127. M Gasser. A connectionist model of sentence generation in a first and second language, TR UCLA-AI-88-13, University of California, 1988.
128. M Gasser. Transfer in a connectionist model of the acquisition of morphology. In: H Baayen, R Schroeder, eds. Yearbook of Morphology 1996. Dordrecht: Foris, 1997.
129. M Gasser. Acquiring receptive morphology: a connectionist model. Proceedings of ACL 94, 1994.
130. C Giles, D Chen, C Miller, H Chen, G Sun, Y Lee. Learning and extracting finite state automata with second-order recurrent neural networks. Neural Comput 4:393–405, 1992.
131. C Giles, C Omlin. Inserting rules into recurrent neural networks. In: S Kung, F Fallside, J Sorenson, eds. Neural networks for Signal Processing II, Proceedings of the 1992 IEEE Workshop. IEEE Press, 1992.
132. C Giles, C Miller, D Chen, G Sun, H Chen, Y Lee. Extracting and learning an unknown grammar with recurrent neural networks. In: J Moody, S Hanson, R Lippmann, eds. Advances in Neural Information Processing Systems 4. Morgan Kaufmann, 1992.
133. C Giles, C. Omlin. Rule refinement with recurrent neural networks. IEEE International Conference on Neural Networks, 1993.
134. C Giles, C Omlin. Extraction, insertion, and refinement of symbolic rules in dynamically driven recurrent neural networks. Connect Sci 5:307–337, 1993.
135. C Giles, B Horne, T Lin. Learning a class of large finite state machines with a recurrent neural network, Neural Networks 8:1359–1365, 1995.
136. C Giles, G Sun, H Chen. Y Lee, D Chen. Second-order recurrent neural networks for grammatical inference. Proceedings of the International Joint Conference on Neural Networks, 1991.
137. F Girosi, T Poggio. Networks and the best approximation property. Biol Cybernet 63: 169–176, 1990.
138. E Gold. Language identification in the limit. Inform Control 10:447–474, 1967.
139. R Golden. A developmental neural model of visual word perception. Cogn Sci 10:241–276, 1986.
140. R Golden. Mathematical Methods for Neural Network Analysis and Design. Cambridge, MA: MIT Press, 1996.
141. S Goonatilake, S Khebbal. Intelligent Hybrid Systems. New York: Wiley, 1994.
142. M Gori, M Maggini, G Soda. Learning finite state grammars from noisy examples using recurrent neural networks. NEURAP'96, 1996.
143. R Hadley. Systematicity in connectionist language learning. Mind Lang 9:247–273, 1994.
144. R Hadley, V Cardei. Acquisition of the active-passive distinction from sparse input and no error feedback, CSS-IS-TR97-01, School of Computer Science and Cognitive Science Program, Vancouver, BC: Simon Fraser University, 1997.

145. S Hanson, J Kegl. PARSNIP: a connectionist network that learns natural language grammar from exposure to natural language sentences. Proceedings of the Ninth Annual Conference of the Cognitive Science Society. Lawrence Erlbaum, 1987.
146. S Hanson, G Drastal, R Rivest. Computational Learning Theory and Natural Learning Systems. Cambridge, MA: MIT Press, 1994.
147. M Hare, J Elman. A connectionist account of English inflectional morphology: evidence from language change. Proceedings of the Fourteenth Annual Conference of the Cognitive Science Society, 1992.
148. M Hare, J Elman. Learning and morphological change. Cognition 56:61–98, 1995.
149. M Hare, J Elman, K Daugherty. Default generalization in connectionist networks. Lang Cogn Processes 10:601–630, 1995.
150. T Harley. Phonological activation of semantic competitors during lexical access in speech production. Lang Cogn Processes 8:291–309, 1993.
151. S Harnad. The symbol grounding problem. Physica D 42:335–346, 1990.
152. S Harnad. Connecting object to symbol in modeling cognition. In: A Clark, R Lutz, eds. Connectionism in Context. Springer, 1992.
153. S Harnad. Problems, problems: the frame problem as a symptom of the symbol grounding problem. Psycoloquy 4, 1993.
154. S Harnad. Grounding symbols in the analog world with neural nets—a hybrid model. Think 2:12–78, 1993.
155. S Harnad. Symbol grounding is an empirical problem: neural nets are just a candidate component. Proceedings of the Fifteenth Annual Meeting of the Cognitive Science Society. Lawrence Erlbaum, 1993.
156. S Harnad. Grounding symbolic capacity in robotic capacity. In: L Steels, R Brooks, eds. The "Artificial Life" Route to "Artificial Intelligence." Building Situated Embodied Agents. Lawrence Erlbaum, 1995.
157. S Harnad. Does the mind piggyback on robotic and symbolic capacity. In: H Morowitz, J Singer, eds. The Mind, the Brain, and Complex Adaptive Systems. Addison-Wesley, 1994.
158. S Harnad. Computation is just interpretable symbol manipulation; cognition isn't. Minds & Mach 4:379–390, 1995.
159. P Haugeland. Artificial Intelligence: The Very Idea. Cambridge, MA: MIT Press, 1985.
160. S Haykin. Neural Networks: A Comprehensive Foundation. New York: Macmillan, 1994.
161. J Henderson. Connectionist syntactic parsing using temporal variable binding. J Psycholinguist Res 23:353–379, 1994.
162. J Henderson, P Lane. A connectionist architecture for learning to parse. Proceedings of 17th International Conference on Computational Linguistics and the 36th Annual Meeting of the Association for Computational Linguistics (COLING-ACL '98), University of Montreal, Canada, 1998.
163. J Henderson. Constituency, context, and connectionism in syntactic parsing. In: M Crocker, M Pickering, C Clifton, eds. Architectures and Mechanisms for Language Processing. Cambridge University Press, to appear.
164. J Hertz, A Krogh, R Palmer. Introduction to the Theory of Neural Computation. Addison–Wesley, 1991.
165. J Hertz. Computing with attractors. In: M Arbib, ed. The Handbook of Brain Theory and Neural Networks. Cambridge: MA: MIT Press, 1995.
166. R Hilborn. Chaos and Nonlinear Dynamics. Oxford: Oxford University Press, 1994.
167. G Hinton. Implementing semantic networks in parallel hardware. In: G Hinton, J Anderson, eds. Parallel Models of Associative Memory. Lawrence Erlbaum, 1981.

168. S Hochreiter, J Schmidhuber. Recurrent neural net learning and vanishing gradient. Proceedings of the Fuzzy-Neuro Workshop, Soest, Germany, 1997.
169. V Honavar, L Uhr. Artificial Intelligence and Neural Networks: Steps Toward Principled Integration. New York: Academic Press, 1994.
170. V Honavar. Toward learning systems that integrate different strategies and representations. In: V Honavar, L Uhr, eds. Artificial Intelligence and Neural Networks: Steps Toward Principled Integration. New York: Academic Press, 1994.
171. V Honavar. Symbolic artificial intelligence and numeric neural networks: towards a resolution of the dichotomy. In: R Sun, L Bookman, eds. Computational Architectures Integrating Neural and Symbolic Processes. Dordrecht: Kluwer, 1995.
172. T Horgan, J Tienson, eds. Connectionism and the Philosophy of Mind. Dordrecht: Kluwer, 1991.
173. B Horne, D Hush. Bounds on the complexity of recurrent neural network implementations of finite state machines. Neural Networks 9:243–252, 1996.
174. T Howells. VITAL—a connectionist parser. Proceedings of the Tenth Annual Conference of the Cognitive Science Society, 1988.
175. G Huang, H Babri. Upper bounds on the number of hidden neurons in feedforward networks with arbitrary bounded nonlinear activation functions. IEEE Trans Neural Netowrks 7:1329–1338, 1996.
176. S Jackson, N Sharkey. Grounding computational engines. Artif Intell Rev 10:65–82, 1996.
177. H Jaeger. From continuous dynamics to symbols. Proceedings of the First Joint Conference on Complex Systems in Psychology: Dynamics, Synergetics, and Autonomous Agents, Gstaad, Switzerland, 1997.
178. A Jain. Parsing complex sentences with structured connectionist networks. Neural Comput 3:110–120, 1991.
179. A Jain, A Waibel. Parsing with connectionist networks. In: M Tomita, ed. Current Issues in Parsing Technology. Kluwer, 1991.
180. A Jain, A Waibel. Incremental parsing in modular recurrent connectionist networks. International Joint Conference on Neural Networks, 1991.
181. D James, R Miikkulainen. SARDNET: a self-organizing feature map for sequences. In: G Tesauro, D Touretzky, T Leen, eds. Advances in Neural Processing Systems 7, 1995.
182. K Jim, C Giles, B Horne. Synaptic noise in dynamically driven recurrent neural networks: convergence and generalization, UMIACS-TR-94-89, Institute for Advanced Computer Studies, University of Maryland, 1994.
183. W Joerding, J Meador. Encoding a priori information in feedforward networks. Neural Networks 4:847–856, 1991.
184. M Jordan, R Jacobs. Modular and hierarchical learning systems. In: M Arbib, ed. The Handbook of Brain Theory and Neural Networks. Cambridge, MA: MIT Press, 1995.
185. M Jordan. Attractor dynamics and parallelism in a connectionist sequential machine. Proceedings of the Eighth Conference of the Cognitive Science Society, 1986.
186. M Jordan. Neural networks. In: A Tucker ed. CRC Handbook of Computer Science. Boca Raton, FL: CRC Press, 1996.
187. JS Judd. Complexity of learning. In: G Smolensky, et al. Mathematical Perspectives on Neural Networks. Hillsdale, NJ: Lawrence Erlbaum, 1996.
188. JS Judd. Time complexity of learning. In: M Arbib, ed. Handbook of Brain Theory and Neural Networks. Cambridge, MA: MIT, Press, 1995.
189. A Karmiloff–Smith. Nature, nurture, and PDP: preposterous developmental predicates? Connect Sci 4:253–270, 1992.
190. A Kawamoto. Distributed representations of ambiguous words and their resolution in a connectionist network. In: S Small, G Cottrell, M Tanenhaus, eds. Lexical Ambiguity Resolution. Morgan Kaufmann, 1989.

191. S Kelso. Dynamic Patterns. Cambridge, MA: MIT Press, 1995.
192. D Kirsch. Foundations of AI: the big issues. Artif Intell 47:3–30, 1991.
193. J Kolen. The origin of clusters in recurrent network state space. Proceedings of the Sixteenth Annual Conference of the Cognitive Science Society, 1994.
194. J Kolen. Fool's gold: extracting finite state machines from recurrent network dynamics. In: J Cowan, G Tesauro, J Alspector, eds. Advances in Neural Information Processing Systems 6. Morgan Kaufmann, 1994.
195. J Kolen. Recurrent networks: state machines or iterated function systems? In: M Mozer, P Smolensky, D Touretzky, J Elman, A Weigend, eds. Proceedings of the 1993 Connectionist Models Summer School. Laurence Erlbaum, 1994.
196. J Kolen, J Pollack. The observer's paradox: apparent computational complexity in physical systems. J of Exp Theor Artif Intell 7:253–269, 1995.
197. K Kukich. Where do phrases come from: some preliminary experiments in connectionist phrase generation. In: G Kempen, ed. Natural Language Generation. Kluwer, 1987.
198. S Kwasny, K Faisal. Connectionism and determinism in a syntactic parser. Connect Sci 2:63–82, 1990.
199. S Kwasny, K Faisal. Symbolic parsing via subsymbolic rules. In: J Dinsmore, ed. The Symbolic and Connectionist Paradigms: Closing the Gap. Lawrence Erlbaum, 1992.
200. S Kwasny, S Johnson, B Kalman. Recurrent natural language parsing. Proceedings of the Sixteenth Annual Meeting of the Cognitive Science Society, Lawrence Erlbaum, 1994.
201. R Lacher, K Nguyen. Hierarchical architectures for reasoning. In: R Sun, L Bookman, eds. Computational Architectures Integrating Neural and Symbolic Processes. Kluwer, 1995.
202. T Lange, M Dyer. High-level inferencing in a connectionist network. Connect Sci 1: 181–217, 1989.
203. T Lange. A structured connectionist approach to inferencing and retrieval. In: R Sun, L Bookman, eds. Computational Architectures Integrating Neural and Symbolic Processes. Kluwer, 1995.
204. S Lawrence, C Giles, S Fong. On the applicability of neural network and machine learning methodologies to natural language processing. UMIACS-TR-95-64. Institute for Advanced Computer Studies, University of Maryland, 1995.
205. S Lawrence, C Giles, S Fong. Natural language grammatical inference with recurrent neural networks. IEEE Trans Knowl Data Eng in press.
206. S Lawrence, A Tsoi, A Back. Function approximation with neural networks and local methods: bias, variance, and smoothness. Australian Conference on Neural Networks, ACNN 96, Australian National University, 1996.
207. S Lawrence, C Giles, S Fong. Can recurrent neural networks learn natural language grammars? Proceedings of the International Conference on Neural Networks, ICNN 96, 1996.
208. S Lawrence, C Giles, A Tsoi. What size network gives optimal generalization? Convergence properties of backpropogation, UMIACS-TR-96-22, Institute for Advanced Computer Studies, University of Maryland, 1996.
209. S Lawrence, C Giles, A Tsoi. Lessons in neural network training: overfitting may be harder than expected. Proceedings of the Fourteenth Annual Conference on Artificial Intelligence, AAAI-97, AAAI Press, 1997.
210. S Lawrence, S Fong, C Giles. Natural language grammatical inference: a comparison of recurrent neural networks and machine learning methods. In: S Wermter, et al., eds. Connectionist, Statistical, and Symbolic Approaches to Learning for Natural Language Processing. Springer, 1996.

211. G Lee, M Flowers, M Dyer. Learning distributed representations for conceptual knowledge and their application to script-based story processing. Connect Sci 2:313–346, 1990.
212. W Lehnert. Possible implications of connectionism. Theoretical Issues in Natural Language Processing. University of New Mexico, 1986.
213. D Levine, M Apariciov. Neural Networks for Knowledge Representation and Inference. Hillsdale, NJ: Lawrence Erlbaum, 1994.
214. T Lin, B Horne, C Giles. How embedded memory in recurrent neural network architectures helps learning long-term temporal dependencies. UMIACS-TR-96-28. Institute for Advanced Computer Studies, University of Maryland, 1996.
215. S Lucas. New directions in grammatical inference. In: Grammatical Inference: Theory, Applications, and Alternatives. IEEE, 1993.
216. J McClelland, D Rumelhart. An interactive activation model of context effects in letter perception: Part 1. An account of basic findings. Psychol Rev. 88:375–407, 1981.
217. J McClelland, A Kawamoto. Mechanisms of sentence processing: assigning roles to constituents. In: D Rumelhart, J McClelland, eds. Parallel Distribution Processing. Cambridge, MA: MIT Press, 1986.
218. T McKenna. The role of interdisciplinary research involving neuroscience in the development of intelligent systems: In: V Honavar, L Uhr, eds. Artificial Intelligence and Neural Networks: Steps Toward Principled Integration. New York: Academic Press, 1994.
219. B MacWhinney, J Leinbach. Implementations are not conceptualizations: revising the verb learning model. Cognition 40 (1991), 121–157.
220. G Marcus. The acquisition of the English past tense in children and multilayered connectionist networks. Cognition 56:271–279, 1997.
221. P Manolios, R Fanelli. First-order recurrent neural networks and deterministic finite state automata. Neural Comput 6:1155–1173, 1994.
222. D Medler. A brief history of connectionism. Neural Comput Surv 1:61–101, 1998.
223. R Miikkulainen, M Dyer. A modular neural network architecture for sequential paraphrasing of script-based stories. TR UCLA-AI-89-02, University of California, 1989.
224. R Miikkulainen, M Dyer. Natural language processing with modular PDP networks and distributed lexicon. Cogn Sci 15:343–399, 1991.
225. R Miikkulainen. Subsymbolic Natural Language Processing: An Integrated Model of Scripts, Lexicon, and Memory. Cambridge, MA: MIT Press, 1993.
226. R Miikkulainen. Integrated connectionist models: building AI systems on subsymbolic foundations. In: V Honavar, L Uhl, eds. Artificial Intelligence and Neural Networks: Steps Toward Principled Integration. New York: Academic Press, 1994.
227. R Miikkulainen. Subsymbolic parsing of embedded structures. In: R Sun, L Bookman, eds. Computational Architectures. Integrating Neural and Symbolic Processes, Kluwer, 1995.
228. R Miikkulainen. Subsymbolic case–role analysis of sentences with embedded clauses. Cogn Sci 20:47–73, 1996.
229. M Minsky, S Papert. Perceptrons. Cambridge, MA: MIT Press, 1969.
230. H Moisl. Connectionist finite state natural language processing. Connect Sci 4:67–91, 1992.
231. M Mozer, S Das. A connectionist symbol manipulator that discovers the structure of context-free languages. In: S Hanson, J Cowan, C Giles, eds. Advances in Neural Information Processing Systems. Morgan Kaufmann, 1993.
232. M Mozer. Neural net architectures for temporal sequence processing. In: A Weigend, N Gershenfeld, eds. Predicting the Future and Understanding the Past. Addison-Wesley, 1994.

233. P Munro, C Cosic, M Tabasko. A network for encoding, decoding, and translating locative prepositions. Connect Sci 3:225–240, 1991.
234. R Nakisa, U Hahn. Where defaults don't help: the case of the German plural system. Proceedings of the Sixteenth Annual Meeting of the Cognitive Science Society. Lawrence Erlbaum, 1996.
235. S Narayanan. Talking the talk *is* like walking the walk: a computational model of verbal aspect. Proceedings of the Annual Conference of the Cognitive Science Society, Stanford, 1997.
236. S Narayanan. Moving right along: a computational model of metaphoric reasoning about events. Proceedings of the National Conference on Artifical Intelligence, AAAI Press, 1999.
237. S Narayanan. Embodiment in language understanding: modeling the semantics of causal narratives. AAAI 1996 Fall Symposium on Embodied Cognition and Action. AAAI Press, 1996.
238. V Nenov, M Dyer. Perceptually grounded language learning: Part 1—a neural network architecture for robust sequence association. Connect Sci 5:115–138, 1993.
239. V Nenov, M Dyer. Perceptually grounded language learning: Part 2—DETE: a neural/procedural model. Connect Sci 6:3–41, 1994.
240. A Newell. Physical symbol systems. Cogn Sci 4:135–183, 1980.
241. L Niklasson, T van Gelder. On being systematically connectionist. Mind Lang 9:288-302, 1994.
242. L Niklasson, N Sharkey. Systematicity and generalization in compositional connectionist representations. In: G Dorffner, ed. Neural Networks and a New Artificial Intelligence. Thomson Computer Press, 1997.
243. A Norton. Dynamics: an introduction. In: R Port, T van Gelder, eds. Mind as Motion. Explorations in the Dynamics of Cognition. Cambridge, MA: MIT Press, 1995.
244. S Olafsson. Some dynamical properties of neural networks. In: R Linggard, D Myers, C Nightingale, eds. Neural Networks for Vision, Speech, and Natural Language. Chapman & Hall, 1992.
245. C Omlin, C Giles. Training second-order recurrent neural networks using hints. In: D Sleeman, P Edwards, eds. Machine Learning: Proceedings of the Ninth International Conference. Morgan Kaufmann, 1992.
246. C Omlin, C Giles. Constructing deterministic finite-state automata in recurrent neural networks. J ACM 43:937–972, 1996.
247. C Omlin, C Giles. Fault-tolerant implementation of finite-state automata in recurrent neural networks. RPI Computer Science Technical Report 95-3, 1995.
248. C Omlin. Stable encoding of large finite state automata in recurrent neural networks with sigmoid discriminants. Neural Comput 8:675–696, 1996.
249. C Omlin, C Giles. Extraction and insertion of symbolic information in recurrent neural networks. In: V Honavar, L Uhr, eds. Artificial Intelligence and Neural Networks: Steps Toward Principled Integration. New York: Academic Press, 1994.
250. C Omlin, C Giles. Extraction of rules from discrete-time recurrent neural networks. Neural Networks 9:41–52, 1996.
251. R Pfeifer, P Verschure. Complete autonomous systems: a research strategy for cognitive science. In: D Dorffner, ed. Neural Networks and a New Artificial Intelligence. Thomson Computer Press, 1997.
252. S Pinker, A Prince. On language and connectionism: analysis of a parallel distributed processing model of language acquisition. Cognition 28:73–193, 1988.
253. S Pinker. The Language Instinct. William Morrow, 1994.
254. T Plate. Holographic recurrent networks. In: C Giles, S Hanson, J Cowan, eds. Advances in Neural Information Processing Systems 5. Morgan Kaufmann, 1993.

255. T Plate. Holographic reduced representations. IEEE Trans Neural Networks 6:623–641, 1995.
256. T Plate. A common framework for distributed representation schemes for compositional structure. In: F Maire, R Hayward, J Diederich, eds. Connectionist Systems for Knowledge Representation and Deduction. Queensland University of Technology, 1997.
257. D Plaut. Semantic and associative priming in a distributed attractor network. Proceedings of the 17th Annual Conference of the Cognitive Science Society. Lawrence Erlbaum, 1995.
258. K Plunkett, V Marchman. U-shaped learning and frequency effects in a multi-layered perceptron: implications for child language acquisition. Cognition 38:43–102, 1991.
259. K Plunkett, V Marchman. From rote learning to system building: acquiring verb morphology in children and connectionist nets. Cognition 48:21–69, 1993.
260. K Plunkett. Language acquisition. In M Arbib, ed. The Handbook of Brain Theory and Neural Networks. Cambridge, MA: MIT Press, 1995.
261. J Pollack. No harm intended: a review of the Perceptrons expanded edition. J Math Psychol 33:358–365, 1989.
262. J Pollack. Implications of recursive distributed representations. In: D Touretzky, ed. Advances in Neural Information Processing Systems 1. Morgan Kaufmann, 1989.
263. J Pollack. Recursive distributed representations. Artif Intell 46:77–105, 1990.
264. J Pollack. Connectionism: past, present, and future. Artif Intell Rev 3:3–20; 1989.
265. J Pollack. The induction of dynamical recognizers. Mach Learn 7:227–252, 1991.
266. S Porat. J Feldman. Learning automata from ordered examples. Mach Learn 7:109–138, 1991.
267. R Port, T van Gelder. Mind as Motion. Explorations in the Dynamics of Cognition. Cambridge, MA: MIT Press, 1995.
268. S Prasada, S Pinker. Similarity-based and rule-based generalizations in inflectional morphology. Lang Cogn Processes 8:1–56, 1993.
269. L Pratt. Experiments on the transfer of knowledge between neural networks. In: S Hanson, et al., eds. Computational Learning Theory and Natural Learning Systems. Cambridge MA: MIT Press, 1994.
270. J Rager. Self-correcting connectionist parsing. In: R Reilly, N Sharkey, eds. Connectionist Approaches to Natural Language Processing, Hillsdale, NJ: Lawrence Erlbaum, 1992.
271. R Reed, R Marks. Neurosmithing: improving neural network learning. In: M Arbib, ed. The Handbook of Brain Theory and Neural Networks. Cambridge, MA: MIT Press, 1995.
272. T Regier. Learning object-relative spatial concepts in the L_0 project. Proceedings of the Thirteenth Annual Meeting of the Cognitive Science Society, 1991.
273. T Regier. Learning perceptually grounded semantics in the L_0 project. Proceedings of the 29th Annual Meeting of the Association for Computational Linguistics, 1991.
274. T Regier. The Human Semantic Potential. Cambridge, MA: MIT Press, 1996.
275. R Reilly. A connectionist model of some aspects of anaphor resolution. Proceedings of the Tenth Annual Conference on Computational Linguistics, 1984.
276. R Reilly, N Sharkey. Connectionist Approaches to Natural Language Processing. Hillside, NJ: Lawrence Erlbaum, 1992.
277. B. Ripley. Pattern Recognition and Neural Networks. Cambridge University Press, 1996.
278. R Rojas. Neural Networks: A Systematic Introduction. Springer, 1996.
279. D Rumelhart, J McClelland. An interactive activation model of context effects in letter perception: Part 2. the contextual enhancement effects and some tests and enhancements of the model. Psychol Rev 89:60–94, 1982.

280. D Rumelhart, J McClelland. On learning the past tenses of English verbs. In: D Rumelhart, J McClelland, eds. Parallel Distributed Processing, vol 2. Cambridge, MA: MIT Press, 1986.
281. D Rumelhart, J McClelland. Parallel Distributed Processing, 2 vols. Cambridge, MA: MIT Press, 1986.
282. D Rumelhart, D Zipser. Feature discovery by competitive learning. In: D Rumelhart, J McClelland, eds. Parallel Distributed Processing. Vol 1. Cambridge, MA: MIT Press, 1986.
283. N Sales, R Evans, I Aleksander. Successful naive representation grounding. Artif Intell Rev 10:83–102, 1996.
284. W Sarle. Neural networks and statistical models. Proceedings of the Nineteenth Annual SAS Users Group International Conference, 1994.
285. F Scarselli, A Tsoi. Universal approximation using feedforward neural networks: a survey of some existing methods, and some new results. Neural Networks 11:15–37, 1998.
286. G Scheler. Generating English plural determiners from semantic representations: a neural network learning approach. In: S Wermter, et al., eds. Connectionist, Statistical and Symbolic Approaches to Learning for Natural Language Processing. Springer, 1996.
287. G Scheler. With raised eyebrows or the eyebrows raised? A neural network approach to grammar checking for definiteness. Proceedings of NEMLAP-2, Ankara, Turkey, 1996.
288. G Scheler. Computer simulation of language acquisition: a proposal concering early language learning in a micro-world. Technical Report FKI-214-95, Technische Universität München, Institut für Informatik, 1995.
289. G Scheler, N Fertig. Constructing semantic representations using the MDL principle. Proceedings of HELNET'97, Montreux, Switzerland, 1997.
290. G Scheler. Learning the semantics of aspect. In: H Somers, D Jones, eds. New Methods in Language Processing. University College London Press, 1996.
291. J Schmidhuber. Learning complex, extended sequences using the principle of history compression. Neural Comput 4:234–242, 1992.
292. J Schmidhuber, S Heil. Sequential neural text compression. IEEE Trans Neural Networks 7:142–146, 1996.
293. M Seidenberg, J McClelland. A distributed, developmental model of word recognition and naming. Psychol Rev 96:523–568, 1989.
294. B Selman. Rule-based processing in a connectionist system for natural language understanding, TR CSRI-168, Computer Systems Research Institute, University of Toronto, 1985.
295. B Selman, G Hirst. A rule-based connectionist parsing system. Proceedings of the Seventh Annual Meeting of the Cognitive Science Society. Lawrence Erlbaum, 1985.
296. B Selman. Connectionist systems for natural language understanding. Artif Intell Rev 3: 23–31, 1989.
297. D Servan–Schreiber, A Cleeremans, J McClelland. Learning sequential structure in simple recurrent networks. In: D Touretzky, eds. Advances in Neural Information Processing Systems 1. Morgan Kaufmann, 1989.
298. D Servan–Schreiber, A Cleeremans, J McClelland. Graded state machines: the representation of temporal contingencies in simple recurrent networks. Mach Learn 7:161–193, 1991.
299. N Sharkey. Connection science and natural language: an emerging discipline. Connect Sci 2: 1990.
300. N Sharkey. Connectionist representation techniques. Artif Intell Rev 5: 143–167, 1991.
301. N Sharkey, ed. Connectionist Natural Language Processing. Kluwer, 1992.
302. N Sharkey, A Sharkey. Adaptive generalisation. Artif Intell Rev 7: 313–328, 1993.

303. N Sharkey, A Sharkey. Separating learning and representations. In: S Wermter, et al., eds. Connectionist, Statistical, and Symbolic Approaches to Learning for Natural Language Processing. Springer, 1996.
304. N Sharkey, S Jackson. Three horns of the representational trilemma. In: V Honavar, L Uhr, eds. Artificial Intelligence and Neural Networks: Steps Toward Principled Integration, New York Academic Press, 1994.
305. L Shastri. Structured connectionist models. In: M Arbib, ed. The Handbook of Brain Theory and Neural Networks. Cambridge, MA: MIT Press, 1995.
306. L Shastri, V Ajjanagadde. From simple associations to systematic reasoning: a connectionist representation of rules, variables, and dynamic bindings using temporal synchrony. Behav Brain Sci 16:417–494, 1993.
307. J Shavlik. Learning by symbolic and neural methods. In: M Arbib, ed. The Handbook of Brain Theory and Neural Networks. Cambridge, MA: MIT Press, 1995.
308. H Siegelman, E Sontag. On the computational power of neural nets. J Comput Syst Sci 50:132–150, 1995.
309. S Small, G Cottrell, L Shastri. Towards connectionist parsing. Proceedings of the National Conference on Artificial Intelligence, 1982.
310. P Smolensky. On the proper treatment of connectionism. Behav Brain Sci 11:1–74, 1988.
311. P Smolensky. The constituent structure of connectionist mental states: a reply to Fodor and Pylyshyn. In: T Horgan, J Tienson, eds. Connectionism and the Philosophy of Mind. Kluwer, 1991.
312. P Smolensky. Connectionism, constituency, and the Language of Thought. In: B Loewer, G Rey, eds. Meaning and Mind. Fodor and his Critics. Blackwell, 1991.
313. P Smolensky. Tensor product variable binding and the representation of symbolic structures in connectionist systems. Artif Intell 46:159–216, 1990.
314. P Smolensky, G Legendre, Y Miyata. Principles for an integrated connectionist/symbolic theory of higher cognition. TR-CU-CS-600-92, Computer Science Department, University of Colorado at Boulder, 1992.
315. P Smolensky. Harmonic grammars for formal languages. In: S Hanson, J Cowan, C Giles, eds. Advances in Neural Information Processing Systems 5. Morgan Kaufmann, 1993.
316. P Smolensky, G Legendre, Y Miyata. Integrating connectionist and symbolic computation for the theory of language. In: V Honavar, L Uhr, eds. Artificial Intelligence and Neural Networks: Steps Toward Principled Integration. New York: Academic Press, 1994.
317. P Smolensky, M Mozer, D Rumelhart. Mathematical Perspectives on Neural Networks. Lawrence Erlbaum, 1996.
318. E Sontag. Automata and neural networks. In: M Arbib, ed. Handbook of Brain Theory and Neural Networks. Cambridge, MA: MIT Press, 1995.
319. J. Sopena. Verbal description of visual blocks world using neural networks, Tr-10, Department de Psicologica Basica, University of Barcelona, 1988.
320. J Sopena. ERSP: a distributed connectionist parser that uses embedded sequences to represent structure. TR UB-PB-1-91. University of Barcelona, Department of Psychology, 1991.
321. A Sperduti. On the computational power of recurrent neural networks for structures. Neural Networks 10:95–400, 1997.
322. R Srihari. Computational models for integrating linguistic and visual information: a survey. Artif Intell Rev 8:349–369, 1994–1995.
323. A Stolcke. Learning feature-based semantics with simple recurrent networks. TR-90-015, International Computer Science Institute, Berkeley, 1990.

324. A Stolcke. Syntactic category formation with vector space grammars. In: Proceedings of the Thirteenth Annual Conference of the Cognitive Science Society. Lawrence Erlbaum, 1991.
325. M St John, J McClelland. Learning and applying contextual constraints in sentence comprehension. Artif Intell 46:217–257, 1990.
326. M St John, J McClelland. Parallel constraint satisfaction as a comprehension mechanism. In: R Reilly, N Sharkey, eds. Connectionist Approaches to Natural Language Processing. Hillsdale, NJ: Lawrence Erlbaum, 1992.
327. G Sun, C Giles, H Chen, Y Lee. The neural network pushdown automaton: model, stack, and learning simulations, UMIACS-TR-93-77, Institute for Advanced Computer Studies, University of Maryland, 1993.
328. R Sun. Beyond associative memories: logics and variables in connectionist networks. Inform Sci 70, 1992.
329. R Sun. Logics and variables in connectionist models: a brief overview. In: V Honavar, L Uhr, eds. Artificial Intelligence and Neural Networks: Steps Toward Principled Integration. New York: Academic Press, 1994.
330. R Sun. A two-level hybrid architecture for structuring knowledge for commonsense reasoning. In: R Sun, L Bookman, eds. Computational Architectures Integrating Neural and Symbolic Processes. Kluwer, 1995.
331. R Sun. Connectionist models of reasoning. In: O Omidvar, C Wilson, eds. Progress in Neural Networks. Vol 5. Ablex Publishing, 1997.
332. R Sun, F Alexandre. Connectionist-Symbolic Integration. From Unified to Hybrid Approaches. Hillsboro, NJ: Lawrence Erlbaum, 1997.
333. R Sun, L Bookman. Computational Architectures Integrating Neural and Symbolic Processes. Kluwer, 1995.
334. W Tabor, C Juliano, M Tanenhaus. Parsing in a dynamical system: an attractor-based account of the interaction of lexical and structural constraints in sentence processing. Lang Cognit Processes 12:211–271, 1997.
335. J Taylor. The historical background. In: E Fiesler, R Beal, eds. Handbook of Neural Computation. Oxford University Press, 1997.
336. S Thorpe. Localized versus distributed representations. In: M Arbib, ed. The Handbook of Brain Theory and Neural Networks. Cambridge, MA: MIT Press, 1995.
337. S Thrun. Explanation-Based Neural Network Learning: A Lifelong Learning Approach. Dordrecht: Kluwer Academic, 1996.
338. S Thrun. Learning one more thing. TR-CMU-CS-94-184, Carnegie Mellon University, 1994.
339. S Thrun. Lifelong learning: a case study. TR-CMU-CS-95-208, Carnegie Mellon University, 1995.
340. S Thrun, L Pratt. Learning To Learn. Dordrecht: Kluwer Academic, 1997.
341. S Thrun. Extracting rules from artificial neural networks with distributed representations. In: G Tesauro, D Touretzky, T Leen, eds. Advances in Neural Information Processing Systems 7. Morgan Kaufmann, 1995.
342. P Tino, B Horne, C Giles. Finite state machines and recurrent neural networks—automata and dynamical systems approaches. UMIACS-TR-95-1, Institute for Advanced Computer Studies, University of Maryland, 1995.
343. D Touretzky. Connectionism and compositional semantics. In: J Barnden, J Pollack, eds. Advances in Connectionist and Neural Computation Theory 1: High-Level Connectionist Models. Ablex, 1991.
344. D Touretzky, D Pomerleau. Reconstructing physical symbol systems. Cogn Sci 18:345–353, 1994.

345. G Towell, J Shavlik. Using knowledge-based neural networks to refine roughly-correct information. In: S Hanson, G Drastal, R Rivest. Computational Learning Theory and Natural Learning Systems. Cambridge, MA: MIT Press, 1994.
346. F Tsung, G Cottrell. Phase-space learning in recurrent networks. Technical Report CS93-285, Dept. of Computer Sciences and Engineering, University of California, San Diego, 1993.
347. T van Gelder. Classical questions, radical answers: connectionism and the structure of mental representations. In: T Horgan, J Tieson, eds. Connection and the Philosophy of Mind. Dordrecht: Kluwer, 1991.
348. T van Gelder. Compositionality: a connectionist variation on a classical theme. Cogn Sci 14:355–384, 1990.
349. T van Gelder. Defining "distributed representation". Connect Sci 4:175–191, 1992.
350. T van Gelder. Why distributed representation is inherently non-symbolic. In: G Dorffner, ed. Konnektionismus in Artificial Intelligence und Kognitionsforschung. Springer, 1990.
351. T van Gelder. The dynamical hypothesis in cognitive science. Behav Brain Sci 21, 1998, 1–14.
352. T van Gelder, R Port. Beyond symbolic: toward a Kama-Sutra of compositionality. In: V Honavar, L Uhr, eds. Artificial Intelligence and Neural Networks: Steps Toward Principled Integration, New York: Academic Press, 1994.
353. J van Leeuwen. Handbook of Theoretical Computer Science. Vol A: Algorithms and Complexity. Elsevier, 1990.
354. J van Leeuwen. Handbook of Theoretrical Computer Science. Vol. B: Formal Models and Semantics. Elsevier, 1990.
355. V Vapnik. Learning and generalization: theoretical bounds. In: M Arbib, ed. The Handbook of Brain Theory and Neural Networks, Cambridge, MA: MIT Press, 1995.
356. P Verschure. Taking connectionism seriously: the vague promise of subsymbolism and an alternative. Proceedings of the Fourteenth Annual Conference of the Cognitive Science Society, 1992.
357. P Verschure. Connectionist explanation: taking positions in the mind-brain dilemma. In: G Dorffner, ed. Neural Networks and New Artificial Intelligence. Thomson Computer Press, 1997.
358. G Wahba. Generalization and regularization in nonlinear learning systems. In: M Arbib, ed. The Handbook of Brain Theory and Neural Networks. Cambridge, MA: MIT Press, 1995.
359. D Waltz, J Pollack. Massively parallel parsing: a strongly interactive model of natural language interpretation. Cogn Sci 9:51–74, 1985.
360. D Wang, B Yuwono. Incremental learning of complex temporal patterns. IEEE Trans Neural Networks 7:1465–1481, 1996.
361. X Wang, E Blum. Dynamics and bifurcation of neural networks. In: M Arbib, ed. The Handbook of Brain Theory and Neural Networks. Cambridge, MA: MIT Press, 1995.
362. R Watrous, G Kuhn. Induction of finite-state languages using second-order recurrent networks. Neural Comput 4:406–414.
363. J Weckerly, J Elman. A PDP approach to processing center-embedded sentences. Proceedings of the Fourteenth Annual Meeting of the Cognitive Science Society, Lawrence Erlbaum, 1992.
364. A Weigend, D Rumelhart. Weight elimination and effective network size. In: S Hansan, et al., eds. Computational Learning Theory and Natural Language Systems, MIT Press, 1994.
365. A Wells. Turing's analysis of computation and theories of cognitive architecture. Cogn Sci 22:269–294, 1998.

366. S Wermter. Learning semantic relationships in compound nouns with connectionist networks. Proceedings of the Eleventh Annual Conference of the Cognitive Science Society, 1989.
367. S Wermter. A hybrid symbolic/connectionist model for noun phrase understanding. In: N Sharkey, ed. Connectionist Natural Language Processing. Kluwer, 1992.
368. S Wermter. Hybrid Connectionist Natural Language Processing. Chapman & Hall, 1995.
369. S Wermter, R Hannuschka. A connectionist model for the interpretation of metaphors. In: G Dorffner, ed. Neural Networks and a New Artificial Intelligence. Thomson Computer Press, 1997.
370. S Wermter, W Lehnert. Noun phrase analysis with connectionist networks. In: R Reilly, N Sharkey. Connectionist Approaches to Natural Language Processing. Hillside, NJ: Lawrence Erlbaum, 1992. (In: Neural Networks and Artificial Intelligence. Thomson computer Press, 1997.)
371. S Wermter, E Riloff, G Scheler. Connectionist, Statistical, and Symbolic Approaches to Learning for Natural Language Processing. Springer, 1996.
372. H White, K Hornik, M Stinchcombe. Multilayer feedforward networks are universal approximators. Neural Networks 2:359–366, 1989.
373. H White, M Stinchcombe. Universal approximation using feedforward networks with non-sigmoid hidden layer activation functions. Proceedings of the International Joint Conference on Neural Networks, Washington DC. IEEE Press, 1989.
374. H White. Artificial Neural Networks: Approximation and Learning Theory. Boston: Blackwell, 1992.
375. H White. Learning in artificial neural networks: a statistical perspective. Neural Comput 1:425–464, 1989.
376. R Williams, D Zipser. A learning algorithm for continually-running fully recurrent neural networks. Neural Comput 1:270–280, 1989.
377. W Winiwarter, E Schweighofer, D Merkl. Knowledge acquisition in concept and document spaces by using self-organizing neural networks. In: S Wermter, E Riloff, G Scheler, eds. Connectionist, Statistical, and Symbolic Approaches to Learning for Natural Language Processing. Springer, 1996.
378. P Wyard, C Nightingale. A single layer higher order neural net and its application to context free grammar recognition. Connect Sci 2:347–370, 1990.
379. Z Zeng, R Goodman, P Smyth. Learning finite state machines with self-clustering recurrent networks. Neural Comput 5:976–990, 1993.
380. Z Zeng, R Goodman, P Smyth. Discrete recurrent neural networks for grammatical inference. IEEE Trans Neural Networks 5:320–330, 1994.

28

Knowledge Representation*

SIMON HAYKIN

McMaster University, Hamilton, Ontario, Canada

The primary characteristics of *knowledge representation* are twofold: (a) what information is actually made explicit; and (b) how the information is physically encoded for subsequent use. By the very nature of it, therefore, knowledge representation is goal-directed. In real-world applications of intelligent machines, it can be said that a good solution depends on a good representation of knowledge [1,9]. So it is with neural networks, representing a special class of intelligent machines. Typically, however, the possible forms of representation from the inputs to internal network parameters are highly diverse, which tends to make the development of a satisfactory solution by means of a neural network a real design challenge.

A major task for a neural network is to learn a model of the world (environment) in which it is embedded and to maintain the model sufficiently consistent with the real world that it achieves the specified goals of the application of interest. Knowlege of the world consists of two kinds of information:

1. Known world state represented by facts about what is and what has been known; this form of knowledge is referred to as *prior information*.
2. Observations (measurements) of the world, obtained by means of sensors designed to probe the environment in which the neural network is supposed to operate. Ordinarily, these observations are inherently noisy, being subject to errors caused by sensor noise and system imperfections. In any

*This article is reproduced, with permission, from S Haykin. *Neural Networks: A Comprehensive Foundation*. Upper Saddle River, NJ: Prentice–Hall.

event, the observations so obtained provide the pool of information from which the examples used to train the neural network are drawn.

Each *example* consists of an input–output pair: an *input signal* and the corresponding *desired response* for the neural network. Thus, a set of examples represents knowledge about the environment of interest. Consider, for example, the *handwritten digit recognition problem*, in which the input consists of an image with black or white pixels, and with each image representing one of ten digits that are well separated from the background. In this example, the direct response is defined by the "identity" of the particular digit the image of which is presented to the network as the input signal. Typically, the set of examples used to train the network consists of a large variety of handwritten digits that are representative of a real-world situation. Given such a set of examples, the design of a neural network may proceed as follows:

1. An appropriate architecture is selected for the neural network, with an input layer consisting of source nodes equal in number to the pixels of an input image, and an output layer consisting of ten neurons (one for each digit). Subset of examples is then used to train the network by means of a suitable algorithm. This phase of the network design is called learning.
2. The recognition performance of the trained network is tested with data that has not been used before. Specifically, an input image is presented to the network, but this time it is not told the identity of the digit to which that particular image belongs. The performance of the network is then addressed by comparing the digit recognition reported by the network with the actual identity of the digit in question. This second phase of the network design is called *generalization*, a term borrowed from psychology.

Herein lies a fundamental difference between the design of a neural network and that of its classic information-processing (IP) counterpart (pattern classifier). In the latter case, we usually proceed by first formulating a mathematical model of environmental observations, validating the models with real data, and then building the design on the basis of the model. In contrast, the design of a neural network is based directly on real data, with the *data set being permitted to speak for itself.* Thus, the neural network provides not only an implicit model of the environment in which it is embedded, but also performs the information-processing function of interest.

The examples used to train a neural network may consist of both *positive* and *negative* examples. For instance, in a passive sonar detection problem, positive examples pertain to input training data that contain the target of interest (e.g., submarine). Now in a passive sonar environment, the possible presence of marine life in the test data is known to cause occasional false alarms. To alleviate this problem, negative examples (e.g., echoes from marine life) are included in the training data to teach the network not to confuse marine life with the target.

In a neural network of specified architecture, knowledge representation of the surrounding environment is defined by the values taken on by the free parameters (i.e., synaptic weights and thresholds) of the network. The form of this knowledge representation constitutes the very design of the neural network; therefore it holds the key to its performance.

The subject of knowledge representation inside an artificial neural network is, however, very complicated. The subject becomes even more compounded when we have multiple sources of information activating the network, and these sources interact with each other. Our present understanding of this important subject is indeed the weakest link in what we know about artificial neural networks. Nevertheless, there are four rules for knowledge representation that are of a general commonsense nature [2]. The four rules are described as follows:

Rule 1. Similar inputs from similar classes should usually produce similar representations inside the network; therefore, they should be classified as belonging to the same category.

There are a plethora of measures for determining the "similarity" between inputs. A commonly used measure of similarity is based on the concept of Euclidian distance. To be specific, let x_i denote an N-by-1 real-valued vector

$$\mathbf{x}_{i1} = [x_{i1}, x_{i2}, \ldots, x_{iN}N]^T \qquad \{1\}$$

for which all of the elements are real; the superscript T denotes matrix *transcription*. The vector $\mathbf{x}_i$ defines a point in an N-dimensional space called *euclidean space* and denoted by $\mathbf{R}^N$. The *euclidean distance* between a pair of N-by-1 vectors x_i and x_j *is* defined by

$$d_{ij} = \|\mathbf{x}_i - \mathbf{x}_j\| = \left(\sum_{n=1}^{N}(x_{in} - x_{jn})^2\right)^{1/2} \qquad \{2\}$$

where x_{in} and x_{jn} are the nth elements of the input vectors $\mathbf{x}_i$ and $\mathbf{x}_j$ respectively. Correspondingly, the similarity between the inputs represented by the vectors $\mathbf{x}_i$ and $\mathbf{x}_j$ is defined as the *reciprocal* of the euclidean distance d_{ij}. The closer the individual elements of the input vectors $\mathbf{x}_i$ and $\mathbf{x}_j$ are to each other, the smaller will the euclidean distance d_{ij} be and therefore, the greater will be the similarity between the vectors $\mathbf{x}_i$ and $\mathbf{x}_j$. Rule 1 states that if the vectors $\mathbf{x}_i$ and $\mathbf{x}_j$ are similar, then they should be assigned to the same category (class).

Another measure of similarity is based on the idea of a *dot product* or *inner product* that is also borrowed from matrix algebra. Given a pair of vectors $\mathbf{x}_i$ and $\mathbf{x}_j$ of the same dimension, their inner product is $\mathbf{x}_i^T\mathbf{x}_j$, written in expanded form as follows

$$\mathbf{x}_i^T\mathbf{x}_j = \sum_{n=1}^{N} x_{in}x_{jn} \qquad \{3\}$$

The inner product $\mathbf{x}_i^T\mathbf{x}_j$ divided by $\mathbf{x}_i\|\|\mathbf{x}_j\|$ is the cosine of the angle subtended between the vectors $\mathbf{x}_i$ and $\mathbf{x}_j$.

The two measures of similarity defined here are indeed intimately related to each other, as illustrated in Fig. 1. The euclidean distance $\|\mathbf{x}_i - \mathbf{x}_j\|$ between the vectors $\mathbf{x}_i$ and $\mathbf{x}_j$ is portrayed as the length of the line joining the tips of these two vectors, and their inner product $\mathbf{x}_i^T\mathbf{x}_j$ is portrayed as the "projection" of the vector $\mathbf{x}_i$ onto the vector $\mathbf{x}_j$. Figure 1 clearly shows that the smaller the euclidean distance $\|\mathbf{x}_i - \mathbf{x}_j\|$ and, therefore, the more similar the vectors $\mathbf{x}_i$ are $\mathbf{x}_j$ are, the larger will be the inner product $\mathbf{x}_i^T\mathbf{x}_j$.

In signal-processing terms, the inner product $\mathbf{x}_i^T\mathbf{x}_j$ may be viewed as a *cross-correlation function*. Recognizing that the inner product is a scalar, we may state that

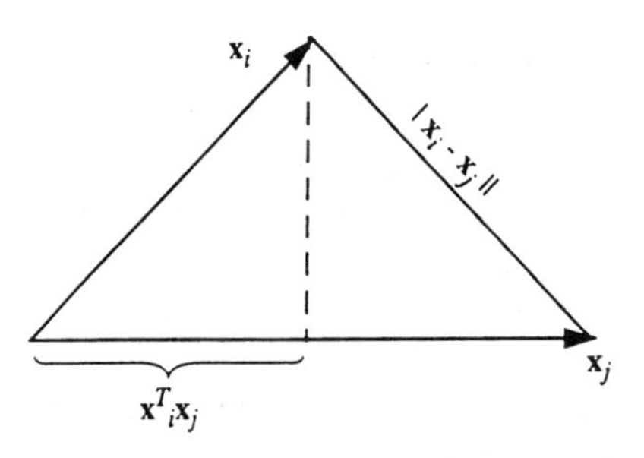

Fig. 1 Illustrating the relation between inner product and euclidean distance as measures of similarity between patterns.

the more positive the inner product $\mathbf{x}_i^T\mathbf{x}_j$ is the more similar (i.e., correlated) the vectors x_i and x_j are to each other. The cross-correlation function is ideally suited for echo location in radar and sonar systems. Specifically, by cross-correlating the echo from a target with a replica of the transmitted signal and finding the peak value of the resultant function, it is a straightforward matter to estimate the arrival-time of the echo. This is the standard method for estimating the target's range (distance).

Rule 2. Items to be categorized as separate classes should be given widely different representations in the network.

The second rule is the exact opposite of Rule 1.

Rule 3. If a particular item is important, then there should be a large number of neurons involved in the representation of that item in the network.

Consider, for example, a radar application involving the detection of a target (e.g., aircraft) in the presence of clutter (i.e., radar reflections from undesirable targets, such as buildings, trees, and weather formations). According to the *Neyman–Pearson criterion*, the probability of detection (i.e., the probability of deciding that a target is present when it is) is maximized, subject to the constraint that the probability of a false alarm (i.e., the probability of deciding that a target is present when it is not) does not exceed a prescribed value [8]. In such an application, the actual presence of a target in the received signal represents an important feature of the input. Rule 3, in effect, states that there should be many neurons involved in making the decision that a target is present when it actually is. By the same token, there should be a very large number of neurons involved in making the decision that the input consists of clutter only when it actually is. In both situtations the larger the number of neurons assures a high degree of accuracy in decision making and tolerance relative to faulty neurons.

Rule 4. Prior information and invariances should be built into the design of a neural network, thereby simplifying the network design by not having to learn them.

Rule 4 is particularly important, because proper adherence to it results in a neural network with a *specialized (restricted) structure*. This is highly desirable for several reasons:

1. Biological visual auditory networks are known to be very specialized.
2. A neural network with specialized structure usually has a much smaller number of free parameters available for adjustment than a fully connected network. Consequently, the specialized network requires a smaller data set for training, learns faster, and often generalizes better.
3. The rate of information transmission through a specialized network (i.e., the network throughput) is accelerated.
4. The cost of building a specialized network is reduced by virtue of its smaller size, compared with its fully connected counterpart.

An important issue that has to be addressed is how to develop a specialized structure by building prior information into its design. Unfortunately, there are as yet no well-defined rules for doing this; rather, we have some ad hoc procedures that are known to yield useful results. To be specific, again consider the example involving the use of a multilayered feedforward network for *handwritten digit recognition* that is a relatively simple human task, but is not an easy machine-vision task, and that has great practical value [4]. The input consists of an image with black or white pixels and ten digits that are well separated from the background. In this example, the prior information is that an image is *two-dimensional* and has a strong *local structure*. Thus, the network is specialized by constraining the synaptic connections in the first few layers of the network to be local; that is, the network is chosen to be locally connected. Additional specialization may be built into the network design by examining the purpose of a *feature detector*, which is to reduce the input data by extracting certain "features" that distinguish the image of one digit from that of another. In particular, if a feature detector is useful in one part of the image, then it is also likely to be useful in other parts of the image. The reason for saying this is that the salient features of a distorted character may be displaced slightly from their position in a typical character. To solve this problem, the input image is scanned with a single neuron that has a local receptive field, and the synaptic weights of the neuron are stored in corresponding locations in a layer called a *feature map*. This operation, illustrated in Fig. 2. Let $\{w_{ij} | i = 0, 1, \ldots, p-1\}$ denote the set of synaptic weights pertaining to neuron j. The *convolution* of the "kernel" represented by this set of synaptic weights and an input pixel denoted by $\{x_i\}$ is defined by the sum

$$y_j = \sum_{i=0}^{p-1} w_{ji} x_i \qquad \{4\}$$

Such a network is sometimes called a *convolutional network*. Thus, the overall operation performed in Fig. 2 is equivalent to the convolution of a small-sized kernel (represented by the set of synaptic weights) of the neuron and the input image, which is then followed by soft-limiting (squashing) performed by the activation function of the neuron. The overall operation is performed in parallel by implementing the feature map in a plane of neurons, the weight vectors of which are constrained to be equal. In other words, the neurons of a feature map are constrained to perform the same mathematical operation on different parts of the image. Such a technique is called *weight-sharing*.[1] Weight-sharing also has a profitable side effect:

[1]It appears that the weight-sharing technique was originally described in Rumelhart et al. [5].

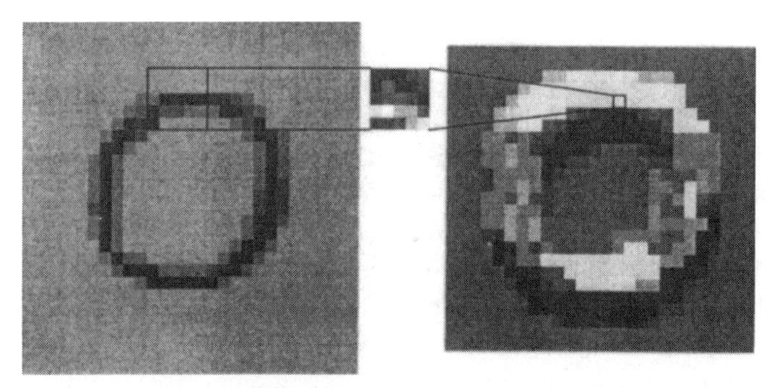

Fig. 2 Input image (left), weight vector (center), and resulting feature map (right). The feature map is obtained by scanning the input image with a single neuron that has a local receptive field, as illustrated: white represents −1, black represents +1. (From Ref. 4.)

The number of free parameters in the network is reduced significantly, because a large number of neurons in the network are constrained to share the same set of synaptic weights.

In summary, prior information may be built into the design of a neural network by using a combination of two techniques: (a) *restricting the network architecture* through the use of local connections, and (b) *constraining the choice of synaptic weights* by the use of weight sharing. Naturally, the manner in which these two techniques are exploited is, in practice, strongly influenced by the application of interest. In a more general context, the development of well-defined procedures for the use of prior information is an open problem. Prior information pertains to one part of Rule 4; the remaining part of the rule involves the issue of invariances, which is considered next.

When an object of interest rotates, the image of the object as perceived by an observer usually changes in a corresponding way. In a coherent radar that provides amplitude as well as phase information about its surrounding environment, the echo from a moving target is shifted in frequency owing to the Doppler effect that arises because of the radial motion of the target in relation to the radar. The utterance from a person may be spoken in a soft or loud voice, and yet again in a slow or quick manner. To build an object recognition system—a radar target recognition system and a speech recognition system for dealing with these phenomena, respectively.—the system must be capable of coping with a range of *transformations* of the observed signal [3]. Accordingly, a primary requirement of pattern recognition is to design a classifier that is invariant to such transformations. In other words, a class estimate represented by an output of the classifier must not be affected by transformations of the observed signal applied to the classifier input.

There are at least three techniques for rendering classifier-type neural networks invariant to transformations [3]:

1. *Invariance by structure.* Invariance may be imposed on a neural network by structuring its design appropriately. Specifically, synaptic connections between the neurons of the network are created such that transformed versions of the same input are formed to produce the same output. Consider, for example, the classification of an input image by a neural

network that is required to be independent of in-plane rotations of the image about its center. We may impose rotational invariance on the network structure as follows. Let w_{ji} be the synaptic weight of neuron j connected to pixel i in the input image. If the condition $w_{ji} = w_{jk}$ is enforced for all pixels i and k that lie at equal distances from the center of the image, then the neural network is invariant to in-plane rotations. However, to maintain rotational invariance, the synaptic weight w_{ji} has to be duplicated for every pixel of the input image at the same radial distance from the origin. This points to a shortcoming of invariance by structure: The number of synaptic connections in the neural network becomes prohibitively large, even for images of moderate size.

2. *Invariance by training.* A neural network has a natural ability for pattern classification. This ability may be exploited directly to obtain transformation invariance as follows. The network is trained by presenting it with a number of different examples of the same object, with the examples being chosen to correspond to different transformations (i.e., different aspect views) of the object. Provided that the number of examples is sufficiently large, and if the network is trained to learn to discriminate the different aspect views of the object, we may then expect the network to correctly generalize transformations other than those shown to it. However, from an engineering perspective, invariance by training has two disadvantages. First, when a neural network has been trained to recognize an object in an invariant fashion relative to known transformations, it is not obvious that this training will also enable the network to recognize other objects of different classes invariantly. Second, the computational demand imposed on the network may be too severe to cope with, especially if the dimensionality of the feature space is high.
3. *Invariant features space.* The third technique of creating an invariant classifier-type neural network is illustrated in Fig. 3. It rests on the basic premise that it is possible to extract *features* that characterize the essential information content of an input data set, and that are invariant to transformations of the input. If such features are used, then, the network as a classifier is relieved from the burden of having to delineate the range of transformations of an object with complicated decision boundaries. Indeed, the only differences that may arise between different instances of the same object are due to unavoidable factors, such as noise and occlusion. The use of an invariant feature space offers three distinct advantages: (a) The number of features applied to the network may be reduced to realistic levels; (b) the requirements imposed on network design are relaxed; and (c) invariance for all objects relative to known transforma-

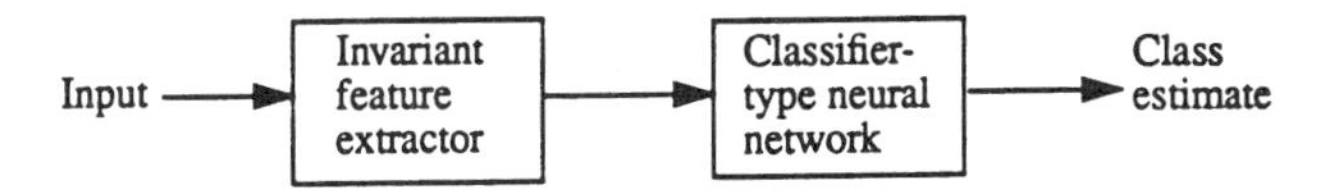

Fig. 3 Block diagram of an invariant feature space-type system.

tions are assured [3]; however, this approach requires prior knowledge of the problem.

In conclusion, the use of an invariant-feature space as described herein may offer the most suitable technique for neural classifiers.

To illustrate the idea of invariant-feature space, consider the example of a coherent radar system used for air surveillance, where the targets of interest include aircraft, weather systems, flocks of migrating birds, and ground objects. The radar echoes from these targets possess different spectral characteristics; moreover, experimental studies show such radar signals can be modeled fairly closely as an *autoregressive (AR) process* of moderate order; an AR model is a special form of regressive model defined by Eq. {5},

$$x(n) = \sum_{i=1}^{M} a_i x(n-i) + e(n) \qquad \{5\}$$

where the $\{a_i | i = 2, 2, \ldots, M\}$ are the *AR coefficients, M* is the *model order,* $x(n)$ is the input, and $e(n)$ is the *error* described as white noise. Basically, the AR model of Eq. {5} is represented by a *tapped-delay–line filter* as illustrated in Fig. 4(a) for $M = 2$. Equivalently, it may be represented by a *lattice filter* as shown in Fig. 4(b), the coefficients of which are called *reflection coefficients.* There is a one-to-

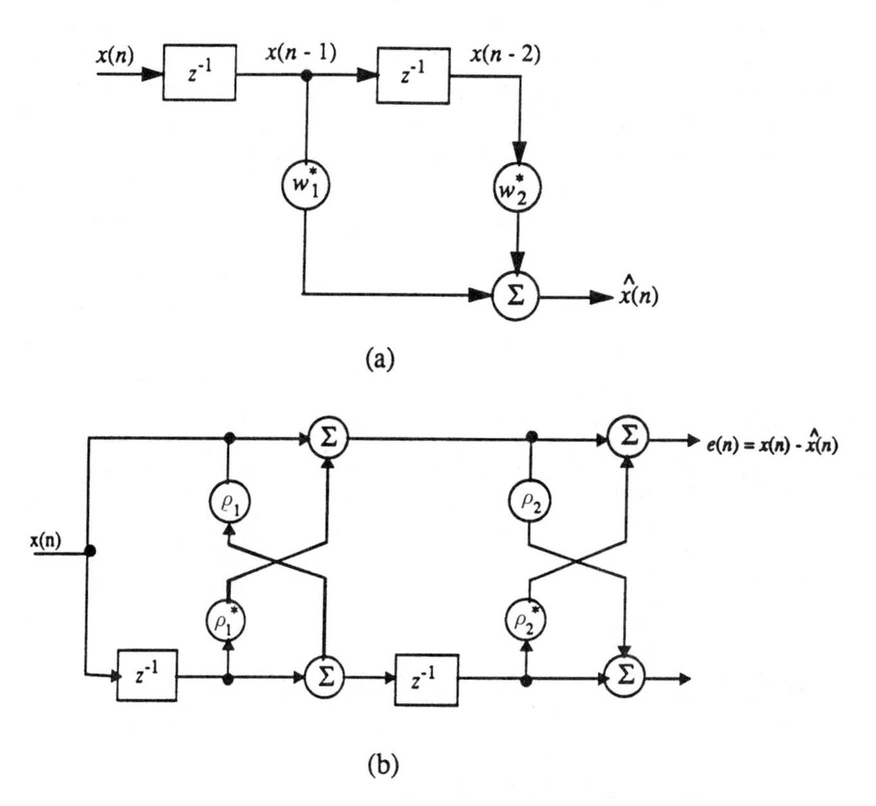

Fig. 4 Autoregressive model of order 2: (a) tapped-delay–line model; (b) lattice filter model (the asterisk denotes complex conjugation).

one correspondence between the AR coefficients of the model in Fig. 4(a) and the reflection coefficients of that in Fig. 4(b). The two models depicted in Fig. 4 assume that the input $x(n)$ is complex valued, as in a coherent radar, in which the AR coefficients and the reflection coefficients are all complex-valued; the asterisk in Fig. 4 signifies *complex conjugation.* For now, it suffices to say that the coherent radar data may be described by a set of *autoregressive coefficients* or, equivalently, by a corresponding set of *reflection coefficients.* The latter set has a computational advantage in that efficient algorithms exist for their computation directly from the input data. The feature-extraction problem, however, is complicated because moving objects produce varying Doppler frequencies that depend on their radial velocities measured relative to the radar, and they tend to obscure the spectral content of the reflection coefficients as feature discriminants. To overcome this difficulty, we must build *Doppler invariance* into the computation of the reflection coefficients. The phase angle of the first reflection coefficient turns out to be equal to the Doppler frequency of the radar signal. Accordingly, Doppler frequency normalization is applied to all coefficients to remove the mean Doppler shift. This is done by defining a new set of reflection coefficients $\{\rho'_{\mathrm{m}}\}$ related to the set of reflection coefficients $\{\rho_{\mathrm{m}}\}$ computed from the input data as follows:

$$\rho'_m = \rho_m e^{-jm\theta}, m = 1, 2, \ldots, M \qquad \{6\}$$

where θ is the phase angle of the first reflection coefficient and M is the order of the AR model. The operation described in Eq. {6} is referred to as *heterodyning.* A set of *Doppler-invariant radar features* is thus represented by the normalized reflection coefficients $\rho'_1, \rho'_2 \ldots, \rho'_{\mathrm{M}}$, with ρ'_1 being the only real-valued coefficient in the set. The major categories of radar targets of interest in air surveillance are weather, birds, aircraft, and ground. The first three targets are moving, whereas the last one is not. The heterodyned spectral parameters of radar echoes from ground have echoes similar in characteristic to those from aircraft. Also, a ground echo can be discriminated from an aircraft echo by virtue of its small Doppler shift. Accordingly, the radar classifier includes a postprocessor, as in Fig. 5, that operates on the classified results (encoded labels) to identify the ground class. Thus the *preprocessor* in Fig. 5 takes care of Doppler shift-invariant feature extraction at the classifier input, whereas the *postprocessor* uses the stored Doppler signature to distinguish between aircraft and ground returns.

A much more fascinating example of knowledge representation in a neural network is found in the biological sonar system of echo-locating bats. Most bats use *frequency-modulated* ("FM" or "chirp") signals for the purposes of acoustic imaging; in an FM signal the instantaneous frequency of the signal varies with time. Specifically, the bat uses its mouth to broadcast short-duration FM sonar

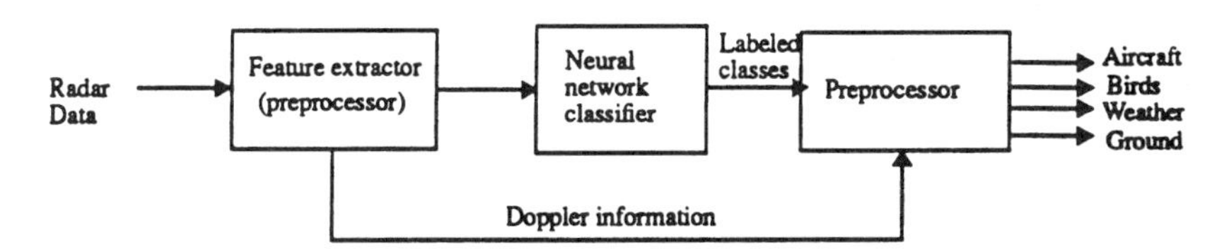

Fig. 5 Doppler shift-invariant classifier of radar signals.

signals and its auditory system as the sonar receiver. Echoes from targets of interest are represented in the auditory system by the activity of neurons that are selective to different combinations of acoustic parameters. There are three principal neural dimensions of the bat's auditory representation [6,7]:

1. *Echo frequency*, which is encoded by "place" originating in the frequency map of the cochlea; it is preserved throughout the entire auditory pathway as an orderly arrangement across certain neurons tuned to different frequencies.
2. *Echo* amplitude, which is encoded by other neurons with different dynamic ranges; it is manifested both as amplitude tuning and as the number of discharges per stimulus.
3. *Echo delay*, which is encoded through neural computations (based on cross-correlation) that produce delay-selective responses; it is manifested as target-range tuning.

The two principal characteristics of a target echo for image-forming purposes are *spectrum* for target shape, and *delay* for target range. The bat perceives "shape" in terms of the arrival time of echoes from different reflecting surfaces (glints) within the target. For this to occur, frequency information in the echo spectrum is converted into estimates of the *time* structure of the target. Experiments conducted by Simmons and co-workers on the large brown bat, *Eptesicus fuscus*, critically identify this conversion process as consisting of parallel time domain and frequency-to-time domain transforms, the converging outputs of which create the common delay or range axis of a perceived image of the target. It appears that the unity of the bat's perception is due to certain properties of the transforms themselves, despite the separate ways in which the auditory time representation of the echo delay and frequency representation of the echo spectrum are initially performed. Moreover, feature invariances are built into the sonar image-forming process such that they make it essentially independent of the motion of the target and the bat's own motion.

Returning to the main theme of this section; namely, that for knowledge representation in a neural network the issue of knowledge representation is intimately related to that of network architecture. Unfortunately, there is no well-developed theory for optimizing the architecture of a neural network required to interact with an environment of interest or for evaluating the way in which changes in the network architecture affect the representation of knowledge inside the network. Indeed, satisfactory answers to these issues are usually found through an exhaustive experimental study, with the designer of the neural network becoming an essential part of the structural learning loop.

REFERENCES

1. S Amarel. On representations of problems of reasoning about actions. In: D Mchie, ed. Machine Intelligence. New York: Elsevier 1968, pp 131–171.
2. JA Anderson. General introduction. In: JA Anderson, E Rosenfeld, eds. Neurocomputing: Foundations of Research. Cambridge, MA: MIT Press, 1988, xiii–x3d.
3. E Barnard, D Casasent. Invariance and neural nets. IEEE Trans Neural Networks 2, 1991, pp 498–508.

4. Y LeCun, B Boser, JA Denker, D Henderson, RE Howard, W Hubbard, LD Jacket. Handwritten digit recognition with a back-propagation network. In: DS Touretzky, ed. Advances in Neural Information Processing Systems 2. San Mateo, CA: Morgan Kaufmann, 1990, pp 396–402.
5. DE Rumelhart, GE Hinton, RJ Williams. Learning internal representations by error propagation. In: DE Rumelhart, JL McClelland, eds.), Vol. 1. Parallel Distributed Processing: Explorations in the Microstructure of Cognition. Cambridge, MA: MIT Press, 1986.
6. JA Simmons. nme-Frequency transforms and images of targets in the sonar of bats. Princeton lectures on Biophysics. NEC Research Institute, Princeton, NJ, 1991.
7. JA Simmons, PA Saillant. Auditory deconvolution in echo processing by bats. Computational Neuroscience Symposium, Indiana University – Purdue University at Indianapolis, 1992, pp 15–32.
8. HL Van Trees. Detection, Estimation and Modulation Theory, Part I. New York: Wiley, 1968.
9. WA Woods. Important issues in knowledge representation. Proceedings of the IEEE 74. 1986, 1322–1334.

29

Grammar Inference, Automata Induction, and Language Acquisition

RAJESH G. PAREKH

Allstate Research and Planning Center, Menlo Park, California

VASANT HONAVAR

Iowa State University, Ames, Iowa

1. INTRODUCTION

How children acquire and use the language of the community is a fundamental problem that has attracted the attention of researchers for several decades. A plausible theory of language acquisition would have to account for a wide range of empirical observations. For example, it would have to explain how children are able to acquire a language on the basis of a finite number of sentences that they encounter during their formative years. Consequently, natural language acquisition has been, and continues to be, a major focus of research [48,75]. The past decade has seen significant theoretical as well as experimental advances in the study of natural language acquisition. Examples of some current developments include *bootstrapping hypotheses* and *constraint-based theories* [9], *optimality theory* [80,89], and *neural theory of language* [27,28]. Empirical evidence from these results argues in favor of language learnability to address some of the key problems encountered in child language acquisition.

Besides obtaining a better understanding of natural language acquisition, interest in studying formal models of language learning stems also from the numerous practical applications of language learning by machines. Research in *instructible robots* [19] *and intelligent software agents and conversational interfaces* [6] is geared toward the design of agents that can understand and execute verbal instructions given in a natural language (such as English) or some restricted subset of a natural language. For example, one might describe the steps needed for cooking dinner in terms of a sequence of instructions: "*Open the microwave door*"—"*Insert the pack-*

aged food"—"*Set the controls*"—"*Push the cook button,*" and so on. Furthermore, the human may want to summarize this entire sequence of actions by the phrase "*Cook dinner*". The robot should learn the foregoing sequence of actions and identify it with the phrase "*Cook dinner*" [19]. A satisfactory solution to this problem will necessarily have to build on recent advances in a number of areas in artificial intelligence, including natural language processing, machine learning, machine perception, robotics, planning, knowledge representation, and reasoning.

Formal language models are extensively used in syntactic or linguistic pattern classification systems [29,30]. The structural interrelations among the linguistic pattern attributes are easily captured by representing the patterns as strings (a collection of syntactic symbols). Learning a set of rules for generating the strings that belong to the same class or category (language) provides for a simple yet adequate syntactic or linguistic classification system. Linguistic pattern recognition has been put to several practical uses, including *speech recognition, discovery of patterns in biosequences, image segmentation, interpretation of electrocardiograms (ECG), handwriting recognition, recognition of seismic signals*, and the like [8,30,57]. The issues and practical difficulties associated with formal language learning models can provide useful insights for the development of language understanding systems. Several key questions in natural language learning, such as the role of prior knowledge, the types of input available to the learner, and the influence of semantic information on learning the syntax of a language, can possibly be answered by studying formal models of language acquisition.

Research in language acquisition has benefited from advances in several disciplines, including cognitive *psychology* [9], *linguistics* [15,16,75], *theoretical computer science* [42,55], *computational learning theory* [45,63], *artificial intelligence* [84], *machine learning* [48,61], and *pattern recognition* [30,57]. Psychological studies of language acquisition have explored the development of the children's cognitive faculties, the nature of stimuli available to them, the types of word combinations they form, and related questions. Studies in linguistics attempt (among other things), to study similarities and differences among the languages to characterize language properties that are universal (i.e., common to all languages). Theoretical computer science has taken a formal view of languages and constructed a hierarchy based on representational power of the different formal languages. Research in computational-learning theory has explored language learnability and has provided insights into questions such as whether it is possible, at least in principle, to learn a language from a set of sentences of the language; whether there are classes of languages that can be efficiently learned by computers and so on. Artificial intelligence has been concerned with the design of languages for communication among intelligent agents, deployment of instructible robots, and development of natural language-understanding systems. Studies in machine learning and pattern recognition have contributed to the development and application of algorithms for language learning.

The language-learning problem concerns the acquisition of the syntax or the grmamar (i.e., rules for generating and recognizing valid sentences in the language) and the semantics (i.e., the underlying meaning conveyed by each sentence) of a target language. Thus far, most of the work on computational approaches to language acquisition has focused on learning the syntax. This component of language acquisition is generally referred to as *grammatical inference*. To make the problem of syntax acquisition well-defined, it is necessary to choose an appropriate class of grammars (or equivalently languages) that is guaranteed to contain the unknown grammar. The classes of grammars such as *regular* or *context-free grammars* that belong to the Chomsky hierarchy of formal grammars [13,42] are often used to model the target grammar. The methods for grammar inference typically identify an unknown grammar (or an approximation of it) from a set of candidate hypotheses (e.g., the set of regular grammars defined over some alphabet). These algorithms are based on the availability of different types of information, such as positive examples, negative examples, the presence of a knowledgable teacher who answers queries and provides hints, and the availability of additional domain-specific prior knowledge.

The remainder of this chapter is organized as follows. The necessary terminology and definitions concerned with language learning are introduced in Sec. 2. A variety of approaches for learning regular grammars are studied in Sec. 3. Algorithms for learning stochastic regular grammars and hidden Markov models are described in Sec. 4. These are motivated by the need for robust methods for dealing with noisy learning scenarios (e.g., when examples are occasionally mislabeled) and the unavailability (in several practical application domains) of suitably labeled negative examples. The limited representational power of regular grammars has prompted researchers to explore methods for inference of more general classes of grammars, such as context-free grammars. Several approaches for learning context-free grammars are reviewed in Sec. 5. Section 6 concludes with a brief discussion of several interesting issues in natural language learning that have been raised by recent research in cognitive psychology, linguistics, and related areas.

2. LANGUAGES, GRAMMARS, AND AUTOMATA

Syntax acquisition is an important component of systems that learn to understand natural language. The syntax of a language is typically represented by grammar rules using which one may generate legal (grammatically correct) sentences of the language, or equivalently, decide whether a given sentence belong to the language or not. *Grammar inference* provides a tool for automatic acquisition of syntax. It is defined as the process of learning a target grammar from a set of labeled sentences (i.e., from examples of grammatically correct and, if available, examples of grammatically incorrect sentences). Formal languages were initially used to model natural languages to better understand the process of natural language learning [13]. Formal languages are associated with recognizing (or equivalently, generating) devices called automata. The task of grammar inference is often modeled as the task of automata induction wherein an appropriate automaton for the target language is learned from a set of labeled examples. A *formal grammar* is a 4-tuple $G = (V_N; V_T; P; S)$ where V_N is the set of nonterminals, V_T is the set of terminals, P is the set of production rules for generating valid sentences of the language, and $S \in V_N$ is a special symbol called the start symbol. The production rules are of the form $\alpha \rightarrow \beta$, where α, β are sentences over the alphabet $(V_N \cup V_T)$ and α contains at least one nonterminal. Valid sentences of the language are sequences of terminal symbols V_T and are obtained by repeatedly applying the production rules as shown in the following:

Example

Consider a simple grammar for declarative English sentences in Fig. 1[1]. These rules can be used to generate the sentence *The boy saw a ferocious tiger*, as shown in Fig. 2. Note that at each step the leftmost nonterminal is matched with the left-hand sides of the rules and is replaced by the right-hand side of one of the matching rules.

The *language* of a grammar is the set of all valid sentences that can be generated using rules of the grammar. Chomsky proposed a hierarchy of formal language grammars based on the types of restrictions placed on the production rules [13,42]. The simplest class of grammars in this hierarchy is the class of *regular grammars* that are recognized by *finite-state automata* (see Sec. 3.A). Regular grammars are limited in their representational power in that they cannot be used to describe languages such as *palindromes*, in which the grammar would be required to memorize specific intrasentential sequences to determine the validity of sentences. This limitation is overcome in the next-higher class of grammars in the hierarchy, the *context-free* grammars. The languages generated by context-free grammars are recognized by *pushdown automata*, which are simply *finite-state automata* augmented with a pushdown stack. Context-free grammars are adequate for several practical natural language-modeling tasks [19,73]. Context-sensitive grammars represent the next level of grammars in the hierarchy and unrestricted grammars (which place no restriction on the form of the production rules) complete the formal language hierarchy.

3. REGULAR GRAMMAR INFERENCE

The regular grammar inference task is defined as follows: Given a finite set of positive examples (sentences belonging to the language of the target grammar) and a finite (possibly empty) set of negative examples (sentences that do not belong to the language of the target grammar), identify a regular grammar G that is equivalent[2] to the target grammar G. An important design choice concerns the representation of the target. Regular grammars can be equivalently represented by a set of production rules, a regular expression, a deterministic finite-state automaton (DFA), or a nondeterministic finite-state automaton (NFA). Most work in regular grammar inference has chosen DFA as the representation of the target regular grammar. This is attributed to the following characteristics of DFAs: they are simple and easy to

$$
\begin{aligned}
S &\rightarrow NP\ VP \\
NP &\rightarrow AR\ N \mid AR\ AP \\
AP &\rightarrow AD\ N \\
VP &\rightarrow V\ NP \\
N &\rightarrow boy \mid tiger \mid fox \\
V &\rightarrow saw \\
AR &\rightarrow a \mid the \\
AD &\rightarrow ferocious \mid sly
\end{aligned}
$$

Fig. 1 Grammar for declarative English sentences.

[1]Note that a rule of the form $\alpha \rightarrow \beta|\gamma$ is a concise way of writing two separate rules $\alpha \rightarrow \beta$ and $\alpha \rightarrow \gamma$.
[2]Two grammars G1 and G2 are equivalent if their languages are exactly the same.

$$
\begin{aligned}
S &\rightarrow NP\ VP \\
&\rightarrow AR\ N\ VP \\
&\rightarrow The\ N\ VP \\
&\rightarrow The\ boy\ VP \\
&\rightarrow The\ boy\ V\ NP \\
&\rightarrow The\ boy\ saw\ NP \\
&\rightarrow The\ boy\ saw\ AR\ AP \\
&\rightarrow The\ boy\ saw\ a\ AP \\
&\rightarrow The\ boy\ saw\ a\ AD\ N \\
&\rightarrow The\ boy\ saw\ a\ ferocious\ N \\
&\rightarrow The\ boy\ saw\ a\ ferocious\ tiger
\end{aligned}
$$

Fig. 2 Generation of the sentence—*The boy saw a ferocious tiger.*

understand; there exists a unique minimum state DFA corresponding to any regular grammar; there exist efficient (polynomial time) algorithms for several operations (such as minimization of a DFA, determining the equivalence of two DFA, determining whether the language of one DFA is a superset of the language of another DFA and the like) that are frequently used in regular grammar inference methods [42].

A. Finite-State Automata

A *deterministic* finite-state automaton (DFA) is a quintuple $A = (Q, \delta, \Sigma, q_0, F)$ where Q is a finite set of states, Σ is the finite alphabet, $q_0 \in Q$ is the start state, $F \subseteq Q$ is the set of accepting states, and δ is the transition function $Q \times \Sigma \rightarrow Q$. $\delta(q, a)$ denotes the state reached when the DFA in state q reads the input letter a. A state $d_0 \in Q$ such that $\forall a \in \Sigma, \delta(d_0, a) = d_0$ is called a dead *state*. The extension of δ to handle input strings (i.e., concatenations of symbols in Σ) is standard and is denoted by δ^*. $\delta^*(q, \alpha)$ denotes the state reached from q on reading the string α. A string α is said to be accepted by a DFA if $\delta^*(q_0, \alpha) \in F$. Strings accepted by the DFA are said to be *positive examples* (or valid sentences) of the language of the DFA. The set of all strings accepted by a DFA A is its language, $L(A)$. Correspondingly, strings not accepted by the DFA are called *negative examples* (or invalid sentences) of the language of the DFA. The language of a DFA is called a regular language. DFA can be conveniently represented using state transition diagrams. Figure 3 shows the state transition diagram for a sample DFA. The start state q_0 is indicated by the

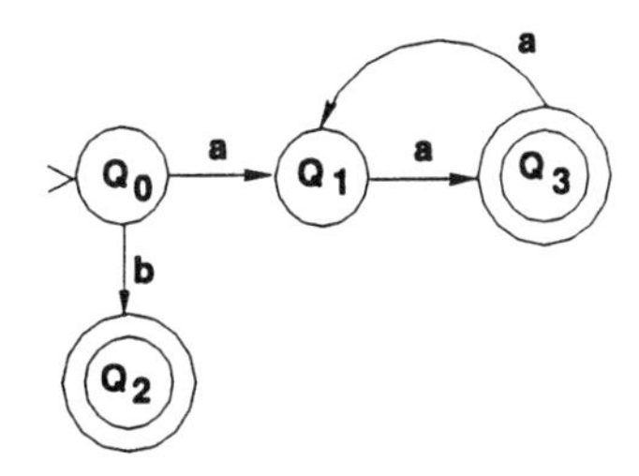

Fig. 3 Deterministic finite-state automaton.

symbol > attached to it. Accepting states are denoted using concentric circles. The state transition $\delta(q_i, a) = q_j$ for any letter $a \in \Sigma$ is depicted by an arrow labeled by the letter a from the state q_i to the state q_j.

A *nondeterministic finite automaton* (NFA) is defined just like the DFA except that the transition function δ defines a mapping from $Q \times \Sigma \rightarrow 2^Q$ (i.e., there can potentially be more than one transition out of a state on the same letter of the alphabet). In general, a finite state automaton (FSA) refers to either a DFA or a NFA. Given any FSA (A'), there exists a minimum state DFA also called the *canonical* DFA (A) such that $L(A) = L(A')$. Without loss of generality we will assume that the target DFA being learned is a canonical DFA. A labeled example $[\alpha, c(\alpha)]$ for A is such that $\alpha \in \Sigma^*$ and $c(\alpha) = +$ if $\alpha \in L(A)$ (i.e., α is a positive example) or $c(\alpha) = -$ if $\alpha \notin L(A)$ (i.e., α is a negative example). Thus $(a, -), (b, +), (aa, +), (aaab, -)$, and $(aaaa; +)$ are labeled examples for the DFA of Fig. 3. Given a set of positive examples denoted by S^+ and a set of negative examples denoted by S^-, we say that A is consistent with the *sample* $S = S^+ \cup S^-$ if it accepts all positive examples and rejects all negative examples.

B. Results in Regular Grammar Inference

Regular grammar inference is a hard problem in that regular grammars cannot be correctly identified from positive examples alone [36]. Moreover, there are no efficient learning algorithms for identifying the minimum state DFA that is consistent with an arbitrary set of positive and negative examples [37]. Efficient algorithms for identification of DFA assume that additional information is provided to the learner. This information is typically in the form of a set of examples S that satisfies certain properties or a knowledgable teacher's responses to queries posed by the learner. Trakhtenbrot and Barzdin have proposed an algorithm to learn the smallest DFA consistent with a *complete-labeled sample* (i.e., a sample that includes all sentences up to a particular length with the corresponding label that indicates whether the sentence is a positive example or a negative example [92]). Oncina and Garcia have defined a *characteristic set* of examples as a representative set that includes information about the states and transitions of the target DFA. They showed that it is possible to exactly identify the target DFA from a characteristic sample [65]. Angluin has described the use of a *minimally adequate teacher* to guide the learner in the identification of the target DFA. A minimally adequate teacher is capable of answering *membership queries* of the form "Does this sentence belong to the target language?" and *equivalence queries* of the form "Is this DFA equivalent to the target?". By using labeled examples together with membership and equivalence queries it is possible to correctly identify the target DFA [3].

C. Search Space

Regular grammar inference can be formulated as a search problem in the space of all finite-state automata (FSA). Clearly, the space of all FSA is infinite. One way to restrict the search space is to map a set of positive examples of the target DFA to a lattice of FSA which is constructed as follows [67]. Initially, the set of positive examples (S^+) is used to define a prefix tree automaton (PTA). The PTA is a DFA with separate paths (modulo common prefixes) from the start state leading to an accepting state for each string in S^+. The PTA accepts only the sentences in S^+

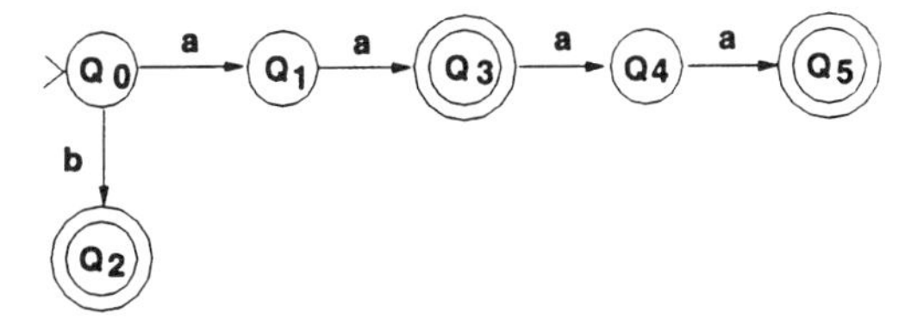

Fig. 4 Prefix tree automaton.

(i.e., the language of the PTA is the set S^+ itself). For example, the PTA corresponding to a set $S^+ = \{b, aa, aaaa\}$ of positive examples is shown in Fig. 4.

The lattice is now defined as the set of all partitions of the set of states of the PTA together with a relation that establishes a *partial order* on the elements. Each element of the lattice (i.e., an individual partition of the set of states of the PTA) is called a *quotient automaton* and is constructed from the PTA by merging together the states that belong to the same block of the partition. For example, the quotient automaton obtained by merging the staes q_0 and q_1 of the PTA is shown in Fig. 5. The corresponding partition of the set of states of the PTA is $\{\{0, 1\}, \{2\}, \{3\}, \{4\}, \{5\}\}$. The quotient automaton is more general than its parent automaton in that its language is a superset of the language of the parent.

The individual partitions in the lattice are partially ordered by the *grammar covers* relationship. One partition is said to cover another if the former is obtained by merging two or more blocks of the latter. For example, the partition $\{\{0, 1, 2\}, \{3\}, \{4, 5\}\}$ covers the partition $\{\{0, 1\}, \{2\}, \{3\}, \{4\}, \{5\}\}$. The important consequence of the grammar covers property is that if a partition covers another, then the language of the FSA representing the former is a superset of the language of the FSA representing the latter. The PTA is the *most specific* element of the lattice. The universal DFA that is obtained by merging all the states of the PTA into a single state is the *most general* element of the lattice. Given the PTA of Fig. 4, the partition corresponding to the *universal DFA* is $\{\{1, 2, 3, 4, 5\}\}$. The modified search space, though finite, is still exponentially large in terms of the number of states of the PTA. Explicit enumeration of the entire search space is thus not practical. Typical search procedures start with either the PTA or the universal DFA and use search operators, such as state merging and state splitting, to generate new elements within the search space.

By mapping the set of positive examples to a lattice of FSA we demonstrated how we could restrict the search space to be finite. How can we guarantee that the target DFA lies within this restricted search space? It can be shown that if the set of

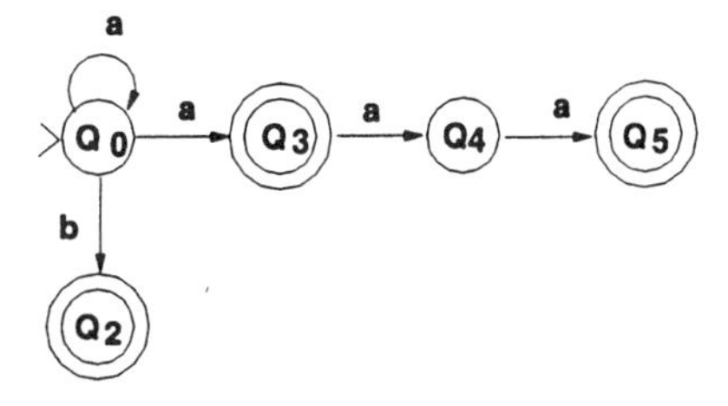

Fig. 5 Quotient automaton.

positive examples provided to the learner is a *structurally complete* set then the lattice constructed is guaranteed to contain the minimum state DFA equivalent to the target [24,67,69]. A structurally complete set for the target DFA is a set of positive examples (S^+) such that for each transition of the target DFA there is at least one string in S^+ that covers the transition and for each accepting state of the target DFA there is at least one string in S^+ that visits the accepting state. For example, the set $S^+ = \{b, aa, aaaa\}$ is structurally complete with respect to the DFA in Fig. 3.

D. Search Strategy

Even though the lattice is guaranteed to contain the target DFA, the size of the lattice is prohibitively large to be efficiently searched using an uninformed strategy such as breadth-first search. Several methods that use additional information provided to the learner have been designed to search the lattice for the target DFA. In the following subsections we summarize three different search methods.

1. Bidirectional Search Using Membership Queries

Parekh and Honavar have proposed a bidirectional search strategy based on the *version space* approach [69,70]. Their method assumes that a structurally complete set of positive examples (S^+) is provided to the learner at the start and a knowledgable teacher is available to answer the membership queries posed by the learner. The lattice of finite state automata defined earlier is implicitly represented using a *version space* [60]. The version space contains all the elements of the lattice that are consistent with the labeled examples and the membership queries. It is represented using two sets of finite-state automata: a set $\boldsymbol{S}$ of *most-specific* FSA that is initialized to the PTA and a set $\boldsymbol{G}$ of *most general* FSA that is initialized to the universal DFA obtained by merging all the states of the PTA. Thus, the initial version space implicitly includes all the elements of the lattice. It is gradually refined by eliminating those elements of the lattice that are inconsistent with the observed data. The version space search proceeds as follows: At each step two automata (one each from the sets $\boldsymbol{S}$ and $\boldsymbol{G}$, respectively) are selected and compared for equivalence. If they are not equivalent, then the shortest string accepted by one automaton, but not the other, is posed as a membership query to the teacher. The refinement of the sets $\boldsymbol{S}$ and $\boldsymbol{G}$ is guided by the teacher's response to the membership query. For example, if the query string is labeled as a negative example by the teacher then FSA in $\boldsymbol{S}$ that accept the negative example are eliminated from $\boldsymbol{S}$. Furthermore, all FSA of $\boldsymbol{G}$ that accept the negative example are progressively specialized by splitting their states until the resulting FSA do not accept the negative example. Thus, effectively all automata accepting the negative example are eliminated from the version space. Membership queries serve to progressively generalize the elements of $\boldsymbol{S}$ and specialize the elements of $\boldsymbol{G}$, thereby moving the two frontiers of the lattice represented by $\boldsymbol{S}$ and $\boldsymbol{G}$ closer to each other. At all times the two frontiers implicitly encompass the elements of the lattice that are consistent with the data observed by the learner. Convergence to the target is guaranteed and is attained when $\boldsymbol{S} = \boldsymbol{G}$. The essence of this bidirectional search of the lattice is captured in Fig. 6. Although this algorithm is guaranteed to converge to the target grammar, the speed of convergence often depends on the actual queries posed. Some query strings have the potential for eliminating huge chunks of the

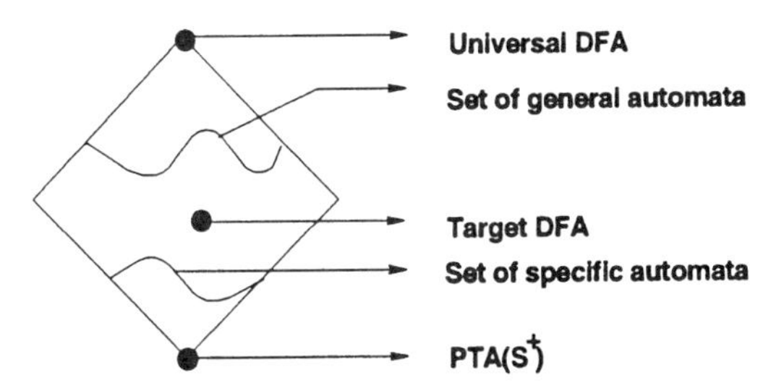

Fig. 6 Bidirectional search of the lattice.

search space. The worst-case time complexity of this method is exponential in the size of the PTA.

2. *Randomized Search*

Genetic algorithms offer an attractive framework for randomized search in large hypothesis spaces [40]. A typical genetic search involves evolving a randomly generated set of individuals (from the hypothesis space) based on the *survival of the fittest* principle of Darwinian evolution. A population of randomly selected elements of the hypothesis space is evolved over a period of time. A suitable representation scheme is chosen to encode the elements of the hypothesis space as *chromosomes*. A fitness function is defined to evaluate the quality of the solution represented by the individual chromosomes. The search progresses over several generations. In each generation a group of fittest (as determined by the fitness function) individuals is selected for genetic reproduction, which is represented by suitably defined operators (such as *mutation* and *crossover*). The genetic operators serve to identify new and potentially interesting areas of the hypothesis space. The offspring are added to the population and the evolution continues. Over time the individuals tend to converge toward high-fitness values (i.e., reasonably good solutions).

Dupont has proposed a framework for regular grammar inference using genetic algorithms [24]. The learner is given a set $S = S^+ \cup S^-$ of labeled examples. A PTA is constructed from the set S^+ as explained in section 3.C. The initial population comprises of a random selection of elements from the set of partitions of the set of states of the PTA. Each element of the initial population is thus a quotient automaton belonging to the lattice of FSA constructed from the PTA. The fitness assigned to each quotient FSA is a function of two variables: the number of states of the FSA and the number of sentences in S^- that are misclassified by the FSA. Individuals with fewer states that make fewer errors are assigned higher fitness. After each generation a subpopulation of individuals is randomly selected for reproduction based on their fitness values. Two genetic operators, *structural mutation and structural crossover* (designed specifically for the problem on hand) are applied to a subpopulation to produce offspring. Structural mutation involves randomly selecting an element from one of the existing blocks of the partition and randomly assigning it to one of the other blocks or creating a new block. Thus, $\{\{0, 1, 3, 5\}, \{2\}, \{4\}\} \rightarrow \{\{0, 1, 5\}, \{2, 3\}, \{4\}\}$ is an example of structural mutation. In structural crossover one block is chosen at random from each parent and both offspring inherit the block obtained by merging the two selected blocks. The remaining blocks for the offspring are obtained by taking the difference[3] of the blocks of the partitions corresponding to the parents that were not chosen before with the common block inherited by both offspring. For example, if the parents are represented by the following partitions {{0}, **{1,4}**, {2, 3, 5}} and {{0}, **{1,3}**, {2}, {4}, {5}} then a single application of the structural crossover might result in the following offspring {{0}, **{1, 3, 4}**, {2, 5}} and {{0}, **{1, 3, 4}**, {2}, {5}}. Here the blocks {1, 4} and {1, 3} are first merged to obtain the common block inherited by both offspring. The difference of the block {0} with {1, 3, 4} is {0}, the difference of the block {2, 3, 5} with {1, 3, 4} is {2, 5} and so on. The offspring produced by the genetic reproduction are valid partitions belonging to the lattice. These are added to the original population. *A fitness-proportionate* selection scheme that assigns high probability to individuals with higher fitness is used to randomly select individuals for the next generation. After a prespecified number of generations the fittest individual from the population is selected and returned by the algorithm as the inferred FSA. Although, this approach is not guaranteed to converge to the target DFA, empirical results have shown that this evolutionary search does result in identifying sufficiently small DFAs that make reasonably fewer errors on the set S^-.

3. Ordered Depth-First Search with Backtracking

The *regular positive and negative inference* (RPNI) algorithm performs an ordered *depth-first* search of the lattice guided by the set of negative examples S^- and in polynomial time identifies a DFA consistent with a given sample $S = S^+ \cup S^-$ [65]. Furthermore, if S happens to be a superset of a *characteristic set* for the target DFA then the algorithm is guaranteed to return a canonical representation of the target DFA. A characteristic set of examples $S = S^+ \cup S^-$ is such that S^+ is *structurally complete* relative to the target and S^- prevents any two states of the PTA of S^+ that are not equivalent to each other from being merged (see [65] for a precise definition).

The algorithm first constructs a PTA for S^+. The states of the PTA are numbered in standard order as follows: The set of strings that lead from the start state to each individual state of the PTA is determined. The strings are sorted in lexicographical order (i.e., $\lambda, a, b, aa, ab, \ldots$). Each state is numbered based on the position of the corresponding string in the sorted list. The states of the PTA shown in Fig. 4 are numbered in standard order. The algorithm initializes the PTA to be the current solution and systematically merges the states of the PTA to identify a more general solution that is consistent with S.

The states of the PTA are systematically merged using a quadratic loop (i.e., at each step i, the algorithm attempts to merge the state q_i with the states $q_0, q_1, \ldots,$ q_{i-1} in order). If the quotient automaton obtained by merging any two states does not accept any example belonging to S^- then the quotient automaton is treated as the current target and the search for a more general solution is continued with the state q_{i+1}. If $\|S^+\|$ and $\|S^-\|$ denote the sums of lengths of each example in S^+ and S^- respectively, then it can be shown that the time complexity of the RPNI algorithm is $\mathbf{O}(\|S^+\| + \|S^-\|.|S^+|^2)$. The interested reader is referred to [23,65] for a detailed description of the algorithm.

[3]The standard set difference operator is used here.

Example

We demonstrate a few steps in the execution of the RPNI algorithm on the task of learning the DFA of Fig. 3. Assume that we are given the *characteristic sample* $S = S^+ \cup S^-$ where $S^+ = \{b, aa, aaaa\}$ and $S^- = \{\lambda, a, aaa, baa\}$. The PTA is depicted in Fig. 4 where the states are numbered in the standard order. The initial partition of the states is {{0}, {1}, {2}, {3}, {4}, {5}}. The blocks 1 and 0 of the partition are merged. The quotient shown in Fig. 7 accepts the negative example *a* and thus the merge is rejected. Next the blocks 2 and 0 of the partition are merged. The quotient automaton shown in Fig. 8 accepts the negative example λ and thus even this merge is rejected. Continuing the merging in this manner it can be shown that the algorithms returns the partition {{0}, {1, 4}, {2}, {3, 5}} which is exactly the target DFA we are trying to learn (see Fig. 3).

E. The L^* Algorithm

The L^* algorithm infers a minimum state DFA corresponding to the target with the help of a *minimally adequate teacher* [3]. Unlike the approaches described so far, the L^* algorithm does not search a lattice of FSA. Instead, it constructs a hypothesis DFA by posing membership queries to the teacher. The learner maintains a table where the rows correspond to the states of the hypothesis DFA and the columns correspond to suffix strings that distinguish between pairs of distinct states of the hypothesis. Individual states are labeled by the strings that lead from the start state to the states themselves. The start state is labeled by the string λ. The states reached from the start state by reading one letter of the alphabet are labeled by *a* and *b* respectively, and so on. Thus, for the DFA in Fig. 3, Q_0 is labeled by the empty string λ, Q_1 is labeled by *a*, Q_2 is labeled by *b*, and Q_3 is labeled by *aa*. For each state the column labels indicate the state that would be reached from that state after reading the string corresponding to the column labels. Membership queries are posed on the string obtained by concatenating the row and column labels. The first column is labeled as λ. The queries λ.λ, *a*.λ, and so on, ask whether the state Q_0 is an accepting state, Q_1 is an accepting state, and so on. The cell representing the intersection of a row *r* and a column *c* is marked either 1 or 0 depending on whether the string *r.c* is accepted or not. Two states are presumed equivalent if all the entries for their corresponding rows are identical. If two states (say q_i and q_j) are presumed equivalent, but the states reached after reading a single letter of the alphabet from these states are not equivalent (i.e., the rows representing these adjacent states are not identical) then the states q_i and q_j cannot be equivalent. A column (representing a

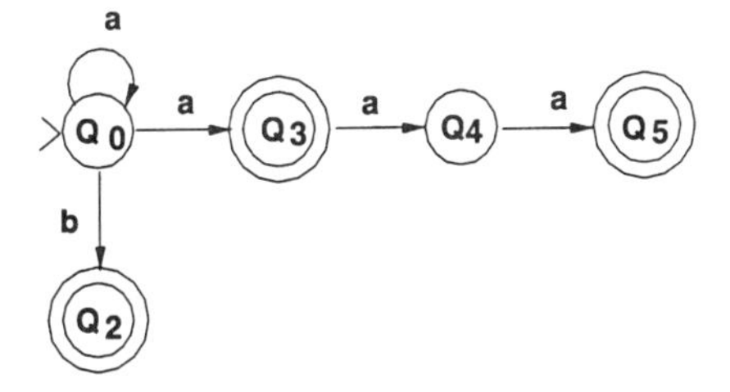

Fig. 7 Quotient automaton obtained by fusing blocks 1 and 0 of the current hypothesis.

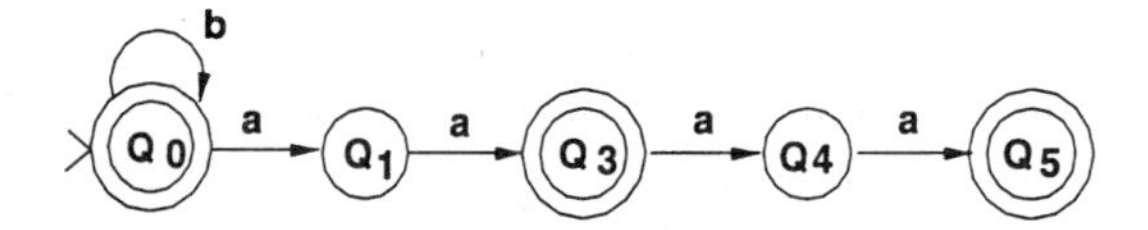

Fig. 8 Quotient automaton obtained by fusing blocks 2 and 0 of the current hypothesis.

distinguishing suffix) is then augmented to the table to distinguish the two states. When pairs of states in the hypothesis DFA cannot be distinguished any further the learner poses an equivalence query asking whether the current hypothesis is equivalent to the target. If the teacher answers *yes* the algorithm terminates and returns the current hypothesis. However, if the teacher answers *no* then it also provides a counterexample that is accepted by the target DFA but not by the current hypothesis or vice versa. The counterexample together with all its prefixes is incorporated into the table as new rows. Intuitively, this modification of the table represents adding new states to the current hypothesis. The learner modifies the hypothesis using the counterexample and additional membership queries and poses another equivalence query. This interaction between the learner and the teacher continues until the teacher's answer to an equivalence query is yes. The algorithm runs in time polynomial in the size of canonical representation of the target DFA and the length of the longest counterexample provided by the teacher and is guaranteed to converge to the unknown target.

The L^* algorithm assumes that the learner has the capacity to reset the DFA to the start state before posing each membership query. In other words, the teacher's response to the membership query indicates whether or not the DFA starting in state q_0 accepts the query string. This assumption might not be realistic in certain situations. For example, consider a robot trying to explore its environment. This environment may be modeled as a finite-state automaton with the different states corresponding to the different situations within which the robot might find itself and the transitions corresponding to the different actions taken by the robot in each situation. Once the robot has made a sequence of moves it may find itself in a particular state (say facing a barrier). However, the robot has no way of knowing from where it started or of retracing its steps to the start state. It has to treat it current state as the start state and explore the environment further. Rivest and Schapire have proposed a method based *on homing sequences* to learn the target DFA in these situations [83]. If it is assumed that each state of the DFA has an output (the output could simply be 1 for an accepting state and 0 for a nonaccepting state) then a homing sequence is defined as a sentence the output of which always uniquely identifies the final state the DFA is in even if the state the DFA started from before processing the sentence is unknown. Rivest & Schapire's algorithm runs N copies of the L^* algorithm (one for each of the N states of the target DFA) in parallel and using homing sequences overcomes the limitation that the start state of the target DFA is not known.

F. Connectionist Methods

Artificial neural networks (ANN) or connectionist networks are biologically inspired models of computation. ANN are networks of elementary processing units that are interconnected by trainable connections. Each processing unit (or *neuron*) computes an elementary function of its inputs. The connections are responsible for transmitting signals between the neurons. Each connection has an associated *strength* or *weight* that prescribes the magnitude by which the signal it carries is amplified or suppressed. ANN are typically represented by weighted, directed graphs. The nodes of the graph correspond to neurons, the edges refer to the connections, and the weights on the edges indicate the strength of the corresponding connections. Each neuron performs its computation locally on the inputs it receives from the neurons to which it is connected. These computations are independent of each other. Thus, ANN have potential for massive parallelism. Typical neural network architectures organize neurons in layers. The *input* layer neurons permit external signals to be input to the network. The network's output is read from the neurons at the *output layer*. Additionally, a network might have one or more *hidden layers* of neurons to facilitate learning.

ANNs have received considerable attention for their ability to successfully model language-learning tasks. The advantages of connectionist methods over the symbolic techniques described earlier include robust behavior in the presence of limited amount of noise, ability to learn from relatively small training sets, and scalability to larger problem sizes [51,52]. Most connectionist language-learning approaches specify the learning task in classic terms (for example; using an automaton to parse grammatically correct sentences of the language) and then implement these using a suitable neural network architecture, such as *recurrent neural networks* (RNN) [62]. A variety of RNN architectures have been investigated for learning grammars from a set of positive and negative examples of the target language (e.g., [17,18,25,26,31,32,34,58,78,85,97]). In the following section we describe the second order recurrent neural network architecture developed by Giles et al. [31].

1. *Second-Order RNN for Regular Grammar Inference*

Recurrent neural networks have feedback connections from the output neurons back to the input that give them the ability to process temporal sequences of arbitrary length. The second-order RNN is depicted in Fig. 9. The network has N recurrent state neurons $S_0, S_1, \ldots, S_{N-1}$, L nonrecurrent input neurons $I_0, I_1, \ldots, I_{L-1}$, and $N^2 \times L$ recurrent real valued connection weights W_{ijk}. The vector of output values of the state neurons is called a *state vector* **S**. **S** is fed back to the network by synchronous time delay lines (i.e., at time $t+1$ the network's output at time t is fed to the network along with the input). The state neuron S_0 is arbitrarily designated as the output neuron and is periodically sampled to read the network's output. The weight W_{ijk} modifies a product of the state neuron S_j and the input neuron I_k. For example, W_{211} would correspond to the weight connecting the state neuron S_2 with the unit computing the product of the state neuron S_1 with the input neuron I_1. This quadratic form directly represents the state transition function $\delta : Q \times \Sigma \rightarrow Q$ and is responsible for the *name second-order* RNN.

Typically, the number of input neurons L is chosen to be $|\Sigma| + 1$ with one neuron per input symbol and the extra neuron used to denote the end of an input string. The input is fed to the network one character at a time. Thus, the string 011 would be presented over four time steps with the activations of the input neurons I_0, I_1, and I_2 being $<100>$, $<010>$, $<010>$, and $<001>$ at $t = 0, 1, 2$, and 3, respectively. The choice of the number of state neurons N is a design parameter.

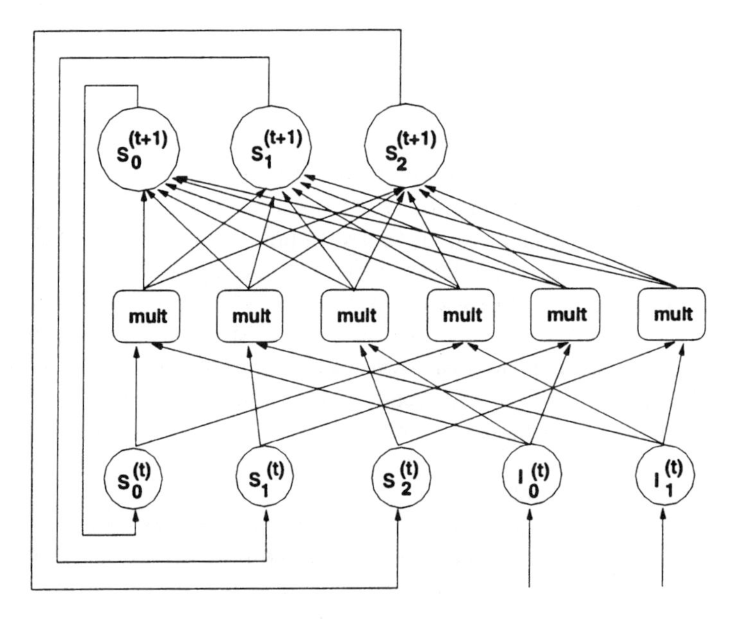

Fig. 9 Recurrent neural network architecture.

Typically N is chosen between 3 and 5. Recently, Siegelmann and Giles have derived a bound on the size of a continuous-valued recurrent network that is needed to correctly classify a given regular set [86]. This bound is easily computable and is tighter than the size of the canonical DFA that is commonly used as the upper bound.

The net input activation $y_{j+1}^{(t+1)}$ of the state neuron S_i is defined by the following expression.

$$y_i^{(t+1)} = \sum_{j=1}^{N} \sum_{k=1}^{L} W_{ijk} S_j^{(t)} I_k^{(t)} + \theta_i$$

where θ_i is the threshold of unit i. Note that the *mult* units in Fig. 9 compute the product $S_j^{(t)} I_k^{(t)}$. The state neurons typically implement the *sigmoid activation function* of the net input.

$$O_i^{(t+1)} = \frac{1}{1 + e^{-y_i^{(t+1)}}}$$

The output of the state neuron S_0 is observed after the complete input string is presented. For a suitably chosen threshold ε (where $0 < \varepsilon < 1/2$), if the output is greater than $1 - \varepsilon$, then the string is declared a positive example and if the output is less than ε then the string is declared a negative example.

The RNN is trained using an iterative *gradient descent* strategy to minimize a selected error function. The squared error is a typical choice for the error function. The error E_p for the sentence p is defined as follows:

$$E_p = \frac{1}{2}(T_p - S_0^{(f)})^2$$

where T_p is the desired label of sentence p (i.e., 1 or 0 depending on wheter p is accepted by the target or not) and $S_0^{(f)}$ is the output of S_0 after the entire sentence is presented. During training an entire input pattern is presented to the network and the final state of the network is determined. In case of a misclassification the network's weights are modified by an amount proportional to the negative gradient of the error function. The weight update equation is derived as follows:

$$\Delta W_{ijk} = -\eta \frac{\partial E_p}{\partial W_{ijk}} = -\eta (T_p - S_o^{(f)}) \frac{\partial S_0^{(f)}}{\partial W_{ijk}}$$

The results of experiments on the Tomita benchmark for regular grammars [91] showed that second-order neural networks converged fairly quickly to compact representations of the target finite-state automata and generalized reasonably well on strings not belonging to the training set [34].

2. *Extracting Finite-State Automata from Trained RNN*

One significant disadvantage of connectionist methods is that the learned models are not as transparent as those for symbolic approaches. Several researchers have studied methods for extracting finite-state automata from a trained RNN [20,33,64,97]. It is observed that RNN develop an internal state representation in the form of clusters in the activation space of the recurrent state neurons. Therefore, a symbolic description of the learned finite state automaton can be extracted from the trained network using clustering techniques. We describe an approach based on partitioning the space of the state neurons [34] and a hybrid method combining symbolic and connectionist approaches [1] for extracting state information from trained RNN.

Giles et al.'s method [34] divides the output range [0, 1] of each recurrent state neuron into q (typically $q = 2$) equal-sized partitions, thereby mapping the outputs of the N state neurons to a set of q^N states of the FSA. The partition P^λ corresponding to the RNN's response to the empty string (i.e., the end of string symbol) is treated as the start state of the FSA. Next the input symbol 0 is fed to the RNN and the output partition P^0 is treated as the state obtained by reading the symbol 0 in the start state P^λ. The partitions in response to the various input strings 1, 00, 01, ... are recorded and the learned FSA is reconstructed. Clearly, the number of states of the learned FSA can be no more than q^N. The extracted DFA is *minimized* [see Ref. 42 for a description of the state minimization procedure]. Experimental results show that the extracted DFA approximates the target DFA fairly closely.

Alquezar and Sanfeliu proposed a method based on *hierarchical clustering* to extract a FSA from the trained RNN [1]. Their algorithm constructs a *prefix tree automaton* (PTA) from the set of positive training examples (see Sec. 3.C). Simultaneously a set of single-point clusters is initialized to the set of activation vectors of the network's state neurons in response to each symbol of each training string. Note that there is a one-to-one correspondence between the states of the PTA and the initial single-point clusters. The PTA is treated as the initial hypothesis FSA. Hierarchical clustering is used to repeatedly merge the two closest clusters. The states of current hypothesis FSA that correspond to the two clusters are also merged. If the resulting automaton is inconsistent with the training set (i.e., it accepts a negative example) then the merge is disallowed. Otherwise, the resulting automaton is treated as the current hypothesis FSA. The next two closest clusters are then considered for merging and so on. This procedure terminates when all further merging results in the hypothesis FSA being inconsistent with some negative example of the training set.

3. *Using Evolutionary Programming to Learn RNN*

Most traditional approaches to training RNN for regular grammar inference use an a priori fixed network architecture where the number of state neurons is often selected in an ad hoc manner. An inappropriate choice of the network architecture can potentially prevent the network from efficiently learning the target grammar. Angeline et al. proposed an evolutionary network induction algorithm for simultaneously acquiring both the network topology and the weight values [2]. Their algorithm, called GNARL (*GeNeralized Acquisition of Recurrent Links*), uses evolutionary programming techniques that rely on *mutation* as the sole genetic reproduction operator (as against genetic algorithms that tend to rely primarily on the crossover operator).

GNARL architectures have a fixed number M of input nodes $I_1, I_2, \ldots I_M$, a fixed number N of output nodes $O_1, O_2, \ldots O_N$, and a variable number of hidden nodes (up to a maximum of h_{max}). Implementational constraints such as no links to an input node, no links from an output node, and at most one connection between any two nodes are enforced to ensure that only plausible network architectures are considered. The algorithm initializes a population of randomly generated networks. The number of hidden neurons, number of network connections, the incident neurons for each network connection, and the values of the connection weights are selected uniformly at random for each network in the initial population. In each generation a subset of the networks in the population is selected for genetic mutation based on their scores on a user provided *fitness function*. The *sum squared* error metric is a commonly used fitness measure. GNARL considers two types of mutation operators: *parametric mutation*, which alters the values of the connection weights, and *structural mutation*, which alters the structure of the network (by adding or deleting nodes or links). The mutation operators are designed to retain the created offspring within a specific locus of the parent (i.e., care is taken to ensure that the network created as a result of mutation is not too different from its parent in terms of both structure and fitness). Experiments on the Tomita benchmark for regular grammars showed that the GNARL networks consistently exhibited better average test accuracy as compared with the second-order recurrent networks described in [95].

G. Incremental Approaches

The problem of regular grammar inference can be made more tractable by assuming that the set of examples provided to the learner is *structurally complete* (see Sec. 3.D.1) or is a *characteristic* sample (see Sec. 3.D.3). In many practical learning scenarios, a sample that satisfies such properties may not be available to the learner at the outset. Instead, a sequence of labeled examples is provided intermittently and the learner is required to construct an approximation of the target DFA based on the information available to it at that time point. In such scenarios, an online or incremental model of learning that is guaranteed to eventually converge to the target DFA in the limit is of interest. Particularly, for intelligent autonomous agents, incremental learning offers an attractive framework for characterizing the behavior of the agents [11].

Parekh et al. have proposed an efficient incremental algorithm for learning regular grammars using membership queries [72]. Their method extends Angluin's *ID* algorithm to an incremental framework. The learning algorithm is intermittently provided with labeled examples and has access to a knowledgable teacher capable of answering membership queries. Based on the observed examples and the teacher's responses to membership queries, the learner constructs a DFA. As in the L^* algorithm, input strings are used to label the individual states. Membership queries are posed on the strings representing the states and the strings representing the adjacent states (obtained from the current state by reading one letter of the alphabet). A set of suffix strings is constructed to distinguish among nonequivalent states. The current hypothesis is guaranteed to be consistent with all the examples observed thus far. If an additional example provided is inconsistent with the learner's current representation of the target DFA then the learner modifies the current representation suitably to make it consistent with the new example. The algorithm is guaranteed to converge to a minimum state DFA corresponding to the target when the information obtained from labeled examples and membership queries is sufficient to enable the learner to identify the different states and their associated transitions. Furthermore, the time and space complexities of this approach are polynomial in the sum of the lengths of the counterexamples seen by the learner.

Two other incremental algorithms for learning DFA are based on only labeled samples and do not require membership queries. Porat and Feldman's incremental algorithm uses a *complete ordered sample* (i.e., all examples up to a particular length provided in strict lexicographical order) [79]. The algorithm maintains a current hypothesis that is updated when a counterexample is encountered and is guaranteed to converge in the limit. It needs only a finite working storage. However, it is based on an ordered presentation of examples and requires a consistency check with all the previous examples when the hypothesis is modified. An incremental version of the *RPNI* algorithm for regular grammar inference was proposed by Dupont [23]. It is guaranteed to converge to the target DFA when the set of examples seen by the learner includes a *characteristic set* for the target automaton as a subset. The algorithm runs in time that is polynomial in the sum of lengths of the observed examples. However, it requires storage of all the examples seen by the learner to ensure that each time the representation of the target is modified, it stays consistent with all examples seen previously.

H. Issues in Efficient Methods for Regular Grammar Inference

The foregoing sections bring out an interesting contrast between the different methods for regular grammar inference. Exact learning of the target DFA requires that the learner be provided with additional information in the form of a characteristic sample or have access to a knowledgable teacher who answer queries. Incremental approaches guarantee convergence in the limit (i.e., when the set of examples seen by the learner satisfies certain properties). Other approaches that attempt to learn the target DFA from only the labeled examples often rely on heuristics and are not guaranteed to converge to the target [29]. An interesting question would be to see whether DFA are *approximately* learnable. Valiant's distribution-independent model of learning also called the *probably approximately correct model* (PAC model) [93] is a widely used model for approximate learning. When adapted to the problem of learning DFA, the goal of a PAC learning algorithm is to obtain from a randomly drawn set of labeled examples, in polynomial time, with high probability, a DFA that is a *good approximation* of the target DFA [76]. A good approximation of the target is defined as one for which the probability of error on an unseen example is less than a prespecified *error parameter* ε. Because on a given run, the randomly drawn examples of the target DFA might not be fully representative of the target, a concession is made to the PAC-learning algorithm to produce a good approximation of the target with high probability (greater than $1-\delta$, where δ is a prespecified *confidence parameter*). For learning DFA, the accepted notion of polynomial time is that the algorithm for approximate learning of DFA must run in time that is polynomial in N (the number of states of the target DFA), $1/\varepsilon$ (the error parameter), $1/\delta$ (the confidence parameter), $|\Sigma|$ (the alphabet size), and m (the length of the longest example seen by the learner) [76].

Angluin's L^* learning algorithm can be adapted to the PAC learning framework to show that DFA can be efficiently PAC learned when the learner is allowed to pose a polynomial number of membership queries [3]. However, PAC learning of DFA is a hard problem in that there exists no polynomial time algorithm that efficiently PAC learns DFA from randomly drawn labeled examples alone [44,77]. The PAC model's requirement of learnability under all conceivable probability distributions is often considered too stringent in practice. Pitt identified the following open research problem: *Are DFA's PAC-identifiable if examples are drawn from the uniform distribution, or some other known simple distribution?* [76].

It is not unreasonable to assume that in a practical learning scenario a learner is provided with simple *representative examples* of the target concept. Intuitively, when we teach a child the rules of *multiplication* we are more likely to first give simple examples such as 3×4, instead of examples such as 1377×428. A *representative set* of examples would enable the learner to identify the target concept exactly. For example, a characteristic set for a DFA would constitute a suitable representative set for the DFA learning problem. The question then is whether we can formalize what simple examples mean. *Komogorov complexity* provides a machine-independent notion of *simplicity* of objects. Intuitively, the Kolmogorov complexity (k) of an object (represented by a binary string x) is the length of the shortest binary program that computes x. Objects that have regularity in their structure (i.e., objects that can be easily compressed) have low Kolmogorov complexity. Consider for example, the string $s_i = 010101\ldots01 = (01)^{500}$. A program to compute this string would be *Print 01 500 times*. On the other hand, consider a totally random string $s_j = 1100111010\,000\ldots01$ of length 500. This string cannot be compressed (as s_i was) which means that a program to compute s_j would be *Print 1100111010000... 01* (i.e., the entire string s_j would have to be specified). The length of the program that computes s_i is shorter than that of the program that computes s_j, thus $K(s_i) \leq K(s_j)$. If there exists a *Turing Machine M* such that it reads a string y as input and produces x as output then y is said to be a program for x. The Kolmogorov complexity of x is thus less than or equal to $|y| + \eta$ (i.e., the length of y plus some constant). Clearly, the Kolmogorov complexity of any string is bounded by its length because there exists a Turing machine that computes the identity function $x \rightarrow x$. Furthermore, there is a *universal* Turing machine U that is capable of simulating all Turing machines. It can be shown that the Kolmogorov complexity of an object relative to U differs from the Kolmogorov complexity of the object relative to any other machine by at most an additive constant. Thus, the Kolmogorov complexity of an object relative to the Universal Turing machine is treated as the Kolmogorov complexity of the object. [The interested reader is referred to Ref. 54 for a complete treatment of Kolmogorov complexity and related topics].

The *Solomonoff–Levin* universal distribution $\mathbf{m}$ assigns high probability to objects that are *simple* [i.e., $\mathbf{m}(x) \approx 2^{-K(x)}$]. If we assume that Kolmogorov complexity is a reasonable notion of simplicity of an object, then universal distribution gives us a way of sampling simple objects. Li and Vitányi proposed the framework of PAC learning under the universal distribution (called the *simple PAC* model) and demonstrated that several concept classes such as *simple* k-reversible DFA and log n-term DNF (i.e., boolean formulas in *disjunctive normal form*) are PAC learnable under this model, while their learnability under the standard PAC model is unknown [53]. Recently, Denis et al. proposed a variant of the simple PAC learning model for which a teacher might intelligently select simple examples based on his or her knowledge of the target. Under this model (called the PACS model) they showed that the entire class of k-reversible DFA and the class of *polyterm* DNF are efficiently PAC learnable. Parekh and Honavar have shown that the class of *simple*-DFA (DFA for which canonical representations have low Kolmogorov complexity) can be efficiently PAC learned under the simple PAC-learning model and the entire class of DFA can be efficiently PAC learned under the PACS model [68,71]. Their approach is based on the fact that for any DFA there exists a characteristic set of *simple* examples. Furthermore, it can be demonstrated that, with very high probability, this set will be drawn when examples are sampled according to the universal distribution. Thus, by using the RPNI algorithm in conjunction with a randomly drawn set of simple examples, it is shown that DFAs are efficiently PAC learnable. This work naturally extends the teachability results of Goldman and Mathias [38] and the equivalent model of learning grammars from characteristic samples developed by Gold [37] to a probabilistic framework [see Ref. 68 for details]. This idea of learning from simple examples holds tremendous potential in difficult learning scenarios. It is of interest to explore the applicability of learning from simple examples (for an appropriately defined notion of *simplicity*) in natural language learning.

4. STOCHASTIC REGULAR GRAMMAR INFERENCE

Often in practical applications of grammar inference, negative examples are not directly available to the learner. Gold proved that regular grammars cannot be learned from positive examples alone [37]. In such cases when only positive examples are available, learning is made more tractable by modeling the target as a stochastic grammar that has probabilities associated with the production rules.

A. Stochastic Finite-State Automata

A stochastic finite-state automaton (SFA) is defined as $A = (Q, \Sigma, q_0, \pi)$ where Q is the finite set of N states (numbered $q_0, q_1, q_2, \ldots, q_{N-1}$); Σ is the finite alphabet; q_0 is the start state; and π is the set of probability matrices. $p_{ij}(a)$ is the probability of transition from state q_i to q_j on observing the symbol a of the alphabet. A vector of

N elements π_f represents the probability that each state is an accepting state. The following normalizing constraint must hold for each state q_i:

$$\sum_{qj \in Q} \sum_{a \in \Sigma} p_{ij}(a) + \pi_f(i) = 1$$

It states that the total probability of each transition out of the state q_i together with the probability that q_i is an accepting state must be 1. The probability $p(\alpha)$ that a string α is generated by the SFA is computed as the sum over all states j of the probability that the SFA arrives at state j after reading α times the probability that j is an accepting state.

$$p(\alpha) = \sum_{q_j \in Q} p_{0j}(\alpha)\pi_f(j)$$

$$p_{ij}(\alpha) = \sum_{q_k \in Q} \sum_{\alpha \in \Sigma} p_{ik}(\beta)p_{kj}(a) \text{ where } \beta\text{a} = \alpha$$

The language generated by an SFA A is called a stochastic regular language and is defined as $L(\text{A}) = \{w \in \Sigma^* | p(w) \neq 0\}$ (i.e., the set of sentences that have a nonzero probability of being accepted by the SFA). Figure 10 shows the state transition diagram and associated probabilities of a stochastic finite state automaton.

1. The Alergia Algorithm for Learning SFA

Carrasco and Oncina have developed an algorithm for the inference of deterministic stochastic finite-state automata (DSFA) [12]. A DSFA is a SFA for which for each state $q_i \in \text{Q}$ and symbol $a \in \Sigma$ there exists at most one state q_j such that $p_{ij}(a) \neq 0$. This algorithm, called *Alergia*, is based on a state-merging approach and is quite similar to the RPNI algorithm for inference of DFA. A *prefix tree automaton* (PTA) is constructed from the given set of positive examples S^+ and its states are numbered in standard order (as described in Sec. 3.D.3). The initial probabilities π and π_f are computed based on the relative frequencies with which each state and transition of the PTA is visited by the examples in S^+ (see the following example).

A quadratic loop merges the states of the PTA in order (i.e., at each step i the algorithm attempts to merge state q_i with the states $q_0, q_1, \ldots, q_{i-1}$ in order). However, unlike the RPNI algorithm for which the state merges were controlled by a set of negative examples S^-, *Alergia* merges states that are considered to be similar in terms of their *transition* behavior (described by the states reached from the current state) and their *acceptance* behavior (described by the number of positive examples that terminate in the current state). The determination of similarity is statistical and is controlled by a parameter α that ranges between 0 and 1. The

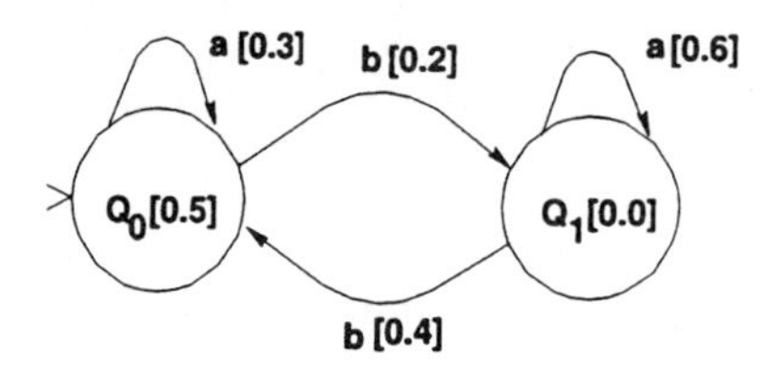

Fig. 10 Deterministic stochastic FSA

probabilities π and π_f are recomputed after each state merge. The algorithm is guaranteed to converge to the target stochastic finite-state automaton in the limit when a complete sample is provided. The worst-case complexity of *Alergia* is cubic in the sum of the lengths of examples in S^+.

Example

Consider the set $S^+ = \{\lambda, a, \lambda, abb, \lambda, bb, a, aa, abab, bb, \lambda, \lambda, a, a, bb\}$. The corresponding PTA is shown in Fig. 11. In the figure, the pair $[x, y]$ beside each state denotes the number of times the state is visited (x) and the number of strings that terminate in the state (y). Similarly, the number $[z]$ by each transition indicates the number of times the arc was used by the strings in S^+. Given this, we compute $\pi_f = \{\frac{5}{15}, \frac{4}{7}, 0, 1, 0, 1, 0, 1, 1\}$. Similarly, $p_{01}(a) = \frac{7}{15}, p_{01}(b) = 0, p_{02}(a) = 0, p_{02}(b) = \frac{3}{15}$, and so on. Intuitively, 5 strings out of 15 that visit state q_0 terminate in q_0. Hence, the probability that q_0 is an accepting state is $\frac{5}{15}$ [i.e., $\pi_f(0) = \frac{5}{15}$]. Similarly, 7 of the 15 strings that visit q_0 take the transition $\delta(q_0, a) = q_1$. Hence, the transition probability, $p_{01}(a) = \frac{7}{15}$ and so on.

The state-merging procedure considers q_0 and q_1 for merging. Assuming that q_0 and q_1 are found to be similar they would be merged together. This introduces nondeterminism because now the new state obtained by merging q_0 and q_1 has two transitions on the letter b (that is, to states q_1 and q_2, respectively). The resulting SFA is determinized by recursively merging the states that cause nondeterminism. This deterministic SFA (obtained by recursive state merging) is shown in Fig. 12. The probabilities π and π_f are recomputed and the state-merging is continued. The procedure terminates when no further state-merging is considered statistically significant and returns the most current DSFA.

B. Hidden Markov Models

Hidden Markov Models (HMM) are the widely used generalizations of stochastic finite-state automata in which both the state transitions and the output symbols are governed by probability distributions. HMMs have been applied successfully in

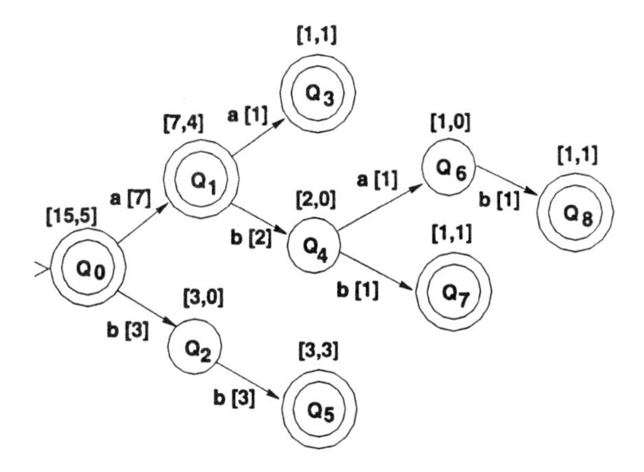

Fig. 11 Prefix tree automaton.

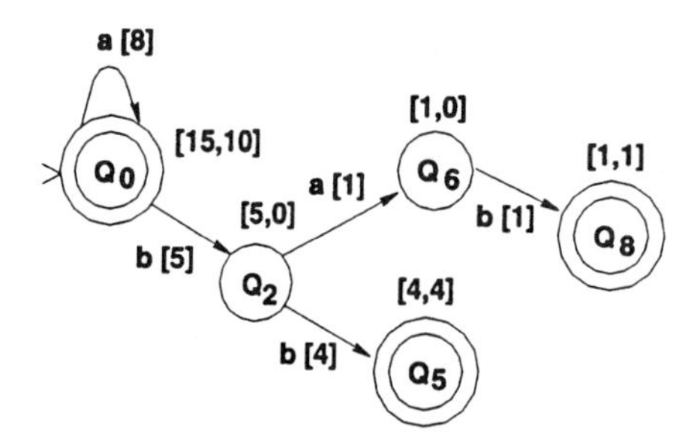

Fig. 12 Deterministic stochastic FSA after merging q_0 and q_1.

speech recognition and cryptography [81]. Formally, a HMM comprises the following elements:

- A finite set of states Q (labeled $q_1, q_2, \ldots, q_N$)
- A finite alphabet Σ (of symbols labeled $\sigma_1, \sigma_2, \ldots \sigma_M$)
- The state transition probability matrix A $(N \times N)$ where a_{ij} represents the probability of reaching state q_j from the state q_i.
- The observation symbol probability matrix B $(N \times M)$ where b_{ij} is the probability of observing the symbol σ_j in the state q_i.
- The initial state probability matrix π $(N \times 1)$ where π_i is the probability of model starting in state q_i.

Note that in the case of HMM, the symbols are associated with individual states instead of the individual transitions, as with SFA. These models satisfy the Markovian property that states that the probability of the model being in a particular state at any instant of time depends only on the state it was in at the previous instant. Moreover, because each symbol is associated with possibly several different states, given a particular symbol, it is not possible to directly determine the state that generated it. It is for this reason that these Markov models are referred to as a hidden. The probabilistic nature of HMM makes them suitable for use in processing temporal sequences. For example, in speech recognition, a separate HMM is created to model each word of the vocabulary. Given an observed sequence of sounds, one then determines the most likely path of the sequence through each model and selects the word associated with the most likely HMM.

Example

Figure 13 shows an HMM with $N = 8$ states and an alphabet $\Sigma = \{a, b, c\}$. The state transition probabilities and the observation symbol probabilities are shown beside the respective transitions and symbols in the figure. If the initial state probability matrix $\pi = \{1, 0, 0, 0, 0, 0, 0, 0\}$ then q_1 is the only initial state. There are two distinct ways of generating the string *bacac* with nonzero probability using this HMM; namely, by using the state sequences $q_1 q_2 q_e q_4 q_5$ and $q_1 q_2 q_6 q_7 q_8$, respectively.

1. *Learning Hidden Markov Models*

Given an N-state HMM with the model parameters $\lambda = (A, B, \pi)$ and an observation sequence $O = O_1 O_2 \ldots O_t$, the probability of this observation sequence $\Pr(O|\lambda)$ can be computed using the *forward–backward* estimation procedure [81]. Also, it is pos-

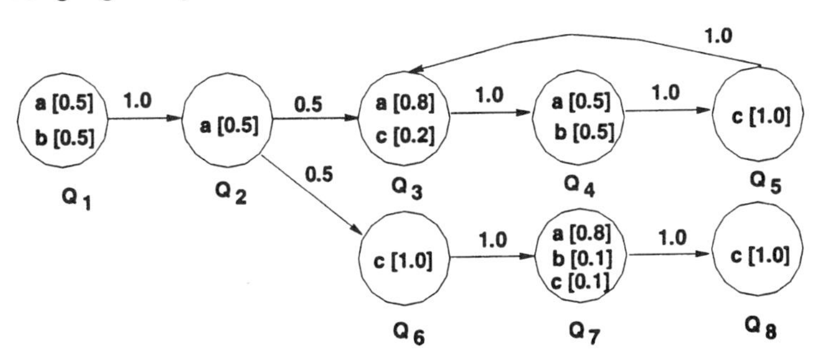

Fig. 13 Hidden Markov model.

sible to determine an optimal state sequence $i_1 i_2 \ldots i_t$ (where $i_k \in Q$; $1 \leq k \leq t$) that most likely produced the observation sequence O using the *Viterbi* algorithm. Perhaps the most critical problem with learning HMMs is the adjustment of the model parameters that would maximize the probability of a single observation sequence. To determine the model parameters an iterative procedure called the *Baum–Welch* algorithm can be used [7].

Stolcke and Omohundro present a more general approach to the HMM learning problem [88]. Their approach is a *Bayesian model-merging* strategy that facilitates learning the HMM structure as well as the model parameters from a given set of positive examples. The first step constructs an initial model comprising of a unique path from the start state q_1 to a final state q_f for each string in the set of positive examples S^+. This is similar to the PTA constructed by the *Alergia* algorithm for learning SFA. The initial probabilities are assigned as follows: The probability of entering the path corresponding to each string from the start state is uniformly distributed (i.e., if there are k paths corresponding to k strings in S^+ then the probability of entering each path is $1/k$). Within each path the probability of observing the particular symbol at each state and the probability of taking the outgoing arc from the state are both set to 1. Next a *state-merging* procedure is used to obtain a generalized model of the HMM. At each step the set of all possible state merges of the current model M_i are considered and two states are chosen for merging, such that the resulting model M_{i+1} maximizes the *posteriori* probability $\Pr(M_{i+1}|S^+)$ of the data. By Bayes' rule, $\Pr(M_{i+1}|S^+) = \Pr(M_{i+1}) \Pr(S^+|M_{i+1})$. $\Pr(M_{i+1})$ represents the *model prior probability*. Simpler models (e.g., models that have fewer states) have a higher prior probability. The data likelihood $\Pr(S^+|M_{i+1})$ is determined using the *Viterbi path* for each string in S^+ (i.e., the optimal state sequence for each string). The state-merging procedure is stopped when no further state merge results in an increase in the posteriori probability.

A brief comparison of *Alergia* with the *Bayesian model-merging* approach is in order. *Alergia* uses a depth-first order of merging states and runs in time that is polynomial in the size of S^+. Moreover, it is guaranteed to converge to the target deterministic stochastic finite automaton in the limit. On the other hand, the Bayesian model-merging procedure for HMM considers all possible state mergings at each step before merging two states and is thus computationally more expensive. Additionally, there is a need for appropriately chosen priors in the Bayesian model-merging procedure. Experimental results have shown that relatively uninformed priors perform reasonably well. Although there is no guarantee for convergence to the target HMM, the Bayesian model-merging procedure has been successfully used in various practical applications. Furthermore, the Bayesian model-merging procedure is also used to learn class based *n-grams* and *stochastic context free grammars* which are more general in terms of their descriptive capabilities than the deterministic stochastic finite automata inferred by *Alergia*.

5. INFERENCE OF CONTEXT-FREE GRAMMARS

Context-Free Grammars (CFG) represent the next level of abstraction (beyond regular grammars) in the Chomsky hierarchy of formal grammars. A CFG is a formal language grammar $G = (V_N, V_T, P, S)$ where the production rules are of the form: $A \rightarrow \alpha$ where $A \in V_N$ and $\alpha \in (V_T \cup V_N)^*$. Most work in inference of context-free grammars has focused on learning the standardized *Chomsky Normal Form* (CNF) of CFG in which the rules are of the form $A \rightarrow BC$ or $A \rightarrow a$ where $A, B, C \in V_N$ and $a \in V_T$. The example grammar for simple declarative English language sentences (see Fig. 1) is in CNF. Given a CFG G and a string $\alpha \in V_T^*$ we are interested in determining whether or not α is generated by the rules of G and if so, how is it derived. A *parse tree* graphically depicts how a particular string is derived from the rules of the grammar. For example, a parse tree for the sentence *The boy saw a ferocious tiger* is shown in Fig. 14. The nonleaf nodes of a parse tree represent the nonterminal symbols. The root of each subtree, together with its children (in order from left to right) represents a single production rule. Parse trees provide useful information about the structure of the string or the way it is interpreted. Often, in learning context-free grammars, parse trees obtained from example sentences are provided to the learner. Parse trees simplify the induction task because they provide the learner with the necessary components of the structure of the target grammar.

A. Inducing Context-Free Grammars

The theoretical limitations that plague regular grammar inference also carry over to the context-free case because the set of context-free languages properly contains the set of regular languages. The problems are further compounded by the fact that several decision problems related to CFG are undecidable. Given two CFG G_1 and G_2 there exists no algorithm that can determine whether G_1 is *more general than* G_2 [i.e., $L(G_1) \supseteq L(G_2)$]. Similarly, there is no algorithm that can answer the question: "*Is* $L(G_1) \cap L(G_2) = \phi$?" [42]. Thus, exact learning of the target grammar is

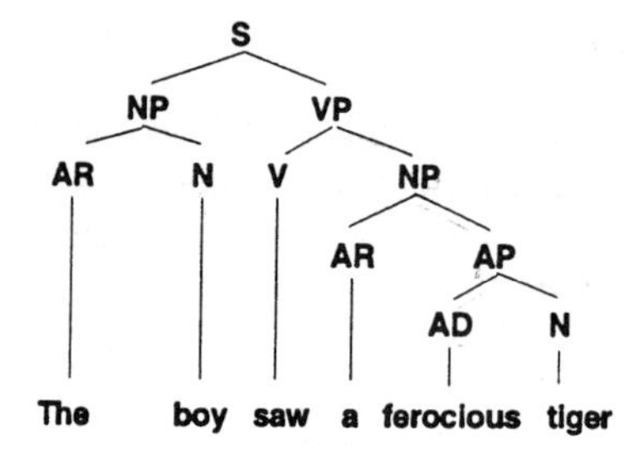

Fig. 14 Parse tree.

seldom attempted in CFG. However, several practically successful approaches have been devised for heuristically learning approximations to the target grammar.

1. *Learning Recursive Transitions Networks*

Recursive Transition Networks are an equivalent representation of CFG. They are essentially simple finite-state diagrams for which the tests on the arcs refer to either terminal symbols or to entire subnetworks [48]. For example, the recursive transition diagram corresponding to the context-free grammar of Fig. 1 is shown in Fig. 15. The first network indicates the start symbol *S* and invokes the subnetworks *NP* and *VP* by specifying them as tests on the arcs. The arcs of the subnetworks to the right of the figure specify individual words from the set of terminals.

One approach to learning recursive transition networks based on Langley's GRIDS algorithm [47] is similar to the state-merging techniques used for learning DFA. Given a set of positive examples, a degenerate one-level transition network is constructed that accepts exactly the sentences that belong to the training set. This initial network is analogous to the prefix tree automaton in the DFA learning problem. For example, if we are given the sentences "The boy saw a sly fox," "A boy followed a ferocious tiger," or "The tiger saw a sly fox," then a sample one-level transition network can be constructed as shown in Fig. 16.

The algorithm identifies patterns that co-occur in the initial network and represents these as new subnetworks that correspond to phrases in the language. Measures such as simplicity of the subnetwork (in terms of nodes and links) are used to select a subnetwork among several competing alternatives. The creation of subnetworks for the foregoing example proceeds as shown in Fig. 17. The subnetwork creation is repeated until further splitting results in a degradation of the performance. Similar subnetworks are then repeatedly merged to give more general transition networks. Sample-merging steps are shown in Fig. 18. Negative examples (if available) or heuristic knowledge can be used to guide the merging process to prevent the creation of overly general networks.

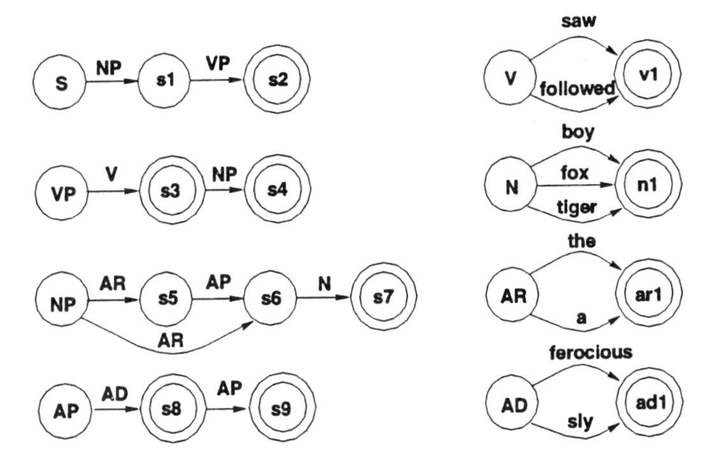

Fig. 15 Recursive transition network.

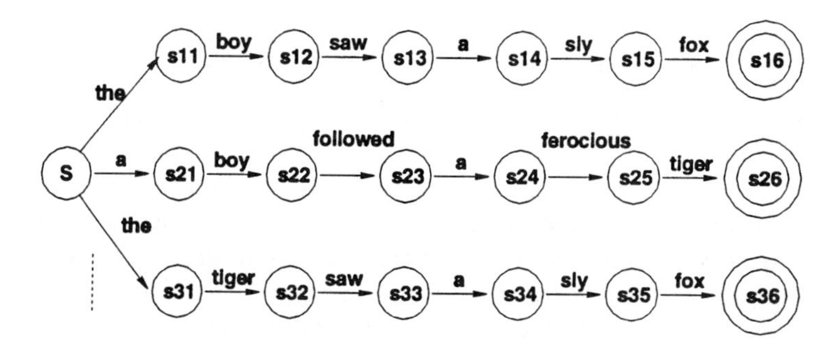

Fig. 16 Initial recursive transition network from positive sentences.

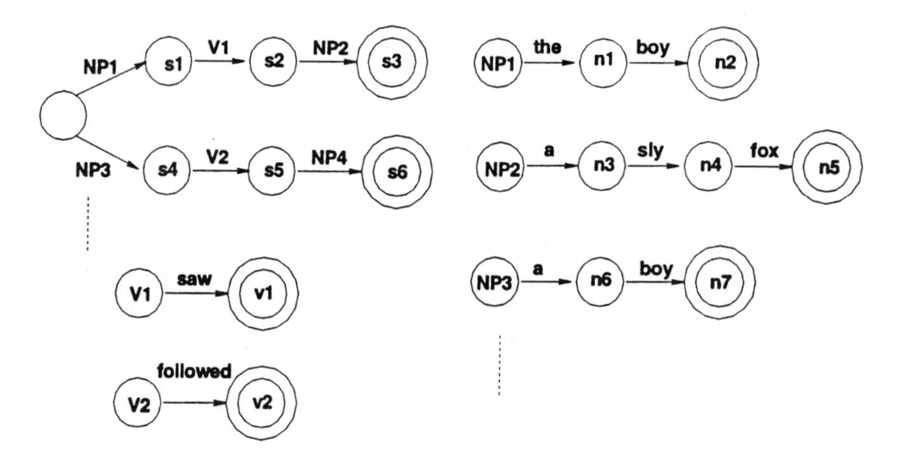

Fig. 17 Creating subnetworks.

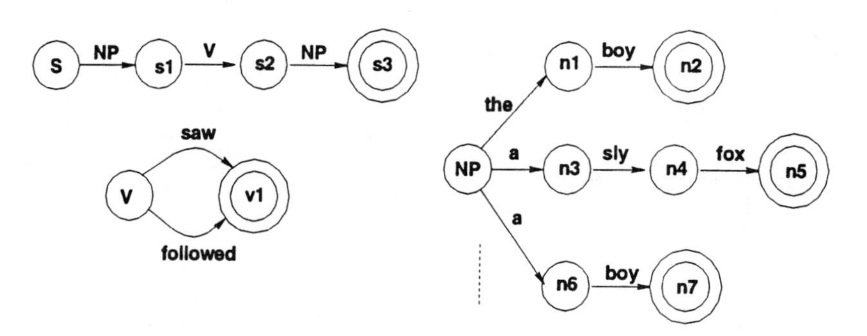

Fig. 18 Merging subnetworks.

2. *Learning CFG Using Version Spaces*

Van Lehn and Ball proposed a version space-based approach for learning CFG [94]. Given a set of labeled examples, a trivial grammar that accepts exactly the set of positive examples can be constructed. Again, this is similar to the PTA constructed by several regular grammar inference algorithms. A version space is defined as the set of all possible generalizations of the trivial grammar that are consistent with the examples. Because the set of grammars consistent with a given presentation is infinite it becomes necessary to restrict the grammars included in the version space. The *reducedness* bias restricts the version space to contain only *reduced* grammars (i.e., grammars that are consistent with the given sentences, but no proper subset of their rules is consistent with the given sentences). A further problem in CFG is that the more general than relation that determines if two grammars G_1 and G_2 are such that $L(G_1) \supseteq L(G_2)$ is undecidable. This problem is circumvented by defining a *derivational version space* (i.e., a version space based on unlabeled parse trees constructed from positive sentences). The idea is to induce partitions on the unlabeled nonterminal nodes of the parse trees. For example, consider a presentation of two sentences *cat* and *black cat*. Two unlabeled parse trees corresponding to these sentences are depicted in Fig. 19. The set of three unlabeled nodes can be partitioned in five different ways (two of which are shown in Fig. 19). Each partition corresponds to a grammar that is consistent with the sentences. The derivational version space is a union of all the grammars obtained from the unlabeled parse trees.

An operation *FastCovers* based on the partitions of the sets of nonterminals determines the partial order among elements of the derivational version space. The algorithm for updating the version space then proceeds as follows. Each new positive example causes the version space to expand by considering the new parse trees corresponding to the example. Each negative example causes the grammars that

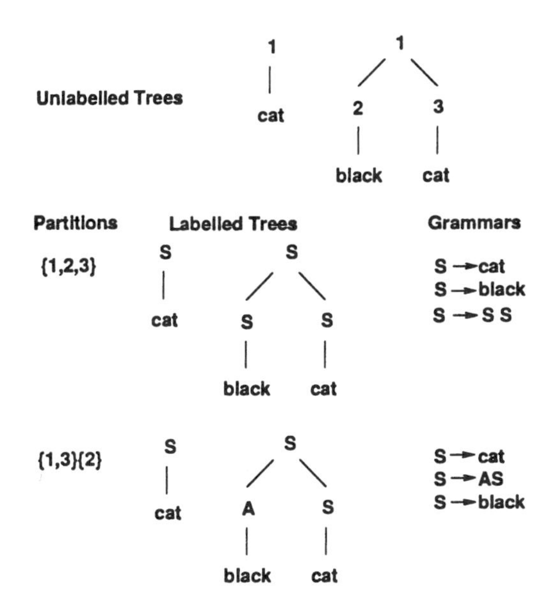

Fig. 19 Derivational version space of CFG.

accept the negative example to be eliminated from the version space. The *FastCovers* relation (which is a variant of the *grammar covers* property described in Sec. 3.C) is used to prune those grammars in the version space that cover the grammar accepting the negative example. This approach completely solves small induction problems and provides an opportunity for incorporation of additional biases to make the learning tractable for larger problems [94].

Giordano studied the version space approach to learning context-free grammars using a notion of *structural containment* to define a partial order on the search space of CFG [35]. This approach provides another mechanism to circumvent the problem that arises because the operation *more general than* is not decidable in CFG. A *structure* generated by a grammar is its derivation tree with all nonterminal labels removed. A grammar G_i is *structurally contained in* (or *structurally equivalent to*) a grammar G_j if the set of structures generated by G_i is contained in (or is equal to) the set of structures generated by G_j. Structural containment implies containment in the sense of languages (i.e., if G_i is structurally contained in G_j then the language of G_i is a subset of the language of G_j). Further, structural containment is polynomial time computable for *uniquely invertible* grammars (i.e., grammars in which the right-hand sides of no two production rules are the same). It can be shown that the set of uniquely invertible grammars is equivalent to the whole set of context-free grammars. Hence, applying the version space strategy for learning uniquely invertible grammars does not restrict the capacity of the inference system in any way.

3. Learning NPDA Using Genetic Search

Lankhorst presented a scheme for learning nondeterministic pushdown automata (NPDA) from labeled examples using genetic search [49]. *Pushdown automata* (PDA) are recognizing devices for the class of context-free grammars (just as FSA are recognizing devices for regular grammars). A PDA comprises of a FSA and a stack. The extra storage provided by the stack enables the PDA to recognize languages, such as *palindromes*, that are beyond the capability of FSA. Lankhorst's method encodes a fixed number of transitions of the PDA on a single chromosome. Each chromosome is represented as a bit string of a fixed length. Standard mutation and crossover operators are used in the genetic algorithm. The fitness of each chromosome (which represents a NPDA) is a function of three parameters: *training accuracy*, *correct prefix identification*, and *residual stack size*. The *training accuracy* measures the fraction of the examples in the training set that are correctly classified. PDA that are able to parse at least a part of the input string correctly are rewarded by the *correct prefix identification* measure. A string is said to be accepted by a PDA if, after reading the entire string, either the FSA is in an accepting state or the stack is empty. The *residual stack size* measure assigns higher fitness to PDAs that leave as few residual symbols on the stack as possible after reading the entire string.

4. Learning Deterministic CFG Using Connectionist Networks

Several researchers have explored connectionist approaches for learning context-free grammars. Das et al. proposed an approach for learning deterministic context-free grammars using recurrent neural network pushdown automata (NNPDA) [21,22]. This model uses a recurrent neural network similar to the one described in Sec. 3.F.1 in conjunction with an external stack to learn a proper subset of deterministic context-free languages. Moisl adopted deterministic pushdown automata (DPDA) as adequate formal models for general natural language processing (NLP) [62]. He showed how a simple recurrent neural network can be trained to implement a finite-state automaton that simulates the DPDA. By using a computer simulation of a parser for a small fragment of the English language, Moisl demonstrated that the recurrent neural network implementation results in a NLP device that is broadly consistent with the requirements of a typical NLP system and has desirable emergent properties. We briefly describe the NNPDA model proposed by Das et al. [21].

The NNPDA consists of a recurrent neural network integrated with an external stack through a hybrid error function. It can be trained to simultaneously learn the state transition function of the underlying pushdown automaton and the actions that are required to control the stack operation. The network architecture is similar to the one described in Sec. 3.F.1. In addition to the *input* and *state* neurons the network maintains a group of *read* neurons to read the top of the external stack and a single nonrecurrent *action* neuron the output of which identifies the stack action to be taken (i.e., *push*, *pop*, or *no-op*). The complexity of this PDA model is reduced by making the following simplifying assumptions. The input and stack alphabets are the same, the *push* operation places the current input symbol on the top of the stack, and ε-transitions are disallowed. These restrictions limit the class of languages learnable under this model to a finite subset of the class of deterministic context free languages. *Third*-order recurrent neural networks are used in this model for which the weights modify a product of the current state, the current input, and the current top of stack to produce the new state and modify the stack. The activation of the state neurons at time $t+1$ is determined as follows.

$$s_i(t+1) = g(\sum_j \sum_k \sum_l w_{ijkl} s_j(t) i_k(t) r_\ell(t))$$

where w_{ijkl} is the third-order weight that modifies the product of the activation of the state neuron S_j, the input I_k, and the activation of the read neuron R_l at time t; g is the standard sigmoid function $g(x) = 1/1 + e^{-x}$. During training, input sequences are presented one at a time, and the activations are propagated until the end of the sequence is reached. At this time, the activation of the designated output neuron is compared with the class label assigned to the particular input sequence (1 or 0). If the sequence is misclassified then the network's weights are modified according to the weight update rule. The weight update rule is designed to minimize a suitably chosen error function E. E is a function of the activation of the designated output neuron and the length of the stack at the end of the presentation of the entire sequence. For positive sequences it is desired that the output activation be close to 1.0 and the stack be empty and for negative sequences it is desired that the output activation be close to 0 and the stack be nonempty. This NNPDA model is capable of learning simple context-free languages such as $a^n b^n$ and *parenthesis*. However, the learning task is computationally intensive. The algorithm does not converge for more nontrivial context-free languages such as $a^n b^n c b^m a^m$.

Das et al. studied the incorporation of prior knowledge in NNPDA to make learning more tractable [22]. Two different types of knowledge could be made available to the model: knowledge that depends on the training data alone and partial knowledge about the automaton being inferred. As explained in Sec. 3.H, presenting simpler examples of the target concept can considerably simplify the learning task. Specifically, in context-free grammars, the learning process benefits from seeing shorter examples early on. Simulation results demonstrated that incremental learning (where the training examples are presented in increasing order by length) achieved a substantial reduction in the training time of the NNPDA algorithm. Additionally, a knowledgable teacher might be asked to indicate for each negative string the character that forces the PDA to a reject state. This information (if available) could be used to stop training the NNPDA on negative examples beyond the rejection point identified by the teacher. Experiments showed that such selective presentation of strings enables the NNPDA algorithm to converge on nontrivial languages such as $a^n b^n c b^m a^m$. The knowledge of the task being solved can be exploited to further assist the NNPDA learning algorithm. Two such methods, including predetermining the initial values of some of the network's weights and presenting structured training examples in which the order of generation of each word of the sentence is indicated by parentheses, improved learning speed [22].

B. Stochastic CFG

The success of HMM in a variety of speech recognition tasks leads one to ask if the more powerful stochastic context free grammars (SCFG) could be employed for speech recognition. The advantages of SCFG lie in their ability to capture the embedded structure within the speech data and their superior predictive power in comparison with regular grammars as measured by prediction entropy [50].

The *Inside–Outside* algorithm can be used to estimate the free parameters of a SCFG. Given a set of positive training sentences and a SCFG for which parameters are randomly initialized, the inside–outside algorithm first computes the most-probable parse tree for each training sentence. The derivations are then used to reestimate the probabilities associated with each rule, and the procedure is repeated until no significant changes occur to the probability values.

Before invoking the inside–outside algorithm one must decide on some appropriate structure for the SCFG. Stolcke and Omohundro's Bayesian model-merging approach can be used for learning both the structure and the associated parameters of the SCFG from a set of training sentences [88]. This approach is analogous to the one for HMM induction described earlier. New sentences are incorporated by adding a top level production from the start symbol S. Model merging involves *merging* of nonterminals in the SCFG to produce a more general grammar with fewer nonterminals and *chunking* of nonterminals where a sequence of nonterminals is abbreviated using a new nonterminal. A powerful beam search (which explores several relatively similar grammars in parallel) is used to search for the grammar with the highest a posteriori model probability.

6. DISCUSSION

Research in computational models of formal language acquisition has received significant attention over the past three decades. Their study is motivated partly by the desire to better understand the process of natural language acquisition and partly by the numerous practical applications of grammar inference. The preceding sections presented several key algorithms for learning regular, stochastic regular, context-free, and stochastic context-free grammars.

Regular grammars represent the simplest class of grammars in the Chomsky hierarchy. The study of regular grammar inference methods is of significant practical interest for several reasons: every *finite* language is regular; a context-free language can often be closely approximated by a regular grammar [30]. Unfortunately, the regular grammar inference problem is known to be difficult (i.e, there exists no efficient algorithmn that can learn the target grammar from an arbitrary set of labeled examples). The regular grammar inference problem can be solved efficiently with the help of *representative samples*, or by drawing examples according to a *helpful* distribution (e.g., simple distributions), or with the help of a knowledgable teacher capable of answering queries. Furthermore, as demonstrated by the results of the recent *Abbadingo One DFA Learning Competition* several efficient heuristic approaches provide satisfactory solutions to the regular grammar inference problem [39,43,46].

Context-free and context-sensitive grammars generate languages that are clearly more expressive than those generated by regular grammars. This has motivated a large body of research on algorithms for learning such grammars. However, existing algorithms for induction of these grammars are generally heuristic. Although some of them have been demonstrated to perform well on small- to medium-sized problems, they do not appear to scale well to larger problem sizes. While context-free and context-sensitive grammars (and their corresponding recognition devices) constitute useful conceptual abstractions for studying fundamental theoretical questions in computer science, and the study of algorithms for learning such languages (or interesting subclasses of such languages) is clearly of significant theoretical interest, their utility in modeling real-world language learning scenarios is open to question. This is because any language that can be generated (and recognized) by a computational device (e.g., a general-purpose computer) with a finite amount of memory can be modeled by a regular language. Sentences that are encountered in practice (e.g., in typical use of natural languages) seldom require infinite memory for generation or recognition. Note that no computer in use today has infinite memory and, hence, is necessarily less powerful than a Turing machine with its infinite tape. But that has never prevented the widespread use of computers for all sorts of practical applications. Similarly, one might argue that algorithms for learning regular grammars are, for the most part, adequate for practical applications of grammar induction.

Although languages generated and recognized by finite state, finite memory computational devices (e.g., a pushdown automaton with a bounded stack) can, in principle, be modeled by regular languages, one might still want to use context-free or context-sensitive languages for reasons of elegance or compactness. Hence, it is of interest to explore efficient algorithms for learning such languages under the assumption of finite-state, bounded memory recognition devices, or a strong *inductive bias* in favor of *simpler* grammars, or under *helpful* distributions.

The different approaches to language acquisition can be grouped into three broad categories: those that make use of purely symbolic representations (e.g., L^*); those that make use of purely numeric representations (e.g., most connectionist networks), and hybrid approaches that combine both symbolic as well as numeric representations. Algorithms that employ purely symbolic representations often lend themselves to rigorous analysis and guarantee of convergence to the unknown target grammar given a noncontradictory, noise-free, representative set of examples. On the other hand, such algorithms can be quite brittle in the presence of noise. Algorithms that employ purely numeric representations are robust in the presence of limited amounts of noise, are able to learn from relatively smaller training sets, and are known to scale well to larger problem sizes [51,52]. However, their convergence properties are generally poorly understood, and the learned models are not as transparent. Even though it is possible to extract a symbolic description of a learned grammar from a connectionist network, the procedures used for doing so are largely heuristic. Against this background, it is of interest to investigate hybrid algorithms that exploit the strengths of both numeric as well as symbolic representations [41].

Research in cognitive psychology, linguistics, and related areas has raised several important issues that must be (at least partially) addressed by any practical system for language acquisition. An in-depth discussion of recent developments in these areas and their implications for models of language learning is beyond the scope of this chapter. However, in what follows, we attempt to briefly highlight some of the key issues.

Mechanisms of language acquisition have been the subject of intense debate for several decades. As in most cognitive phenomena, the opinions that have been voiced span a wide spectrum. At one end of the spectrum is the behaviorist view, advocated by Skinner, that suggests a *tabula rasa* approach to learning [87]. In this view, language acquisition is driven collectively by *stimuli* experienced by an individual, the *response* evoked by the stimuli, and the *reinforcing stimuli* that follow the response. For example, when a child feels thirsty—the stimulus of *milk-deprivation*, it might respond by saying "*milk*" (essentially by a process of trial and error) and the parents might provide a reinforcing stimulus in the form of a bottle of milk. Thus, it was argued that language originates from physical need and is a means to a physical end. This behaviorist view equates learning with essentially the formation of (typically but not necessarily direct) stimulate–response associations. Note that *semantics* play a central role in this model of language learnng.

At the other end of the spectrum is the view advocated by Chomsky that language is *not* a cultural artifiact that can be learned [14]. This claim is based on what Chomsky has called *the argument from the poverty of stimulus* which states that the language stimuli received by the child are insufficient for acquiring abstract grammatical rules solely by inductive learning [15]. Language is thus innate in the biological makeup of the brain. In other words, language is a human *instinct* [75]. Thus, children must be born with a *Universal Grammar* or a system of principles, conditions, and rules that are elements or properties of all human languages [15]. Chomsky's subsequent writings [16] argue that the innate *universal grammar* ir organized into modular subsystems of principles concerning *government*, *binding*, and *thematic roles*. Principles or, alternatively, particular lexical items may have parameters associated with them for which values are set according to language experience [56].

It is perhaps worth noting that Chomsky's arguments of innateness of language appear to be restricted to language *syntax*, whereas the behaviorist view of language acquisition is primarily concerned with language *semantics*. If we restrict our attention to syntax acquisition, results in machine learning and computational learning theory appear to lend credence to Chomsky's view. Almost all learning algorithms either explicitly or implicitly use appropriate *representational* and *inductive* biases [59,61]. The representational bias of the algorithm determines the space of candidate hypotheses (or sets of grammars in the case of language acquisition) that are considered by the learner. For learning to succeed, the hypothesis space must be *expressive enough* to contain the target grammar. On the other hand, because learning typically involves searching the hypothesis space to identify an unknown target on the basis of data or other information available in the learner's environment, if it is *overly expressive*, it can increase the complexity of the learning task. Inductive bias of the algorithm tends to focus or bias the search for the unknown target hypothesis. For instance, an algorithm for regular language acquisition has a representational bias that allows it to consider only regular grammars as candidate solutions. In addition, such an algorithm might have a strong inductive bias that directs its search in favor of, say, *simpler* rules.

Most efficient algorithms that have been devised for language learning rely on the availability of negative examples of the unknown target grammar. Although this may be a reasonable assumption to make in several practical applications of language acquisition or grammatical inference (e.g., syntactic approaches to pattern recognition), it has been argued that the apparent lack of *negative examples* presents a language *learnability dilemma*. Wexler and Manzini point out that children seldom see negative examples of a grammar and are rarely corrected when they utter grammatically incorrect sentences [96]; Consequently, they often overgeneralize from the limited number of sentences they hear. The lack of negative examples prevents any direct recovery from overgeneralizations of the target grammar. Against this background, several mechanisms (such as *constraint sampling* and use of the *uniqueness principle*) have been suggested for recovering from such overgeneralizations [74].

It is tempting to think of Chomsky's Universal Grammar (if in fact it exists) as determining the representational bias of a natural language learner. The inductive bias of the learner might be enforced by the innately determined range of values that the various parameters are allowed to take. It might also be a result of the process of brain development in children. In this context, Prince and Smolensky [80] have recently developed a grammatical formalism, called *optimality theory*, that brings together certain connectionist computational principles into the essentially symbolic theory of universal grammar. In this formalism, possible human languages share a common set of universal constraints on well-formedness. These constraints are highly general, hence conflicting. Consequently, some of them must be violated in *optimal* (i.e., grammatical) structures. The differences in the world's languages emerge by different priority rankings of the universal constraints: each ranking is a language-particular grammar, a means of resolving the inherent conflicts among the universal constraints.

In this chapter, we have studied language learning primarily from the point of view of acquiring syntax. The learning of *semantics* or the meanings of words and sentences is central to natural language acquisition. An excellent overview of empirical methods for natural language processing (including techniques for speech recognition, syntactic parsing, semantic processing, information extraction, and machine translation) is available [10]. Thompson et al. have designed a system for automatically acquiring a semantic lexicon from a corpus of sentences paired with representations of their meanings [90]. This system learns to parse novel natural language sentences into a suitable semantic representation (such as a logical database query), and it successfully learns useful lexicons for a database interface in four different natural languages. Current understanding of semantic development in children, as summarized by Pan and Gleason [66], is surprisingly consistent with the essential tenets of the behaviorist view advocated by Skinner. It is believed that children typically learn meanings of words and phrases by identifying suitable mappings between them and perceptual stimuli. In summary, it seems plausible that something akin to a universal grammar and its interaction with the rest of the brain is what enables children to become fluent in any language during their childhood. Thus, language acquisition has to necessarily rely on not only mechanisms for learning language-specific grammars (which might involve tuning the parameters of the universal grammar), but also processes for inducing the meaning of the words and sentences in the language. This argues for a model of language learning that accounts for both syntactic as well as semantic components of the language. Recent research results offer some intriguing preliminary insights on the development of formal models of how children learn language [27,28]. Regier proposed a computational model of how some lexical items describing spatial relations might develop in different languages [82]. His system includes a simple model of the visual system that is common to all human beings and thus must be the source from which all visual concepts arise. Using conventional backpropagation techniques his system was able to learn spatial terms from labeled example movies for a wide range of languages. Stolcke's Bayesian model-merging approach [88] was able to learn semantics on the same domain (of labeled example movies) in the form of a simple stochastic attributed grammar. Bailey proposed a computational model to address the issue of how a child makes use of the concepts that he or she has learned [5,4]. His system learns to produce verb labels for actions and also carries out the actions specified by the verbs it has learned.

Against this background, we expect to see further advances in models of language acquisition that tie together the syntactic, semantic, and progmatic aspects of language.

ACKNOWLEDGMENT

This research was partially supported by grants from the National Science Foundation (IRI-9409580 and IRI-9643299) and the John Deere Foundation to Vasant Honavar.

REFERENCES

1. R Alquezar, A Sanfeliu. A hybrid connectionist-symbolic approach to regular grammar inference based on neural learning and hierarchical clustering. In: RC Carrasco, J Oncina, eds. Proceedings of the Second ICGI-94, Lecture Notes in Artificial Intelligence 862. Springer-Verlag, 1994, pp 203–211.
2. PJ Angeline, GM Saunders, JB Pollack. An evolutionary algorithm that constructs recurrent neural networks. IEEE Trans Neural Networks 5:54–65, 1994.
3. D Angluin. Learning regular sets from queries and counterexamples. Inform Comput 75:87–106, 1987.
4. D Bailey. When push comes to shove: a computational model of the role of motor control in the acquisition of action verbs. PhD dissertation, Unviersity of California at Berkeley, Berkeley, CA, 1997.

5. D Bailey, J Feldman, S Narayanan, G Lakoff. Modeling embodied lexical development. In: Proceedings of the 19th Cognitive Science Society Conference, Stanford University Press, 1997.
6. G Ball, D Ling, D Kurlander, J Miller, D Pugh, T Skelly, A Stankosky, D Thiel, M Van Dantzich, T Wax. Lifelike computer characters: the persona project at Microsoft Research. In: J Bradshaw, ed. Software Agents. MIT Press, Cambridge, MA, 1997.
7. LE Baum, T Petrie, G Soules, N Weiss. A maximization technique occuring in the statistical analysis of probabilistic functions in Markov chains. Ann Math Stat 41:164–171, 1970.
8. A Brazma, I Jonassen, I Eidhammer, D Gilbert. Approaches to automatic discovery of patterns in biosequences. J Comput Biol 5:277–303, 1998.
9. MR Brent. Advances in the computational study of language acquisition. Cognition 61:1–38, 1996.
10. E Brill, R Mooney. An overview of empirical natural language processing. AI Mag 18(4):13–24, 1997.
11. D Carmel, S Markovitch. Learning models of intelligent agents. In Proceedings of the AAAI-96 (vol 1). AAAI Press/MIT Press, 1996, pp 62–67.
12. RC Carrasco, J Oncina. Learning stochastic regular grammar by means of a state merging method. In: RC Carrasco, J Oncina, eds. Proceedings of the Second ICGI-94, Lecture Notes in Artificial Intelligence 862. Springer-Verlag, 1994, pp 139–152.
13. N Chomsky. Three models for the description of language. PGIT 2(3):113–124, 1956.
14. N Chomsky. Review of BF Skinner. Verbal Behavior. Language 35:26–58, 1959.
15. N Chomsky. Reflections on Language. Temple Smith, London, 1976.
16. N Chomsky. Some Concepts and Consequences of the Theory of Government and Binding, MIT Press, Cambridge, MA, 1982.
17. A Cleeremans, D Servan–Schreiber, J McClelland. Finite state automata and simple recurrent networks. Neural Comput 1:372–381, 1989.
18. D Clouse, C Giles, B Horne, G Cottrell. Time-delay neural networks: representation and induction of finite state machines. IEEE Trans Neural Networks 8:1065–1070, 1997.
19. C Crangle, P Suppes. Language and Learning for Robots. CSLI Lecture Notes: No. 41. CLSI Publications, Stanford, CA, 1994.
20. S Das, R Das. Induction of discrete-state machine by stabilizing a simple recurrent network using clustering. Comput Sci Inform 21(2):35–40, 1991.
21. S Das, C Giles, GZ Sun. Learning context free grammars: capabilities and limitations of a recurrent neural network with an external stack memory. Proceedings of the Fourteenth Annual Conference of the Cognitive Science Society, 1992, pp 791–795.
22. S Das, C Giles, GZ Sun. Using prior knowledge in a NNPDA to learn context free languages. In: C Giles, S Hanson, J Cowan, eds. Advances in Neural Information Processing Systems, 5. 1993, pp 65–72.
23. P Dupont. Incremental regular inference. In: L Miclet, C Higuera, eds. Proceedings of the Third ICGI-96, Lecture Notes in Artificial Intelligence 1147, Montpellier, France, 1996. Springer, pp 222–237.
24. P Dupont, L Miclet, E Vidal. What is the search space of the regular inference? Proceedings of the Second International Colloquium on Grammatical Inference (ICGI'94), Alicante, Spain, 1994, pp 25–37.
25. J Elman. Finding structure in time. Cogn Sci 14:179-211, 1990.
26. J Elman. Distributed representation, simple recurrent networks, and grammatical structure. Mach Learn 7:195–225, 1991.
27. J Feldman, G Lakofi, D Bailey, S Narayanan, T Regier, A Stolcke. Lzero: the first five years. Artif Intell Rev 10:103–129, 1996.
28. JA Feldman. Real language learning. In: V Hanavar, G Slutzki, eds. Proceedings of the Fourth ICGI-98 Lecture Notes in Artificial Intelligence 1433, Ames, Iowa, 1998. Springer, pp. 114–125.

29. K Fu. Syntactic Pattern Recognition and Applications. Prentice-Hall, Englewood Cliffs, NJ, 1992.
30. KS Fu, TL Booth. Grammatical inference: introduction and survey (part 1). *IEEE Trans Syst Man Cybern* 5:85–111, 1975.
31. C Giles, D Chen, H Miller, G Sun. Second-order recurrent neural networks for grammatical inference. Proc Int Joint Conf Neural Networks 2:273–281, 1991.
32. C Giles, B Horne, T Lin. Learning a class of large finite state machines with a recurrent neural network. Neural Networks 8:1359–1365, 1995.
33. C Giles, C Omlin. Extraction, insertion and refinement of symbolic rules in dynamically-driven recurrent neural networks. Connect Sci (special issue on Architectures for Integrating Symbolic and Neural Processes) 5:307–337, 1993.
34. C Giles, C Miller, D Chen, H Chen, G Sun, Y Lee. Learning and extracting finite state automata with second-order recurrent neural networks. Neural Comput 4:393–405, 1992.
35. J Giordano. Inference of context-free grammars by enumeration: structural containment as an ordering bias. In: RC Carrasco, J Oncina, eds. Proceedings of the Second ICGI-94, Lecture Notes in Artificial Intelligence 862. Springer-Verlag, 1994, pp 212–221.
36. EM Gold. Language identification in the limit. Inform Control 10:447–474, 1967.
37. EM Gold. Complexity of automaton identification from given data. Inform Control 37:302–320, 1978.
38. S Goldman, H Mathias. Teaching a smarter learner. In: Proceedings of the Workshop on Computational Learning Theory (COLT'93). ACM Press, 1993, pp 67–76.
39. C de la Higuera, J Oncina, E Vidal. Identification of DFA: data-dependent vs data-independent algorithms. In: L Miclet, C Higuera, eds. Proceedings of the third ICGI-96, Lecture Notes in Artificial Intelligence 1147. Springer, Montpellier, France, 1996, pp 313–326.
40. JH Holland. Adaptation in Natural and Artificial Systems. University of Michigan Press, Ann Arbor. MI, 1975.
41. V Honavar. Toward learning systems that integrate multiple strategies and representations. In: V Honavar, L Uhr, eds. Artificial Intelligence and Neural Networks: Steps Toward Principled Integration. Academic Press: New York, 1994, pp 615–644.
42. J Hopcroft, J Ullman. Introduction to Automata Theory, Languages, and Computation. Addison–Wesley, 1979.
43. H Juillé, J Pollack. A sampling based heuristic for tree search applied to Grammar Induction. Proceedings of the Fifteenth National Conference on Artificial Intelligence, Madison, Wisconsin, 1998.
44. M Kearns, LG Valiant. Cryptographic limitations on learning boolean formulae and finite automata. In: Proceedings of the 21st Annual ACM Symposium on Theory of Computing. ACM, New York, 1989, pp 433–444.
45. M Kearns, U Vazirani. An Introduction to Computational Learning Theory. MIT Press, Cambridge, MA, 1994.
46. K Lang, B Pearlmutter, R Price. Results of the Abbadingo One learning competition and a new evidence-driven state merging algorithm. In: V Honavar, G Slutzki, eds. Proceedings of the Fourth ICGI-98, Lecture Notes in Artificial Intelligence 1433, Ames, Iowa, 1998. Springer, pp. 1-12.
47. P Langley. Simplicity and representation change in grammar induction. Robotics Laboratory, Stanford Laboratory. 1994. Unpublished manuscript.
48. P Langley. Elements of Machine Learning. Morgan Kauffman, Palo Alto, CA, 1995.
49. M Lankhorst. A genetic algorithm for induction of nondeterministic pushdown automata. Technical Report CS-R 9502, University of Groningen, The Netherlands, 1995.
50. K Lari, SJ Young. The estimation of stochastic context-free grammars using the inside–outside algorithm. Comp Speech Lang 4:35–56, 1990.

51. S Lawrence, S Fong, C Giles. Natural language grammatical inference: a comparison of recurrent neural networks and machine learning methods. In: S Wermter, E Rilofi, G Scheler, eds. Connectionist, Statistical, and Symbolic Approaches to Learning for Natural Language Processing, Lecture Notes in Artificial Intelligence 1040. Springer-Verlag, 1996, pp 33–47.
52. S Lawrence, C Giles, S Fong. Natural language grammatical inference with recurrent neural networks. IEEE Trans Knowledge and Data Eng. IEEE Press, (To appear).
53. M Li, P Vitányi. Learning simple concepts under simple distributions. SIAM J Comput 20:911–935, 1991.
54. M Li, P Vitányi. An Introduction to Komogorov Complexity and its Applications. 2nd edn. Springer Verlag, New York, 1997.
55. JC Martin. Introduction to Languages and The Theory of Computation, McGraw–Hill, New York, 1991.
56. JM Meisel. Parameters in acquisition. In: P Fletcher, B MacWhinney, eds. A Handbook of Child Language Acquisition. Blackwell, 1985, pp 10–35.
57. L Miclet, J Quinqueton. Learning from examples in sequences and grammatical inference. In: G Ferrate, et al., eds. Syntactic and Structural Pattern Recognition. NATO ASI Ser vol F45, 1986, pp 153–171.
58. C Miller, C Giles. Experimental comparison of the effect of order in recurrent neural networks. Int J Pattern Recogn Artif Intell 7:849–872, 1993.
59. T Mitchell. The need for biases in learning generalizations. Technical Report CBM-TR-117, Department of Computer Science, Rutgers University, New Bruskwick, NJ, 1980.
60. T Mitchell. Generalization as search. Artif Intell 18:203–226, 1982.
61. T Mitchell. Machine Learning. McGraw–Hill, New York, 1997.
62. H Moisl. Connectionist finite state natural language processing. Connect Sci 4(2):67–91, 1992.
63. BK Natarajan. Machine Learning: A Theoretical Approach. Morgan Kauffman, Palo Alto, CA, 1991.
64. C Omlin, C Giles. Extraction of rules from discrete-time recurrent neural networks. Neural Networks 9:41–52, 1996.
65. J Oncina, P Garcia. Inferring regular languages in polynomial update time. In: N Pérez, et al., eds. Pattern Recognition and Image Analysis. World Scientific, 1992, pp 49–61.
66. B Pan, J Gleason. Semantic development: learning the meaning of words. In: J Gleason, ed. The Development of Language. 4th ed. Allyn and Bacon, Boston, MA, 1997, pp 122–158.
67. T Pao, J Carr. A solution of the syntactic induction–inference problem for regular languages. Comp Lang 3:53–64, 1978.
68. R Parekh. Constructive Learning: Inducing Grammars and Neural Networks. PhD dissertation, Iowa State University, Ames, IA, 1998.
69. RG Parekh, VG Honavar. Efficient learning of regular languages using teacher supplied positive examples and learner generated queries. Proceedings of the Fifth UNB Conference on AI, Fredricton, Canada, 1993, pp 195–203.
70. RG Parekh, VG Honavar. An incremental interactive algorithm for regular grammar inference. In: L Miclet, C Hiiguera, eds. Proceedings of the Third ICGI-96, Lecture Notes in Artificial Intelligence 1147, Montpellier, France, Springer, 1996, pp 238–250.
71. RG Parekh, VG Honavar. Learning DFA from simple examples. In: Proceedings of the Eighth International Workshop on Algorithmic Learning Theory (ALT'97), Lecture Notes in Artificial Intelligence 1316, 1997, Sendai, Japan, Springer, pp 116–131. Also presented at the Workshop on Grammar Inference, Automata Induction, and Language Acquisition (ICML'97), Nashville, TN. July 12, 1997.
72. R Parekh, C Nichitiu, V Honavar. A polynomial time incremental algorithm for regular grammar inference. In: V Honavar, G Slutzki, eds. Proceedings of the Fourth ICGI-98, Lecture Notes in Artificial Intelligence 1433, Ames, Iowa, 1998. Springer, pp. 37–49.

73. S Pinker. Formal models of language learning. Cognition 7:217–283, 1979.
74. S Pinker. Productivity and conservatism in language acquisition. In: W Demopoulos, A Marras, eds. Language Acquisition and Concept Learning. Ablex Publishing, Norwood, NJ, 1986.
75. S Pinker. The Language Instinct. William Morrow & Co, 1994.
76. L Pitt. Inductive inference, DFAs and computational complexity. In: Analogical and Inductive Inference, Lecture Notes in Artificial Intelligence 397. Springer-Verlag, 1989, pp 18–44.
77. L Pitt, MK Warmuth. Reductions among prediction problems: on the difficulty of predicting automata. In: Proceedings of the 3rd IEEE Conference on Structure in Complexity Theory, 1988, pp 60–69.
78. JB Pollack. The induction of dynamical recognizers. Mach Learn 7;123–148, 1991.
79. S Porat, J Feldman. Learning automata from ordered examples. Mach Learn 7:109–138, 1991.
80. A Prince, P Smolensky. Optimality: from neural networks to universal grammar. Science 1997 March 14, 275: 1604–1610.
81. LR Rabiner, BH Juang. An introduction to hidden Markov models. IEEE Acoust Speech Signal Processing Mag 3:4–16, 1986.
82. T Regier. The Human Semantic Potential. MIT Press, Cambridge, MA, 1996.
83. RL Rivest, RE Schapire. Inference of finite automata using homing sequences. Inform Comput 103:299–347, 1993.
84. S Russell, P Norvig. Artificial Intelligence: A Modern Approach. Prentice–Hall, Englewood Cliffs, NJ, 1995.
85. D Servan–Schreiber, A Cleeremans, J McClelland. Graded state machines: the representation of temporal contingencies in simple recurrent networks. Mach Learn 7:161–193, 1991.
86. H Siegelmann, C Giles. The complexity of language recognition by neural networks. Neurocomputing 15:327–345, 1997.
87. BF Skinner. Verbal Behavior. Appleton–Century–Crofts, New York, 1957.
88. A Stolcke, S Omohundro. Inducing probabilistic grammars by Bayesian model merging. In: RC Carrasco, J Oncina, eds. Proceedings of the Second ICGI–94, Lecture Notes in Artificial Intelligence 862. Springer-Verlag, 1994, pp 106–118.
89. B Tesar, P Smolensky. Learnability in optimality theory. Technical Report JHU-Cog-Sci-96-2, Johns Hopkins University, Cognitive Science Department, 1996.
90. C Thompson, R Mooney. Semantic lexicon acquisition for learning natural language interfaces. Technical Report TR AI98-273, Artificial Intelligence Laboratory, University of Texas, Austin, TX, 1998.
91. M. Tomita. Dynamic construction of finite-automata from examples using hill-climbing. Proceedings of the 4th Annual Cognitive Science Conference, 1982, pp 105–108.
92. B Trakhtenbrot, Y Barzdin. Finite Automata: Behavior and Synthesis. North-Holland Publishing, Amsterdam, 1973.
93. L Valiant. A theory of the learnable. Commun ACM 27:1134–1142, 1984.
94. K Vanlehn, W Ball. A version space approach to learning context-free grammars. Mach Learn 2:39–74, 1987.
95. RL Watrous, GM Kuhn. Induction of finite-state automata using second-order recurrent networks. In: Advances in Neural Information Processing, 4. Morgan Kaufmann, Palo Alto, CA, 1992.
96. K Wexler, R Manzini. Parameters and learnability in binding theory. In: T Poeper, E Williams, eds. Parameter Setting, D Reidel, Dordrecht; 1987. pp 41–76.
97. Z Zeng, R Goodman, P Smyth. Learning finite state machines with self-clustering recurrent networks. Neural Comput 5:976–990, 1993.

30

The Symbolic Approach to ANN-Based Natural Language Processing

MICHAEL WITBROCK

Lycos Inc., Waltham, Massachusetts

1. INTRODUCTION

Symbolic systems have been widely used for natural language processing (NLP); connectionist systems less so. The advantages of symbol systems are made clear elsewhere in this handbook: well-defined semantics, straightforward handling of variable binding and recursion, and well-developed systems for manipulating the symbols. The advantages of connectionist systems, also discussed in nearby chapters, are the potential speedup that can be derived from their inherent parallelism; the graceful degradation, in the face of imperfect input, that can result from the use of continuous representations, and in distributed representations, the "natural" inferences that can result from pattern completion and from representating semantic distance as vector distance.

As Sun [1] has pointed out, these qualities of symbolic and connectionist systems, to some extent, are complementary. This chapter discusses NLP systems that attempt to take advantage of the strengths of both approaches by implementing symbol systems in a neural network framework. The argument for taking advantage of symbol processing, in even neural–net-based language-processing systems, was perhaps stated most succinctly by Touretzky, one of the pioneers of this approach: "Connectionists need to get over the idea that mechanisms sophisticated enough to handle real natural language will emerge from simple gradient descent learning algorithms" [2].

Although this chapter attempts to concentrate on techniques that might be applied to the reader's NLP research or applications, some material has been included for historical perspective. The coverage is by no means exhaustive, and readers are encouraged to refer to other chapters in the handbook, and to the literature, as a palliative for its failings.

2. IMPLEMENTATION OF COMPUTATIONAL DEVICES

Much of the success, to date, of Artificial Intelligence (AI) has rested on its adoption of Newell and Simon's *physical symbol hypothesis* [3,4]. This hypothesis holds that cognition can be understood, at least in the abstract, in terms of the manipulation of discrete symbols representing physical objects. This approach is particularly attractive for Natural Language Processing (NLP), because at least at some levels of analysis, words are self-evidently symbolic.

Artificial Neural Networks (ANN), on the other hand, attempt to approximate the hardware substrate of human cognition, using units that compute by applying some simple function to activation values passed from other units over weighted connections. Representations in such systems are less obviously symbolic.

Following the resurgence of popularity of computatiaon using ANNs around the time of the publication of Rumelhart and McClelland's *Parallel Distributed Processing* collection [5], it was important to discover whether the fruits of decades of symbolic AI research could be translated into this newly popular computation model.

Perhaps the earliest concerted set of demonstrations that it was possible to do purely symbolic computation with a connectionist network were provided by Touretzky and Hinton's Distributed Connectionist Production System (DCPS) [6–8] and Touretzky's BoltzCONS [7,9,10]. In its simplest form, the BoltzCONS system represented stacks and binary trees in a Boltzmann Machine [11,12] by explicitly building a representation of LISP cons-cells, which were, in turn, used to construct more complex data structures in the conventional LISP manner.

A. The BoltzCONS System

Although it is difficult to imagine a system such as BoltzCONS being put to practical use, it is worth describing in a little detail, both because it was historically important, and because two of the techniques used—building reliable distributed representations and using sets of units as working memory buffers—are generally applicable. Two main techniques were used in representing the LISP cons-cells in BoltzCONS:

1. Providing the equivalent of an address by using a distributed identification tag.
2. Using a "pullout" network to allow accurate retrieval of an individual cons-cell from the collection stored in memory.

Each cons-cell was represented by a triple of symbols (TAG, CAR, CDR) chosen from a 25-symbol vocabulary. One possibility for representing such triples would have been to use a collection of 25^3 units, each corresponding to one of the 25^3 possible (TAG, CAR, CDR) combinations. This possibility is unattractive for two reasons: a single unit's failure could make it impossible to represent a cons-cell, and, more importantly, the number of units required by such a representation increases so quickly with the size of data structures that even modest representational ambitions—a tree with 200 nodes, for example—would require millions of units.

Instead, a reliable distributed representation of the triples was adopted. Storage of a triple, (F A A) for example, in BoltzCONS consists of the activation of a set of units, each of which is activated to represent that triple or any other member of a particular subset of other triples. Figure 1 shows the "receptive field" of one of these units, together with a subset of the cons-cell representations that activate it.

A cons-cell is considered to be stored in working memory if it is in the receptive field of sufficiently many active units. With 2000 units constituting the memory, and 25 symbols to be represented, each possible triple would fall within the receptive field of 28 units, on average. An analysis by Touretzky and Hinton [8] showed that such a memory can store between 3 and 25 cons-cells without excessive interference, depending on their similarity. If cons-cells are chosen at random, about 20 can be stored. The representation, therefore, is fairly efficient, for it can reliably store a fairly sizable set of symbol relations in a relatively small set of binary units. It is also robust, as the loss of a single unit decreases the activation of the symbol by only 1/28 of its total value.

To enable symbolic manipulation of the stored cons-cells, BoltzCONS uses a "pullout" network that allows the single symbol most closely satisfying externally imposed constraints to be extracted from the working memory. This network contains exactly as many units as the working memory store, connected one-to-one to the working memory units through gated connections. Unlike the working memory, however, each unit in the cons-pullout network has inhibitory connections to every other. The weights of this winner-take-all arrangement are set so that approximately 28 active units can be on at any time—the number required to represent a single cons-cell. Additional bidirectional connections lead from the cons-pullout network to TAG, CAR, and CDR distributed winner-take-all networks, each of which can represent exactly 1 of the 25 symbols at a time. A cons-cell is retrieved from memory by placing, or "clamping", symbols in some subset of these TAG, CAR, and CDR

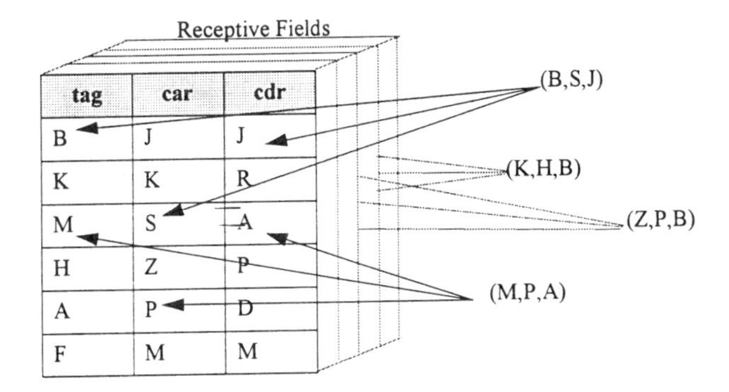

Fig. 1 The receptive field of a unit used in the distributed representation of BoltzCONS. On the right are several triples representing Lisp cons-cells. The unit represented will be active if either of the triples (B,S,J) or (M,P,A) are stored in memory, because the unit has tag, car, and cdr inputs corresponding to the first, second, and third elements of the triples. The triples (K,H,B) and (Z,P,B), on the other hand, do not activate the unit, because at least one of their elements does not find a match in the tag, car, or cdr receptors of the unit shown. These two triples are represented by activating other units in the memory. Approximately 28 units are activated to represent each triple in working memory, as described in the text.

networks, and using simulated annealing [13] between the units composing the TAG, CAR, CDR, and CONS-PULLOUT networks to find activation patterns in the remaining, unclamped, symbol networks that are most consistent with the contents of working memory. With a little extra machinery, this entire arrangement provides for the storage or deletion as well as the retrieval of reliable symbolic representations.

Touretzky [9] used LISP code to push and pop a stack stored in the BoltzCONS working memory and to construct and traverse a tree. Remembering that the TAG symbol represents the "address" of a cons-cell, it should be readily apparent how this could be done with the sort of (TAG, CAR, CDR) triples that BoltzCONS can represent.

B. Distributed Connectionist Production System (DCPS)

Touretzky and Hinton had also applied the techniques used in BoltzCONS to the highly symbolic problem of representing and applying rules [6,8]. Because this task is not particular to NLP, the operation of DCPS will not be described in great detail here. What is particularly interesting about it, from the NLP point of view, is its use of a "Bind-Space" to perform variable binding. As in BoltzCONS, the basic memory element in DCPS is a triple of symbols chosen from a 25-symbol vocabulary. A typical rule, such as: **(=x T P)(=x W U) → +(W M =x) –(=x A =x)** consists of two triples on the left-hand side that must match with working memory contents and may contain a variable in the first position, and any number of right-hand size clauses that specify changes to working memory and may, as in the example given, contain the same variable in any position. The left-hand side of such a rule, or "production" specifies preconditions, and the right-hand side actions to be taken if the preconditions are met. Production systems with arbitrarily large memories are computationally equivalent to Turing machines. In DCPS, the rules are represented by a Rule-Space containing mutually inhibitory cliques of 40 mutually excitatory units. Activation of a clique represents activation of the corresponding rule. Rules do not share units.

To perform rule matching, DCPS maintains two Clause Spaces, which, similar to the pullout network in BoltzCONS, are each responsible for extracting a single triple each from among those stored in working memory. These two extracted triples correspond to the two preconditions in a rule, and the units with receptive fields for any possible match to the corresponding precondition are directly connected by bidirectional excitatory links to the units, in rule space, for the corresponding rule. Because the form of the rules is stored in a hard-wired pattern of connections, DCPS is not a stored program computer.[1] During rule matching, simulated annealing is used to find a pair of triples in the clause spaces that are supported both by the pattern of activation in working memory, and by the existence of a rule in the rule space that allows the two triples to support each other.

[1] Although, as with a Turing machine, it is possible to principle to simulate a stored program computer in a system such as DCPS by encoding an interpreter for that computer's instructions in the hard-wired rules, one would not be advised to make the attempt.

Variable binding works similarly. A "bind space" consists of 25 cliques representing the 25 symbols in the alphabet.[2] There are connections from the units representing each symbol to a randomly chosen subset of the units in the two clause spaces that have that symbol in the first element of their receptive fields. In this way, two triples being considered by the two clause spaces during simulated annealing can support each other through the units in bind space, if they share a first symbol in common.

Having settled on a variable binding for the preconditions of a rule, DCPS then proceeds to use the bound value in the right-hand side "actions" as follows. During rule firing, gated connections from the rule units to the working memory units corresponding to RHS triples open, allowing working memory units to be turned on or off. In RHS triples with variables, the connections run to the working memory units corresponding to every instantiation of the variable, but these connections are gated by second-order connections to the bind cliques for each instance. This means that only the working memory units corresponding to a RHS triple with its variable bound to the symbol that is active in bind-space are affected by the gating operation.

The DCPS and BoltzCONS subsystems, described in the foregoing, were combined into a complete symbol-processing system [7], with DCPS providing the control for alterations of tree structures stored as cons-cells in BoltzCONS's working memory. Using this system, Touretzky was able to build a toy NLP system that reliably converted the parse trees of SVO sentences to passive form by switching the subject and direct object, and inserting "be" before the verb and "by" after it.

C. Lessons from BoltzCONS and DCPS

BoltzCONS and DCPS were important pieces of work, in that they demonstrated solidly that connectionist systems could manipulate symbols, and also that they could control that manipulation. As a practical matter, though, their use of simulated annealing to search for consistent variable bindings, to match rules, and to search memory was a barrier to their extension and application to more practical tasks; Boltzman machines are remarkably slow. The systems also had the disadvantage, perhaps, of too closely implementing purely symbolic processes—they achieved, with quite an array of connectionist hardware, exactly the same effect as a few instructions and a few bytes of memory on a digital computer. They made no attempt to take advantage of the ability of distributed representations to represent and process similar objects similarly.

The feasibility of connectionist symbol processing having been demonstrated, the question of its usefulness remains. In the remainder of the chapter, the methods described will still be essentially symbolic, but the focus will be more on solving realistic problems and on representing complex data structures in ways that also take advantage of the properties of connectionist representations.

[2]Unlike those in the Rule Space, these cliques are used for a distributed representation, with each symbol in the receptive field of three cliques, but this is not a central point.

3. LOCALIST SYSTEMS

Although the fault tolerance and the representational efficiency of distributed representations are important benefits of connectionist systems, an important body of research has concentrated on purely localist representations. In these representations, each represented object, such as a concept, a word, or a role assignment, is represented by a unique unit. These units are connected together by various links with specific functions. Usually these units and links constitute a fixed network architecture, in which calculations are performed by activating a subset of the units and feeding activation through the links until a result is obtained. For symbolic natural language processing, however, several systems have been designed that dynamically construct networks to solve a particular instance of a linguistic task, such as attachment. In this section, both of these kinds of localist systems will be considered.

A. Static Localist Systems

Perhaps the best way to introduce these systems is by giving a simple example. Consider the preposition attachment problem presented by the sentence fragment "the car with the dog in the garage." Clearly the first prepositional phrase "with the dog" attaches to the first noun phrase "the car," but there is, at least arguably, some doubt about whether the second prepositional phrase (PP) "in the garage" should attach to "the car" or to "the dog." Touretzky [2] describes a network that aims to solve such attachment problems, as outlined in Fig. 2.

Although a black box associator is used to assess the degree of semantic relatedness between the noun phrases (NPs) that are candidates for attachment, an explicitly localist network is used to make the final decision about the previous noun phrase to which PP 2 should be attached. The presence of a candidate PP 2

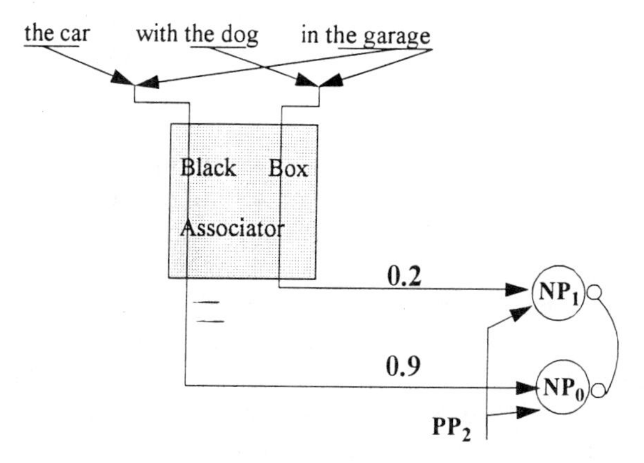

Fig. 2 A static localist network for solving preposition attachment problems: The strength of semantic relation between the third noun phrase "the garage" (NP 2) and each of the first and second is assessed by a black box associator. These association strengths are then used to modulate the activities of two units corresponding to the phrases, NP 0 and NP 1, to which the preposition "in" might attach. The two localist units correspond directly to the cases of the second preposition being attached to NP 0 or NP 1. Competition between the two attachments is achieved by means of the inhibitory link. (From Ref. 2.)

partially activates units for attachments involving the two preceding NPs, which also receive activation from the corresponding output of the associator. A single choice of NP for the attachment is enforced by the mutually inhibitory connection between the two units. The paper that describes this system also shows a similar, but more elaborate, network that can handle a sequence of up to four prepositional phrases, and that enforces the no-crossing rule and the preference for left attachment [2].

Even before Touretzky's application of localist networks to simple parsing tasks, researchers had considered them as the basis of a solution to the problem of disambiguating natural language. Small and co-workers [14], for example, outlined a network using spreading activation and mutual inhibition to perform lexical sense disambiguation for sequentially presented words. This system was elaborated by Cottrell [15] who described not only the lexical sense resolution process, but also a system for parsing by relaxation in a collection of subnetworks generated from productions representing a grammar. Essentially, Cottrell's system produced a large network containing, as subnetworks, every possible instantiation of the productions in the grammar, down to the level of "NOUN" or "VERB" nodes, for example. These units were, in turn, connected to the lexical disambiguation networks. The parse was represented by the active subnetwork after the sentence had been completely presented and the network had reached a stable state. The use of relaxation to find a consistent interpretation of a system's input is a consistent feature of localist networks, and was also used in Waltz and Pollack's parser, described in the next section. A further feature of Cottrell's system was the use of binding units to represent syntactic roles of constituents. The fact that the first Noun Phrase in a certain sentence is the subject, for example, was represented by the activity in a Concept1–Subject binding node, in competition with, among others, a Concept1–Agent node.

B. Dynamically Configured Localist Systems

There is a tradition in AI of representing syntactic structures using trees, and of representing semantics using a network of interconnected frames. With the development of the connectionist framework of doing computation using networks of simple, interconnected units, it was natural that the former structures should be imposed on the latter computing devices. Under this model, networks of connectionist units are dynamically assembled into structures representing, for example, competing parses or competing semantic interpretations for a sentence. These networks are essentially similar to the static localist networks described earlier, but do not require a network to represent every possible parse for every possible input. Even though neither learning nor claims of biological plausibility are possible for such networks, they do allow the connectionist techniques of spreading activation, mutual inhibition, ready integration of multiple knowledge sources, and constraint satisfaction by relaxation to be applied when useful in an otherwise more or less conventional parsing framework.

A very early model of this type was Waltz and Pollack's "phenomenologically plausible" parser [16]. This system used spreading activation to share support among elements of a consistent interpretation of syntactic, semantic, lexical, contextual, and pragmatic cues to a sentence's meaning, and lateral inhibition to allow inconsistent interpretations to compete with each other. In interpreting the sentence "Mary hit the mark," for example, the network[3] in Fig. 3 can reach two stable interpretations, depending on whether previous context has activated the "Gambling" or "Archery" context units. When the "Archery" context is active, the network has settled into the interpretation "Mary succeeded in striking the target"; in the "Gambling" context, the interpretation is "Mary gave another card to her card-sharking victim."

The network is initially constructed by adding a set of possible syntactic networks, connected internally by excitatory connections and to each other by inhibitory connections, to a static network representing semantic knowledge about each of the terms in the sentence and to context units that can be activated to retain topic information from previous sentences. Next, the appropriate context units are activated, and all others are set to zero, and activation values are repeatedly computed for the units in the network, until a stable activity pattern is found. In this way, semantic, syntactic, and contextual knowledge sources are integrated to find a consistent interpretation. In an elaboration of the system, described in the same paper, the unitary context units shown here were replaced by sets of units representing contextual microfeatures, combinations of which could be activated to produce a "priming context" representing a variety of concepts, even somewhat novel ones.

For garden path sentences, such as "the astronomer married a star," the network exhibited the same "double-take" that humans experience, first attempting the CELESTIAL-BODY interpretation of "star," but finally settling on the MOVIE-STAR interpretation.

4. REPRESENTATIONS

The DCPS, BoltzCONS, and the localist systems, described in the foregoing, all depended on customized representations designed by the experimentor for their operation. For ease of application, and ultimately, to take advantage of learned representational correspondences, it was important that more general representational schemes be developed. In the following, a technique that uses tensor products to allow a wide variety of structures will be described, followed by the powerful technique of symbol recirculation, which allows representations to be learned in the course of network operation. After symbol recirculation has been introduced, systems that apply and extend the technique will be described.

A. Representing Symbolic Structures with Tensor Products

Instead of a custom distributed representation, such as the one used in DCPS and BoltzCONS, Smolensky [17] suggested the use of tensor products as a general way of representing value-variable bindings. The tensor, or Kronecker product [18] $A \oplus B$ of two matrices A and B is produced by replacing each element of A with its scalar value multipled by the matrix B. If A and B are two vectors, the tensor product is the matrix of all products of elements of A and B.

For this representation to be used requires that symbolic structures be decomposed into a set of role–filler bindings. Both fillers and roles in Smolensky's formulation are represented by sparse vectors that may be either local (only one element of

[3]This example was constructed for the purpose of illustrating this chapter, and might have been handled slightly differently by Waltz and Pollack's actual system.

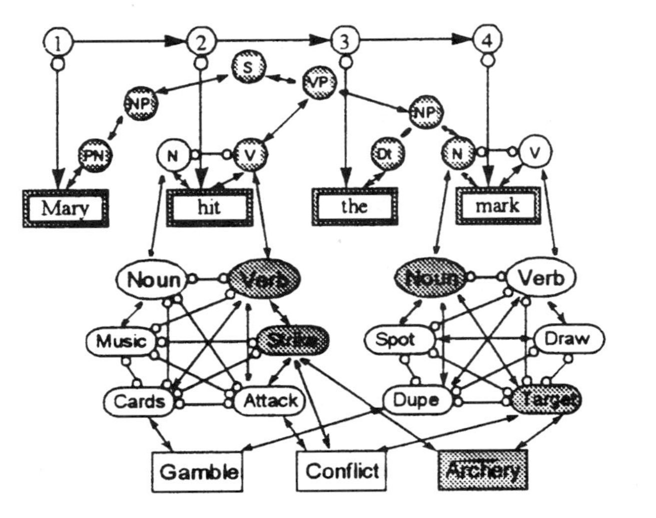

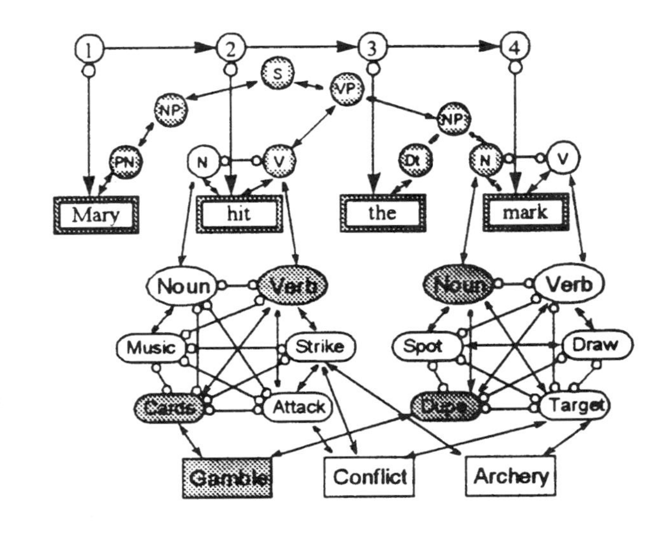

Fig. 3 In Waltz and Pollack's model, spreading activation and lateral inhibition in a network constructed at run-time are used to combine syntactic, semantic, and contextual cues to interpret sentences. Here, the sentence "Mary hit the mark" is given two interpretations following the fashion of the original system. In the interpretation at left, Mary has succeeded at archery, on the right, she has given another card to her card-sharking victim.

the vector is nonzero) or distributed. Tensor product representations are purely local if both vectors are local, semilocal if one is local and the other is distributed, and fully distributed if both vectors are distributed. To store multiple role–filler pairs in a tensor product memory, one must simply add together the products for each role–filler pair $M = \sum \vec{f} \oplus \vec{r}$. To extract a filler from M, one can use a self-addressing unbinding procedure: $\vec{f}_i \approx M \cdot \vec{r}_i$. A symmetrical procedure can be used to extract roles given fillers. A detailed discussion of the expected intrusion of stored pairs into others during retrieval is given in Smolensky's paper.

As the tensor product operation can take place between arbitrary arrays, it is clear that iterated structures can be stored. If M_i is a superposition of tensor product representations of role–filler pairs that represents some case frame f_i, then $N = \vec{r} \otimes M_i$ is a representation of that frame filling some role r. A number of the more successful previous connectionist representations were analyzed in these terms [17], and also according to whether some of the elements of the tensor product were discarded to form the final representation. Following Touretzky's pattern, Smolensky also showed that tensor product representations can be used to implement the symbolic operations push, pop, car, cdr, and cons. In a later paper [19], this parallel development was completed with the demonstration of TPPS, a tensor product-based version of DCPS.

B. Recursion and Compositionality

An important feature of natural language is that it is recursive. A noun phrase can, for example, be constructed from a noun phrase, a preposition, and another noun phrase. One way of handling this property is simply to set a fixed maximum depth for recursion, and to represent each level explicitly. Although this would work in most practical applications,[4] it is unsatisfying, because it guarantees that there are sentences that a system operating in this way cannot parse, but that would be identified as grammatical, if perhaps odd, by a human. Finding a connectionist implementation of recursion was important in establishing the credibility of using neural networks for NLP [20]. One approach to achieving such recursion is Pollack's Recursive Autoassociative Memory, or RAAM [21,22].

RAAMs are multilayer backpropagation networks that can be trained to store a given collection of trees, stacks, and other similar structures. RAAMs are trained as autoassociators, attempting to reproduce their input exactly on their output units. The number of hidden units is smaller than the number of input and output units, with the result that the activity of the hidden units is a reduced description of the input pattern. It is these hidden unit activation patterns that are used to represent recursive structures. Bit patterns representing leaf symbols are concatenated with copies of hidden activity patterns representing trees and used to build more complex trees.[5] For example, Fig. 4 shows how the tree ((A B) C) is represented by a network trained to exactly reproduce the following input patterns on its output {A.B, R(A.B).C}, where the dot represents concatenation, and R(A.B) is the hidden unit activation pattern obtained when the pattern A.B is presented to the input. This latter pattern changes slowly during training,

[4]And is, in fact, the basis for Jain's PARSEC system, described later in the chapter.
[5]Stacks are just right-branching trees, and can be handled in the same manner.

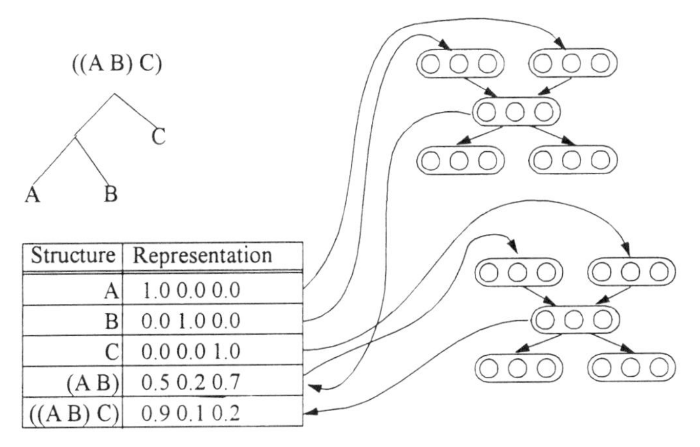

Structure	Representation
A	1.0 0.0 0.0
B	0.0 1.0 0.0
C	0.0 0.0 1.0
(A B)	0.5 0.2 0.7
((A B) C)	0.9 0.1 0.2

Fig. 4 Representing a simple tree using a RAAM: A single autoassociator network is trained to represent both the subtree (A B) and the full tree ((A B) C). Fixed representations for the terminal symbols A, B, and C are initially stored in a global symbol memory, shown at the left of the diagram. To train (A B), the representations for both A and B are simultaneously presented as both inputs and targets for the autoassociator. After weight update, the representation of (A B) from the hidden units of the network is stored back in the symbol memory. That representation, along with the representation for C, is then presented to the network as input and target, and the resulting hidden representation is stored back in the symbol memory as the representation of ((A B) C). The process of training hidden representations for (A B) and ((A B) C) and recirculating these representations to the symbol memory is repeated until stable representations are formed. These final structural representations are effective because, if placed on the hidden units of the network and propagated forward, the representations of their child-components will appear on the network outputs.

causing the input–output (I/O) patterns in the training set to gradually converge to a set of patterns representing both terminal symbols and subtrees. This technique of using neural network training and symbol recirculation to build subsymbolic representations is generally useful, and it was also used to learn subsymbolic representations of lexical items by backpropagating corrections for output errors into their representations on the network inputs in Miikkulainen and Dyer's FGREP system [23–25]. In RAAM, the final hidden unit representations can be thought of, in a sense, as pointers, because placing the representation on the hidden units and propagating forward to the outputs permits one to retrieve the object to which the representations point.

RAAM networks are recursive. Provided that tight enough training tolerances can be reached, there is no limit to the depth of the tree they can represent. They also exhibit representational compositionality, as the fixed-length representation vectors of, for example, the symbols BLUE and COMPUTER, can be combined into a single vector of equal length representing BLUE-COMPUTER. Because this combination and dimensionality reduction is performed identically for all pairs of symbols presented to the RAAM input, the representation for BLUE-COMPUTER should be related to that of its components BLUE and COMPUTER in a way that other networks can learn to exploit.

RAAM networks have become a popular tool for representing symbolic structures. Chan and Franklin [26], for example, used two RAAM networks, together with a simple recurrent (Elman) network [27] and a feedforward network, to build a system capable of reading sentences and producing tree structured representations of their semantics. An overview of this system is given in Fig 5.

Although precise details of the syntactic representation are not given in Chan and Franklin's paper, we may suppose that the parse trees for the training sentences are represented in the system by triples, such as ({subject},{verb},{object}) or (NIL,NIL,JOHN). Some of the elements of these triples, such as {subject},{verb}, and {object}, are RAAM representations of other triples, and some, such as JOHN or NIL are 20-element random lexical vectors. Once the RAAM has learned to store and reproduce the syntactic trees corresponding to the training sentences, its hidden representations for these sentences are used as a training target for the recurrent parsing network. The parsing network receives, on its input, successive 20-element lexical vectors representing words in the sentence, and the recirculated hidden unit representation resulting from the previous word presentation. As each word is presented, the network attempts to reproduce the RAAM syntactic representation for the whole sentence on its output units.

The semantics are also represented by triples, such as (GO,{thing},{path}) in which the some of the items, in this case {thing} and {path}, are RAAM representations of other triples, and some, such as GO, are 20-element vector representations of objects, drawn from a semantic lexicon. Once the semantic and syntactic RAAM networks have been trained, a further "categorization" or transfer network is trained to convert the syntactic representation of a sentence into the corresponding semantic representation. This network is also explicitly given the lexical vector of the sentence verb as input, although it is not clear whether this extra hint is necessary for the system to operate.

After training on 100 sentences, Chan and Franklin's system successfully produced semantic structures corresponding to 34 out of 60 test sentences.

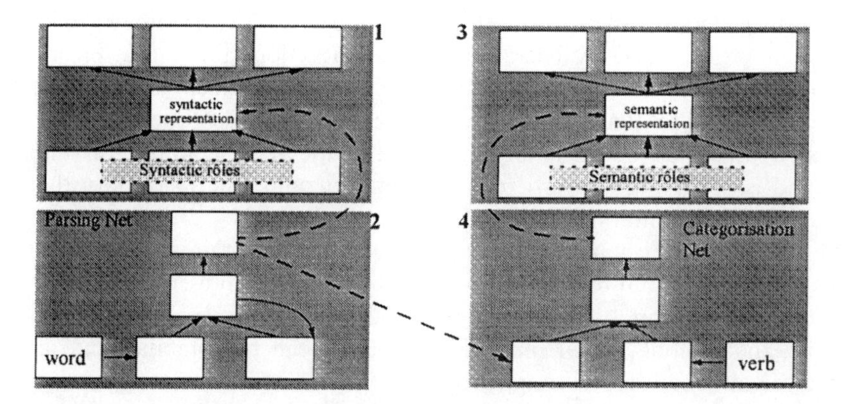

Fig. 5 Chan and Franklin constructed a complete sentence-understanding system from three types of backpropagation networks. The networks are numbered in the diagram in the order in which they were trained. Two RAAM networks (numbered 1 and 3) were used to encode semantic and syntactic tree structures, a recurrent network (2) was used to parse into the syntactic RAAM representation, and a simple feedforward "categorization" network (4) was used to convert syntactic to semantic RAAM representations.

C. Manipulating Transformations: The DUAL System

An even more involved symbol recirculation scheme was used in the DUAL system [28], which was intended to support symbolic structures sufficiently well to represent the background knowledge needed to understand simple stories. As with the FGREP system mentioned before, symbols in DUAL are represented by pairs of symbol names and patterns of activation, and are stored in a global symbol memory. These activation patterns are retrieved for use in the input–output patterns of networks doing mapping tasks and, during learning, are modified by backpropagation into the inputs before being stored back into the symbol memory. In the course of learning mapping tasks, the symbol representations are gradually modified to improve their usefulness for the tasks in which they are applied.

The additional representational technique used in DUAL, but not present in preceding symbol recirculation systems, rests on the observation that both neural network weight sets and the patterns of activation used on those networks' inputs and outputs are simply vectors of floating point numbers. Consequently, the entire weight matrix for a network preforming some mapping can be manipulated as an input or output pattern by another, larger network. It can also be treated as the representation of a symbol and stored in a FGREP style symbol memory.

In DUAL, this observation is exploited by using such a weight matrix-manipulating autoassociative network as a long-term memory (LTM) that stores a complete semantic network. The weight patterns stored in this memory each represent the entire set of (link, link-destination) pairs[6] associated with a particular node in the network. Any of these weight patterns, once output by the LTM, can be loaded as the weight matrix of a short-term memory (STM), which outputs link-targets when a pattern representing a link type is placed on its input.

That the long-term memory in DUAL is autoassociative is exploited in another way. Instead of storing the complete STM weight matrices corresponding to each node in the semantic network in the external global symbol memory, the system stores only a "handle" consisting of the smaller, hidden unit activity pattern that appears in LTM when a symbol's weight matrix is placed on its input. By placing one of these handles directly on the LTM hidden units, the entire weight matrix representing the links from the corresponding symbol can be reconstructed. These handles are also used to represent link destinations on the outputs of the STM networks. The operation of the system as a whole is described by Fig. 6, which shows the roles of the DUAL system's global symbol memory, LTM, and STM with loadable weights, in representing a fragment of a semantic network in which "node i" is connected to "node 1" by way of a "type 3" link. Although handles for nodes converge to a final representation over time as the systems learns, link types in DUAL are represented by hand-coded, orthogonal activity patterns.

Training takes place by first choosing random patterns for the node handles, and then training STM networks for each node. The set of these STM weights is then used as training data for the LTM autoassociator, which produces, after training has converged, an initial handle for each STM symbol on its hidden units. These handles are placed in the global symbol memory, and used to retrain the STMs for each

[6]Which may be thought of as role–filler pairs associated with a frame.

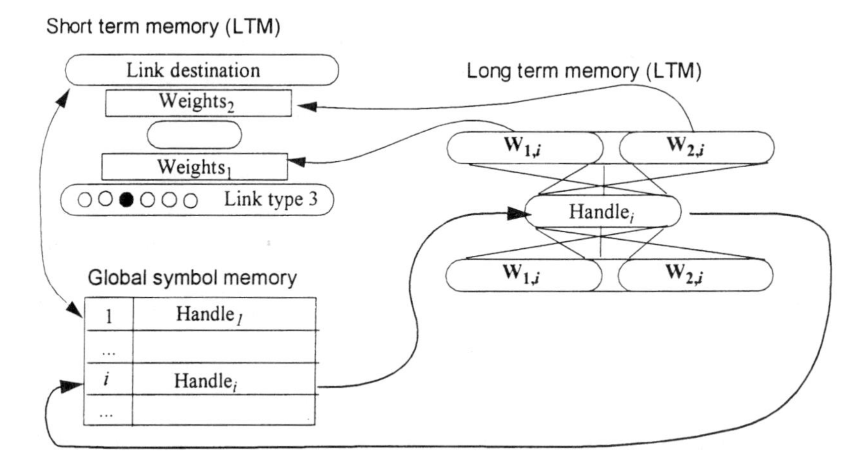

Fig. 6 In DUAL, weight matrices and activation patterns are interchangeable. Long-term memory (LTM) represents a semantic network by storing weight matrices that, when loaded into the short-term memory network, perform link-type, link-destination mappings for some particular node *i*. The link-destinations, in turn, are represented by "handles" consisting of the hidden unit activity pattern that forms in the LTM network when that node's STM weight matrix is placed on its input. These handles are associated with node names by storing them in a global symbolic memory. In the example, part of the network in the LTM connects node *i* to node 1 through link type 3.

node. The LTM is then reloaded with the new weight matrices, and the process is repeated until a stable set of handles for the nodes in the network is learned.

Although the DUAL networks have a great deal of generality and representational power, the training process is time-consuming, even for relatively small semantic networks, and there is no guarantee of convergence to a stable set of handles. As computation becomes steadily less expensive, however, this sort of representation should become an appealing subject for further research. In the meantime, it is possible that giving the system hints, for example, by using a microfeature representation for the initial node handles, could speed convergence.

Ultimately, the intention of systems such as DUAL is to allow constituent and recursive structures to be represented without the need to manually construct any part of those representations. Dyer et al. point out that symbol recirculation systems such as DUAL, while not themselves providing a solution to the variable binding problem, can provide the representations that variable-binding systems, such as that found in Touretzky and Hinton's DCPS and Dolan and Smolensky's tensor product networks, manipulate.

5. FRAMES

Frames are a versatile means for representing symbolic knowledge. They consist of collections of named slots, representing roles of a concept, and fillers for those roles. Conventionally, frames have been represented as records with atomic symbols as the role names, and the fillers have been represented by either atomic symbols or by pointers to other frames. In computational linguistics, frames have been widely used to represent syntactic or semantic structures. A frame representing a simple English noun phrase, for example, might be represented by a frame with slots for the head-noun, a pointer to a frame representing adjectival modifiers, and a flag to represent definiteness.

Because frames are a widespread and useful tool for dealing with language, it comes as no surprise that connectionist researchers have made an effort to cast them into the connectionist framework, and to extend their usefulness.

A. Dynamically Updatable Concept Structures

One such early effort was Touretzky and Geva's Dynamically Updatable Concept Structures (DUCS) [29]. In DUCS, binary activity patterns representing names and fillers are combined in selectors, each of which contains a randomly chosen subset of the tensor product of the slot and filler names.[7] The units that make up these selectors, role units, and filler units are symmetrically connected, in a Hopfield and Tank-style network, so that activity in the selector units can combine with role activity to activate filler units, or with activity in filler units to activate roles.

Activity in multiple selector units is accumulated in a concept buffer that can store multiple role bindings at once. This concept buffer, in turn, interacts with a concept memory that stores buffer contents by activating a unit for every pair of active buffer units. Because of the bidirectional connectivity pattern, this allows completion of concept buffer patterns. This, in turn, activates the selectors, and completes the frame, filing in fillers for roles, and possibly correcting roles. For example, if the role "nose" were filled with "gray-curved-thing" for a parrot, in an animal frame, the role name could be corrected to "bleak." This form of role name completion, enabled by the use of distributed representations instead of atomic symbols, is not easily achieved in conventional, nonconnectionist frame systems.

Moreover, the DUCS system allows reduced descriptions of whole frames—in this case bitslices—to be used as fillers, allowing compositionality, as long as the reduced description retains sufficient detail to allow retrieval of the whole concept.

B. A Sequential Frame Filler

In Jain's more elaborate "Connectionist Architecture for Sequential Symbolic Domains" [30,31], a network is trained, using backpropagation, to control the assignment of sequentially presented words to slots in frames. Although the lexical representation is subsymbolic, the network in Jain's system operates in an essentially symbolic mode, copying lexical representations from one part of the network to another using gating units. There are two stages to the assignment of inputs to slots. In the first phase, the words are assigned to roles in phrase "blocks," and, in the second, these blocks are assigned to roles, such as agent and recipient, in a case frame. This second assignment represents a parse of the input data.

The input to Jain's system is sequential activation of units representing each of the words in a sentence. These word units have connections to a set of feature units, representing both hand-designed semantic and syntactic features (such as plural or

[7]The same size subset of role bits is used for every filler bit, so the selection is not completely random. Units within the selectors are also mutually inhibitory.

animate) and an ID part, that enables words with identical features, such as "John" and "Mark" to be distinguished. As connections from these ID units are not learned, a convenient form of generalization is obtained from this design.

The general architecture of the system is shown in Fig. 7. To enable phrase parsing, gating units connect the feature units to three noun blocks and two verb blocks. When these gating units operate, they may copy feature patterns into preposition, determiner, adjective 1 and 2, or noun roles within a noun block, or adverb, auxiliary, or verb roles in a verb block. Because it fills a fixed number of fixed roles, the phrase-parsing system does not allow the representation of general recursive structure, but it can represent a fairly large class of sentence forms.

The phrase-parsing network is trained independently of the case frame assignment network described in the next paragraph. A set of hidden units in the phrase network takes input from the feature units, from the current contents of the phrase blocks, and from the gating units, and sends outputs to the gating units, determining which phrase block role (or roles) the current input word should occupy. During training the target slot assignments for all words in a sentence are presented as target training signals for the units in the phrase blocks, and a variant of the backpropagation training algorithm for recurrent networks is used to update all weights except those from word ID units. To accommodate role assignment ambiguity, the network must initially learn to assign a word, while it is present on the input, to all possible slots, and later overwrite or empty out those assignments that are contradicted by subsequent input.

Assignments of the phrase blocks to case roles in the sentence are represented by role groups. These groups represent relations between noun blocks and other noun blocks (prepositional attachment), between verb blocks and other verb blocks (subordinate clause) and between noun blocks and verb blocks (agent, patient, recipient, relative clause). The patient relatiton is, for example, represented in a three by

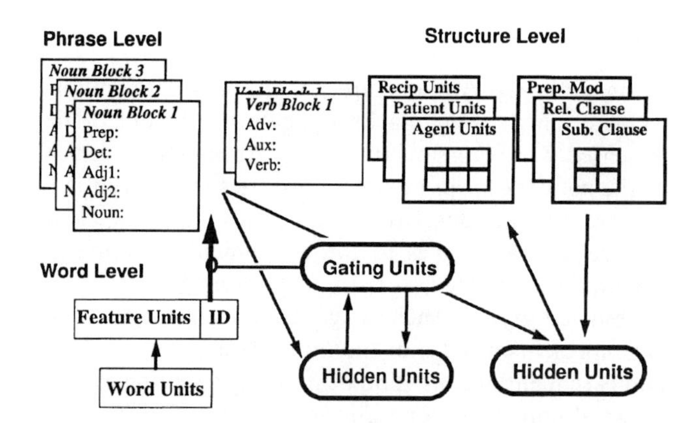

Fig. 7 In Jain's connectionist frame-filling system, sequentially presented words are first assigned to roles in phrase blocks by a recurrent gating network. After this mapping is learned, the contents of the phrase blocks can be used as input to a role assignment network that identifies the syntactic relations between phrases.

two grid of units, with unit X,Y active if noun block X is the patient of verb block Y. After the phrase block assignment network has been fully trained, the contents of the input word feature units and of the phrase block role units are used as input to a set of hidden units that feed into the case role groups. The hidden units also have recurrent inputs from the case role groups, and the network is trained in the same manner as the phrase network, by using the role assignments for the whole sentence as targets, and training network connections using a backprop variant as words are presented.

This system learned to parse its small set of training sentences correctly and to generalize to similar sentences outside its training set. It was later extended as PARSEC [32], which could handle interclause relations, such as "X is a subordinate clause of Y," or "X is a relative clause attached to Y," and identify the mood of sentences. Perhaps more enticingly, when used to parse the output of a speech recognizer for use in machine translation, it was able, following the simple addition of another input, to use the output of a pitch tracker to distinguish syntactically identical questions and statements. Given only a lexicon and hand-labeled parses for 240 sentences, PARSEC was able to learn a parser for a conference registration domain that achieved almost double the coverage on 117 test sentences obtained by the best of three hand-designed parsers. Later, the system was further extended [33] to allow the tagging of output parses with arbitrary linguistic features and, by substantially modifying the basic architecture, to perform shift-reduce parsing, enabling it to handle arbitrarily embedded tree structures.

6. FINAL NOTES

In the introduction, the potential advantages of symbolic and connectionist NLP were reviewed. Ideally, one would wish for a system that combined the advantages of both—from symbolic processing one would draw recursion, compositionality, and easy association of values with variables or fillers with frame roles, and from connectionist processing one would draw subsymbolic representations with a natural sense of semantic distance, constraint satisfaction by relaxation, and graceful degradation in the face of input error. Full NLP systems with these characteristics remain as the subject of future research. What the systems discussed in this chapter provide are a set of useful demonstrations of the possibilities of such a combination, and a set of useful techniques that a reader may apply in building systems for research or application.

The last decade of work on connectionist symbol processing systems has made important advances. Demonstrations are now available that one can perform general computations with these systems, using them as a stored program computer, and that one can represent a large class of data structures including frames and trees, while taking advantage of the semantics of distributed representation. There have even been some large-scale connectionist symbol-processing systems, such as Jain's PARSEC system, that outperformed conventional systems applied to the same tasks.

Much work remains for the future. It is not immediately apparent how any of the systems described here can be extended to build comprehensive systems able to manipulate the hundreds of thousands of lexemes and millions of concepts expressed by natural languages. Perhaps a fruitful line of attack will lie in the combination of connectionism with the statistical techniques that have shown such success in the past few years.[8] Perhaps faster computers and improved network-training techniques will provide partial solutions. Given the promise of the systems that have been developed so far, improvements to connectionist symbol-processing should be well worth the research effort.

REFERENCES

1. Sun R. Integrating Rules and Connectionism for Robust Commonsense Reasoning. John Wiley & Sons, New York, 1994.
2. Touretzky DS. Connectionism and PP Attachment. In: Touretzky D, Hinton G, Sejnowski T, eds. Proceedings of the 1988 Connectionist Models Summer School. Morgan Kaufmann, San Mateo, CA, 1988, pp 325–332.
3. Newell A, Simon HA, Human Problem Solving. Prentice-Hall, Englewood Cliffs, NJ, 1972.
4. Newell A. Physical symbol systems. Cogn Sci 4: 135–183, 1980.
5. Rumelhart DE, McClelland JL, the PDP Research Group, eds. Parallel Distributed Processing: Explorations in the Microstructure of Cognition. MIT Press, Cambridge, MA, 1986.
6. Touretzky DS, Hinton GE. Symbols among neurons: details of a connectionist inference architecture. Proceedings of the 9th IJCAI, IJCAI-85, Los Angeles. Morgan Kaufmann, San Mateo, CA, 1985, pp 238–243.
7. Touretzky DS. Representing and transforming recursive objects in a neural network, or "trees do grow on Boltzmann machines". Proceedings of the 1986 IEEE International Conference on Systems, Man and Cybernetics, Atlanta, GA, 1986, pp 238–243.
8. Touretzky DS, Hinton GE. A distributed connectionist production system. Cogn Sci 12:423–466, 1988.
9. Touretzky DS. BoltzCONS: reconciling connectionism with the recursive nature of stacks and trees. Proceedings of the Eighth Annual Conference of the Cognitive Science Society, Amherst, 1986, pp 522–530.
10. Touretzky DS. BoltzCONS: dynamic symbol structures in a connectionist network. Artif Intell 46:5–46, 1990.
11. Fahlman SE, Hinton GE, Sejnowsky TJ. Massively parallel architectures for AI: Netl, Thistle and Boltzmann machines. Proceedings of AAAI-83, Washington, DC, 1983, pp 109–113.
12. Ackley DH, Hinton GE, Sejnowsky TJ. A learning algorithm for Boltzman machines. Cogn Sci 9:147–169, 1985.
13. Kirkpatrick S, Gelatt CD, Vecchi MP. Optimisation by simulated annealing. Science 220:671–680, 1983.
14. Small S, Cottrell G, Shastri L. Toward connectionist parsing. AAAI Proceedings of the National Conference on Artificial Intelligence, Pittsburgh, 1982.
15. Cottrell G. Connectionist parsing. Proceedings of the 7th Annual Meeting of the Cognitive Science Society, Irvine, CA, 1985, pp 201–211.
16. Waltz D, Pollack J. Phenomenologically plausible parsing. AAAI 84: Proceedings of the National Conference on Artificial Intelligence, Austin, TX, 1984, pp 335–339.
17. Smolensky P. On variable binding and the representation of symbolic structures in connectionist systems. Technical Report CU-CS-355-87, Department of Computer Science, University of Colorado at Boulder, 1987.

[8]An excellent introduction to this area is given in Charniak's book [34]. More detail can be found elsewhere in these volumes.

18. Pearson C. Handbook of Applied Mathematics. Selected Results and Methods. Van Nostrand Reinhold, New York, 1974, p 933.
19. Dolan C, Smolensky P. Implementing a connectionist production system using tensor products. In: Touretzky D. Hinton G, Sejnowski T, eds. Proceedings of the 1988 Connectionist Models Summer School, Morgan Kaufmann, San Mateo, CA, 1988, pp 265–272.
20. Fodor J, Pylyshyn Z. Connectionism and cognitive architecture: a critical analysis. Cognition 28:3–71, 1988.
21. Pollack J. Recursive auto-associative memory. Neural Networks 1:122, 1988.
22. Pollack J. Recursive distributed representations. Artif Intell 46:77–105, 1990.
23. Dyer M. Symbolic neuroengineering for natural language processing: a multilevel research approach. University of California Los Angeles Artificial Intelligence Laboratory. Technical Report UCLA-AI-88-14, 1988.
24. Miikkulainen R, Dyer M. Encoding input/output representations in connectionist cognitive systems. In: Touretzky D, Hinton G, Sejnowski T, eds. Proceedings of the 1988 Connectionist Models Summer School, Morgan Kaufmann, San Mateo, CA, 1988, pp 347–356.
25. Miikkulainen R. A neural network model of script processing and memory. UCLA AI Lab Technical Report UCLA-AI-90-02, 1990.
26. Chan S, Franklin J. A neural network model for acquisition of semantic structures. 1994 International Symposium on Speech, Image Processing and Neural Networks, Hong Kong, 13–16 April 1994, pp 221–224.
27. Elman J. Distributed representations, simple recurrent networks, and grammatical structure. Mach Learn 7:195–225, 1991.
28. Dyer M, Flowers M, Wang Y. Distributed symbol discovery through symbol recirculation: toward natural language processing in distributed connectionist networks. In: Reilly R, Sharkey N, eds. Connectionist Approaches to Natural Language Processing. Lawrence Erlbaum, Hillsdale, NJ, 1992.
29. Touretzky D, Geva S. A distributed connectionist representation for concept structures. Ninth Annual Conference of the Cognitive Science Society, Seattle, WA, 1987, pp 155–164.
30. Jain A. A connectionist architecture for sequential symbolic domains. Technical Report, Carnegie Mellon University School of Computer Science CMU-CS-89–187, 1989.
31. Jain A. Parsing complex sentences with structured connectionist networks. Neural Comput 3:110–120, 1991.
32. Jain A, Waibel A, Touretzky D. PARSEC: a structural connectionist-parsing system for spoken language. ICASSP 1992, International Conference on Acoustics, Speech and Signal Processing, vol 1, 1992, pp 205–208.
33. Buø F, Polzin T, Waibel A. Learning complex output representations in connectionist parsing of spoken language. ICASSP 1994, International Conference on Acoustics, Speech, and Signal Processing, 1984, pp I365–I368.
34. Charniak E. Statistical Language Learning. MIT Press, Cambridge, MA, 1993.

31

The Subsymbolic Approach to ANN-Based Natural Language Processing

GEORG DORFFNER

Austrian Research Institute for Artificial Intelligence and University of Vienna, Vienna, Austria

1. INTRODUCTION

This chapter provides an overview of subsymbolic connectionist approaches (i.e., subsymbolic approaches based on artificial neural networks, ANN) and models in natural language processing (NLP). *Subsymbolic* mainly refers to the exploitation of distributed representations of linguistic and conceptual knowledge and the replacement of classic symbol manipulation by the distributed processes in densely connected artificial neural networks. Such an approach, partly due to its focus on learning and self-organization, and on downscaling of predefined representations, is currently limited in terms of practical applicability. In other words, hardly any existing purely subsymbolic approach to language processing can immediatley be turned into practical applications, such as speech recognition systems or database query systems (as many symbolic and hybrid systems can—compare Chap. 30). Instead the focus is on descriptive adequacy of human language processes in linguistic terms—often, such models strive to define novel concepts for linguistics, challenging fundamental aspects of modern linguistics such as rule-governed behavior and generative grammar.

Nevertheless, subsymbolic approaches show considerable potential for improving natural language processing systems relative to their capability of performing in a natural way as compared with human language understanding. The examples introduced in this chapter are meant to highlight the major contributions of connectionism to such an improvement. The larger part of this chapter is structured along linguistic subfields, such as morphology, syntax, and semantics. Because the aim of this book is to introduce text processing, rather than speech processing, connectionist work in phonology is left out of this chapter. This overview will be complemented by a discussion of the term "subsymbolic" itself, as well as of recent progress toward a novel, situated, and grounded view of language.

Section 2 briefly analyzes the term subsymbolic, its different uses in literature, the basic strengths and weaknesses of subsymbolic models, and the role of symbols in language.

Section 3 gives an overview over the most common neural network architectures used in subsymbolic models: Perceptrons, multilayer perceptrons, recurrent perceptrons, and competitive learning. It discusses the major contributions of those architectures, the role and limits of distributed representations, rule-governed versus rule-following behavior, and the learning of internal representations.

Section 4 discusses connectionist models of morphology. At the center lies the so-called past tense debate, initiated by the famous model by Rumelhart and McClelland and the subsequent criticisms by Pinker and others. It can be viewed as a prototypical case for the discussion of the virtues and limits of subsymbolic language models. It highlights the role of similarity and frequency of input patterns in developing rule-like behavior, and puts forward a major hypothesis raised by connectionism: namely, that regular as well as irregular forms are treated by a single process, whereas classic linguistics would advocate a dual-process theory.

Section 5 focuses on aspects of language syntax. Because the basic properties of recurrent neural networks in implementing formal automata is discussed in a different chapter (see Chap. 29, on automata induction), this section will focus on the work by Elman and Pollack, and on developments of that work. By training recurrent perceptrons on sequential language inputs, Elman has shown that grammatical categories can be learned on the basis of a word's typical position in a sentence. Furthermore, he demonstrates how center-embedding of clauses can be reflected by the dynamical properties of a recurrent network operating in a continuous state space. This section also discusses the concepts behind recursive autoassociative memory for learning distributed representations of syntactical trees.

Section 6 deals with how semantic content can enter a subsymbolic model. Two classes of models have to be distinguished. The first class focusses on explicit hand-wired representations of semantic contents and the subsymbolic mapping between sentences and those representations. The second class exploits the concept of representational grounding and focusses on the mapping between words and semantic categories, learned on the basis of perceptual features. Somewhere in between lies the model of the so-called sentence gestalt—a distributed representation of sentential content in a recurrent network based on the task of answering questions about this content. Work combining semantic and syntactic components is also introduced.

Section 7 summarizes the main contributions and commonalities of the different connectionist models discussed in the previous sections, and discusses their possible practical applicability.

2. WHEN IS A MODEL SUBSYMBOLIC?

The term *subsymbolic* is quite loosely used throughout the literature. Often it is synonymous with "connectionist," excluding only models that assign predefined meaning to all of its components (units)—sometimes referred to as "structured" or "localist" connectionist models [1,2]. Often it refers to approaches making use of *distributed representations*, be it through explicit definition (e.g., through setting up feature vectors to represent conceptual knowledge), or through learned representations in the hidden layer of a feedforward network. A few authors have tried to define subsymbolic more stringently. Smolensky [3], in his seminal paper on connectionism, located subsymbolic models on a lower level than the conceptual one, meaning that what is expressed by these models is more fine-grained than a conceptual (symbolic) description. In particular he states (p. 17) that

> [I]n the symbolic paradigm, the context of a symbol is manifest around it and consists of other symbols; in the subsymbolic paradigm, the context of a symbol is manifest inside it and consists of subsymbols.

An incarnation of this view is the concept of *harmonic grammar* [4,5], in which two levels are posited: a lower level consisting of distributed connectionist representations, and a higher conceptual level of local representations. Both levels interact through jointly maximizing their harmony.

Dorffner [6] has pointed out that what is referred to as "symbol" in classic artificial intelligence is very similar to symbols in semiotics, that is, a sign that is arbitrary (its shape does not reflect its meaning) and discrete (see also [7]). With connectionist models in mind one can now argue that there is no real evidence that cognitive systems need anything like arbitrary signs to represent the world. Instead, the notion is replaced by *concepts*—special connectionist states in state space, which are *not* arbitrary and not necessarily fully discrete. Signs (and among them, symbols), according to this view, enter cognition only when and where they are actively used by a cognitive agent to refer to something, for instance, in language. Therefore, I [6] suggest that it would often be more appropriate to call the models "subconceptual," whereas a focus on symbols is specific to words and word use.

For the purpose of this chapter, *subsymbolic* mainly refers to models that contain at least some process, the components of which cannot be clearly interpreted in a conceptual way. This mainly concerns the process transferring input to output representations through a large number of connections, with or without hidden units. Examples are networks deriving phonological or morphological variants of word forms without a single component clearly identifiable as a symbolic rule (see Sec. 4), and networks developing distributed representations of syntactic form (see Sec. 5) or semantic content (see Sec. 6) of sentences.

3. THE MOST COMMON NETWORK TYPES

The core architecture for the majority of subsymbolic language models is the multilayer perceptron (MLP; Fig. 1) (i.e., a feedforward neural network with at least one hidden layer, trained by error backpropagation [8]). A few models use single-layer perceptrons without hidden units, which are identical with linear discriminators. The important function of perceptrons is the *associative mapping* between two linguistic entities, for instance, between one word form and a corresponding derived form, between a sentence and its meaning, or between a word and its pronunciation. Through the concerted spread of activations in a large number of connections, one strength of this type of process lies in its ability to acquire rule-like (or *rule-following*) behavior from training on examples, without actually explicitly implementing any of the rules (i.e., without being *rule-governed*). Another frequent use of multilayer perceptrons is their ability to learn *internal representations* in the hidden layer based on the task-specific training. In other words, input patterns are recoded into hidden layer patterns taking similarities in the input and the task at hand (similarities in the required output) into account.

A major property of MLPs trained by backpropagation is the dependency of their performance on two important aspects:

1. The similarities of input, as well as target output patterns
2. The frequency of input patterns

Training with backpropagation extracts important statistical properties of the input pattern distribution. Frequency refers to both *type* (how many different input patterns or classes of input patterns) and *token* (the number of occurrences of single-input pattern instances). Because the network attends to similarities in the input, frequency does not only play a role for single patterns themselves. Classes of input patterns are defined through similarity and thus frequent patterns can influence rare but similar patterns as well. Many authors put forward strong claims that this property is the right kind of process for language modeling. On the other hand, many authors have seen this as a disadvantage of connectionist models, for there are many apparent counterexamples for which neither similarity nor frequency seems to be the decisive parameter for learning. The examples in the following sections will highlight the pros and cons of this central property of connectionist models. For instance, Sec. 4.B presents a few examples for the ability of networks to learn rule-based behavior that is low in frequency.

Because many aspects of language involve the processing of sequential input, another frequently used class of networks are *simple recurrent networks* (SRNs, see Fig. 1). At their core, those networks are multilayer perceptrons, but they possess one or more additional input layers, receiving feedback from other layers in the network at the previous time step. Two very common architectures are frequently named after their original authors; namely, the Jordan-type SRN, with a "state layer" receiving feedback from the net's output layer [9]; and the Elman-type SRN [10], with a "context layer" receiving input from the net's hidden layer (or from the central hidden layer, if there is more than one [11]). Both context and state layers represent a type of memory for the network used to process input, depending on its temporal position in a sequence. Recurrent networks can implement a mapping between a sequence of inputs and a single output or a sequence of outputs. In addition, they can be used to predict future elements of the sequence, or they can be used to learn internal representations about the entire sequence.

Owing to their basic nature as MLPs, backpropagation training can be applied to recurrent networks as well. A suitable extension is backpropagation through time [8], where the recurrent network is seen as being unfolded into a feedforward network by considering several subsequent time steps. Through this learning algorithm, longer time dependencies can be learned more efficiently [see, for instance, Ref. 12].

A third class of models is based on unsupervised learning and categorization, with competitive learning [13,14] and its variants such as adaptive resonance theory [15] or self-organizing feature maps [16]. The function of these networks is to learn internal, often categorical, representations even without teaching feedback through the environment.

4. CONNECTIONIST MODELS OF MORPHOLOGY

Similar to phonology—which I am not discussing in more detail here—morphological processes are usually modeled in a rule-based framework in classic linguistic theory. Based on lexical information stored with each morpheme—the smallest unit with semantic content—a set of rules are applied to derive inflected or derived forms given syntactic and semantic context. Only in some cases analogical processes are assumed to be at work, such as in the inflection of so-called strong verbs into their past tense forms in English (e.g., sing–sang, ring–rang, and such).

Most of the connectionist work on morphology is focussed on linguistic theory and aspects interesting from a cognitive science point of view (e.g., plausible replication of child language acquisition data). For engineering applications, at least at first sight, there do not seem to be many advantages of using a connectionist approach to interpret or generate word forms over a rule-based one. Nevertheless, it is worth taking a brief look at the major discussions to show the basic workings of rule-following behavior in connectionist networks on one hand, and to highlight possible virtues of subsymbolic models reflecting some of the facets that give language its flexibility and creativity in many cases, on the other.

A. The Past Tense Debate

The formation of the English past tense has become the most prototypical example demonstrating the possible contribution of connectionism to aspects of linguistics, the main differences between connectionist and classic theory, and the possible shortcomings of the connectionist approach. This is why this part of morphology deserves a more detailed discussion.

Compared with many other languages, morphology in modern English is reduced to a rather simple core set of inflections. Therefore, it is rather surprising that English morphology—and here even a more reduced subset, the formation of past tense forms of verbs—has become such a central case for connectionism in linguistics. Some authors [e.g., 17] have argued that the anglocentric view that has been applied to the field of morphology in connectionism has led to a variety of misconceptions and limitations of the models. Nevertheless, the English past tense shows complex and rich enough phenomena to have sprouted two quite controversial theories.

The discussion was started with the seminal work by Rumelhart and McClelland [18] who demonstrated that a simple feedforward neural network (a perceptron without hidden units) could be trained to successfully map phonological representations of English verbs onto phonological representations of their corresponding past tense. They used a training set that starts small and is gradually extended to mimic the increasing exposure of children to words. The major arguments around this work can be summarized as follows:

- The network achieves the infinitive–past tense mapping without the incorporation of explicit rules.
- Regular forms (e.g., walk–walked), irregular forms with subregularities (e.g., sing–sang), as well as completely irregular forms (e.g., go–went) can be modeled by one single process.

- The network learning behavior reflects important properties of child language acquisition, in particular, so-called U-shaped curves of learning performance, where—owing to the "discovery" of a rule and its subsequent overgeneralization—performance can temporarily decrease before increasing again (Fig. 2).

The first achievement highlights one of the major contributions of subsymbolic connectionist models to language processing, which also characterizes many subsequent models. Through learning and self-organization in distributed weight spaces, processes that are classically modeled through explicit rules can be implemented without such rules.

The second aspect has come to be known as the "single versus dual process hypothesis" argument, discussion of which has continued to this day. As one of the major critics of the original work [18]—Pinker—and his colleagues [19] have frequently contended, mappings in morphology (and other subfields of linguistics, as well) are best viewed by the interplay of two distinct processes: a rule-based process applied to regularly derived forms, and rote learning (a look-up table) for irregular forms. Only recently, classic linguists seem to have admitted that an associative network can take the processing part for some irregular forms [20] (as cited in [21]).

The third aspect touches on one of the most attractive properties of connectionist models; namely, the capability of reproducing intricate patterns of human learning. Because of their incremental and adaptive nature, neural networks are able to show improvements and changes during learning that have remarkable similarity with changes observed in humans—in this case, children in their early language acquisition phase. U-shaped curves are but one important example.

Plunkett and Marchman [22] have argued that the size of these U-shapes might have been overestimated in the work by Rumelhart and McClelland [18]. They trained a multilayer perceptron including a hidden layer, demonstrating so-called micro–U-shaped curves more similar to what is exhibited in child learning, using a more uniform training set. By this they also address a major criticism by Pinker and Prince [19]; namely, that the careful selection of initially small and subsequently increasing training sets [18] is highly implausible.

Another major criticism [19], further elaborated [23], is that similarity between inputs cannot be the major cue used to determine the past tense form of a verb. They constructed examples, such as "Alcatraz out-Sing-Singed Sing-Sing" (meaning that Alcatraz became a better Sing-Sing than Sing-Sing itself) where they can show that test subjects would hardly ever choose to say "out-Sing-Sang" despite the large similarity to the irregular verb "sing." Also, homophonous words with different ways of forming their past tense (such as "break–broke" and "brake–braked") are not accounted for by the model. They further argue that, contrary to what connectionists suggest, the inclusion of semantic features in the input would not solve the problem, but again make false predictions (e.g., that "hit," "strike," and "slap" are semantically closely related, but have quite different past tense formation rules).

Kim et al. [23] offer an alternative theory for the class of verbs that have been derived from nouns (denominal verbs). This theory states that denominal verbs cannot be irregular, because nouns do not have markers for past tense, and thus the verb cannot inherit it. With this they explain, for instance, why people say "He flied out to center field" (meaning that a baseball player hit a fly ball and was put out) rather than "flew out." Harris [24], and subsequently Daugherty et al. [25], convincingly argue that connectionist models can easily account for this phenomenon by resorting to *semantic distance* between a derived verb and the verb exhibiting the irregular past tense. Furthermore, they demonstrate that the connectionist model goes further than the rule-based scheme in that it rather naturally predicts preferences, which do not always have to be all-or-none. For instance, it is mentioned [25] that some people do, in fact, prefer "flew out," because the semantic distance to the prototype of "to fly" can be rather low. As a result, despite a long list of phenomena not accounted for by the original model, connectionism can stand up to scrutiny and rather naturally explain a variety of phenomena based on similarity and frequency. They can explain why "Alcatrax out-Sing-Sang Sing-Sing" or even "He broke the car in front of the elk" (meaning that he slowed it down) are rather unlikely but can nevertheless occur to be interpreted[1] given the proper context. The subsymbolic approach never completely rules out any hypothesis.

B. Frequency and Default

Marcus et al. [26] bring forward another argument against the connectionist hypothesis that a single, associative, process accounts for all morphological transformations. They argue, and prove with experimental data, that in German plural formation the *–s* ending is the default plural, even though it is rather rare (only about 7.2% of noun types, and 1.9% of noun tokens take that plural in standard German). Furthermore, they argue that a connectionist account would predict that the default is always the most frequent case, such as -ed in English past tense. Thus, they conclude, German plural is the "exception that proves the rule." This is in line with another major criticism of connectionist models; that is, that they are merely based on statistical distributions in the training set.

Hare and Elman [27], while concentrating on replicating the historical development of English morphology, demonstrate that a network can learn a default rule even if it is not the most frequent one (both in type and token frequency). Multilayer perceptrons divide the input space through a number of hyperplanes (one per hidden unit) into potentially complex decision regions. Such decision regions can be unbounded after learning, giving the network the capability to generalize into regions not richly covered by training data. Whereas in engineering applications of neural networks (e.g., classification in medicine) this property is unwanted, in language models this can often lead to plausible extrapolations (by treating everything that is *not* covered by explicit training data as the default case).

Similarly, Indefrey and Goebel [28], while comparing feedforward with recurrent networks, show that connectionist models can successfully learn low-frequency rules. Their example is weak German noun declension, which occurs in only about 1% of noun types. Their results are another demonstration that connectionist models, although basically implementing a statistical approach, nevertheless are able to represent low-frequency mechanisms and thus are not limited by grossly unequal input type and token distributions.

---

[1]Note, however, that none of the presented models deal with interpreting (understanding) word forms, but only with their generation.

Sanchez et al. [17], however, have pointed out that the focus on English (and in some cases, German) morphology might give a biased view not appropriate for other languages. For instance, they argue, languages with more than one default class might pose major, still unsolved, problems for connectionist models.

C. Other Work in Morphology

Gupta and MacWhinney [29] present a model on German articles and noun inflection combining a categorization network (competitive learning) and two multilayer perceptrons (Fig. 3). It is trained on a number of words reflecting the frequency of standard German. German has a a set of six articles and several ways of modification (umlaut and suffixes) for expressing plural forms. The authors [29] argue that categorization—in this case, categorization of nouns into gender—plays an important role in linguistic learning. A single article, together with case and number information, does not uniquely determine the gender of a noun. Therefore, the categorization network learns to distinguish between 14 different situations (combinations of article, case, and number), and accumulates these categories into a distributed representation of gender (in the layer entitled "co-occurrence"). This representation is used in the two MLPs learning to output the correct article, as well as the correct modification (e.g., suffix or umlaut) to produce the inflected form. Additional input is the phonology of the noun.

Lee [30] analyzed simple recurrent networks (see Sec. 3) used for the task of producing different types of noun inflections, such as suffixation, assimilation, prefixation, and reversal. He was able to show that the network reflects many, although not all, preferences for some mechanisms over others that can be observed across the world's languages. For instance, suffixation is much more widespread than prefixation, reflected by the fact that the network learns the former much more easily than the latter. A mechanism such as reversal of phonemes is hardly ever seen in any language, and the network has great difficulties learning it. Thus, the network can explain some kinds of *universals* in human language, demonstrating that connectionism might indeed be on the right track to represent human language capabilities.

Similar results were achieved by Gasser [31], who also trained recurrent neural nets to learn different types of morphological processes. While emphasizing on the receptive nature of the morphological process (i.e., a human learner learns to understand word forms before learning to produce them), he can also show that processes observed in human languages are favored by the network over processes that have not been observed (and thus appear to be difficult, or even impossible, to learn for humans).

5. GRAMMAR AND THE PROCESSING OF SYNTACTIC STRUCTURE

Syntax is generally believed to be a—if not the—most central aspect of language. Classic natural language processing systems, therefore, focus mainly on parsing—the analysis of symbol strings into a syntactic structure given by a grammar. It would seem that this process is the prototypical case for symbol manipulation, thus leaving little or no room for subsymbolic approaches to show any strengths. However, this is not true. Quite a large body of connectionist literature deals with aspects such as

1. The implementation of formal grammars in recurrent neural networks

2. The problem of learning a grammar based on examples
3. The distributed representation of syntactic structures
4. The view of grammar as a dynamical system

The first, and partly the second, aspects are dealt with elsewhere in this book (see Chap. 29). Therefore, I will concentrate mainly on the remaining ones.

A. Finding Structure in Time

The title of this section is that of an influential paper by Elman [10], who demonstrated that a simple recurrent neural network (SRN) can learn interesting facts about grammatical structure based on the presentation of word strings (simple sentences) alone. The network he used is a multilayer perceptron with hidden layer feedback—a network often named "Elman network," as mentioned in Sec. 3 (see Fig. 1). The input layer was fed with local representations of words, one at a time, forming simple sentences, such as "boy eat cookie" or "man smash glass."[2] The task of the network was to predict the following word in the sentence (e.g., "cookie" after having seen "boy eat"). Thus, the target for backpropagation learning (or backpropagation through time) [see also Ref. 12] was the subsequent word in the training sequence, which consisted of a concatenation of a large number of simple sentences.

The task the network is trained on cannot be learned perfectly, as in most cases it cannot uniquely be decided which word follows next.[3] At best, the network can give probabilities for each word to follow—which, for the chosen local representations, is indeed possible. For this task, however, it must be able to extract information about the underlying structure in the given sentences. Elman visualizes this rather elegantly by looking at the hidden layer activations. For each word occurring in the training sequence he calculates the average vector of hidden unit activations. The set of these vectors is then subjected to a hierarchical cluster analysis depicting the relative euclidean distances between patterns in a hierarchical tree, with patterns closest to each other forming binary subtrees (Fig. 4).

This clustering reveals a quite interesting and useful grouping of the patterns. First, it is important to observe that these average hidden unit patterns can be seen as an internal representation of the words, because they are the average patterns activated by these words. These representations are distributed, even though the original input representation is a local one, showing similarity relations that are reflected in the cluster tree. This tree shows that the internal representations of most words that we would call nouns form a natural group, and those that we would call verbs form another. It seems that the network has discovered grammatical categories without having been explicitly told about them. This makes much sense, for the concept of grammatical category helps in solving the task of predicting the subsequent word in a sentence. Even though one cannot say with certainty what word will follow the sequence "boy eat," one can easily make the prediction that what follows the sequence "noun verb" will be a noun again (viewed against the background of

[2]Note that in these simple examples, details such as inflectional endings or articles were left out for simplicity.

[3]In technical terms, the process is non-Markovian in that the current state—expressed through the activation pattern in the context layer—does not uniquely determine how the sequence continues.

Elman's simple sentences). After all, classic grammar theory is defined over those categories and not over specific words (except in some cases of auxiliary words).

I deliberately said that the network "seemed" to have learned about grammatical categories. A closer look at the internal representations reveals that the clusters formed in the hidden layer only partially correspond to "noun" and "verb":

- The clusters are formed solely on the basis of the given training sequences and the given task, thus expressing only syntactic information. Traditionally, "noun" and "verb" also play a role in semantic decisions; for instance, in that nouns most often refer to objects or states, whereas verbs most often refer to transitions or processes. Nothing of this kind can have entered the network's internal representations. Even the apparent subgrouping in semantic categories, such as "animate" and "inanimate" (see Fig. 4) is a pure syntactic distinction (e.g., some verbs such as "eat" follow only nouns that refer to animates).
- The clusters do not seem to be perfect. In some cases (e.g., the word "see" in Fig. 4) words are far away from the main clusters to which we would assign them. It is easily possible that this is not simply an error, but reflects details about the given training sentences, and thus enables the network to perform its task better than if those words had developed representations closer to those of the others.

From this we see that the internal representations are clearly domain- and task-specific. They do not exactly correspond to any category a linguist would apply in syntactic analysis, but they are close to optimal for the task that must be performed. This is one major strength of this subsymbolic approach: instead of relying on representations defined and given by the programmer, which might be inadequate, misleading, or incomplete relative to the given task, the network is able to develop internal representations that are optimally[4] suited to the task at hand.

Chater and Conkey [12] have elaborated a bit more on these hidden layer representations. They show that an even more plausible clustering can be obtained when viewing patterns according to the predicted word, or by clustering the difference vectors that change the hidden layer patterns at each time step. They also demonstrate that this clustering relatively directly reflects the statistical dependencies between words in the sentences, when directly calculating the conditional probabilities for subsequent words given a sequence of preceding words.

B. State Space Transitions

The hidden layer patterns clustered in Fig. 4 can be given another interpretation, which was subsequently explored by Elman himself [11]: the current activation pattern in the hidden layer, which is fed back to the context layer before processing the next input, can be seen as a state[5] in a high-dimensional state space. The formation of a new state, based on the old state and the current input, can be seen as a state

[4]They are optimal only provided that the network architecture is suited for the task and that the learning algorithm finds an optimal solution.

[5]For this reason, it would be more appropriate to call Elman's layer a "state layer," as Jordan [9] has done it for his network.

transition (Fig. 5). By looking at this state space and the resulting state transitions one obtains a formal *automaton* similar to automata used in traditional formal language theory [32,38]. The biggest difference is that hidden layer vectors span a *continuous* state space, where states bear a euclidean distance, and thus a similarity, measure to one another. In other words, states can be more or less similar, a property that is readily exploited by the network. Although usually this state space is bounded (through the minimum and maximum activations of the hidden units), it is nevertheless potentially *infinite* relative to the number of states.

This last observation is important for the computational complexity of the automata that can be implemented with recurrent networks. For natural language, it is well known that at least context-free grammars are necessary to describe its complexity. Such grammars, expressed as automata (usually so-called pushdown automata) need an infinite number of states or a stack mechanism (as opposed to regular languages describable with finite-state automata). Pushdown automata, which would be necessary for processing context-free languages, exploit a stack mechanism permitting the return to the point the parse left off before processing an embedded structure. Miikkulainen [33] introduces a model that implements a stack in a connectionist way. This model will be described in more detail in Sec. 6. Another approach using an explicit but connectionist version of a stack is that of Das et al. [34].

On the other hand, humans have natural limitations to understand sentences with an arbitrary number of recursive embeddings (an aspect of language that makes it context-free, rather than regular). This limitation is usually attributed to memory limitations. Weckerly and Elman [35] present simulations in which SRNs show limitations similar to humans, but that are due to the precision in which hidden units can exploit the infinite state space, and not to some arbitrary memory restrictions (e.g., a stack of limited length). Thus, the network appears to express human language in a more natural way than traditional parsers.

This more "natural" way of representing grammar is worth a closer look. On one hand Ref. [35] distinguishes between center-embedded sentences such as "The man the boy the woman saw heard left" and right-bracketing structures like "The woman saw the boy that heard the man that left." Similar to the network, humans are much better at interpreting the second kind of sentence. Given classic complexity theory, this is not surprising for right-bracketing (or "tail-recursive") structures are describable by finite-state automata and are thus regular, whereas center-embedding truly needs the full power of context-free grammars. SRNs also appear to have fewer problems in representing and learning regular languages than context-free ones. In fact, it is not even clear whether they are at all capable of representing, let alone learning, context-free languages, as they are defined classically.[6] However, given the analysis [35] and related ones, one can argue that they do not need to represent them

[6]Quite a few references deal with demonstrations and proofs of second-order recurrent networks as being able to represent regular languages [32,37,38]. Reference [32], in turn, proves that there exists a first-order recurrent neural network that is equivalent to a Turing machine. The architecture they use in their proof is quite complex, and thus it is still unclear what the complexity of SRNs as described here would be. Substantially fewer results exist for the learnability (e.g., when gradient descent algorithms such as back-propagation are used).

in order to model natural language, as long as they cover the ranges humans are able to cover easily—which they seem to, in their simple examples.

Yet another facet has been added [35] to this analysis. As is also well-known, humans are able to interpret center-embedded sentences much more easily if there is a semantic bias (i.e., if the nouns and verbs involved constrain each other such that possible confusion is reduced). A sample sentence they give would be "The claim the horse he entered in the race at the last minute was a ringer was absolutely false." All involved nouns and verb phrases constrain each other (e.g., only horses can be entered, not claims) such that there is much less confusion as to who did what. The SRN reproduces these results. Returning to the network as a continuous-state automaton again, a possible explanation is this.[7] We observed that states possess a similarity structure (distance measure). In the SRN, new states are derived based on the previous state together with the new input. Both similar states and similar (or identical) inputs can give rise to similar state transitions (given through the input-to-hidden connections and the nonlinear activation function of the hidden units). Through error backpropagation, similar outputs (in their case, the predicted subsequent word) can also have an influence on the state. This means that if at a certain point in the input sentence similar outputs are predicted, the resulting states will be close together. If dissimilar outputs are predicted, states will be farther apart. Words such as "woman" and "man" have a high probability of being followed by the same set of words, whereas words such as "claim" and "horse" are likely to be followed by quite different words. As a result, state transitions in the first case will all cover very much the same portion of state space, whereas in the latter they will move in different directions. This is roughly visualized in Fig. 6. As a result, state transitions in the former case will have greater difficulties in "staying on track" (i.e., in following the path of a complete automaton which describes the sentence) than in the latter.

Similar results have been obtained [11]. An SRN is trained with simple sentences including center embeddings, very similar to [10]. In this case, Elman explicitly analyzes the resulting state transitions in the hidden layer. Because this layer has a large number of units, depiction of such state transitions becomes difficult. One method to at least roughly visualize the trajectories is to apply a principal component analysis to the resulting patterns and depicting the two or three most relevant components. Figure 7 depicts such trajectories for the three sentences

1. Boy walks.
2. Boy sees boy.
3. Boy chases boy.

One can clearly see that the trajectories for sentences 2 and 3 take quite similar paths, since the expectations after the verbs "sees" and "chases" are similar, in that a particular object is expected, whereas sentence 1 deviates from that, resulting in a state similar to the end state of the other two sentences (from then on expectations are similar again). Elman also demonstrates that the network can successfully learn to handle person and number agreement between noun and verb, even if the two words are separated by an embedded sentence.

[7] This goes a bit beyond how these authors [35] interpret their own results.

This work on viewing syntactic processes as trajectories in state space has subsequently lead Elman to view language as a dynamic process [36]. Combined with other results on the dynamic nature of grammar learning [e.g., 37,38] this should lead to a quite novel way of viewing and modeling syntax in the near future.

Finally, Elman has pointed out that his network succeeded in learning complex embedded sentences more easily if, prior to that learning task, simple clauses without embeddings are learned. He referred to this effect as "the importance of starting small" [39], and it could become a major cornerstone for SRN training in more elaborate applications to language processing.

C. Recursive Autoassociative Memory

In one of the most influential cricitisms of connectionism, Fodor and Pylyshyn [40] have argued that a connectionist representation can never represent complex structures built combinatorially from primitive constituents. This is, as they argue, because unit activations in a network form only sets without any labeled interrelations between them. When more than one representation is active at the same time, there is no way of telling which concept plays which constituent role in the relations. Therefore, they conclude, connectionism is inadequate for modeling cognitive processes, including, most conspicuously, language.

In a simplistic view of distributed representations this seems true. If one pattern of activations represents the relation—say, "ask"—, another pattern represents the agent, and a third one the recipient, then simultaneous activation of all three patterns would lead to a representation expressing no information about the binding of the constituents to their roles (i.e., who asked whom). This, incidentally, apparently applies to both distributed and local representations. For this reason, it should be impossible to represent any syntactic structure of a sentence if it involves more than one constituent. However, there is much more to distributed representations than this simplistic view suggests.[8] This has been nicely demonstrated by Pollack [41] using a network architecture called *recursive autoassociative memory (RAAM)*.

A RAAM consists of a multilayer perceptron with one hidden layer, and input and output layers of identical size, divided into two or more equally sized groups. The size of the hidden layer is identical with the size of one of those groups (see Fig. 8 for an example of two groups). The purpose of the network is to recursively develop internal distributed representations of syntactic trees by training the net to reproduce the input at the output, an autoassociative task. In the case of two input and output groups, arbitrary binary trees can be encoded. Consider the example in Fig. 8. By first letting the two input groups represent the two leaf nodes A and B of the simple binary tree, and training the RAAM to autoassociatively reproduce A and B at the output, the hidden layer develops a compressed and truly distributed pattern (independent of whether the original representations of A and B were local or distributed), which could be considered as representing the combination of A and B. This representation is of the same size as the original leaf node representations and can thus be used subsequently to encode it together with node C as the next binary substructure of the tree. Provided one does not reach the capacity limit of the net-

[8]An elegant solution to this so-called binding problem in a network using local representations has been presented [1].

work, one can encode arbitrary binary trees with this procedure, obtaining a compressed pattern that represents the entire tree.

To see that the resulting pattern indeed represents the entire tree, one just has to look at the network and its processing steps. Any hidden layer pattern can be seen as being situated between an encoder (the input-to-hidden connections) and a decoder (the hidden-to-output connections). Once the network has been successfully trained, the transformation can be executed uniquely in both directions: A tree can be encoded stepwise into a compressed pattern, and any compressed pattern can be decoded (using only hidden-to-output connections) into the original patterns it contains. Van Gelder [42] speaks about "functional compositionality" in this context: the resulting distributed pattern representing a compositional structure is functionally equivalent to the classic concatenative structure (a symbolic representation of the tree through concatenating symbols), but is not in itself concatenative (i.e., the constituents cannot be easily identified by splitting the representation into parts[9]).

The RAAM clearly demonstrates that it is *not* true that distributed representations cannot express information about complex relations and their constituents. The way it achieves this is not by providing one unit for each possible combination of constituent and role in a relation (the only alternative acknowledged [40]), but by superimposing patterns onto each other through nonlinear functions. To visualize this, Sharkey [43] suggested looking at what a single hidden unit is doing (Fig. 9). Consider a constituent A (for simplicity, represented locally) paired with another constituent B in one case, and with C in another. Owing to the nonlinearity in the hidden unit (in this case, a sigmoid function), the contribution of A to the unit's activation is different when it is paired with B than when it is paired with C. The activation thus uniquely encodes the fact that it is the pair A/B (or A/C, respectively) that is being represented. Looking at the entire hidden layer, there will be hidden units largely activated by A alone, by B (or C) alone, or by the specific pairing. If this pattern is then processed together with yet another constituent D, nonlinearity ensures that the resulting pattern is not simply the sum of all constituents, but that it preserves their specific ordering.

Such distributed representations are only useful, if (a) they can be further processed in a connectionist network, and (b) they show strengths of their own, rather than being just a reimplementation of the classic symbol-processing mechanism.

With respect to (a), several authors have shown that using these representations as inputs and outputs to an additional multilayer perceptron, transformation of syntactic structures can indeed be learned, with some capacity of generalization. Chalmers [44], for instance, has used two independent RAAMs to encode syntactic trees of English sentences in the active and passive voice, respectively. The additional MLP could then successfully learn to transfer an active sentence into its passive version.

With respect to (b), rather little discussion can be found in the literature, at least for RAAM-like architectures processing purely syntactic structures (for

[9]To be fair, one must add that this view is relative to what we humans can perceive as being compositional through concatenation. Imagine we had a visual system that performs exactly the decoding process of a RAAM; in this case we could again easily identify the constituents.

approaches involving semantic interpretations, see Sec. 6.A). Taken at face value, the proposal [41] must be seen as a kind of implementationalism, proving that connectionist networks are capable of representing compositional structure, but providing no motivation for doing so in the connectionist rather than the symbolic way. In a sense, models such as Chalmers [44] cannot be seen as truly subsymbolic, because all they do is reimplement a symbolic mechanism.

One strength, nevertheless, is the neural network's ability to learn from examples and to generalize based on similarities. Distributed RAAM representations do show interesting similarities between related structures, and between structures containing related constituents, and thus give rise to generalization properties in further processing. It remains to be seen whether this kind of generalization bears any properties not easily implemented with traditional rule-based approaches.

We [45] have argued that another major strength of distributed representations in the style of RAAM could be their ability to express a "soft" variant of compositionality. Not every concept is unambiguously identified as being compositional, that is, consisting of constituents, and the roles played by the constituents can be as fuzzy as object concepts themselves. Thus, we conclude, connectionism can go farther than simply reimplementing a classic syntactic mechanism.

A major weakness of RAAM is its lack of *systematicity* [46]. Take the problem addressed in Ref. 44, for instance. From a system that plausibly handles human language one would expect that once sentences such as "John gave Mary a book" can be transformed into their passive version "Mary was given a book by John," it should automatically be able to also transform the following sentences:

- "Mary gave John a book"
- "John gave his sister a teddy bear"
- "John gave Mary, who he met just yesterday, a book"

Hadley [46] distinguishes between three types of systematicity (meaning that parts of the sentences can systematically be replaced with other parts, without influencing the performance):

1. Weak systematicity: the model is able to generalize to novel sentences only if every word has occurred in every permissible position in the training set.
2. Quasi-systematicity: the model is able to generalize if any embedded sentence is isomorphic to the main learned sentences, and if every word occurs in every permissible position on at least one level,
3. Strong systematicity: the model can generalize to words in positions where it has never occurred in the training set.[10]

Now, connectionist models using RAAM representations have only shown, at best, weak systematicity. As Clark [47] puts it, representations from RAAM are not freely *usable* because they are bound to the contexts with which they have been trained, and thus any process working on them cannot be (strongly) systematically extended to new contexts and structures. Traditional symbolic approaches have no problem with this. Therefore, connectionism still bears the burden of proof that all

---

[10]Note that [46] does not even require arbitrary embeddings for this kind of systematicity.

types of systematicity—as long as they are plausible[11]—can be successfully addressed.

Another criticism on RAAM (or similar distributed representations encoding compositional structure) was raised by Fodor and McLaughlin [48], saying that the distributed representations encode structural information, but that the constituents contained therein cannot be tokened. By this they mean that any process working on the basis that structure X contains constituent A (no matter where) cannot be easily implemented. Sharkey [43] showed that this is not true, at least not to the extent that Fodor and McLaughlin [48] suggest.

D. Relation Between SRNs and RAAM

Use of a hidden layer pattern as input, as is done in the RAAM training procedure, is very similar to the feedback of the hidden layer to the context layer in an Elman-type SRN. The main difference between the SRN and the RAAM, as described in the foregoing, is that, in the SRN the feedback is fixed, whereas in the RAAM the pattern can be fed back to any of the groups in the input layer. A further difference is that the RAAM is trained autoassociatively.

There is also a varient of RAAM—the sequential RAAM, or SRAAM [41]—that bears an even closer relation to Elman-type SRNs. An SRAAM does not assume trees with a fixed number of successors in each node as the basic structure, but a sequential structure. Every tree can be transformed into a sequence if additional symbols to represent parentheses (to indicate subtrees) are used. An SRAAM accepts elements from this sequence, one by one, and transforms them into a compressed representation, which could be termed a (connectionist) stack. In Fig. 8, the right part of the input could be seen as the stack, continuously receiving feedback from the hidden layer. The size of the hidden layer no longer needs to be equal to the size of the representation of a leaf in the tree, and trees are no longer restricted to equal valence (equal number of successors for each node). Now the network becomes almost identical with the Elman-type SRN, except that it is still trained on an autoassociative task, including the stack pattern itself as a target output pattern.

Kwasny and Kalman [49] have illustrated this relation between SRAAMs and SRNs. They constructed sequences from parse trees, given by a simple context-free grammar, and trained an SRAAM to represent these. Subsequently they subjected all hidden layer patterns that occur during processing any of the trained sequences to a cluster analysis, the same way Elman [10] did, with the average hidden patterns corresponding to words. The resulting clusters (similar to Fig. 4) represent groups of points in state space. Each encoding process in the SRAAM can now be followed step by step by looking at into which cluster the resulting hidden layer pattern falls. By identifying an entire cluster with one state, a finite-state automaton can be abstracted, describing the state transitions during "parsing" (a similar procedure was applied to recurrent nets for sequence processing by, for instance, Giles et al. [50]). Because cluster analysis results in a hierarchical structure of clusters and subclusters, one can choose the granularity of the automaton. The choice of different

[11]Humans also have their limits in processing embedded and arbitrarily transformed structures.

granularities results in automata accepting more or fewer sentences from the original grammar [49].

We see that the workings of RAAM (in this case, SRAAM) can be interpreted in the same way as that of SRNs. Continuous, high-dimensional state spaces are exploited to represent the intricacies of complex syntactic structures, at the same time expressing similarities between similar structures through euclidean measures.

6. SEMANTIC INTERPRETATION OF LANGUAGE

The thrust of literature on models incorporating semantic representations can be roughly divided into two strands:

1. Work involving explicit representations of semantic content, to which a string of words must be mapped
2. Work dealing with the question of how meaningful structures can come about through immediate experiences in the world (i.e., how meaning can be *grounded* in perceptual experiences)

I will discuss these two strands in turn.

A. Semantic Interpretation of Sentences and Texts

A variety of subsymbolic approaches deal with the semantic interpretation of language by answering questions such as the following:

How can words and phrases in a sentence be mapped onto semantic case-frame representations of propositions?
How can missing information be filled in from experience?
How can novel words be interpreted?
How can pronoun references be resolved in texts with several sentences?

Most work in this area exploits distributed representations—either programmed explicitly or learned in the hidden layer of a neural network—to form the basis for generalizations (e.g., filling in a missing concept can be viewed as a generalization based on similarities to previously learned sentences). Furthermore, many papers deal with the application of recurrent network structures, similar to the ones described in Sec. 3. What distinguishes these approaches from the foregoing ones is that they explicitly incorporate semantic information, thus making the internal structures richer than those of networks learning on word sequences alone. A major common theme behind this work is the emphasis on a neural network as a *constraint satisfaction* process, in which semantic interpretation is the result from resolving a large number of cues into a coherent representation.

The paper by McClelland and Kawamoto [51] was one of the first attempts to use a connectionist model to map sentences (surface structures) onto semantic representations (deep structures). The authors set the scene for later work by introducing semantic case frames [52] as the major medium to represent semantic content. For instance, the sentence "John gave Mary flowers" can be interpreted as involving the concept GIVE, which has three roles: AGENT, RECIPIENT, OBJECT, which are filled with the concepts JOHN, MARY, and FLOWERS. They use distributed semantic feature representations for the words in their grammatical categories (such as verb, subject, object) as input, and distributed slot–filler pairs according to the semantic roles as output of a single-layer perceptron. They demonstrate that the network can fill in missing roles, solve lexical ambiguities, and express different shades of meaning.

St. John and McClelland [53] extended this concept of sentence to case–frame mapping by using recurrent networks based on the fact that sentences are basically sequential inputs of varying length. The model is depicted in Fig. 10. This network can be seen as actually consisting of two networks: a recurrent one with Jordan-type feedback connections from the "output" layer, and a regular multilayer perceptron. The task of the recurrent network is to learn distributed representations based on sequential input in the layer entitled "sentence gestalt." This name suggests that, similar to the RAAM, the representations holistically represent the syntactic as well as semantic information in the sentence. Because this representation cannot be given external feedback, it itself serves as the input layer to the MLP, the task of which is to extract information again, by answering "queries" about the sentence. For this, the additional input layer is fed with case roles with empty fillers, which have to be filled by the network. Errors from the final output to the network are propagated all the way back to the original input, via the sentence gestalt.

St. John [54] uses the same concept to encode representations of entire texts, or stories. Consequently, the resulting representations are called "story gestalts." The architecture is identical with that in Fig. 10, except that entire propositions are used as input. However, "no attempt was made to produce naturally sounding text." Instead of sentences consisting of sequences of words, propositions divided into agents, predicates, recipients, and so on, with a unit for each possible filler were represented at the input. Thus, an important step of text understanding—that of parsing sentences into propositions (the goal of the previous work [53]) was bypassed.[12] Stories consisting of several propositions were generated based on six different scripts, such as eating at a restaurant or going to the beach.

The model exhibited several important properties:

- *Pronoun resolution*: When pronouns (i.e., unspecific fillers of roles in the input propositions) were used, the query-answering network could successfully replace them with the concrete filler, taking cues like recency or gender into account.
- *Inference*: Missing propositions (e.g., who left the tip in the restaurant) could be filled in based on previously learned more complete stories (e.g., that it is always the person who pays the bill who leaves the tip). Note that this is one of the original motivations of using scripts in traditional text understanding. In the connectionist network, scripts are implicitly represented in the distributed story gestalt.
- *Revision*. The network can both predict the final proposition (e.g., that the person who chooses the restaurant will leave the tip) and revise this prediction when new information comes in during the course of the story (e.g., that another person pays the bill).

[12]The same can be said about Ref. [51].

- *Generalization*: The model is able to generalize to novel texts composed of familiar pieces, in a plausible way. For instance, the network would "misread" novel concepts in familiar contexts, just as humans do.

Schulenburg [55] addresses the foregoing problem [54] of using preprocessed propositions rather than continuous text by combining the models of sentence gestalts [53] and story gestalts [54]. Furthermore, he argues that the context of a sentence in a text influences its interpretation. He extends the model by also using the story gestalt as an input for the sentence-processing network. The overall architecture (called SAIL2) is depicted in Fig. 11. A further difference to St. John [54] is that Schulenburg [55] uses representations for entire questions, rather than single predicates in the query network.

Miikkulainen and Dyer [56] propose another connectionist approach to text understanding using recurrent networks and distributed representations. They emphasize the need for modular systems and for distributed input representations that are not handcrafted [51]. For this they propose the FGREP module that develops meaningful representations while solving the task of assigning semantic roles to syntactic constituents. The feedforward version of FGREP consists of a multilayer perceptron that maps a set of word representations in syntactic positions (such as subject or "with-phrase") onto a set of identical word representations as fillers of semantic roles (such as agent or instrument). This is still very similar to [51]. The major difference is that word representations start with random patterns and are adapted during learning. For this, the error used for adapting weights is further propagated back to the input and used to adapt the representations. As a result, words used in similar contexts develop similar representations. Those word representations are stored in a global lexicon and can subsequently be used for further processing.

Also, a recurrent variant of FGREP is proposed [56] to avoid the fact that the foregoing scheme requires quite a bit of preprocessing (similar to the preprocessed text [54]). Furthermore, to distinguish between different instances of the same word (e.g., the word *man* in "the man who helped the boy blamed the man who hit the boy"), they extend word representations by a unique pattern, which they call the ID-part of the representation. The whole pattern then consists of an ID-part and a content part developed by FGREP. Note that this highlights an essential difference to approaches, such as [54] or [10], that use local representations of words. Here, words are not input as arbitrary symbols, but as a pattern that expresses their content through their previous use. Thus, these patterns actually do not really represent words, but concepts that form the underlying meaning of those words. The separate global lexicon establishes the mapping between the words and the corresponding concepts.

From recurrent FGREP modules, a larger model—DISPAR—was constructed, which takes sentences of a story as the input and learns to output paraphrases of these stories. This architecture is depicted in Fig. 12. It was successfully trained on a number of script-based stories, showing similar performance to [54].

Miikkulainen [33] introduces a model (SPEC) combining SRNs, RAAM and FGREP to map sentences including embeddings onto semantic case roles. Its architecture is depicted in Fig. 13. One of the main arguments is that most other networks for sentence processing cannot generalize to novel structure. This is related to both the discussion on systematicity and the discussion on SRNs and context-free grammars. As a solution, Miikkulainen suggests an SRN combined with a stack, realized as a sequential RAAM that pushes the current hidden unit representation whenever an embedding is detected. This representation is popped when the embedding has been processed, and the previous analysis by the SRN can be continued. An additional network, called the "segmenter," is used to detect the embedded structure by detecting a relative pronoun such as "who" and to adjust the hidden layer representation on one hand, and to activate one of three control units (for the actions *push*, *pop*, and *output*), on the other. The latter aspect demonstrates that a subsymbolic model consisting of several modules can easily "take care of its own control" [33].

A comment on processing embedded structures is in place here. To detect an embedding and thus to issue a push action on the stack it needs a special marker, such as a relative pronoun. This is especially important because no current neural network approach is capable of backtracking once the analysis has gone the wrong way. Put differently, the subsymbolic models described here can only handle sentences for which, at any point, it can be decided unambiguously whether the analysis should be continued on the same level or whether a recursive application of the analysis (either directly in state space, or through the application of a stack mechanism) should be performed. Garden path sentences like the famous "The horse raced past the barn fell" are still beyond the scope of such models.

B. Grounding Meaning in Perception

A variety of authors have investigated how meaning—and here, mostly meaning of single-content words or short phrases is meant—can come about as a result from experiencing the world through sensors. The main observation is that meaning can be acquired only through perception of the world, through subsequent categorization or other buildup of internal representations, and through associating word forms with these representations. In this sense, explicit design of meaningful representation is replaced by self-adapting structures in a neural network that interacts with its environment. This process of tying representations and thus meaning to perception is usually called "grounding." Offering the basic ingredients for grounding representations (and thus solving the "symbol grounding problem" [57]) is another main contribution of connectionism to cognitive and language modeling.

The basic architecture of many of the models in this realm consists of two components to recognize or categorize two types of perceptual input (e.g., visual and acoustic), and associative connections between them to map word forms (e.g., presented acoustically) to meaning (e.g., a visually induced category; Fig. 14). None of these models claim to entirely solve the problem of perception itself (which would require an intricate treatment of such complex processes as visual or auditory recognition). Therefore, sensory stimuli in these models are extremely simple, in most cases consisting of simple bit patterns, distorted by noise to simulate the stochastic nature of the stimuli.

Cottrell et al. [58], for instance, introduce a model that uses an autoassociative network on each side to derive internal representations of acoustic stimuli on one hand and images on the other. Two more backpropagation networks then learn to map these two representations onto each other. In a subsequent paper [59], the authors extend the approach to short sentences and action sequences by using simple recurrent nets.

Chauvin [60] and Plunkett et al. [61] realize both categorization in the two perceptual components and the associative mapping between them through one large autoassociative multilayer perceptron, with two separate and one common hidden layer (Fig. 15). This model is able to reflect important properties of early language acquisition in the child, such as over- and underextensions of meaning, and the comprehension-before-production asymmetry. The former refers to the observation that children often assign meanings to words that are either more restricted than their adult uses (e.g., "ball" referring only to a particular stuffed ball), or more extended (e.g., "ball" referring to all round objects). The latter refers to the observation that children usually can comprehend many more words than they can produce. Another modeled aspect [61] is the so-called "vocabulary spurt"—a phase in which the rate of learning dramatically increases after some time of relatively little progress. An explanation based on dynamical systems theory is given [39] for this phenomenon.

Two pieces of work [62–64] model the same associative process between words and their meanings using variants of competitive learning. Schyns [62] uses a self-organizing feature map [16] as the basic component for building categorical representations based on visual stimuli. A bidirectional associative network can subsequently map word representations to those categories. Dorffner et al. [63] achieve categorization through soft competitive learning, exploiting mechanisms from adaptive resonance theory [15] to achieve stable representations and in some cases fast learning of new representations. The latter accounts for the often-observed fact that children can learn words after a very low number of exposures, provided the stimuli are sufficiently novel or perspicuous. While being able to simulate observations similar to Plunkett et al. [61], both these models put forward the following additional arguments:

- Categorization—and thus concept formation—although being strongly influenced by language, also happens independently of language (prelinguistic categories) and thus requires a separate module. In the MLP approach [61], this modularity is also reflected through separate hidden layers and particular learning schemes, but word learning and categorization are more intricately linked than in Refs. [62] or [63].
- Human categorization exhibits important properties, such as prototype or basic level effects [65], which can be nicely simulated in competitive learning schemes. Whereas the authors [61] also analyze the performance of their model relative to prototypes and more borderline cases of categories, both Refs. [62] and [63] more explicitly discuss this apparently important property of categories as the basis for language meaning.

Incidentally, the comparison between models based on multilayer perceptrons and models based on competitive learning points to potentially important extensions of connectionist models of language processing. Although MLPs exhibit interesting dynamical properties for speed and progress of learning, competitive learning appears to provide solutions for common problems with MLPs usually referred to as "catastrophic interference" [66] and the "stability-plasticity dilemma" [13]. These problems deal with the observation that language learning apparently is more stable and allows for less interference between newly acquired and previously learned representations than MLPs usually exhibit [for a discussion see Ref. 64]. While this is true for all aspects of language discussed in this chapter so far, it has become most evident in the discussion surrounding the acquisition of grounded meaning.

The models of grounding meaning described so far are rather simple, deal only with some content words, and use a rather artificial perceptual input (except [58], where real images are used). Furthermore, it is evident that

- Meaning cannot be reduced to categories based on a single mode of perceptual input. Multimodal sensors, context, action processes when dealing with objects, and such, all play a decisive role in building meaningful representations.
- When using rather artificial quasi-featural input patterns (as is done, for instance, in [61] or [64]), the difference between "grounded" models and the ones using explicit preprogrammed representations (see Sec. 6.A) is rather ecumenical and mainly serves theoretical discussion.

The model introduced by Nenov and Dyer [67]—called DETE—goes a decisive step beyond this apparent simplicity. DETE consists of a large number of modules, both connectionist and algorithmic, aimed at simulating, rather complex, aspects of grounded language learning including the following:

The acquisition of short sentences referring to action sequences
The acquisition of simple tenses such as past and future
The learning of the ability to answer simple queries
The ability to acquire different languages that "carve out" different categorical meanings from the sensory environment.

The last point on this list in particular emphasizes the main contribution of models focusing on the grounding of meaning. This contribution is concerned with the observation that knowledge—be it linguistic or nonlinguistic—is not predefined and objective, but must be seen in relation to a cognitive agent. Connectionism, through mechanisms and architectures for learning based on sensory stimuli, lends itself perfectly for what could be called a *constructivist* view of cognition [68,69]. In technical terms, this means that language can be appropriately understood and modeled only when the relativity of meaning can be accounted for. In DETE, this is highlighted by showing that different languages can refer to different conceptual representations of the same environment. By taking this view further it becomes conceivable that such a model would also account for the relativity of language in robots or other embodied devices interacting with a physical environment [70].

The main connectionist module in DETE for processing stimuli, both sensory and linguistic, is a component termed Katamic memory [71]. This is a network inspired by the temporal nature of dendritic trees in biological neurons designed to be able to learn, recognize, and predict sequences of patterns. It was shown [72] that Katamic memories can be visualized as a kind of time-delay neural networks [73] with intricate random connection patterns.

Inputs to DETE consist of simple sensory patterns on a visual field reflecting still and moving objects of different size and shape. Words are also presented as physical stimuli (a property DETE shares with the model [63]); namely, in terms of simple sequential phonological patterns. Based on these stimuli, DETE can successfully learn to map sentences such as "Red ball will move up hit wall" to corresponding two-dimensional action sequences in its visual field and answer questions such as "What size [does object have]?" or "What moves?" Although the claim is not to model the visual system as such, DETE reflects several details known from neurophysiology, such as the separate processing of shape, motion, and color. To solve the so-called binding problem in bringing these separate pieces of information together, synchrony in pulse sequences emanated by the units in the network is used. Figure 16 gives a rough overview of the DETE model.

A model introduced by Hutchins and Hazlehurst [74] takes another interesting and important step in understanding the grounding of meaning, viewing language as a social phenomenon involving several individuals learning from each other and adapting such that language serves its purpose in transmitting information between the agents. Although similar approaches exist in the domain of autonomous robotics [75], the model [74] is one of the few connectionist attempts to model language as an interagent process, rather than one involving only one passive agent.

The main idea is to use autoassociative MLPs to extract distributed representations from visual stimuli, the hidden layer of which receives a target input from the hidden layer of a different agent (Fig. 17). This hidden layer representation is viewed as an external symbol (word) that must be learned to be shared with others. Therefore, hidden layer patterns from one individual is used as a target pattern for the hidden layer of another individual. Note the similarity to the network in Fig. 15. The major difference is that here two (or more) separate agents—possibly with different goals and experiences—learn their internal representations in an interactive scheme. Thus, not only the influence of language on meaningful representations is modeled, but also the influence of language as a social phenomenon that must work in a society of agents.

The simulations show that a population of agents can develop a shared language (a lexicon of symbols), even if no preassigned mappings between word and meanings are given. Again, subjectivity of internal representations plays a major role in this setting. In addition, the model can deal with the limits of words to convey private meanings to other agents, and provide a plausible account for the stable development of shared lexica.

In summary, grounded models of language learning, although still in their infancy, promise to provide exciting new routes that can lead both to a better understanding of natural language and to improved natural language processing systems.

7. SUMMARY: THE MAIN CONTRIBUTIONS FROM SUBSYMBOLIC MODELS

Although, as we have seen, most work on subsymbolic models of language processing focuses on questions interesting from a linguistic or cognitive science point of view, there are several main contributions that could enter practical natural language processing systems in the near future. These include:

1. The concept of *distributed representations* offers an elegant way of dealing with associative processes and similarity based generalization in language. Such representations are, in principle, capable of expressing information about arbitrarily complex compositional syntactic, as well as semantic, structures. Therefore, the obvious problems with systematicity notwith-

standing, it can be expected that connectionist systems exploiting these representations show a natural and more robust behavior than do classic systems, when dealing with everyday language.
2. Several models demonstrate that subsymbolic models form an intricate system of *constraint satisfaction* taking syntactic, semantic, and contextual information into account. The constraints involved are "soft ones" (i.e., they can be relaxed in favor of other constraints). Thus, the systems always produce the most likely hypothesis given the current input and previous experience, never completely ruling out any hypothesis or relying on all-or-none decisions. This, too, contributes to the robustness of the systems.
3. In the domain of syntax, we have seen that connectionist models appear to offer a quite novel theory based on continuous state transitions and dynamic systems theory. In many respects, this theory appears to reflect the capabilities and limits of human syntactic processes more naturally and adequately than traditional generative models.
4. At first sight, it might not seem particularly relevant for a natural language processing system to reproduce human behavior as naturally as possible (including humans' weaknesses), as long as it produces useful interpretations. However, any computer system has limited resources, both for memory and processing. If, now, a system is able to exploit these resources in a way such that it maximizes its performance for aspects in which humans are strong, while sacrificing performance wherever humans have difficulties, those resources might be used in a much more fruitful way. Consider traditional symbolic language processing. That approach is unsurpassed in processing arbitrarily complex recursive syntactic structure, even including hundreds of embeddings. However, they often fail in processing creative novel uses of words or phrases. Subsymbolic systems promise to provide a shift away from the former toward the latter. In this view, many of the results in morphology or syntax described in the foregoing might prove very useful for practical systems in the near future.

One property conspicuously missing from foregoing list is a model's *adaptivity*—usually brought forward as another strength of connectionist models. By adaptivity one would understand the capability of the system to learn from experience, to improve its performance while performing, and thus to adapt to new situations. Although the term "learning" occurs in almost every paper on connectionist language models, one must say that the issue of true adaptivity (in a practical sense) is still largely unresolved. Most models, especially the ones relying on multilayer perceptrons, are *not* able to continuously adapt their performance to new language stimuli without seriously interfering with previously learned processing. This has become known as *catastrophic forgetting* or *catastrophic interference* [e.g., 66], and it has not been sufficiently addressed in connectionist literature [for arguments, see also 64].

8. CONCLUSION

This chapter has given an overview of some major subsymbolic approaches to natural language processing, focussing on aspects such as morphology, syntax, and semantics. It is clear that the list of references given here must be incomplete given the vast literature in this field. Nevertheless, the chosen examples should have highlighted the most important contributions of connectionism to language processing. Although for the time being most models focus on issues interesting from a cognitive science point of view, there is a realistic prospect of engineering applications to subsymbolic natural language processing in the future. In particular, connectionism promises to contribute to robust and natural systems that deal with language in the way humans do.

ACKNOWLEDGMENTS

Most of the references given here were used in a seminar on connectionism and language acquisition at the University of Vienna, jointly held with Prof. Dressler, Dept of Linguistics. I thank him and the participants in the seminar for valuable comments on the papers.

The Austrian Research Institute for Artificial Intelligence is supported by the Austrian Federal Ministry of Science and Transport.

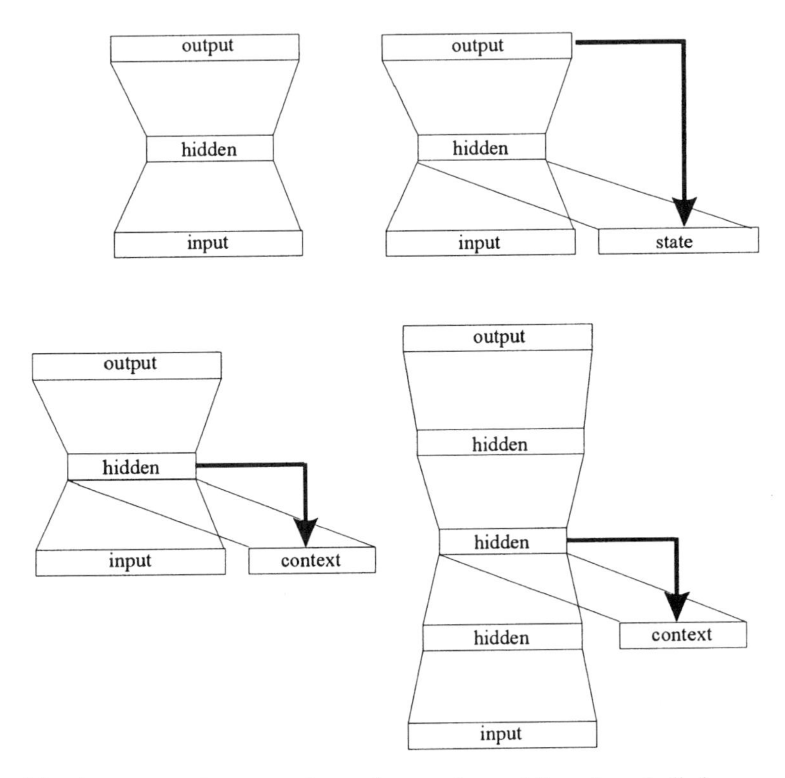

Fig. 1 Four major types of neural networks used for subsymbolic language models: a multilayer perceptron (upper left), a Jordan-type recurrent network (upper right), an Elman-type recurrent network (lower left), and an extended Elman-type recurrent network with additional hidden layers. Trapezoids between layers depict full feedforward connections from the bottom to the top. Arrows depict one-to-one recurrent copy connections. These networks are usually trained by backpropagation.

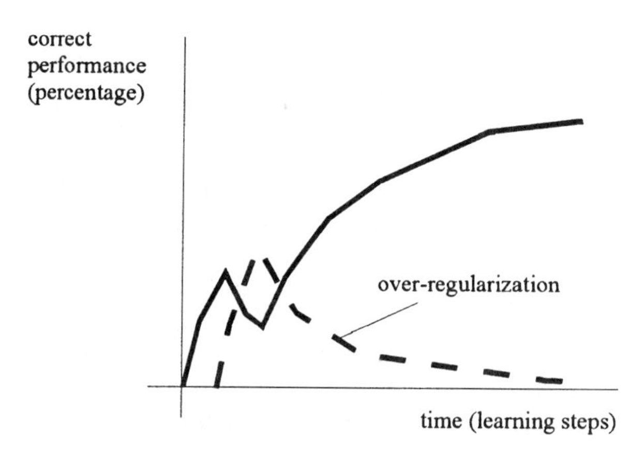

Fig. 2 Learning curves of the past-tense model: The network performance initially rises, meaning that the relatively small training set is memorized by rote learning. As soon as the regular productive rule is discovered (induced by extending the training set, now containing many regular past-tense forms), the performance drops owing to overregularization (e.g., go–goed). Only after continuous learning with both regular and irregular verbs does overregularization decrease and performance increase again. This behavior is called a "U-shaped" curve. (From Ref. 18.)

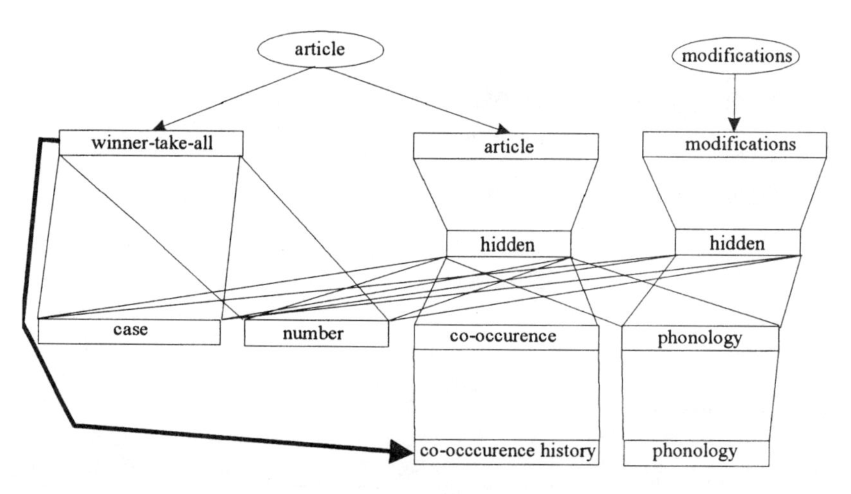

Fig. 3 The model for learning the correct German article and noun inflections: It consists of three networks: an unsupervised competitive learning (left), an MLP to learn the association between past case/number/article co-occurrences and the correct article (middle), and an MLP to learn the association between noun phonology and the modifications needed for inflection. The recurrent connections lead to cumulative patterns expressing all different categories of co-occurrences detected by the unsupervised winner-take-all network, thus forming a distributed representation of a noun's gender. (From Ref. 29.)

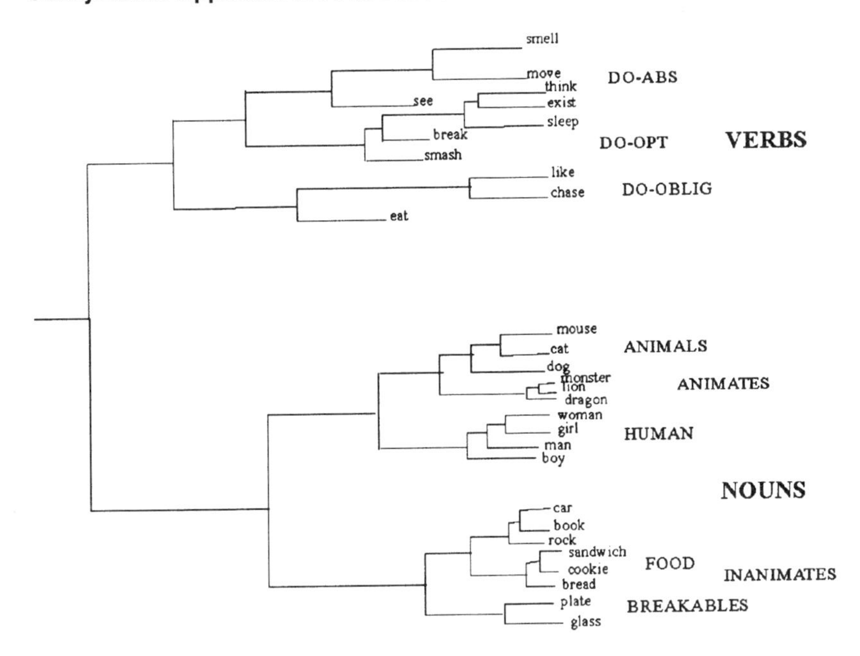

Fig. 4 Results of a cluster analysis of hidden unit representations in the current network after learning to predict words in simple sentences, as described in [10]. Neighboring words (leaf nodes in the tree) indicate proximity in the state space of the hidden layer. The results show that the internal representations of the words naturally group themselves into verbs and nouns, and into subclusters, such as words referring to animates or food. (From Ref. 10.)

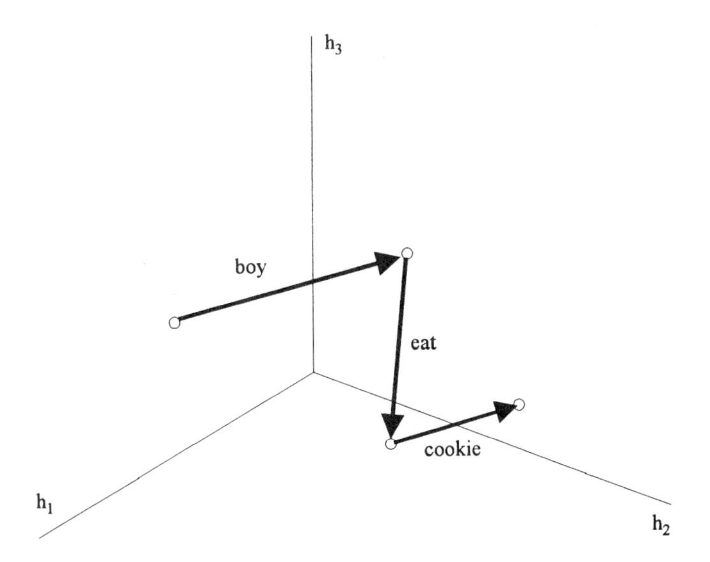

Fig. 5 Depiction of the changes in hidden layer representations as transitions in the state space spanned by the activations of the hidden units (for illustrative purposes, only three hidden units are assumed). These transitions can be seen as the arcs of a state automaton, with the exception that the state space is continuous and, thus, there exists a distance measure between nodes in the automaton.

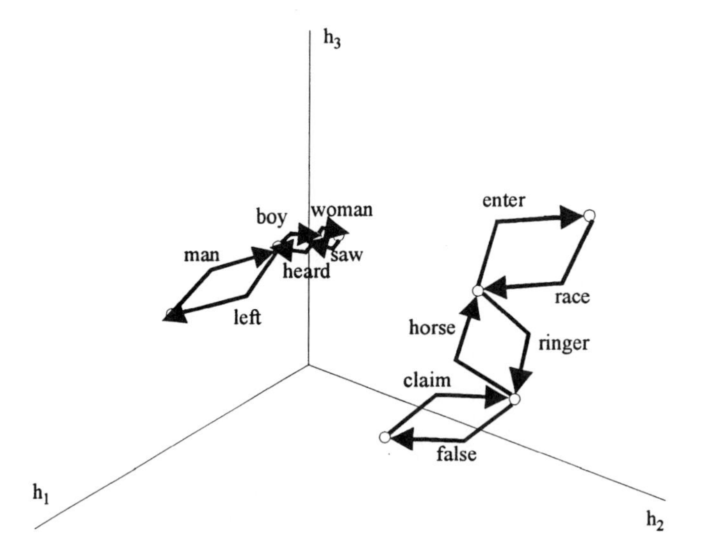

Fig. 6 An illustration in state space of the phenomenon that semantically similar words lead to confusion in embeddings more easily than unrelated words. It is assumed that more similar words lead to more similar transitions and thus to states that are closer in state space, which can be perturbed by errors or noise more easily [35].

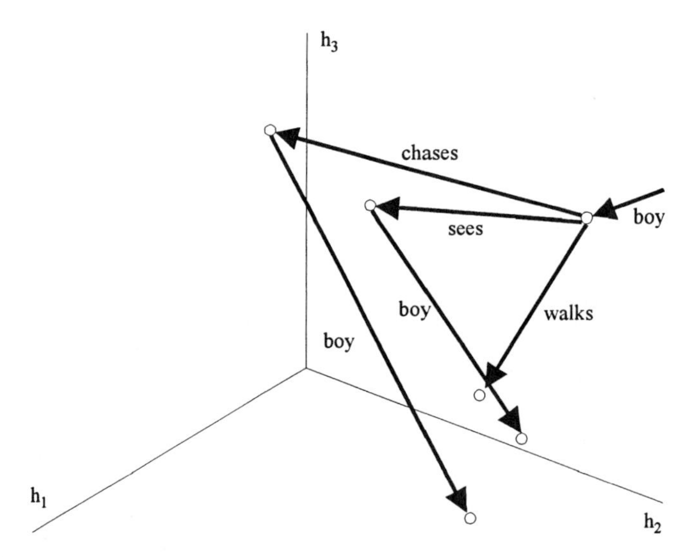

Fig. 7 Resulting state transitions in the network described [11]: Three different sentences are depicted, which all end in approximately the same region (indicating a successful parse). Depending on the transitivity of the verb, however (meaning that another noun is expected or not), the transition either directly moves to the end state or deviates to parse the additional input. A sentence such as "boy chases" would thus be considered as ungrammatical. Because the hidden layer in this case has more than three units, here h_1 through h_3 refer to principal components of the high-dimensional state space. (From Ref. 11.)

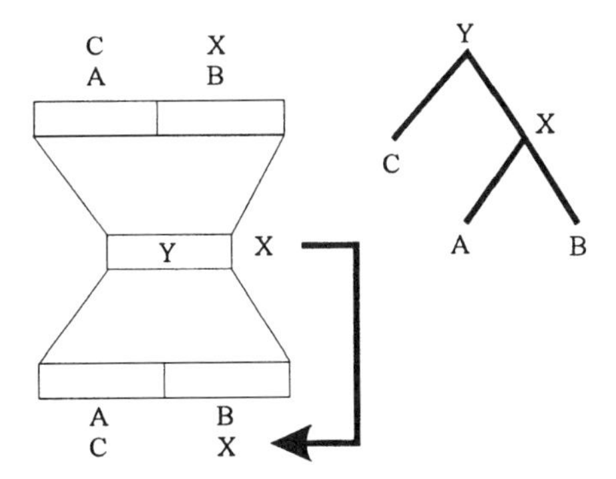

Fig. 8 A recursive autoassociative memory (or RAAM): In its simplest form, a RAAM consists of an autoassociative MLP with a hidden layer of exactly half the size of input and output. With this network, compact internal representations of binary trees (such as the one depicted in the upper right) can be learned through backpropagation training. For this, two representations of nodes in the tree (starting with leaf nodes) are autoassociatively mapped onto themselves, leading to a compact representation of the node joining the two nodes (e.g., X, joining A and B). This representation can then be used recursively to form the representation of the next higher node (e.g., Y), and so on. (From Ref. 41.)

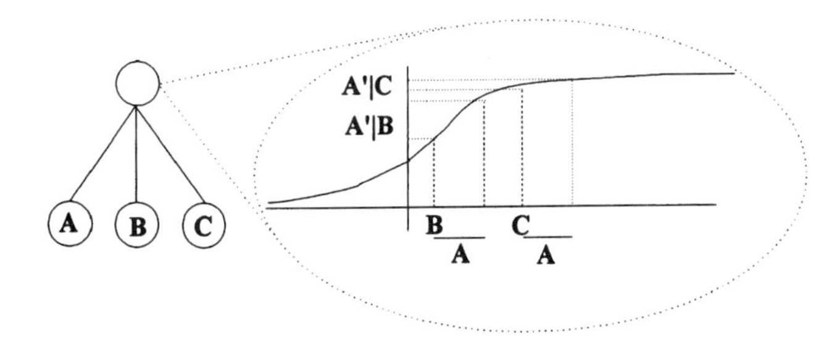

Fig. 9 An illustration of how nonlinear hidden units can encode the binding of different roles and fillers: When A is input together with B, the information supplied by A (A′| B) is different than if A is input together with C (A′| C), owing to the nonlinearity of the transfer function. Thus the resulting activation implicitly bears information about the specific binding (A and B vs. A and C) (From Ref. 43.)

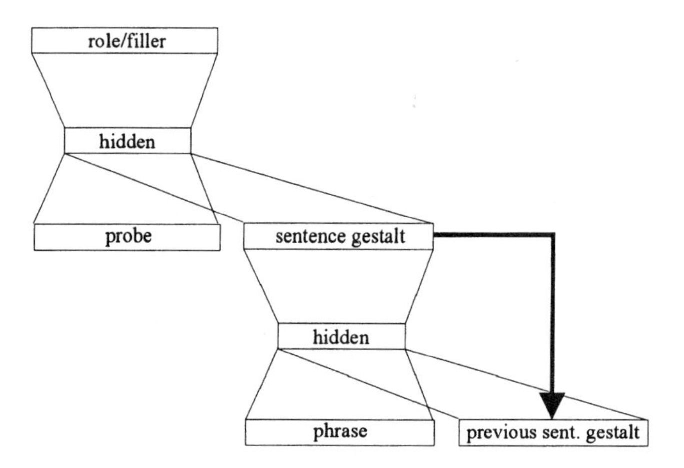

Fig. 10 The SAIL2 network learning to understand sentences [53]: The network receives sequences of phrases as the input and learns through answering queries ("probe") about the sentences with outputting roles and fillers. Errors are backpropagated all the way from the upper network to the lower one. The "output" layer of the lower network (which is actually a hidden layer) thus develops internal representations about the sentence's content, called sentence gestalts. These sentence gestalts are used recurrently as context input in the processing of the sequences of inputs. (From Ref. 53.)

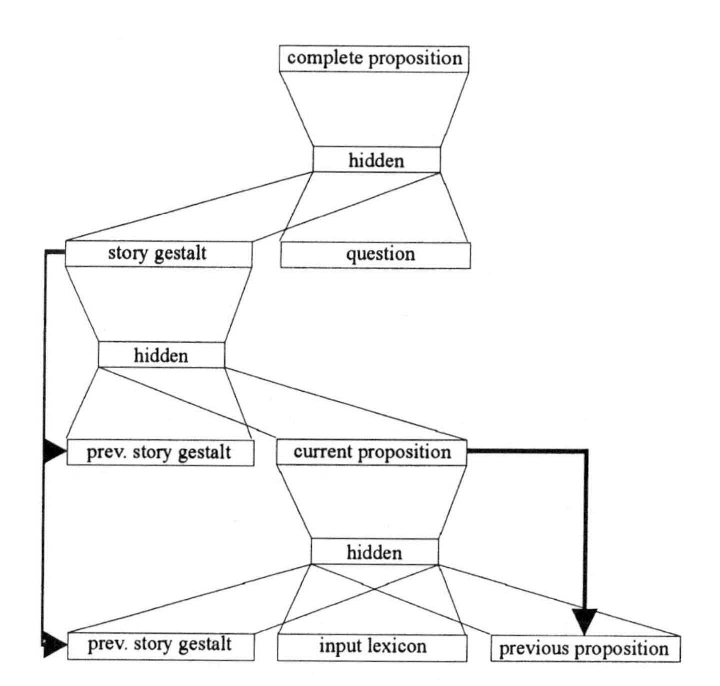

Fig. 11 The network capable of processing entire stories [54]: This is an extension to the network-learning sentence gestalts [53] (see Fig. 10). Input is a sequence of words that are mapped onto a distributed representation of the current proposition (sentence). These representations are input to a further pair of networks learning to build representations through answering questions about the story. This results in internal representations about the entire story ("story gestalts"). Both story gestalts and current proposition representations are recurrently used as further input to the network. Errors are backpropagated all the way through the three separate networks. (From Ref. 54.)

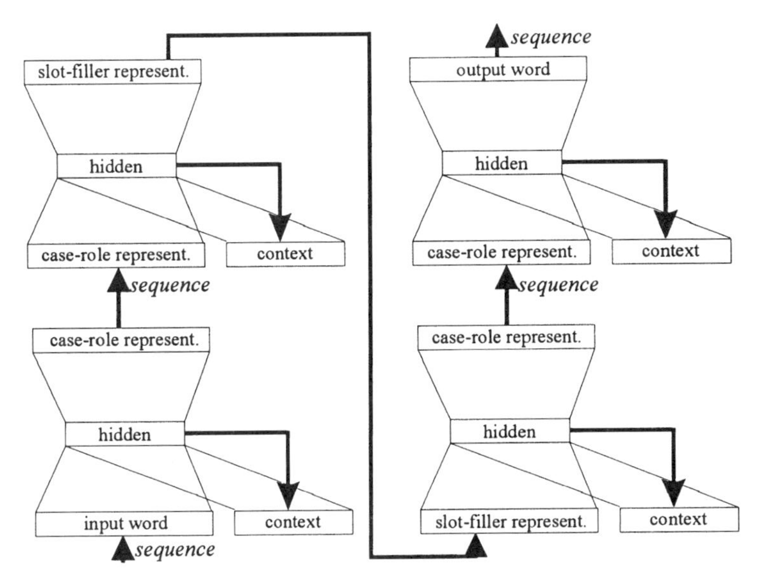

Fig. 12 The DISPAR model [56]: The model consists of two parts, each in turn consisting of two recurrent FGREP modules. The left part first processes words into sentence representations, which are subsequently processed into representations of entire stories. The right part, while working in a similar fashion, generates paraphrases of the stories by turning the story representations into a sequence of words. Although the tasks of the four networks are separable, they are trained in combination such that representations can influence each other. (From Ref. 56.)

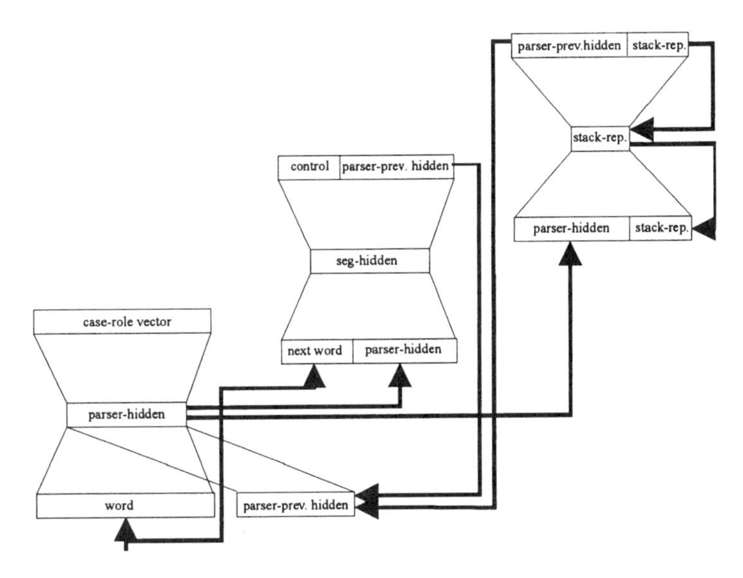

Fig. 13 The SPEC model [33]: This model is able to process arbitrary sentences into case–role assignments and to generalize to novel syntactic structure including embedded sentences. The model consists of three interconnected networks: the parser, the segmenter, and the stack. The stack is a connectionist version of a classic stack (as discussed in Sec. 5) similar to sequential RAAM. When a relative pronoun is detected the network can store the current hidden layer representation and later restore it to continue to parse of the higher-level sentence. To detect this situation (i.e., to segment the sentence into clauses), is the role of the segmenter (middle network). (From Ref. 33.)

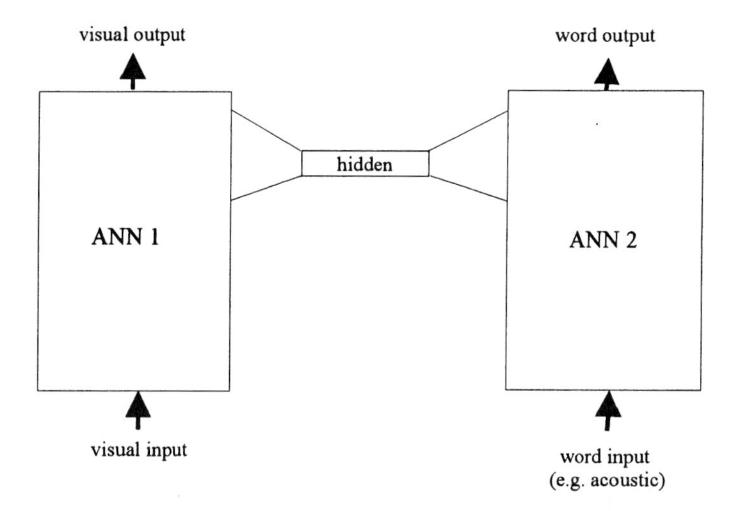

Fig. 14 A generic model for grounded word learning: Two neural networks (ANN 1 and ANN 2) learn to build internal representations based on two different input modalities (e.g., a visual stimulus representing an object and an acoustic stimulus representing a spoken word). These internal representations are subsequently linked with each other to build a mapping between the word and its meaning.

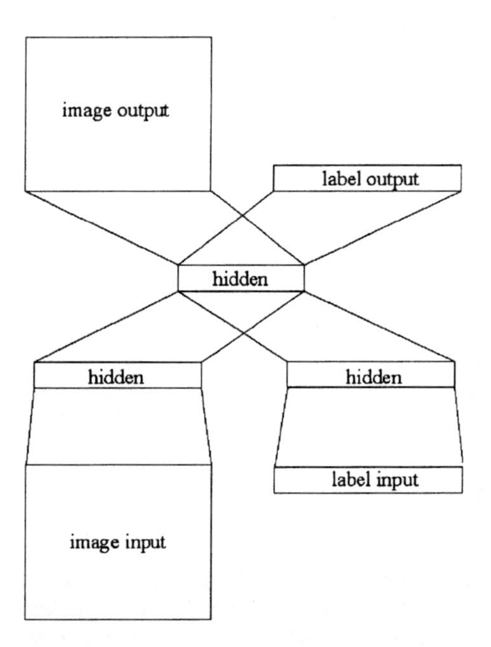

Fig. 15 The word-learning model [61]: Simple bit patterns as image input and local representations of words ("labels") are autoassociatively mapped onto themselves using an MLP and backpropagation. Comprehension in the trained network means inputting a label without image input and associating an image with it. Naming (production) means inputting an image without label and associating the corresponding label at the output. (From Ref. 61.)

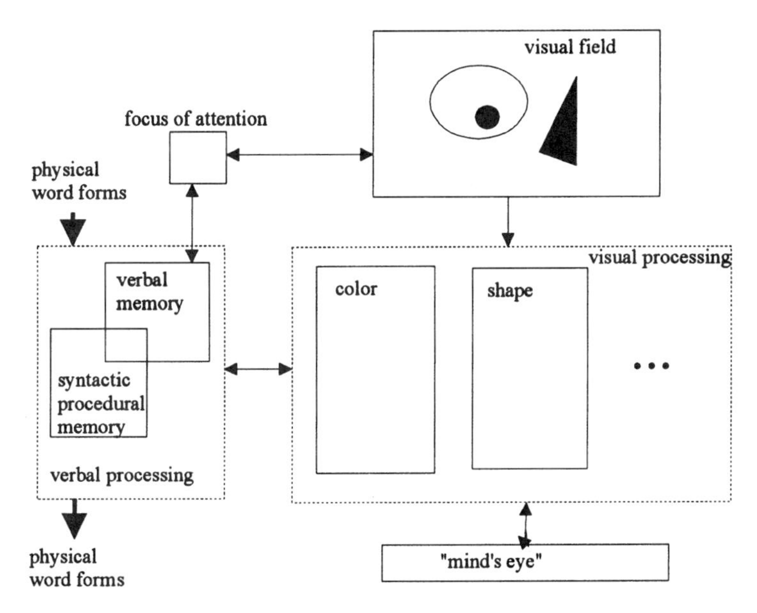

Fig. 16 A simplified depiction of the DETE model [67]: The model exemplifies grounded language learning by processing physical (visual and acoustic) stimuli and thus associating simple sentences with visual scenes. It consists of a large number of modules responsible for tasks such as focus of attention, visual processing divided into aspects such as shape, color, and others, and memorizing scenes and word sequences. The model is also able to predict the continuation of sequences by using an internal "mind's eye," which is necessary to express events in the future. The main network types to process sequential input are so-called Katamic memories. (From Ref. 67.)

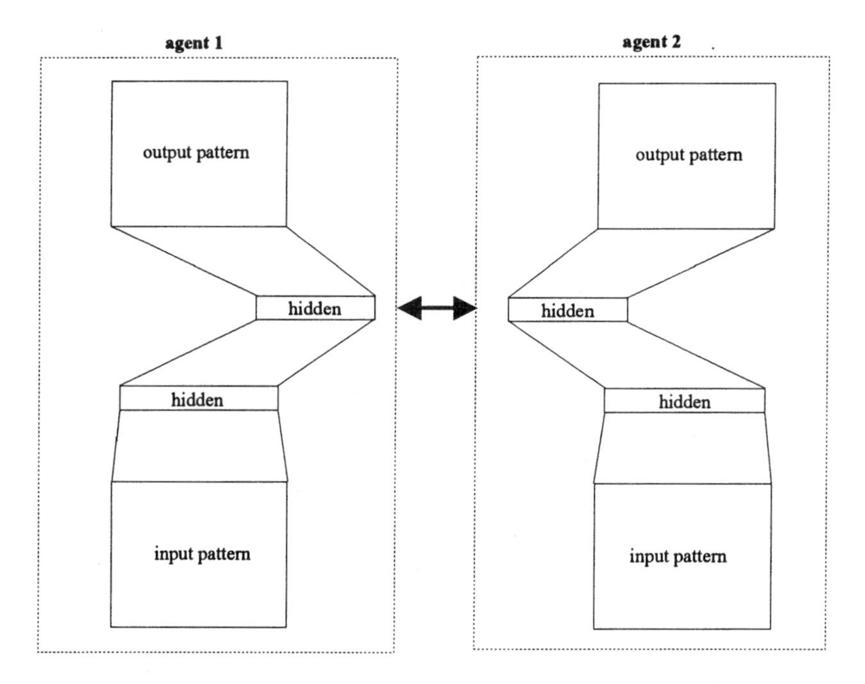

Fig. 17 The model developing a shared lexicon between several individuals [74]: Each large box represents an individual consisting of an autoassociative MLP learning to map visual input patterns (simple scenes) onto themselves. The patterns in one of the hidden layers are viewed as external symbols (words), which refer to the scenes and which must be learned to be shared with other individuals. For this, hidden layer patterns of one individual are used as a target pattern for the hidden layer of another. (From Ref. 74.)

REFERENCES

1. Shastri L. Ajjanagadde V. From simple associations to systematic reasoning: a connectionist representation of rules, variables and dynamic bindings using temporal synchrony. Behav Brain Sci 16:417–451, 1993.
2. Diederich J. Connectionist recruitment learning. In: Kodratoff Y, ed. Proceedings of the 8th European Conference on Artificial Intelligence (ECAI-88), Pitman, London, 1988, pp 351–356.
3. Smolensky P. On the proper treatment of connectionism. Behav Brain Sci 11:1–74, 1988.
4. Legendre G, Miyata Y, Smolensky P. Harmonic grammar—a formal multi-level connectionist theory of linguistic well-formedness: theoretical foundations. Proceedings of the Twelfth Annual Conference of the Cognitive Science Society. Lawrence Erlbaum, Hillsdale, NJ, 1990, pp 388–395.
5. Legendre G, Miyata Y, Smolensky P. Harmonic grammar—a formal multi-level connectionist theory of linguistic well-formedness: an application. Proceedings of the Twelfth Annual Conference of the Cognitive Science Society. Lawrence Erlbaum, Hillsdale, NJ, 1990, pp 884–891.
6. Dorffner G. Radical connectionism—a neural bottom-up approach to AI. In: Dorffner G, ed. Neural Networks and a New Artificial Intelligence. International Thomson Computer Press, London, 1997.

7. Dorffner G, Prem E, Trost H. Words, symbols, and symbol grounding. Österreichisches Forschungsinstitut für Artificial Intelligence, Vienna, Technical Report TR-93-30, 1993.
8. Rumelhart DE, Hinton GE, Williams RJ. Learning internal representations by error propagation. In: Rumelhart DE, McClelland JL. Parallel Distributed Processing, Explorations in the Microstructure of Cognition, vol 1: Foundations. MIT Press, Cambridge, MA, 1986, pp 282–317.
9. Jordan MI. Serisl order: a parallel distributed processing approach. ICS-UCSD. Report No. 8604, 1986.
10. Elman JL. Finding structure in time. Cogn Sci 14:179–212, 1990.
11. Elman JL. Distributed representations, simple recurrent networks, and grammatical structure. Mach Learn 7:195–226, 1991.
12. Chater N, Conkey P. Finding linguistic structure with recurrent neural networks. Proceedings of the Fourteenth Annual Conference of the Cognitive Science Society. Lawrence Erlbaum, Hillsdale, NJ, 1992, pp 402–407.
13. Grossberg S. Studies of Mind and Brain. Reidel Press, Boston, 1982.
14. Rumelhart DE, Zipser D. Feature discovery by competitive learning. Cogn Sci 9:75–112, 1985.
15. Grossberg S. Competitive learning: from interactive activation to adaptive resonance. Cogn Sci 11:23–64, 1987.
16. Kohonen T. Self-Organizing Maps. Springer, Berlin, 1995.
17. Sanchez Miret F, Koliadis A, Dressler WU. Connectionism vs. rules in diachronic morphology, University of Vienna, Dept. of Linguistics. Folia Linguistica Historica 18:149–182, 1998.
18. Rumelhart DE, McClelland JL. On learning the past tenses of English verbs. In: McClelland JL, Rumelhart DE. Parallel Distributed Processing, Explorations in the Microstructure of Cognition, vol 2: Psychological and Biological Models. MIT Press, Cambridge, MA, 1986.
19. Pinker S, Prince A. On language and connectionism: analysis of a parallel distributed processing model of language acquisition. Cognition 28:73–193, 1988.
20. Pinker S. Rules of language. Science 253:530–534, 1991.
21. Daugherty K, Seidenberg M. Rules or connections? The past tense revisited. Proceedings of the Fourteenth Annual Conference of the Cognitive Science Society. Lawrence Erlbaum, Hillsdale, NJ, 1992, pp 259–264.
22. Plunkett K, Marchman V. U-shaped learning and frequency effects in a multi-layered perceptron: implications for child language acquisition. Cognition 38:43–102, 1991.
23. Kim JJ, Pinker S, Prince A, Prasada S. Why no mere mortal has ever flown out to center field. Cogn Sci 15:173–218, 1991.
24. Harris C. Understanding English past-tense formation: the shared meaning hypothesis. Proceedings of the Fourteenth Annual Conference of the Cognitive Science Society. Lawrence Erlbaum, Hillsdale, NJ, 1992, pp 100–105.
25. Daugherty KG, MacDonald MC, Petersen AS, Seidenberg MS. Why no mere mortal has ever flown out to center field but people often say they do. Proceedings of the Fifteenth Annual Conference of the Cognitive Science Society 1993. Lawrence Erlbaum, Hillsdale, NJ, 1993, pp 383–388.
26. Marcus GF, Brinkmann U, Clahsen H, Wiese R, Woest A, Pinker S. German inflection: the exception that proves the rule, Proceedings of the Fifteenth Annual Conference of the Cognitive Science Society 1993, Lawrence Erlbaum, Hillsdale, NJ, 1993, 670–675.
27. Hare M, Elman J. A connectionist account of English inflectional morphology: evidence from language change. Proceedings of the Fourteenth Annual Conference of the Cognitive Science Society, Lawrence Erlbaum, Hillsdale, NJ, 1992, pp 265–270.

28. Indefrey P, Goebel R. The learning of weak noun declension in German: children vs. artificial network models. Proceedings of the Fifteenth Annual Conference of the Cognitive Science Society 1993. Lawrence Erlbaum, Hillsdale, NJ, 1993, pp 575–580.
29. Gupta P, MacWhinney B. Integrating category acquisition with inflectional marking: a model of the German nominal system. Proceedings of the Fourteenth Annual Conference of the Cognitive Science Society. Lawrence Erlbaum, Hillsdale, NJ, 1992, pp 253–258.
30. Lee C-D. Explaining language universals in connectionist networks: the acquisition of morphological rules. Proceedings of the Fifteenth Annual Conference of the Cognitive Science Society 1993. Lawrence Erlbaum, Hillsdale, NJ, 1993, pp 653–658.
31. Gasser M. Acquiring receptive morphology: a connectionist model. Proceedings of the 32nd Annual Meeting of the Association for Computational Linguistics. ACL, 1994.
32. Siegelmann HT, Sontag ED. On the computational power of neural nets. J Comput Syst Sci 50:132–150, 1995.
33. Miikkulainen R. Subsymbolic case–role analysis of sentences with embedded clauses. Cogn Sci 20:47–73, 1996.
34. Das S, Giles C, Sun G. Learning context-free grammars: capabilities and limitations of a recurrent neural network with an external stack memory. Proceedings of the Fourteenth Annual Conference of the Cognitive Science Society. Lawrence Erlbaum, Hillsdale, NJ, 1992, pp 791–796.
35. Weckerly J, Elman J. A PDP approach to processing center-embedded sentences. Proceedings of the Fourteenth Annual Conference of the Cognitive Science Society. Lawrence Erlbaum, Hillsdale, NJ, 1992, pp 414–419.
36. Elman JL. Language as a dynamic system. In: Port RF, van Gelder TJ, eds. Mind as Motion. MIT Press, Cambridge, MA, 1995, pp 195–226.
37. Pollack JB. The induction of dynamical recognizers. [Special Issue on Connectionist Approaches to Language Learning.] Mach Learn 7:227–252, 1991.
38. Casey M. How discrete-time recurrent neural networks work. Institute for Neural Computation, University of California, San Diego, INC-9503, 1995.
39. Elman JL, Bates EA, Johnson MH, Karmiloff-Smith A, Parisi D, Plunkett K. Rethinking Innateness. MIT Press, Cambridge, MA, 1996.
40. Fodor JA, Pylyshyn ZW. Connectionism and cognitive architecture: a critical analysis. Cognition 28:3–71, 1988.
41. Pollack JB. Recursive distributed representations. [Special Issue on Connectionist Symbol Processing]. Artif Intell 46:77–106, 1990.
42. van Gelder T. Compositionality: a connectionist variation on a classical theme. Cogn Sci 14:208–212, 1990.
43. Sharkey NE. The causal role of the constituents of superpositional representations. In: Trappl R, ed. Cybernetics and Systems '92. World Scientific Publishing, Singapore, 1992, pp 1375–1382.
44. Chalmers DJ. Syntactic transformations on distributed representations. Connect Sci 2:53–62, 1990.
45. Dorffner G, Rotter M. On the virtues of functional connectionist compositionality. In: Neumann B, ed. Proceedings of the Tenth European Conference on Artificial Intelligence (ECAI92), Wiley, Chichester, UK, 1992, pp 203–205.
46. Hadley R. Compositionality and systematicity in connectionist language learning. Proceedings of the Fourteenth Annual Conference of the Cognitive Science Society. Lawrence Erlbaum, Hillsdale, NJ, 1992, pp 659–664.
47. Clark A. The presence of a symbol. Connect Sci 4:193–206, 1993.
48. Fodor JA, McLaughlin B. Connectionism and the problems of systematicity: why Smolensky's solution doesn't work. Cognition 35:183–204, 1990.

49. Kwasny SC, Kalman BL. Tail-recursive distributed representations and simple recurrent networks. Connect Sci 7:61–80, 1995.
50. Giles CL, Miller CB, Chen D, Chen HH, Sun GZ, Lee YC. Learning and extracting finite state automata with second-order recurrent neural networks. Neural Comput 4:393–405, 1992.
51. McClelland JL, Kawamoto AH. Mechanisms of sentence processing: assigning roles to constituents of sentence. In: Rumelhart DE, McClelland JL, eds. Parallel Distributed Processing, Explorations in the Microstructure of Cognition, vol 2: Psychological and Biological Models. MIT Press, Cambridge, MA, 1986, pp 272–326.
52. Fillmore CJ. The case for case. In: Bach E, Harms RT, eds. Universals in Linguistic Theory. Holt, Rinehart, Winston, New York, 1968.
53. St John MF, McClelland JL. Applying contextual constraints in sentence comprehension. In: Touretzky D, ed. 1988 Connectionist Models Summer School. Morgan Kaufmann, Los Altos, CA, 1988, pp 338–346.
54. St John MF. The story gestalt: a model of knowledge-intensive processes in text comprehension. Cogn Sci 16:271–306, 1992.
55. Schulenburg D. Using context to interpret indirect requests in a connectionist model of NLU. Proceedings of the Fifteenth Annual Conference of the Cognitive Science Society 1993. Hillsdale, NJ, Erlbaum, 1993, pp 895–899.
56. Miikkulainen R, Dyer MG. Natural language processing with modular PDP networks and distributed lexicon. Cogn Sci 15:343–400, 1991.
57. Harnad S. The symbol grounding problem. Physica D 42:335–346, 1990.
58. Cottrell GW, Bartell B, Haupt C. Grounding meaning in perception. In: 'Marburger H, ed. Proceedings of the German Workshop on AI (GWAI-90), Springer, Berlin, 1990.
59. Bartell B, Cottrell GW. A model of symbol grounding in a temporal environment. Proceedings of the International Joint Conference on Neural Networks, Seattle, WA, 1991.
60. Chauvin Y. Toward a connectionist model of symbolic emergence. Proceedings of the Eleventh Annual Conference of the Cognitive Science Society. Lawrence Erlbaum, Hillsdale, NJ, 1989, pp 580–587.
61. Plunkett K, Sinha C, Moller M, Strandsby O. Symbol grounding or the emergence of symbols? Vocabulary growth in children and a connectionist net. Connect Sci 4:293–312, 1992.
62. Schyns PG. A modular neural network model of conept acquisition. Cogn Sci 15:461–508, 1991.
63. Dorffner G, Hentze M, Thurner G. A connectionist model of categorization and grounded word learning. In: Koster C, Wijnen F, eds. Proceedings of the Groningen Assembly on Language Acquisition (GALA '95), 1996.
64. Dorffner G. Categorization in early language acquisition—accounts from a connectionist model. Österreichisches Forschungsinstitut für Artificial Intelligence, Vienna, Technical Report TR-96-16, 1996.
65. Rosch E. Principles of categorization. In Rosch E, Lloyd BB, eds. Cognition and Categorization. Lawrence Erlbaum, Hillsdale, NJ, 1978.
66. McCloskey M, Cohen NJ. Catastrophic interference in connectionist networks: the sequential learning problem. In: Bower G, ed. The Psychology of Learning and Motivation, vol 24. New York: Academic Press, 1989.
67. Nenov VI, Dyer MG. Perceptually grounded language learning: part 2—DETE: a neural/procedural model. Connect Sci 6: 3–42, 1994.
68. Maturana HR, Varela FJ. Autopoiesis and Cognition. Reidel, Dordrecht, 1980.
69. von Glasersfeld E. The Construction of Knowledge. Intersystems Publications, Seaside, CA, 1988.

70. Dorffner G, Prem E. Connectionism, symbol grounding, and autonomous agents. Proceedings of the Fifteenth Annual Conference of the Cognitive Science Society. Lawrence Erlbaum, Hillsdale, NJ, 1993, pp 144–148.
71. Nenov VI, Dyer MG. perceptually grounded language learning: part 1—a neural network architecture for robust sequence association. Connect Sci 5:115–138, 1993.
72. Dorffner G. Why connectionism and language modeling need DETE. Connect Sci 6:115–118, 1994.
73. Waibel A. Consonant recognition by modular construction of large phonemic time-delay neural networks. In: Touretzky D, ed. Advances in Neural Information Processing Systems. Morgan Kaufmann, Los Altos, CA, 1989, pp. 215–223.
74. Hutchins E, Hazlehurst B. How to invent a lexicon: the development of shared symbols in interaction. In: Gilbert N, Conte R, eds. Artificial Societies: The Computer Simulation of Social Life. UCL Press, London, 1995.
75. Steels L. A self-organizing spatial vocabulary. Artif Life J 2, 1996.

32

The Hybrid Approach to ANN-Based Natural Language Processing

STEFAN WERMTER

University of Sunderland, Sunderland, England

1. MOTIVATION FOR HYBRID SYMBOLIC–CONNECTIONIST PROCESSING

In recent years, the field of hybrid symbolic–connectionist processing has seen a remarkable development [2,9,18,20,36–38,48,55,59,61]. Currently, it is still an open issue whether connectionist[1] or symbolic approaches alone will be sufficient to provide a general framework for processing natural language [11,27,51,56]. However, because human language capabilities are based on real neural networks in the brain, artificial neural networks (also called connectionist networks) provide one essential starting point for modeling language processing. On the other hand, human language capabilities include rule-based reasoning that is supported well by symbolic processing. Given this general situation, the motivation for examining hybrid connectionist models of natural language processing (NLP) comes from different directions.

From the perspective of cognitive neuroscience, a symbolic interpretation of a connectionist architecture is desirable, because the brain has a neuronal structure and has the capability to perform symbolic reasoning. This leads to the question of how different processing mechanisms can bridge the large gap between, for instance, acoustic or visual input signals and symbolic reasoning for language processing. The brain uses a complex specialization of different structures. Although much of the functionality of the brain is not yet known in detail, its architecture is highly specialized and organized at various levels of neurons, networks, nodes, cortex areas, and their respective connections. Furthermore, different cognitive processes are not homogeneous, and it is to be expected that they are based on different representa-

[1]Sometimes connectionist approaches are also called subsymbolic approaches, together with other statistical approaches.

tions [54]. Therefore, there is evidence from the brain that multiple architectural representations may also be involved in language processing.

From the perspective of knowledge-based natural language processing, hybrid symbolic–connectionist representations are advantageous, becaues different mutually complementary properties can be combined. Symbolic representations have the advantages of easy interpretation, explicit control, fast initial coding, dynamic variable binding, and knowledge abstraction. On the other hand, connectionist representations show advantages for gradual analog plausibility, learning, robust fault-tolerant processing, and generalization to similar input. As these advantages are mutually complementary, a hybrid symbolic connectionist architecture can be useful if different processing strategies have to be supported.

The development of hybrid symbolic–connectionist architectures is still a new research area, and there is no general theory of hybrid architectures [6]: "The development of additional constraints for defining appropriate mappings between hybrid models and the resultant types of hybrid models remain areas for future research." However, the question of an appropriate architecture for a given task is very important [54]: "We need to consider *architectures*, which thus occupy a clearly more prominent place in this area of research compared with other areas in AI."

In general, there are different global possibilities for a symbolic–connectionist integration. Based on the argument from cognitive neuroscience, symbolic connectionist integration relies on a symbolic interpretation of a connectionist architecture and connectionist processing. There are just connectionist representations for different tasks that can be interpreted symbolically. On the other hand, based on the argument from knowledge-based natural language processing, symbolic–connectionist integration relies on a combination of different symbolic and connectionist representations. However, these interpretation approaches and representation approaches are closely related and, in general, there is a continuum of possible connectionist–symbolic architectures.

2. TYPES OF SYMBOLIC–CONNECTIONIST INTEGRATION FOR LANGUAGE PROCESSING

Although there has been quite a lot of work in the field of hybrid and connectionist natural language processing, most work has concentrated on a single individual system, rather than on types of hybrid systems, their interpretation, and communication principles within various architectures. Previous characterizations of architectures have covered certain specific connectionist architectures, for instance, recurrent networks [31,39], or they have covered expert systems or knowledge-based systems [24,37,55]. In contrast, here we will concentrate on various types of hybrid connectionist natural language processing.

In Fig. 1 there is an overview of different possibilities for integration in natural language processing. Continuous connectionist representations are represented by a circle, discrete symbolic representations by a square. Symbolic interpretations of connectionist representations are shown as squares with dotted lines.

Connectionist structure architectures are the first type of symbolic–connectionist architectures. They can model higher cognitive functions and rely solely on connectionist representations. Symbolic knowledge arises by an interpretation process of

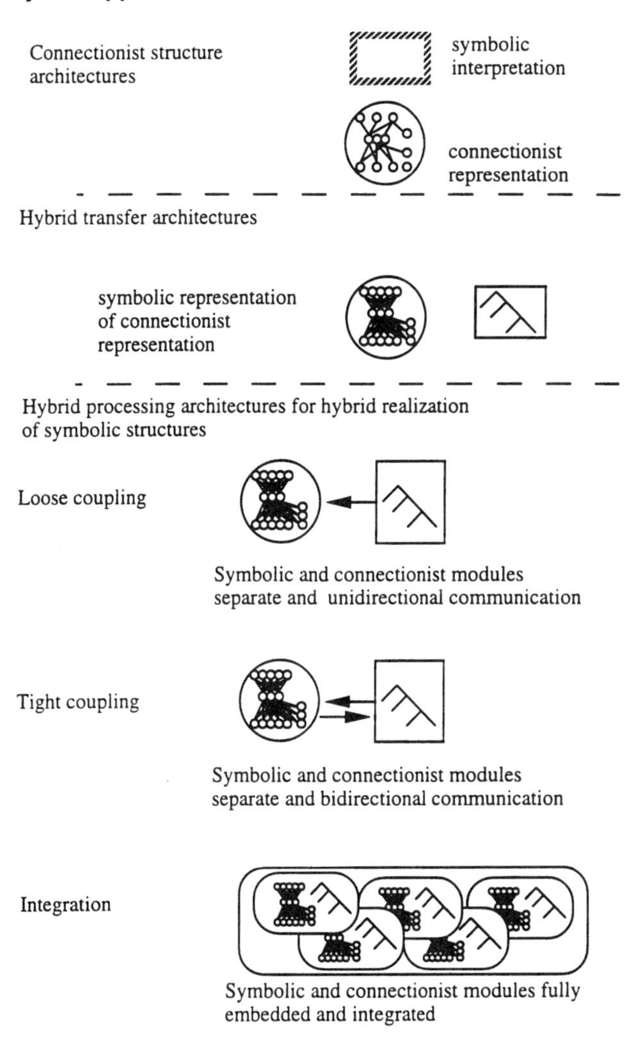

Fig. 1 Overview of various types of symbolic–connectionist integration.

the connectionist representations. Often specific knowledge of the task is built into the connectionist structure architecture.

Hybrid transfer architectures transfer symbolic representations into connectionist representations or vice versa. Using a transfer architecture it is possible to insert or extract symbolic knowledge into or from a connectionist architecture. The main processing is performed by connectionist representations, but there are automatic procedures for transferring from connectionist representations to symbolic representations or vice versa. Hybrid transfer architectures differ from connectionist structure architectures by the automatic transfer into and from symbolic representations. While certain units in connectionist structure architectures may be interpreted symbolically by an observer, only hybrid transfer architectures allow the knowledge transfer into rules, automata, grammars.

Hybrid transfer architectures transfer symbolic and connectionist representations, but the symbolic and connectionist knowledge is not yet applied at the same time to combine the complementary advantages of each representation for a given task. Such a combination of symbolic and connectionist representations is possible in *hybrid processing architectures*, which contain both symbolic and connectionist modules appropriate to the task. Here, symbolic representations are not just initial or final representations. Rather, they are combined and integrated with connectionist representations in many different ways.

Connectionist and symbolic modules in hybrid processing architectures can be loosely coupled, tightly coupled, or completely integrated (see Fig. 1). *A loosely coupled hybrid architecture* has separate symbolic and connectionist modules. The control flow is sequential in the sense that processing has to be finished in one module before the next module can begin. Only one module is active at any time, and the communication between modules is unidirectional.

A *tightly coupled hybrid architecture* contains separate symbolic and connectionist modules, and control and communication are by common internal data structures in each module. The main difference between loosely and tightly coupled hybrid architectures is common data structures that allow bidirectional exchanges of knowledge between two or more modules. Processing is still in a single module at any given time, but the output of a connectionist module can have direct influence on a symbolic module (or vice versa) before it finishes its global processing. In this way feedback between two modules is possible and the communication can be bidirectional.

In a *fully integrated hybrid architecture* there is no discernible external difference between symbolic and connectionist modules, because the modules have the same interface, and they are embedded in the same architecture. The control flow may be parallel, and the communication between symbolic and connectionist modules is by messages. Communication may be bidirectional between many modules, although not all possible communication channels have to be used. This is the most advanced of the *hybrid processing* architectures. In the remainder of this article we will show detailed examples of each of these types of symbolic–connectionist architectures.

3. CONNECTIONIST STRUCTURE ARCHITECTURES

In this section we describe principles of connectionist structure architectures. Knowledge of some task is built into the connectionist structure architecture, and a symbolic interpretation is assigned by an observer. Much early work on structured connectionism can be traced back to work by Feldman and Ballard, who provided a general framework of structured connectionism [15]. This framework was extended for natural language processing in many different directions, for instance for parsing [12,13,46], language acquisition [14,25], explanation [8], and semantic processing [26,47].

More recent work along these lines focuses on the so-called *NTL, neural theory of language*, which attempts to bridge the large gap between neurons and cognitive behavior [16,49]. The NTL framework is challenging, for it tries to study neural processing mechanisms for high conceptual cognitive processing such as embodied semantics, reasoning, metaphor interpretation.

To attack such a challenging task it is necessary to focus on different processing mechanisms and certain representative problems. One such problem is verbal aspect, which describes the temporal character of events (such as past tense or progressive form). Narayanan has developed and implemented a computational model of verbal aspect, and he argues that the semantics of aspect is grounded in sensory–motor primitives [41]. This model was developed around processing mechanisms similar to schemas and Petri-networks that could also be transferred into structured connectionist networks. In related work Bailey developed a computational motor-control model of children's acquisition of verb semantics for words such as *push*, *pull* [1]. This work supports the claim that the semantics of verb actions is grounded in sensory–motor primitives. Statistic Bayesian model merging is used for learning the task of verb semantics. Both models focus on the computational level, but they have close relations to neurally plausible models or recruitment of neuron-like elements.

Although these two related models for embodied event modeling can be viewed at a computational level, there is more neurally inspired work for event modeling and temporal processing at a connectionist level as well [e.g., 21,49]. For instance, schemas from a computational level can be modeled using structured connectionist terms from the SHRUTI system, which was previously developed by Shastri and Ajjanagadde [21]. A particularly interesting recent work is a model of rapid memory formation in the Hippocampal System [49]. Focusing on events such as "John gave Mary a book on Tuesday," the aim is to find how a neural code can be transferred rapidly into a structural code so that it can be retrieved later. This SMRITI system is one of few connectionist systems that take temporal neural processing seriously at a neurobiological level. It relates the function of the system to the human hippocampal system which is taken to be central for memory formation and event encoding.

In general, within the NTL framework, structured connectionist networks are seen as central for modeling systems of natural language processing. However, beyond this level of structured connectionist networks there is a computational level that provides interpretations at a symbolic level and links to a conceptual linguistic level. At the computational level well-known techniques such as Petri Nets, statistical model merging, and feature structures are used for symbolic interpretation. On the other hand, below the level of structured connectionist networks, there is a computational neurobiology level that links connectionist networks with the biological level. Because of this large challenging task of linking biological, neurobiological, connectionist, computational, and cognitive levels, it will be interesting to see what a complete detailed NTL architecture that contains structured connectionist networks will look like in the future.

We will now focus on a few examples of different structured connectionist architectures that have been developed in recent years. One example of a connectionist structure architecture is provided as part of the SCAN project [59]. One task within this framework is structural disambiguation of a given sentence or phrase. In general many different structural interpretations of phrases are possible and only the integration of different constraints allows the system to make a correct structural disambiguation decision. For instance, there are syntactic locality constraints that realize a preference for a local attachment of constituents. On the other hand, there are also semantic plausibility constraints that have to be considered for the structural disambiguation. Different constraints may differ in strength and have to be integrated.

Such an integration of competing constraints of varying strengths can be represented well in a localist connectionist network, since the explicit nodes and connections in the network can be used to model these competing constraints (Fig. 2). There is one noun node for each head noun of a phrase group, one semantic node for the semantic plausibility of a semantic relation between two head nouns, and one locality node for the syntactic locality of a head noun. Excitatory connections link the semantic nodes and the noun nodes; competing semantic nodes are connected by inhibitory connections. Furthermore, locality nodes provide activation through excitatory connections for the semantic nodes and the locality nodes are inhibited by the semantic nodes. When using a localist network, phrases such as "conference on hydrodynamics in the outer atmosphere" and "conference on hydrodynamics in room 113" could be assigned different structural interpretations depending on the attachment of the last constituent.

This localist network represents the possible static connections for the structural disambiguation task. The relevant semantic knowledge can be learned in distributed connectionist plausibility networks. Although the syntactic representation in the localist network can be applied for different domains, the semantic representations have to be adapted to different domains. Therefore, it is important that the knowledge acquisition effort is restricted to the semantic representations; this is one motivation for learning semantic relations in plausibility networks. There is one plausibility network for each semantic node. Each plausibility network has a feedforward structure and receives the semantic representations of the two connecting

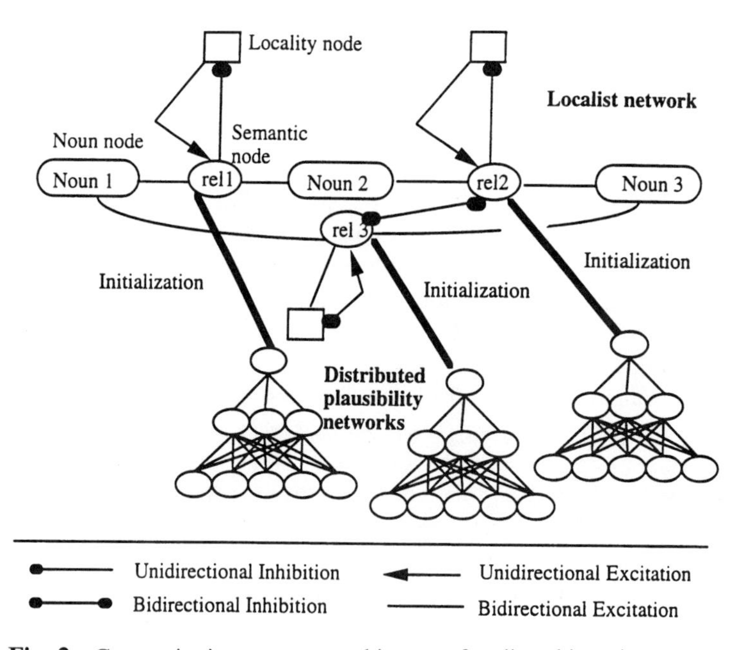

Fig. 2 Connectionist structure architecture for disambiguation.

nouns as input. The output of such a network consists of a plausibility value that indicates the likelihood of the semantic relations.

In this connectionist structure architecture, then, we have distributed plausibility networks that provide bottom-up activation for the plausibility of semantic relations within a localist connectionist network. After initialization, activation spreads within the localist network. In each cycle the activation of a particular node is updated. The new activation of a node is calculated from the activation of the previous cycle plus the activation over excitatory connections minus the activation over inhibitory connections. This spreading activation process in the localist network stops after the different possibly competing constraints have been integrated. The final activation of the nodes determines the structural disambiguation of a phrase.

Another recent connectionist structure architecture is CONSYDERR [53]. The task of this system is to make inferences, as they are required, for instance, in natural language understanding systems. The system has an interesting architecture because it consists of a structured connectionist architecture with two different components. The localist level represents inference knowledge at a more abstract concept level, whereas the distributed level represents the details of the inferencing process at the feature level (Fig. 3).

Inferences are represented either directly within the localist component (A→B) or indirectly by flow of activation within the distributed component. Each node within the overall architecture is connected in a structured meaningful way to other nodes in the architecture. Each node has a certain activation, receives activation, and sends activation in each cycle. Each cycle of communication between the nodes in the localist and distributed components is divided into three phases. First, the activations of the nodes are computed in the localist inference component. Next, activation is spread top-down from the localist nodes to the nodes in the distributed component (phase 1). Then, the activations are updated in the distributed compo-

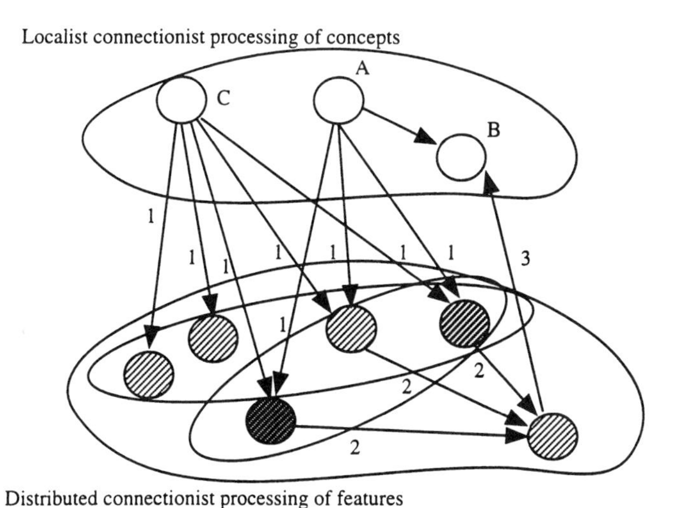

Fig. 3 Connectionist structure architecture for inferencing.

nent (phase 2), and finally, these activations are propagated upward to the localist component again (phase 3).

In general, this architecture is a good representative of connectionist structure architectures. Each node has a meaning; that is, each node represents a concept or a feature, and each node is selectively connected to other nodes. Even the flow of activation is structured according to a three-phase cycle. Furthermore, concepts are organized in the localist concept component, whereas more detailed features for these concepts are organized in the distributed feature component. Each node in this architecture can be given a symbolic interpretation based on its current activation.

As we have seen, connectionist structure architectures can also contain distributed components. In such a case, there are several modular parts within the overall connectionist architecture. The individual modules provide a kind of structure within this architecture, even though some internal elements within a module may constitute distributed representations. A good example of a connectionist structure architecture of this modular-distributed category was developed for automatic feature acquisition. Connectionist architectures can provide an automatic formation of distributed representations for input and output [e.g., 10,11,38] that have the potential to increase robustness and facilitate learning. One way of building distributed input representations is by determining them dynamically during the learning process. A distributed connectionist representation of a symbol is held in a global lexicon and is updated automatically based on its relations to the distributed representations of other lexical entries. This update of the representation has been called symbolic recirculation [10,38] because the representations are constantly changed and recirculated within the architecture during learning.

Symbolic recirculation can be used within a single network, for instance, a feedforward network, but symbolic recirculation can also be used within a larger architecture. We focus first on the recirculation principle within a single network, and then go on to describe a larger connectionist structure architecture with symbolic recirculatioan. For determining the input representation dynamically, the error is computed not only for the output layer and hidden layers, but also for the input layer. Then the new representation values of the input units are computed based on their values in the preceding cycle plus the current change, which is based on the error within the input layer.

During this automatic feature formation process, all the communication is directed by a central lexicon, which contains the learned distributed representation of each symbol. In Fig. 4 we show a generalization for a connectionist structure architecture with six networks. Input and output of these networks are symbols that are represented as distributed vectors. For each learning step within each network it is possible to change the input representation and, therefore, change the distributed representation of the respective symbols in the global lexicon. This architecture has been used for parsing simple sentences and for providing case role and script role representations. It is a connectionist structure architecture, because only connectionist networks are used for the assignment of case and script roles. Furthermore, it is possible to interpret the distributed representations symbolically.

There are many more connectionist architectures that allow a symbolic interpretation; for instance, at the output layers of their network. There are modular network architectures for case role assignment in gestalt networks [52], confluent

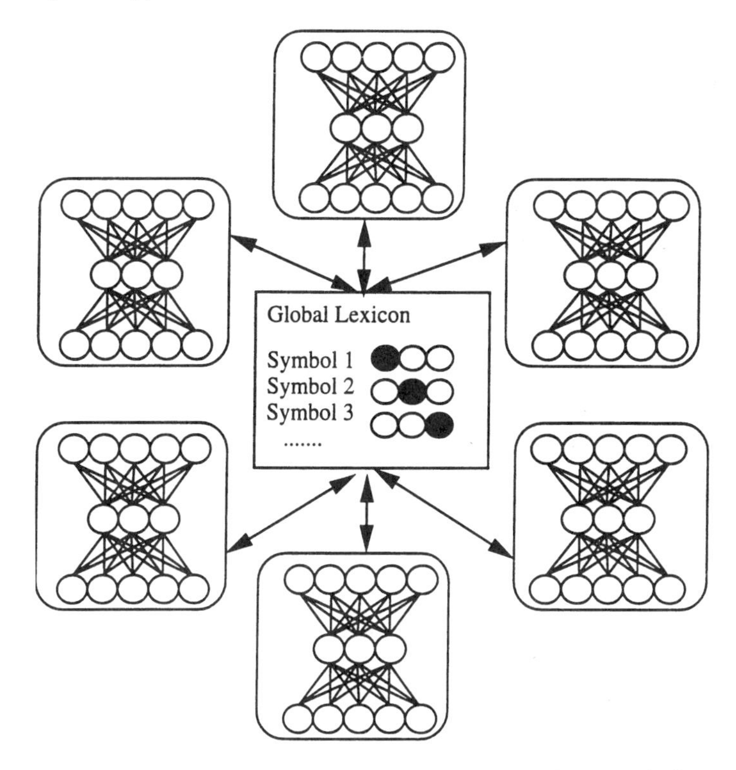

Fig. 4 Connectionist structure architecture using symbolic recirculation for script understanding.

dual RAAM networks for translation [5], the mixture of expert networks for hierarchical control [28], and the two-tier architecture for corpus analysis [3]. All this recent work contains various forms of modularity and structuring within an overall connectionist architecture. Furthermore, symbolic interpretation of the results is possible, at least at certain nodes.

4. HYBRID TRANSFER ARCHITECTURES

Hybrid transfer architectures transfer symbolic representations into connectionist representations or vice versa. By using a transfer architecture, it is possible to insert or extract symbolic knowledge into or from a connectionist architecture. Hybrid transfer architectures differ from connectionist structure architectures by this automatic transfer into and from symbolic representations. Although certain units in connectionist structure architectures may be interpreted symbolically, only hybrid transfer architectures allow the knowledge transfer into or from rules, automata, grammars, and such.

There are many ways to insert or extract symbolic representations into or from connectionist representations. The range of possibilities depends on the form of the symbolic representations (e.g., context-free rules, automata), and the connectionist representations (e.g., use of weights, activations). In this section, we will show two representative approaches for extracting knowledge from connectionist networks.

We have chosen these approaches because they have been widely explored during the last few years and they are relatively straightforward. At the end of this section we will refer to additional approaches using hybrid transfer architectures.

One major problem for hybrid transfer architectures is that the architecture itself has to support the transfer. A good example is the work on activation-based automata extraction from recurrent networks [19,42,43]. First, simple finite state automata are used for generating training examples of a given regular grammar. These training sequences are used to train a second-order connectionist network as an acceptor for the regular grammar. This network has been designed particularly to meet the constraints of symbolic automata behavior. For a given input and a given state, the network determines the next state. So this network structure directly supports the underlying symbolic transfer to and from the network by integrating architectural constraints from symbolic automata.

Figure 5 shows the connectionist architecture. The network receives the representation of the input $I_k(t)$ and the representation of the current state $S_k(t)$ at time t. Then the output representation $S_j(t+1)$ is the state at the next time step. For the extraction of symbolic representations from a trained connectionist network, the output unit activations are partitioned into q discrete intervals. Therefore, there are q^n partitions for n units. The presentation of all input patterns generates a search tree of transitions of an automaton where one node corresponds to one partition. Then, a minimization algorithm can be used, that reduces this deterministic automaton. The extracted rules have the form: If the state is x and the input is y, then go to state z. In general, this architecture is a hybrid transfer architecture where symbolic automata are extracted from connectionist networks.

The activation-based extraction of automata is just one example of a hybrid transfer architecture, and there are several further possibilities. For instance, a weight-based transfer between symbolic rules and feedforward networks has been

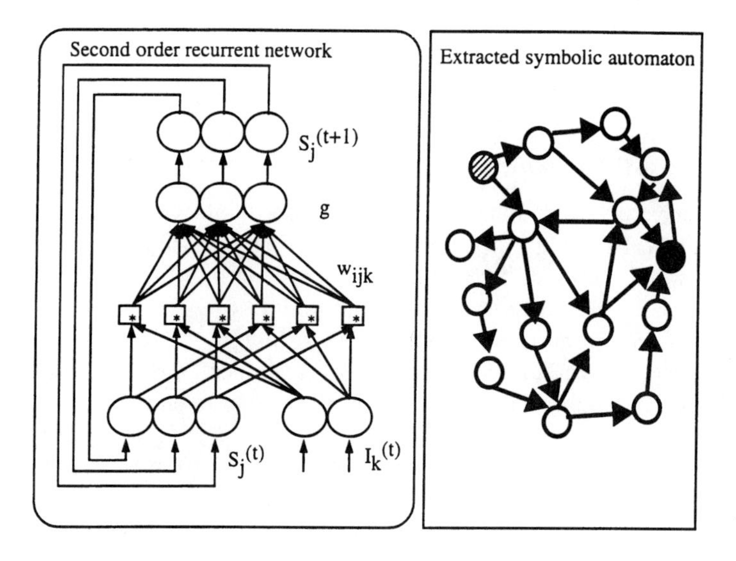

Fig. 5 Hybrid transfer architecture for automata extraction from a recurrent network.

extensively examined in knowledge-based artificial neural networks (KBANN) [7,50]. This weight-based transfer uses the weights, rather than the activations, as the main knowledge source for induction and extraction. Whereas an activation-based transfer is based on the activations of certain units, a weight-based transfer focuses on a more detailed weight level. Therefore, the architectures are fairly simple, in this case, three-layer feedforward networks.

The underlying assumption of this approach is that simple rule knowledge is often available, and that this knowledge should be used for initializing connectionist networks [50]. Then connectionist learning can be used to refine this knowledge. Ultimately the symbolic knowledge can be extracted from the connectionist network, yielding an improved set of interpretation rules. Therefore, this KBANN approach is a good example for a hybrid transfer architecture.

Figure 6 shows a simple set of rules and the corresponding network architecture. Thick connections are associated with high weights, thin connections are low weights. That way initial domain knowledge can be integrated into the connectionist architecture. After a training period the weights can be transferred again to symbolic rules.

In contrast to activations, there are many more weights in a connectionist network, so that heuristics for grouping, eliminating, and clustering of weights are particularly important. For extraction, the weights associated with a connectionist unit are clustered and replaced by the average of the weights of the cluster. If there are clusters that do not contribute to the functionality of a unit (because the weights are too small), these clusters are eliminated. In a final step, the weights are written as rules. Here it is assumed that units can only have two states. Typical simple rules could be "if 2 of A B G then J." This so-called N-of-M-method assumes that *N* of *M* units are equally important for the conclusion. The rules in this KBANN approach are still fairly simple: propositional, binary, acyclic, without recursion and variables. However, simple transfers of connectionist and symbolic knowledge can be performed in a general manner, as for example, in simple forms of natural language processing, but also for many other domains.

In further work, activation-based transfer between context-free rules and feedforward networks is described in a symbolic manipulator for context-free languages [40], and another form of weight-based insertion of symbolic rules has been proposed

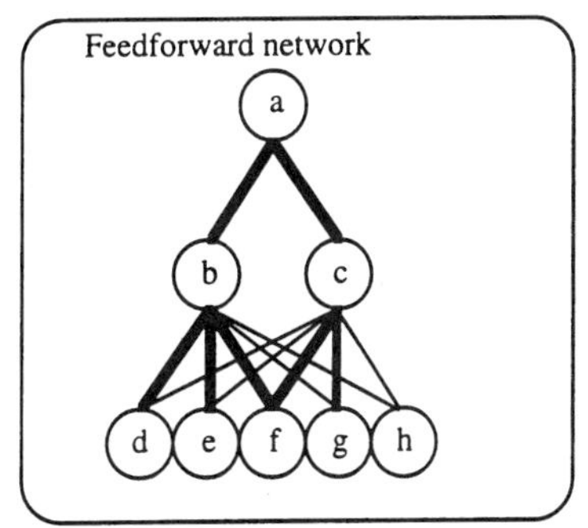

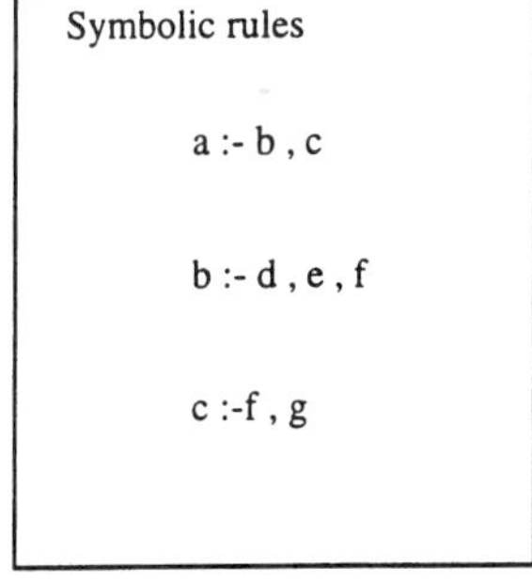

Fig. 6 Hybrid transfer architecture for rule extraction and insertion from and to a feedforward network.

for recurrent networks [33]. Extraction of symbolic rules can also be based on multiplicative networks that control the association of conditions and actions [34,35].

5. HYBRID PROCESSING ARCHITECTURES

Hybrid transfer architectures do not apply symbolic and connectionist knowledge simultaneously to combine the complementary advantages of each representation in solving a given task. Hybrid tranfer architectures just transfer symbolic and connectionist representations into each other. However, a combination of mutual advantages of symbolic and connectionist representations may be advantagous for "hybrid tasks." This combination of symbolic and connectionist representations during task processing is performed in *hybrid processing architectures*, which contain symbolic and connectionist modules.

A. Loosely Coupled Hybrid Processing Architectures

As we have outlined, connectionist and symbolic modules in hybrid processing architectures can be loosely coupled, tightly coupled, or completely integrated (see Fig. 1). In a *loosely coupled hybrid architecture* processing has to be completed in one module before the next module can begin. For instance, the WP model [44,57] for phenomenologically plausible parsing used a symbolic chart parser to construct syntactic localist networks that were also connected to semantic networks. Although the network itself may be viewed as a purely connectionist architecture, the overall architecture is a hybrid processing architecture. First a symbolic parser is used for network building, and after the chart parser has finished its processing the connectionist spreading activation in localist networks takes place.

Another architecture for which the division of symbolic and connectionist work is even more loosely coupled has been described in a model for structural parsing within the SCAN framework [59]. First, a chart parser is used to provide a structural tree representation for a given input; for instance, for phrases or sentences (Fig. 7). This symbolic structural tree representation is used to extract important head nouns and their relations. In a second step, these triples of "noun relationship noun" are used as input for several feedforward networks which produce a plausibility measure of the relations. Based on this connectionist output, another symbolic restructuring component changes the original tree representation if the semantic feedforward networks indicate that this is necessary.

This system has a loosely coupled hybrid processing architecture, as there is a clear division between symbolic parsing, connectionist semantic analysis, and symbolic restructuring. Only if the preceding module has finished completely, will the subsequent module start its processing. The architecture and interfaces are relatively simple to realize owing to the sequential control between symbolic and connectionist processing. On the other hand, this simple sequential sequence of symbolic and connectionist processing does not support feedback mechanisms.

There are several other loosely coupled hybrid processing architectures. For instance, in the SICSA system, connectionist recurrent networks and symbolic case role parsing are combined for semantic analysis of database queries [4]. In BerP, connectionist feedforward networks for speech processing and symbolic dialogue understanding are combined in a spoken language dialogue system [30,65]. In

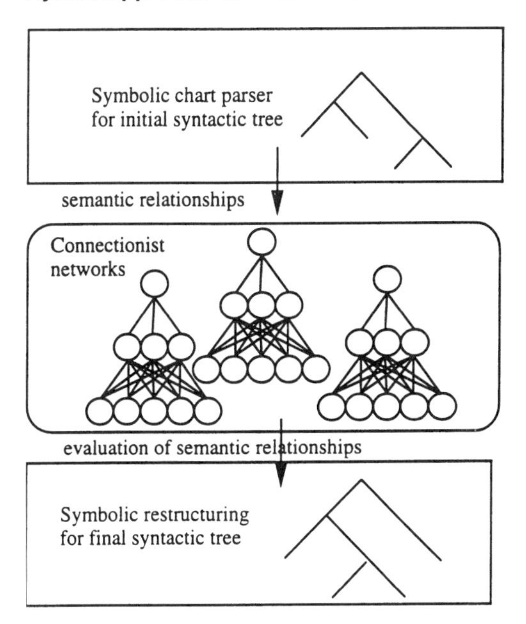

Fig. 7 Loosely coupled symbolic chart parser and connectionist plausibility networks.

FeasPar, connectionist feedforward networks for feature assignment are combined with symbolic search for finding better analysis results. As in the WP model, in all these loosely coupled hybrid processing architectures the connectionist module has to finish before the symbolic module can start (and vice versa).

B. Tightly Coupled Hybrid Processing Architectures

A *tightly coupled hybrid architecture* allows multiple exchanges of knowledge between two or more modules. Processing is still a single module at a time, but the result of a connectionist module can have a direct influence on a symbolic module (or vice versa) before it finishes its global processing. For instance, CDP is a system for connectionist deterministic parsing [32]. Although the choice of the next action is performed in a connectionist feedforward network, the action itself is performed in a symbolic module (Fig. 8). During the process of parsing, control is switched back and forth between these two modules, but processing is confined to a single module at a time. Therefore, a tightly coupled hybrid architecture has the potential for feedback to and from modules.

In other tightly coupled hybrid processing architectures [22,23], the control changes between symbolic marker passing and connectionist similarity determination. In ProPars two connectionist feedforward networks are loosely coupled with a symbolic module for syntactic shift reduce parsing [45]. These tightly coupled hybrid processing architectures interleave symbolic and connectionist processing at the module level and allow the control switch between these different modules. Therefore, this tight coupling has the potential for more powerful interactions. On the other hand, the development of such architectures needs more dynamic and more

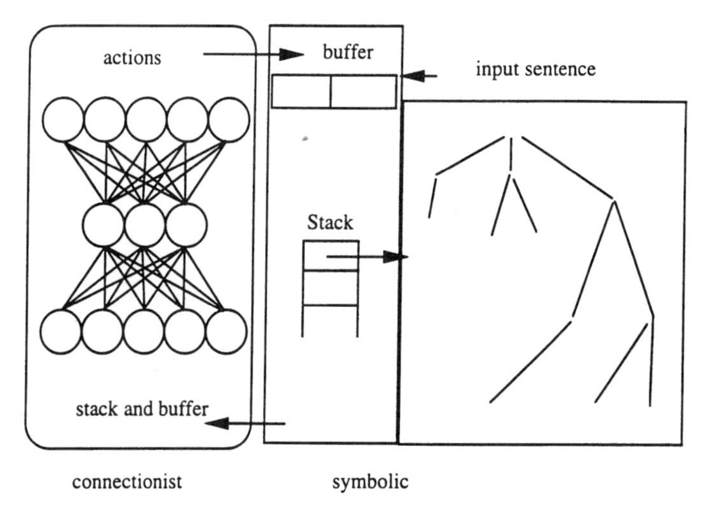

Fig. 8 Tightly coupled symbolic deterministic parser and connectionist control network.

complex interfaces to support the dynamic control between symbolic and connectionist modules.

C. Fully Integrated Hybrid Processing Architectures

The modules in a *fully integrated hybrid architecture* all have the same interface. They are embedded in the same architecture, and externally there is no way to distinguish a symbolic from a connectionist module. The control flow may be parallel, and the communication between symbolic and connectionist modules is through messages. This is the most integrated and interleaved version of the hybrid processing architectures.

One example of an integrated hybrid architecture is SCREEN, which was developed for exploring integrated hybrid processing for spontaneous language analysis. To give the reader an impression of the state of the art of current hybrid language technology we focus on this architecture in slightly more detail. Here, we focus primarily on the architectural principles of this approach, whereas other task-related details can be found elsewhere [58,60,62,63]. One main architectural motivation is the use of a common interface between symbolic and connectionist modules that are externally indistinguishable. This allows incremental and parallel processing involving many different modules. Besides this architectural motivation there are also other task-oriented motivations of exploring learning and fault-tolerance in a hybrid connectionist architecture for robust processing of faulty spoken language.

SCREEN consists of many modules, which can be grouped into six organizational parts (Fig. 9). The input to the system consists of word hypotheses from a speech recognizer which, typically, provides many word hypotheses for a given signal segment. Each word hypothesis is assigned a plausibility value, a start time, and an end time. Such word hypotheses have to be connected to word hypothesis sequences, and this subtask is performed in the word hypothesis construction component. Because this is a constructive problem and because symbolic operations perform

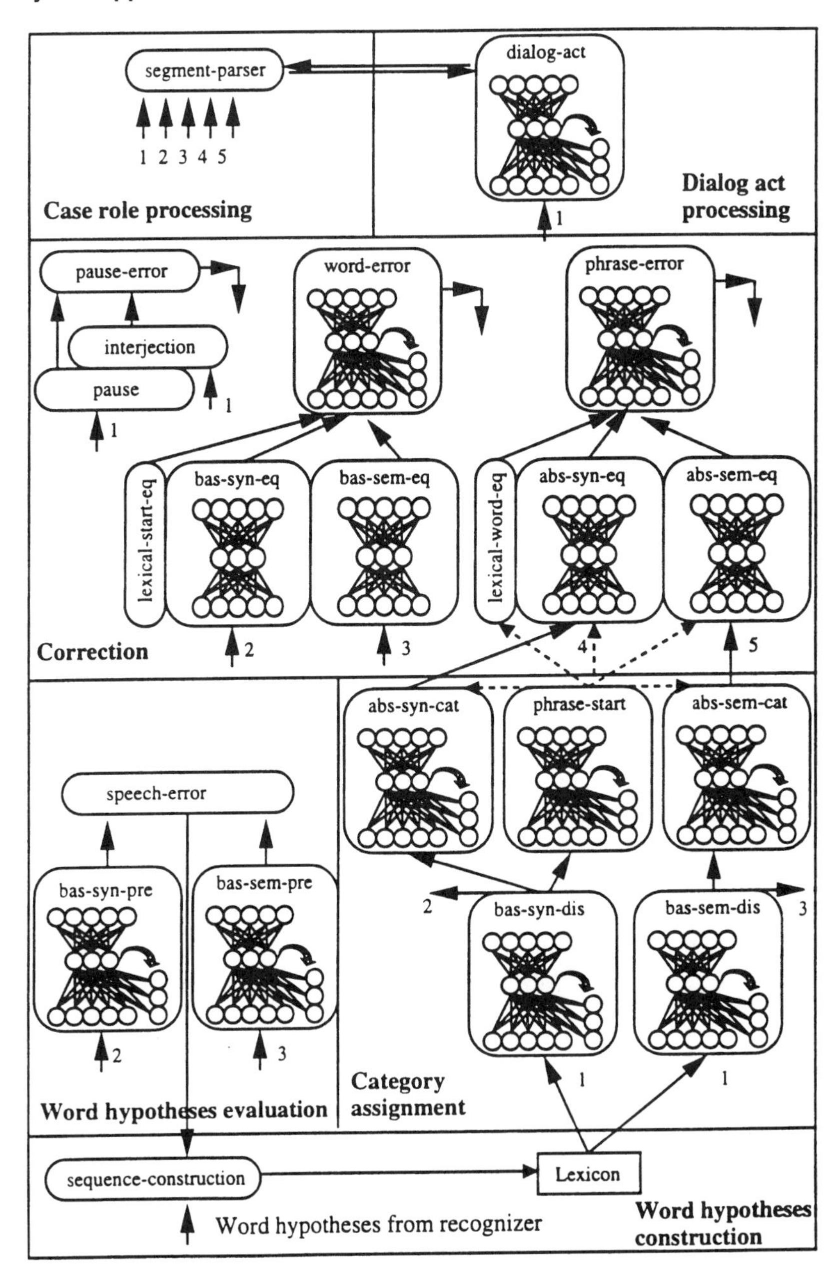

Fig. 9 SCREEN: an integrated hybrid connectionist processing architecture for spoken language analysis. The following abbreviations are used: abs(tract), bas(ic), dis(ambiguation), cat(egorization), syn(tactic), sem(antic), pre(diction), eq(uality). So, for instance, bas–sem–dis is the module for basic semantic disambiguation.

well on such constructive problems, we used a symbolic mechanism for performing this task incrementally on incoming word hypotheses.

During this process of word hypothesis sequence construction it is possible to evaluate many candidate sequences using the word hypothesis evaluation component. Depending on the plausibility of a sequence of semantic and syntactic categories, word hypothesis sequences with a high plausibility are retained while sequences with a lower plausibility are eliminated. Since this subproblem relies on the continuous plausibility of syntactic or semantic category sequences and since connectionist networks support such plausibility evaluations, we used connectionist networks to evaluate these sequences.

The retained word hypothesis sequences are analyzed incrementally in the component for category assignment. In particular, this component performs a syntactic and semantic category assignment. Each word hypothesis within each sequence is assigned the most plausible basic syntactic, abstract syntactic, basic semantic, and abstract semantic categories. Basic syntactic categories describe a word hypothesis sequence at a level of noun, adjective, verb, determiner, and so forth, whereas abstract syntactic categories describe a sequence at a phrase level as noun phrase, verb phrase, prepositional phrase, or other. Similarly, there are basic semantic categories and abstract semantic categories that will depend on the specific domain. For instance, in a corpus for scheduling meetings basic semantic categories could be "meet," "select," "animate object," or other, whereas abstract semantic categories could be "action", "object," and such. Furthermore, the word hypothesis sequences are separated into phrases to support various other tasks.

All these category assignments can be learned in recurrent connectionist networks based on examples. The use of connectionist networks is desirable to reduce the knowledge acquisition effort. Furthermore, the data-driven learning from connectionist networks allows the category assignment component to learn regularities from the data, which otherwise may not be manually encoded. Such unforeseeable regularities are particularly important when working with noisy spoken language data. These were in fact the reasons why we used connectionist networks for category assignment.

Spoken language may contain "errors," such as interjections, like "oh," "eh," repetitions, corrections, false starts, and all kinds of syntactic and semantic irregularities. Although it is impossible to predict all possible irregularities, there are some classes of "error" that occur relatively often. Therefore, we model the elimination of interjections, word, and phrase errors, in the correction component, because these are the most frequent mistakes. Some error detections can be supported by lexical comparisons, for instance, the lexical equality of two subsequent words may support the hypothesis that there is a repetition (the the meeting ...). Such lexical comparisons can be encoded directly in a symbolic manner and, therefore, they are realized in symbolic modules. On the other hand a comparison between two real-valued vectors of basic syntactic categories could be realized in a connectionist module.

So far we have dealt with words and phrase sequences. However, it is also important to understand complete sentences and turns, where a turn is the whole sequence of utterances before the next speaker begins. Boundaries between the utterances in a turn have to be detected to trigger different frames for different utterances. Furthermore, the analyzed constituents have to be transferred to case roles to be ready for later retrieval. All this is performed in a case role parser.

Because much explicit control is necessary to pull out the required knowledge and fill the case frame incrementally from the underlying modules, this component is realized symbolically.

In addition to the syntactic and semantic knowledge from the category assignment component, there is also knowledge at the dialogue level that plays an important role. Each utterance can be assigned to a certain dialogue act, for instance, that a certain utterance is a request. As with syntactic and semantic category assignment, this knowledge can be learned in connectionist networks in the dialogue processing component.

Before we present an example of processing in SCREEN, we summarize the categories for our flat analysis in Table 1. The categories belong to five different knowledge sources: a basic syntactic and a basis semantic word description, an abstract syntactic and an abstract semantic phrase description, and a dialogue act description.

SCREEN is an integrated architecture, as it does not crucially rely on a single connectionist or symbolic module. Rather, there are many connectionist and symbolic modules, but they have a common interface, and they can communicate with each other in many directions. For instance, Fig. 9 gives an impression of the many communication paths between various modules using arrows and numbers. From a module-external point of view it does not matter whether the internal processing within a module is connectionist or symbolic. In fact, during the design of the system, several modules, for instance, in the correction component, were replaced by their symbolic or connectionist counterparts. This architecture, therefore, exploits a full integration of symbolic and connectionist processing at the module level. Furthermore, the modules can run in parallel and produce an analysis in an incremental manner.

Now we will give an example of how a simple spoken sentence is processed. Figure 10 shows the system environment of SCREEN. Word hypothesis sequences are depicted horizontally. They are ordered in a decreasing manner according to their combined acoustic, syntactic, and semantic plausibility; that is, the top word hypothesis sequence represents the highest ranked word hypothesis sequence at that time step. Word hypotheses are processed incrementally, and for each new possible word hypothesis there is an evaluation of whether the best n possible word hypothesis sequences can be continued.

In analyzing the incremental word hypotheses it is possible to stop processing (Stop), to start (Go), and to continue processing in different steps (single, multiple step). Furthermore, we can store the current internal state of all categories. By using the left vertical scroll bar we can inspect further lower ranked word hypothesis sequences, using the lower horizontal bar we can view arbitrarily long word hypothesis sequences.

The example sentence in German is *sag das doch gleich* (literal translation: say that yet immediately). This sentence is processed incrementally and the state after 1010 ms is shown in Fig. 10. Each word hypothesis in a sequence is shown with the activations of the highest-ranked category assignments for that word hypothesis, as computed by the underlying connectionist networks (for a list of abbreviations see Table 1). In general, for each word hypothesis six boxes are shown: syntactic basic category (upper left), semantic basic category (lower left), abstract syntactic category

Table 1 Categories for the Flat Analysis[a]

Basic syntax	Basic semantics	Abstract syntax	Abstract semantics	Dialogue act
noun	**sel**ect	**v**erb **g**roup	**act**ion	**acc**ept
ad**j**ective	**sug**gest	**n**oun **g**roup	**aux**-action	**query**
verb	**meet**	**a**dverbial **g**roup	**agent**	**rej**ect
adverb	**utter**	**p**repositional **g**roup	**obj**ect	**re**quest-**c**omment
p**r**eposition	**is**	**c**onjunction **g**roup	**recip**ient	**r**equest-**s**uggest
conjunction	**have**	**m**odus **g**roup	**instr**ument	**state**
prono**un**	**move**	**s**pecial **g**roup	**manner**	**sug**gest
determiner	**aux**iliary	**i**nterjection **g**roup	**time-at**	**misc**ellaneous
nu**m**eral	**quest**ion		**time-from**	
interjection	**phys**ical		**time-to**	
participle	**anim**ate		location-**at**	
other	**abstr**act		location-**from**	
pause (/)	**here**		location-**to**	
	sou**r**ce		**conf**irmation	
	destination		**neg**ation	
	location		**quest**ion	
	time		**misc**ellaneous	
	negative evaluation (**no**)			
	positive evaluation (**yes**)			
	unspecific (**nil**)			

[a]Abbreviations are depicted in bold face.

(upper middle), abstract semantic category (lower middle), dialogue act category (upper right), and plausibility of word hypothesis sequence (lower right).

For instance, if we focus on the first word hypothesis "sag" (say) of the third word hypothesis sequence "sag das" (say that), the word hypothesis "sag" has the basic syntactic category V (verb), the basic semantic category of an utterance (UTTER), the abstract syntactic category of a verbal group (VG), the abstract semantic category of an action (ACT), the dialogue act category miscellaneous (MISC), and a particular combined acoustic–syntactic–semantic plausibility value (PLAUS). After 2760 ms we can see the final analysis in Fig. 11. The higher-ranked word hypothesis sequences are also the desired sequences. The two lower word hypothesis sequences show an additional undesired "hier" (here) at the end of the sequence.

Although we cannot go into all the details of the whole hybrid system, we believe it is important to demonstrate the applicability of hybrid techniques in real-world environments of language processing. For more details the interested reader is referred to [58,60,62,63]. During the development of the SCREEN system we have taken advantage of the integrated hybrid processing architecture which provides the same interface independently of whether it contains a connectionist or symbolic module. For instance, in the correction part or the case role part of the architecture, we have interchanged symbolic and connectionist versions of the

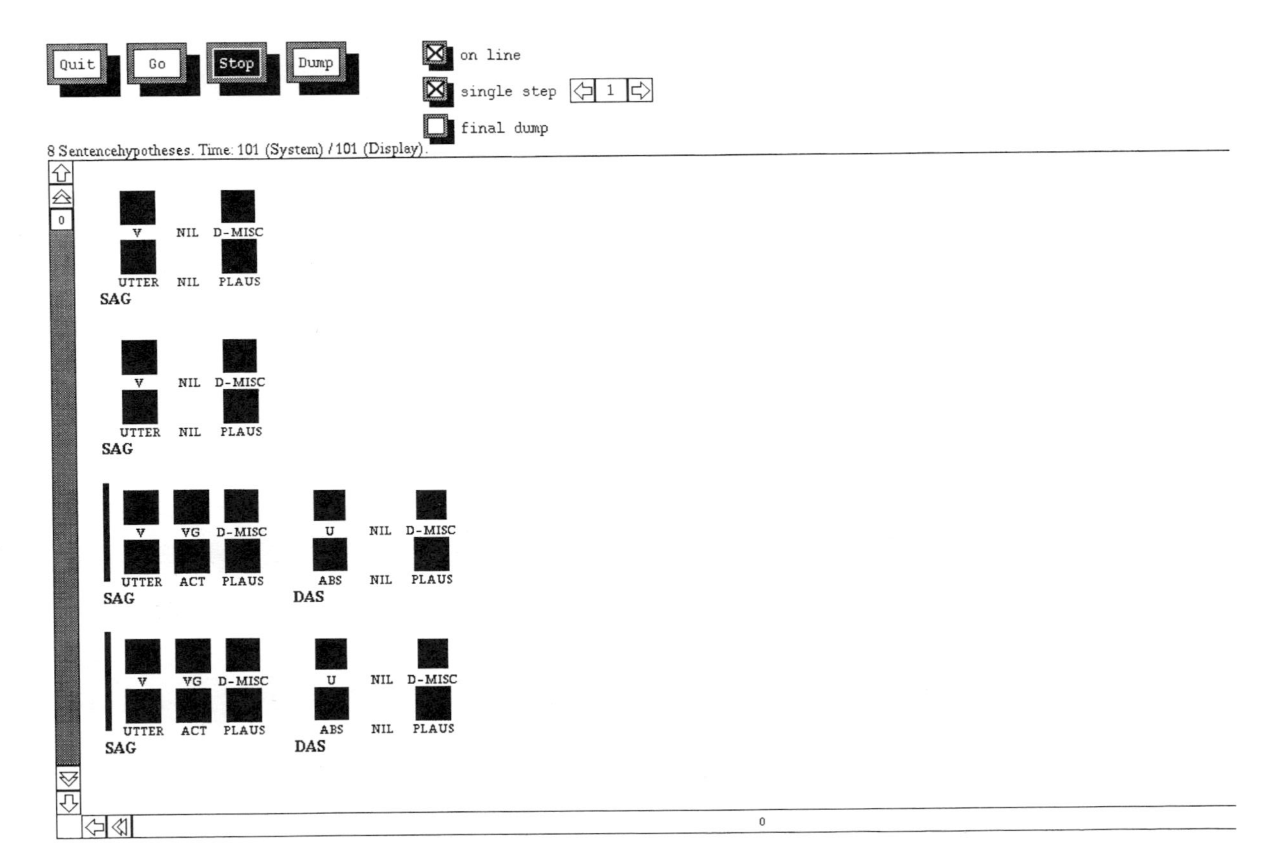

Fig. 10 SCREEN snapshot 1 for sentence *sag das doch gleich* (literal translation: say that yet immediately). See Table 1 for abbreviations.

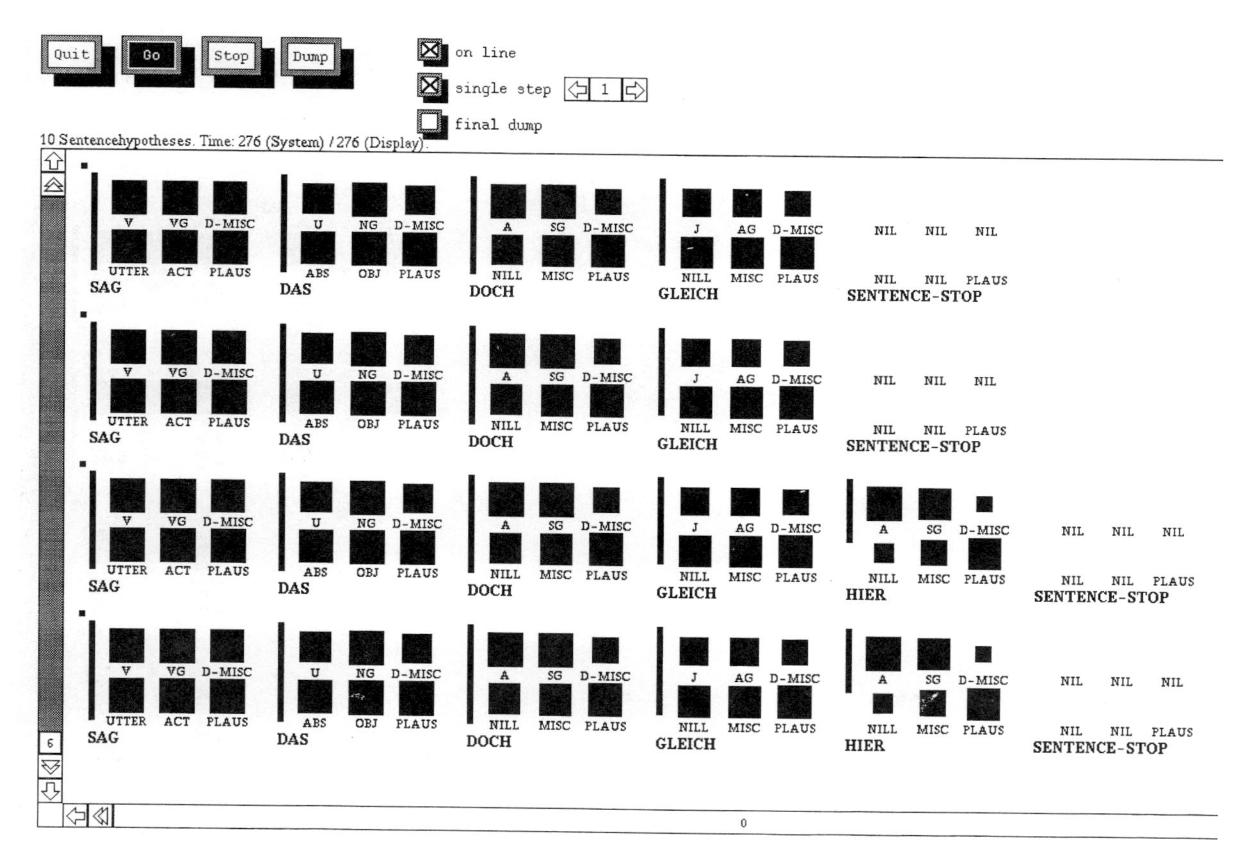

Fig. 11 SCREEN snapshot 2 for sentence *sag das doch gleich* (literal translation: say that yet immediately). See Table 1 for abbreviations.

same module. Our goal was to use connectionist techniques wherever possible owing to their advantages for robustness and learning capabilities, but to use also symbolic techniques wherever convenient. Even though all modules have a symbolically interpretable interface, the connectionist or symbolic techniques are hidden within the modules.

Other integrated architectures can be found in PARSEC, a system for language analysis in the conference registration domain [29], which uses Jordan networks to trigger symbolic transformations within the modules, and CONNCERT [29], which has been developed for technical control problems using a mixture of expert networks controlled by symbolic supervisors.

6. SUMMARY AND CONCLUSIONS

Recently, there has been an increasing interest in hybrid symbolic–connectionist interpretation, combination, and integration. We have provided an overview of the foundations of hybrid connectionist architectures and a classification of architectures. Furthermore, we have described some representative and current examples of such hybrid connectionist structures. From this perspective it is clear that hybrid symbolic–connectionist techniques have become an important class of artificial intelligence techniques. Hybrid techniques can be used successfully for those problems where the tasks require different discrete and analogue modes of computation and representation. Natural language processing is one important area for hybrid architectures, because natural language processing can be supported well by symbolic structure manipulations as well as connectionist gradual plausibility, learning, and robustness.

Hybrid architectures constitute a continuum from connectionist structure architectures with symbolic interpretation to hybrid transfer architectures and hybrid processing architectures with symbolic and connectionist representations. Today, hybrid connectionist architectures provide technology for realizing larger real-world systems for natural language processing. For instance, the SCREEN system uses a hybrid processing architecture including many different modules at speech, syntax, semantics, and dialogue processing levels. While early work on connectionist and hybrid architectures often had to focus on small tasks and small architectures to gain initial knowledge about the possibilities of combination and integration, we now begin to see the potential of hybrid connectionist systems being used for larger real-world tasks in natural language processing.

ACKNOWLEDGMENTS

This chapter was written during a research stay at ICSI, Berkeley. I would like to thank Jerry Feldman, Lokendra Shastri, Srini Narayanan, and the members of the NTL research group at ICSI, Berkeley for their support and discussions. I would like to thank the members of the hybrid connectionist language group at Hamburg University (J Chen, S Haack, M Loechel, M Meurer, U Sauerland, M Schrattenholzer, and V Weber) for their cooperation as well as many discussions. I am also grateful to H Moisl who provided detailed comments on a previous draft. This research was funded by Grant DFG We1468/4-1.

REFERENCES

1. D Bailey, J Feldman, S Narayanan, G Lakoff. Modelling embodied lexical development. Proceedings of the Meeting of the Cognitive Science Society, Stanford, 1997.
2. JA Barnden, KJ Holyoak, eds. Advances in Connectionist and Neural Computation Theory. Vol 3. Ablex Publishing, Norwood, NJ, 1994.
3. L Bookman. Trajectories Through Knowledge Space. Kluwer, Boston, 1994.
4. Y Cheng, P Fortier, Y Normandin. A system integrating connectionist and symbolic approaches for spoken language understanding. Proceedings of the International Conference on Spoken Language Processing, Yokohama, 1994, pp 1511–1514.
5. L Chrisman. Learning recursive distributed representations for holistic computation. Connect Sci 3:345–365, 1991.
6. R Cooper, B Franks. How hybrid should a hybrid model be? Proceedings of the ECAI Workshop on Combining Symbolic and Connectionist Processing, Amsterdam, 1994, pp 59–66.
7. MW Craven, JW Shavlik. Using sampling and queries to extract rules from trained neural networks. Proceedings of the 11th International Conference on Machine Learning, Rutgers University, New Brunswick, NJ; 1994, pp 37–45.
8. J Diederich, DL Long. Efficient question answering in a hybrid system. Proceedings of the International Joint Conference on Neural Networks, Singapore, 1992.
9. G Dorffner. Neural Networks and a New AI. Chapman & Hall, London, 1996.
10. MG Dyer. Distributed symbol formation and processing in connectionist networks. J Exp Theor Artif Intell, 2:215–239, 1990.
11. MG Dyer. Symbolic neuroengineering for natural language processing: a multilevel research approach. In: JA Barnden, JB Pollack, eds. Advances in Connectionist and Neural Computation Theory, vol 1: High Level Connectionist Models. Ablex Publishing, Norwood, NJ, 1991, pp 32–86.
12. M Fanty. Context-free parsing in connectionist networks. Technical Report 174, University of Rochester, Rochester, NY, 1985.
13. M Fanty. Learning in structured connectionist networks. Technical Report 252, University of Rochester, Rochester, NY, 1988.
14. JA Feldman. Structured connectionist models and language learning. Artif Intell Rev 7:301–312, 1993.
15. JA Feldman, DH Ballard. Connectionist models and their properties. Cogn Sci 6:205–254, 1982.
16. JA Feldman, G Lakoff, DR Bailey, S Narayanan, T Regier, A Stolcke. L_0—the first five years of an automated language acquisition project. AI Rev 8, 1996.
17. L Fu. Rule generation from neural networks. IEEE Trans Syst Man Cybernet 24:1114–1124, 1994.
18. SI Gallant. Neural Network Learning and Expert Systems. MIT Press, Cambridge, MA, 1993.
19. CL Giles, CW Omlin. Extraction, insertion and refinement of symbolic rules in dynamically driven recurrent neural networks. Connect Sci 5:307–337, 1993.
20. S Goonatilake, S Khebbal. Intelligent Hybrid Systems. Wiley, Chichester, 1995.
21. D Grannes, L Shastri, S Narayanan, J Feldman. A connectionist encoding of schemas and reactive plans. Proceedings of the Meeting of the Cognitive Science Society, Stanford, 1997.
22. J Hendler. Developing hybrid symbolic/connectionist models. In: JA Barnden, JB Pollack, eds. Advances in Connectionist and Neural Computation Theory. Vol 1: High Level Connectionist Models. Ablex Publishing, Norwood, NJ, 1991, pp 165–179.
23. JA Hendler. Marker passing over microfeatures: towards a hybrid symbolic connectionist model. Cogn Sci 13:79–106, 1989.

24. M Hilario. An overview of strategies for neurosymbolic integration. Proceedings of the Workshop on Connectionist–Symbolic Integration: From Unified to Hybrid Approaches, Montreal, 1995, pp 1–6.
25. S Hollbach Weber, A Stolcke. L_0: a testbed for miniature language acquisition. Technical Report TR-90-010, International Computer Science Institute, Berkeley, CA, 1990.
26. S Hollbach Weber. A structured connectionist approach to direct inferences and figurative adjective noun combination. Technical Report 289, University of Rochester, Rochester, NY, 1989.
27. V Honavar, L Uhr. Artificial Intelligence and Neural Networks: Steps Toward Principled Integration. Academic Press, Cambridge, MA, 1994.
28. RA Jacobs, MI Jordan, AG Barto. Task decomposition through competition in a modular connectionist architecture: the what and where vision tasks. Cogn Sci 15:219–250, 1991.
29. AN Jain. Parsec: a connectionist learning architecture for parsing spoken language. Technical Report CMU-CS-91-208, Carnegie Mellon University, Pittsburgh, PA, 1991.
30. D Jurafsky, C Wooters, G Tajchman, J Segal, A Stolcke, E Fosler, N Morgan. The Berkeley Restaurant Project. Proceedings of the International Conference on Speech and Language Processing, Yokohama, 1994, pp 2139–2142.
31. SC Kremer. A theory of grammatical induction in the connectionist paradigm. PhD dissertation, University of Alberta, Edmonton, 1996.
32. SC Kwasny, KA Faisal. Connectionism and determinism in a syntactic parser. In: N Sharkey, ed. Connectionist Natural Language Processing. Lawrence Erlbaum, Hillsdale, NJ, 1992, pp 119–162.
33. R Maclin. Learning from instruction and experience: methods for incorporating procedural domain theories into knowledge-based neural networks. PhD dissertation, University of Wisconsin–Madison, 1995. [Also appears as UW Technical Report CS-TR-95-1285].
34. C McMillan, M Mozer, P Smolensky. Dynamic conflict resolution in a connectionist rule-based system. Proceedings of the International Joint Conference on Artificial Intelligence, Chambéry, France, 1993, pp 1366–1371.
35. C McMillan, MC Mozer, P Smolensky. The connectionist scientist game: rule extraction and refinement in a neural network. Proceedings of the 13th Annual Conference of the Cognitive Science Society, 1991, pp 424–430.
36. LR Medsker. Hybrid Neural Network and Expert Systems. Kluwer, Boston, 1994.
37. LR Medsker. Hybrid Intelligent Systems. Kluwer, Boston, 1995.
38. R Miikkulainen. Subsymbolic Natural Language Processing. MIT Press, Cambridge, MA, 1993.
39. MC Mozer. Neural net architectures for temporal sequence processing. In: A Weigend, N Gershenfeld, eds. Time Series Prediction: Forecasting the Future and Understanding the Past. Addison–Wesley, Redwood City, CA, 1993, pp 243–264.
40. MC Mozer, S Das. A connectionist symbol manipulator that discovers the structure of context-free languages. In: SJ Hanson, JD Cowan, CL Giles, eds. Advances in Neural Information Processing Systems 5. Morgan Kaufmann, San Mateo, CA, 1993, pp 863–870.
41. S Narayanan. Talking the talk is like walking the walk: a computational model of verbal aspect. Proceedings of the Meeting of the Cognitive Science Society, Stanford, 1997.
42. CW Omlin, CL Giles. Extraction and insertion of symbolic information in recurrent neural networks. In: V Honavar, L Uhr, eds. Artificial Intelligence and Neural Networks: Steps Towards Principled Integration. Academic Press, San Diego, 1994, pp 271–299.
43. CW Omlin, CL Giles. Extraction of rules from discrete-time current neural networks. Neural Networks, 9:41–52, 1996.

44. JB Pollack. On connectionist models of natural language processing. PhD dissertation. Technical Report MCCS-87-100, New Mexico State University, Las Cruces, NM, 1987.
45. T Polzin. Parsing spontaneous speech: a hybrid approach. Proceedings of the ECAI Workshop on Combining Symbolic and Connectionist Processing, Amsterdam, 1994, pp 104–113.
46. JE Rager. Self-correcting connectionist parsing. In: RG Reilly, NE Sharkey, eds. Connectionist Approaches to Natural Language Processing. Lawrence Erlbaum, Hillsdale, NJ, 1992.
47. T Regier. The acquisition of lexical semantics for spatial terms: a connectionist model of perceptual categorization. Technical Report, International Computer Science Institute, 1992.
48. RG Reilly, NE Sharkey. Connectionist Approaches to Natural Language Processing. Lawrence Erlbaum, Hillsdale, NJ, 1992.
49. L Shastri. A model of rapid memory formation in the hippocampal system. Proceedings of the Meeting of the Cognitive Science Society, Stanford, 1997.
50. J Shavlik. A framework for combining symbolic and neural learning. In: V Honavar, L Uhr, eds. Artificial Intelligence and Neural Networks: Steps Towards Principled Integration. Academic Press, San Diego, 1994, pp 561–580.
51. P Smolensky. On the proper treatment of connectionism. Behav Brain Sci 11:1–74, 1988.
52. MF St John, JL McClelland. Learning and applying contextual constraints in sentence comprehension. Artif Intell 46:217–257, 1990.
53. R Sun. Integrating Rules and Connectionism for Robust Commonsense Reasoning. Wiley, New York, 1994.
54. R Sun. Hybrid connectionist–symbolic models: a report from the IJCAI 95 workshop on connectionist-symbolic integration. Artif Intell Mag 1996.
55. R Sun, F Alexandre. Proceedings of the Workshop on Connectionist–Symbolic Integration: From Unified to Hybrid Approaches. McGraw-Hill, Montreal, 1995.
56. R Sun, LA Bookman. Computational Architctures Integrating Neural and Symbolic Processes. Kluwer, MA, 1995.
57. DL Waltz, JB Pollack. Massively parallel parsing: a strongly interactive model of natural language interpretation. Cogn Sci 9:51–74, 1985.
58. S Wermter. Hybride symbolische und subsymbolische Verarbeitung am Beispiel der Sprachverarbeitung. In: I Duwe, F Kurfess, G Paass, S Vogel, eds. Herbstschule Konnektionismus und Neuronale Netze. GMD, Sankt Augustin, 1994.
59. S Wermter. Hybrid Connectionist Natural Language Processing. Chapman & Hall, London, 1995.
60. S Wermter, M Löchel. Learning dialog act processing. Proceedings of the International Conference on Computational Linguistics, Copenhagen, Denmark, 1996.
61. S Wermter, E Riloff, G Scheler. Connectionist, Statistical and Symbolic Approaches to Learning for Natural Language Processing. Springer, Berlin, 1996.
62. S Wermter, V Weber. SCREEN: learning a flat syntactic and semantic spoken language analysis using artificial neural networks. J Artif Intell Res 6:35–86, 1997.
63. S Wermter, M Meurer. Building lexical representations dynamically using artificial neural networks. Proceedings of the International Conference of the Cognitive Science Society, Stanford, 1997.
64. A Wilson, J Hendler. Linking symbolic and subsymbolic computing. Connect Sci 5:395–414, 1993.
65. CC Wooters. Lexical modeling in a speaker independent speech understanding system. Technical Report TR-93-068, International Computer Science Institute, Berkeley, 1993.

33

Character Recognition with Syntactic Neural Networks

SIMON LUCAS

University of Essex, Colchester, England

1. INTRODUCTION

This chapter gives a brief overview of syntactic neural networks (SNNs) (see Sec. 2), and demonstrates two complementary ways in which they can be applied to problems in optical character recognition (OCR).

At the character recognition level, a special simplified SNN is used to build a statistical language model of the chain-codes of character images, which are subsequently used in recognition. This method, described in Sec. 3, yields a character classifier that can be trained on over 30,000 characters per second, and can recognise over 1,000 characters per second, on an entry level PC, yet which is nearly as accurate as the best of other approaches.

In nearly all practical applications of character recognition, character images do not exist in a vacuum, but exist in relation to other characters, combining with these to form words, either in the sense of some natural language, or in the sense of symbol sequences that constitute postcodes or social security numbers, and such. Furthermore, there are constraints on these "words" that can also be exploited. Hence, the image of a hand-written form is typically highly redundant. The state of the art in isolated digit or character recognition is that machines can now do this about as reliably as people in the absence of context. Future OCR research, therefore, will focus on how best to recognise cursive script, and how best to apply contextual knowledge to optimally interpret the character and word hypotheses produced by low-level classifiers.

Section 4 describes how an SNN can be constructed to store dictionary or lexicon-type knowledge in an extremely efficient manner. This is especially appropriate for massive dictionaries, in which the number of entries exceeds tens of thousands. The example used is the United Kingdom (UK) Postcode dictionary, which contains about 1.4 million entries. Efficient retrieval given uncertain character hypotheses is achieved by lazy evaluation of the SNN. Interestingly, the retrieval speed actually gets faster as the size of the dictionary increases. The method is much better than using a trie structure (the conventional approach) for this size of dictionary, both in terms of speed and memory requirements. Furthermore, the SNN structure can also incorporate other types of constraint, giving a comprehensive and uniform solution to the optimal interpretation of character images.

2. SYNTACTIC NEURAL NETWORKS

Syntactic neural networks (SNNs) [1,2] are a class of artificial neural architecture based on context-free grammars (CFGs). The SNN has a well-defined modular structure. The local inference machines (LIMs) parse simple grammar fragments, and the way in which these are interconnected dictates the class of language that the network as a whole may learn and recognise. The architecture is derived from a matrix-parsing algorithm, also known as the CKY (after Cocke, Kasami, and Younger) algorithm [3,4]. Matrix parsing is appropriate for highly ambiguous grammars, but much less efficient than LR parsing for the LR-grammars used to define programming languages, for example. When the parse-matrix is stored in sparse form, these algorithms are known as chart-parsers [5–7]. Figure 1 shows the general network structure (macrostructure) for dealing with the general class of context-free grammar. By selectively pruning this general structure, SNNs can be constructed specifically to deal with various subsets of CFG, such as regular or strictly hierarchical (see Fig. 4). This represents an interesting alternative to the more conventional approach of applying neural networks to symbol–sequence learning problems using a nonhierarchical recurrent neural net [8–11], the structure of which is well adapted only for learning regular grammars.

The schematic structure of the LIM is shown in Fig. 2. The weights (or rewrite rules) can be made common to all LIMs, or unique to each LIM. The LIM can be designed to handle stochastic or nonstochastic grammars, and can operate with a fixed grammar, or be designed to infer a grammar. There are various ways in which this can be achieved—the method used in the following rapid retrieval system is incredibly simple (but useful). More interesting approaches are possible (e.g., by using forward–backward function modules [12] the system can be made to infer more complex grammars, in which case the system is performing a kind of backpropagation through time/structure [13]).

The focus of this chapter, however, is how special cases of SNN architecture that, at first glance, might appear trivial can be put to good use in practical OCR problems. The first application is for high-speed OCR. For this, the images are first mapped to strings and then partially parsed by an SNN with a fixed, trivial grammar. The statistics of the partial parses are then computed to make classifications. The second application is dictionary search, for which a strictly hierarchical SNN is used to store a dictionary (by inferring a strictly hierarchical context-free grammar) which can then be searched very efficiently by means of a lazy evaluation procedure.

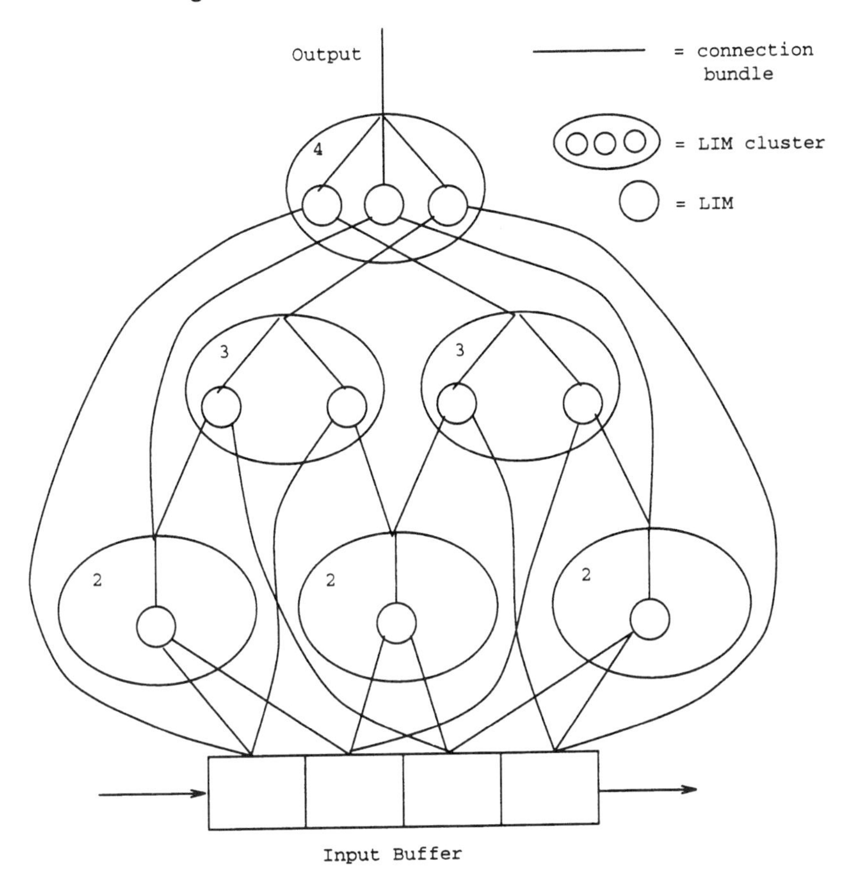

Fig. 1 Architecture of SNN for general context-free grammars.

3. OCR

This section describes how a simplified form of SNN can be used for high-speed OCR. A more complete description of this work is available [14]. We first look at how the images are mapped into strings, and then go on to describe how the SNN models the strings. In fact, the resulting model is little more than a statistical n-gram model [15], although the results in Table 4 show how it outperforms a conventional n-gram model. Also, it is closely related to n-tuple classifiers [16]; henceforth, we refer to the method as the sn-tuple system for scanning n-tuple, because essentially, the system behaves as a probabilistic n-tuple classifier scanned over the entire chain-code of the image.

A. Mapping Images to Strings: Chain-Coding Procedure

Before applying the sn-tuple method we map each image to a string (Fig. 3). Here we use a simple chain-coding procedure to accomplish this. The initial procedure uses just four direction vectors, and retains *all* the information in the original binary image—in fact, it is a type of lossless compression algorithm for this data and, incidentally, generally compresses the images by a ratio of about 50:1.

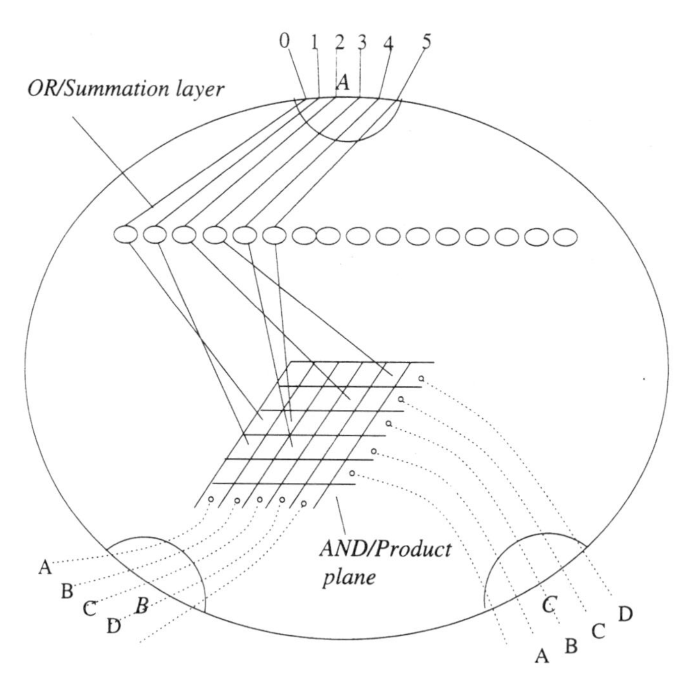

Fig. 2 Schematic of the local inference machine.

The procedure first detects all the edge transitions in the images, assigning four possible codes according to whether the transition is in a direction from top to bottom, or left to right, and whether it is going from white to black or black to white as shown in Fig. 3c. These edges are then followed in specific directions, and connected up into strings by a simple finite-state machine. The starting point for each chain code is found by scanning the image from top to bottom, left to right.

Most images consist of more than one chain code, but the sn-tuples are applied here to model just one string per pattern. We map a set of strings for a given image to a single string by removing the positional information (i.e., the X and Y start coordinates), and then concatenating the strings together.

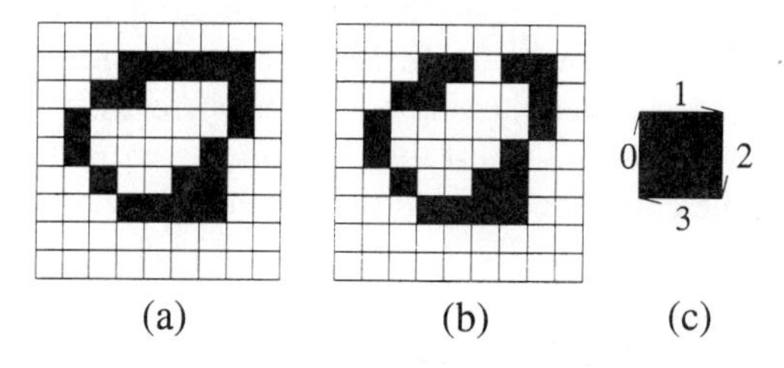

Fig. 3 The sensitivity of the chain-coding algorithm to minor changes in the bitmap. The two images differ only in a single pixel, but (a) the first one produces the strings: (3 1 11111222322233330303001010) (4 1 233221211010100333), whereas (b), the second one produces the string (3 1 11232332212110101003011222322233330303001010). The first two numbers are the X and Y start positions of the chain, and the rest is the chain code itself. (c) shows the coding of the edge transitions as movement symbols 0, 1, 2, and 3.

B. Problems with Chain Coding and Possible Solutions

In fact, minor variations in the bitmap can cause rather dramatic changes in the chain code, as illustrated in Fig. 3. Here we see the omission of a single pixel changing two codes into one, and actually completely altering the symbol set of the coding of the interior edge. The sort of arbitrary sensitivity shown in Fig. 3 is enough to put one off applying this strategy further, but Vidal et al. [17] showed that grammatical inference (language learning) techniques such as the Error Correcting Grammatical Inference algorithm (ECGI) were able to cope with such variations to a reasonable degree, and in a follow-up paper [18] the ECGI was shown to outperform the other techniques under test.

The chain code representation used here may also be criticised for not possessing rotational invariance characteristics. However, the degree to which rotational invariance is a necessary or desirable property in OCR systems is arguable, and total invariance to rotation is certainly undesirable, for this would lead to confusion between 6s and 9s for example. The easiest way to build the required degree of invariance into the sn-tuple OCR method reported here would be to build a new enlarged training set by transforming all the training data by a number of distinct rotation angles (e.g., +/–5°, +/–10°). Some investigation was made into this [19], and expanding the training set in this way increases the accuracy from 97.6% (see results in following section) to 98.3%.

C. Mapping 4-Direction to 8-Direction Strings

The results quoted in Sec. 3.E show a clear advantage in using 8-direction, rather than 4-direction chain codes, in increased recognition accuracy, increased speed, and reduced memory requirements. The following system was used to map each 4-direction string y_4 to an 8-direction string y_8:

$$y_8(i) = f(y_4(i^*2 - 1), y_4(1^*2))\forall(i^*2) \leq \|y_4\| \qquad \{1\}$$

where the function f: $\Sigma_4 \times \Sigma_4 \rightarrow \Sigma_8$ is defined for the entries in Table 1 and undefined elsewhere. It is apparent from Eq. {1} that the 8-directional string y_8 is half the length of the original string y_4. Thus, using strings over Σ_8 not only results in more accurate recognition, as shown later, but it is also twice as fast (the cost of transforming the string $y_4 \rightarrow y_8$ is negligible).

Table 1 Rewrite Rules for Mapping 4-Direction to 8-Direction Chain-Codes

Rule
0 → 00
1 → 01|10
2 → 11
3 → 12|21
4 → 22
5 → 23/32
6 → 33
7 → 30|03

D. N-Tuple Classifiers and the Scanning N-Tuple

In standard n-tuple classifiers [16] the d-dimensional (discrete) input space is sampled by m n-tuples. The range of each dimension in the general case is the alphabet $\Sigma = \{0, \ldots, \sigma - 1\}$ but most n-tuple methods reported in the literature are defined over a binary input space where $\sigma = 2$ and $\Sigma = \{0, 1\backslash\}$.

Each n-tuple defines a fixed set of locations in the input space. Let the set of locations defining the jth n-tuple be $n_j = \{a_{j1}, a_{j2}, \ldots, a_{jn} | 1 \leq a_{ji} \leq d\}$ where each a_{ji} is chosen as a random integer in the specified range. This mapping is normally the same across all classes. For a given d-dimensional input pattern $\mathbf{x} = x(1) \ldots x(d)$ an address $b_j(\mathbf{x})$ may be calculated for each n-tuple mapping n_j as shown in Eq. {2}.

$$b_j(\mathbf{x}) = \sum_{k=1}^{n} x(a_{jk}) \times \sigma^{k-1} \tag{2}$$

These addresses are used to access memory elements, where there is a memory n_{cj} for each class c in the set of all classes C and n-tuple mapping n_j. We denote the value at location b in memory n_{cj} as $n_{cj}[b]$. The set of all memory values for all the n-tuple mappings for a given class we denote M_c, the model for a class c. The size of the address space of each memory n_{cj} is σ^n.

During training the value at location $n_{cj}[b]$ is incremented each time a pattern of class c addresses location b. During recognition, there are three established ways of interpreting the value at each address: binary, frequency weighted, and probabilistic [20]. Only the probabilistic version will be derived here, for the case of the scanning n-tuple, but results are also quoted later for the standard binary n-tuple classifier.

The difference between the n-tuple and scanning n-tuple (sn-tuple) is that whereas each n-tuple samples a set of fixed points in the input space, each sn-tuple defines a set of relative offsets between its input points. Each sn-tuple is then scanned over the entire input space. The input space is now a variable length string $\mathbf{y}$ of length $\|\mathbf{y}\|$ rather than a fixed-length array.

We redefine sn-tuple address computation as follows for offset o relative to the start of the string as shown in Eq. {3}.

$$b_{jo}(\mathbf{y}) = \sum_{k=1}^{n} y(o + a_{jk}) \times \sigma^{k-1} \tag{3}$$

The probability that this address is accessed by a pattern from class c is given in Eq. {4} where $N_{cj} = \sum_{b=0}^{\sigma^n - 1} n_{cj}[b]$.

$$P(b_{jo}(\mathbf{y}) | M_{cj}) = \frac{n_{cj}[b_{jo}(\mathbf{y})]}{N_{cj}} \tag{4}$$

From this we calculate the probability of the whole string given sn-tuple model M_{cj}, under the assumption that the n-tuples at different offsets in the string are statistically independent:

$$P(\mathbf{y} | M_c) = \prod_{o=1}^{\|\mathbf{y}\| = \max(a_{jk} \forall k \in \{1, \ldots, n\})} P(b_{jo}(\mathbf{y}) | M_{cj}) \tag{5}$$

As pointed out by Rohwer [20], the assumption of statistical independence between n-tuples is unrealistic, but there exists as yet no superior alternative. Note that Eq. {5} is very similar to the equation for the probability of a sequence given the statistical n-gram language models used in speech recognition (as mentioned earlier), except that now we are able to model long-range correlations as well as short-range ones. Table 4 clearly shows the benefit of this.

The probability of a stringy given all the sn-tuple models of class c is given in Eq {6}

$$P(\mathbf{y}|M_c) = \prod_{j=1}^{m} P(\mathbf{y}|M_{cj}) \qquad \{6\}$$

Subsequent pattern classification proceeds according to Bayes theorem. If the prior class probabilities are assumed equal, which we do here, then the maximum likelihood decision is to assign the pattern to the class c for which $P(\mathbf{y}|M_c)$ is a maximum.

The algorithm for training the scanning n-tuple is as follows. All the memory contents are initialised to zero. For each pattern (string) of each class and for each mapping n_j, we scan the mapping along the string $\mathbf{y}$ from beginning to end by adjusting an offset o. In each case we increment the value at memory $n_{cj}[b_{jo}(\mathbf{y})]$ the address calculation $b_{jo}(\mathbf{y})$ is defined in Eq. {3}.

E. Results

Experimental results are reported for two freely available (and widely used) databases of handwritten digits: Essex and Cedar. In the original data sets, the images vary greatly in size, and the position of the character within a bitmap map also vary. To work well, n-tuple methods first require that the image be size and position normalised. For these experiments, a bounding box for each original character image was automatically computed; the image within this box was then scaled to fit on a 16×20 grid and then submitted directly to the n-tuple classifier, or chain-coded and then fed to the sn-tuple classifier. Best results for the standard binary n-tuple system were obtained using 40 8-tuples. Best results for the sn-tuple were obtained using 4 5-tuples, with evenly spaced offsets, spaced 2, 3, 4, and 5 places apart, respectively.

Table 2 shows the test set recognition accuracy for a standard binary n-tuple system, and a probabilistic sn-tuple system. Note that the sn-tuple-8 achieves superior performance in every case.

Table 2 Recognition Rates (%) on the Essex and CEDAR Test Data[a]

Method	Essex data	CEDAR data
n-Tuple	90.6	91.8
sn-Tuple-4	89.3	96.3
sn-Tuple-8	91.4	97.6

[a] All methods scored very close to 100% on both training sets.

Table 3 Results of Ten Experiments[a] Comparing Performance of Binary n-Tuple and Probabilistic n-Tuple Applied to the Image, and Scanned Across the Chain-code (String)

	Fixed (image)	Scanning (chain-code)
Bin	86.0% (0.37)	92.1% (0.41)
Prob	92.5% (0.27)	96.5% (0.24)

[a]Mean with standard deviation in parentheses.

To give some idea of statistical significance, further experiments were conducted based on a random sampling method. For each entry in Table 3, ten experiments were performed. For each experiment, a training set and disjoint test set of 500 characters per class for each set were created by randomly sampling the CEDAR training set. The results show the recognition accuracy obtained when applying both the binary and probabilistic n-tuple systems both to the image, or as scanning n-tuples to a chain code of the image. The advantages of adopting both a probabilistic interpretation and a scanning mode of application are clear. The reason the results are poorer than those for the CEDAR set in Table 2 is that fewer samples were used for training (500 versus approximately 2000 per class). We also tested a conventional n-gram model [15] using this random sampling method on the same sample set. Best results obtained were 92.5% with $n = 6$. The sn-tuple gives superior performance because it is able to model longer-range constraints owing to its use of nonconsecutive offsets.

F. Recognition Accuracy Versus *n* and Offset Gap

The test-set recognition accuracy has been measured for various values of n ranging from 1 to 6, and for a number of different gaps between each relative offset, ranging also from 1 to 6. These results are summarised in Table 4. This table shows that a standard bigram-modelling approach would score 67.2% on average, whereas we might expect 79.8% from a standard trigram model. Note the advantage of using nonconsecutive offsets; a trigram model with offsets placed every fourth position scores 90.7%—an improvement of more than 10% over the standard trigram model.

Table 4 Variation of Recognition Accuracy with Size of n-Tuple and Length of Offset Gap[a]

		n					
		1	2	3	4	5	6
Gap	1	48.2 (0.45)	67.2 (0.53)	79.8 (0.62)	87.0 (0.40)	90.8 (0.28)	92.452 (0.267)
	2	—	74.3 (0.44)	86.4 (0.49)	91.8 (0.42)	94.6 (0.29)	94.618 (0.279)
	3	—	77.4 (0.67)	89.9 (0.42)	93.8 (0.38)	95.2 (0.24)	94.584 (0.254)
	4	—	77.9 (0.49)	90.7 (0.37)	94.0 (0.44)	95.0 (0.21)	94.082 (0.215)
	5	—	78.4 (0.40)	90.4 (0.39)	93.0 (0.37)	93.9 (0.28)	93.024 (0.324)
	6	—	77.3 (0.61)	89.0 (0.50)	91.6 (0.30)	92.3 (0.27)	91.456 (0.425)

[a]Each table entry shows the mean of ten experiments with standard deviation in parentheses.

A composite sn-tuple system was designed using this table of data by using four 5-tuples with offset gaps of 2, 3, 4, and 5, respectively.

G. Speed

In this section we investigate the claimed speed performance of the sn-tuple system. The timing results are in CPU seconds on a Pentium PC running at 66 Mhz with 24 Mb of RAM, running under Windows 95. The code was developed using Borland C++ 4.52. To measure the time taken for each operation, a subset of 1000 images (100 of each class zero to nine) was extracted from the CEDAR training set. These figures are shown in Table 5. From this we can work out the average time per operation per image.

Once we have a database of chain-code strings, we can then train the system at a rate of up to 33,000 characters per second (cps), and subsequently recognise 1200 cps. The significance of this is that it allows us to perform many experiments in a short space of time to determine a close to optimal system configuration, in terms of the number of inputs and gap between each input in each sn-tuple, and the total number of sn-tuples. Optimal performance is often achieved with three or four sn-tuples. Thus, if we include the time taken to preprocess the images and a setup involving four sn-tuples, then we can still work at training speeds of 500 cps and recognition speeds of 200 cps.

4. THE SNN-BASED DICTIONARY SEARCH SYSTEM

Contextual knowledge is of critical importance in most pattern-recognition applications. Within a given application, there may be a need to apply many different types of knowledge to maximise the system accuracy. Examples of the type of knowledge that can be applied to improve recognition accuracy include dictionary, grammatical, arithmetic, logical, or database-type knowledge [see Ref. 21 for discussion of the importance of context in document image analysis].

This section gives an overview of the SNN-based rapid retrieval system, recently developed by the author [22,23]. The original system was developed to perform best-first retrieval from massive dictionaries given uncertain inputs, and has since been extended to provide a complete and uniform method for incorporating most kinds of contextual knowledge in the pattern-recognition process [24].

Note that dictionary constraints are much tighter than stochastic grammar models. Therefore, this method provides better contextual constraints than

Table 5 Time (CPUs) for Processing 1000 Images (Each One 16×20)[a]

Operation	Time (CPU seconds)
Read images from disk	2.69
Chain-code images	1.89
Train sn-tuple on chain codes	0.02
Classify chain codes with sn-tuple	0.83

[a]The sn-tuple timings are *per sn-tuple*.

HMMs, for example, if a complete dictionary exists and any generalisation is undesirable. In natural lexicons this is not so—people generate and understand new words with ease. In artificial lexicons of postcodes, telephone numbers, vehicle licence numbers, or social security numbers, however, the best strategy (if we have the computational resources to facilitate it) is to store and search an absolutely faithful copy of the dictionary. The SNN rapid retrieval system has proved to be an excellent method of applying contextual knowledge to pattern recognition systems, to provide an optimal interpretation of the uncertain outputs of low-level pattern classifiers.

The overall structure of the system is shown in Fig. 4. Each of the shaded circles is called a Local Inference Machine (LIM) and is responsible for inferring and parsing local fragments of grammar (i.e., here, each LIM has its own grammar). The idea of using a grammar to represent a lexicon is not new; for example, Dodd [25] applied the inside–outside grammar to estimate a stochastic context-free grammar for the spelling of English words. Here, the aim is quite different, however; we wish to infer a grammar to represent a finite set of words exactly, and not generalise at all. This makes the grammatical inference rather trivial, but nonetheless useful.

Figure 5 illustrates the postcode recognition problem, with some sample characters and their probability estimates as assigned by a simple pattern classifier and sorted into order such that the most likely characters are on the top ('_' indicates the space character).

A. Storing the Data

In general, the training of an SNN corresponds to a form of grammatical inference, because a type of language is being learned. However, we do not wish the system to generalise *at all*, and the grammatical inference is trivial (but useful). Hence, the operation of the LIMs during training is very simple: if a pattern (pair of symbols)

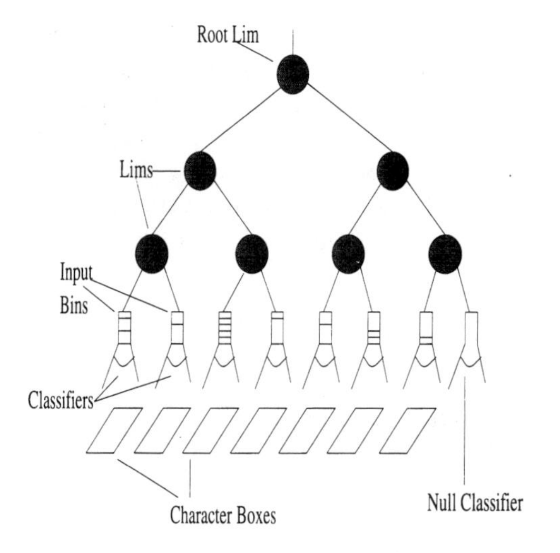

Fig. 4 Macrostructure of the strictly hierarchical syntactic neural network, as applied to the UK postcode recognition problem.

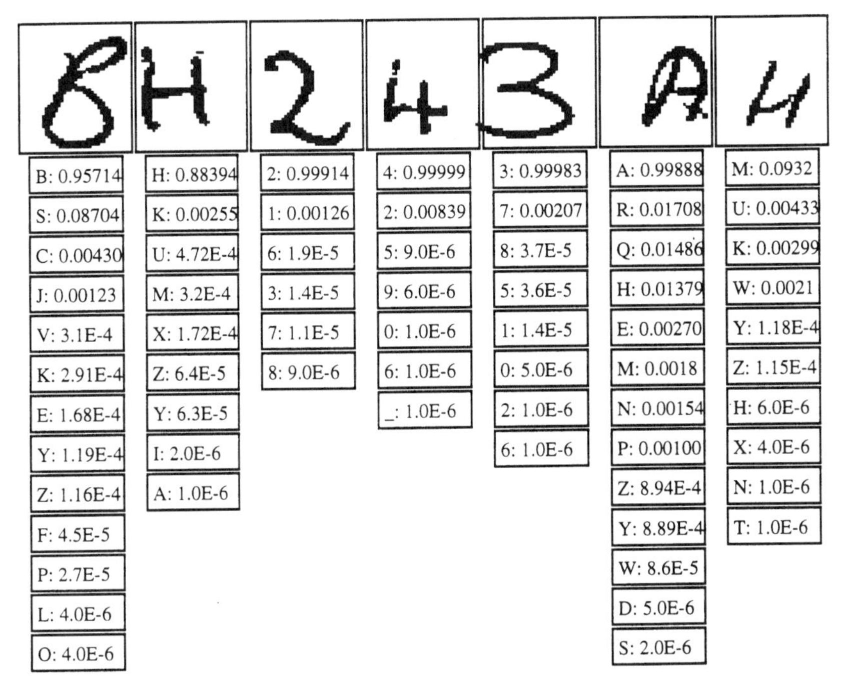

Fig. 5 A sample postcode written in boxes. The associated classifier outputs are shown in the stacks of symbol–probability pairs below each character.

has been seen before, then the corresponding nonterminal is passed on to the output; or else a new nonterminal is allocated and passed on to the output. An SNN can be built for a dictionary of over 1 million postcode strings in approximately 1 min on a 200 Mhz PC.

B. Retrieving the Data

The data retrieval corresponds to finding the maximum likelihood parse of the uncertain input sets. Note that, because of the structure of the SNN employed here and the nongeneralising nature of the LIMs, each root-level nonterminal rewrites to a single postcode, and each postcode has a single parse.

The lazy retrieval method is based entirely at the LIM level. Each LIM, on request for the next most likely symbol, explores its current candidate set, returns the most likely valid symbol, and recomputes the current candidate set.

The operation of the LIM can be decomposed into two parts: the lazy multiplier and the validity checker. The lazy multiplier returns the next best pair each time it is asked for one, and the validity checker simply checks that this pair is valid (i.e., has been seen by this LIM during training). If the pair is valid, it is returned, else the lazy multiplier produces the next one, and so on until it produces a pair that is valid.

The operation of higher-level LIMs is identical, the only difference being the context in which they operate (i.e., when the lazy multiplier asks for the next best symbol on a particular input line, it comes from another LIM, and not one of the input bins. The result is that the root node returns only legal sequences, and does so in strictly best-first order.

C. Example: The Postcode Recognition Problem

In the UK, there are approximately 1.4 million postcodes, each of which is up to seven characters long. In the original work [23], we assumed that the location of each character could be reliably identified, but that the identity would be highly uncertain. This amounts to catering for substitution errors, but not insertion or deletion errors. This has since been extended to cope with insertion and deletion errors caused by poor segmentation [26], but this latter work is still in prototype. Here we look at only the original system in which inputs were assumed to be reliably segmented.

Figure 5 illustrates the postcode recognition problem, with some sample characters and their confidence estimates as assigned by a simple pattern classifier and sorted into order such that the most likely characters are on the top ('_' indicates the space character). The actual character recogniser used to generate the outputs was a multilayer perceptron with 40 hidden units working on radial mean features [27].

D. SNN Retrieval Method Versus Trie

A conventional method for solving this problem is to store the dictionary in a special tree type of data structure called a trie [28–30]. This method is reasonably efficient in terms of space, and if the inputs are presented appropriately (in best-first order) the structure can be searched quite efficiently.

The SNN method improves on this in that it tends to be more efficient in terms of space and is able to offer the best or the n-best alternatives more rapidly than the trie. The graph in Fig. 6 illustrates this by plotting the negative log probability of each sequence against its order of retrieval. Retrievals from the SNN are in strictly best-first order, whereas this is clearly not so for the trie structure (because the most efficient way to search the trie, and the method used here, is depth-first).

Table 6 shows the desirable properties of the SNN method; namely, the high-speed and low-memory requirements for large dictionaries. Also, the SNN data structure (i.e., strictly hierarchical context-free grammar) is symmetrical. This means that the SNN is able to cope with errors equally, regardless of where they occur in the input string. The trie, however, is very good at dealing with errors at the end of the string, but bad at dealing with errors at the start of the string.

Table 6 Average Time (CPUs) to Retrieve the Most Likely Postcode from the SNN and Trie, and Memory Requirements for Various Dictionary Sizes.

Postcodes (approx)	Trie		SNN	
	Time	Memory	Time	Memory
10,000	0.0055 s	0.9 Mb	0.0226 s	1.2 Mb
100,000	0.0038 s	5.6 Mb	0.0019 s	1.5 Mb
1,400,000	0.0163 s	32.4 Mb	0.0006 s	1.6 Mb

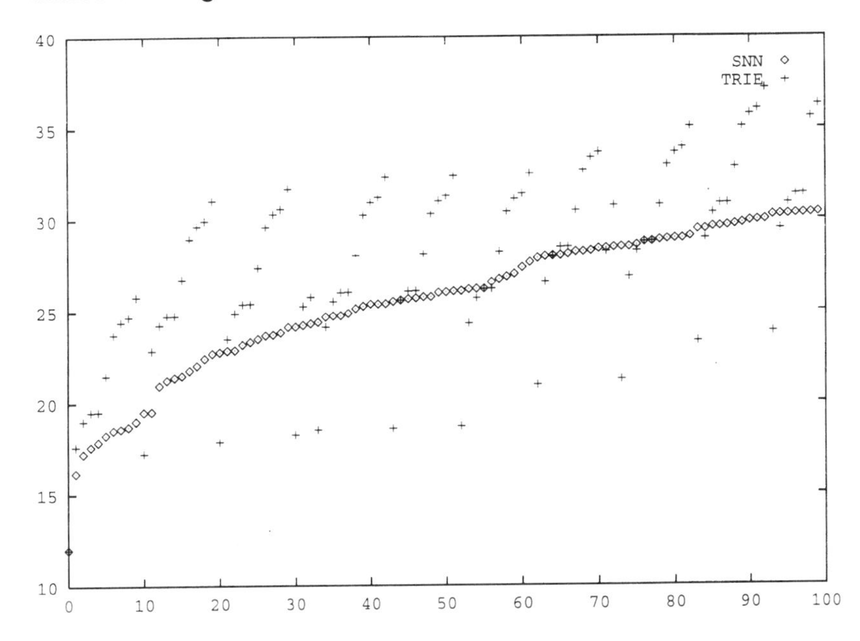

Fig. 6 The negative logarithm of the postcode probability is shown plotted against the retrieval order for SNN and Trie.

E. Why Retrieval Can Become Faster for Larger Dictionaries

In Table 6, the SNN retrieval time is shown as becoming faster as the dictionary becomes larger. During retrieval, each LIM calls on its children for their best alternatives so far. As it processes each pair, it checks that it is valid (i.e., has occurred in training). Only valid pairs are passed on to its output. The likelihood of a pair being valid is proportional to the density of the LIM, which is defined as the number of valid pairs divided by the total number of pairs. Typically, as more strings are seen from a given domain (e.g., such as postcodes or vehicle licence numbers), the density of each LIM increases. For the postcode example, the density of the root LIM is 0.0011 for 1000 postcodes (Table 7) but this rises to 0.13 for 1.4 million postcodes (Table 8).

Table 7 Density Figures for the Network with 1076 Postcodes Loaded In

Lim	Outputs	Left inputs	Right inputs	Density
Root: 0:8	1076	1063	915	0.0011
Lim: 0:4	1063	160	74	0.09
Lim: 0:2	160	22	32	0.23
Lim: 2:4	74	13	14	0.40
Lim: 4:8	915	171	20	0.27
Lim: 4:6	171	10	19	0.9
Lim: 6:8	20	20	1	1.0

Table 8 Density Figures for the Network with 1,399,114 Postcodes Loaded In

Lim	Outputs	Left inputs	Right inputs	Density
Root: 0:8	1,399,114	2681	3994	0.13
Lim: 0:4	2681	184	130	0.11
Lim: 0:2	184	23	33	0.24
Lim: 2:4	130	19	23	0.29
Lim: 4:8	3994	201	21	0.95
Lim: 4:6	201	11	21	0.87
Lim: 6:8	21	21	1	1.0

F. Future Work

The SNN dictionary search system offers an extremely efficient way for storing and searching massive dictionaries (i.e., dictionaries with millions of entries). The efficiency of the SNN depends on how dense each LIM is. For dictionaries such as UK postcodes and UK vehicle licence numbers (which are both essentially composed of two parts) the binary branching structure of the strictly hierarchical SNN offers a natural match with the structure of the data, and this certainly aids the efficiency of the system. There is some interesting work yet to be done in optimising the system structure for other types of dictionaries.

The main focus of the current work on these types of dictionary-search systems, however, is in coping with poorly segmented input. The next version of the system will take as input a graph of character hypotheses, rather than a sequence. Use of this more-general structure as input to the retrieval system will invoke a higher computational cost than the current sequential input. Initial estimates indicate that this will offer best-first retrieval times from massive dictionaries in the order of tens of milliseconds per retrieval. Although this is significantly slower than the current system, this would still be fast enough for many practical applications.

5. CONCLUSIONS

This chapter has given a brief overview on two recent applications of SNNs to character recognition: one for performing the low-level recognition, and one for an optimal contextual interpretation of the low-level character hypotheses, given dictionary-type constraints. The low-level OCR system offers perhaps the greatest speed of any reported system, with a comparable level of accuracy, for which the retrieval system offers much better memory and speed efficiency than conventional trie structures when the dictionaries are large (more precisely, when the LIMs become dense).

In each of these applications, very simple noniterative training schemes are used. These are incapable of learning recursive grammars, but offer excellent speed when that facility is unnecessary. An interesting aspect of the work is that it demonstrates what can be accomplished by learning very simple structures (i.e, the statistics of partial parses, or strictly hierarchical CFGs).

REFERENCES

1. S Lucas. Connectionist architectures for syntactic pattern recognition. PhD dissertation; University of Southampton, 1991.
2. S Lucas, R Damper. Syntactic neural networks. Connect Sci 2:199–225, 1990.
3. T Kasami. An efficient recognition and syntax analysis algorithm for context-free languages. Technical Report AF-CRL-65-758, Air Force Cambridge Research Laboratory, Bedford, MA, 1965.
4. D Younger. Recognition and parsing of context-free languages in time n^3. Inform Control 10:189–208, 1967.
5. R Kaplan. A general syntactic processor. In: Natural Language Processing. R Rustin, ed. New York: Algorithmics Press, 1973, pp 193–241.
6. M Kay. The mind system. In: Natural Language Processing. R Rustin, ed. New York: Algorithmics Press, 1973, pp 193–241.
7. H Thompson, G Ritchie. Implementing natural language parsers. In: Artificial Intelligence: Tools, Techniques and Applications. T O'Shea, M Eisenstadt, eds. New York: Harper & Row, 1984.
8. JL Elman. distributed representations, simple recurrent networks and grammatical structure. Mach Learn 7:195–225, 1991.
9. JB Pollack. The induction of dynamical recognisers. Mach Learn 7:227–252, 1991.
10. C Giles, G Sun, H Chen, Y Lee, D Chen. Higher order recurrent neural networks and grammatical inference. In: Advances in Neural Information Processing Systems, 2. D Touretzky, ed. San Mateo, CA. Morgan Kaufmann, 1990.
11. Y Bengio, P Frasconi. Input–output HMMs for sequence processing. IEEE Trans Neural Networks 7:1231–1249, 1996.
12. S Lucas. Forward–backward building blocks for evolving neural networks with intrinsic learning behaviours. In: Lecture Notes in Computer Science (1240): Biologial and Artificial Computation: From Neuroscience to Technology. Berlin: Springer-Verlag, 1997, pp 723–732.
13. P Frasconi, M Gori, A Sperduti. On the efficient classification of data structures by neural networks. Proceedings of the International Joint Conference on Artificial Intelligence (IJCAI-97), 1997.
14. S Lucas, A Amiri. Statistical syntactic methods for high performance OCR. IEE Proc Vis Image Signal Processing 143:23–30, 1996.
15. F Jelinek, R Mercer, L Bahl. The development of an experimental discrete dictation recognizer. Proc IEEE 73:1616–1624, 1985.
16. I Aleksander, T Stonham. Guide to pattern recognition using random-access memories. IEE Proc Comput Digit Tech 2:29–40, 1979.
17. E Vidal, H Rulot, J Valiente, G Andreu. Recognition of planar shapes through the error-correcting grammatical inference algorithm (ECGI). Proceedings of IEE Colloquium Grammatical Inference: Theory, Applications and Alternatives, 22–23 April 1993, Digest No. 1993/092. London: IEE, 1993.
18. S Lucas, E Vidal, A Amiri, S Hanlon, A Amengual. A comparison of syntactic and statistical techniques for off-line OCR. In: Lecture Notes in Artificial Intelligence (862): Grammatical Inference and Applications. Berlin: Springer-Verlag, 1994, pp 168–179.
19. S Lucas. Improving scanning n-tuple classifiers by pre-transforming training data. In: Progress in Handwriting Recognition. A Downton, S Impedovo, eds. London: World Scientific, 1997, pp 345–352.
20. R Rohwer, M Morciniec. The theoretical and experimental status of the n-tuple classifier. Technical Report NCRG/4347: Neural Computing Research Group, Aston University, UK, 1995.

21. J Schurmann, N Bartneck, T Bayer, J Franke, E Mandler, M Oberlander. Document analysis—from pixels to content. Proc IEEE 80:1101–1119, 1992.
22. S Lucas. Rapid retrieval from massive dictionaries by lazy evaluation of a syntactic neural network. Proceedings of IEEE International Conference on Neural Networks, Perth, Western Australia: IEEE, 1995, pp 2237–2242.
23. S Lucas. Rapid best-first retrieval from massive dictionaries. Pattern Recogn Lett 17:1507–1512, 1996.
24. A Downton, L Du, S Lucas, A Badr. Generalized contextual recognition of hand-printed documents using semantic trees with lazy evaluation. In: Proceedings of the Fourth International Conference on Document Analysis and Recognition. H Baird, A Denkel, Y Nakano, eds. Ulm, Germany: IEEE, 1997, pp 238–242.
25. L Dodd. Grammatical inference for automatic speech recognition: an application of the inside/outside algorithm to the spelling of English words. In: Proceedings of Speech '88. W Ainsworth, J Holmes, eds. Edinburgh: Institute of Acoustics, 1988, pp 1061–1068.
26. S Lucas. Spatially aware rapid retrieval system. Proceedings of IEE Third European Workshop on Handwriting Analysis and Recognition, London: IEE, 1998, pp 15/1–15/6.
27. K Yamamoto, S Mori. Recognition of hand-printed characters by an outermost point method. Pattern Recogn 12:229–236, 1980.
28. R de la Briandais. File searching using variable length keys. Proc West Joint Comput Conf 15:295–298, 1959.
29. D Knuth. The Art of Computer Programming, vol 3; Sorting and Searching. Reading, MA: Addison–Wesley, 1973.
30. C Wells, L Evett, P Whitby, R Whitrow. Fast dictionary look-up for contextual word recognition. Pattern Recogn 23:501–508, 1990.

34

Compressing Texts with Neural Nets*

JÜRGEN SCHMIDHUBER

IDSIA, Lugano, Switzerland

STEFAN HEIL

Technical University of Munich, Munich, Germany

1. INTRODUCTION

Text compression is important [e.g., 1]. It is cheaper to communicate compressed text files instead of original text files. Moreover, compressed files are cheaper to store. For such reasons various text-encoding algorithms have been developed. A text-encoding algorithm takes a text file and generates a shorter compressed file from it. The compressed file contains all the information necessary to restore the original file, which can be done by calling the corresponding decoding algorithm. Unlike image compression, text compression requires loss-free compression. The most widely used text compression algorithms are based on Lempel–Ziv techniques [e.g., 2]. Lempel–Ziv compresses symbol strings sequentially, essentially replacing substrings by pointers to equal substrings encountered earlier. As the file size goes to infinity, Lempel–Ziv becomes asymptotically optimal in a certain information theoretical sense [3].

The average ratio between the lengths of original and compressed files is called the average compression ratio. We cite a statement from Held's book [4], where he refers to text represented by 8 bits per character:

> *"In general, good algorithms can be expected to achieve an average compression ratio of 1.5, while excellent algorithms based upon sophisticated processing techniques will achieve an average compression ratio exceeding 2.0."*

This paper will show that neural networks may be used to design "excellent" text compression algorithms.

Section 2 describes the basic approach combining neural nets and the technique of predictive coding. Section 3 focuses on details of a neural predictor of conditional probabilities. In addition, Sec. 3 describes three alternative coding techniques to be used in conjunction with the predictor. Section 4 presents comparative simulations. Section 5 discusses limitations and extensions.

2. BASIC APPROACH

We combine neural nets, standard statistical compression methods such as Huffman–Coding [e.g., 4] and Arithmetic Coding [e.g., 5], and variants of the "'principle of history compression" [6,7]. The main ideas of the various alternatives will be explained in Sec. 3.

All our methods are instances of a strategy known as "predictive coding" or "model-based coding." We use a neural predictor network P, which is trained to approximate the conditional probability distribution of possible characters, given previous characters. P's outputs are fed into coding algorithms that generate short codes for characters with low information content (characters with high predicted probability) and long codes for characters conveying a lot of information (highly unpredictable characters).

Why not use a lookup table instead of a network? Because lookup tables tend to be inefficient. A lookup table requires k^{n+1} entries for all the conditional probabilities of k possible characters, given n previous characters. In addition, a special procedure is required for dealing with previously unseen character combinations. In contrast, the size of a neural net typically grows in propostion to n^2 (assuming the number of hidden units grows in proportion to the number of inputs), and its inherent "generalization capability" takes care of previously unseen character combinations (hopefully, by coming up with good predicted probabilities).

We will make the distinction between *on-line* and *off-line* variants of our approach. With *off-line* methods, P is trained on a separate set F of training files. After training, the weights are frozen and copies of P are installed at all machines functioning as message receivers or senders. From then on, P is used to encode and decode unknown files without being changed any more. The weights become part of the code of the compression algorithm. The storage occupied by the network weights does not have to be taken into account to measure the performance on unknown files—just as the code for a conventional data compression algorithm does not have to be taken into account.

The *on-line* variants are based on the insight that even if the predictor learns *during* compression, the modified weights need not be sent from the sender to the receiver across the communication channel—as long as the predictor employed for decoding uses exactly the same initial conditions and learning algorithm as the predictor used for encoding (this observation goes back to Shannon). Because on-line methods can adapt to the statistical properties of specific files, they promise signficantly better performance than off-line methods. But there is a price to pay: on-line methods tend to be computationally more expensive.

Section 4 will show that even off-line methods can sometimes achieve excellent results. We will briefly come back to on-line methods in the final section of this paper.

3. OFF-LINE METHODS

In what follows, we will first describe the training phase of the predictor network P (a strictly layered feedforward net trained by backpropagation [8,9]. The training phase is based on a set F of training files. Then we will describe three working off-line variants of "compress" and "uncompress" functions based on P. All methods are guaranteed to encode and decode arbitrary unknown text files without loss of information.

A. The Predictor Network *P*

Assume that the alphabet contains k possible characters $z_1, z_2, \ldots z_k$. The (local) representation of z_i is a binary k-dimensional vector $r(z_i)$ with exactly one nonzero component (at the ith position). P has nk input units and k output units. n is called the "time-window" size. We insert n default characters z_0 at the beginning of each file. The representation of the default character, $r(z_0)$, is the k-dimensional zero-vector. The mth character of file f (starting from the first default character) is called c_m^f.

For all $f \in F$ and all possible $m > n$, P receives as input

$$r(c_{m-n}^f) \circ r(c_{m-n_{+1}}^f) \circ \ldots \circ r(c_{m-1}^f)$$

where $\circ$ is the concatenation operator for vectors. P produces as output P_m^f, a k-dimensional output vector. Using backpropagation [8,9], P is trained to minimize

$$\frac{1}{2}\sum_{f\in F}\sum_{m>n} \|r(c_m^f) - P_m^f\|^2 \quad \{1\}$$

Equation {1} is minimal if P_m^f always equals

$$E(r(c_m^f)|c_{m-n}^f, \ldots, c_{m-1}^f) \quad \{2\}$$

the conditional expectation of $r(c_m^f)$, given $r(c_{m-n}^f) \circ r(c_{m-n+1}^f) \circ \ldots \circ r(c_{m-1}^f)$. Because of the local character representation, this is equivalent to $(P_m^f)_i$ being equal to the conditional probability

$$Pr(c_m^f = z_i|c_{m-n}^f, \ldots, c_{m-1}^f) \quad \{3\}$$

for all f and for all appropriate $m > n$, where $(P_m^f)_j$ denotes the jth component of the vector to P_m^f.

For instance, assume that a given "context string" of size n is followed by a certain character in one-third of all training exemplars involving this string. Then, given the context, the predictor's corresponding output unit will tend to predict a value of 0.3333.

In practical applications, the $(P_m^f)_i$ will not always sum up to 1. To obtain outputs satisfying the properties of a proper probability distribution, we normalize by defining

$$P_m^f(i) = \frac{(P_m^f)_i}{\sum_{j=1}^k (P_m^f)_j} \tag{4}$$

B. Method 1

With the help of a copy of P, an unknown file f can be compressed as follows: again, n default characters are inserted at the beginning. For each character c_m^f $(m > n)$, the predictor emits its output P_m^f based on the n previous characters. There will be a k such that $c_m^f = z_k$. The estimate of

$$Pr(c_m^f = z_k | c_{m-n}^f, \ldots, c_{m-1}^f)$$

is given by $P_m^f(k)$. The code of c_m^f, the bitstring *code* (c_m^f), is generated by feeding $P_m^f(k)$ into the Huffman Coding algorithm (see the following); *code* (c_m^f) is written into into compressed file.

1. Huffman Coding

With a given probability distribution on a set of possible characters, Huffman Coding [e.g., 4] encodes characters by bitstrings as follows. Characters correspond to terminal nodes of a binary tree to be built in an incremental fashion. The probability of a terminal node is defined as the probability of the corresponding character. The probability of a nonterminal node is defined as the sum of the probabilities of its sons. Starting from the terminal nodes, a binary tree is built as follows:

Repeat as long as possible:
Among those nodes that are not children of any nonterminal nodes created earlier, pick two with lowest associated probabilities. Make them the two sons of a newly generated nonterminal node.

The branch to the "left" son of each nonterminal node is labeled 0. The branch to its "right" son is labeled 1. The code of a character c, *code(c)*, is the bitstring obtained by following the path from the root to the corresponding terminal node. Obviously, if $c \neq d$, then *code*(c) cannot be the prefix of *code*(d). This makes the code uniquely decipherable. Note that characters with high associated probability are encoded by short bitstrings. Characters with low associated probability are encoded by long bitstrings.

The probability distribution on the characters is not required to remain fixed. This allows for use of "time-varying" conditional probability distributions as generated by our neural predictor.

2. How to Decode

The information in the compressed file is sufficient to reconstruct the original file. This is done with the "uncompress" algorithm, which works as follows: again, for each character c_m^f $(m > n)$, the predictor (sequentially) emits its output P_m^f based on the n previous characters, where the c_l^f with $n < l < m$ were obtained sequentially by feeding the approximations $c_l^f(k)$ of the probabilities

$$Pr(c_l^f = z_k | c_{l-n}^f, \ldots, c_{l-1}^f)$$

into the inverse Huffman Coding procedure. The latter is able to correctly decode c_l^f from $code(c_l^f)$. Note that to correctly decode some character, we first need to decode all previous characters.

C. Method 2

Similar to method 1, but with Arithmetic Coding (see following) replacing the non-optimal Huffman Coding (a comparison of alternative coding schemes will be given in Sec. 3.E).

1. *Arithmetic Coding*

The basic idea of Arithmetic Coding is as follows: a message is encoded by an interval of real numbers from the unit interval [0; 1]. The output of Arithmetic Coding is a binary representation of the boundaries of the corresponding interval. This binary representation is incrementally generated during message processing. Starting with the unit interval, for each observed character the interval is made smaller, essentially in proportion to the probability of the character. A message with low information content (and high corresponding probability) is encoded by a comparatively large interval, for which the precise boundaries can be specified with comparatively few bits. A message with high information content (and low corresponding probability) is encoded by a comparatively small interval, the boundaries of which require comparatively many bits to be specified. Although the basic idea is elegant and simple, additional technical considerations are necessary to make Arithmetic Coding practicable [see Ref. 5 for details].

D. Method 3

This section presents another alternative way of "predicting away" redundant information in sequences. Again, we preprocess input sequences by a network that tries to predict the next input, given previous inputs. The input vector corresponding to time step t of sequence p is denoted by $x^p(t)$. The network's real-valued output vector is denoted by $y^p(t)$. Among the possible input vectors, there is one with minimal euclidean distance to $y^p(t)$. This one is denoted by $z^p(t)$. $z^p(t)$ is interpreted as the deterministic vector-valued prediction of $x^p(t+1)$.

It is important to observe that all information about the input vector $x^p(t_k)$ (at time t_k) is conveyed by the following data: the time t_k, a description of the predictor and its initial state, and the set

$$\{(t_s, x^p(t_s)) with\ 0 < t_s \leq t_k, z^p(t_s - 1) \neq x^p(t_s)\}$$

In what follows, this observation will be used to compress text files.

1. *Application to Text Compression*

As with methods 1 and 2, the "time-window" corresponding to the predictor input is sequentially shifted across the unknown text file. The P_m^f, however, are used in a different way. The character z_i for which the representation $r(z_i)$ has minimal euclidean distance to P_m^f is taken as the predictor's deterministic prediction (if there is more than one character with minimal distance to the output, then we take the one with lowest ASCII value). If P_m^f does not match the prediction, then it is stored in a second file, together with a number indicating how many characters were processed since the last nonmatching character [6,7]. Expected characters are simply ignored: they represent redundant information. To avoid confusions between unexpected numbers from the original file and numbers indicating how many correct predictions went by since the last wrong prediction, we introduce an escape character to mark unexpected number characters in the second file. The escape character is also used to mark unexpected escape characters. Finally, we apply Huffman Coding (as embodied by the UNIX function *pack*) to the second file and obtain the final compressed file.

The "uncompress" algorithm works as follows: we first unpack the compressed file by inverse Huffman Coding (as employed by the UNIX function *unpack*). Then, starting from n default characters, the predictor sequentially tries to predict each character of the original file from the n previous characters (deterministic predictions are obtained similar to the foregoing compression procedure). The numbers in the unpacked file contain all information about which predictions are wrong, and the associated characters tell us how to correct wrong predictions: if the unpacked file indicates that the current prediction is correct, it is fed back to the predictor input and becomes part of the basis for the next prediction. If the unpacked file indicates that the current prediction is wrong, the corresponding entry in the unpacked file (the correct character associated with the number indicating how many correct predictions went by since the last unexpected character) replaces the prediction and is fed back to the predictor input where it becomes part of the basis for the next prediction.

E. Comparison of Methods 1, 2, and 3

With a given probability distribution on the characters, Huffman Coding guarantees minimal expected code length, provided all character probabilities are integer powers of 1/2. In general, however, Arithmetic Coding works slightly better than Huffman Coding. For sufficiently long messages, Arithmetic Coding achieves expected code lengths arbitrarily close to the information-theoretic lower bound. This is true even if the character probabilities are not powers of 1/2 [see e.g., 5].

Method 3 is of interest if typical files contain long sequences of predictable characters. Among the foregoing methods, it is the only one that explicitly encodes strings of characters (as opposed to single characters). It does not make use of all the information about conditional probabilities, however.

Once the current conditional probability distribution is known, the computational complexity of Huffman Coding is $O(k \log k)$. The computational complexity of Arithmetic Coding is $O(k)$. So is the computational complexity of method 3. In practical applications, however, the computational effort required for all three variants is negligible in comparison with the effort required for the predictor updates.

4. SIMULATIONS

Our current computing environment prohibits extensive experimental evaluations of the foregoing three methods. On an HP 700 workstation, the training phase for the predictor is quite time-consuming, taking days of computation time. Once the predictor is trained, the method still tends to be on the order of 1000 times slower than standard methods. In many data transmission applications, communication is not expensive enough to justify this in absence of specialized hardware (given the current state of workstation technology). This leads us to recommend special neural net hardware for our approach. The software simulations presented in this section, however, will show that "neural" compression techniques can sometimes achieve excellent compression ratios.

We applied our off-line methods to German newspaper articles (simulations were run and evaluated by the second author). We compared the results to those obtained with standard-encoding techniques provided by the operating system UNIX, namely "pack," "compress," and "grip." The corresponding decoding algorithms are "unpack," "uncompress," and "gunzip," respectively. "pack" is based on Huffman Coding [e.g., 4], whereas "compress" and "gzip" are based on the asymptotically "optimal" Lempel–Ziv technique [2]. It should be noted that "pack," "compress," and "gzip" ought to be classified as on-line methods—they adapt to the specific text file they see. In contrast, the competing "neural" methods ran off-line, owing to time limitations. Therefore, our comparison was unfair in the sense that it was biased against the "neural" methods. See Sec. 5, however, for on-line "neural" alternatives.

The training set for the predictor was given by a set of 40 articles from the *newspaper Münchner Merkur*, each containing between 10,000 and 20,000 characters. The alphabet consisted of $k = 80$ possible characters, including upper-case and lower-case letters, ciphers, interpunction symbols, and special German letters such as "ö", "ü", "ä." P had 430 hidden units. A "true" unit with constant activation 1.0 was connected to all hidden and output units. The learning rate was 0.2. The training phase consisted of 25 sweeps though the training set, taking 3 days of computation time on an HP 700 station. Why just 25 sweeps? On a separate test set, numbers of sweeps between 20 and 40 empirically led to acceptable performance. Note that a single sweep actually provides many different training examples for the predictor.

The test set consisted of 20 newspaper articles (from the same newspaper), each containing between 10,000 and 20,000 characters. The test set did not overlap with the training set. Table 1 lists the average compression ratios and the corresponding variances. Our methods achieved excellent performance. Even method 3 led to an excellent compression ratio, although it does not make use of all the information about the conditional probabilities. The best performance was obtained with method

Table 1 Average compression ratios (and corresponding variances) of various compression algorithms tested on short German text files (< 20,000 bytes) from the unknown test set from *Münchner Merkur*.

Method	Av. compression ratio	Variance
Huffman Coding (UNIX: pack)	1.74	0.0002
Lempel–Ziv Coding (UNIX: compress)	1.99	0.0014
Method 3, $n = 5$	2.20	0.0014
Improved Lempel–Ziv (UNIX: gzip –9)	2.29	0.0033
Method 1, $n = 5$	2.70	0.0158
Method 2, $n = 5$	2.72	0.0234

2, which outperformed the strongest conventional competitor, the UNIX "gzip" function based on the asymptotically optimal Lempel–Ziv algorithm. Note that variance goes up (but always remains within acceptable limits) as compression performance improves.

The hidden units were actually necessary to achieve good performance. A network without hidden units was not able to achieve average compression ratios exceeding 2.0. The precise number of hidden units do not appear to be very important, though. A network with 300 hidden units achieved performance similar to the foregoing network.

How does a neural net trained on articles from *Münchner Merkur* perform on articles from other sources? Without retraining the neural predictor, we applied all competing methods with ten articles from another German newspaper (the *Frankenpost*). The results are given in Table 2. The *Frankenpost* articles were harder to compress for all algorithms. But relative performance remained comparable.

Note that we used quite a small time-window ($n = 5$). In general, larger time windows will make more information available to the predictor. In turn, this will improve the prediction quality and increase the compression ratio. Therefore, we expect to obtain even bette results for $n > 5$ and for recurrent predictor networks (note that recurrent nets are less limited than the time window approach—in principle, they can emit predictions based on all previous characters). Another reason for optimism is given by a performance comparison with three human subjects who had to predict characters (randomly selected from the test files) from n preceding characters. With $n = 5$, the humans were able to predict 52% of all characters, while our predictor predicted 49% (the character with the highest predicted probability was taken as the prediction). With $n = 10$, humans were able to predict about 59% of all characters. With $n = 15$, humans were able to predict about 63% of all characters. We expect that P will remain close to human performance for $n > 5$. More training data, however, are required to avoid overfitting.

5. DISCUSSION

Our results show that neural networks are promising tools for loss-free data compression. It was demonstrated that even off-line methods based on small time windows can lead to excellent compression ratios. We have hardly begun, however, to exhaust the potential of the basic approach.

Table 2 Average compression ratios and variances for the *Frankenpost*. The neural predictor was not retrained.

Method	Av. compression ratio	Variance
Huffman Coding (UNIX: pack)	1.67	0.0003
Lempel–Ziv Coding (UNIX: compress)	1.71	0.0036
Method 3, $n = 5$	1.99	0.0013
Improved Lempel–Ziv (UNIX: gzip -9)	2.03	0.0099
Method 1, $n = 5$	2.25	0.0077
Method 2, $n = 5$	2.20	0.0112

A disadvantage of the off-line technique is that it is off-line: the predictor does not adapt to the specific text file it sees. Instead, it relies on regularities extracted during the training phase, and on its ability to generalize. This tends to make it language-specific. English texts or C-code should be compressed with a predictor different from the one used for German texts (unless one takes the effort and trains the predictor on texts from many different sources).

As mentioned in Sec. 2, this limitation is not essential. It is straightforward to construct on-line variants of all three methods described in the previous sections. With these on-line variants, the predictor continues to learn during compression. A typical on-line variant proceeds like this: both the sender and the receiver start with exactly the same initial predictor. Whenever the sender sees a new character, it encodes it using its current predictor. The code is sent to the receiver who decodes it. Both the sender and the receiver use exactly the same learning protocol to modify their weights (for instance: after processing every 1,000th character, take the last 10,000 symbols to retrain the predictor). The modified weights need not be sent from the sender to the receiver and do not have to be taken into account to compute the average compression ratio. Especially with long unknown text files, the on-line variant should make a big difference. Initial experiments with on-line variants of methods 2 and 3 led to additional significant improvements of the compression ratio.

The main disadvantage of both on-line and off-line variants, however, is their computational complexity. Our current off-line implementations are clearly slower than conventional standard techniques, by a factor of about 1000 (but we did not attempt to optimize our systems for speed). And the complexity of an on-line method is typically even worse than the one of the corresponding off-line method (the precise slow-down factor depends on the nature of the learning protocol). For this reason, especially the promising on-line variants can be recommended only if special neural net hardware is available. Note, however, that there are many commercial data compression applications that rely on specialized electronic chips.

There are a few obvious directions for future experimental research: (a) Use larger time windows or recurrent nets—they seem to be promising even for off-line methods (see the last paragraph of Sec. 4). (b) Thoroughly test the potential of on-line methods. Both (a) and (b) should greatly benefit from fast hardware.

Finally, we mention that there are additional interesting applications of neural predictors of conditional probabilities. [See Ref. 10 for a method that uses a predictor of conditional probabilities to modulate the sequence processing strategy of a separate recurrent network *R*]. This can greatly improve *R*'s ability to detect correlations between events separated by long time lags. [See Ref. 11 for a method that uses predictors of conditional probabilities to develop factorial codes of environmental input patterns—codes with the property that the code components are statistically independent; see Refs. 12 and 13 for applications]. This can be useful in conjunction with statistical classifiers that assume statistical independence of their input variables.

ACKNOWLEDGMENT

Thanks to David MacKay for directing our attention towards Arithmetic Coding. Thanks to Gerhard Weiss for useful comments. This research was supported in part by a DFG fellowship to J Schmidhuber.

REFERENCES

1. TC Bell, JG Cleary, IH Witten. Text Compression. Prentice Hall, Englewood Cliffs, NJ, 1990.
2. J Ziv, A Lempel. A universal algorithm for sequential data compression. IEEE Trans Inform Theory, IT-23(5):337–343, 1997.
3. A Wyner, J Ziv. Fixed data base version of the Lempel–Ziv data compression algorithm. IEEE Trans Inform Theory, 37:878–880, 1991.
4. G Held. Data Compression. John Wiley & Sons, New York, 1991.
5. IH Witten, RM Neal, JG Cleary. Arithmetic coding for data compression. Commun ACM 30:520–540, 1987.
6. JH Schmidhuber. Learning unambiguous reduced sequence descriptions. In: JE Moody, SJ Hanson, RP Lippman, eds. Advances in Neural Information Processing Systems 4. San Mateo, CA: Morgan Kaufmann, 1992, pp 291–298.
7. JH Schmidhuber. Learning complex, extended sequences using the principle of history compression. Neural Comput 4:234–242, 1992.
8. PJ Werbos. Beyond Regression: New Tools for Prediction and Analysis in the Behavioral Sciences. PhD dissertation, Harvard University, Cambridge, MA, 1974.
9. DE Rumelhart, GE Hinton, RJ Williams. Learning internal representations by error propagation. Parallel Distributed Processing, vol 1. MIT Press, Cambridge, MA, 1986, pp 318–362.
10. JH Schmidhuber, MC Mozer, D Prelinger. Continuous history compression. In: H Hüning, S Neuhauser, M Raus, W Ritschel, eds. Proceedings International Workshop on Neural Networks, RWTH Aachen, Augustinus, 1993, pp 87–95.
11. JH Schmidhuber. Learning factorial codes by predictability minimization. Neural Comput 4:863–879, 1992.
12. JH Schmidhuber, D Prelinger. Discovering predictable classifications. Neural Comput 5:625–636, 1993.
13. S Lindstädt. Comparison of two unsupervised neural network models for redundancy reduction. In: MC Mozer, P Smolensky, DS Touretzky, JL Elman, AS Weigend, eds. Proc. of the 1993 Connectionist Models Summer School, Hillsdale, NJ: Erlbaum Associates, 1993, pp. 308–315.

35

Neural Architectures for Information Retrieval and Database Query

CHUN-HSIEN CHEN

Chang Gung University, Kwei-Shan, Tao-Yuan, Taiwan, Republic of China

VASANT HONAVAR

Iowa State University, Ames, Iowa

1. MOTIVATION FOR HYBRID SYMBOLIC–CONNECTIONIST PROCESSING

Artificial neural networks (ANN) offer an attractive computational model for a variety of applications in pattern classification, language processing, complex systems modelling, control, optimization, prediction, and automated reasoning for a variety of reasons, including potential for massively parallel, high-speed processing, resilience in the presence of faults (failure of components), and noise. Despite a large number of successful applications of ANN in aforementioned areas, their use in complex symbolic computing tasks (including storage and retrieval of records in large databases, and inference in deductive knowledge bases) is only beginning to be explored [1–4].

This chapter explores the application of neural networks to noise-tolerant information retrieval and database query processing. Database query entails a process of table lookup which is used in a wide variety of computing applications. Examples of such lookup tables include: *routing tables*, used in routing of messages in communication networks; *symbol tables*, used in compiling computer programs written in high-level languages; and *machine-readable lexicons*, used in natural language processing (NLP). In such tables, every table entry is an associated input–output (I/O) ordered pair. As the number of table entries and the occurrence of partially specified inputs increase, the speed of locating an associative table entry can become a severe bottleneck in large-scale information-processing tasks that involve extensive associative table lookup. Many applications require the lookup mechanism or query-processing system to be capable of retrieving items based on *partial matches* or retrieval of *multiple* records matching the specified query criteria.

This capability is computationally rather expensive in many current computer systems. The ANN-based approach to database query processing that is proposed in this chapter exploits the fact that a table lookup task can be viewed at an abstract level in terms of *associative pattern matching and retrieval*, which can be efficiently realized using neural associative memories [5,6].

The remainder of the chapter is organized as follows: The rest of Sec. 1 briefly introduces the basic ANN architecture we use. Section 2 reviews the neural associative memory [5], which is capable of massively parallel best match, exact match, and partial match and recall. Section 3 develops an ANN design for a high-speed query system for text-based, machine-readable lexicon by taking advantage of the capability of the neural memory for massive parallel pattern matching and retrieval. Section 4 compares the performance of the proposed ANN-based query processing system with that of several commonly used techniques. Section 5 concludes with a summary.

A. Artificial Neural Units

A typical computing unit (node) in an ANN has n input and m output connections, each of which has an associated weight. The node computes the weighted sum on the inputs, compares the sum to its node threshold, and produces its output based on an activation function. A commonly used activation function is threshold function. The resulting output is sent along the output connections to other nodes. The output of such a node used in this chapter is defined by:

$$y = f(s) \qquad \text{and} \qquad s = \sum_{i=1}^{n} w_i x_i - \theta$$

where x_i is the value of input i, w_i is the associated weight on input connection i, θ is the node threshold, y is the output value, and f is the activation function. Figure 1 shows such a node.

The types of activation functions used by an ANN affect its expressive power, computational capabilities, and performance (in terms of speed). The threshold function we use is *binary hardlimiter* f_H which is defined as follows:

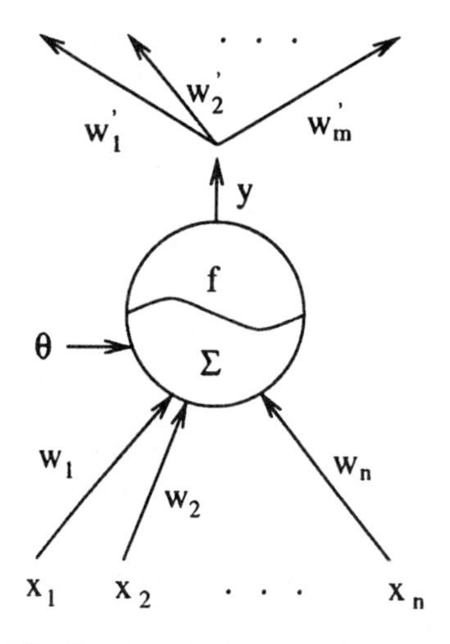

Fig. 1 A typical computing unit of an ANN.

$$f_H(s) = \begin{cases} 1 \text{ if } s \geq 0 \\ 0 \text{ otherwise} \end{cases}$$

The simplicity allows simple and efficient hardware implementation of such a threshold function.

B. Perceptrons

A one-layer perceptron has n input neurons, m output neurons and one layer of interconnections. The output y_i of output neuron i, $1 \leq i \leq m$, is given by $y_i = f_H(\sum_{j=1}^{n} w_{ij}x_j - \theta_i)$; w_{ij} denotes the weight on the connection from input neuron j to output neuron i, θ_i is the threshold of output neuron i, and x_j is the value of the input neuron j. Such a one-layer perceptron can implement only linearly separable functions from $\mathbf{R}^n$ to $\{0,1\}^m$[7]. We can see the connection weight vector $w_i = < w_{i1}, \ldots, w_{in} >$ and the node threshold θ_i as defining a linear hyperplane H_i which partitions the n-dimensional pattern space into two half-spaces.

A two-layer perceptron has one layer of k hidden neurons (and, hence, two layers of interconnections with each hidden neuron being connected to each of the input as well as output neurons). In this chapter, every hidden neuron and output neuron in the two-layer perceptron uses binary hardlimiter function f_H as activation function and produces binary outputs; its weights are restricted to values from $\{-1,0,1\}$, and it uses integer thresholds. It is known that such a two-layer perceptron can realize arbitrary binary mappings [5].

2. A NEURAL ASSOCIATIVE MEMORY FOR INFORMATION RETRIEVAL

Most database systems store (symbolic) data in the form of structured records. When a *query* is made, the database system searches and retrieves records that match the user's query criteria, which typically only partially specify the contents of records to be retrieved. Also, there are usually multiple records that match a query (e.g., retrieving the lexical specifications of the words matching the partially specified input pattern **ma?e** from a machine-readable lexicon, where the symbol **?** means the letter at that position is unavailable). Thus, query processing in a database can be viewed as an instance of the task of recall of multiple stored patterns given a partial specification of the patterns to be recalled. A neural associative memory that is capable of massively parallel best match, exact match, and partial match and recall of binary (bipolar) patterns was proposed [5]. This section summarizes the relevant properties of the neural memory developed by these authors [5]. It also briefly discusses how to represent symbolic information in terms of binary codings to facilitate symbolic information manipulation on the proposed neural memory, which operates on binary (bipolar) values.

A. A Neural Associative Memory Capable of Exact Match, Best Match, and Partial Match

The neural associative memory used here is based on a two-layer perceptron. Given a set U of k bipolar input vectors $u_1 \ldots, u_k$ of dimension n and a set V of k desired binary output vectors $v_1, \ldots, v_k$ of dimension m, such an associative memory module can be synthesized using one-shot learning as follows. The memory module (Fig. 2)

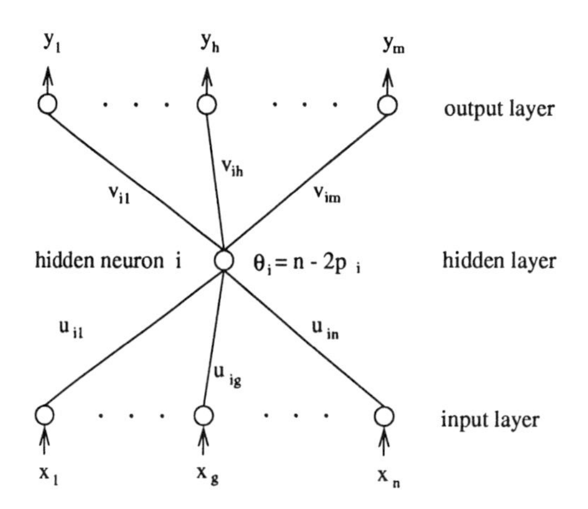

Fig. 2 The settings of connection weights and hidden node threshold in the proposed neural associative memory module for an associative ordered pair $(u_i v_i)$. The threshold for each of the output neurons is set to 1. p_i (in Hamming distance) is the adjustable precision level for this associative pair.

has n input, k hidden, and m output neurons. For each associative ordered pair (u_i, v_i), where $1 \leq i \leq k$, a hidden neuron i is created with threshold $n - 2p_i$, where $p_i \in \mathbf{N}$ is the adjustable precision level (in Hamming distance) for that associative pair. The connection weight from input neuron g to hidden neuron i is u_{ig} and that from hidden neuron i to output neuron h is v_{ih}, where u_{ig} is the gth component of bipolar vector u_i and v_{ih} is the hth component of binary vector v_i. The threshold for each of the output neurons is set to 1. The activation function at hidden and output nodes is binary hardlimiter function f_H.

The computation in the associative memory can be viewed as a two-stage associative process: *identification* and *recall.* Because input is bipolar, the weights on the first-layer connections of the associative memory are either 1 or -1. During identification, a bit of an input pattern that is wrongly on or off (relative to the stored pattern) contributes -1 to the activation of the corresponding hidden neuron. Each hidden neuron sums up the contributions to its activation from its first-layer connections, compares the result with its threshold and produces output value 1 if its activation exceeds or equals its threshold. If one of the hidden neurons is turned on, one of the stored memory patterns will be *recalled* by that hidden neuron and its associated second-layer connections. Note that an input pattern is matched against all the stored memory patterns in *parallel.*

B. Associative Recall from Partially Specified Input

This section summarizes a mathematical model of associative recall from a partially specified bipolar input pattern and its neural network realization. The model assumes that the unavailable components of a bipolar input pattern have a default value of 0. Thus, a bipolar partial input pattern is *completed* by filling in a 0 for each of the unavailable components.

Let $\dot{u}$ be a partially specified n-dimensional bipolar pattern with the values of some of its components being unavailable; *bits*$(\dot{u})$, a function that counts the number of components with known values ($+1$ or -1) of partial pattern $\dot{u}$; $pad0(\dot{u})$, a function that pads the unavailable bits of bipolar partial pattern $\dot{u}$ with 0s; and $u \odot v$, a binary predicate that tests whether '*u is a partial pattern of v*', where '*u is a partial pattern of v*' means that the values of available bits of u are same as those of their corresponding bits in v.

Let $\dot{D}_H(\dot{u}, \dot{v})$ denote the Hamming distance between two bipolar partial patterns $\dot{u}$ and $\dot{v}$ which have same corresponding unavailable components. If $bits(\dot{u}) = j$, $pad0(\dot{u})$ is called a *padded j-bit partial pattern* derived from partially specified pattern $\dot{u}$. Define $\dot{U}^j_{Pi} = \{\dot{u} |\ bits(\dot{u}) = j\ \&\dot{u} \odot u_i\}$, $1 \le j \le n$ & $1 \le i \le k$, that is, $\dot{U}^j_{Pi}(p_i)$ is the set of partial patterns, with j available bits, of bipolar full pattern u_i. Define $\ddot{U}^j_{Pi}(p_i) = \{pad0(\ddot{u}) | \exists \dot{u},\ \dot{u} \in \dot{U}^j_{Pi}\ \&\ \dot{D}_H(\ddot{u}, \dot{u}) \le \lfloor j/n \rfloor \times p_i\}$, $1 \le j \le n$ & $1 \le i \le k$, that is $\ddot{U}^j_{Pi}(p_i)$ is the set of padded j-bit partial patterns having Hamming distance less than or equal to $\lfloor j/n \rfloor \times p_i$ to any one of the padded j-bit partial patterns of bipolar full pattern u_i.

Let $p_i = p,\ 1 \le i \le k$; $\ddot{U}^{c\sim n}_{Pi}(p) = \cup^n_{j=c} \ddot{U}^j_{Pi}(p)$; $\ddot{U}^{c\sim n}_{P}(p) = \cup^k_{i=1} \ddot{U}^{c\sim n}_{Pi}(p)$; and $\ddot{U}^h_{Pi}(p) \cap \ddot{U}^h_{Pj}(p) = \emptyset$ for $i \ne j$, $c \le h \le n$ and $1 \le i,\ j \le k$. Let f_P denote the function of recall from padded bipolar partial pattern. Then f_P is defined as follows:

$$f_P : \ddot{U}^{c\sim n}_P(p) \to V$$

$$f_P(x) = v_i;\ \text{if } x \in \ddot{U}^{c\sim n}_{Pi}(p),\ 1 \le i \le k$$

f_P is a partial function and is extended to a full function $\hat{f}_P$ for recall from padded bipolar partial pattern using associative memory as follows:

$$\hat{f}_P : \ddot{\mathbf{B}}^{c\sim n} \to (V \cup \{\langle 0^m \rangle\})$$

$$\hat{f}_P(x) = \begin{cases} f_p(x) \textbf{ if } x \in \ddot{U}^{c\sim n}_P(p) \\ < 0^m > \textbf{ if} x \in (\ddot{\mathbf{B}}^{c\sim n} - \ddot{U}^{c\sim n}_P(p)) \end{cases}$$

where $\ddot{\mathbf{B}}^{c\sim n}$ is the universe of n-dimensional vectors each component of which is 1, 0, or -1 and that have at least c nonzero components. It is assumed that the m-dimensional null pattern $< 0^m >$ is excluded from V.

The neural associative memory designed for recall from a fully specified input pattern can be used for associative recall from a partially specified input only by adjusting the thresholds of the hidden neurons as follows. Multiply the threshold of each hidden neuron by the ratio of the number of available components of a partially specified input pattern to that of a complete pattern. That is, reduce the threshold θ_i of each hidden neuron i from $n - 2p$ to $(n - 2p) \times n_a/n = n_a - 2(p \times n_a/n)$, where $n_a \le n$ is the number of available bits of a partially specified input pattern.

Note that p is the precision level set for every memory pattern for recall from a partial pattern, n_a equals the number of available bits of a partial input pattern, and $p \times n_a/n$ is the new precision level. In the interest of efficiency for hardware realization, it is desirable to use $\lfloor p \times n_a/n \rfloor$ as the new precision level (i.e., to use $n_a - 2\lfloor p \times n_a/n \rfloor$ as the new θ_i value).

C. Multiple Associative Recalls

The information-retrieval process in aforementioned neural associative memory contains two stages: *identification* and *recall*. During identification of an input pattern, the first-layer connections perform similarity measurements and sufficiently activate zero or more hidden neurons so that they produce high outputs of value 1. The actual choice of hidden neurons to be turned on is a function of the first-layer weights, the input pattern, and the threshold settings of the hidden neurons. During recall, if only one hidden neuron is turned on, one of the stored memory patterns will be *recalled* by that hidden neuron along with its associated second-layer connections. Without any additional control, if multiple hidden neurons are enabled, the corresponding output pattern will be a superposition of the output patterns associated with each of the enabled individual hidden neurons. With the addition of appropriate control circuitry, this behavior can be modified to yield sequential recall of more than one stored pattern.

Multiple recalls are possible if some of the *associative partitions* realized in the neural associative memory are not *isolated* [see Ref. 5 for details]. An input pattern (a vertex of an n-dimensional bipolar hypercube) located in a region of overlap between several associative partitions is close enough to the corresponding partition centers (stored memory patterns) and, hence, can turn on more than one hidden neuron, as explained in the following.

Suppose we define $\dot{U}_i^n(p_i) = \{u | u \in \dot{\mathbf{B}}^n \ \& \ D_H(u, u_i) \leq p_i\}$, $1 \leq i \leq k$, where $\dot{\mathbf{B}}^n$ is the universe of n-dimensional bipolar vector, $D_H(u, v)$ denotes the Hamming distance between two bipolar patterns u and v, and p_i is the specified precision level (in Hamming distance) for the stored memory pattern u_i; that is, $\dot{U}_i^n(p_i)$ is the set of n-dimensional bipolar patterns that have Hamming distance less than or equal to p_i away from the given n-dimensional bipolar memory pattern u_i. To keep things simple, let us assume $p_i = p$, $1 \leq i \leq k$. Suppose we define f_M as follows:

$$f_M : \dot{U}^n(p) \rightarrow (2^V - \varnothing)$$
$$f_M(x) = \{v_i | x \in \dot{U}_i^n(p), 1 \leq i \leq k\}$$

where $\dot{U}^n(p) = \dot{U}_1^n(p) \cup \dot{U}_2^n(p) \ldots \cup \dot{U}_k^n(p)$, $\dot{U}_i(p) \cap \dot{U}_j(p) \neq \varnothing$ for some $i \neq j$, and 2^V is the *power set* of V (i.e., the set of all subsets of V). The output of f_M is a set of binary vectors that corresponds to the set of patterns that should be recalled given a bipolar input vector x; f_M is a partial function and is extended to a full function $\hat{f}_M$ to describe multiple recall in the associative memory module as follows:

$$\hat{f}_M : \dot{\mathbf{B}}^n \rightarrow (2^V \cup \{< 0^m >\} - \varnothing)$$
$$\hat{f}_M(x) = \begin{cases} f_M(x) \text{ if } x \in \dot{U}^n(p) \\ (< 0^m >\} \text{ if } x \in (\dot{\mathbf{B}}^n - \dot{U}^n(p)) \end{cases}$$

f_M can be further extended to function f_{MP} for dealing with recall from a partially specified bipolar input pattern. f_{MP} is defined as follows:

$$f_{MP} : \ddot{U}_P^{c \sim n}(p) \rightarrow (2^V - \varnothing)$$
$$f_{MP}(x) = \{v_i | x \in \ddot{U}_{Pi}^{c \sim n}(p), 1 \leq i \leq k\}$$

where $\ddot{U}_{Pi}^h(p) \cap \ddot{U}_{Pj}^h(p) \neq \varnothing$ for some hs and $i \neq j$s, $c \leq h \leq n$ and $1 \leq i, j \leq k$.

f_{MP} is a partial function and is extended to a full function $\hat{f}_{MP}$ for multiple recalls from padded bipolar partial patterns in associative memory as follows:

$$\hat{f}_{MP} : \ddot{\mathbf{B}}^{c \sim n} \rightarrow (2^V \cup \{< 0^m >\} - \emptyset)$$

$$\hat{f}_{MP}(x) = \begin{cases} f_{MP}(x) \text{ if } x \in \ddot{U}_P^{c \sim n}(p) \\ < 0^m > \text{ if } x \in (\ddot{\mathbf{B}}^{c \sim n} - \ddot{U}_P^{c \sim n}(p)) \end{cases}$$

D. Information Retrieval in Neural Associative Memories

The foregoing neural associative memories operate on binary (bipolar) values. Because it is hard and error-prone for humans to reason with binary codings, we use symbolic representations when the neural associative memories are used for information storage and retrieval. The translation from symbolic reprsentations to binary codings can be done automatically, and this is not discussed here. In general, symbolic information retrieval (lookup) from a table can be viewed in terms of a binary random mapping f_B, which is defined as follows. Let U be a set of k distinct binary input vectors $u_1, \ldots, u_k$ of dimension n, and let V be a set of k binary output vectors $v_l, \ldots, v_k$ of dimension m. Then:

$$f_B : U \rightarrow V$$

$$f_B(u_i) = v_i \text{ for } 1 \leq i \leq k$$

Let $|Z|$ denote the cardinality of set Z. The binary vector u_i, where $1 \leq i \leq k$, represents an ordered set of r binary-coded symbols from symbol sets $\Gamma_1, \Gamma_2, \ldots, \Gamma_r$, respectively (i.e., $\exists \alpha_1 \in \Gamma_r, s.t.\ u_i = \alpha_1.\alpha_2.\ldots.\alpha_r$, where . denotes the concatenation of two binary codes). The binary vector v_i, where $1 \leq i \leq k$, represents an ordered set of t symbols from symbol sets $\Delta_1, \Delta_2, \ldots, \Delta_t$, respectively. In the context of Sec. 3, $r = L$, every Γ_i denotes the set of ASCII-coded English letters, $t = 1$, and Δ_1 is the set of M-bit lexical (record) pointers, where $1 \leq i \leq L.f_B$ defines a symbolic mapping function f_S such that

$$f_S : \Gamma_1 \times \ldots \times \Gamma_r \rightarrow \Delta_1 \times \ldots \times \Delta_t$$

In this case, the I/O mapping of a symbolic function f_S (information retrieval from a symbolic table, given a query criterion) can be viewed in terms of the binary (bipolar) mapping operations of f_B, which is realized by the proposed neural associative memory.

3. QUERY PROCESSING USING NEURAL ASSOCIATIVE MEMORIES

This section describes the use of the neural associative memory described in the previous section to implement a high-speed database query system. An ANN-based query system for a text-based machine-readable lexicon used in natural language processing (NLP) is presented to illustrate the key concepts. In the analysis, interpretation, and generation of natural languages, the lexicon is one of the central components of many NLP applications. As the quantity of entries of a lexical database increases, the speed of lexical access can become a bottleneck in real-time, large-scale machine processing of text. The number of words a native English speaker knows is estimated to be between 50,000 and 250,000 [8]. In the proposed ANN-based query system for text-based NLP lexicon, such a bottleneck can be alleviated by taking advantage of the capability of the proposed neural associative memory for massively parallel pattern matching and retrieval.

A. Realization of Lexical Access for Machine-Readable Lexicons Using a Neural Associative Memory

Basically, the lexical specification for a word in a lexicon contains phonological, morphosyntactic, syntactic, semantic, and other fields [9]. Each field may contain several subfields. In a lexical database that realizes a machine-readable lexicon for real-time NLP, the lengths of the fields and subfields are usually fixed to allow efficient random access to them. This is where a computational lexicon is distinguished from a dictionary in which the format of lexical entries is mostly irregular and, hence, the access of the lexical fields for a word (lexeme) is sequential. Typically, a dictionary contains much free text including definitions, examples, cross-reference words, and others.

Generally, there are two basic conceptions about the form of the items that serve as access keys in a lexicon. One is the *minimal listing hypothesis* [10], which lists only lexemes and results in a *root lexicon.* A lexeme may have several variants (e.g., in English, the words **produces**, **produced**, **producing**, **producer**, **productive**, and **production** are variants of the lexeme **produce**, and the words **shorter**, **shortest**, and **shortly** are variants of the lexeme short). The other is the *full-listing hypothesis* that lists all possible words of a language and results in a *full-form lexicon.* A root lexicon is more compact and requires a rule system to process the variants of lexemes, whereas a full-form lexicon is more computationally efficient in terms of lexical access and more user-friendly in terms of lexicon editing and extension [9]. Therefore, a hybrid of the two conceptions is often adopted in many computational lexicon applications. In the following, we use the term *access key* to stand for either word or lexeme in a computational lexicon no matter whether it is a root or full-form lexicon.

There are several models of lexical access in a computational lexicon. Our ANN-based query system for an NLP lexicon is based on the search model of lexical access (indirect access) [9,11]. In such a model, a text-based computational lexicon that associates every access key with its lexical specification contains two organizations: one is called the *master file*, which stores entries of lexical specifications, and the other is called the *access file*, which consists of pairs (<access key>, <lexical pointer>). The access keys are somewhat organized to allow location of a desired access key and its associated lexical pointer efficiently. The lexical pointers point to the lexical specifications of their corresponding entries in the master file. The process of lexical access in the search model is similar to that of locating a book in a library. To locate a book from a collection of shelves (the master file) in a library, the book catalog (the access file) is searched using author name(s) or book title, or both, to find the call number (a pointer indicating the location) of a desired book.

A noise-tolerant neural associative memory that can efficiently realize the process of search and retrieval of desired lexical pointers for a text-based machine-readable English lexicon is designed as follows. Suppose English letters of the lexicon are represented using 8-bit ASCII codes (extended to 8 bits by padding each 7-bit ASCII code with a leading 0). Assume the maximal length of an English word be L letters. Because each letter is represented by an 8-bit ASCII code, we need $8L$ input neurons in the ANN memory. Each binary bit x_b of the ASCII input is converted into a bipolar bit x_p by expression $x_p = 2x_b - 1$ before it is fed into the ANN memory to execute a query. Let the output (the lexical pointer) be represented as an M-bit binary vector that can access at most 2^M lexical specifications in the lexical database. So, the ANN memory uses M output neurons.

For every associative ordered pair of an access key and a lexical pointer, a hidden neuron is used in the ANN memory. Suppose there are k such pairs. Then, the ANN memory uses k hidden neurons. In the ANN memory, every access key is represented by padding its corresponding English word with trailing **spaces**. The ASCII code for the special symbol **space** is $20^{16} = 0010\ 0000_2$. For example, if an English word has j letters ($j \leq L$), then the first j letters of its corresponding access key are from the English word and the last $L - j$ letters of the access key are all **spaces**. The reason for such padding will become obvious in the coming examples. During storage of an associated pair, the connection weights are set as explained in Sec. 2.A. During recall, the thresholds of hidden neurons are adjusted for each query as outlined in Sec. 2.B (where, for each query, the value of n_a can be set either by a centralized check on the number of letters of the input access key, or by distributed circuitry embedded in input neurons). The precision level p is set at 0 for this associative ANN memory.

1. *Examples of Query Processing in the Proposed ANN Memory*

The following examples illustrate how the proposed ANN memory for an NLP lexicon retrieves desired lexical pointers by processing a query that may contain a partially specified input (target access key).

- **Example 1** (exact match): Suppose the lexical pointer of the word **product** is to be retrieved from the ANN memory. Then, the first seven letters of the target access key to be searched are **p**, **r**, **o**, **d**, **u**, **c**, and **t**, and the last $L - 7$ letters are **spaces**. In this case, no letter of the target access key is unavailable. Therefore, in the ANN memory, the threshold set at all hidden neurons is $L \times 8 = 8L$. Suppose a hidden neuron i is used for the association of this access key and its associated lexical pointer. When the target access key is presented to the ANN memory, only hidden neuron i has net sum of 0 and other hidden neurons have net sum less than 0 (see Sec. 2 for details). So, hidden neuron i is activated to recall the desired lexical pointer using the weights on the second-layer connections associated with hidden neuron i.
- **Example 2** (partial match of type 1): Suppose the lexical pointer(s) of the word(s) matching the pattern **product*** is to be retrieved from the ANN memory, where the symbol * means the trailing English letters starting from that position are unavailable. In this case, the last L–7 letters of the target access key are viewed as unavailable; only the first seven letters are available, and its first seven letters are **p**, **r**, **o**, **d**, **u**, **c**, and **t**. Therefore, in the ANN memory, only the first $7 \times 8 = 56$ neurons have input value either 1 or -1, the other input neurons are fed with 0, and the threshold set at all hidden neurons is $7 \times 8 = 56$. Suppose, in the lexicon, **product**, **production**, **productive**, **productively**, **productiveness**, and **productivity** are the words the first seven letters of which match the pattern **product***. In this case, six

hidden neurons are used for the associations of these six access keys and their lexical pointers, respectively, in the ANN memory. When the partially specified target access key is presented to the ANN memory, only these six hidden neurons have net sum of 0, and other hidden neurons have net sum less than 0. So, these six hidden neurons are activated one at a time to sequentially recall the associated lexical pointers using the weights on the second-layer connections associated with these six hidden neurons.

- **Example 3** (partial match of type 2): Suppose the lexical pointer(s) of a noisy seven-letter word **pro??ct** is to be retrieved from the ANN memory, where the symbol ? means the English letter at that position is unavailable. In this case, two of the letters (the fourth and fifth letters) of the target access key are viewed as unavailable, its first three, sixth and seventh letters are **p**, **r**, **o**, **c**, and **t**, respectively, and the last $L-7$ letters are **spaces**. Therefore, in the ANN memory, the input neurons representing the fourth and fifth input letters are fed with 0, other input neurons have input value either 1 or -1, and the threshold set at all hidden neurons is $8(L-2)$. Suppose, in the lexicon, **product**, **project**, and **protect** are the only seven-letter words that match the pattern **pro??ct**. Therefore, three hidden neurons are used for the associations of these three access keys and their lexical pointers, respectively, in the ANN memory. When the partially specified target access key is presented to the ANN memory, only these three hidden neurons have net sum of 0 and other hidden neurons have net sum less than 0. So, these three hidden neurons are activated one at a time to sequentially recall the associated lexical pointers using those weights on the second-layer connections that are, respectively, associated with them.

The large number of hidden neurons in such an ANN module poses a problem (because of the large fan-out for input neurons and large fan-in for output neurons) in the hardware realization of such an ANN module using electronic circuits. One solution to this problem is to divide the whole module into several submodules that contain the same number of input, hidden, and output neurons. These submodules are linked together by a shared input and output bus (Fig. 3). Such a bus topology also makes it possible to easily expand the size of the ANN memory. The one-dimensional array structure shown in Fig. 3 can be easily extended to two- or three-dimensional array structures.

2. *The Implementation of Case-Insensitive Pattern Matching*

It is rather straightforward to modify the proposed ANN-based query system to make it case-insensitive. The following shows ASCII codes of English letters, which are denoted in hexadecimal and binary codes.

$$A = 41_{16} = 0100\ 0001_2, \ldots, Z = 5\mathrm{A}_{16} = 0101\ 1010_2$$
$$a = 61_{16} = 0110\ 0001_2, \ldots, z = 7\mathrm{A}_{16} = 0111\ 1010_2$$

The binary codes for the capital case and small case of every same English letter differ only at the third bit counted from the left-hand side. If that bit is viewed as "don't care" (or unavailable), this query system will be case-insensitive. This effect can be achieved by treating the corresponding input value as though it was unavailable.

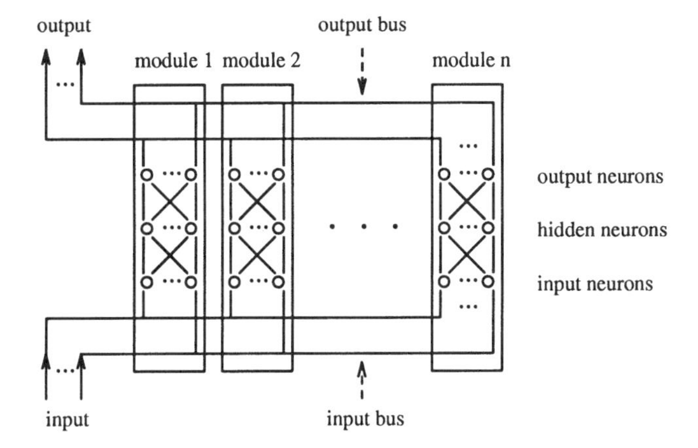

Fig. 3 A modular design of the proposed ANN memory for easy expansion. This one-dimensional array structure can be easily extended to two- or three-dimensional array structures.

4. COMPARISON WITH OTHER DATABASE QUERY-PROCESSING TECHNIQUES

This section compares the anticipated performance of the proposed neural architecture for database query processing with other approaches that are widely used in present computer systems. Such a performance comparison takes into account the performance of hardware that is used in these systems and the process used for locating data items. It is assumed that the systems have comparable I/O performance, which is not discussed here. First the performance of the proposed neural network is estimated, based on current CMOS technology for realizing neural networks. Next, the operation of conventional database systems is examined, and their performance is estimated and compared with that of the proposed neural architecture.

A. Performance of Current Electronic Realization for Neural Networks

Many electronic realizations of artificial neuron networks (ANNs) have been reported [12–21]. ANNs are implemented using mostly CMOS-based analog, digital, and hybrid electronic circuits. The analog circuit, which consists mainly of processing elements for multiplication, summation, and thresholding, is popular for the realization of ANNs because compact circuits capable of high-speed, asynchronous operation can be achieved [22]. A measured propagation delay of 104 ns (*nanosecond*) from the *START* signal until the result is latched by using a digital synapse circuit containing an 8-bit memory, an 8-bit subtractor, and an 8-bit adder has been reported [20]. A hybrid analog–digital design with 4-bit binary synapse weight values and current-summing circuits to achieve a network computation time of less than 100 ns between the loading of the imput and the latching of the result in a comparator was adopted [12]. Also adopted [16] is a hybrid analog–digital design with 5-bit (4 bits + sign) binary synapse weight values and current-summing circuits to implement a feedforward neural network with two connection layers, and a network computation time of less than 20 ns. The throughput was achieved [13] at the rate of 10 MHz (delay = 100 ns) in a Hamming Net pattern classifier using analog circuits. The first-layer and second-layer subnetworks of the proposed neural architecture for database query processing are very similar to the first-layer subnetwork of a Hamming Net, respectively, and the neural architecture with two connection layers in the proposed ANN is exactly same as that implemented [16], except that these authors [16] use discretized inputs, 5-bit synaptic weights, and a sigmoid-like activation function. The proposed ANN uses bipolar inputs, weights in $\{-1, 0, 1\}$ and binary hardlimiter as activation function. Hence, the computation delay of the proposed ANN can be expected to be at worst on the order of 100 ns and at best 20 ns given the current CMOS technology for realizing ANNs.

The development of specialized hardware for implementation of ANNs is still in its early stages. Conventional CMOS technology, which is currently the main technology for VLSI implementation of ANNs, is known to be slow [23,24]. Other technologies, such as BiCMOS, NCMOS [23], pseudo-NMOS logic, standard N-P domino logic, and quasi-N-P domino logic [24], may provide better performance for the realization of ANNs. Thus, the performance of the hardware implementation of ANNs is likely to improve with technological advances in VLSI.

B. Analysis of Query Processing in Conventional Computer Systems

In database systems implemented on conventional computer systems, given the value for a key, a record is located efficiently by using key-based organizations including *hashing*, *index-sequential access files*, and *B-trees* [25]. Such a key-based organization usually contains two data structures: *index files(s)* and *master file*. In an index file, every key is somewhat organized and usually associated with a record pointer that points to a corresponding record in the master file, which is typically stored in secondary storage devices such as hard disks for very large databases. Conventionally, estimated cost of locating a record is based on the number of physical block accesses of secondary storage devices [25], because the access latency with current cost-effective disk systems is close to 5–10 ms and every one of the repetitive search steps that together facilitate locating a desired record pointer from index files (loaded on main memory) takes only several CPU clocks. The clock cycle of cost-effective CPUs is about 2–10 ns. But, as the number of entries in an index file increases, the speed of locating a desired record pointer can become a bottleneck in very large databases.

The following compares the anticipated performance of the proposed neural associative memory with other approaches that are widely used in present computer systems for locating a record pointer associated with a given key. In the following analysis, it is assumed that all program and index files for processing queries using current computer systems are preloaded into the main memory. The effect of data dependency among instructions that offsets pipeline and superscalar effects and thus much reduces the average performance of current computer systems is not considered here.

To simplify the comparison, it is assumed that each instruction on a conventional computer takes τ ns on an average. For instance, on a relatively cost-effective 100-MIPS processor, a typical instruction would take 10 ns (The MIPS measure for speed combines clock speed, effect of caching, pipelining, and superscalar design into a single figure for speed of a microprocessor). Similarly, we will assume that a single identification and recall operation cycle by a neural associative memory takes α ns. Assuming hardware implementation based on current CMOS VLSI technology, α is close to 20–100 ns. Table 1 summarizes from the following analysis the estimated performance of the proposed neural associative memory and other techniques commonly used in conventional computer systems for locating a desired record pointer. Table 2 summarizes the capabilities of the proposed neural associative memory and other techniques commonly used in conventional computer systems for exact match and partial match of the types mentioned in Sec. 3.

1. Analysis of Locating a Record Pointer Using Hashing Functions

Hashing structure is the fastest of all key-based search for locating a record pointer for a single record. However, although it is effective in locating a single record by exact match (e.g., example 1 of Sec. 3), it is inefficient or unable to locate several related records in response to a single query (e.g., the partially specified inputs in examples 2 or 3 of Sec. 3). Let us consider the time needed for locating a record pointer using a hash function in current computer systems. Commonly used hash functions are based on multiplication, division, and addition operations [26,27]. In hardware implementation, addition is faster than multiplication, which is far faster than division. Assume that computing a hashing function on a key with a length of L bytes (characters) takes $\lceil 8L/n \rceil$ cycles using a processor with an n-bit data bus and every cycle takes h instructions. Then, the estimated computation time for locating a record pointer is $\lceil 8L/n \rceil h\tau$. Other overheads of computing a hashing function in such systems includes the time for handling the potential problem of collisions in hash functions. If a single-CPU 100-MIPS processor with a 32-bit data bus is used, it is expected that the total computation time for locating a record pointer will typically be in excess of 100 ns (If $L = 15$ and $h = 5$, the total computation time is $\lceil 8 \times 15/32 \rceil \times 5 \times 10$ ns $=$ 200 ns).

Table 1 A Comparison of the Estimated Performance of the Proposed Neural Associative Memory with That of Other Techniques Commonly Used in Conventional Computer Systems for Locating a Record Pointer in Key-Based Organizations[a]

Method	Estimated time (ns)
Hashing	$\lceil 8L/n \rceil h\ \tau$
Index search	$(M-1)\lceil 4L/n \rceil b\ \tau$
ANN memory	α
k-d-tree (partial match)	$O(N^{(K-J)/K})$

[a]The comparison assumes that the value of the key is given, the data structures and programs are loaded in the main memory of the computer systems used, index search occurs in a balanced binary tree of $(2^M - 1)$ records, and partial match occurs in a *k-d*-tree of M records. L is the total number of bytes of a key, n is the data bus width of the computer system used, h is the average number of executed instructions in a hashing cycle, τ is the average time delay for executing an instruction, b is the average number of executed instructions in a comparison cycle for every n bits in a binary search cycle, α is the time delay of the proposed neural memory, K is the number of index fields used in the *k-d*-tree, and J is the number of index fields specified in a query criterion.

Table 2 A Comparison of the Capabilities of the Proposed Neural Associative Memory with Those of Other Techniques Commonly Used in Conventional Computer Systems for Exact Match and Partial Match of the Types Mentioned in Sec. 3

Method	Exact match	Partial match (type 1)	Partial match (type 2)
Hashing	Efficient	Unable	Unable
Index search	Efficient	Efficient	Inefficient
ANN memory	Efficient	Efficient	Efficient
k-*d*-tree	Satisfactory	Satisfactory	Inefficient

2. *Analysis of Locating a Record Pointer Using Index Search*

The perfectly balanced binary search tree is another popular, efficient data structure used in conventional database systems to locate a single record by exact match (e.g., example 1 of Sec. 3) or several related records by partial match (e.g., example 2, but not example 3 of Sec. 3). Assume every nonterminal node in a perfectly balanced binary search tree links two child subtrees and there are $(2^M - 1)$ nodes in the tree. Assume the length of the index key is L bytes (characters). The average number of nodes visited for locating a desired key would be $(2^M(M-1)+1)/(2^M-1) \approx M-1$. On average, every visit takes $\lceil (1/2 \times 8L)/n \rceil = \lceil 4L/n \rceil$ comparison cycles for a processor with an n-bit data bus. Suppose every comparison cycle takes b instructions. Then, the estimated computation time for locating a desired record pointer is $(M-1)\lceil 4L/n \rceil b\tau$. If $L = 15$, a 100-MIPS processor with a 32-bit data bus is used, the comparison cycle for every 32 bits takes five instructions on average, and there are $2^{16} - 1 = 65,535$ records (the number of words a native English speaker knows is estimated to be between 50,000 and 250,000 [8]), then the overhead for locating a desired record pointer is about $(16-1) \times \lceil 4 \times 15/32 \rceil \times 5 \times 10$ ns $= 1500$ ns. Note that this is only the cost for locating a record pointer for a single record. The cost for locating several record pointers of related records using user-entered data in an index file containing multiple index fields is examined in next section.

3. *The Cost of Partial-Match Queries*

One of the most commonly used data structures for processing partial-match queries on multiple index fields is *k-d-tree* [28]. It can provide approximately satisfactory performance for locating a single record by exact match or several related records by partial match. In the worst case, the number of visited nodes in an ideal *k-d-tree* of N nodes (one for each record stored) for locating the desired record pointers for a partial-match query is

$$\frac{(J+2)2^{K-J-1} - 1}{2^{K-J} - 1}\left[(N+1)^{(K-J)/K} - 1\right] \approx O(N^{(K-J)/K})$$

where K is the number of index fields used to construct the *k-d-tree*, and J out of K index fields are explicitly specified by a user query. For typical values of N, K, and J, the performance of such systems is far worse than that of the proposed ANN-based model according to the foregoing equation.

5. SUMMARY AND DISCUSSION

Artificial neural networks, because of their inherent parallelism and potential for noise tolerance, offer an attractive paradigm for efficient implementations of a broad range of information-processing tasks. In this chapter, we have explored the use of artificial neural networks for query processing in large databases. The use of the proposed approach was demonstrated using the example of a query system for a text-based machine-readable lexicon used in natural language processing. The performance of a CMOS hardware realization of the proposed neural associative memory for a database query-processing system was estimated and compared with that of other approaches that are widely used in conventional databases implemented on current computer systems. The comparison shows that ANN architectures for query processing offer an attractive alternative to conventional approaches, especially for dealing with partial-match queries in large databases. With the need for real-time response in language translation and with the explosive growth of the Internet as well as gradually increased use of large networked databases over the Internet, efficient architectures for high-speed associative table lookup, message routing, and database query processing have assumed great practical significance. Extensions of the proposed ANN architcture for syntax analysis is in progress [29].

ACKNOWLEDGMENT

This research was partially supported by the National Science Foundation through the grant IRI-9409580 to Vasant Honavar.

REFERENCES

1. S Goonatilake, S Khebbal, eds. Intelligent Hybrid Systems. Wiley, London, 1995.
2. V Honavar, L Uhr, eds. Artificial Intelligence and Neural Networks: Steps Toward Principled Integration. Academic Press, New York, 1994.
3. D Levine, M Aparicio IV, eds. Neural Networks for Knowledge Representation and Inference. Lawrence Erlbaum, Hillsdale, NJ, 1994.
4. R Sun, L Bookman, eds. Computational Architectures Integrating Symbolic and Neural Processes. Kluwer Academic, Norwell, MA, 1994.
5. C Chen, V Honavar. Neural associative memories for content as well as address-based storage and recall: theory and applications. Connect Sci 7:293–312, 1995.
6. C Chen, V Honavar. A neural network architecture for high-speed database query processing system. Microcomp Appl 15:7–13, 1996.
7. M Minsky, S Papert. Perceptrons: An Introduction to Computational Geometry. MIT Press, Cambridge, MA, 1969.
8. J Aitchison. Words in the Mind. An Introduction to the Mental Lexicon. Basil Blackwell, Oxford, 1987.
9. J Handke. The Structure of Lexicon: Human Versus Machine. Mouton de Gruyter, Berlin, 1995.
10. B Butterworth. Lexical representation. In: Language Production, vol 2: Development, Writing and Other Language Processes. B Butterworth, ed. Academic Press, London, 1983, pp. 257–294.
11. KI Forster. Accessing the Mental Lexicon. In: New Approaches to Language Mechanisms. R Walker, RJ Wales, eds., North-Holland, Amsterdam, 1976, pp 257–287.

12. HP Graf, D Henderson. A reconfigurable CMOS neural network. ISSCC Dig. Tech. Papers, San Francisco, CA, 1990, pp 144–145.
13. D Grant, et al. Design, implementation and evaluation of a high-speed integrated Hamming neural classifier. IEEE J Solid-State Circuits 29:1154–1157, 1994.
14. A Hamilton, et al. Integrated pulse stream neural networks: results, issues, and pointers. IEEE Trans Neural Networks 3:385–393, 1992.
15. JB Lont, W Guggenbühl. Analog CMOS implementation of a multilayer perceptron with nonlinear synapses. IEEE Trans Neural Networks 3:457–465, 1992.
16. P Masa, K Hoen, H Wallinga. 70 input, 20 nanosecond pattern classifier. Proceedings of IEEE International Joint Conference on Neural Networks, vol 3, Orlando, FL, 1994.
17. LW Massengill, DB Mundie. An analog neural network hardware implementation using charge-injection multipliers and neuron-specific gain control. IEEE Trans Neural Networks 3:354–362, 1992.
18. G Moon, et al. VLSI implementation of synaptic weighting and summing in pulse coded neural-type cells. IEEE Trans Neural Networks, 3:394–403, 1992.
19. ME Robinson, et al. A modular CMOS design of a Hamming network. IEEE Trans Neural Networks 3:444–456, 1992.
20. K Uchimura, et al. An 8G connection-per-second 54 mW digital neural network with low-power chain-reaction architecture. ISSCC Dig. Tech. Papers. San Francisco, CA, 1992, pp 134–135.
21. T Watanabe, et al. A single 1.5-V digital chip for a 10 6 synapse neural network. IEEE Trans Neural Networks 4:387–393, 1993.
22. SM Gowda, et al. Design and characterization of analog VLSI neural network modules. IEEE J Solid-State Circ 28:301–313, 1993.
23. R Kumar. NCMOS: a high performance CMOS logic. IEEE J Solid-State Circ 29:631–633, 1994.
24. F Lu, H Samueli. A 200-MHz CMOS pipelined multiplier–accumulator using a quasi-domino dynamic full-adder cell design. IEEE J Solid-State Circ. 28:123–132, 1993.
25. JD Ullman. Principles of Databases and Knowledge-Base Systems. vol 1, Chap 6, Computer Science Press, Maryland, 1988.
26. T Kohonen. Content-Addressable Memories, 2nd ed. Springer-Verlag, Berlin, 1987.
27. R Sedgewick. Algorithms, 2nd ed. Addison–Wesley, Reading, MA, 1988.
28. JL Bently. Multidimensional binary search trees used for associative searching. Commun ACM 18:507–517, 1975.
29. C Chen, V Honavar. A neural network architecture for syntax analysis. IEEE Trans Neural Networks 10:94–114, 1999.

36

Text Data Mining

DIETER MERKL

Vienna University of Technology, Vienna, Austria

1. INTRODUCTION

In recent years we have witnessed an ever-increasing flood of written information culminating in the advent of massive digital libraries. Powerful methods for organizing, exploring, and searching collections of textual documents are needed to deal with that information. The classic way of dealing with textual information, as developed in the information retrieval community, is defined by means of keyword-based document representations. These methods may be enhanced with proximity search functionality and keyword combination using Boole's algebra. Other approaches rely on document similarity measures, based on a vector representation of the various texts. What is still missing, however, are tools providing assistance for explorative search in text collections.

Exploration of document archives may be supported by organizing the various documents into taxonomies or hierarchies. Parenthetically, we should note that such an organization has been in use by librarians for centuries. To achieve such a classification a number of approaches are available. Among the oldest and most widely used is statistics, especially cluster analysis. The use of cluster analysis for document classification has a long tradition in information retrieval research and its specific strength and weaknesses are well explored [1,2].

The field of artificial neural networks in a wide variety of applications has attracted renewed interest, which is at least partly due to increased computing power available at reasonable prices. There is general agreement that the application of artificial neural networks may be recommended in areas that are characterized by (a) noise, (b) poorly understood intrinsic structure, and (c) changing characteristics.

Each of these characteristics is present in text classification. Noise is present because no completely satisfying way of representing text documents has yet been found. Second, poorly understood intrinsic structure is due to the nonexistence of an authority that knows the contents of each and every document. In other words, in document classification we cannot rely on the availability of a person who is able to describe the decisions needed for automating the classification process in an algorithmic fashion based on that person's knowledge of the contents of each document. Hence, approaches based on learning techniques have their potential in document classification. Finally, the changing characteristics of document collections are due to the collections being regularly enlarged to include additional documents. As a consequence, the contents of the document collection changes constantly. Learning techniques are advantageous in such an environment compared with algorithmic or knowledge-based approaches.

From the wide range of proposed artificial neural network (ANN) architectures we consider unsupervised models as especially well suited for the exploration of text collections. This is because in a supervised environment, one would have to define proper input–output-mappings anew each time the text archive changes, and such changes can be expected to occur quite frequently. By *input–output-mapping* we refer to the manual assignment of documents to classes, which is, obviously, only possible when one assumes the availability of considerable insight in the structure of the text archive. On the other hand, in an unsupervised environment, it remains the task of the artificial neural network to uncover the structure of the document archive. Hence, the unrealistic assumption of being able to provide proper input–output-mappings is unnecessary in an unsupervised environment. A number of successful applications of unsupervised neural networks in the area of text archive exploration have already been reported in the literature [3–10].

One of the most versatile and successful unsupervised neural network architectures is the *self-organizing map* [11]. It is a general unsupervised tool for ordering high-dimensional statistical data in a way such that similar input items are grouped spatially close to one another. To use the self-organizing map to cluster text documents, the various texts have to be represented by means of a histogram of word occurrences. With this data, the network is capable of performing the classification task in a completely unsupervised fashion.

From an exploratory data analysis point of view, however, this neural network model may have its limitations because it commonly uses a two-dimensional plane as the output space for input data representation. In such a restricted low-dimensional space it does not come as a surprise that at least some relations between various input items are visualized in a rather limited fashion. The limitations arise because the self-organizing map performs a mapping from a very high-dimensional input space represented by the various words used to describe the documents to a two-dimensional output space. Such a mapping, obviously, cannot mirror the similarity between the various documents exactly. As a result, one might come up with an erroneous perception of the similarities inherent in the underlying document collection. A recently proposed neural network architecture that overcomes these limitations, while still relying on the self-organizing map as the general underlying model is the *hierarchical feature map* [12]. A specific benefit of this model is the hierarchical organization, which imposes a hierarchical structure on the underlying document archive. By making use of this feature, the establishment of a reliable document taxonomy is made feasible, enabling a straightforward exploration of document similarities.

The remainder of this discussion is organized as follows. In Sec. 2 we give a very short description of the text representation used for the experiments. Because it is a common feature-based text representation, we feel that the quick review provided in this section is sufficient. In Sec. 3 we give a brief description of the neural network architectures used for document space exploration. These networks are referred to as *topology-preserving neural networks* because of their inherent ability to map similar input patterns onto adjacent regions of the network. In Sec. 4 we give a detailed exposition of a series of experiments in text archive exploration based on an experimental text collection consisting of the manual pages describing various software components. In particular, we provide a comparison of the effects of using either self-organizing maps or hierarchical feature maps. Finally, we give some conclusions in Sec. 5.

2. TEXT REPRESENTATION

Generally, the task of text clustering aims to discover the semantic similarities between various documents. In the spirit of Fayyad et al. and Mannila [13–16], we regard clustering as one of the essential techniques in the data-mining process for discovery of useful data patterns. Because the various documents that compose text archives do not lend themselves to immediate analysis, some preprocessing is necessary.

To enable further analyses the documents have, in the first instance, to be mapped onto some representation language. One of the most widely used representation languages is still single-term full-text indexing. In such an environment, the documents are represented by feature vectors $\mathbf{x} = (\xi_1, \xi_2, \ldots, \xi_n)^T$. Thereby the ξ_i, $1 \leq i \leq n$, refer to terms[1] extracted from the varoius documents contained in the archive. The specific value of ξ_i corresponds to the importance of this feature in describing the particular document at hand. Numerous strategies for assigning degrees of importance to features in any particular document are available [17]. Without loss of generality, we may assume that importance is represented as a scalar in the range of [0, 1], where zero means that a particular feature is unimportant for describing the document. Any gradation from 0 to 1 is proportional to the increased importance of the feature in question.

In what is probably the most basic text representation (i.e., binary single-term indexing), the importance of a specific feature is represented by one of two numerical values. Hence, the notion of "importance" is reduced to whether or not a particular word (i.e., document feature) is contained in the document. A value of 1 represents the fact that the corresponding feature was extracted from the document at hand. On the other hand, a value of 0 means that the corresponding feature is not contained in that document.

[1]In the remainder of this discussion we will use the words *term* and *keyword* interchangeably to refer to entities that are selected to represent the contents of documents. Collectively, these terms represent the feature space to describe the document collection.

Such a feature-based document representation is known as the vector-space model for information retrieval. In this model, the similarity betwen two text documents corresponds to the distance between their vector representations [18,19].

In the experiments described later we will rely on a binary feature-based document representation. This representation is used as the input to an artificial neural network. There are, however, other document representation techniques available. The utilization of a feature-based document representation may lead to a very high-dimensional feature space. This feature space is generally not free from correlations. As a consequence, one might be interested in the transformation of the original document representation into a lower-dimensional space. Statistical techniques that can be used to perform such a transformation include multidimensional scaling [20] and principal component analysis [21]. The former is the underlying principle of *Latent Semantic Indexing* [22,23]. The latter has been described and successfully applied to text corpora analysis in [24]. The feature space for document representation may be reduced by means of an autoassociative multilayer perceptron [25]. In such a network configuration the activations of the hidden layer units, which are numerically fewer that those in the input and output layer, represent an approximation to the principal components of the input data. These activations are then used for self-organizing map training instead of the original input patterns, yielding substantially reduced training time while still enabling comparable results as far as text archive organization is concerned. Recently, several papers have been published on the utilization of the self-organizing map for large-scale document representation [4] based on the seminal work of Ritter and Kohonen [26] and subsequent interactive exploration [3,6].

3. TOPOLOGY PRESERVING NEURAL NETWORKS

A. Competitive Learning Basics

Competitive learning [27], or *winner-takes-all,* as it is termed quite often, may be considered as the basis of a number of unsupervised learning strategies. In its most rudimentary form, a competitive learning network consists of k units with weight vectors $\mathbf{m}_k$ of dimension equal to the input data, $\mathbf{m}_k \in \mathcal{R}^n$. During learning, the unit with its weight vector closest to the input vector $\mathbf{x}$ is adapted in such a way that the weight vector resembles the input vector more closely after the adaptation. To determine the distance between the vectors, any distance (or similarity) metric may be selected, although the most common choices are marked by the euclidean vector norm and the inner product of the vectors. The unit with the closest weight vector is dubbed the *winner* of the selection process. This learning strategy may be implemented by gradually reducing the difference between weight vector and input vector. The actual amount of difference reduction at each learning step may be guided by means of a so-called learning-rate α in the interval [0,1]. When given such a learning environment, the various weight vectors converge toward the mean of the set of input data represented by the unit in question.

Obviously, the term *winner-takes-all* refers to the fact that only the winner is adapted, whereas all other units remain unchanged. As a severe limitation of this basic learning strategy consider a situation where there are some units which, because of random initialization of their weight vectors, are never selected as the winner and whose weight vectors are consequently never adapted. Strictly speaking, such units may be referred to as *dead units*, as they do not contribute to the learning process and thus do not contribute to input data representation. This possibility has led to the development of a number of learning rules and network architectures that overcome this limitation by enlarging the set of units that are affected by adaptation at each learning step.[2] Apart from the winner, adaptation is performed with units in some defined vicinity around the winner. This type of learning rule may be referred to as *soft competitive learning*, representatives of which are described in the following subsections.

B. Self-Organizing Maps

The *self-organizing map* as proposed [11] and described thoroughly [28] is one of the most widely used unsupervised artificial neural network models. It consists of a layer of input units each of which is fully connected to a set of output units. These output units are arranged in some topology where the most common choice is represented by a two-dimensional grid.

Input units receive the input patterns $\mathbf{x}, \mathbf{x} \in \mathcal{R}^{\mathbf{n}}$, and propagate them as they are onto the output units. Each of the output units i is assigned a weight vector $\mathbf{m}_i$. These weight vectors have the same dimension as the input data, $\mathbf{m}_i \in \mathcal{R}^n$.

During each learning step, the unit c with the highest activity level relative to a randomly selected input pattern $\mathbf{x}$ is adapted in such a way that it will exhibit an even higher activity level at future presentations of $\mathbf{x}$. Commonly, the activity level of a unit is computed as the euclidean distance between the input pattern and that unit's weight vector. Hence, the selection of the winner c may be written as given in Eq. (1).

$$c : \|\mathbf{x} - \mathbf{m}_c\| = \min_i\{\|\mathbf{x} - \mathbf{m}_i\|\} \qquad \{1\}$$

Adaptation takes place at each learning iteration and constitutes a gradual reduction of the difference between the respective components of the input vector and the weight vector. The amount of adaptation is guided by a learning-rate parameter α that gradually decreases in the course of learning. This decrease ensures large adaptation steps at the beginning of the learning process where the weight vectors have to be tuned from their random initialization toward the actual requirements of the input space. Furthermore, the ever smaller adaptation steps toward the end of the learning process enable a fine-tuned input space representation.

As an extension to standard competitive learning, units in a time-varying and gradually decreasing neighborhood around the winner are also adapted. During the learning steps of the self-organizing map a set of units around the winner is tuned toward the currently presented input pattern enabling a spatial arrangement of the input patterns such that similar inputs are mapped onto regions close to each other in the grid of output units. Thus, the training process of the self-organizing map results in a topological ordering of the input patterns. According to Ripley [29] we may thus refer to the self-organizing map as a neural network model performing a spatially smooth version of k-means clustering.

---

[2]We have to admit that this is a rather sloppy formulation as far as historical chronology is concerned. In fact the *self-organizing map*, the neural network model to be described in the next subsection, appeared earlier in the literature as the basic *winner-takes-all* learning rule.

The neighborhood of units around the winner may be described implicitly by means of a neighborhood kernel h_{ci} taking into account the distance between unit i under consideration and unit c, the winner of the current learning iteration. This neighborhood kernel assigns scalars in the range of [0, 1] that are used to determine the amount of adaptation, ensuring that nearby units are adapted more strongly than units farther away from the winner. A gaussian may be used as the neighborhood kernel. It is common practice that at the start of the learning process, the neighborhood kernel is selected large enough to cover a wide area of the output space. The spatial width of the neighborhood kernel is reduced gradually during the learning process so that, toward the end of the learning process, just the winner itself is adapted. This strategy enables the formation of large clusters at the beginning and fine-grained input discrimination toward the end of the learning process.

Combining these principles of self-organizing map training, we may write the learning rules as given in Eq. (2). Please note that we use a discrete time notation with t denoting the current learning iteration. The other parts of this expression are α, representing the time-varying learning-rate, h_{ci} representing the time-varying neighborhood kernel, $\mathbf{x}$ representing the current input pattern, and, finally, $\mathbf{m}_i$ denoting the weight vector assigned to unit i.

$$\mathbf{m}_i(t+1) = \mathbf{m}_i(t) + \alpha(t) \times h_{ci}(t) \times [\mathbf{x}(t) - \mathbf{m}_i(t)] \tag{2}$$

A simple graphical representation of a self-organizing map's architecture and its learning process is provided in Fig. 1. In this figure, the output space consists of a square of 36 units, depicted as circles. One input vector $\mathbf{x}(t)$ is randomly chosen and mapped onto the grid of output units. In a subsequent step the winner is selected, which is depicted as the black node in the figure. The weight vector of the winner, $\mathbf{m}_c(t)$, is now moved toward the current input vector. This movement is symbolized in the input space in Fig. 1. As a consequence of the adaptation, this unit c will produce an even higher activation to input pattern $\mathbf{x}$ at the next learning iteration, $t+1$, because the unit's weight vector, $\mathbf{m}_c(t+1)$, is now nearer to the input pattern $\mathbf{x}$ in terms of the input space. Apart from the winner, adaptation is also performed with neighboring units. Units that are subject to adaptation are depicted as shaded nodes in the figure. The shading of the various nodes corresponds to the amount of adaptation and, thus, to the spatial width of the neighborhood kernel. Generally,

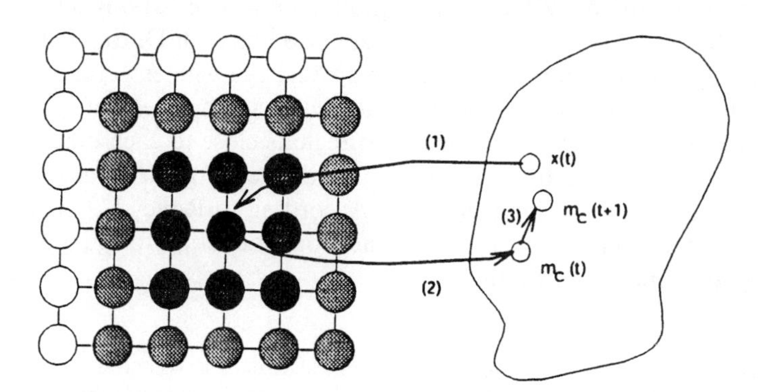

Fig. 1 Architecture of a self-organizing map.

units in close vicinity of the winner are adapted more strongly, and consequently, they are depicted with a darker shade in the figure.

C. Hierarchical Feature Maps

The self-organizing map is a neural network model capable of arranging high-dimensional input data within its (usually) two-dimensional output space in such a way that the similarity of the input data is mirrored as faithfully as possible in terms of the topographical distance between the respective winning units. The utilization of this model is thus especially well suited in application areas where one is highly dependent on a convenient data representation and visualization for subsequent exploration of the input data space. The exploration of text collections exemplifies this type of application.

For such exploration, however, we have to note a deficiency of the self-organizing map. This deficiency is the inability of the standard model to represent cluster boundaries explicitly. A substantial amount of research addressing the representation of cluster boundaries has emerged as a natural consequence. Without going into greater detail, we may distinguish three different approaches to overcoming this deficiency. The first focuses on improved visualization of the training result. Examples of this line of research are given elsewhere [30–32]. A second approach addresses cluster boundary detection by means of adaptive network architectures. More precisely, these models rely on incrementally growing and splitting architectures, where the final shape of the network is determined by the specific requirements of the input space. Several studies [33–36] describe this line of research. The major problem with these models is that they require that a few more parameters be specified before network training. Consequently, the utilization of these models is more demanding as far as the necessary experience of the users is concerned [5]. The third and final approach to cluster boundary uses individual self-organizing maps assembled into a neural network with a layered architecture [12,37].

Relative to our goal of text data mining we consider the *hierarchical feature map* [12,38–40] as the most promising among the approaches to cluster boundary recognition. The reason is that this model imposes a hierarchical structure on the underlying input data that resembles the organizational principle in use by librarians for centuries for text corpora organization. Moreover, the time needed to train the network is substantially shorter than for self-organizing maps, thereby providing a highly convenient environment for text analysis and exploration.

The key idea of the hierarchical feature map is to arrange a number of self-organizing maps in a hierarchy such that for each unit on one level of the hierarchy a two-dimensional self-organizing map is added to the next level. The resulting architecture may thus be characterized as having the shape of a pyramid, as depicted in Fig. 2.

The training of the hierarchical feature map is performed sequentially from the first layer downward in the hierarchy. The maps on each layer are trained according to the standard learning rule of self-organizing maps as outlined in the foregoing. As soon as the first-layer map has reached a stable state, training continues with the maps forming the second layer. Within the second layer, each map is trained only with that subset of the original input data that was mapped onto the corresponding unit of the first-layer map. Moreover, the dimension of the input patterns may be reduced at the transition from one layer to the next by omitting that portion of the

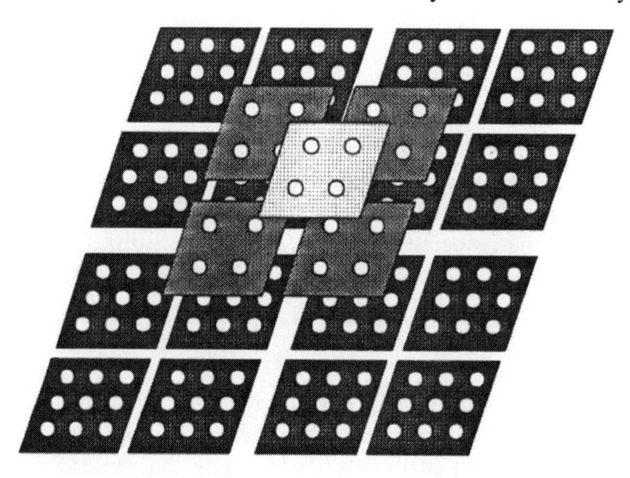

Fig. 2 Architecture of a hierarchical feature map.

document representation that is equal in the patterns mapped onto the same unit. In other words, the keywords that are found to represent each document mapped onto the same unit may be removed, and the next layer is trained only with the now shorter, remaining portion of the document representation. The rationale behind this reduction is that the omitted vector components represent features that are already learned by the higher-layer map. This characteristic is of especial importance when dealing with text as the underlying input data because, first, text documents are represented in a very high-dimensional feature space, and, second, some of these features are common to each text belonging to a particular subject. Owing to this dimension reduction, the time needed to train the maps is also reduced. Training of the second layer is completed when each map has reached a stable state. Analogously, the same training procedure is used to train the third and any subsequent layers of the hierarchical feature map.

A valuable property of the hierarchical feature map is the substantial speed-up of the training process as compared with conventional self-organizing maps. An explanation that goes beyond the obvious dimension reduction of the input data emerges from an investigation of the general properties of the self-organizing training process. In self-organizing maps, the units that are subject to adaptation are selected by means of the neighborhood kernel. It is common practice that, at the start of the training process, almost the whole map is affected by the presentation of an input vector. With this strategy, the map is forced to establish initial clusters of similar input items at the outset of learning. By reducing, the width of the neighborhood kernel in the course of learning, the training process is able to learn ever finer distinctions within the clusters while the overall topology of cluster arrangement is maintained. The flipside of the coin, however, is that units along the boundary between two clusters tend to be occasionally modified as belonging to either one of these clusters. This interference is the reason for the time-consuming self-organizing process. Such an interference is dramatically reduced in hierarchical feature maps. This reduction is due to the architcture of this neural network. The topology of the high-level categories is represented in the first layer of the hierarchy. Each of its subcategories are then independently organized within separate maps at lower levels within the hierarchy. These maps, in turn, are free from having to represent the overall structure, as this structure is already determined by the architecture of the hierarchical feature map. As a consequence, a set of documents that is mapped onto the same unit in any of the layers is further separated in the following layer of the hierarchical feature map. Assuming that such a unit represents a meaningful cluster of documents (i.e., a set of documents covering the same subject matter), this cluster is further organized in the next layer of the artificial neural network. In summary, much computational effort is saved because the overall structure of clusters is determined by the architecture of the artificial neural network, rather than by its learning rule.

4. A VOYAGE THROUGH DOCUMENT SPACES

In this section we will describe some of the results we achieved when using unsupervised neural networks for the exploration of a document archive. In particular, we will give samples from text archive organization by means of self-organizing maps and hierarchical feature maps.

A. An Experimental Document Collection

Throughout the remainder of this work we will use the various manual pages of the NIH Class Library [41], *NIHCL*, as a sample document archive. The *NIHCL* is a collection of classes developed in the C++ programming language. The class library covers classes for storing and retrieving arbitrarily complex data structures on disk, generally useful data types such as **String**, **Time**, and **Date**, and, finally, a number of container classes such as, for example, **Set**, **Dictionary**, and **OrderedCltn**. More general information concering the *NIHCL* may be found elsewhere [42].

The selection of the *NIHCL* is motivated by the fact that software libraries represent a convenient application arena for information retrieval systems. The reason is that much of the information about a particular software component is available in textual description organized as manual pages. Moreover, the results of the classification process may easily be evaluated because the semantics of the software component (i.e., its functionality) is well known.

The full text of the various manual pages describing the classes was accessed and indexed to generate a binary vector-space representation of the documents. The indexing process identified 489 distinct content terms, and each component is thus represented by a 489-dimensional feature vector. These vectors are subsequently used as the input data to the artificial neural network. Note that we do not make use of the metainformation contained in the manual entries, such as references to *base class*, *derived classes*, and *related classes*. Such information is not generally available in document archives, and thus, for the sake of general applicability, we excluded this type of metainformation from our work. Such information might well be of importance, however, if one were specifically interested in an application fine-tuned to this particular document archive.

B. A Map of the Document Collection

A typical result from the application of self-organizing maps to this type of data is represented in Fig. 3. In this case we have used a 10 × 10 self-organizing map to represent the document collection. Generally, the graphic representation may be interpreted as follows. Each output unit of the self-organizing map is represented by means of either a dot or a class name. The class name appears where the respective unit is the winner for that specific input pattern. By contrast, a dot marks units that have not won the competition for an input pattern. These units, however, have an important role in determining the spatial range of the various data clusters.

For convenience of discussion we have marked some of the interesting regions manually in Fig. 3. In the upper right part of the map we recognize the arrangement of all classes performing file I/O; these classes are designated by the "**OIO**" part of their respective class names. The region itself is organized in such a way that a class performing an input operation is mapped neighboring its output counterpart (e.g. **OIOifd** and **OIOofd**). Just below the file I/O region we find a large area of the map consisting of the various classes representing data structures. Within this larger region we have marked the area comprising classes that allow an access to their elements through a key-attribute (i.e., **Dictionary**, **IdentDict**, and **KeySortCltn**). Within this area we also find the classes that actually implement this type of access (i.e., **Assoc**, **AssocInt**, and **LookupKey**). Finally, we want to shift the attention toward the left-hand side of the map where the classes representing data types are located. Again, their arrangement reflects their mutual similarity. For instance, note the placement of class **Random**, a random number generator producing numbers of **Float** data type. This class is mapped onto a unit neighboring the basic numerical data types **Float** and **Integer**. Many more interesting areas may be located in this representation. They cannot be covered in any detail here, so the reader is referred to experimental results [9] that are compared with those obtained from statistical cluster analysis.

C. A Taxonomy of Documents

In Fig. 4 we show a typical result from organizing the *NIHCL* document archive by means of the hierarchical feature map. Owing to the hierarchical arrangement of a

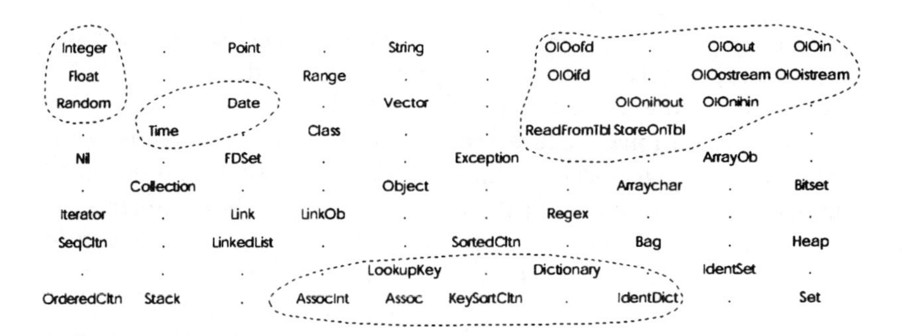

Fig. 3 *NIHCL* presented in a 10 × 10 self-organizing map.

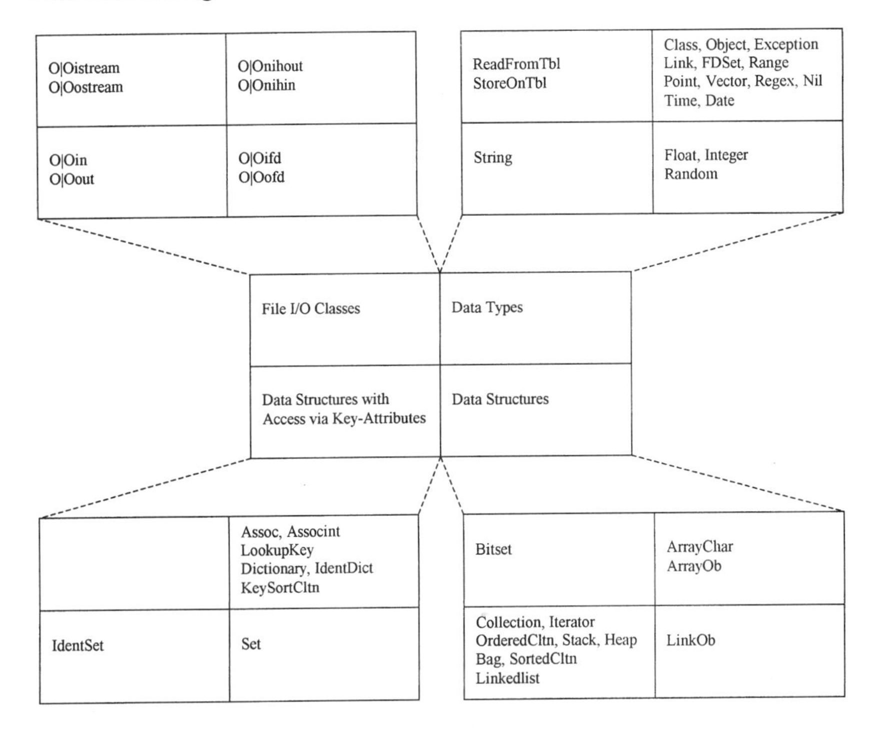

Fig. 4 *NIHCL* presented in a hierarchical feature map.

number of independent self-organizing maps, the inherent hierarchical organization of the text archive is mirrored in almost perfect fashion. In the top-layer map, shown in the center of Fig. 4, the four main groups of classes are clearly separated. Each group, in its turn, is further detailed within its own self-organizing map in the second layer of the hierarchy.

More precisely, we find the classes performing file I/O in the left upper map of the second layer depicted in Fig. 4. Within this map, the various classes are arranged in pairs according to their respective functionality. In the right upper map we recognize the classes representing data types. The large number of entries in the right upper unit of this particular map is because the library's general base classes (i.e., **Object**, **Class**, and **Exception**) are also considered as "data types." These classes, however, are clearly separated in the third layer. We refrain from its graphic representation here for space considerations. In the lower left map of the second layer we find mostly the data structures that allow access through a key attribute. As with the self-organizing map, they are represented together with the classes that actually perform the key–attribute-based access. Finally, the remaining data structures are located in the lower right map of the second layer. Here again, we find a unit that represents a large number of classes. In this case, the base class for all container classes (i.e., **Collection**) and the class providing the uniform means of access for all container classes (i.e., **Iterator**) are mapped onto this unit as well. Again, these classes are clearly separated in the next layer of the hierarchy.

D. A Comparison of Both Approaches

In the previous subsections we have described the results of both neural network architectures without favoring either one. We feel, however, that the results from the hierarchical feature map are preferable for a number of reasons. First, imagine Fig. 3 without the lines indicating the cluster boundaries as shown in Fig. 5. Such a representation may be considered as the standard form of visualization with self-organizing maps. In such a representation, a number of erroneous perceptions concerning the mutual similarity of various documents in the collection might occur because it is often difficult to demarcate region boundaries of similar documents. Consider, for instance, the borderline between the file I/O classes and the data types. A casual user might conclude that the classes **OIOofd** and **String** are as similar as the classes **String** and **Point** are because both pairs are mapped onto equidistant units in the upper middle of the map. The similarity of the latter pair is obvious in that both are data types. The former pair, though, consists of rather unrelated classes, the one being a data type, the other a file I/O class. To draw the borderline between clusters some insight into the underlying document archive is definitely necessary.

Conversely, with the hierarchical feature map this separation into distinct clusters comes for free because of the architecture of the neural network. A hierarchical organization of text collections is especially appealing because it emerges naturally. The inherent hierarchical organization of text collections arises from their decomposition into various topics that are dealt with within the documents. This is, obviously, the underlying organizational principle used by librarians to structure the contents of a library. In this sense, we rely on the user's experience in coping with hierarchically arranged data spaces.

Finally, a strong argument in favor of the hierarchical feature map emerges from the time needed to train the network. With the data set and the network dimensions as described in the foregoing, the time needed to finish the training process is about 1 h for the self-organizing map and about 10 min for the hierarchical feature map [43]. This timing was done on an otherwise idle SPARC-20 workstation. These characteristics scale-up quite nicely, as indicated with a larger document collection [44] both in terms of the number of documents and in terms of the dimension of the feature space used to describe the various texts.

Integer	.	Point	.	String	.	OIOofd	.	OIOout	OIOin
Float	.	.	Range	.	.	OIOifd	.	OIOostream	OIOistream
Random	.	Date	.	Vector	.	.	OIOnihout	OIOnihin	.
.	Time	.	Class	.	.	ReadFromTbl	StoreOnTbl	.	.
Nil	.	FDSet	.	.	Exception	.	.	ArrayOb	.
.	Collection	.	.	Object	.	.	Arraychar	.	Bitset
Iterator	.	Link	LinkOb	.	.	Regex	.	.	.
SeqCltn	.	LinkedList	.	.	SortedCltn	.	Bag	.	Heap
.	.	.	.	LookupKey	.	Dictionary	.	IdentSet	.
OrderedCltn	Stack	.	AssocInt	Assoc	KeySortCltn	.	IdentDict	.	Set

Fig. 5 *NIHCL* without manually provided cluster boundaries.

5. CONCLUSION

In this chapter we have described how unsupervised neural networks can be used to reveal the relative similarity of documents contained in a text archive. More precisely, we contrasted the results obtained from self-organizing maps with those from hierarchical feature maps. In a nutshell, both models proved to be successful in organizing the various documents according to their relative similarities. As far as an explicit and intuitive visualization is concerned, however, we noted a specific benefit of hierarchical feature maps. This benefit is marked by the inherent hierarchical organization of the underlying input data space. Such a hierarchical organization seems to be well suited to an application area, such as text archive exploration, in which a corresponding organization according to the various topics that the documents deal with may be found. Librarians have made use of exactly this property of text archives for centuries to organize the contents of libraries. Hierarchical feature maps were, moreover, found to require much shorter training times than self-organizing maps. Both of the neural network models described in this chapter assist the user in uncovering the inherent structure of a text archive. They thus represent useful tools for an interactive exploration of document collections.

ACKNOWLEDGMENTS

Part of this work was done while the author was visiting the Department of Computer Science at the Royal Melbourne Institute of Technology.

REFERENCES

1. G Salton. Automatic Text Processing: The Transformation, Analysis, and Retrieval of Information by Computer. Addison–Wesley, Reading, MA, 1989.
2. P Willet. Recent trends in hierarchic document clustering: a critical review. Inform Process Manage 24, 1988.
3. T Honkela, S Kaski, K Lagus, T Kohonen. Newsgroup exploration with WEBSOM method and browsing interface. Helsinki University of Technology, Laboratory of Computer and Information Science, Technical Report A32, Espoo, Finland, 1996.
4. T Honkela, V Pulkki, T Kohonen. Contextual relations of words in Grimm tales analyzed by self-organizing maps. Proceedings International Conference on Artificial Neural Networks (ICANN'95), Paris, France, 1995.
5. M Köhle, D Merkl. Visualizing similarities in high dimensional input spaces with a growing and splitting neural network. Proceedings International Conference on Artificial Neural Networks (ICANN'96), Bochum, Germany, 1996.
6. K Lagus, T Honkela, S Kaski, T Kohonen. Self-organizing maps of document collections: a new approach to interactive exploration. Proceedings International Conference on Knowledge Discovery and Data Mining (KDD-96), Portland, OR, 1996.
7. X Lin, D Soergel, G Marchionini. A self-organizing semantic map for information retrieval. Proceedings ACM SIGIR International Conference on Research and Development in Information Retrieval (SIGIR'91), Chicago, IL, 1991.
8. D Merkl. Structuring software for reuse: the case of self-organizing maps. Proceedings International Joint Conference on Neural Networks (IJCNN'93), Nagoya, Japan, 1993.
9. D Merkl. A connectionist view on document classification. Proceedings Australasian Database Conference (ADC'95), Adelaide, Australia, 1995.

10. D Merkl. Content-based software classification by self-organization. Proceedings IEEE International Conference on Neural Networks (ICNN'95), Perth, Australia, 1995.
11. T Kohonen. Self-organized formation of topologically correct feature maps. Biol Cybern 43, 1982.
12. R Miikkulainen. Script recognition with hierarchical feature maps. Connect Sci 2, 1990.
13. UM Fayyad, G Piatetsky–Shapiro, P Smyth. From data mining to knowledge discovery: an overview. In: Advances in Knowledge Discovery and Data Mining. UM Fayyad, G Piatetsky–Shapiro, P Smyth, R Uthurusamy, eds., AAAI Press, Menlo Park, CA, 1996.
14. UM Fayyad, G Piatetsky–Shapiro, P Smyth. The KDD process for extracting useful knowledge from volumes of data. Commun ACM 39(11), 1996.
15. H Mannila. Data mining: machine learning, statistics, and databases. Proceedings International Conference on Scientific and Statistical Database Management, Stockholm, Sweden, 1996.
16. H Mannila. Methods and problems in data mining. Proceedings International Conference on Database Theory, Delphi, Greece, 1997.
17. G Salton, C Buckley. Term weighting approaches in automatic text retrieval. Inform Process Manage 24(5), 1988.
18. G Salton, MJ McGill. Introduction to Modern Information Retrieval. McGraw–Hill, New York, 1983.
19. HR Turtle, WB Croft. A comparison of text retrieval models. Comput J 35(3), 1992.
20. TF Cox, MAA Cox. Multidimensional Scaling. Chapman & Hall, London, 1994.
21. IT Jolliffe. Principal Component Analysis. Springer-Verlag, Berlin, 1986.
22. S Deerwester, ST Dumais, GW Furnas, TK Landauer, R Hashman. Indexing by latent semantic analysis. J Am Soc Inform Sci 41(6), 1990.
23. BT Bartell, GW Cottrell, RK Belew. Latent semantic indexing is an optimal special case of multidimensional scaling. Proceedings ACM SIGIR International Conference on Research and Development in Information Retrieval (SIGIR'92), Copenhagen, Denmark, 1992.
24. T Bayer, I Renz, M Stein, U Kressel. Domain and language independent feature extraction for statistical text categorization. Proceedings Workshop on Language Engineering for Document Analysis and Recognition, Sussex, UK, 1996.
25. D Merkl. Content-based document classification with highly compressed input data. Proceedings 5th International Conference on Artificial Neural Networks (ICANN'95), Paris, France, 1995.
26. H Ritter, T Kohonen. Self-organizing semantic maps. Biol Cybern 61, 1989.
27. DE Rumelhart, D Zipser. Feature discovery by competitive learning. In: Parallel Distributed Processing: Explorations in the Microstructure of Cognition, vol I: Foundations, (DE Rumelhart, JL McClelland, the PDP Research Group, eds.), MIT Press, Cambridge, MA, 1986.
28. T Kohonen. Self-Organizing Maps. Springer-Verlag, Berlin, 1995.
29. BD Ripley. Pattern Recognition and Neural Networks. Cambridge University Press, Cambridge, UK, 1996.
30. M Cottrell, E de Bodt. A Kohonen map representation to avoid misleading interpretations. Proceedings European Symposium on Artificial Neural Networks (ESANN'96), Brugge, Belgium, 1996.
31. D Merkl, A Rauber. On the similarity of eagles, hawks, and cows: visualization of similarity in self-organizing maps. Proceedings International Workshop Fuzzy-Neuro-Systems '97, Soest, Germany, 1997.
32. A Ultsch. Self-organizing neural networks for visualization and classification. In: Information and Classification: Concepts, Methods, and Applications. O Opitz, B Lausen, R Klar, eds. Springer-Verlag, Berlin, 1993.

33. J Blackmore, R Miikkulainen. Incremental grid growing: encoding high-dimensional structure into a two-dimensional feature map. Proceedings IEEE International Conference on Neural Networks, San Francisco, CA, 1993.
34. J Blackmore, R Miikkulainen. Visualizing high-dimensional structure with the incremental grid growing network. Proceedings International Conference on Machine Learning, Lake Tahoe, NV, 1995.
35. B Fritzke. Kohonen feature maps and growing cell structures: a performance comparison. In: Advances in Neural Information Processing Systems 5. CL Gibs, SJ Hanson, JD Cowan, eds., Morgan Kaufmann, San Mateo, CA, 1993.
36. B Fritzke. Growing cell structures: a self-organizing network for unsupervised and supervised learning. Neural Networks 7(9), 1994.
37. W Wan, D Fraser. Multiple Kohonen self-organizing maps: supervised and unsupervised formation with application to remotely sensed image analysis. Proceedings Australian Conference on Neural Networks, Brisbane, Australia, 1994.
38. R Miikkulainen. Trace feature map: a model of episodic associative memory. Biol Cybern 66, 1992.
39. R Miikkulainen. Subsymbolic Natural Language Processing: an Integrated Model of Scripts, Lexicon, and Memory. MIT Press, Cambridge, MA, 1993.
40. R Miikkulainen. Script-based inference and memory retrieval in subsymbolic story processing. Appl Intell 5, 1995.
41. KE Gorlen. NIH Class Library Reference Manual. National Institutes of Health, Bethesda, MD, 1990.
42. KE Gorlen, S Orlow, P Plexico. Abstraction and Object-Oriented Programming in C++. John Wiley & Sons, New York, 1990.
43. D Merkl. Exploration of text collections with hierarchical feature maps. Proceedings ACM SIGIR International Conference on Research and Development in Information Retrieval (SIGIR'97), Philadelphia, PA, 1997.
44. D Merkl, E Schweighofer. The exploration of legal text corpora with hierarchical neural networks: a guided tour in public international law. Proceedings International Conference on Artificial Intelligence and Law, Melbourne, Australia, 1997.

37

Text and Discourse Understanding: The DISCERN System

RISTO MIIKKULAINEN

The University of Texas at Austin, Austin, Texas

1. INTRODUCTION

The subsymbolic approach to natural language processing (NLP) captures a number of intriguing properties of human-like information processing, such as learning from examples, context sensitivity, generalization, robustness of behavior, and intuitive reasoning. Within this new paradigm, the central issues are quite different from (even incompatible with) the traditional issues in symbolic NLP, and the research has proceeded without much in common with the past. However, the ultimate goal is still the same: to understand how humans process language. Even if NLP is being built on a new foundation, as can be argued, many of the results obtained through symbolic research are still valid, and they could be used as a guide for developing subsymbolic models of natural language processing.

This is where DISCERN (DIstributed SCript processing and Episodic memoRy Network) [1], a subsymbolic neural network model of script-based story understanding, fits in. DISCERN is purely a subsymbolic model, but at the high level it consists of modules and information structures similar to those of symbolic systems, such as scripts, lexicon, and episodic memory. At the highest level of natural language processing such as text and discourse understanding, the symbolic and subsymbolic paradigms have to address the same basic issues. Outlining a subsymbolic approach to those issues is the purpose of DISCERN.

In more specific terms, DISCERN aims (a) to demonstrate that distributed artificial neural networks can be used to build a large-scale natural language processing system that performs approximately at the level of symbolic models; (b) to show that several cognitive phenomena can be explained at the subsymbolic level using the special properties of these networks; and (c) to identify central issues in subsymbolic NLP and to develop well-motivated techniques to deal with them. To the extent that DISCERN is successful in these areas, it constitutes a first step toward building text and discourse understanding systems within the subsymbolic paradigm.

2. THE SCRIPT-PROCESSING TASK

Scripts [2–4] are schemas of often-encountered, stereotypic event sequences, such as visiting a restaurant, traveling by airplane, and shopping at a supermarket. Each script divides further into tracks, or established minor variations. A script can be represented as a causal chain of events with a number of open roles. Script-based understanding means reading a script-based story, identifying the proper script and track, and filling its roles with the constituents of the story. Events and role fillers that were not mentioned in the story, but are part of the script can then be inferred. Understanding is demonstrated by generating an expanded paraphrase of the original story, and by answering questions about the story.

To see what is involved in the task, let us consider an example of DISCERN input–output behavior. The following input stories are examples of the fancy-restaurant, plane-travel, and electronics-shopping tracks:

1. John went to MaMaison. John asked the waiter for lobster. John left a big tip.
2. John went to LAX. John checked in for a flight to JFK. The plane landed at JFK.
3. John went to Radio-Shack. John asked the staff questions about CD-players. John chose the best CD-player.

DISCERN reads the orthographic word symbols sequentially, one at a time. An internal representation of each story is formed, where all inferences are made explicit. These representations are stored in the episodic memory. The system then answers questions about the stories:

Q: What did John buy at Radio-Shack?
A: John bought a CD-player at Radio-Shack.
Q: Where did John take a plane to?
A: John took a plane to JFK.
Q: How did John like the lobster at MaMaison?
A: John thought the lobster was good at MaMaison.

With the question as a cue, the appropriate story representation is retrieved from the episodic memory and the answer is generated word by word. DISCERN also generates full paraphrases of the input stories. For example, it generates an expanded version of the restaurant story:

John went to MaMaison. The waiter seated John. John asked the waiter for lobster. John ate a good lobster. John paid the waiter. John left a big tip. John left MaMaison.

The answers and the paraphrase show that DISCERN has made a number of inferences beyond the original story. For example, it inferred that John ate the lobster and the lobster tasted good. The inferences are not based on specific rules, but are statistical and learned from experience. DISCERN has read a number of similar stories in the past and the unmentioned events and role bindings have occurred in most cases. They are assumed immediately and automatically on reading the story and have become part of the memory of the story. In a similar fashion, human readers often confuse what was mentioned in the story with what was only inferred [5–7].

Several issues can be identified from the foregoing examples. Specifically, DISCERN has to (a) make statistical, script-based inferences and account for learning them from experience; (b) store items in the episodic memory in a single presentation and retrieve them with a partial cue; (c) develop a meaningful organization for the episodic memory, based on the stories it reads; (d) represent meanings of words, sentences, and stories internally; and (e) organize a lexicon of symbol and concept representations based on examples of how words are used in the language and form a many-to-many mapping between them. Script processing constitutes a good framework for studying these issues, and a good domain for developing an approach toward general text and discourse understanding.

3. APPROACH

Subsymbolic models typically have very little internal structure. They produce the statistically most likely answer given the input conditions in a process that is opaque to the external observer. This is well suited to the modeling of isolated low-level tasks, such as learning past tense forms of verbs or word pronunciation [8,9]. Given the success of such models, a possible approach to higher-level cognitive modeling would be to construct the system from several submodules that work together to produce the higher-level behavior.

In DISCERN, the immediate goal is to build a complete, integrated system that performs well in the script-processing task. In this sense, DISCERN is very similar to traditional models in artificial intelligence. However, DISCERN also aims to show how certain parts of human cognition could actually be built. The components of DISCERN were designed as independent cognitive models that can account for interesting language-processing and memory phenomena, many of which are not even required in the DISCERN task. Combining these models into a single, working system is one way of validating them. In DISCERN, the components are not just models of isolated cognitive phenomena; they are sufficient constituents for generating complex high-level behavior.

4. THE DISCERN SYSTEM

DISCERN can be divided into parsing, generating, question answering, and memory subsystems, each with two modules (Fig. 1). Each module is trained in its task separately and in parallel. During performance, the modules form a network of networks, each feeding its output to the input of another module.

The sentence parser reads the input words one at a time and forms a representation of each sentence. The story parser combines the sequence of sentences into an internal representation of the story, which is then stored in the episodic memory. The story generator receives the internal representation and generates the sentences of the paraphrase one at a time. The sentence generator outputs the sequence of words for each sentence. The cue former receives a question representation, built by

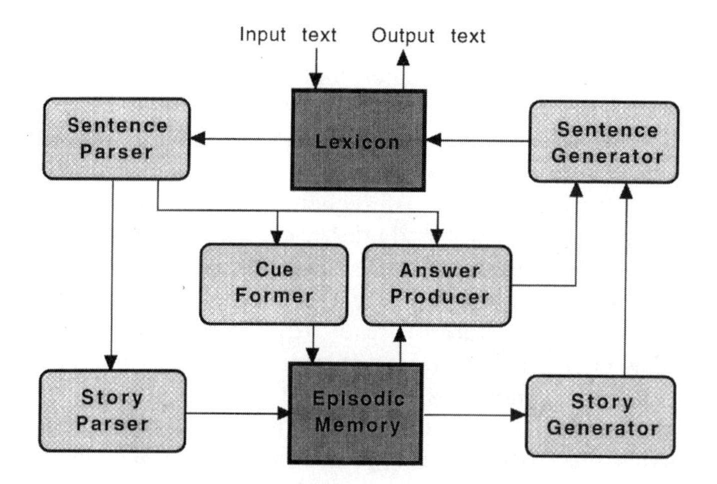

Fig. 1 The DISCERN architecture: The system consists of parsing, generating, question answering, and memory subsystems, two modules each. A dark square indicates a memory module, a light square stands for a processing module. The lines indicate pathways carrying distributed word, sentence, and story representations during the performance phase of the system. The modules are trained separately with compatible I/O data. (From Ref. 1.)

the sentence parser, and forms a cue pattern for the episodic memory, which returns the appropriate story representation. The answer producer receives the question and the story and generates an answer representation, which is output word by word by the sentence generator. The architecture and behavior of each of these modules in isolation is outlined in the following.

5. LEXICON

The input and output of DISCERN consist of distributed representations for orthographic word symbols (also called lexical words). Internally, DISCERN processes semantic concept representations (semantic words). Both the lexical and semantic words are represented distributively as vectors of gray-scale values between 0.0 and 1.0. The lexical representations are based on the visual patterns of characters that make up the written word; they remain fixed throughout the training and performance of DISCERN. The semantic representations stand for distinct meanings and are developed automatically by the system while it is learning the processing task.

The lexicon stores the lexical and semantic representations and translates between them (Fig. 2; [10]). It is implemented as two feature maps [11,12], one lexical and the other semantic. Words for which lexical forms are similar, such as **LINE** and **LIKE**, are represented by nearby units in the lexical map. In the semantic map, words with similar semantic content, such as **John** and **Mary** or **Leone's** and **MaMaison** are mapped near each other. There is a dense set of associative interconnections between the two maps. A localized activity pattern representing a word in one map will cause a localized activity pattern to form in the other map, representing the same word. The output representation is then obtained from the weight vector of the most highly active unit. The lexicon thus transforms a lexical input

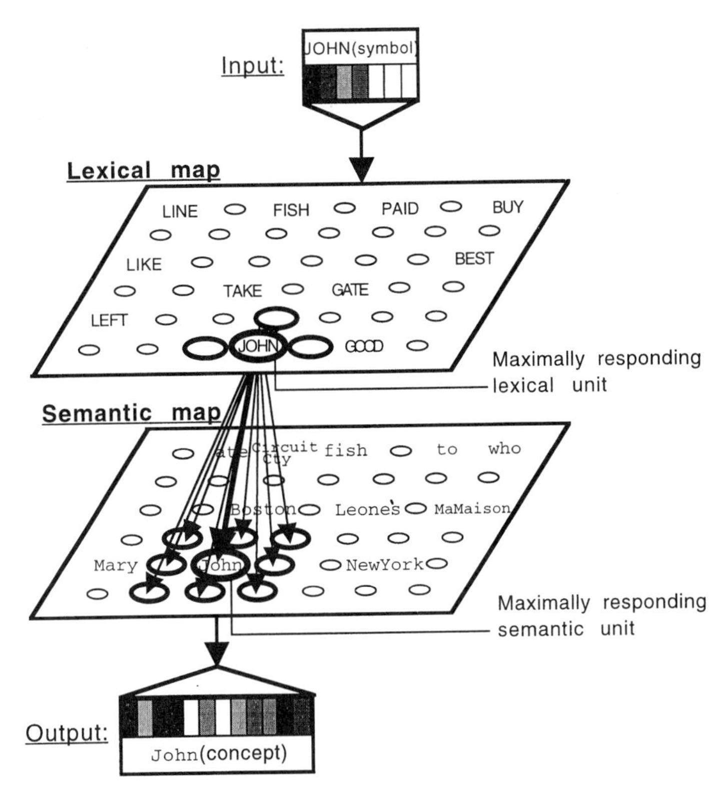

Fig. 2 The lexicon: The lexical input symbol **JOHN** is translated into the semantic representation of the concept **John**. The representations are vectors of gray-scale values between 0.0 and 1.0, stored in the weights of the units. The size of the unit on the map indicates how strongly it responds. Only a small part of each map, and only a few strongest associative connections of the lexical unit **JOHN** are shown in this figure.

vector into a semantic output vector and vice versa. Both maps and the associative connections between them are organized simultaneously, based on examples of co-occurring symbols and meanings.

The lexicon architecture facilitates interesting behavior. Localized damage to the semantic map results in category-specific lexical deficits similar to human aphasia [13,14]. For example, the system selectively loses access to restaurant names, or animate words, when that part of the map is damaged. Dyslexic performance errors can also be modeled. If the performance is degraded, for example, by adding noise to the connections, parsing and generation errors that occur are quite similar to those observed in human deep dyslexia [15]. For example, the system may confuse **Leone's** with **MaMaison**, or **LINE** with **LIKE**, because they are nearby in the map and share similar associative connections.

6. FGREP PROCESSING MODULES

Processing in DISCERN is carried out by hierarchically organized pattern-transformation networks. Each module performs a specific subtask, such as parsing a sentence or generating an answer to a question. All these networks have the same basic architecture: they are three-layer, simple-recurrent backpropagation networks [16], with the extension called FGREP that allows them to develop distributed representations for their input–output words.

The FGREP mechanism (Forming Global Representations with Extended back Propagation) [17] is based on a basic three-layer backward error propagation network, with the I/O representation patterns stored in an external lexicon (Fig. 3). The input and output layers of the network are divided into assemblies (i.e., groups of units). The assemblies stand for slots in the I/O representation, such as the different case roles of the sentence representation, or the role bindings in the story representation. Each input pattern is formed by concatenating the current semantic lexicon entries of the input words; likewise, the corresponding target pattern is formed by concatenating the lexicon entries of the target words. For example, the target sentence **John went to MaMaison** would be represented at the output of the sentence parser network as a single vector, formed by concatenating the representations for **John**, **went**, and **MaMaison** (see Fig. 7a).

Three types of FGREP modules are used in the system: nonrecurrent (the cue former and the answer producer), sequential input (the parsers), and sequential output modules (the generators). In the recurrent modules the previous hidden layer serves as sequence memory, remembering where in the sequence the system currently is and what has occurred before (see Fig. 3). In a sequential input network, the input changes at each time step, while the target pattern stays the same. The network learns to form a stationary representation of the sequence. In a sequential output network, the input is stationary, but the teaching pattern changes at each step. The network learns to produce a sequential interpretation of its input.

The network learns the processing task by adapting the connection weights according to the standard online backpropagation procedure [18]. The error signal is propagated to the input layer, and the current input representations are modified as

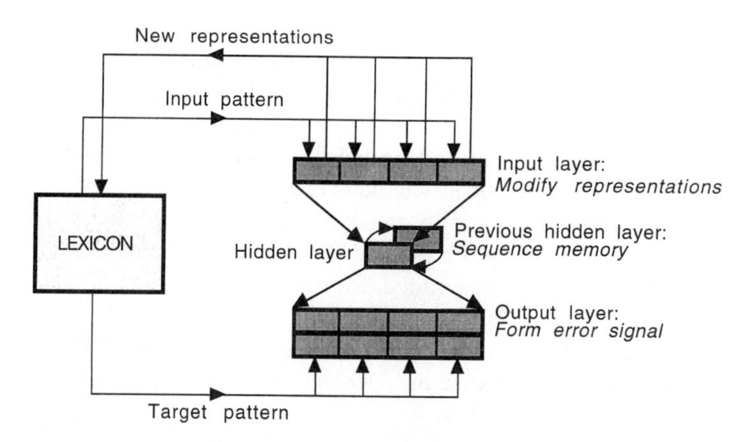

Fig. 3 The FGREP-module: At each I/O presentation, the representations at the input layer are modified according to the backpropagation error signal, and replace the old representations in the lexicon. For sequential input or output, the hidden layer pattern is saved after each step in the sequence, and used as input to the hidden layer during the next step, together with the actual input. (From Ref. 1).

if they were an extra layer of weights. The modified representation vectors are put back in the lexicon, replacing the old representations. Next time the same words occur in the input or output, their new representations are used to form the input–output patterns for the network. In FGREP, therefore, the required mappings change as the representations evolve, and backpropagation is shooting at a moving target.

The representations that result from this process have a number of useful properties for natural language processing: (a) Because they adapt to the error signal, they end up coding information most crucial to the task. Representations for words that are used in similar ways in the examples become similar. Thus, these profiles of continuous activity values can be claimed to code the meanings of the words as well. (b) As a result, the system never has to process very novel input patterns, because generalization has already been done in the representations. (c) The representation of a word is determined by all the contexts in which that word has been encountered; consequently, it is also a representation of all those contexts. Expectations emerge automatically and cumulatively from the input word representations. (d) Individual representation components do not usually stand for identifiable semantic features. Instead, the representation is holographic: word categories can often be recovered from the values of single components. (e) Holography makes the system very robust against noise and damage. Performance degrades approximately linearly as representation components become defective or inaccurate.

After a core set of semantic representations have been developed in the FGREP process, it is possible to extend the vocabulary through a technique called cloning. Each FGREP representation stands for a unique meaning and constitutes a semantic prototype. In cloning, several distinct copies, or instances, are created from the same prototype. For example, instances such as **John**, **Mary**, and **Bill** can be created from the prototype **human**. Such cloned representations consist of two parts: the content part, which was developed in the FGREP process and encodes the meaning of the word, and the ID part, which is unique for each instance of the same prototype. They all share the share meaning and, therefore, the system knows how to process them, but at the same time, the ID part allows the system to keep them distinct. The ID + content technique can be applied to any word in the training data, and in principle, the number of instances per word is unlimited. This allows us to approximate a large vocabulary with only a small number of semantically different representations (which are expensive to develop) at our disposal.

7. EPISODIC MEMORY

The episodic memory in DISCERN [19,20] consists of a hierarchical pyramid of feature maps organized according to the taxonomy of script-based stories (Fig. 4). The highest level of the hierarchy is a single feature map that lays out the different script classes. Beneath each unit of this map there is another feature map that lays out the tracks within the particular script. The different role bindings within each track are separated at the bottom level. The map hierarchy receives a story representation vector as its input and classifies it as an instance of a particular script, track, and role binding. The hierarchy thereby provides a unique memory representation for each script-based story as the maximally responding units in the feature maps at the three levels.

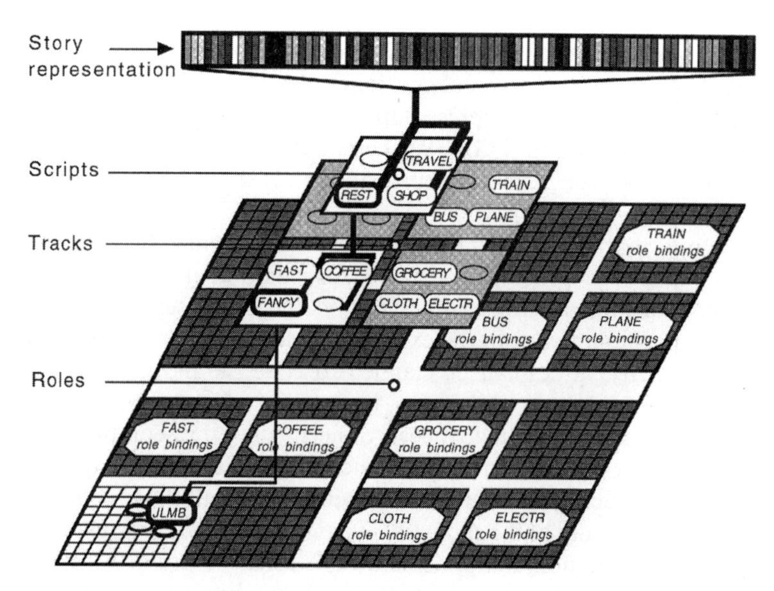

Fig. 4 The hierarchical feature map classification of script-based stories: Labels indicate the maximally responding unit for the different scripts and tracks. This particular input story representation is classified as an instance of the restaurant script (top level) and fancy-restaurant track (middle level), with role bindings customer = **John**; food = **lobster**; restaurant = **MaMaison**; tip = **big** (i.e., unit JLMB, bottom level).

Whereas the top and the middle level in the hierarchy serve only as classifiers, selecting the appropriate track and role-binding map for each input, at the bottom level a permanent trace of the story must be created. The role-binding maps are trace feature maps, with modifiable lateral connections (Fig. 5). When the story representation vector is presented to a role-binding map, a localized activity pattern forms as a response. Each lateral connection to a unit with higher activity is made excitatory, whereas a connection to a unit with lower activity is made inhibitory. The units within the response now "point" toward the unit with highest activity, permanently encoding that the story was mapped at that location.

A story is retrieved from the episodic memory by giving it a partial story representation as a cue. Unless the cue is highly deficient, the map hierarchy is able to recognize it as an instance of the correct script and track and form a partial cue for the role-binding map. The trace feature map mechanism then completes the role binding. The initial response of the map is again a localized activity pattern; because the map is topological, it is likely to be located somewhere near the stored trace. If the cue is close enough, the lateral connections pull the activity to the center of the stored trace. The complete story representation is retrieved from the weight vectors of the maximally responding units at the script, track, and role-binding levels.

Hierarchical feature maps have a number of properties that make them useful for memory organization: (a) The organization is formed in an unsupervised manner, extracting it from the input experience of the system. (b) The resulting order reflects the properties of the data, the hierarchy corresponding to the levels of variation, and the maps laying out the similarities at each level. (c) By dividing the data

MLMS						JLMB	MLMB
		MLLS					
JLLS						JSMB	
JLMS							MSMB
MSMS						MSLB	JSLB
	JSMS						
	MSLS	JSLS				MLLB	

(a)

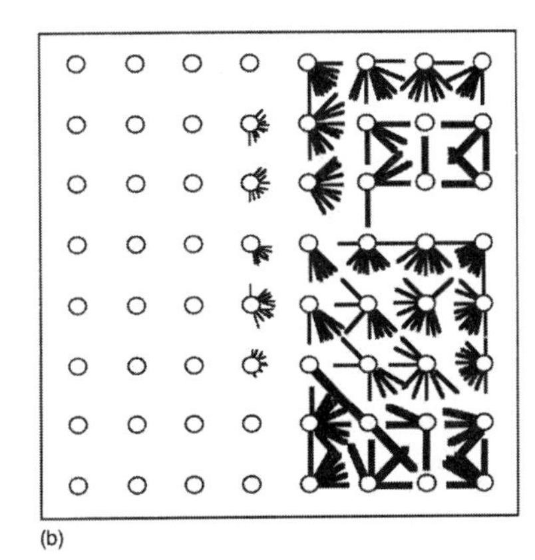

(b)

Fig. 5 The trace feature map for fancy-restaurant stories: (a) The organization of the map. Each story in the test set was mapped on one unit in the map, labeled by the role bindings (where **J** = **John**, **M** = **Mary**, **L** = **lobster**, **S** = **steak**, **M** = **Mamaison**, **L** = **Leone's**, **B** = **big**, and **S** = **small**, for size of tip). The activation of the units (shown in gray-scale) indicates retrieval of the JLMB story. (b) Lateral connections after storing first JLMB and then MLLB. Line segments indicate excitatory lateral connections originating from each unit, length and width proportional to the magnitude of the weight. Inhibitory connections are not shown. The latter trace has partially obscured the earlier one. (From Ref. 1.)

first into major categories and gradually making finer distinctions lower in the hierarchy, the most salient components of the input data are singled out and more resources are allocated for representing them accurately. (d) Because the representation is based on salient differences in the data, the classification is very robust, and usually correct even if the input is noisy or incomplete. (e) Because the memory is based on classifying the similarities and storing the differences, retrieval becomes a reconstructive process [21,22] similar to human memory.

The trace feature map exhibits interesting memory effects that result from interactions between traces. Later traces capture units from earlier ones, making later traces more likely to be retrieved (see Fig. 5). The extent of the traces determines memory capacity. The smaller the traces, the more of them will fit in the map, but more accurate cues are required to retrieve them. If the memory capacity is exceeded, older traces will be selectively replaced by newer ones. Traces that are unique, that is, located in a sparse area of the map, are not affected, no matter how old they are. Similar effects are common in human long-term memory [23,24].

8. DISCERN HIGH-LEVEL BEHAVIOR

DISCERN is more than just a collection of individual cognitive models. Interesting behavior results from the interaction of the components in a complete story-processing system. Let us follow DISCERN as it processes the story about John's visit to MaMaison (Fig. 7). The lexical representations for each word are presented to the lexical map of the lexicon, which produces the corresponding semantic representation as its output (see Fig. 2). These are fed one at a time to the sentence parser, which gradually forms a stationary case-role representation of each sentence at its

```
[ parsing input story: ]                                     (comment by DISCERN)
JOHN WENT TO MAMAISON .                                      (sequence of input words)
|John went _ _ _ MaMaison|                                   (sentence parser output)
JOHN ASKED THE WAITER FOR LOBSTER .
|John asked waiter _ lobster _|
JOHN LEFT A BIG TIP .
|John left waiter big tip _|                                 (figure 7a)

[ into internal representation: ]
|$restaurant $fancy John lobster MaMaison good big|          (story parser output; figure 7b)
[ storing into episodic memory: ]
[ image units (0,0), (0,0), (6,6): ]                         (location in the memory; figure 4)
|$restaurant $fancy John lobster MaMaison good big|          (episodic memory representation)
[ generating paraphrase: ]
|John went _ _ _ MaMaison|                                   (story generator output)
JOHN WENT TO MAMAISON .                                      (sequence of output words)
|waiter seated _ _ John _|
THE WAITER SEATED JOHN .
|John asked waiter _ lobster _|
JOHN ASKED THE WAITER FOR LOBSTER .
|John ate _ good lobster _|
JOHN ATE A GOOD LOBSTER .
|John paid waiter _ _ _|
JOHN PAID THE WAITER .
|John left waiter big tip _|
JOHN LEFT A BIG TIP .
|John left _ _ _ MaMaison|
JOHN LEFT MAMAISON .

[ parsing question: ]
WHAT DID JOHN EAT AT MAMAISON ?
|John ate _ _ what MaMaison|                                 (sentence parser output; figure 7c)

[ into cue: ]
|$restaurant $fancy John (?) MaMaison good (?)|              (cue former output; figure 7d)
[ retrieving from episodic memory: ]
[ image units (0,0), (0,0), (6,6): ]                         (location in the memory; figures 4 and 5a)
|$restaurant $fancy John lobster MaMaison good big|          (episodic memory output; figure 7b)
[ generating answer: ]
|John ate _ good lobster _|                                  (answer producer output; figure 7e)
JOHN ATE A GOOD LOBSTER .
```

Fig. 6 A listing of DISCERN output as it processes the example story: Comments printed out by DISCERN are enclosed in brackets; otherwise all words stand for activity patterns, with "_" indicating the blank pattern. Output layers consisting of several assemblies are enclosed between bars.

output layer (Fig. 7a). After a period is input, ending the sentence, the final case-role pattern is fed to the input of the story parser.

In a similar manner, the story parser receives a sequence of sentence case-role representations as its input, and forms a stationary slot-filler representation of the whole story at its output layer (see Fig. 7b). This is a representation of the story in terms of its role bindings, and constitutes the final result of the parse. The story representation is fed to the episodic memory, which classifies it as an instance of a particular script, track, and role binding, and creates a trace in the appropriate role-binding map (see Figs. 4 and 5b).

The generator subsystem reverses the parsing process. The story generator network receives the story representation as its input and generates a sequence of sentence case-role representations. Each of these is fed to the sentence generator, which outputs the semantic representations of the output words one at a time. Finally, the lexicon transforms these into lexical words.

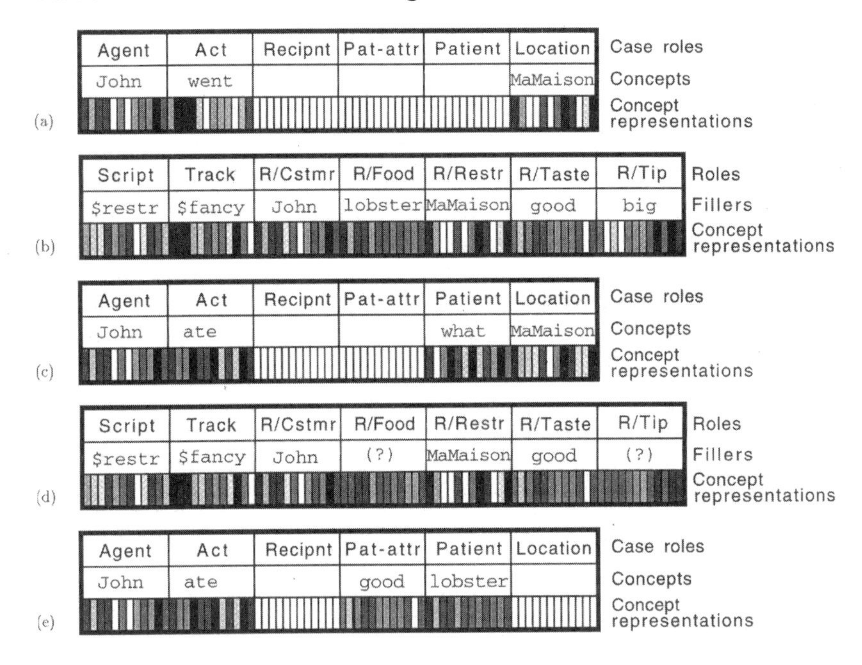

Fig. 7 Sentence and story representations: (a) Case-role representation of the sentence **John went to MaMaison**. The concept representations in each case-role correspond to the concept representations in the lexicon. (b) Representation of the story by its role bindings. The assemblies are data-specific: their interpretation depends on the pattern in the script slot. The role names R/... are specific for the restaurant script. (c) Case-role representation of the question **What did John eat at MaMaison?** questions are represented as sentences, but processed through a different pathway (see Fig. 1). (d) Memory cue. Most of the story representation is complete, but the patterns in Food and Tip slots indicate averages of all possible alternatives. (e) Case-role representation of the answer **John ate a good lobster**. (From Ref. 1.)

The sentence parser and the sentence generator are also trained to process question sentences and answer sentences. The cue former receives the case-role representation of the question (see Fig. 7c), produced by the sentence parser, and generates an approximate story representation as its output (see Fig. 7d). This pattern is fed to the episodic memory, which classifies it as an instance of a script, track, and role binding. The trace feature map settles to a previously stored memory trace (see Fig. 5a), and the complete story representation (see Fig. 7b) is retrieved from the weights of the maximally responding units at the three levels of the hierarchy (see Fig. 4). The answer producer receives the complete story representation, together with the case-role representation of the question, and generates a case-role representation of the answer sentence (see Fig. 7e), which is then output word by word by the sentence generator.

DISCERN was trained and tested with an artificially generated corpus of script-based stories consisting of three scripts (restaurant, shopping, and travel), with three tracks and three open roles each.

The complete DISCERN system performs very well: at the output, 98% of the words are correct. This is rather remarkable for a chain of networks that is nine modules long and consists of several different types of modules.

A modular neural network system can operate only if it is stable, that is, if small deviations from the normal flow of information are automatically corrected. It turns out that DISCERN has several built-in safeguards against minor inaccuracies and noise. The semantic representations are distributed and redundant, and inaccuracies in the output of one module are cleaned up by the module that uses the output The memory modules clean up by categorical processing: a noisy input is recognized as a representative of an established class and replaced by the correct representation of that class. As a result, small deviations do not throw the system off course, but rather, the system filters out the errors and returns to the normal course of processing, which is an essential requirement for building robust natural language processing models.

DISCERN also demonstrates strong script-based inferencing [25]. Even when the input story is incomplete, consisting of only a few main events, DISCERN can usually form an accurate internal representation of it. DISCERN was trained to form complete story representations from the first sentence on, and because the stories are stereotypical, missing sentences have little effect on the parsing process. Once the story representation has been formed, DISCERN performs as if the script had been fully instantiated. Questions about missing events and role-binding are answered as if they were part of the original story. If events occurred in an unusual order, they are recalled in the stereotypical order in the paraphrase. If there is not enough information to fill a role, the most likely filler is selected and maintained throughout the paraphrase generation. Such behavior automatically results from the modular architecture of DISCERN and is consistent with experimental observations on how people remember stories of familiar event sequences [5–7].

In general, given the information in the question, DISCERN recalls the story that best matches it in the memory. An interesting issue is: what happens when DISCERN is asked a question that is inaccurate or ambiguous, that is, one that does not uniquely specify a story? For example, DISCERN might have read a story about John eating lobster at MaMaison, and then about Mary doing the same at Leone's, and the question could be **Who ate lobster?** Because later traces are more prominent in the memory, DISCERN is more likely to retrieve the Mary-at-Leone's story in this case (see Fig. 5b). The earlier story is still in the memory, but to recall it, more details need to be specified in the question, such as **Who ate lobster at MaMaison?** Similarly, DISCERN can robustly retrieve a story even if the question is slightly inaccurate. When asked **How did John like the steak at MaMaison?**, DISCERN generates the answer **John thought lobster was good at MaMaison**, ignoring the inaccuracy in the question, because the cue is still close enough to the stored trace. DISCERN does recognize, though, when a question is too different from anything in the memory, and should not be answered. For **Who ate at Mcdonald's?** the cue vector is not close to any trace, the memory does not settle, and nothing is retrieved. Note that these mechanisms were not explicitly built into DISCERN, but they emerge automatically from the physical layout of the architecture and representations.

9. DISCUSSION

There is an important distinction between scripts (or more generally, schemas) in symbolic systems, and scripts in subsymbolic models such as DISCERN. In the symbolic approach, a script is stored in memory as a separate, exact knowledge structure, coded by the knowledge engineer. The script has to be instantiated by searching the schema memory sequentially for a structure that matches the input. After instantiation, the script is active in the memory and later inputs are interpreted primarily in terms of this script. Deviations are easy to recognize and can be taken care of with special mechanisms.

In the subsymbolic approach, schemas are based on statistical properties of the training examples, extracted automatically during training. The resulting knowledge structures do not have explicit representations. For example, a script exists in a neural network only as statistical correlations coded in the weights. Every input is automatically matched to every correlation in parallel. There is no all-or-none instantiation of a particular knowledge structure. The strongest, most probable correlations will dominate, depending on how well they match the input, but all of them are simultaneously active at all times. Regularities that make up scripts can be particularly well captured by such correlations, making script-based inference a good domain for the subsymbolic approach. Generalization and graceful degradation give rise to inferencing that is intuitive, immediate, and occurs without conscious control, as script-based inference in humans. On the other hand, it is very difficult to recognize deviations from the script and to initiate exception-processing when the automatic mechanisms fail. Such sequential reasoning would require intervention of a high-level "conscious" monitor, which has yet to be built in the connectionist framework.

10. CONCLUSION

The main conclusion from DISCERN is that building subsymbolic models is a feasible approach to understanding mechanisms underlying natural language processing. DISCERN shows how several cognitive phenomena may result from subsymbolic mechanisms. Learning word meanings, script processing, and episodic memory organization are based on self-organization and gradient-descent in error in this system. Script-based inferences, expectations, and defaults automatically result from generalization and graceful degradation. Several types of performance errors in role binding, episodic memory, and lexical access emerge from the physical organization of the system. Perhaps most significantly, DISCERN shows how individual connectionist models can be combined into a large, integrated system that demonstrates that these models are sufficient constituents for generating sequential, symbolic, high-level behavior.

Although processing simple script instantiations is a start, there is a long way to go before subsymbolic models will rival the best symbolic cognitive models. For example, in story understanding, symbolic systems have been developed that analyze realistic stories in depth, based on higher-level knowledge structures such as goals, plans, themes, affects, beliefs, argument structures, plots, and morals [e.g., 4,26–28]. In designing subsymbolic models that would do that, we are faced with two major challenges [1]: (a) how to implement connectionist control of high–level-processing strategies (making it possible to model processes more sophisticated than a series of reflex responses); and (b) how to represent and learn abstractions (making it possible to process information at a higher level than correlations in the raw input data).

Progress in these areas would constitute a major step toward extending the capabilities of subsymbolic natural language processing models beyond those of DISCERN.

ACKNOWLEDGMENTS

This research was supported in part by the Texas Higher Education Coordinating Board under grant ARP–444.

Note

Software for the DISCERN system is available in the World Wide Web at URL http://www.cs.utexas.edu/users/nn/pages/software/software.html. An interactive X11 graphics demo, showing DISCERN in processing example stories and questions, can be run remotely under the World Wide Web at: http://www.cs.utexas.edu/users/nn/pages/research/discern/discern.html.

REFERENCES

1. R Miikkulainen. Subsymbolic Natural Language Processing: An Integrated Model of Scripts, Lexicon, and Memory. MIT Press, Cambridge, MA, 1993.
2. RE Cullingford. Script application: computer understanding of newspaper stories. PhD dissertation, Yale University, New Haven, CT, 1978.
3. MG Dyer, RE Cullingford, S Alvarado. Scripts. In: Encyclopedia of Artificial Intelligence. SC Shapiro, ed. Wiley, New York, 1987, pp 980–994.
4. RC Schank, RP Abelson. Scripts, Plans, Goals, and Understanding: An Inquiry into Human Knowledge Structures. Erlbaum, Hillsdale, NJ, 1977.
5. GH Bower, JB Black, TJ Turner. Scripts in memory for text. Cogn Psychol 11:177–220, 1979.
6. AC Graesser, SE Gordon, JD Sawyer. Recognition memory for typical and atypical actions in scripted activities: tests for the script pointer + tag hypothesis. J Verbal Learn Verbal Behav 18:319–332, 1979.
7. AC Graesser, SB Woll, DJ Kowalski, DA Smith. Memory for typical and atypical actions in scripted activities. J Exp Psychol Hum Learn Mem 6:503–515, 1980.
8. DE Rumelhart, JL McClelland. On learning past tenses of English verbs. In: Parallel Distributed Processing: Explorations in the Microstructure of Cognition. Vol 2: Psychological and Biological Models. DE Rumelhart, JL McClelland, eds. MIT Press, Cambridge, MA, 1986, pp 216–271.
9. TJ Sejnowski, CR Rosenberg. Parallel networks that learn to pronounce English text. Complex Syst 1:145–168, 1987.
10. R Miikkulainen. Dyslexic and category-specific impairments in a self-organizing feature map model of the lexicon. Brain Lang 59:334–366, 1997.
11. T Kohonen. The self-organizing map. Proc IEEE 78:1464–1480, 1990.
12. T Kohonen. Self-Organizing Maps. Springer, New York, 1995.
13. A Caramazza. Some aspects of language processing revealed through the analysis of acquired aphasia: The lexical system. Ann Rev Neurosci 11:395–421, 1988.
14. RA McCarthy, EK Warrington. Cognitive Neuropsychology: A Clinical Introduction. Academic Press, New York, 1990.
15. M Coltheart, K Patterson, JC Marshall, eds. Deep Dyslexia. 2nd ed. Routledge and Kegal Paul, London, 1988.
16. JL Elman. Finding structure in time. Cogn Sci 14:179–211, 1990.

17. R Miikkulainen, MG Dyer. Natural language processing with modular neural networks and distributed lexicon. Cogn Sci 15:343–399, 1991.
18. DE Rumelhart, GE Hinton, RJ Williams. Learning internal representations by error propagation. In: Parallel Distributed Processing: Explorations in the Microstructure of Cognition. Vol 1: Foundations. DE Rumelhart, JL McClelland, eds. MIT Press, Cambridge, MA, 1986, pp 318–362.
19. R Miikkulainen. Script recognition with hierarchical feature maps. Connect Sci 2:83–101, 1990.
20. R Miikkulainen. Trace feature map: a model of episodic associative memory. Biol Cybern 66:273–282, 1992.
21. JL Kolodner. Retrieval and Organizational Strategies in Conceptual Memory: A Computer Model. Erlbaum, Hillsdale, NJ, 1984.
22. MD Williams, JD Hollan. The process of retrieval from very long-term memory. Cogn Sci 5:87–119, 1981.
23. AD Baddeley. The Psychology of Memory. Basic Books, New York, 1976.
24. L Postman. Transfer, interference and forgetting. In: Woodworth and Schlosberg's Experimental Psychology. 3rd ed. JW Kling, LA Riggs, eds. Holt, Rinehart and Winston, New York, 1971, pp 1019–1132.
25. R Miikkulainen. Script-based inference and memory retrieval in subsymbolic story processing. Appl Intell 5:137–163, 1995.
26. S Alvarado, MG Dyer, M Flowers. Argument representation for editorial text. Knowledge Based Syst 3:87–107, 1990.
27. MG Dyer. In-Depth Understanding: A Computer Model of Integrated Processing for Narrative Comprehension. MIT Press, Cambridge, MA, 1983.
28. JF Reeves. Computational morality: a process model of belief conflict and resolution for story understanding. PhD dissertation, University of California, Los Angeles, CA, 1991.

Index

Note: Grammatical theories are indexed alphabetically by full name. Named corpora are similarly indexed under their full names. Names of software systems are generally in all uppercase characters.